Phases of Matter and their Transitions

Phases of Matter and their Transitions

Concepts and Principles for Chemists, Physicists, Engineers, and Materials Scientists

Gijsbertus de With

WILEY-VCH

Author

Gijsbertus de With
Professor Emeritus, Department of Chemical Engineering and Chemistry
Eindhoven University of Technology
Helix Building
PO Box 513
5600MB Eindhoven
The Netherlands

Cover image designed by Martijn de With (2023)

Library of Congress Card No.: applied for

British Library Cataloguing-in-Publication Data
A catalogue record for this book is available from the British Library.

Bibliographic information published by the Deutsche Nationalbibliothek
The Deutsche Nationalbibliothek lists this publication in the Deutsche Nationalbibliografie; detailed bibliographic data are available on the Internet at http://dnb.d-nb.de.

Print ISBN: 978-3-527-35031-5
ePDF ISBN: 978-3-527-83693-2
ePub ISBN: 978-3-527-83692-5

Typesetting: Straive, Chennai, India
Printing and Binding: CPI Group (UK) Ltd, Croydon, CR0 4YY

C9783527350315_191023

Nothing has such power to broaden the mind as the ability to investigate systematically and truly all that comes under thy observation in life.

Marcus Aurelius (121–180)

Progress is made by trial and failure; the failures are generally a hundred times more numerous than the successes, yet they are usually left unchronicled.

William Ramsay (1852–1916)

If you thought that science was certain? Well, that is just an error on your part.

Richard Feynman (1918–1988)

Measure what is measurable and make measurable what is not so.

Galileo Galilei (1564–1642)

Contents

Preface

At some point in time, I got intrigued by the fact that attention for transitions between phases is more limited than for transformations within one phase. For example, in many books on the chemistry, physics, or thermodynamics of solids, melting is mentioned in passing, often merely stating the rather large change in properties between solids and liquids accompanying this discontinuous transition. This interest led to the present book. Discussing phase transitions requires knowledge of the phases themselves as well as various other disciplines. The physical chemistry, or chemical physics, of phases of matter and their transitions is wide, and therefore choices had to be made. This is always a difficult process, considering the interests of both readers and the author, coherency of the material covered, and size of the book.

As I always felt that a book, as far as is reasonably possible, should be more or less self-contained, the text is divided into three parts. Extending the metaphor of "stones" and "house" by Poincaré[1] (1854–1912), the first part (Chapters 2–7) deals with the various disciplines involved, the "tools" or *Basics*, supported by some further background (Appendices A–E). In the second part (Chapters 8–11), we discuss the "stones," mostly "bricks" nowadays, or the *Phases* themselves. These parts are necessarily compact and selective. In the third part (Chapters 12–19), the "houses" or the *Transitions* are dealt with. In the same vein, the houses are part of the "city" of materials science, many cities form the "country" of natural science, what used to be called natural philosophy, and many countries constitute the "world" of science(s). Clearly, some of the same types of bricks are used for different houses, meaning that some topics could be considered to belong to either gases, liquids, or solids. The choice is evidently somewhat arbitrary. The topic at hand is also inevitably connected with some mathematics, and, contrary to present popular approaches, the mathematics required is presented concisely in Appendix A. This choice attempts to avoid entangling physical and mathematical issues and to make the more general use of these mathematical topics clear, quite apart from where they are really used in this book. Evidently, these "bricks" are all imported from the "city" of mathematics.

Given the "three-fold way," it is still clear that not all aspects of each of the three ways can be discussed, not even superficially. Making these choices is a matter of personal choice as well as one of what is of broad interest. The contents section makes the choices made

1 "Le savant doit ordonner ; on fait la science avec des faits comme une maison avec des pierres ; mais une accumulation de faits n'est pas plus une science qu'un tas de pierres n'est une maison." or "The man of science must work with method. Science is built up of facts, as a house is built of stones; but an accumulation of facts is no more a science than a heap of stones is a house."

clear: the focus is on thermodynamics and properties of dielectric behavior. Although the book focuses on phase transitions (i.e. changes from one phase to another), reviewers of the content outline pointed out that the inclusion of some aspects of phase transformations (i.e. structural changes within one phase) would be desirable. To remain in line with the focus on dielectric behavior, the phase transformation part deals with ferroelectrics and liquid crystals, as these are the most closely related to dielectric behavior (and, for ferroelectrics, within my primary interest). What to choose to discuss for liquids is particularly precarious, and I opted to include, what I now call, structural models. No apology is offered for that, as I am convinced that these models are there to stay for quite some time.

Another difficult choice is how to present the matter. There are essentially two ways to go. The first option, which may be more Anglo-Saxon in style, is to demonstrate concepts through examples of applications and thereafter generalize, that is, use an inductive approach. The second option, which may be more Franco-Roman in style, is to postulate concepts and thereafter specialize in specific examples of application, that is, use a deductive style. In view of the "Zeitgeist," the spirit of times, most books nowadays use the former approach. Nevertheless, I have chosen the second approach, as that is closer to my heart. To alleviate possible concerns, I also opted to include examples at many stages of the formal development to keep readers connected to practical issues. Nevertheless, it is important to keep in mind, as I learned already during my scientific education from *Inorganic Chemistry* by Phillips and Williams, that it is necessary "to distinguish quite clearly between the usefulness of a model in predicting or correlating certain types of experimental observation and the general validity of the picture that this model seems to convey. Usefulness and validity often go hand in hand, but they do not always do so". For several topics, different approaches, not necessarily compatible but all describing the facts reasonably well, exist, indicating that neither all is clear nor that all authors agree. Wherever appropriate, I pointed to that. As illustration, I refer for thermodynamics to Maugin, who jokingly (but I think not 100%) quoted, what he called a faked true common citation: Some many thermodynamicists, some many thermodynamics. For glasses, a field recognized to be loaded with uncertainty, Weitz has even joked that: There are more theories of the glass transition than there are theorists who propose them. This clearly suffices to underline that not all is certain.

In many cases, the relevant derivations are given in some detail, so that readers need not retrieve the original source right away if they are trying to follow the road towards the result instead of looking only for the results. This is another, at least partial, explanation for the inclusion of what I called *Basics* (Chapters 2–7, Appendices A–E).

Each chapter contains at the end a section on *Further Reading* with a list of books recommended. For each of them, I made a small comment, hopefully helping readers assess what they can expect. The fact that a book is on the list implies that I like that book, and most of the time I refrain from further recommendations.

To some, the *Basics* as described in Chapters 2–7 may seem, to quote Pippard (1920–2008) in the introduction of his *Classical Thermodynamics*, to be "concentrated too much on the dry bones and too little on the flesh." He continued with, "but I would ask such critics to concede at least that the bones have an austere beauty of themselves." I cannot agree more. On a more personal note, I remember that as a youngster of about 22 years, I was highly impressed by *The Principles of Quantum Mechanics* by Dirac (1902–1984), not only because of the clarity but also because of the conciseness and beauty of presentation.

Not all of the text is completely new, as I used some, generally heavily rewritten, parts of my earlier books. Still, the content is essentially new, with a focus on what the title promises. The concept of justification vignettes for dealing with more detailed or advanced arguments, as used before, is retained, as is the concept of examples, these being essentially solved problems. As I know from experience (in lecturing and when studying new topics), full answers to problems invite the reader, after a short time, to accept the implicit invitation to have a look at these solutions. Therefore, as a compromise, I provided in an appendix the final answers to selected problems and sometimes their full solutions.

On a somewhat different tone, Kuhn (1921–1996) has advocated that scientific progress is made by large leaps. Sometimes it is even advocated that a single person is mainly responsible for a certain topic, in one sense or another. Sometimes that seems true; for example, for the theory of special relativity, it is often stated that it was largely single-handedly created by Einstein (1879–1955). Actually, many of the relevant ingredients were already there, but Darrignol made clear that Einstein was most likely the first to take the principle of relativity and the constancy of the speed of light at face value. To quote another example, classical mechanics is often said to be based on the three laws formulated by Newton (1642–1727), implying that they were the result of his effort only, but in reality, many predecessors and successors contributed: Galilei (1564–1642), Huygens (1629–1695), Borelli (1608–1679), and so on. In fact, what is now known as Newton's first law, the principle of inertia, was formulated first by Galilei and reformulated by Huygens. The second law started with Galilei, was reformulated by Huygens and Newton, while the form $f = ma$ as we know it now was first given by Euler (1750), as made clear by Maltese and Maronne and Panza. Another example is d'Alembert's principle (1743), first formulated in the form we know it by Lagrange (1788). For the history of mechanics, I refer to Dugas (1897–1957) and Dijksterhuis (1892–1965), while a critical assessment of the basis of mechanics was given by Mach (1838–1916). Highly readable in this connection is also the chapter on classical mechanics in the book by Lindsay and Margenau, which contains quite relevant parts on other parts of physics. For a brief review of energy and inertia, admittedly from a physicist's point of view, see a paper by von Laue and the highly relevant papers by Hecht referred to in the Introduction.

For those who object that this is a rather physical picture, I plead guilty. As an antidote, I like to refer to the concept of an atom, which is a good example of a converging chemical idea over time (see the books by Leicester or Partington). Accepted by chemists long before it was accepted by physicists, it is now a concept generally accepted by almost all scientists. Well known is probably that the physical chemist Ostwald (1853–1932), after being an ardent opponent for quite some time, in a later stage of his long career nevertheless accepted the idea.[2] On the contrary, Mach, the influential physicist, thinker, and contemporary of Ostwald, never accepted the concept.

In general, history is not always kind to authors, but many labels referring to scientists, although historically incorrect, have been so frequently used that it would be foolish to change them. Nevertheless, one should realize that labels indicating a responsible scientist more often than not are incorrect, or at least largely incomplete.

Finally, errors are hopefully minor, but, as Callen (1919–1993) stated in the introduction to his book *Thermodynamics and an Introduction to Thermostatistics*, "no error, numerical

2 The story is told by Ostwald himself in his reissued autobiography, translated by Jack and Scholz.

or textual, is truly minor to the student reader." I am sure that despite all efforts, the text will not be free of errors. The responsibility is entirely mine, and any comments, corrections, or indications of omissions will be appreciated.

Writing is done for a large part outside regular office hours, and I am, again, indebted to my wife Ada for her patience and forbearance. Although writing is done alone, improving written text is not. I would like to express my gratitude to my former PhD students, Dr. Gökhan Kaçar (Trakya University, Edirne, Turkey) and Dr. Bernette Oosterlaken (Leyden-Jar Technologies, Eindhoven) for helpful comments and suggestions on several sections. Further, to Prof. E. Hecht (formerly at Adelphi University, New York), Dr. K. Trachenko (Queen Mary, London), Prof. Erik Nies (KU Leuven), Prof. Helmut Cölfen, (University of Konstanz), Dr. Liesbeth Janssen, C. Laudicina, MSc, ir. Peter Koets, Dr. Mark Vis, and Prof. Dick Broer (all TU/e, Eindhoven). Finally, I would like to thank the staff of Wiley-VCH, in particular Dr. Frank Weinreich (initial Commissioning Editor), Dr. Sarah Higginbotham (final Commissioning Editor), Satvinder Kaur Sandhu (Managing Editor), and Judit Anbu Hena Daniel (Content Refinement Specialist) for their efforts during the production of this book.

Having said all this, I hope that the aim, purpose, and approach of the book have been made clear.

March 2023

Gijsbertus de With
TU Eindhoven

References

Callen, H.B. (1985). *Thermodynamics and an Introduction to Thermostatistics*. New York: John Wiley & Sons.

Darrigol, O. (2006). *The Genesis of the Theory of Relativity*, Ch. 1 in *Einstein, 1905–2005: Poincaré Seminar 2005, Progress* in *Mathematical Physics*, Vol. 47 (ed. T. Damour, O. Darrigol, B. Duplantier, and V. Rivasseau), p. 1. Basel: Birkhäuser Verlag.

Dijksterhuis, E.J. (1950, 1977, 3e). *De Mechanisering van het Wereldbeeld* (in Dutch), 3e. Amsterdam: Meulenhoff; (1961). *The Mechanization of the World Picture*. Oxford: Clarendon Press.

Dirac, P.A.M. (1958). *The Principles of Quantum Mechanics*, 4e. Oxford: Oxford University Press.

Dugas, R. (1955). *A History of Mechanics*. Neuchâtel, Switzerland: Éditions du Griffon (also Dover, 1988).

Jack, R.S. and Scholz, F. (2017). *Wilhelm Ostwald, The Autobiography*. Cham, Switzerland: Springer.

Kuhn, T.S. (1996). *The Structure of Scientific Revolutions*, 3e. Chicago: University of Chicago Press.

von Laue, M. (1949). *Inertia and Energy*, Ch. II.19 in *Albert Einstein Philosopher-Scientist* (ed. P.A. Schilpp), p. 501. Evanston: Ill., Library of Living Philosophers; (1969). 3e. New York: MJF Books.

Leicester, H.M. (1956). *The Historical Background of Chemistry*. New York: John Wiley & Sons (also Dover, 1971).

Lindsay, R.B. and Margenau, H. (1936). *Foundations of Physics*. New York: John Wiley & Sons (also Dover, 1957).

Mach, E. (1912). *Die Mechanik in Ihrer Entwicklung Historisch-Kritisch Dargestellt*, 7e, F.A. Brockhaus, Leipzig (1921, 8e and 1933, 9e); (1960). *The Science of Mechanics*. 6e. LaSalle, IL: Open Court Publishing Company (after the 7th German ed. of 1912, with revisions through the 9th German ed.).

Maltese, G. (2000). On the relativity of motion in Leonhard Euler's science. *Arch. Hist. Exact Sci.* 54: 319.

Maronne, A. and Panza, M. (2014). Euler, reader of Newton: mechanics and algebraic analysis. *Adv. Hist. Stud.* 3: 12.

Maugin, G.A. (1999). *The Thermomechanics of Nonlinear Irreversible Behaviors*. Singapore: World Scientific Publishing Company.

Partington, J.R. (1957). *A short History of Chemistry*. New York: MacMillan & Co. (also Dover, 1989).

Phillips, C.S.G. and Williams, R.J.P. (1965). *Inorganic Chemistry, Principles and Non-metals*, Vol. 1. New York: Oxford University Press.

Pippard, A. (1966). *Classical Thermodynamics*. Cambridge: Cambridge University Press.

Poincaré, H. (1902). *La Science et l'Hypothèse*. Paris: Flammarion; (1952). *Science and Hypothesis*. New York: Dover.

Weitz, D.A. (2008). In an interview by K. Chang in the *New York Times* (29 July). www.nytimes .com/2008/07/29/science/29glass.html.

List of Frequently Used Symbols and Abbreviations

(X)	dimension dependent on property, (–) dimensionless property
ϑ	kinetic energy (J)
Λ	thermal wavelength (m)
Ξ	grand partition function (–)
Φ	potential energy (J)
Ψ	wave function (–)
	voltage (V)
Ω	external potential energy (J)
	grand potential (J)
	atomic volume (m^3)
α	constant (X)
	activity coefficient (–)
	thermal expansivity (K^{-1})
	polarizability (C^2 m^2 J^{-1})
β	constant (X)
	$1/kT$ (J^{-1})
	activity coefficient (–)
γ	constant (X)
	activity coefficient (–)
δ	solubility parameter ($MPa^{1/2}$)
	Dirac delta function (–)
ε	small scalar (X)
	energy (J)
ε_j	strain component j (–)
η	(bulk) viscosity (Pa s)
	packing density (–)
	internal variable (X)
κ	compressibility (m^2 N^{-1})
θ	characteristic temperature
λ	parameter (X)
	wavelength (m)
μ	dipole moment (C m)
ν	frequency (s^{-1})

ξ	internal variable (X)
ρ	number density (m^{-3})
	mass density ($g\,m^{-3}$)
	radius of curvature (m)
ρ'	mass density ($kg\,m^{-3}$)
σ	(hard-sphere) diameter (m)
σ_j	stress component j (X)
τ	characteristic time (s)
ϕ	(pair) potential energy (J)
	electrical potential ($V\,m^{-1}$)
	volume fraction (–)
χ	susceptibility (X)
ω	circular frequency (s^{-1})
$\boldsymbol{D}$	dielectric displacement ($A\,s\,m^{-2}$)
$\boldsymbol{E}$	electric field ($V\,m^{-1}$)
$\boldsymbol{F}$	force (N)
$\boldsymbol{P}$	polarization ($A\,s\,m^{-2}$)
$\boldsymbol{Q}$	quadrupole moment ($C\,m^2$)
$\boldsymbol{e}$	unit vector (m)
$\boldsymbol{f}$	force (N)
$\boldsymbol{p}$	generalized momentum (X)
$\boldsymbol{q}$	generalized coordinate (X)
$\boldsymbol{r}$	coordinate (m)
$\boldsymbol{u}$	displacement (m)
$\boldsymbol{v}$	velocity ($m\,s^{-1}$)
$\boldsymbol{x}$	coordinate (m)
$\mathcal{A}$	affinity
$\mathcal{H}$	Hamilton function
$\mathcal{L}$	Lagrange function
$\mathcal{M}$	molar mass
$\hbar$	Hamilton density
ℓ	Lagrange density
A	area (m^2)
	generalized force (X)
	constant (X)
B	constant (X)
	generalized force (X)
B_i	virial coefficients
C	constant (X)
	heat capacity
D	diffusivity ($m^2\,s^{-1}$)
E	energy (J)
	electric field ($V\,m^{-1}$)
F	Helmholtz energy (J)
	force (N)
	Faraday constant ($C\,mol^{-1}$)

G	Gibbs energy (J)
H	enthalpy (J)
I	moment of inertia (kg m^2)
J	flux (x s^{-1} m^{-2})
K	equilibrium constant (X)
	bulk modulus (N m^{-2})
L	length (m)
M	molecular weight
N	number of molecules (particles) (–)
N_i	number of molecules of type i (–)
N_A	Avogadro's number (mol^{-1})
P	pressure (MPa)
	probability (–)
	polarization (m^3)
Q	configurational partition function (–)
R	gas constant (J K^{-1} mol^{-1})
	radius (m)
	distance (m)
S	entropy (J K^{-1})
T	temperature (K)
U	(internal) energy (J)
V	potential energy (J)
	volume (m^3)
W	work (J)
Z	canonical partition function (–)
a	generalized displacement (X)
	constant (X)
	spring constant (N m^{-1})
b	second virial coefficient (m^3 mol^{-1})
	constant (X)
	generalized displacement (X)
c	constant (X)
	inverse spring constant (compliance) (m N^{-1})
e	unit charge (C)
f	(volume) fraction (–)
	specific Helmholtz energy (J kg^{-1})
	force (N)
	spring constant (N m^{-1})
	rate constant (X)
g	specific Gibbs energy (J kg^{-1})
	density of states (J^{-1})
h	Planck's constant
j	flux (X m^{-2} s^{-1})
k	Boltzmann's constant (J K^{-1})
	spring constant (N^{-1})
	rate constant (X)

l	length (m)
m	Mie constant (–)
	mass (kg)
n	Mie constant (–)
	material constant (X)
	number of moles (–)
	number density (m^{-3})
n_α	number of moles of component α (–)
p_j	probability of state j (–)
q	charge (C)
r	distance (m)
	rate ($X\,s^{-1}$)
s	specific entropy ($J\,K^{-1}\,kg^{-1}$)
t	time (s)
u	(pair) potential energy (J)
v	volume (m^3)
w	regular solution parameter (J)
x_α	mole fraction of component α (X)
z	single-particle partition function (–)
	coordination number (–)
z_j	valency of particle j (–)

Main Abbreviations

AAD	average absolute difference
AB	acid–base
AO	atomic orbital
BCC	body-centered cubic
BZ	Brillouin zone
CM	continuum mechanics
CMC	critical micelle concentration
CN	coordination number
CNT	classical nucleation theory
CWT	capillary wave theory
DFT	density functional theory
DoF	degrees of freedom
DoS	density of states
EN	electronegativity
EoS	equation of state
FCC	face-centered cubic
FVT	free volume theory
GAE	Gibbs adsorption equation
HCP	hexagonal closed packed
HNC	hyper-netted chain equation
HS	hard sphere

IR	infrared
MC	Monte Carlo
MCT	mode coupling theory
MD	molecular dynamics
MO	molecular orbital
MT	macroscopic thermodynamics
LCAO	linear combination of AOs
LJ(D)	Lennard-Jones(-Devonshire)
NFE	nearly free electron
NMR	nuclear magnetic resonance
NRD	neutron-ray diffraction
OZ	Ornstein–Zernike
PM	particle mechanics
PME	principle of maximum entropy
PoCS	principle of corresponding states
PY	Percus–Yevick equation
(Q)HA	(quasi-) harmonic approximation
QM	quantum mechanics
RCP	random close packed
RDF	radial distribution function
RS	regular solution
RVE	representative volume element
SC	simple cubic
SCF	self-consistent field
SLS	significant liquid structure
ST	statistical thermodynamics
STP	standard temperature and pressure
SW	square well
T	tetragonal
TB	tight binding
TEM	transmission electron microscopy
TST	transition state theory
UV	ultraviolet
VFT	free volume theory
XRD	X-ray diffraction
YBG	Yvon–Born–Green equation
lhs	left-hand side
rhs	right-hand side
vdW	van der Waals
$\cong$	approximately equal
$\approx$	very approximately equal
$\equiv$	identical
$\sim$	proportional to
$\Leftrightarrow$	corresponds with
$\mathcal{O}(x)$	order of magnitude x

Superscripts

E	excess
∞	infinite dilution
$\circ$	standard
*	pure substance
‡	activated complex
id	ideal

Subscripts

f	formation reaction
fus	fusion
g	glass
mix	mixing
m	molar
mel	melting
r	reaction in general
sol	solution
sub	sublimation
trs	transition
vap	vaporization

SI Units, Physical Constants, and Conversion Factors

In 2019, the Système International d'Unités or International System of Units (SI) was revised to comprise a coherent system of units of measurement in terms of defined physical constants.

Basic SI Units

Quantity	Unit	Symbol
Length	Metre	m
Mass	Kilogram	kg
Time	Second	s
Electric current	Ampere	A
Temperature	Kelvin	K
	°C	$t/°\mathrm{C} = T/\mathrm{K} - 273.15$
Amount of substance	Mole	mol
Luminosity	Candela	cd

Derived SI Units

Force	Newton	$\mathrm{N = kg\,m\,s^{-2}}$
Work, energy, heat	Joule	$\mathrm{J = N\,m}$
Power	Watt	$\mathrm{W = J\,s^{-1}}$
Pressure	Pascal	$\mathrm{Pa = N\,m^{-2}}$
Frequency	Hertz	$\mathrm{Hz = s^{-1}}$
Electrical charge	Coulomb	$\mathrm{C = A\,s}$
Electrical potential	Volt	$\mathrm{V = J\,C^{-1}}$
Electrical resistance	Ohm	$\Omega = \mathrm{V\,A^{-1}}$

SI Defining Physical Constants

(… indicating that more digits are in the definition)

Constant	Symbol	Value
Avogadro's number	N_A	$6.022\,141\ldots \times 10^{23}$ mol^{-1}
Elementary charge	e	$1.602\,177\ldots \times 10^{-19}$ C
Boltzmann's constant	k	$1.380\,649 \times 10^{-23}$ J K^{-1}
		$8.617\,337$ eV K^{-1}
Planck's constant	h	$6.626\,070\ldots \times 10^{-34}$ J s
	$\hbar = h/2\pi$	$1.054\,572\ldots \times 10^{-34}$ J s
Speed of light	c	$2.997\,925\ldots \times 10^{10}$ m s^{-1}
Hyperfine transition frequency of Cs	–	$9.192\,632\ldots \times 10^{9}$ Hz
Luminous efficacy (of 540 THz radiation)	K_{cd}	683 lm W^{-1}

Other Constants

Electron rest mass	m_e	$9.109\,384 \times 10^{-31}$ kg
Proton rest mass	m_p	$1.672\,622 \times 10^{-27}$ kg
Neutron rest mass	m_n	$1.674\,927 \times 10^{-27}$ kg
Gas constant	$R = kN_A$	$8.314\,463$ J mol^{-1} K^{-1}
Standard acceleration of gravity	g	$9.806\,65$ m s^{-2}
Faraday's constant	$F = eN_A$	$9.648\,533 \times 10^{4}$ C mol^{-1}
Permeability of vacuum	μ_0	$1.000\,000 \times 4\pi \times 10^{-7}$ N A^{-2}
Permittivity of vacuum	$\varepsilon_0 = 1/\mu_0 c^2$	$8.854\,187 \times 10^{-12}$ C^2 N^{-1} m^{-2}

Conversion Factors for Non-SI Units

1 dyne	=	10^{-5} N
1 bar	=	10^{5} Pa
1 atm	=	1.013 25 bar
1 mm Hg (torr)	=	1/760 atm
1 D	=	$3.336\,641 \times 10^{-30}$ C m
1 L	=	1 dm^3 = 10^{-3} m^3
1 eV	=	$1.602\,176\ldots \times 10^{-19}$ J
1 erg	=	10^{-7} J
1 eV/particle	=	96.485 332 kJ mol^{-1}
1 Å	=	10^{-10} m
1 cm^{-1}	=	$1.986\,302 \times 10^{-23}$ J
hc/k	=	1.438 777 cm K
1 cal$_{th}$ (refers to the *thermochemical calorie*)	≡	4.184 J
1 u (Da) (unified atomic mass unit, dalton)	=	$1.660\,539 \times 10^{-27}$ kg

Prefixes

femto, f	=	10^{-15}
pico, p	=	10^{-12}
nano, n	=	10^{-9}
micro, μ	=	10^{-6}
milli, m	=	10^{-3}
kilo, k	=	10^{3}
mega, M	=	10^{6}
giga, G	=	10^{9}
tera, T	=	10^{12}

Greek Alphabet

Α, α	Alpha
Β, β	Beta
Γ, γ	Gamma
Δ, δ	Delta
Ε, ε	Epsilon
Ζ, ς	Zeta
Η, η	Eta
Θ, θ, ϑ	Theta
Ι, ι	Iota
Κ, κ	Kappa
Λ, λ	Lambda
Μ, μ	Mu
Ν, ν	Nu
Ξ, ξ	Xi
Ο, ο	Omicron
∏, π	Pi
Ρ, ρ	Rho
Σ, σ	Sigma
Τ, τ	Tau
Υ, υ	Upsilon
Φ, φ, ϕ	Phi
Χ, χ	Chi
Ψ, ψ	Psi
Ω, ω	Omega

Standard Values

Until 1982, *standard temperature and pressure* (STP) were defined as a temperature of 273.15 K (0 °C) and an absolute pressure of exactly 1 atm ($P_{atm}{}^{\circ}$ = 101.325 kPa), but since

1982, STP are defined as a temperature of 273.15 K (0 °C) and an absolute pressure of exactly 10^5 Pa ($P_{bar}{}^{\circ}$ = 100 kPa, 1 bar). A reference temperature of 25 °C is used as well (*standard ambient temperature and pressure*, SATP). Good practice always incorporates the reference conditions (including relative humidity).

For the standard molar volume of gases, the four most important choices are the following:

$V_m = RT/P_{atm}{}^{\circ} = R \times 273.15/101.325 = 22.414\ \text{dm}^3\ \text{mol}^{-1}$ at 0 °C and 1 atm,
$V_m = RT/P_{bar}{}^{\circ} = R \times 273.15/100.000 = 22.711\ \text{dm}^3\ \text{mol}^{-1}$ at 0 °C and 1 bar,
$V_m = RT/P_{atm}{}^{\circ} = R \times 298.15/101.325 = 24.466\ \text{dm}^3\ \text{mol}^{-1}$ at 25 °C and 1 atm,
$V_m = RT/P_{bar}{}^{\circ} = R \times 298.15/100.000 = 24.790\ \text{dm}^3\ \text{mol}^{-1}$ at 25 °C and 1 bar.

Thermal Energy *kT* at 298 K

$207.2\ \text{cm}^{-1} = 4.116 \times 10^{-21}\ \text{J} = 2.569 \times 10^{-2}\ \text{eV} = 2.479\ \text{kJ mol}^{-1} = 0.5925\ \text{kcal mol}^{-1}$.

Crystallographic Directions and Planes

Specific plane (hkl), set of equivalent planes {hkl}

Specific direction <hkl>, set of equivalent directions [hkl]

Summary of Notation

Scalar a
Vector $\boldsymbol{a}$, component a_j
Unit vector $\boldsymbol{e}$, $\boldsymbol{n}$
Column matrix $\mathbf{a}$, element a_j
Row matrix $\mathbf{a}^{\mathrm{T}}$ (transpose of column matrix)
Matrix $\mathbf{A}$, element A_{ij}
Tensor $\mathbb{A}$, component A_{ij}
Scalar product $\boldsymbol{a} \cdot \boldsymbol{b} = \mathbf{a}^{\mathrm{T}}\mathbf{b} = \Sigma_j\ a_j b_j$
Vector product $\boldsymbol{a} \times \boldsymbol{b} = \varepsilon_{ijk}\boldsymbol{e}_i a_j b_k$ with alternator $\varepsilon_{ijk} = ½(i-j)(j-k)(k-i)$
Triple product $\boldsymbol{a} \cdot (\boldsymbol{b} \times \boldsymbol{c}) = \Sigma_{i,j,k}\ \varepsilon_{ijk} a_i b_j c_k$
Gradient $\nabla \boldsymbol{a} = \boldsymbol{e}_k \partial a_j / \partial r_k$
Divergence $\nabla \cdot \boldsymbol{a} = \Sigma_j\ \partial a_j / \partial r_j$
Gauss theorem $\int \nabla \cdot \mathbb{A}\, \mathrm{d}V = \int \mathbb{A} \cdot \boldsymbol{n}\, \mathrm{d}S$, $\int \nabla a\, \mathrm{d}V = \int \boldsymbol{n}a\, \mathrm{d}S$, $\int \nabla \cdot \boldsymbol{a}\, \mathrm{d}V = \int \boldsymbol{n} \cdot \boldsymbol{a}\, \mathrm{d}S$, and $\int \nabla \times a\, \mathrm{d}V = \int \boldsymbol{n} \times \boldsymbol{a}\, \mathrm{d}S$
Fourier transform $F[f(\boldsymbol{r}, t)] = \tilde{f}(\boldsymbol{q}, \omega) = \int_{-\infty}^{+\infty} f(\boldsymbol{r}, t)\ \exp[\mathrm{i}(\boldsymbol{q} \cdot \boldsymbol{r} - \omega t)\, \mathrm{d}\boldsymbol{r}\mathrm{d}t$
Inverse transform $F^{-1}[\tilde{f}(\boldsymbol{q}, \omega)] = f(\boldsymbol{r}, t) = (2\pi)^{-4} \int_{-\infty}^{+\infty} \tilde{f}(\boldsymbol{q}, \omega)\ \exp[-\mathrm{i}(\boldsymbol{q} \cdot \boldsymbol{r} - \omega t)\, \mathrm{d}\boldsymbol{q}\mathrm{d}\omega$
Operator A, a
Volume V, pressure P, number molecules N, number of moles n
Internal energy $U(S, V, N)$ with natural variables S, V, and N
Enthalpy $H(S, P, N) = U + PV$ with natural variables S, P, and N
Entropy $S(U, V, N)$ with natural variables U, V, and N
Helmholtz energy $F(T, V, N) = U - TS$ with natural variables T, V, and N
Gibbs energy $G(T, P, N) = F + PV = U - TS + PV$ with natural variables T, P, and N
Chemical potential μ, (absolute) activity $\lambda = \exp(\beta\mu)$ with $\beta = 1/kT$
Mole fraction x_j of the component j
(Relative) Activity $a_j = f_j x_j$ with activity coefficient f_j or $a_j = \gamma_j x_j$ with activity coefficient γ_j
Superscript (α) for bulk phase α, (σ) for surface phase
Superscript* for pure state,$^\circ$ for reference state
Number of complexions $W = \exp(S/k)$
Canonical partition function $Z = \exp(-\beta F)$, configurational partition function Q
Thermal wavelength $\Lambda = (h^2/2\pi mkT)^{1/2}$ with mass m and Planck's constant h
Grand canonical partition function $\Xi = \exp(PV)$

1 Introduction

The change in appearance and properties of substances upon applying various stimuli is a well-known phenomenon. For example, probably the most important substance on earth, water, is known to exist in three forms or *phases*: solid (ice), liquid (water), and gas (water vapor), depending on the temperature applied. Although ubiquitous, the transition processes between these phases are generally complex. In this book, we deal with (some of) the chemical physics of substances and their phase transitions.

As we need many concepts to deal with these transitions and the transition processes inevitably refer to the phases themselves, the book is divided into three parts. By and large, in the first part we deal with the theoretical tools needed and in the second part with the phases themselves, while in the third part we discuss the transitions. As many basic concepts and assumptions are used throughout, we briefly review them in this chapter. First, the concepts of matter, energy, and charge and their conservation laws are introduced, followed by some considerations about macroscopic and microscopic descriptions. Thereafter, the essential differences between solids, liquids, and gases are outlined, followed by a description of the approach to the various topics dealt with in this book. The final section deals with the units and notations used.

1.1 Constituents of Matter

We first briefly review the constituents of matter as far as they are relevant to our topic. For that, we accept its discrete nature. Basically, the history boils down to the fact that matter was considered, by chemists from the 1780s and later by physicists as well, to consist of a set of immutable and indestructible *elements* that could be divided into *metals*, *nonmetals*, and *gases*. The elements were supposed to have constant properties, such as mass, and consisted of *atoms* that got labels like H (hydrogen), O (oxygen), C (carbon), N (nitrogen), and which were systematically organized in the *periodic table of elements*. Elements can bond to other elements, resulting in *compounds* that were classified as *salt* (combining a metal with a nonmetal), *acid* (combining hydrogen with a nonmetal), and *base* (combining water with a metal), while the particular combination of elements in the substance was (is) addressed as the *composition*. The smallest entity that behaves chemically like the whole is a *molecule*. Molecules have a relatively simple, fixed ratio of atoms, as in H_2O (water). In other cases, the

Phases of Matter and their Transitions: Concepts and Principles for Chemists, Physicists, Engineers, and Materials Scientists, First Edition. Gijsbertus de With.

stoichiometry, that is, the ratio(s) of the number of elements in a compound, varies considerably. The largest variation is found in compounds where there is a small energy difference between the various defect states, often transition metal oxides; for example, for TiO_x the range is $0.92 < x < 1.26$ [1]. It appeared somewhat later that some elements that are gaseous at room temperature are also compounds and are actually mostly diatomic molecules, for example, H_2, O_2, and N_2.

All these results were made possible by the (reproducible) *preparation* and *isolation*, often followed by *purification*, of substances, whereafter the study of their *physical properties* (response to stimuli that do not change the composition) and *chemical properties* (response to stimuli that do change the composition) became possible. One of the most important early results is the discovery that at constant temperature and pressure, certain amounts of the gaseous elements, labeled *mole*, contain the same number of molecules, a number known as the *number of Avogadro*.

Nowadays, all elementary entities are often counted in moles. The *molecular weight* of a substance translates this amount into kilograms of the substance.

As is well known in the meantime, atoms appear not to be indestructible but to be composed of other particles that together determine the properties of atoms. These are the *nuclei*, composite particles containing singly positively charged *protons* and neutral *neutrons*, collectively denoted as *nucleons*, and singly negatively charged *electrons*, which all three were for some time considered elementary particles. Still, later, it appeared that these particles were composed of a zoo of (still more) elementary particles. For our field of interest, employing energies relevant to life and terrestrial phenomena, foremost atoms, electrons, and protons are relevant. Adding an electron to an atom (with the related energy characteristic, the *electron affinity*) or removing an electron from an atom (with the related energy characteristic, the *ionization potential*) results in a charged atom called an *ion*. The numbers of protons and neutrons in the nucleus determine the *mass* of the nucleus, while the number of protons determines its *charge*. For a neutral atom, that charge is compensated by an equal number of electrons. The number of neutrons varies for most elements while having the same number of protons; each variety is called an *isotope*. For example, the isotope deuterium (D) of hydrogen (H) has one proton and one neutron, while H has one proton. Neutrons play a relatively minor role, but still, via their effect on mass, they can influence some properties of elements and molecules significantly (Table 1.1).

Table 1.1 Some properties of normal water (H_2O) and heavy water (D_2O).

	ρ (g cm^{-3}) at 25 °C	T_{mel} (°C)	T_n (°C)	γ (mN m^{-1}) at 25 °C	ρ_{max} at °C	η (mPa s) at 25 °C	$\Delta_{fus}H$ (cal mol^{-1})	$\Delta_{vap}H$ (cal mol^{-1})
H_2O	1.00	0.00	100.00	72.75	4.0	10.09	1436	10 480
D_2O	1.1079	3.82	101.42	67.8	11.6	12.6	1510	10 740

ρ, density; T_{mel}, melting point; T_n, boiling point; γ, surface tension; ρ_{max}, maximum density; η, viscosity; $\Delta_{fus}H$, enthalpy of fusion; $\Delta_{vap}H$, enthalpy of vaporization.
Source: Adapted from Prutton and Maron [2].

1.2 Matter and Energy: Interaction and Change

Let us now turn to the three conservation laws that appear to be fundamental: *conservation of mass, energy, and charge.* In this section, we deal first with energy, while in Section 1.3, we discuss mass and charge.

Surprisingly (or may be, not), the definition of energy is treated as a stepchild in many treatments of physical phenomena. The usual way to introduce energy is to refer to the kinetic energy (KE) of particles in mechanics, introduce thereafter the potential energy (PE), and show that their sum, that is, energy,[1] is conserved, and from there on more or less assume that energy conservation is always the case, including effects of heat, radiation, and so forth. It sounds a bit like this drink is brown and sold in bottles, and there are many competitors on the market. All true, but not really defining the drink. Using the abovementioned process of describing energy avoids the real question: What *is* energy?

Only recently a serious attempt was made by Hecht [4] as a result of a longer running activity to provide a real definition. As there is not much point in rephrasing what is a clear exposition, we will largely use his words ("…" for exact quotes, '…' for approximate quotes). In this connection, we will frequently need the concept of a *particle*, a not necessarily microscopic entity with mass m (to be discussed later) and an inaccessible internal structure that can move in space with velocity $\boldsymbol{v}$. Such a particle is located at one point in space at one time, that is, it is considered small as compared to the total system considered.

The point of departure is that "on a very basic level, there is matter, interaction and change." "*Matter is that which interacts and, as a result, manifests observable physical properties that are amenable to being measured.*" Matter interacts via equal and oppositely directed forces (interaction pairs): "there is no such thing as a single force. If matter did not interact, it could not be known and could not be said to exist. All matter interacts in one way or another, either via long-range or short-range forces. *Force is the agent of all observable physical change.* Nothing determinable happens without the action of force."

So, physical change occurs because of interactions. "*Energy is a measure of the capacity of matter interacting with matter to effectuate physical change. It is the conserved scalar measure of the extent of change that has already occurred, and/or change that can yet occur.*" Thus, while force is that which changes the state of a system, 'energy is a measure of the extent of that change, whether it has already happened or is yet to happen. Energy is associated with all matter that exists and is a measure of the change resulting from matter interacting with matter; it is not an entity in and of itself, independent of matter.' "If matter did not interact, energy in all its manifestations would vanish and the concept would be meaningless." 'Due to the symmetrical nature of interactions, a change in one entity is accompanied by an energy-equivalent change in the entity with which it interacts.' "Accordingly, we have the principle of *Conservation of Energy*: *the total energy associated with any system can be accounted for at all times.*"

---

1 Initially I intended to insert here brief remarks about the history of the concept of energy, but refrained from that, not for not being a professional historian, but mainly because a short story probably will do injustice to players in this game: Young, Joule, the brothers William and James Thomson, Mayer, Clausius, von Helmholtz, Mach, Planck, and a few more scientists and, not to forget, engineers like father and son Lazare and Sadi Carnot, Clapeyron, Rankine, Gouy, Stodola, and Zeuner [3].

As an example, consider a stretched spring that has a quantity of energy stored (via internal interactions) by an applied force that is the agent of change. The spring in its energized state, characterized by the amount of energy stored, has the capacity to effectuate in the future a variety of changes, and "the energy stored is a measure of the ability of the spring to produce change (internally and externally) via the action of the forces it can exert."

1.3 Mass and Charge

We now have to turn to the concepts of mass and charge, and for that, we will have to use one or two results from special relativity theory. We use the formulation in which the mass m is considered speed independent (for a brief review, see Appendix B). For a single free particle, the energy $E = \gamma mc^2$, where $\gamma = (1 - v^2/c^2)^{-1/2}$ with $v = (\mathbf{v} \cdot \mathbf{v})^{1/2}$ the speed of the particle and c is the speed of light. So, when $v = 0$, $\gamma = 1$ and the energy becomes the *energy of rest* $E_0 = mc^2$. The difference between E and E_0 is the *energy of motion* or the organized KE ϑ of the particle moving as a whole and given by $\vartheta = \gamma mc^2 - mc^2$. Introducing the momentum $\mathbf{p} = \gamma m\mathbf{v}$ leads to $E = [c^2p^2 - (mc^2)^2]^{1/2} \cong mc^2 + p^2/2m$, and in its approximate form, apart from the constant mc^2, in agreement with the classical result. For a composite body containing many particles with total mass M, the object's total energy, measured in its center-of-mass frame (where it is motionless), is the rest energy $E_0 = Mc^2$. It encompasses the individual rest energies, kinetic energies, and potential energies of all the interacting particles that constitute the object. Note therefore that, as $E_0 = Mc^2$, $M \neq \Sigma_j m_j$ and the mass of a composite object as a whole change with its energy content.

Although seldom formally defined, mass is often considered to be a quantity of matter, generally simply assumed to be synonymous with the "amount of material," a measure presumably having to do with the type and number of particles constituting the material object. In view of the results of the theory of special relativity, this interpretation no longer holds. Another often-used definition is in terms of inertia. Using $\vartheta = \gamma mc^2 - mc^2$, we see that as v goes to c, γ approaches infinity and ϑ of an object possessing mass approaches infinity. Hence, it will take an infinite amount of energy for such an object to be propelled up to c, which is the maximum velocity. The well-known Michelson–Morley and Kennedy–Thorndike experiments confirmed that the speed of light is constant within 3×10^{-5}. More recently, Brecher [5] showed from observations of regularly pulsating X-ray sources in binary star systems that the speed of light is independent of the velocity of the source within 2×10^{-9}. Comparing frequencies of a cryogenic sapphire oscillator and a hydrogen maser (Figure 1.1a), Wolf et al. [6] confirmed that Lorentz transformations are obeyed with an overall uncertainty of $\leq 8 \times 10^{-7}$, again indicating a constant c. Bertozzi [7] showed similarly that c is the maximum speed for electrons (Figure 1.1b). Apparently, energy is the determinant of inertia so that mass cannot be defined in terms of inertia. Finally, there is the interpretation that mass produces gravity and is a measure thereof. In special relativity theory, both m and c are constants, and thus mc^2 as well. As $E = \gamma mc^2$, mass is derived from energy, and mass (and momentum) cannot be defined in terms of gravity.

So, then, what *is* mass? It seems that energy is a more fundamental concept than mass. After all, everything that has mass always has energy, but not everything that has energy always has mass (e.g. a photon). Considering $E_0 = Mc^2$, we learned that "*the mass of a*

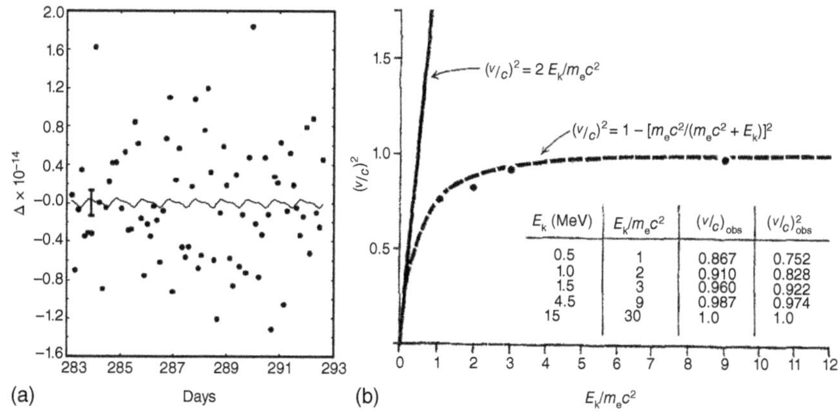

Figure 1.1 Relativity. (a) Constant relative frequency difference Δ between an oscillator and maser indicating that Lorentz transformations are obeyed with an uncertainty of $\leq 8 \times 10^{-7}$. Source: Adapted from Wolf et al. [6]. (b) The quantity v^2/c^2 versus $E_k/m_e c^2$ for various kinetic energies E_k for an electron with mass m_e showing the limiting velocity c [7].

composite system is a property of its matter, the internal motions thereof, and the interactions therein." The mass of any composite entity seems to represent an account of the entity's history: "*The mass of any system at rest is a measure of the amount of energy that went into creating that object as it exists at that moment.*" If a system cannot exist at rest, that is, at $\boldsymbol{p} = \mathbf{0}$, that system has no mass.

Traditionally, PE is the energy said to be stored as a result of a configurational change in a system of interacting parts. PE Φ is a tremendously useful bookkeeping device, but it is not a quantity possessing measurable physicality. Only changes in Φ are defined, and these are defined as the work done on a system by conservative forces. Such work is measurable as it is being done. However, once work is done, it no longer exists and is no longer measurable. So, PE is not an empirical quantity as it is not itself amenable to measurement.

A composite system (at rest) possesses (excess) energy when it can undergo spontaneous change. To return to our example, a stretched spring is said to contain elastic energy; when released, it can experience self-initiated change. If a physical property of a system of interacting parts undergoes a change (temperature, volume, etc.), that property can be associated with a resulting difference in energy. Because the interaction energy, that is, the PE, is associated with the system's internal energy, it is conceptually subsumed into the rest energy E_0. Because $E_0 = Mc^2$, whenever the conceptual quantity known as the PE of a system increases or decreases, the mass of that system actually, sometimes measurably, increases or decreases.

The change in PE is thus a convenient way to keep track of the energy stored via interactions within a system, but which is actually stored as mass: $\Delta m = \Delta\Phi/c^2$, which is at least in principle, if not always in practice, measurable. One atomic mass unit (Da) corresponds to 931.494 MeV/c^2. As an example, consider ^{56}Fe, with 91.8%, the most abundant isotope of Fe, containing 26 protons with mass $m_p = 1.007\,276$ Da, 30 neutrons with mass $m_n = 1.008\,665$ Da, and 26 electrons with $m_e = 0.000\,548\,6$ Da. The actual mass of a nucleus M of atomic number Z and mass number A is given by

$$M = Zm_p + (A - Z)m_n - \delta m$$

where M is less than the mass of the sum of the parts by the *mass defect* δm, corresponding to a *binding energy* $c^2\delta m$: the minimum energy that must be added to the system to dissociate it, to relocate its constituent parts, at rest and far from each other. For ^{56}Fe, we thus have $26 \times m_p + 30 \times m_n + 26 \times m_e = 56.4634$ Da, while its actual mass is 55.9349 Da. Hence, $\delta m = 0.5285$ Da $= 492.3$ MeV/c^2 or $\delta m/M = 0.92\%$ or $\delta m/(A+Z) = 8.8$ MeV/c^2 per nucleon [8]. For most of physics and chemistry, however, δm is much smaller. Consider, for example, the ^{1}H atom containing one proton and one electron, for which $\delta m = 1.461 \times 10^{-8}$ Da or $c^2\delta m = 13.61$ eV.

Thus, a change in the PE of a system of interacting objects corresponds to a change in mass, which is, in principle, measurable. However, for energies relevant to life and terrestrial phenomena, the mass deficit δm is normally immeasurably small, but the corresponding energy $c^2\delta m$ can be measured, and we accept conservation of mass (which is thus not exactly true) and introduce PE to satisfy conservation of energy (which is always exactly true).

Summarizing so far, there is matter interacting via forces that effectuate all change, and that interaction produces observable change. As a result of this interaction, a material system possesses a quantity of energy. Energy is the capacity of a system, acted upon by forces, to experience a specific amount of change, change that has already been imparted to it, or change that it can impart to itself and/or to its environment in the future. Work, heat, and electromagnetic radiation correspond to energy in transit and are means of transferring energy from one system to another.

Like mass, charge is velocity independent. This is shown by considering the H_2 molecule. Such a molecule contains two protons and two electrons. As the electrons move around the protons, they move on average faster than the protons. If a velocity dependence of charges is present, this would lead to an effective charge on the molecule. King [9–11] has shown that within one-part of 10^{20}, the charges of electrons and protons in H_2 are equal. Similar experiments for He, also containing two protons (and two neutrons), but with the protons moving much faster than in H_2 as they are contained in the nucleus, yielded the same result. Hence, these (and other) molecules are neutral, and charge is exactly conserved in special relativity theory and therefore in classical theory.

1.4 Macroscopic and Microscopic Approaches

Although a molecular image of matter nowadays seems to prevail, a *macroscopic description* is rather useful, in science as well as in technology. In the nineteenth century, the debate over whether atoms or molecules exist was still ongoing, and in such a case, one way to make progress is to avoid the issue. The macroscopic approach led to two related branches of science. The first extended particle mechanics to continuous media and is called *continuum mechanics*. The second considered the relationship between work and heat and led to the discipline we now call *thermodynamics*. At present, these two branches are often treated in a highly coupled way called *thermomechanics* [12]. For thermodynamics, the description is given in macroscopic or thermodynamic variables (pressure, volume, temperature, electric field, etc.), which [13]

- are few in number,
- involve no special assumptions concerning the structure of matter,

- are suggested more or less directly by our sensory perceptions, and
- can, in general, be directly measured.

The result is one of the most secure, if not the most secure, parts of science.

Somewhat parallel, a more mechanistic description based on rather simplified models of atoms and molecules, for example, as hard spheres, developed. In this *microscopic description*,

- assumptions are made concerning the structure of matter, for example, the existence of molecules is assumed;
- many quantities must be specified, whereas
- these quantities specified are not suggested by our sense perceptions; and
- they generally cannot be measured directly.

In this picture the few directly measurable properties that constitute the macroscopic description are in reality averages over a period of time of a large number of microscopic characteristics. Both approaches have their pros and cons, but both are useful, albeit typically for rather different types of problems.

1.5 Gases, Liquids, and Solids

Let us now survey the differences and similarities between gases, liquids, and solids, and the associated energetics. Starting with solids, two classes of solids can be distinguished: crystalline and amorphous materials. As is well known, the basic feature of crystalline solids is order. *Crystalline* solids can be further divided into single-crystalline or polycrystalline materials, in both of which a regularly ordered structure exists at the atomic scale (Figure 1.2a). This structure is maintained, at least in principle, throughout the whole material in a *single-crystalline* material, whereas in a *polycrystalline* material, regions of different crystallographic orientations exist. These regions are referred to as *grains*, and the boundaries between them are *grain boundaries*. Studies using X-ray diffraction have clearly revealed the long-range atomic order in these materials. In *amorphous* solids, there is no long-range order (Figure 1.2b), although the local coordination of a specific molecule[2] in the amorphous state may not be that different from the coordination of the same type of molecule in the corresponding crystalline state (if it exists). A *glass* is any thermodynamically nonequilibrium (amorphous or crystalline) solid where a state of order or disorder is frozen in, corresponding to an equilibrium state, which is possible for a higher temperature or any other value of the external parameters. Such a definition thus includes, for example, frozen-in disordered crystals.

From the observation that the structure of a solid is, in essence, maintained with increasing temperature up to the melting point,[3] it follows that the PE U_{pot} is more important than the KE U_{kin} because a high value for the PE more or less immobilizes the molecules in space. Therefore, we have

$$U_{pot} \gg U_{kin} \text{ for solids}$$

2 Although we denote the basic entities as molecules for convenience, the term is generally also supposed to include free atoms and ions as well, whenever appropriate.

3 We "forget" here for convenience of phase transformations (Section 1.6).

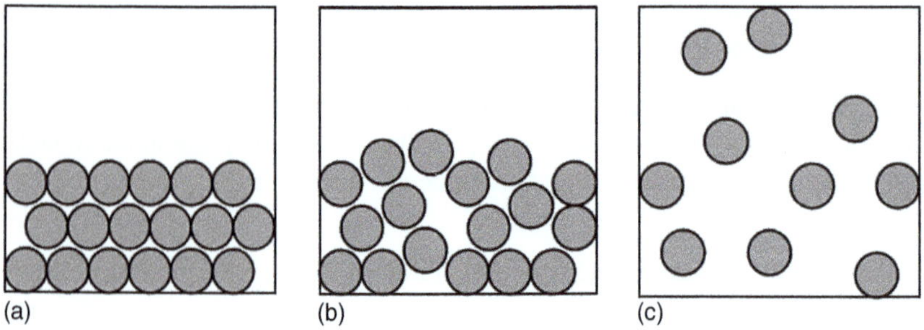

Figure 1.2 Schematic of structure and coordination of (a) a (crystalline) solid, (b) a liquid or amorphous solid, and (c) a gas. While solids and liquids have comparable values of density ρ, thermal expansion coefficient α, and compressibility κ, liquids and gases have more comparable values of viscosity η, shear modulus G, and diffusion coefficient D.

This makes a regular spatial array of molecules the most suitable *reference configuration* for modeling a crystalline solid. This regularity can be described globally by the concept of *lattices* (long-range order) and locally by a well-defined *coordination number* (short-range order). Other aspects, such as the KE of the molecules or defects in the regularity of the structure, can be considered to first order as perturbations of this regularity. This implies that relatively simple models for particular features of the solid state – that is, models that ignore many details – can already describe the physical phenomena of solids reasonably well.

For gases, the molecules move through space almost independently of each other, as evidenced by the wide applicability of the perfect gas law $PV = nRT$, with the pressure P, the volume V, the number of moles n, the gas constant R, and the temperature T. Here, order is nearly absent, and the reference configuration can be described as random (Figure 1.2c). This is exactly the reverse of the situation for a solid. Thus, we conclude that the PE is small as compared with the KE, and we have

$$U_{\text{pot}} \ll U_{\text{kin}} \text{ for gases}$$

The influence of intermolecular interactions can be considered first order as a perturbation, leading to some coordination of molecules with each other, so that also for gases, relatively simple models can provide a good clue for understanding.

Liquids do have some properties akin to those of solids and some other properties more like those of gases; that is, they have a *dual nature*. For example, their thermodynamic properties like density ρ, thermal expansivity α, and compressibility κ are typically not too different from those of the corresponding solid. As a rule of thumb, the specific volume increases by only 5–15% upon melting (water is a well-known exception). On the other hand, fluids (liquids and gases) have fluidity in common, although a liquid has a meniscus, whereas a gas has no such thing, and the viscosity η of liquids is higher than that of gases. This indicates that the movement of molecules in fluids is relatively easy when compared to solids, as also reflected by similar values of the shear modulus G – commonly assumed to be (near) zero – and diffusion coefficient D for liquids and gases. As might be expected, the situation for liquids with respect to energy is somewhere in the middle, and both the PE and the KE play an important role. Therefore, we have

$$U_{\text{pot}} \approx U_{\text{kin}} \text{ for liquids}$$

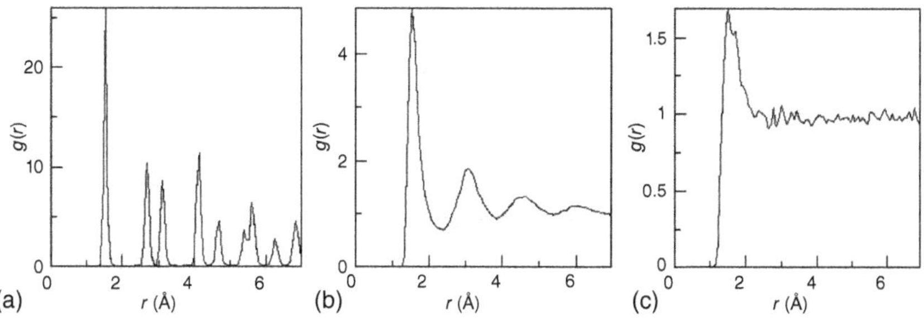

Figure 1.3 The pair correlation function $g(r)$. (a) A solid with a regular array of molecules leading to long-range order and well-defined coordination shells. (b) A liquid with similar density as the solid but having a random dense packing of molecules, leading only to short-range order in which the coordination number is of prime relevance. (c) A gas with a random packing of molecules and some rather limited ordering due to weak mutual attraction and exclusive volume of the molecules.

This implies that neither an ordered nor a fully disordered configuration is present in a liquid. The choice of a reference configuration becomes accordingly (much) more troublesome. Although long-range order is absent, short-range order is present, and the concept of coordination number is still valuable for liquids. However, because of the approximately equal importance of U_{pot} and U_{kin}, relatively simple models of liquids are usually less reliable than those for either solids or gases.

To summarize, while the dominant feature of a solid is order and that of a gas is disorder, the liquid is somewhere in between, both in structural and energetic aspects. In Figure 1.3, their static structure is illustrated by the *pair correlation function* $g(r)$, which describes the probability to find a molecule at a distance r from a reference molecule at the origin, scaled so that its average value is unity. For crystalline solids, only at discrete distances are other molecules present, whereas for gases, the probability of finding another molecule rapidly becomes constant with increasing r. For liquids, the situation is intermediate, as evidenced by some structure in $g(r)$ for small r and the limiting behavior of $g(r) = 1$ for large r. The coordination number is, however, similar to the coordination number in the corresponding solid.

For future reference, we note that often a threefold classification of fluids is made:

- *Simple fluids* are fluids of (more or less) spherical molecules such as Ar, CH_4, and N_2 for which angular dependence of the interaction potential is (almost) absent.
- *Normal fluids* are fluids without a specific interaction such as hydrogen bonding. Examples are C_6H_6, CCl_4, and cyclo-C_6H_{12}.
- *Complex fluids* if these specific interactions are present. Examples are H_2O, NH_3, and CH_3COOH.

Justification 1.1 ***Kinetic Versus Potential Energy*** Let us consider the magnitude of U_{kin} and U_{pot} a bit more quantitatively, and use from thermodynamics that $(\partial U/\partial V)_T = T(\partial P/\partial T)_V - P$ with $U = U_{kin} + U_{pot}$ the (internal) energy. As for a perfect gas $PV_m = RT$ (with $V_m = V/n$) and $U = U_{kin} = 3RT/2$, we have for a perfect gas $(\partial U/\partial V)_T = 0$. For a van der Waals (vdW) fluid (Section 8.2), using the fact that the limit of the vdW energy for $V \to \infty$ approaches the energy for the perfect gas, one obtains, by integrating

Table 1.2 Argon thermodynamic essentials.

State	T (K)	V_m (l mol^{-1})	U_{pot} (kJ mol^{-1})	U_{kin} (kJ mol^{-1})	U (kJ mol^{-1})
Gas	273	22.4	−0.006 16	3.41	3.40
Vapor at T_n	87.3	6.79	−0.0203	1.09	1.07
Fluid at T_{cri}	151	0.0746	−1.85	1.88	0.0325
Liquid at T_n	87.3	0.0287	−4.81	1.09	−3.72
Solid at T_n	87.3	0.0248	−8.40	1.09	−7.31

Source: Adapted from Pryde [14] and Rice [15].

$(\partial U/\partial V)_T$, $U = U_{pot} + U_{kin} = -a/V_m + 3RT/2$, where a is one of the vdW constants, so that $U_{pot} = -a/V_m$ and $U_{kin} = 3RT/2$. Table 1.2 shows the magnitudes of U_{pot} and U_{kin} as calculated from the vdW equation for Ar in various states, where $a = 138.1\ \text{J mol}^{-1}$. The change from $|U_{pot}| > U_{kin}$ for the solid and liquid at the normal boiling temperature T_n via $|U_{pot}| \cong U_{kin}$ for the fluid at the critical temperature T_{cri} to $|U_{pot}| < U_{kin}$ for the gas at any temperature is evident. The experimental value for U_{pot} for the liquid state at T_n is $U_{pot} = -5.84\ \text{kJ mol}^{-1}$.

In the course of history, it also became clear that most substances can manifest themselves, given appropriate conditions, as gas, liquid, and solid, and this phase behavior is schematically shown in Figure 1.4. At sufficiently low temperature and sufficiently high pressure, a vapor (V) condenses to a liquid (L) or a solid (S), and this leads to the coexistence of vapor and liquid (V–L coexistence line), vapor and solid (V–S coexistence line), or liquid and solid (L–S coexistence line). Although, as far as is known, the fusion (L–S) curve continues to exist with increasing temperature,[4] the vaporization (V–L) curve ends at the *critical temperature* T_{cri} or critical point (CP). Above the CP, condensation is impossible, and the vapor is often addressed as gas (G). Moreover, above the critical isotherm, the difference between liquids and gases ceases to exist. Along the line A–CP (Figure 1.4), the liquid is in equilibrium with (an infinitesimal amount of) gas, while along the line B–CP the vapor is in equilibrium with (an infinitesimal amount of) liquid. These lines are conveniently addressed as *saturation lines*. Another special point is the *triple point* (TP, T_{tri}), the intersection of the V–L/V–S and L–S coexistence lines, where the vapor, liquid, and solid phases are in equilibrium. Although the typical values for $P(T_{tri})$ are $\approx 10^{-3}$ bar, those for $P(T_{cri})$ are ≈ 50 bar, but both show a wide range: for example, $P(T_{tri})$ for CO_2 is 5.2 bar and for neo-pentane $\cong 10^{-7}$ bar, and $P(T_{cri})$ for He is 2.3 bar and for Hg $\cong$ 1510 bar. The TP is more sensitive to intermolecular interactions than the CP. To illustrate this, consider the isomers *n*-octane and 2,2,3,3-tetramethylbutane with T_{cri} of 569 and 568 K, but with T_{tri} of 216 and 374 K, respectively. The *normal melting point*, T_{mel}, that is, the melting temperature at 1 bar, is generally only marginally different from T_{tri}, for example, for water, $T_{tri} = 0.01\ °\text{C}$ (at 0.0061 bar) and $T_{mel} = 0\ °\text{C}$ (at 1 bar). Often pseudo-spherical molecules

4 For the most extensively investigated liquid, Ar, no change in the character of the solid–liquid transition is observed up to $P = 12$ kbar at $T = 330$ K [17].

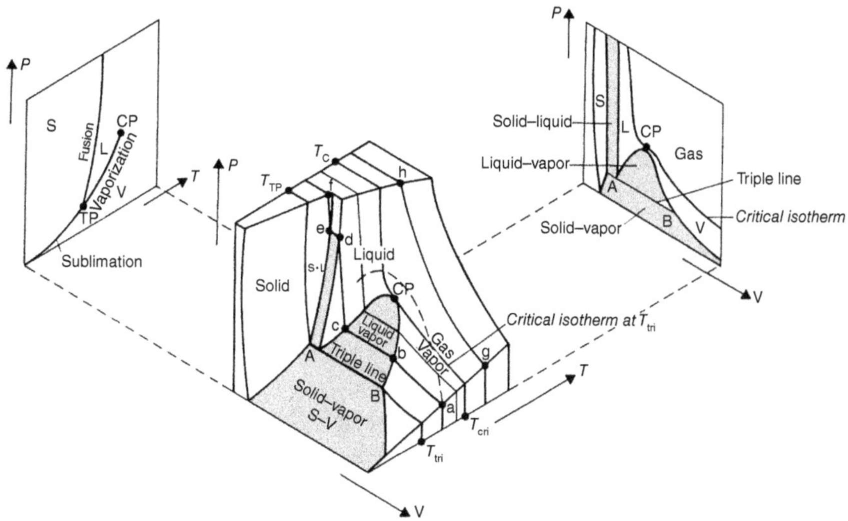

Figure 1.4 Phase relationships for a simple fluid between the solid (S), liquid (L), and vapor (V) phases. The triple point (TP) and critical point (CP) are indicated. As in the figure, some authors use the word "gas" for above-critical and "vapor" for below-critical conditions. Source: Finn [16]/Taylor & Francis.

have an unusually high melting point; for example, for adamantane ($C_{10}H_{16}$), $T_{mel} = 543$ K and for C_2Cl_6, $T_{mel} = 456$ K. Finally, we have the *normal boiling point* T_n, that is, the boiling temperature at 1 bar.

1.6 What to Expect?

As said, we will deal in Part 2 with solids, liquids, and gases and (some of) their properties as such, before entering the arena of transitions in Part 3. But first, to be self-contained as far as possible and using the metaphor of Poincaré as introduced in the Preface, we review in Part 1 the "tools" we need. We discuss in Chapter 2 *classical particle mechanics* (PM) and in Chapter 3 *quantum mechanics* (QM). Both disciplines are based on the discrete (or molecular) nature of substances, that is, on a mechanistic, *microscopic picture*. As an application of this picture, we discuss *intermolecular interactions* in Chapter 4. But before this microscopic picture was fully accepted, a purely phenomenological, *macroscopic picture* already emerged from engineering considerations, ignoring largely, if you are a purist completely, microscopic considerations. Not taking thermal effects into account (apart from the temperature dependence of properties), the approach is normally called *continuum mechanics* (CM) and a brief introduction is given in Chapter 5. Incorporating thermal phenomena led to a field nowadays known as *macroscopic* (or *phenomenological* or *classical*) *thermodynamics* (MT), which is reviewed in Chapter 6, including its extension to *linear irreversible thermodynamics*. Often, an integrated approach to classical and statistical thermodynamics is advocated [18, 19]. In this book, we review thermodynamics as an individual discipline, while its merger with *microscopic* or *statistical thermodynamics* (ST) appears naturally in Chapter 7.

In Part 2, we discuss the “bricks”: the salient aspects of gases, liquids, and solids. In Chapter 8, we overview the macroscopic description of *gases* as given by the various equations of state and outline the microscopic description as embodied in the kinetic theory of gases and the virial approach. For *liquids*, we introduce in Chapter 9 the various approaches used in general, after which we mainly follow a structural approach. There are at least three distinguishable types of *solids* – inorganic, metallic, and polymeric materials – which we describe in Chapter 10. The importance of *interfaces* – the “cement” – warrants a separate discussion (Chapter 11).

In Part 3, we turn to the “house”: the transitions and transformations between phases. Here, we distinguish between *phase transitions* (the change from one phase to another) and *phase transformations* (the structural change within one phase), the latter obviously being most important for solids. We also distinguish between continuous and discontinuous phase transitions. In Chapter 12, we discuss *general aspects*. *Discontinuous phase transitions* are divided into transformations between gases and liquids (Chapter 13) and between liquids and solids (Chapter 14). In Chapter 15, we deal with *continuous phase transitions*, the description of which is an example of an intimate mixture between molecular and mesoscopic arguments. Two phase transformations are discussed: one in the liquid state, for *liquid crystals* (Chapter 16), and one in the solid state, for *ferroelectrics* (Chapter 17), followed by the *glass transition* (Chapter 18). Finally, Chapter 19 deals with *irreversibility* and the *return to equilibrium*. The appendices provide elements of mathematics, special relativity theory, the lattice gas model, electrostatics, and probability, as well as some physical property data and answers to selected problems.

Overall, the focus in Parts 2 and 3 is on structure and thermodynamic aspects. As not all physical properties can be discussed, we opted to deal with dielectric behavior for all three phases to provide some continuity. For the basics, in Part 1, an axiomatic approach is followed. Generally, we refrain from discussing experimental methods and focus on the underlying theoretical concepts. Problems are mostly limited to the somewhat more complex types, ignoring the “plug-in” types. Each chapter contains specific references and a section titled “Further Reading,” referring to a small set of recommended relevant books, that are referred to by author(s) and year.

1.7 Units and Notation

To avoid unnecessary semantic discussions, one better agrees on the meaning of the basic concepts used. This is the purpose of Sections 1.1–1.4. It is also expedient to know what to expect, and this is what is described in Section 1.6. Of course, one also must be able to measure properties quantitatively, and for that, a system of units for quantities is required. Restricting ourselves here to the Système International (SI), there are seven *base units*. The relevant ones for our purpose are for length (*meter*, m), for mass (*kilogram*, kg), for time (*second*, s), for temperature (*kelvin*, K or °C where $t/°\text{C} = T/\text{K} - 273.15$), and for amount of substance (*mole*, mol). Originally, standards were defined for these units, but since 2019, all base units are defined in terms of fundamental (physical) constants. *Derived units* are defined in terms of base units. Some important derived units for our considerations are the unit for force, the *newton* (N) defined as $\text{N} = \text{kg m s}^{-2}$, the unit for pressure, the *pascal* (Pa) defined as $\text{Pa} = \text{N m}^{-2}$, and the unit for energy, the *joule* (J) defined as $\text{J} = \text{N m}$. A brief description of SI units, physical constants, and conversion factors is given in the front matter. This is not to say that other units will not be used, as these are frequently used in the literature. For a short or extensive review of measurement, see Ref. [20] or [21]. For the use of preferred units and symbols in the physical sciences, we refer to [22].

There remains the matter of notation, for which the conventions used should be clear. We use in (macroscopic and microscopic) thermodynamic discussions for summations over entities like molecules and states, a lowercase Latin index, say i or j, and for a summation over chemical components generally (but not always) a lowercase Greek index, say α or β. Furthermore, a superscript * is used for a pure compound, for example, the partial volume V_α^* of component α, and a superscript ° for a reference state, for example, the pressure $P°$, conventionally taken as 1 bar.

With respect to mathematical notation, scalars are addressed via an italic letter, say a, and vectors by an italic bold face letter, say $\boldsymbol{a}$ (direct notation) with components a_i (index notation). Column matrices are also labeled by, say a_i, or by a roman bold face letter, say $\mathbf{a}$ (matrix notation). Similarly, matrices are addressed by an italic letter with two subscripts, say A_{ij}, or by $\mathbf{A}$. The column $\mathbf{a}$ is used as a shorthand for a collective of N quantities, that is, $\mathbf{a} = \mathbf{a}_1\mathbf{a}_2\ldots\mathbf{a}_N$. For example, for N molecules each with coordinates $\boldsymbol{r}_i$, where $\boldsymbol{r}_i = (x_i,y_i,z_i)$, we denote the coordinates collectively by $\mathbf{r} = \boldsymbol{r}_1\boldsymbol{r}_2\ldots\boldsymbol{r}_N = x_1y_1z_1x_2y_2z_2\ldots x_Ny_Nz_N$, or in a multidimensional integral, we write $\int \mathrm{d}\mathbf{r}$, where $\mathrm{d}\mathbf{r} = \mathrm{d}\boldsymbol{r}_1\mathrm{d}\boldsymbol{r}_2\ldots\mathrm{d}\boldsymbol{r}_N = \mathrm{d}x_1\mathrm{d}y_1\mathrm{d}z_1\mathrm{d}x_2\mathrm{d}y_2\mathrm{d}z_2\ldots\mathrm{d}x_N\mathrm{d}y_N\mathrm{d}z_N$. Denoting the set b_i by $\mathbf{b}$ and the set a_i by $\mathbf{a}$, we can therefore write $c = \Sigma_i b_i a_i = \mathbf{b}^\mathrm{T}\mathbf{a}$, using the transpose $\mathbf{b}^\mathrm{T}$ of $\mathbf{b}$. This allows us to write the derivatives of a function $f(a_i)$ given by $b_i = \partial f/\partial a_i$ collectively as $\mathbf{b} = \partial f/\partial \mathbf{a}$ or of a set of functions $f_i(a_j)$ given by $B_{ij} = \partial f_i/\partial a_j$ as $\mathbf{B} = \partial \mathbf{f}/\partial \mathbf{a}$. We, therefore, distinguish between a vector $\boldsymbol{a}$ and its matrix representation $\mathbf{a}$. For a function, the notation $y = y(\mathbf{x})$ is often used, but if clarity is required, we use $y = f(\mathbf{x})$. The inner product c of two vectors $\boldsymbol{a}$ and $\boldsymbol{b}$ is in direct notation $c = \boldsymbol{ab}(= \Sigma_i b_i a_i)$ and in matrix notation $c = \mathbf{a}^\mathrm{T}\mathbf{b}$. As we try to be pragmatic, we use direct, index, and matrix notations, whenever convenient. Generally, direct and matrix notations are the most compact, but in case of possible confusion, index notation, which is unambiguous, is given. Finally, for a reminder of some mathematics, we refer to Appendix A and to the front matter for a summary of the notation used.

References

1 Gusev, A.I. and Valeeva, A.A. (2003). *Phys. Solid State* 45: 1242.
2 Prutton, C.F. and Maron, S.H. (1951). *Fundamental Principles of Physical Chemistry*, Revised edition. New York: MacMillan Company.
3 Bejan, A. (2006). *Advanced Engineering Thermodynamics*, 3e, p. 26–34. Hoboken: John Wiley & Sons.
4 Hecht, E. (2019). *Am. J. Phys.* 87: 495 and references therein.
5 Brecher, K. (1977). *Phys. Rev. Lett.* 39: 1051.
6 Wolf, P., Bize, S., Clairon, A. et al. (2003). *Phys. Rev. Lett.* 90: 060402.
7 Bertozzi, W. (1964). *Am. J. Phys.* 32: 551.
8 Fewell, M.P. (1995). *Am. J. Phys.* 63: 653.
9 King, J.G. (1960). *Phys. Rev. Lett.* 5: 562.
10 Dylla, H.F. and King, J.G. (1973). *Phys. Rev. A* 7: 1224.

11 Marinelli, M. and Morpurgo, G. (1984). *Phys. Lett. B* 137: 439. This paper contains a brief summary on charge conservation.
12 Ziegler, H. (1983). *Introduction to Thermomechanics*, 2e. Amsterdam: North-Holland.
13 Zemansky, M.W. and Dittman, R.H. (1981). *Heat and Thermodynamics*, 6e. Auckland: McGraw-Hill Book Comp.
14 Pryde, J.A. (1966). *The Liquid State*. London: Hutchinson University Library.
15 Rice, O.K. (1946). *J. Chem. Phys.* 14: 321.
16 Finn, C.B.P. (1993). *Thermal Physics*. London: Chapman and Hall.
17 Rowlinson, J.S. and Swinton, F.L. (1982). *Liquids and Liquid Mixtures*, 3e. London: Butterworth.
18 Swendsen, R.H. (2012). *An Introduction to Statistical Mechanics and Thermodynamics*. Oxford: Oxford University Press.
19 Scott Shell, M. (2015). *Thermodynamics and Statistical Mechanics*. Cambridge: Cambridge University Press.
20 Himbert, M.E. (2009). *Eur. Phys. J. Spec. Top.* 172: 25.
21 Treese, S.A. (2018). *History and Measurement of the Base and Derived Units*. Cham, Switzerland: Springer International Publishing.
22 Cohen, E.R., Cvitaš, T., Frey, J.G. et al. (2007). *Quantities, Units and Symbols in Physical Chemistry*, 3e. RSC Publishing.

Further Reading

Berry, R.S., Rice, S.A., and Ross, J.R. (2000). *Physical Chemistry*, 2e. Oxford: Oxford University Press. A comprehensive and high-level modern text on physical chemistry in general.
Eyring, H., Henderson, D., and Jost, W. (ed.) (1971). *Physical Chemistry, an Advanced Treatise*, Vol. I–XI. New York: Academic Press. This set of books contains many useful parts for all aspects of physical chemistry, despite being somewhat dated.
Partington, J.R., *Advanced Treatise on Physical Chemistry*; Vol. 1 (1949). *Fundamental Principles–The Properties of Gases*; Vol. 2 (1951). *The Properties of Liquids*; Vol. 3 (1952). *The Properties of Solids*; Vol. 4 (1953). *Physico-Chemical Optics*; Vol. 5 (1954). *Molecular Spectra and Structure, Dielectrics and Dipole Moments*. London: Longmans. This book series contains a wealth of old data and concepts.

2

Classical Mechanics

The microscopic description of the motion of particles can be done by either classical mechanics or quantum mechanics. In this chapter, a brief review of the first "tool" – classical particle mechanics (PM) based on Hamilton's principle – is provided. The presentation is abstract but efficient and useful in quantum mechanics as well as in statistical thermodynamics [1]. We first introduce the notion of a particle, generalized coordinates, and the basic mechanics principles. Thereafter, we deal with Hamilton's principle, the Lagrange equations, and their consequences. Finally, we introduce Hamilton's equations and an important consequence of all of these, the virial theorem.

2.1 Frames, Particles, and Coordinates

In mechanics we distinguish between *kinematics* – the description of motion and change without considering the cause – and *kinetics* – why things move and change. Here, we discuss kinematics, and in the remaining sections kinetics. To describe phenomena, we need a frame of reference. A basic assumption of nonrelativistic mechanics is that *space–time* is *isotropic* and *homogeneous* with *zero curvature*, that is, it is a three-dimensional (3D) *Euclidean space* with a *universal time*, independent of relative motion. It follows that a so-called *inertial frame* can be found in which the laws of mechanics do not depend explicitly on the position and orientation of the system in space and are independent of the time origin. A freely moving particle (one that experiences no external force) moves with constant velocity in such a frame. This is the *law of inertia*. Thus, any frame I′ moving with a constant velocity $\boldsymbol{V}$ with respect to the inertial frame I is also an inertial frame. Galilei's *Principle of Relativity*[1] (PoR) states that all laws of nature are identical in all inertial frames, that is, the laws are form invariant or *covariant* with respect to transformations of coordinates and time from one inertial frame to another. This suggests the use of coordinate-independent quantities like scalars, vectors, and tensors. Some quantities involved are *covariant quantities*; they have values that do depend on the frame used, for example, position and velocity, but within that frame these values can be used in the same form of law as for any other frame. *Invariant quantities* have values unaffected by relative

1 Almost always if equations, laws, or principles are indicated by their (real or alleged) discoverer, the family name is used, with Galileo Galilei (1564–1642) being an exception. Therefore, I use Galilei's relativity principle instead of the often-used Galileo's relativity principle.

Phases of Matter and their Transitions: Concepts and Principles for Chemists, Physicists, Engineers, and Materials Scientists, First Edition. Gijsbertus de With.

motion, for example, mass and charge. To obtain covariant equations, one applies in classical nonrelativistic physics a *Galilei transformation*, in which spatial positions are considered relative, but time difference is a parameter identical to all observers. This implies that a particle at position $\boldsymbol{r}$ at time t in frame I can also be described by $\boldsymbol{r}'$ at t' in I′ with $\boldsymbol{r} = \mathbb{R} \cdot \boldsymbol{r}' + \boldsymbol{V}t + \boldsymbol{r}_0$ and $t = t' + t_0$, with $\mathbb{R}$ a constant orthogonal tensor ($\mathbb{R} : \mathbb{R} = \mathbb{1}$), $\boldsymbol{V}$ and $\boldsymbol{r}_0$ constant vectors, and t_0 a scalar constant. It also implies that interactions are instantaneous.

A (*material*) *particle* is considered to be a point mass without (accessible) internal structure that can move in space. Obviously, this idealization can only be used to advantage if the dimensions of the entity are small as compared to the overall size of the system studied and the internal structure of the entity is not changing.

The (dynamical) *state* of such a particle appears to be determined by its Cartesian *position vector* $\boldsymbol{r}$ and the rate of change of $\boldsymbol{r}$, that is, the *velocity vector* $\boldsymbol{v} = \mathrm{d}\boldsymbol{r}/\mathrm{d}t \equiv \dot{\boldsymbol{r}}$. For N particles, we need to specify N position vectors, collectively denoted by $\mathbf{r}^{\mathrm{T}} = (\boldsymbol{r}_1, \boldsymbol{r}_2, \ldots, \boldsymbol{r}_N)$, and we have $3N$ coordinates.[2] Similarly, $\mathbf{v}^{\mathrm{T}} = (\boldsymbol{v}_1, \boldsymbol{v}_2, \ldots, \boldsymbol{v}_N)$. The number of independent coordinates necessary to specify uniquely the position of all particles in the system is called the number of *Degrees of Freedom* (DoFs). For a system of N particles with constraints, the number of DoFs does not equal $3N$ and changes in, for example, velocity $\mathrm{d}\boldsymbol{v}_j$, are not independent. *Holonomic constraints* depending only on $\mathbf{r}$ and t but not on $\dot{\mathbf{r}}$ and expressed by $k = 3N–M$ equations ($M < 3N$)

$$C_k(\mathbf{r}, t) = C_k(\boldsymbol{r}_1, \ldots, \boldsymbol{r}_j, \ldots, t) = 0$$

can be taken into account by using Lagrange multipliers. If done so, the system itself is called *holonomic*. We do not discuss explicitly constraints for which we refer to the literature. Within an abstract framework we find it therefore convenient to describe the motion of systems of N particles with N *generalized coordinates* $\mathbf{q}$, free of internal constraints, and their associated *generalized velocities* $\dot{\mathbf{q}}$. Similar to Cartesian coordinates, the coordinates $\mathbf{q}$ are used to describe the instantaneous configuration of the system, while the velocities $\dot{\mathbf{q}}$ specify the instantaneous motion of the system. Specifying $\mathbf{q}$ and $\dot{\mathbf{q}}$ simultaneously determines the state of the system completely.

In general, the Cartesian coordinates $\mathbf{r}$ are nonlinear functions of the generalized coordinates $\mathbf{q}$, which can be expressed in matrix notation as

$$\mathbf{r} = \mathbf{r}(\mathbf{q}, t) \quad \text{or} \quad \mathrm{d}\mathbf{r} = \frac{\partial \mathbf{r}}{\partial \mathbf{q}}\mathrm{d}\mathbf{q} \equiv \mathbf{A}\mathrm{d}\mathbf{q} + \frac{\partial \mathbf{r}}{\partial t}$$

where $\mathbf{r}^{\mathrm{T}} = [x_1, y_1, z_1, x_2, \ldots, z_3, t]$. Note that the matrix $\mathbf{A}$ is orthogonal, that is, $\mathbf{A}^{-1} = \mathbf{A}^{\mathrm{T}}$. For the velocity $\mathbf{v}$ we obtain

$$\mathbf{v} = \dot{\mathbf{r}} = \frac{\mathrm{d}\mathbf{r}}{\mathrm{d}t} = \frac{\partial \mathbf{r}}{\partial \mathbf{q}}\frac{\mathrm{d}\mathbf{q}}{\mathrm{d}t} = \mathbf{A}\dot{\mathbf{q}} \quad \text{so that} \quad \mathbf{A} = \frac{\partial \mathbf{r}}{\partial \mathbf{q}} = \frac{\partial \dot{\mathbf{r}}}{\partial \dot{\mathbf{q}}}$$

If $C_k = C_k(\mathbf{r})$ and $\mathbf{r} = \mathbf{r}(\mathbf{q})$, that is, if both functions are explicitly independent of time, the system is called *scleronomic*; otherwise, it is called *rheonomic*.

Example 2.1 The H_2O molecule is nonlinear with an O—H bond length of 0.0958 nm and an included angle of ≅105°. Since the molecule has $N = 3$ atoms, we have $3N = 9$ coordinates. We can describe the geometry of the molecule using as coordinates the bond lengths

2 Note that $\boldsymbol{r}$ denotes a vector and $\mathbf{r}$ is a (column) matrix. So, for N particles we have $\mathbf{r}^{\mathrm{T}} = (\boldsymbol{r}_1, \boldsymbol{r}_2, \ldots, \boldsymbol{r}_N)$.

r_1 and r_2 and the bond angle ϕ. Of the $3N-3$ other coordinates, three refer to translation and three to the rotation of the molecule as a whole.

2.2 From Newton to Hamilton

The basis of classical mechanics is formed by what we now call Newton's laws, formulated by him in his *Philosophiæ Naturalis Principia Mathematica* (1687). We consider a system of interacting particles and denote with a symbol with no subscript the quantity for the system, while a subscript j indicates that quantity for a specific particle j. For example, $\boldsymbol{f}$ denotes the total force on the system and m_1 is the mass of particle 1. The notation $\boldsymbol{f}_{12}$ indicates the force on particle 1 exerted by particle 2. Summarizing, in modern terms *Newton's laws* read [2]

- If $\boldsymbol{f} = 0$, $\boldsymbol{v} = \text{constant}$
- $\boldsymbol{f} = \mathrm{d}(m\boldsymbol{v})/\mathrm{d}t \equiv \dot{\boldsymbol{p}}$
- $\boldsymbol{f}_{12} + \boldsymbol{f}_{21} = 0$

Although called laws, they are essentially a summary of a set of ideas comprising a postulate, an experimental summary, and a physical assumption. We introduce the laws and some follow-up here, mainly as background to Hamilton's principle.

Newton's laws immediately lead to three consequences for an isolated system, namely the conservation of *linear momentum* $\boldsymbol{p} = m\boldsymbol{v}$, of *angular momentum* $\boldsymbol{l} = \boldsymbol{r} \times \boldsymbol{p}$ and, if the force $\boldsymbol{f}$ is *conservative*, that is, related as $\boldsymbol{f} = -\nabla\Phi(\boldsymbol{r})$ to a *potential energy* Φ, the conservation of energy $E = \vartheta + \Phi$ with $\vartheta = \frac{1}{2}mv^2$ the *kinetic energy*. In the latter case, the system is also called *conservative*. The derivation is left as an exercise. Here, we only comment on a few general aspects. First, we note that in case we have $\mathrm{d}\boldsymbol{p}/\mathrm{d}t = 0$ and $\mathrm{d}\boldsymbol{l}/\mathrm{d}t = 0$, the equations for conservation of linear and angular momenta become the force equilibrium $\boldsymbol{f} = 0$ and the moment (torque) equilibrium $\boldsymbol{r} \times \boldsymbol{f} = 0$, respectively. Second, in nonrelativistic PM, m is normally constant and $\mathrm{d}(m\boldsymbol{v})/\mathrm{d}t = m\mathrm{d}\boldsymbol{v}/\mathrm{d}t = m\boldsymbol{a} = \boldsymbol{f}$, an expression first given by Euler[3] (1752). Although for constant m the first law follows from the second law, its essence is that it expresses the existence of the law of inertia. The second law is basically a summary of experimental facts. The third law is an assumption (not generally true) and implies that the forces are equal, oppositely directed, and lie along a line joining the particles. It is implied that the force for one particle is the sum of the interactions with all other particles. Third, the laws introduce the concepts of force and mass, as discussed in Chapter 1. It may be useful to recall that a force is an interaction that results in a change of state, here acceleration, of the body to which it is applied and noting that the mass of a particle is a numerical measure of its reluctance to being accelerated. The third law provides a way to measure mass. Consider just two particles 1 and 2, interacting only with each other, so that $\boldsymbol{f}_{12} + \boldsymbol{f}_{21} = 0$. Equivalently, $m_2\boldsymbol{a}_2 + m_1\boldsymbol{a}_1 = 0$ or $m_2/m_1 = -\boldsymbol{a}_1/\boldsymbol{a}_2$. If we take the mass of particle 1 as unit mass, the mass of particle 2 can be expressed as a multiple of m_1.

Further, we note that Newton in his *Principia* also formulated the *law of gravity*, nowadays reading $f = GmM/r^2$, where m is the mass of the particle under consideration and M is the mass of the earth. The gravity force f is directed along the connection line between the centers of mass of the earth and the particle located at distance r from each other. From the determination of the mean density of the Earth ($5.48\,\mathrm{g\,cm^{-3}}$) by Cavendish in 1798, the

3 This seems straightforward but remember that the mathematics was developed concurrently.

gravitation constant $G \cong 6.7 \times 10^{-11}$ m^3 kg^{-1} s^{-2} can be calculated, as first done by Cornu and Baille[4] in 1873. The presently accepted value $G = 6.674 \times 10^{-11}$ m^3 kg^{-1} s^{-2}. This raises the question whether the gravitational mass (as used in the gravitation law) is the same as the inertial mass (as used in the second law). Newton assumed so and his assumption, named the *equivalence principle*, has been confirmed experimentally with increasing accuracy over the years, recently up to 1 in 10^{14} [4].

Almost always, the most economic description of physical laws is in the form of an extremum principle, and this is also the case with classical mechanics. An outline of the train of thoughts on how to reach such a principle is now given. In statics, for a system of particles, mechanical equilibrium is described by $\boldsymbol{f} = 0$, where $\boldsymbol{f} = \Sigma_j \boldsymbol{f}_j$ denotes the sum of all forces operating on a particle j. Now consider first the forces. Apart from the *external* and *internal forces*, here together labeled $\boldsymbol{f}_j^{(a)}$, some constraints for the constituting particles can be present, represented by *constraint* or *reaction forces* $\boldsymbol{f}_j^{(c)}$. In that case we have in static equilibrium $\boldsymbol{f} = \boldsymbol{f}^{(a)} + \boldsymbol{f}^{(c)} = 0$ or $\boldsymbol{f}^{(a)} \equiv \Sigma_j \boldsymbol{f}_j^{(a)} = -\Sigma_j \boldsymbol{f}_j^{(c)} \equiv -\boldsymbol{f}^{(c)}$. Generally, if we displace particles by $\mathrm{d}\boldsymbol{r}_j$, the (real) work done on the system is $\mathbf{f}^{\mathrm{T}}\mathrm{d}\mathbf{r} = \Sigma_j \boldsymbol{f}_j \cdot \mathrm{d}\boldsymbol{r}_j$. Using the calculus of variations, we can also define for an arbitrary, small, and momentary variation of displacement $\delta\mathbf{r}$, consistent with the forces and constraints, the *virtual work* $\mathbf{f}^{\mathrm{T}}\delta\mathbf{r}$. Obviously, for a system in static equilibrium we have $\mathbf{f}^{\mathrm{T}}\delta\mathbf{r} = 0$ because the $\delta\mathbf{r}$ are arbitrary. Often, we can assume that, to maintain the constraints, the constraint forces perform no work, and if we do assume workless constraints, we have $(\mathbf{f}^{(c)})^{\mathrm{T}}\delta\mathbf{r} = 0$, resulting in the *Principle of Virtual Work* (PoVW) $(\mathbf{f}^{(a)})^{\mathrm{T}}\delta\mathbf{r} = 0$: a static system with workless constraints is in equilibrium if no virtual work is done by the forces.

For dynamics, the force equilibrium $\boldsymbol{f} = \Sigma_j \boldsymbol{f}_j = 0$ becomes $\Sigma_j(\boldsymbol{f}_j - m_j\boldsymbol{a}_j) = 0$. The acceleration term $-m_j\boldsymbol{a}_j$ is also taken to be a force and called the *inertial force*. Using the same argumentation as mentioned earlier, we obtain *d'Alembert's principle* (1743)

$$\textstyle\sum_j \left(\boldsymbol{f}_j^{(a)} - m_j\boldsymbol{a}_j\right) \cdot \delta\boldsymbol{r}_j = 0$$

as formulated by Lagrange (1788). In this way the PoVW applies to dynamic systems as well. This is not yet an extremum principle, but a further step can be made for conservative forces with $\boldsymbol{f}_j = -\partial\Phi/\partial\boldsymbol{r}_j$ for which we can write $\Sigma_j \boldsymbol{f}_j \cdot \delta\boldsymbol{r}_j = -\delta\Phi$. However, this cannot be done for the acceleration term. To arrive nevertheless at a single principle, we integrate d'Alembert's expression over time. Using partial integration for the acceleration term and $\delta\dot{\mathbf{r}} = \mathrm{d}(\delta\mathbf{r})/\mathrm{d}t$, we have

$$\int_{t_1}^{t_2} m_j\ddot{\boldsymbol{r}}_j \cdot \delta\boldsymbol{r}_j\,\mathrm{d}t = m_j\dot{\boldsymbol{r}}_j\delta\boldsymbol{r}_j\Big|_{t_1}^{t_2} - \int_{t_1}^{t_2} m_j\dot{\boldsymbol{r}}_j \cdot \delta\dot{\boldsymbol{r}}_j\,\mathrm{d}t = -\int_{t_1}^{t_2} \delta\vartheta_j\,\mathrm{d}t$$

with $\vartheta_j = \tfrac{1}{2}m\dot{\boldsymbol{r}}_j \cdot \dot{\boldsymbol{r}}_j$. Hence, with $\vartheta = \Sigma_j\vartheta_j$, the result is *Hamilton's principle* (1834)

$$\int_{t_1}^{t_2} \delta\vartheta\,\mathrm{d}t - \int_{t_1}^{t_2} \delta\Phi\,\mathrm{d}t = \delta\int_{t_1}^{t_2} (\vartheta - \Phi)\,\mathrm{d}t = \delta\int_{t_1}^{t_2} \mathscr{L}\,\mathrm{d}t = \delta S = 0 \tag{2.1}$$

where $\mathscr{L}$ is the *Lagrange function* and S the *action*. This is a true extremum principle, which states that the actual solution for a mechanical problem is the one for which the action is stationary, which in practice is often minimal. It is the most used formulation for the mechanics of holonomic systems comprising discrete particles.

4 So, Cavendish did *not* measure G. For the history of G, see [3].

More formulations along these lines are possible. Just consider for the moment a non-holonomic constraint given by $\mathbf{dr} = \mathbf{x}^{\mathrm{T}}\mathbf{dq} + \mathbf{y}\mathrm{d}t$ with $\mathbf{x} = \mathbf{x}(\mathbf{q},t)$ and $\mathbf{y} = \mathbf{y}(\mathbf{q},t)$. This expression would be integrable if it were a holonomic constraint. It will be evident that the variation of $\mathbf{r}$ (which implies $\delta t = 0$) is $\delta\mathbf{r} = \mathbf{x}^{\mathrm{T}}\delta\mathbf{q}$. Now consider that $\dot{\mathbf{r}} = \mathbf{x}^{\mathrm{T}}\dot{\mathbf{q}} + \mathbf{y}$ and consider another variation $\delta_1\dot{\mathbf{r}}$, defined for fixed time and configuration and which thus implies $\delta t = 0$ and $\delta\mathbf{r} = 0$. Evidently, $\delta_1\dot{\mathbf{r}} = \mathbf{x}^{\mathrm{T}}\delta\dot{\mathbf{q}}$ and thus we have

$$\sum_j \left(\boldsymbol{f}_j^{(\mathrm{a})} - m_j\boldsymbol{a}_j\right) \cdot \delta_1\dot{\boldsymbol{r}}_j = 0$$

This is the *Principle of Virtual Power* (PoVP) or *Jordain's principle* (1909). We will not pursue this topic further but refer to Pars [5] for details. The PoVP will be employed, though, in Chapter 5, dealing with continuum mechanics.

Problem 2.1 ***Conservation of Linear and Angular Momenta, and Energy*** Derive that for an isolated particle system, $\dot{\boldsymbol{p}} = 0$, $\dot{\boldsymbol{l}} = 0$, and $\dot{E} = 0$.

2.3 Hamilton's Principle and Lagrange's Equations

As indicated, Hamilton's principle [6, 7], which we take as the basic equation for arbitrary holonomic systems, states that, using a certain function called the *Lagrange function* $\mathscr{L}$ (and to which we come in some detail later), the *action* S is defined by

$$S \equiv \int_{t_1}^{t_2} \mathscr{L}\,\mathrm{d}t = \int_{t_1}^{t_2} \mathscr{L}(\mathbf{q}, \dot{\mathbf{q}}, t)\,\mathrm{d}t \tag{2.2}$$

has an extremum (is minimal) for the actual path of $\mathbf{q}(t)$, given $\mathbf{q}(t_1)$ and $\mathbf{q}(t_2)$ at times t_1 and t_2. Since $\mathscr{L}$ is a scalar, it is immaterial in which coordinates it is expressed. In particular, it may be expressed in generalized coordinates as done here.

Evaluating the action between fixed times t_1 and t_2, using variation calculus with as variables $\mathbf{q}$ and $\dot{\mathbf{q}}$ (Section A.8), results in $\delta S = S(\mathbf{q} + \delta\mathbf{q}, \dot{\mathbf{q}} + \delta\dot{\mathbf{q}}, t) - S(\mathbf{q}, \dot{\mathbf{q}}, t) = 0$ or

$$\delta S = \delta \int_{t_1}^{t_2} \mathscr{L}(\mathbf{q}, \dot{\mathbf{q}}, t)\,\mathrm{d}t = \int_{t_1}^{t_2} \delta\mathscr{L}(\mathbf{q}, \dot{\mathbf{q}}, t)\,\mathrm{d}t = \int_{t_1}^{t_2} \left[\left(\frac{\partial\mathscr{L}}{\partial\mathbf{q}}\right)^{\mathrm{T}} \delta\mathbf{q} + \left(\frac{\partial\mathscr{L}}{\partial\dot{\mathbf{q}}}\right)^{\mathrm{T}} \delta\dot{\mathbf{q}}\right] \mathrm{d}t = 0$$

Partial integration of the second term of this expression using $\delta\dot{\mathbf{q}} = \mathrm{d}(\delta\mathbf{q})/\mathrm{d}t$ yields

$$\int_{t_1}^{t_2} \left(\frac{\partial\mathscr{L}}{\partial\dot{\mathbf{q}}}\right)^{\mathrm{T}} \frac{\mathrm{d}}{\mathrm{d}t}\delta\mathbf{q}\,\mathrm{d}t = \left(\frac{\partial\mathscr{L}}{\partial\dot{\mathbf{q}}}\right)^{\mathrm{T}} \delta\mathbf{q}\Big|_{t_1}^{t_2} - \int_{t_1}^{t_2} \frac{\mathrm{d}}{\mathrm{d}t}\left(\frac{\partial\mathscr{L}}{\partial\dot{\mathbf{q}}}\right)^{\mathrm{T}} \delta\mathbf{q}\,\mathrm{d}t$$

The boundary term vanishes since $\delta\mathbf{q}$ vanishes at the boundary. Combining results in

$$\delta S = \int_{t_1}^{t_2} \left(\frac{\partial\mathscr{L}}{\partial\mathbf{q}} - \frac{\mathrm{d}}{\mathrm{d}t}\frac{\partial\mathscr{L}}{\partial\dot{\mathbf{q}}}\right)^{\mathrm{T}} \delta\mathbf{q}\,\mathrm{d}t = 0 \quad \text{or} \quad \frac{\mathrm{d}}{\mathrm{d}t}\frac{\partial\mathscr{L}}{\partial\dot{\mathbf{q}}} - \frac{\partial\mathscr{L}}{\partial\mathbf{q}} = 0 \tag{2.3}$$

where the last step can be made since the variations $\delta\mathbf{q}$ are arbitrary. These so-called *Lagrange equations (of motion)*, given the Lagrange function $\mathscr{L}$, describe the behavior of a holonomic system. Using this description, $\mathscr{L}$ is defined to within an additive constant and a multiplicative constant, but also within a total time derivative of any function $f(\mathbf{q},t)$, as $\int_{t_1}^{t_2}(\mathrm{d}f/\mathrm{d}t)\,\mathrm{d}t = f(\mathbf{q}, t_2) - f(\mathbf{q}, t_1)$, which gives zero on variation [7].

For a free (or noninteracting) particle, it follows from the homogeneity of space that $\mathscr{L}$ depends only on the velocity $\boldsymbol{v} = \dot{\boldsymbol{r}}$, that is, $\mathscr{L} = \mathscr{L}(\boldsymbol{v})$, while from the isotropy it follows that $\mathscr{L}$ depends only on the magnitude of $\boldsymbol{v}$, that is, $\mathscr{L} = \mathscr{L}(v^2)$. Hence $\mathscr{L} \sim v^2 = |\boldsymbol{v}\cdot\boldsymbol{v}|$ or $\mathscr{L} = ½mv^2$, representing the kinetic energy, and where $\boldsymbol{v}$ is the velocity of the particle with respect to the inertial frame and the proportionality factor m is the *mass*. For a set of free particles, $\mathscr{L}$ represents the total *kinetic energy* ϑ so that $\mathscr{L} = \vartheta = ½\Sigma_j m_j v_j{}^2 = ½\mathbf{v}^{\mathrm{T}}\mathbf{M}'\mathbf{v}$, where the mass matrix $\mathbf{M}'$ is defined by

$$\mathbf{M}' = \begin{pmatrix} m_1 & 0 & 0 & 0 & .. & 0 \\ 0 & m_1 & 0 & 0 & .. & 0 \\ 0 & 0 & m_1 & 0 & .. & 0 \\ 0 & 0 & 0 & m_2 & .. & 0 \\ .. & .. & .. & .. & .. & 0 \\ 0 & 0 & 0 & 0 & 0 & m_N \end{pmatrix} \tag{2.4}$$

Transforming to generalized coordinates, the kinetic energy ϑ can be expressed as

$$\vartheta = ½\mathbf{v}^{\mathrm{T}}\mathbf{M}'\mathbf{v} = ½\dot{\mathbf{q}}^{\mathrm{T}}\mathbf{A}^{\mathrm{T}}\mathbf{M}'\mathbf{A}\dot{\mathbf{q}} \equiv ½\dot{\mathbf{q}}^{\mathrm{T}}\mathbf{M}(\mathbf{q})\dot{\mathbf{q}}$$

with $\mathbf{M} = \mathbf{A}^{\mathrm{T}}\mathbf{M}'\mathbf{A}$, the generalized mass matrix. In general, therefore, ϑ must be considered as a function of $\mathbf{q}$ and $\dot{\mathbf{q}}$, that is, $\vartheta = \vartheta(\mathbf{q}, \dot{\mathbf{q}})$.

Consider now a system of particles A where the particles do interact with each other but not with other bodies, as described by a potential energy function $\Phi(\mathbf{q}_{\mathrm{A}})$. In this case, to balance kinetic and potential energy, we take $\mathscr{L} = \vartheta(\mathbf{q}_{\mathrm{A}}, \dot{\mathbf{q}}_{\mathrm{A}}) - \Phi(\mathbf{q}_{\mathrm{A}})$. If interaction with the environment is allowed, we use a compound system composed of the original system A plus a system B having a given motion, which as compound system is not interacting with the outside. So we have $\mathscr{L} = \vartheta_{\mathrm{A}}(\mathbf{q}_{\mathrm{A}}, \dot{\mathbf{q}}_{\mathrm{A}}) + \vartheta_{\mathrm{B}}(\mathbf{q}_{\mathrm{B}}, \dot{\mathbf{q}}_{\mathrm{B}}) - \Phi(\mathbf{q}_{\mathrm{A}},\mathbf{q}_{\mathrm{B}})$. Since the $\mathbf{q}$s are given as functions of time t and omitting ϑ_{B} as this is a total derivative of a function of t, we obtain $\mathscr{L} = \vartheta_{\mathrm{A}}(\mathbf{q}_{\mathrm{A}}, \dot{\mathbf{q}}_{\mathrm{A}}) - \Phi[\mathbf{q}_{\mathrm{A}},\mathbf{q}_{\mathrm{B}}(t)]$. Therefore, $\mathscr{L} = \vartheta - \Phi$ remains valid, with as only change that Φ has become time dependent and the system is no longer conservative. For a conservative system, using as generalized coordinates Cartesian coordinates, substitution in the Lagrange equations results in matrix and direct notations in, respectively,

$$\frac{\mathrm{d}\mathbf{M}'\mathbf{v}}{\mathrm{d}t} = \mathbf{M}'\frac{\mathrm{d}\mathbf{v}}{\mathrm{d}t} = -\frac{\mathrm{d}\Phi}{\mathrm{d}\mathbf{r}} = -\nabla\Phi \equiv \mathbf{f} \quad \text{and} \quad \frac{\mathrm{d}\boldsymbol{p}}{\mathrm{d}t} = \sum\nolimits_j \frac{\mathrm{d}\boldsymbol{p}_j}{\mathrm{d}t} = \sum\nolimits_j \boldsymbol{f}_j \equiv \boldsymbol{f}$$

Here, $\boldsymbol{f}$ is the *force*, $\boldsymbol{f}_j$ is the force on particle j, $\boldsymbol{p}$ is the *linear momentum*, and $\boldsymbol{p}_j = m_j\boldsymbol{v}_j$ is the linear momentum of particle j, while $\mathrm{d}\boldsymbol{p}/\mathrm{d}t$ is addressed as the *inertial force*. We thus regained *Newton's second law*: a system of particles is in equilibrium if the force $\boldsymbol{f}_j$ on each particle j is equal to its rate of change of linear momentum $\mathrm{d}\boldsymbol{p}_j/\mathrm{d}t$. From this result, we infer *Newton's first law*: $\mathbf{v}$ = constant if $\mathbf{f} = 0$. Hence, for a conservative system, the force $\mathbf{f} = -\nabla\Phi(\mathbf{r})$ depends only on $\mathbf{r}$ and the work done when moving a particle from one point to another is independent of the path taken.

If we introduce the *generalized force* $\mathbf{Q} = -\partial\Phi/\partial\mathbf{q}$ and $\mathbf{f}$ is conservative, $\mathbf{Q}$ is also conservative because

$$\mathbf{f} = -\frac{\partial\Phi}{\partial\mathbf{r}} = \left(\frac{\partial\mathbf{q}}{\partial\mathbf{r}}\right)^{\mathrm{T}}\left(-\frac{\partial\Phi}{\partial\mathbf{q}}\right) \equiv (\mathbf{A}^{-1})^{\mathrm{T}}\mathbf{Q} = \mathbf{A}\mathbf{Q} \quad \text{or} \quad \mathbf{Q} = \mathbf{A}^{-1}\mathbf{f} = \mathbf{A}^{\mathrm{T}}\mathbf{f}$$

Using the *Lagrange function* $\mathscr{L} = \vartheta - \Phi$ and realizing that $\partial\Phi/\partial\dot{\mathbf{q}} = 0$, we may write

$$\frac{\mathrm{d}}{\mathrm{d}t}\left(\frac{\partial\vartheta}{\partial\dot{\mathbf{q}}}\right) + \frac{\partial\Phi}{\partial\mathbf{q}} = 0 \quad \text{or equivalently} \quad \frac{\mathrm{d}}{\mathrm{d}t}\left(\frac{\partial\vartheta}{\partial\dot{q}_j}\right) + \frac{\partial\Phi}{\partial q_j} = 0 \tag{2.5}$$

which are *Lagrange's equations of motion* for a conservative system describing a system of N particles with $3N$ second-order equations.

Example 2.2 ***The Harmonic Oscillator*** In a harmonic oscillator, a force f acts on a mass m which is linearly related to the extension x, that is, $f = -kx$, where k is the spring constant. The force f can be obtained from the potential energy $\Phi = \frac{1}{2}kx^2$ via $f = -\partial\Phi/\partial x$. If the momentum and velocity are given by $p = mv$ and v, respectively, the kinetic energy $\vartheta = p^2/2m = \frac{1}{2}mv^2$ and Newton's second law reads $f = \dot{p} = \mathrm{d}(mv)/\mathrm{d}t = m\ddot{x}$. Combining leads to $m\ddot{x} = -kx$. If we take as generalized coordinate the Cartesian coordinate, the Lagrange function reads $\mathscr{L} = \vartheta - \Phi = \frac{1}{2}mv^2 - \frac{1}{2}kx^2$. Inserting $\mathrm{d}(\partial\mathscr{L}/\partial\dot{x})/\mathrm{d}t = m\ddot{x}$ and $\partial\mathscr{L}/\partial x = -kx$ in the Lagrange equations $\mathrm{d}(\partial\mathscr{L}/\partial\dot{x})/\mathrm{d}t - (\partial\mathscr{L}/\partial x) = 0$, yields again $m\ddot{x} = -kx$. Defining the circular frequency $\omega = (k/m)^{1/2}$, the solution of this differential equation is $x = x_0\exp(-\mathrm{i}\omega t)$ (or, equivalently, $x = x_0\cos(\omega t + \phi_0)$ with x_0 the *amplitude* and ϕ_0 the *phase*).

2.4 Conservation Laws

In principle, there exist $3N - 1$ functions of $\mathbf{q}$ and $\dot{\mathbf{q}}$, called *constants* (or *integrals*) *of the motion*, whose values remain constant during the motion. The most important and only ones that are additive over subsystems can be derived from the isotropy and homogeneity of space and time.

Let us consider *homogeneity of time*. Due to this homogeneity, the Lagrange function of a closed system is explicitly independent of time. Hence, $\mathscr{L} = \mathscr{L}(\mathbf{q}, \dot{\mathbf{q}})$ and

$$\frac{\mathrm{d}\mathscr{L}}{\mathrm{d}t} = \left(\frac{\partial\mathscr{L}}{\partial\mathbf{q}}\right)^{\mathrm{T}}\dot{\mathbf{q}} + \left(\frac{\partial\mathscr{L}}{\partial\dot{\mathbf{q}}}\right)^{\mathrm{T}}\ddot{\mathbf{q}} = \left(\frac{\mathrm{d}}{\mathrm{d}t}\frac{\partial\mathscr{L}}{\partial\dot{\mathbf{q}}}\right)^{\mathrm{T}}\dot{\mathbf{q}} + \left(\frac{\partial\mathscr{L}}{\partial\dot{\mathbf{q}}}\right)^{\mathrm{T}}\ddot{\mathbf{q}} = \frac{\mathrm{d}}{\mathrm{d}t}\left[\left(\frac{\partial\mathscr{L}}{\partial\dot{\mathbf{q}}}\right)^{\mathrm{T}}\dot{\mathbf{q}}\right]$$

where in the second step the Lagrange equations are used. So, we obtain

$$\frac{\mathrm{d}}{\mathrm{d}t}\left[\left(\frac{\partial\mathscr{L}}{\partial\dot{\mathbf{q}}}\right)^{\mathrm{T}}\dot{\mathbf{q}} - \mathscr{L}\right] = 0 \quad \text{or} \quad \left(\frac{\partial\mathscr{L}}{\partial\dot{\mathbf{q}}}\right)^{\mathrm{T}}\dot{\mathbf{q}} - \mathscr{L} \equiv E$$

We thus have as one constant of the motion E, called the (*canonical*) *energy* of the system. This relation is also valid for systems in a constant external field since in that case $\partial\mathscr{L}/\partial t = 0$ is still valid. Because $\mathscr{L} = \vartheta(\mathbf{q},\dot{\mathbf{q}}) - \Phi(\mathbf{q})$ with ϑ a homogeneous quadratic function of $\dot{\mathbf{q}}$, Euler's theorem for homogenous functions results in

$$\left(\frac{\partial\mathscr{L}}{\partial\dot{\mathbf{q}}}\right)^{\mathrm{T}}\dot{\mathbf{q}} = \left(\frac{\partial\vartheta}{\partial\dot{\mathbf{q}}}\right)^{\mathrm{T}}\dot{\mathbf{q}} = 2\vartheta \quad \text{or} \quad E = \vartheta(\mathbf{q}, \dot{\mathbf{q}}) + \Phi(\mathbf{q})$$

Consequently, for these systems, the canonical energy E equals the *mechanical energy*, that is, the sum of ϑ and Φ, and such systems are called *conservative*.

Let us now consider *homogeneity of space* and apply an infinitesimal displacement d$\boldsymbol{\varepsilon}$ to the particles. For a parallel displacement, where every particle is moved by the same amount d$\boldsymbol{\varepsilon}$, the variation in $\mathcal{L}$ should be zero and is given by

$$\delta\mathcal{L} = \left(\frac{\partial\mathcal{L}}{\partial\mathbf{r}}\right)^{\mathrm{T}}\delta\mathbf{r} = \left(\frac{\partial\mathcal{L}}{\partial\mathbf{r}}\right)^{\mathrm{T}}\mathrm{d}\boldsymbol{\varepsilon} = 0$$

Because d$\boldsymbol{\varepsilon}$ is arbitrary, $\partial\mathcal{L}/\partial\mathbf{r} = 0$ and, using the Lagrange equations (2.3), the result is

$$\frac{\partial\mathcal{L}}{\partial\mathbf{r}} = \frac{\mathrm{d}}{\mathrm{d}t}\frac{\partial\mathcal{L}}{\partial\dot{\mathbf{r}}} = 0$$

Hence, for a closed system, the *linear momentum* vector $\boldsymbol{p} \equiv \Sigma_j\partial\mathcal{L}/\partial\boldsymbol{v}_j = \Sigma_j m_j\boldsymbol{v}_j \equiv \Sigma_j\boldsymbol{p}_j$ is conserved during the motion. It will be clear that $\boldsymbol{p}$ is additive over the individual particles j, independent of whether interaction is present or not. The interpretation follows from the expression $\partial\mathcal{L}/\partial\boldsymbol{r}_1 = -\partial\Phi/\partial\boldsymbol{r}_1 = \boldsymbol{f}_1$, which determines the force on particle 1. Since $\partial\mathcal{L}/\partial\mathbf{r} = 0$, the total force $\boldsymbol{f} = \Sigma_j\boldsymbol{f}_j$ is zero. In particular, for two particles $\boldsymbol{f}_{12} + \boldsymbol{f}_{21} = 0$, where $\boldsymbol{f}_{12}$ ($\boldsymbol{f}_{21}$) denotes the force on particle 1 (2) by particle 2 (1). This represents *Newton's third law*, implying that the force by particle 1 exerted on particle 2 is equal to, but oppositely directed, to the force exerted by particle 2 on particle 1, both lying along the line joining the particles. From this result one can show that the total energy E is given by $E = \frac{1}{2}\mu\boldsymbol{V}^2 + E_{\mathrm{int}}$, with total mass $M = \Sigma_j m_j$ and $\boldsymbol{V} = \boldsymbol{p}/M$ the velocity of the *center of mass*, the latter being given by $\boldsymbol{R} = \Sigma_j m_j\boldsymbol{r}_j/M$, and E_{int} the energy of the system being at rest as a whole (Problem 2.3).

Let us finally consider *isotropy of space* and use an infinitesimal angle of rotation d$\boldsymbol{\phi}$ around a vector $\boldsymbol{\phi}$. The variation in radius vector $\boldsymbol{r}$ for any particle is $\delta\boldsymbol{r} = \mathrm{d}\boldsymbol{\phi}\times\boldsymbol{r}$, while the variation in velocity reads $\delta\boldsymbol{v} = \mathrm{d}\boldsymbol{\phi}\times\boldsymbol{v}$. For a rotation where each particle is rotated by the same amount d$\boldsymbol{\phi}$, $\delta\mathcal{L}$ should be zero and in direct notation reads

$$\begin{aligned}\delta\mathcal{L} &= \sum\nolimits_j\left[\frac{\partial\mathcal{L}}{\partial\boldsymbol{r}_j}\cdot\delta\boldsymbol{r}_j + \left(\frac{\partial\mathcal{L}}{\partial\dot{\boldsymbol{r}}_j}\right)\cdot\delta\dot{\boldsymbol{r}}_j\right] = \sum\nolimits_j(\dot{\boldsymbol{p}}_j\cdot\delta\boldsymbol{r}_j + \boldsymbol{p}_j\cdot\delta\dot{\boldsymbol{r}}_j)\\ &= \sum\nolimits_j[\dot{\boldsymbol{p}}_j\cdot\mathrm{d}\boldsymbol{\phi}\times\boldsymbol{r}_j + \boldsymbol{p}_j\cdot\mathrm{d}\boldsymbol{\phi}\times\dot{\boldsymbol{r}}_j] = 0\end{aligned}$$

or, taking d$\boldsymbol{\phi}$ outside after permutation of factors,

$$\mathrm{d}\boldsymbol{\phi}\cdot\sum\nolimits_j(\boldsymbol{r}_j\times\dot{\boldsymbol{p}}_j + \dot{\boldsymbol{r}}_j\times\boldsymbol{p}_j) = \mathrm{d}\boldsymbol{\phi}\cdot\frac{\mathrm{d}}{\mathrm{d}t}\sum\nolimits_j(\boldsymbol{r}_j\times\boldsymbol{p}_j) = 0$$

Because d$\boldsymbol{\phi}$ is arbitrary, we have $\mathrm{d}\Sigma_j(\boldsymbol{r}_j\times\boldsymbol{p}_j)/\mathrm{d}t = 0$ and therefore for a closed system the *angular momentum* vector $\boldsymbol{l} \equiv \Sigma_j\boldsymbol{r}_j\times\boldsymbol{p}_j$ is conserved during the motion. Like the momentum vector $\boldsymbol{p}$, it is additive over the particles of the system.

For a set of interacting particles, three types of forces can be distinguished. First, forces acting alike on all particles due to long-range external influences, for example, gravity or an externally imposed electromagnetic field. We call these forces resulting from such an external field, *volume* (or *body*) *forces*, and indicate them for a particle j by $\boldsymbol{f}_j^{(\mathrm{vol})}$. Second, forces applied to a particle due to short-range external forces, for example, interactions with walls of an enclosure or when a weight is resting on a solid. We call these *surface* (or *contact*) *forces* and indicate them by $\boldsymbol{f}_j^{(\mathrm{sur})}$. Volume and surface forces are collectively called *external* (or *applied* or *impressed*) *forces*, here labeled $\boldsymbol{f}_j^{(\mathrm{ext})}$, that is, $\boldsymbol{f}_j^{(\mathrm{ext})} = \boldsymbol{f}_j^{(\mathrm{vol})} + \boldsymbol{f}_j^{(\mathrm{sur})}$. Third, we have forces due to the presence of the other particles, that is, *internal forces*, indicated here by $\boldsymbol{f}_j^{(\mathrm{int})}$.

Let $\boldsymbol{f}_{ij}^{(\mathrm{pp})}$ denote the force on particle i due to particle–particle interaction with particle j. According to Newton's third law, we have $\boldsymbol{f}_{ij}^{(\mathrm{pp})} = -\boldsymbol{f}_{ji}^{(\mathrm{pp})}$. The resultant internal force acting on particle j is $\boldsymbol{f}_j^{(\mathrm{int})} = \Sigma_{i\neq k}\boldsymbol{f}_{ik}^{(\mathrm{pp})}$, and the total force on particle j is

$$\boldsymbol{f}_j = \boldsymbol{f}_j^{(\mathrm{vol})} + \boldsymbol{f}_j^{(\mathrm{sur})} + \boldsymbol{f}_j^{(\mathrm{int})} = \boldsymbol{f}_j^{(\mathrm{ext})} + \sum\nolimits_{j\neq k}\boldsymbol{f}_{jk}^{(\mathrm{int})}$$

The system of particles is in equilibrium if the force $\boldsymbol{f}_j$ on each particle j in the system is equal to its rate of change of linear momentum $\mathrm{d}\boldsymbol{p}_j/\mathrm{d}t$, also known as the *inertial force*. Obviously, it holds that

$$\boldsymbol{f} = \sum\nolimits_j \boldsymbol{f}_j = \sum\nolimits_j \mathrm{d}\boldsymbol{p}_j/\mathrm{d}t = \mathrm{d}\boldsymbol{p}/\mathrm{d}t \quad \text{or} \quad \boldsymbol{f} = \sum\nolimits_j \boldsymbol{f}_j = 0$$

for quasi-static problems where dynamic effects can be neglected and $\mathrm{d}\boldsymbol{p}/\mathrm{d}t = 0$.

The energy of a system with coordinates $\boldsymbol{r}_{\mathrm{sys}}$ can be changed by applying external (reservoir) forces $\boldsymbol{f}_{\mathrm{res}}$, delivering work $\mathrm{d}W_{\mathrm{sys}} = \boldsymbol{f}_{\mathrm{res}}\cdot\mathrm{d}\boldsymbol{r}_{\mathrm{sys}}$ on the system. Using Newton's second law $\mathrm{d}\boldsymbol{p}/\mathrm{d}t = \boldsymbol{f}$ for a system with mass m and velocity $\boldsymbol{v}$ as a whole, we have

$$\mathrm{d}\int \dot{\boldsymbol{p}}\cdot\mathrm{d}\boldsymbol{r} = \mathrm{d}\int m\dot{\boldsymbol{v}}\cdot\mathrm{d}\boldsymbol{r} = \mathrm{d}\int m\dot{\boldsymbol{v}}\cdot\boldsymbol{v}\,\mathrm{d}t = \mathrm{d}(\tfrac{1}{2}mv^2) = \mathrm{d}E_{\mathrm{kin}} \quad \text{and}$$

$$\begin{aligned}\mathrm{d}\int \boldsymbol{f}\cdot\mathrm{d}\boldsymbol{r} &= \mathrm{d}\int\left(\boldsymbol{f}_{\mathrm{res}}^{(\mathrm{vol})} + \boldsymbol{f}_{\mathrm{res}}^{(\mathrm{sur})} + \boldsymbol{f}_{\mathrm{sys}}^{(\mathrm{int})}\right)\cdot\mathrm{d}\boldsymbol{r}_{\mathrm{sys}}\\ &= \mathrm{d}\int \boldsymbol{f}_{\mathrm{res}}^{(\mathrm{vol})}\cdot\mathrm{d}\boldsymbol{r}_{\mathrm{sys}} + \mathrm{d}\int \boldsymbol{f}_{\mathrm{res}}^{(\mathrm{sur})}\cdot\mathrm{d}\boldsymbol{r}_{\mathrm{sys}} + \mathrm{d}\int \boldsymbol{f}_{\mathrm{sys}}^{(\mathrm{int})}\cdot\mathrm{d}\boldsymbol{r}_{\mathrm{res}} \equiv \mathrm{d}E_{\mathrm{fie}} + \mathrm{d}W_{\mathrm{sys}} + \mathrm{d}W_{\mathrm{res}}\end{aligned} \tag{2.6}$$

Thus, we obtain $\mathrm{d}E_{\mathrm{kin}} = \mathrm{d}E_{\mathrm{fie}} + \mathrm{d}W_{\mathrm{res}} + \mathrm{d}W_{\mathrm{sys}}$, known as the *conservation of (mechanical) energy*. For stationary systems in a constant force field, $\mathrm{d}E_{\mathrm{kin}} = \mathrm{d}E_{\mathrm{fie}} = 0$, and $\mathrm{d}W_{\mathrm{sys}} = -\mathrm{d}W_{\mathrm{res}} = -\mathrm{d}E_{\mathrm{int}}$ with the *internal energy* $E_{\mathrm{int}} = \vartheta + \Phi$ where ϑ and Φ now refer only to the kinetic and potential energy of the particles of the system. Internal energy is a concept central to thermodynamics and indicated normally by U. Motions of the collection of all particles of the body that leave the distances between particles unchanged are called *rigid body motions*. Obviously, according to Newton's third law, the work due to internal forces vanishes for rigid body motion.

Problem 2.2 ***Rotating Mass*** An object of mass m moves in a plane with speed v at a constant distance r from the center of rotation, also taken as the origin of the coordinate system. Show that:

(a) the angular speed $\omega = v/r$ and acceleration $\alpha = \dot{v}/r$,
(b) the angular momentum $l = mvr = I\omega$ with $I = mr^2$ the moment of inertia,
(c) the torque $q = \dot{l} = mr^2\alpha = I\dot{\omega}$,
(d) the kinetic energy $\vartheta = \tfrac{1}{2}I\omega^2 = l^2/2I$.
(e) Derive the equation of motion $m\ddot{r} = mr\dot{\theta}^2 - \Phi(r)$ for this system if the potential energy with respect to the center of rotation is $\Phi(r)$.

Problem 2.3 ***Separation of Energy*** Show that the total energy E of a system of particles for an inertial frame is given by $E = \tfrac{1}{2}MV^2 + E_{\mathrm{int}}$ with total mass $M = \Sigma_j m_j$, $\boldsymbol{V} = \boldsymbol{p}/M$ the velocity of the center of mass, given by $\boldsymbol{R} = \Sigma_j m_j\boldsymbol{r}_j/M$, and E_{int} the energy of the system being at rest as a whole.

2.5 Hamilton's Equations

The abovementioned formalism may be put in another form by using the *canonically conjugate* variable for $\mathbf{q}$, the *generalized momentum* $\mathbf{p}$, defined by

$$\mathbf{p} \equiv \partial\mathcal{L}/\partial\dot{\mathbf{q}} = \mathbf{M}\dot{\mathbf{q}} \tag{2.7}$$

so that for Lagrange's equation $\mathrm{d}(\partial\mathcal{L}/\partial\dot{\mathbf{q}})/\mathrm{d}t - \partial\mathcal{L}/\partial\mathbf{q} = 0$, we can write $\dot{\mathbf{p}} = \partial\mathcal{L}/\partial\mathbf{q}$. Defining the *Hamilton function*[5] $\mathcal{H}$ by the Legendre transform (Section A.5)

$$\mathcal{H}(\mathbf{p},\mathbf{q},t) \equiv \mathbf{p}^{\mathrm{T}}\dot{\mathbf{q}} - \mathcal{L}(\mathbf{q},\dot{\mathbf{q}},t)$$

we obtain for its differential

$$\begin{aligned}\mathrm{d}\mathcal{H} &= \mathbf{p}^{\mathrm{T}}\mathrm{d}\dot{\mathbf{q}} + \dot{\mathbf{q}}^{\mathrm{T}}\mathrm{d}\mathbf{p} - \left(\frac{\partial\mathcal{L}}{\partial\mathbf{q}}\right)^{\mathrm{T}}\mathrm{d}\mathbf{q} - \left(\frac{\partial\mathcal{L}}{\partial\dot{\mathbf{q}}}\right)^{\mathrm{T}}\mathrm{d}\dot{\mathbf{q}} - \frac{\partial\mathcal{L}}{\partial t}\mathrm{d}t\\ &= \left(\mathbf{p} - \frac{\partial\mathcal{L}}{\partial\dot{\mathbf{q}}}\right)^{\mathrm{T}}\mathrm{d}\dot{\mathbf{q}} + \dot{\mathbf{q}}^{\mathrm{T}}\mathrm{d}\mathbf{p} - \left(\frac{\partial\mathcal{L}}{\partial\mathbf{q}}\right)^{\mathrm{T}}\mathrm{d}\mathbf{q} - \frac{\partial\mathcal{L}}{\partial t}\mathrm{d}t = \dot{\mathbf{q}}^{\mathrm{T}}\,\mathrm{d}\mathbf{p} - \dot{\mathbf{p}}^{\mathrm{T}}\mathrm{d}\mathbf{q} - \frac{\partial\mathcal{L}}{\partial t}\mathrm{d}t\end{aligned}$$

where in the last step $\mathbf{p} = \partial\mathcal{L}/\partial\dot{\mathbf{q}}$ and $\dot{\mathbf{p}} = \partial\mathcal{L}/\partial\mathbf{q}$ are used. We thus obtain

$$\dot{\mathbf{q}} = \frac{\partial\mathcal{H}}{\partial\mathbf{p}}, \qquad \dot{\mathbf{p}} = -\frac{\partial\mathcal{H}}{\partial\mathrm{q}} \qquad \text{and} \qquad \frac{\partial\mathcal{H}}{\partial t} = -\frac{\partial\mathcal{L}}{\partial t} \tag{2.8}$$

of which the first two are known as *Hamilton's* equations of motion and, because of their symmetry and simplicity, are also often called the *canonical* equations of motion. In this way, the system is described by $6N$ first-order equations instead of the $3N$ second-order equations. The motion of set of particles can thus be described by a path, a so-called *trajectory*, in a $6N$-dimensional space with coordinates $\mathbf{q}$ and $\mathbf{p}$, the so-called *phase space*. As at each point in phase space (or *phase point*), the direction of a trajectory is uniquely given by Hamilton's equations, it follows that through each phase point can pass one and only one trajectory. Hence trajectories do not cross. Any (physical) *observable*, a quantity determined by the (dynamical) state of the system, is represented by a *phase function* $X(\mathbf{q},\mathbf{p})$ or $X(\mathbf{q},\mathbf{p},t)$. Phase space and phase functions are essential concepts in classical statistical mechanics.

For an arbitrary phase function $X(\mathbf{q},\mathbf{p},t)$, we have

$$\begin{aligned}\frac{\mathrm{d}X}{\mathrm{d}t} &= \left(\frac{\partial X}{\partial\mathbf{q}}\right)^{\mathrm{T}}\frac{\mathrm{d}\mathbf{q}}{\mathrm{d}t} + \left(\frac{\partial X}{\partial\mathbf{p}}\right)^{\mathrm{T}}\frac{\mathrm{d}\mathbf{p}}{\mathrm{d}t} + \frac{\partial X}{\partial t} = \left(\frac{\partial X}{\partial\mathbf{q}}\right)^{\mathrm{T}}\left(\frac{\partial\mathcal{H}}{\partial\mathbf{p}}\right) - \left(\frac{\partial X}{\partial\mathbf{p}}\right)^{\mathrm{T}}\left(\frac{\partial\mathcal{H}}{\partial\mathbf{q}}\right) + \frac{\partial X}{\partial t}\\ &\equiv \{X,\mathcal{H}\} + \partial X/\partial t\end{aligned} \tag{2.9}$$

where $\{X,\mathcal{H}\}$ is the so-called *Poisson bracket*. For the components $\dot{q}_j$ and $\dot{p}_j$ of $\dot{\mathbf{q}}$ and $\dot{\mathbf{p}}$, we obtain, respectively, the pleasingly symmetric expressions

$$\dot{q}_j = \{q_j,\mathcal{H}\} \quad \text{and} \quad \dot{p}_j = \{p_j,\mathcal{H}\} \tag{2.10}$$

as $\partial q_j/\partial q_k = \partial p_j/\partial p_k = \delta_{jk}$ and $\partial q_j/\partial p_k = \partial p_j/\partial q_k = 0$ because, by definition, the various q_js and p_js are independent variables. Note that a Poisson bracket can be written down generally, for example, for q_i and p_i. In that case, we have

$$\{q_i,p_j\} = \sum_k \left(\frac{\partial q_i}{\partial q_k}\frac{\partial p_j}{\partial p_k} - \frac{\partial q_i}{\partial p_k}\frac{p_j}{\partial q_k}\right) = \delta_{ij} \quad \text{and} \quad \{q_i,q_j\} = \{p_i,p_j\} = 0 \tag{2.11}$$

5 Introduced by Hamilton (1805–1865) and labeled by him as $\mathcal{H}$ in honor of Huygens (1629–1695).

where q_i and p_j are said to *commute* if $i \neq j$. For a conservative system, the Poisson bracket $\{t,\mathcal{H}\} = 1$, so that time t and energy $\mathcal{H}$ can also said to be canonically conjugate. Formally, for a change of the Hamilton function $\mathcal{H}$ with time t, we have

$$\frac{\mathrm{d}\mathcal{H}}{\mathrm{d}t} = \left(\frac{\partial \mathcal{H}}{\partial \mathbf{q}}\right)^T \frac{\mathrm{d}\mathbf{q}}{\mathrm{d}t} + \left(\frac{\partial \mathcal{H}}{\partial \mathbf{p}}\right)^T \frac{\mathrm{d}\mathbf{p}}{\mathrm{d}t} + \frac{\partial \mathcal{H}}{\partial t} = \{\mathcal{H},\mathcal{H}\} + \frac{\partial \mathcal{H}}{\partial t} = \frac{\partial \mathcal{H}}{\partial t}$$

where Hamilton's equations are used. If the Lagrange function $\mathcal{L}$ is explicitly independent of time, that is, $\mathcal{L} = \mathcal{L}(\mathbf{q}, \dot{\mathbf{q}})$, this is also true for Hamilton's function $\mathcal{H}$. Hence, for a conservative system, $\partial\mathcal{H}/\partial t = 0$, and $\mathcal{H}$ is constant, that is, $\mathcal{H}(\mathbf{q},\mathbf{p}) = E$, where E is that constant. To interpret the constant E, we consider $\mathcal{H}$ and obtain

$$\mathcal{H} = \mathbf{p}^\mathrm{T}\dot{\mathbf{q}} - \mathcal{L} = \left(\frac{\partial \mathcal{L}}{\partial \dot{\mathbf{q}}}\right)^\mathrm{T} \dot{\mathbf{q}} - \mathcal{L} = \left(\frac{\partial \vartheta}{\partial \dot{\mathbf{q}}}\right)^\mathrm{T} \dot{\mathbf{q}} - \mathcal{L} = 2\vartheta - \vartheta + \Phi = \vartheta + \Phi$$

As seen before, the constant E equals the (mechanical) energy but now expressed in terms of position and momentum as independent coordinates. Since $\mathcal{H}$ is a function of $\mathbf{q}$ and $\mathbf{p}$, we must express ϑ also in terms of $\mathbf{p}$. Using $\dot{\mathbf{q}} = \mathbf{M}^{-1}\mathbf{p}$, we obtain

$$\vartheta = \tfrac{1}{2}\dot{\mathbf{q}}^\mathrm{T}\mathbf{M}\dot{\mathbf{q}} = \tfrac{1}{2}(\mathbf{M}^{-1}\mathbf{p})^\mathrm{T}\mathbf{M}(\mathbf{M}^{-1}\mathbf{p}) = \tfrac{1}{2}\mathbf{p}^\mathrm{T}\mathbf{M}^{-1}\mathbf{p}$$

The Hamilton function thus becomes $\mathcal{H} = \vartheta + \Phi = \tfrac{1}{2}\mathbf{p}^\mathrm{T}\mathbf{M}^{-1}\mathbf{p} + \Phi(\mathbf{q})$. Note that the generalized force $\mathbf{Q} = \mathbf{A}^\mathrm{T}\mathbf{f}$ is only equal to $\dot{\mathbf{p}}$ if $\mathbf{M}$ is independent of $\mathbf{q}$ which generally is not the case. However, the definition used for $\mathbf{Q}$ conserves the qualification conservative, because the work $\mathrm{d}W = \mathbf{f}^\mathrm{T}\mathrm{d}\mathbf{r} = \mathbf{Q}^\mathrm{T}\mathrm{d}\mathbf{q}$. An alternative, and possibly expected, definition $\mathbf{Q}' = \partial\mathcal{L}/\partial\mathbf{q}$ [7] does lead to $\dot{\mathbf{p}} = \mathbf{Q}'$, but $\mathbf{Q}'$ is not conservative when $\mathbf{f}$ is conservative and $\mathrm{d}W \neq \mathbf{Q}'^\mathrm{T}\mathrm{d}\mathbf{q}$.

Finally, we prove *Liouville's theorem*, which implies that the value of a volume element $\mathrm{d}\mathbf{q}\mathrm{d}\mathbf{p}$ in phase space remains invariant in time (although its shape changes). If we write $\mathbf{p} = \mathbf{p}(t)$, $\mathbf{p}' = \mathbf{p}(t + \mathrm{d}t)$, $\mathbf{q} = \mathbf{q}(t)$ and $\mathbf{q}' = \mathbf{q}(t + \mathrm{d}t)$, we have

$$\mathbf{q}' = \mathbf{q} + \dot{\mathbf{q}}\mathrm{d}t = \mathbf{q} + \frac{\partial \mathcal{H}}{\partial \mathbf{p}}\mathrm{d}t \quad \text{and} \quad \mathbf{p}' = \mathbf{p} + \dot{\mathbf{p}}\mathrm{d}t = \mathbf{p} - \frac{\partial \mathcal{H}}{\partial \mathbf{q}}\mathrm{d}t$$

The Jacobian determinant $J = \partial(\mathbf{q}',\mathbf{p}')/\partial(\mathbf{q},\mathbf{p})$ is given by

$$J = \begin{vmatrix} 1 + \dfrac{\partial^2 \mathcal{H}}{\partial \mathbf{p}\partial \mathbf{q}}\mathrm{d}t & \dfrac{\partial^2 \mathcal{H}}{\partial \mathbf{p}^2}\mathrm{d}t \\ -\dfrac{\partial^2 \mathcal{H}}{\partial \mathbf{q}^2}\mathrm{d}t & 1 + \dfrac{\partial^2 \mathcal{H}}{\partial \mathbf{q}\partial \mathbf{p}}\mathrm{d}t \end{vmatrix} = 1 + \mathcal{O}(\mathrm{d}t)^2$$

so that $\mathrm{d}J/\mathrm{d}t = 0$ or $J = 1$. Concisely, $\mathrm{d}\mathbf{q}'\mathrm{d}\mathbf{p}' = \mathrm{d}\mathbf{q}\mathrm{d}\mathbf{p}$. This implies that for the probability density $\rho(\mathbf{q},\mathbf{p},t)$ of phase points we have $\dot{\rho} = \mathrm{d}\rho/\mathrm{d}t = 0$, and thus

$$\frac{\partial \rho}{\partial t} = -\left[\left(\frac{\partial \rho}{\partial \mathbf{q}}\right)^\mathrm{T} \dot{\mathbf{q}} + \left(\frac{\partial \rho}{\partial \mathbf{p}}\right)^\mathrm{T} \dot{\mathbf{p}}\right] = \left(\frac{\partial \mathcal{H}}{\partial \mathbf{q}}\right)^\mathrm{T} \frac{\partial \rho}{\partial \mathbf{p}} - \left(\frac{\partial \mathcal{H}}{\partial \mathbf{p}}\right)^\mathrm{T} \frac{\partial \rho}{\partial \mathbf{q}} = \{\mathcal{H},\rho\} \tag{2.12}$$

where both $\dot{\rho} = 0$ and $\partial\rho/\partial t = \{\mathcal{H},\rho\}$ are referred to as *Liouville's equation*. The latter expression is also written as $\partial\rho/\partial t = -\boldsymbol{L}\rho$, with $\boldsymbol{L}\cdot = \{\cdot,\mathcal{H}\}$ the *Liouville operator*.

Example 2.3 ***The Harmonic Oscillator Revisited*** We treat once more the oscillator with kinetic energy $\vartheta = \tfrac{1}{2}mv^2 = p^2/2m$ and potential energy $\Phi = \tfrac{1}{2}kx^2$. From Hamilton's equation, we get right away $m\dot{v} = m\ddot{x} = -kx$, resulting in the solution indicated in Example 2.2. From Hamilton's $\dot{x} = \partial\mathcal{H}/\partial p$, we have the identity $\dot{x} = p/m = m\dot{x}/m = \dot{x}$.

Problem 2.4 ***Hamilton and Poisson*** Verify Eqs. (2.8), (2.9), (2.10), and (2.11).

Problem 2.5 ***Hamilton and Action*** Show using $S = \int \mathscr{L}(\mathbf{q}, \dot{\mathbf{q}})\,\mathrm{d}t$, Eq. (2.2), that for a conservative system $\partial S/\partial t = -\mathscr{H}$.

2.6 Hamilton's Principle for Continuous Systems

So far, we discussed the relations for discrete particle systems. In this section, we introduce some relations for continuous material systems. While in the discrete particle case the particles were labeled by an index j, in the continuum case the material particles were labeled by their coordinates $\boldsymbol{r}$. Thus, for a property f_j, the summation over particles $\Sigma_j f_j$ is replaced by integration over the volume of the continuous body $\int f(\boldsymbol{r})\mathrm{d}V$. For example, the kinetic energy for a particle with mass m_j and coordinate $\boldsymbol{q}_j$ reads $\frac{1}{2}\Sigma_j m_j \dot{\boldsymbol{q}}_j^2$ which becomes for a continuous body $\frac{1}{2}\int \rho[\dot{\boldsymbol{u}}(\boldsymbol{r})]^2\mathrm{d}V$, where $\theta = \frac{1}{2}\rho[\dot{\boldsymbol{u}}(\boldsymbol{r})]^2$ denotes the *kinetic energy density*, ρ is the mass density, and $\boldsymbol{u}$ is the displacement. Similarly, the Lagrange function $\mathscr{L}$ becomes an integral over the *Lagrange density* ℓ given in direct notation by

$$\mathscr{L} = \int \ell(\boldsymbol{u}, \dot{\boldsymbol{u}}, \boldsymbol{r}, t)\,\mathrm{d}V$$

Because the displacement $\boldsymbol{u}$ and displacement rate $\dot{\boldsymbol{u}}$ depend on the coordinate $\boldsymbol{r}$, the variation $\delta\ell = (\partial\ell/\partial\boldsymbol{u})\cdot\delta\boldsymbol{u} + (\partial\ell/\partial\dot{\boldsymbol{u}})\cdot\delta\dot{\boldsymbol{u}} + (\partial\ell/\partial\boldsymbol{u}')\cdot\delta\boldsymbol{u}'$, where $\boldsymbol{u}' = \partial\boldsymbol{u}/\partial\boldsymbol{r}$, so that

$$\delta S = \int_{t_1}^{t_2}\int \left(\frac{\partial\ell}{\partial\boldsymbol{u}}\cdot\delta\boldsymbol{u} + \frac{\partial\ell}{\partial\dot{\boldsymbol{u}}}\cdot\delta\dot{\boldsymbol{u}} + \frac{\partial\ell}{\partial\boldsymbol{u}'}\cdot\delta\boldsymbol{u}'\right)\mathrm{d}V\,\mathrm{d}t = 0$$

Partial integration of the second and third terms leads to, respectively,

$$-\int_{t_1}^{t_2}\int\left[\left(\frac{\mathrm{d}}{\mathrm{d}t}\frac{\partial\ell}{\partial\dot{\boldsymbol{u}}}\right)\cdot\delta\boldsymbol{u}\right]\mathrm{d}V\,\mathrm{d}t \quad \text{and} \quad -\int_{t_1}^{t_2}\int\left[\left(\frac{\mathrm{d}}{\mathrm{d}\boldsymbol{r}}\frac{\partial\ell}{\partial\boldsymbol{u}'}\right)\cdot\delta\boldsymbol{u}\right]\mathrm{d}V\,\mathrm{d}t$$

and consequently, as the variations $\delta\boldsymbol{u}$ are arbitrary, the Lagrange equations become

$$\frac{\partial\ell}{\partial\boldsymbol{u}} - \frac{\mathrm{d}}{\mathrm{d}t}\frac{\partial\ell}{\partial\dot{\boldsymbol{u}}} - \frac{\mathrm{d}}{\mathrm{d}\boldsymbol{r}}\frac{\partial\ell}{\partial\boldsymbol{u}'} = 0$$

In the calculus of variations, the *functional derivative* of any functional $F = \int f[\eta, \dot{\eta}, t]\mathrm{d}V$ is normally defined as $\delta F/\delta\eta = \partial f/\partial\eta - \mathrm{d}(\partial f/\partial\dot{\eta})/\mathrm{d}t$. In discrete particle classical mechanics, however, if a functional $F = \int f[\eta, \dot{\eta}, r, t]\mathrm{d}V$ is used with the argument η dependent on the spatial coordinates r and time t, one normally defines the functional derivative as $\delta F/\delta\eta = \partial f/\partial\eta - \mathrm{d}(\partial f/\partial\eta')/\mathrm{d}r$ with $\eta' = \partial\eta/\partial r$, that is, one leaves the time t out of the definition. Accepting the latter definition, we have

$$\delta\mathscr{L} = \int\left(\frac{\partial\ell}{\partial\boldsymbol{u}}\cdot\delta\boldsymbol{u} - \frac{\mathrm{d}}{\mathrm{d}\boldsymbol{r}}\frac{\partial\ell}{\partial\boldsymbol{u}'}\cdot\delta\boldsymbol{u} + \frac{\delta\ell}{\delta\dot{\boldsymbol{u}}}\cdot\delta\dot{\boldsymbol{u}}\right)\mathrm{d}V = \int\left(\frac{\delta\mathscr{L}}{\delta\boldsymbol{u}}\cdot\delta\boldsymbol{u} + \frac{\delta\mathscr{L}}{\delta\dot{\boldsymbol{u}}}\cdot\delta\dot{\boldsymbol{u}}\right)\mathrm{d}V$$

where the step $\partial\ell/\partial\dot{\boldsymbol{u}} = \delta\mathscr{L}/\delta\dot{\boldsymbol{u}}$ can be made because ℓ does not depend on the gradient of $\dot{\boldsymbol{u}}$. This is the same result as obtained from applying the δ variation to $\mathscr{L}$, ignoring the dependence on $\boldsymbol{u}'$, and demonstrates the usefulness of the functional derivative. Similarly, the Lagrange equations become

$$\frac{\delta\mathscr{L}}{\delta\boldsymbol{u}} - \frac{\mathrm{d}}{\mathrm{d}t}\frac{\delta\mathscr{L}}{\delta\dot{\boldsymbol{u}}} = 0$$

This expression is also similar to the expression for the discrete case, thereby explaining why the second definition of the functional derivative is preferred.

Similar as the Lagrange function $\mathscr{L}$ can be transformed to the Hamilton function $\mathscr{H} = \boldsymbol{p}\cdot\boldsymbol{q} - \mathscr{L}$ by introducing the momentum $\boldsymbol{p} = \partial\mathscr{L}/\partial\dot{\boldsymbol{q}}$, the Lagrange density ℓ can be transformed to the *Hamilton density* $\hbar$ via $\mathscr{H} = \int \hbar\,\mathrm{d}V = \int(\boldsymbol{\pi}\cdot\boldsymbol{u} - \ell)\mathrm{d}V$ by defining the *momentum density* $\boldsymbol{\pi} = \partial\ell/\partial\dot{\boldsymbol{u}} = \delta\mathscr{L}/\delta\dot{\boldsymbol{u}}$. Variation of $\mathscr{H}$ yields

$$\begin{aligned}\delta\mathscr{H} &= \int\left(\boldsymbol{\pi}\cdot\delta\dot{\boldsymbol{u}} + \dot{\boldsymbol{u}}\cdot\delta\boldsymbol{\pi} - \frac{\partial\ell}{\partial\boldsymbol{u}}\cdot\delta\boldsymbol{u} - \frac{\partial\ell}{\partial\dot{\boldsymbol{u}}}\cdot\delta\dot{\boldsymbol{u}}\right)\mathrm{d}V\\ &= \int\left[\left(\boldsymbol{\pi} - \frac{\partial\ell}{\partial\dot{\boldsymbol{u}}}\right)\cdot\delta\dot{\boldsymbol{u}} + \dot{\boldsymbol{u}}\cdot\delta\boldsymbol{\pi} - \dot{\boldsymbol{\pi}}\cdot\delta\boldsymbol{u}\right]\mathrm{d}V\\ &= \int(\dot{\boldsymbol{u}}\cdot\delta\boldsymbol{\pi} - \dot{\boldsymbol{\pi}}\cdot\delta\boldsymbol{u})\,\mathrm{d}V \end{aligned} \tag{2.13}$$

where in the second step we used $\partial\mathscr{L}/\partial\boldsymbol{u} = \mathrm{d}(\delta\mathscr{L}/\delta\dot{\boldsymbol{u}})/\mathrm{d}t = \dot{\boldsymbol{\pi}}$. We conclude that

$$\dot{\boldsymbol{u}} = \frac{\delta\hbar}{\delta\boldsymbol{\pi}} \quad\text{and}\quad \dot{\boldsymbol{\pi}} = -\frac{\delta\hbar}{\delta\boldsymbol{u}}$$

If $\mathscr{L}$ is conserved, that is, time independent, $\mathscr{H}$ is conserved and

$$\begin{aligned}\delta\mathscr{H} &= \delta\int\left(\frac{\partial\ell}{\partial\dot{\boldsymbol{u}}}\cdot\dot{\boldsymbol{u}} - \ell\right)\mathrm{d}V = \delta\int(2\theta - \theta + \phi)\,\mathrm{d}V\\ &= \delta\int(\theta+\phi)\,\mathrm{d}V = \delta\int e\,\mathrm{d}V = \delta E = 0\end{aligned}$$

where θ is the *kinetic energy density*, ϕ is the *potential energy density*, $e = \theta + \phi$ is the *energy density*, and E is the (total) energy. We thus again obtain that the actual displacements of the system are those for which the virtual work of the applied forces vanishes.

2.7 The Virial Theorem

Even without considering the details of the dynamics of a system of particles, we can connect the average kinetic energy with the average potential energy. This is achieved via a useful theorem, first introduced by Clausius [8], of which we present here a short derivation. First, note that for a bounded function $G(t)$ with time t as variable, the time average $\overline{g}$ of $g = \mathrm{d}G/\mathrm{d}t$ is zero, since

$$\overline{g} = \lim_{t\to\infty} t^{-1}\int_0^t(\mathrm{d}G/\mathrm{d}\tau)\,\mathrm{d}\tau = \lim_{t\to\infty}[G(t) - G(0)]/t = 0 \tag{2.14}$$

Second, consider for a molecule j enclosed in a finite volume, the bounded function $G = \boldsymbol{p}_j\cdot\boldsymbol{q}_j$ [9]. Differentiating, we have $g = \boldsymbol{p}_j\cdot\dot{\boldsymbol{q}}_j + \dot{\boldsymbol{p}}_j\cdot\boldsymbol{q}_j$. Inserting Hamilton's equations $\dot{\boldsymbol{q}}_j = \partial\mathscr{H}/\partial\boldsymbol{p}_j$ and $\dot{\boldsymbol{p}}_j = -\partial\mathscr{H}/\partial\boldsymbol{q}_j$ in Eq. (2.14), the result is $\overline{\boldsymbol{q}_j\cdot(\partial\mathscr{H}/\mathrm{d}\boldsymbol{q}_j)} = \overline{\boldsymbol{p}_j\cdot(\partial\mathscr{H}/\mathrm{d}\boldsymbol{p}_j)}$. From $\vartheta_j = \boldsymbol{p}_j^{\,2}/2m$, we have $\boldsymbol{p}_j\cdot(\partial\mathscr{H}/\mathrm{d}\boldsymbol{p}_j) = 2\vartheta_j$. Summing over all particles and defining the *virial* $\mathscr{V}_j \equiv \tfrac{1}{2}\,\boldsymbol{q}_j\cdot(\partial\mathscr{H}/\mathrm{d}\boldsymbol{q}_j)$, we obtain the *virial theorem*:

$$\vartheta_j = \mathscr{V}_j \quad\text{or}\quad \overline{\vartheta} \equiv \sum\nolimits_j\overline{\vartheta_j} = \mathscr{V} \equiv \sum\nolimits_j\mathscr{V}_j = \tfrac{1}{2}\sum\nolimits_j\boldsymbol{q}_j\cdot(\partial\mathscr{H}/\partial\boldsymbol{q}_j) \tag{2.15}$$

Originally, using Cartesian coordinates, Clausius defined the virial as $-\tfrac{1}{2}\Sigma_j\overline{\boldsymbol{f}_j\cdot\boldsymbol{r}_j}$, which follows from inserting $\mathscr{H} = \vartheta(\boldsymbol{p}) + \Phi(\boldsymbol{r})$ and $\boldsymbol{f}_j = -\partial\Phi/\partial\boldsymbol{r}_j$ in the virial $\mathscr{V}_j$.

Frequently, Φ is a homogeneous function of degree n in $\boldsymbol{r}$, say $\Phi = ar^n$, and from Euler's theorem (Section A.3) we have $\sum_j(\partial\Phi/\partial\boldsymbol{r}_j)\cdot\boldsymbol{r}_j = n\Phi$, or combining with Eq. (2.15), $2\overline{\vartheta} = n\overline{\Phi}$. Since the total energy $U = \overline{\mathcal{H}} = \overline{\vartheta} + \overline{\Phi}$, we have $\overline{\Phi} = 2U/(n+2)$ and $\overline{\vartheta} = nU/(n+2)$. For a harmonic potential $n = 2$ and $\overline{\vartheta} = \overline{\Phi} = ½U$, so that the total energy is on average equally divided over the kinetic and potential energy. For the Coulomb or gravitational potential $n = -1$, therefore $\overline{\vartheta} = -½\,\overline{\Phi} = -U$, and since ϑ must be ≥ 0, this implies $U \leq 0$. For such a system to stay together, we must have $U < 0$, and accordingly, the charges have to move as otherwise $\vartheta = 0$ and thus $U = 0$.

2.8 Final Remarks

For our discussion of phases and their transitions, the basic concepts of classical mechanics discussed here more than suffice. In fact, the most important concepts are possibly Hamilton's equations, which are used in classical statistical mechanics, and the Poisson brackets, as used for quantum mechanics, for which we have collected here all results required. We note that the Lagrange function is not necessarily $\mathcal{L} = \vartheta - \Phi$ for all types of problems; for example, in special relativity or electromagnetism, the concept of a minimum principle can be used to advantage with a different expression for $\mathcal{L}$, and in these cases, one uses often the term *Principle of Least Action* instead of Hamilton's principle. Anyway, for holonomic systems, Hamilton's principle and Newton's laws are equivalent [6]. As on molecular level all systems are holonomic, we do not bother whether the PoVP (allowing non-holonomic systems) and Newtons' laws are fully equivalent as well. The matter is still under debate [5, 6, 10]. However, in many cases using a single scalar function, either $\mathcal{L}$ or $\mathcal{H}$, instead of a set of vectors, like $\boldsymbol{v}$, $\boldsymbol{a}$, and $\boldsymbol{f}$, eases an abstract development considerably. Apart from the books mentioned in the section "Further Reading", Refs. [11–13] provide modern introductions.

References

1 Tolman, R.C. (1938). *The Principles of Statistical Mechanics*. Oxford: Oxford University Press.

2 Goldstein, 1981.

3 Clotfelter, B.E. (1987). *Am. J. Phys.* 55: 210; Falconer, I. (1999). *Meas. Sci. Technol.* 10: 470; Lally, S.P. (1999). *Phys. Teach.* 37: 34; Moreno González, A. (2001). *Sci. Educ.* 10: 515.

4 Touboul, P., Metris, G., Rodrigues, M. et al. (2020). *Comptes Rendus Physique* 21: 139.

5 Pars, L.A. (1965). *A Treatise on Analytical Dynamics*. London: Heinemann.

6 Lanczos, 1970.

7 Landau and Lifshitz, 1976.

8 Clausius, R.J.E. (1870). *Phil. Mag. Ser. 4* 40: 122; (1871). *Ann. Phys.* 141: 124.

9 Corben, H.C. and Stehle, P. (1960). *Classical Mechanics*, 2e. New York: John Wiley & Sons.

10 Wang, L.-S. and Pao, Y.-H. (2003). *Am. J. Phys.* 71: 72.

11 Gregory, R.D. (2006). *Classical Mechanics*. Cambridge: Cambridge University Press.

12 Deriglazov, A. (2010). *Classical Mechanics, Hamiltonian and Lagrangian Formalism*. Berlin: Springer.

13 Likharev, K.K. (2017). *Classical Mechanics, Lectures Notes and Classical Mechanics, Problems with Solutions*. Bristol: IOP Publishing.

Further Reading

Goldstein, H. (1981). *Classical Mechanics*, 2e. Amsterdam: Addison-Wesley. A well-known, generally very clear exposition.

ter Haar, D. (1961). *Elements of Hamiltonian Mechanics*. Amsterdam: North-Holland. Contains a similar treatment as Goldstein, but often more concise.

Lanczos, C. (1970). *The Variational Principles of Mechanics*, 4e. Toronto: University of Toronto Press (also Dover, 1986). Highly readable and highly recommended.

Landau, L.D. and Lifshitz, E.M. (1976). *Mechanics*, 3e. Oxford: Butterworth-Heinemann. A classic text dealing mainly with principles, collisions, small oscillations and spinning rigid bodies.

3

Quantum Mechanics

After reviewing classical particle mechanics (PM), in this chapter we discuss the second "tool," quantum mechanics (QM). We first describe the kinematics and the kinetics. Subsequently, we deal with the interpretation and some exactly solvable and approximately solvable problems.

3.1 Quantum Concepts

For molecular phenomena, classical PM is in principle inadequate, which led to the discovery of QM. Here, we provide a necessarily compact outline of basic quantum theory only and, similar as for PM and later on thermodynamics, we refrain from comparison with experiment. The basic postulate is that on a molecular level all information for a system, that is, the molecule or molecules at hand, is contained in a "state vector" and that any observable property can be extracted from the "state vector" by applying the appropriate operator [1–3]. This abstract formulation (Sections 3.1.1–3.1.3) will be helpful as background for Chapter 19 and can be skipped by readers more interested in the subsequent more down-to-earth sections on interpretation and some practical examples (Section 3.2 and further).

3.1.1 Fundamental Quantum Kinematics

A (quantum) state is an abstract quantity, dependent on the number of degrees of freedom (DoF) d of the molecular system, which can be described by a vector in a d-dimensional (dD) complex vector space $\mathbf{H}$, conventionally addressed as *Hilbert space*. Such a vector is called a *ket* and labeled as $|\ldots\rangle$. Abstract implies that these vectors are defined by their formal properties. Let $\alpha, \beta, \ldots$ be complex numbers and $|a\rangle, |b\rangle, \ldots$ be kets. Then, by definition, the product $\alpha|a\rangle$ and the sum $|c\rangle = \alpha|a\rangle + \beta|b\rangle$ are also vectors in the same space. Adding kets is denoted as *superposition of states*. Because kets are complex quantities, one needs to define to every ket $|a\rangle$ in a one-to-one way a dual vector, labeled $\langle b|$ and denoted as *bra*.[1] The *inner product* is then defined by

$$\langle a|b\rangle \equiv \langle b|a\rangle^*$$

1 Here I use, what is called, the Dirac ⟨bra|c|ket⟩ notation. The association with brackets is obvious but the meaning of the bra $\langle\Psi|$ and ket $|\Psi\rangle$ is much deeper. The basic idea for these vectors is similar to that of a

Phases of Matter and their Transitions: Concepts and Principles for Chemists, Physicists, Engineers, and Materials Scientists, First Edition. Gijsbertus de With.

where the asterisk denotes the complex conjugate and implying that $\langle c| = \alpha^*\langle a| + \beta^*\langle b|$. For the inner product, the *Schwartz inequality*

$$|\langle a|b\rangle|^2 \leq \langle a|a\rangle\langle b|b\rangle$$

applies, where the *norm* $\langle a|a\rangle$ of $|a\rangle$ is by definition positive. The norm is used to normalize the state vectors via requiring that $\langle a|a\rangle = 1$. Moreover, for any ket $|d\rangle$, we have $\langle d|c\rangle = \alpha\,\langle d|a\rangle + \beta\,\langle d|b\rangle$.

A *basis* in **H** is defined by a denumerable set[2] of kets $\{|k\rangle\}$, $k = 1, 2, \ldots, d$ that spans **H** in the sense that any ket $|w\rangle$ can be expressed as

$$|w\rangle = \Sigma_k c_k |k\rangle$$

Assuming an orthonormal basis with property $\langle k|k'\rangle = \delta_{kk'}$, we have $c_k = \langle k|w\rangle$ and thus $|w\rangle = \Sigma_k\, |k\rangle\langle k|w\rangle$. Here $\delta_{kk'}$ is the *Kronecker delta* having the property $\delta_{kk'} = 1$ if $k = k'$ and $\delta_{kk'} = 0$ if $k \neq k'$. From $\langle k|k'\rangle = \delta_{kk'}$, we conclude also that a state is determined up to a complex phase factor $\exp(\mathrm{i}\phi)$, since for $|l\rangle = \mathrm{e}^{\mathrm{i}\phi}|k\rangle$, we have

$$\langle l|l'\rangle = \langle k|\mathrm{e}^{-\mathrm{i}\phi}\mathrm{e}^{\mathrm{i}\phi}|k'\rangle = \langle k|k'\rangle = \delta_{kk'} \tag{3.1}$$

Therefore, in fact, only the direction of a ket $|w\rangle$ in **H** determines the state.[3]

A *subspace* $\mathbf{H}_\mathrm{s}$ of **H** is spanned by a subset of $\{|k\rangle\}$. For the 1D subspace $\mathbf{H}_k$, spanned by $|k\rangle$ only, a *projection operator* P_k operating on a general state $|w\rangle$ in **H** yields the vector $|k\rangle$ via $P_k|w\rangle = c_k|k\rangle$. Introducing the *outer product* $|k\rangle\langle k|$, we have

$$P_k = |k\rangle\langle k|$$

so that $c_k = \langle k|w\rangle$. The sum $\Sigma_k P_k = \Sigma_k\, |k\rangle\langle k|$ is thus the total projection on **H** and therefore just $\mathit{1}$, where $\mathit{1}$ represents the unity operator. Hence, we have the *completeness relation*

$$\mathit{1} = \Sigma_k P_k = \Sigma_k |k\rangle\langle k|$$

indicating that the basis spans the complete space **H**. This relation is also called the *resolution of the identity*. Once a projection is made, another projection does not change the projected vector anymore and therefore $P^2 = P$ (*idempotency*).

In general, an operator X is defined by $|a\rangle\langle b|$, and we have

$$X|w\rangle = (|a\rangle\langle b|)|w\rangle = |a\rangle\langle b|w\rangle \quad \text{and} \quad \langle w|X = \langle w|(|a\rangle\langle b|) = \langle w|a\rangle\langle b|$$

The product of $X = |a\rangle\langle b|$ and $Y = |c\rangle\langle d|$ is given by

$$XY = (|a\rangle\langle b|)(|c\rangle\langle d|) = |a\rangle\langle b|c\rangle\langle d| = \langle b|c\rangle|a\rangle\langle d|$$

If A is a linear operator,[4] we write in the k-representation, $A|k\rangle = \mathit{1}A|k\rangle = \Sigma_{k'}\, |k'\rangle\langle k'|A|k\rangle$, where $\langle k'|A|k\rangle$ is the expansion coefficient of $A|k\rangle$ in the set $\{|k'\rangle\}$.

---

vector in 3D Euclidean space. Any such vector $\boldsymbol{v}$ can be described by three mutually independent base vectors $\boldsymbol{e}_j$ via $\boldsymbol{v} = \Sigma_j c_j \boldsymbol{e}_j$, where the set $\{c_j\}$ is the *representation* of $\boldsymbol{v}$. For any two vectors $\boldsymbol{a}$ and $\boldsymbol{b}$, there exists a scalar $c = \boldsymbol{a}\cdot\boldsymbol{b}$ (inner product) and vector $\boldsymbol{c} = \boldsymbol{a}\times\boldsymbol{b}$ (outer product). The essential point is that c and $\boldsymbol{c}$ are independent of the basis $\boldsymbol{e}_j$ chosen, but their representations obviously not (Section A.7). For further details, see, for example [1–3].

2 If confusion is possible whether a set is meant or not, I denote a set by {…}. A prime indicates another member of the same set.

3 It is thus actually what mathematically is called a *ray*, although generally the word vector is used.

4 Linearity implies $A(c_1 f + c_2 g) = c_1 Af + c_2 Ag$, with c_1, c_2 constants and f, g functions.

For BA, we then have

$$BA|k\rangle = 1B1A|k\rangle = \Sigma_{k'}\Sigma_{k''}|k'\rangle\langle k'|B|k''\rangle\langle k''|A|k\rangle \tag{3.2}$$

representing matrix multiplication (Section A.9) since $\langle k|BA|k'\rangle = \Sigma_{k''}\langle k|B|k''\rangle\langle k''|A|k'\rangle$ and the *matrix*[5] A contains the *matrix elements* $\mathrm{A}_{kk'} = \langle k|A|k'\rangle$. Since $A = 1A1$, we can write $A = \Sigma_k\Sigma_{k'}|k\rangle\langle k|A|k'\rangle\langle k'|$, where the diagonal elements $\langle k|A|k\rangle$ of the representation $\mathrm{A} = \langle k|A|k'\rangle$ are denoted as the *expectation values* of the operator A for the basis $\{|k\rangle\}$. Further, we have the *trace* and *determinant*, given by (Section A.9)

$$\mathrm{trA} \equiv \Sigma_k \langle k|A|k\rangle \quad \text{and} \quad \mathrm{detA} = \det(\langle k'|A|k\rangle) \tag{3.3}$$

For an operator A, the *transpose* A^{T} and *complex conjugate* A^* are defined by

$$A^{\mathrm{T}} = \Sigma_k\Sigma_{k'}|k\rangle\langle k'|A|k\rangle\langle k'| \quad \text{and} \quad A^* = \Sigma_k\Sigma_{k'}|k\rangle\langle k|A|k'\rangle^*\langle k'|$$

respectively. When $A^{\mathrm{T}} = A$ holds, the operator is *symmetric*, while when $A = (A^*)^{\mathrm{T}} \equiv A^\dagger$ is true, it is called *Hermitian*.[6] We thus have $\langle k|A|k'\rangle^* = \langle k'|A^\dagger|k\rangle$, and hence for $A|w\rangle = |w'\rangle$ it holds that $\langle w'| = \langle w|A^\dagger$. A convenient and often used transcription of the trace of an operator is $\mathrm{tr}\,|a\rangle\langle b| = \sum_n\langle n|a\rangle\langle b|n\rangle = \sum_n\langle b|n\rangle\langle n|a\rangle = \langle b|a\rangle$.

From these definitions it follows that $(AB)^\dagger = B^\dagger A^\dagger$. If A and B are Hermitian, the product AB is Hermitian; if A commutes with B, that is, if for the *commutator*

$$[A,B] \equiv AB - BA \tag{3.4}$$

the relation $[A,B] = 0$ is true. If an operator U satisfies

$$UU^\dagger = U^\dagger U = 1$$

it is called *unitary* and $\Sigma_{k'}\langle k|U|k'\rangle\langle k''|U|k'\rangle^* = \Sigma_{k'}U_{kk'}U^*_{k''k'} = \delta_{kk''}$, where the $U_{kk'}$ are the matrix elements $\langle k|U|k'\rangle$ of U for the k-basis used. These matrices are important because they transform the basis $\{|k\rangle\}$ for **H** to another one, say $\{|r\rangle\}$. If we have

$$|r\rangle = \Sigma_k|k\rangle\langle k|r\rangle \quad \text{and} \quad |k\rangle = \Sigma_r|r\rangle\langle r|k\rangle$$

the *transformation functions* $\langle k|r\rangle$ are elements of a unitary matrix, since $\Sigma_k\langle r|k\rangle\langle k|r'\rangle = \delta_{rr'}$ and $\Sigma_r\langle k|r\rangle\langle r|k'\rangle = \delta_{kk'}$. Hence, for two bases $\{|a_j\rangle\}$ and $\{|b_j\rangle\}$ with $U_{ba} = \Sigma_j|b_j\rangle\langle a_j|$ and $U_{ba}{}^\dagger = U_{ab}$, we have $U_{ba}|a_j\rangle = |b_j\rangle$ and $U_{bc}U_{ca} = U_{ba}$. A unitary transformation of an operator A is defined by $A' = UAU^\dagger$. For two operators A and B, we have

$$\mathrm{trAB} = \mathrm{trBA} \quad \text{and} \quad \det(\mathrm{AB}) = (\mathrm{detA})(\mathrm{detB}) \quad \text{and hence}$$

$$\mathrm{trA}' = \mathrm{tr\,UAU}^\dagger = \mathrm{trA} \quad \text{and} \quad \mathrm{detA}' = \det(\mathrm{UAU}^\dagger) = \mathrm{detA}$$

If for $A|n\rangle = |b\rangle$ the transformed ket $|b\rangle$ is given by $|b\rangle = a_n|n\rangle$, we have an *eigenvalue equation* with *eigenvalue* a_n. Taking the inner product $\langle n'|A|n\rangle = a_n\langle n'|n\rangle = a_n\delta_{n'n}$, we see that $A_{n'n} - a_n\delta_{n'n} = 0$ or $\mathrm{A} - 1\mathbf{a} = 0$, where $1\mathbf{a}$ denotes a diagonal matrix containing the eigenvalues $\mathbf{a}$ of A (the matrix representing the operator A) and having a solution only if

5 Recall the notation: scalar X (italic), vector $\boldsymbol{X}$ (bold italic), column matrix $\mathbf{X}$ (bold roman) of a set of elements X_i or $\boldsymbol{X}_i$ (Section 1.4). Here X denotes an operator, and X is its matrix representation.

6 For a Hermitian operator A, it holds that $\int f^*\mathrm{A}g\,\mathrm{d}\tau = \int g(\mathrm{A}f)^*\mathrm{d}\tau$, where the integration is over all of the domain of the functions f and g.

$\det A \neq 0$. A unitary transformation can bring an arbitrary matrix $A' = A_{n'n}'$ in diagonal form $A = a_n\delta_{n'n}$ by choosing U as defined by $A = UA'U^\dagger$, so that

$$U(A' - 1\mathbf{a})U^\dagger = UA'U^\dagger - U1\mathbf{a}U^\dagger = A - 1\mathbf{a} = 0$$

From $A|n\rangle = a_n|n\rangle$, we obtain what is called the *spectral decomposition*

$$A = \Sigma_n\, |n\rangle a_n\langle n|$$

Multiple eigenvectors can have the same eigenvalue in which case they are called *degenerate*. The various eigenvectors of A are orthogonal or, if degenerate, can be chosen so. The eigenvalues of a Hermitian matrix are real.

So far, we have restricted ourselves to denumerable sets $\{|k\rangle\}$. We now extend the notation to non-denumerable sets $\{|q\rangle\}$ without bothering about the mathematical subtleties as required, for example, in discussing convergence. For these aspects we refer to the literature. The analogue of the Kronecker symbol $\delta_{nn'}$ for discrete sets on the domain $(-\tfrac{1}{2}L,+\tfrac{1}{2}L)$ is for continuous sets on the same domain the *Dirac function* $\delta(k\text{–}k')$, defined by (Section A.14)

$$\int \delta(k - k')f(k')\mathrm{d}k' = f(k)$$

Hence, analogues of some previously introduced expressions are:

$$\langle k|k'\rangle = \delta_{kk'} \quad \Leftrightarrow \quad \langle q|q'\rangle = \delta(q - q') \qquad \text{orthonormality}$$

$$\textstyle\sum_k |k\rangle\langle k| = 1 \quad \Leftrightarrow \quad \int |q\rangle\langle q'|\mathrm{d}q' = 1 \qquad \text{resolution of identity}$$

$$|w\rangle = \textstyle\sum_k |k\rangle\langle k|w\rangle \quad \Leftrightarrow \quad |w\rangle = \int |q\rangle\langle q|\,w\rangle\mathrm{d}q \qquad \text{change of representation}$$

3.1.2 Operators and their Representation

Measurable quantities, or *observables*, are described by operators. Note that, since measurable quantities are real, operators representing observables, must be Hermitian.

The simplest operator is possibly the position operator $\mathcal{r}$, which determines the position of the particle at hand. Here we have $\mathcal{r}\,|r\rangle = r|r\rangle$, where, as often, we label the ket $|r\rangle$ with its eigenvalue r. Since classical translations in orthogonal directions commute, we assume this remains true on a molecular level, and hence we obtain, using the component $\mathcal{r}_j$ of $\mathcal{r}$, the commutation relation $[\mathcal{r}_j, \mathcal{r}_k] = 0$. To describe translation, we assume an operator $\mathcal{J}(\mathrm{d}r)$ in such a way that $\mathcal{J}(\mathrm{d}r)|r\rangle = |r + \mathrm{d}r\rangle$. Making use of the resolution of identity, we obtain

$$\mathcal{J}(\mathrm{d}r)|a\rangle = \mathcal{J}(\mathrm{d}r)\int |r\rangle\langle r|a\rangle\,\mathrm{d}r = \int |r + \mathrm{d}r\rangle\langle r|a\rangle\,\mathrm{d}r = \int |r\rangle\langle r - \mathrm{d}r|a\rangle\,\mathrm{d}r$$

where the last step can be made since r is just a label. We have four requirements [4]:

- Normalized states remain normalized, that is, $\langle a|a\rangle = \langle a|\mathcal{J}^\dagger(\mathrm{d}r)\,\mathcal{J}(\mathrm{d}r)|a\rangle$, or equivalently, $\mathcal{J}^\dagger(\mathrm{d}r)\,\mathcal{J}(\mathrm{d}r) = 1$;
- Translation is additive, that is, $\mathcal{J}(\mathrm{d}r'')\,\mathcal{J}(\mathrm{d}r') = \mathcal{J}(\mathrm{d}r'' + \mathrm{d}r')$;
- A reversed translation is described by the inverse operator, that is, $\mathcal{J}(-\mathrm{d}r) = \mathcal{J}^{-1}(\mathrm{d}r)$; and
- Translation is continuous, that is, $\lim_{\mathrm{d}r\to 0} \mathcal{J}(\mathrm{d}r) = 1$.

From the continuity condition one can conjecture that $\mathcal{J}(\mathrm{d}r) = \mathit{1} - \mathrm{i}K\mathrm{d}r$, where K is a Hermitian operator. Classically, translation is inferred by momentum, so that one could think that K should be the momentum operator $\mathcal{p}$ associated with $\mathcal{r}$. However, as $K\mathrm{d}r$ has to be dimensionless, K has dimension m^{-1} and it appears that it is appropriate to take $K = \mathcal{p}/\hbar$, where $2\pi\hbar = h$ represents Planck's constant, similar as in the de Broglie relation $\mathbf{p} = \hbar\mathbf{k}$ with $\mathbf{k}$ the wave vector. Hence, $\mathcal{J}(\mathrm{d}r) = \mathit{1} - \mathrm{i}\hbar^{-1}\mathcal{p}\,\mathrm{d}r$. For a finite translation $\mathcal{J}(\Delta r)|r\rangle = |r + \Delta r\rangle$, considered as a repetitive infinitesimal translation $\mathrm{d}r = \Delta r/N$, we consequently obtain

$$\mathcal{J}(\Delta r) = \lim_{N\to\infty}[\,\mathit{1} - \mathrm{i}\hbar^{-1}\mathcal{p}\Delta x/N]^N = \exp(-\mathrm{i}\hbar^{-1}\mathcal{p}\Delta r) \tag{3.5}$$

The momentum and coordinate operator obey, given in components, the commutation relation $[\mathcal{r}_j, \mathcal{p}_k] = \mathrm{i}\hbar\delta_{jk}$. Similar to $[\mathcal{r}_j, \mathcal{r}_k] = 0$, the relations $[\mathcal{p}_j, \mathcal{p}_k] = 0$ hold.

Summarizing the above, we have the three commutation relations

$$[\mathcal{r}_j, \mathcal{r}_k] = 0, \quad [\mathcal{r}_j, \mathcal{p}_k] = \mathrm{i}\hbar\delta_{jk}, \quad [\mathcal{p}_j, \mathcal{p}_k] = 0 \tag{3.6}$$

These fundamental quantum relations show a close resemblance to their classical (Poisson bracket) counterparts.

So far, we have employed the abstract state description with kets and bras, somewhat comparable to the description of vectors in direct notation.[1] For vectors, a representation is made by choosing a coordinate set, and we do something similar here. If we have the position operator $\mathcal{r}$ that has as eigenvalue the position r, that is, $\mathcal{r}\,|r\rangle = r|r\rangle$, normalized so that $\langle r|r'\rangle = \delta(r - r')$, we can write

$$|a\rangle = \int |r\rangle\langle r|a\rangle \mathrm{d}r$$

Using these position eigenkets as base vectors, we have what is called the *coordinate representation*. The coefficient $\langle r|a\rangle$ represents the state $|a\rangle$ in the coordinate representation and is usually denoted as the *wave function* $\psi_a(r)$. This implies that for expressions like $\langle b|a\rangle$ and $\langle b|\mathcal{A}|a\rangle$, we obtain

$$\langle b|a\rangle = \int \psi_b^*(r)\psi_a(r)\,\mathrm{d}r \quad \text{and} \quad \langle b|\mathcal{A}|a\rangle = \int \psi_b^*(r)\mathcal{A}\psi_a(r')\,\mathrm{d}r\mathrm{d}r'$$

respectively. In case $\mathcal{A} = f(\mathcal{r})$, $\langle b|\mathcal{A}|a\rangle$ reduces to $\langle b|f(\mathcal{r})|a\rangle = \int \psi_b{}^*(r)f(r)\psi_a(r)\,\mathrm{d}r$.

To obtain an explicit expression for the momentum operator $\mathcal{p}_r$ in the coordinate representation, we consider $\mathcal{J}(\mathrm{d}r) = 1 - \mathrm{i}\hbar^{-1}\mathcal{p}_r\,\mathrm{d}r$ acting on $|a\rangle$,

$$1 - \mathrm{i}\hbar^{-1}\mathcal{p}_r\mathrm{d}r|a\rangle = \int \mathcal{J}(\mathrm{d}r)|r\rangle\langle r|a\rangle\,\mathrm{d}r = \int |r + \mathrm{d}r\rangle\langle r|a\rangle\,\mathrm{d}r =$$

$$\int |r\rangle\langle r - \mathrm{d}r|a\rangle \mathrm{d}r = \int |r\rangle\left[\langle r|a\rangle - \frac{\partial\langle r|a\rangle}{\partial r}\mathrm{d}r\right]\mathrm{d}r \quad \text{resulting in}$$

$$\mathcal{p}_r|a\rangle = \int |r\rangle\left[-\mathrm{i}\hbar\frac{\partial\langle r|a\rangle}{\partial r}\right]\mathrm{d}r \quad \text{or} \quad \langle r|\mathcal{p}_r|a\rangle = -\mathrm{i}\hbar\frac{\partial\langle r|a\rangle}{\partial r}$$

Substituting $|a\rangle = |r'\rangle$, we obtain $\langle r|\mathcal{p}_r|r'\rangle = -\mathrm{i}\hbar\partial\delta(r - r')/\partial r'$. Similarly, substituting $\langle r| = \langle b|$, we have

$$\langle b|\mathcal{p}_r|a\rangle = \int \langle b|r\rangle\left[-\mathrm{i}\hbar\frac{\partial\langle r|a\rangle}{\partial r}\right]\mathrm{d}r = \int \psi_b^*(r)\left[-\mathrm{i}\hbar\frac{\partial}{\partial r}\right]\psi_a(r)\,\mathrm{d}r$$

so that in coordinate representation $\mathcal{p}_r(\cdot) = -\mathrm{i}\hbar(\partial(\cdot)/\partial r$. Therefore, we have $\mathcal{p}_r^2(\cdot) = (-\mathrm{i}\hbar)^2\partial^2(\cdot)/\partial r^2$ so that, since $\mathscr{H} = \mathbf{p}^2/2m + V$, with V the potential energy, $H(\cdot) = -(\hbar^2/2\,m)\,\nabla^2(\cdot) + V(\cdot)$.

If we use, instead of the base kets $|r\rangle$, the eigenvectors $|p\rangle$ of the momentum operator p from the eigenvalue equation $p|p\rangle = p|p\rangle$ as base kets, a similar discussion leads to the *momentum representation*. As we will not use this representation, we refrain from details and refer to the literature.

3.1.3 Fundamental Quantum Kinetics

Generally, a quantum state evolves from time t_0 to another state at time t, that is, $|a,t_0\rangle \to |a,t\rangle$. From the continuity of states, we have $\lim_{t\to t_0}|a,t\rangle = |a,t_0\rangle$. This evolution is described by a *time-evolution operator* $U(t,t_0)$, so that we have $|a,t\rangle = U(t,t_0)|a,t_0\rangle$. Similar as for the translation operator, we require that [4]:

- Normalized states remain normalized, that is, $\langle a,t|a,t\rangle = \langle a,t_0|a,t_0\rangle = 1$. This implies that $U(t,t_0)$ is unitary or $U^\dagger(t,t_0)U(t,t_0) = \mathit{1}$.
- For a time series $t_2 > t_1 > t_0$, we have $U(t_2,t_0) = U(t_2,t_1)U(t_1,t_0)$, and
- Time is continuous, that is, $\lim_{\mathrm{d}t\to 0} U(t+\mathrm{d}t,t) = \mathit{1}$.

For an infinitesimal time interval $\mathrm{d}t$, we have $|a,t+\mathrm{d}t\rangle = U(t+\mathrm{d}t,t)|a,t\rangle$. From the continuity of time we surmise that $U(t+\mathrm{d}t,t) = \mathit{1} - \mathrm{i}A\mathrm{d}t$ with A an Hermitian operator. Realizing that A should have dimension s^{-1} and that the Planck–Einstein expression connecting energy E and (circular) frequency ω reads $E = \hbar\omega$, one could expect that $A = H/\hbar$. Accepting this, we thus have $U(t+\mathrm{d}t,t) = \mathit{1} - \mathrm{i}\hbar^{-1}H\mathrm{d}t$. Consequently,

$$U(t+\mathrm{d}t,t_0) - U(t,t_0) = \left(-\mathrm{i}\frac{H}{\hbar}\mathrm{d}t\right)U(t,t_0) \quad \text{or} \quad \mathrm{i}\hbar\frac{\partial}{\partial t}U(t,t_0) = HU(t,t_0) \tag{3.7}$$

representing the *Schrödinger equation* for the time-evolution operator $U(t,t_0)$. Since this operation can be applied to any ket $|a,t\rangle$, we also have $\mathrm{i}\hbar\,\partial|a,t\rangle/\partial t = H|a,t\rangle$. For the time dependence of H we can distinguish the following three cases:

- H is independent of time. In this case, the solution is $U(t,t_0) = \exp[-\mathrm{i}\hbar^{-1}H(t-t_0)]$.
- H is dependent on time but $H(t_1)H(t_2) = H(t_2)H(t_1)$. One can show that the solution is then $U(t,t_0) = \exp[-\mathrm{i}\hbar^{-1}\int H(t')\mathrm{d}t']$.
- H is dependent on time but $H(t_1)H(t_2) \neq H(t_2)H(t_1)$. Now the solution, known as the *Dyson series* and which we do not discuss further, becomes

$$U(t,t_0) = \mathit{1} + \sum\nolimits_{n=1}^{n=\infty}(-\mathrm{i}/\hbar)^n \int_{t_0}^{t_1}\int_{t_0}^{t_2}\cdots\int_{t_0}^{t_{n-1}} H(t_1)H(t_2)\ldots H(t_n)\,\mathrm{d}t_1\mathrm{d}t_2\ldots\mathrm{d}t_3$$

As the only requirement for A is that it should have the dimension t^{-1}, the question is legitimate whether Eqs. (3.5) and (3.7) should contain both the same constant $\hbar$. In fact, one can show that a classical relation like $\mathbf{p} = m\dot{\mathbf{x}}$ cannot be obtained as the limit of the corresponding quantum expression, if this constant is not the same.

To assess the time dependence for a general ket, we need to evaluate first the time dependence of the base kets. If the base kets are the eigenvectors $|a\rangle$ of an operator A that commutes with H, then the eigenvectors of A are also eigenvectors for H and we have $H|a\rangle = E_a|a\rangle$. Hence, taking $t_0 = 0$ so that $|a,t_0\rangle = |a,0\rangle \equiv |a\rangle$, we have

$$\mathrm{e}^{-\mathrm{i}Ht/\hbar} = \sum\nolimits_a\sum\nolimits_{a'}|a'\rangle\langle a'|\mathrm{e}^{-\mathrm{i}Ht/\hbar}|a\rangle\langle a| = \sum\nolimits_a|a\rangle\mathrm{e}^{-\mathrm{i}E_at/\hbar}\langle a|$$

We thus have $|a,t\rangle = \exp(-\mathrm{i}\hbar^{-1}Ht)|a\rangle = \exp(-\mathrm{i}\hbar^{-1}E_at)|a\rangle$, which implies that in the expansion $|b,t\rangle = \Sigma_a c_a(t)|a\rangle$, the expansion coefficient $c_a(t) = c_a(0)\exp(-\mathrm{i}\hbar^{-1}E_at)$. The same arguments apply for all operators that commute with H, and the energy is specified uniquely once this set of operators is specified. The set of kets common to all those operators form the *basis*.

Applying $|a,t\rangle = U(t)|a\rangle$, where $U(t) \equiv U(t,0)$, to the expectation value of any operator B with respect to any state $|b\rangle$ we obtain, using $c_a = \langle a,t|b\rangle$,

$$\begin{aligned}\langle b|B|b\rangle &= \sum\nolimits_{a'}\sum\nolimits_{a}\langle b|a',t\rangle\langle a',t|B|a,t\rangle\langle a,t|b\rangle| \\ &= \sum\nolimits_{a'}\sum\nolimits_{a} c_{a'}^* c_a \langle a'|U^\dagger(t)BU(t)|a\rangle \\ &= \sum\nolimits_{a'}\sum\nolimits_{a} c_{a'}^* c_a \langle a'|B|a\rangle|a\rangle\, \mathrm{e}^{-\mathrm{i}(E_a - E_{a'})/\hbar}\end{aligned}$$

From this one infers that if $|b\rangle$ is an eigenstate of H, the expectation values are stationary, that is $\langle b,t|a,t\rangle = \langle b|a\rangle$. If $|b\rangle$ is an arbitrary state, the expectation value oscillates with the *Bohr frequency* $\omega_{a'a} = (E_a{'} - E_a)/\hbar$.

Let us now express the dynamics in the coordinate representation. Recall that $\langle x|a\rangle$ in the coordinate representation reads as $\Psi_a \equiv \Psi_a(r)$. Moreover, $H(\cdot) = -\mathrm{i}\hbar^2\nabla^2(\cdot) + V(\cdot)$ so that the *time-dependent Schrödinger equation* becomes

$$\mathrm{i}\hbar\frac{\partial}{\partial t}\langle r|a,t\rangle = \langle r|H|a,t\rangle \quad \text{or} \quad \mathrm{i}\hbar\frac{\partial}{\partial t}\Psi_a = H\Psi_a \tag{3.8}$$

Using the energy eigenstates, $\langle r|a,t\rangle = \exp(-\mathrm{i}\hbar^{-1}E_a t)\langle r|a\rangle = \exp(-\mathrm{i}\hbar^{-1}E_a t)\Psi_a$, we have

$$-\frac{\hbar^2}{2m}\nabla^2\langle r|a\rangle + V\langle r|a\rangle = E_a\langle r|a\rangle \quad \text{or} \quad -\frac{\hbar^2}{2m}\nabla^2\Psi_a + V\Psi_a = E_a\Psi_a \tag{3.9}$$

abbreviated as $H\Psi = E\Psi$, and known as the *time-independent Schrödinger equation*.

From the time-evolution operator U, we have $|a,t\rangle = U|a\rangle$. The inner product reads $\langle b,t|a,t\rangle = \langle b|a\rangle$ and thus remains unchanged. For an expectation value, we have in the so-called *Schrödinger picture*,

$$\langle b,t|A|a,t\rangle = \left[\langle b|U^\dagger(t)\right] A \left[U(t)|a\rangle\right] \equiv \langle b,t|A_\mathrm{S}|a,t\rangle$$

in which the state kets are changing and the operators are constant. We can also write

$$\langle b,t|A|a,t\rangle = \langle b|[U^\dagger(t)A_\mathrm{S}U(t)]|a\rangle \equiv \langle b|A_\mathrm{H}|a\rangle$$

to obtain the *Heisenberg picture*, in which the state kets are constant but the operators are changing. It will be clear that the expectation value does not depend on the choice made. Let us now evaluate time evolution in the Heisenberg picture. Here we have at $t_0 = 0$, $A_\mathrm{H}(0) = A_\mathrm{S}$, with A_H evolving in time, while $|a,t\rangle_\mathrm{H} = |a\rangle_\mathrm{H}$ remains constant in time. Differentiating $A_\mathrm{H} = U^\dagger(t)A_\mathrm{S}U(t)$, we obtain

$$\begin{aligned}\frac{\mathrm{d}A_\mathrm{H}}{\mathrm{d}t} &= -\frac{\partial U^\dagger}{\partial t}A_\mathrm{S}U + U^\dagger A_\mathrm{S}\frac{\partial U}{\partial t} \\ &= -(\mathrm{i}\hbar)^{-1}U^\dagger H(UU^\dagger)A_\mathrm{S}U + (\mathrm{i}\hbar)^{-1}U^\dagger A_\mathrm{S}(UU^\dagger)HU \\ &= (\mathrm{i}\hbar)^{-1}[A_\mathrm{H}, U^\dagger HU] = (\mathrm{i}\hbar)^{-1}[A_\mathrm{H}, H_\mathrm{H}] = (\mathrm{i}\hbar)^{-1}[A_\mathrm{H}, H]\end{aligned} \tag{3.10}$$

The last step results if $H_\mathrm{H} = U^\dagger HU$, that is, if H and U commute, which is true for time-independent Hamilton operators since in that case $U(t,t_0{=}0) \equiv U(t) = \exp(-\mathrm{i}\hbar^{-1}Ht)$. We have to add the term $\partial A_\mathrm{S}/\partial t$ to the right-hand side of Eq. (3.10) if A is an explicit function of t. Equation (3.10) is known as the *Heisenberg equation of motion*. The resemblance to the time evolution of a classical quantity using the Poisson brackets of classical mechanics, Eq. (2.27), will be clear.

Table 3.1 Summary of the three quantum-mechanical pictures.

	Schrödinger picture	Dirac picture	Heisenberg picture
State vector	$\lvert a,t\rangle_S = U\lvert a\rangle$	$\lvert a,t\rangle_I = U_0\lvert a\rangle$	$\lvert a,t\rangle_H = \lvert a\rangle$
Operator	$A_S(t) = A_S(0)$	$A_I(t) = U_0^{\dagger}A_S(0)\,U_0$	$A_H(t) = U^{\dagger}A_S(0)U$

In case the Hamilton operator can be written as $H = H_0 + H'$, where H' can be considered as a perturbation or interaction, we can also only transfer the time dependence of the unperturbed configuration from the state vectors to the operators. In this *Dirac* or *interaction picture*, we write for the state vector $\lvert a\rangle_I = U_0\lvert a\rangle$ with $U_0 = \exp(-iH_0t/\hbar)$, while for an operator we have $A_I = U_0^{\dagger}A_S U_0$. The Schrödinger equation in the Dirac picture thus becomes $i\hbar\,\partial\lvert a,t\rangle_I/\partial t = H_1\lvert a,t\rangle_I$ with $H_I = U_0^{\dagger}H'U_0$. Differentiating A_I, we now obtain $dA_I/dt = (i\hbar)^{-1}[A_I, H_0]$. In case there is no interaction, $\lvert a\rangle_I$ is time independent and we regain the Heisenberg picture of the unperturbed system. Table 3.1 summarizes the three pictures described.

3.2 Interpretation and Some Exact Solutions

So, in the Schrödinger representation, the state of a system is given by the *wave function* $\widetilde{\Psi}(\mathbf{r},t) \equiv \Psi(\mathbf{r})\exp(-i\hbar^{-1}Et)$, where $\mathbf{r}$ stands for the full set of coordinates, E for energy, and t for time. This function and its gradient must be single valued, finite, and continuous for all values of its arguments. A function satisfying these conditions is said to be *well behaved*. It also must have a finite integral of its quadrate if integrated over the complete range of coordinates rendering *normalization* possible. As noted, the only possible values which a measurement of an observable with operator A can yield are the *eigenvalues a* of the *eigenvalue equation* $A\widetilde{\Psi}(\mathbf{r},t) = a\widetilde{\Psi}(\mathbf{r},t)$ with $\widetilde{\Psi}(\mathbf{r},t)$ as the *eigenfunction*. For the coordinate operator $\mathcal{r}(i) = \mathbf{r}(i)$, the eigenvalues ξ_i of a single particle i are the values for which the equation

$$\mathbf{r}\widetilde{\Psi}(\mathbf{r},t) = \xi\widetilde{\Psi}(\mathbf{r},t) \qquad \text{or} \qquad (\mathbf{r}-\xi)\,\widetilde{\Psi}(\mathbf{r},t) = 0$$

which is an ordinary algebraic equation, possesses solutions. It is evident that for this expression to be true either $\mathbf{r} = \xi$ or $\widetilde{\Psi}(\mathbf{r},t) = 0$. This corresponds exactly to the definition of the Dirac δ-function (Section A.14), which is thus the eigenfunction $\delta(\mathbf{r} - \xi)$ associated with the operator $\mathcal{r}$ for the eigenvalue ξ.

We also recall that the energy E for conservative systems is given by the Hamilton function $\mathcal{H}$ in terms of the coordinates $\mathbf{r}$ and conjugated momenta $\mathbf{p}$, that is, $E = \mathcal{H}(\mathbf{p},\mathbf{r}) = \vartheta(\mathbf{p}) + V(\mathbf{r})$. The Hamilton operator H is formed accordingly (using a symmetrized function of $\mathbf{r}$ and $\mathbf{p}$ if necessary). The eigenfunctions Ψ_j of H for the *Schrödinger equation* $H\Psi_j = E\Psi_j$ form a complete set, implying that any arbitrary well-behaving function Φ that depends on the same variables in the same domain and obeys the same boundary conditions can be expressed as a sum in terms of these eigenfunctions, that is, $\Phi = \Sigma_j c_j\Psi_j$ with the c_j's being constants. Completeness also applies for the eigenfunctions of any operator, for example, the momentum operator $-i\hbar\nabla$. Here, we have for a free particle,

$$-i\hbar\nabla\Psi = \boldsymbol{p}\Psi \quad \text{with} \quad \text{solution} \quad \Psi = C\exp(i\boldsymbol{k}\cdot\boldsymbol{r}) = C\exp(i\boldsymbol{p}\cdot\boldsymbol{r}/\hbar) \tag{3.11}$$

Using the de Broglie relation $\boldsymbol{p} = \hbar\boldsymbol{k}$ with $k = |\boldsymbol{k}| = 2\pi/\lambda$, where $\boldsymbol{p}$ is the momentum and $\boldsymbol{k}$ is the wave vector, we can also write the wave function for a free particle as $\exp(i\boldsymbol{k}\cdot\boldsymbol{r} - i\omega t)$ with the circular frequency $\omega = 2\pi\nu$, and, as usual, λ and ν representing the wavelength and frequency, respectively. Normalization of this Ψ can be done in two ways. First, by periodic box normalization, where we assume that the particle is contained in an arbitrary box of volume L^3 and that the potential at $x = y = z = 0$ equals that at $x = y = z = L$. We obtain, using $\langle\Psi_k|\Psi_l\rangle = \delta_{\boldsymbol{kl}} = \delta_{kx,lx}\delta_{ky,ly}\delta_{kz,lz}$, $C = L^{-3/2}$. Second, by δ-function normalization, where we employ a particular representation of the Dirac δ-function (Section A.14), namely

$$\delta(x) = \lim_{g\to\infty} (\sin(gx)/\pi x)$$

The function $\sin(gx)/\pi x$ has a value of g/π at $x = 0$, oscillates with decreasing amplitude and period $2\pi/g$ with increasing $|x|$, and is normalized independent of the value of g. Hence,

$$\int_{-\infty}^{+\infty} \exp[i(\boldsymbol{k}-\boldsymbol{l})\cdot\boldsymbol{r}]\,d\boldsymbol{r} = \lim_{g\to\infty}\int_{-g}^{+g} \exp[i(\boldsymbol{k}-\boldsymbol{l})\cdot\boldsymbol{r}]\,d\boldsymbol{r} = \lim_{g\to\infty}\frac{2\sin \boldsymbol{g}\cdot(\boldsymbol{k}-\boldsymbol{l})}{\boldsymbol{k}-\boldsymbol{l}} = 2\pi\delta(\boldsymbol{k}-\boldsymbol{l})$$

Consequently, $C = 2\pi^{-3/2}$ using $\exp(i\boldsymbol{k}\cdot\boldsymbol{r})$ or $C = h^{-3/2}$ using $\exp(i\boldsymbol{p}\cdot\boldsymbol{r})$.

When a system is in an arbitrary state Φ, the expected mean of a sequence of measurements of the observable with operator A is given by the *expectation value*

$$\langle A\rangle \equiv \int \Phi^* A\Phi\,d\mathbf{r} = \langle\Phi|A|\Phi\rangle \tag{3.12}$$

to be interpreted as $\langle A\rangle = \sum_i \rho_i a_i$, where a_i denotes the measured value and ρ_i its probability of occurrence. If we expand Φ in terms of Ψ_ks, we have $\Phi = \sum_k c_k\Psi_k$, and the expectation value becomes

$$\langle A\rangle = \langle\textstyle\sum_k c_k\Psi_k|A|\sum_l c_l\Psi_l\rangle = \sum_{k,l} c_k c_l\langle\Psi_k|A|\Psi_l\rangle = \sum_k c_k^2 a_k$$

Hence, when the system is in a state Φ, a measurement of the observable A yields the value a_k with a probability $c_k{}^2$, where c_k is the expansion coefficient (or *probability amplitude*) in the expansion of Φ in terms of Ψ_ks, or $\rho_k = c_k{}^2$. The probability amplitude may be expressed in terms of Φ and Ψ_k as

$$\langle\Psi_k|\Phi\rangle = \langle\Psi_k|\textstyle\sum_l c_l\Psi_l\rangle = c_k \tag{3.13}$$

If Φ is one of the eigenfunctions Ψ_l of A, we obviously have $\rho_k = |\langle\Psi_k|\Psi_l\rangle|^2 = \delta_{kl}$. In this case, the expectation value $\langle A\rangle$ is thus equal to the eigenvalue a_l. The abovementioned considerations lead directly to the interpretation of the wave function. Consider the probability that a measurement of the position of a single particle i will give the value ξ. The eigenfunction corresponding to the coordinate operator $r = \boldsymbol{r}$ for the eigenvalue ξ has shown to be $\Psi_\xi = \delta(\boldsymbol{r} - \xi)$. From $\rho_k = c_k{}^2$ and Eq. (3.12), we obtain

$$\rho_\xi = |\langle\delta(\boldsymbol{r}-\xi)|\Phi(\boldsymbol{r})\rangle|^2 = |\Phi(\xi)|^2$$

Hence, the probability of finding the particle i in a narrow range $d\boldsymbol{r}$ located at $\boldsymbol{r} = \xi$ is given by the square of its wave function evaluated at ξ. Generalization to many particles is straightforward.

An important aspect is that conjugated variables, such as the position coordinate $\boldsymbol{r}$ and momentum $\boldsymbol{p}$, cannot be determined precisely at the same time or, in other words, if one of the conjugated variables is exactly known, the other is fully undetermined. More generally, if we define the operator $\delta A \equiv A - \langle A \rangle$ and write $|a\rangle = \delta A|\cdot\rangle$ where $|\cdot\rangle$ denotes any ket, we have

$$\delta A \delta B = \tfrac{1}{2}(\delta A \delta B - \delta A \delta B) + \tfrac{1}{2}(\delta A \delta B + \delta A \delta B) \equiv \tfrac{1}{2}[\delta A, \delta B] + \tfrac{1}{2}[\delta A, \delta B]_+$$

The Schwartz inequality yields for any pair of operators A and B are

$$\langle (\delta A)^2 \rangle \langle (\delta B)^2 \rangle \geq |\langle \delta A \delta B \rangle|^2$$

Using the expression for $\delta A \delta B$, the right-hand side becomes

$$|\langle \delta A \delta B \rangle|^2 = \tfrac{1}{4}|\langle [\delta A, \delta B] \rangle|^2 + \tfrac{1}{4}|\langle [\delta A, \delta B]_+ \rangle|^2 \quad \text{or} \quad |\langle \delta A \delta B \rangle|^2 \geq \tfrac{1}{4}|\langle [\delta A, \delta B] \rangle|^2$$

since $[\delta A, \delta B]$ equals $[A, B]$ and $\langle [\delta A, \delta B]_+ \rangle|^2 \geq 0$. For any pair of conjugated variables a_j and b_k, obeying the commutator $[a_j, b_k] = \mathrm{i}\hbar\delta_{jk}$, we thus have

$$\delta a \delta b \geq \hbar/2 \tag{3.14}$$

which are known as the Heisenberg *uncertainty relations*. The implication for the pair coordinate $\boldsymbol{r}$ and momentum $\boldsymbol{p}$ is that a precisely localized particle has an indeterminate momentum and vice versa. For energy E and time t, a similar relation holds.[7] In a stationary state at *any* time t, the energy is *exactly* E. In general, any process that shortens the lifetime, broadens the energy level. Since transitions to other states are required for a change in time, there is a limited residence time δt, and this leads to a small uncertainty in energy δE given by $\delta E \delta t = \tfrac{1}{2}\hbar$.

Another important point is that particles also possess *spin angular momentum*, for which no classical analogue is available, but which obeys the same quantization rules as orbital angular momentum. Although introduced here in an ad hoc way, the concept follows from relativistic QM, as shown first by Dirac [1]. For an electron, the eigenvalue for the spin operator is either $\tfrac{1}{2}\hbar$ or $-\tfrac{1}{2}\hbar$, with the eigenfunctions generally denoted by α and β, respectively.

To conclude, we note that the number of exact solutions in QM for realistic systems is limited. Virtually the only realistic one is the hydrogen atom, at least if a non-relativistic Hamilton operator is used (Section 10.6.1). Fortunately, several model systems can be solved exactly. Therefore, as has been emphasized many times before, to obtain (even approximate) solutions to the Schrödinger equation, one needs drastic approximations. In the next sections, we indicate the solutions of a few, exactly solvable single-particle problems and thereafter the basics of many-particle solutions.

3.2.1 The Particle in a Box

For a particle in a one-dimensional (1D) box, the potential energy $V(x)$ is given by

$$V(x) = 0 \ \text{ for } \ 0 < x < l \qquad \text{and} \qquad V(x) = \infty \ \text{ otherwise} \tag{3.15}$$

The kinetic energy operator $T(x)$ and Schrödinger equation read, respectively,

$$T(x) = -\frac{\hbar^2}{2m}\frac{\mathrm{d}^2}{\mathrm{d}x^2} \qquad \text{and} \qquad -\frac{\hbar^2}{2m}\frac{\mathrm{d}^2}{\mathrm{d}x^2}\Psi = E\Psi$$

7 In non-relativistic QM, time is a parameter, not a variable, and the derivation of the uncertainty relation for E and t, for which we refer to the literature, proceeds differently.

where m is the mass of the particle. The solutions for the wave functions are

$$\Psi_n = \sqrt{2/w}\,\sin(2\pi nx/l)$$

where $n = 1, 2, \ldots$ is a positive integer, the so-called *quantum number*, arising since only for discrete wavelengths the solution obeys the boundary conditions. These wavelengths are given by $\lambda = 2\,l/1, 2\,l/2, \ldots, 2\,l/n$ with allowed energies

$$E_n = \frac{h^2}{8m}\left(\frac{n^2}{l^2}\right) = \frac{\hbar^2}{2m}\left(\frac{2\pi\mathbf{n}}{2l}\right)^2 = \frac{\hbar^2}{2m}\left(\frac{\pi\mathbf{n}}{l}\right)^2 = \frac{(\hbar\mathbf{k})^2}{2m} = \frac{\mathbf{p}^2}{2m}$$

where $\boldsymbol{k}$ is the wave vector and $\boldsymbol{p}$ is the momentum. Note that in the ground state with $n = 1$ the energy is still not zero. For a three-dimensional (3D) box with potential energy $V(\boldsymbol{r}) = 0$ for $\boldsymbol{0} < \boldsymbol{r} < \mathbf{w}$ and $V(\boldsymbol{r}) = \infty$ otherwise, the energy levels are

$$E_{\mathbf{n}} = \frac{h^2}{8m}\left(\frac{n_1^2}{l_1^2} + \frac{n_2^2}{l_2^2} + \frac{n_3^2}{l_3^2}\right) \tag{3.16}$$

If one of the dimensions of the box is equal to another, the energy levels become *degenerate*, that is, there are two (or more) states with the same energy. For example, if $w_1 = w_2 = w_3$, the energies for states $\mathbf{n} = (1,2,2)$, $\mathbf{n} = (2,1,2)$, and $\mathbf{n} = (2,2,1)$ are the same and the system is said to be threefold degenerate. Using periodic boundary conditions $V(\boldsymbol{0}) = V(\mathbf{w}) = const.$, instead of the fixed box boundary conditions given by Eq. (3.15), a similar procedure will lead to the same expression for $E_{\mathbf{n}}$ but with the factor 8 replaced by 2. In this case, $\mathbf{n} = (n_1,n_2,n_3) \in (\pm1,\pm2,\ldots)$ and the wave functions $\exp(\mathrm{i}\boldsymbol{p}\cdot\boldsymbol{r}/\hbar) = \exp(\mathrm{i}\boldsymbol{k}\cdot\boldsymbol{r})$ represent running waves.

3.2.2 The Harmonic Oscillator

For a 1D harmonic oscillator, located at the origin, the potential energy $V(x)$ with force constant a and the kinetic energy operator $T(x)$ are given by

$$V(x) = \tfrac{1}{2}ax^2 \quad \text{and} \quad T(x) = -\frac{\hbar^2}{2m}\frac{\mathrm{d}^2}{\mathrm{d}x^2}$$

respectively, so that the Schrödinger equation reads

$$\left[-\frac{\hbar^2}{2m}\frac{\mathrm{d}^2}{\mathrm{d}x^2} + \tfrac{1}{2}ax^2\right]\Psi = E\Psi$$

The solutions for the wave functions are

$$\Psi_n(\xi) = \left(\frac{\sqrt{\alpha/\pi}}{2^n n!}\right)^{1/2} H_n(\xi)\exp(-\tfrac{1}{2}\xi^2)$$

$$\text{with}\quad \xi = \sqrt{\alpha}x, \qquad \alpha = \hbar^{-1}\sqrt{ma}, \qquad n = 0, 1, 2, \ldots$$

where $H_n(x)$ are known as *Hermite functions* defined by

$$H_n(x) = (-1)^n \exp(x^2)\mathrm{d}^n[\exp(-x^2)]/\mathrm{d}x^n$$

Explicitly the first few Hermite functions are

$$H_0(x) = 1 \quad H_1(x) = 2x \quad H_2(x) = 4x^2 - 2 \quad H_3(x) = 8x^3 - 12x$$

$$H_4(x) = 16x^4 - 48x^2 + 12 \qquad H_5(x) = 32x^5 - 160x^3 + 120x, \ldots$$

The energy levels are given by

$$E_n = \hbar\omega(n + ½) \quad \text{with circular frequency } \omega = \sqrt{a/m}$$

For a 3D oscillator with potential

$$V(x, y, z) = ½a_x(x - x_0)^2 + ½a_y(y - y_0)^2 + ½a_z(z - z_0)^2$$

where a shift in origin to x_0, y_0 and z_0 is applied, the energy levels are given by

$$E_{\mathbf{n}} = \hbar\omega_x(n_x + ½) + \hbar\omega_y(n_y + ½) + \hbar\omega_z(n_z + ½) \tag{3.17}$$

Again, if one of the frequencies is equal to another, the levels are degenerate. For example, if $\omega_x = \omega_y = \omega_z$, for states $\mathbf{n} = (1,1,2)$, $\mathbf{n} = (1,2,1)$ and $\mathbf{n} = (2,1,1)$ the energy levels are given by $E_{\mathbf{n}} = 11\hbar\omega/2$ and the level is a threefold degenerate. Finally, we note that the presence of the zero-point energy $½\hbar\omega$ for the ground state $n = 0$ is in accord with the uncertainty relations. If the oscillator had no zero-point energy, it would have zero momentum and be located exactly at the minimum of $V(x)$. The necessary uncertainties in position and momentum thus give rise to the zero-point energy. Finally, for completeness, we mention that the selection rule for transitions between levels is given by $\Delta n = \pm 1$.

3.2.3 The Rigid Rotator

For the rigid two-particle (atom) rotator, there is no potential in the conventional sense, but there is the constraint that the distance between the two particles of the rotator should remain constant. Hence, the Schrödinger equation reads

$$-\frac{\hbar^2}{2\mu}\nabla^2\Psi = E\Psi$$

Here, μ indicates the reduced mass of the rotator given by $1/\mu = 1/m_1 + 1/m_2$, with m_1 and m_2 the masses of the two particles. The constraint is taken into account as follows. Rigid rotation is realized if we suppose that the two rigidly connected particles rotate freely over a sphere. For this problem, it is convenient to think in polar coordinates. In these coordinates, the radial variable or radius r is fixed, but the angular coordinates θ and ϕ are free. The *Laplace operator* ∇^2, acting on a scalar x, results in

$$\nabla^2 x = \frac{\partial^2 x}{\partial r^2} + \frac{2}{r}\frac{\partial x}{\partial r} + \frac{1}{r^2}\frac{\partial^2 x}{\partial \theta^2} + \frac{\cot\theta}{r^2}\frac{\partial x}{\partial \theta} + \frac{1}{r^2\sin^2\theta}\frac{\partial^2 x}{\partial \varphi^2} \quad \text{or}$$

$$\nabla^2 x = r^{-1}\frac{\partial^2}{\partial r^2}rx + r^{-2}\Lambda^2 \quad \text{with} \quad \Lambda^2 = \frac{\partial^2 x}{\partial \theta^2} + \cot\theta\frac{\partial x}{\partial \theta} + \frac{1}{\sin^2\theta}\frac{\partial^2 x}{\partial \phi^2}$$

The *Legendre operator* Λ^2 is the angular part of ∇^2. The constraint $r = const.$ implies that we can skip the radial part of ∇^2. The Schrödinger equation thus becomes

$$-\frac{\hbar^2}{2I}\Lambda^2\Psi(\theta,\phi) = E\Psi(\theta,\phi)$$

where $I = \mu r^2$ is the moment of inertia. The solutions for this equation are the *spherical harmonics*, conventionally indicated by $Y_{lm}(\theta,\phi)$, and characterized by the two quantum numbers l and m [5]. The number $l = 0, 1, 2, \ldots$ indicates the overall angular momentum, given by $[l(l+1)]^{1/2}\hbar$, while the number $m = -l, -l+1, \ldots, l-1, l$ indicates the angular momentum in a specified direction, often taken as z, and given by $m\hbar$. The energy expression is independent of the quantum number m and becomes

$$E_l = \frac{\hbar^2}{2I} l(l+1) \quad \text{with} \quad l = 0, 1, 2 \ldots \tag{3.18}$$

The abovementioned notation is the conventional one in QM for the solution of the angular part of the Laplace operator. In spectroscopy though, the symbol for the quantum number l is usually replaced by J and we will do so here as well. Each energy level is thus characterized by the quantum number J corresponding to $2J + 1$ quantum states, that is, the degeneracy is $2J + 1$. Again, for completeness, we mention that the selection rule for transitions between levels is $\Delta J = \pm 1$.

3.2.4 Many Particles

Finally, we focus our attention on systems with a Hamilton operator with additive properties. For many-particle systems, the Hamilton operator H is a function of the coordinates and the momenta of all the particles $i, j, \cdots$, that is, $H = H(i, j, \ldots)$. In a number of cases, however, the Hamilton operator is the sum (or can be approximated as the sum) of the operators $h(i)$ for single particles i, that is, $H = \Sigma_i h(i)$. These single-particle operators satisfy the (single-particle) Schrödinger wave equation

$$h(i)\phi_k(i) = \varepsilon_k \phi_k(i)$$

where $\phi_k(i)$ and ε_k denote single-particle wave functions and eigenvalues (particle energies), respectively. The total wave function Ψ for the N-particle system can be taken as the product of the individual particle wave functions $\Psi(i) = \prod_k \phi_k(i)$ so that the N-particle Schrödinger equation reads

$$H\Psi = \sum_i h(i) \prod_k \phi_k(i) = \sum_i \varepsilon_i \prod_k \phi_k(j) = E\Psi$$

and the total energy is given by $E = \sum_k \varepsilon_k$. However, we recall that the individual particles in a system are indistinguishable. This implies that $|\phi_i(k)\phi_j(l)|^2 = |\phi_i(l)\phi_j(k)|^2$ or that either $\phi_i(k)\phi_j(l) = \phi_i(l)\phi_j(k)$ or $\phi_i(k)\phi_j(l) = -\phi_i(l)\phi_j(k)$. Thus Ψ must be either symmetrical or anti-symmetrical with respect to the exchange of particle coordinates including spin. Electrons are particles with a half-integer spin, the so-called *fermions* (or Fermi–Dirac [FD] particles), for which the wave function must be anti-symmetric in the coordinates of all electrons. A direct consequence, usually referred to as *Pauli's exclusion principle*, is that each system state can be occupied with only one electron. For an electron with spin ½, this implies either spin up (½$\hbar$) or with spin down (−½$\hbar$). Making allowance for Pauli's principle, a many-electron wave function can be expanded in anti-symmetrized products of one-electron wave functions (or *spin orbitals*) ϕ_j, each ϕ_j being the product of a spatial part (or *orbital*) and a spin function ($\sigma_j = \alpha$ or β). For this the *permutation operator* $P_r{}'$ is used, which permutes the coordinates $\boldsymbol{r}$ of the wave function. A permutation is either odd or even, dependent on the number of binary switches in coordinates $|P_r{}'|$ that is required to realize that permutation. A completely anti-symmetrized wave function $\Psi^{(A)}$ is obtained by combining all $N!$ permutations $P_r{}'[\Psi]$ of the wave function $\Psi = \Psi(\boldsymbol{r}_1, \boldsymbol{r}_2, \ldots, \boldsymbol{r}_N)$, that is, by taking

$$P\Psi \equiv (N!)^{-1} \sum_{P_r{}'} (-1)^{|P_r{}'|} P_r{}' \left[\Psi(\boldsymbol{r}_1, \boldsymbol{r}_2, \ldots, \boldsymbol{r}_N)\right] \equiv (N!)^{-1} \sum_{P_r} P_r \left[\Psi(\mathbf{r})\right]$$

So, we have $(-1)^{|P_r'|} = -1$ or 1 for an odd or even permutation, respectively. For future use, we include the prefactor $(-1)^{|P_r'|}$ in the operator P_r' by writing P_r. A convenient form for such an antisymmetrized product is the *Slater determinant*:

$$\Psi^{(A)} \equiv |\phi_a(1)\,\phi_b(2)\ldots\phi_n(N)| \equiv \begin{vmatrix} \phi_a(1) & \phi_a(2) & \ldots & \phi_a(N) \\ \phi_b(1) & \phi_b(2) & \ldots & \phi_b(N) \\ \ldots & \ldots & \ldots & \ldots \\ \phi_n(1) & \phi_n(2) & \ldots & \phi_n(N) \end{vmatrix}$$

where (i) denotes the coordinates of an electron including spin and N is the number of electrons. Similarly, for integer spin particles, the so-called *bosons* (or Bose–Einstein [BE] particles), the wave function must be symmetric in all the coordinates of the particles and system states may contain any number of particles. This symmetry can be realized by taking the *permanent* of individual particle wave functions indicated by

$$\Psi^{(S)} = ||\phi_a(1)\,\phi_b(2)\ldots\phi_n(N)||$$

The permanent is constructed similarly as the determinant but upon expanding one takes all signs as positive. Hence $(+1)^{|P_r'|} = 1$ always. To obtain a normalized wave function Ψ, if all ϕs are orthonormal, one includes a normalization factor $C = (N!)^{-1/2}$

$$\Psi = C\,|\phi_a(1)\,\phi_b(2)\ldots\phi_n(N)| \quad \text{or} \quad \Psi = C\,||\phi_a(1)\,\phi_b(2)\ldots\phi_n(N)||$$

for fermions and bosons, respectively. Here, we labeled the system state Ψ explicitly by all its particle coordinates (i) with $i = 1, \ldots, N$ and individual particle states ϕ_j can show up more than once. If we label Ψ with all individual particle states ϕ_j with $j = a, \ldots, n$, we allow n_j particles in each state ϕ_j and C becomes $C = [N!/\Pi_j(n_j!)]^{-1/2}$. Note that for FD particles, C still yields $C = (N!)^{-1/2}$ because all n_j are either 1 or 0. The factor C is often incorporated in the definition of the anti-symmetrized product function so that we write for the wave function $\Psi = |\phi_a(1)\;\phi_b(2)\ldots\phi_n(N)|$ or even, assuming a fixed order of coordinates, $\Psi = |\phi_a\;\phi_b\ldots\phi_n|$. The same convention is used for the symmetrized product function. A straightforward calculation shows that for the Hamilton operator $H = \Sigma_i h(i)$ the N-particle system eigenvalue is still

$$E = \sum\nolimits_i \varepsilon_i$$

When $H \neq \Sigma_i h(i)$, the energy E is no longer the sum of the particle energies and the product function and determinant (or permanent) no longer yield the same energy. In that case approximation methods must be used.

3.3 Approximate Quantum Mechanics Solutions

Since for many particle systems the Schrödinger equation cannot be solved exactly, we need approximations. In this section, we describe some of them.

3.3.1 The Born–Oppenheimer Approximation

Let us separate the electronic motion in order to arrive at the concept of potential energy between molecules. In a molecule, both electrons and nuclei are considered as particles.

For the moment, we denote the coordinates of all the electrons by **r** and all the coordinates of the nuclei by **R**. Neglecting relativistic terms, spin interactions, and the possibility of explicit time dependence, the Hamilton operator $H(\mathbf{r},\mathbf{R})$ consists of three terms representing the kinetic energy of the nuclei $T^{(n)}(\mathbf{R})$, that of the electrons $T^{(e)}(\mathbf{r})$ and the potential energy of interaction $V(\mathbf{r},\mathbf{R})$ (appendix VIII of [6]):

$$H(\mathbf{r},\mathbf{R}) = T^{(n)}(\mathbf{R}) + [T^{(e)}(\mathbf{r}) + V(\mathbf{r},\mathbf{R})] = T^{(n)}(\mathbf{R}) + H^{(e)}(\mathbf{r},\mathbf{R})$$

where the labels (n) and (e) indicate the nuclei and electrons, respectively. A further step can be made by adopting the *Born–Oppenheimer approximation*

$$\Psi(\mathbf{r},\mathbf{R}) = \Psi^{(e)}(\mathbf{r};\mathbf{R})\Psi^{(n)}(\mathbf{R})$$

where the separate factors are obtained from

$$H^{(e)}(\mathbf{r};\mathbf{R})\Psi^{(e)}(\mathbf{r};\mathbf{R}) = E(\mathbf{R})\Psi^{(e)}(\mathbf{r};\mathbf{R}) \quad \text{and}$$
$$[T^{(n)}(\mathbf{R}) + E(\mathbf{R})]\,\Psi^{(n)}(\mathbf{R}) = E_{\text{tot}}\Psi^{(n)}(\mathbf{R}) \tag{3.19}$$

As Eq. (3.19) is usually solved for a fixed configuration of nuclei, so that the nuclear coordinates enter the electronic wave function parametrically, we write $\Psi^{(e)}(\mathbf{r};\mathbf{R})$. We see that in this approximation $E(\mathbf{R})$, the electronic energy of the electrons moving in the field of the fixed nuclei *plus* the repulsion of the fixed nuclei, can be considered as the *effective potential energy* for the motion of the nuclei. The approximation thus allows us to discuss the electronic structure and nuclear motion separately. In fact, $E(\mathbf{R})$ is the potential energy employed in all our discussions on intermolecular interactions. Adding the energy of nuclear motion to E, we obtain the total energy E_{tot}.

3.3.2 The Variation Principle

Let us now focus on the electronic Schrödinger equation [7]. Dropping the label (e) and the nuclear coordinates **R**, Eq. (3.19) reads

$$H(\mathbf{r})\Psi_j(\mathbf{r}) = E_j\Psi_j(\mathbf{r}) \tag{3.20}$$

where the index j refers to a specific electronic wave function Ψ_j with associated energy E_j. Indicating from now on the coordinates of electron i with $\boldsymbol{r}_i$, in general a many-electron wave function $\Psi(\boldsymbol{r}_1,\ldots,\boldsymbol{r}_N)$ can be expanded as

$$\Psi(\boldsymbol{r}_1,\ldots,\boldsymbol{r}_N) = \sum\nolimits_k c_k\Phi_k(\boldsymbol{r}_1,\ldots,\boldsymbol{r}_N)$$

where c_ks are the coefficients of the expansion and Φ_ks are the members of a proper complete set. For example, for many-electron systems, Φ_k may be a properly anti-symmetrized product function of single-electron functions or Slater determinant:

$$\Phi_k(\boldsymbol{r}_1,\ldots,\boldsymbol{r}_N) = |\phi_l(\boldsymbol{r}_1)\phi_m(\boldsymbol{r}_2)\ldots\phi_p(\boldsymbol{r}_N)|$$

where k is a collective index representing $l,\ldots,p$. If we collect the functions Φ_k and the coefficients c_k in a column matrix, we may write $\Psi = \mathbf{\Phi}^{\mathrm{T}}\mathbf{c}$. Now the Hamilton operator acting on Ψ yields a new function Ψ'. Expanding Ψ' similarly as Ψ, we have

$$H\Psi = \Psi' \quad \text{or} \quad H\mathbf{\Phi}^{\mathrm{T}}\mathbf{c} = \mathbf{\Phi}^{\mathrm{T}}\mathbf{c}'$$

Multiplying the last equation by $\boldsymbol{\Phi}^*$ and integrating over all coordinates results in

$$\left(\int \boldsymbol{\Phi}^* H \boldsymbol{\Phi}^{\mathrm{T}} \mathrm{d}\mathbf{r}\right) \mathbf{c} = \left(\int \boldsymbol{\Phi}^* \boldsymbol{\Phi}^{\mathrm{T}} \mathrm{d}\mathbf{r}\right) \mathbf{c}' \quad \text{or} \quad H_{ij}c_j = S_{ij}c_j' \quad \text{or} \quad \mathrm{H}\mathbf{c} = \mathbf{S}\mathbf{c}'$$

where $H_{ij} = \mathrm{H} = \langle \Phi_i | H | \Phi_j \rangle$ and $S_{ij} = \mathbf{S} = \langle \Phi_i | \Phi_j \rangle$ are referred to as *Hamilton* and *overlap matrix elements*, respectively. For an orthonormal set, $\mathbf{S} = \mathbf{1}$ and $\mathrm{H}\mathbf{c} = \mathbf{c}'$.

Consider now the Schrödinger equation, Eq. (3.20). Applying the same procedure as before, that is, multiplying by $\boldsymbol{\Phi}^*$ and integrating over all coordinates, we obtain

$$\mathrm{H}\mathbf{c} = E\mathbf{S}\mathbf{c} \qquad (\text{or if } \mathbf{S} = \mathbf{1}, \mathrm{H}\mathbf{c} = E\mathbf{c})$$

In practice, we use a truncated set of n basis functions. Let us write for this case $\mathrm{H}^{(n)}\mathbf{c}^{(n)} = E^{(n)}\mathbf{S}^{(n)}\mathbf{c}^{(n)}$. This equation will have eigenvalues $E_1{}^{(n)}$, $E_2{}^{(n)}$, etc. One can show, if we use a truncated set with $n+1$ functions, that

$$E_i^{(n)} > E_i^{(n+1)} > \cdots > E_i^{(\infty)} = E_0$$

where $i = 1, \ldots, n$ and E_0 is the exact solution. This result implies that by admitting one more function to an n-truncated complete set, an improved estimate is made for the cigenvalues. This is true for the ground state as well as all excited states. The conventional *variation theorem* arises by taking $n = 1$ so that $\mathbf{c}$ drops out and we have $H_{11}{}^{(1)} - E_1{}^{(1)}S_{11} = 0$ or (omitting the superscript and subscripts)

$$E = \frac{\langle \Phi | H | \Phi \rangle}{\langle \Phi | \Phi \rangle} \geq E_0 \tag{3.21}$$

where E_0 is the energy of the first state of the corresponding symmetry, usually the ground state. According to this theorem, an upper-bound E can be obtained from an approximate wave function. Only in cases where Φ is an exact solution the equality sign holds. In all other cases, the quotient in the middle results in a larger value than the exact value E_0. This implies that one can introduce one or more parameters $\boldsymbol{\alpha}$ in the electronic wave function so that it reads $\Phi(\mathbf{r};\boldsymbol{\alpha})$ and minimize the left-hand side of Eq. (3.21) with respect to the parameters $\boldsymbol{\alpha}$. In this way the best approximate wave function, given a certain functional form, is obtained.

A direct but important consequence of the variation theorem is the *Hellmann–Feynman theorem* [8], which says that, if we have the Hamilton operator $H(\mathbf{p},\mathbf{r};\boldsymbol{\alpha})$ dependent on the parameters $\boldsymbol{\alpha}$, the energy derivative $\partial E/\partial \boldsymbol{\alpha} = \langle \Psi | \partial H(\boldsymbol{\alpha})/\partial \boldsymbol{\alpha} | \Psi \rangle$, to be compared to the classical expression $\partial E/\partial \boldsymbol{\alpha} = \partial \mathscr{H}(\boldsymbol{\alpha})/\partial \alpha$. The theorem finds an important application in the calculation of intermolecular forces via $\mathbf{F} = -\partial E/\partial \boldsymbol{\alpha}$. Also, the virial theorem remains valid in QM. Using that the Hamilton operator H is a Hermitian operator, one can show that the expectation value $\langle [H,A] \rangle = 0$ for any linear, time-independent operator A. For $A = p{\cdot}r = -\mathrm{i}\hbar[\mathbf{r}{\cdot}(\partial(\cdot)/\partial \mathbf{r})]$, we obtain $[H,A] = \mathrm{i}\hbar[\mathbf{r}{\cdot}(\partial V/\partial \mathbf{r})] - 2\mathrm{i}\hbar T$. Here V is the potential energy operator. Using $\langle [H,A] \rangle = 0$ leads to the QM *virial theorem* given by

$$2\langle T \rangle = \langle (\partial V/\partial \boldsymbol{r}){\cdot}\boldsymbol{r} \rangle$$

Approximate wave functions do not necessarily obey the virial theorem, but one can scale these wave functions in such a way that they *do obey* the virial theorem. This is done by introducing a scaling parameter η for all coordinates in the system. For example, for a diatomic molecule, this means that the bond distance $R \to \eta R$ and the electron coordinates $\boldsymbol{r} \to \eta \boldsymbol{r}$. Using $w = \eta R$, the kinetic and potential energy become $\langle T \rangle = \eta^2 \langle T(w) \rangle \equiv \eta^2 \vartheta(w)$

and $\langle V\rangle = \eta\langle V(w)\rangle \equiv \eta V(w)$, and thus $\langle H\rangle = E(\eta,R) = \eta^2\vartheta(w) + \eta V(w)$. Considering R as a fixed parameter, the optimum value is given by $\mathrm{d}E(\eta,R)/\mathrm{d}w = 0$, resulting in the expression $\eta(w) = -[V + w(\mathrm{d}V/\mathrm{d}w)]/[2\vartheta + w(\mathrm{d}\vartheta/\mathrm{d}w)]$. At equilibrium $\mathrm{d}\vartheta/\mathrm{d}w = \mathrm{d}V/\mathrm{d}w = 0$, $\eta^* = -V/2\vartheta$ and the scaled energy becomes $E^* = \eta^{*2}\vartheta + \eta^* V = -V^2/4\vartheta$. To obtain the full energy function, one substitutes $w(R)$, as solved from $R = w/\eta(w)$, in $E(w)$ to yield $E(R)$.

Example 3.1 ***The Helium Atom*** As an example of the variation theorem, we discuss the He atom with nuclear charge $Z = 2$ and 2 electrons. The Hamilton operator for He in atomic units[8] is

$$H = -\tfrac{1}{2}\left(\nabla_1^{\ 2} + \nabla_2^{\ 2}\right) - Z(1/r_1 + 1/r_2) + 1/r_{12} = K + N + I$$

where the first term indicates the kinetic energy of electrons 1 and 2, the second term the nuclear attraction for electrons 1 and 2, and the third term the electron–electron repulsion. As a normalized trial wave function, we use a product of two simple exponential functions, namely $\Psi = \phi(1)\phi(2) = (\eta^3/\pi)\exp[-\eta(r_1 + r_2)]$, where the parameter η, acting as an effective nuclear charge, is introduced. The expectation value of the Hamilton operator is $\langle\Psi|H|\Psi\rangle$. The kinetic energy integral results in $K = \eta^2$, the nuclear attraction integral yields $N = -2Z\eta$, and the electron–electron repulsion integral is $I = 5\eta/8$ [7]. The total energy is thus given by $\langle\Psi|H\,|\Psi\rangle = (K+N)+I = \eta^2 - 2Z\eta + 5\eta/8$, at minimum when

$$\partial\langle\Psi|H|\Psi\,\rangle/\partial\eta = 0 = 2\eta - 2Z + 5/8 \quad \text{or when} \quad \eta = (2Z - 5/8)/2 = 27/16 = 1.688$$

The energy is $-2.848\,E_\mathrm{h}$, to be compared with the experimental value of $-2.903\,E_\mathrm{h}$.

We now return to the original linear expansion

$$\Psi = \sum\nolimits_k c_k\Phi_k \quad \text{or} \quad \Psi = \mathbf{\Phi}^\mathrm{T}\mathbf{c}$$

Multiplying $H\Psi = E\Psi$ by Ψ^*, integrating over all electron coordinates and using the above-mentioned expansion results in

$$\mathbf{c}^\dagger\mathrm{H}\mathbf{c} = E\mathbf{c}^\dagger\mathrm{S}\mathbf{c} \quad \text{or} \quad \mathbf{c}^\dagger(\mathrm{H} - E\mathbf{S})\mathbf{c} = 0 \quad \text{or} \quad E = \mathbf{c}^\dagger\mathrm{H}\mathbf{c}/\mathbf{c}^\dagger\mathrm{S}\mathbf{c}$$

where † denotes the Hermitian transpose. Minimizing with respect to c_i yields

$$\partial[\mathbf{c}^\dagger(\mathrm{H} - E\mathbf{S})\mathbf{c}]/\partial\mathbf{c} = \mathbf{c}^\dagger(\mathrm{H} - E\mathbf{S}) + (\mathrm{H} - E\mathbf{S})\mathbf{c} = 0$$

Because the **c**s are independent, the determinant of $(\mathrm{H} - E\mathbf{S})$ must vanish or $\det(\mathrm{H} - E\mathbf{S}) = 0$, corresponding to the generalized eigenvalue equation $\mathrm{H}\mathbf{c} - E\mathbf{S}\mathbf{c} = \mathbf{0}$. This equation can be transformed to the conventional eigenvalue equation by applying a unitary transformation to the basis functions. Since $\det(\mathbf{S}) \neq 0$, we can take

$$\mathbf{\Phi}' = \mathbf{S}^{-1/2}\mathbf{\Phi} \quad \text{to obtain} \quad \langle\mathbf{\Phi}'|\mathbf{\Phi}'\rangle = 1 \quad \text{and} \quad \mathrm{H}' = \langle\mathbf{\Phi}'|H|\mathbf{\Phi}'\rangle = \mathbf{S}^{-1/2}\mathrm{H}\mathbf{S}^{-1/2}$$

---

8 In the *atomic unit system*, the length unit is the *Bohr radius*, $1\ a_0 = (4\pi\varepsilon_0)\hbar^2/me^2 = 0.5292 \times 10^{-10}$ m, and the energy unit is the *Hartree*, $1\ E_\mathrm{h} = me^4/(4\pi\varepsilon_0\hbar)^2 = 4.360 \times 10^{-18}$ J = 27.11 eV. Here $\hbar = h/2\pi$, where h is Planck's constant, m and e are the mass and the charge of an electron, respectively, and ε_0 is the permittivity of vacuum. In this system, $e^2/(4\pi\varepsilon_0) = 1$ and $\hbar^2/2m = \tfrac{1}{2}$. One also uses the *Rydberg* with 2 Ry = 1 E_h. In the Hartree convention the kinetic energy reads $-\tfrac{1}{2}\nabla^2$ and the electrostatic potential $1/r$, while in the Rydberg convention they read $-\nabla^2$ and $2/r$, respectively. Expressing results in atomic units renders them independent of the numerical values of the fundamental constants used.

In this way, we are led to the conventional eigenvalue problem $\mathrm{H}'\mathbf{c} - E\mathbf{c} = \mathbf{0}$, which can be solved in the usual way to find the eigenvalues. The lowest one approximates the energy of the ground state. Substituting the eigenvalues back in the secular equation results in the coefficients $\mathbf{c}$ from which the approximate wave function can be constructed. In case the functions $\boldsymbol{\Phi}$ are already orthonormal, $\mathbf{S} = \mathbf{1}$ and thus $\mathbf{S}^{-1/2} = \mathbf{1}$, and we arrive directly at the standard eigenvalue problem.

Since the eigenfunctions are all orthonormal or can be made to be so, the approximate wave functions are, like the exact ones, orthonormal. Moreover, they provide stationary solutions. As usual with eigenvalue problems we can collect the various eigenvalues E_k in a diagonal matrix $\mathbf{E}$ where each eigenvalue takes a diagonal position. The eigenvectors can be collected in a matrix $\mathbf{C} = (\mathbf{c}_1|\mathbf{c}_2|\ldots)$. In this case, we can summarize the problem by writing

$$\mathrm{H}\mathbf{C} = \mathbf{ESC} \quad \text{or equivalently} \quad \mathbf{E} = \mathbf{C}^{\dagger}\mathrm{H}\mathbf{C} \quad \text{and} \quad \mathbf{C}^{\dagger}\mathbf{SC} = \mathbf{1} \tag{3.22}$$

which shows that the matrix $\mathbf{C}$ brings H and $\mathbf{S}$ simultaneously to a diagonal form, $\mathbf{E}$ and $\mathbf{1}$, respectively. This approach is widely used nowadays for the calculation of molecular wave functions.

3.3.3 The Hartree–Fock Method

Even by using the Born–Oppenheimer approximation, an exact solution for molecules exists only for the H_2^+ ion molecule [9]. No exact solutions are known for bonding in many-electron molecules or for solids. The description of bonding in molecules and solids relies heavily on orbital theories. As stated before, making allowance for Pauli's principle, a many-electron wave function can be expanded in *Slater determinants* containing *spin orbitals* ϕ_j, each ϕ_j being the product of a spatial part (or *orbital*) χ_j and a spin function $\sigma_j = \alpha$ or $\sigma_j = \beta$ [7, 10]. Sometimes the overbar notation is used whereby $\chi\alpha \rightarrow \chi$ and $\chi\beta \rightarrow \overline{\chi}$.

An electronic wave function must also be an eigenfunction of the total spin operator S^2. The determinants, constructed from the same orbitals, can be grouped in such a way that each group is an eigenfunction of S^2. Such a group is called a *configuration* Δ. Approximating the wave function by more than one configuration is denoted by *configuration interaction* (CI). Hence, we can write

$$\Psi_i^{(e)}(\mathbf{r};\mathbf{R}) = \sum\nolimits_j C_{ij}\Delta_j$$

If all possible configurations for a given set of spin orbitals are included, one speaks of complete CI, otherwise of a limited CI. Choosing spin orbitals that are spread throughout a molecule (and belong to one of the irreducible representations of the symmetry group of the molecule), one obtains what is called *molecular orbital* (MO) theory. In solids this becomes the crystal orbital theory although it is usually called the *tight binding* (TB) *approximation*. Within this choice, one frequently tries to describe the system with only one configuration. For a closed shell ground state this configuration contains only one determinant.

$$\Psi_i^{(e)}(\mathbf{r};\mathbf{R}) = \Delta_i = |\phi_a(1)\phi_b(2)\ldots\phi_n(N)|$$

for which the total energy $E = E_{\text{ele}} + E_{\text{nuc}}$ in E_{h} is given by

$$E = E_{\text{ele}} + E_{\text{nuc}} = \left[2\sum\nolimits_i h_i + \sum\nolimits_{i,j}(2J_{ij} - K_{ij})\right] + ½\sum\nolimits_{\text{A,B}}\frac{Z_{\text{A}}Z_{\text{B}}}{R_{\text{AB}}} \quad \text{where} \tag{3.23}$$

$$h_i = \langle\phi_i| - ½\nabla^2(1) - \sum\nolimits_A \frac{Z_A}{r_{1A}}|\phi_i\rangle,$$

$$J_{ij} = \langle\phi_i^*(1)\phi_j^*(2)|\frac{1}{r_{12}}|\phi_i(1)\phi_j(2)\rangle, \quad \text{often abbreviated to } \langle ij|ij\rangle \quad \text{and}$$

$$K_{ij} = \langle\phi_i^*(1)\phi_j^*(2)|\frac{1}{r_{12}}|\phi_j(1)\phi_i(2)\rangle, \quad \text{often abbreviated to } \langle ij|ji\rangle.$$

In these equations, r_{1A} denotes the distance between electron 1 and nucleus A with charge Z_A while r_{12} and R_{AB} denote the distances between electrons 1 and 2 and nuclei A and B, respectively. The first term in Eq. (3.23) indicates the sum over the contributions of the electrons in the field of the nuclei, the second the Coulomb repulsion between electrons, the third the exchange interaction between electrons (due to the anti-symmetry of the wave function), and the last term the nuclear repulsion between the various nuclei A and B.

Applying the variation principle to the electronic energy expression to determine an approximate product wave function or a one-determinant wave function under the constraint $\langle\phi_i|\phi_j\rangle = \delta_{ij}$ leads to the *Hartree* or *Hartree–Fock self-consistent field* (HF-SCF) equations, respectively. For the latter we have

$$f(1)\phi_i(1) = \varepsilon_i\phi_i(1) \quad i = 1, \ldots, N \quad \text{with} \tag{3.24}$$

$$f(1) = -½\nabla^2(1) - \sum\nolimits_A \frac{Z_A}{r_{1A}} + \int \frac{\rho(2)}{r_{12}}\,d\mathbf{r}_2 - \int \frac{\rho_{Xi}(1,2)}{r_{12}}\,d\mathbf{r}_2 \quad \text{where} \tag{3.25}$$

$$\rho(1) = \sum n_j\phi_i^*(1)\phi_i(1) \quad \text{and}$$

$$\rho_{Xi}(1,2) = \sum\nolimits_j n_j \frac{\phi_i^*(1)\phi_j^*(2)\phi_j(1)\phi_i(2)}{\phi_i^*(1)\phi_i(1)}$$

Here the number n_j is the occupation number, 1 for an occupied orbital and 0 for an unoccupied one. The *Fock operator* $f(j)$ is an effective one-electron operator, describing the behavior of the electrons in the field of the others. The first two terms in Eq. (3.25) represent the kinetic energy and nuclear attraction energy, respectively. The last two terms describe the Coulomb repulsion and exchange energy, both thought to be due to a charge density, $\rho(1)$ and $\rho_{Xi}(1,2)$, respectively. The latter quantity is nonlocal and different for each spin orbital. The *Hartree–Fock* (HF) equations are (pseudo-)eigenvalue equations where the eigenvalues (or Lagrange multipliers) ε_i are to be interpreted as orbital energies. The orbital energies ε_i are given by Eq. (3.24), applying the Fock operator $f(j)$ upon each spin orbital ϕ_j. The result is

$$\varepsilon_i = h_i + \sum\nolimits_j (2J_{ij} - K_{ij})$$

Note that the total electronic energy E_{ele} *does not equal* $2\Sigma_j\,\varepsilon_j$.

Only for atoms Eq. (3.24) can be solved numerically to any desired accuracy. In molecules and solids, the orbital χ_j is usually approximated by a linear combination of basis functions ξ_k. If the basis functions are largely localized at the various atomic sites, they are conventionally addressed as atomic orbitals, so that one speaks of a *linear combination of atomic orbitals* (LCAO). Hence for a spin orbital ϕ_j,

$$\phi_j = \sigma_j\chi_j = \sigma_j\sum\nolimits_k c_{jk}\xi_k$$

The set of basis functions ξ_k is usually called the *basis set*. A *minimal basis set* is a set, which employs only one basis function of the proper symmetry per occupied orbital for each atom. For example, for H_2 we use one basis function, say $1s_A$, at nucleus A and one, say $1s_B$, at nucleus B. Similarly, for CO we have at each atom the 1s, 2s, $2p_x$, $2p_y$, and $2p_z$ functions. For large sets with functions of sufficient flexibility, the results approach the results which would be obtained from a numerical solution of Eq. (3.24). The latter is called the *HF limit*. Restricting ourselves to closed shell configurations as before, each orbital is doubly occupied, and we may write $\boldsymbol{\phi} = \mathbf{C}\boldsymbol{\xi}$. Given a set of basis functions ξ_k (equivalently $\boldsymbol{\xi}$) the optimum coefficients c_{jk} are obtained, similar as for Eq. (3.22), from

$$\mathbf{FC} - \mathbf{ESC} = \mathbf{0} \tag{3.26}$$

The *overlap integrals* S_{rs} and *Fock integrals* F_{rs}, taking the place of the Hamilton matrix elements, are given by

$$S_{rs} = \int \xi_r^* \xi_s \, d\mathbf{r} \quad \text{and} \quad F_{rs} = \int \xi_r^* f(1) \xi_s \, d\mathbf{r}$$

where the integration is over all space and $f(1)$ again indicates the Fock operator.

The evaluation of Fock integrals assumes the LCAO coefficients to be known and involves the calculation of many one- and two-electron integrals. The former aspect makes the LCAO-MO-SCF method an iterative procedure. The number of integrals increases with the fourth power of the basis set size, and thus various calculational schemes have been applied. Schemes containing (semi-)empirical parameters, to be gauged against experimental data, are indicated as *semi-empirical* calculations [11]. Ab-initio calculations, that is, calculations without empirical parameters, can be done with a maximum number of atoms that can be handled, which is largely limited by the computing power and memory available. Although Slater-type (exponential) orbitals (STOs) are preferred theoretically for basis sets, Gauss-type orbitals (GTOs) have considerable numerical advantages so that expansion of STOs in GTOs has become common [12].

The HF theory is general, can be applied to molecules as well as to solids, but it does not represent an exact solution. It takes into account the exchange correlation, that is, electrons with the same spin repel each other, and one obtains in this approximation a diminished probability of finding another electron near the reference electron. This diminished probability for one electron being in the neighborhood of another is the so-called *exchange hole* or *Fermi hole*. Integrating over all space this hole contains exactly one electron. On the other hand, the Coulomb correlation, that is, the repulsion of electrons with different spins, is not properly taken into account although each electron is moving in the average field of the others. One can show that a diminished probability near and an enhanced probability somewhat further away from the reference electron should be present, which, when integrated over all space, yields exactly zero electron. However, within the HF framework such a *Coulomb hole*, corresponding to the exchange hole, is absent. The difference between the HF limit energy E_{HF} and the exact energy E_0 is called the *correlation energy* [13, 14].

Example 3.2 ***The Minimal Basis Set H_2^+ Molecule Ion and H_2 Molecule*** The simplest molecule is H_2^+, containing two nuclei (protons) A and B separated by the internuclear distance R, and one electron, while the next simplest is H_2, containing two electrons indicated by 1 and 2. Within the MO model the simplest description of these molecules

Table 3.2 Bonding data for the H_2^+ and H_2 molecules.

	D (E_h)	D (eV)	R (a_0)	R (Å)	ϑ	$-V$	$-V/2\vartheta$
H_2^+: $\eta = 1.000$	0.06483	1.76	2.493	1.319	0.3827	0.9475	1.238
H_2^+: $\eta = 1.238$	0.08651	2.35	2.000	1.058	0.5865	1.1730	1.000
Exp.	0.10264	2.79	1.997	1.057			
H_2: $\eta = 1.000$	0.0991	2.70	1.603	0.848	0.7927	1.8914	1.193
H_2: $\eta = 1.193$	0.1282	3.49	1.385	0.732	1.1282	2.2564	1.000
Exp.	0.1646	4.46	1.401	0.741			

D dissociation energy, R bond distance, ϑ kinetic energy, V potential energy.
Data from [15].

is obtained by using a minimal basis set of a 1s function $\xi_A = \pi^{-1/2}\exp[-(\boldsymbol{r} - \boldsymbol{r}_A)])$ for atom A as well as for atom B (ξ_B). The overlap integral S for the basis functions given by $S = \langle\xi_A|\xi_B\rangle = \langle\xi_B|\xi_A\rangle$. From these two basis functions we can construct the symmetrical MO $\chi_1 = c_1\xi_A + c_2\xi_B$ and the anti-symmetrical MO $\chi_2 = c_3\xi_A - c_4\xi_B$. In this case, the expansion coefficients c_i are determined fully by the symmetry of the molecule and the normalization condition $\langle\chi_1|\chi_1\rangle = \langle\chi_2|\chi_2\rangle = 1$. This leads for χ_1 to $c_1 = c_2 = [2(1+S)]^{-1/2}$ and for χ_2 to $c_3 = c_4 = [2(1-S)]^{-1/2}$. In general, the coefficients c_i must be carried through the calculation to determine them via Eq. (3.24). From the spatial orbitals χ we can form four spin orbitals ϕ by multiplication with the spin functions α or β. Hence, either in full or using the overbar notation,

$$\phi_1 = \chi_1\alpha \text{ or } \chi_1, \quad \phi_2 = \chi_1\beta \text{ or } \overline{\chi}_1, \quad \phi_3 = \chi_2\alpha \text{ or } \chi_2, \quad \phi_4 = \chi_2\beta \text{ or } \overline{\chi}_2$$

The spin orbitals ϕ_1 and ϕ_2 are degenerate, have lower energy, and describe bonding. Similarly, ϕ_3 and ϕ_4 are also degenerate and describe the first excited state.

For H_2^+, with Hamilton operator $H = -\frac{1}{2}\nabla^2 - (1/r_A + 1/r_B) + 1/R$, E becomes

$$E = -1/2 + (h_{AA} + h_{AB})/(1 + S) \quad \text{where}$$

$$h_{AA} = \langle\xi_A| - 1/r_A + 1/R|\xi_A\rangle \quad \text{and} \quad h_{AB} = \langle\xi_A| - 1/r_B + 1/R|\xi_B\rangle$$

The full expression for E is somewhat lengthy [7, 15], and we omit it here. In Table 3.2, the result for the dissociation energy $D = E - E(\text{H}) = E - 0.5$ is given, together with the experimental data, indicating an underestimate of D by about 37%.

The HF ground state for H_2 is given by $\Psi = |\chi_1\overline{\chi}_1|$ with the Hamilton operator $H = -\frac{1}{2}(\nabla_1^2 + \nabla_2^2) - \Sigma_{i=1,2,j=A,B}(1/r_{ij} + 1/r_{ij}) + 1/r_{12} + 1/R$. For two electrons the determinant can be split into a spatial and spin part according to

$$\Psi = [\xi_A(1) + \xi_B(1)][\xi_A(2) + \xi_B(2)][\alpha(1)\beta(2) - \beta(1)\alpha(2)]/\sqrt{2}[2(1+S)]$$

leading, after a somewhat lengthy, straightforward calculation, to the energy

$$E = -1 + 2(h_{AA} + h_{AB})/(1 + S) + J - 1/R \quad \text{where}$$

$$J = \langle\chi_1(1)\chi_1(2)|1/r_{12}|\chi_2(1)\chi_2(2)\rangle$$

The total energy E of the H_2 molecule is thus twice the energy of the H_2^+ molecule ion except for the term J, correcting for the electron–electron repulsion, and for the term $-1/R$, correcting for counting the nuclear repulsion twice. The full expression, again being lengthy [7, 15], is omitted. The results and experimental data given in Table 3.2 show an underestimate of $D = E - 2E(\text{H}) = D - 1$ by about 43%.

The approximate wave functions as discussed earlier do not obey the virial theorem, which might have been expected as no real variation is done. As stated in Section 3.3.2, one can scale the wave function by introducing a parameter η, so that the basis functions become $\xi_X = (\eta^3/\pi)^{1/2}\exp[-\eta(\boldsymbol{r} - \boldsymbol{r}_X)]$. Using $\eta^* = -V/2\vartheta$ and $E^* = -V^2/4\vartheta$ leads to the data given in Table 3.2 which show a considerable improvement in D and R when compared with experiment. Nevertheless, accuracy is not really in the chemical range ($\cong$0.05 eV or $\cong$5 kJ mol^{-1}). An elaborate review on bonding in these molecules in given in [16].

3.3.4 Perturbation Theory

Another way to approximate wave functions is by perturbation theory. If we use the same matrix notation again, we may partition the equation $\text{H}\mathbf{c} = E\mathbf{c}$ to [7]

$$\begin{pmatrix} \text{H}_{11} & \text{H}_{12} \\ \text{H}_{21} & \text{H}_{22} \end{pmatrix} \begin{pmatrix} \mathbf{a} \\ \mathbf{b} \end{pmatrix} = E \begin{pmatrix} \mathbf{a} \\ \mathbf{b} \end{pmatrix}$$

Solving for $\mathbf{b}$ from $\text{H}_{21}\mathbf{a} + \text{H}_{22}\mathbf{b} = E\mathbf{b}$ we obtain

$$\mathbf{b} = (E1_{22} - \text{H}_{22})^{-1}\text{H}_{21}\mathbf{a}$$

which upon substitution in $\text{H}_{11}\mathbf{a} + \text{H}_{12}\mathbf{b} = E\mathbf{a}$ yields

$$\text{H}_{\text{eff}}\mathbf{a} = E\mathbf{a} \quad \text{with} \quad \text{H}_{\text{eff}} \equiv \text{H}_{11} + \text{H}_{12}(E1_{22} - \text{H}_{22})^{-1}\text{H}_{21}$$

Let us take the number of elements in $\mathbf{a}$ equal to 1. In this case the coefficient drops out and we have $E = \text{H}_{\text{eff}}(E)$. Since H_{eff} depends on E we have to iterate to obtain a solution and we do so by inserting as a first approximation $E = \text{H}_{11}$ in H_{eff}. By writing $(E1_{22} - \text{H}_{22})^{-1} = \{(E1_{22} - \text{H}_{22}{}^{(\text{d})})[1 - \text{H}_{22}{}^{(\text{nd})}/(E1_{22} - \text{H}_{22}{}^{(\text{d})})\}^{-1}$, where $\text{H}_{22}{}^{(\text{d})}$ and $\text{H}_{22}{}^{(\text{nd})}$ denote the diagonal and non-diagonal part of H_{22}, and using $(1 - \text{X})^{-1} \cong 1 + \text{X}$, we obtain second order in off-diagonal elements

$$E = H_{11} + \sum\nolimits_{m \neq 1} \frac{H_{1m}H_{m1}}{H_{11} - H_{mm}} \tag{3.27}$$

This is a general form of perturbation analysis since the basis is entirely arbitrarily; in particular, it is not necessary to assume a complete set, and so far it has not been necessary to divide the Hamilton operator into a perturbed and unperturbed part.

The conventional *Rayleigh–Schrödinger perturbation* equations result if we do divide the Hamilton operator into an unperturbed part H_0 and perturbed part H' or

$$H = H_0 + \lambda H'$$

where λ is the order parameter to be used for classifying orders, for example, a term in λ^n being of nth order. The perturbation may be regarded as switched on by changing λ from 0 to 1, assuming that the energy levels and wave functions are continuous functions of λ. Furthermore, we assume that we have the exact solutions of

$$H_0\Phi_k = E_k^{(0)}\Phi_k$$

and we use these solutions Φ_k in Eq. (3.27). The matrix elements become

$$H_{11} = \langle \Phi_1 | H_0 + \lambda H' | \Phi_1 \rangle = E_1^{(0)} + \lambda \langle \Phi_1 | H' | \Phi_1 \rangle$$

$$H_{1m} = \langle \Phi_1 | H_0 + \lambda H' | \Phi_m \rangle = \lambda \langle \Phi_1 | H' | \Phi_m \rangle$$

Since label 1 is arbitrary, we replace it by k to obtain the final result:

$$E_k = E_k^{(0)} + \langle \Phi_k | H' | \Phi_k \rangle + \sum\nolimits_{m \neq k} \frac{\langle \Phi_k | H' | \Phi_m \rangle \langle \Phi_m | H' | \Phi_k \rangle}{E_k^{(0)} - E_m^{(0)}} \tag{3.28}$$

where λ has been suppressed. This expression is used among others for the calculation of the scattering of radiation with matter (first-order term) and the van der Waals forces between molecules (first- and second-order terms).

Example 3.3 ***The Helium Atom Revisited*** To illustrate perturbation theory, we discuss the He atom again. We consider the electron–electron repulsion as the perturbation and as zeroth-order wave function we use the product function $\Psi = \phi(1)\phi(2) = (Z^3/\pi)\exp[-Z(r_1 + r_2)]$ with the nuclear charge Z instead of the variation parameter η. For the kinetic energy integral K we obtained in Example 3.1 $K = Z^2$, and for the nuclear attraction integral N we derived $N = -2Z^2$. The zeroth-order energy ε_0 is $\varepsilon_0 = K + N = Z^2 - 2Z^2 = -Z^2$. For the electron–electron repulsion integral I, which here represents the first-order energy ε_1, we obtained $I = 5Z/8$. The total energy $\varepsilon = \varepsilon_0 + \varepsilon_1 = -Z^2 + 5Z/8$ or $-2.750\,E_h$, to be compared with the variation result of $-2.848\,E_h$ and the experimental value $-2.905\,E_h$. With perturbation theory using $\eta = 1$, the ionization potential IP $= E(\text{He}^+) - E(\text{He}) = -2 + 11/4 = 0.7500\,E_h$ while for variation theory using $\eta = 27/16$ it is $-2 + (27/16)^2 = 0.8477\,E_h$, to be compared with experimental value $0.9033\,E_h$.

A slightly different line of approach is to start with the time-dependent Schrödinger equation right away. This is particularly useful for time-dependent perturbations. In this case we divide the Hamilton operator directly into $H = H_0 + \lambda H'$ and suppose that the solutions of $H_0\Phi_n = E_n\Phi_n$ are available. Using the same notation as before and allowing the coefficients c_k to be time dependent, we obtain

$$\Psi = \sum\nolimits_k c_k(t)\Phi_k(\mathbf{r}) \exp(-\mathrm{i}E_k t/\hbar) \quad \text{or} \quad \Psi = \tilde{\mathbf{\Phi}}^{\mathrm{T}}(\mathbf{r}, t)\mathbf{c}(t)$$

where $\tilde{\mathbf{\Phi}}_k(\mathbf{r}, t) = \Phi_k(\mathbf{r}) \exp(-\mathrm{i}E_k t/\hbar)$. Insertion in the Schrödinger equation leads to

$$H\tilde{\mathbf{\Phi}}^{\mathrm{T}}(\mathbf{r}, t)\mathbf{c}(t) = \mathrm{i}\hbar\frac{\partial}{\partial t}\tilde{\mathbf{\Phi}}^{\mathrm{T}}(\mathbf{r}, t)\mathbf{c}(t) \quad \text{or omitting the arguments for } \tilde{\mathbf{\Phi}} \text{ and } \mathbf{c},$$

$$H\tilde{\mathbf{\Phi}}^{\mathrm{T}}\mathbf{c} = \mathrm{i}\hbar\tilde{\mathbf{\Phi}}^{\mathrm{T}}\frac{\mathrm{d}}{\mathrm{d}t}\mathbf{c} + \mathrm{i}\hbar\left(\frac{\partial}{\partial t}\tilde{\mathbf{\Phi}}^{\mathrm{T}}\right)\mathbf{c} = \mathrm{i}\hbar\tilde{\mathbf{\Phi}}^{\mathrm{T}}\dot{\mathbf{c}} + \mathrm{i}\hbar[(\mathrm{i}\hbar)^{-1}\mathbf{E}\tilde{\mathbf{\Phi}}^{\mathrm{T}}\mathbf{c}]$$

with $\mathbf{E}$ a diagonal matrix containing all eigenvalues E_k. Evaluating using

$$H\tilde{\mathbf{\Phi}}^{\mathrm{T}}\mathbf{c} = (H_0 + \lambda H')\tilde{\mathbf{\Phi}}^{\mathrm{T}}\mathbf{c} = \mathbf{E}\tilde{\mathbf{\Phi}}^{\mathrm{T}}\mathbf{c} + \lambda H'\tilde{\mathbf{\Phi}}^{\mathrm{T}}\mathbf{c} \quad \text{leads to} \quad \lambda H'\tilde{\mathbf{\Phi}}^{\mathrm{T}}\mathbf{c} = \mathrm{i}\hbar\tilde{\mathbf{\Phi}}^{\mathrm{T}}\dot{\mathbf{c}}$$

Multiplying by $\tilde{\mathbf{\Phi}}^*$ and integrating over spatial coordinates yields

$$\lambda\tilde{\mathrm{H}}'\mathbf{c}(t) = \mathrm{i}\hbar\dot{\mathbf{c}}(t) \quad \text{with} \quad \tilde{H}'_{jk} = \exp[\mathrm{i}(E_j - E_k)\,t/\hbar]\,\langle \Phi_j | H' | \Phi_k \rangle$$

Finally, integrating over time, we obtain the result

$$\mathbf{c}(t) = (\mathrm{i}\hbar)^{-1}\lambda \int_0^t \widetilde{\mathrm{H}}'\mathbf{c}(t')\,\mathrm{d}t' \tag{3.29}$$

which is an equation that can be solved formally by iteration. Further particular solutions depend on the nature of the Hamilton operator and the boundary conditions.

We leave the perturbation H' undetermined but assume that at $t = 0$ the time-dependent effect is switched on, that is, $\lambda = 1$, and that at that moment the system is in a definite state, say j. In that case $c_j(0) = 1$ and $c_k(0) = 0$ for $k \neq j$, or equivalently, $c_j = \delta_{jk}$. We thus write to first order

$$c_j(t) = (\mathrm{i}\hbar)^{-1} \int_0^t \sum\nolimits_k \widetilde{H}_{jk}{}'c_k(t')\,\mathrm{d}t' \quad \Rightarrow \quad (\mathrm{i}\hbar)^{-1} \int_0^t H_{jk}{}' \exp[\mathrm{i}\hbar^{-1}(E_j - E_k)\,t']\,\mathrm{d}t'$$

Using the abbreviation $(E_j - E_k)/\hbar = \omega_{jk}$ and evaluating leads to

$$c_j(t) = -H_{jk}{}'(\hbar\omega_{jk})^{-1}\,[\exp(\mathrm{i}\omega_{jk}t) - 1]$$

If we may assume that $H_{jk}{}'$ varies but weakly (or not at all) with t. The probability that the system is in state I at the time t is given by $|c_j(t)|^2$, and this evaluates to

$$|c_j(t)|^2 = 4\left|H_{jk}{}'\right|^2(\hbar\omega_{jk})^{-2}\sin^2 \tfrac{1}{2}\omega_{jk}t$$

Introducing the final density of states $g_k = \mathrm{d}n_k/\mathrm{d}E$ of states k with nearly the same energy as the initial state j, the transition probability w to state k per unit time reads

$$w = \frac{1}{t}\int |c_k(t)|^2 g_k\,\mathrm{d}E_k = \frac{1}{t\hbar}4\left|H_{jk}{}'\right|^2 g_k \int_{-\infty}^{+\infty} \frac{\sin^2 \frac{1}{2}\omega_{jk}t}{\omega_{jk}^2}\,\mathrm{d}\omega_{jk} = \frac{2\pi}{\hbar}\left|H_{jk}{}'\right|^2 g_k \tag{3.30}$$

where the last but one step can be made if $H_{jk}{}'$ and g_k vary slowly with energy and the last step as the integral evaluates to $\frac{1}{2}\pi t$. This result is known as *Fermi's golden rule*.

In fact, the energy available is slightly uncertain due to the Heisenberg uncertainty relations. Therefore, the state i is actually one out of a group having nearly the same energy and labeled A. Similarly, j is one out of a set B with nearly the same energy. The occupation probability p_i of each of the states i may change in time. If a system in state i within a set of states A makes a transition to any state within a set of states B, the *one-to-many jump rate* w_{AB} is given by Eq. (3.30). Since we are interested in the *one-to-one jump rate* ν_{ij} from any state i in group A to every state j in group B, we make the *accessibility assumption*, that is, all states within a (small) energy range ΔE, the *accessibility range*, are equally accessible. It can be shown that the exact size of ΔE is unimportant for macroscopic systems (Section 7.1.2). Hence,

$$\nu_{ij} = w_{\mathrm{AB}}/g_{\mathrm{B}}\delta E = 2\pi\left|H_{ij}{}'\right|^2/\hbar\Delta E$$

Because $H_{ji}{}' = \left(H_{ij}{}'\right)^*$, we obtain *jump rate symmetry* $\nu_{ji} = \nu_{ij}$. This result is used in conjunction with the master equation to show that entropy always increases.

3.3.5 The Density Operator

So far, we have discussed only systems containing a small number of particles in which the separate energy levels and associated eigenfunctions dominate and for which the number of micro-states is relatively limited. Macroscopic systems, however, contain many particles, resulting in a really large number of micro-states (Example 7.1). Moreover, the spacing of the energy levels becomes so small that it is irrelevant to speak about the individual levels and states (Justification 7.2). To cope with such a situation, we need a further concept, namely the density operator, which summarizes the available information about a macro-quantum system concisely.

For a single system, or a set of systems occupying all the same state, denoted as a *pure ensemble*, only one state described by a ket, say $|a\rangle$, is relevant. Consider now a single system for which many states are relevant. This single system can be described by an ensemble of systems in which each system j occupies a different state $|j\rangle$ from the set $\{|j\rangle\}$ with probability w_j and is referred to as a *mixed ensemble*. Alternatively, one has actually many systems j, each in state $|j\rangle$. For example, in an ensemble of electrons, all electrons could have the same spin state, say $+\frac{1}{2}\hbar$ (pure ensemble) or $-\frac{1}{2}\hbar$ (pure ensemble), or half of the electrons with $+\frac{1}{2}\hbar$ and the other half $-\frac{1}{2}\hbar$ (mixed ensemble), and so on. Obviously, $\Sigma_j w_j = 1$. In the case all N accessible states have the same weight, $w_j = 1/N$, and we denote the ensemble as *random*. From a measurement of an observable A, represented by an operator A, we find for the ensemble average[9] $\langle A\rangle \equiv \Sigma_j w_j\ \langle j|A|j\rangle$ in which probability plays twice a role: first, in the quantum expectation value $\langle j|A|j\rangle$, and, second, via the weight w_j indicating how many systems are in state j. Changing base vectors, we have

$$\langle A\rangle = \sum\nolimits_j w_j \sum\nolimits_{j'} \sum\nolimits_{j''} \langle j|j'\rangle\langle j'|A|j''\rangle\langle j''|j\rangle$$

$$= \sum\nolimits_j w_j \left(\sum\nolimits_{j'} \sum\nolimits_{j''} \langle j''|j\rangle\langle j|j'\rangle\right) \langle j'|A|j''\rangle$$

Defining the *density operator* $D \equiv \Sigma_j w_j|j\rangle\langle j|$, its matrix representative, the *density matrix*,[10] reads $\mathrm{D} = \langle j''|D|j'\rangle = \Sigma_j w_j\ \langle j''|j\rangle\langle j|j'\rangle$. Using D, the ensemble average is

$$\langle A\rangle = \sum\nolimits_{j'} \sum\nolimits_{j''} \langle j''|D|j'\rangle\langle j'|A|j''\rangle = \sum\nolimits_{j''} \langle j''|DA|j''\rangle = \mathrm{tr}(\mathrm{DA})$$

It is easy to show that $\mathrm{tr\,D} = 1$. For a pure ensemble $w_j = 1$ for a single j, while for a random ensemble $w_j = 1/N$ for all j, that is,

$$\mathrm{D} = \begin{pmatrix} 0 & . & 0 \\ & \cdot & \cdot \\ . & 1 & . \\ 0 & . & 0 \end{pmatrix} \quad \text{and} \quad \mathrm{D} = N^{-1} \begin{pmatrix} 1 & & . & 0 \\ & 1 & & . \\ . & & 1 & \\ 0 & . & & 1 \end{pmatrix}$$

respectively. For the pure ensemble $\mathrm{D}^2 = \mathrm{D}$ and thus $\mathrm{tr\,D}^2 = 1$, while for the random ensemble $\mathrm{tr\,D}^2 = N^{-2}\mathrm{tr\,D} = N^{-1}$. In fact, one can show that $\mathrm{tr\,D}^2$ is maximal for a pure ensemble. For the time evolution of D, consider $D(t) = \Sigma_j w_j|j(t)\rangle\langle j(t)|$. For an undisturbed ensemble w_j remains constant and time evolution follows from the time-evolution state of the ket $|j\rangle \equiv |j(0)\rangle$ to $|j(t)\rangle$. Using the Schrödinger equation, $i\hbar\partial\,|j\rangle/\partial t = H|j\rangle$, we obtain the *von Neumann equation*

$$i\hbar\frac{\partial D}{\partial t} = \sum\nolimits_j w_j[H|j\rangle\langle j| - |j\rangle\langle j|H] = [H, D] \equiv -LD \tag{3.31}$$

---

9 Note that we use $\langle A\rangle$ for both the quantum and statistical expectation value of A. Sometimes $\langle\langle A\rangle\rangle \equiv \Sigma_j w_j\ \langle j|A|j\rangle$ is used, but $\langle A\rangle$ is conventional and confusion is unlikely to occur.

10 Le Bellac [3] rightfully calls this the *state operator* as it does not only provide information on the density of the states but also about their phases. However, the designation density operator, actually even the designation density matrix for the density operator, is generally accepted.

which is the equivalent of the *Liouville equation* in classical mechanics with L the *von Neumann operator*. Although similar to the equation of state in the Heisenberg picture, we describe here time evolution in the Schrödinger picture (note the difference in sign). In thermal equilibrium we expect $\partial D/\partial t = 0$. Hence, in view of Eq. (3.31), H and D can be simultaneously diagonalized. The simultaneous eigenvectors of H and D represent the state vector and, in this representation, the so-called *number representation*, the various D_{jj}, the matrix elements of D, represents the probability of the system to be in state $|j\rangle$ with energy E_j. In fact, the p_js used in Chapter 7 are the probabilities D_{jj} in this representation.

Further, suppose we have a Hilbert space $\mathbf{H} = \mathbf{H}_1\mathbf{H}_2$, a property A that depends only on $\mathbf{H}_1$, that is, with operator A_1, 1_2, and that we want to obtain a density operator D_1 in such a way that $\langle A\rangle = \text{tr}\, \mathrm{D}_1\mathrm{A}$. To achieve this, we write

$$\langle A_1 1_2\rangle = \text{tr}A_1 1_2 \mathrm{D} = \sum\nolimits_{m_1 m_2 n_1 n_2} A_{n_1 m_1}\delta_{n_2 m_2} D_{m_1 m_2, n_1 n_2}$$
$$= \sum\nolimits_{n_1 m_2} A_{n_1 m_1} \sum\nolimits_{n_2} D_{m_1 n_2, n_1 n_2} \equiv \sum\nolimits_{n_1 m_2} A_{n_1 m_1} D^{(1)}_{m_1 n_1} = \text{tr}_2 \mathrm{AD}_1$$

The *reduced density operator* D_1 with elements $D^{(1)}_{m_1 n_1}$ is the partial trace tr_2 over states of system 2 of the (total) density operator D.

Problem 3.1 ***Hermiticity and Real Eigenvalues*** Show that a Hermitian operator has real eigenvalues and orthogonal eigenfunctions.

Problem 3.2 ***Energy Dispersion*** Show that $(\Delta H)^2 = \langle (H - \langle H\rangle)^2\rangle = 0$.

Problem 3.3 ***Energy for an Additive Hamilton Operator*** Show that a two-electron determinant wave function can be factorized in a spatial and a spin factor and verify that for the operator $H = \Sigma_i h(i)$ the total energy $E = \Sigma_k \varepsilon_k$.

Problem 3.4 ***Energy for a Non-additive Hamilton Operator*** Calculate the energy for a two-electron product and determinant wave function using the Hamilton operator $H = h(1) + h(2) + g(12)$. Show that the total energy $E \neq \Sigma_k \varepsilon_k$.

Problem 3.5 ***The Hellmann–Feynman Theorem*** Prove the Hellmann–Feynman theorem.

3.4 Final Remarks

For our discussion of phases and their transitions, the concepts discussed here are amply sufficient. In fact, the existence of energy levels is the most important aspect for the basis of statistical thermodynamics. However, to discuss irreversibility (Chapter 19), a more elaborate background is helpful, and we have collected here all results required.

References

1 Dirac, 1958.

2 Gottfried and Yan, 2003.

3 Le Bellac, 2006.
4 Sakurai and Napolitano, 2011.
5 Pauling, L. and Wilson, E.B. (1935). *Introduction to Quantum Mechanics*. New York: McGraw-Hill Book Comp.
6 Born, M. and Huang, K. (1954). *Dynamical Theory of Crystal Lattices*. Oxford: Oxford University Press.
7 Pilar, 1968.
8 Hellmann, H. (1937). *Einführung in die Quantenchemie*, p. 285. Leipzig: Franz Deuticke; Feynman, R.P. (1939). *Phys. Rev.* 56: 340; Güttinger, P. (1932). *Z. Phys.* 73: 169; Pauli, W. (1933). *Handbuch der Physik*, Vol. 24, p. 162. Berlin: Springer Verlag.
9 Mitin, A.V. (2015). arXiv:1508.01359 (accessed 09 April 2021).
10 Seitz, F. (1940). *Modern Theory of Solids*. New York: McGraw-Hill Book Comp.
11 Harrison, W.A. (1980). *Electronic Structure and the Properties of Solids*. San Francisco: W.H. Freeman & Comp. (also Dover, 1989).
12 Szabo, A. and Ostlund, N.S. (1996). *Modern Quantum Chemistry*, rev. ed. New York: Dover.
13 Sutton, A. (1993). *Electronic Structure of Materials*. Oxford: Clarendon Press.
14 Pettifor, D.G. (1995). *Bonding and Structure of Molecules and Crystals*. Oxford: Clarendon Press.
15 McQuarrie, 2008.
16 Nordholm, S. and Bacskay, G.B. (2020). *Molecules* 25: 2667.

Further Reading

Atkins, P.W. and Friedman, R.S. (2010). *Molecular Quantum Mechanics*, 5e. Oxford: Oxford University Press. A thorough discussion of the application of quantum mechanics to chemical topics.

Le Bellac, M. (2006). *Quantum Physics*. Cambridge: Cambridge University Press. A modern, compact and relatively high-level treatment of quantum mechanics.

Dirac, P.A.M. (1958). *The Principles of Quantum Mechanics*, 4e. Oxford: Oxford University Press. A classic and somewhat abstract, but highly recommended book.

Gottfried, K. and Yan, T.-M. (2003). *Quantum Mechanics: Fundamentals*, 2e. New York: Springer. This book does exactly what the titles says, and it does so at an accessible, intermediate level.

McQuarrie, D.A. (2008). *Quantum Chemistry*. Sausalito, CA: University Science Books. A modern treatment using a relatively slow rate and similar in scope as Pilar.

Pilar, F.L. (1968). *Elementary Quantum Chemistry*. London: McGraw-Hill Book Comp. Despite its age, a highly readable and valuable book, in particular for those parts dealing with the basics including Hartree-Fock theory.

Sakurai, J.J. and Napolitano, J.J. (2011). *Modern Quantum Mechanics*, 2e. San Francisco: Addison-Wesley. A modern treatment but less compact than Le Bellac.

Schiff, L.I. (1955). *Quantum Mechanics*, 2e. New York: McGraw-Hill Book Comp. Another classic still relevant at a similar level as Gottfried and Yan.

4

Intermolecular Interactions

All bonding and molecular interaction characteristics are determined by quantum mechanics. Intermolecular interactions, however, may be described to quite some extent via classical considerations. In this chapter, we use the "tools" as dealt with so far and describe long-range electrostatic intermolecular interactions using the multipole expansion to which we add induction and dispersion interactions. Short-range repulsion, due to orbital overlap, is treated in an empirical way. Finally, we discuss some often-used model potentials, hydrogen bonding, and three-body interactions.

4.1 The Semi-classical Approach

In principle, molecules and their interactions should be described by the Schrödinger equation. As this formidable problem cannot be solved exactly, we use approximate methods, and in this chapter we describe some of them for intermolecular interactions. From the *Born–Oppenheimer approximation* (Section 3.3.1), we know that an (*effective*) *potential energy* $E(\mathbf{r})$ for the motion of the nuclei with coordinates $\mathbf{r}$ exists.[1] Emphasizing that $E(\mathbf{r})$ acts as a potential energy employed in all our discussions on intermolecular interactions, we label it as $\Phi(\mathbf{r})$ from now on.

Experiments together with theory have made it clear that all interaction potentials between atoms have a *potential energy curve* as shown in Figure 4.1, which shows for short distances a strong repulsive part and for longer distances an attractive part, whose the magnitude strongly depends on the type of bonding.[2] For a poly-atomic molecule the potential energy curve becomes a *potential energy surface*, or in more pictorial language, a *potential energy landscape*. Such a surface contains one absolute minimum (or, dependent on symmetry, several equivalent ones) where the curvature is positive in all directions. It also contains several relative minima connected via pathways containing a saddle point, where the curvature is negative along the pathway and positive perpendicular to that pathway [1]. However, repulsion and attraction remain, and we need to explain the origins of both contributions.

1 Note again that r is a scalar, $\boldsymbol{r}$ a vector, and $\mathbf{r}$ a (column matrix of a) set of elements r_i or $\boldsymbol{r}_i$.
2 In fact, since gases do condense in liquids, at intermediate distances attraction must be present, but since liquids and solids have a finite but small compressibility, at shorter distances repulsion must set in.

Phases of Matter and their Transitions: Concepts and Principles for Chemists, Physicists, Engineers, and Materials Scientists, First Edition. Gijsbertus de With.

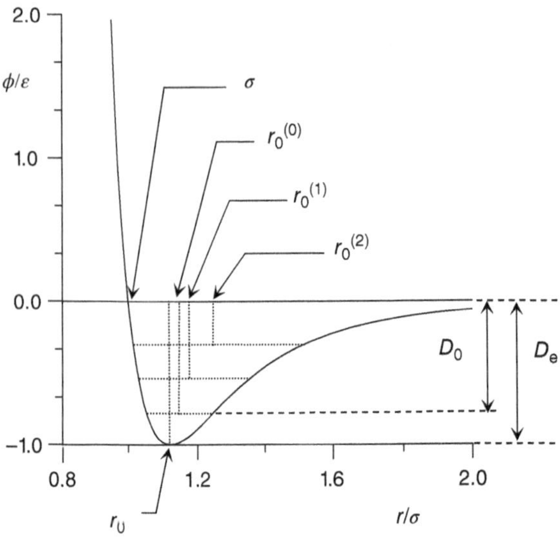

Figure 4.1 The Mie pair potential for a diatomic molecule in units of ε and σ for $n = 12$ and $m = 6$. The equilibrium distance is r_0, practically equal to $r_0^{(0)}$ for the vibrational ground state at $T = 0$ K. This distance increases with T due to the asymmetry of the potential and the increase in the population of excited vibrational states, as indicated by $r_0^{(1)}$ and $r_0^{(2)}$. The *equilibrium energy* is D_e while the *dissociation energy* is $D_0 = D_e - \frac{1}{2}h\nu$ (vibrational state spacing $h\nu$ exaggerated).

The balance between the long-range attraction Φ_{att} and the short-range repulsion Φ_{rep} determines whether a complex of molecules is stable or not. The total potential energy Φ being given by $\Phi = \Phi_{att} + \Phi_{rep}$, the *equilibrium distances* $\mathbf{r}_0$ are given by the solution of $\partial\Phi/\partial\mathbf{r} = \mathbf{0}$, while the *force constant matrix*, describing the harmonic vibrations, reads $\mathbf{K}(\mathbf{r}_0) = \partial^2\Phi/\partial\mathbf{r}\partial\mathbf{r}'$ or $K_{ij}(\mathbf{r}_0) = \partial^2\Phi/\partial_i\partial r_j$.

Before we discuss the interactions, let us introduce the moments of a charge distribution. The charge distribution $\rho(\boldsymbol{r})$ of a molecule is described by either a continuum distribution or by a discrete set of discrete charges q_i at positions $\boldsymbol{r}_i$. Both nuclei and electrons contribute to the distribution, and we assume that $\rho(\boldsymbol{r})$ is given. The zeroth moment of this distribution is a scalar and is given by

$$q = \int_V \boldsymbol{r}^0 \rho(\boldsymbol{r})\,\mathrm{d}\boldsymbol{r} = \int_V \rho(\boldsymbol{r})\,\mathrm{d}\boldsymbol{r} \qquad \text{or} \qquad q = \sum\nolimits_i \boldsymbol{r}_i^0 q_i = \sum\nolimits_i q_i$$

It is essentially the *charge* q of the molecule, with the coulomb [C] as unit for q. A molecule with $q = 0$ is called *neutral*. The first moment obviously becomes

$$\boldsymbol{\mu} = \int_V \boldsymbol{r}\rho(\boldsymbol{r})\,\mathrm{d}\boldsymbol{r} \qquad \text{or} \qquad \boldsymbol{\mu} = \sum\nolimits_i q_i \boldsymbol{r}_i$$

It is a vector called the *dipole moment* with components μ_j, the magnitude of which is given by $\mu = |\boldsymbol{\mu}| = q\delta$, where δ is the distance between the centers of gravity of the positive and negative charges $+q$ and $-q$ in the molecule. The unit for $\boldsymbol{\mu}$ is [C m], but for historical reasons, the Debye [D] = 3.336×10^{-30} [C m] is often used. Conventionally, $\boldsymbol{\mu}$ is directed from $-q$ to $+q$. The quantity $\delta \neq 0$ renders the dipole moment a vector and leads to strong orientational forces. The *ideal dipole moment* is obtained by decreasing the distance δ to zero, while the charge q increases in such a way that the product $q\delta$ remains μ. Higher order moments, such as the *quadrupole moment* tensor $\mathbb{R}$ with components $R_{ij} = R_{ji}$, can be defined similarly

(Justification 4.1), but we will use mainly q and $\boldsymbol{\mu}$. Only the first non-zero moment is independent of the location of the origin, which is usually taken as the center of gravity of the molecule. A full set of moments would be equivalent to the original distribution.

In Sections 4.1.1–4.1.4, we deal with the intermolecular interactions semi-classically. For readers interested in the general picture only, these sections will suffice, but Section 4.2 provides an introduction to the quantum-mechanical derivations.

4.1.1 Electrostatic Interaction

The *Coulomb energy* w of a charge q_1 in the presence of another charge q_2 is

$$w(r) = q_1\phi_2 = \mathscr{C}q_1q_2r^{-1} \quad \text{with} \quad \phi_2 \equiv \mathscr{C}q_2r^{-1} \quad \text{and} \quad \mathscr{C} \equiv (4\pi\varepsilon_0)^{-1}$$

Here ϕ_2 represents the *potential* associated with the *force* $-\nabla w = \mathscr{C}q_1q_2\boldsymbol{r}/r^3$ between the charges q_1 and q_2 with $r = |\boldsymbol{r}|$ their distance and ε_0 the permittivity of vacuum. Denoting here the charges by the subscript i and the molecule by the superscript (i), molecule 1 is represented by the charge distribution $\rho_1 = \sum_j q_j^{(1)}$ and the potential ϕ_2 due to the charge distribution of molecule 2 by $\phi_2(\boldsymbol{r}) = \sum_k q_k^{(2)}/s_k$, where $s_k = |\boldsymbol{s}_k| = |\boldsymbol{r} - \boldsymbol{r}_k|$ is the distance of charge q_k to position $\boldsymbol{r}$ (Figure 4.2). The total interaction energy $w_{12} = \rho_1\phi_2 = \sum_j q_j^{(1)}\phi_2(\boldsymbol{r}_i) = \sum_j q_j^{(1)}\sum_k(q_k^{(2)}/s_{jk})$ with $s_{jk} = |\boldsymbol{s}_{jk}| = |\boldsymbol{r}_j - \boldsymbol{r}_k|$. After expanding the function ϕ_2, evaluating all the derivatives, and collecting terms, a procedure with considerable understatement often called "a lengthy but straightforward calculation" (Justification 4.1), one obtains the desired result. Note that, since this process takes place at constant T and V (Section 6.1), the interaction w_{12} represents the Helmholtz energy.

Justification 4.1 ***The Multipole Expansion*** To find the proper expression for the electrostatic interaction, let us consider an arbitrary charge distribution ρ of point charges q_j at position $\boldsymbol{r}_j$ from the origin O located at the center of mass (Figure 4.2). The potential energy ϕ at a certain point P located at $\boldsymbol{r}$ outside the sphere containing all charges is

$$\phi(\boldsymbol{r}) = \mathscr{C}\sum\nolimits_j(q_j/s_j)$$

where $s_j = |\boldsymbol{s}_j| = |\boldsymbol{r} - \boldsymbol{r}_j|$ is the distance of charge q_j to P. Developing $1/s_j$ in a Taylor series with respect to $\boldsymbol{r}$, we may write $1/s_j = 1/r + \boldsymbol{r}_j\cdot(\nabla r^{-1})_\text{O} + \cdots$, and therefore (using direct notation, Section A.12; note that $\boldsymbol{xx} = x_kx_l$ and $\boldsymbol{x}\cdot\boldsymbol{x} = x^2 = \Sigma_jx_j^2$ for any vector $\boldsymbol{x}$),

$$\begin{aligned}\phi(\boldsymbol{r}) &= \mathscr{C}\sum\nolimits_j q_jr^{-1} + \mathscr{C}\sum\nolimits_j q_j\boldsymbol{r}_j\cdot(\nabla r^{-1})_\text{O} + {}^1\!/\!_2\mathscr{C}\sum\nolimits_j q_j\boldsymbol{r}_j\cdot(\nabla\nabla r^{-1})_\text{O}\cdot\boldsymbol{r}_j + \cdots\\ &= \mathscr{C}qr^{-1} - \boldsymbol{\mu}\cdot(\mathscr{C}\nabla r^{-1})_\text{P} + {}^1\!/\!_2\mathbb{Q}:(\mathscr{C}\nabla\nabla r^{-1})_\text{P} - \cdots\\ &= q\phi - \boldsymbol{\mu}\cdot\phi' + {}^1\!/\!_2\mathbb{Q}:\phi'' - \cdots = q\phi - \boldsymbol{\mu}\cdot\phi' + \mathbb{R}:\phi''/3 - \cdots\end{aligned} \tag{4.1}$$

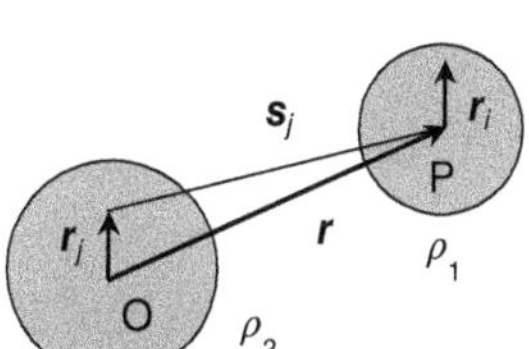

Figure 4.2 The interaction $\rho_1\phi_2$ of two charge distributions ρ_1 and ρ_2 with common origin O.

where $r = |\mathbf{r}|$ is the distance of the point P to the origin O of ρ. In the second line, we defined the *total charge* $q = \sum_j q_j$, the *dipole moment* $\boldsymbol{\mu} = \sum_j q_j\mathbf{r}_j$, and the second moment $\mathbb{Q} = \sum_j q_j\mathbf{r}_j\mathbf{r}_j$. The minus sign in the second line for the second, fourth, and so on terms arises because the differentiation is now made at P, and not at the origin of the potential O. This follows since for an arbitrary vector $\boldsymbol{x}$ it holds that $\nabla x^{-1} = -\boldsymbol{x}/x^3$, and therefore $(\nabla x^{-1})_\mathrm{P} = -(\nabla x^{-1})_\mathrm{O}$. In the third line, we abbreviated $\mathscr{C} r^{-1}$ by ϕ, $(\mathscr{C}\nabla r^{-1})_\mathrm{P}$ by ϕ', and so on. Since Laplace's equation $\nabla^2 r^{-1} = \mathscr{C}^{-1}\phi'' : \mathbb{I} = 0$ holds, the relation $\mathbb{Q}' : \phi'' \equiv (\mathbb{Q} - \zeta\mathbb{I}) : \phi'' = \mathbb{Q} : \phi'' - \zeta\phi'' : \mathbb{I} = \mathbb{Q} : \phi''$ holds for any ζ. Taking $\zeta = \mathbb{Q} : \mathbb{I}/3 = \mathrm{tr}\mathbb{Q}/3$ reduces $\mathbb{Q}$ to $2\mathbb{R}/3$ with $\mathbb{R} = ½[3\mathbb{Q}-\mathrm{tr}(\mathbb{Q})\mathbb{I}]$ referred to as the *quadrupole moment*. Equation (4.1) is usually called the *multipole expansion* of the potential. The *interaction energy* w of a charge q at $\boldsymbol{r}$ with the potential ϕ of a charge distribution is $w = q\phi(\boldsymbol{r})$. Hence, to calculate the interaction energy w_{12} between two charge distributions ρ_1 and ρ_2, separated by $\boldsymbol{r}$ (Figure 4.2), we express this energy as the energy of $\rho_1 = \sum_i q_i^{(1)}$ in the field of $\rho_2 = \sum_j q_j^{(2)}$. Hence, $w_{12} = \rho_1\phi_2$ is

$$\begin{aligned} w_{12} &= \sum\nolimits_i q_i^{(1)}\phi_2(\boldsymbol{r}_i) = \sum\nolimits_i q_i^{(1)}\left(\phi_2 + \boldsymbol{r}_i^{(1)}\cdot\phi_2' + ½\boldsymbol{r}_i^{(1)}\cdot\phi_2''\cdot\boldsymbol{r}_i^{(1)} + \cdots\right) \\ &= q_1\phi_2 + \boldsymbol{\mu}_1\cdot\phi_2' + \mathbb{R}_1 : \phi_2''/3 + \cdots \end{aligned} \tag{4.2}$$

If we take the center of mass P of ρ_1 as common origin, the potential ϕ_2 due to ρ_2 is given by the last line of Eq. (4.1), and substitution in Eq. (4.2) results in

$$\begin{aligned} &w_{12} = A\phi + \boldsymbol{B}\cdot\phi' + \mathbb{\Gamma} : \phi'' + \cdots \qquad \text{where} \\ &A = q_1q_2, \quad \boldsymbol{B} = q_1\boldsymbol{\mu}_2 - \boldsymbol{\mu}_1q_2, \quad \mathbb{\Gamma} = q_1\mathbb{R}_2/3 - \boldsymbol{\mu}_1\boldsymbol{\mu}_2 + \mathbb{R}_1q_2/3, \ldots \end{aligned} \tag{4.3}$$

This is the general electrostatic interaction energy expression in terms of multipole moments of the charge distributions with respect to their own center of mass [2–4]. For further evaluation, we need again $\nabla(1/r) = -\boldsymbol{r}/r^3$ and $-\mathbb{T} \equiv \nabla\nabla(1/r) = (3\boldsymbol{rr} - r^2\mathbb{I})/r^5$ with $\mathbb{I}$ the unit tensor, derived using $r^2 = x^2 + y^2 + z^2$ and leading to Eq. (4.4).

The total result up to and including the dipole–dipole interaction is

$$w_{12}(r) = \frac{\mathscr{C}\, q_1q_2}{r} + \frac{\mathscr{C}}{r^3}[(\boldsymbol{r}\cdot\boldsymbol{\mu}_1)q_2 - (\boldsymbol{r}\cdot\boldsymbol{\mu}_2)q_1] + \frac{\mathscr{C}}{r^5}[r^2\boldsymbol{\mu}_1\cdot\boldsymbol{\mu}_2 - 3(\boldsymbol{\mu}_1\cdot\boldsymbol{r})(\boldsymbol{\mu}_2\cdot\boldsymbol{r})] + . \tag{4.4}$$

or explicitly (see Figure 4.3 for notation, introducing the abbreviation $\boldsymbol{\omega} = \theta, \phi$)

$$\begin{aligned} w_{12}(r, \boldsymbol{\omega}_1, \boldsymbol{\omega}_2) = {} & \frac{\mathscr{C}}{r}(q_1q_2) + \frac{\mathscr{C}}{r^2}(q_2\mu_1\cos\theta_1 - q_1\mu_2\cos\theta_2) \\ & - \frac{\mathscr{C}\,\mu_1\mu_2}{r^3}[2\cos\theta_1\cos\theta_2 - \sin\theta_1\sin\theta_2\cos(\phi_1 - \phi_2)] + \ldots \end{aligned} \tag{4.5}$$

In this expression $q_1 = \sum_j q_j^{(1)}$ is the *charge* of molecule 1 and $\boldsymbol{\mu}_1 = \sum_j q_j^{(1)}\boldsymbol{r}_j^{(1)}$ is the *dipole moment* with $\mu_1 = |\boldsymbol{\mu}_1|$ and similar expressions for q_2 and μ_2.

The first term (proportional to q_1, q_2, and $1/r$) is the (charge–charge) *Coulomb interaction* $W_\text{c–c}$ between the molecules at distance r. At long distance r, this interaction term will dominate. Obviously, it is zero if one of the molecules is neutral.

The second term (proportional to q_1, μ_2, q_2, μ_1, and $1/r^2$; Figure 4.3a) is the interaction between the charge of molecule 2 with the dipole moment of molecule 1 and the interaction between the charge of molecule 1 with the dipole moment of molecule 2. This is the *charge–dipole (Coulomb) interaction*, decreasing as $1/r^2$, as compared to $1/r$ for the

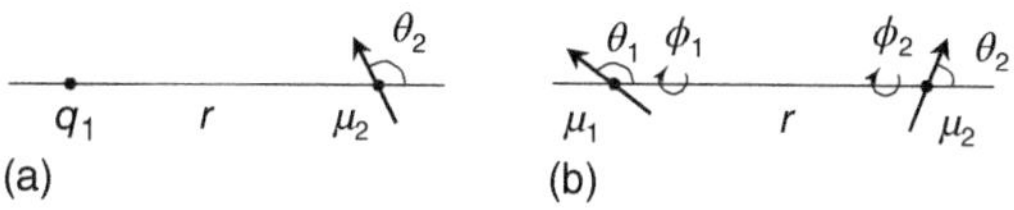

Figure 4.3 (a) The interaction at a distance r of a point charge q_1 with a dipole μ_2 (in which the charges q are separated by a distance δ). (b) The interaction of a dipole with a dipole at distance r.

charge–charge Coulomb interaction. Because the charge–dipole interaction is orientation dependent and molecules move generally rather rapidly, an orientational average (overbar) must be made (Problem 7.10). For the moment, we just quote the result, valid for sufficiently high temperatures,

$$w_{\text{c-d}}(r) = \overline{w_{\text{c-d}}(r,\theta)} = -\frac{\mathscr{C}^2}{6kT}\left(\frac{q_1^2\mu_2^2}{r^4}\right) - \frac{\mathscr{C}^2}{6kT}\left(\frac{q_2^2\mu_1^2}{r^4}\right) \tag{4.6}$$

The third term (proportional to μ_1, μ_2, and $1/r^3$; Figure 4.3b) represents the *dipole–dipole (Coulomb) interaction* between the dipole moments of molecule 1 and molecule 2 and is often addressed as the *Keesom interaction* (it also includes the charge–quadrupole interaction, which we neglect). This interaction is also orientation dependent and for the orientational average one obtains, again for sufficiently high temperature and using $\boldsymbol{\omega} = (\theta,\phi)$ with θ and ϕ the orientation angles,

$$w_{\text{d-d}}(r) = \overline{w_{\text{d-d}}(r,\boldsymbol{\omega}_1,\boldsymbol{\omega}_2)} = -\frac{\mathscr{C}^2}{3kT}\left(\frac{\mu_1^2\mu_2^2}{r^6}\right) \equiv -\frac{C_{\text{K}}}{r^6} \tag{4.7}$$

For neutral molecules, the dipole–dipole interaction is the leading term. Note (again) that the expressions given represent the work done to set up the charge distributions at constant T and V and thus represent Helmholtz energies f. As $w \sim 1/T$, the internal energy expressions u are, using $u = \partial(f/T)/\partial(1/T)$, given by $u = 2w$.

Problem 4.1 ***Permanent Dipole Interactions*** In a simple model to estimate the charge–dipole interaction (Figure 4.3a), the distances between the charge q_1 of molecule 1 and the charges of the dipole moment of molecule 2, $+q_2$ and $-q_2$, respectively, are given by

$$r(q_1,+q_2) = r + \tfrac{1}{2}\delta\cos\theta_2 \quad \text{and} \quad r(q_1,-q_2) = r - \tfrac{1}{2}\delta\cos\theta_2$$

Assuming that $r \gg \delta$, using Coulomb's law and the binomial expansion, show that for a fixed direction, the interaction $W_{\text{c-d}} = -\mathscr{C}\,r^{-2}q_1\mu_2\cos\theta_2$ where $\mathscr{C} = 1/4\pi\varepsilon_0$. Show similarly that for the dipole–dipole interaction (Figure 4.3b), the in-line orientation energy (orientation parallel to the connection line) and the parallel orientation energy (orientation perpendicular to the connection line) are, respectively,

$$w_{\text{d-d}}(r) = \pm 2\mathscr{C}\,r^{-3}\mu_1\mu_2 \quad \text{and} \quad w_{\text{d-d}}(r) = \pm\mathscr{C}\,r^{-3}\mu_1\mu_2$$

Here, $\pm$ indicates whether the orientation is in the same way (attraction) or in the opposite way (repulsion) or antiparallel (attraction) or parallel (repulsion). How large should r be as compared to δ in order to have an error less than 5%?

Problem 4.2 ***The Multipole Expansion*** Verify Eq. (4.3). Alternatively, derive Eq. (4.3) up to second order by developing s_{ij}^{-1} in $w_{12} = \mathscr{C}\,\Sigma_{i,j}\,q_i^{(1)}s_{ij}^{-1}\Sigma_j\,q_j^{(2)}$, where $s_{ij} = |\boldsymbol{r} + \boldsymbol{r}_j - \boldsymbol{r}_i|$, in a double Taylor expansion.

4.1.2 Induction Interaction

Apart from the electrostatic interactions between permanent charges (c) and dipoles (d), a charge in molecule 1 will also induce a dipole moment in molecule 2 (as will a dipole in molecule 1). The induced dipole (id) in molecule 2 is proportional to the field generated by the charge (or the dipole) of molecule 1 with as proportionality factor the (isotropic[3]) *polarizability*[4] α_2 of molecule 2, that is, $\boldsymbol{\mu}_2 = \alpha_2 \boldsymbol{E}_1$.

The field of a charge q_1 is given by

$$\boldsymbol{E}_{\text{c}}(\boldsymbol{r}) = -\frac{\partial \phi_{\text{c}}(r)}{\partial \boldsymbol{r}} = -\frac{\partial}{\partial \boldsymbol{r}}\frac{\mathscr{C} q_1}{r} = \frac{q_1 \boldsymbol{r}}{4\pi\varepsilon_0 r^3}$$

The interaction between the field $\boldsymbol{E}_{\text{c}}$ of molecule 1 and the induced dipole moment $\boldsymbol{\mu}_{\text{id}}$ in molecule 2 can be calculated as

$$w_{\text{c-id}}(r) = -\int_0^{\boldsymbol{E}_{\text{c}}} \boldsymbol{\mu}_{\text{id}} \cdot \text{d}\boldsymbol{E} = -\int_0^{\boldsymbol{E}_{\text{c}}} \alpha_2 \boldsymbol{E}_1 \cdot \text{d}\boldsymbol{E}_1 = -\tfrac{1}{2}\mathscr{C}^2 \alpha_2 q_1^2 / r^4$$

and is a contribution to the total intermolecular interaction.

The field of a dipole moment $\boldsymbol{\mu}_1$ is given by

$$\boldsymbol{E}_{\text{d}}(\boldsymbol{r}) = -\frac{\partial \phi_{\text{d}}(r)}{\partial \boldsymbol{r}} = -\frac{\partial}{\partial \boldsymbol{r}}\frac{\mathscr{C}\boldsymbol{\mu}_1 \cdot \boldsymbol{r}}{r^3} = \frac{1}{4\pi\varepsilon_0 r^5}[3(\boldsymbol{\mu}_1 \cdot \boldsymbol{r})\boldsymbol{r} - r^2 \boldsymbol{\mu}_1]$$

The interaction between the field $\boldsymbol{E}_{\text{d}}$ of molecule 1 and the induced dipole moment $\boldsymbol{\mu}_{\text{id}}$ of molecule 2 can be calculated similarly as

$$w_{\text{d-id}}(r) = \overline{w_{\text{d-id}}(r,\theta)} = -\int_0^{\boldsymbol{E}_{\text{d}}} \boldsymbol{\mu}_{\text{id}} \cdot \text{d}\boldsymbol{E} = -\mathscr{C}^2 \alpha_2 \mu_1^2 / r^6 \equiv -C_{\text{D}}/r^6 \tag{4.8}$$

using $[3(\boldsymbol{\mu}\cdot\boldsymbol{r})\boldsymbol{r} - r^2\boldsymbol{\mu}]^2 = (1 + 3\cos^2\theta)\mu^2 r^4$ and $\overline{\cos^2\theta} = 1/3$. This interaction is denoted as the *Debye interaction* and is always attractive. Since the Helmholtz energy w is independent of temperature (at least, as derived here for an isotropic polarizability α), it also represents the internal energy u. Both interactions are due to the field of molecule 1 (originating either from its charge or dipole moment) and the polarizability of molecule 2. For the total interaction, we must add the interaction between the field of molecule 2 and the polarizability of molecule 1.

Problem 4.3 ***Induced Dipole Interactions*** Verify the expressions for the energy $w_{\text{c-id}}$ and the energy $w_{\text{d-id}}$.

Problem 4.4 ***The Bohr Atom*** In a Bohr atom, an electron orbits a proton at a fixed distance a_0 at which the Coulomb energy $\mathscr{C} e^2/a_0 = 2h\nu$, where h is Planck's constant and ν is the orbiting frequency. Show that for a Bohr atom, by balancing the forces required for the displacement of the plane of the orbit from the nucleus, $\alpha = 4\pi\varepsilon_0 a_0^3$.

---

3 Isotropy implies equal response for all directions. When the polarization becomes anisotropic, these interactions become orientation dependent and therefore (in principle) temperature dependent.

4 The unit of polarizability is $\text{C}^2\ \text{m}^2\ \text{J}^{-1}$. Writing $\alpha' = \alpha/(4\pi\varepsilon_0)$ renders the dimension of α' to be m^3. Since, according to electrostatics, the polarizability of a homogeneously conducting sphere of radius r in vacuum as well as the polarizability of a Bohr atom with radius r is given by $4\pi\varepsilon_0 r^3$, the molecular radius r can be estimated from $r^3 = \alpha'$. Hence, α' is known as the *polarizability volume*. For example, for H_2O $\alpha' = 1.49\ \text{Å}^3$ or $r = 1.13\ \text{Å}$, while $\sigma/2$ as calculated from the vdW constant $b = 2\pi N_{\text{A}}\sigma^3/3 = 0.0305\ \text{l}\,\text{mol}^{-1}$ yields $\sigma/2 = 1.45\ \text{Å}$, or $r/(\sigma/2) \cong 0.8$.

4.1.3 Dispersion Interaction

Although the motion of electrons in a molecule is rather fast, at any moment the charge distribution of a molecule will have an instantaneous dipole moment. The field associated with this dipole moment also induces an instantaneous dipole moment in a neighboring molecule. Both induced dipoles interact, and this leads to the *London dispersion interaction*, which is relatively weak but omnipresent and (in a vacuum) always attractive. An introductory, full quantum-mechanical treatment is given in Section 4.2. However, a relatively simple model, due to London,[5] using the Drude model for atoms, yields good insight into the nature of the interaction. Moreover, it yields the proper expression apart from the exact numerical pre-factor.

In the *Drude model*, one assumes that a molecule can be considered as a set of particles (electrons) with charge e_j and mass m_j. Each of these particles is harmonically and isotropically bound to its equilibrium position. The configuration and notation as used for a single electron are sketched in Figure 4.4. For the in-line configuration, the dynamics of this electron are thus given by those of a harmonic oscillator, for which, from quantum mechanics (Section 3.2.2), we know that u is

$$u = \hbar\omega_0(n + {}^1\!/\!_2) \qquad \text{with} \qquad \omega_0 = \sqrt{a/m}$$

where n denotes the quantum number $n = 0, 1, 2, 3, \ldots$ and ω_0 the circular frequency of the oscillator with a the force constant with which the electron with mass m is bound to the nucleus. Without any interaction, the total energy is given by

$$u = u_1(n_1) + u_2(n_2) = \hbar\omega_0(n_1 + {}^1\!/\!_2) + \hbar\omega_0(n_2 + {}^1\!/\!_2) \quad \text{or} \quad u_0 = u_1(0) + u_2(0) = \hbar\omega_0$$

for the ground state with $n = 0$. At any moment, electron 1 of atom 1 exerts a force on electron 2 of atom 2, and for a distance r, using Eq. (4.5), the interaction is given by

$$w_{zz} = -2\mathscr{C}\,\mu_z\mu_z/r^3 = -2\mathscr{C}\,z_1z_2e^2/r^3$$

so that the resulting total energy u is

$$u = \left({}^1\!/\!_2 a z_1^2 + {}^1\!/\!_2 m\dot{z}_1^2\right) + \left({}^1\!/\!_2 a z_2^2 + {}^1\!/\!_2 m\dot{z}_2^2\right) - (2\mathscr{C}\,z_1z_2e^2/r^3)$$

This result can be reduced to the sum of the energies of two independent harmonic oscillators with different frequencies. This is accomplished by the transformation

$$z_+ = (z_1 + z_2)/\sqrt{2} \quad \text{and} \quad z_- = (z_1 - z_2)/\sqrt{2} \quad \text{leading to}$$

$$u = \left[{}^1\!/\!_2(a - 2\mathscr{C}\,e^2/r^3)z_+^2 + {}^1\!/\!_2 m\dot{z}_+^2\right] + \left[{}^1\!/\!_2(a + 2\mathscr{C}\,e^2/r^3)z_-^2 + {}^1\!/\!_2 m\dot{z}_-^2\right]$$

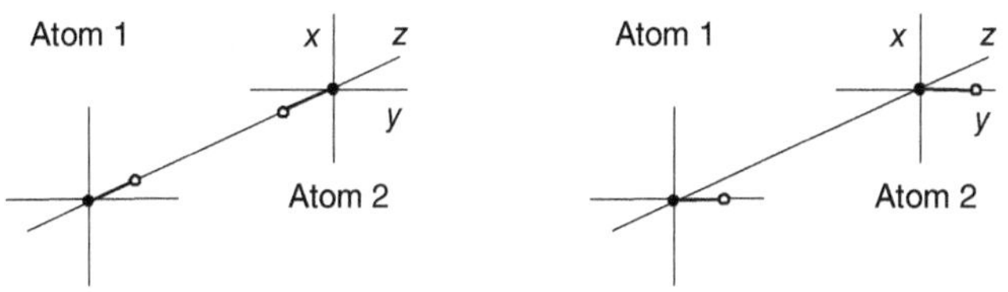

Figure 4.4 Schematic of the momentary interaction between the electrons (o) bonded to the nucleus (•) of two Drude atoms for an in-line orientation (a) and a parallel orientation (b).

5 For an overview by London himself, see [5]. The term "dispersion forces" was coined by him to indicate the analogy of the expressions derived to those which appear in the dispersion formula for the polarizability of a molecule when acted upon by an alternating electric field.

Hence, the solution becomes

$$u = u_+(n_+) + u_-(n_-) = \hbar\omega_+(n_+ + ½) + \hbar\omega_-(n_- + ½)$$

with the new frequencies and force constants

$$\omega_+ = \sqrt{a_-/m}, \omega_- = \sqrt{a_+/m}, a_- = a - (2\mathscr{C} e^2/r^3) \quad \text{and} \quad a_+ = a + 2\mathscr{C} e^2/r^3$$

The new ground-state energy becomes

$$u_0{}' = u_+(0) + u_-(0) = ½\hbar\omega_+ + ½\hbar\omega_- \cong \hbar\omega_0 - (\mathscr{C}^2 e^4 \hbar\omega_0/2a^2 r^6)$$

where the last step is made by expanding the square roots up by the binomial theorem to second order. Finally, we link the force constant a to the polarizability α using the force eE of the electric field E as compensated by the restoring force az, that is, $z = eE/a$. The induced dipole moment is given by $\mu = ez = e^2E/a$ as well as $\mu = \alpha E$, so that $\alpha = e^2/a$. The final expression for the in-line interaction energy becomes

$$w_{zz} = u_0{}' - u_0 = -\mathscr{C}^2\alpha^2\hbar\omega_0/2r^6 \equiv -\mathscr{C}^2\alpha^2 I/2r^6$$

For the two parallel vibrations, perpendicular to the joining axis, of which the solutions are degenerated, a similar calculation with $w_{xx} = \mathscr{C}\ x_1x_2e^2/r^3$ and $w_{yy} = \mathscr{C}\ y_1y_2e^2/r^3$ leads to one quarter of the abovementioned expression for each orientation, so that the total interaction becomes

$$w_{\text{dis}} = w_{zz} + w_{xx} + w_{yy} = -3\mathscr{C}^2 I\alpha^2/4r^6 \equiv -C_{\text{L}}/r^{-6}$$

The dispersion energy is thus proportional to the polarizability α squared, a characteristic energy $I \equiv \hbar\omega_0$, and the reciprocal sixth power of the distance r and represents a purely quantum-mechanical effect due to a difference in zero-point energy. For dissimilar molecules, a similar but more complex calculation leads to

$$C_{\text{L}} = -\frac{3}{2}\mathscr{C}^2\frac{I_1I_2}{I_1 + I_2}\alpha_1\alpha_2 \tag{4.9}$$

Apart from the "binomial approximation," the result is exact for the Drude model of molecules with only one characteristic frequency. For real molecules, the energies I should be chosen in accordance with the strongest absorption frequencies. In the absence of information for I, ionization energies I_0 are often taken. This works reasonably well for He ($I/I_0 \cong 1.20$) and H_2 ($I/I_0 \cong 1.09$), but for Ne, Ar, Kr, Xe, N_2, Cl_2, and CH_4, a mean value of $I/I_0 \cong 2.3$ (with a standard deviation in the mean of 0.2) is required to match the experimental data [6]. In general, although estimates for α are fairly reliable (Section 8.5.3), estimates for I are less trustworthy. Hence, alternatives to estimating C_{L} have been proposed (Section 4.3).

Problem 4.5 *London Interactions* Derive the expression for C_{L} for similar and dissimilar molecules.

4.1.4 The Total Interaction

A comparison of the values of the various contributions to the total interaction is useful, and values as estimated for several small molecules are given in Table 4.1. Here we limit

Table 4.1 Data and vdW energy coefficients (10^{-79} J m^6) for several molecules at 300 K.

Molecule	$d/2$ (pm)	μ (D)	α' (Å^3)	I (eV)	C_{Keesom}	C_{Debye}	C_{London}	C_{vdW}	C_{EoS}
Ar	188	0	1.63	15.8	0	0	50.4	50.4	45.3
Kr	201	0	2.46	14.0	0	0	102	102	
Xe	216	0	4.0	12.1	0	0	233	233	
N_2	—	0	1.77	15.6	0	0	58.7	58.7	55.3
O_2	180	0	1.58	12.1	0	0	36.3	36.3	46.0
HI	—	0.42	5.45	10.4	0.50	1.92	371	374	349
HBr	190	0.80	3.61	11.7	6.59	4.62	183	194	207
HCl	180	1.08	2.63	12.8	21.9	6.13	106	134	157
HF	—	1.91	0.51	15.8	214	3.72	4.94	223	
CCl_4	280	0	10.5	11.1	0	0	1520	1520	
$CHCl_3$	255	1.04	8.8	11.4	19	19	1058	1096	1632
H_2O	140	1.85	1.48	12.6	189	10.1	33.2	232	176
NH_3	180	1.47	2.22	10.9	75.2	9.59	64.5	149	164
CH_4	200	0	2.60	14.4	0	0	116	116	103
C_6H_6	265	0	11.0	11.0	0	0	1520	1520	
$(CH_3)_2CO$	—	2.87	6.33	10.1	2400	104	486	2990	
CH_3OH	210	1.71	3.23	10.9	138	18.9	136	293	651

The parameter $d/2$ refers to the vdW radius as estimated from packing considerations [10]. The value of C_{EoS} is calculated from the vdW Equation of State (EoS) $(P + a/V_{\text{m}}^2)(V_{\text{m}} - b) = RT$ according to $C_{\text{EoS}} = 9ab/4\pi^2 N_{\text{A}}^3$, assuming hard sphere particles with diameter σ or co-volume $b = 2\pi N_{\text{A}}\sigma^3/3$ plus an attractive interaction $-a/V_{\text{m}}$ or $a = 2\pi N_{\text{A}}^2 C_{\text{EoS}}/3\sigma^3$.
Source: Data taken from [7–9].

ourselves to the Keesom, Debye, and London contributions, collectively denoted as the *van der Waals interactions*, since in many cases the interaction between neutral molecules is the most important, and three-body interactions are neglected anyway (Section 4.4). The total attractive (internal) energy $\phi_{\text{vdW}} = 2w_{\text{d-d}} + w_{\text{d-id}} + w_{\text{dis}}$ between two molecules is then

$$\varphi_{\text{vdW}} = -\mathscr{C}^2 \left(\frac{2\mu_1^2\mu_2^2}{3kT} + \mu_1^2\alpha_2 + \mu_2^2\alpha_1 + \frac{3}{2}\alpha_1\alpha_2 \frac{I_1 I_2}{I_1 + I_2} \right) \frac{1}{r^6} \equiv -\frac{C_6}{r^6} \tag{4.10}$$

As for an interaction proportional to $1/T$, $u = 2f$, the first term had to be multiplied by a factor 2. For this calculation, we used the simple Drude model for the London interaction, with the characteristic energy I estimated as the ionization potential. From this comparison, it becomes clear that, in many cases, the London dispersion interaction is dominating. Only for highly polar molecules does the Keesom interaction contribute significantly, for example, H_2O and HF. As an aside, we note that incorporating higher order terms in the analysis of the previous sections leads to $\phi_{\text{vdW}} = -C_6/r^6 - C_8/r^8 + \ldots$, where the quantities C_j all can be expressed in terms of the moments of the charge distribution and polarizabilities of the molecules. Nevertheless, frequently, the r^{-6} approximation is still made, and we do so here.[6]

To obtain the total interaction, we need to add the repulsion acting at short distances between molecules. This interaction, due to orbital overlap, is often called the *Born repulsion*. From quantum mechanics, one can rationalize an exponential dependence, although in practice it is often described by a power law. Hence, we have

$$\phi_{\text{rep}} = b\ \exp(-r/\rho) \qquad \text{or} \qquad \phi_{\text{rep}} = b/r^n$$

where (b,ρ) or (b,n) are parameters. The parameter r/ρ is typically between 12 and 16, while n generally ranges from 9 to 15. A value of $n = 12$ is often taken for mathematical convenience. The total interaction energy is thus given by

$$\phi = \phi_{\text{rep}} + \phi_{\text{vdW}}$$

If we use $\phi = Ar^{-n} - Br^{-m}$, $\text{d}\phi/\text{d}r = 0$ for $r_0 = (nA/mB)^{1/(n-m)}$, so that $\phi = -Br_0{}^{-m}(1 - m/n)$. Hence, using the experimental value for r_0 and $m = 6$, repulsion can be effectively accounted for by n only.

Finally, we note that, using the virial theorem $\overline{\vartheta} = \frac{1}{2}\overline{\mathbf{q} \cdot (\partial\mathscr{H}/\partial\mathbf{q})}$, for any system, a separation of the energy $E(r)$ in kinetic and potential energy parts is possible whenever $\text{d}E(r)/\text{d}r$ is available, either experimentally or theoretically [11–13]. For example, for the Lennard-Jones (LJ) potential $\phi_{\text{LJ}} = \phi_{\text{rep}} + \phi_{\text{att}} = 4\varepsilon(\sigma/r)^{12} - 4\varepsilon(\sigma/r)^6$ (Section 4.3), we have $3\phi_{\text{att}} - 6\phi_{\text{rep}} = \vartheta$. More generally, for two nuclei the electronic energy, $E = \vartheta + \phi$ with ϑ the kinetic energy of the electrons and ϕ the electrostatic potential energy, the force on a nucleus is $-\text{d}E(r)/\text{d}r$, and the equilibrating force to keep a nucleus fixed is $\text{d}E(r)/\text{d}r$. Hence, omitting the overbar, $2\vartheta = -\phi - r(\text{d}E(r)/\text{d}r)$, which combined with $E = \vartheta + \phi$, results in $\vartheta = -E - r(\text{d}E/\text{d}r)$ and $\phi = 2E + r(\text{d}E/\text{d}r)$.

4.2 The Quantum Approach

In Sections 4.1.1–4.1.4, we discussed the interaction between two charge distributions, adding ad hoc induction and dispersion contributions. Actually, the interaction should be described by quantum mechanics, and this is usually done by using second-order perturbation theory, in which the multipole expansion is used as a quantum-mechanical operator [4], as generalized by McLachlan [14]. First, we derive an expression for the (frequency-dependent) polarizability and thereafter reconsider the dispersion energy. Finally, we deal with some aspects of higher-order terms.

To calculate the polarizability due to an (oscillating) electric field, we use time-dependent perturbation theory (Section 3.3.4) with the Hamilton operator $H = H_0 + H' = H_0 + Vf(t)$.

6 Electromagnetic radiation has a finite velocity c, and thus a second molecule feels the potential of the first molecule at a distance r only after a time r/c. This *retardation effect* is important when $r \gg \lambda$, where λ is the wavelength related to the characteristic frequency via $\lambda = c/\nu = hc/I$ and leads to $\phi \sim r^{-7}$. For $I \cong 12.5$ eV (typically), $\lambda \cong 100$ nm. For $r = \lambda$ (0.3λ, 0.1λ), the (Drude model-based) correction factor to $\phi \sim r^{-6}$ is $\cong 0.35$ (0.72, 0.95), so that retardation becomes important for r larger than, say, 25 nm. The effect is important for the mutual interaction between macroscopic bodies and between a single particle and a solid surface at distances larger than about 5 nm [7].

We assume that the wavelength of the radiation is large and that we may neglect the gradient and higher derivatives. We also assume that the field is gradually switched on so that $H' = 2V\cos\omega t \exp\varepsilon t = V\{\exp[(\varepsilon+\mathrm{i}\omega)t]+\exp[(\varepsilon-\mathrm{i}\omega)t]\}$, where ε is small and eventually set to zero. From Eq. (3.28) we have $c_k(t) = (\mathrm{i}\hbar)^{-1}\int_0^t V_{kn}f(\tau)\exp(\mathrm{i}\omega_{kn}\tau)\mathrm{d}\tau$, with $V_{kn} = \langle k|V|n\rangle$. Substitution of H' leads, after integration and using $\hbar\omega_{kn} = E_k - E_n$ where E_k is the eigenvalue (energy) of state $|k\rangle$, to

$$c_k(t) = V_{kn}\frac{\exp(\varepsilon+\mathrm{i}\omega_{kn}+\mathrm{i}\omega)t}{\hbar(\omega_{kn}+\omega-\mathrm{i}\varepsilon)} + \frac{\exp(\varepsilon+\mathrm{i}\omega_{kn}-\omega)t}{\hbar(\omega_{kn}-\omega-\mathrm{i}\varepsilon)} \quad \text{and by setting } \varepsilon = 0 \text{ to}$$

$$c_k(t) = V_{kn}\frac{\exp[\mathrm{i}(\omega_{kn}+\omega)t]}{\hbar(\omega_{kn}+\omega)} + \frac{\exp[\mathrm{i}(\omega_{kn}-\omega)t]}{\hbar(\omega_{kn}-\omega)}$$

Evaluating the dipole moment in the ground state $|0\rangle$ in the presence of the perturbation $H' = 2V\cos\omega t = -\boldsymbol{\phi}'\cdot\hat{\boldsymbol{\mu}}\cos\omega t$, where $\hat{\boldsymbol{\mu}}$ denotes the operator for $\boldsymbol{\mu}$ and $-\boldsymbol{\phi}'$ represents the electric field, results to first order, using $|0\rangle + \Sigma_k c_k|k\rangle$, in

$$\boldsymbol{\mu}(t) = \langle 0|\hat{\boldsymbol{\mu}}|0\rangle + \sum\nolimits_k' c_k(t)\langle 0|\hat{\boldsymbol{\mu}}|k\rangle\exp(\mathrm{i}\omega_{0k}t) + \text{c.c.}$$

where $\langle 0|\hat{\boldsymbol{\mu}}|0\rangle$ is the permanent dipole moment and c.c. denotes the complex conjugate. Substituting $c_k(t)$ leads, after some calculation, to

$$\begin{aligned}\boldsymbol{\mu}(t) &= \langle 0|\hat{\boldsymbol{\mu}}|0\rangle + \sum\nolimits_k'\left\{V_{0k}\langle 0|\hat{\boldsymbol{\mu}}|k\rangle\left[\frac{\exp(\mathrm{i}\omega t)}{\hbar(\omega_{0k}+\omega)} + \frac{\exp(-\mathrm{i}\omega t)}{\hbar(\omega_{0k}-\omega)}\right] + \text{c.c.}\right\}\\ &= \langle 0|\hat{\boldsymbol{\mu}}|0\rangle - \boldsymbol{\phi}'\cos\omega t\sum\nolimits_k'\frac{2\omega_{0k}\langle 0|\hat{\boldsymbol{\mu}}|k\rangle\langle k|\hat{\boldsymbol{\mu}}|0\rangle}{\hbar\left(\omega_{0k}^2+\omega^2\right)}\end{aligned}$$

where we assumed a non-degenerate state so that the sine terms vanished. Comparing with $\boldsymbol{\mu} = \langle 0|\hat{\boldsymbol{\mu}}|0\rangle - \boldsymbol{\alpha}(\omega)\cdot\boldsymbol{\phi}'\cos\omega t$ shows that the sum in the second term represents the polarizability $\boldsymbol{\alpha}(\omega)$. Taking $\omega = 0$, one obtains the static polarizability $\boldsymbol{\alpha}(0)$ of which the spherical average $\alpha = \alpha(0)$ is obtained by calculating $\alpha = ⅓\mathrm{tr}\boldsymbol{\alpha}(0)$. Finally, using the *Unsöld approximation*, where only one excited state is assumed to contribute mainly and thus only one (average) characteristic energy $I = \hbar\omega_0$ is used, leads to $\Sigma_k\langle 0|\hat{\boldsymbol{\mu}}|k\rangle\langle k|\hat{\boldsymbol{\mu}}|0\rangle = \langle 0|\hat{\boldsymbol{\mu}}\hat{\boldsymbol{\mu}}|0\rangle = 3\alpha I/2$. For real molecules, I should be chosen in accordance with their strongest absorption frequency.

Now turning to the dispersion energy, we note that the multipole expansion

$$w_{12} = A\phi + \boldsymbol{B}\cdot\boldsymbol{\phi}' + \mathbb{\Gamma} : \boldsymbol{\phi}'' + \cdots \qquad \text{with}$$

$$A = q_1q_2,\quad \boldsymbol{B} = \hat{\boldsymbol{\mu}}_1q_2 - q_1\hat{\boldsymbol{\mu}}_2,\quad \mathbb{\Gamma} = q_1\mathbb{R}_2/3 - \hat{\boldsymbol{\mu}}_1\hat{\boldsymbol{\mu}}_2 + \hat{\mathbb{R}}_1q_2/3, \ldots \tag{4.11}$$

is considered an operator. The zero-order wave functions are the product functions of the noninteracting molecules. If we denote the electronic ground states in molecules 1 and 2 by k and m, respectively, and similarly the excited states by l and n, the result of second-order perturbation theory using terms up to the dipole–dipole interaction is

$$\phi_{\text{vdW}} = w_{\text{C}} + w_{\text{ind}} + w_{\text{dis}}, \qquad w_{\text{C}} = \langle k|w_{12}|k\rangle\langle m|w_{12}|m\rangle,$$

$$\begin{aligned}w_{\text{ind}} &= \sum\nolimits_l'(\hbar\omega_{kl})^{-1}|\langle k|\hat{\boldsymbol{\mu}}_1|l\rangle\langle m|\hat{\boldsymbol{\mu}}_2|m\rangle\boldsymbol{\phi}'|^2\\ &\quad+\sum\nolimits_n'(\hbar\omega_{mn})^{-1}|\langle k|\hat{\boldsymbol{\mu}}_1|k\rangle\langle m|\hat{\boldsymbol{\mu}}_2|n\rangle\boldsymbol{\phi}'|^2 \qquad \text{and}\end{aligned}$$

$$w_{\text{dis}} = \sum\nolimits_{l,n}'\frac{|\langle k|\hat{\boldsymbol{\mu}}_1|l\rangle\langle m|\hat{\boldsymbol{\mu}}_2|n\rangle\boldsymbol{\phi}''|^2}{\hbar\omega_{kl}+\hbar\omega_{mn}} \tag{4.12}$$

where the prime in the summation excludes $k = l$ and $m = n$. The quantity $\langle k|\hat{\boldsymbol{\mu}}_1|l\rangle$ is the matrix element of the dipole operator of molecule 1 between states k and l. The first-order term w_C represents the *Coulomb* (or *electrostatic*) interaction, but for neutral molecules, the A and $\boldsymbol{B}$ terms in Eq. (4.11) are zero. If the molecules are polar ($\boldsymbol{\mu}_1,\boldsymbol{\mu}_2 \neq 0$ and therefore $\mathbb{T} \neq 0$), we have the *Keesom* (or *orientation*) interaction. The second-order term contains the *Debye* (or *induction*) interaction w_{ind} and the *London* (or *dispersion*) interaction w_{dis}. The term ϕ_{ind} is only non-zero if at least one of the molecules has a permanent dipole moment $\langle k|\hat{\boldsymbol{\mu}}|k\rangle$. The second-order term w_{dis} in Eq. (4.12) is always present and, after some calculation, reduces for the spherical average[7] using $\text{tr}(|\phi''|^2) = 6\mathscr{C}^2/r^{-6}$, to

$$w_{\text{dis}} = \frac{2}{3}\sum\nolimits'_{l,n} \frac{|\langle k|\hat{\boldsymbol{\mu}}_1|l\rangle|^2|\langle m|\hat{\boldsymbol{\mu}}_2|n\rangle|^2}{\hbar\omega_{kl} + \hbar\omega_{mn}}\frac{\mathscr{C}^2}{r^6} \equiv -\frac{C_{\text{L}}}{r^6} \tag{4.13}$$

Since $\hbar\omega_{kl} = (E_k - E_l) < 0$, $\hbar\omega_{mn} = (E_m - E_n) < 0$ and the enumerator >0, the London interaction is always negative and therefore attractive.

Although nowadays ab initio calculations based on this equation are feasible for realistic systems, approximations are still generally made. There are two (approximate) routes to evaluate Eq. (4.13). The first route is to use again the Unsöld approximation $\Sigma_k\langle 0|\hat{\boldsymbol{\mu}}|k\rangle\langle k|\hat{\boldsymbol{\mu}}|0\rangle = 3\alpha I/2$, leading directly to the London estimate, Eq. (4.9). Alternatively, one uses the identity, first used by Casimir and Polder [15],

$$\frac{1}{A+B} = \frac{2}{\pi}\int_0^\infty \frac{AB}{(A^2+x^2)(B^2+x^2)}\,\text{d}x \quad \text{for } A > 0 \text{ and } B > 0 \tag{4.14}$$

Applying this relation to $(\hbar\omega_{kl} + \hbar\omega_{mn})^{-1}$ with ground state $|0\rangle$, the result is [4, 14]

$$C_{\text{L}} = \frac{3\hbar\mathscr{C}^2}{\pi}\int_0^\infty \alpha_1(\text{i}\omega)\alpha_2(\text{i}\omega)\,\text{d}\omega \quad \text{with} \quad \alpha_j(\text{i}\omega) = 2\sum\nolimits_l \frac{|\langle 0|\hat{\boldsymbol{\mu}}_j|l\rangle|^2\hbar\omega_{0l}}{\hbar^2(\omega_{0l}^2+\omega^2)} \tag{4.15}$$

To avoid difficulties with the poles occurring in $\alpha(\omega)$ when $\omega = \omega_n$ upon integration, $\alpha(\text{i}\omega)$ is introduced as this is a continuously decreasing function of ω. The physical meaning of $\alpha(\text{i}\omega)$ is not clear, and it is best to consider it a mathematical convenience. The (spherically averaged) polarizability $\alpha(\omega)$ with frequency ω can be empirically well described by

$$\alpha(\omega) = \frac{e^2}{m}\sum\nolimits_l \frac{f_l}{\omega_{0l}^2 - \omega^2 - \text{i}\omega\gamma_{0l}} \cong \frac{\alpha(0)\omega_{0l}^2}{\omega_{0l}^2 - \omega^2}$$

Here e and m are the electron charge and mass, ω_{0l} and γ_{0l} are the characteristic frequency and damping constant for a transition from state $|0\rangle$ to $|l\rangle$, and f_l is the associated *oscillator strength*. In the last step, a single-term approximation is made, meanwhile neglecting absorption. Inserting the single-term approximation in Eq. (4.15) again leads to the London expression. A clear advantage of the latter formulation is that, once $\alpha_j(\text{i}\omega)$ is established, $C_{\text{L},ij}$ can be calculated straightforwardly.

Similar considerations for the dipole–quadrupole interaction lead to [4]

$$-\frac{C_8}{r^8} \quad \text{with} \quad C_8 = \mathscr{C}^2\frac{15}{8}\frac{I_1I_2}{(I_1+I_2)}(\alpha_1\xi_2 + \xi_2\alpha_2)$$

---

7 For a spherical average T_{sph} of a second-rank tensor $\mathbb{T}$, we have $T_{\text{sph}} = ⅓\text{tr}\mathbb{T}$. If $\mathbb{T}$ is given as the dyadic product $\mathbb{T} = \boldsymbol{rr}$ of a vector $\boldsymbol{r}$, this reduces to $(r_{\text{sph}})^2 = ⅓\text{tr}(\boldsymbol{rr}) = ⅓\boldsymbol{r}\cdot\boldsymbol{r} = ⅓r^2 = x^2 = y^2 = z^2$.

where ξ denotes the (spherically averaged) quadrupole polarizability. The coefficient C_8 reduces for Drude atoms with a single excitation energy to [16]

$$C_8 = \mathscr{C}^2 \frac{45}{8} I_1 I_2 \alpha_1 \alpha_2 \left(\frac{\alpha_1 I_1}{2I_1 + I_2} + \frac{\alpha_2 I_2}{I_1 + 2I_2} \right)$$

Estimates for the various coefficients thus can be made.

Having shown that the interaction at 0 K is given by $w(0)$, McLachlan also calculated the interaction $w(T)$ at temperature T. The line-of-thought uses statistical thermodynamics as well as several advanced concepts. For a clear formulation, we refer to the original papers [14], although many details have to be filled in by the reader. Here we quote only the general, spherically averaged result

$$w = -6kT\mathscr{C}^2 r^{-6} \sum\nolimits'_{n=0} \alpha_1(\mathrm{i}\omega_n)\alpha_2(\mathrm{i}\omega_n)$$

where $\omega_n = (2\pi/\beta\hbar)n$ $(n = 0, \pm 1, \pm 2, \ldots)$ and the prime indicates that the $n = 0$ term should be taken half. Assuming that a molecule has one typical absorption frequency ω_α and a dipole moment μ with typical orientational frequency ω_μ, the frequency-dependent response can be modeled by [17]

$$\alpha(\omega) = \alpha_{\text{abs}}(\omega) + \alpha_{\text{ori}}(\omega) = \alpha\left[1 - (\omega/\omega_\alpha)^2\right]^{-1} + \mu^2\{3kT[1 - \mathrm{i}(\omega/\omega_\mu)]\}^{-1}$$

Using this expression for $\alpha(\mathrm{i}\omega)$, the $n = 0$ term therefore yields

$$w(r)_{\omega=0} = -3kT\mathscr{C}^2 \left[\left(\alpha_1 + \mu_1^2/3kT\right)\right]\left[\left(\alpha_2 + \mu_2^2/3kT\right)\right]/r^{-6}$$

describing the Keesom and Debye interactions. At room temperature, where $\hbar\omega_\alpha \gg kT$, the α contribution is usually much larger than the μ contribution. Moreover, the points $\mathrm{i}\omega_n$ are close together, and we may replace the sum by an integral using

$$kT\sum\nolimits_{n=1,..} = \hbar\int_{\omega_1}^{\infty} \mathrm{d}\omega \cong \hbar\int_0^{\infty} \mathrm{d}\omega \quad \text{so that} \quad w(r)_{\omega\geq 1} = -\frac{3\hbar\mathscr{C}^2}{\pi r^6}\int_0^{\infty} \alpha_1(r)\alpha_2(r)\mathrm{d}\omega$$

Using once more Eq. (4.14), the result can be shown to be the London equation. These results generalize $w(0)$ and provide a useful tool, including, in principle, all multipole terms and retardation [14]. It is also possible to include the effect of a medium (by considering the polarizabilities as excess polarizabilities) and three-body effects. For details, we refer to the original paper [18].

Problem 4.6 ***Dispersion Energy*** Derive Eq. (4.13) (see footnote 7). Note that $\langle i|\hat{\boldsymbol{\mu}}|i\rangle$ behaves like a vector and $[\nabla\nabla(1/r)]_0$ like a second-rank tensor. Further, use $\nabla r^n = nr^{n-1}\nabla r = nr^{n-2}\mathbf{r}$ and $\nabla\mathbf{r} = \mathbb{I}$.

4.3 Model Interactions

The above given treatment provides an insightful description of intermolecular interactions, but often a simpler expression is required. Almost invariably, these expressions are called potentials, although they are in fact energies. Any simplified potential should contain an attractive and a repulsive part, but the simplest potential used is the *hard-sphere* (HS) *potential* containing only repulsion and given by

$$\phi_{\text{HS}} = \infty \text{ for } r < \sigma \qquad \text{and} \qquad \phi_{\text{HS}} = 0 \text{ for } r \geq \sigma$$

where σ is the HS diameter. Although highly artificial, this potential is nevertheless often used in simple models and as a reference potential.

The simplest way to include attractive interaction to the HS potential is to add a region with constant attraction. This leads to the *square-well* (SW) *potential*

$$\phi_{SW} = \infty \text{ if } r < \sigma, \quad \phi_{SW} = -\varepsilon \text{ if } \sigma \leq r \leq R_{SW}\sigma \quad \text{and} \quad \phi_{SW} = 0 \text{ if } r > R_{SW}\sigma$$

where, as before, σ is the HS diameter and ε and R_{SW} are the depth and range of the SW, respectively. Typically, the value of $R_{SW} \cong 1.5$–2.0, and although the potential is simple, it will appear to be versatile. Another option is to add, instead of a constant attraction over a finite range, a vdW-like attraction $-cr^{-m}$, with c and m constants, resulting in what is known as the *Sutherland potential*.

A more realistic expression is the *Mie potential* given by

$$\phi_{Mie} = br^{-n} - cr^{-m} \tag{4.16}$$

where b, c, n, and m are parameters and $n > m$. An often-used choice, mainly for mathematical convenience, is the 12-6 or LJ *potential*[8] with $n = 12$ and $m = 6$. For this potential, alternative forms for Eq. (4.16) are

$$\phi_{LJ} = 4\varepsilon[(\sigma/r)^{12} - (\sigma/r)^6] = \varepsilon\left[(r_0/r)^{12} - 2(r_0/r)^6\right]$$

Here, the parameter ε describes the depth of the potential energy curve, while r_0 is the position of the minimum. For the LJ 12-6 potential, we thus have $r_0 = 2^{1/6}\sigma$, $\phi_{LJ}(r_0) = -\varepsilon$ and $\phi_{LJ}(\sigma) = 0$. The parameter σ can be seen as the diameter of the molecule. The LJ potential describes the interaction energy qualitatively well (Figure 4.1). Typical values of parameters for the LJ potential for some molecules are given in Sections F.1–F.2. It will be clear that slightly different values will be obtained depending on which properties and procedures are used for their evaluation.

For quasi-spherical molecules, such as SF_6, CF_4, and $C(CH_3)_4$, both the attractive and repulsive forces decrease more rapidly than would be expected from the LJ potential. This implies a higher value for both exponents 12 and 6. In fact, we can write the Mie potential, using ε and r_0, as

$$\phi_{Mie} = \frac{nm\varepsilon}{n-m}\left[n(r_0/r)^n - m(r_0/r)^m\right]$$

The potential for quasi-spherical molecules can be described by $(n,m) = (28,7)$, where the precise choice is largely dictated by mathematical convenience. An alternative for quasi-spherical molecules is the *Kihara potential* [19], which describes the interaction in a similar way as the LJ potential, but replaces σ and r with $\sigma - \delta$ and $r - \delta$, where δ is the hard core diameter. We thus have $\phi_K(r < \delta) = \infty$ and $\phi_K(\sigma) = 0$. Typical values are $\delta/\sigma = 0.084$ for Ar, 0.194 for N_2, 0.655 for C_6H_6, and 0.900 for n-C_6H_{14}. Its shape as a function of r resembles that of the Mie 28-7 potential, taking δ equal to ¼ of the vdW radius [20, 21]. For some data, see Section F.7.

While the power–law relationship with $m = 6$ can be rationalized for the attractive part of the potential, the repulsion is, on quantum-mechanical grounds, expected to be better described by an exponential. This leads to the *Buckingham* or *exp-6 potential*:

$$\phi_B = \varepsilon\left[\frac{6}{\alpha-6}\,e^{-\alpha(r-r_0)/r_0} - \frac{\alpha}{\alpha-6}\left(\frac{r_0}{r}\right)^{-6}\right] \quad \text{and} \quad \phi_B = \infty \text{ for } r < r_{max}$$

8 Although I generally refrain from historical remarks, I must make an exception here: John E. Jones married Kathleen Lennard in 1925 and added his wife's name to his own to become Lennard–Jones.

where ε, r_0, and α are parameters, representing dissociation energy, equilibrium distance, and (scaled) force constant, respectively. Because this potential shows a maximum at a small value of r, the condition $\phi_B = \infty$ for $r < r_{max}$ is added. Another often-used expression, with ε, r_0, and α as parameters, is the *Morse potential* [22]:

$$\phi_M = \varepsilon \left[e^{-2\alpha(r-r_0)} - 2\,e^{-\alpha(r-r_0)} \right] = \varepsilon \left[1 - e^{-\alpha(r-r_0)} \right]^2 - \varepsilon$$

Finally, we mention the *Yukawa* (or *screened Coulomb*) *potential* given by the expression $\phi_Y = (\varepsilon/r)\exp[-\alpha\,(r - r_0)]$. Combining two such potentials leads to the *double Yukawa potential* given by

$$\phi_{DY} = \varepsilon \left\{ e^{-\alpha(r-r_0)} - e^{-\beta(r-r_0)} \right\} / r$$

This is a flexible potential for which many analytical results can be obtained [23].

For estimating the interaction between unlike molecules, several combining rules exist. The repulsion is often modeled as for hard spheres, that is, one uses for spheres i with diameter σ_{ii} the combining rule $\sigma_{ij} = [(½\sigma_{ii})^3 + (½\sigma_{jj})^3]^{1/3} \cong (\sigma_{ii} + \sigma_{jj})/2$, usually denoted as the *Lorentz rule*. For attraction, the London expression $-C_L/r^6$ is combined with the attractive part of the LJ potential $-4\varepsilon_{jj}\sigma_{jj}{}^6/r^6$, resulting in

$$\gamma_{ij} \equiv \frac{3}{2}\frac{I_i I_j}{I_i + I_j}\alpha_i'\alpha_j' = 4\varepsilon_{ij}\sigma_{ij}^6 \quad \text{so that} \quad I_j = \frac{4}{3}\frac{\gamma_{jj}}{\alpha_j'^2} = \frac{4}{3}\frac{4\varepsilon_{jj}\sigma_{jj}^6}{\alpha_j'^2} \equiv \frac{4}{3}\chi_j$$

After some manipulation, we obtain

$$\frac{\alpha_1'\alpha_2'}{\gamma_{12}} = ½\left(\frac{\alpha_1'^2}{\gamma_{11}} + \frac{\alpha_2'^2}{\gamma_{22}}\right) \quad \text{and} \quad \varepsilon_{12} = ½\frac{\alpha_1'\alpha_2'}{\sigma_{12}^6}\frac{\chi_1\chi_2}{\chi_1 + \chi_2} \tag{4.17}$$

Since reliable information on I is scarce, this route is essentially a clever way, as proposed by Kohler [24, 25], to circumvent the need for an estimate of I.

If we take $\sigma_{ii} = \sigma_{jj}$ and $I_i = I_j$, we obtain the geometric mean $\varepsilon_{ij} = (\varepsilon_{ii}\varepsilon_{jj})^{1/2}$, usually denoted as the *Berthelot rule*. However, while the Lorentz rule is quite acceptable, the Berthelot rule is only approximate. Some algebra leads to

$$\frac{\varepsilon_{12}}{(\varepsilon_{11}\varepsilon_{22})^{1/2}} = \frac{\left(\sigma_{11}^6\sigma_{22}^6\right)^{1/2}}{[(\sigma_{11} + \sigma_{22})/2]^6}\frac{(\chi_1\chi_2)^{1/2}}{(\chi_1 + \chi_2)/2}$$

showing that, since that the geometric mean $\leq$ the arithmetic mean, $(\varepsilon_{11}\varepsilon_{22})^{1/2}$ always overestimates ε_{12}. A sensitivity analysis for ε_{12} using the Lorentz–Berthelot (LB) rule shows that the ratio σ_{11}/σ_{22} has a particular large influence. Still, if intermolecular interactions are fitted to the experimental data, the Berthelot rule is frequently used. We note that the success of Berthelot's rule is fortuitous and due to partial canceling of errors.

It is therefore no surprise that many other combining rules are in use. For example, using the geometric mean rule for the attractive interaction leads for the LJ potential to $\varepsilon_{ij}\sigma_{ij}{}^6 = (\varepsilon_{ii}\sigma_{ii}{}^6\varepsilon_{jj}\sigma_{jj}{}^6)^{1/2}$. Expressing the repulsion as a sum of the deformation energies for each molecule, $U_{ij}{}^{rep}(r_i + r_j) = ½[U_{ii}{}^{rep}(2r_i) + U_{jj}{}^{rep}(2r_j)]$, at separations r_i and r_j with equal slope of the potential energy curve so that $\left[dU_{ii}^{rep}(r)/dr\right]_{r=2r_i} = \left[dU_{jj}^{rep}(r)/dr\right]_{r=2r_j}$, one obtains for the LJ potential

$$\left(\varepsilon_{ij}\sigma_{ij}^{12}\right)^{1/13} = ½\left[\left(\varepsilon_{ii}\sigma_{ii}^{12}\right)^{1/13} + \left(\varepsilon_{jj}\sigma_{jj}^{12}\right)^{1/13}\right]$$

For simple molecules, it appears that this combining rule [26] provides a much more accurate estimate than the Berthelot rule [27]. For related views, see [28, 29].

Problem 4.7 ***Force Constants*** Show that the force constant $a = \partial^2\phi/\partial x^2|_{r=r0}$ for the Morse and Kihara potentials is given by $a = 2\alpha^2\varepsilon$ and $a = 72\varepsilon/(r_0 - \delta)^2$, respectively.

Problem 4.8 ***The Universal Bonding Curve*** Another well-known potential is the so-called universal bonding curve [30]:

$$\phi_{\text{uni}}(r) = \varepsilon g(x), \qquad g(x) = -(1 + x + 0.05x^3)\exp(-x), \qquad x = (r - r_0)/l$$

Show that the force constant a for this potential is $a = \varepsilon/l^2$, neglecting the x^3 term.

Problem 4.9 ***Berthelot's Rule*** Show, using Eq. (4.17), that only if $\sigma_{11} = \sigma_{22}$ and $I_1 = I_2$, that $\varepsilon_{12} = (\varepsilon_{11}\varepsilon_{22})^{1/2}$. Also estimate ε_{12} using $\varepsilon_{12} = (\varepsilon_{11}\varepsilon_{22})^{1/2}$ and Eq. (4.17), taking a typical value of $\alpha_1{}'/\alpha_2{}' = 2$ and σ_{11}/σ_{22} and $\varepsilon_{11}/\varepsilon_{22}$ ranging from 1 to 2. Using this, assess the Berthelot rule.

Problem 4.10 ***Cohesive Energy*** Show, assuming $\varepsilon_{\text{coh}} = -C/r^6$, that the average value $\langle\varepsilon_{\text{coh}}\rangle$ for a fluid is given by

$$\langle\varepsilon_{\text{coh}}\rangle = -\frac{4\pi C\rho}{N-1}\left(\frac{1}{3\sigma^3} - \frac{1}{R^3}\right) \cong -\frac{4\pi C\rho}{3\sigma^3(N-1)}$$

Assume a random distribution of N molecules over a volume, that is, a uniform number density $\rho = N/V$, by integrating ε_{coh} for a reference molecule over a spherical shell for the smallest distance of approach σ to some macroscopic distance R. By taking this interaction over all pairs of molecules, show that U_{coh} reads

$$U_{\text{coh}} = -2\pi N_A^2 C/3\sigma^3 V_m$$

Estimate the value of U_{coh} for CCl_4 with mass density $\rho' = 1.59\,\text{g cm}^{-3}$ and molecular mass $\mathcal{M} = 153.8\,\text{g mol}^{-1}$. Do so first by estimating σ from the polarizability volume α' and C from Eq. (4.10), and second from the LJ parameters ε and σ, leading to $C \equiv \gamma = 4\varepsilon\sigma^6$. Comment on the agreement with the experimental value $U_{\text{coh}} = 32.8\,\text{kJ mol}^{-1}$.

4.4 Refinements

In this section, we introduce hydrogen bonding and three-body interactions and deal briefly with how far empirical potentials are accurate.

4.4.1 Hydrogen Bonding

All the abovementioned interactions are generally valid. There is, however, a special bond, designated as *hydrogen bond*, which is a fairly directional, attractive interaction between two electronegative species A and B that form a link A–H···B via the covalently bonded H atom and the lone pair of electrons (or polarizable π electrons) of B. The Lewis acid A withdraws some charge from the H atom, leading to a partial charge δ^+ that interacts with

δ^- of the Lewis base B. More generally, if a molecule accepts some charge from another charge-donating molecule, one denotes this as a *donor–acceptor* or *acid–base* (AB) *interaction* [31, 32].

The hydrogen bond can also be considered a delocalized bond in which orbitals from the A and B species as well as the 1s orbital of the H atom are involved and which accommodates four electrons (two electrons from the A—H bond and two from the lone pair of atom B). These four electrons occupy the two lowest molecular orbitals, leading to overall energy lowering. Based on orbital overlap, the hydrogen bond strength decreases rapidly with distance. Nevertheless, despite the orbital overlap picture, one considers the hydrogen bond nowadays to be largely electrostatic, with for water only a small amount of charge transfer. When hydrogen bonding is active, it often dominates over the other vdW interactions. Charge transfer may be more important in other solvents and for heavy metals. For example, Na_2SO_4 is soluble in H_2O but not in NH_3, but for AgCl, the reverse is true.

Two types of hydrogen bonds can be distinguished: an *intramolecular* bond between A and B atoms within the same molecule and an *intermolecular* bond between A and B atoms from different molecules. An example of the former is found in ortho-hydroxybenzoic acid, while the latter occurs, for example, in CH_3COOH and in H_2O. For F, O, and N atoms, the interaction strength decreases from the FH to the OH to the NH bond. While the O—H···O hydrogen bond plays an important role in water, bonds of the type N—H···O=C occur in, for example, urea and proteins, playing an important role in biological systems.

The hydrogen atom in the O—H···O bond has two nonequivalent minimum energy positions near the two O atoms, separated by an energy barrier. For example, in water, the O—H bond (≅0.097 nm) is shorter than the H···O distance (≅ 0.176 nm). For strong hydrogen bonds, the bond length is usually shorter than the sum of the vdW radii. The enthalpy of dissociation $\Delta_{\text{dis}}H$ ranges from 10 to 40 kJ mol^{-1} per bond; for formic and acetic acids in the gas phase at about 400 K, for example, $\Delta_{\text{dis}}H \cong 30\,\text{kJ mol}^{-1}$ per bond. In view of the electrostatic nature of the hydrogen bond, it is no surprise that in solution $\Delta_{\text{dis}}G$ scales with the permittivity ε_r of the solvent as $\Delta_{\text{dis}}G^{(\text{liq})} = \Delta_{\text{dis}}G^{(\text{gas})}/\varepsilon_r$. Hence, $\Delta_{\text{dis}}H^{(\text{liq})} = [\Delta_{\text{dis}}H^{(\text{gas})} + T\Delta_{\text{dis}}G^{(\text{gas})}(\text{d}\ln\varepsilon_r/\text{d}T)]\Delta_{\text{dis}}H^{(\text{gas})}/\varepsilon_r \cong [1 + T(\text{d}\ln\varepsilon_r/\text{d}T)]\Delta_{\text{dis}}H^{(\text{gas})}/\varepsilon_r$, consistent with the experimental data for, for example, acetic acid [33]. Hydrogen bonds provide multiple options for bonding in water, which leads to a residual entropy for ice of $R\ln(3/2)$ and the anomalous behavior of water, for example, a low compressibility κ, high permittivity ε_r, high normal boiling point T_n, and minimum density ρ at 4 °C.

Clearly, the infrared (IR) frequencies of the OH bond are influenced by the presence of a hydrogen bond. For example, for gaseous water, the bond is characterized by a wave number of ≅ 3756 cm^{-1}, whereas for liquid water, the wave number decreases to ≅ 3453 cm^{-1}. In a simple model for collinear O—H···O bonds, the O—H···O bond energy $\phi_{\text{H-bond}}$ can be described by one Morse potential for the O—H bond with distance r and one for the H···O bond with distance $R - r$ plus the interaction between the two O atoms, $u(R)$. This expression contains several parameters: the bond dissociation energies ε_0, the force constant a_0 for these bonds unperturbed and their associated equilibrium distances r_0, the distance between the two O atoms R, and, finally, the parameters in the function $u(R)$. The values for these parameters can be obtained from the condition $(\partial\phi_{\text{H-bond}}/\partial r)_{\text{eq}} = 0$. The IR shifts $\Delta\nu = \nu_{\text{H-bond}} - \nu_0$, with ν_0 the unperturbed bond frequency, are obtained from the force constant $a = (\partial^2\phi_{\text{H-bond}}/\partial r^2)_{\text{eq}}$ and are to be compared with the experimental data.

Hydrogen bonding leads to some phenomena deviating from expected, normal behavior; for example, when hydrogen bonding is present, significant deviations occur from Trouton's rule (Section 13.1) and Eötvös' rule (Section 11.2). Hydrogen bonding also affects other properties considerably, but two examples for the normal boiling point T_n will suffice. Ethanol CH_3CH_3OH and dimethyl ether CH_3OCH_3 are of similar size and have the same number of electrons; moreover, their dipole moments are not very different. Nevertheless, their boiling points are 78.5 and −24.8 °C, respectively. Pentane C_5H_{12} and butanol C_4H_9OH are also of similar size and have the same number of electrons. In this case, the dipole moment of butanol aids to increase the intermolecular interaction, and their boiling points are 36.3 and 117 °C, respectively. The melting point T_{mel} is similarly affected. For example, at room temperature, $B(OH)_3$ forms flaky crystals with the planar molecules bound stably in layers with three adjacent molecules in the same plane, leading to $T_{mel} = 185\,°C$, while BF_3 with a rather similar molecular weight has $T_{mel} = -127\,°C$.

The field of hydrogen bonding is rich and elaborate. Classical reviews have been given by Pimentel [34] and Joesten and Schaad [35], while Grabowski [36] provides a more recent review. Another extensive source is the series by Schuster et al. [37].

4.4.2 Three-Body Interactions

So far, we have been discussing intermolecular interactions in terms of two-body potentials. However, there is no reason why a third body would not influence the interactions between two bodies, apart from contributing to the sum of the two-body potentials for all three bodies. The dipolar *three-body interaction* is probably [38] the most important and has been studied by Axilrod and Teller [39–41] using third-order perturbation theory. Using the Drude model with a single excitation energy results in

$$\phi_{123} = \gamma_{123}(1 + 3\cos\theta_1 \cos\theta_2 \cos\theta_3)/r_{12}^3 r_{23}^3 r_{31}^3 \quad \text{with}$$

$$\gamma_{123} = \frac{3I_1I_2I_3(I_1 + I_2 + I_3)\alpha_1\alpha_2\alpha_3}{2(I_1 + I_2)(I_2 + I_3)(I_3 + I_1)} \quad \text{or} \quad \gamma_{111} = 3\mathscr{C}\alpha C_L/4$$

where the last step can be made when the molecules are identical. Further, θ_1 denotes the internal angle between the vector with length r_{12} between particles 1 and 2 and r_{13} the length of the vector between particles 1 and 3. The angles θ_2 and θ_3 have a similar meaning, and the sign of this contribution depends on the geometry of the triangle formed. For acute triangles, $\phi_{123} > 0$, while for obtuse triangles, $\phi_{123} < 0$. The significance of the three-body interaction has been estimated rather differently by various authors. On the one hand, Rowlinson [42] indicates a too negative potential energy by "10–15% of the total potential energy of a normal liquid," later alleviating this [43] to 5–10% for a typical configuration in a liquid, where most close triplets of molecules form acute-angled triangles but also indicating substantial cancelation of multibody terms. Dymond and Alder [44], on the other hand, indicate a contribution "… as little as 1–2%," supported by Berendsen [45], who stated that the contribution is probably negligible. The matter is further reviewed in [46, 47], and we will pay attention to three-body interactions only occasionally.

4.4.3 Accurate Empirical Potentials

Although model potentials are useful, they are idealized. Figure 4.5a shows a typical set of curves for Ar [48] fitted on second virial coefficients over a temperature range of

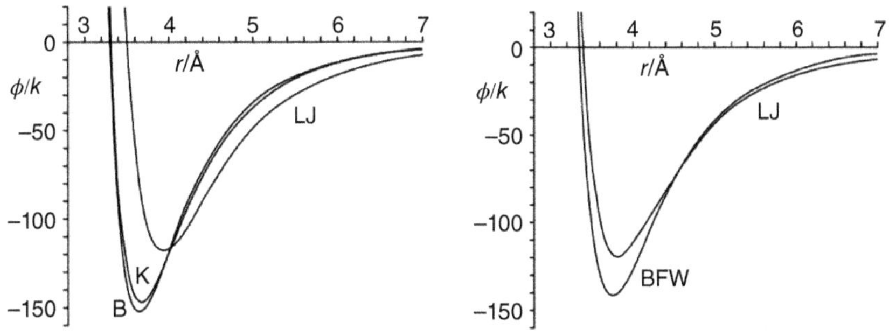

Figure 4.5 (a) Comparison of the Lennard–Jones (LJ), Kihara (K), and Buckingham (B) potentials for Ar showing the difference in minimum and depth of the various potentials. (b) Comparison of the LJ and Barker–Fisher–Watts potential (BFW) for Ar showing that the LJ potential is shallower at its minimum and has a longer range than the BFW potential.

−190 to 600 °C. It will be clear that both depth and location of the potential vary considerably with the type of potential. Similar differences are obtained for other molecules. A rather complete evaluation of Ar dimers has been given by Parson et al. [49]. Only in a few cases are accurate empirical potentials available, the inert gases being such a case. Barker and co-workers [50] represented the Ar potential as a sum of pair potentials and the Axilrod–Teller three-body interaction, resulting in a 15-parameter equation (!). For fitting, they used the thermodynamic and transport data of the gas, the low-temperature properties of the solid, the internal energy and self-diffusion coefficient of the liquid, and the equation of state. This provided excellent agreement with the vibration spectrum of bound Ar dimers and with molecular beam scattering cross-section determinations [49]. It appeared also to describe the solid-state internal energy, the elastic constants, the phonon curves, and the solid–fluid coexistence curve well. Calculations of the third virial coefficient with this potential showed the contribution of the three-body interaction to be quite significant and to bring the theoretical results in good agreement with the experiment data. Figure 4.5b shows a comparison with the optimum LJ potential, indicating that the latter is shallower at its minimum and has a longer range than the optimum potential. It seems that the Kihara and Buckingham potentials are doing better in this sense. Nevertheless, simple pair potential calculations for the condensed phase perform quite well in practice, and one might wonder why they do not perform very badly. Barker et al. [50] concluded that the three-body effects partially cancel out in the condensed phase and that three-body effects are taken into account in an average way using a fitted two-body potential. Their study shows that, for Ar at least, a single potential derived from bulk data can describe all properties.

4.5 Final Remarks

This concludes our discussion on intermolecular interactions. Israelachvili [10] provides a good introduction. Comprehensive overviews are given by Stone [4] and Kaplan [51], while Margenau and Kestner [16] and Hirschfelder et al. [3] are classic references. Mahanty and Ninham [52] discuss dispersion forces from a more macroscopic point of view.

References

1 Murrell, J.N. and Laidler, K.J. (1968). *Trans. Far. Soc.* 64: 371; Wales, D.J. and Berry, R.S. (1992). *J. Chem. Soc. Far. Trans.* 88: 543.
2 Böttcher, C.J.F. (1973). *Theory of Dielectric Polarization*, 2e, Vol. 1. Amsterdam: Elsevier.
3 Hirschfelder, Curtiss and Bird, 1964.
4 Stone, 1996 and 2013.
5 London, F. (1937). *Trans. Far. Soc.* 33: 8.
6 Pitzer, K.S. (1959). *Adv. Chem. Phys.* 2: 59.
7 Atkins, P.W. (2002). *Physical Chemistry*, 7e. Oxford: Oxford University Press.
8 Butt, H.-J., Graf, K., and Kappl, M. (2006). *Physics and Chemistry of Interfaces*, 2e. Weinheim: Wiley-VCH.
9 Yildirim Erbil, H. (2006). *Surface Chemistry of Solid and Liquid Interfaces*. Oxford: Blackwell Publ. Ltd.
10 Israelachvili, J., 1991 and 2011.
11 Slater, J.C. (1933). *J. Chem. Phys.* 1: 687.
12 Löwdin, P.O. (1959). *J. Mol. Spectr.* 3: 46.
13 Rioux, F. (1998). *J. Chem. Ed.* 75: 538; (2001). *Chem. Ed.* 6, 288.
14 McLachlan, A.D. (1963). *Proc. Roy. Soc. A* 271: 387; *Proc. Roy. Soc. A* 274: 80.
15 Casimir, H.B.G. and Polder, D. (1948). *Phys. Rev.* 73: 360.
16 Margenau and Kestner, 1969 and 1971.
17 von Hippel, A.R. (1954). *Dielectrics and Waves*. Cambridge, MA: MIT Press.
18 McLachlan, A.D. (1966). *Disc. Far. Soc.* 40: 239.
19 Kihara, T. (1954). *Rev. Mod. Phys.* 25: 831.
20 Hamann, S.D. and Lambert, J.A. (1954). *Austr. J. Chem.* 7: 1.
21 Hildebrand, J. and Scott, R.L. (1962). *Regular Solutions*. Englewood Cliffs: Prentice-Hall.
22 Morse, P.M. (1929). *Phys. Rev.* 20: 57.
23 Rowlinson, J.S. (1989). *Physica A* 156: 15; Tang, Y., Tong, Z., and Lu, B.C.-Y. (1997). *Fluid Phase Eq.* 134: 21; Montes, J., Robles, M., and López de Haro, M. (2016). *J. Chem. Phys* 144: 084503.
24 Kohler, F. (1957). *Monatsh. Chem.* 88: 857.
25 Munn, R.J. (1961). *Trans. Far. Soc.* 57: 187.
26 Kong, C.L. (1973). *J. Chem. Phys.* 59: 2464.
27 Delhommele, J. and Millié, P. (2001). *Mol. Phys.* 99: 619.
28 Boda, D. and Henderson, D. (2008). *Mol. Phys.* 106: 2367.
29 Al-Matar, A.K. and Rockstraw, D.A. (2006). *Dirasat Eng. Sci.* 33: 27.
30 Rose, J.H., Smith, J.R., Guinea, F. et al. (1984). *Phys. Rev. B* 29: 2963.
31 Gutmann, V. (1978). *The Donor-Acceptor Approach to Molecular Interactions*. New York: Plenum Press.
32 Berg, J.C. (1993). *Wetting by Solutions*, Ch. 2 in *Wettability* (ed. J.C. Berg), p. 75. New York: Marcel Dekker.
33 Davies, M., Jones, P., Patnaik, D. et al. (1951). *Trans. Chem. Soc.*, paper 278, 1249.
34 Pimentel, G.C. and McLellan, A.L. (1960). *The Hydrogen Bond*. New York: W.H. Freeman and Co.
35 Joesten, M.D. and Schaad, J.L. (1974). *Hydrogen Bonding*. New York: Dekker.
36 Grabowski, S.J. (2006). *Hydrogen Bonding – New Insights*. Dordrecht: Springer.

37 Schuster, P., Zundel, G., and Sandorfy, C. (1976). *The Hydrogen Bond*, Vol. I-III. Amsterdam: North-Holland.
38 Scott, R.L. (1971). *Introduction*, Ch. 1 in *Physical Chemistry*, Vol. VIIIA (ed. H. Eyring, D. Henderson, and W. Jost), p. 1. New York: Academic Press.
39 Axilrod, B.M. (1949). *J. Chem. Phys.* 17: 299; *Ibid.* 19: 71.
40 Axilrod, B.M. and Teller, E. (1943). *J. Chem. Phys.* 11: 299.
41 Midzuno, Y. and Kihara, T. (1956). *J. Phys. Soc. Japan.* 11: 1045.
42 Rowlinson, J.S. (1965). *Disc. Far. Soc.* 40: 19.
43 Rowlinson, J.S. and Swinton, F.L. (1982). *Liquids and Liquid Mixtures*, 3e. London: Butterworth Scientific.
44 Dymond, J.H. and Alder, B.J. (1968). *Chem. Phys. Lett.* 2: 54.
45 Berendsen, H.J.C. (2007). *Simulating the Physical World*, p. 187. Cambridge: Cambridge UP.
46 Elrod, M.J. and Saykally, R.J. (1994). *Chem. Rev.* 94: 1975.
47 Reilly, A.M. and Tkatchenko, A. (2015). *Chem. Sci.* 6: 3289.
48 Sherwood, A.E. and Prausnitz, J.M. (1964). *J. Chem. Phys.* 41: 429.
49 Parson, J.M., Siska, P.E., and Lee, Y.T. (1972). *J. Chem. Phys.* 56: 1511.
50 Barker, J.A., Fisher, R.A., and Watts, R.O. (1971). *Mol. Phys.* 21: 657.
51 Kaplan, 2006.
52 Mahanty, J. and Ninham, B.W. (1976). *Dispersion Forces*. New York: Academic Press.

Further Reading

Hirschfelder, J.O., Curtiss, C.F., and Bird, R.B. (1954). *Molecular Theory of Gases and Liquids*. New York: John Wiley & Sons; corrected ed. (1964). A classic book containing a wealth of information on theory and experiments.

Israelachvili, J. (1991). *Intermolecular and Surface Forces*, 2e. Academic Press; (2011). Elsevier, Amsterdam, 3e. A highly readable and recommended introduction.

Kaplan, I.G. (2006). *Intermolecular Interactions – Physical Picture, Computational Methods and Model Potentials*. Chichester: John Wiley & Sons. An intermediate treatment, providing extensive background.

Maitland, G.C., Rigby, M., Smith, E.B., and Wakeham, W.A. (1981). *Intermolecular Forces – Their Origin and Determination*. Oxford: Clarendon Press. A broad overview dealing with many phenomena and experimental methods.

Margenau, H. and Kestner, N.R. (1969). *Theory of Intermolecular Forces*. Oxford: Pergamon (1971). 2e. A now somewhat outdated book containing detailed analytical theory.

Parsegian, V.A. (2006). *Van der Waals Forces: A Handbook for Biologists, Chemists, Engineers and Physicists*. Cambridge: Cambridge University Press. A useful book containing solutions for many models involving van der Waals interactions.

Stone, A.J. (1996). *The Theory of Intermolecular Forces*. Oxford: Clarendon Press; (2013). 2e. A clear and modern treatment, highly recommended.

5

Continuum Mechanics

Having discussed elements of the microscopic approach, namely classical and quantum particles and their interactions, we turn now to the next "tool," a macroscopic approach called (conventional) continuum mechanics (CM). We describe first the nature of the continuum, then kinematics including the distinction between material and spatial coordinates, kinetics, and the associated conservation relations.

5.1 The Nature of the Continuum

At the macroscopic level, a *continuum description* is used where for every point of the material a response – as described by the constitutive behavior and the accompanying phenomenological parameters – is prescribed. The response may be the same (different) for different directions, in which case we call the material *isotropic* (*anisotropic*). The response may also be the same (different) at different locations, and in this case, we call the material *homogeneous* (*inhomogeneous*). Single crystals are anisotropic and homogeneous, whereas composites are typically both anisotropic and inhomogeneous. Macroscopic properties are the result of the collective behavior of molecular and structural features that make up a *representative volume element* (RVE). This often, but not always, results in an isotropic and homogeneous response. So, although in a continuum description one speaks of material points of infinitely small size, one actually means such an RVE. In line with that, the concept of a *field* – a continuous function of coordinates and time for the quantity at hand – is used.

Let us try to be a little more precise [1] and take an arbitrary volume element $\Delta v = \int \mathrm{d}V$ at position $\boldsymbol{r}$. Indicating the integral of the relevant parameter y over this volume by $\Delta y = \int y\,\mathrm{d}V$, the space average of y is given by $\Delta y/\Delta v$. If we denote the volume of the RVE by Δv^* and the average over the RVE by y^*, in continuum theory it is thus assumed that the identification of the physical quantity y^* with the field quantity y is allowed, that is,

$$y^* \equiv \lim_{\Delta v \to \Delta v^*} (\Delta y/\Delta v) \cong \lim_{\Delta v \to 0} (\Delta y/\Delta v) \equiv y(\boldsymbol{r}, t)$$

A similar statement can be made about time dependence. Indicating the integral of the relevant parameter z over an arbitrary time interval $\Delta t = \int \mathrm{d}t$ with $\Delta z = \int z\,\mathrm{d}V$, the time average of z is given by $\Delta z/\Delta t$. If the relaxation time of the parameter z is τ, the necessary

Phases of Matter and their Transitions: Concepts and Principles for Chemists, Physicists, Engineers, and Materials Scientists, First Edition. Gijsbertus de With.

minimum time for measuring z of the RVE is Δt^*, for which it holds that $\Delta t^* \gg \tau$. Denoting the average over Δt^* by z^*, the following transition should be allowed:

$$z^* \equiv \lim_{\Delta t \to \Delta t*} (\Delta z/\Delta t) \cong \lim_{\Delta t \to 0} (\Delta z/\Delta t) \equiv z(\boldsymbol{r}, t)$$

Any field property in continuum theory should be interpreted accordingly. Hence, for each point of a continuum the properties of an RVE, centered at the coordinates of that point, must be given. This type of point is usually denoted as *material points* or just *particles* in continuum theory (obviously with a different interpretation as the corpuscular particles discussed in Chapter 1) and the continuum consists of a set of these particles. For continuum theory to be physically meaningful, the size of the RVE should be large in comparison with the underlying microstructural and molecular features of the substance but small compared to the length of the macroscopic variation over the structure. Obviously, the above implies that continuum theory is useful for the macroscopic description of a structure but must be supplemented by models, based on (micro)-structural information or measurements, which yield a prescription for the behavior of the phenomenological parameters.

5.2 Kinematics

Kinematics describes the motion and deformation of a continuous body with respect to a reference frame without paying attention to the origin of this motion and deformation or to the nature of the body. A general motion of a body is composed of a global and a local part. The *global part* or *rigid body motion* is the part where the distance between any two particles of the body does not change with time at all, leading only to a different position and/or orientation of the body in space [1]. It describes the motion as translation of the total mass of the body located at the center of mass and rotation around an axis through the center of mass of the body as a whole. The *local part* is the part where the distances between particles do change and leads to a *deformation*, characteristic of changes in shape and/or volume of the body.

5.2.1 Material and Spatial Coordinates

For the description of motion and deformation of a continuous body, we use a right-handed Cartesian axis coordinate system fixed in Euclidean space with unit vectors $\boldsymbol{e}_1$, $\boldsymbol{e}_2$, and $\boldsymbol{e}_3$. Fixed points in this *reference coordinate system* are known as *spatial points* or just *points*. It is the coordinate system used by an observer at rest. As discussed in Section 5.1, the continuous body itself is supposed to be an assembly of *material points* or *particles*, the collective motion of which describes the motion and deformation and to which the properties averaged over an RVE centered at the coordinates of the particle can be ascribed. To eliminate confusion as far as possible, spatial points will be referred to as points and material points as particles. It is convenient to introduce another coordinate system, in parallel orientation to the reference system that accompanies the particle considered during its motion. This *accompanying coordinate system* is the system used by an observer moving along with the particle.

Suppose that a particle P of the body is located at point $\boldsymbol{x}$ at time t_0. The configuration of all particles of the body at time t_0 is the undeformed, *reference configuration*. At a later time t, the particle considered will be at point $\boldsymbol{r}$, and the configuration of all particles of the body at time t is known as the *current configuration*. The mapping $\boldsymbol{r} = \boldsymbol{r}(\boldsymbol{x}, t)$ for all particles describes the motion and deformation of the body. It is known as the *Lagrange* or *material description* and considers the position $\boldsymbol{r}$ as a function of time for one and the same particle P, labeled with $\boldsymbol{x}$ in the reference configuration.

If we assume that the mapping $\boldsymbol{r} = \boldsymbol{r}(\boldsymbol{x}, t)$ is single valued and sufficiently differentiable, inversion leads to $\boldsymbol{x} = \boldsymbol{x}(\boldsymbol{r}, t)$, referred to as the *Euler* or *spatial description*. This description considers the occupation of a point $\boldsymbol{r}$ by a particle that was in the reference configuration at position $\boldsymbol{x}$. This implies that at the same point different particles are considered at different times. The components of $\boldsymbol{r}$ are called the *spatial coordinates*, and the components of $\boldsymbol{x}$ are the *material coordinates*.

A simple analogy of these two descriptions is provided by the motion of traffic. If you are a driver (moving observer), you move with your car (particle $\boldsymbol{x}$) and you will describe the motion in Lagrange terms. However, if you are a traffic controller (observer at rest), you are fixed in space (point $\boldsymbol{r}$), you will see the cars pass and describe the motion in Euler terms.

Consider now an arbitrary tensor $\mathbb{c}$. In the material description for an observer using the accompanying coordinate system, $\boldsymbol{r} = \boldsymbol{r}(\boldsymbol{x}, t)$, the coordinates $\boldsymbol{x}$ are fixed and the time derivative of $\mathbb{c}$ is $\mathrm{d}\mathbb{c}/\mathrm{d}t \equiv \dot{\mathbb{c}} = (\partial \mathbb{c}/\partial t)_{\boldsymbol{x}}$. For example, the *velocity* is given by $\boldsymbol{v} = (\partial \boldsymbol{r}/\partial t)_{\boldsymbol{x}}$. In the spatial description for an observer using the reference coordinate system, $\boldsymbol{x} = \boldsymbol{x}(\boldsymbol{r}, t)$, we obtain for the time derivative $\mathrm{d}\mathbb{c}/\mathrm{d}t = (\partial \mathbb{c}/\partial t)_{\boldsymbol{r}} + (\partial \mathbb{c}/\partial \boldsymbol{r}) \cdot (\partial \boldsymbol{r}/\partial t) = \partial \mathbb{c}/\partial t + \boldsymbol{v}\,\nabla \mathbb{c}$. From now on, we use for the time derivative in the material description $\mathrm{d}\mathbb{c}/\mathrm{d}t$ and in the spatial description $\partial \mathbb{c}/\partial t$. As both expressions denote the same change, the time derivative of $\mathbb{c}$ for a material point, the *material derivative*, is given by

$$\frac{\mathrm{d}\mathbb{c}}{\mathrm{d}t} \equiv \dot{\mathbb{c}} = \frac{\partial \mathbb{c}}{\partial t} + \boldsymbol{v} \cdot \nabla \mathbb{c} \tag{5.1}$$

CM can be formulated in either spatial or material coordinates. We use mainly spatial coordinates and refer to the literature for further details, for example, Fung or Malvern [2]. The treatise by Haupt [3] deals, apart from with CM, also with the theory of materials.

5.2.2 General Deformations

To describe a general motion, let us consider a particle in the reference configuration at time $t_0 = 0$. A line element at $\boldsymbol{x}$ is given by $\mathrm{d}\boldsymbol{x}$. At a later time t, the particle will be at point $\boldsymbol{r}$ and in this deformed configuration $\mathrm{d}\boldsymbol{x}$ becomes $\mathrm{d}\boldsymbol{r}$. The deformation is described by $\mathrm{d}\boldsymbol{r} = \mathbb{F} \cdot \mathrm{d}\boldsymbol{x}$, where $\mathbb{F} = \partial \boldsymbol{r}/\partial \boldsymbol{x}$ is the *deformation gradient*, or by $\mathrm{d}\boldsymbol{x} = \mathbb{F}^{-1} \cdot \mathrm{d}\boldsymbol{r}$, where $\mathbb{F}^{-1} = \partial \boldsymbol{x}/\partial \boldsymbol{r}$ is used. The length squared of $\mathrm{d}\boldsymbol{x}$ and $\mathrm{d}\boldsymbol{r}$ are given by

$$\mathrm{d}\boldsymbol{x} \cdot \mathrm{d}\boldsymbol{x} = \mathrm{d}\boldsymbol{r} \cdot \mathbb{F}^{\mathrm{T}} \cdot \mathbb{F} \cdot \mathrm{d}\boldsymbol{r} \equiv \mathrm{d}\boldsymbol{r} \cdot \mathbb{C} \cdot \mathrm{d}\boldsymbol{r} \quad \text{and}$$
$$\mathrm{d}\boldsymbol{r} \cdot \mathrm{d}\boldsymbol{r} = \mathrm{d}\boldsymbol{x} \cdot \mathbb{F}^{-\mathrm{T}} \cdot \mathbb{F}^{-1} \cdot \mathrm{d}\boldsymbol{x} \equiv \mathrm{d}\boldsymbol{x} \cdot \mathbb{B} \cdot \mathrm{d}\boldsymbol{x}$$

where the tensors $\mathbb{C} = \mathbb{F}^{\mathrm{T}} \cdot \mathbb{F}$ and $\mathbb{B} = \mathbb{F}^{-\mathrm{T}} \cdot \mathbb{F}^{-1}$ are measures of the pure deformation.

For a particle in the reference configuration, using the material description, the *displacement* vector $\boldsymbol{u}$ can be introduced by [1]

$$\boldsymbol{x} + \boldsymbol{u} = \boldsymbol{r} + \boldsymbol{c}$$

where $\boldsymbol{u}$ is considered as a function of the material coordinates $\boldsymbol{x}$. The vector $\boldsymbol{c}$ is a constant vector, specifying the origin $\mathrm{O_s}$ of the spatial coordinate system relative to the origin $\mathrm{O_m}$ of the material coordinate system. If the reference configuration in material coordinates $\boldsymbol{x}$ and spatial coordinates $\boldsymbol{r}$ at $t = t_0$ coincide, $\mathbf{c} = 0$. If for all particles $\boldsymbol{x}$ the displacement $\boldsymbol{u}(\boldsymbol{x})$ is known, a complete material description of the deformation is given. Thus, taking $\mathbf{c} = 0$, it follows that

$$\mathbb{F} = \mathbb{I} + \partial \boldsymbol{u}/\partial \boldsymbol{x} \quad \text{and} \quad \mathbb{F}^{-1} = \mathbb{I} - \partial \boldsymbol{u}/\partial \boldsymbol{r}$$

where $\mathbb{I}$ is the unit tensor and $\partial \boldsymbol{u}/\partial \boldsymbol{x}$ and $\partial \boldsymbol{u}/\partial \boldsymbol{r}$ are the *material* and *spatial displacement gradients*, respectively. The tensors $\mathbb{C} = \mathbb{F}^{\mathrm{T}} \cdot \mathbb{F}$ and $\mathbb{B} = \mathbb{F}^{-\mathrm{T}} \cdot \mathbb{F}^{-1}$ thus can be written as

$$\mathbb{C} = \mathbb{I} + \left[\left(\frac{\partial \boldsymbol{u}}{\partial \boldsymbol{x}}\right) + \left(\frac{\partial \boldsymbol{u}}{\partial \boldsymbol{x}}\right)^{\mathrm{T}}\right] + \left(\frac{\partial \boldsymbol{u}}{\partial \boldsymbol{x}}\right)^{\mathrm{T}} \cdot \left(\frac{\partial \boldsymbol{u}}{\partial \boldsymbol{x}}\right) \quad \text{and}$$

$$\mathbb{B} = \mathbb{I} - \left[\left(\frac{\partial \boldsymbol{u}}{\partial \boldsymbol{r}}\right) + \left(\frac{\partial \boldsymbol{u}}{\partial \boldsymbol{r}}\right)^{\mathrm{T}}\right] + \left(\frac{\partial \boldsymbol{u}}{\partial \boldsymbol{r}}\right)^{\mathrm{T}} \cdot \left(\frac{\partial \boldsymbol{u}}{\partial \boldsymbol{r}}\right)$$

We can calculate the increase in length of the line element from

$$\mathrm{d}\boldsymbol{r} \cdot \mathrm{d}\boldsymbol{r} - \mathrm{d}\boldsymbol{x} \cdot \mathrm{d}\boldsymbol{x} = \mathrm{d}\boldsymbol{x} \cdot 2\mathbb{L} \cdot \mathrm{d}\boldsymbol{x} = \mathrm{d}\boldsymbol{r} \cdot 2\mathbb{E} \cdot \mathrm{d}\boldsymbol{r}$$

where $\mathbb{L} = ½(\mathbb{C} - \mathbb{I})$ is the *material* (or Lagrange) *strain tensor* and $\mathbb{E} = ½(\mathbb{I} - \mathbb{B})$ is the *spatial* (or Euler) *strain tensor*. Therefore,

$$\mathbb{L} = ½\left[\left(\frac{\partial \boldsymbol{u}}{\partial \boldsymbol{x}}\right) + \left(\frac{\partial \boldsymbol{u}}{\partial \boldsymbol{x}}\right)^{\mathrm{T}}\right] + ½\left(\frac{\partial \boldsymbol{u}}{\partial \boldsymbol{x}}\right)^{\mathrm{T}} \cdot \left(\frac{\partial \boldsymbol{u}}{\partial \boldsymbol{x}}\right) \equiv \boldsymbol{\varepsilon}_{\mathrm{L}} + ½\left(\frac{\partial \boldsymbol{u}}{\partial \boldsymbol{x}}\right)^{\mathrm{T}} \cdot \left(\frac{\partial \boldsymbol{u}}{\partial \boldsymbol{x}}\right) \quad \text{and} \tag{5.2}$$

$$\mathbb{E} = ½\left[\left(\frac{\partial \boldsymbol{u}}{\partial \boldsymbol{r}}\right) + \left(\frac{\partial \boldsymbol{u}}{\partial \boldsymbol{r}}\right)^{\mathrm{T}}\right] - ½\left(\frac{\partial \boldsymbol{u}}{\partial \boldsymbol{r}}\right)^{\mathrm{T}} \cdot \left(\frac{\partial \boldsymbol{u}}{\partial \boldsymbol{r}}\right) \equiv \boldsymbol{\varepsilon}_{\mathrm{E}} - ½\left(\frac{\partial \boldsymbol{u}}{\partial \boldsymbol{r}}\right)^{\mathrm{T}} \cdot \left(\frac{\partial \boldsymbol{u}}{\partial \boldsymbol{r}}\right) \tag{5.3}$$

related by $\mathbb{L} = \mathbb{F}^{\mathrm{T}} \cdot \mathbb{E} \cdot \mathbb{F}$ and $\mathbb{E} = \mathbb{F}^{-\mathrm{T}} \cdot \mathbb{L} \cdot \mathbb{F}^{-1}$. If $|\partial \boldsymbol{u}/\partial \boldsymbol{x}| \ll 1$, one has from $\partial \boldsymbol{u}/\partial \boldsymbol{x} = (\partial \boldsymbol{u}/\partial \boldsymbol{r})^{\mathrm{T}} \cdot (\partial \boldsymbol{r}/\partial \boldsymbol{x}) = (\partial \boldsymbol{u}/\partial \boldsymbol{r})^{\mathrm{T}} \cdot \mathbb{F} = (\partial \boldsymbol{u}/\partial \boldsymbol{r})^{\mathrm{T}} \cdot [\mathbb{I} + (\partial \boldsymbol{u}/\partial \boldsymbol{x})]$ that $|\partial \boldsymbol{u}/\partial \boldsymbol{r}| \ll 1$. Therefore, neglecting second-order quantities, we have $(\mathbb{L} \cong \boldsymbol{\varepsilon}_{\mathrm{L}}) \cong (\mathbb{E} \cong \boldsymbol{\varepsilon}_{\mathrm{E}})$, all equal to the *small displacement gradient* (SDG) *strain tensor* $\boldsymbol{\varepsilon} \equiv ½[(\partial \boldsymbol{u}/\partial \boldsymbol{x}) + (\partial \boldsymbol{u}/\partial \boldsymbol{x})^{\mathrm{T}}]$, in short the *strain tensor*. Note that the symbol $\boldsymbol{\varepsilon}$ is deviating from our usual notation for tensors.

5.2.3 The Small Displacement Gradient Approximation

For many formal calculations involving solids, the Lagrange description is the most useful, whereas for fluids the Euler description is more fruitful. As indicated earlier, if the gradient of the deformation is small, the difference between these two descriptions is negligible, and we can use the SDG approximation, which is the most often used continuum description for solids [4]. Although $\boldsymbol{r} = \boldsymbol{r}(\boldsymbol{x}, t)$ can be inverted to $\boldsymbol{x} = \boldsymbol{x}(\boldsymbol{r}, t)$, in practice this process is often quite complex. Example 5.1 illustrates the essential difference in functional dependence for the velocity for a simple one-dimensional (1D) motion.

Example 5.1 ***Materials and Spatial Velocity*** Consider the 1D motion $r = x(1 + tx)^{-1}$. We obtain $v(x, t) = (\partial r/\partial t)_x = -x^2(1 + tx)^{-2}$. Solving x from $r = x(1 + tx)^{-1}$ yields $x = r(1 - tr)^{-1}$. Substituting x in $v(x, t)$ leads to $v(r, t) = -r^2$, illustrating the difference in functional dependence between $v(x, t)$ and $v(r, t)$.

Consider now the deformation of a line element $\mathrm{d}\boldsymbol{x}$ that in the reference state has a length $\mathrm{d}l_0$ and a fixed direction $\boldsymbol{m}$, so that $\mathrm{d}\boldsymbol{x} = \boldsymbol{m}\,\mathrm{d}l_0$. After deforming the body, $\mathrm{d}l_0$ changed to $\mathrm{d}l$ and its direction to $\boldsymbol{n}$, or $\mathrm{d}\boldsymbol{r} = \boldsymbol{n}\,\mathrm{d}l$. So, $\mathrm{d}\boldsymbol{r} = \mathbb{F} \cdot \mathrm{d}\boldsymbol{x}$ can be written as

$$\boldsymbol{n}\,\mathrm{d}l = \mathbb{F} \cdot \boldsymbol{m}\,\mathrm{d}l_0 \quad \text{or} \quad \lambda \boldsymbol{n} = \mathbb{F} \cdot \boldsymbol{m}$$

where $\lambda = \mathrm{d}l/\mathrm{d}l_0$ is the *stretch* describing the elongation ratio of the vectors $\mathrm{d}\boldsymbol{x}$ and $\mathrm{d}\boldsymbol{r}$. Note that the stretch must be positive. It follows, since $|\boldsymbol{n}| = 1$, that

$$\lambda^2 = \boldsymbol{m} \cdot \mathbb{F}^{\mathrm{T}} \cdot \mathbb{F} \cdot \boldsymbol{m} = \boldsymbol{m} \cdot \mathbb{C} \cdot \boldsymbol{m} = \boldsymbol{m} \cdot (\mathbb{I} + 2\mathbb{L}) \cdot \boldsymbol{m} \tag{5.4}$$

This expression represents a *representation of an ellipsoid* for which the length of a vector in the direction $\boldsymbol{m}$ from the center of the ellipsoid to a particular point at the surface is given by $\lambda^2(\boldsymbol{m})$. It also leads to the *relative* or *unit extension* $\in(\boldsymbol{m}) = (\mathrm{d}l\text{–}\mathrm{d}l_0)/\mathrm{d}l_0 = \lambda(\boldsymbol{m}) - 1$ [4]. In particular, if the material vector in the reference configuration is parallel to the unit vector $\boldsymbol{e}_1$, the stretch $\lambda(1)$ and unit extension $\in(1)$ are given by

$$\lambda^2(1) = C_{11} = 1 + 2L_{11} \quad \text{and} \quad \in(1) = \sqrt{C_{11}} - 1 = \sqrt{1 + 2L_{11}} - 1$$

In the SDG approximation, the unit extension reduces to

$$\in(1) = \sqrt{1 + 2L_{11}} - 1 \cong \sqrt{1 + 2\varepsilon_{11}} - 1 \cong \varepsilon_{11}$$

Let us now consider how the stretch varies with time. Because $\boldsymbol{m}$ has a fixed direction, the time derivative of λ^2 is given by [5]

$$\mathrm{d}\lambda^2/\mathrm{d}t = 2\lambda\dot{\lambda} = \boldsymbol{m} \cdot \dot{\mathbb{C}} \cdot \boldsymbol{m} = \boldsymbol{m} \cdot (\dot{\mathbb{F}}^{\mathrm{T}} \cdot \mathbb{F} + \mathbb{F}^{\mathrm{T}} \cdot \dot{\mathbb{F}}) \cdot \boldsymbol{m} \quad \text{where}$$

$$\dot{\mathbb{F}}^{\mathrm{T}} \cdot \mathbb{F} = \left(\frac{\partial \boldsymbol{v}}{\partial \boldsymbol{x}}\right)^{\mathrm{T}} \cdot \mathbb{F} = \left(\frac{\partial \boldsymbol{r}}{\partial \boldsymbol{x}}\right)^{\mathrm{T}} \cdot \left(\frac{\partial \boldsymbol{v}}{\partial \boldsymbol{r}}\right)^{\mathrm{T}} \cdot \mathbb{F} = \mathbb{F}^{\mathrm{T}} \cdot \left(\frac{\partial \boldsymbol{v}}{\partial \boldsymbol{r}}\right)^{\mathrm{T}} \cdot \mathbb{F}$$

Using the latter expression and its transpose, together with $\lambda\boldsymbol{n} = \mathbb{F} \cdot \boldsymbol{m}$, yields

$$\dot{\lambda}/\lambda = \boldsymbol{n} \cdot \mathbb{D} \cdot \boldsymbol{n} \quad \text{where} \quad \mathbb{D} = \tfrac{1}{2}\left[(\partial \boldsymbol{v}/\partial \boldsymbol{r}) + (\partial \boldsymbol{v}/\partial \boldsymbol{r})^{\mathrm{T}}\right] \tag{5.5}$$

so that, maybe somewhat surprising, the relative rate of elongation $\dot{\lambda}/\lambda$ can be expressed entirely in the spatial coordinates $\boldsymbol{r}$ using the *deformation rate* $\mathbb{D}$. For a rigid body, obviously, $\mathbb{D} = 0$.

The angle changes for general deformations also follow from the material strain tensor $\mathbb{L} = \tfrac{1}{2}(\mathbb{C} - \mathbb{I})$ [4]. Let $\mathrm{d}\boldsymbol{x}'$ and $\mathrm{d}\boldsymbol{x}''$ be two material vectors with unit vectors $\boldsymbol{m}'$ and $\boldsymbol{m}''$ having lengths $\mathrm{d}l_0'$ and $\mathrm{d}l_0''$, respectively, in the reference configuration. In the current configuration, these vectors become $\mathrm{d}\boldsymbol{r}'$ and $\mathrm{d}\boldsymbol{r}''$ with unit vectors $\boldsymbol{n}'$ and $\boldsymbol{n}''$ having lengths $\mathrm{d}l'$ and $\mathrm{d}l''$, respectively. In that case, we have

$$\cos(\boldsymbol{n}', \boldsymbol{n}'') = \boldsymbol{n}' \cdot \boldsymbol{n}'' = \frac{\mathrm{d}\boldsymbol{r}' \cdot \mathrm{d}\boldsymbol{r}''}{\mathrm{d}l'\mathrm{d}l''} = \frac{\mathrm{d}\boldsymbol{x}' \cdot \mathbb{C} \cdot \mathrm{d}\boldsymbol{x}''}{\mathrm{d}l'\mathrm{d}l''} = \frac{\boldsymbol{m}' \cdot \mathbb{C} \cdot \boldsymbol{m}''}{\lambda(\boldsymbol{m}')\lambda(\boldsymbol{m}'')}$$

where $\mathrm{d}\boldsymbol{x}' = \boldsymbol{m}'\mathrm{d}l_0'$, $\mathrm{d}\boldsymbol{x}'' = \boldsymbol{m}''\mathrm{d}l_0''$, $\mathrm{d}l' = \lambda(\boldsymbol{m}')\mathrm{d}l_0'$ and $\mathrm{d}l'' = \lambda(\boldsymbol{m}'')\mathrm{d}l_0''$ have been used. From $\cos(\boldsymbol{n}', \boldsymbol{n}'')$ in the reference configuration the angle change can be calculated by subtraction. For example, if the material vectors $\mathrm{d}\boldsymbol{x}'$ and $\mathrm{d}\boldsymbol{x}''$ are in the reference configuration parallel to $\boldsymbol{e}_1$ and $\boldsymbol{e}_2$, the angle change $\Delta\theta_{12} = \pi/2 - \theta_{12}$, where

$$\cos(\theta_{12}) = \frac{C_{12}}{\lambda(1)\lambda(2)} = \frac{C_{12}}{\sqrt{C_{11}C_{22}}} = \frac{2L_{12}}{\sqrt{(1 + 2L_{11})(1 + 2L_{22})}}$$

In the SDG approximation, the above expression reduces to

$$\cos(\theta_{12}) \cong \frac{2\varepsilon_{12}}{\sqrt{(1 + 2\varepsilon_{11})(1 + 2\varepsilon_{22})}} \cong \frac{2\varepsilon_{12}}{(1 + \varepsilon_{11})(1 + \varepsilon_{22})} \cong 2\varepsilon_{12}$$

and, since θ_{12} is small, via $\cos(\theta_{12}) = \sin[(\pi/2) - \theta_{12}] \cong (\pi/2) - \theta_{12}$, to $\Delta\theta_{12}/2 = \varepsilon_{12}$.

In summary, for describing general deformations both the tensor $\mathbb{C}$ and $\mathbb{L}$ can be used. In the SDG approximation $\mathbb{C} \cong \mathbb{I} + 2\boldsymbol{\varepsilon}$ and $\mathbb{L} \cong \boldsymbol{\varepsilon}$, leading to the unit extension $\in(\boldsymbol{n}) = \boldsymbol{n} \cdot \boldsymbol{\varepsilon} \cdot \boldsymbol{n}$ and the angle change between two initially orthogonal line elements $\frac{1}{2}\theta(\boldsymbol{n}, \boldsymbol{m}) = \boldsymbol{n} \cdot \boldsymbol{\varepsilon} \cdot \boldsymbol{m}$. In the SDG approximation, the deformations are thus completely defined by the strain tensor $\boldsymbol{\varepsilon}$. The components such as $\varepsilon_{11}, \ldots$, are called the *normal strains* and the components ε_{12}, …, are the *shear strains*. We note that since the angle of shear $\theta(\boldsymbol{n}, \boldsymbol{m})$ is equal to $2\boldsymbol{n} \cdot \boldsymbol{\varepsilon} \cdot \boldsymbol{m}$, frequently also the *conventional* or *engineering shear strains* $\gamma_{ij} = 2\varepsilon_{ij}$ $(i \neq j)$ are introduced, while ε_{ij} denote the *tensorial shear strains*.

Because the strain tensor is symmetric ($\boldsymbol{\varepsilon} = \boldsymbol{\varepsilon}^{\mathrm{T}}$), it has all properties of a symmetric tensor. In particular, the tensor can be brought at principal axes with the diagonal elements ε_{11}, ε_{22}, and ε_{33} as the only non-zero elements. They are conventionally called the *principal values*, while the corresponding axes are the *principal axes* (Section A.10).

Using the decomposition of the symmetric tensor $\boldsymbol{\varepsilon}$ in an *isotropic strain* $\frac{1}{3}(\mathrm{tr}\,\boldsymbol{\varepsilon})\mathbb{I}$ and a *deviatoric strain* $\boldsymbol{\varepsilon}'$, we can write $\boldsymbol{\varepsilon} = \frac{1}{3}(\mathrm{tr}\,\boldsymbol{\varepsilon})\mathbb{I} + \boldsymbol{\varepsilon}'$. While the latter term represents the change in shape (at constant volume), the former term describes the change in volume (at constant shape). The trace of the strain tensor, $\mathrm{tr}\,\boldsymbol{\varepsilon} = \nabla \cdot \boldsymbol{u}$, equals the *dilatation* $\Delta V/V_0$ with ΔV the volume change and V_0 the initial volume.

Finally, we will also need the strain rate $\mathrm{d}\boldsymbol{\varepsilon}/\mathrm{d}t = \dot{\boldsymbol{\varepsilon}}$, given by

$$2\dot{\boldsymbol{\varepsilon}} = \frac{\mathrm{d}}{\mathrm{d}t}\left[\left(\frac{\partial \boldsymbol{u}}{\partial \boldsymbol{x}}\right)^{\mathrm{T}} + \left(\frac{\partial \boldsymbol{u}}{\partial \boldsymbol{x}}\right)\right] \cong \frac{\mathrm{d}}{\mathrm{d}t}\left[\left(\frac{\partial \boldsymbol{u}}{\partial \boldsymbol{r}}\right)^{\mathrm{T}} + \left(\frac{\partial \boldsymbol{u}}{\partial \boldsymbol{r}}\right)\right] = \left[\left(\frac{\partial \boldsymbol{v}}{\partial \boldsymbol{r}}\right)^{\mathrm{T}} + \left(\frac{\partial \boldsymbol{v}}{\partial \boldsymbol{r}}\right)\right] = 2\mathbb{D}$$

In the SDG approximation, $\dot{\boldsymbol{\varepsilon}}$ and $\mathbb{D}$ are thus equal, and the strain is given by

$$\boldsymbol{\varepsilon} = \int \dot{\boldsymbol{\varepsilon}}\,\mathrm{d}t \cong \int \mathbb{D}\,\mathrm{d}t$$

5.3 Balance Equations

Let us consider an arbitrary extensive property C, either a scalar or a vector (tensor) component, and its *specific value* c, the value of C per unit of mass. If C is distributed over a volume V with density ρ, the expressions for C and the rate for C are

$$C = \int \rho c\,\mathrm{d}V \text{ and } \dot{C} \equiv \frac{\mathrm{d}C}{\mathrm{d}t} = \frac{\mathrm{d}}{\mathrm{d}t}\int \rho c\,\mathrm{d}V$$

Changes in C can result from two contributions. The first is the (in- and out-) flow of material due to transport by diffusion. The second is the production (or elimination) of a certain amount of C within the volume by chemical reactions [6].

Let us use spatial coordinates and note that in this case the volume and bounding surface are fixed in space and not changing. This we indicate by using V° and A°, the latter having the outward normal $\boldsymbol{n}^\circ$ (Figure 5.1a). As the coordinates are fixed, the rate of change of C becomes $\dot{C} = \int (\partial \rho c/\partial t)\,\mathrm{d}V^\circ$. For the transport term we write $-\int \boldsymbol{J}_c^\circ \cdot \boldsymbol{n}^\circ\,\mathrm{d}A^\circ$, where $\boldsymbol{J}_c^\circ = \rho c \boldsymbol{v}_c$ represents the rate of transport of c per unit time and per unit area. Counting inflow positive, this term has a negative sign because $\boldsymbol{n}^\circ$ is the outward normal. We label the production term per unit time and per unit volume as π_c°. The rate of C thus becomes

$$\dot{C} = \int \frac{\partial \rho c}{\partial t}\,\mathrm{d}V^\circ = -\int \boldsymbol{J}_c^\circ \cdot \boldsymbol{n}^\circ \mathrm{d}A^\circ + \int \pi_c^\circ \mathrm{d}V^\circ = -\int \nabla \cdot \boldsymbol{J}_c^\circ \mathrm{d}V^\circ + \int \pi_c^\circ \mathrm{d}V^\circ \quad (5.6)$$

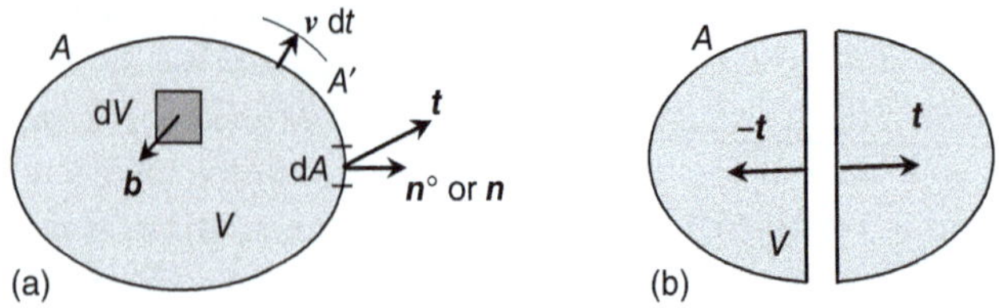

Figure 5.1 Schematic picture of a continuous body. (a) Deformation from A to A', the normal vector $\boldsymbol{n}$, and the acting external forces $\boldsymbol{b}$ and $\boldsymbol{t}$. (b) A cut and associated stress vectors $\boldsymbol{t}$ and $-\boldsymbol{t}$.

where in the last step the divergence theorem is used. Because the volume is arbitrary, the integrands on the left- and right-hand side are equal and Eq. (5.6) reduces to

$$\partial\rho c/\partial t = -\nabla\cdot\boldsymbol{J}_c{}^\circ + \pi_c{}^\circ = -\nabla\cdot\rho c\boldsymbol{v}_c + \pi_c{}^\circ \tag{5.7}$$

which is the *general balance equation in spatial coordinates*. If we take for C the mass M of the volume, $c = 1$, $\boldsymbol{J}_c{}^\circ \to \boldsymbol{J}^\circ = \rho\boldsymbol{v}$, and $\pi_c{}^\circ \to \pi^\circ = 0$ because mass is conserved, so that we have

$$\partial\rho/\partial t = -\nabla\cdot\boldsymbol{J}^\circ = -\nabla\cdot\rho\boldsymbol{v} \tag{5.8}$$

which is known as the (spatial) *continuity equation*, valid for an observer at rest.

Let us do the same exercise again, but now use material coordinates. In this case, the mass $\mathrm{d}M = \rho\mathrm{d}V$ of the material point or RVE is fixed and we label the volume by V, the surface by A, and the outward normal by $\boldsymbol{n}$. Introducing the rate $\boldsymbol{J}_c \equiv \boldsymbol{J}_c{}^\circ - \rho c\boldsymbol{v} = \rho c(\boldsymbol{v}_c - \boldsymbol{v})$, conservation of mass is expressed by $\boldsymbol{J} = \boldsymbol{J}^\circ - \rho\boldsymbol{v} = 0$. The rate of change of C is now $\dot{C} = \int \rho\dot{c}\,\mathrm{d}V$ and, adding the transport and production terms, we have

$$\dot{C} = \int \rho\dot{c}\,\mathrm{d}V = -\int \boldsymbol{J}_c\cdot\boldsymbol{n}\,\mathrm{d}A + \int \pi_c\,\mathrm{d}V = -\int \nabla\cdot\boldsymbol{J}_c\,\mathrm{d}V + \int \pi_c\,\mathrm{d}V$$

where in the last step the divergence theorem is used. Because $\mathrm{d}V$ is arbitrary, this expression reduces similarly to the *general balance equation in material coordinates*

$$\rho\dot{c} = -\nabla\cdot\boldsymbol{J}_c + \pi_c = -\nabla\cdot\rho c(\boldsymbol{v}_c - \boldsymbol{v}) + \pi_c \tag{5.9}$$

valid for an observer moving with the accompanying system. If we take for C again the mass M of the volume, we have $c = 1$ and $\pi_c \to \pi = 0$, and from the material derivative $\mathrm{d}(\rho c)/\mathrm{d}t = \partial(\rho c)/\partial t + \boldsymbol{v}\cdot\nabla\rho c$, we obtain the (material) *continuity equation*

$$\dot{\rho} = -\rho\nabla\cdot\boldsymbol{v} \tag{5.10}$$

Moreover, using $\nabla\cdot\rho c\boldsymbol{v} = \rho c(\nabla\cdot\boldsymbol{v}) + \boldsymbol{v}\cdot\nabla\rho c = \rho c(\nabla\cdot\boldsymbol{v}) + \boldsymbol{v}\cdot c\nabla\rho + \boldsymbol{v}\cdot\rho\nabla c$ and twice $\mathrm{d}(\cdot)/\mathrm{d}t = \partial(\cdot)/\partial t + \boldsymbol{v}\cdot\nabla(\cdot)$, it appears that $\pi_c{}^\circ = \pi_c$, so that the source term is identical for the spatial and material description. Finally, inserting the material continuity equation $\mathrm{d}\rho/\mathrm{d}t = -\rho\nabla\cdot\boldsymbol{v}$ in the material derivative $\mathrm{d}\rho c/\mathrm{d}t = \partial\rho c/\partial t + \boldsymbol{v}\cdot\nabla\rho c$, we obtain a relation between the material and spatial variation of c given by

$$\rho\dot{c} = \partial\rho c/\partial t + \nabla\cdot\rho c\boldsymbol{v} \tag{5.11}$$

So far, we have considered the particle or RVE as containing only one chemical component. Generally, more components are present (Section 6.1.7), and for a volume element $\mathrm{d}V$, having mass $\mathrm{d}M$ and n components α with mass $\mathrm{d}M_\alpha$, we have for the (*mass*) *density* $\rho(\boldsymbol{x}, t) = \mathrm{d}M/\mathrm{d}V = \Sigma_\alpha\rho_\alpha$ with $\rho_\alpha = \mathrm{d}M_\alpha/\mathrm{d}V$. We assume that each of these components may be produced within the RVE by a chemical reaction from other components with as constraint dM = 0. For each component we have

$$\partial\rho_\alpha/\partial t + \nabla\cdot \boldsymbol{J}_\alpha{}^\circ = \pi_\alpha{}^\circ \quad \text{and} \quad \dot{\rho}_\alpha + \nabla\cdot\boldsymbol{J}_\alpha + \rho_\alpha\nabla\cdot\boldsymbol{v} = \pi_\alpha \tag{5.12}$$

where $\boldsymbol{J}_\alpha{}^\circ = \rho_\alpha \boldsymbol{v}_\alpha$ and $\boldsymbol{J}_\alpha = \rho_\alpha(\boldsymbol{v}_\alpha - \boldsymbol{v})$. Equation (5.12b) is obtained by inserting $\partial\rho_k/\partial t = \mathrm{d}\rho_k/\mathrm{d}t - \boldsymbol{v}\cdot\nabla\rho_k$ in Eq. (5.12a) and $\boldsymbol{J}_\alpha$ describes the flux for each component with respect to the center of gravity of the RVE. Summing over α, noting that $\rho\boldsymbol{v} = \Sigma_\alpha\rho_\alpha\boldsymbol{v}_\alpha$ and that $\Sigma_\alpha\pi_\alpha{}^\circ = \Sigma_\alpha\pi_\alpha = 0$, we regain the continuity Eqs. (5.8) and (5.10). For later use we introduce for any component α the weight fraction $w_\alpha = \rho_\alpha/\rho$, so that $\Sigma_\alpha w_\alpha = 1$. Choosing $c = w_\alpha$, we have from Eq. (5.9), more concisely than Eq. (5.12b),

$$\rho\dot{w}_\alpha = -\nabla\cdot\boldsymbol{J}_\alpha + \pi_\alpha \tag{5.13}$$

Obviously, if no reaction occurs, $\pi_\alpha = 0$, while if no transport occurs, $\boldsymbol{J}_\alpha = 0$.

5.4 Kinetics

We now turn to kinetics, dealing with the forces acting on a body [5]. To that purpose we consider again a volume V of a continuum with a regular surface A (Figure 5.1a). A volume element dV contains a mass dM, and dA is a surface element with exterior normal $\boldsymbol{n}$.

Like for a system of corpuscular particles, there are two types of external force for a continuous body. First, the *body* (or *volume*) *force* acting on all volume elements of the body. The specific body force $\boldsymbol{b}$ is the body force per unit of mass of which the most well-known one is the specific weight, due to the acceleration of gravity. It is assumed that $\boldsymbol{b}$ is continuous and that couples acting on the volume element dV are excluded. Second, the *surface force* continuously distributed over the surface A with couples acting on a surface element dA excluded. Here, it is convenient to refer to the *traction*, that is, the force per unit area. The traction $\boldsymbol{t}(\boldsymbol{r}, \boldsymbol{n})$ depends on the position $\boldsymbol{r}$ and on the orientation of the surface element dA, indicated by the exterior normal $\boldsymbol{n}$. The volume and surface forces for a continuous body are the analogues of $\boldsymbol{f}_j{}^{(\text{ext})}$ for the case of corpuscular particles. We also need the *inertial force*, which is, like $\boldsymbol{b}$, distributed over the volume of the entire body. The specific inertial force is the force per unit mass and is equal to the acceleration $\boldsymbol{a} = \dot{\boldsymbol{v}}$. This contribution is analogous to d$\boldsymbol{p}_j$/dt for corpuscular particles. Besides the external and inertial forces, there are also the *internal forces*, the analogues of the forces $\boldsymbol{f}_j{}^{(\text{int})} = \Sigma_{i\neq k}\boldsymbol{f}_{ik}{}^{(\text{pp})}$ for corpuscular particles. The precise expressions will be obtained later.

Now, consider an arbitrary body divided into two arbitrary parts with a common boundary. To assess the effect of the interaction between these parts, Cauchy proposed (1823) that this interaction may be replaced by a continuous distribution of surface forces $\boldsymbol{t}(\boldsymbol{r},\boldsymbol{n})$ acting on both sides of the common boundary with normal vector $\boldsymbol{n}$. Thus, if we cut the body into two pieces along this boundary (Figure 5.1b), the configuration can be maintained only if we apply at the same time extra forces at the cut, which compensate for the missing forces from the other part. The extra forces on both parts are equal in magnitude but of opposite direction since their sum must be zero. Using a small surface element ΔA of the cut with the orientation $\boldsymbol{n}$ and a force $\Delta\boldsymbol{f}$ acting on that element, we can define the *stress vector* $\boldsymbol{t}$ by

$$\boldsymbol{t} = \boldsymbol{t}(\boldsymbol{r},\boldsymbol{n}) = \lim_{\Delta A\to 0}(\Delta\boldsymbol{f}/\Delta A)$$

The above indicated process can be done for any point on the cut, and since the cut itself is arbitrary, a stress vector can be defined for any surface element at any point within the body. Hence, if we mention the stress at a point $\boldsymbol{r}$ in the body, we mean the components of the stress vector on a certain plane. We used again the symbol $\boldsymbol{t}$ since the stress vector is an external surface force for the detached part. In the literature, the terms stress vector and traction are often used interchangeably. We use traction for external force and stress vector for internal force.

5.4.1 The Principle of Virtual Power

We now turn to the next step, namely, how the forces determine motion and deformation. In Section 2.2, we introduced the principle of virtual power (PoVP)

$$\sum_j \left(\boldsymbol{f}_j^{(a)} - m_j \boldsymbol{a}_j\right) \cdot \delta_1 \dot{\boldsymbol{r}}_j = 0$$

(we omit the subscript 1 as well as the superscript (a) from now on). To be able to apply this principle, we recall that the work done on a particle (material point) is $\boldsymbol{f} \cdot \delta\boldsymbol{r}$, or equivalently the power applied is $\boldsymbol{f} \cdot \dot{\boldsymbol{r}} = \boldsymbol{f} \cdot \boldsymbol{v}$. Let us assume that the surface and volume forces do not depend on the deformation, neither in direction, nor in magnitude. For the potential of the surface forces, we take $\phi_{\text{sur}} = -\boldsymbol{t} \cdot \boldsymbol{u}$, with $\boldsymbol{u}(\boldsymbol{r})$ the displacement, so that a force $\boldsymbol{f}_{\text{sur}} = \boldsymbol{t}$ acts on surface element $\mathrm{d}A$. Similarly, for the potential of the volume forces we use $\phi_{\text{vol}} = -\rho\boldsymbol{b} \cdot \boldsymbol{u}$, so that a force $\boldsymbol{f}_{\text{vol}} = \rho\boldsymbol{b}$ acts on volume element $\mathrm{d}V$. Further, we have to consider the inertial forces due to the kinetic energy density θ leading to $\delta\theta = \rho\dot{\boldsymbol{v}} \cdot \delta\boldsymbol{v}$. Finally, there are the internal forces $\boldsymbol{f}_{\text{int}}$ acting between the various particles. We thus have for the PoVP

$$P_{\text{acc}}^* = \int \rho\dot{\boldsymbol{v}} \cdot \delta\boldsymbol{v}\,\mathrm{d}V = P_{\text{vol}}^* + P_{\text{sur}}^* + P_{\text{int}}^* = \int \rho\boldsymbol{b} \cdot \delta\boldsymbol{v}\,\mathrm{d}V + \int \boldsymbol{t} \cdot \delta\boldsymbol{v}\,\mathrm{d}A + \int \boldsymbol{f}_{\text{int}} \cdot \delta\boldsymbol{v}\,\mathrm{d}V$$

To complement the PoVP for a continuous body, we need the reaction principle, another name for Newton's third law. We noted that it is always possible to detach a system from its surroundings if we introduce at the same time external, compensating surface forces. The system may then be moved as if it were rigid. The *reaction principle* for a continuous body can now be stated as

$$P_{\text{int}}^* = 0$$

or the virtual power of the internal forces vanishes for a rigid body motion.

Consider now an arbitrary rigid body motion of a volume element $\mathrm{d}V$ described by the virtual velocity field $\delta\boldsymbol{v}$. On account of the rigid body motion, the virtual power of the internal forces $P_{\text{int}}{}^*$ is zero. Hence in this case the PoVP states that

$$P_{\text{acc}}^* = P_{\text{vol}}^* + P_{\text{sur}}^* \quad \text{or} \quad \int \rho\boldsymbol{a} \cdot \delta\boldsymbol{v}\ \mathrm{d}V = \int \rho\boldsymbol{b} \cdot \delta\boldsymbol{v}\ \mathrm{d}V + \int \boldsymbol{t} \cdot \delta\boldsymbol{v}\ \mathrm{d}A \tag{5.14}$$

where $\boldsymbol{a} = \dot{\boldsymbol{v}}$ is used. Expression (5.14) holds for arbitrary virtual rigid motion.

5.4.2 Linear Momentum

Applying Eq. (5.14) to rigid body translations, the velocity $\delta\boldsymbol{v}$ is constant throughout the body and can be dropped from the equation, leading to

$$\int \rho\boldsymbol{a}\,\mathrm{d}V = \int \rho\boldsymbol{b}\,\mathrm{d}V + \int \boldsymbol{t}\,\mathrm{d}A \tag{5.15}$$

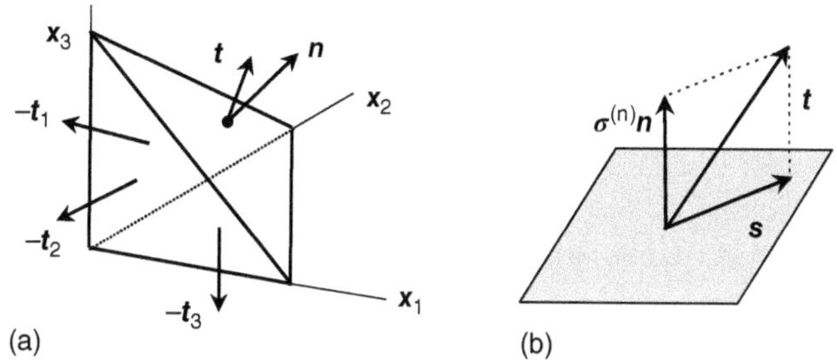

Figure 5.2 (a) Cauchy's tetrahedron; (b) stress vector $\boldsymbol{t}$, normal stress vector $\sigma^{(n)}\boldsymbol{n}$, and shear stress vector $\boldsymbol{s}$ at a plane with a normal vector n.

The total linear momentum of the body is defined by

$$\boldsymbol{p} = \int \rho \boldsymbol{v} \, \mathrm{d}V$$

so that, since the mass density is conserved, the left-hand side of Eq. (5.15) represents the material derivative of the total linear momentum $\boldsymbol{p}$. Consequently, Eq. (5.15) represents the *theorem of linear momentum*: the material derivative of the linear momentum $\boldsymbol{p}$ is equal to the sum of the external forces $\boldsymbol{b}$ and $\boldsymbol{t}$.

Let us now apply this theorem to a small tetrahedron with volume ΔV whose three edges are parallel to the coordinate axes $\boldsymbol{e}_j$ (Figure 5.2a). If ΔA is the area of the oblique face with exterior normal $\boldsymbol{n}$, the areas of the other three faces are $\Delta A_j = (\boldsymbol{e}_j \cdot \boldsymbol{n})\Delta A$. The stress vector on ΔA is denoted by $\boldsymbol{t}$, and the stress vectors on the other faces by $-\boldsymbol{t}_j$, as they are pointing in the $-j$ direction. The stress vector of a surface element is thus defined in such a way that a stress vector pointing in the positive j direction is $\boldsymbol{t}_j$. If the tetrahedron is sufficiently small, it holds that

$$\int \rho \boldsymbol{b} \, \mathrm{d}V \cong \rho \boldsymbol{b} \ \Delta V, \qquad \int \rho \boldsymbol{a} \, \mathrm{d}V \cong \rho \boldsymbol{a} \ \Delta V \quad \text{and} \quad \int \boldsymbol{t} \, \mathrm{d}A \cong \boldsymbol{t} \, \Delta A - \sum\nolimits_j \boldsymbol{t}_j \, \Delta A_j$$

Using $\Delta A_j = (\boldsymbol{e}_j \cdot \boldsymbol{n})\Delta A$, we thus obtain from the theorem of linear momentum

$$\boldsymbol{t} \cong \sum\nolimits_j (\boldsymbol{t}_j)(\boldsymbol{e}_j \cdot \boldsymbol{n}) + \rho(\boldsymbol{b} - \boldsymbol{a})(\Delta V/\Delta A)$$

Taking the limit to an infinitesimal tetrahedron ($\Delta V/\Delta A \to 0$), the result is

$$\boldsymbol{t} = \sum\nolimits_j (\boldsymbol{t}_j)(\boldsymbol{e}_j \cdot \boldsymbol{n}) = \left(\sum\nolimits_j \boldsymbol{t}_j \boldsymbol{e}_j\right) \cdot \boldsymbol{n} \equiv \boldsymbol{\sigma} \cdot \boldsymbol{n} \tag{5.16}$$

where we have used the fact that $\Sigma_j \boldsymbol{t}_j \boldsymbol{e}_j$ is a sum of three tensor products and therefore a tensor itself, which we label $\boldsymbol{\sigma}$ and call the *Cauchy stress tensor* (again, the symbol used is conventional but deviating from our usual tensor notation). Obviously, $\boldsymbol{t}$ is a linear function of $\boldsymbol{n}$. Using the definition of $\boldsymbol{\sigma}$ one finds that

$$\sigma_{ij} = \boldsymbol{e}_i \cdot \boldsymbol{\sigma} \cdot \boldsymbol{e}_j = \boldsymbol{t}_i \cdot \boldsymbol{e}_j$$

So, σ_{ij} is the component of the stress vector $\boldsymbol{t}_i$ acting on a surface element with normal vector $\boldsymbol{e}_j$. A component is positive if it points in the positive direction for a plane with an outward normal also pointing in the positive direction. The components σ_{ij} $(i = j)$ are the *normal stresses*, whereas the components σ_{ij} $(i \neq j)$ are the *shear stresses*.

5.4.3 Angular Momentum

We now turn our attention to rigid body rotations around an origin O. The virtual velocity for a rigid body rotation is given by

$$\delta \boldsymbol{v} = \delta(\boldsymbol{\omega} \times \boldsymbol{r})$$

where $\delta\boldsymbol{\omega}$ denotes the virtual angular velocity and $\boldsymbol{r}$ is the position vector. Inserting the above equation in the PoVP for a virtual rigid body motion, Eq. (5.14), and noting that $\delta\boldsymbol{\omega}$ can be dropped since it is independent of position, leads to

$$\int \rho(\boldsymbol{r} \times \boldsymbol{a})\,\mathrm{d}V = \int \rho(\boldsymbol{r} \times \mathbf{b})\,\mathrm{d}V + \int (\boldsymbol{r} \times \boldsymbol{t})\,\mathrm{d}A$$

Further $\dot{\boldsymbol{r}} \times \boldsymbol{v} = \boldsymbol{v} \times \boldsymbol{v} = 0$, so that $\boldsymbol{r} \times \boldsymbol{a} = \boldsymbol{r} \times \dot{\boldsymbol{v}} = \mathrm{d}(\boldsymbol{r} \times \boldsymbol{v})/\mathrm{d}t$, and using $\boldsymbol{t} = \boldsymbol{\sigma} \cdot \boldsymbol{n}$ yields

$$\int \rho[\mathrm{d}(\boldsymbol{r} \times \boldsymbol{v})/\mathrm{d}t]\,\mathrm{d}V = \int \rho(\boldsymbol{r} \times \mathbf{b})\,\mathrm{d}V + \int (\boldsymbol{r} \times \boldsymbol{\sigma} \cdot \boldsymbol{n})\,\mathrm{d}A \tag{5.17}$$

The total angular momentum of the body is given by

$$\boldsymbol{l} = \int \rho(\boldsymbol{r} \times \boldsymbol{v})\,\mathrm{d}V$$

so that, since the mass is conserved, the left-hand side of Eq. (5.17) is equal to the material derivative of the total angular momentum. Consequently, Eq. (5.17) represents the *theorem of angular momentum*: the material derivative of the angular momentum $\boldsymbol{l}$ with respect to the origin O is equal to the sum of the moments of the external forces $\boldsymbol{b}$ and $\boldsymbol{t}$ with respect to O.

5.4.4 Cauchy's Equations of Motion

So far, we have obtained from the PoVP, conservation of linear and angular momentum, represented by

$$\int \rho(\boldsymbol{a} - \mathbf{b})\,\mathrm{d}V = \int \boldsymbol{\sigma} \cdot \boldsymbol{n}\,\mathrm{d}A \quad \text{and} \quad \int \rho\boldsymbol{r} \times (\boldsymbol{a} - \mathbf{b})\,\mathrm{d}V = \int \boldsymbol{r} \times (\boldsymbol{\sigma} \cdot \boldsymbol{n})\,\mathrm{d}A \tag{5.18}$$

We now derive in a compact way the equations of motion. By taking the outer product of Eq. (5.18b) with an arbitrary constant vector $\mathbf{c}$ and using the vector relation $\boldsymbol{x} \times \boldsymbol{y} \times \boldsymbol{z} = (\boldsymbol{yz}\text{-}\boldsymbol{zy})\boldsymbol{x}$ for arbitrary $\boldsymbol{x}$, $\boldsymbol{y}$, and $\boldsymbol{z}$, this equation becomes, after deleting the vector $\mathbf{c}$,

$$\int \rho[\boldsymbol{r}(\boldsymbol{a} - \mathbf{b}) - (\boldsymbol{a} - \mathbf{b})\boldsymbol{r}]\,\mathrm{d}V = \int [\boldsymbol{r}(\boldsymbol{\sigma} \cdot \boldsymbol{n}) - (\boldsymbol{\sigma} \cdot \boldsymbol{n})\boldsymbol{r}]\,\mathrm{d}A$$

To transform the surface integral to a volume integral, we use a somewhat different form of the Gauss theorem, which states, for an arbitrary vector $\boldsymbol{z}$, an arbitrary tensor $\mathbb{Z}$ and an outward normal $\boldsymbol{n}$ to the surface element $\mathrm{d}A$, that

$$\int \boldsymbol{z}(\mathbb{Z}^{\mathrm{T}} \cdot \boldsymbol{n})\,\mathrm{d}A = \int \boldsymbol{z}(\nabla \cdot \mathbb{Z})\,\mathrm{d}V + \int (\nabla \boldsymbol{z})\mathbb{Z}\,\mathrm{d}V$$

reducing for a constant vector $\boldsymbol{z}$ to the form $\int \mathbb{Z}^{\mathrm{T}} \cdot \boldsymbol{n}\,\mathrm{d}A = \int \nabla \cdot \mathbb{Z}\,\mathrm{d}V$ and for $\mathbb{Z} = \mathbb{I}$ to the form $\int \boldsymbol{z}\boldsymbol{n}\,\mathrm{d}A = \int \nabla \boldsymbol{z}\,\mathrm{d}V$ or, using $\mathrm{tr}(\boldsymbol{x}\boldsymbol{y}) = \boldsymbol{x} \cdot \boldsymbol{y}$ and $\mathrm{tr}(\nabla \boldsymbol{x}) = \nabla \cdot \boldsymbol{x}$, to $\int \boldsymbol{z} \cdot \boldsymbol{n}\,\mathrm{d}A = \int \nabla \cdot \boldsymbol{z}\,\mathrm{d}V$, which is the standard form.[1] For $\boldsymbol{z} = \boldsymbol{r}$ and $\mathbb{Z}^T = \boldsymbol{\sigma}^T$, we obtain

$$\int [\rho(\boldsymbol{a} - \boldsymbol{b}) - \nabla \cdot \sigma]\,\mathrm{d}V = 0 \qquad \text{(from the standard form) and}$$

$$\int \left\{ \boldsymbol{r}[\rho(\boldsymbol{a} - \boldsymbol{b}) - \nabla \cdot \sigma] - [\rho(\boldsymbol{a} - \boldsymbol{b}) - \nabla \cdot \sigma]\boldsymbol{r} + \boldsymbol{\sigma}^{\mathrm{T}} - \boldsymbol{\sigma} \right\} \mathrm{d}V = 0$$

Since the volume element is arbitrary, the integrands are zero and we arrive at *Cauchy's equations of motion*:

$$\nabla \cdot \sigma + \rho \boldsymbol{b} = \rho \boldsymbol{a} \qquad \text{and} \qquad \boldsymbol{\sigma} = \boldsymbol{\sigma}^{\mathrm{T}} \tag{5.19}$$

Problem 5.1 ***Linear Momentum*** Show that the theorem of linear momentum, valid for an observer at rest, reads

$$\int \frac{\partial \rho \boldsymbol{v}}{\partial t}\,\mathrm{d}V = \int \rho \boldsymbol{b}\,\mathrm{d}V + \int (\sigma - \rho \boldsymbol{v}\boldsymbol{v}) \cdot \boldsymbol{n}\,\mathrm{d}A$$

Problem 5.2 ***Angular Momentum*** Show that the theorem of angular momentum, valid for an observer at rest, reads

$$\int \boldsymbol{x} \times (\mathrm{d}\rho \boldsymbol{v}/\mathrm{d}t)\,\mathrm{d}V = \int \rho \boldsymbol{x} \times \boldsymbol{b}\,\mathrm{d}V + \int \boldsymbol{x} \times (\sigma - \rho \boldsymbol{v}\boldsymbol{v}) \cdot \boldsymbol{n}\,\mathrm{d}A$$

5.5 The Stress Tensor

As discussed, the stress vector $\boldsymbol{t}$ acting on an infinitesimal plane with area $\mathrm{d}A$, position vector $\boldsymbol{r}$, and outer normal $\boldsymbol{n}$ is given by $\boldsymbol{t}(\boldsymbol{r}, \boldsymbol{n}) = \boldsymbol{\sigma} \cdot \boldsymbol{n}$. This relation applies up to and including the boundary of a body. So, for a free boundary, one has $\boldsymbol{\sigma} \cdot \boldsymbol{n} = 0$, and on those parts of a boundary where a traction $\bar{\boldsymbol{t}}$ is prescribed, one has $\boldsymbol{\sigma} \cdot \boldsymbol{n} = \bar{\boldsymbol{t}}$. Often the normal stress is required, that is, the component of the stress vector normal to $\mathrm{d}A$. The length $\sigma^{(\mathrm{n})}$ of this stress component is given by the projection of the stress vector $\boldsymbol{t}$ on the normal vector $\boldsymbol{n}$ and thus equal to (Figure 5.2b)

$$\sigma^{(\mathrm{n})} = \boldsymbol{n} \cdot \boldsymbol{t} = \boldsymbol{n} \cdot \sigma \cdot \boldsymbol{n}$$

The normal stress vector is accordingly $\sigma^{(\mathrm{n})}\boldsymbol{n}$. The component of the stress vector $\boldsymbol{s}$ in the plane $\mathrm{d}A$ is known as the shear component and, according to the Pythagorean theorem, its length $\sigma^{(\mathrm{s})} = [\boldsymbol{t}^2 - (\sigma^{(\mathrm{n})}\boldsymbol{n})^2]^{1/2}$. Using vector subtraction, the shear stress vector $\boldsymbol{s}$ can also be written as

$$\boldsymbol{s} = \boldsymbol{t} - \sigma^{(\mathrm{n})}\boldsymbol{n} \qquad \text{or equivalently as} \qquad \boldsymbol{s} = \boldsymbol{n} \times (\boldsymbol{t} \times \boldsymbol{n}) \qquad \text{with} \qquad |\boldsymbol{s}| = \sigma^{(\mathrm{s})}$$

---

1 The standard form looks less general than the form used here, but in fact they are equivalent. This can be proven using $(\boldsymbol{z} \cdot \boldsymbol{x})(\mathbb{Z} \cdot \boldsymbol{y})$ for $\boldsymbol{z}$ in the standard form with $\boldsymbol{x}$ and $\boldsymbol{y}$ arbitrary vectors. Using for arbitrary tensor $\mathbb{A}$ the transpose relation $\boldsymbol{x} \cdot (\mathbb{A}^{\mathrm{T}} \cdot \boldsymbol{y}) = \boldsymbol{y} \cdot (\mathbb{A} \cdot \boldsymbol{x})$ and the relations $(\boldsymbol{u}\boldsymbol{v}) \cdot \boldsymbol{x} = (\boldsymbol{v} \cdot \boldsymbol{x})\boldsymbol{u}$, $(\nabla \boldsymbol{u})^{\mathrm{T}} \cdot \boldsymbol{x} = \nabla(\boldsymbol{u} \cdot \boldsymbol{x})$, $(\nabla \cdot \mathbb{A}) \cdot \boldsymbol{x} = \nabla \cdot (\mathbb{A} \cdot \boldsymbol{x})$, and $\nabla \cdot (\phi \boldsymbol{x}) = \boldsymbol{x} \cdot \nabla \phi + \phi \nabla \cdot \boldsymbol{x}$ for arbitrary $\boldsymbol{u}$, $\boldsymbol{v}$, and ϕ, one can derive $[(\boldsymbol{z} \cdot \boldsymbol{x})(\mathbb{Z} \cdot \boldsymbol{y})] \cdot \boldsymbol{n} = \boldsymbol{x} \cdot [\boldsymbol{z}(\mathbb{Z}^{\mathrm{T}} \cdot \boldsymbol{n})] \cdot \boldsymbol{y}$ and $\nabla \cdot [(\boldsymbol{z} \cdot \boldsymbol{x})(\mathbb{Z} \cdot \boldsymbol{y})] = \boldsymbol{x}[(\nabla \boldsymbol{z}) \cdot \mathbb{Z} + \boldsymbol{z}(\nabla \cdot \mathbb{Z})] \cdot \boldsymbol{y}$, which substituted in the standard Gauss theorem, and deleting the arbitrary $\boldsymbol{x}$ and $\boldsymbol{y}$ yields the form used here [7].

The stress tensor being symmetric, it can be decomposed into an isotropic part $\frac{1}{3}(\mathrm{tr}\boldsymbol{\sigma})\mathbb{I}$ and a deviator $\boldsymbol{\sigma}'$ according to $\boldsymbol{\sigma} = \frac{1}{3}(\mathrm{tr}\boldsymbol{\sigma})\mathbb{I} + \boldsymbol{\sigma}'$. The invariant $\sigma = \frac{1}{3}\mathrm{tr}\boldsymbol{\sigma}$ is the *mean normal stress* and its negative $P = -\frac{1}{3}\mathrm{tr}\boldsymbol{\sigma}$ the *pressure*. Further, the stress tensor $\boldsymbol{\sigma}$ can be brought at *principal axes* so that only diagonal elements, the *principal stresses*, remain.

5.6 Mechanical Energy

We now turn to the theorem of mechanical energy, the discussion of which results in an expression for the power of the internal forces [5].

Taking the scalar product of both sides of the equilibrium equation $\nabla \cdot \boldsymbol{\sigma} + \rho\,\boldsymbol{b} = \rho\dot{\boldsymbol{v}}$ with $\boldsymbol{v}$ and integrating over the volume, we obtain

$$\begin{aligned}\int \rho\boldsymbol{v}\cdot\dot{\boldsymbol{v}}\,\mathrm{d}V &= \int \rho\boldsymbol{b}\cdot\boldsymbol{v}\,\mathrm{d}V + \int (\nabla\cdot\boldsymbol{\sigma})\cdot\boldsymbol{v}\,\mathrm{d}V \\ &= \int \rho\boldsymbol{b}\cdot\boldsymbol{v}\,\mathrm{d}V + \int \nabla\cdot(\boldsymbol{\sigma}\cdot\boldsymbol{v})\,\mathrm{d}V - \int \boldsymbol{\sigma}\cdot(\nabla\cdot\boldsymbol{v})\,\mathrm{d}V\end{aligned}$$

On account of the symmetry of $\boldsymbol{\sigma}$ and the deformation rate $\mathbb{D} = [\nabla\cdot\boldsymbol{v} + (\nabla\cdot\boldsymbol{v})^{\mathrm{T}}]/2$, we can replace $\boldsymbol{\sigma}\cdot(\nabla\cdot\boldsymbol{v})$ by $\boldsymbol{\sigma}:\mathbb{D}$. Moreover, we apply once more the Gauss theorem, in this case to $\int \nabla\cdot(\boldsymbol{\sigma}\cdot\boldsymbol{v})\,\mathrm{d}V$. Thus, we obtain

$$\int \rho\boldsymbol{v}\cdot\dot{\boldsymbol{v}}\,\mathrm{d}V = \int \rho\boldsymbol{b}\cdot\boldsymbol{v}\,\mathrm{d}V + \int (\boldsymbol{\sigma}\cdot\boldsymbol{v})\cdot\boldsymbol{n}\,\mathrm{d}A - \int \boldsymbol{\sigma} : \mathbb{D}\,\mathrm{d}V \tag{5.20}$$

The left-hand side represents the material derivative of the kinetic energy given by

$$\vartheta = \tfrac{1}{2}\int \rho\boldsymbol{v}\cdot\boldsymbol{v}\,\mathrm{d}V \quad \text{so that} \quad P_{\mathrm{acc}} = \dot{\vartheta} = \int \rho\boldsymbol{v}\cdot\dot{\boldsymbol{v}}\,\mathrm{d}V$$

represents the power of the inertial forces. The first two terms on the right-hand side

$$P_{\mathrm{vol}} = \int \rho\boldsymbol{b}\cdot\boldsymbol{v}\,\mathrm{d}V \quad \text{and} \quad P_{\mathrm{sur}} = \int (\boldsymbol{\sigma}\cdot\boldsymbol{v})\cdot\boldsymbol{n}\,\mathrm{d}A = \int \boldsymbol{t}\cdot\boldsymbol{v}\,\mathrm{d}A$$

represent the power of the volume and surface forces. Now the PoVP also contains a term representing the power of the internal forces P_{int}, so that the third term

$$P_{\mathrm{int}} = -\int \boldsymbol{\sigma} : \mathbb{D}\,\mathrm{d}V$$

is interpreted as the power of the internal forces, consistent with the reaction principle $P_{\mathrm{int}} = 0$ for rigid body motion with $\mathbb{D} = 0$. Note that the minus sign appears because the internal forces $\boldsymbol{f}_{\mathrm{int}}$ are characterized by the external traction $\boldsymbol{t}$. In this way, Eq. (5.20) can be seen as the global form of the theorem of *conservation of mechanical energy*: the power of the inertial forces P_{acc} is equal to the sum of the power of the external forces $P_{\mathrm{ext}} = P_{\mathrm{vol}} + P_{\mathrm{sur}}$ and internal forces P_{int} or $P_{\mathrm{acc}} = P_{\mathrm{vol}} + P_{\mathrm{sur}} + P_{\mathrm{int}}$. Conservation of energy can thus be derived from the PoVP by interpreting Eq. (5.20) as the PoVP applied to the real state of motion considered as a particular case of virtual motion. The volume V does not remain rigid, and this accounts for the difference between Eqs. (5.20) and (5.14), where in the former $\boldsymbol{v}$ replaces $\delta\boldsymbol{v}$. Consequently, $P_{\mathrm{int}}^{*} = -\int \boldsymbol{\sigma} : \delta\mathbb{D}\,\mathrm{d}V$ and the final form of the PoVP becomes

$$\begin{aligned}P_{\mathrm{vol}}^{*} + P_{\mathrm{sur}}^{*} + P_{\mathrm{int}}^{*} &= \int \rho\boldsymbol{b}\cdot\delta\boldsymbol{v}\,\mathrm{d}V + \int \boldsymbol{t}\cdot\delta\boldsymbol{v}\,\mathrm{d}A - \int \boldsymbol{\sigma} : \delta\mathbb{D}\,\mathrm{d}V \\ &= P_{\mathrm{acc}}^{*} = \int \rho\dot{\boldsymbol{v}}\cdot\delta\boldsymbol{v}\,\mathrm{d}V\end{aligned} \tag{5.21}$$

stating that, *at any time t, the virtual power of the external and internal forces is equal to the virtual power of the inertial forces for any virtual state of motion.*

To show that this interpretation is consistent with previous results, we consider that

$$P^*_{\text{int}} = -\int \boldsymbol{\sigma} : \delta\mathbb{D}\,\mathrm{d}V = -\int (\boldsymbol{\sigma}\cdot\boldsymbol{n})\cdot\delta\boldsymbol{v}\,\mathrm{d}A + \int (\nabla\cdot\boldsymbol{\sigma})\cdot\delta\boldsymbol{v}\,\mathrm{d}V$$

where $\boldsymbol{\sigma} : \delta\mathbb{D} = \boldsymbol{\sigma} : \delta\nabla\boldsymbol{v} = \nabla\cdot(\boldsymbol{\sigma}\cdot\delta\boldsymbol{v}) - (\nabla\cdot\boldsymbol{\sigma})\cdot\delta\boldsymbol{v}$ and the Gauss theorem have been used.

To find the solution $\boldsymbol{\sigma}(\boldsymbol{r})$, one must know the *boundary conditions* on the surface A. At any point, either $\boldsymbol{t} = \bar{\boldsymbol{t}}$ (on part A_t of A) or $\boldsymbol{v} = \bar{\boldsymbol{v}}$ (on part A_v of A) is prescribed. The traction boundary condition is thus $\boldsymbol{\sigma}\cdot\boldsymbol{n} = \bar{\boldsymbol{t}}$ on A_t and the velocity boundary condition is $\boldsymbol{v} = \bar{\boldsymbol{v}}$ on A_v. Since these conditions apply per component, A_t and A_v may overlap. When velocities are prescribed one speaks of *external constraints*. In addition, there may be *internal constraints*, for example, incompressibility. A velocity field is *kinematically admissible* if it is continuous, sufficiently differentiable, and obeys the constraints. A stress field satisfying the equilibrium condition $\nabla\cdot\boldsymbol{\sigma} + \rho\boldsymbol{b} = \rho\boldsymbol{a}$ and the traction boundary condition $\bar{\boldsymbol{t}} = \boldsymbol{\sigma}\cdot\boldsymbol{n}$ is called *dynamically admissible.*

Because $\delta\boldsymbol{v} = 0$ on A_v we restrict the surface integration to A_t and write

$$\int (\boldsymbol{\sigma}\cdot\boldsymbol{n})\cdot\delta\boldsymbol{v}\,\mathrm{d}A = \int (\boldsymbol{\sigma}\cdot\boldsymbol{n})\cdot\delta\boldsymbol{v}\,\mathrm{d}A_t, \qquad \text{leading to}$$

$$P^*_{\text{ext}} + P^*_{\text{int}} = P^*_{\text{acc}} \;\text{ or }\; \int [\nabla\cdot\boldsymbol{\sigma} + \rho(\boldsymbol{b}-\boldsymbol{a})]\cdot\delta\boldsymbol{v}\,\mathrm{d}V = \int \left(\boldsymbol{\sigma}\cdot\boldsymbol{n} - \bar{\boldsymbol{t}}\right)\cdot\delta\boldsymbol{v}\,\mathrm{d}A_t$$

The variations $\delta\boldsymbol{v}$ being arbitrary, we regain the equilibrium and boundary conditions

$$\nabla\cdot\boldsymbol{\sigma} + \rho(\boldsymbol{b}-\boldsymbol{a}) = 0 \text{ in } V, \qquad \boldsymbol{\sigma}\cdot\boldsymbol{n} = \bar{\boldsymbol{t}} \text{ on } A_t \qquad \text{and} \qquad \boldsymbol{v} = \bar{\boldsymbol{v}}\,(\delta\boldsymbol{v} = 0) \text{ on } A_v$$

and this warrants our interpretation for $\boldsymbol{f}_{\text{int}}$.

5.7 Final Remarks

In principle, many (kinematically admissible) strain and (dynamically admissible) stress tensors can be defined, but the only ones properly allowed are those for which the product of stress and strain increment yields the work done [8]. Limiting ourselves here to spatial coordinates $\boldsymbol{r}$, the work W_{int} of the internal forces can be expressed as the integral of the power P_{int}. Since we have $P_{\text{int}} = \dot{W}_{\text{int}} = -\int \boldsymbol{\sigma} : \mathbb{D}\,\mathrm{d}V$, in the SDG approximation we obtain $P_{\text{int}} \cong -\int \boldsymbol{\sigma} : \dot{\boldsymbol{\varepsilon}}\mathrm{d}V$ and thus $W_{\text{int}} = \int P_{\text{int}}\,\mathrm{d}t = -\int\int \boldsymbol{\sigma} : \dot{\boldsymbol{\varepsilon}}\mathrm{d}t\mathrm{d}V = -\int\int \boldsymbol{\sigma} : \mathrm{d}\boldsymbol{\varepsilon}\mathrm{d}V$. The work done on the system $\mathrm{d}W_{\text{mec}}$ is given by the negative work of the internal forces, that is, $\mathrm{d}W_{\text{mec}} = -\mathrm{d}W_{\text{int}} = -\mathrm{d}E_{\text{int}}$, where E_{int} is *internal energy* containing the potential and kinetic energies of the molecules constituting the solid.

We will often use *isotropic materials* (Section 5.1) for which the constitutive relations are functions of the invariants of the state parameters only. In particular, if a constitutive relation connects two symmetric second-order tensors, for example, a stress s_{ij} and a strain t_{ij}, its form according to the Hamilton–Cayley theorem (Section A.10) is

$$s_{ij} = C_{ij} + \sum\nolimits_{kl} C_{ijkl}t_{kl} + \sum\nolimits_{klmn} C_{ijklmn}t_{kl}t_{mn}$$

where the material properties are represented by C_{ij}, C_{ijkl}, and C_{ijklmn}. In general, the components of these tensors are orientation dependent, and it follows that the equation represents the description of *anisotropic materials*. For an isotropic material, the relation itself must be isotropic and of the form

$$s_{ij} = a\left(t_{(1)}, t_{(2)}, t_{(3)}\right)\delta_{ij} + b(t_{(1)}, t_{(2)}, t_{(3)})t_{ij} + c(t_{(1)}, t_{(2)}, t_{(3)})\sum\nolimits_k t_{ik}t_{kj}$$

where $t_{(1)}$, $t_{(2)}$, and $t_{(3)}$ are the basic invariants of the tensor t_{ij}. Because $t_{(1)}$, $t_{(2)}$, and $t_{(3)}$ are of order 1, 2, and 3, respectively, this allows us to discuss approximations up to a certain order. In a first-order approximation $c = 0$, b is a constant, say $b = 2\mu$ to be consistent with literature, and a is a linear function of $t_{(1)}$, say $\kappa + \lambda t_{(1)}$, and we obtain

$$s_{ij} = (\kappa + \lambda t_{(1)})\delta_{ij} + 2\mu\, t_{ij} \quad \text{or} \quad s_{ij} = \lambda t_{(1)}\delta_{ij} + 2\mu\, t_{ij} = \lambda\delta_{ij}\sum\nolimits_k t_{kk} + 2\mu\, t_{ij}$$

if in the reference state there is no response, that is, if $s_{ij} = 0$ when $t_{ij} = 0$ and $\kappa = 0$. The scalars λ and μ are (still temperature dependent) material constants. For example, a gas is isotropic, and the simplest choice one can make for the stress tensor is $\mathbb{P} = P\mathbb{I}$ with pressure P. Higher order approximations can be made in a similar way.

For our discussion of phases and phase transitions, the equations of motion $\nabla \cdot \boldsymbol{\sigma} + \rho\mathbf{b} = \rho\boldsymbol{a}$ and $\boldsymbol{\sigma} = \boldsymbol{\sigma}^{\mathrm{T}}$ and the associated basic concepts will do. Their main use will be in the treatment of (linear) irreversible thermodynamics.

References

1 Kuiken, G.D.C. (1994). *Thermodynamics of Irreversible Processes*. Chichester: John Wiley & Sons.

2 Fung, 1965; Malvern, 1969.

3 Haupt, P. (2002). *Continuum Mechanics and Theory of Materials*, 2e. Berlin: Springer.

4 Teodosiu, C. (1982). *Elastic Models of Crystal Defects*. Berlin: Springer.

5 Ziegler, 1983.

6 Gyarmati, 1970.

7 Chadwick, P. (1976). *Continuum Mechanics, Concise Theory and Problems*. George Allen & Unwin Ltd. (also Dover, 1999, enlarged and corrected).

8 McVean, D.B. (1968). *Z. Angew. Math. Phys.* 19: 157.

Further Reading

Fung, Y.C. (1965). *Foundations of Solid Mechanics*. Englewood Cliffs, NJ: Prentice-Hall. A solid book on the basics and extensive applications for solids but does not deal with thermodynamics.

Gyarmati, I. (1970). *Non-equilibrium Thermodynamics*. Berlin: Springer. In addition to what the title suggests, this little book also contains a compact and clear introduction to continuum mechanics. The thermodynamic part largely deals with a variational treatment.

Malvern, L.E. (1969). *Introduction to the Mechanics of a Continuous Medium*. Englewood Cliffs, NJ: Prentice-Hall. A highly readable classic containing basic theory, and a number of applications on fluids and solids.

Ziegler, H. (1983). *Introduction to Thermomechanics*, 2e. Amsterdam: North-Holland. A nice book with an amalgamation of continuum mechanics and thermodynamics, dealing with basics, mainly in the SDG approximation, some applications and the orthogonality principle.

6

Macroscopic Thermodynamics

In most of macroscopic and much of microscopic science we need another “tool”: thermodynamics. We first discuss macroscopic thermodynamics (MT) and its extension to continuum systems, and thereafter linear irreversible thermodynamics.

6.1 Classical Thermodynamics

Thermodynamic considerations are basic to many models and theories in science and technology.[1] In this section, the phenomenological aspects of thermodynamics are briefly described, but less briefly as in the summary given by Clausius[2]:

- Die Energie der Welt ist constant.
- Die Entropie der Welt strebt einem Maximum zu.

We use a relatively abstract formulation [9] and omit discussions on the link with experiments for which we refer to [10].

6.1.1 The Four Laws

Thermodynamics (i.e. MT) is based on four well-known basic laws. To describe them, we start with introducing a few concepts. For the sake of analysis, the part of the physical world that is under consideration, the *system*, is separated from the rest, the *surroundings* or *environment*. It is normally taken as *stationary* (no overall translation and rotation). The *thermodynamic state* of the system is assumed to be controlled completely by a set

1 Thermodynamics extends into all branches of *natural science*. Hence, its generally applicable basic concepts, indispensable for expressing our *observations* of events [1, 2], should be introduced without any resort to a specific branch of natural science. The goal of natural science is the description and understanding of *reproducible events*, and reproducibility distinguishes natural science from other intellectual endeavor. A complete description includes, therefore, explicitly or implicitly instructions that can be put into reality, sufficient for the reproduction of the event. This requires *measurables*, defined by a prescription of their measurement. In physical *theory*, we also use *conceptuals* whose values are obtained by a mathematical procedure on measurables.
2 The energy of the world is constant. The entropy of the world strives to a maximum [3]. A compact summary, but as Planck [4] remarked, not to be taken literally as “energy and entropy of the world have no meaning because they admit of no accurate definition.” Similar short remarks are given in [5, 6], while Buchdahl [7] discussed the matter in some detail, as did Bridgman [8].

Phases of Matter and their Transitions: Concepts and Principles for Chemists, Physicists, Engineers, and Materials Scientists, First Edition. Gijsbertus de With.

of macroscopic, independent, *external variables*[3] $\boldsymbol{a}$ (or a_j), which can be arbitrarily kept constant and arbitrarily changed without affecting others, and one "extra" variable related to the thermal condition of the system. For the latter, one has several choices, one of which is the most appropriate given the conditions and to which we return later. Together, these variables constitute the set of *state variables*. When a new set of state variable values is applied, the system adapts itself via internal processes, collectively characterized by the *internal variables* $\boldsymbol{b}$ (or b_j). When the properties of a system do not change with time at an observable rate given certain constraints and the state so obtained is independent of the path between the initial and the final state, the system is said to be in *equilibrium*. *Thermodynamics* is concerned with the equilibrium states available to systems, the transitions between them, and the effect of external influences upon the systems and transitions [11].

With each external variable a having a rate $\dot{a} = \mathrm{d}a/\mathrm{d}t$, a length scale L_a and a time scale $t_a = a/\dot{a}$ is associated. Therefore, not all of the internal variables b are relevant for a process. For example, if a certain b has a maximum gradient $b' = \partial b/\partial x$, the relevant length scale L_b associated with b is given by $L_b = b/b'$. When $L_b \gg L_a$, the gradient can be neglected, and the induced change can be considered as uniform over L_a. Similarly, if b changes with a time derivative $\dot{b}$, the associated time scale is $t_b = b/\dot{b}$. If $t_b \ll t_a$, the variable b is always in equilibrium with a and is denoted as *relaxed*, while if $t_b \gg t_a$, the variable b does not change significantly and is addressed as *frozen*. Thus, for the process at hand only the internal variables with $t_b \approx t_a$ are the most relevant. Obviously, for a specific process the considerations depend on both $\boldsymbol{b}$ (controlled by the system) and $\boldsymbol{a}$ (controlled by the experimenter). *Extensive* (*intensive*) variables are variables for which the magnitude for a system containing two noninteracting similar subsystems is additive (the same as for the subsystems). Thus, volume is an extensive property, while pressure is an intensive property.

Depending on the problem, one assumes a specific type of boundary between the system and the surroundings: an *adiabatic* (contrary *diathermal*) boundary allowing changes in the system by changes in external variables *only* (that is, by work, see below), a *flexible* (contrary *rigid*) boundary allowing all changes except matter exchange, or a *permeable* (contrary *impermeable*) boundary allowing also matter exchange. Systems separated by an adiabatic, rigid boundary are *isolated*, while systems separated by a diathermal, flexible boundary are said to be *closed*, but in *thermal contact*. Finally, systems separated by diathermal flexible permeable boundaries are labeled as *open*. The considerations given above for $\boldsymbol{b}$ and $\boldsymbol{a}$ also apply to the boundaries of the system.[4]

A basic premise, usually tacitly assumed but made explicit in, for example, [16], is that *an isolated system in an arbitrary initial state within a finite fixed volume will attain a unique state of equilibrium*, and for which the label "minus first law" has been coined [17] (Chapter 19). If any two separate systems, each in equilibrium, are brought in thermal contact, the two systems will gradually adjust themselves until they do reach mutual

3 External means externally controllable but belonging to the system. Note that in this chapter, $\boldsymbol{a}$ and $\boldsymbol{b}$ are the external and internal variables, while in Chapter 5, they denote the acceleration and volume force.

4 These concepts imply idealization, sometimes specified but often not, and therefore necessarily fall short of a complete description of nature. In any well-designed experiment, imperfections have only a small influence on the result. Moreover, we try to estimate their magnitude and correct for these. The uncertainty of this estimate is part of the experimental error. The use of idealized concepts is justified on the assumption that the effect of deviations from ideality decreases with the magnitude of these deviations, an assumption not necessarily always correct [2, 12]. Obviously, system and isolation are coupled: a system is a part of the world that can be isolated arbitrarily, and isolation can be discerned only by observation of a system and its environment.

equilibrium. The *zeroth law*[5] states that *if two systems are both in thermal equilibrium with a third system, they are also in thermal equilibrium with each other*, that is, they have a common value for a (measurable) state variable θ, called *(empirical) temperature*, which we take as positive. Temperature is therefore a possible candidate for the "thermal" coordinate. If the first system is much smaller than the second system, the state of the second – the *reservoir* or *temperature bath* or *thermostat* – will only change negligibly in comparison with the state of the first one if thermal contact is made. We may then regard the third system, which is much smaller than the first, as a measuring device for the temperature, that is, as a *thermometer*.

With each variable a_j, also indicated as *(generalized) displacement*, a dependent (measurable) variable is associated, generally denoted as *(generalized) force* A_j. A force A is a quantity in the surroundings that, when multiplied with a change in displacement $\mathrm{d}a$, yields the associated *work transfer* (*work* for short) $đW = A\mathrm{d}a$. Displacement and force are denoted as *conjugated variables*.[6] The work $đW(t)$, with time t as a parameter, generally depends on the path between the initial and final states. The line integral $\int đW = \int \dot{W}\,\mathrm{d}t = \int (\mathrm{d}W/\mathrm{d}t)\,\mathrm{d}t = \int (\mathrm{d}W/\mathrm{d}a)(\mathrm{d}a/\mathrm{d}t)\,\mathrm{d}t = W$ is thus not a *state function*, that is, a function of state variables only, contrary to $\int \mathrm{d}a = \Delta a$, where a is a state variable. Hence, $đW$ is in general not a total differential (Section A.1), and we use the notation $đW$ instead of $\mathrm{d}W$. When A and a are collinear, work is done on the system and counted positive. For several coordinates $\mathrm{d}a_j$, the work $đW$[7] is

$$đW = \sum\nolimits_j A_j \mathrm{d}a_j = \boldsymbol{A} \cdot \mathrm{d}\boldsymbol{a}$$

The most familiar example of work is the mechanical work for an inviscid fluid $đW_{\mathrm{mec}} = -P\mathrm{d}V$, where P and V denote the pressure and volume of the fluid, respectively.[8] To elaborate a bit, in general, upon transfer of energy from a reservoir to a system by applying work via coordinate a, both reversible work $đW_{\mathrm{rev}} = A^{(\mathrm{q})}\mathrm{d}a$ and dissipative work $đW_{\mathrm{dis}} = A^{(\mathrm{d})}\mathrm{d}a$ are present so that $đW = A\mathrm{d}a = (A^{(\mathrm{q})} + A^{(\mathrm{d})})\mathrm{d}a$, and where $A^{(\mathrm{q})}$ and $A^{(\mathrm{q})}$ are the *quasi-conservative* and *dissipative force*, respectively. Consider now an adiabatic super-system (system + reservoir) and assume that no dissipative effects occur in the reservoir, so that from the conservation of mechanical energy for a stationary

5 The peculiar numbering of the laws of thermodynamics results from the fact that the importance of the zeroth law was realized only after the first law was established. Its status is not without dispute though, as it can be considered a consequence of a measurement similar to force and therefore not a law [2]. Frivol additions are the fourth law stating that no experimental apparatus works the first time it is set up, and the fifth law stating that no experiment gives quite the expected numerical result [15].
6 For changing the state of a system, we must establish *interaction*, the characteristic feature of which is the loss of one independent variable (Section 1.2). If a_1 and a_2 were independent before interaction, they are no longer afterwards, so that $F(a_1,a_2) = 0$. Since we can replace any quantity by a monotonically increasing function, we may always transform the interaction condition to $a_1 + a_2 =$ constant or $\mathrm{d}a_1 + \mathrm{d}a_2 = 0$. This is usually, but not always, done. After having realized the interaction, we have three cases: ($\mathrm{d}a_1 > 0$, $\mathrm{d}a_2 < 0$), ($\mathrm{d}a_1 < 0$, $\mathrm{d}a_2 > 0$), or ($\mathrm{d}a_1 = \mathrm{d}a_2 = 0$). A *(generalized) force* f is defined by stating that these three cases indicate $f_1 < f_2$, $f_1 > f_2$ or $f_1 = f_2$. The third case represents *equilibrium*. To measure f, we have to choose a standard, an appropriate calibration so that $f_2\mathrm{d}a_1$ represents energy, and equilibrium between the system and standard must be established, at least with respect to the interaction involved [1, 2, 12]. The same considerations apply for temperature.
7 A significant objection against the use of the differential $\mathrm{d}z(x,y)$ of a variable z, where x and y are functions of time t, has been raised by Menger [16]. Essentially his criticism boils down to that $\mathrm{d}z$ describes progress of z with t, as given by $\mathrm{d}z/\mathrm{d}t = (\partial z/\partial x)(\mathrm{d}x/\mathrm{d}t) + (\partial z/\partial y)(\mathrm{d}y/\mathrm{d}t)$, without saying so.
8 We use here for work a local, system-based definition [17] in a Galilean rest frame (Section 2.1) with such a low velocity of the piston that P_{sys} remains essentially uniform throughout the system.

super-system (Section 2.4) $đW_{\mathrm{sys}} + đW_{\mathrm{res}} = 0$. Using one coordinate system for system and reservoir, $da_{\mathrm{res}} = da_{\mathrm{sys}}$, $A_{\mathrm{res}} = -A_{\mathrm{sys}}$, $đW_{\mathrm{sys}} = A^{(q)}_{\mathrm{res}} da_{\mathrm{sys}}$, and $đW_{\mathrm{res}} = -(A^{(q)}_{\mathrm{sys}} + A^{(d)}_{\mathrm{sys}}) da_{\mathrm{res}}$, and we obtain $A^{(d)}_{\mathrm{sys}} da_{\mathrm{res}} \equiv đW_{\mathrm{dis}} = (A^{(q)}_{\mathrm{sys}} - A^{(q)}_{\mathrm{res}}) da_{\mathrm{sys}}$. Experimentally, we have for $(A^{(q)}_{\mathrm{sys}} - A^{(q)}_{\mathrm{res}}) > 0$, $da_{\mathrm{sys}} > 0$ and for $(A^{(q)}_{\mathrm{sys}} - A^{(q)}_{\mathrm{res}}) < 0$, $da_{\mathrm{sys}} < 0$, so that $đW_{\mathrm{dis}} \geq 0$ always. For the limiting case of reversible work only, where $đW_{\mathrm{dis}} = 0$ or $A^{(q)}_{\mathrm{res}} = -A^{(q)}_{\mathrm{sys}}$, we have $đW_{\mathrm{sys}} = A^{(q)}_{\mathrm{res}} da_{\mathrm{sys}} = -A^{(q)}_{\mathrm{sys}} da_{\mathrm{sys}}$. For P-V work, to agree with the sign convention of work, we take $A^{(q)}_{\mathrm{sys}} = P$ and $da_{\mathrm{sys}} = dV$, so that $đW_{\mathrm{sys}} = -PdV$. For further details on work, see [18].

When work is done on an adiabatically enclosed system having an initial state with coordinates $\boldsymbol{a}_{\mathrm{ini}}$, it appears experimentally that the final state with coordinates $\boldsymbol{a}_{\mathrm{fin}}$ is independent of the type of work and its process of delivery. This leads to a new (conceptual) quantity, the state function U called *internal energy*, for which we thus have $\Delta U = U(\boldsymbol{a}_{\mathrm{fin}}) - U(\boldsymbol{a}_{\mathrm{ini}}) = \Delta W^{(\mathrm{a})}$ as well as $dU = dW^{(\mathrm{a})}$ with $dW^{(\mathrm{a})}$ the *adiabatic work*. Work thus refers to energy transfer and the unit for U is *joule* (1 J = 1 N m) or *calorie* (1 cal = 4.184 J). Using other enclosures as well, the *first law* states that the change in internal energy dU from one state to another is given by [6, 9, 19–21]

$$dU = đQ + đW = đQ + \sum\nolimits_j A_j \, da_j = đQ + \boldsymbol{A} \cdot d\boldsymbol{a} \tag{6.1}$$

where $đQ$ refers to the *heat transfer* (*heat* for short) during the process[9] and is interpreted as work done by the microscopic forces associated with the internal variables $\boldsymbol{b}$ of the system. Heat depends on the path between the two states, and when heat enters a system, it is counted as positive. Thus, both $đQ$ and $đW$ depend on the path between the initial and final states, but dU is independent of the path and depends only on the initial and final states. Hence, dU is a total differential and U a state variable. Although $đQ$ is defined mechanically by Eq. (6.1), in practice heat is measured calorimetrically. In summary, the *first law*, Eq. (6.1), can be stated succinctly as follows: *for an isolated system, the internal energy is constant.*

So far, each state is characterized by a set of state variables $\boldsymbol{a}$ and the (empirical) temperature θ. A process or system at constant θ is denoted as *isothermal* while at constant P (or V) they are called *isobaric* (or *isochoric*). If $đQ = 0$, the process and the system are called *adiabatic*, while if $đW = 0$ we refer to the process involved as pure *heating* or *cooling*. If $đQ = đW = 0$, the system is *isolated*. In a *quasi-static process* "infinitely" slow rates $\dot{\boldsymbol{a}}$ are applied, but the system is not necessarily fully in equilibrium and hence dissipation may occur. For example, in compressing a system with a piston having friction, dissipation occurs. A *reversible process* is a quasi-static process in which no dissipation occurs and during the entire process $t_{\boldsymbol{b}} \approx t_{\mathbf{a}}$ for the largest $t_{\boldsymbol{b}}$, so that the system is at any moment fully in equilibrium, and the process can be controlled in both directions. Clearly, reversible processes are quasi-static, although quasi-static processes are only reversible if dissipation is absent.[10]

---

9 We neglect volume forces (Chapter 2 and/or Chapter 5). Note that only by defining "adiabatic" as the inability to influence the system in the absence of forces, a definition of heat in terms of the set $\boldsymbol{a}$ becomes feasible. For an elaborate discussion, see, for example, [9, 20].

10 A quasi-static process can be described by a path in the $(\boldsymbol{A}, \boldsymbol{a})$ state space. The internal variables of the system, collectively indicated by $\boldsymbol{b}$, control the direction and rate of the processes, externally driven by $\boldsymbol{a}$, so that any change should actually be written as $d\boldsymbol{A} = (\partial \boldsymbol{A}/\partial \boldsymbol{b}) \cdot (\partial \boldsymbol{b}/\partial \boldsymbol{a}) \cdot (d\boldsymbol{a}/dt)dt$, keeping the criticism of Menger [16] in mind. Limiting to reversible processes, the system is also at any moment in equilibrium but has no dissipation, resulting in $d\boldsymbol{A} = (\partial \boldsymbol{A}/\partial \boldsymbol{a}) \cdot (d\boldsymbol{a}/dt)dt = (\partial \boldsymbol{A}/\partial \boldsymbol{a}) \cdot d\boldsymbol{a}$.

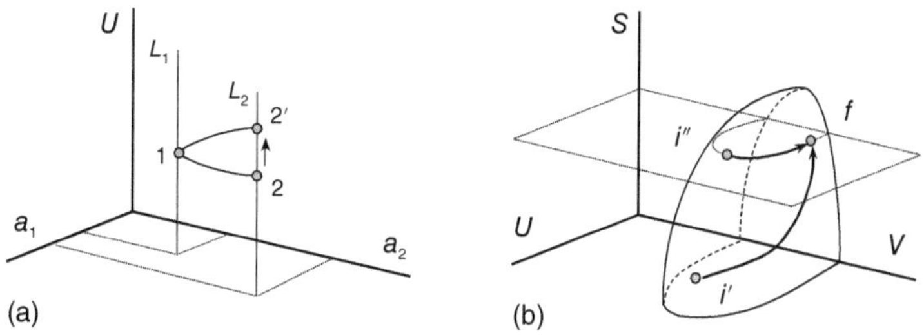

Figure 6.1 (a) If in energy space $U(a_1,a_2)$ points 1 and 2 are isentropic and point 2′ with energy $U_{2'} > U_2$ would be isentropic with point 2, a reversible path from 2′ via 1 to 2 would decrease U adiabatically, in contradiction with the second law. (b) The entropy surface $S(U,V)$ with a quasi-static path from initial state i′ to final state f and a reversible path from initial state i″ to final state f.

Experimentally, it appears that adiabatic work is always positive, that is, $\mathrm{d}W^{(a)} \geq 0$, or since $đW = đW_{\text{rev}} + đW_{\text{dis}}$, $đW_{\text{dis}}{}^{(a)} \geq 0$ for a cyclic process with $\boldsymbol{a}_{\text{fin}} = \boldsymbol{a}_{\text{ini}}$. This implies $U_{\text{fin}} \geq U_{\text{ini}}$ or, equivalently, that there are states in the neighborhood of any state that are *inaccessible* by an adiabatic process. Hence, on a line L defined by $\boldsymbol{a}$ = constant in $(U,\boldsymbol{a})$ space, since $U = U(\boldsymbol{a},\theta)$, we can take $\theta_{\text{fin}} > \theta_{\text{ini}}$. As any two points on L cannot have the same value for U, we have $(\partial U/\partial\theta)_{\mathbf{a}} > 0$; hence θ $(U,\boldsymbol{a})$ is single-valued and U may replace θ as a coordinate. We do so here, and U is therefore our present choice for the "extra," thermal variable of state (Figure 6.1a).

Consider now a reversible adiabatic process, so that $\mathrm{d}U = đW_{\text{rev}}$ and label states that can be reached by that process as *isentropic*. Denote again by L a line $\boldsymbol{a}$ = constant in $(U,\boldsymbol{a})$ space. On such a line only one and no other isentropic point can be present, as otherwise $\mathrm{d}W^{(a)} < 0$ would be possible. By moving L parallel to itself, the intersection with the hyperplane $U = const.$ generates a connected region in that plane. Similarly, an isentropic point on L will generate a coherent isentropic hyperplane. In fact, each point on L will generate such plane, which can be labeled by $\varphi(U,\boldsymbol{a}) = const.$, as these planes cannot intersect. This new (conceptual) quantity φ is a state function called (*empirical*) *entropy*. To move from one value of φ to another, the process must violate either the adiabaticity or reversibility condition, as assumed for isentropic surfaces. To describe a general process without constraints, define $\boldsymbol{R} = (1, -\boldsymbol{A}^{(q)})$ with $\boldsymbol{A}^{(q)}$ the quasi-conservative part of $\boldsymbol{A}$, $\boldsymbol{r} = (U,\boldsymbol{a})$ and $\mathrm{d}\boldsymbol{r} = (\mathrm{d}U,\mathrm{d}\boldsymbol{a})$, so that we can write for an isentropic process $\mathrm{d}U - \boldsymbol{A}^{(q)}\cdot\mathrm{d}\boldsymbol{a} = \boldsymbol{R}\cdot\mathrm{d}\boldsymbol{r} = 0$. Furthermore, all curves satisfying $\boldsymbol{R}\cdot\mathrm{d}\boldsymbol{r} = 0$ lie on isentropic surfaces $\varphi = const.$ and obey $\mathrm{d}\varphi = (\nabla\varphi)\cdot\mathrm{d}\boldsymbol{r} = 0$. Hence, $\boldsymbol{R}$ and $\nabla\varphi$ are parallel, and therefore $\boldsymbol{R}\cdot\mathrm{d}\boldsymbol{r}$ can be written as $\boldsymbol{R}\cdot\mathrm{d}\boldsymbol{r} = \mu\,\mathrm{d}\varphi$ with as proportionality factor $\mu(\boldsymbol{r})$ a scalar function. For nonadiabatic irreversible processes $\mathrm{d}U = đQ + đW$, so that in that case $\mu\,\mathrm{d}\varphi = \boldsymbol{R}\cdot\mathrm{d}\boldsymbol{r} = \mathrm{d}U - \boldsymbol{A}^{(q)}\cdot\mathrm{d}\boldsymbol{a} = đQ + đW_{\text{dis}} \geq đQ$. One can show, by a somewhat complex procedure [9, 20, 21], that $\mu\mathrm{d}\varphi = T_{\text{ext}}\mathrm{d}S$ with T_{ext} the (*thermodynamic*) *temperature* of the environment supplying the heat $đQ$ and S the (*thermodynamic*) *entropy*. Both T and S are state functions; T is independent of the scale used for θ, and one can show that for all equilibrium states $T \geq 0$ with $T = 0$ the *absolute zero*, while S is independent of the scale used for φ. Hence, from now on we will use T, S, or U as thermal state variables. The unit for T is *kelvin* (K) with 1 K = 1 °C, as used for the *Celsius* scale, where 0.01 °C ≡ 273.16 K is defined by the reliably and reproducibly measurable triple point T_{tri} of water. Thus,

$$x\,^\circ\mathrm{C} \equiv (273.15 + x)\ \mathrm{K}.$$

In summary, the *second law* states that there exists a state function $S = S^{(r)} + S^{(i)}$, called *entropy* and composed of a reversible part $S^{(r)}$ and irreversible part $S^{(i)}$, such that for a transition between two states [9, 19, 20]

$$dS \equiv dS^{(r)} + dS^{(i)} = đQ/T_{ext} + dS^{(i)} \quad \text{with} \quad dS^{(i)} \geq 0 \tag{6.2}$$

where T_{ext} is the external temperature just outside to the volume element considered (Problem 2.1). The second law, Eq. (6.2), thus expresses that *for a closed adiabatic system the entropy can increase only or remains constant at most* [6, 9, 19–22]. The condition $dS \geq đQ/T_{ext}$ is referred as the *Clausius(–Duhem) inequality*.

Since for any system the energy U is a characteristic of the state, the entropy S is a function of U and $\boldsymbol{a}$, and we write $S = S(U,\boldsymbol{a})$.[11] The entropy of a composite system is usually taken to be *additive*, that is, the sum of the entropies of the constituent subsystems, *analytical*, that is, continuously differentiable (Section A.3) and *monotonically increasing with energy*. Additivity is true if the range of interactions between particles is small as compared to the size of the system, that is, if only short-range intermolecular interactions are present. Conventionally, additivity is taken to imply *extensivity*, that is, if the extensive parameters are multiplied by λ, the entropy will obey $S(\lambda U,\lambda \boldsymbol{a}) = \lambda S(U,\boldsymbol{a})$, and the entropy is thus taken as a homogeneous function of degree 1 of the extensive parameters (Section A.3). However, this is not necessarily true but only if the effect of the surfaces of the system (surface energy) as well as the interaction of the system with the walls of the enclosure (e.g. by adsorption) can be neglected. Analyticity and monotonicity are also not always true. For example, for a phase transition it may be impossible to express the entropy in a Taylor series around the transition temperature and it may show a jump (discontinuous transition) or a discontinuity (continuous transition) in derivatives at that temperature. Normally, we accept analyticity and monotonicity.

Therefore, according to the second law, for given U and $\boldsymbol{a}$, the equilibrium state is obtained when S is maximized by the microscopic processes associated with the internal variables $\boldsymbol{b}$ of the system or $dS(U,\boldsymbol{a}) = 0$ and $d^2S(U,\boldsymbol{a}) < 0$ (Figure 6.1b). If the equality in Eq. (6.2) holds, the process is *reversible* ($dS^{(i)} = 0$), otherwise it is *irreversible* ($dS^{(i)} > 0$). Evidently, if for a process $dS = 0$, it is *isentropic*. A *natural process* (contrary to an *unnatural process*) is a process occurring in nature, which proceeds toward equilibrium, and in this sense a reversible process can be seen as the limiting case between natural and unnatural processes.

The entropy $S = S(U,\boldsymbol{a})$ is a *fundamental equation*: once S is known as a function of its *natural variables* U and $\boldsymbol{a}$, all other thermodynamic properties can be calculated. The description $S = S(U,\boldsymbol{a})$ is called the *entropy representation*. Since S is a single-valued continuously increasing function of U, the equation $S = S(U,\boldsymbol{a})$ can be inverted to $U = U(S,\boldsymbol{a})$ without ambiguity. For a given entropy S the equilibrium state is thus reached when U is minimized by the internal processes of the system or $dU(S,\boldsymbol{a}) = 0$ and $d^2U(S,\boldsymbol{a}) > 0$. The description $U = U(S,\boldsymbol{a})$ is the *energy representation*

11 The entropy S depends on the choice of the set $\boldsymbol{a}$ and therefore different choices lead to different values for S. Hence, S is not only a *conceptual property*, but also what one may call, a *contextual property*. Moreover, equilibrium depends on the choice for the set $\boldsymbol{a}$, and different choices lead to different equilibration times. Consider as an example a cup into which hot water is poured. Thermal equilibrium between cup and water is reached within, say, a minute. After about half an hour, cup and water will bet at the room temperature, whereas after a few days, the water will have evaporated. The first two situations indicate the difference in the time scale of thermal equilibration, while the last two conditions illustrate the incorporation of another variable, namely the number of molecules.

with natural variables S and $\boldsymbol{a}$, and $U(S,\boldsymbol{a})$ is also a fundamental equation. We obtain from[12] $\mathrm{d}U = (\partial U/\partial S)\mathrm{d}S + (\partial U/\partial \boldsymbol{a})\cdot \mathrm{d}\boldsymbol{a}$ the *(thermodynamic) temperature* (of the system) $T = \partial U/\partial S$ and the *(quasi-conservative) forces* (of the system) $\boldsymbol{A} = \partial U/\partial \boldsymbol{a}$.

We only mention the Nernst–Simon formulation of what is called the *third law* [23, 24]: for any isothermal process involving only phases in internal equilibrium or phases in frozen metastable equilibrium, provided the process does not disturb this frozen equilibrium,

$$\lim_{T\to 0} \Delta S = 0 \tag{6.3}$$

In Planck's formulation [4], the law states that not only the entropy differences ΔS, but also the entropy S of the components approach zero for $T \to 0\,\mathrm{K}$, allowing the introduction of an "absolute" entropy, if the convention is adopted that nuclear states and isotopic mixing do not affect the process [25]. According to the third law, $T = 0\,\mathrm{K}$ cannot be reached; the *unattainability of absolute zero* can be considered as an equivalent statement under certain conditions [19, 24]. For a general discussion, we refer to the literature [26, 27] and references therein.[13]

6.1.2 Quasi-Conservative and Dissipative Forces

At this point it may be useful to review the foregoing. Systems can be described by the coordinates $\boldsymbol{a}$ and forces $\boldsymbol{A} = \boldsymbol{A}^{(q)} + \boldsymbol{A}^{(d)}$, with $\boldsymbol{A}^{(q)}$ and $\boldsymbol{A}^{(d)}$ the quasi-conservative and dissipative forces, respectively. Using the energy change $\mathrm{d}U$, heat $đQ$, work $đW = đW_{\mathrm{rev}} + đW_{\mathrm{dis}} = (\boldsymbol{A}^{(q)} + \boldsymbol{A}^{(d)})\cdot \mathrm{d}\boldsymbol{a}$, reversible entropy change $\mathrm{d}S^{(r)} = đQ/T_{\mathrm{ext}}$, and irreversible entropy change $\mathrm{d}S^{(i)} \geq 0$, we have as main results the first and second law

$$\mathrm{d}U = đQ + đW = đQ + \boldsymbol{A}\cdot \mathrm{d}\boldsymbol{a} = T\,\mathrm{d}S + \boldsymbol{A}^{(q)}\cdot \mathrm{d}\boldsymbol{a} \quad \text{and}$$
$$\mathrm{d}S = \mathrm{d}S^{(r)} + \mathrm{d}S^{(i)} = đQ/T_{\mathrm{ext}} + \mathrm{d}S^{(i)} \quad \text{with} \quad \mathrm{d}S^{(i)} \geq 0$$

respectively, where T and T_{ext} represent the temperature of the system and just outside the system, respectively. Only for reversible systems we can associate $T\mathrm{d}S$ with $T_{\mathrm{ext}}\mathrm{d}S^{(r)} = đQ$ and in this case $đW_{\mathrm{rev}} = \boldsymbol{A}^{(q)}\cdot \mathrm{d}\boldsymbol{a}$ is the *reversible work*.[14] For irreversible systems, $T_{\mathrm{ext}}\mathrm{d}S > đQ$ and the work $đW$ contains a dissipative part. Writing

$$đW = \mathrm{d}U - đQ = \mathrm{d}U - T_{\mathrm{ext}}\,\mathrm{d}S^{(r)} = \boldsymbol{A}\cdot \mathrm{d}\boldsymbol{a} + T\,\mathrm{d}S^{(r)} + T\,\mathrm{d}S^{(i)} - T_{\mathrm{ext}}\,\mathrm{d}S^{(r)}$$

and solving this expression for $T\mathrm{d}S^{(i)}$ leads to

$$T\mathrm{d}S^{(i)} = đW_{\mathrm{dis}} + (1 - T/T_{\mathrm{ext}})\,đQ \geq 0$$

with $đW_{\mathrm{dis}} = đW - đW_{\mathrm{rev}} = \boldsymbol{A}^{(d)}\cdot \mathrm{d}\boldsymbol{a}$ the *dissipative work*, dependent on the internal variables $\boldsymbol{b}$. Hence, if T_{ext} and T differ finitely, both $đW_{\mathrm{dis}}$ and $đQ$ contribute to $T\mathrm{d}S^{(i)}$. If T_{ext} and T differ infinitesimally, $T_{\mathrm{ext}} = T + \mathrm{d}T$ and the thermal contribution vanishes to first order so that $T\mathrm{d}S^{(i)} = đW_{\mathrm{dis}}$. So, for $T_{\mathrm{ext}} = T + \mathrm{d}T$, we have

$$\begin{aligned} \mathrm{d}U &= đW + đQ = \boldsymbol{A}\cdot \mathrm{d}\boldsymbol{a} + T\mathrm{d}S^{(r)} = \boldsymbol{A}^{(q)}\cdot \mathrm{d}\boldsymbol{a} + (\boldsymbol{A}^{(d)}\cdot \mathrm{d}\boldsymbol{a} + T\mathrm{d}S^{(r)}) \\ &= \boldsymbol{A}^{(q)}\cdot \mathrm{d}\boldsymbol{a} + (T\mathrm{d}S^{(i)} + T\mathrm{d}S^{(r)}) = \boldsymbol{A}^{(q)}\cdot \mathrm{d}\boldsymbol{a} + T\mathrm{d}S = đW_{\mathrm{rev}} + T\mathrm{d}S \end{aligned} \tag{6.4}$$

12 No subscripts of variables in the partial derivatives imply that the various state functions U, S, and so on are expressed in their natural parameters, that is, $U = U(S,\mathbf{a})$ and $\partial U/\partial S = [\partial U(S,\mathbf{a})/\partial S]_{\mathbf{a}}$, etc.

13 The small booklet by Nernst [28], the originator of all this, is still delightful to read.

14 For reversible processes, heat $đQ = T\mathrm{d}S^{(r)}$ can be seen as thermal work (compare $đW_{\mathrm{rev}} = \boldsymbol{A}\cdot \mathrm{d}\boldsymbol{a}$).

known as the *Gibbs equation*. Equilibrium is obtained when U is minimal at constant S and $\boldsymbol{a}$ or when S is maximal at constant U and $\boldsymbol{a}$, and both $S(U,\boldsymbol{a})$ and $U(S,\boldsymbol{a})$ are fundamental equations. However, note that in a nonequilibrium state and for dissipative processes, the quasi-conservative forces $\boldsymbol{A}^{(q)}$ are not the only forces, and that more independent variables are required. For these situations the internal variables $\boldsymbol{b}$ relax codependent on the external variables $\boldsymbol{a}$, but in equilibrium all $\boldsymbol{b}$s have reached their optimum values and $\mathbf{b}^{(\text{equ})} = \boldsymbol{b}(\boldsymbol{a})$. In these processes the $\boldsymbol{b}$s do not contribute to $đW$ but via $\boldsymbol{A}^{(d)}(\boldsymbol{b})$ to $\mathrm{d}S^{(i)}$ and for a closed system they can be omitted.[10] This is obviously not true for a closed system in a force gradient or an open system but suggests that a nonequilibrium system can be treated as a set of connected subsystems, each in *local equilibrium*. Local equilibrium can exist for systems where the macro-changes are much slower than the micro-changes so that at any point $\boldsymbol{r}$ and time t a representative volume element (RVE, subsystem) $v = \partial V/\partial M$ (with $\mathrm{d}M$ the mass of the RVE) can be defined having a local T and P. Defining similarly $u = \partial U/\partial M$ and $s = \partial S/\partial M$, the relation $u = T\mathrm{d}s - P\mathrm{d}v$ holds [29], so that the conventional thermodynamic formulation still applies. We return to this topic in Section 6.4.

For the force $\boldsymbol{A}^{(q)}$, the adjective "quasi-" stems from the fact that $\boldsymbol{A}^{(q)}$ is dependent on the temperature while the adjective "conservative" indicates that U acts as a potential. Generally, the adjective "quasi-conservative" is not added; using only the volume V for $\mathbf{a}$, one refers to $\boldsymbol{A}^{(q)} = -P$ as pressure. We will do so likewise hereafter and use $\boldsymbol{A}^{(q)} = -P$, $\boldsymbol{a} = V$, and write $U = U(S,V)$ so that the pressure P and temperature T become $P = -\partial U/\partial V$ and $T = \partial U/\partial S$. The Gibbs equation accordingly becomes

$$\mathrm{d}U = (\partial U/\partial S)\mathrm{d}S + (\partial U/\partial V)\mathrm{d}V = T\mathrm{d}S - P\mathrm{d}V$$

6.1.3 Equations of State

Expressed in its natural variables S and V, $U(S,V)$ is a fundamental equation, just like $S(U,V)$. Again, once it is known, all other thermodynamic properties of the system can be calculated. If U is not expressed in its natural variables but, for example, as $U = U(T,V)$, we have slightly less information. In fact it follows from $T = \partial U/\partial S$ that $U = U(T,V)$ represents a differential equation for $U(S,V)$. Functional relationships such as

$$T = T(S, V) \quad \text{and} \quad P = P(S, V)$$

expressing a certain intensive state variable as a single-valued function of the remaining extensive state variables are called *equations of state* (EoS) and this intensive variable is thus described by a *state function*. If all the EoSs are known, the fundamental equation can be inferred to within an arbitrary constant. To that purpose we note that, since U is an extensive quantity, $U(S,V)$ can be taken as homogeneous of degree 1 and thus, using Euler's theorem (Section A.3), that

$$U = \frac{\partial U}{\partial S}S + \frac{\partial U}{\partial V}V = TS - PV \tag{6.5}$$

Given $T = T(S,V)$ and $P = P(S,V)$, the energy U can be calculated. A similar argumentation for S leads to $S = (1/T)U + (P/T)V$. Only in some special cases, for example, if $\mathrm{d}S(U,V) = f(U)\mathrm{d}U + g(V)\mathrm{d}V$, the differential can be integrated directly to obtain the fundamental equation. Equation (6.5) is often referred to as an *Euler equation*.

6.1.4 Mechanical and Thermal Equilibrium

Consider an isolated system consisting of two subsystems only capable of exchanging heat ($\mathrm{d}V_i = 0$). We now ask under what conditions thermal equilibrium occurs, and to this purpose we consider arbitrary but small variations in energy in both subsystems. The first law says $\mathrm{d}U = \mathrm{d}U_1 + \mathrm{d}U_2 = 0$. If the system is in *thermal equilibrium*, since S is maximal, we also have from the second law $\mathrm{d}S = \mathrm{d}S_1 + \mathrm{d}S_2 = 0$. Because

$$\mathrm{d}S = \frac{\partial S_1}{\partial U_1}\mathrm{d}U_1 + \frac{\partial S_2}{\partial U_2}\mathrm{d}U_2 = \left(\frac{\partial S_1}{\partial U_1} - \frac{\partial S_2}{\partial U_2}\right)\mathrm{d}U_1 = \left(\frac{1}{T_1} - \frac{1}{T_2}\right)\mathrm{d}U_1 = 0$$

and $\mathrm{d}U_1$ is arbitrary, we obtain $T_1 = T_2$ in agreement with the zeroth law. Starting from a nonequilibrium state and using $\mathrm{d}S \geq 0$, we have for $T_1 > T_2$, $\mathrm{d}U_1 < 0$, and energy (heat) flows from subsystem 1 to subsystem 2. For $T_1 < 0$, we have $\mathrm{d}U_1 > 0$ so in all cases heat flows from high to low temperature, in agreement with intuition [11].

The procedure outlined above is in fact the general method to deal with basic equilibrium thermodynamic problems: one considers an isolated system consisting of two subsystems of which one is the object of interest and the other represents the environment. For the total isolated system, $\mathrm{d}U = 0$ and in equilibrium $\mathrm{d}S = 0$. Using the *closure relations* – the coupling relations between the displacements a_i of the subsystem of interest and of the environment – a solution can be obtained.

Considering hydrostatic equilibrium along these lines, we assume again an isolated system consisting of two (homogeneous) subsystems but now capable of exchanging heat and work. For this system to be in equilibrium we write again $\mathrm{d}S = \mathrm{d}S_1 + \mathrm{d}S_2 = 0$. However, the entropy S is now a function of U and the various a_i. Being interested in hydrostatic equilibrium, we take just the volume V for each subsystem. From the Gibbs equation $\mathrm{d}U = T\,\mathrm{d}S - P\,\mathrm{d}V$ we obtain

$$\mathrm{d}S = (1/T)\mathrm{d}U + (P/T)\mathrm{d}V$$

and thus for the equilibrium configuration

$$\mathrm{d}S = \mathrm{d}S_1 + \mathrm{d}S_2 = (1/T_1)\mathrm{d}U_1 + (P_1/V_1)\mathrm{d}V_1 + (1/T_2)\mathrm{d}U_2 + (P_2/V_2)\mathrm{d}V_2 = 0$$

Because the total system is isolated, we have the closure relations $\mathrm{d}U = \mathrm{d}U_1 + \mathrm{d}U_2 = 0$ and $\mathrm{d}V = \mathrm{d}V_1 + \mathrm{d}V_2 = 0$. Therefore, since $\mathrm{d}U_1$ and $\mathrm{d}V_1$ are arbitrary, we regain thermal equilibrium $T_1 = T_2$ and

$$P_1/T_1 = P_2/T_2 \quad \text{or using } T_1 = T_2, \quad P_1 = P_2$$

The latter condition corresponds to *hydrostatic equilibrium*. Hence, in hydrostatic equilibrium the temperature and the pressure of the subsystems are equal.

6.1.5 Auxiliary Functions

The Gibbs equation for a closed system is given by

$$\mathrm{d}U = T\,\mathrm{d}S - P\,\mathrm{d}V$$

and yields the dependent variable U as a function of the independent, natural variables S and V. From consideration of the second law, the criterion for equilibrium for a closed system with fixed composition is that $\mathrm{d}S(U,V) = 0$ or S is a maximum at constant U. Equivalently, as stated earlier, $\mathrm{d}U(S,V) = 0$ or U is a minimum at constant S. Both descriptions are complete but not very practical, since it is difficult to keep the entropy and energy for a system constant, to say the least. Therefore, auxiliary functions having more practical natural variables are introduced.

If we write the energy as $U(S,V)$, the *Helmholtz energy* F is the Legendre transform[15] of U with respect to $T = \partial U/\partial S$, which is obtained from

$$F \equiv U - TS \tag{6.6}$$

After differentiation and combining with $\mathrm{d}U = T\mathrm{d}S - P\mathrm{d}V$, the expression for F yields

$$\mathrm{d}F = -S\mathrm{d}T - P\mathrm{d}V$$

The natural variables for the Helmholtz energy F are thus T and V, and the equilibrium condition becomes $\mathrm{d}F(T,V) = 0$. A second transform of U, now with respect to $P = -\partial U/\partial V$, is the *enthalpy* H, defined by

$$H \equiv U + PV \quad \text{and resulting in } \mathrm{d}H = T\,\mathrm{d}S + V\,\mathrm{d}P \tag{6.7}$$

with natural variables S and P, and the equilibrium condition $\mathrm{d}H(S,P) = 0$. Similarly, the *Gibbs energy* G is the Legendre transform of U with respect to both $T = \partial U/\partial S$ and $P = -\partial U/\partial V$ (or of H with respect to $T = \partial H/\partial S$) and given by

$$G \equiv H - TS = U + PV - TS \quad \text{resulting in} \quad \mathrm{d}G = -S\mathrm{d}T + V\mathrm{d}P \tag{6.8}$$

with natural variables T and P. Consequently, the equilibrium condition becomes $\mathrm{d}G(T,P) = 0$. It follows that

$$T = \frac{\partial U}{\partial S} = \frac{\partial H}{\partial S} \quad \text{and} \quad S = -\frac{\partial F}{\partial T} = -\frac{\partial G}{\partial T}$$

$$-P = \frac{\partial U}{\partial V} = \frac{\partial F}{\partial V} \quad \text{and} \quad V = \frac{\partial H}{\partial P} = \frac{\partial G}{\partial P}$$

The functions $U(S,V)$, $H(S,P)$, $F(T,V)$, and $G(T,P)$ thus act as *thermodynamic potentials*, similar as the potential energy in mechanics, and are also fundamental equations. The derivatives with respect to their natural, independent variables yield the conjugate, dependent variables. A stable equilibrium state is reached when these potentials are minimized for a given set of their natural variables, that is, when $\mathrm{d}X = 0$ and when $\mathrm{d}^2X > 0$, where X is any of the four functions U, H, F, or G. Processes can only occur when $\mathrm{d}X < 0$. The advantage of using F and G is obvious: while it is possible to control experimentally either the set (V,T) or (P,T), control of either the set (V,S) or (P,S) is possible, in principle, by extracting heat, but at least impractical and possibly insufficient. As an aside, we note that Legendre transforms can also be applied to the entropy yielding *Massieu functions*, acting as potentials as well, for example, $X(V,T) = S - U/T$ or $Y(P,T) = S - U/T - PV/T$. It will be clear that X is related to F and Y to G. In practice, although invented before the transforms of the energy, these functions are nowadays limitedly used in MT.

15 A clear discussion of Legendre transforms is given by Callen [11]. See also Section A.5. I use here F for the Helmholtz energy, contrary to the IUPAC recommendation A, and use A for area.

6.1.6 Some Derivatives and their Relationships

We also need some derivatives of the potentials, frequently denoted as *response functions*, and their mutual relationships. Consider the following relations:

$$đQ = T\mathrm{d}S = \mathrm{d}U + P\mathrm{d}V \quad \text{and} \quad đQ = T\mathrm{d}S = \mathrm{d}H - V\mathrm{d}P$$

The *heat capacities*, defined by $C_A = (\partial Q/\partial T)_A$ ($A = V$ or P) are thus given by

$$C_V = T(\partial S/\partial T)_V = (\partial U/\partial T)_V \quad \text{and} \quad C_P = T(\partial S/\partial T)_P = (\partial H/\partial T)_P$$

Other derivatives that occur frequently are the *compressibilities* κ_B ($B = T$ or S) and the *thermal expansion coefficient* or *expansivity* $\alpha \equiv \alpha_P$ reading

$$\kappa_T = -(1/V)(\partial V/\partial P)_T, \quad \kappa_S = -(1/V)(\partial V/\partial P)_S \quad \text{and} \quad \alpha_P = (1/V)(\partial V/\partial T)_P$$

The inverse of the compressibility, $K = \kappa^{-1}$ and called *bulk modulus*, is also often used. Moreover, if $\mathrm{d}\phi = X\mathrm{d}x + Y\mathrm{d}y$ is a *total differential*, we can use the *Maxwell relations* $(\partial X/\partial y)_x = (\partial Y/\partial x)_y$. From the expressions for $\mathrm{d}U$, $\mathrm{d}H$, $\mathrm{d}F$, and $\mathrm{d}G$ we obtain

$$(\partial T/\partial V)_S = -(\partial P/\partial S)_V \qquad (\partial V/\partial S)_P = (\partial T/\partial P)_S$$

$$(\partial P/\partial T)_V = (\partial S/\partial V)_T \qquad (\partial S/\partial P)_T = -(\partial V/\partial T)_P$$

which can be used to reduce any thermodynamic quantity to a set of measurable quantities. Only three of the quantities C_V, C_P, κ_T, κ_S, and α are independent because

$$\kappa_T - \kappa_S = TV\alpha^2/C_P \quad \text{and} \quad C_P - C_V = TV\alpha^2/\kappa_T \tag{6.9}$$

For condensed matter one often assumes that $C_P \cong C_V$, but $C_P - C_V$ can amount up to $0.2C_V$ [30]. According to the third law, for any variable a, $(\partial S/\partial a)_T$ and $(\partial S/\partial a)_P \to 0$ for $T \to 0$ K, it follows from the appropriate Maxwell relation that $(\partial a/\partial T)_{Z'}$ and $(\partial a/\partial P)_{Z'} \to 0$. Therefore, when $T \to 0$ K, C_P, C_V, and $\alpha \to 0$, as has been amply verified [31]. Also, for the surface tension γ, $\mathrm{d}\gamma/\mathrm{d}T \to 0$. For ^{4}He, the only substance that remains liquid at 1 bar when $T \to 0$ K, this has been verified experimentally [32]. Finally, using $S = -\partial F/\partial T$ and $S = -\partial G/\partial T$, we have the *Gibbs–Helmholtz equations*

$$U = F + TS = F - T\frac{\partial F}{\partial T} = \frac{\partial(F/T)}{\partial(1/T)} \text{ and } H = G + TS = G - T\frac{\partial G}{\partial T} = \frac{\partial(G/T)}{\partial(1/T)} \tag{6.10}$$

6.1.7 Chemical Content

The content of a system is defined by the amount of its *components*, that is, the amount of moles n_α of chemical species α in the system. A *mixture* is a system with more than one component. A homogeneous part of a mixture, that is, a part with uniform properties, is addressed as a *phase*. A multicomponent phase is labeled a *solution* with as majority component the *solvent*, while *solute* refers to the minority component. A system consisting of a single phase is called *homogeneous*, while one consisting of more than one phase is labeled *heterogeneous*. A phase can be *isotropic* or *anisotropic* (properties independent or dependent on direction within that phase).

For chemical problems, where a change in composition is involved, the fundamental equation becomes $U = U(S,V,n_\alpha)$ or $S = S(U,V,n_\alpha)$. This leads to

$$\mathrm{d}U = \frac{\partial U}{\partial S}\,\mathrm{d}S + \frac{\partial U}{\partial V}\,\mathrm{d}V + \sum\nolimits_\alpha \frac{\partial U}{\partial n_\alpha}\,\mathrm{d}n_\alpha = T\,\mathrm{d}S - P\,\mathrm{d}V + \sum\nolimits_\alpha \mu_\alpha\,\mathrm{d}n_\alpha$$

The partial derivative $\mu_\alpha = \partial U/\partial n_\alpha$, called the *chemical potential*, is the conjugate intensive variable associated with the extensive variable n_α in the *chemical work* $đW_{\text{che}} = \Sigma_\alpha \mu_\alpha \text{d}n_\alpha$. Equilibrium is now obtained for closed systems, starting out of equilibrium, and open systems, where the amount of components change, when $\text{d}U(S,V,n_\alpha) = 0$. Note that in open systems heat and work are ambiguous terms [8, 33].

For obtaining a potential in terms of T and V, we use a Legendre transformation of U with respect to $T = \partial U/\partial S$ using the Helmholtz energy $F = U - TS$ and obtain

$$\text{d}F(T,V,n_\alpha) = -S\text{d}T - P\text{d}V + \sum\nolimits_\alpha \mu_\alpha \text{d}n_\alpha$$

Hence, $F = F(T,V,n_\alpha)$ and for the equilibrium condition we have $\text{d}F(T,V,n_\alpha) = 0$.

Applying a Legendre transformation of U with respect to $T = \partial U/\partial S$ and $P = -\partial U/\partial V$ using the Gibbs energy $G = U - TS + PV$ leads to

$$\text{d}G = \text{d}(U - TS + PV) = -S\,\text{d}T + V\,\text{d}P + \sum\nolimits_\alpha \mu_\alpha\,\text{d}n_\alpha$$

Therefore, $G = G(T,P,n_\alpha)$ and the equilibrium condition becomes $\text{d}G(T,P,n_\alpha) = 0$.

Finally, we note that we can apply a Legendre transformation to U with respect to $T = \partial U/\partial S$ and $\mu_\alpha = \partial U/\partial n_\alpha$ by using $\Omega = U - TS - \Sigma_\alpha \mu_\alpha n_\alpha = F - \Sigma_\alpha \mu_\alpha n_\alpha$ leading to

$$\text{d}\Omega = -S\,\text{d}T - P\,\text{d}V + \sum\nolimits_\alpha n_\alpha\,\text{d}\mu_\alpha$$

The so-called *grand potential* Ω is mainly used in statistical thermodynamics (Section 7.1.2). Since $G = U - TS + PV = \Sigma_\alpha \mu_\alpha n_\alpha$, we identify the grand potential as

$$\Omega = F - G = -PV \tag{6.11}$$

For chemical solutions, it is useful to discuss the situation with respect to extensive quantities $Z = Z(P,T,n_\alpha)$ in terms of *partial (molar) quantities* $Z_\alpha \equiv (\partial Z/\partial n_\alpha)_{P,T,n'}$, withe n' indicating all components except component α.[16] Generally, for Z we have

$$\text{d}Z = (\partial Z/\partial P)\text{d}P + (\partial Z/\partial T)\text{d}T + \sum\nolimits_\alpha Z_\alpha\,\text{d}n_\alpha \tag{6.12}$$

Since the extensive quantity $Z(P,T,n_\alpha)$ is a homogeneous function of degree 1 in n_α, we have by Euler's theorem

$$Z = \sum\nolimits_\alpha Z_\alpha n_\alpha \tag{6.13}$$

So, we have for dZ on the one hand Eq. (6.12), and on the other hand differentiating Eq. (6.13) $\text{d}Z = \Sigma_\alpha Z_\alpha \text{d}n_\alpha + \Sigma_\alpha n_\alpha \text{d}Z_\alpha$. Hence,

$$(\partial Z/\partial P)\text{d}P + (\partial Z/\partial T)\text{d}T = \sum\nolimits_\alpha n_\alpha\,\text{d}Z_\alpha \quad \text{or} \quad \sum\nolimits_\alpha n_\alpha\,\text{d}Z_\alpha = 0 \ (\textit{const.}\ T, P) \tag{6.14}$$

This is the (*general*) *Gibbs–Duhem equation*, which puts a constraint on possible changes of a system.

Taking for Z the volume V, the partial volume is given by $V_\alpha = (\partial V/\partial n_\alpha)_{P,T,n'}$ and the total volume becomes at constant T and P, using Euler's theorem,

$$V_\text{m} = x_1 V_1 + x_2 V_2$$

---

16 I use Z_m for a *molar quantity* but often the subscript m is omitted if the meaning is clear from the context. For arbitrary values of n_α, $Z = n_\alpha Z_\alpha$ holds exactly for any Z but it is sometimes approximated by $Z = n_\alpha Z_\alpha^*$, in particular for the volume V. Note that for a pure component $Z_\alpha = Z_\alpha^*$.

with *mole fraction* $x_\alpha = n_\alpha/\sum_\alpha n_\alpha = n_\alpha/n$ and $n \equiv \sum_\alpha n_\alpha$. Similarly, for the Gibbs energy G, we have the partial Gibbs energy $G_\alpha = (\partial G/\partial n_\alpha)_{P,T,n'}$, and since this quantity equals the chemical potential defined by $\mu_\alpha \equiv (\partial G/\partial n_\alpha)_{P,T,n'}$, the total Gibbs energy becomes

$$G_m = x_1 G_1 + x_2 G_2 = x_1\mu_1 + x_2\mu_2 \tag{6.15}$$

Note that the partial quantity G_α and chemical potential μ_α are equivalent, but that for the Helmholtz energy F one would have $\mu_\alpha = (\partial F/\partial n_\alpha)_{V,T,n'} \neq (\partial F/\partial n_\alpha)_{P,T,n'} = F_\alpha$.

For the Gibbs energy G we further obtain at constant P and T

$$x_1 \mathrm{d}\mu_1 + x_2 \mathrm{d}\mu_2 = 0$$

This is the *Gibbs–Duhem equation* most frequently used and of which another form is

$$\mathrm{d}\mu_1 = -(x_2/x_1)\mathrm{d}\mu_2 \quad \text{or} \quad \mu_1 - \mu_1^* = -\int_0^{\mu_2} \frac{x_2}{1-x_2}\mathrm{d}\mu_2 = -\int_0^{x_2} \frac{x_2}{1-x_2}\frac{\mathrm{d}\mu_2}{\mathrm{d}x_2}\mathrm{d}x_2$$

where $\mu_1{}^*$ is the chemical potential of the pure component 1. This relation can be used for a consistency check on the experimental data. If the molar Gibbs energy G_m is known, we can obtain μ_1 and μ_2 by differentiation with respect to x_2, since

$$\frac{\partial G_m}{\partial x_2} = \frac{\partial}{\partial x_2}(x_1\mu_1 + x_2\mu_2) = \mu_2 - \mu_1 \tag{6.16}$$

and solving μ_1 and μ_2 from Eqs. (6.15) and (6.16) results in

$$\mu_1 = G_m + x_2\frac{\partial G_m}{\partial x_1} = G_m - x_2\frac{\partial G_m}{\partial x_2} \quad \text{and} \quad \mu_2 = G_m + x_1\frac{\partial G_m}{\partial x_2} = G_m - x_1\frac{\partial G_m}{\partial x_1}$$

Thus, if the dependence of G_m as a function of x_2 (or x_1 for that matter) is known (Figure 6.2a), the behavior of μ_1 and μ_2 can be obtained by differentiation. It will be clear that a similar reasoning applies to all partial quantities, for example, V_m.

Furthermore, consider a system with c components and p phases at constant P and T. For such a system only $c-1$ components can be varied independently for each phase at the given P and T, leading to $p(c-1)+2$ independent variables. However, in equilibrium, we also have $\mu_j^{(1)} = \ldots = \mu_j^{(p)}$, $j = 1 \ldots c$, leading to $c(p-1)$ restrictions. The total number of independent variables, the number of *degrees of freedom* f, which can be changed without changing the number of phases is given by the *phase rule* $f = p(c-1)+2-c(p-1) = c-p+2$.

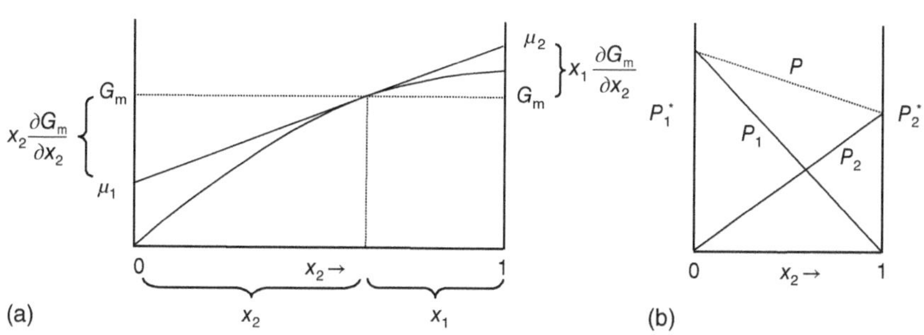

Figure 6.2 (a) The calculation of the chemical potentials μ_1 and μ_2 from the Gibbs energy G_m. Although illustrated here for μ_1 and μ_2 using G_m, this so-called *Bakhuis Roozeboom construction* can be done for any partial quantity remembering that μ_α equals G_α. (b) The pressure diagram for a perfect mixture showing the total pressure P and the partial pressures P_1 and P_2.

Finally, we note that, besides the number of moles n_α, we will also use the number of molecules $N_\alpha = N_A n_\alpha$ with N_A Avogadro's number. For composition per unit volume, we use, apart from the mole fraction $x_\alpha = N_\alpha / \sum_\alpha N_\alpha$, the *molarity* (or *molar density*) $c_\alpha \equiv n_\alpha / V$ with V the total volume of the solution. Since $\rho V = (M/V)V\,\Sigma_\alpha n_\alpha \mathcal{M}_\alpha$ with ρ the mass density[17] of the solution, M the total mass and $\mathcal{M}_\alpha$ the molar mass, we have $c_\alpha = \rho n_\alpha / \Sigma_\alpha n_\alpha \mathcal{M}_\alpha$. Using x_α again, we obtain

$$\frac{c_\alpha}{x_\alpha} = \frac{\rho n}{\sum_\alpha n_\alpha \mathcal{M}_\alpha} \qquad \text{or} \qquad \frac{c_2}{x_2} = \frac{\rho(n_1 + n_2)}{n_1 \mathcal{M}_1 + n_2 \mathcal{M}_2} = \frac{\rho n}{n_1 \mathcal{M}_1 + n_2 \mathcal{M}_2}$$

for a binary system with index 1 referring to the solvent and index 2 to the solute. For dilute solutions, $n_{\alpha \neq 1} \to 0$ and we obtain $c_\alpha / x_\alpha = \rho / \mathcal{M}_1$ $(c_2/x_2 = \rho/\mathcal{M}_1)$. If c_α is expressed in mol l^{-1}, labeled as M, one often speaks of *concentration*, for example, for $c_\alpha = 0.1$ mol l^{-1}, one writes 0.1 M. Experimentally, it is expedient to use as the other independent variables P and T, so that differentiation with respect to T implies constant P and vice versa. From c_α we obtain the relation $\partial c_\alpha / \partial T = -c_\alpha \alpha_P$ with α_P the thermal expansivity and therefore c_α is a function of T. This renders a direct T-dependency calculation using c_α as independent variable often cumbersome.

6.1.8 Chemical Equilibrium

At constant P and T, the Gibbs energy in a chemically reacting system varies with composition as $\mathrm{d}G = \Sigma_\alpha \mu_\alpha \mathrm{d}n_\alpha$. For a reaction to be able to occur $\mathrm{d}G = \Sigma_\alpha \mu_\alpha \mathrm{d}n_\alpha < 0$ while at equilibrium[18] $\mathrm{d}G = \Sigma_\alpha \mu_\alpha \mathrm{d}n_\alpha = 0$. Let us consider the reaction

$$a\mathrm{A} + b\mathrm{B} \leftrightarrow c\mathrm{C} + d\mathrm{D} \tag{6.17}$$

where $a, \ldots, d$ denote the stoichiometric coefficients of components A, …, D. It will be convenient to rearrange this expression to

$$0 = c\mathrm{C} + d\mathrm{D} - a\mathrm{A} - b\mathrm{B} \qquad \text{or, even more compact, to} \qquad 0 = \sum\nolimits_\alpha \nu_\alpha \mathrm{X}$$

with a positive value for the coefficient ν_α when α is a product (C, D) and a negative value when α is a reactant (A, B). We define a factor of proportionality $\mathrm{d}\xi(t)$ in such a way that $\mathrm{d}n_\alpha = \nu_\alpha \mathrm{d}\xi$. Starting at time $t = 0$ with initially $n_\alpha(0)$ moles of each species, the changes of the number of moles of each species in the time interval $\mathrm{d}t$ are

$$\mathrm{d}n_\alpha = \nu_\alpha \frac{\mathrm{d}\xi}{\mathrm{d}t} \mathrm{d}t \equiv \nu_\alpha \dot{\xi}\, \mathrm{d}t \qquad \text{or} \qquad n_\alpha = n_\alpha(0) + \int \mathrm{d}n_\alpha = n_\alpha(0) + \nu_\alpha \Delta\xi$$

where $\dot{\xi}$ is the *extent of reaction*. Defining the *affinity* $\mathcal{A} \equiv -\Sigma_\alpha(\nu_\alpha \mu_\alpha)$, this leads to

$$\mathrm{d}G = \sum\nolimits_\alpha \mu_\alpha \mathrm{d}n_\alpha = \sum\nolimits_\alpha (\mu_\alpha \nu_\alpha)\dot{\xi}\, \mathrm{d}t = -\mathcal{A}\,\dot{\xi}\, \mathrm{d}t \leq 0$$

From $\mathrm{d}G \leq 0$, we conclude that a reaction can occur only if the *chemical rate of work*

$$\mathcal{A}\,\dot{\xi} > 0 \tag{6.18}$$

17 A *density* is a quantity expressed per unit volume, for example, using the same symbol, the number density $\rho = N/V = N_A \Sigma_\alpha n_\alpha / V$. If confusion is possible, I use $\rho' = M/V$ for mass density.

18 Chemical equilibrium conditions can also be derived similarly as those for thermal equilibrium and mechanical equilibrium.

while at equilibrium $\mathscr{A} = 0$. So, for a single reaction, $\mathscr{A}$ and $\dot{\xi}$ must have the same sign or be zero. Since $\nu_\alpha = \mathrm{d}n_\alpha/\mathrm{d}\xi$, the affinity $\mathscr{A}$ is related to the fundamental equations, the most important ones being, using $\mu_\alpha = \partial U/\partial n_\alpha = \partial G/\partial n_\alpha = -T\,\partial S/\partial n_\alpha$,

$$\mathscr{A} = -\sum\nolimits_\alpha \nu_\alpha \mu_\alpha = -\frac{\partial U(S,V,n_\alpha)}{\partial \xi} = -\frac{\partial G(T,P,n_\alpha)}{\partial \xi} = T\frac{\partial S(U,V,n_\alpha)}{\partial \xi}$$

For n reactions labeled $k = 1, \ldots, n$, we similarly have $\mathscr{A}_k$ and $\dot{\xi}_k$. In this case, $\mathrm{d}G \le 0$ leads to $\mathscr{A} \cdot \dot{\xi} = \sum_k^n \mathscr{A}_k \dot{\xi}_k \ge 0$, so that, if the reactions are coupled, we can have $\sum_l^m \mathscr{A}_l \dot{\xi}_l \le 0$ for m reactions with $m < n$ as long as $\mathscr{A} \cdot \dot{\xi} \ge 0$.

Justification 6.1 ***The Chemical Potential*** To calculate $\mathscr{A}$, the functional behavior of μ should be known. For a *perfect gas* the EoS is $PV = nRT$, and with $V = \partial G/\partial P$ we obtain, for arbitrary number of moles n,

$$G = G^\circ(T) + nRT\ln(P/P^\circ)] = n[\mu^\circ(T) + RT\ln(P/P^\circ)] \equiv n\mu(T,P) \qquad \text{and}$$
$$F = G - PV = n[\mu^\circ(T) - RT + RT\ln(nRT/VP^\circ)]$$

where P° is a reference pressure (1 bar) and μ° the value for $\mu = G_\mathrm{m} = G/n$ if $P = P^\circ$.

For *perfect gas mixtures*, because at constant T and V and sufficiently low density each species behaves experimentally largely independent of the other species, we use that each component α contributes independently from the others to F, that is, $F_{T,V} = \Sigma_\alpha (F_j)_{T,V}$, sometimes labeled as *Gibbs' theorem*. Hence,

$$F_\mathrm{mix} = \sum\nolimits_\alpha n_\alpha \left[\mu_\alpha^\circ(T) - RT + RT\ln(n_\alpha RT/VP^\circ)\right] \qquad \text{and}$$
$$G_\mathrm{mix} = \sum\nolimits_\alpha n_\alpha \left[\mu_\alpha^\circ(T) + RT\ln(x_\alpha P/P^\circ)\right] = \sum\nolimits_\alpha n_\alpha \left[\mu_\alpha^\circ(T) + RT\ln(P_\alpha/P^\circ)\right]$$

where $P_\alpha \equiv x_\alpha P$ is referred to as the *partial pressure*. The total pressure is given by $P = \Sigma_\alpha P_\alpha$, known as *Dalton's law of partial pressures*.[19] To calculate $\Delta_\mathrm{mix}G$ for a binary gas mixture with partial pressure P_1 and P_2, we subtract $n_1\mu_1$ and $n_2\mu_2$, initially at P_1' and P_2', respectively, from G_mix to obtain

$$\Delta_\mathrm{mix}G = n_1 RT\ln\left(P_1/P_1'\right) + n_2 RT\ln\left(P_2/P_2'\right)$$

In the special case that $P_1' = P_2' = P'$, that is, when the original pressures of each gas P_1' and P_2' are equal to the final pressure P', we have $P_\alpha/P_\alpha' = P_\alpha/P' = x_\alpha$ and

$$\Delta_\mathrm{mix}G = RT(n_1\ln x_1 + n_2\ln x_2) \qquad \text{and} \qquad \Delta_\mathrm{mix}S = -R(n_1\ln x_1 + n_2\ln x_2)$$

A *perfect solution* is defined as having the same $\Delta_\mathrm{mix}G$ as a perfect gas mixture for the whole composition range. Using a superscript * for pure substances, we have

$$\Delta_\mathrm{mix}\mu = RT(x_1\ln x_1 + x_2\ln x_2) \qquad \text{or} \qquad \mu_\alpha(T,P) = \mu_\alpha^*(T,P) + RT\ln x_\alpha \tag{6.19}$$

Equation (6.19) forms the basis for all further treatments of solutions. In practice, perfect solution behavior is often obeyed only for a limited range of concentrations, and we label

---

19 Note that the unfortunate but conventional name partial pressure refers *not* to a partial quantity. *Amagat's law* $V_{T,P} = \Sigma_\alpha x_\alpha V_\alpha^*$ is comparable to Dalton's law but considerably less accurate.

such a solution as an *ideal solution*. For perfect and ideal solutions (within the range of validity) and the perfect gas we have, respectively, the expressions

$$\mu_\alpha^{\rm L}(T,P) = \mu_\alpha^*(T,P) + RT\ln x_\alpha \quad \text{and} \quad \mu_\alpha^{\rm V}(T,P) = \mu_\alpha^\circ(T) + RT\ln P_\alpha$$

leading to, using the equilibrium condition $\mu_\alpha{}^{\rm L} = \mu_\alpha{}^{\rm V}$,

$$\mu_\alpha^* + RT\ln x_\alpha = \mu_\alpha^\circ + RT\ln P_\alpha \quad \text{or} \quad K_{{\rm H},\alpha} = P_\alpha/x_\alpha = \exp\left[\left(\mu_\alpha^* - \mu_\alpha^\circ\right)/RT\right]$$

The parameter $K_{{\rm H},\alpha}$ is independent of composition. For the solute we thus have $P_2 = K_{{\rm H},2}x_2$, known as *Henry's law* (1803!), stating that the partial pressure P_2 of solute 2 is proportional to its mole fraction in the liquid phase. For the solvent, the limiting situation $x_1 = 1$ yields $K_{{\rm H},1} = P_1{}^*$. Hence approximately $P_1 = P_1^* x_1$, known as *Raoult's law* (1887), stating that the partial pressure P_1 of solvent 1 is directly proportional to the vapor pressure $P_1{}^*$ of the pure solvent.

To describe deviations from perfect solution behavior one uses the (*relative*) *activity* $a_\alpha = \gamma_\alpha x_\alpha$, where the *activity coefficient* $\gamma_\alpha = \gamma_\alpha(T,x)$ is introduced to keep the same formal expression for μ_α as for perfect solutions. Hence, $\mu_\alpha = \mu_\alpha{}^* + RT \ln x_\alpha$ becomes $\mu_\alpha = \mu_\alpha{}^* + RT \ln a_\alpha$. Two conventions for γ_α are in use:

- Convention I: $\gamma_\alpha \to 1$ for $x_\alpha \to 1$ and for $x_\alpha \to 0$. This convention is usually employed if the solute is fully soluble in the solvent. Because in a dilute solution of component β in α, molecules of type α are mainly surrounded by molecules of type α, component α behaves largely as if the component was pure. The reference state is accordingly the pure component with chemical potential $\mu_\alpha{}^*$.
- Convention II: $\gamma_1 \to 1$ for $x_1 \to 1$ and $\gamma_2 \to 1$ for $x_2 \to 0$. This convention is usually employed if the solute is only partially soluble. Because in a dilute solution of component 2 in 1, molecules of type 2 hardly interact with each other but mainly with molecules of type 1, the activity coefficient approaches a constant value, characteristic for the 1–2 interaction. Since we would like to have $\gamma_2 \to 1$ for $x_2 \to 0$, the reference state for the solute is taken as the pure solute in a hypothetical state extrapolated to infinite dilution using Henry's law.

The abovementioned formulation leads immediately to the conventional chemical description. Let us introduce for component α the (*absolute*) *activity* $\lambda_\alpha = \exp(\mu_\alpha/RT)$ and define, considering again the reaction given by Eq. (6.17), the *reaction product*

$$\lambda_{\rm C}{}^c\lambda_{\rm D}{}^d/\lambda_{\rm A}{}^a\lambda_{\rm B}{}^b \quad \text{or} \quad \prod\nolimits_\alpha \lambda_\alpha{}^{\nu_\alpha}$$

using the compact notation introduced before. The equilibrium condition $\mathscr{A} = -\Sigma_\alpha \nu_\alpha \mu_\alpha = 0$ can then be written as $\prod_\alpha \lambda_\alpha^{\nu_\alpha} = 1$.

Now we distinguish between gases (α) and condensed phases (β). This allows us to write $\prod_\alpha \lambda_\alpha^{\nu_\alpha} \prod_\beta \lambda_\beta^{\nu_\beta} = 1$, where the first product contains all the terms relating to gaseous species and the second to the condensed species. Now note that the chemical potential μ_α of a gaseous component α is given by $\mu_\alpha = \mu_\alpha{}^\circ + RT \ln y_\alpha/P^\circ$, where $\mu_\alpha{}^\circ$ is the chemical potential in the standard state, $y_\alpha = \gamma_\alpha x_\alpha P$ is the (*relative*) *activity* (for gases often denoted as *fugacity*), γ_α is the (*mole fraction*) *activity coefficient*, P is the total pressure, and P° is the standard pressure (1 bar, for the details of changing from 1 atm to 1 bar, see [34]). Hence, for a gas $\lambda_\alpha = \exp(\mu_\alpha/RT) = \lambda_\alpha{}^\circ y_\alpha/P^\circ$, where $\lambda_\alpha{}^\circ$ is the value of λ_α for $P = P^\circ$. For gases at low pressures (hence activity coefficient $\gamma_\alpha = 1$), the activity becomes the *partial pressure* P_α, given by $P_\alpha = x_\alpha P$. For pure condensed phases, $\mu_\beta = \mu_\beta{}^\circ + \int_{P_0}^{P} V_\beta \mathrm{d}P \cong \mu_\beta{}^\circ$, because the integral, the *Poynting term*, is small. Hence, $\lambda_\beta \cong \lambda_\beta{}^\circ$. In total, we have

$$\prod\nolimits_\alpha (\lambda_\alpha{}^\circ)^{\nu_\alpha} x_\alpha^{\nu_\alpha} (P^\circ)^{-\nu_\alpha} \prod\nolimits_\beta (\lambda_\beta{}^\circ)^{\nu_\beta} = 1 \quad \text{or}$$

$$\prod\nolimits_\alpha x_\alpha^{\nu_\alpha} = K_P \quad \text{with} \quad K_P \equiv \prod\nolimits_\alpha (\lambda_\alpha{}^\circ)^{-\nu_\alpha} (P^\circ)^{\nu_\alpha} \prod\nolimits_\beta (\lambda_\beta{}^\circ)^{-\nu_\beta}$$

The (*pressure*) *equilibrium constant* K_P is related to the standard affinity

$$-\mathscr{A} = c\mu_\mathrm{C}{}^\circ + d\mu_\mathrm{D}{}^\circ - a\mu_\mathrm{A}{}^\circ - b\mu_\mathrm{B}{}^\circ \quad \text{or, more generally,} \quad \mathscr{A} = -\sum\nolimits_\alpha \nu_\alpha \mu_\alpha{}^\circ$$

via $\mathscr{A} = RT \ln K_P$. Since $\mathscr{A}$ is constant at constant temperature, the value of K_P is constant at constant temperature, which explains the name. Slightly more general, in terms of the mole fraction, we have

$$\prod\nolimits_\alpha (n_\alpha)^{\nu_\alpha} \prod\nolimits_\alpha (\gamma_\alpha)^{\nu_\alpha} = K_n \quad \text{with} \quad K_n \equiv \prod\nolimits_\alpha (\lambda_\alpha^\circ)^{-\nu_\alpha} = P^{-\nu} K_P$$

where K_n is the (*mole fraction*) *equilibrium constant* and $\nu = \Sigma_\alpha \nu_\alpha$. As many types of equilibrium constants are used [19], one should be careful to use them consistently.

Problem 6.1 ***Entropy*** Show that in Eq. (6.2), $T_\mathrm{ext}đQ$ instead of, possibly expected, $TđQ$ should be used by considering an isolated system composed of a (sub)system of interest and a reservoir.

Problem 6.2 ***The Perfect Gas Entropy*** For a perfect gas defined by $PV = nRT$, show that $đQ/T$ is a state function and that $(\partial U/\partial V)_T = 0$ or $U = U(T)$. Show, for a classical monatomic gas with $U = cnRT$, $c = 3/2$, n is the number of moles, and the other symbols have their usual meaning, that the fundamental equation $S = S(U,V,n)$ is given by

$$S = (n/n^\circ)S^\circ + nR \ln[(U/U^\circ)^c (V/V^\circ)(n^\circ/n)^{c+1}] \quad \text{with} \quad S^\circ = (c+1)n^\circ R - n^\circ(\mu/T)^\circ$$

where the superscript ° refers to reference values. First, use the Gibbs–Duhem equation for $\mathrm{d}(\mu/T)$ to show that μ is given by $\mu/T = (\mu/T)^\circ - R\ln[(U_\mathrm{m}/U_\mathrm{m}{}^\circ)^c (V_\mathrm{m}/V_\mathrm{m}{}^\circ)]$. Thereafter, use the Euler equation $S = (1/T)(U + PV - \mu N)$.

Problem 6.3 ***The Grüneisen Parameter*** Prove Eq. (6.9). Show that $C_P/C_V = \kappa_T/\kappa_S = 1 + \gamma\alpha T$ with $\gamma = \alpha V/\kappa_T C_V = \alpha V/\kappa_S C_P$ the (thermodynamic) *Grüneisen parameter*, and that $\gamma = 2/3$ for a perfect gas ($c = 3/2$).

Problem 6.4 ***Blackbody Radiation*** For blackbody radiation in a volume V, the pressure $P = U/3V$ and the total energy $U(T)$ being a function of temperature T only. Using $\mathrm{d}U = T\mathrm{d}S - P\mathrm{d}V$ and $(\partial S/\partial V)_T = (\partial P/\partial T)_V$, show that $U/V = aT^4$ with a the Stefan–Boltzmann constant. Also show that $S/V = 4aT^3/3$ and that $S = CU^{3/4}V^{1/4}$ with $C = 4a^{1/3}/3$.

Problem 6.5 ***Chemical Equilibrium*** Derive the chemical equilibrium condition along the lines as done for hydrostatic and thermal equilibrium by using Lagrange undetermined multipliers (Section A.4).

6.2 The Local State and Internal Variables

Many problems in thermodynamics can be solved by the formulation as outlined in the previous sections. However, we also need to consider dissipative and/or inhomogeneous systems and their time dependence. This calls for a field formulation like used for continuum mechanics (CM). Recall that in CM a system is a collective of (continuum) particles or RVEs. Such an RVE is a macroscopic element when considering microscopic mechanisms, but an infinitesimal element considering the whole system. We now assume that equilibrium is present in each RVE, though not necessarily each RVE is in the same equilibrium state, that is, we assume *local equilibrium* for each RVE, but transport of quantities between RVEs can occur. Local equilibrium requires that $\partial T_{\text{system}}/\partial t \ll \partial T_{T\text{-meter}}/\partial t$ and that $\partial P_{\text{system}}/\partial t \ll \partial P_{P\text{-meter}}/\partial t$, so that T and P are useful local concepts. Consequently, U and S are useful concepts, and the usual thermodynamical equations apply. Changes within the RVE occur through changes in internal variables of the RVE and by transport from or to neighboring RVEs. Chemical reactions occur only within the RVE, while diffusion is responsible for transport. In this way the full machinery of CM can be used.

As in CM we use rates, we have to do likewise for thermodynamics and replace increments $\mathrm{d}\mathbf{a}$ by *rates* $\mathrm{d}\boldsymbol{a}/\mathrm{d}t = \dot{\boldsymbol{a}}$, which are to be interpreted as material derivatives. The work becomes the power $P_{\text{ext}} = \boldsymbol{A} \cdot \dot{\boldsymbol{a}}$ and with the heat supply per unit time $đQ/\mathrm{d}t = Q^*$, we have for the first law [35]

$$\dot{U} = P_{\text{ext}} + Q^* = \boldsymbol{A} \cdot \dot{\boldsymbol{a}} + Q^* = (\boldsymbol{A}^{(\mathrm{q})} + \boldsymbol{A}^{(\mathrm{d})}) \cdot \dot{\boldsymbol{a}} + Q^*$$

with U the internal energy. Similarly, the second law becomes

$$T\dot{S} = T(S^{*(\mathrm{r})} + S^{*(\mathrm{i})}) \geq Q^* \quad \text{where} \quad S^{*(\mathrm{r})} = Q^*/T \quad \text{and} \quad S^{*(\mathrm{i})} \geq 0$$

Combining the abovementioned yields, the *Gibbs equation* in rate form

$$\dot{U} = \boldsymbol{A}^{(\mathrm{q})} \cdot \dot{\boldsymbol{a}} + \boldsymbol{A}^{(\mathrm{d})} \cdot \dot{\boldsymbol{a}} + Q^* = \boldsymbol{A}^{(\mathrm{q})} \cdot \dot{\boldsymbol{a}} + TS^{*(\mathrm{i})} + TS^{*(\mathrm{r})} = \boldsymbol{A}^{(\mathrm{q})} \cdot \dot{\boldsymbol{a}} + T\dot{S}$$

Here $\mathbf{A}^{(\mathrm{q})} \cdot \dot{\mathbf{a}}$ represents the quasi-conservative part of the power and $Q^* = TS^{*(\mathrm{r})}$ the heat input. The *dissipation rate* $TS^{*(\mathrm{i})}$ or *dissipation function* $\boldsymbol{\Phi}$ is given by

$$\Phi(\boldsymbol{a}, T, \dot{\boldsymbol{a}}, \dot{T}) = TS^{*(\mathrm{i})} = \boldsymbol{A}^{(\mathrm{d})} \cdot \dot{\boldsymbol{a}} \geq 0$$

Extensive quantities, such as U and S, when expressed per unit mass are referred to as *specific quantities*. For example, we have the *specific energy* $u = \partial U/\partial M$ where $\mathrm{d}M$ is the mass of the RVE, so that $U = \int \rho\, u\, \mathrm{d}V$ with ρ the mass density and ρu the *energy density*.[20] The specific energy $u = u(\boldsymbol{a},s)$ is thus a function of $\boldsymbol{a}$ and s. Similar expressions apply for the entropy, the Helmholtz energy, and the dissipation function.

Generally, therefore, we obtain for the quasi-conservative stress $\boldsymbol{A}^{(\mathrm{q})}$, the specific entropy s and specific dissipation function ϕ, the expressions

$$\boldsymbol{A}^{(\mathrm{q})} = \rho[\partial f(\boldsymbol{a},T)/\partial \boldsymbol{a}], \quad s = -\partial f(\boldsymbol{a},T)/\partial T \quad \text{and} \quad \phi = Ts^{*(\mathrm{i})} \tag{6.20}$$

which are assumed to be valid for any (continuum) particle[21] (or RVE, Section 5.1). Equivalently, we have $\boldsymbol{A}^{(\mathrm{q})} = \rho\partial u(\boldsymbol{a},s)/\partial \boldsymbol{a} = \rho T\partial s(\boldsymbol{a},u)/\partial \boldsymbol{a}$. As an example, consider the

20 For a molar quantity $z_{\mathrm{m},\alpha}$, the corresponding specific quantity $z_\alpha = z_{\mathrm{m},\alpha}/\mathcal{M}_\alpha$ with $\mathcal{M}_\alpha$ the molar mass.

21 This is the *principle of local action*, which states that in determining the property of a given material point, the behavior outside an arbitrary neighborhood of that point may be discarded. If only the coordinates of one material point are involved, the material is called *simple*. Still, its behavior may depend

mechanical case. In Chapter 5, we have shown that, using the deformation rate $\mathbb{D}$ as external variable, the conjugated force is the mechanical stress $\boldsymbol{\sigma}$. Thus, the *power density* is given by $\rho l = \boldsymbol{A}\cdot\dot{\mathbf{a}} = \boldsymbol{\sigma}{:}\mathbb{D}$ ($\cong \boldsymbol{\sigma}{:}\dot{\boldsymbol{\varepsilon}}$ with $\boldsymbol{\varepsilon}$ the SDG strain), so that the specific dissipation power is given by $l^{(\mathrm{d1})} = \rho^{-1}\boldsymbol{\sigma}^{(\mathrm{d})}{:}\mathbb{D}$.

6.2.1 The Behavior of Internal Variables

In the classical theory, the variables $\boldsymbol{a}$ can be varied at will, at least in principle and in many cases also in practice. If such a state variable is varied, the system reacts with a change in associated force, and the work performed in this way enters the first law. One example is given by a cylinder closed with a piston and containing a gas. If we decrease the volume of the gas by moving the piston, the pressure increases. Similarly, if we deform an elastic solid, a stress occurs due to the prescribed strain. However, the state of many systems is also determined by variables that cannot be prescribed at will, and which before we called *internal variables*[22] and labeled $\boldsymbol{b}$. So, they are in principle measurable, but not directly controllable. Although internal variables cannot be influenced directly, this does not mean that they cannot be influenced at all. Consider, as an example, the following one-dimensional (1D) mechanical models.

Example 6.1 ***The Maxwell and Voigt Model*** Figure 6.3a shows a (black) box containing a *Maxwell element*, that is, a spring in series with a dashpot. The elongation of the dashpot is α, while the total elongation of the element is ε, so that for the spring the elongation $\varepsilon - \alpha$ remains. Hence ε and α are the independent variables for the element. However, the only variable that can be influenced directly is the total elongation ε. The force on the element σ is the same for the spring and the dashpot, is determined by ε and α, and in fact decreases slowly at constant elongation ε. While the state functions like u and s depend on ε, α, and T, the force depends not only on ε, α, and T, but also on $\dot{\varepsilon}$ and $\dot{\alpha}$. The total

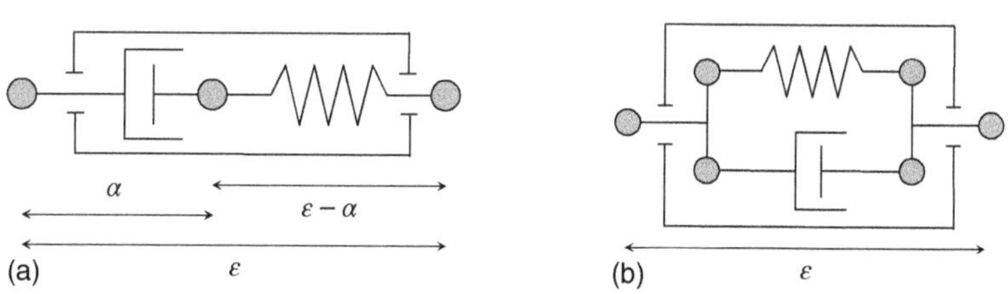

Figure 6.3 (a) The Maxwell element with a spring and dashpot in series. (b) The Voigt element with as spring and dashpot in parallel.

---

on the derivatives at that point, leading to *gradient theories*. If the material behavior depends on the coordinates of two (or more) material points, it is addressed as *nonlocal*.

22 First introduced, it seems, by P.W. Bridgman [8], although some French authors claim that in the work of Duhem the idea is already implicated. Bridgman wrote: "I believe that in general the analysis of such systems will be furthered by the recognition of a new type of large-scale thermodynamic parameter of state, namely the parameter of state which can be measured but not controlled … These parameters are measurable, but they are not controllable, which means that they are coupled to no external forces variable which might provide the means of control. And not being coupled to a force variable, they cannot take part in mechanical work." See also [36].

elongation is an external variable, and the elongation of the dashpot is an internal variable. Similarly, a *Voigt element* contains a spring and damper connected in parallel (Figure 6.3b). For this element $\varepsilon = \alpha$ so that, although dissipation occurs, it has no internal variable.

Although simple models like those indicated above cannot describe material behavior with sufficient generality, the concept of internal variable is nevertheless extremely useful. Just like the external variables $\boldsymbol{a}$, the internal variables $\boldsymbol{b}$ have thermodynamic forces associated with them, but the internal variables in conjunction with these forces do not contribute to the work in the first law. Writing the energy u, entropy s, and Helmholtz energy f as $u(\boldsymbol{a},\boldsymbol{b},s)$, $s(\boldsymbol{a},\boldsymbol{b},u)$, and $f(\boldsymbol{a},\boldsymbol{b},T)$, we associate with $\boldsymbol{b}$ the *internal forces* $\boldsymbol{B}$, the quasi-conservative parts of which are defined by

$$\boldsymbol{B}^{(\mathrm{q})} = \rho(\partial f/\partial \boldsymbol{b})$$

They are state functions like $\boldsymbol{A}^{(\mathrm{q})}$, and depend on $\boldsymbol{a}$, $\boldsymbol{b}$, and T. The dissipative parts $\boldsymbol{B}^{(\mathrm{d})}$ also depend on $\dot{\boldsymbol{a}}$ and $\dot{\boldsymbol{b}}$. Their contribution to the specific power of dissipation is

$$l^{(\mathrm{d2})} = \rho^{-1}\boldsymbol{B}^{(\mathrm{d})} \cdot \dot{\boldsymbol{b}}$$

which should be added to $l^{(\mathrm{d1})}$. Because the internal forces do not appear in the expression for the total specific power, we conclude that for an arbitrary process

$$\rho^{-1}\boldsymbol{B} \cdot \dot{\boldsymbol{b}} = \rho^{-1}(\boldsymbol{B}^{(\mathrm{q})} + \boldsymbol{B}^{(\mathrm{d})}) \cdot \dot{\boldsymbol{b}} = 0 \qquad \text{or} \qquad \boldsymbol{B}^{(\mathrm{q})} = -\boldsymbol{B}^{(\mathrm{d})}$$

if the various internal variables $\boldsymbol{b}$ are independent.

While the rates $\dot{\mathbf{a}}$ can be controlled by the experimenter, the rates $\dot{\mathbf{b}}$ are controlled by the system itself and only indirectly by the rates $\dot{\mathbf{a}}$. Therefore, to describe $\dot{\mathbf{b}}$, one needs an extra equation, typically an *evolution equation* like [37]

$$\dot{\boldsymbol{b}} = \boldsymbol{x}(\boldsymbol{a}, \boldsymbol{b}) + \boldsymbol{y}(\boldsymbol{a}, \boldsymbol{b}) \cdot \dot{\boldsymbol{a}} \tag{6.21}$$

where $\boldsymbol{x}$ and $\boldsymbol{y}$ are functions of the variables $\boldsymbol{a}$ and $\boldsymbol{b}$. Often we are able to select $\boldsymbol{b}$ in such a way that $\boldsymbol{y}(\boldsymbol{a},\boldsymbol{b}) = \boldsymbol{0}$, and in this case an instantaneous change in $\boldsymbol{a}$ does not cause an instantaneous change in $\boldsymbol{b}$. For the mechanical case, this corresponds to the fact that instantaneous strains are either elastic or zero. For example, for the Maxwell element the evolution equation describes the rate of change of the damper length $\dot{\alpha}$ as the ratio of the damper force σ and damper constant η. In fact, this equation describes the viscous behavior of the damper and reads explicitly $\dot{\alpha} = \sigma(\varepsilon, \alpha)/\eta$.

Elaborating a bit on the character of the internal variables, we recall that a (continuum) particle is an RVE of which the local internal energy u is the global internal energy of the RVE. Similarly, the strain ε may be viewed as the global kinematical description of the RVE. With such an RVE a characteristic time scale $\tau_{\mathrm{m}} = \varepsilon/\dot{\varepsilon}$ is associated. A *relaxation type* internal variable is characterized by a relaxation time $\tau_\alpha = (\alpha - \alpha^{(\mathrm{equ})})/\dot{\alpha}$. If $\tau_\alpha/\tau_{\mathrm{m}} \ll 1$, the corresponding internal variables take their equilibrium values $\alpha^{(\mathrm{equ})}$ (for u, ε, and the other α's constant), while for $\tau_\alpha/\tau_{\mathrm{m}} \gg 1$ the internal variable is said to be *frozen in* (on the time scale of the applied rate) and can be neglected. For $\tau_\alpha/\tau_{\mathrm{m}} \cong 1$, the internal variable does change on the time scale of the applied rate. A *rate-independent type* internal variable does not involve a characteristic time, but has a limiting value, say, in mechanics the yield strength σ_{y}. As long as $\sigma_{\mathrm{y}}/\sigma \gg 1$, the variable may be considered as frozen in, but σ_{y} follows σ as soon as $\sigma_{\mathrm{y}}/\sigma_{\mathrm{y}} > 1$.

In fact, the extent of reaction ξ as discussed in Section 6.1.8 is such an internal variable: in a closed system, ξ cannot be influenced directly but only indirectly by changing T and P.

The power $\mathscr{A} \cdot \dot{\xi}$ mimics the expression for $\boldsymbol{B}^{(\mathrm{d})} \cdot \dot{\boldsymbol{b}}$, and in both cases the power is positive semi-definite, that is, $\boldsymbol{B}^{(\mathrm{d})} \cdot \dot{\boldsymbol{b}} \geq 0$ and $\mathscr{A} \cdot \dot{\xi} \geq 0$.

Problem 6.6 ***The Maxwell Element*** Consider a Maxwell element. Assuming $\sigma = \eta\dot{\alpha}$ for the dissipative element and $\sigma = E(\varepsilon - \alpha)$ for the elastic element, write down the expression for the Helmholtz energy F and dissipation function Φ and derive the time dependence.

Problem 6.7 ***The Voigt Element*** Consider a Voigt element. Assuming $\sigma = \eta\dot{\alpha}$ for the dissipative element and $\sigma = E\varepsilon$ for the elastic element, write down the expression for the Helmholtz energy F and dissipation function Φ and derive its time dependence.

Problem 6.8 ***The Standard Linear Body*** Derive the time dependence for the Helmholtz energy F and dissipation function Φ for a *standard linear body* in which a spring and Maxwell element are connected in parallel using the same assumptions as in Problems 6.6 and 6.7.

6.2.2 The Local State

So, for the RVE we have two sets of equations that describe its behavior. The first, the EoS set Eq. (6.20), describes the quasi-conservative forces $\boldsymbol{A}^{(\mathrm{q})}$. The second, the evolution equation set Eq. (6.21), describes the rate of change of the internal variables $\boldsymbol{b}$. Elimination of the variables $\boldsymbol{b}$ via time integration of the evolution equation and substitution in the EoS yields a functional for $\boldsymbol{A}$ in t and $\boldsymbol{a}$.[23] This functional can be used to describe the material behavior, although the link with experiment might be cumbersome. However, in internal variable thermodynamics one keeps $\boldsymbol{b}$ explicitly, as one wants to describe the system with a limited set of variables for which all relevant thermodynamic equations can be used. Moreover, an evolution equation is reasonable anyway and leads to a simpler mathematical implementation.

We use again a division between system and environment with temperature T^* and forces $\boldsymbol{A}^*$, where the latter is taken sufficiently large so that it is always in equilibrium. The *equilibrium state* of the RVE is described by the set $\mathbf{E} = \{u,\boldsymbol{a}\}$, but the description of a thermodynamic process in the RVE requires the internal variables $\boldsymbol{b}$ and we introduce a *constrained equilibrium state* space, described by the set $\mathbf{C} = \{u,\boldsymbol{a},\boldsymbol{b}\}$, in which $\boldsymbol{b}$ is kept fixed. Moreover, we need the *contact temperature* Θ [38], which is independent of u for nonequilibrium states but becomes identical to T^* in equilibrium. It is the thermal analogue of the nonequilibrium kinematical variables. For example, the equilibrium pressure $P^{(\mathrm{equ})}$ of a gas in an adiabatic cylinder closed by a piston is measured by, say, a spring installed in the piston rod (Figure 6.4a). The pressure P is defined by the zero of the time derivative of the volume V of the system

$$(P - P^{(\mathrm{equ})})\dot{V} \geq 0 \qquad (Q^* = 0 \quad \text{and} \quad A_j \neq P = 0)$$

where Q^* is the heat exchange rate. Similarly, we write

$$(\Theta^{-1} - T^{-1})Q^* \geq 0 \qquad (\boldsymbol{A} \cdot \boldsymbol{a} = 0)$$

---

23 The *principle of determinism* states that, although materials can have memories, they cannot have foresight. Hence, a property of a body is determined by the history of that body so that, in principle, its behavior is described by a functional of the external variables.

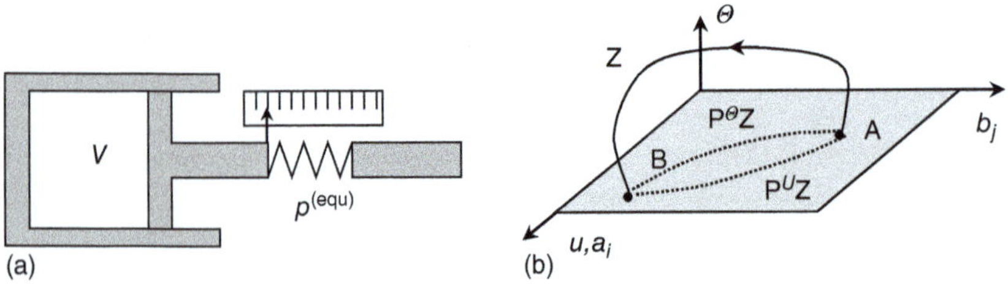

Figure 6.4 (a) The dynamical pressure as illustrated by a gas in a cylinder. (b) Schematic of the constrained equilibrium state space $\mathbf{C} = \{u,a_i,b_j\} \equiv \{(u,a_i),b_j\}$ and the extended state space $\mathbf{N} = \{u,a_i,b_j,\boldsymbol{\Theta};T^*,A_k^*\} \equiv \{(u,a_i), b_j,\boldsymbol{\Theta}\}$ depicting the transition from state A to state B via the process Z. While the process Z is described in extended state space, characterized by the energy u, the set of external and internal variables a_i and b_j as well as $\boldsymbol{\Theta}$, T^*, and A_k^*, the accompanying process P^UZ (or $P^{\Theta}Z$) is described in constrained equilibrium state space, characterized by the energy u and temperature $T^{(U)}$ (or u^{Θ} and $\boldsymbol{\Theta}$) and the set of external and internal variables a_i and b_j.

Therefore, we may envisage not only a constrained equilibrium state space described by the set $\mathbf{C} = \{u,\mathbf{a},\mathbf{b}\}$, but also an *extended state space* described by the set $\mathbf{N} = \{u,\boldsymbol{a},\boldsymbol{b},\Theta;T^*,\boldsymbol{A}^*\} = \{u,\boldsymbol{a},\boldsymbol{b},\Theta\}$. A reversible process can be represented by a path in **E**, while an irreversible process requires a path in **N** (Figure 6.4b). Consider the transition from equilibrium state A to B and let these two states be connected by an irreversible process. With each point of the path in the space **N** we can associate a point in the space **C**. In a "Gedankenexperiment" this can be done in two ways.

The first way to isolate the RVE is by *adiabatic closure* for which we isolate the system so that no heat or work is exchanged with the environment. For such a process du = 0. After isolation, the temperature of the RVE will change from the contact temperature Θ to the equilibrium temperature $T^{(U)}$, as calculated from $T^{-1} = \partial s(u,\boldsymbol{a},\mathbf{b})/\partial u$. As the process can be described by the projection P^U of the path in space **N** on space **C** for which the energy of the RVE is kept constant, we may write

$$\mathrm{P}^U\{u,\boldsymbol{a},\mathbf{b},\Theta\} = \{u,\boldsymbol{a},\mathbf{b},T^{(U)}\}$$

Within **C**, the variables u, $\boldsymbol{a}$, and $\mathbf{b}$ have the same value as in the nonequilibrium state, but the temperature changed from Θ to $T^{(U)}$ (Figure 6.4b).

The second way is by *thermal equilibration* in which the RVE is heated or cooled, but no work is exchanged with the environment. In this case, we keep Θ constant, so that after equilibration u changes to $u^{(\Theta)}$, corresponding to the temperature Θ. This process is described by the projection P^{Θ} of the path in space **N** on space **C** given by

$$\mathrm{P}^{\Theta}\{u,\boldsymbol{a},\mathbf{b},\Theta\} = \{u^{(\Theta)},\boldsymbol{a},\mathbf{b},\Theta\}$$

Within **C**, the variables Θ, $\boldsymbol{a}$, and $\mathbf{b}$ have the same values as in the nonequilibrium state but the energy changed from u to $u^{(\Theta)}$ (Figure 6.4b).

Now an entropy in the extended state space can be defined. Using $s^{(\mathrm{equ})} = s^{(\mathrm{equ})}(u^{(\mathrm{equ})},\boldsymbol{a},\boldsymbol{b})$ and remembering that $\boldsymbol{b}^{(\mathrm{equ})} = \mathbf{b}(u,\boldsymbol{a})$, at equilibrium we have

$$T^{(\mathrm{equ})}\dot{s}^{(\mathrm{equ})} = \dot{u}^{(\mathrm{equ})} - \boldsymbol{A}^{(\mathrm{equ,q})}\cdot\dot{\boldsymbol{a}}$$

We now take for the P^{Θ} associated equilibrium state with $s^{(\Theta)}(u^{(\Theta)},\boldsymbol{a},\boldsymbol{b},\Theta)$ and for the P^U associated equilibrium state with $s^{(U)}(u,\boldsymbol{a},\boldsymbol{b},T^{(U)})$, respectively,

$$\Theta\dot{s}^{(\Theta)} = \dot{u}^{(\Theta)} - \boldsymbol{A}^{(\Theta,\mathrm{q})}\cdot\dot{\boldsymbol{a}} \quad \text{and} \quad T^{(U)}\dot{s}^{(U)} = \dot{u} - \boldsymbol{A}^{(U,\mathrm{q})}\cdot\dot{\boldsymbol{a}}$$

Muschik [38] showed that, for open as well as inhomogeneous systems and irrespective of whether T^* or Θ is used, the Clausius inequality $\int_A^B \rho s^{*(i)}\,dt \geq 0$ applies. This corroborates the existence of a nonequilibrium entropy. This nonequilibrium (as well as the equilibrium) entropy is nonunique, as should be clear from $s = s(u,\boldsymbol{a},\boldsymbol{b})$, which is evidently dependent on the choice of the parameters $\boldsymbol{a}$ and $\boldsymbol{b}$.

The approximation of the *local state* proclaims that it is sufficient for the description of a nonequilibrium process to use s and T of the associated states in state space **C**. Since either T or u can be projected identically from **N** to **C** but not both, P^U and P^Θ define different local states. Nevertheless, normally P^U is used, so that $u = u^{(\mathrm{equ})}$, but in conjunction with $\Theta = T^*$. Finally, note that the entropy of a nonequilibrium state $s > s^{(U)}$ since the change from Θ to $T^{(U)}$ occurs adiabatically.

6.3 Field Formulation

Up to now we described the behavior of an RVE of constant mass $\mathrm{d}M$, given by $\rho\mathrm{d}V$ with ρ its mass density and $\mathrm{d}V$ its volume, with a homogeneous distribution of quantities. However, for continuum considerations we want to be able to use gradients of parameters over the system and hence consider the situation anew [9, 35, 37].

6.3.1 The First Law

The (internal) energy of the system U is obtained by integration of the specific energy u times the mass density ρ over the volume of the system resulting in

$$U = \int \rho\, u\, \mathrm{d}V$$

The power of the external mechanical forces $\boldsymbol{f}$ is (Section 5.4, using $\boldsymbol{f}$ instead of $\boldsymbol{b}$)

$$P_{\mathrm{ext}} = \int \rho \boldsymbol{f} \cdot \boldsymbol{v}\, \mathrm{d}V + \int (\boldsymbol{\sigma} \cdot \boldsymbol{n}) \cdot \boldsymbol{v}\, \mathrm{d}A = P_{\mathrm{vol}} + P_{\mathrm{sur}}$$

Further we need the heat supply. Neglecting any volume mechanism by radiation, the heat supply per unit time Q^* is due to conduction across the surface A.[24] If $\boldsymbol{q}$ denotes the heat flow vector per unit time across a surface element $\mathrm{d}A$ with unit outward normal $\boldsymbol{n}$, Q^* is given by

$$Q^* = -\int \boldsymbol{q} \cdot \boldsymbol{n}\, \mathrm{d}A$$

Finally, we need the kinetic energy ϑ or, actually, its material derivative given by

$$\dot{\vartheta} = \int \rho \boldsymbol{v} \cdot \dot{\boldsymbol{v}}\, \mathrm{d}V$$

Conservation of energy in global rate form is now written as

$$\dot{\vartheta} + \dot{U} = P_{\mathrm{ext}} + Q^* \qquad \text{or, explicitly, as}$$

24 For extension with an applied electromagnetic field, see [33, 39]. Such a contribution does not appear in the dissipation function in the conventional formulation used here, as there is no state variable that can couple with this "force." For a critical discussion, see [40].

$$\int \rho(\boldsymbol{v}\cdot\dot{\boldsymbol{v}}+\dot{u})\,\mathrm{d}V = \int \rho\boldsymbol{f}\cdot\boldsymbol{v}\,\mathrm{d}V + \int(\boldsymbol{\sigma}\cdot\boldsymbol{v}-\boldsymbol{q})\cdot\boldsymbol{n}\,\mathrm{d}A$$

Applying the divergence theorem to the last integral yields

$$\int \rho(\boldsymbol{v}\cdot\dot{\boldsymbol{v}}+\dot{u})\,\mathrm{d}V = \int(\rho\boldsymbol{f}\cdot\boldsymbol{v}+\boldsymbol{\sigma}\cdot\nabla\boldsymbol{v}+\boldsymbol{v}\cdot\nabla\cdot\boldsymbol{\sigma})\,\mathrm{d}V - \int\nabla\cdot\boldsymbol{q}\,\mathrm{d}V$$

reducing, using the equation of motion $\rho\,\dot{\boldsymbol{v}} = \rho\boldsymbol{f}+\nabla\cdot\boldsymbol{\sigma}$ and the symmetry $\boldsymbol{\sigma}=\boldsymbol{\sigma}^{\mathrm{T}}$, to

$$\int \rho\dot{u}\,\mathrm{d}V = \int\boldsymbol{\sigma}:\mathbb{D}\,\mathrm{d}V - \int\nabla\cdot\boldsymbol{q}\,\mathrm{d}V$$

with $\mathbb{D} = ½[(\partial\boldsymbol{v}/\partial\boldsymbol{r})+(\partial\boldsymbol{v}/\partial\boldsymbol{r})^{\mathrm{T}}]$ the deformation rate. Noting that this result is valid for any volume, we obtain for a closed system the first law in local form

$$\rho\,\dot{u} = \boldsymbol{\sigma}:\mathbb{D}-\nabla\cdot\boldsymbol{q} \qquad \text{or, more generally,} \qquad \rho\dot{u} = \boldsymbol{A}\cdot\dot{\boldsymbol{a}}-\nabla\cdot\boldsymbol{q} \tag{6.22}$$

if we realize that $\boldsymbol{\sigma}$ is only one of the forces $\boldsymbol{A}$ and $\mathbb{D}$ only one of the rates $\dot{\boldsymbol{a}}$. For example, if the amount of component α for the RVE is changing by diffusion, we must add $\Sigma_\alpha\boldsymbol{f}_\alpha\cdot\boldsymbol{J}_\alpha$ with $\boldsymbol{J}_\alpha = \rho_\alpha(\boldsymbol{v}_\alpha-\boldsymbol{v})$ the rate and $\boldsymbol{f}_\alpha$ the volume force, so that we have $\rho\dot{u} = \boldsymbol{\sigma}:\mathbb{D}+\Sigma_\alpha\boldsymbol{f}_\alpha\cdot\boldsymbol{J}_\alpha-\nabla\cdot\boldsymbol{q}$. Equation (6.22) states that the material derivative of the internal energy equals the power of the applied forces plus the rate of heat supply.

6.3.2 The Second Law

In a similar way as for the energy U, the entropy S of the system is obtained from

$$S = \int\rho s\,\mathrm{d}V = S^{(\mathrm{r})}+S^{(\mathrm{i})} \qquad \text{and} \qquad \dot{S} = \int\rho\dot{s}\,\mathrm{d}V = S^{*(\mathrm{r})}+S^{*(\mathrm{i})} \geq S^{*(\mathrm{r})}$$

From the vector $\boldsymbol{q}/T$ with T the local temperature, the rate of entropy supply per unit time across area A due to heat transfer[25] is, using the divergence theorem, given by

$$S^{*(\mathrm{r})} = -\int(\boldsymbol{q}/T)\cdot\boldsymbol{n}\,\mathrm{d}A = -\int\nabla\cdot(\boldsymbol{q}/T)\,\mathrm{d}V$$

To obtain the local form of the second law we use the global form $\dot{S}\geq S^{*(\mathrm{r})}$ and obtain

$$\int\rho\dot{s}\,\mathrm{d}V \geq -\int\nabla\cdot(\boldsymbol{q}/T)\mathrm{d}V \qquad \text{or} \qquad \rho\dot{s}\geq\rho s^{*(\mathrm{r})} = -\nabla\cdot(\boldsymbol{q}/T)$$

Thus, the rate of entropy change $\rho\dot{s}$ is never less than the rate of entropy supply $\rho s^{*(\mathrm{r})}$.

To obtain $\phi = Ts^{*(\mathrm{i})}$, we first transform the Gibbs equation $T\mathrm{d}S = \mathrm{d}U-\boldsymbol{A}^{(\mathrm{q})}\cdot\mathrm{d}\boldsymbol{a}-\boldsymbol{B}^{(\mathrm{q})}\cdot\mathrm{d}\boldsymbol{b}$, using $S = sM$, … with M the total mass. After substitution of the specific quantities, the Euler relation $TS = U+\boldsymbol{A}^{(\mathrm{q})}\cdot\boldsymbol{a}+\boldsymbol{B}^{(\mathrm{q})}\cdot\boldsymbol{b}$ and dividing by M, the result is

$$\rho T\dot{s} = \rho\dot{u}-\boldsymbol{A}^{(\mathrm{q})}\cdot\dot{\boldsymbol{a}}-\boldsymbol{B}^{(\mathrm{q})}\cdot\dot{\boldsymbol{b}} \tag{6.23}$$

Using the first law $\rho\dot{u} = \boldsymbol{A}\cdot\dot{\boldsymbol{a}}-\nabla\cdot\boldsymbol{q}$ and the equality $\boldsymbol{B}^{(\mathrm{q})} = -\boldsymbol{B}^{(\mathrm{d})}$, Eq. (6.23) yields

$$\rho\dot{s} = \frac{1}{T}(\boldsymbol{A}^{(\mathrm{d})}\cdot\dot{\boldsymbol{a}}+\boldsymbol{B}^{(\mathrm{d})}\cdot\dot{\boldsymbol{b}}-\nabla\cdot\boldsymbol{q}) = \frac{1}{T}(\boldsymbol{A}^{(\mathrm{d})}\cdot\dot{\boldsymbol{a}}+\boldsymbol{B}^{(\mathrm{d})}\cdot\dot{\boldsymbol{b}})-\nabla\cdot\left(\frac{\boldsymbol{q}}{T}\right)+\boldsymbol{q}\cdot\nabla\frac{1}{T}$$

25 In case not only heat transfer occurs at the surface, the entropy flow contains also other terms. In essence, all terms in the general balance equation (Section 5.3), written for the entropy as $\rho\mathrm{d}s/\mathrm{d}t = -\nabla\cdot\rho s(\boldsymbol{v}_s-\boldsymbol{v})+\pi_s$, that can be collected in the divergence term contribute to the entropy flow.

As $\rho\dot{s} \geq -\nabla \cdot (\boldsymbol{q}/T)$ and $\dot{s} = s^{*(\mathrm{r})} + s^{*(\mathrm{i})}$, we obtain the *Clausius–Duhem inequality*

$$\rho\phi \equiv \rho T s^{*(\mathrm{i})} = \boldsymbol{A}^{(\mathrm{d})} \cdot \dot{\boldsymbol{a}} + \boldsymbol{B}^{(\mathrm{d})} \cdot \dot{\boldsymbol{b}} + \boldsymbol{q} \cdot T\nabla(1/T) = \rho(l^{(\mathrm{d1})} + l^{(\mathrm{d2})} + l^{(\mathrm{d3})}) \geq 0 \qquad (6.24)$$

The terms $l^{(\mathrm{d1})}$ and $l^{(\mathrm{d2})}$ represent the power of the external and internal forces for which the quantities $\boldsymbol{A}^{(\mathrm{d})}$ and $\boldsymbol{B}^{(\mathrm{d})}$ are the forces while the quantities $\dot{\boldsymbol{a}}$ and $\dot{\boldsymbol{b}}$ are the corresponding rates. For our general topic, an important example for external work is diffusion (Example 6.3) and for internal work chemical reactions. In the local state approximation, the RVE is an open system for which the terms heat and work are ambiguous and must be labeled explicitly (see in particular [41], but also [8, 33]). The term $l^{(\mathrm{d3})}$ reflects an additional entropy production due to heat flow across the RVE under consideration. It does not arise for a homogeneous situation and $T\nabla(1/T)$ can be interpreted as the force with $\boldsymbol{q}$ as the corresponding rate (Example 6.2). Although the second law only requires that $\phi \geq 0$, it is often assumed that $l^{(\mathrm{d1})} + l^{(\mathrm{d2})} \geq 0$ and $l^{(\mathrm{d3})} \geq 0$ apply separately. Generally, we do so as well in this book.

Let us now elaborate a bit on chemical reactions before we redo the above in specific terms for reaction–diffusion in the presence of surface and volume forces. In the local state approximation, chemical reactions take place within the RVE and in Section 6.1.8 we introduced for a set of n coupled reactions comprising ω components the affinities $\mathscr{A}_j$ and extents of reaction ξ_j. We obtained for the dissipation in a closed system $\sum_j^n \overline{\mathscr{A}}_j\dot{\xi}_j = -\sum_j^n \sum_\alpha^\omega \overline{\mu}_\alpha \overline{\nu}_{\alpha j}\dot{\xi}_j$ with $\overline{\mu}_\alpha$ the chemical potential per mole, $\overline{\nu}_{\alpha j}$ the stoichiometric coefficient and $\mathrm{d}_j n_\alpha = \nu_{\alpha j}\xi_j$ the change in moles for component in reaction j.[26] Using $\mathscr{M}_\alpha$ for the molar mass of component α, we define $\mu_\alpha = \overline{\mu}_\alpha/\mathscr{M}_\alpha$ as the chemical potential per unit mass and $\nu_{\alpha j} = \overline{\nu}_{\alpha j}\mathscr{M}_\alpha/a_j$ as a modified stoichiometric coefficient with a_j a normalization constant. Because the mass of the RVE is constant, we have $\sum_\alpha^\omega \nu_{\alpha j} = 0$ and use $a_j = \sum_\pi^\omega \overline{\nu}_{\pi j}\mathscr{M}_\pi$, where the index π only runs over the reaction products. So, $\overline{\mathscr{A}}_j = -\sum_\alpha^\omega \overline{\mu}_\alpha \overline{\nu}_{\alpha j} = -a_j \sum_\alpha^\omega \mu_\alpha \nu_{\alpha j} \equiv a_j \mathscr{A}_j$. For the RVE we have $\mathrm{d}M = \sum_\alpha \mathrm{d}M_\alpha = \sum_\alpha \rho_\alpha \mathrm{d}V = \rho \mathrm{d}V$ with

$$\mathrm{d}M_\alpha = \sum\nolimits_j^n \mathrm{d}_j M_\alpha = \sum\nolimits_j^n \nu_{\alpha j} J_j \mathrm{d}t$$

where $\mathrm{d}_j M_\alpha$ is the change in mass for component α in reaction j and J_j is the rate of reaction j. With $M_\alpha = n_\alpha \mathscr{M}_\alpha$, we have $\overline{\mu}_\alpha \mathrm{d}_j n_\alpha = \mu_\alpha \mathrm{d}_j M_\alpha$ and

$$\sum\nolimits_j^n \overline{\mathscr{A}}_j \dot{\xi}_j = \sum\nolimits_j^n a_j \mathscr{A}_j (\overline{\nu}_{\alpha j})^{-1}(\mathrm{d}_j n_\alpha/\mathrm{d}t) = \sum\nolimits_j^n \mathscr{A}_j (\nu_{\alpha j})^{-1}(\mathrm{d}_j M_\alpha/\mathrm{d}t) = \sum\nolimits_j^n \mathscr{A}_j J_j$$

For an open system, we obtain from $\sum_\alpha^\omega \mu_\alpha \mathrm{d}M_\alpha$ for the total chemical power $\sum_\alpha^\omega \mu_\alpha (\partial \dot{M}_\alpha/\partial V)\mathrm{d}V = \rho \sum_\alpha^\omega \mu_\alpha \dot{w}_\alpha$, where the factor $\rho\dot{w}_\alpha$ is given by the balance equation $\rho\dot{w}_\alpha = -\nabla \cdot \boldsymbol{J}_\alpha + \sum_j \nu_{\alpha j} J_j$ with $\boldsymbol{J}_\alpha$ the transport rate for component α (Eq. (5.13)). Thus, we have $\rho \sum_\alpha^\omega \mu_\alpha \dot{w}_\alpha = -\sum_\alpha^\omega \mu_\alpha \nabla \cdot \boldsymbol{J}_\alpha - \sum_j \mathscr{A}_j J_j$ with $\mathscr{A}_j = -\sum_\alpha \mu_\alpha \nu_{\alpha j}$.

Let us now redo the derivation given above for diffusion–reaction for which the entropy and energy rate are given by, respectively,

$$\rho T\dot{s} = \rho\dot{u} - \boldsymbol{\sigma}^{(\mathrm{q})} : \mathbb{D} - \rho\sum\nolimits_\alpha \mu_\alpha \dot{w}_\alpha \quad \text{and} \quad \rho\dot{u} = -\nabla \cdot \boldsymbol{q} + \boldsymbol{\sigma} : \mathbb{D} + \sum\nolimits_\alpha \boldsymbol{f}_\alpha \cdot \boldsymbol{J}_\alpha \qquad (6.25)$$

Inserting $\rho \sum_\alpha^\omega \mu_\alpha \dot{w}_\alpha$ and Eq. (6.25b) in Eq. (6.25a) yields

$$\rho T\dot{s} = -\nabla \cdot \boldsymbol{q} + \boldsymbol{\sigma}^{(\mathrm{d})} : \mathbb{D} + \sum\nolimits_\alpha \boldsymbol{f}_\alpha \cdot \boldsymbol{J}_\alpha + \sum\nolimits_\alpha \mu_\alpha \nabla \cdot \boldsymbol{J}_\alpha + \sum\nolimits_j \mathscr{A}_j J_j$$

26 To be as clear as possible, I use here overbars to indicate molar quantities, in consonance with [33].

which can be converted, after dividing by T and using $x\nabla \cdot \mathbf{z} = \nabla \cdot x\mathbf{z} - \mathbf{z} \cdot \nabla x$, to

$$\rho\dot{s} = -\nabla \cdot (\mathbf{q}/T) + \mathbf{q} \cdot \nabla(1/T) + T^{-1}\boldsymbol{\sigma}^{(\mathrm{d})} : \mathbb{D} + T^{-1}\sum\nolimits_{\alpha}\mathbf{f}_{\alpha} \cdot \mathbf{J}_{\alpha}$$
$$+ \sum\nolimits_{\alpha}\nabla \cdot (\mu_{\alpha}\mathbf{J}_{\alpha}/T) - \sum\nolimits_{\alpha}\mathbf{J}_{\alpha} \cdot \nabla(\mu_{\alpha}/T) - T^{-1}\sum\nolimits_{j}\mathscr{A}_{j}J_{j}$$

Comparing this expression with the entropy balance equation $\rho\dot{s} = -\nabla \cdot \mathbf{J}_s + \pi_s$ and the definition $\rho\dot{s} = \rho(s^{*(\mathrm{r})} + s^{*(\mathrm{i})})$, we see that $-\nabla \cdot \mathbf{J}_s = \rho s^{*(\mathrm{r})}$ and $\pi_s = \rho s^{*(\mathrm{i})}$ with

$$\rho s^{*(\mathrm{r})} = -\nabla \cdot \left[(\mathbf{q}/T) - \sum\nolimits_{\alpha}(\mu_{\alpha}\mathbf{J}_{\alpha}/T)\right] = -\nabla \cdot \mathbf{J}_s \quad \text{and}$$

$$\rho s^{*(\mathrm{i})} = \frac{1}{T}\left[\boldsymbol{\sigma}^{(\mathrm{d})} : \mathbb{D} + \mathbf{q} \cdot T\nabla\frac{1}{T} + \sum\nolimits_{\alpha}\mathbf{J}_{\alpha} \cdot \left(\mathbf{f}_{\alpha} - T\nabla\frac{\mu_{\alpha}}{T}\right) + \sum\nolimits_{j}\mathscr{A}_{j}J_{j}\right] = \pi_s \tag{6.26}$$

These equations describe diffusion–reaction in general. Note that the entropy flux $\mathbf{J}_s$ now also contains a transport term besides the heat term (conform footnote 25). Finally, we note that, although the divergence transformation used above may seem arbitrary, it is Galilei invariant, while the result yields $s^{*(\mathrm{i})} = 0$ for equilibrium and $\mathrm{d}S^{(\mathrm{i})}/\mathrm{d}t \geq 0$ after integration of $s^{*(\mathrm{i})}$, all as it should be.

6.4 The Linear Approximation in Non-equilibrium Thermodynamics

For irreversible processes the dissipation density $\rho\phi$ is given by Eq. (6.24). We now change notation (to one more conventional in this field) and denote the dissipative forces collectively by $\mathbf{X} \equiv \mathbf{X}^{(\mathrm{d})} = (T\mathbf{A}^{(\mathrm{d})}, T\mathbf{B}^{(\mathrm{d})}, T\nabla T^{-1})$ and the rates, associated with the state variables $\mathbf{y} \equiv (\mathbf{a},\mathbf{b},u)$, by $\mathbf{J} \equiv (\dot{\mathbf{a}}, \dot{\mathbf{b}}, \mathbf{J}_u)$. A generic expression is thus $\rho\phi = \mathbf{J} \cdot \mathbf{X}$.

Now consider the entropy s. At equilibrium, the entropy has a maximum s_0 and we calculate s somewhat remote from equilibrium using the equilibrium state as the reference state by developing s in a Taylor series [42] in the state variables $\mathbf{y}$

$$\rho s = \rho s_0 + \left(\frac{\partial \rho s}{\partial \mathbf{y}}\right)_0 \cdot \mathrm{d}\mathbf{y} + ½\mathrm{d}\mathbf{y} \cdot \left(\frac{\partial^2 \rho s}{\partial \mathbf{y}\partial \mathbf{y}'}\right)_0 \cdot \mathrm{d}\mathbf{y}' + \cdots$$

For the equilibrium state the first derivatives $(\partial\rho s/\partial\mathbf{y})_0$ are zero. Moreover, defining $\mathbf{y}_0 = \mathbf{0}$ and $\rho\Delta s = \rho s - \rho s_0$, we have $\mathrm{d}\mathbf{y} = \mathbf{y} - \mathbf{y}_0 = \mathbf{y}$, and we write for a "small" deviation from equilibrium for $\rho\Delta s$ a quadratic expression in the state variables $\mathbf{y}$ given by

$$\rho\Delta s = ½\mathbf{y} \cdot \left(\frac{\partial^2 \rho s}{\partial \mathbf{y}\partial \mathbf{y}'}\right)_0 \cdot \mathbf{y}' \equiv ½\mathbf{y} \cdot \mathbb{G} \cdot \mathbf{y}' \quad \text{or} \quad \rho s^{*(\mathrm{i})} = \dot{\mathbf{y}} \cdot \mathbb{G} \cdot \mathbf{y}'$$

where the last step can be made because $s^{*(\mathrm{r})}$ is independent of $\dot{\mathbf{y}}$. Realizing that in this formulation $\mathbb{G} \cdot \mathbf{y} = \partial\rho\Delta s/\partial\mathbf{y} = \partial\rho s^{*(\mathrm{i})}/\partial\dot{\mathbf{y}}$ corresponds with $\mathbf{X}/T$, we obtain

$$\rho\phi = \rho T s^{*(\mathrm{i})} = \dot{\mathbf{y}} \cdot T\mathbb{G} \cdot \mathbf{y}' \equiv \mathbf{J} \cdot \mathbf{X} \quad \text{and} \quad \mathbf{y} = T^{-1}\mathbb{G}^{-1} \cdot \mathbf{X}$$

consistent with $\phi = \mathbf{J} \cdot \mathbf{X}$ as defined earlier. Now one expects that $\mathbf{J}$ is a function of $\mathbf{y}$, say $\mathbf{J} = \mathbf{J}(\mathbf{y})$, that can be developed in $\mathbf{y}$ resulting in $\mathbf{J} = (\partial\mathbf{J}/\partial\mathbf{y}) \cdot \mathbf{y} + \ldots \equiv \mathbb{C} \cdot \mathbf{y}$. Inserting $\mathbf{y} = T^{-1}\mathbb{G}^{-1} \cdot \mathbf{X}$ yields the *phenomenological equations*

$$\mathbf{J} = \mathbb{C} \cdot T^{-1}\mathbb{G}^{-1} \cdot \mathbf{X} \equiv \mathbb{L} \cdot \mathbf{X} \quad \text{and} \quad \rho T s^{*(\mathrm{i})} = \mathbf{J} \cdot \mathbf{X} = \mathbf{X} \cdot \mathbb{L}^{\mathrm{T}} \cdot \mathbf{X} = \mathbf{J} \cdot \mathbb{L}^{-1} \cdot \mathbf{J} \tag{6.27}$$

Based on microscopic arguments, Onsager showed that $\mathbb{L} = \mathbb{L}^{\mathrm{T}}$, in the literature known as the *Onsager reciprocal relations*. They apply provided the rate $\mathbf{J}$ is chosen as $\dot{\mathbf{y}}$ and the force

$\boldsymbol{X}$ is chosen as $\partial\rho T\Delta s/\partial\boldsymbol{y}$. Several attempts have been made to derive the reciprocal relations also by phenomenological arguments. May be not completely watertight, but nevertheless rather convincing arguments are given in Justification 6.2.

Justification 6.2 ***Phenomenological Arguments for*** $\mathbb{L} = \mathbb{L}^T$ One approach to rationalize the Onsager relations phenomenologically has been given by Gyarmati [43]. In a linear theory, we should be able to write not only $s^{*(\mathrm{i})} = s^{*(\mathrm{i})}(\boldsymbol{J})$ but also $s^{*(\mathrm{i})} = s^{*(\mathrm{i})}(\boldsymbol{X})$, leading in a similar way as for $\rho Ts^{*(\mathrm{i})}(\boldsymbol{J})$ to $\rho Ts^{*(\mathrm{i})}(\boldsymbol{X}) = \boldsymbol{X}\cdot\mathbb{Q}\cdot\boldsymbol{X}$. However, the rate is also given by $\rho Ts^{*(\mathrm{i})} = \boldsymbol{J}\cdot\boldsymbol{X} = \mathbb{C}\cdot\boldsymbol{y}\cdot\boldsymbol{X} = \mathbb{C}\cdot T^{-1}\mathbb{G}^{-1}\cdot\boldsymbol{X}\cdot\boldsymbol{X} = \mathbb{L}\cdot\boldsymbol{X}\cdot\boldsymbol{X} = \boldsymbol{X}\cdot\mathbb{L}^{\mathrm{T}}\cdot\boldsymbol{X}$. Comparing with $\rho Ts^{*(\mathrm{i})} = \boldsymbol{X}\cdot\mathbb{Q}\cdot\boldsymbol{X}$ and realizing that $\mathbb{Q} = \mathbb{Q}^{\mathrm{T}}$, we have $\mathbb{L} = \mathbb{L}^{\mathrm{T}}$.

Even more convincing arguments are provided by Woods [9]. His basic argument is that upon time reversal (Section 19.2) for an infinitesimal, macroscopic time interval $\mathrm{d}t$, the variables X have a definite time parity, that is, $X(-\mathrm{d}t) = \eta X(\mathrm{d}t)$ with $\eta = 1$ or $\eta = -1$ ($\overline{X} = \eta X$ for brevity), but that the rate $s^{*(\mathrm{i})}$ does not change sign, that is, $s^{*(\mathrm{i})}(\mathrm{d}t) = s^{*(\mathrm{i})}(-\mathrm{d}t)$. Therefore, taking just two state variables labeled α and β, we have for the entropy production $s^{*(\mathrm{i})} = X_\alpha L_{\alpha\beta} X_\beta = \frac{1}{2} X_\alpha (L_{\alpha\beta} + L_{\beta\alpha}) X_\beta$ and $s^{*(\mathrm{i})}(-\mathrm{d}t) = \frac{1}{2}\overline{X}_\alpha[\overline{L}_{\alpha\beta} + \overline{L}_{\beta\alpha}]\overline{X}_\beta = \frac{1}{2}\eta_\alpha\eta_\beta X_\alpha[\overline{L}_{\alpha\beta} + \overline{L}_{\beta\alpha}]X_\beta$. Consequently,

$$L_{\alpha\beta} + L_{\beta\alpha} = \eta_\alpha\eta_\beta(\overline{L}_{\alpha\beta} + \overline{L}_{\beta\alpha}) \tag{6.28}$$

Now consider

$$J_\alpha = L_{\alpha\alpha}X_\alpha + L_{\alpha\beta}X_\beta \quad \text{and} \quad J_\beta = L_{\beta\alpha}X_\alpha + L_{\beta\beta}X_\beta \tag{6.29}$$

Because the forces are independent, we can choose J_α and X_β to be the forces, so that

$$X_\alpha = L_{\alpha\alpha}{}' J_\alpha + L_{\alpha\beta}{}' X_\beta \quad \text{and} \quad J_\beta = L_{\beta\alpha}{}' X_\alpha + L_{\beta\beta}{}' X_\beta$$

As Eqs. (6.28) and (6.29) refer to the same physical phenomenon, solving leads to

$$L_{\alpha\alpha}{}' = L_{\alpha\alpha}{}^{-1}, \quad L_{\alpha\beta}{}' = -L_{\alpha\alpha}{}^{-1}L_{\alpha\beta}, \quad L_{\beta\alpha}{}' = L_{\alpha\alpha}{}^{-1}L_{\beta\alpha} \quad \text{and}$$
$$L_{\beta\beta}{}' = L_{\alpha\alpha}{}^{-1}(L_{\alpha\alpha}L_{\beta\beta} - L_{\alpha\beta}L_{\beta\alpha})$$

after some algebra. Now $L_{\alpha\beta} + L_{\beta\alpha} = \eta_\alpha\eta_\beta(\overline{L}_{\alpha\beta} + \overline{L}_{\beta\alpha})$ is independent of the choice of forces and using J_α and X_β as forces results in $L_{\alpha\beta}{}' + L_{\beta\alpha}{}' = \eta_\alpha{}'\eta_\beta\left(\overline{L}_{\alpha\beta}{}' + \overline{L}_{\beta\alpha}{}'\right)$ with $\eta_\alpha{}' = -\eta_\alpha$. Substitution of $L_{\alpha\alpha}{}'$, $L_{\alpha\beta}{}'$, $L_{\beta\alpha}{}'$ and $L_{\beta\alpha\beta}$ results in

$$-L_{\alpha\beta} + L_{\beta\alpha} = \eta_\alpha\eta_\beta(\overline{L}_{\alpha\beta} - \overline{L}_{\beta\alpha}) \tag{6.30}$$

Subtracting Eq. (6.28) from Eq. (6.30) delivers the reciprocal relations $L_{\alpha\beta} = \eta_\alpha\eta_\beta\overline{L}_{\beta\alpha}$. As the choice of α and β is arbitrary, the result is general.

Because $\phi = \rho Ts^{*(\mathrm{i})}$ can be split in several ways into (a sum of) conjugated $\boldsymbol{J}$s and $\boldsymbol{X}$s, there is a certain freedom in choice of $\boldsymbol{J}$ and $\boldsymbol{X}$, but for any proper division of ϕ in $\boldsymbol{J} = \dot{\boldsymbol{y}}$ and $\boldsymbol{X} = \partial\rho T\Delta s/\partial\boldsymbol{y} = \rho\partial\phi/\partial\dot{\boldsymbol{y}}$ with $\boldsymbol{y}$ a state variable, the symmetry relation $\mathbb{L} = \mathbb{L}^{\mathrm{T}}$ remains valid. Woods [9] showed that the reciprocal relations are valid for any linear transformation between the $\boldsymbol{J}$s and $\boldsymbol{X}$s that preserve the bilinear form $\boldsymbol{J}\cdot\boldsymbol{X}$, as long as the new $\boldsymbol{J}$s and $\boldsymbol{X}$s have a definite time parity (as introduced in Justification 6.2). We omit the derivation here as it is somewhat long, but this result eases the realization of a proper formulation considerably. From these remarks it is hopefully also clear that the division of ϕ in $\boldsymbol{J}$s and $\boldsymbol{X}$s is only a division between dependent and independent variables, convenient for theory but possibly also realizable in practice.

Using only two rates we have from $\rho\phi = \boldsymbol{X}\cdot\mathbb{L}\cdot\boldsymbol{X} \geq 0$, $\rho\phi = L_{00}X_0{}^2 + (L_{01}+L_{10})X_0X_1 + L_{11}X_1{}^2 \geq 0$, which by completing the square, can be written alternatively as

$$\rho\phi = \left[L_{00} - (L_{01}+L_{10})^2/4L_{11}\right]X_0{}^2 + L_{11}\{X_1 + [(L_{01}+L_{10})/2L_{11}]\,X_0\}^2 \geq 0 \qquad \text{or}$$
$$\rho\phi = L_{00}\{X_0 + [(L_{01}+L_{10})/2L_{00})]X_1\}^2 + [L_{11} - (L_{01}+L_{10})^2/4L_{00}]X_1{}^2 \geq 0$$

This is equivalent to $L_{jj} \geq 0$ and $L_{ii}L_{jj} \geq (L_{ji}+L_{ji})^2/4$, which applies for any i and j if the processes are independent.

The formulation $\boldsymbol{X} = \lambda\rho[\partial\phi(\boldsymbol{J})/\partial\boldsymbol{J}]$, where λ is chosen such that $\boldsymbol{J}\cdot\boldsymbol{X} = \rho\phi$ or $\lambda = ½$, is valid for small deviations from equilibrium. For this formulation, Ziegler in particular [35] advocated that the expression for $\boldsymbol{X}$ is also valid for larger deviations from equilibrium, that is, for non-quadratic dissipation functions in $\boldsymbol{J}$, a postulate known as the *orthogonality principle*. If we consider the dissipation function ϕ as a surface in a space, where the rates $\boldsymbol{J}$ act as coordinate axes, the forces $\boldsymbol{X}$ are perpendicular to surfaces of constant values for ϕ. To be consistent in this more general case, the constant $\lambda = ½$ must be replaced by $\lambda = \phi/[\partial\phi(\boldsymbol{J})/\partial\boldsymbol{J}]$. For homogeneous dissipation functions of degree r we obtain $\lambda = 1/r$ and, hence, for $r = 2$, like $\rho\phi = \boldsymbol{J}\cdot\mathbb{L}^{-1}\cdot\boldsymbol{J}$, $\lambda = ½$ as before. However, the orthogonality principle is not universally accepted.[27] Fortunately, the "small-deviation-from-equilibrium" theory, labeled as *linear irreversible thermodynamics*, is often sufficiently accurate. Within this approach many useful results, based on different force and rate choices and their associated transformation relations, can be obtained. Possibly the most important one is the *Curie principle*, stating that for isotropic media, such as a (regular) fluid and an amorphous or polycrystalline solid, forces of a different tensorial nature do not couple. For these and many other aspects, we refer to de Groot and Mazur [33], Haase [41], Gyarmati [44], and Kuiken [39]. The (relatively simple) phenomenon of heat conduction and the (more complex) one of mass diffusion are discussed in the coming sections.

Summarizing, from a thermodynamic point of view the "only" remaining task for given material is to determine the Helmholtz function f and the dissipation function ϕ, and solve the resulting equations, which is generally, however, a not-so-easy task.

Example 6.2 ***Heat Conduction*** Probably pure heat conduction is the simplest irreversible phenomenon, at least if we limit the discussion to rigid (anisotropic), non-polarizable bodies without diffusion and internal reactions. In this case, the dissipation density $\rho\phi = \rho T\,s^{*(i)} = \rho l^{(d3)} = \boldsymbol{q}\cdot T\nabla(1/T) = \boldsymbol{q}\cdot[-(1/T)\nabla T]$. Identifying the rate as $\boldsymbol{J} = \boldsymbol{q}$ and the force as $\boldsymbol{X} = -(\nabla T)/T$, the linear approximation yields $\boldsymbol{J} = \mathbb{L}\cdot\boldsymbol{X}$. Hence,

$$\rho\phi = \boldsymbol{J}\cdot\boldsymbol{X} = \boldsymbol{J}\cdot\mathbb{L}^{-1}\cdot\boldsymbol{J} = \boldsymbol{q}\cdot\mathbb{L}^{-1}\cdot\boldsymbol{q}$$

27 Note that the dissipation function ϕ is *convex* guaranteeing that it is positive semi-definite ($\phi \geq 0$), conform the second law. The orthogonality principle in velocity space can be inverted to one in force space, provided ϕ is a function of the (internal and external) rates alone [35], that is, $\phi = \phi(\boldsymbol{J})$. Both are only valid for *purely dissipative processes* where gyroscopic forces – forces for which the power is always zero, such as Lorentz forces – are absent. This restriction is often tacitly assumed in thermodynamics and does not constitute a serious restriction.

The dissipation density $\rho\phi$ being homogeneous of degree $r = 2$, $\lambda = ½$. The force $\boldsymbol{X}$ is given either by inverting $\boldsymbol{J} = \mathbb{L} \cdot \boldsymbol{X}$ or by

$$\boldsymbol{X} = \xi\rho\partial\phi/\partial\boldsymbol{J} = \mathbb{L}^{-1} \cdot \boldsymbol{q} \quad \text{so that} \quad \boldsymbol{q} = \mathbb{L} \cdot \boldsymbol{X} = -(\mathbb{L}/T) \cdot \nabla T = -\boldsymbol{\kappa} \cdot \nabla T$$

which is *Fourier's law of thermal conduction* with *thermal conductivity*[28] $\boldsymbol{\kappa} = \mathbb{L}/T$. Now $\boldsymbol{q}$ is not the time derivative of a state variable so that the identification $\boldsymbol{J} = \boldsymbol{q}$ is questionable, but the parity argument of Woods shows that for each component j of X we have $X_j(t) = X_j(-t)$, so that $\mathbb{L}_{ij} = \overline{\mathbb{L}}_{ji}$. Hence, $\phi(t) = \phi(-t)$ and $q_j(t) = q_j(-t)$, as it should be. Finally, the reciprocal relations show that $\boldsymbol{\kappa}$ is symmetrical, even if not required by the (lattice) symmetry of the material.

Example 6.3 ***Mass Diffusion*** Possibly the most important application of irreversible thermodynamics is diffusion. In the absence of chemical reactions, volume forces, and viscous phenomena, so that we can assume mechanical equilibrium with[29] $\boldsymbol{\sigma} = \boldsymbol{\sigma}^{(q)}$, the terms $\boldsymbol{\sigma}^{(d)}{:}\mathbb{D}$, $\Sigma_i A_i J_j$ and $\Sigma_\alpha \boldsymbol{f}_\alpha \cdot \boldsymbol{J}_\alpha$ in Eq. (6.26) are absent. In this case, we identify, limiting ourselves to binary systems, as forces $\boldsymbol{X}_0 = T\nabla(1/T)$ and $\boldsymbol{X}_\alpha = T\nabla(-\mu_\alpha/T)$ with $\alpha = 1, 2$ and associated rates $\boldsymbol{J}_0 = \boldsymbol{q}$ and $\boldsymbol{J}_\alpha = \rho_\alpha(\boldsymbol{v}_\alpha - \boldsymbol{v})$. The linear relations therefore become

$$\boldsymbol{J}_k = L_{k0}T\nabla(1/T) + L_{k1}T\nabla(-\mu_1/T) + L_{k2}T\nabla(-\mu_2/T) \qquad \text{for} \qquad k = 0,1,2$$

Generally, the dissipation function is given by $\rho\phi = \sum_j^n \boldsymbol{J}_j \cdot \boldsymbol{X}_j$ but, because $\boldsymbol{J}_\alpha = \rho(\boldsymbol{v}_\alpha - \boldsymbol{v})$ and thus $\sum_\alpha^\omega \boldsymbol{J}_\alpha = \boldsymbol{0}$, it becomes $\rho\phi = \sum_j^{n-1} \boldsymbol{J}_j \cdot (\boldsymbol{X}_j - \boldsymbol{X}_n)$. Substituting $\boldsymbol{J}_k = \Sigma_k L_{kj}\boldsymbol{X}_j$ in $\Sigma_k \boldsymbol{J}_k = 0$ leads to $\Sigma_k L_{kj} = 0$ and, since $L_{kj} = L_{jk}$ to $\Sigma_k L_{jk} = 0$ and $\Sigma_{k,j} L_{kj} = 0$. For isothermal conditions of the binary case considered here, we thus have $L_{11} + L_{12} = 0$ and $\boldsymbol{J}_1 = -\boldsymbol{J}_2 = L_{11}\boldsymbol{X}_1{}'$ with $\boldsymbol{X}_1{}' = \boldsymbol{X}_1 - \boldsymbol{X}_2 = -\nabla\mu_1 + \nabla\mu_2$. Using the Gibbs–Duhem equation $x_1 \mathrm{d}\overline{\mu}_1 + x_2 \mathrm{d}\overline{\mu}_2 = w_1 \mathrm{d}\mu_1 + w_2 \mathrm{d}\mu_2 = 0$, we obtain $\boldsymbol{X}_1{}' = -w_2{}^{-1}\nabla\mu_1 = -w_1{}^{-1}\nabla\mu_2$. In isothermal binary diffusion there is thus only one phenomenological constant, which agrees with the idea that binary diffusion is a single phenomenon. A similar analysis for a ternary system results in three independent coefficients [33, 42].

The description in $T\nabla(1/T)$ and $T\nabla(-\mu_\alpha/T)$ can be converted into one using ∇T and ∇x_α with x_α the mole fraction of component α. As $\overline{\mu}_\alpha = \overline{\mu}_\alpha{}^\circ + RT \ln a_\alpha$ with activity $a_\alpha = \gamma_\alpha x_\alpha$ and mole fraction activity coefficient γ_α, we obtain

$$\nabla(-\overline{\mu}_\alpha/T) = -\Xi_\alpha \nabla T - (R/x_\alpha)\Gamma_\alpha \nabla x_\alpha$$

where $\Xi_\alpha = -\overline{\mu}_\alpha{}^\circ/T^2 + T^{-1}(\partial\overline{\mu}_\alpha{}^\circ/\partial T)$ and $\Gamma_\alpha = 1 + \partial \ln \gamma_\alpha/\partial \ln x_\alpha$ is the so-called *thermodynamic factor*. Limiting to unary systems, the rates for $k = 0,1$ become

$$J_k = (-L_{k0}/T - L_{k1}T\Xi_1)\nabla T - L_{k1}(R/a_1)T\Gamma_1\nabla x_1 \equiv -D_{k0}\nabla T - D_{k1}\nabla x_1$$

defining the diffusion coefficients D_{ij}. Note that in this formulation $D_{01} \neq D_{10}$ as ∇T and ∇x are not proper variables. Often the cross (or Dufour and Soret) effects are negligible and, if so and $\nabla T = 0$, we have *Fick's law of mass diffusion* $J_1 = -D_{11}\nabla x_1$ with *diffusivity* D_{11}. Similarly, if $\nabla x_1 = 0$, D_{00} represents the thermal conductivity $\boldsymbol{\kappa}$.

---

28 Some authors write this expression as $\boldsymbol{\kappa} = \boldsymbol{L}/T^2$ because they use $\rho s^{*(i)} = \boldsymbol{J} \cdot \boldsymbol{X}$ instead of $\rho T s^{*(i)} = \boldsymbol{J} \cdot \boldsymbol{X}$.

29 This relation is based on the experimental fact that for materials for which viscous effects can be neglected, mechanical equilibrium is much faster reached than chemical equilibrium. Note that both stress tensors are defined according to their outward normal so that for the traction $\boldsymbol{t} = \boldsymbol{t}'$ holds.

6.5 Final Remarks

The field of thermodynamics is extensive, and many books have been written on both the reversible approach, providing presentations with rather different emphasis, and the irreversible approach, for which the various books deal with a wide range of phenomena and with different representations and their transformations. Although we presented here an abstract outline, thermodynamics is deeply rooted in experiments and, perhaps remarkable, authors do not agree on certain aspects, while new aspects are still discovered. This is true for both classical and irreversible thermodynamics. For all this, we refer to the literature and books in the Further Reading list. Finally, for a mixed technical-historical treatment of thermodynamics, we refer to Bailyn (1994).

References

1 Redlich, O. (1962). *J. Phys. Chem. Ref. Data.* 66: 585.
2 Redlich, O. (1968). *Rev. Mod. Phys.* 40: 556.
3 Clausius, R. (1865). *Pogg. Ann. Phys. Chem.* 125: 353.
4 Planck, M. (1945). *Treatise on Thermodynamics*, 3e. after the 7th German ed. Dover (originally Longmans, Green & Co., 1926).
5 Buchdahl, H.A. (1975). *Twenty Lectures on Thermodynamics*. Rushcutters Bay, N.S.W: Pergamon Press (Australia).
6 Landsberg, P.T. (1961). *Thermodynamics with Quantum Statistical Illustrations*. New York: Wiley-Interscience.
7 Buchdahl, 1966.
8 Bridgman, P.W. (1941). *The Nature of Thermodynamics*. Harvard University Press (also Harper and Brothers, 1961); In spite of its age, still a very much to-the-point booklet.
9 Woods, 1986. This reference is highly recommended.
10 McGlashan, 1979.
11 Callen, 1960; Waldram, 1985; Bailyn, 1994. There are too many books on macroscopic thermodynamics to enumerate, but the Further Reading list indicates which ones I like.
12 Redlich, O. (1972). *Thermodynamics: Fundamentals, Applications*. Amsterdam: Elsevier.
13 Bazarov, I. (1964). *Thermodynamics*. Oxford: Pergamon Press.
14 Brown, M.R. and Uffink, J. (2001). *Stud. Hist. Phil. Mod. Phys.* 32: 525.
15 Kuhn, T.S. (1977). *The Essential Tension*. Chicago: University of Chicago Press.
16 Menger, K. (1950). *Am. J. Phys.* 18: 89 and 19: 476.
17 de Voe, H. (2013). *J. Chem. Ed.* 90: 591; Gislason, J. and Craig, N.C. (2005). *J. Chem. Therm.* 37: 954; (2007). *J. Chem. Ed.* 84: 499; Schmidt-Rohr, K. (2014). *J. Chem. Ed.* 91: 402.
18 Hatsopoulos, G.N. and Keenan, J.H. (1965). *Principles of General Thermodynamics*. New York: Wiley.
19 Haase, R. (1971). *Survey of Fundamental Laws*, Ch. 1 in *Thermodynamics, Physical Chemistry: An Advanced Treatment*, Vol. 1 (ed. H. Eyring, D. Henderson, and W. Jost), p. 1. New York: Academic Press.
20 Pippard, 1964.

21 Wilson, A.H. (1957). *Thermodynamics and Statistical Mechanics*, (1960) reprinted ed. Cambridge: Cambridge University Press.

22 Mello, P.R. and Rodriguez, R.F. (2013). *Eur. J. Phys.* 34: 83.

23 Guggenheim, 1967.

24 Fowler, R.H. and Guggenheim, E.A. (1939). *Statistical Thermodynamics*. London: Cambridge University Press.

25 Denbigh, 1981.

26 Belgiorno, F. (2002). *J. Phys. A* 36: 8165 and 8195.

27 Wreszinski, W.F. and Abdalla, E. (2009). *J. Stat. Phys.* 134: 781.

28 Nernst, W. (1926). *The New Heat Theorem*. London: Methuen (also Dover, 1969).

29 Eckart, C. (1940). *Phys. Rev.* 58: 267.

30 Smith, N.O. (1965). *J. Chem. Ed.* 42: 654.

31 Wilks, J. (1961). *The Third Law of Thermodynamics*. London: Oxford University Press; Eyring, H., Henderson, D., and Jost, W. (ed.) (1971). *The Third Law of Thermodynamics*, Ch. 6 in *Thermodynamics, Physical Chemistry: an Advanced Treatment*, p. 437. New York: Academic Press.

32 Donnelly, R.J. and Barenghi, C.F. (1998). *J. Phys. Chem. Ref. Data.* 27: 1217.

33 de Groot and Mazur, 1962.

34 Freeman, R.D. (1985). *J. Chem. Ed.* 62: 681.

35 Ziegler, H. (1983). *Introduction to Thermomechanics*, 2e. Amsterdam: North-Holland.

36 Bridgman, P.W. (1950). *Phys. Rev.* 22: 56.

37 Maugin, G.A. (1992). *The Thermomechanics of Plasticity and Fracture*. Cambridge: Cambridge University Press.

38 Muschik, W. (1979). *J. Non-Equilib. Thermodyn.* 4: 277 and 377.

39 Kuiken, 1994.

40 Lavenda, B.H. (1978). *Thermodynamics of Irreversible Processes*. London: MacMillan Press (also Dover, 1997).

41 Haase, 1969.

42 de Groot, S.R. (1951). *Thermodynamics of Irreversible Processes*. North-Holland Amsterdam.

43 Gyarmati, I. (1961). *Period. Polytech.* 5: 219.

44 Gyarmati, I. (1970). *Non-equilibrium Thermodynamics*. Berlin: Springer.

Further Reading

Bailyn, M. (1994). *A Survey on Thermodynamics*. New York: AIP Press. A mixed technical-historical text, giving a clear and detailed background on many macro- and micro-thermodynamical concepts.

Buchdahl, H. (1966). *The Concepts of Classical Thermodynamics*. London: Cambridge University Press. A nice booklet, deceptively looking "difficult" but actually highly readable with a clear discussion on basics.

Callen, H. (1960). *Thermodynamics*, (1985). *Thermodynamics and an Introduction to Thermostatistics*. New York: John Wiley & Sons. The first, now classic, textbook employing the postulational approach. The second edition, enlarged with statistical thermodynamics, omits some topics contained in the first.

de Groot, S.R. and Mazur, P. (1962). *Non-equilibrium Thermodynamics*. Amsterdam: North-Holland Publ. Comp. (also Dover, 1984). A classic in the field for the linear approximation.

Denbigh, K. (1981). *The Principles of Chemical Equilibrium*, 4e. London: Cambridge University Press. Clearly written textbook dealing with both macroscopic and elementary statistical topics.

Demirel, Y. (2014). *Nonequilibrium Thermodynamics*, 3e. Amsterdam: Elsevier. An elaborate modern text covering a broad range of topics and providing many examples.

DeVoe, H. (2001). *Thermodynamics and Chemistry*. Upper Saddle River, NJ: Prentice Hall. Highly readable and focusing, conform the title, entirely on chemistry.

Guggenheim, E.A. (1967). *Thermodynamics*, 5e. Amsterdam: North-Holland. After a compact and strict introduction of basics, dealing with physical–chemical phenomena while being rigorous.

Haase, R. (1969). *Thermodynamics of Irreversible Processes*. Reading, MA: Addison-Wesley Publ. Comp. A detailed exposition on the basics and many applications.

Honig, J.M. (2007). *Thermodynamics*, 3e. London: Academic Press. A modern treatise, similar in scope and approach as Guggenheim, but having a rather different style.

Kuiken, G.D.C. (1994). *Thermodynamics of Irreversible Processes*. Chichester: John Wiley & Sons. A clearly written and appealing book containing the basics and applications to diffusion and rheology.

McGlashan, M.L. (1979). *Chemical Thermodynamics*. London: Academic Press. An illuminating and precise book that emphasizes the link with experiments.

Pippard, A.B. (1957). *The Elements of Classical Thermodynamics*. Cambridge: Cambridge University Press, corrected ed. (1964). Another classic booklet dealing mainly with basic aspects.

Waldram, J.R. (1985). *The Theory of Thermodynamics*. Cambridge: Cambridge University Press. Provides a broad overview on both macroscopic and statistical thermodynamics.

Woods, L.C. (1975). *The Thermodynamics of Fluid Systems*. Oxford: Oxford University Press, reprinted and corrected ed. (1986). A gem of a book dealing with classical, internal variable and statistical thermodynamics of (plasma-like) fluids, but, along the lines, elaborating on many issues.

7

Microscopic Thermodynamics

In the description of matter, the link between microscopic considerations and concepts to macroscopic aspects is given by the last "tool," statistical thermodynamics (ST). In this chapter, we introduce and outline ST in a concise way. First, we deal with the basic formalism and apply it to noninteracting molecules. Thereafter, the effects of intermolecular interactions and intramolecular contributions are discussed. We conclude with some general aspects.

7.1 Basics of Statistical Thermodynamics

While classical thermodynamics provides relations between macroscopic properties, ST relates the macroscopic properties of a system, a set of molecules, to the microscopic phenomena, that is, to molecular behavior.

7.1.1 Preliminaries

Let us first describe a system by classical particle mechanics (PM). If the system is a single molecule, one may proceed as follows. In PM, the system is described by *Hamilton's equations* (Section 2.4). Each system can be characterized by a set of n generalized coordinates and the associated momenta ($2n$ degrees of freedom). A state of a molecule can thus be depicted as a point in a $2n$-dimensional Cartesian space whose axes are labeled by the allowed momenta and coordinates of the molecule. This space is called the (molecule) *phase space* or μ-space. If we have a system containing many non or weakly interacting molecules, such as in a volume of gas, the collective of points describes the gas in μ-space, with each point describing a molecule. Similar considerations hold in quantum mechanics (QM). However, in many cases, we need the behavior of a macroscopic system, that is, the collective behavior of many, strongly interacting molecules. Now four problems arise:

- The size of the macroscopic system, containing many particles and thus a large number of degrees of freedom (DoFs), renders a general solution (both in PM and in QM) highly unlikely to be found.
- The initial conditions of such systems are unknown so that, even if a general solution is possible in principle, a particular solution cannot be obtained.

Phases of Matter and their Transitions: Concepts and Principles for Chemists, Physicists, Engineers, and Materials Scientists, First Edition. Gijsbertus de With.

- Even if the initial conditions are known, they have a limited accuracy. Note that, for example, Avogadro's number is "only" known with an accuracy of about 10^{-7}. Since it appears that the relevant equations are extremely sensitive to small changes in initial conditions, this rapidly leads to chaotic behavior (Section 19.2).
- It is difficult to incorporate interaction between molecules and with the environment.

To overcome these problems, one uses the *ensemble* – a large (virtual) collection of identical systems with the same Hamilton function but different initial conditions. By taking a $2nN$-dimensional space, where N is the total number of molecules in the system, we can make a similar representation as before. The axes are labeled with the allowed momenta and coordinates of all the molecules. This enlarged space is called the (gas) *phase space* or Γ-space. To each system in the ensemble there corresponds a representative *phase point* in the $2nN$-dimensional Γ-space. Since the system evolves in time as described by Hamilton's equations, the phase point will describe a path as a function of time in Γ-space, and this path is called a *trajectory*. Although a trajectory can visit any point in phase point, its direction at each phase point is determined by Hamilton's equations, so that trajectories do not cross. The ensemble can be depicted as a swirl of points in Γ-space, and the macroscopic state of a system is described by the average behavior of this swirl. Since the macroscopic system considered has a large number of DoFs, the density of phase points can be considered as continuous in many cases; this density is denoted by $\rho(\mathbf{q}, \mathbf{p})$, where $\mathbf{q}$ and $\mathbf{p}$ denote the collective of coordinates and momenta, respectively (Section 2.1). Obviously, $\rho(\mathbf{q}, \mathbf{p}) \geq 0$ and we take it normalized, that is, $\int \rho(\mathbf{q}, \mathbf{p})\,\mathrm{d}\mathbf{q}\mathrm{d}\mathbf{p} = 1$. One can prove from the equations of motion that this density is constant if we follow along with the swirl, a general theorem called *Liouville's theorem* (Section 2.4). A function $F(\mathbf{q}, \mathbf{p})$ in phase space representing a certain property is called a *phase function*, the Hamilton function $\mathscr{H}$ providing an important example. To calculate the average of $F(\mathbf{q}, \mathbf{p})$ over the ensemble, one uses the phase average

$$\langle F\rangle_\Gamma = \int_\Gamma F(\mathbf{q},\mathbf{p})\rho(\mathbf{q},\mathbf{p})\mathrm{d}\mathbf{q}\mathrm{d}\mathbf{p}$$

We discuss ST using a quantum description, since that approach is, arguably, conceptually simpler, and in a later stage make the transition to classical ST. However, the ensemble language of classical statistical mechanics, as systematically employed by Gibbs [1], Tolman [2], and Phillies [3], is frequently used in literature, even when dealing with quantum systems.

The first question is how to characterize a macroscopic state in terms of molecular parameters. It might be useful to recall that a macroscopic system is considered to possess a *thermodynamic* or *macrostate*, characterized by a limited set of macroscopic parameters, such as the volume V and the temperature T. However, such a system contains many particles, for example, atoms, molecules, electrons, and photons. A macroscopic system is thus also a quantum system, that is, an object that contains energy and particles and is identified by a large set of distinct *quantum* or *microstates* of the system. Identical particles – for example, the electrons of a molecule – are objects that have access to the same subset of microstates. From these considerations, we conclude that *each macrostate contains many microstates* and describe the macroscopic state of the system by the fraction p_j of each possible quantum state j in the macroscopic system. With each quantum state j, an energy E_j and a number of particles N_j are associated. The probability of a system being in state j

can in classical terms be interpreted as the fraction of systems in the ensemble in state j. In that case, in view of the uncertainty relations, j refers to a volume in Γ-space of size h^{nN}, labeled by a particular set $(\mathbf{q}, \mathbf{p})$. The simple label j implicates the complete description of the microstate, while the distribution over microstates j describes the macroscopic system, and thus the whole hides a considerable complexity.

Further, it should be noted that the number of quantum states increases rapidly with increasing energy. For large systems, the states of the system will form a quasi-continuum that can be described by the *density of states* g, indicating the number of quantum states $\mathrm{d}n$ for a certain energy range ΔE, given by

$$g(E)\,\Delta E = (\mathrm{d}n/\mathrm{d}E)\,\Delta E$$

where ΔE is the *accessibility range*, an energy range small as compared to E, but large when compared to the Heisenberg uncertainty δE. For a macroscopic system, g is astronomically large, related to the temperature (Problem 7.3), and increases extremely rapidly with the size of the system, as shown in Example 7.1 for translation.

Example 7.1 ***The Density of States for Translation*** First, we calculate the volume G of a hypersphere with radius R and of dimension D, given by $AR^D = A(R^2)^{D/2}$. To calculate A, we consider the integral

$$I = \int \mathrm{e}^{-R^2}(\mathrm{d}G/\mathrm{d}R^2)\,\mathrm{d}R^2 = \tfrac{1}{2}AD\int \mathrm{e}^{-R^2}(R^2)^{(D/2)-1}\mathrm{d}R^2 = \tfrac{1}{2}AD\Gamma(D/2)$$

with $\Gamma(t)$ the gamma function. The volume can also be written as $G = \int h(R^2 - \mathbf{x}^\mathrm{T}\mathbf{x})\mathrm{d}\mathbf{x}$, with $h(t)$ the Heaviside function and $\mathbf{x}^\mathrm{T}\mathbf{x} = \Sigma_j x_j^2$. So, $\mathrm{d}G/\mathrm{d}R^2 = \int \delta(R^2 - \mathbf{x}^\mathrm{T}\mathbf{x})\mathrm{d}\mathbf{x}$, using the Dirac function $\delta(t) = \mathrm{d}h(t)/\mathrm{d}t$ (Section A.14). Therefore, the integral I also becomes

$$I = \int\int \exp(-R^2)\delta(R^2 - \mathbf{x}^\mathrm{T}\mathbf{x})\mathrm{d}R^2\mathrm{d}\mathbf{x} = \int \exp(-\mathbf{x}^\mathrm{T}\mathbf{x})\mathrm{d}\mathbf{x} = \pi^{D/2}$$

Combining yields $A = 2\pi^{D/2}/D\Gamma(D/2) = \pi^{D/2}/\Gamma(\tfrac{1}{2}D+1)$. Consider now N particles of mass m in a three-dimensional (3D) box with edges a so that the energy E can be written as $E = (h^2/8ma^2)\mathbf{n}^\mathrm{T}\mathbf{n} = (h^2/8ma^2)\Sigma_j n_j^2$, where $\mathbf{n} = n_j$ covers the set $n_{1x}, n_{1y}, \ldots, n_{Nz}$ with $n_j \in (\pm1, \pm2, \pm3, \ldots)$. The degeneracy of energy levels is given by the number of ways the integer $\mathbf{n}^\mathrm{T}\mathbf{n} = 8ma^2E/h^2$ can be written as a sum of squares of integers. To estimate the degeneracy, consider a D-dimensional space with the axes labeled with $\mathbf{n}$. The number of lattice points inside a hypersphere with radius squared $\mathbf{n}^\mathrm{T}\mathbf{n} = 8ma^2E/h^2$ provides the number of energy levels. Equating $\mathbf{n}^\mathrm{T}\mathbf{n}$ to R^2, we have

$$G = A\left(\frac{8ma^2E}{h^2}\right)^{D/2} = \frac{\pi^{D/2}}{\Gamma(\tfrac{1}{2}D+1)}\left(\frac{8ma^2E}{h^2}\right)^{D/2} = \frac{2^D}{\Gamma(\tfrac{1}{2}D+1)}\left(\frac{2\pi ma^2E}{h^2}\right)^{D/2}$$

with $D = 3N$. The final step is to realize that we need only one "octant" of the hypersphere, so we have to divide by 2^D, and that the states are indistinguishable, so we have to divide by $N!$ (Section 7.3). The volume G and density of states $g(E)$ thus become

$$G = (2\pi ma^2E/h^2)^{D/2}/N!\Gamma(\tfrac{1}{2}D+1) \quad \text{and} \quad g(E) = \mathrm{d}G(E)/\mathrm{d}E \approx E^{(3N/2)-1}$$

For $N = 1$, using $E = 3kT/2$, $T = 300$ K, $m = 10^{-22}$ g, and $a = 10^{-2}$ m, we obtain $g \cong 10^{30}$, which is a large number. For N particles, using again $T = 300$ K, $m = 10^{-22}$ g, and $a = 10^{-2}$ m, but now with $N = 6\times10^{23}$ and $E = 3NkT/2$, we obtain $g \cong 10^N$, which is an extremely large number. Finally, the expression for g shows that it increases extremely rapidly with the number of particles in the system.

7.1.2 Entropy and Partition Functions

Since in thermodynamics equilibrium is reached at maximum entropy $S(U, V, N)$, given energy U, volume V, and number of particles N, it seems natural in ST to start with defining the entropy. For the statistical representation of S various expressions can be given.[1] Since the macroscopic system is characterized by the p_js, we will have $S = S(p_j;U, V, N)$, or $S(p_j)$ for short.[2] Probably the most direct way to obtain S for a system having n states is to require the following properties [4]:

- $S(p_j) \leq S(1/n)$. This statement implies that equiprobability for states with the same energy yields the maximum entropy and thus the equilibrium state. In fact, this is the basic postulate of ST.
- $S(p_j,0) = S(p_j)$. This statement requires that if a state cannot be occupied it, will not contribute to the entropy.
- $S = S_A + S_B$ for noninteracting systems A and B. This condition delivers the usual additivity of entropy. For interacting systems, the condition changes to $S = S_A + S_B{}^{(A)} \leq S_A + S_B$, where $S_B{}^{(A)}$ denotes the conditional entropy of system B, that is, the entropy of system B given the entropy of system A.

These requirements have a unique solution (Justification 7.1) apart from a multiplicative constant, usually chosen in such a way that the statistical entropy thus defined corresponds to the conventional thermodynamic entropy, and this constant appears to be Boltzmann's constant k. The *(Gibbs) entropy* S is then defined by

$$S \equiv -k\sum\nolimits_j p_j \ln p_j \tag{7.1}$$

Justification 7.1 ***The Gibbs Entropy*** To show that, given the three properties indicated above, Eq. (7.1) represents the unique solution for the entropy, we argue as follows [4, 5]. First, we set

$$L(n) \equiv S(1/n, \ldots, 1/n)$$

as an abbreviation. Using properties 1 and 2, we have

$$L(n) = S(1/n, \ldots, 1/n, 0) \leq S(1/(n+1), \ldots, 1/(n+1)) = L(n+1)$$

so that $L(n)$ is a nondecreasing function of n. Now, taking m and r as positive integers, consider m mutually independent systems $A_1, A_2, \ldots, A_m$, each containing r equally likely states. So, we have $S(A_k) = S(1/r,\ldots,1/r) = L(r)$ for $1 \leq k \leq m$. Using property 3, since the systems are independent, we have

$$S(A_1A_2 \ldots A_m) = \Sigma_k S(A_k) = mL(r)$$

1 Formally, the statistical representation of the S should be denoted by another symbol, for example, $\langle S\rangle$, see Chapter 19. Here, considering equilibrium mainly, we identify right away $\langle S\rangle$ with S.

2 The QM probability p_j for a single system corresponds in PM to the fraction of systems of the ensemble being in a volume of phase space of size h^{nN}, where n is the number of DoFs and N the number of molecules and labeled by $(\mathbf{q}, \mathbf{p})$, see Section 7.3.

Since the combined (product) system $A_1A_2...A_m$ contains r^m equally likely events, we have for its entropy also $L(r^m)$ and therefore

$$L(r^m) = mL(r)$$

This relation holds for any other pair of integers m and r. As a next step, take three arbitrary integers r, s, and n, and a number m determined by

$$r^m \leq s^n \leq r^{m+1} \quad \text{or} \quad m \ln r \leq n \ln s \leq (m+1) \ln r \quad \text{or}$$

$$m/n \leq \ln s/\ln r \leq (m+1)/n \tag{7.2}$$

Since $L(n)$ is monotonous, we have also

$$L(r^m) \leq L(s^n) \leq L(r^{m+1}) \quad \text{or} \quad mL(r) \leq nL(s) \leq (m+1)L(r) \quad \text{or}$$

$$m/n \leq L(s)/L(r) \leq (m+1)/n \tag{7.3}$$

By combining Eqs. (7.2) and (7.3), we have

$$| L(s)/L(r) - \ln s/\ln r | \leq 1/n \tag{7.4}$$

Equation (7.4) is independent of m, and since n can be arbitrarily large, we obtain

$$L(s)/L(r) = \ln s/\ln r$$

which implies, since r and s are arbitrary, that $L(n) = \lambda \ln n$, with $\lambda \geq 0$ a nonnegative constant as $L(n)$ is monotonous.

So far, the special case $p_k = 1/n$ $(1 \leq k \leq n)$ is proved, and now we turn the case where p_k can be any rational number. Suppose we have $p_k = g_k/g$ with $g = \Sigma_k g_k$, where the g_k are all positive integers. Further, suppose that we have a system A consisting of n states with probabilities $p_1, p_2,..., p_n$. We also have a specially devised system B, dependent on A and consisting of n groups of states such that the kth group contains g_k states. If state A_k is realized in system A, then in system B all the g_k events of the kth group have the same probability $1/g_k$, while all the events of the other groups have probability zero. In this case, the system B reduces to a system of g_k equally probable states with conditional entropy $S_B{}^{(k)} = S(1/g_k,...,1/g_k) = L(g_k) = \lambda \ln g_k$, implying that

$$S_B^{(A)} = \sum\nolimits_k S_B^{(k)} = \lambda \sum\nolimits_k p_k \ln g_k = \lambda \sum\nolimits_k p_k \ln p_k + \lambda \ln g$$

Consider now the product system AB containing A_kB_l states $(1 \leq k \leq n, 1 \leq l \leq g)$. A state in this scheme is only possible if B_l belongs to the kth group, so that the number of possible states A_kB_l for a given k is g_k. The total number of states in system AB is then $\Sigma_k g_k = g$, and the probability of each possible state A_kB_l is $p_k/g_k = 1/g$. Hence, the system AB consists of g equally likely states and thus $S = L(g) = \lambda \ln g$. We now use property 3, $S = S_A + S_B{}^{(A)}$, to obtain

$$\lambda \ln g = S_A + \lambda \sum\nolimits_k p_k \ln p_k + \lambda \ln g \quad \text{or} \quad S_A = -\lambda \sum\nolimits_k p_k \ln p_k \tag{7.5}$$

Since the entropy must be continuous, Eq. (7.5) is valid not only for the rational numbers $p_1, p_2,..., p_n$ but also for any value of its arguments. Note that, although we used a specially devised system B, the result is independent of system B. The constant λ can be identified with Boltzmann's constant k by calculating the pressure for a perfect gas, Eq. (7.18), leading to $\lambda = k$, which we use from now on. Finally, we mention that other ways to introduce the entropy expression can be found in [1–3, 6].

A short route to obtain the relevant thermodynamic expressions runs as follows. We consider an *open system*[3] with volume V to be in contact with a thermal bath characterized by the temperature T, and a particle bath characterized by the chemical potential μ. Therefore, the conjugate variables, pressure P, energy U, and number of particles N, fluctuate. Alternatively, we may assume that the system is in contact with many other, similar systems. For such a system we may assume that we are always near equilibrium. The probabilities of the system should be normalized, that is,

$$\sum\nolimits_i p_i = 1 \tag{7.6}$$

In addition, we require that the average energy $\langle H\rangle$ of the system[4] is constant, that is,

$$\sum\nolimits_i p_i E_i = \langle H\rangle$$

and that the average number of particles $\langle N\rangle$ of the system[5] is constant, that is,

$$\sum\nolimits_i p_i N_i = \langle N\rangle$$

Conform macroscopic thermodynamics, we seek the maximum of the entropy $S = -k\,\Sigma_i\, p_i \ln p_i$, given the abovementioned constraints. This maximum can be obtained using the Lagrange method of undetermined multipliers (Section A.4). Taking these multipliers as $-k\alpha$, $-k\beta$, and $-k\gamma$, the maximum is obtained from

$$\frac{\partial}{\partial p_i}\left[S(p_i) - k\alpha\left(\sum\nolimits_i p_i - 1\right) - k\beta\left(\sum\nolimits_i p_i E_i - \langle H\rangle\right) - k\gamma\left(\sum\nolimits_i p_i N_i - \langle N\rangle\right)\right] = 0$$

and after some calculation this expression leads to

$$p_i^* = \exp[-(1+\alpha+\beta E_i+\gamma N_i)]$$

The normalization condition, Eq. (7.6), yields

$$\sum\nolimits_i p_i^* = \exp[-(1+\alpha)]\sum\nolimits_i \exp[-(\beta E_i+\gamma N_i)] = 1$$

so that we may define

$$\Xi \equiv \sum\nolimits_i \exp[-(\beta E_i+\gamma N_i)] = \exp(1+\alpha)$$

For the probabilities we thus obtain

$$p_i^* \equiv p_i^*(\mu, V, T) = \exp[-(\beta E_i+\gamma N_i)]/\Xi$$

The function Ξ is the macrocanonical or grand canonical partition function, often just called the *grand partition function*. We find for the entropy, meanwhile using the normalization condition $\Sigma_i\, p_i^* = 1$,

$$S = k\beta\langle H\rangle + k\gamma\langle N\rangle + k\ln\,\Xi$$

We compare this expression with the thermodynamic expression for the entropy

$$S = (U - \mu N + PV)/T$$

3 For simplicity of notation, we assume only one type of particle.

4 We write $\langle H\rangle$ since E represents the numerical value of the energy, while the Hamilton operator H represents the energy expression (Section 3.1.2).

5 Similar to H representing the energy operator, N represents the number operator. For all purposes in this book, it can be taken as the number of particles.

If we identify the average microscopic energy $\langle H \rangle$ with the internal energy U and the average number of particles $\langle N \rangle$ with the macroscopic number N, we obtain

$$\beta = 1/kT, \quad \gamma = -\mu/kT \quad \text{and} \quad \ln \Xi = PV/kT$$

so that we finally have the solution, known as the *Gibbs distribution*,

$$p_i^* = \exp[(\mu N_i - E_i)/kT]/\Xi \tag{7.7}$$

Since we are only using the values of p_i corresponding to the maximum entropy (or close to it), in the sequel we omit the asterisk and just write p_i. For future reference, we also write the expressions[6] for p_i and Ξ as

$$p_i = \lambda^{N_i} \exp(-\beta E_i)/\Xi \quad \text{and}$$

$$\Xi = \sum\nolimits_i \lambda^{N_i} \exp(-\beta E_i) \quad \text{with} \quad \lambda = \exp(-\gamma) = \exp(\beta\mu)$$

where λ is the *absolute activity* (Section 6.1.8). The grand partition function is directly related to the grand potential Ω, since we have $-PV = \Omega$ and thus $\Omega = -kT \ln \Xi$. The thermodynamic properties can be calculated in the usual thermodynamic way by realizing that the natural variables for Ξ are T, V, and μ, that is, $\Omega = \Omega(T, V, \mu)$. Hence, we have $S = -\partial\Omega/\partial T$, $P = -\partial\Omega/\partial V$, and $N = -\partial\Omega/\partial\mu$.

If the number of particles is fixed at, say N, we have a *closed system*, the constraint characterized by γ is removed, and one can take $\gamma = 0$. Therefore,

$$p_i \equiv p_i(N, V, T) = \exp(-E_i/kT)/Z_N \quad \text{with} \quad Z_N = \sum\nolimits_i \exp(-E_i/kT) \tag{7.8}$$

Here, Z_N denotes the canonical partition function for N particles, often labeled as just Z and addressed as the *partition function*. The distribution is known as the *Boltzmann distribution*, and p_i is the *Boltzmann probability*. For the entropy, we have in this case

$$S = k\beta\langle H \rangle + k \ln Z$$

which upon comparison with

$$S = (U + PV - \mu N)/T = (U - F)/T$$

results again in $\beta = 1/kT$, as expected, and $F = -kT \ln Z$. From the Helmholtz energy F we obtain the internal energy U, entropy S, pressure P, and chemical potential μ:

$$U = F + TS = \frac{\partial(F/T)}{\partial(1/T)} = kT^2\left(\frac{\partial \ln Z}{\partial T}\right)_V \quad S = -\frac{\partial F}{\partial T} = k \ln Z + kT\left(\frac{\partial \ln Z}{\partial T}\right)_V \tag{7.9}$$

$$P = -\left(\frac{\partial F}{\partial V}\right)_T = kT\left(\frac{\partial \ln Z}{\partial V}\right)_T \quad \mu = \left(\frac{\partial F}{\partial N}\right)_{T,V} = -kT\left(\frac{\partial \ln Z}{\partial N}\right)_{T,V} \tag{7.10}$$

We note that using Z_N, we can write $\Xi = \Sigma_N \lambda^N Z_N$, where the original index i has been replaced by the pair (N, j) with j the index used in the expression for Z_N.

Finally, if the energy is also fixed, we have an *isolated system*, the constraint characterized by β is also removed and one can take $\beta = 0$ as well. In this case, the probability p_i describes the *microcanonical distribution*

$$p_i \equiv p_i(N, V, E) = 1/W \tag{7.11}$$

6 Although β is identified as $1/kT$, for convenience the symbol β is still frequently used.

where $W = \Sigma_i(1)$ is the number of accessible states for the macroscopic system at fixed internal energy U.[7] In this case, the entropy expression $S(p_i) = -k\,\Sigma_i\, p_i \ln p_i$ becomes the *Boltzmann entropy*

$$S = k \ln W$$

Writing this expression as $-TS = -kT \ln W$ renders the expressions for the relation between the potentials ($\Omega = -PV = U - TS - \mu N$, $F = U - TS$, $-TS$) and the partition functions (Ξ, Z, W) all similar. Once having calculated the p_is, the *mean* or *average value* for any property X of the system can be calculated using

$$\langle X\rangle = \Sigma_i p_i X_i$$

where X_i denotes the value in microstate i. This expression is valid for all three distributions discussed, and adopting this view we can write for the entropy $S = -\,k\,\Sigma_i p_i \ln p_i = -\,k\langle \ln p\rangle$.

To show that the entropy so defined (for situations close to equilibrium) can be identified as the thermodynamic entropy (originally defined for equilibrium situations only), we calculate $\mathrm{d}S$ for a closed system using $\ln p_i = -E_i/kT - \ln Z$ to obtain

$$\mathrm{d}S = -k\sum\nolimits_i \mathrm{d}(p_i \ln p_i) = -k\sum\nolimits_i (\ln p_i + 1)\,\mathrm{d}p_i = \sum\nolimits_i E_i\,\mathrm{d}p_i/T \tag{7.12}$$

Here, we have used the definition of the probability p_i and twice the fact that $\sum_i \mathrm{d}p_i = 0$. Hence, for reversible heat flow, we have $đQ = T\,\mathrm{d}S = \Sigma_i\, E_i\,\mathrm{d}p_i$. The total energy of the system is given by $\langle H\rangle = \Sigma_i\, E_i\, p_i$ and its differential by $\mathrm{d}\langle H\rangle = \Sigma_i(E_i\,\mathrm{d}p_i + p_i\,\mathrm{d}E_i)$. From thermodynamics we know that the internal energy U is given by the first law of thermodynamics $\mathrm{d}U = đW + đQ$, where $đW$ and $đQ$ denote the increment in work and heat, respectively. If we again identify U with $\langle H\rangle$, we can associate $\Sigma_i\, p_i\,\mathrm{d}E_i$ with $đW$ since $\Sigma_i\, E_i\mathrm{d}p_i$ corresponds to $T\mathrm{d}S = đQ$. Furthermore, from Eq. (7.12), one can show that $\partial S/\partial U = 1/T$ (Problem 7.1).

Finally, one might wonder what the partition function represents. In fact, it just provides the number of *complexions*, that is, the number of microstates that are thermally accessible for the system compatible with the constraints applied. This is particularly clear for the microcanonical ensemble ($S = k\,\ln W$). Actually, W can always be interpreted likewise. For a further discussion, see Chapter 19.

7.1.3 Fluctuations

The difference between the results from the grand partition function Ξ and the partition function Z is usually small for macroscopic systems, that is, fluctuations in the number of molecules and in energy are generally not terribly important. This can be seen as follows. First, note that we can add another term Yy to the Euler equation $U = TS - PV + \mu N$, leading to the extra constraint $\Sigma_j p_j y_j - \langle y\rangle = 0$ to the entropy expression [8]. Second, generalizing by denoting the properties by a_j ($a_1 = U, a_2 = N, \ldots$), the associated conjugated multipliers by $x_j = -\beta A_j$ ($x_1 = \beta, x_2 = \gamma, \ldots$) with A_j auxiliary variables ($A_1 = -1, A_2 = \mu, \ldots$), we can

7 The uncertainty relations tell us that the energy E for any system has an uncertainty δE. Moreover, exactly defined energies are not realizable experimentally and cover a range ΔE. Typically, it holds that $\Delta E \gg \delta E$. Actually, we have to write $W = g(E)\Delta E$ with $g(E) = \mathrm{d}n/\mathrm{d}E$ the density of states and ΔE the *accessibility range* for the energy E, in which we can take $g(E)$ as a constant. However, $\ln(\Delta E)$ is usually completely negligible as compared to $\ln g(E)$, even if $\Delta E = \delta E$ [7], and therefore omitted.

write $\Xi = \text{tr}\exp(-\Sigma_j x_j a_j) = \text{tr}\exp(-\mathbf{x}^\mathrm{T}\mathbf{a})$, where tr (trace) denotes the summation over the probabilities for the properties a_j. Differentiating $\ln \Xi$ with respect to x_j, one obtains

$$\partial \ln \Xi/\partial \mathbf{x} = -\langle \mathbf{a} \rangle \quad \text{and} \quad \partial^2 \ln \Xi/\partial \mathbf{x}\partial \mathbf{x}' = -\partial\langle \mathbf{a}\rangle/\partial \mathbf{x}' = C_{\mathbf{aa}'} \tag{7.13}$$

where $C_{\mathbf{aa}'} \equiv \langle \mathbf{a}^\mathrm{T}\mathbf{a}'\rangle$ is the *correlation matrix* [3]. Equivalently, using the definition $\delta\mathbf{a} \equiv \mathbf{a} - \langle \mathbf{a}\rangle$, $C_{\delta\mathbf{a}\delta\mathbf{a}'} = \langle \mathbf{a}^\mathrm{T}\mathbf{a}'\rangle - \langle \mathbf{a}^\mathrm{T}\rangle\langle \mathbf{a}'\rangle$. Because Ξ is symmetric in $\mathbf{x}$ and $\mathbf{a}$, similar expressions result upon differentiating with respect to a_j. Applying these results for $A_j = N_j$ and $A_k = N_k$, or $x_j = -\beta\mu_j$ and $x_k = -\beta\mu_k$, we obtain

$$\partial\langle N_j\rangle/\partial\mu_k = \partial\langle N_k\rangle/\partial\mu_j = \beta\left(\langle N_j N_k\rangle - \langle N_j\rangle\langle N_k\rangle\right)$$

which reduces, if $j = k$, to $\partial\langle N\rangle/\partial\mu = \beta(\langle N^2\rangle - \langle N\rangle^2) \equiv \beta\sigma_N{}^2$, where the *variance* $\sigma_X{}^2 \equiv C_{\delta X\delta X} = \langle \delta X^2\rangle = \langle X^2\rangle - \langle X\rangle^2$ is used. For $\langle N\rangle$ we also have

$$\left(\frac{\partial\langle N\rangle}{\partial\mu}\right)_{V,T} = \left(\frac{\partial N}{\partial\mu}\right)_{V,T} = \left(\frac{\partial N}{\partial P}\right)_{V,T}\left(\frac{\partial P}{\partial\mu}\right)_{V,T} \quad \text{and} \quad \left(\frac{\partial P}{\partial\mu}\right)_{V,T} = \frac{N}{V}$$

Since the derivative $(\partial N/\partial P)_{V,T}$ is at constant V, we can write $(\partial N/\partial P)_{V,T} = V[\partial(N/V)/\partial P]_{V,T} = V[\partial(N/V)/\partial P]_{N,T} = NV[\partial(1/V)/\partial P]_{N,T} = N\kappa_T$, where the second step can be made because N/V is an intensive quantity. Therefore, we have $(\partial N/\partial\mu)_{V,T} = N^2\kappa_T/V$. Hence, the relative fluctuation in number of molecules in the system σ_N/N is given by

$$(\sigma_N/N)^2 = kT\kappa_T/V \tag{7.14}$$

and is therefore negligible for large N. A property X for which $\sigma_X \ll \langle X\rangle$ is sometimes called a *normal property*. The fluctuation in energy for an open system can be derived similarly as (Problem 7.5)

$$\sigma_E^2 = kT^2 C_V + (\partial\langle H\rangle/\partial\langle N\rangle)\,\sigma_N^2$$

It remains to be discussed, of course, whether ensemble fluctuations represent fluctuations of a single system in time. For this discussion, see Chapter 19.

Problem 7.1 ***Work and the Grand Canonical Distribution*** Show, using Eq. (7.12), that $\partial S/\partial U = 1/T$, and verify from the expression for $\mathrm{d}U$ and the statistical expression for $\mathrm{d}S$ that $đW$ in the grand canonical ensemble is given by

$$đW = \Sigma_i p_i \mathrm{d}E_i + \mu\Sigma_i N_i \mathrm{d}p_i \quad \text{corresponding} \quad đW = -P\mathrm{d}V + \mu\mathrm{d}N.$$

Problem 7.2 ***The Two-State Model*** A simple model in statistical mechanics is the two-state model with (obviously) two states: state 1 with energy 0 and state 2 with energy ε.

(a) Give the expression for the partition function Z.
(b) For $T = \varepsilon/k$, with k Boltzmann's constant, calculate the probabilities p_1 and p_2.
(c) Calculate the internal energy U and heat capacity C_V, and sketch the $C_V(T)$ behavior.

Problem 7.3 ***Temperature and the Density of States*** Show that T is related to $g(E)$ for a system with energy E_s via $1/kT = \partial\,\mathrm{ln}g(E)/\partial E|_{E=Es}$.

Problem 7.4 ***Entropy and the Microcanonical Distribution*** Show in detail for macrosystems, although we have $S = k\ln[g(E)\Delta E]$, that S is essentially independent of ΔE.

Consider to that purpose the value of ΔE based on the quantum uncertainty relations δE and a typical experimental uncertainty ΔE.

Problem 7.5 ***The Correlation Function and Energy Fluctuations*** Derive Eq. (7.13). Also show that $(\partial P/\partial \mu)_{V,T} = N/V$ is generally valid. Derive $\sigma_E{}^2 = kT^2 C_V$ for a closed system similarly as $(\sigma_N/N)^2 = kT\kappa_T/V$ is derived for the open system. For an open system, show that $\sigma_E{}^2 = kT^2 C_V + (\partial\langle H\rangle/\partial\langle N\rangle)\sigma_N{}^2$.

7.2 Noninteracting Particles

Let us apply the statistical mechanics ideas to noninteracting particles. The simplest model is that of a perfect gas consisting of particles without internal structure and without any mutual interaction contained in a box. To evaluate the partition function, we first describe the partition function for a single such particle. Thereafter, we extend the result to many particles and in a later section to many interacting particles.

7.2.1 Single Particle

From QM we know that for a particle of mass m in a 3D box with edge l and volume $V = l^3$, we have for the energy levels (Section 3.2.1)

$$\varepsilon_{\mathbf{n}} = \frac{\hbar^2}{8ml^2}(2\pi\mathbf{n})^2 = \frac{\hbar^2}{2m}\mathbf{k}^2 = \frac{\mathbf{p}^2}{2m}$$

where $2\pi\hbar = h =$ Planck's constant and $\mathbf{n} = (n_1, n_2, n_3)$ with $(n_1, n_2, n_3) \in (\pm 1, \pm 2, \pm 3, \ldots)$. The momentum $\mathbf{p} = \hbar\mathbf{k}$ of such a particle is given by $\mathbf{k} = 2\pi\mathbf{n}/2l$, where $\mathbf{k}$ is the wave vector. The single-particle partition function z is thus [2, 3, 6]

$$z = \sum\nolimits_{\mathbf{n}} \exp(-\beta\varepsilon_{\mathbf{n}}) = \sum\nolimits_{n_1}\sum\nolimits_{n_2}\sum\nolimits_{n_3} \exp\left[-\frac{\beta h^2}{8ml^2}\left(n_1^2 + n_2^2 + n_3^2\right)\right]$$

The spacing of energy levels is close so that the summation may be replaced by an integration, resulting, using standard integrals (Section A.16), in

$$z \cong \int_{\mathbf{n}=0}^{\mathbf{n}=\infty} \exp(-\beta\varepsilon_{\mathbf{n}})\mathrm{d}\mathbf{n} = \left[\int_{n=0}^{n=\infty} \exp\left(-\frac{\beta h^2 n^2}{8ml^2}\right)\mathrm{d}n\right]^3 = V\left(\frac{2\pi mkT}{h^2}\right)^{3/2} \equiv \frac{V}{\Lambda^3} \tag{7.15}$$

with the *thermal wavelength* $\Lambda \equiv (h^2/2\pi mkT)^{1/2}$.

7.2.2 Many Particles

In some cases, such as for a perfect gas, the total energy E_j in state j can be written as the sum of the energies ε_{ji} of particles i. In other cases, the interaction energy between particles is negligible as compared with the total energy, and in both cases we have,

$$E_j = \varepsilon_{j1} + \varepsilon_{j2} + \ldots \varepsilon_{jN} = \sum\nolimits_i \varepsilon_{ji}$$

The partition function becomes [2, 3, 8]

$$Z_N = \sum\nolimits_j \mathrm{e}^{-\beta E_j} = \sum\nolimits_j \mathrm{e}^{\sum_i -\beta\varepsilon_{ji}} = \sum\nolimits_j\prod\nolimits_i \mathrm{e}^{-\beta\varepsilon_{ji}} = \prod\nolimits_i\sum\nolimits_j \mathrm{e}^{-\beta\varepsilon_{ji}} = z^N \tag{7.16}$$

The total partition function Z_N for N independent particles is thus the product of the single-particle partition functions z.

For low density the size of the particles can be neglected, but for high(er) density V must be corrected to $V - V_{\text{excl}}$, with V_{excl} the *excluded volume*. It appears that even for hard sphere atoms this is an involved problem for which the Carnahan-Starling approximation is a practical solution (Section 8.2.1).

If the gas consists of molecules instead of atoms, we must realize that molecules have internal motion with an associated energy. To a good first approximation, the total energy can be taken as the sum of energies of the contributing mechanisms. For polyatomic molecules, these mechanisms would be (overall) translation, (overall and internal) rotation, vibration, and electronic transitions. So, we similarly have

$$\varepsilon_j = \varepsilon_{j\kappa} + \varepsilon_{j\lambda} + \cdots \quad \text{and} \quad z = z_\kappa z_\lambda \ldots$$

where each of the indices κ, λ,... indicates a mechanism with the total single partition function z being product of the partition functions z_α for each of the mechanisms.

Finally, we note that the states of an N-particle system with the same energy are indistinguishable, which leads to a correction for Z_N. If we consider particles as indistinguishable, for each of the N particles we could choose another one, leading to overcounting by $N!$ and this leads to an extra factor $1/N!$ in the partition function.[8] The real reason originates from the indistinguishability of states, that is, states differing in the exchange of two identical particles are to be considered as one and the same, required in both QM and PM, although not completely clear why.[9] For QM, after a somewhat complex, approximate reasoning, this leads indeed to the factor $1/N!$ (Justification 7.2). Here, we accept that result and replace the expression $Z_N = z^N$ by

$$Z_N = z^N/N! \tag{7.17}$$

where z is the total single-particle partition function. If we apply this approximation, we say that we use *Boltzmann statistics*. We will see that the factor $N!$ does not influence the pressure P (i.e. the equation of state, EoS) but does influence the entropy S (Problem 7.6).

7.2.3 Pressure and Energy

The Helmholtz energy F is related to the partition function Z via

$$F = -kT \ln Z = -kT \ln(z^N/N!)$$

So, let us calculate the mechanical EoS for a perfect gas $P = -\partial F/\partial V$ as a function of the independent variables. We find, using the Stirling approximation[10] for $\ln N!$ and $z = V/\Lambda^3$,

$$F = -kT \ln(z^N/N!) = -NkT\,[\ln V + \ln(\Lambda^{-3}) - \ln(N/e)]$$

$$P = -\frac{\partial F}{\partial V} = NkT\frac{\partial \ln V}{\partial V} + \frac{\partial \ldots}{\partial V} = \frac{NkT}{V}\frac{\partial V}{\partial V} \quad \text{or with } N = N_{\text{A}}, \quad P = \frac{RT}{V_{\text{m}}} \tag{7.18}$$

In this last expression, we recognize the well-known perfect gas law.

8 This correction is overcorrecting the situation. In the double sum in Eq. (7.16), the contribution ij ($i \neq j$) is counted twice because ji represents the same physical configuration while double counting is not done for ii. This shows right away that the correction is only valid for non-condensed systems.

9 Classically, the topic is still discussed. For a discussion, see [9] and references quoted therein.

10 The Stirling approximation for factorials reads $\ln x! = x \ln x - x + \frac{1}{2}\ln(2\pi x) + \ldots$. This approximation is excellent even for $x = 3$, the difference with the exact value being about −2.7%. Often, the term $\frac{1}{2}\ln(2\pi x)$ is neglected. Although the latter approximation is considerably less accurate, for $x = 50$ it deviates only about −2% from the exact value. Since typically much larger numbers are used, the approximation $\ln x! = x \ln x - x$, or alternatively $x! = (x/e)^x$, is usually quite sufficient. A curious improvement is due to Gosper [10], who replaced $\ln(2\pi x)$ by $\ln[(2x+\frac{1}{3})\pi]$, resulting in a very good approximation (−0.132% error for $x = 2$ and −0.002% error for $x = 20$).

Because from the Helmholtz function $F = -kT \ln Z$, given as a function of its natural variables T and V, all other thermodynamic expressions can be derived, the same is true for the partition function Z. For the entropy S and energy U we use, respectively, $S = -\partial F/\partial T$ and $U = F + TS$ (Eq. (7.17)). We can also calculate U directly from Z using the Gibbs–Helmholtz expression $U = \partial(F/T)/\partial(1/T)$. The internal energy U is entirely kinetic and given for structureless molecules by

$$U = 3NkT/2$$

and hence the heat capacity is given by $C_V = (\partial U/\partial T)_V = 3Nk/2$. Since we have the general thermodynamic relation $C_P - C_V = TV\alpha_P{}^2/\kappa_T$, we obtain $C_P = 5Nk/2$.

Problem 7.6 ***The Sackur–Tetrode Equation*** Show that the entropy S for a perfect gas is given by the *Sackur–Tetrode* equation

$$S = Nk \ln(ze^{5/2}/N) \quad \text{with} \quad z = V/\Lambda^3 = (2\pi mkT/h^2)^{3/2}V \tag{7.19}$$

Also show that the chemical potential $\mu = kT \ln(N/z)$.

Problem 7.7 ***Fermi–Dirac and Bose–Einstein Statistics*** According to QM, a wave function should be either antisymmetric (Fermi-Dirac (FD) particles) or symmetric (Bose-Einstein (BE) particles) in its coordinates (Section 3.2.4). Show, for N weakly interacting FD particles, so that we may assume the existence of q single-particle states ($q \geq N$) with energy ε_j, that the number of system states is $W_{\text{FD}} = q!/(q-N)!N!$, leading to the *Fermi-Dirac distribution* $N_j/(1-N_j) = \lambda \exp(-\beta\varepsilon_j)$, where λ is the absolute activity and N_j is the number of particles in state j. Similarly, show that for BE particles $W_{\text{BE}} = (q+N-1)!/N!(q-1)! \cong (q+N)!/N!\ (q)!$ if $q \gg N$, leading to the *Bose–Einstein distribution* $N_j/(1+N_j) = \lambda \exp(-\beta\varepsilon_j)$. Also show that for $q \gg N$, both distributions reduce to the *Maxwell–Boltzmann distribution* $N_j = \lambda \exp(-\beta\varepsilon_j)$ and $W_{\text{MB}} = q^N/N!$.

Problem 7.8 ***Fluctuations*** Show for a perfect gas, using $U = \langle H\rangle$, that $\sigma_U{}^2 = (2U/3N)^{1/2}$ and thus that fluctuations vanish in the *thermodynamic limit* ($N \to \infty$ and $V \to \infty$, but N/V remains finite).

7.3 The Semi-classical Approximation

The scheme for structureless independent particles outlined so far requires the calculation of:

- the quantum energy levels ε_j;
- the partition function, say Z (or Ξ or W), either exact via summation or approximate via integration;
- the corresponding thermodynamic potential, say F (or Ω or S); and
- the remaining properties (e.g. $S, P \ldots$) using macroscopic thermodynamics.

The scheme also applies to other situations, but a few remarks are appropriate. We said "independent" particles; however, interaction is important, and we deal with the effect of interactions in Section 7.4. We also said "structureless" particles, but molecules do contain internal structure, the effect of which is discussed in Section 7.5. Here we elaborate a bit on the required energy expression. In many cases, instead of using the energy E_{QM} as calculated from the Schrödinger equation $H\Psi = E\Psi$, one uses the PM expression $E_{\text{PM}} = \mathcal{H}(\mathbf{q}, \mathbf{p})$, where $\mathcal{H}$ is the Hamilton function. In fact, ST was largely developed before the introduction of QM [1]. The classical development leads to the partition function:

$$Z_{\text{PM}} = \int \dots \int \exp(-\beta\mathcal{H})\,\mathrm{d}\mathbf{q}\,\mathrm{d}\mathbf{p} \qquad \text{with} \qquad \mathcal{H} = \vartheta(\mathbf{p}) + \Phi(\mathbf{q})$$

where $\mathbf{q} = \mathbf{q}_1 \dots \mathbf{q}_N = q_{1x}q_{1y}q_{1z}q_{2x} \dots q_{Nz}$, $\mathbf{p} = \mathbf{p}_1 \dots \mathbf{p}_N = p_{1x}p_{1y}p_{1z}p_{2x} \dots p_{Nz}$, $\vartheta(\mathbf{p})$ and $\Phi(\mathbf{q})$ represent the kinetic and potential energy, respectively, and $\beta = 1/kT$. Here, $\mathbf{q}_j$ is the (generalized) coordinate and $\mathbf{p}_j$ is the (generalized) momentum of particle j. The choice of coordinates is to some extent arbitrary, since for conjugated coordinates we have $\mathrm{d}\mathbf{q}\mathrm{d}\mathbf{p} = \mathrm{d}\mathbf{q}'\mathrm{d}\mathbf{p}'$. The connection between Z_{QM} and Z_{PM} has been made [11], but the process is somewhat complex, and we illustrate the result here by an example. Since the factor between Z_{PM} and Z_{QM} should depend only on fundamental constants, this is in principle also sufficient. Consider a single particle with three DoFs with as energy expression $\varepsilon = \mathcal{H}(\mathbf{q}, \mathbf{p}) = \mathbf{p}^2/2m$, so that z_{PM} becomes

$$z_{\text{PM}} = \int_0^l \mathrm{d}\mathbf{q} \int_0^\infty \exp(-\beta\mathcal{H})\mathrm{d}\mathbf{p} = V\int_0^\infty \exp(-\mathbf{p}^2/2mkT)\mathrm{d}\mathbf{p} = V(2\pi mkT)^{3/2} \quad (7.20)$$

We see that z_{PM} is similar to z_{QM} (Eq. 7.15) except for a factor h^{-3}, which is missing in the classical expression. So, for each DoF an extra factor h^{-1} is required to match the quantum expression. This result appears to be general: if we calculate the partition function z_{1D} using a classical expression for the energy, we need an extra factor h^{-1} for every DoF (Justification 7.2) so that

$$z_{\text{1D}} = l(2\pi mkT/h^2)^{1/2} \qquad \text{or} \qquad z_{\text{1D}} = l/\Lambda \qquad \text{with} \qquad \Lambda = (h^2/2\pi mkT)^{1/2}$$

Because Λ has dimension "length", z_{1D} is dimensionless. As long as $z_{\text{1D}} \gg 1$, this DoF can be described by classical mechanics. As soon as $z_{\text{1D}} \cong 1$, a full quantum description for that particular DoF is required. This applies also to many particles if l is taken to be the mean particle distance $l = (V/N)^{1/3}$. To apply classical mechanics to a gas thus requires that $\Lambda^3 \ll V/N$. Since $\mu = kT\ln(N/z)$ with $z = V/\Lambda^3$ (Problem 7.6) this implies that $\exp(\beta\mu) \gg 1$. For translation it can be shown that a semi-classical description is nearly always sufficient. This is not generally true for the internal contributions. It is thus possible that classical evaluation is sufficient for one mechanism, while a quantum evaluation is necessary for another. If energies are additive, this presents no special problems.

Example 7.2 ***The Harmonic Oscillator*** A single harmonic oscillator provides a good demonstration for the factor h. In QM, the energy for an oscillator with spring constant a and mass m is given by $E_n = \hbar\omega(n + ½)$ with $\omega = \sqrt{a/m}$ so that the energy difference ΔE between two successive states is $\Delta E = \hbar\omega$. In classical mechanics, the total energy can be written as $\mathcal{H}(p, q) = p^2/2m + ½m\omega^2q^2$, where p is the momentum and q is the coordinate, or as $p^2/\alpha^2 + q^2/\beta^2 = 1$ with $\alpha = (2mE)^{1/2}$ and $\beta = (2E/m\omega^2)^{1/2}$. If we plot constant energy curves in μ-space, which is in this case a 2D space, we obtain ellipses. The area enclosed by such an ellipse is given by the integral $I = \oint p\,\mathrm{d}q$, where the integration is over one period,

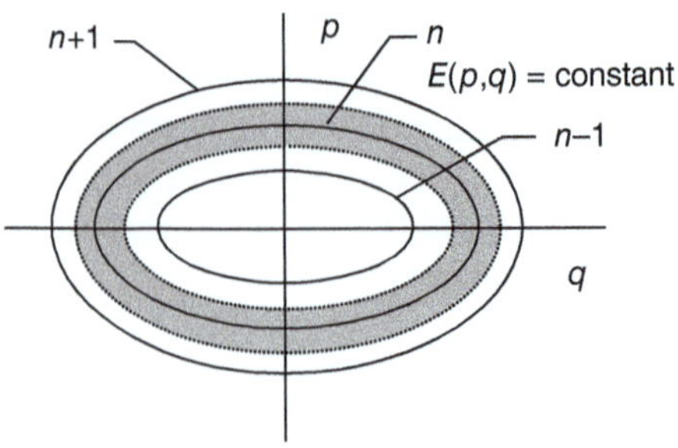

Figure 7.1 Phase space for a 1D oscillator indicating phase space between $n-1$ and $n+1$.

or $I = \pi\alpha\beta = 2\pi E/\omega$. Alternatively, we have $q = q_0 \sin(\omega t)$, $p = m\dot{q} = m\omega q_0 \cos(\omega t)$ and $I = \int_0^{2\pi/\omega} p\dot{q}\,\mathrm{d}t = \pi m\omega q_0^{\,2} = 2\pi E/\omega$. Let us draw ellipses with energies corresponding to $n-1$, n, and $n+1$ (Figure 7.1). The area between two successive ellipses is the area in classical phase space associated with one quantum state and corresponds to $2\pi\Delta E/\omega = 2\pi\hbar\omega/\omega = h$. For large n, the classical energy $\mathscr{H}(p, q)$ is almost constant in the phase region between $n - ½$ and $n + ½$, so that we can approximate the expression $\int\int \exp[-\mathscr{H}(p, q)/kT]\,\mathrm{d}p\mathrm{d}q$ by $h\,\Sigma_n \exp(-E_n/kT)$. For $T \to \infty$ and $n \to \infty$, the argument becomes exact.

Overall, the result for N particles leads to the introduction of the factor $(h^{3N}N!)^{-1}$ so that we have in total, using the Hamilton function $\mathscr{H}$ and potential energy Φ,

$$Z_N = (h^{3N}N!)^{-1} \int \ldots \int \mathrm{e}^{-\beta\mathscr{H}}\,\mathbf{dqdp} = (\Lambda^{3N}N!)^{-1} \int \ldots \int \mathrm{e}^{-\beta\Phi}\mathbf{dq} \equiv (\Lambda^{3N}N!)^{-1} Q_N \quad (7.21)$$

This *semi-classical partition function* is the most frequently encountered expression for the partition function. We have seen that in Eq. (7.20) for the single-particle partition function, the integrations over the momenta and position coordinates are separate. The factor V, due to integration over the position coordinates, is called the *configurational part*, while the factor Λ, due to the integration over the momenta, is the *kinetic part*. Because in energy expressions in classical mechanics the kinetic energy is generally only dependent on $\mathbf{p}$ and the potential energy is generally only dependent on $\mathbf{q}$, this separation is also generally possible. For many particles this separation is shown in the last part of Eq. (7.21). For N independent particles, the configurational part is deceptively simple ($Q_N = V^N$), but we will see that this integration becomes the major problem for interacting particles[11] (Section 7.4).

Finally, we note that the expectation value for the canonical average value of a property $X(\mathbf{q}, \mathbf{p})$ in the semi-classical approximation becomes

$$\langle X \rangle = (Z_N)^{-1} \int \ldots \int X(\mathbf{q}, \mathbf{p})\mathrm{e}^{-\beta\mathscr{H}}\,\mathbf{dqdp} = (Q_N)^{-1} \int \ldots \int X(\mathbf{q})\mathrm{e}^{-\beta\Phi}\mathbf{dq} \quad (7.22)$$

where the last step can be made if X depends only on the coordinates $\mathbf{q}$. If we just want to count the relative number of molecules with coordinates $\mathbf{q}$ and $\mathbf{p}$ in infinitesimal elements $\mathbf{dq}$ and $\mathbf{dp}$, we take $X = 1$ and obtain $\mathrm{d}N/N = \exp(-\beta\mathscr{H})\,\mathbf{dqdp}/Z_N$, often referred to as the *Maxwell–Boltzmann* (MB) *distribution*.

11 The symbolism for Z_N and Q_N is not uniformly used. In some cases, the factor $1/N!$ is included in Q_N, whereas in other cases the reverse designation for Z_N and Q_N is used.

Justification 7.2 ***The Introduction of Planck's Constant h and the Factorial N!*** First, we note that, for the Hamilton operator H operating on wave function Ψ and corresponding to the classical Hamilton function $\mathscr{H}$ [6, 11, 12],

$$\exp(H)\,|\Psi\rangle = (1 + H + \ldots)\,|\Psi\rangle \cong (1 + \mathscr{H} + \ldots)\,|\Psi\rangle = \exp(\mathscr{H})\,|\Psi\rangle$$

Second, a state $|\Psi\rangle$ for N particles can be expanded in any complete set of the proper symmetry, and we use here the antisymmetrized eigenstates of the momentum operator $\mathbf{p}$ using the δ-function normalization (Section 3.2)

$$|\Psi\rangle = (N!h^{3N/2})^{-1}\int_{-\infty}^{+\infty} A(\mathbf{p})\sum\nolimits_{P_\mathbf{r}} P_\mathbf{r}\left[\left|\exp\left(\mathrm{i}\mathbf{p}^\mathrm{T}\mathbf{r}\right)\right\rangle\right]\,\mathrm{d}\mathbf{p}$$

with the permutation operator $P_\mathbf{r}$ for (Cartesian) coordinates $\mathbf{r}$ (Section 3.2) and $A(\mathbf{p})$ the corresponding coefficient. Note that $|\exp(\mathrm{i}\mathbf{p}^\mathrm{T}\mathbf{r})\rangle$ denotes the product function $\Pi_i\,|\exp(\mathrm{i}\mathbf{p}_i{}^\mathrm{T}\mathbf{r}_i)\rangle$ of the individual eigenfunctions $|\exp(\mathrm{i}\mathbf{p}_i{}^\mathrm{T}\mathbf{r}_i\rangle$ for particle i. Third, we will need for an arbitrary function $F(\mathbf{p})$ of $\mathbf{p}$, the relation

$$\int_{-\infty}^{+\infty} F(\mathbf{p})\,\mathrm{d}\mathbf{p} = (N!)^{-1}\int_{-\infty}^{+\infty} \sum\nolimits_{P_\mathbf{p}} P_\mathbf{p}[F(\mathbf{p})]\,\mathrm{d}\mathbf{p}$$

Using bra(c)ket notation, the partition function $Z_N = \Sigma_j \exp(-\beta E_j)$ is written as

$$Z_N = \sum\nolimits_j \langle\Psi_j \mid \Psi_j\rangle \exp(-\beta E_j) = \sum\nolimits_j \langle\Psi_j \mid \exp(-\beta H) \mid \Psi_j\rangle$$

with Ψ_j the energy eigenfunction corresponding to state j. We see that this result can be interpreted as the sum of the diagonal elements of the *density matrix* with as elements $\langle\Psi_i\,|\exp(-\beta H)|\,\Psi_j\rangle$ (Section 2.5). We expand Ψ_j in the aforementioned antisymmetrized eigenfunctions of the momentum operator $\mathbf{p}$ to obtain

$$\begin{aligned}&\left\langle\Psi_j\,|\exp(-\beta H)|\,\Psi_j\right\rangle\\ &= (N!h^{3N/2})^{-2}\int_{-\infty}^{+\infty}\int_{-\infty}^{+\infty}\left\langle A(\mathbf{p})\sum\nolimits_{P_\mathbf{r}} P_\mathbf{r}\left[\mathrm{e}^{\mathrm{i}\mathbf{p}^\mathrm{T}\mathbf{r}}\right]\left|\mathrm{e}^{-\beta H}\right| A(\mathbf{p}')\sum\nolimits_{P_\mathbf{r}} P_\mathbf{r}\left[\mathrm{e}^{\mathrm{i}\mathbf{p}'^\mathrm{T}\mathbf{r}}\right]\right\rangle \mathrm{d}\mathbf{p}\,\mathrm{d}\mathbf{p}'\end{aligned}$$

The partition function Z_N becomes

$$\begin{aligned}Z_N = \sum\nolimits_j \langle\Psi_j|\mathrm{e}^{-\beta H}|\Psi_j\rangle = \frac{1}{(N!h^{3N/2})^2}\sum\nolimits_j\int_{-\infty}^{+\infty}\int_{-\infty}^{+\infty}\left\langle A_j(\mathbf{p})\sum\nolimits_{P_\mathbf{r}} P_\mathbf{r}\left[\mathrm{e}^{\mathrm{i}\mathbf{p}_j^\mathrm{T}\mathbf{r}}\right]\left|\mathrm{e}^{-\beta H}\right|\ldots\right.\\ \left.\ldots A_j(\mathbf{p}')\sum\nolimits_{P_\mathbf{r}} P_\mathbf{r}\left[\mathrm{e}^{\mathrm{i}\mathbf{p}_j'^\mathrm{T}\mathbf{r}}\right]\right\rangle \mathrm{d}\mathbf{p}_j\mathrm{d}\mathbf{p}_j'\end{aligned}$$

Now, using $P_\mathbf{r} = P_\mathbf{p}$ for $\exp(\mathrm{i}\mathbf{p}'^\mathrm{T}\mathbf{r})$, $N!F(\mathbf{p}) = \sum_{P_\mathbf{p}} P_\mathbf{p}[F(\mathbf{p})]$, $\Sigma_j A_j(\mathbf{p})A_j(\mathbf{p}') = \delta(\mathbf{p}'-\mathbf{p})$, and $\exp(H)\,|\Psi\rangle \cong \exp(\mathscr{H})|\Psi\rangle$, we obtain

$$\begin{aligned}Z_N &= \frac{1}{N!^2h^{3N}}\sum\nolimits_j\int_{-\infty}^{+\infty}\int_{-\infty}^{+\infty}\left\langle A_j(\mathbf{p})\sum\nolimits_{P_\mathbf{r}} P_\mathbf{r}\left[\mathrm{e}^{\mathrm{i}\mathbf{p}_j^\mathrm{T}\mathbf{r}}\right]\left|\mathrm{e}^{-\beta H}\right|\ldots\right.\\ &\qquad \left.\ldots\left|\sum\nolimits_{P_\mathbf{p}} P_\mathbf{p}\left[A_j(\mathbf{p}')\,\mathrm{e}^{\mathrm{i}\mathbf{p}_j'^\mathrm{T}\mathbf{r}}\right]\right.\right\rangle \mathrm{d}\mathbf{p}_j\mathrm{d}\mathbf{p}_j'\\ &= \frac{1}{N!h^{3N}}\sum\nolimits_j\int_{-\infty}^{+\infty}\int_{-\infty}^{+\infty}\left\langle A_j(\mathbf{p})\sum\nolimits_{P_\mathbf{r}} P_\mathbf{r}\left[\mathrm{e}^{\mathrm{i}\mathbf{p}_j^\mathrm{T}\mathbf{r}}\right]\left|\mathrm{e}^{-\beta H}\right| A_j(\mathbf{p}')\,\mathrm{e}^{\mathrm{i}\mathbf{p}_j'^\mathrm{T}\cdot\mathbf{r}}\right\rangle \mathrm{d}\mathbf{p}_j\mathrm{d}\mathbf{p}_j'\\ &= \frac{1}{N!h^{3N}}\int_{-\infty}^{+\infty}\int_{-\infty}^{+\infty}\sum\nolimits_j A_j(\mathbf{p})A_j(\mathbf{p}')\left\langle\sum\nolimits_{P_\mathbf{r}} P_\mathbf{r}\left[\mathrm{e}^{\mathrm{i}\mathbf{p}_j^\mathrm{T}\mathbf{r}}\right]\left|\mathrm{e}^{-\beta H}\right|\,\mathrm{e}^{\mathrm{i}\mathbf{p}_j'^\mathrm{T}\mathbf{r}}\right\rangle \mathrm{d}\mathbf{p}_j\mathrm{d}\mathbf{p}_j'\end{aligned}$$

$$= \frac{1}{N!h^{3N}} \int_{-\infty}^{+\infty} \left\langle e^{i\mathbf{p}^T\mathbf{r}} \left| e^{-\beta H} \right| e^{i\mathbf{p}^T\mathbf{r}} \right\rangle \, d\mathbf{p} \quad \text{or}$$

$$Z_{CM} \cong \frac{1}{N!h^{3N}} \int_{-\infty}^{+\infty} \cdots \int_{-\infty}^{+\infty} \exp(-\beta\mathcal{H})\, d\mathbf{r} d\mathbf{p}$$

The classical partition function Z_{PM} is thus $(N!h^{3N})^{-1}$ times the phase integral over $\exp(-\beta\mathcal{H})$. This analysis not only shows the origin of the factor h^{-3N} but also of the factor $(N!)^{-1}$. Extending the analysis further – in principle a bit but in practice considerably (detailed discussions given in [6, 12]) – by introducing second-order terms in $\exp(H)|\Psi\rangle \cong \exp(\mathcal{H})|\Psi\rangle$ leads to the first-order quantum correction on the classical result, which reads, using $\langle\ldots\rangle$ as the canonical average,

$$Z_{QM} = Z_{PM}\left(1 - \frac{\beta^2\hbar^2}{24}\sum_{a=1}^{3N}\frac{\Phi_{aa}}{m_a} + \ldots\right) \quad \text{with} \quad \Phi_{aa} \equiv \left\langle \frac{\partial^2\Phi(\mathbf{r})}{\partial x_a^2} \right\rangle$$

so that the Helmholtz energy becomes

$$\beta F_{QM} = \beta F_{PM} + \frac{\beta^2\hbar}{24}\sum_{a=1}^{3N}\frac{\Phi_{aa}}{m_a} + \ldots = \beta F_{PM} + \left(\frac{298.2}{T}\right)^2 \sum_{a=1}^{3N}\left(\frac{\overline{\omega}_a}{1015}\right)^2 + \ldots$$

The quantity Φ_{aa} is an effective force constant so that $\omega_a = (\Phi_{aa}/m_a)^{1/2}$ is the corresponding angular frequency. At room temperature for vibrations with wave number[13] $\overline{\omega} \sim 1000\ \text{cm}^{-1}$ or more, the correction becomes significant.

Problem 7.9 ***Angle-averaged Potentials*** For arbitrary molecules, the interaction is dependent on their distance r and both their orientations, indicated by $\boldsymbol{\omega} = (\theta, \phi)$, where θ and ϕ are the angles for the polar coordinate system ($0 \le \theta \le \pi$ and $0 \le \phi \le 2\pi$). The differential element $d\boldsymbol{\omega}$ is $d\boldsymbol{\omega} = \sin\theta\, d\theta\, d\phi$. Using the multipole expression for $W(r)$ for a *fixed* distance r, expanding the exponentials, and integrating term by term, show that,

$$W(r) = -1/2\beta\overline{W^2(r,\boldsymbol{\omega})} \quad \text{for} \quad \beta W \ll 1 \quad \text{with}$$

$$\overline{W^n(r,\boldsymbol{\omega})} = \int W^n(r,\boldsymbol{\omega}_1,\boldsymbol{\omega}_2)\, d\boldsymbol{\omega}_1 d\boldsymbol{\omega}_2 \Big/ \!\! \int d\boldsymbol{\omega}_1 d\boldsymbol{\omega}_2$$

Discuss why the terms with n odd are identically zero. Show also that the angle-averaged internal energy, with the subscript $\boldsymbol{\omega}$ indicating integration over $\boldsymbol{\omega}$ only, is

$$U(r) = \langle W(r,\boldsymbol{\omega})\rangle_{\boldsymbol{\omega}} = -\beta\overline{W^2(r,\boldsymbol{\omega})}$$

Problem 7.10 ***Multipole Interactions*** Show, by using the result of Problem 7.9, that the charge–dipole contribution to the intermolecular interaction is given by

$$W_{c\text{-}d}(r) = \overline{W_{c\text{-}d}(r,\theta)} = -\mathcal{C}^2\left[\left(q_1^2\mu_2^2/6kT\right) + \left(q_2^2\mu_1^2/6kT\right)\right] r^{-4} \tag{4.7}$$

Similarly show that the dipole-induced dipole contribution is given by

$$W_{d\text{-}id}(r) = \overline{W_{d\text{-}id}(r,\theta)} = -\mathcal{C}^2\left(\alpha_1\mu_2^2 + \alpha_2\mu_1^2\right) r^{-6} \tag{4.4}$$

and that the dipole–dipole contribution is given by

$$W_{d\text{-}d}(r) = \overline{W_{d\text{-}d}(r,\boldsymbol{\omega}_1,\boldsymbol{\omega}_2)} = -\mathcal{C}^2\left(\mu_1^2\mu_2^2/3kT\right) r^{-6} \tag{4.10}$$

7.4 Interacting Particles

In Sections 7.2–7.3, we discussed the partition function z of a single particle with mass m, momentum $\mathbf{p}_1$, and (Cartesian) coordinate $\mathbf{r}_1$ enclosed in a box with edge l and volume $V = l^3$. The energy E, given by the Hamilton function $\mathscr{H}$, and partition function z are

$$E = \mathscr{H}(\mathbf{p}_1, \mathbf{r}_1) = {\mathbf{p}_1}^2/2m \quad \text{and} \quad z = V\,(2\pi mkT/h^2)^{3/2} = (l/\Lambda)^3$$

respectively, where $\Lambda = (h^2/2\pi mkT)^{1/2}$ is the *thermal wavelength.*

Let us now evaluate the partition function for two interacting particles, characterized by the set of variables $(\mathbf{p}, \mathbf{r}) = (\mathbf{p}_1, \mathbf{r}_1, \mathbf{p}_2, \mathbf{r}_2)$ and their potential energy of interaction $\phi(\mathbf{r}_1, \mathbf{r}_2)$ [8, 13]. In this case, the energy is given by

$$E = \mathscr{H}(\mathbf{p}, \mathbf{r}) = \mathscr{H}(\mathbf{p}_1, \mathbf{r}_1, \mathbf{p}_2, \mathbf{r}_2) = {\mathbf{p}_1}^2/2m + {\mathbf{p}_2}^2/2m + \phi(\mathbf{r}_1, \mathbf{r}_2)$$

If we assume that ϕ is spherically symmetrical and thus only depends on the distance r between the two particles, the potential energy ϕ can be written as $\phi(\mathbf{r}_1, \mathbf{r}_2) = \phi(r)$ with $r = |\mathbf{r}_2 - \mathbf{r}_1|$. Thus, we obtain for the partition function[12]

$$\begin{aligned} Z &= h^{-3N}(N!)^{-1} \iint \exp[-\beta\mathscr{H}(\mathbf{p}, \mathbf{r})]\,\mathrm{d}\mathbf{p}\mathrm{d}\mathbf{r} \\ &= h^{-6}(2!)^{-1} \int \exp\left(-\beta{\mathbf{p}_1}^2/2m\right)\,\mathrm{d}\mathbf{p}_1 \times \int \exp\left(-\beta{\mathbf{p}_2}^2/2m\right)\,\mathrm{d}\mathbf{p}_2 \\ &\quad \times \iint \exp(-\beta\phi(r)\,\mathrm{d}\mathbf{r}_1\mathrm{d}\mathbf{r}_2 \\ &= (2!)^{-1}\Lambda^{-3} \times \Lambda^{-3} \times V \int \exp[-\beta\phi\,(r)]\,4\pi r^2\,\mathrm{d}r \equiv (2!)^{-1}\Lambda^{-6}Q_{12} \end{aligned}$$

The factor Q_{12} is usually addressed as the *(two-particle) configuration integral.*

This result can be easily generalized to N particles, characterized by $(\mathbf{p}, \mathbf{r}) = (\mathbf{p}_1, \mathbf{r}_1, \mathbf{p}_2, \mathbf{r}_2, \ldots, \mathbf{p}_N, \mathbf{r}_N)$ and $\Phi(\mathbf{r}) = \Phi(\mathbf{r}_1, \mathbf{r}_2, \ldots, \mathbf{r}_N)$. The total energy is

$$E = \mathscr{H}(\mathbf{p}, \mathbf{r}) = \mathscr{H}(\mathbf{p}_1, \mathbf{r}_1, \mathbf{p}_2, \mathbf{r}_2, \ldots, \mathbf{p}_N, \mathbf{r}_N) = \sum\nolimits_i \mathbf{p}_i^2/2m + \Phi(\mathbf{r}_1, \mathbf{r}_2, \ldots, \mathbf{r}_N)$$

Assuming again spherically symmetrical interactions between the particles, we have

$$\Phi(\mathbf{r}) = \Phi(\mathbf{r}_1, \mathbf{r}_2, \ldots, \mathbf{r}_N) = \Phi(r_{12}, \ldots, r_{1N}, r_{23}, \ldots, r_{2N}, \ldots, r_{N-1,N})$$

where r_{ij} is the distance between particles i and j. The partition function becomes

$$\begin{aligned} Z &= h^{-3N}\,(N!)^{-1} \iint \exp(-\beta\mathscr{H})\,\mathrm{d}\mathbf{p}\mathrm{d}\mathbf{r} \\ &= (N!\Lambda^{3N})^{-1} \times \int \exp[-\beta\Phi\,(\mathbf{r})]\,\mathrm{d}\mathbf{r}_1 \ldots \mathrm{d}\mathbf{r}_N \equiv (N!\Lambda^{3N})^{-1}Q \end{aligned}$$

with Q the *configuration integral.* Our "only" remaining task is thus to evaluate this many-dimensional integral, but in general this is a rather difficult task. In Section 8.2, we discuss the simplest possible situation, that is, for intermediate pressure gases. Even for that case, it will become clear that complications rapidly set in.

---

12 Using the transformation $[\mathbf{r} \equiv \mathbf{r}_1 - \mathbf{r}_2, \mathbf{R} \equiv (\mathbf{r}_1 + \mathbf{r}_2)/2]$, the Jacobian determinant J (Section A.7) is easily calculated as $J = 1$. Hence, $\mathrm{d}\mathbf{r}_1\mathrm{d}\mathbf{r}_2 = J\mathrm{d}\mathbf{r}\mathrm{d}\mathbf{R} = \mathrm{d}\mathbf{r}\mathrm{d}\mathbf{R}$. Integration over $\mathbf{R}$ yields the volume V, whereas for spherically symmetrical polar coordinates, $\mathrm{d}\mathbf{r}$ reduces to $4\pi r^2\mathrm{d}r$.

7.5 Internal Contributions

So far, we have limited the discussion to molecules without internal DoFs. In Section 7.2.2, we saw that, if energy contributions are additive, the overall partition function Z can be factorized, for example, $Z = N!^{-1}z_{\text{tra}}z_{\text{vib}}z_{\text{rot}}z_{\text{ele}}$, where z_{tra}, z_{vib}, z_{rot}, and z_{ele} denote the translation, vibration, rotation, and electronic partition function, respectively. In this section, we address the effect of vibrations, rotations, and electronic transitions in molecules. The pattern how to calculate thermodynamic properties will be clear by now: find the energy expression, evaluate the partition function, and calculate from the partition function the thermodynamic properties.

7.5.1 Vibrations

The simplest molecule with internal structure is a diatomic molecule. The vibrations of such a molecule can be described by the harmonic oscillator (HO) model for two masses m_1 and m_2 with equilibrium distance r, spring constant a, and reduced mass $1/\mu = 1/m_1 + 1/m_2$. The energy levels of this oscillator using the circular frequency ω, as given in Section 3.2.2 and with selection rule for transitions $\Delta n = \pm 1$, are

$$\varepsilon_n = \hbar\omega(n + ½) \quad \text{with} \quad n = 0, 1, 2, \ldots, \quad \omega = (a/\mu)^{1/2}$$

For the average energy $U = \langle \varepsilon \rangle$ we find [6, 8], with $\beta = 1/kT$,

$$U = z_{\text{vib}}^{-1}\sum\nolimits_n \exp[-\beta\hbar\omega(n + ½)]\hbar\omega\,(n + ½) \quad \text{with} \quad z_{\text{vib}} = \sum\nolimits_n \exp[-\beta\hbar\omega\,(n + ½)]$$

Since z_{vib} is given as a geometric series, we can evaluate the partition function as

$$z_{\text{vib}} = \exp(-½\beta\hbar\omega)[1 - \exp(-\beta\hbar\omega)]^{-1} = [\exp(+½\beta\hbar\omega) - \exp(-½\beta\hbar\omega)]^{-1} \tag{7.23}$$

This expression cannot be used for too high temperature for two reasons. First, at higher temperature T, anharmonicity occurs and the HO model no longer applies. Second, for z_{vib} we assumed that $n \to \infty$, but in reality, n is finite. Knox [14] estimated that one should have $kT < 0.1D$, where D is the dissociation energy. For a typical value of D = 4 eV, we obtain $T < 4600$ K, and this is therefore not a problem in practice.

Since $\sum_n \exp[-\beta\hbar\omega\,(n + ½)]\,\hbar\omega\,(n + ½) = -\partial z_{\text{vib}}/\partial\beta$, we can write

$$U = -\frac{\partial z_{\text{vib}}/\partial\beta}{z_{\text{vib}}} = -\frac{\partial \ln z_{\text{vib}}}{\partial\beta} = ½\hbar\omega + \frac{\hbar\omega}{\exp(\beta\hbar\omega) - 1} \equiv \hbar\omega(\overline{n} + ½)$$

with $\overline{n}$ the *mean occupation number*. The Helmholtz energy F is given by

$$F = -kT \ln z_{\text{vib}} = ½\hbar\omega + kT \ln[1 - \exp(-\beta\hbar\omega)]$$

so that the entropy S is given by

$$S = -\frac{\partial F}{\partial T} = -k \ln[1 - \exp(-\beta\hbar\omega)] + \frac{k(\beta\hbar\omega)\exp(-\beta\hbar\omega)}{1 - \exp(-\beta\hbar\omega)}$$

The energy can also be calculated from $U = F + TS$. The heat capacity C_V reads

$$C_V = \frac{\partial U}{\partial T} = \frac{k(\beta\hbar\omega)^2\exp(\beta\hbar\omega)}{[\exp(\beta\hbar\omega) - 1]^2}$$

Table 7.1 Vibration, rotation, and dissociation data for several diatomic molecules.

Mol.	θ_{vib} (K)	$\tilde{\omega}$ (cm^{-1})	$\tilde{x}\tilde{\omega}$ (cm^{-1})	θ_{rot} (K)	$\tilde{B}$ (cm^{-1})	$\tilde{D}$ (cm^{-1})	$\tilde{\alpha}$ (cm^{-1})	$r_{n=0}$ (Å)	$D_{n=0}$ (kJ mol^{-1})
H_2	6336	4403.56	123.86	87.55	60.8530	$4.71\cdot10^{-2}$	3.0622	0.7414	432.1
HF	5954	4138.32	89.88	30.15	20.9557	$2.15\cdot10^{-3}$	0.798	0.9168	566.2
HCl	4303	2990.94	52.819	15.24	10.5934	$5.32\cdot10^{-4}$	0.3072	1.2746	427.8
HBr	3811	2648.97	45.218	12.18	8.4649	$3.46\cdot10^{-4}$	0.2333	1.4144	362.6
HI	3322	2309.01	39.644	9.366	6.510	$2.07\cdot10^{-4}$	0.1689	1.6092	294.7
CO	3122	2169.89	13.288	2.779	1.9313	$6.12\cdot10^{-6}$	0.0175	1.1283	1070.2
NO	2740	1904.20	14.075	2.406	1.6719	$5.4\cdot10^{-6}$	0.0171	1.1508	626.8
N_2	3393	2358.57	14.324	2.875	1.9982	$5.76\cdot10^{-6}$	0.0173	1.0977	941.6
O_2	2274	1580.19	11.98	2.080	1.4456	$4.84\cdot10^{-6}$	0.0159	1.2075	493.6
F_2	1319	916.64	11.236	1.281	0.8902	$3.3\cdot10^{-6}$	0.1385	1.4119	154.6
Cl_2	805.3	559.72	2.675	0.3511	0.2440	$1.86\cdot10^{-7}$	0.00149	1.9879	239.2
Br_2	468.1	325.321	1.0774	0.1181	0.0821	$2.09\cdot10^{-8}$	0.00319	2.2811	190.1
I_2	308.6	214.502	0.6147	0.05377	0.03737	$4.25\cdot10^{-9}$	0.00011	2.6663	148.8

Data from [21]. All $g_0 = 1$, except $g_0(NO) = 2$ and $g_0(O_2) = 3$.

At low and high temperatures, the expression for C_V expands to, respectively,

$$C_V = k(\beta\hbar\omega)^2 \exp(-\beta\hbar\omega) + \dots \quad \text{and} \quad C_V = k[1 - (\beta\hbar\omega)^2/12 + \dots]$$

The behavior of the harmonic oscillator is characterized by ω. Equivalently, we use the *characteristic (vibration) temperature* θ_{vib} given $\theta_{vib} = \hbar\omega/k$. When $T \gg \theta_{vib}$, the behavior can be classified as classical, and the energy becomes $U \cong kT$. When $T \ll \theta_{vib}$, expansion of the expression results in $U \cong {}^1\!/_2\hbar\omega$. When $T \cong \theta_{vib}$, the full expression must be used. Since θ_{vib} is typically hundreds of degrees, this implies that in most cases the quantum expression for the partition function must be applied.

More accurately, molecules behave like anharmonic oscillators and anharmonicity changes the energy expression and selection rule for an HO to

$$\varepsilon_n/hc = \tilde{\omega}(n + {}^1\!/_2) - \tilde{x}\tilde{\omega}(n + {}^1\!/_2)^2 \quad \text{and} \quad \Delta n = \pm1, \pm2, \dots$$

where ε_n/hc is the energy in cm^{-1}, $\tilde{\omega}$ is the vibrational wave number,[13] and $\tilde{x}$ is the *anharmonicity constant*. For several diatomic molecules, parameter values including r_0 and D_0 (Figure 4.1) are given in Table 7.1. We note that the equilibrium distance r depends on the vibrational state, that is, on n, and that we should write r_n.

For a N-atomic molecule we can describe the vibrational behavior by $3N - 6$ *normal coordinates* ($3N - 5$ for a linear molecule), each of which can be modeled as a harmonic oscillator.[14] In this case, these normal coordinates are independent, and hence the behavior can be described as the sum of the behaviors of the individual normal coordinates

13 The energy is often characterized by the frequency $\nu = \varepsilon/h$ in s^{-1} (with rotational parameter $B = h/8\pi^2 I$) or by the wave number $\tilde{\omega} = \varepsilon/hc$ in cm^{-1} (with rotational parameter $\tilde{B} = h/8\pi^2 Ic$).

14 Any N-atom molecule has $3N - 6$ internal DoFs. For linear molecules, this number reduces to effectively to $3N - 5$ because rotation around the length axis, although possible in principle, will require an energy $\gg kT$ in view of the small moment of inertia around that axis, and thus the molecule will remain in

(Justification 7.3). In fact, they are an example of generalized coordinates. For example, for H_2O the three vibrational temperatures are 2294, 5262, and 5404 K, respectively, while they read 960, 960, 1932, and 3380 K for CO_2. For an extensive discussion, see the classic reference on vibrational spectra [16].

Justification 7.3 ***Normal Coordinates*** The kinetic energy ϑ and potential energy Φ of a molecule are, respectively,

$$2\vartheta = \sum\nolimits_i m_i \dot{r}_i^2 = \dot{\mathbf{r}}^\mathrm{T}\mathbf{M}'\dot{\mathbf{r}}$$

with $\mathbf{M}'$ a diagonal matrix containing the masses m_i of the atoms, see Eq. (2.4), and

$$2\Phi = \sum\nolimits_{i,j} B_{ij} r_i r_j = \mathbf{r}^\mathrm{T}\mathbf{B}\mathbf{r}$$

with r_i or $\mathbf{r}$ the coordinates of the atoms and $B_{ij} = \partial^2\Phi/\partial r_i \partial r_j$. Using $\mathbf{r} = \mathbf{A}\mathbf{q}$, with $\mathbf{q}$ the generalized coordinates, we easily obtain $2\vartheta = \dot{\mathbf{q}}^\mathrm{T}\mathbf{A}^\mathrm{T}\mathbf{M}'\mathbf{A}\dot{\mathbf{q}} \equiv \dot{\mathbf{q}}^\mathrm{T}\mathbf{M}\dot{\mathbf{q}}$. Next, we diagonalize Φ using $\mathbf{r} = \mathbf{A}\mathbf{q}$ to obtain $\mathbf{q}^\mathrm{T}\mathbf{A}^\mathrm{T}\mathbf{B}\mathbf{A}\mathbf{r} - \mathbf{q}^\mathrm{T}\boldsymbol{\Lambda}'\mathbf{q} = \mathbf{q}^\mathrm{T}(\mathbf{A}^\mathrm{T}\mathbf{B}\mathbf{A} - \boldsymbol{\Lambda}')\mathbf{q} = 0$, where $\boldsymbol{\Lambda}'$ is a diagonal matrix containing the eigenvalues λ_i of $\mathbf{B}$. This equation has solutions only if $\det(\mathbf{A}^\mathrm{T}\mathbf{B}\mathbf{A} - \boldsymbol{\Lambda}') = 0$. For the dynamics we use Lagrange's equations

$$\frac{\mathrm{d}}{\mathrm{d}t}\frac{\partial\vartheta}{\partial\dot{\mathbf{q}}} + \frac{\partial\Phi}{\partial\mathbf{q}} = \mathbf{M}\ddot{\mathbf{q}} + \boldsymbol{\Lambda}'\mathbf{q} = 0$$

Solving leads to $\ddot{\mathbf{q}} + \mathbf{M}^{-1}\boldsymbol{\Lambda}'\mathbf{q} = 0$ or $\ddot{\mathbf{q}} + \boldsymbol{\Lambda}\mathbf{q} = 0$ with $\boldsymbol{\Lambda} \equiv \mathbf{M}^{-1}\boldsymbol{\Lambda}'$. If the eigenvalue $\lambda_i > 0$, the solution is a harmonic oscillation described by $q_i = \alpha_i\cos(\lambda_i^{1/2}t + \delta_i)$ with α_i the amplitude and δ_i the phase shift. If the eigenvalue $\lambda_i = 0$, the solution is a translation described by $q_i = \alpha_i t + \delta_i$, while if $\lambda_i < 0$ an imaginary solution obtained.

As an example for the relation $\mathbf{r} = \mathbf{A}\mathbf{q}$, take water ($H_2O$). In Example 2.1, we used generalized coordinates directly related to bond stretching and bending. Here, we use for convenience slightly different coordinates (for notation, see Figure 7.2), where we take into

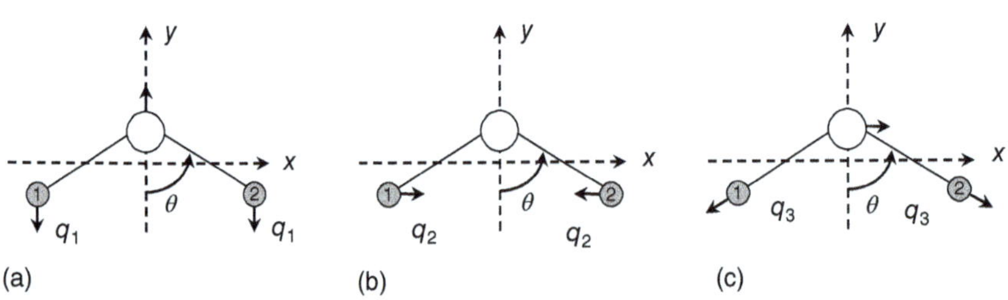

Figure 7.2 Normal modes for water: (a) mode 1 or symmetric stretching (3652 cm^{-1}), (b) mode 2 or bending (1595 cm^{-1}), and (c) mode 3 or asymmetric stretching (3756 cm^{-1}). As the molecule is planar, only x and y displacements occur. The H—O—H bond angle is 2θ.

---

its ground state. The same argument applies for rotation of atoms. Moreover, from quantum mechanics for an axially symmetric system the allowed wave functions have zero angular momentum about the symmetry axis. Hence, spherical nuclei cannot rotate and nonspherical nuclei require an energy far greater than kT, leaving in all cases only translation [15].

account that overall, no translation or rotation should occur. For mode 1 we have $y_1 = -q_1$ and $y_2 = -q_1$. Overall rotation is absent but to avoid overall translation we must require $-m_\text{H}q_1 - m_\text{H}q_1 + m_\text{O}y_\text{O} = 0$ or $y_\text{O} = (2m_\text{H}/m_\text{O})q_1$. For mode 2, we have $x_1 = q_2$ and $x_2 = -q_2$. In this case, overall translation and rotation is avoided. Mode 3 is slightly more complex and leads to $x_1 = -q_3\sin\theta$, $x_2 = -q_3\sin\theta$, $y_1 = -q_3\cos\theta$ and $y_2 = q_3\cos\theta$. These displacements should be compensated to avoid overall translation and rotation by requiring that $-m_\text{H}x_1 - m_\text{H}x_2 + m_\text{O}x_\text{O} = -2q_3m_\text{H}\sin\theta + m_\text{O}x_\text{O} = 0$ or $x_\text{O} = (2m_\text{H}\sin\theta/m_\text{O})q_3$. Hence, adding the displacements of the three modes, we have in total

$$\begin{pmatrix} x_\text{O} \\ y_\text{O} \\ x_1 \\ y_1 \\ x_2 \\ y_2 \end{pmatrix} = \begin{pmatrix} 0 & 0 & 2m_\text{H}\sin\theta/m_\text{O} \\ 2m_\text{H}/m_\text{O} & 0 & 0 \\ 0 & 1 & -\sin\theta \\ -1 & 0 & -\cos\theta \\ 0 & -1 & -\sin\theta \\ -1 & 0 & \cos\theta \end{pmatrix} \begin{pmatrix} q_1 \\ q_2 \\ q_3 \end{pmatrix}$$

The kinetic energy becomes $\vartheta = \left(m_\text{H} + 2m_\text{H}^2/m_\text{O}\right)\dot{q}_1^2 + m_\text{H}\dot{q}_2^2 + \left(m_\text{H} + 2m_\text{H}^2\sin^2/m_\text{O}\right)\theta\,\dot{q}_3^2$ or $\vartheta \equiv \gamma_1\dot{q}_1^2 + \gamma_2\dot{q}_2^2 + \gamma_3\dot{q}_3^2$, while the potential energy reads $\Phi = a_1q_1^2 + a_2q_2^2 + a_3q_3^2$ with a_j the force constants. Altogether, we have the Hamilton function $\mathscr{H} = \vartheta + \Phi = \sum_{j=1}^{j=3}\left(\gamma_j\dot{q}_j^2 + a_jq_j^2\right)$, representing three independent harmonic oscillators.

7.5.2 Rotations

The rotation behavior of molecules is somewhat more complex than their vibration behavior. We start with homonuclear and heteronuclear diatomic molecules, and thereafter deal briefly with polyatomic molecules and internal rotation.

The energy levels of the diatomic rigid rotator (RR), as given in Section 3.2.3, are

$$\varepsilon_J = BJ(J+1) \text{ with } J = 0, 1, 2, \ldots \text{ and } B = h^2/8\pi^2 I \equiv k\theta_\text{rot}$$

with $\Delta J = \pm 1$ as the selection rule for transitions. Here, the moment of inertia $I = \mu r^2$ with the reduced mass $1/\mu = 1/m_1 + 1/m_2$, where m_1 and m_2 are the masses of the particles. The constant B is the *rotational constant*, alternatively expressed as the *characteristic (rotation) temperature* $\theta_\text{rot} = B/k$. Each of the energy levels J has a $(2J+1)$ degeneracy. At low temperatures, only the first few terms are contributing, and we have [6, 8, 13]

$$z_\text{rot} = \sum\nolimits_J (2J+1)\exp[-\beta BJ(J+1)] = 1 + 3\exp\left(-\frac{2\theta_\text{rot}}{T}\right) + 5\exp\left(-\frac{6\theta_\text{rot}}{T}\right) + \ldots$$

At high temperatures, the summation can be approximated by integration and

$$z_\text{rot} \cong \int_0^\infty (2J+1)\exp[-\beta BJ(J+1)]\,\text{d}J = \int_0^\infty \exp[-\beta BJ(J+1)]\,\text{d}[J(J+1)] = kT/B \tag{7.24}$$

By treating J as a continuous variable, the most probable rotational state for a diatomic molecule at high temperatures is given by $J_\text{max} \cong (T/2\theta_\text{rot})^{1/2} - 1/2$.

Actually, we have to divide z_{rot} by the symmetry number σ, denoting the number of ways the molecule can be rotated into a configuration indistinguishable from the original configuration. Obviously, $\sigma = 1$ for AB and $\sigma = 2$ for AA molecules.[15] A more complete analysis using the Euler–McLaurin summation expression results in

$$z_{rot} = (T/\sigma\theta_{rot})\left[1 + \theta_{rot}/3T + (\theta_{rot}/T)^2/15 + 4(\theta_{rot}/T)^3/15 + \dots\right]$$

which is, using the given truncation, good to within 1% for $\theta_{rot} < T$. Replacing the summation by integration is thus only justified at sufficiently elevated temperature, say for $T > 5\theta_{rot}$. For a lower T one must sum z_{rot} term by term.

More accurately, diatomic molecules do not behave as rigid rotors and the rotational energy is more appropriately described by

$$\varepsilon_J/hc = \widetilde{B}J(J+1) - \widetilde{D}J^2(J+1)^2, \quad \widetilde{B} = h/8\pi^2 Ic, \quad \widetilde{D} = 4\widetilde{B}^3/\widetilde{\omega}^2$$

with rotational constant[13] $\widetilde{B}$, vibrational wave number $\widetilde{\omega}$, and *centrifugal distortion constant* $\widetilde{D}$. Moreover, the rotational constants depend on n, the so-called *vibration–rotation interaction*, which leads to $\widetilde{B}_n = \widetilde{B}_e - \widetilde{\alpha}_e(n + 1/2)$ with $\widetilde{\alpha}_e$ a constant. Typical parameter values for several diatomic molecules are given in Table 7.1.

If a polyatomic molecule is linear, the same expression as for diatomic molecules applies, of course, with adjusted moments of inertia. For a nonlinear polyatomic molecule, a similar but more complex reasoning leads to

$$z_{rot} = (8\pi^2 kT/h^2)^{3/2}(I_A I_B I_C)^{1/2}\sqrt{\pi}/\sigma = \sqrt{T/\theta_A}\sqrt{T/\theta_B}\sqrt{T/\theta_C}\sqrt{\pi}/\sigma$$

where I_A, I_B, and I_C are the principal moments of inertia (the principal components of the tensor defined by $I_{xx} = \Sigma_j m_j(y_j^2 + z_j^2)$ and $I_{xy} = \Sigma_j m_j x_j y_j$). Equivalently, one uses the rotational temperatures θ_A, θ_B, and θ_C. The symmetry number has the same meaning as before. As examples, we quote $\sigma(H_2O) = 2$, $\sigma(NH_3) = 3$, $\sigma(CH_4) = 12$, $\sigma(CH_3Cl) = 3$, $\sigma(C_6H_6) = 12$, and $\sigma(C_6H_5Cl) = 2$. Also, for polyatomic molecules this semi-classical partition function

15 The origin of the symmetry factor lies in the symmetry of the overall wave function (for details see, e.g. [13]). In brief, for a molecule containing identical nuclei, the total wave function Ψ, that is, the electronic, vibrational, rotational, translational, and nuclear wave functions, must be either symmetric (if the nuclei have integral spins, bosons) or antisymmetric (if they have half-integral spins, fermions) under the interchange of the two identical nuclei. This interchange can be seen as the result of (i) an inversion of all the particles (electrons and nuclei) through the origin, and thereafter (ii) an inversion of only the electrons back through the origin. This two-step process is equivalent to an exchange of the nuclei. The translational and vibrational wave functions are not affected by an inversion, and the electronic wave function of the ground state is nearly always symmetric. Therefore, the overall symmetry is determined by the nuclear spin and rotational wave functions. For a heteronuclear diatomic molecule, $z_{nuc}z_{rot} = \rho_A\rho_B\,\Sigma_{J=0,1,2,\dots}(2J+1)\exp[-J(J+1)\theta_{rot}/T] \equiv \rho_A\rho_B\Sigma_{all}$, where $\rho_j = 2I_j+1$ is the statistical weight associated with the nuclear spin I_j and we have abbreviated the sum by Σ_{all}. The factor $\rho_A\rho_B$ adds $R(\ln\rho_A + \ln\rho_B)$ to the entropy, a quantity that is constant for almost all processes since the magnetic interactions between the nuclei are so small that ordering only occurs below $T_{ord} \lesssim 10^{-5}$ K, and for $T \gtrsim T_{ord}$ the weights always remain the same. However, for a homonuclear diatomic molecule where each nucleus has statistical weight ρ, the nuclear spins combine to $\frac{1}{2}\rho(\rho+1)$ symmetric or *ortho states* and $\frac{1}{2}\rho(\rho-1)$ antisymmetric or *para states*. If Ψ is symmetric, the para (ortho) nuclear states must combine with the odd J, antisymmetric (even J, symmetric) rotational states, while if Ψ is antisymmetric, the para (ortho) states must combine with symmetric (antisymmetric) rotational states. Hence, $z_{nuc}z_{rot} = \frac{1}{2}\rho(\rho\pm1)\Sigma_{even} + \frac{1}{2}\rho(\rho\mp1)\Sigma_{odd}$, where the upper sign applies if Ψ is symmetric and the lower sign if it is antisymmetric. For $T \gtrsim 5\theta_{rot}$, $\Sigma_{even} \cong \Sigma_{odd} \cong \frac{1}{2}\Sigma_{all}$, and we have $z = \frac{1}{2}\rho^2\Sigma_{all}$, where the factor $\frac{1}{2}$ is interpreted as the *symmetry number*. Similar considerations apply for more complex molecules.

is generally valid for $T \gtrsim 5\theta_{rot}$, but since θ_{rot} is generally small, in this case this provides seldom a problem. For an extensive discussion, see the classic references [17, 18], and for the vibration–rotation interaction, see [19].

Internal rotations, such as the transition from the staggered to eclipsed conformation in CH_3–CH_3, also contribute. For molecules for which the associated barrier is relatively large, say $>10kT$, the two groups only liberate with respect to each other, and the harmonic oscillator approximation can be used. If the barrier to internal rotation is relatively low, say $<1kT$, one could consider the internal rotation as a free rotation around a single axis for which the partition function is $z' = (kT/B')^{1/2}\sigma'$ with $B' = h^2/8\pi^2I'$ and I' the moment of inertia for rotation around the internal rotation axis. The number σ' is the symmetry number for internal rotation, for example, for acetone $\sigma' = 3$ for each CH_3 group. The total partition function becomes [20]

$$z = (8\pi^3kT/h^2)^{n/2}(I_AI_BI_CI_aI_b \dots I_j)^{1/2}/\pi\sigma_{tot}$$

where $\sigma_{tot} = \sigma\sigma'$, $I_a, I_b, \dots, I_j$ are the moments of inertia of j internal rotors, and $n = 3 + j$ is the total number of DoFs. For example, for acetone, $\sigma_{tot} = 2 \times 3^2$ and $n = 3 + 2 = 5$. In the case of ethane, however, the barrier is $\cong$12.1 kJ mol^{-1}, while at 300 K $N_AkT \cong 2.5$ kJ mol^{-1}, and thus neither the harmonic oscillator, nor the free rotator approximation may be used. Calculating z is involved [22, 23], but an accurate approximation [24] is

$$z = (\pi r/\theta)^{1/2} \exp(-r/2\theta)[\hbar\omega/(2 + 16r)]I_0(r/2\theta) \tag{7.25}$$

$$I_\nu(x) = \exp(x)/(2\pi x)^{1/2}\Sigma_k[(-1)^k\Gamma(\nu + k + ½)/(2x)^k k!\Gamma(\nu - k + ½)] \quad \text{for} \quad |x| > 3.75$$

$$I_\nu(x) = \Sigma_k[(x/2)^{\nu+2k}/k!(k + \nu)!] \quad \text{for} \quad |x| < 3.75$$

with $\Gamma(\cdot)$ the gamma function. From Eq. (7.25), F, U, and so on can be calculated as usual.

7.5.3 Electronic Transitions

Generally, the spacing between electronic states in molecules is large compared with kT. This suggests that transitions from the ground state to an excited state are rare and only the ground state needs to be considered. However, this is an oversimplification.

For simplicity, we consider the hydrogen atom having energy $E_n = R(1 - n^{-2})$ with $n = 1,2,\dots,\infty$ and R the Rydberg constant. With degeneracy factor $g_n = 2n^2$, we obtain

$$Q = \sum_{n=1}^{n=\infty} g_n \exp[-\beta R(1 - n^{-2})]$$

For $n = 1$, $E_n = 0$ and the exponential equals one. For $n = 2$, $E_n = 10.20$ eV, while kT at 298 K equals 2.569×10^{-2} eV, so that the exponential reads e^{-397} or 10^{-172}. For higher values of n, the energies rapidly approach the ionization limit of 13.60 eV, and the exponential term approaches e^{-529} or 10^{-230}. Although these are small values, there is an infinite number of them, and Q diverges. Since the probability for state n is $p_n = \exp(-\beta E_n)/Q$, this would imply that no bound state is ever observed whatever the temperature. The resolution of this paradox is that the range of $\boldsymbol{r}$ and therefore n is limited. For an atom in a volume $V = l^3$, the radius r_m of the level m that maximally can be occupied should be smaller than $l/2$, or, $r_m = m^2a_0 \le l/2$ with a_0 the Bohr radius. Taking for l a macroscopic size, say, $l = 1$ m, the maximum quantum number $m \approx 10^5$. Approximating all terms but the first with the limiting term 10^{-230} results in

$$Q \approx 1 + 10^{-230}\sum_{n=2}^{n=10^5} 2n^2 = 1 + 2(n^3/3 + n^2/2 + n/6 - 1)\Big|_2^{10^5} \approx 1 + 10^{-215}$$

Therefore, Q remains finite, rather close to one, and the probability for the atom to be in a bound state non-zero. Since all molecules have such high lying electronic states, the *Rydberg states*, the argument is generally valid. If the atom is in outer space somewhere, where the electron can move (almost) any distance without interference, this is still true. Consider the radius of the known universe,[16] $R_U \approx 10^{26}$ m. Doing the same calculation, we find that $n \approx 10^{18}$ and $Q \approx 1 + 10^{-176}$, still not significantly different from unity. Of course, for this case, $T \ll 298$, say 2.7 K, the temperature of the cosmic microwave background radiation, rendering the results again closer to one.[17] More importantly, general relativity must be accounted for. However, all possible solutions essentially boil down to having a limiting "container" size [28].

Hence, the electronic partition function is generally closely approximated by

$$z_{ele} = g_0 \exp(-\varepsilon_0/kT) + g_1 \exp(-\varepsilon_1/kT) + \dots \cong g_0 \exp(-\varepsilon_0/kT)$$

where g_i indicates the electronic state degeneracy, for the ground state often $g_0 = 1$. Taking the zero level for energy at the electronic ground state, the approximation $z_{ele} = g_0$ is in most cases sufficiently accurate for both atoms and molecules. One exception is formed by systems where the electronic ground state is split by spin–orbit coupling. For example, for the F atom the ground state is split in a lower $^3P_{3/2}$ level with $g_0 = 4$ and an upper $^2P_{1/2}$ level with $g_1 = 2$. The gap between these levels is 404 cm^{-1}, while the gap with the next electronic level is much larger than $kT = 207$ cm^{-1} at 298 K. Another example is the NO molecule with an unpaired electron in a π^* orbital giving rise to a $^2\Pi_{1/2}$ ($g_0 = 2$) and a $^2\Pi_{3/2}$ ($g_1 = 2$) level with a gap of 121 cm^{-1}. Again, the gap to the first excited level, in this case a vibration level, is much larger than kT.

Problem 7.11 *C_V, Anharmonicity, and Rotation* Estimate for a diatomic molecule the contribution to C_V due to anharmonicity (assuming $x = 0.01$) and centrifugal distortion. Show that the rotation contribution is $C_V = R[1 + (\theta_{rot}{}^2/45T^2) + \dots)]$ and to S by $S = R[1 - \ln(\sigma\theta_{rot}/T) - (\theta_{rot}{}^2/90T^2) + \dots]$.

7.6 Some General Aspects

In this section, we discuss some general and useful aspects not easily fitting elsewhere.

7.6.1 Mode or Average?

The presented line of thought takes the mode of the distribution as representing the average behavior, while one might think that one should calculate the average behavior directly. Apart from the fact that macroscopic thermodynamics also calculates the mode (i.e. maximizes S), the distribution is extremely sharp, and mode and average are virtually the same. Essentially, one writes $\Sigma_j \exp(-\beta\varepsilon_j) \cong \int \exp(-\beta\varepsilon)\, dn(\varepsilon) = \int \exp(-\beta\varepsilon) g(\varepsilon)\, d\varepsilon$ with $g(\varepsilon) = dn(\varepsilon)/d\varepsilon$. The sharpness is due to the compensating effects of the rapidly increasing

16 The radius of the universe is somewhat ambiguous as there is a difference between the radius of the observable universe, $R_U \approx 4.4 \times 10^{26}$ m [25, 26], and the whole universe, estimates of which depend heavily on the curvature of space assumed.

17 We used the arguments of Strickler [27]. For other solutions, see [28] and references given therein.

value of the density of states $g(\varepsilon)$ and the rapidly decreasing value of the Boltzmann factor $\exp(-\beta\varepsilon)$ (Problem 7.12). In the *thermodynamic limit* ($N \to \infty$ and $V \to \infty$, but N/V finite), mode and average are identical [29]. The fact that mode and average are virtually the same also leads to the (surprising) maximum-term method (Justification 7.4), which allows one to calculate the value of a logarithmic sum by taking just the largest term. A completely rigorous calculation of average values is the Darwin–Fowler method (see, e.g. [30]).

Justification 7.4 ***The Maximum-Term Method*** Surprisingly, in ST the logarithm of a sum often can be approximated by the logarithm of its maximum term. Consider the typical expression

$$X = \sum\nolimits_{n=0}^{n=N} t_n \quad \text{with} \quad t_n = N!x^n/n!(N-n)!$$

Here $x = \mathcal{O}(1)$ but >0 and $N = \mathcal{O}(10^{20})$, where $\mathcal{O}(x)$ indicates the order of magnitude of x. The sum X can be evaluated exactly as $X = (1+x)^N$. The largest term for X with index n^* is also the largest term for $\ln X$. Since n^* and $N-n^*$ are both large, we can use Stirling's approximation $\ln m! \cong m \ln m - m$, and applying $\partial \ln t_n/\partial n = 0$ leads to $n^*/(N-n^*) = x$ or $n^* = xN/(1+x)$. The maximum term becomes $t_{n^*} = (1+x)^N$, identical to the sum itself! This is too good to be true and, in fact, accidental. Using $\ln m! \cong m \ln m - m + ½\ln 2\pi m$ leads with the same procedure to the implicit function $x = n^* e^A/(N-n^*)$ with $A = (N-2n)/2n(N-n) \cong (1-x)/2n$, or employing $e^A \cong 1+A$, to $n^* \cong xN/(1+x) - (1-x)/2(1+x)$ and $\ln t_{n^*} = \ln(2\pi Nx)^{-½}(1+x)^{N+1}$. Expanding $\ln t_n$ about n^* to second order in $n-n^*$, we obtain, with either the simple or extended Stirling approximation,

$$\ln t_n = \ln t_{n^*} + ½\left(\frac{\partial^2 \ln t_n}{\partial n^2}\right)_{n=n^*}(n-n^*)^2 + \ldots \quad \text{or} \quad \ln t_n = \ln t_{n^*} - \frac{(1+x)^2(n-n^*)^2}{2xN} + \ldots$$

where the linear term is missing because $(\partial \ln t_n/\partial n)_{n=n^*} = 0$. Note from this expression that the extended Stirling approximation leads to an improved value of n^* and t_{n^*}, while the expression for the ratio t_n/t_{n^*} is identical for both approximations. Note also that $\ln t_{n^*}$ can be written as $\ln t_{n^*} = N\ln(1+x) - ½\ln N + ½\ln[(1+x)^2/2\pi x] = N\ln(1+x) + \mathcal{O}(\ln N)$, so that if we are only interested in terms of order $N\ln N$, we can use the simple Stirling approximation. Summarizing, using the Stirling approximation, we have

$$t_n = t_{n^*}\exp[-(n-n^*)^2/2\sigma^2] \quad \text{where} \quad \sigma = x^{1/2}N^{1/2}/(1+x)$$

In all cases, this is thus a *very* sharp Gaussian distribution centered about n^*, since the relative standard deviation $\sigma/n^* = (xN)^{-1/2} = \mathcal{O}(N^{-1/2}) = \mathcal{O}(10^{-10})$. Higher-order terms yield a negligible contribution, and we have $\ln X \cong \ln t_{n^*}$ and, consequently, $\langle n \rangle \cong n^*$.

7.6.2 Fluctuations and Other Ensembles

We now discuss the difference in behavior of conjugated variables in microscopic thermodynamics as compared to macroscopic thermodynamics. While in the latter case both members of a pair of conjugated variables can be fixed, in the former they behave complementarily. For example, if the temperature is prescribed via a temperature bath (closed system), the energy of the system will fluctuate, whereas if the energy is fixed (isolated system), the temperature will fluctuate. The same applies to other pairs of conjugated variables, although for macroscopic systems fluctuations are normally small. Fluctuations resulting from our statistical considerations must not be confused with the intrinsic uncertainty of QM, as described by the Heisenberg inequalities (Eq. (3.13)), in which both members of a conjugated pair of variables can have an uncertainty. Our statistical fluctuations obviously add to the quantum uncertainty but are usually much larger.

Other ensembles can be defined by using other constraints to the energy expression. For example, using constant P instead of constant V, we can calculate directly $G(P) = -kT\ln\Delta$, where $\Delta = \Sigma_i \exp[-\beta(E_i + PV)]$ using $H + PV$ as energy expression [6, 8, 12] instead of calculating $G(P) = F(P) + PV(P)$ after solving V from $P = -\partial F(V)/\partial V$. Using an obvious notation for independent variables, the partition function Δ refers to the *NPT* or *pressure ensemble*, while Z refers to the *NVT* or canonical ensemble. Actually, the procedure is similar as for the grand canonical (or μVT) ensemble, where $H + \mu N$ results in constant chemical potential conditions. Although, strictly speaking, $H + PV$ and similar expressions do not represent internal energy, in ST they are still often referred to as Hamiltonians.

7.6.3 Equipartition of Energy

In physical–chemical discussions, one uses with some regularity the "equipartition" theorem. Using semi-classical statistics this theorem follows if the energy expression $\mathcal{H}$ can be separated in a sum of independent terms. While for the kinetic energy of a mass m we generally have $\vartheta = p^2/2m$, the potential energy is typically given by $\phi(q) = \alpha q^n$, where p and q denote the momentum and coordinate, respectively. For example, for cyclobutanone one of the normal coordinates q describes the so-called "ring-puckering," for which the potential energy $\phi(q)$ can be written as $\phi(q) \cong \alpha q^4$ with α a constant. So, let us consider the average $\langle x^n \rangle$, with x either p or q.

Writing down the expression for the average with as usual $\beta = 1/kT$, we have

$$\langle x^n \rangle = z^{-1} \int x^n \exp(-\beta\phi)\,\mathrm{d}x = z^{-1} \int x^n \exp(-\beta x^n)\,\mathrm{d}x$$

$$\text{with} \quad z = \int \exp(-\beta\phi)\,\mathrm{d}x = \int \exp(-\beta x^n)\,\mathrm{d}x$$

The integrals are standard (Section A.16), and using the results for $n = 2$, we obtain

$$\langle x^2 \rangle = (4\beta)^{-1}(\pi/\beta)^{1/2}/2^{-1}(\pi/\beta)^{1/2} = 1/2\beta = kT/2$$

Thus, for a quadratic dependence on the coordinate, we obtain $\tfrac{1}{2}kT$ per DoF. This is the conventional "equipartition" of energy. More generally, consider

$$\int x_i \frac{\partial \mathcal{H}}{\partial x_j} \mathrm{e}^{-\beta\mathcal{H}}\,\mathrm{d}\omega = \iint x_i \frac{\partial \mathcal{H}}{\partial x_j} \mathrm{e}^{-\beta\mathcal{H}}\,\mathrm{d}x_j \mathrm{d}\omega^{(j)}$$

where $\mathrm{d}\omega = \mathrm{d}x_j \mathrm{d}\omega^{(j)}$ with $\mathrm{d}\omega^{(j)} = \mathrm{d}x_1 \ldots \mathrm{d}x_{j-1}\mathrm{d}x_{j+1} \ldots \mathrm{d}x_N$ is an element of phase space. By partial integration, we obtain

$$-\beta^{-1} \iiint x_i \mathrm{d}(\mathrm{e}^{-\beta\mathcal{H}})_j \mathrm{d}\omega^{(j)} = -\beta^{-1} \int \left\{ x_i \mathrm{e}^{-\beta\mathcal{H}} - \int \mathrm{e}^{-\beta\mathcal{H}}\,\mathrm{d}x_i \right\} \mathrm{d}\omega^{(j)}$$

$$= \beta^{-1} \iint \mathrm{e}^{-\beta\mathcal{H}}\,\mathrm{d}x_i \mathrm{d}\omega^{(j)} = \beta^{-1} \int \mathrm{e}^{-\beta\mathcal{H}}\,\mathrm{d}\omega \frac{\mathrm{d}x_i}{\mathrm{d}x_j} \quad \text{or} \quad \left\langle x_i \frac{\partial \mathcal{H}}{\partial x_j} \right\rangle = kT\delta_{ij} \tag{7.26}$$

Here the term $x_i \exp(-\beta\mathcal{H})$ vanishes because at the boundary $\mathcal{H} = \infty$ and $\mathrm{d}x_i/\mathrm{d}x_j = \delta_{ij}$ as x_i and x_j are independent coordinates. This result, due to Tolman [31], represents the (generalized) *equipartition theorem*: If $\mathcal{H} = \alpha x^n$, $x(\partial\mathcal{H}/\partial x) = n\mathcal{H}$, and $\langle x(\partial\mathcal{H}/\partial x)\rangle = kT/n$. The theorem applies as long as the partition function can be approximated by an integral. As stated before, this is generally true for translation, for sufficiently high T true for rotation and seldom true for vibration. If we nevertheless apply the theorem to the harmonic oscillator with $\mathcal{H} = p^2/2m + \alpha q^2$, we have a contribution ½kT from the kinetic energy and also ½kT from the potential energy, leading to kT per oscillator.

7.6.4 The Gibbs–Bogoliubov Inequality

Suppose we have a Hamilton expression consisting of a reference part $\mathcal{H}_0$ and a perturbation part $\mathcal{H}_1$ and we write, using λ as a bookkeeping label, $\mathcal{H} = \mathcal{H}_0 + \lambda\mathcal{H}_1$. The configurational partition function Q becomes

$$Q = \int \exp(-\beta\mathcal{H})\,\mathrm{d}\mathbf{r} = Q_0\left(Q_0^{-1}\int \exp[-\beta(\mathcal{H}_0 + \mathcal{H}_1)]\,\mathrm{d}\mathbf{r}\right) = Q_0\langle\exp(-\beta\mathcal{H}_1)\rangle_0$$

where Q_0 is the configurational partition function of the reference system and $\langle\cdots\rangle_0$ denotes the canonical average for that reference system. Using $\exp(x) = 1 + x + x^2/2! + \ldots$ for $\langle\exp(\mathcal{H}_1)\rangle_0$, we find $F = -kT\ln Q = -kT[\ln Q_0 + \ln(1 - \beta\langle\mathcal{H}_1\rangle_0 + ½\beta^2\langle\mathcal{H}_1^2\rangle_0 - \ldots)]$. Expanding the logarithm via $\ln(1+x) = x - x^2/2 + \ldots$, the result is the λ *expansion*

$$-\beta F_\lambda = -\beta F_0 - \lambda\beta\langle\mathcal{H}_1\rangle_0 + ½\lambda^2\beta^2\left[\left\langle\mathcal{H}_1^2\right\rangle_0 - \langle\mathcal{H}_1\rangle_0^2\right] + \ldots$$

Here, $\langle f\rangle_\lambda$ is the canonical average given λ. Hence, for the reference system we have $\langle f\rangle_0$, while the complete system we use $\langle f\rangle_1$. Since $\langle\mathcal{H}_1^2\rangle_0 - \langle\mathcal{H}_1\rangle_0^2 = \langle\mathcal{H}_1 - \langle\mathcal{H}_1\rangle_0\rangle_0^2$, the λ^2 term is always positive. Therefore, $F_0 - \langle\mathcal{H}_1\rangle_0$ is an upper bound for F. We can make further progress noting that for $q \geq 0$, $q\ln q - q + 1 \geq 0$ [18] and write $q = f/g$. For arbitrary distribution functions f and g with $\int f\mathrm{d}\mathbf{r} = \int g\mathrm{d}\mathbf{r} = 1$, we have

$$\int g[(f/g)\ln(f/g) - (f/g) + 1]\,\mathrm{d}\mathbf{r} \geq 0$$

since $g(r) \geq 0$. Identifying the reference system ($\lambda = 0$) with $f = \exp[-\beta(\mathcal{H}_0 - F_0)]$ and the actual system ($\lambda > 0$) with $g = \exp[-\beta(\mathcal{H} - F_\lambda)]$, we obtain [32]

$$-\beta(\langle\mathcal{H}_0\rangle_0 - F_0) \geq -\beta(\langle\mathcal{H}\rangle_0 - F_\lambda) \quad \text{or} \quad F_\lambda \leq F_0 + \lambda\langle\mathcal{H}_1\rangle_0$$

Reversing the roles of f and g, we similarly obtain $F_\lambda \geq F_0 + \lambda\langle\mathcal{H}_1\rangle_1$. In total, we thus have, what is called the *Gibbs–Bogoliubov inequality* [33]

$$F_0 + \lambda\langle\mathcal{H}_1\rangle_1 \leq F_\lambda \leq F_0 + \lambda\langle\mathcal{H}_1\rangle_0 \tag{7.27}$$

Figure 7.3 shows that $\langle\mathcal{H}_1\rangle_0$ and $\langle\mathcal{H}_1\rangle_\lambda$ provide an upper and lower bound on $F_\lambda - F_0$. Usually, only $\langle\mathcal{H}_1\rangle_0$ is considered. If $\mathcal{H}_1$ contains parameters ξ, $\langle\mathcal{H}_1\rangle_0$ can be minimized with respect to ξ to obtain the optimum result. Some consideration makes it clear that the "flatter" $\mathcal{H}_1$, the smaller $\langle\mathcal{H}_1\rangle_0$ will be. In case we have a pairwise additive potential

18 If $y(q) = q\ln q - q + 1$, $y' = \ln q$, and from $y' = 0$, we see that $y(q)$ has one extremum at $q = 1$. From $y'' = 1/q \geq 0$, this extremum is a minimum.

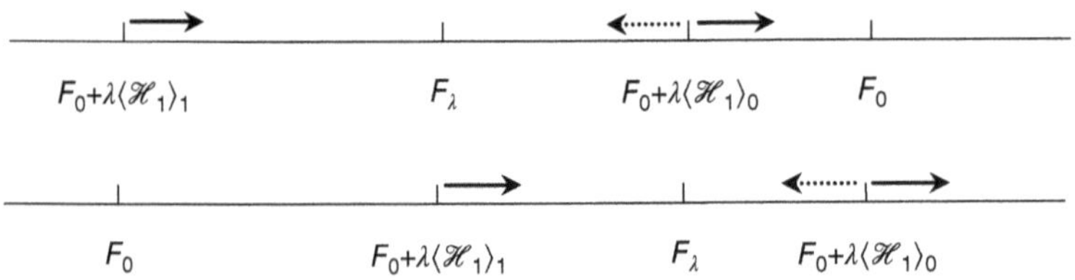

Figure 7.3 Terms in the Gibbs–Bogoliubov inequality for $\langle \mathscr{H}_1 \rangle_0 < 0$ (upper) and $\langle \mathscr{H}_1 \rangle_0 > 0$ (lower). Arrow → indicates the effect of F'' while arrow <···. represents the effect of minimization.

$\Phi = ½\Sigma_{ij}\phi_{ij}$, which can be split as $\phi(r) = \phi_0(r) + \phi_1(r)$ with $\phi_0(r)$ a reference part and $\phi_1(r)$ the perturbation, $\langle \mathscr{H}_1 \rangle_0$ can be simplified and to first order

$$F_\lambda = F_0 + 2\pi\lambda N\rho \int g_0(r)\phi_1(r)r^2\,\mathrm{d}r + \ldots$$

with g_0 the distribution function of the reference system. As an example of the use of the Gibbs–Bogoliubov inequality, we sketch a proof of the *Henderson theorem* [34], which states that under rather mild conditions, two pair potentials that give rise to the same correlation function $g(r)$ (Section 9.2) cannot differ by more than a constant. Suppose we have two systems with Hamilton function $\mathscr{H}_j$ and potential ϕ_j, leading to $g_j(r)$, so that for the Helmholtz energies we have $F_1 \leq F_2 + \langle \mathscr{H}_2 - \mathscr{H}_1 \rangle_1$. Equality only holds if $\mathscr{H}_2 - \mathscr{H}_1$ is independent of all DoFs, implying that ϕ_1 and ϕ_2 can only differ by a constant. If ϕ_1 and ϕ_2 differ by more than a constant, the same holds for $\mathscr{H}_1$ and $\mathscr{H}_2$ and the equality cannot hold, or $F_1 < F_2 + ½N\rho \int [\phi_2(\mathbf{r}) - \phi_1(\mathbf{r})]g_1(\mathbf{r})\mathrm{d}\mathbf{r}$. This also holds for systems 1 and 2 interchanged so that $F_2 < F_1 + ½N\rho \int [\phi_1(\mathbf{r}) - \phi_2(\mathbf{r})]g_2(\mathbf{r})\mathrm{d}\mathbf{r}$. Inserting $g_1(\mathbf{r}) = g_2(\mathbf{r})$ and adding both equations leads to $0 < 0$. Hence the initial assumption must be wrong. The uniqueness proof for ϕ is simple, but its existence proof is complex [34]. Basically, the Henderson theorem states that, if $g(r)$ is a two-particle correlation function for any admissible N-particle distribution, an effective pair potential reproducing $g(r)$ always exists.

Problem 7.12 ***The Sharpness of the Energy Distribution*** Consider as system X a particle in a 3D box with density of states (DoS) $g_X(\varepsilon_X) \sim \varepsilon_X^{1/2}$. Take as weakly coupled reservoir R also such a system, so that the total energy becomes $\varepsilon = \varepsilon_X + \varepsilon_R$. Maximize the total DoS $g = g_X g_R$ using $\partial g/\partial\varepsilon_X = 0$ and show that $\varepsilon_X = ½\varepsilon$. Is this the expected result, and is the associated energy distribution sharp? Next, consider a many-particle system with N_X particles with DoS $g_X \sim E_X^{\alpha}$ with $\alpha = (3N_X/2) - 1$. Couple this system to a many-particle reservoir with N_R particles with DoS $g_R \sim E_R^{\beta}$ with $\beta = (3N_R/2) - 1$. Show that in this case that g is sharply peaked for large N_X and N_R with mode $E_X = \alpha E/(\alpha + \beta)$, where $E = E_X + E_R$.

References

1 Gibbs, 1902.
2 Tolman, 1938.
3 Phillies, 2000.

4 Landsberg, P.T. (1978). *Thermodynamics and Statistical Mechanics.* Oxford, UK: Oxford University Press (also Dover, 1990).

5 Khinchin, A.I. (1957). *Mathematical Foundations of Information Theory.* New York: Dover.

6 Landau and Lifshitz, 1980.

7 Mayer, J.E. and Mayer, M.G. (1940). *Statistical Mechanics.* New York: Wiley.

8 Fowler and Guggenheim, 1939.

9 Paillusson, F. (2018). *Mol. Phys.* 116: 3196; Koutsoyiannis, D. (2013). *Stud. Hist. Phil. Mod. Phys.* 44: 480; Ainsworth, P.M. (2012). *Phil. Sci.* 79: 542.

10 Gosper, R.W. (1978). *Proc. Natl. Acad. Sci. U.S.A.* 75: 40.

11 Wigner, E.P. (1932). *Phys. Rev.* 40: 749; Kirkwood, J.G. (1933). *Phys. Rev.* 44: 31 and (1934). 45: 116; Green, H.S. (1951). *J. Chem. Phys.* 19: 955.

12 Hill, T.L. (1956). *Statistical Mechanics.* London: McGraw-Hill Book Comp. (also Dover, 1987).

13 McQuarrie, 1973.

14 Knox, J.H. (1978). *Molecular Thermodynamics.* New York: Wiley-Interscience.

15 Gearhart, C.A. (1996). *Am. J. Phys.* 64: 995.

16 Wilson, E.B., Decius, J.C., and Cross, P.C. (1955). *Molecular Vibrations.* New York: McGrawHill Book Comp.

17 Townes, C.H. and Schawlow, A.L. (1955). *Microwave Spectroscopy.* New York: McGraw-Hill Publ. Comp.

18 Wollrab, J.E. (1967). *Rotational Spectra and Molecular Structure.* New York: Academic Press.

19 Allen, H.C. and Cross, P.C. (1963). *Molecular Vib-Rotors.* New York: John Wiley & Sons.

20 Halford, J.O. (1934). *J. Chem. Phys.* 2: 694.

21 McQuarrie, D.A. (2008). *Quantum Chemistry*, 2e. Sausalito, CA: University Science Books.

22 Pitzer, K.S. (1937). *J. Chem. Phys.* 5: 469; (1942). 10: 428; (1946). 14: 239; (1949). 17: 1064; (1954). *J. Phys. Chem.* 60: 466.

23 Lucas, K. (2007). *Molecular Models of Fluids.* Cambridge: Cambridge University Press.

24 McClurg, R.B., Flagan, F.C., and Goddard, III, W.A. (1997). *J. Chem. Phys.* 106: 6675; McClurg, R.B. (1999). *J. Chem. Phys.* 111: 7164.

25 Halpern, P. and Tomasello, N. (2016). *Adv. Astrophys.* 1: 135.

26 https://en.wikipedia.org/wiki/Observable_universe (accessed 23 August 2015).

27 Strickler, S.J. (1966). *J. Chem. Ed.* 43: 364.

28 Blinder, S.M. (1995). *J. Math. Phys.* 36: 1208; Blinder, S.M. (1996). *J. Math. Chem.* 19: 43.

29 Ruelle, D. (1969). *Statistical Mechanics: Rigorous Results.* New York: Benjamin, Inc. (also 1999, Imperial College Press, London).

30 Fowler, R.H. (1936). *Statistical Mechanics*, 2e. London: Cambridge University Press; Schrödinger, E. (1952). *Statistical Thermodynamics*, 2e. Cambridge: Cambridge University Press.

31 Tolman, R.C. (1918). *Phys. Rev.* 11: 262.

32 Falk, H. (1963). *Physica* 29: 1114.

33 Friedman, H.L. (1985). *A Course on Statistical Mechanics.* Englewood Cliffs, NJ: Prentice-Hall.

34 Henderson, R.L. (1974). *Phys. Lett. A* 49: 197; Chayes, J.T., Chayes, L., and Lieb, E.L. (1984). *Commun. Math. Phys.* 93: 57; Chayes, J.T. and Chayes, L. (1984). *J. Stat. Phys.* 36: 471.

Further Reading

Chandler, D. (1987). *Introduction to Modern Statistical Mechanics*. Oxford: Oxford University Press. A concise introduction to phase transitions, Monte Carlo simulations and non-equilibrium systems.

Fowler, R.H. and Guggenheim, E.A. (1939). *Statistical Thermodynamics*. London: Cambridge University Press. A classic focusing on physical-chemical topics and discussions not found easily elsewhere.

Gibbs, J.W. (1902). *Elementary Principles in Statistical Mechanics*. New Haven: Yale University Press. A clearly pre-quantum era but must-read at some point book by the main founder of statistical mechanics.

Landau, L.D. and Lifshitz, E.M. (1980). *Statistical Physics*, 3e, Vol. 1. Oxford: Pergamon Press. Advanced and condensed treatment including thermodynamics, well-known for physical insight.

McQuarrie, D.A. (1973). *Statistical Mechanics*. New York: Harper and Row. A standard reference focusing on the basic aspects including liquids, meanwhile providing many details.

Phillies, G.D.J. (2000). *Elementary Lectures in Statistical Mechanics*. New York: Springer. Excellent introduction, while still treating more sophisticated topics such as cluster expansions and correlation functions, including Mori-Zwanzig and linear response theory.

Reichl, L.E. (2009). *A Modern Course in Statistical Physics*. Weinheim: Wiley-VCH. Balanced modern and relatively high-level treatment of equilibrium and non-equilibrium aspects.

Swendsen, R.H. (2012). *Introduction to Statistical Mechanics and Thermodynamics*. Oxford: Oxford University Press. A modern and highly readable introduction along classical lines.

Tolman, R.C. (1938). *The Principles of Statistical Mechanics*. Oxford: Oxford University Press (also Dover, 1979). A classical treatise containing not only what the title promises, but also containing useful discussions about classical and quantum mechanics.

8

Gases

After having discussed in Chapters 2–7 the "tools," we now turn to the "bricks," our first type being gases. We start with a microscopic discussion based on the kinetic theory of gases, followed by a discussion on the virial expansion. Thereafter, we turn to a macroscopic description and outline some equations of state and the principle of corresponding states. Finally, we discuss dielectric behavior.

8.1 Basic Kinetic Theory of Gases

In Chapter 7, we introduced the probabilities to realize the relative number of molecules $\mathrm{d}N/N$ with coordinates $\mathbf{q}$ and $\mathbf{p}$ and the associated canonical partition function Z_N, which led to what is called the *Maxwell–Boltzmann* (MB) *distribution*

$$\mathrm{d}N/N = \exp(-\beta\mathscr{H})\,\mathrm{d}\mathbf{q}\mathrm{d}\mathbf{p}/Z_N \qquad \text{with}$$

$$Z_N = (h^{3N}N!)^{-1}\int\cdots\int \mathrm{e}^{-\beta\mathscr{H}}\,\mathrm{d}\mathbf{q}\,\mathrm{d}\mathbf{p} = (\Lambda^{3N}N!)^{-1}\int\cdots\int \mathrm{e}^{-\beta\Phi}\,\mathrm{d}\mathbf{q} \equiv (\Lambda^{3N}N!)^{-1}Q_N$$

Here, we use $\beta = 1/kT$ with Boltzmann's constant k and temperature T, the Hamilton function $\mathscr{H} = \vartheta + \Phi$ with kinetic energy ϑ and potential energy Φ, Planck's constant h, thermal wavelength $\Lambda = h/l \equiv h/(2\pi mkT)^{1/2}$ with mass m, and the configuration integral Q_N. For noninteracting particles, $\Phi = 0$, and we obtained the *perfect gas Equation of State* (EoS) $PV = nRT$ with pressure P, volume V, number of moles n, and the gas constant $R = N_\mathrm{A}k$. We also indicated that, generally, evaluation of Q_N is the problem. However, one can learn still a great deal more from the kinetic part alone and that is what we will pursue here. Hence, for the moment we take $\Phi = 0$ and consider independent particles with velocity $\boldsymbol{v}$ or momentum $\boldsymbol{p}$. The MB distribution then reads $\mathrm{d}N/N = B(\boldsymbol{p}) = \exp(-\beta\vartheta)\mathrm{d}\boldsymbol{p}/l^3$ or $\mathrm{d}N/N = B(\boldsymbol{v}) = \exp(-\beta\vartheta)m^3\mathrm{d}\boldsymbol{v}/l^3$.

First, we note that we use integrals with an integrand containing $\exp(-cx^2)$. To evaluate such an integral, one uses in general tables of integrals (Section A.16), but as they are so essential, we briefly discuss them first. We start by noting that

$$I = \int_{-\infty}^{+\infty} \exp(-cx^2)\,\mathrm{d}x = \sqrt{\pi/c}$$

Phases of Matter and their Transitions: Concepts and Principles for Chemists, Physicists, Engineers, and Materials Scientists, First Edition. Gijsbertus de With.

Differentiation of I yields somewhat more complex integrals, for example,

$$\mathrm{d}I/\mathrm{d}c = -\int_{-\infty}^{+\infty} x^2 \exp(-cx^2)\,\mathrm{d}x = \mathrm{d}(\sqrt{\pi/c})/\mathrm{d}c = -\sqrt{\pi/c}/2c$$

Proceeding similarly with $J = \int_0^{+\infty} x \exp(-cx^2)\,\mathrm{d}x = 1/2c$, one can generalize to

$$\int_{-\infty}^{+\infty} x^{2n} \exp(-cx^2)\,\mathrm{d}x = (-1)^n \mathrm{d}^n I/\mathrm{d}c^n \quad \text{with} \quad I = (\pi/c)^{1/2} \quad \text{and}$$

$$\int_0^{+\infty} x^{2n+1} \exp(-cx^2)\,\mathrm{d}x = (-1)^n \mathrm{d}^n J/\mathrm{d}c^n \quad \text{with} \quad J = 1/2c$$

We note that for even exponents $\int_0^{+\infty} x^{2n} \exp(-cx^2)\,\mathrm{d}x = ½ \int_{-\infty}^{+\infty} x^{2n} \exp(-cx^2)\,\mathrm{d}x$ and that for $c = \beta/2m$, the result $(\pi/c)^{1/2}$ equals $l = (2\pi mkT)^{1/2}$ [1].

Let us now calculate some averages. Using $\mathcal{H} = \vartheta = p^2/2m = ({p_x}^2 + {p_y}^2 + {p_z}^2)/2m = \vartheta_x + \vartheta_y + \vartheta_z$, the average momentum $\langle p \rangle$ of a molecule evidently becomes

$$\langle p \rangle = \int p\mathrm{e}^{-\beta\vartheta}\,\mathrm{d}\mathbf{p}/l^3 = 4\pi \int_0^{+\infty} p\mathrm{e}^{-\beta\vartheta} p^2 \mathrm{d}p/l^3 = \sqrt{8mkT/\pi}$$

so that the *average speed* becomes $\langle v \rangle = \langle p \rangle/m = (8kT/\pi m)^{1/2}$. Similarly, one can calculate the average square momentum $\langle p^2 \rangle$ leading to

$$\langle p^2 \rangle = \int p^2 \mathrm{e}^{-\beta\vartheta}\,\mathrm{d}\mathbf{p}/l^3 = 4\pi \int_0^{\infty} p^2 \mathrm{e}^{-\beta\vartheta} p^2 \mathrm{d}p/l^3 = 3kTm$$

Therefore, the *root-mean-square speed* $\langle v^2 \rangle^{1/2} = \langle p^2 \rangle^{1/2}/m = (3kT/m)^{1/2}$. The most probable or modal momentum $\tilde{p}$ is determined by $\mathrm{d}B/\mathrm{d}p = 0$. This leads to

$$\frac{\mathrm{d}B}{\mathrm{d}p} = \frac{\mathrm{d}}{\mathrm{d}p} 4\pi \int_0^{+\infty} (\mathrm{e}^{-\beta\vartheta} p^2)\,\mathrm{d}p/l^3 = 0 \quad \text{or} \quad \mathrm{d}(\mathrm{e}^{-\beta\vartheta} p^2)/\mathrm{d}p = 0$$

and results in $\tilde{p} = (2mkT)^{1/2}$ or in the *modal speed* $\tilde{v} = (2kT/m)^{1/2}$. Comparing these characteristics, we see that the root-mean-square speed $\langle v^2 \rangle^{1/2} = (3\pi/8)^{1/2} \langle v \rangle = 1.085 \langle v \rangle$, about 10% larger than the average speed $\langle v \rangle$, while the modal speed $\tilde{v} = (2\pi/8)^{1/2} \langle v \rangle = 0.886 \langle v \rangle$, about 10% smaller than $\langle v \rangle$.

Another often required result is the flux of molecules toward a surface. Generally, if for a gas with number density $\rho = N/V$, a fraction $\mathrm{d}\rho$ of the molecules has a velocity component u toward a surface with area A, the number of these molecules reaching the surface per second becomes $Au\mathrm{d}\rho$. Taking for the surface the x–y plane the *flux* – the rate of approach of molecules per unit surface area per unit time – becomes $\mathrm{d}f = u\mathrm{d}\rho = p_z \mathrm{d}\rho/m$. We further consider that $\mathrm{d}\rho = \rho B(\mathbf{p})\mathrm{d}\mathbf{p}$. Therefore,

$$m\mathrm{d}f = \rho p_z \mathrm{e}^{-\beta\vartheta}\,\mathrm{d}\mathbf{p}/l^3 = \rho \left(\mathrm{e}^{-\beta\vartheta_x} \mathrm{d}p_x/l\right) \left(\mathrm{e}^{-\beta\vartheta_y} \mathrm{d}p_y/l\right) \left(p_z \mathrm{e}^{-\beta\vartheta_z} \mathrm{d}p_z/l\right)$$

or, for the rate solely in the z-direction by integrating over the x- and y-directions,

$$m\mathrm{d}f_z = \rho p_z \mathrm{e}^{-\beta\vartheta_z}\,\mathrm{d}p_z/l = \rho \left(p_z \mathrm{e}^{-\beta\vartheta_z} \mathrm{d}p_z/l\right) \quad \text{as} \quad \int_{-\infty}^{+\infty} \exp^{-\beta\vartheta_j} \mathrm{d}p_j/l = 1$$

Integration over the z-direction leads to

$$mf = \rho \int_0^{\infty} p_z \mathrm{e}^{-\beta\vartheta_z}\,\mathrm{d}p_z/l = \rho \cdot ½(2mkT)/l = \rho(mkT/2\pi)^{1/2}$$

We thus have for the flux of molecules toward a surface $f = \rho(kT/2\pi m)^{1/2} = \rho\langle v\rangle/4$, or using the perfect gas law, $f = P_{\text{vap}}/l$, where P_{vap} is the vapor pressure.

It is also of interest to calculate the average distance a molecule travels between collisions. We first somewhat heuristically derive the distribution of the collision distances. Assuming isotropy, we take the probability that a collision occurs in an interval dr as dr/λ, so that $1 - \mathrm{d}r/\lambda$ is the probability that no collision occurs in the interval dr. If $P(r)$ is the probability that no collision occurs in an interval of length r, the probability that no collisions occur in an interval of length $r + \mathrm{d}r$ is $P(r + \mathrm{d}r) = P(r)(1 - \mathrm{d}r/\lambda)$. Rearranging and taking the limit $\mathrm{d}r \to 0$ leads to $\mathrm{d}P/\mathrm{d}r = -P/\lambda$, and taking into account that $P(0) = 1$, to the solution $P(r) = \exp(-r/\lambda)$. Therefore, the probability that a collision occurs in the interval $r \to r + \mathrm{d}r$ is $\mathrm{d}P = P(r)\mathrm{d}(r/\lambda)$, and the average distance between collisions or *mean free path* $\int_0^{+\infty} r\,\mathrm{d}P = \int_0^{+\infty} r\,P(r)\mathrm{d}(r/\lambda) = \lambda$.

To obtain an expression for λ, consider a gas of hard sphere (HS) molecules of type 1 (2) having mass m_1 (m_2), number density ρ_1 (ρ_2), and collision diameter σ_1 (σ_2). For each type, the velocity distribution is given by the MB distribution $B(\mathbf{v})$. Consider now the average relative speed $v_{\text{rel}} = |\mathbf{v}_2 - \mathbf{v}_1|$ between particles of type 1 and 2 given by

$$\langle v_{\text{rel}}\rangle = \int\int B(\mathbf{v}_1)B(\mathbf{v}_2)\,|\mathbf{v}_2 - \mathbf{v}_1|\,\mathrm{d}\mathbf{v}_1\mathrm{d}\mathbf{v}_2$$

The first step to make is to convert to center-of-mass coordinates. Defining the *total mass* $M = m_1 + m_2$ and *reduced mass* $1/\mu = 1/m_1 + 1/m_2$, we can use the *center-of-mass velocity* $\boldsymbol{V} = m_1\mathbf{v}_1 + m_2\mathbf{v}_2/M$ and *relative velocity* $\mathbf{v} = \mathbf{v}_2 - \mathbf{v}_1$. The Jacobian determinant J for this transformation is $J = 1$. Thus, we obtain

$$\langle v_{\text{rel}}\rangle = (\beta M/2\pi)^{3/2}(\beta\mu/2\pi)^{3/2}\int\int \mathbf{v}\mathrm{e}^{-\frac{1}{2}\beta MV^2}\,\mathrm{e}^{-\frac{1}{2}\beta\mu v^2}\,\mathrm{d}\mathbf{v}\mathrm{d}\boldsymbol{V} = (8kT/\pi\mu)^{1/2}$$

Next, to calculate the collision frequency f_{12} between particles of type 1 and 2, we can now assume that particle 1 is at rest and particle 2 moves with speed $\mathbf{v}_{\text{rel}}$. The latter particle sweeps out a cylinder of radius $\sigma_{12} = \frac{1}{2}(\sigma_1 + \sigma_2)$. The number of particles of type 2 colliding with the one of type 1 per unit of time is given by $\rho_2\pi\sigma_{12}{}^2\langle v_{\text{rel}}\rangle$, so that the total number of collisions between particles of type 1 and 2 per second is $f_{12} = \rho_1\rho_2\pi\sigma_{12}{}^2\langle v_{\text{rel}}\rangle$. Now reducing to one type of particles, we have $f = \frac{1}{2}\rho^2\pi\sigma^2\langle v_{\text{rel}}\rangle = \frac{1}{2}\rho^2\pi\sigma^2(16kT/\pi m)^{1/2}$, where the factor ½ is introduced to avoid double counting. Finally, we need the expression for $\langle v_{\text{rel}}\rangle$ in terms of the average speed $\langle v\rangle$. Using $\langle v_{\text{rel}}{}^2\rangle = \langle(\mathbf{v}_2 - \mathbf{v}_1)^2\rangle = \langle v_2{}^2\rangle + \langle v_1{}^2\rangle$, we have $\langle v_{\text{rel}}{}^2\rangle^{1/2} = 2^{1/2}\langle v^2\rangle^{1/2}$ for identical masses. Because $\langle v\rangle \sim \langle v^2\rangle^{1/2}$, we also have $\langle v_{\text{rel}}\rangle = 2^{1/2}\langle v\rangle$. Hence, for a gas of one type of particles, the mean free path is $\lambda = \langle v\rangle/f = \langle v_{\text{rel}}\rangle/2^{1/2}f = (2^{1/2}\pi\rho\sigma^2)^{-1}$.

All this leads directly to an expression for the transport coefficient of any property $Z(z)$. Using a Taylor expansion, the value of Z for a plane located at $z_0 - z$ with respect to a reference plane of area dO located at $z = z_0$ is $Z(z) = Z_0 - (\mathrm{d}Z/\mathrm{d}z)_0 z = Z_0 - (\mathrm{d}Z/\mathrm{d}z)_0 r\cos\theta$, with θ the polar angle (Figure A.1c). For a volume element d$\boldsymbol{r}$, the product $\langle v\rangle\mathrm{d}\boldsymbol{r}$ is the distance per unit time traveled by a molecule within d$\boldsymbol{r}$, so that the number of collisions per unit time is $\langle v\rangle\mathrm{d}\boldsymbol{r}/\lambda$. Of the molecules scattered within d$\boldsymbol{r}$, a fraction $(\cos\theta)/4\pi r^2$ reaches the area dO. Finally, only the fraction $\exp(-r/\lambda)$ of molecules having at least path length r reaches dO. Therefore, the total flux is

$$J_Z = \int \frac{\langle v\rangle Z}{\lambda}\frac{\cos\theta}{4\pi r^2}\mathrm{e}^{-r/\lambda}\mathrm{d}\boldsymbol{r} = C\int_0^{2\pi}\mathrm{d}\phi\int_0^{\pi}\cos\theta\sin\theta\,\mathrm{d}\theta\int_0^{\infty}Z(r)\mathrm{e}^{-r/\lambda}\mathrm{d}r$$

Table 8.1 Diffusivity D, viscosity η, and thermal conductivity κ based on kinetic theory.

Law	J_Z	Z	Z'		ξ_{CC}
Fick	$J = -Dm\rho'$	ρm	$\rho' m$	$D = ⅓\langle v\rangle\lambda = \xi r(\pi\sigma^2)^{-1}(\rho m)^{-1}$	3/8
Newton	$F = -\eta v'$	$\rho m\langle v\rangle$	$\rho m\langle v\rangle'$	$\eta = ⅓\langle v\rangle\lambda\rho m = \xi r(\pi\sigma^2)^{-1}$	5/16
Fourier	$q = -\kappa T'$	$\rho\langle\varepsilon\rangle = \rho C_V T$	$\rho\langle\varepsilon\rangle' = nc_V T'$	$\kappa = ⅓\langle v\rangle\lambda\rho c_V = \xi r(\pi\sigma^2)^{-1}c_V/m$	25/32

J, molecular flux; F, force (momentum flux); q, heat flux; $\xi = 2/3\pi$, $r = l/2^{1/2} = (\pi mkT)^{1/2}$.
Source: Hirschfelder et al. [2]/John Wiley & Sons.

Table 8.2 Molecular diameter σ_η from viscosity η and σ_D from diffusivity D at 0 °C and 1 atm.

Gas	η (μPa s)	D (cm² s⁻¹)	σ_η (Å)	σ_D (Å)	Gas	η (μPa s)	D (cm² s⁻¹)	σ_η (Å)	σ_D (Å)
Ne	209.9	0.156	3.64	3.47	O_2	191.8	0.187	3.61	3.35
Ar	296.7	0.452	2.58	2.42	CH_4	103.0	0.206	4.14	3.79
N_2	166.3	0.185	3.75	3.48	CO_2	136.6	0.0974	4.63	4.28

Source: Hirschfelder et al. [2]/John Wiley & Sons.

where $C = \langle v\rangle/4\pi\lambda$. Inserting the Taylor expansion, noting that the integral over Z_0 vanishes and evaluating the integral over $(\mathrm{d}Z/\mathrm{d}z)_0 = Z'$, leads to

$$J_Z = -C\int_0^{2\pi}\mathrm{d}\phi\int_0^{\pi}\cos^2\theta\sin\theta\,\mathrm{d}\theta\int_0^{\infty}Z'\mathrm{e}^{-r/\lambda}r\,\mathrm{d}r = -C\cdot 2\pi\cdot ⅔\cdot Z'\lambda^2 = -⅓\langle v\rangle\lambda Z'$$

In Table 8.1, we summarize the results for diffusivity D, viscosity η, and thermal conductivity κ. More rigorous theory for HS molecules [2, 3] yields the same functional dependence but with different coefficients (ξ_{CC} in Table 8.1). Calculated values for D, η, and κ show reasonable agreement with the experimental data using ξ_{CC}. Because molecules are not HSs, we expect that σ as determined from, say, D and η differ. Actually, the agreement is also quite reasonable (Table 8.2). For other potentials than the HS potential, the more rigorous calculations require extensive numerical evaluation. For a *Sutherland potential* $\phi_S = -\varepsilon(\sigma/r)^{-6}$ and $kT \approx \varepsilon$, a first-order evaluation yields $\sigma^2 = \sigma_0{}^2(1 + S/T)$ with $S_\eta = 0.1722\varepsilon/k$ and $S_D = 0.1667\varepsilon/k$.

Finally, to provide some numbers, consider nitrogen (N_2) at 293 K and 1 atm. Using σ = 3.64 Å, we have $\langle v\rangle \cong 471\ \mathrm{m\,s^{-1}}$, $\lambda \cong 678$ Å, $f \cong 8.7\times10^9\ \mathrm{s^{-1}}$ and the mean spacing $l = \rho^{-1/3} \cong 34.2$ Å. Clearly for these conditions, $\lambda \gg l$. In fact, the mean free path λ is useful to characterize the difference between phases. To do this we note that the average volume per molecule in the gas phase $v_{\mathrm{mol}} = V/N$ is also given by $v_{\mathrm{mol}} = 2^{1/2}\pi\sigma^2\lambda$. In solids, using an FCC lattice, the volume per molecule is $v_0 = 2^{-1/2}\sigma^3$. Experimentally, it appears that at T_{cri} the mean volume $v_{\mathrm{cri}} \cong 4v_0$ or $v_{\mathrm{mol}} = 2\cdot2^{1/2}\sigma^3 = 2^{1/2}\pi\sigma^2\lambda_{\mathrm{cri}}$ or $\lambda_{\mathrm{cri}}/\sigma = 2/\pi$. Hence, for gases $\lambda > \sigma$, for liquids $\lambda \cong \sigma$ and for solids $\lambda < \sigma$, which is another way to characterize the difference between gases, liquids, and solids than that given in Chapter 1. For further details on kinetic gas theory, see [4–6].

8.2 The Virial Expansion

So far, we neglected intermolecular interactions, which led to the perfect gas EoS. Adding contributions from intramolecular degrees of freedom (DoFs), this provides a useful description of nearly perfect (or low-density) gases, proof being that the standard entropy $S_m{}^\circ$ can be calculated in precise agreement with the experimental data.[1] We need, of course, also a description for nonperfect gases. To that purpose, we recall that in Section 7.4, the interaction potential $\Phi(\mathbf{r})$ was introduced in the partition function Z, leading to the configuration integral Q. We now discuss the *virial expansion* to evaluate Q and the EoS.

The virial theorem (Section 2.7) is given by $\overline{\vartheta} = \mathscr{V} = -½\Sigma_j \overline{\boldsymbol{f}_j \cdot \boldsymbol{r}_j}$. According to the equipartition theorem (Section 7.7), the average kinetic energy per atom is $3kT/2$, where T is the absolute temperature and k is Boltzmann's constant. Thus, for N atoms $\overline{\vartheta} = 3NkT/2$. For the force on particle j, neglecting long-range forces, we have $\boldsymbol{f}_j = \boldsymbol{f}_j{}^{(\text{sur})} + \boldsymbol{f}_j{}^{(\text{int})}$. With P the pressure, the instantaneous force on an area of wall $\text{d}A$ is $\Sigma_j \boldsymbol{f}_j{}^{(\text{sur})} = \boldsymbol{n}P\text{d}A$, where $\boldsymbol{n}$ is the *inward* unit vector normal to the surface. Thus, we have $\sum_j \overline{\boldsymbol{f}_j^{(\text{sur})} \cdot \boldsymbol{r}_j} = \sum_j \int \boldsymbol{r}_j \cdot \boldsymbol{f}_j^{(\text{sur})}\,\text{d}A = \sum_j P \int \boldsymbol{n} \cdot \boldsymbol{r}_j\,\text{d}A = -\sum_j P \int \nabla \cdot \boldsymbol{r}_j\,\text{d}V = -3PV$, where Gauss's theorem is used. Further, replacing $\boldsymbol{f}_j{}^{(\text{int})}$ by $-\partial\Phi/\partial\boldsymbol{r}_j$ and the time average $\sum_j \overline{\boldsymbol{f}_j^{(\text{int})} \cdot \boldsymbol{r}_j}$ by the phase average $\sum_j \int \boldsymbol{r}_j \cdot (\partial\Phi/\partial\boldsymbol{r}_j)\text{e}^{-\beta\mathscr{H}}\,\text{d}V$, the total result is $PV = NkT - ⅓\sum_j \int \boldsymbol{r}_j \cdot (\partial\Phi/\partial\boldsymbol{r}_j)\text{e}^{-\beta\mathscr{H}}\,\text{d}V$. Remembering that $P = -\partial F/\partial V = kT\partial\ln Z/\partial V$, partial integration shows that the integral term corresponds to $kT\partial\ln Q/\partial V$.

The simplest way to evaluate Q is to assume that the potential Φ is a sum of pair potentials $\Phi(\boldsymbol{r}) = \Sigma_{i<j}\phi(r_{ij}) = ½\Sigma_{i,j}\phi(r_{ij})$, so that

$$Q = \int \ldots \int \exp[-\beta\Phi(\mathbf{r})]\,\text{d}\boldsymbol{r}_1 \ldots \text{d}\boldsymbol{r}_N = \int \ldots \int \Pi_{ij} \exp[-½\beta\phi(r_{ij})]\,\text{d}\boldsymbol{r}_1 \ldots \text{d}\boldsymbol{r}_N$$

The trick is now to introduce the *Mayer function* $f_{ij} = \exp[-\beta\phi(r_{ij})] - 1$, resulting in

$$\begin{aligned} Q &= \int \cdots \int \prod\nolimits_{ij} (1 + f_{ij})^{1/2} \text{d}\boldsymbol{r}_1 \ldots \text{d}\boldsymbol{r}_N \\ &= \int \cdots \int \left(1 + ½\sum\nolimits_{ij} f_{ij} - ⅛\sum\nolimits_{ij}\sum\nolimits_{kl\neq ij} f_{ij}f_{kl} + \cdots\right) \text{d}\boldsymbol{r}_1 \ldots \text{d}\boldsymbol{r}_N \\ &\cong \int \cdots \int \left(1 + ½\sum\nolimits_{ij} f_{ij}\right) \text{d}\boldsymbol{r}_1 \ldots \text{d}\boldsymbol{r}_N = V^N + ½N(N-1)V^{N-2}Q_{12} + \cdots \quad \text{with} \end{aligned}$$

$$Q_{12} \equiv \int\int f_{12}\,\text{d}\boldsymbol{r}_1\text{d}\boldsymbol{r}_2 = \int \text{d}\boldsymbol{r}_1 \int f(r)4\pi r^2 \text{d}r = 4\pi V \int f(r)r^2\,\text{d}r \equiv 2Vb_2$$

and where $|N(N-1)b_2/V| \ll 1$ is assumed. The final result is

$$Q = V^N[1 + N(N-1)V^{-1}b_2 + \cdots] \quad \text{with} \quad b_2 \equiv 2\pi \int f(r)r^2\,\text{d}r$$

Using $P = -\partial F/\partial V = kT\partial\ln Z/\partial V = kT\,\partial[\ln(N!\Lambda^{3N})^{-1}Q]/\partial V$, we obtain $PV/nRT = 1 + B_2/V + \cdots$ with $B_2 = -b_2$ or $\beta P = \rho + B_2\rho^2 + \cdots$ with $\rho = N/V$ and $\beta = 1/kT$. In general, the

---

1 Such calculations are based on translation, harmonic vibration, rigid rotation, and the third law, that is, the residual entropy $S_{\text{res}} = 0$, see, for example, [7–9]. For molecules like CO with only a small dipole moment, (almost) random ordering in the lattice occurs, resulting in $W = 2^N$ possible configurations at 0 K and thus to $S_{\text{res}} = k\ln W = 5.8\ \text{J K}^{-1}\ \text{mol}^{-1}$, approximately agreeing with experiment, $S_{\text{res}} = 4.5\ \text{J K}^{-1}\ \text{mol}^{-1}$. The remaining difference can be explained if the characteristic temperature below which no reorientation can take place is below the melting temperature [10, 11].

expression for ϕ is complex so that analytical evaluation of b_2 is almost never possible and we have to resort to numerical integration.

Example 8.1 ***The Hard-Sphere Second Virial Coefficient*** A potential for which analytical evaluation of b_2 is possible is the HS potential

$$\phi = \infty \text{ for } r < \sigma \quad \text{and} \quad \phi = 0 \text{ otherwise}$$

Splitting the integration for b_2 from 0 to ∞ in 0 to σ and σ to ∞, the integral becomes

$$B_2 = -b_2 = 2\pi \int_0^\sigma (\mathrm{e}^{-\beta\infty} - 1)r^2 \,\mathrm{d}r + 2\pi \int_\sigma^\infty (\mathrm{e}^{-\beta 0} - 1)r^2 \,\mathrm{d}r = 2\pi\sigma^3/3$$

From Example 8.1 we note that $b_2 < 0$, and this appears generally to be the case for attractive interactions. Often the quantity $b \equiv B_2 = -b_2 = 2\pi\sigma^3/3$ is used as a reference value in more extended calculations, and we do so likewise.

Now there is a problem with the abovementioned derivation. We assumed that $[N(N-1)/V]b \ll 1$. Clearly, because b is of the order of V_m, this is not the case. Another approach is thus required, and the shortest route is via the grand canonical partition function.

Recall that the grand canonical partition function reads

$$\Xi = \exp(\beta PV) = \sum_N \lambda^N Z_N \quad \text{where} \quad \lambda = \exp(\beta\mu)$$

and with Z_N the canonical partition function including all energies belonging to a state containing N particles. For consistency, Z_0 is defined as $Z_0 \equiv 1$, while Z_1 represents the situation containing one particle and is thus $Z_1 = Vz_\mathrm{int}/\Lambda^3$, where z_int represents the internal contributions. We will neglect the latter and thus use $Z_1 = Q_1/\Lambda^3 = V/\Lambda^3$, since for a single classical particle, the configurational partition function equals the volume. The semi-classical partition function thus becomes

$$Z_N = (N!\Lambda^{3N})^{-1} Q_N$$

where Q_N is the configurational partition function for N particles, and Ξ becomes

$$\Xi = \sum_N (\lambda/N!\Lambda^3)^N Q_N = \sum_N (N!)^{-1} \xi^N Q_N \quad \text{with} \quad \xi = \lambda/\Lambda^3 \tag{8.1}$$

Let us now assume that we may expand the pressure in a power series in ξ, that is,

$$\beta P = \sum_j b_j \xi^j \tag{8.2}$$

so that the grand partition function, using $\exp(x) = 1 + x + x^2/2 + \cdots$, becomes

$$\Xi = \exp(\beta PV) = 1 + Vb_1\xi + \left(Vb_2 + \tfrac{1}{2}V^2{b_1}^2\right)\xi^2 + \left(Vb_3 + V^2 b_1 b_2 + \tfrac{1}{6}V^3{b_1}^3\right)\xi^3 + \cdots$$

Comparing terms with $\Xi = 1 + Q_1\xi + Q_2\xi^2/2! + Q_3\xi^3/3! + \cdots$, Eq. (8.1), yields

$$Vb_1 = Q_1 \quad \rightarrow \quad b_1 = Q_1/V = 1$$

$$Vb_2 + V^2{b_1}^2/2 = Q_2/2 \quad \rightarrow \quad b_2 = \left(Q_2 - {Q_1}^2\right)/2V$$

$$Vb_3 + V^2 b_1 b_2 + V^3{b_1}^3/6 = Q_3/6 \quad \rightarrow \quad b_3 = \left(Q_3 - 3Q_1Q_2 + 2{Q_1}^3\right)/6V$$

From the expression for the average number of particles

$$\langle N\rangle = \lambda\frac{\partial \ln \Xi}{\partial \lambda} = \xi\frac{\partial \ln \Xi}{\partial \xi} = \xi\frac{\partial (PV/kT)}{\partial \xi} = \xi V\frac{\partial}{\partial \xi}\sum_j b_j\xi^j = V\sum_j jb_j\xi^j$$

we obtain the number density

$$\rho = \langle N\rangle/V = b_1\xi + 2b_2\xi^2 + 3b_3\xi^3 + \cdots \tag{8.3}$$

Inverting this series, we write

$$\xi = a_1\rho + a_2\rho^2 + a_3\rho^3 + \cdots \tag{8.4}$$

and substitute this expression in the one for ρ, Eq. (8.3), to obtain

$$\rho = b_1(a_1\rho + a_2\rho^2 + a_3\rho^3 + \cdots) + 2b_2\left(a_1{}^2\rho^2 + 2a_1a_2\rho^3 + \cdots\right) + 3b_3\left(a_1{}^3\rho^3 + \cdots\right) + \cdots$$

so that the coefficient a_i become

$$a_1b_1 = 1 \quad \rightarrow \quad a_1 = b_1 = 1$$

$$b_1a_2 + 2b_2a_1{}^2 = 0 \quad \rightarrow \quad a_2 = -2b_2$$

$$b_1a_3 + 4b_2a_1a_2 + 3b_3a_1{}^3 = 0 \quad \rightarrow \quad a_3 = 8b_2{}^2 - 3b_3$$

Now these coefficients are used in the series for ξ, Eq. (8.4), and we obtain

$$\xi = b_1\rho - 2b_2\rho^2 + \left(8b_2{}^2 - 3b_3\right)\rho^3 + \cdots \tag{8.5}$$

Finally, substitution of Eq. (8.5) in the expression for βP, Eq. (8.2), results in

$$\beta P = \sum_j b_j\xi^j = b_1\xi + b_2\xi^2 + b_3\xi^3 + \cdots = \rho - b_2\rho^2 + \left(4b_2{}^2 - 2b_3\right)\rho^3 + \cdots$$

Comparing this result with the virial expansion $\beta P = \rho + B_2\rho^2 + B_3\rho^3 + \cdots$, we obtain

$$B_2 = -b_2 \text{ and } B_3 = 4b_2{}^2 - 2b_3$$

For B_2 we thus have $B_2 = -(Q_2 - Q_1{}^2)/2V$ or

$$B_2 = -\frac{1}{2V}\int [\exp(-\beta\phi) - 1]\,\mathrm{d}\mathbf{r}_1\mathrm{d}\mathbf{r}_2 = -\frac{1}{2V}\int f(r)\,\mathrm{d}\mathbf{r}_1\mathrm{d}\mathbf{r}_2 = -\frac{1}{2}\int f(r)\,\mathrm{d}\mathbf{r}$$

The result for B_2 using the grand canonical partition function thus agrees with the result from the canonical partition function. This is not true for B_n with $n \geq 3$. More importantly, in the former case B_2 is the coefficient of a small term, while in the latter case it is the coefficient of a large term, rendering the expansion doubtful. It also shows that B_2 is strictly dependent only on two-body interactions. This applies also to the higher-order terms: B_n is only dependent on the n-body interactions (and the lower-order ones). Note that this derivation is made without making any assumption for the potential, as also is the case for two other methods [12, 13].

The excess Helmholtz energy $F^{\mathrm{E}} = F - F^{\mathrm{idg}}$ is $-\beta F^{\mathrm{E}}/V = B_2^{\mathrm{F}}\rho^2 + B_3^{\mathrm{F}}\rho^3 + \cdots$ with the perfect gas Helmholtz energy $\beta F^{\mathrm{idg}} = N\ln(\rho/\rho^\circ)$ and ρ° a reference density. The coefficients $B_n{}^{\mathrm{F}}$ are obtained by integration from the coefficients B_n, and we obtain

$$B_n = -(n-1)B_n{}^{\mathrm{F}}$$

Table 8.3 Virial coefficients of the hard-sphere fluid in terms of $B_2 = b = 2\pi\sigma^3/3$.

Term	ρ Expansion	η Expansion	Term	ρ Expansion	η Expansion
B_3	$5/8\ b^2$	10	B_7	$0.013\,023\,54(91)\ b^6$	53.34
B_4	$0.286\,949\,5\ldots\ b^3$	18.36	B_8	$0.004\,183\,2(11)\ b^7$	68.54
B_5	$0.110\,252(1)\ b^4$	28.22	B_9	$0.001\,309\,4(13)\ b^8$	85.81
B_6	$0.038\,881\,98(91)\ b^5$	39.82	B_{10}	$0.000\,403\,5(15)\ b^9$	105.8

Source: Adapted from Clisby and McCoy [16].

8.2.1 Some Further Remarks

A systematic extension of the virial expansion can be realized. This is usually done via a graphical technique, known as the cluster expansion and for which we refer to the literature [14, 15]. The calculation of high-order virial coefficients is rather complex, even for a simple potential like the HS potential. While B_3 and B_4 for the HS potential can be calculated analytically, the values for the higher-order coefficients are obtained from simulations (Table 8.3). For more realistic potentials, however, usually only B_2 or, may be, B_3 and B_4 are available.

From Table 8.3, it will be clear that the convergence of the virial expansion is not rapid. This is even made clearer if we introduce as reference volume the close-packed volume, that is, the volume per molecule for a FCC lattice with volume V_0 for N atoms, given by $V_0/N = v_0 = \rho_0{}^{-1} = \sigma^3/2^{1/2}$ for atoms with a diameter σ. Introducing the *reduced density* $\eta = N(4\pi/3)(\sigma/2)^3/V = \rho\pi\sigma^3/6 = \rho b/4 = 2^{1/2}\pi V_0/6V$, mostly called the *packing fraction*,[2] the virial expansion can also be expressed in η instead of ρ. Hence,

$$\begin{aligned} Z &= 1 + b\rho + 0.625b^2\rho^2 + 0.28695b^3\rho^3 + 0.1103b^4\rho^4 + \cdots \\ &= 1 + 4\eta + 10\eta^2 + 18.36\eta^3 + 28.22\eta^4 + \cdots \end{aligned}$$

with $Z = P/\rho kT$. For HS, $Z \geq 1$ and continuously increases with ρ. A comparison of the virial results including B_{10} with those of computer simulations [17] shows that they begin to deviate at $\rho\sigma^3 \cong 0.5$ or $\eta \cong 0.26$.

A few problems exist with the virial expansion; already mentioned are the availability and complexity of virial coefficient calculations and the slow convergence. Moreover, the virial expansion only converges in some non-zero region, although in that region convergence occurs for a wide class of potentials. Essentially, $B_n \sim V_0{}^{n-1}$ and if $\rho \to \rho_0$ the expansion diverges. Perhaps more important is that Monte Carlo computer simulations by Hoover and Ree in 1968 [18] have shown that for the HS system, a freezing transition occurs at $\eta_{\text{fre}} = 0.494 \pm 0.002$ from a fluid disordered phase to a metastable solid, while the ordered closed-packed phase shows a melting transition at $\eta_{\text{mel}} = 0.545 \pm 0.002$. The melting temperature T_{mel} and pressure P_{mel} are related by $P_{\text{mel}}v_0 = (8.27 \pm 0.13)kT_{\text{mel}}$ or $P_{\text{mel}}v_* = (6.12 \pm 0.10)kT_{\text{mel}}$.[3] The virial expansion fails to reveal all this. The range

2 For the packing fraction, we thus have $0 \leq \eta \leq \eta_{\text{CP}} \equiv 2^{1/2}\pi/6 \cong 0.741$.

3 Note that two definitions of molecular volume exist. Here, we use both $v_0 = V_0/N$ for a volume V_0 with N atoms and $v_* = 4\pi(\sigma/2)^3/3$ with σ the diameter of the atom. Hence, for a lattice $v_0 = v_*/\eta_{\text{lat}}$ with η_{lat} the packing fraction of the lattice. For example, for an FCC lattice, $\eta_{\text{FCC}} = 2^{1/2}\pi/6 \cong 0.741$ and $v_0 = v_*/\eta_{\text{FCC}} = [4\pi(\sigma/2)^3/3]/(2^{1/2}\pi/6) = \sigma^3/2^{1/2} = \sigma^3/\gamma$, with γ a constant characteristic for the lattice.

$\eta = 0.494 - 0.545$ ($V/V_0 = 1.50 - 1.36$) is much lower than the packing fraction for random close-packed structures, $\eta_{\text{RCP}} \cong 0.637$, or close-packed structures, $\eta_{\text{CP}} \equiv 2^{1/2}\pi/6 \cong 0.741$. Hence, a regular solid phase is favored thermodynamically at much lower density than at which the transition becomes necessary for geometrical reasons. The at-present best estimates $\eta_{\text{fre}} = 0.492$ and $\eta_{\text{mel}} = 0.543$ [19] are remarkably close to those of Hoover and Ree.

If B_2 to B_6 are rounded to the integers 4, 10, 18, 28, and 40, we obtain $B_n = (n^2 + n - 2)$. Assuming that this expression is applicable for $n > 6$ as well, summing the series $S = \Sigma_{n=2}(n^2 + n - 2)\eta^{n-1} = \Sigma_{n=0}(n^2 + 3n)\eta^n$ using $S_0 = \Sigma_{n=0}\eta^n$, $S_1 = \text{d}S_0/\text{d}\eta$, and $S_2 = \text{d}S_1/\text{d}\eta$, we obtain for $Z_{\text{CS}} = 1 + S$ the *Carnahan–Starling* (CS) expression [20]

$$Z_{\text{CS}} = (1 + \eta + \eta^2 - \eta^3)/(1 - \eta)^3 \tag{8.6}$$

This surprising procedure describes the HS EoS well by matching simulation results [21] with an average absolute deviation (AAD) of 1.5×10^{-3} [19] over the stable range $0 \leq \eta \leq 0.49$. Another simple but accurate description is the cubic EoS [22]

$$Z = 1 + 4\eta/(1 - 1.216\eta) + 5.696\eta^2/(1 - 1.216\eta)^2$$

with an AAD of 8.5×10^{-4} [19]. An excellent description is a [5,4] Padé approximant[4] using the HS virial coefficients B_2 to B_{10} [16], and given by [17]

$$Z = \frac{1 + 0.804\,09b\rho + 0.466\,62b^2\rho^2 + 0.141\,98b^3\rho^3 + 0.029\,739b^4\rho^4 + 0.002\,819\,2b^5\rho^5}{1 - 0.195\,91b\rho + 0.037\,534b^2\rho^2 - 0.060\,058b^3\rho^3 + 0.012\,303b^4\rho^4}$$

with an AAD of 6.0×10^{-5} [19]. Using more recent simulation data [21] results in an AAD of less than 5.0×10^{-5} [19] and a modified Padé EoS of even less than 1.0×10^{-6} [19]. Based on the virial coefficients, simpler Padé approximants can be easily constructed. A comparison of a variety of HS EoSs is made in [19, 25, 26]. We will not pursue this topic further, but refer the reader to, for example, [15, 27–29].

In summary, the virial approach is useful for gases at intermediate density, but a first-principles calculation of the coefficients requires extensive numerical work. Because the virial expansion cannot describe the gas–liquid transition, EoSs as discussed in the next section remain important for many technical problems.

Problem 8.1 ***The Square-Well Potential*** Show that for the square-well (SW) potential, the virial coefficient B_2 and the Boyle temperature T_{B}, where $B_2(T_{\text{B}}) = 0$, are given by, respectively,

$$B_2 = 2\pi\sigma^3 \left[R_{\text{SW}}{}^3 - \left(R_{\text{SW}}{}^3 - 1\right)\exp\beta\varepsilon\right]/3 \quad \text{and} \quad T_{\text{B}} = (\varepsilon/k)/\ln\left[R_{\text{SW}}{}^3/\left(R_{\text{SW}}{}^3 - 1\right)\right]$$

Show that B_2 increases continuously with T without showing a maximum as experimentally observed and discuss the reason why. Also show that for $\beta\varepsilon \ll 1$, the truncated virial expansion $\beta P/\rho = 1 + B_2\rho$ reduces to the van der Waals equation.

---

4 In numerical analysis it is well known that a truncated power series of a function is an unsatisfactory approximation. A more sophisticated method is the use of Padé approximants [23] which represent the function by a ratio of polynomials. The coefficients are found by expanding the ratio and requiring the coefficients to represent the first k Taylor coefficients correctly. For example, $f(x) \cong c_0 + c_1x + \cdots c_kx^k$ is approximated by the $[n,m]$ approximant $f(x) \cong (a_0 + a_1x + \cdots a_nx^n)/(1 + b_1x + \cdots b_mx^m)$. Obviously, $n + m + 1$ should equal $k + 1$. A matrix recipe is given by Ree et al. [24].

Problem 8.2 ***The Sutherland Potential*** Consider the Sutherland potential with $\phi(r) = \infty$ for $r < \sigma$ and $\phi(r) = -\varepsilon r^{-n}$ for $r \geq \sigma$. Show that B_2 is only finite if $n > 3$. Use the expansion for the exponential.

8.3 Equations of State

To recapitulate, for not too-low temperature and not too-high pressure, the pressure P of a gas is described by the well-known perfect gas EoS $P = nRT/V$. All gases obey this equation in the low-pressure limit, as illustrated in Figure 8.1, where the behavior of the *compression factor* $Z = PV/RT$ is shown for H_2, N_2, and CH_4. Typically, the accuracy for P according to the perfect gas EoS is ≈1% for $P < 10$ bar for $T > 273$ K. However, for higher pressures, deviations from the perfect gas law occur and the EoS becomes material specific (Figures 8.1a and 8.2).

In Section 8.2, we discussed the virial equation of state, which was initially given as an empirical description by Kamerlingh Onnes (1901) using an expansion in $1/V$. Obviously for the perfect gas $Z = 1$, but for a real gas Z is

$$Z(T,V) = \sum_{j=0} B_{j+1} V^{-j} \quad \text{or} \quad Z(T,V) = 1 + B_2 V^{-1} + B_3 V^{-2} + B_4 V^{-3} + \cdots \tag{8.7}$$

The jth virial coefficient B_j is a function of T only, except B_1, which obviously equals 1. The temperature dependence of B_2 can be empirically described by $B_2 = \alpha - \beta\exp(\gamma/T)$. Some data are given in Table 8.4. For a perfect gas $\mathrm{d}Z/\mathrm{d}(1/V) = 0$, but for a real gas we have $\mathrm{d}Z/\mathrm{d}(1/V) \cong B_2$. Alternatively, one can expand Z in P, and obtain

$$Z(T,P) = \sum_{j=0} B_{j+1}' P^j \quad \text{or} \quad Z(T,P) = 1 + B_2' P + B_3' P^2 + B_4' P^3 + \cdots$$

In this case, $\mathrm{d}Z/\mathrm{d}P \cong B_2'$. The coefficients B_j' can be expressed in terms of the coefficients B_j by inverting the series (for general conversion relations, see [30]). The first three coefficients read

$$B_2' = B_2/RT, \quad B_3' = \left(B_3 - B_2^2\right)/(RT)^2 \quad \text{and} \quad B_4' = \left(B_4 - 3B_2B_3 + 2B_2^3\right)/(RT)^3$$

If only B_2 is available, one would be inclined to use $Z = 1 + B_2/V$. However, the expression $Z = 1 + B_2P/RT$, obtained from substituting $B_2' = B_2/RT$ in $Z = 1 + B_2'P$, is often slightly

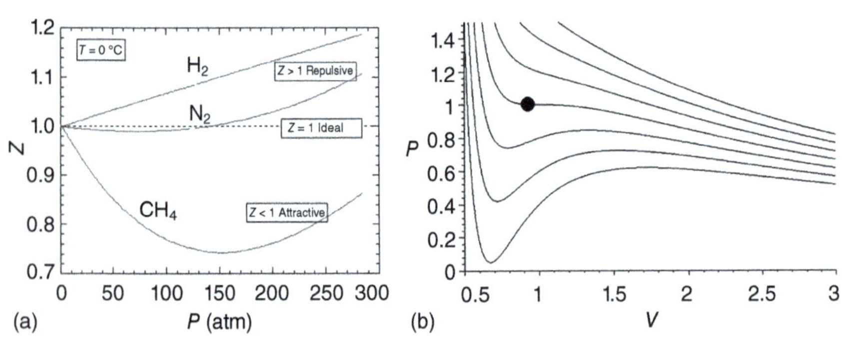

Figure 8.1 (a) Approaching the perfect gas EoS for gases as exemplified by H_2, N_2, and CH_4. The limiting value of $PV = 2271$ Pa m^3 mol^{-1} at the triple point of water, $T = 273.16$ K. This yields $R = 8.314$ Pa m^3 mol^{-1} K^{-1} or 8.314 J mol^{-1} K^{-1}. (b) The van der Waals EoS isotherms in the PV plane for $T_{red} = 0.85, 0.90, 0.95, 1.00, 1.05, 1.10$, and 1.15. The critical point is indicated by •.

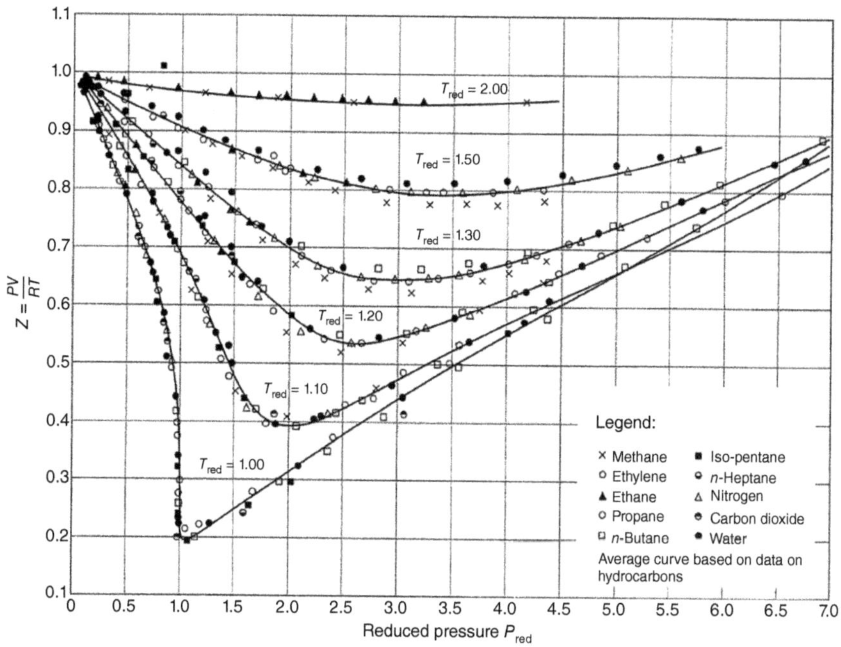

Figure 8.2 Compression factor Z as a function of the reduced pressure for various fluids. Source: Adapted from Sen [62].

Table 8.4 Values for α, β, and γ in $B_2 = \alpha - \beta\exp(\gamma/T)$ for some simple molecules.

Molecule	α ($cm^3\ mol^{-1}$)	β ($cm^3\ mol^{-1}$)	γ (K)	Range of validity (K)
Ar	154.2	119.3	105.1	80–1024
N_2	185.4	141.8	88.7	75–700
H_2O	33.0	15.2	1300.7	293–1248
CH_4	206.4	159.5	133.0	110–600
CH_3OH	14	3.958	1840	280–400
C_6H_6	571.4	384.4	526.6	290–600

Source: Kaye et al. [42]/National Physical Laboratory.

more accurate ($P < \frac{1}{2}P_{cri}$). Such gases are denoted as "*slightly imperfect*" [31]. In this approximation for Ar, a negligible error occurs for $P < 10$ bar and an error of $\approx 2\%$ up to 100 bar. On the contrary, if B_3 is available, the expression $Z = 1 + (B_2/V) + (B_3/V^2)$ is much more accurate than $Z = 1 + B_2'P + B_3'P^2$ ($P < \frac{3}{4}P_{cri}$).

It will be clear that the virial EoS describes only the behavior of the gas. Already in 1873, van der Waals (vdW) emphasized the continuity between the gas and the liquid state and advocated the EoS named after him [32]. The *van der Waals equation* describes both liquids and gases, and reads

$$(P + n^2a/V^2)(V - nb) = nRTa \quad \text{or} \quad (P + a\rho^2)(1 - b\rho) = \rho RT$$

with $\rho = nN_A/V$ the number density for n moles. The parameters a and b are material specific. The intuitive interpretation for a is that there is attraction between the molecules, so that molecules in the bulk tend to pull a molecule near the wall away from the wall. The measured pressure is thus less than the perfect gas pressure. As the pressure depends on force and frequency of the colliding molecules with the wall and both are proportional to n/V, we have $a(n/V)^2$ resulting in $P + a(n/V)^2$. For b, one considers that a molecule has certain radius, corresponding to a certain *excluded volume* (or *co-volume*) b, so that the volume available becomes $V - b$. Considering only binary interactions, for a rigid particle with radius $\sigma/2$ the volume that cannot be occupied by another particle is $4\pi\sigma^3/3$ and one finds $2b = 4\pi\sigma^3/3$. Figure 8.1b illustrates the vdW behavior.[5] For low densities, the vdW expression reduces to the perfect gas law, but for high densities sinuous curves are obtained representing an unstable situation. The interpretation is that the region of the sinuous curves corresponds to the separation of the fluid into liquid and vapor, and that the equation breaks down here. Maxwell argued that a horizontal line should be inserted to make the areas below and above that line equal (Section 12.2), obtaining in this way a general resemblance to the actual isotherms of a real fluid. The vdW EoS is not very accurate but shows qualitatively the correct behavior, combined with simple mathematics [36].

Other two-parameter equations of state exist, for example, the *Berthelot* (B) and *Dieterici* (D) *equations*, given for 1 mol by, respectively,

$$(P + a/TV^2)(V - b) = RT \quad \text{and} \quad P(V - b) = RT\exp(-a/RTV)$$

Like the vdW equation, they are not very accurate. For many details on these (and some other) EoSs, see [37]. The Berthelot EoS holds best at low pressure [38]. Changing V_{cri} from $3b$ to $4b$ [39] and expanding the result in a virial equation leads to the modified Berthelot EoS $PV = RT[1 + (9/128)(PT_{cri}/P_{cri}T)(1 + 6(T_{cri}/T)^2)]$ describing Z and B_2 rather accurately ($P < 0.6P_{cri}$). Even better accuracy is obtained by empirically changing 9/128 to 1/7 and 6 to 15/2 [40]. The most successful two-parameter EoS, however, appears to be the Redlich–Kwong (RK, [41]) equation reading

$$P = \frac{RT}{V - b} - \frac{a}{T^{1/2}V(V + b)}$$

Overall, it represents both the gaseous and liquid regions rather accurately, combining simplicity and accuracy over wide ranges of temperature and pressure but predicts a monotonically increasing heat capacity C_P, while experimentally at high pressure the C_P decreases with increasing density. A modified RK (mRK) equation [43] uses the CS expression (Section 8.2) for the repulsion instead of $1/(V - b)$. While the RK equation behaves generally much better than the vdW equation, the mRK equation improves on enthalpy, density, pressure, and fugacity. However, a concise study on vapor–liquid equilibria, densities, pressure, and vaporization enthalpy using the CS expression instead of the vdW expression for the repulsion in the vdW, D, and RK EoSs shows, as might be expected, that these two-parameter equations are not satisfactory for all properties of fluids if one set of parameters is used [44].

5 The need for modifications was at the time "in the air" for already quite some time. The expression $P(1 - b\rho) = \rho RT$ was proposed by Clausius in 1857 [33]. For brief historical reviews, see [34, 35]. Further, note that in the kinetic theory of gases, the excluded volume is neglected, but that molecules do have a collision cross-section.

Well-known cubic EoSs, using three parameters, are the Redlich–Kwong–Soave (RKS, [45]) and Peng–Robinson (PR, [46]) equations. The general form is

$$\frac{P}{RT\rho} = Z = \frac{1}{1-b\rho} - \frac{a\alpha(T)\rho}{RTf(\rho)}$$

In this notation, the vdW equation uses $\alpha(T) = f(\rho) = 1$ and the RK equation uses $\alpha(T) = T^{-1/2}$ and $f(\rho) = 1 + b\rho$. The RKS equation differentiates between various normal liquids by introducing the acentric factor ω (Section 8.4.1) in the expression for $\alpha(T)$, namely $\alpha(T) = [1 + m(1 - T_{\text{red}}{}^{1/2})]^2$ with $m = 0.480 + 1.574\omega - 0.176\omega^2$, in conjunction with $f(\rho) = 1 + b\rho$. The PR equation also modifies $f(\rho)$ by using $f(\rho) = 1 + 2b\rho - b^2\rho^2$, but with a slightly modified $m = 0.374\,64 + 1.542\,26\omega - 0.269\,92\omega^2$. It appears that the RKS EoS is better suited for small values of ω, while the PR is better for larger values of ω. Introducing temperature dependent a and b values in the MRK EoS leads to further improvement [47]. More complex EoSs are the Beattie–Bridgeman (BB, [48]) and Benedict–Webb–Rubin (BWR, [49]) equations and their modifications, where the vdW repulsive term is replaced by polynomial and exponential expressions, the form being chosen mainly to be easily integrable to yield the Helmholtz energy. The original BB and BWR equations use five and eight parameters, respectively, but to realize accuracy at high density they have been extended considerably.

A compromise between accuracy and complexity is the general cubic EoS [50] of which the vdW equation is the prototype, although cubic EoSs have limited flexibility [51]. This equation reads for 1 mol (omitting the subscript m)

$$P = \frac{RT}{V-b} - \frac{\theta(V-\eta)}{(V-b)(V^2+\kappa V+\lambda)}$$

with θ, η, κ, λ, and b as parameters. The first three virial coefficients are given by

$$B_2 = b - \frac{\theta}{RT}, \quad B_3 = bB_2 + \frac{\theta(\kappa+\eta)}{RT}, \quad \text{and} \quad B_4 = bB_3 - \frac{\theta(\kappa^2+\kappa\eta-\lambda)}{RT} \tag{8.8}$$

The vdW equation is obtained if $\eta = b$, $\theta = a$, and $\kappa = \lambda = 0$. An often-used class is given by $\eta = b$, $\theta = a(T)$, $\kappa = (\varepsilon + \sigma)b$, and $\lambda = \varepsilon\sigma b^2$. The parameters ε and σ have the same value for all substances in the class, while $a(T)$ and b are substance dependent, for example, for the vdW equation $a(T) = a$ and $\varepsilon = \sigma = 0$.

Most EoSs can be developed in a power series in $1/V$. For example, for the vdW equation we obtain $B_2 = b - a/RT$, $B_3 = b^2$, and $B_4 = b^3$, conforming to Eq. (8.8). A similar procedure may be applied to other EoSs. For more complex EoSs, such as the RK EoS, obtaining B_2, B_3, … can become quite involved. Experiments show that $B_2 < 0$ at low temperatures, becomes positive at higher temperatures, reaches a maximum, and then decreases. The vdW, B, D, and RK equations all give $B_2 < 0$ at low temperatures, change to $B_2 > 0$ at higher temperatures, but do not predict a maximum.

Often an EoS for mixtures is required for which parameters for the interaction between components α and β are introduced. For example, for the virial approach, $Z(T,V,n)$ generalizes for n_α (n_β) moles of component α (β) to

$$Z(T,V,n_\alpha,n_\beta) = (n_\alpha + n_\beta) + \left(n_\alpha{}^2B_{\alpha\alpha} + 2n_\alpha n_\beta B_{\alpha\beta} + n_\beta{}^2B_{\beta\beta}\right)/V + \cdots$$

with the second virial cross-coefficient $B_{\alpha\beta}$. When the *principle of corresponding states* (PoCS) (Section 8.4) is obeyed, one can estimate $B_{\alpha\beta}$ from $B_{\alpha\alpha}$ and $B_{\beta\beta}$ [52]. Often one assumes combining rules: the *Berthelot rule* $a_{\alpha\beta} = (a_\alpha a_\beta)^{1/2}$ and the *Lorentz rule*

Table 8.5 The second virial coefficient $B_{\alpha\beta}$ for some mixtures.

Molecule	Exp. $B_{\alpha\beta}$ ($cm^3\ mol^{-1}$)	PoCS $B_{\alpha\beta}$ ($cm^3\ mol^{-1}$)	Average $B_{\alpha\beta}$ ($cm^3\ mol^{-1}$)	T (K)
Ar-Kr	208	213	232	116.5
Ar-CH_4	233	239	255	107.1
Kr-CH_4	297	297	297	118.6

Source: McGlashan [54]/with permission of Elsevier.

$b_{\alpha\beta} = ½(b_\alpha{}^{1/3} + b_\beta{}^{1/3})^3$, generally approximated by $b_{\alpha\beta} = (b_\alpha + b_\beta)/2$. For polyatomic molecules this fails, largely due to Berthelot's rule [53], which accordingly is modified to $B_{\alpha\beta} = (1 - k_{\alpha\beta})(B_{\alpha\alpha}B_{\beta\beta})^{1/2}$, with the parameter $k_{\alpha\beta} \ll 1$. In the absence of data, one often takes $B_{\alpha\beta} = ½(B_{\alpha\alpha} + B_{\beta\beta})$, but that estimate is usually not very accurate. Table 8.5 provides some examples.

Having an EoS also allows us to determine the thermodynamics by integration. If the EoS is explicit in V, the Helmholtz energy $F = -\int P\,\mathrm{d}V + C = \int (P/\rho^2)\,\mathrm{d}\rho + C$ with C a constant can be used. Consider, for example, the virial expansion in $1/V$ truncated after the third term and written as $P = RT\rho + B_2\rho^2 + B_3\rho^3$. We obtain

$$F = \int (P/\rho^2)\,\mathrm{d}\rho + C = RT\ln\rho + B_2(\rho - 1) + ½B_3(\rho^2 - 1) + F^\circ$$

where in the last step the constant C has been replaced by $F^\circ - B_2 - ½B_3$ with F° the Helmholtz energy at unit density. Entropy S, energy U, specific heat C_V, and its derivative $(\mathrm{d}C_V/\mathrm{d}\rho)_T$ follow by differentiation. If the EoS is explicit in P, the Gibbs energy $G = \int V\,\mathrm{d}P + C'$ is used with C' another constant. Similarly, using the virial expansion in P truncated in the same way, we have $V = RT/P + B_2' + B_3'P$ and obtain

$$G = \int V\,\mathrm{d}P + C = RT\ln P + B_2'(P - 1) + ½B_3'(P^2 - 1) + G^\circ$$

again leading by differentiation to S, U, C_P, and $(\mathrm{d}C_P/\mathrm{d}P)_T$. Finally, for many more details, we refer to reviews on the EoS for fluids [55–57].

8.4 The Principle of Corresponding States

The critical point (CP) is characterized by $(\partial P/\partial V)_T = (\partial^2 P/\partial V^2)_T = 0$ and this leads for the vdW equation to $V_{cri} = 3b$, $T_{cri} = 8a/27Rb$ and $P_{cri} = a/27b^2$. Defining the reduced quantities $V_{red} = V/V_{cri}$, $T_{red} = T/T_{cri}$ and $P_{red} = P/P_{cri}$, we can write the vdW EoS in a nondimensional form, usually called the *reduced vdW EoS*, as given by

$$\left(P_{red} + 3/V_{red}{}^2\right)(3V_{red} - 1) = 8T_{red}$$

To estimate a and b, values of the critical constants (Section F.10) can be used, preferably using P_{cri} and T_{cri} as V_{cri} is difficult to determine accurately [58]. Another way of determining a and b is fitting P to the vdW equation for a range of temperatures, or using α_P and κ_T, evidently yielding different values for a and b.

Table 8.6 Critical compression factor and surface tension for several simple molecules.

Molecule	Z_{cri}	$\gamma_0 V_{cri}^{2/3}/T_{cri}$ (erg K^{-1} mol$^{-2/3}$)	Molecule	Z_{cri}	$\gamma_0 V_{cri}^{2/3}/T_{cri}$ (erg K^{-1} mol$^{-2/3}$)
Ne	0.305	4.05	N_2	0.292	4.48
Ar	0.292	4.25	O_2	0.292	4.4
Kr	0.290	—	CO	0.294	—
Xe	0.288	4.2	CH_4	0.289	4.39

Source: Guggenheim [31]/with permission of Elsevier.

All two-parameter EoSs can be scaled similarly. The reduced vdW EoS provides an example for the PoCS, an experimental fact that asserts that for a group of similar substances the EoS can be written as [59–61]

$$\Omega(P_{red}, T_{red}, V_{red}) = 0$$

where Ω is the same function for all the substances of the group.[6] The (approximate) validity of the PoCS for gases is indicated by the fact that $Z_{cri} = P_{cri}V_{cri}/RT_{cri}$ is approximately constant for a set of spherical, nonpolar molecules at 0.293 ± 0.005, where $\pm$ indicates the sample standard deviation (Table 8.6). Note that for the vdW fluid $Z_{cri} = 3/8 = 0.375$. Perhaps more convincing is a plot of Z versus P_{red} (Figure 8.2), which shows that the curves for a certain fixed T_{red} nicely coincide.

Paynter [63] wrote the vdW equation as $P_{red} = [1 + (\rho_{red} - 1)^3 M]T_{red} - A\rho^2(1 - T_{red})$ with $M = 1/(1 - b_{red}\rho_{red})$ and $A = a_{red}$. In the conventional formulation $a_{red} = 3$ and $b_{red} = ⅓$, but they may be chosen otherwise. Using $M' = |\rho_{red} - 1|M$ and $A' = A = a_{red}$, he obtained a significant improvement, both in agreement with experiment and in compliance with modern theory of continuous transitions. His choice of $a_{red} = 7$ and $b_{red} = ¼$ reproduces Figure 8.2 well and yields $Z_{cri} = 1/(4 - b_{red}) = 0.267$.

Another example of the PoCS is given in Figure 8.3a, where the second virial coefficient B/V_{cri} versus T/T_{cri} is shown for four simple fluids. The curve can be described well down to $T/T_{cri} = 0.5$ by the square-well potential with a hard-core diameter σ and a depth ε over the range 1.0σ–1.5σ given by [31]

$$B/V_{cri} = 0.440 + 1.40[1 - \exp(0.75T_{cri}/T)]$$

with $⅔\pi N_A\sigma^3 = 0.447V_{cri}$ and $\varepsilon = 0.936kT_{cri}$.[7] As still another example, Figure 8.3b shows the curve of T/T_{cri} versus ρ/ρ_{cri} for various liquids, quite well fitted by [31]

$$(\rho_L + \rho_V)/2\rho_{cri} = 1 + 3(1 - T/T_{cri})/4 \quad \text{and} \quad (\rho_L - \rho_V)/\rho_{cri} = 7(1 - T/T_{cri})^{1/3}/2 \quad (8.9)$$

where ρ_L and ρ_V represent the density of the liquid and the vapor, respectively. Equation (8.9a) represents the *law of rectilinear diameters* or *law of Cailletet–Mathias*,

6 Su [62] used $V_{red}' = VP_{cri}/RT_{cri}$ and thus $\Omega(P_{red}, T_{red}, V_{red}') = 0$. For substances with different Z_{cri}, the vdW PoCS is exact at the CP and (largely) in error for $P \to 0$, and vice versa for the Su PoCS.

7 McGlashan [54] excluded Ne and, using slightly different values, obtained $B/V_{cri} = 0.597 - 0.462 \exp(0.7002T_{cri}/T)$, $(\rho_L + \rho_V)/2\rho_{cri} = 1 + 0.797(1 - T/T_{cri})$ and $(\rho_L - \rho_V)/\rho_{cri} = 3.524(1 - T/T_{cri})^{0.326}$.

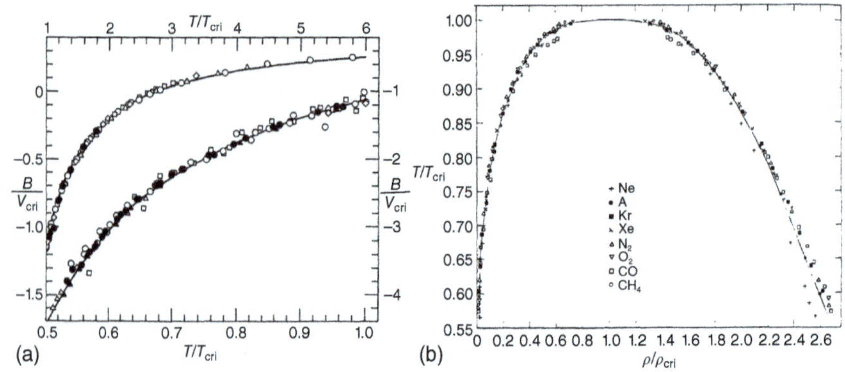

Figure 8.3 (a) The curve of B/V_{cri} versus T/T_{cri} for Ar, Kr, Xe, and CH_4. The left-hand and upper scales relate to the upper curve, while the right-hand and lower scales relate to the lower curve. (b) The curve of T/T_{cri} versus ρ/ρ_{cri} for some fluids. Source: Guggenheim [31]/with permission of Elsevier.

stating that the average density of gas and liquid is a linear function of temperature.[8] Equation (8.9b) shows a discontinuity at T_{cri}, in agreement with critical phenomena theory. Below $T \cong 0.65T_{cri}$ its inaccuracy becomes large and for $T = 0.565T_{cri}$, $\rho_V = 0$, although this is below $T_{tri} = 0.58T_{cri}$. The accuracy of $\rho_L + \rho_V$ and $\rho_L - \rho_V$ always remains good [31]. For other liquids the exponents and coefficients vary slightly [66].

Considerably below $T \cong 0.65T_{cri}$, it is best to consider the saturated vapor pressure P_{sat} instead of ρ_V. If the PoCS applies, we expect that P_{sat}/P_{cri} is a common function of T/T_{cri} and Figure 8.4a shows this indeed to be the case. The data can be well described from T_{tri} to T_{cri} by the reduced Clapeyron–Clausius equation $\ln(P_{sat}/P_{cri}) = A - BT_{cri}/T$ with $A = 5.29$ and $B = 5.31$. The small difference between A and B indicates that the fitted line does not go through the CP, as probably expected because the Clapeyron equation has theoretical significance only in the low-temperature regime.

The PoCS can be rationalized [61] assuming that (i) internal modes are independent of the density; (ii) quantum effects play no role, in particular, translation can be described classically; and (iii) the intermolecular potential has a similar shape for the molecules within the class of molecules considered. For a two-parameter intermolecular potential of the form $\phi(r) = \varepsilon f(r/\sigma)$ – a so-called *conformal potential* – distances are scaled by σ, while the energy is scaled by ε. Consequently, we can scale the pressure P according to $P^* = \sigma^3 P/\varepsilon$, the volume V according to $V^* = V/\sigma^3$, and the temperature T according to $T^* = kT/\varepsilon$. Finally, the configuration partition function Q scales according to $Q^* = Q/\sigma^{3N}$. In this way, the pressure becomes

$$P = kT(\partial \ln Z/\partial V)_T = kT(\partial \ln Q/\partial V)_T \quad \rightarrow \quad P^*(T^*, V^*) = T^*(\partial \ln Q^*/\partial V^*) \qquad (8.10)$$

In this dimensionless representation, the P–V curves of a set of molecules should coincide for the same T^*, as illustrated in Figure 8.2, where for the various T^*s the corresponding reduced temperature T_{red} has been used.

8 From the linear relation $(\rho_L + \rho_V)/2 = a - bT$, we have at $T = 0$, $\rho_V = 0$, and $\rho_L = \rho_S$, with ρ_S the density of the solid at $T = 0$, and thus $a = \rho_S/2$. At $T = T_{cri}$, $\rho_L = \rho_V = \rho_{cri}$ and $(\rho_L + \rho_V)/\rho_{cri} = \rho_S/\rho_{cri} + (\rho_S/\rho_{cri} - 2)(T/T_{cri})$. Jüptner [64] obtained for 33 liquids $\rho_S/\rho_{cri} = 3.998$ from the intercept and $\rho_S/\rho_{cri} = 3.984$ from the slope. Evidently deviations from linearity do occur, for example, for CCl_4 Young [65] obtained $(\rho_L + \rho_V)/2\rho_{cri} = 1.958 - 1.041(T/T_{cri}) + 0.08292(T/T_{cri})^2$, leading to $\rho_S/\rho_{cri} = 3.916$.

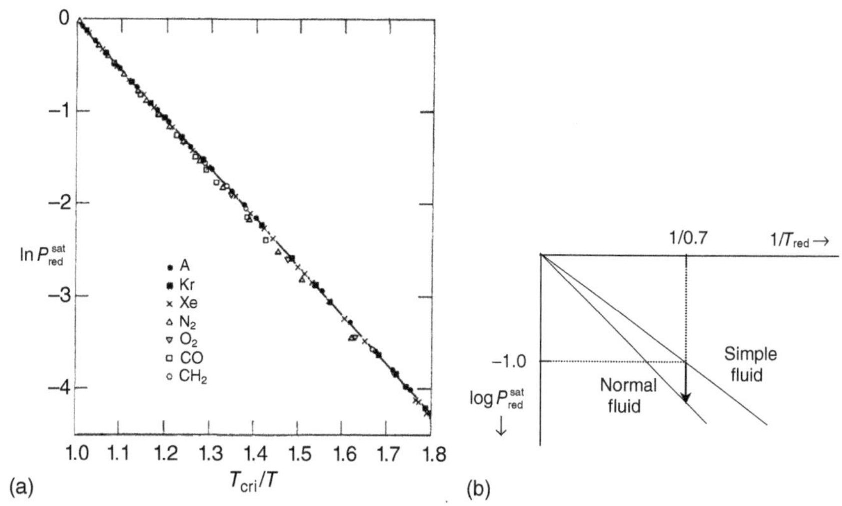

Figure 8.4 (a) The reduced vapor pressure P_{sat}/P_{cri} versus T_{cri}/T for various fluids. Source: Guggenheim [31]/with permission of Elsevier. (b) The acentric factor ω as the deviation of the slope of the vapor pressure of a normal fluid from the slope for a simple fluid. Note that log means logarithm to the base 10.

8.4.1 The Extended Principle

The conventional corresponding state approach deals with two parameters, for example, using the compression factor $Z = Z(T_{red}, P_{red})$. For simple, (more or less) spherical molecules this approach works reasonably well, as evidenced by the "universal" value of the compression factor Z. In fact, any EoS that contains only two parameters can be mathematically reduced likewise. The reduction is essentially based on dimensional analysis with force, length, and temperature as basic dimensions. In the case of categories of molecules for which extra dimensions are important, one does not expect the principle to apply. A significant improvement in describing experimental results can be obtained by extending the concept to three parameters, of which the best-known approaches are by Riedel and Pitzer. Both approaches use the pressure P versus the temperature T. Riedel [67] used the parameter $\alpha = (\mathrm{d}\ln P/\mathrm{d}\ln T)_{T=T\mathrm{cri}}$ taken along the saturation curve. Pitzer et al. [68] used a parameter that can be obtained if we plot the logarithm of the saturated vapor pressure in reduced units versus $1/T_{red}$. In this representation, we obtain approximately straight lines, that is, $\mathrm{d}\log P_{red}^{sat}/\mathrm{d}(1/T_{red}) = S$. The two-parameter principle predicts that the slope S is the same for all substances. If we plot the vapor pressures of simple fluids (spherical molecules such as Ar, CH_4, …), they indeed all fall more or less on the same line that appears to go through $\log P_{red}^{sat} = -1.0$ at $T_{red} = 0.7$ (Figure 8.4a). However, other, less spherical molecules do have different slopes. These facts are used to define the so-called *acentric factor* $\omega = -\log P_{red}^{sat}(T_{red} = 0.7) = -1.0$ (Figure 8.4b). This apparently arbitrary definition was chosen because of the ease and precision with which this quantity can be determined (contrary to Riedel's α, which might be difficult to determine precisely). Since at $T/T_{cri} = 0.7$, the vapor pressure of simple fluids is close to $0.1P_{cri}$, the acentric factor is essentially zero for these compounds. The relation of Pitzer's ω to Riedel's α reads $\alpha = 5.808 + 4.93\omega$, and therefore we limit ourselves further on to ω. For values of ω, see Sections F.8 and F.11 [69].

Table 8.7 The acentric factor ω for the first 20 aliphatic alkanes C_nH_{2n+2}.

n	ω	n	ω	n	ω	n	ω
1	0.011	6	0.300	11	0.537	16	0.718
2	0.099	7	0.350	12	0.576	17	0.753
3	0.152	8	0.399	13	0.618	18	0.800
4	0.200	9	0.445	14	0.644	19	0.845
5	0.252	10	0.490	15	0.685	20	0.865

Source: Adapted from Poling et al. [69].

The simplest correlation [70] for normal fluids correlates Z with ω and reads

$$Z_{\rm cri} = P_{\rm cri}V_{\rm cri}/RT_{\rm cri} = 0.2905 - 0.0787\omega$$

Abbott [71] showed that, if we assume slight imperfection,

$$Z = 1 + (B_2P/RT) = 1 + (B^*P_{\rm red}/T_{\rm red})B \quad \text{with} \quad B^* = B_2P_{\rm cri}/RT_{\rm cri} = B^{(0)} + \omega B^{(1)}$$

and $B^{(0)}$ and $B^{(1)}$ functions of $T_{\rm red}$, for $T_{\rm red} > 0.6$ and $\omega < 0.4$, reasonably fitted by

$$B^{(0)} = 0.083 - \left(0.422/T_{\rm red}^{1.6}\right) \quad \text{and} \quad B^{(1)} = 0.139 - \left(0.172/T_{\rm red}^{4.2}\right)$$

We thus have $Z = Z(T_{\rm red}, P_{\rm red}, \omega)$. More generally, the compression factor Z can be accurately described by $Z = Z^{(0)} + \omega Z^{(1)}$, where $Z^{(0)}$ and $Z^{(1)}$ are complex, extensively tabulated functions of $T_{\rm red}$ [72]. A reasonable fit for $\ln P_{\rm red} = f^{(0)} + f^{(1)}\omega$ reads

$$f^{(0)} = 5.927\,14 - (6.096\,48/T_{\rm red}) - 1.288\,62\ln T_{\rm red} + 0.1693\,47T_{\rm red}^6$$

$$f^{(1)} = 15.2518 - (15.6875/T_{\rm red}) - 13.4721\ln T_{\rm red} + 0.43577T_{\rm red}^6$$

For nonpolar fluids above 1 bar, the accuracy for Z is about 2–3%, but for more polar molecules and low pressures, the calculated pressure is typically too low and the deviations are larger [69], perhaps up to 10%. It should be noted that these correlations typically apply only to normal fluids, as defined in Chapter 1.

For a homologous series, ω often shows a systematic trend, as illustrated in Table 8.7 for the first 20 aliphatic alkanes. Using data for $n = 2$–13, $\omega = 0.049n$, where n is the number of carbon atoms.[9] Further discussion on the PoCS is given in [73, 74].

Problem 8.3 ***Boyle and Joule–Thomson Temperature*** Show that for the vdW fluid the Boyle temperature $T_{\rm B}$, where $B_2(T_{\rm B}) = 0$, is given by $T_{\rm B} = a/Rb$. Also show that the Joule–Thomson inversion temperature $T_{\rm JT}$ for which the $\mu_{\rm JT} \equiv (\partial T/\partial P)_H = V(\alpha T - 1)/C_P$ changes sign, hence cooling changes to heating and vice versa, and where $\partial(B_2/T)/\partial T = 0$, is given by $T_{\rm JT} = 2T_{\rm B}$.

---

9 Using $n = 2$–20, $\omega = 0.046n$. Accounting for the slight nonlinearity, $\omega = 5.37 \times 10^{-2}n - 5.18 \times 10^{-4}n^2$.

Problem 8.4 ***Virial Coefficients and Critical Points*** Show that the second and third virial coefficients for the EOSs indicated are:

Berthelot:	$B_2 = b - a/RT^2$	$B_3 = b^2$
Dieterici:	$B_2 = b - a/RT$	$B_3 = b^2 - ab/RT + a^2/2R^2T^2$

Further, show that the CPs and the values for Z_{cri} are:

Berthelot:	$P_{cri} = (aR/216b^3)^{1/3}$	$V_{cri} = 3b$	$T_{cri} = (a/27Rb)^{1/2}$	$Z_{cri} = 2/e^2$
Dieterici:	$P_{cri} = a/4e^2b^2$	$V_{cri} = 2b$	$T_{cri} = a/4Rb$	$Z_{cri} = 3/8$

Problem 8.5 ***The van der Waals Equation*** An equivalent procedure to obtain the vdW parameters is to use $(V - V_{cri})^3 = 0$, since at the critical point $V = V_{cri}$ and the vdW equation is a cubic equation in V. Compare the expansion of this equation with the standard vdW equation in its polynomial form $V^3 - \alpha V^2 + \beta V - \gamma = 0$. Show that $3V_{cri} = b + RT_{cri}/P_{cri}$, $3V_{cri}^2 = a/P_{cri}$, and $V_{cri}^3 = ab/P_{cri}$. Show that from the last two equations it follows that $a = 3P_{cri}V_{cri}^2$ and $b = V_{cri}/3$. Also show that substitution of the expression for b in the first of these three equations allows solution for V_{cri}, which then can be eliminated from a and b to obtain $V_{cri} = 3RT_{cri}/8P_{cri}$, $a = 27R^2T_{cri}^2/64P_{cri}$, and $b = RT_{cri}/8P_{cri}$.

Problem 8.6 ***van der Waals Thermodynamics*** Show, by integrating the pressure P over the volume V, that the vdW expression $P = RT/(V - b) - a/V^2$ yields the Helmholtz energy $F(T,V) - F^\circ(T,V) = -RT\ln[(V - b)/V^\circ] - a/V$, the entropy $S(T, V) - S^\circ(T, V) = R\ln[(V - b)/V^\circ]$, and the internal energy $U(T,V) - U^\circ(T,V) = -a/V$, where F°, S°, and U° refer to the corresponding perfect gas properties. Also show that $\rho_{vdW}/\rho_{idg} = 1 + (3PT_{cri}/8P_{cri}T)[(9T_{cri}/8T) - (1/3)]$, where $\rho = N/V$ is the number density, and estimate its value for Ar, N_2, and O_2 at 0 °C and 1 atm.

Problem 8.7 ***Carnahan–Starling Thermodynamics*** Show that, using the CS EoS, the Helmholtz energy F_{CS} of the HS fluid is

$$\frac{\beta F_{CS}}{N} = \int_0^\eta \frac{\beta P}{\rho\eta'}\,d\eta' = \ln\left(\frac{\rho\Lambda^3}{e}\right) + \frac{\eta(4-3\eta)}{(1-\eta)^2}$$

Show also that the compressibility $\kappa_{T,CS}$ and chemical potential μ_{CS} read, respectively,

$$\frac{\rho\kappa_{T,CS}}{\beta} = \frac{(1-\eta)^4}{2\eta(4-\eta)+(1-\eta)^4} \quad \text{and} \quad \beta\mu_{CS} = \ln(\rho\Lambda^3) + \frac{8\eta - 9\eta^2 + 3\eta^3}{(1-\eta)^3}$$

Problem 8.8 ***The van der Waals and Carnahan–Starling Hard-Sphere Fluid*** Consider a reasonably dense packed fluid. The vdW repulsion is $\beta P_{vdW} = \rho/(1 - b\rho)$ with $b = 4v_0$, while the CS approximation is given by Eq. (8.6). Show that the vdW and CS expressions agree to second order in ρ but differ in higher order. For what value of η these expressions diverge? Compare your answer with the value for η for Ar using as HS diameter $\sigma = 3.4$ Å at the triple point (85 K, $V_m = 28.39$ cm^3 mol^{-1}).

8.5 Transition State Theory

Transition state theory (TST) is an important follow-up from statistical mechanics, used throughout in chemistry and physics, which we illustrate for chemical reactions in the gas phase. But first, we recall some chemical kinetics concepts.

8.5.1 Chemical Kinetics Basics

For the description of reactions, some terms and concepts are required, which we briefly iterate here. A typical reaction may be written as

$$\nu_A\mathrm{A} + \nu_B\mathrm{B} \leftrightarrow \nu_X\mathrm{X} + \nu_Y\mathrm{Y} \tag{8.11}$$

where A and B represents the reactants (R) and X and Y represents the products (P). The *stoichiometric sum* is defined by $\Sigma_j\nu_j = \Sigma_P\nu_P - \Sigma_R\nu_R$. Thus, for Eq. (8.11), $\Sigma_j\nu_j$ becomes $(\nu_X + \nu_Y) - (\nu_A + \nu_B)$. Progress of the reaction as a function of time t is given by the *extent of reaction* $\xi(t)$, defined by

$$\xi(t) = [n_j(t) - n_j(0)]/\nu_j$$

where n_j and ν_j denote the number of moles and the reaction coefficient for component j, respectively. The *rate of reaction* is defined by

$$r \equiv \frac{1}{V}\frac{\mathrm{d}\xi}{\mathrm{d}t} = \frac{1}{\nu_j V}\frac{\mathrm{d}n_j}{\mathrm{d}t} \quad \text{or} \quad r = \frac{1}{\nu_j}\frac{\mathrm{d}c_j}{\mathrm{d}t} \quad \text{for the volume } V \text{ constant}$$

So, for Eq. (8.11) we have, using the conventional notation [Z] for the molarity c_Z,

$$r = -\frac{1}{\nu_A}\frac{\mathrm{d[A]}}{\mathrm{d}t} = -\frac{1}{\nu_B}\frac{\mathrm{d[B]}}{\mathrm{d}t} = \frac{1}{\nu_X}\frac{\mathrm{d[X]}}{\mathrm{d}t} = \frac{1}{\nu_Y}\frac{\mathrm{d[Y]}}{\mathrm{d}t}$$

The *forward* rate r and the *reverse rate* r_{rev} can often be described by

$$r = k_{for}[\mathrm{A}]^\alpha[\mathrm{B}]^\beta \quad \text{and} \quad r = k_{rev}[\mathrm{X}]^\gamma[\mathrm{Y}]^\delta \tag{8.12}$$

Generally, the *reaction order* for component Z is defined by $\partial \ln r/\partial \ln[\mathrm{Z}]$ leading for Eq. (8.12a) to the total order $n = \alpha + \beta$ with α and β the *partial reaction order*. But r is not necessarily described by a power law. For example, using P and Q as constants, it may also be described by

$$r = P[\mathrm{Z}]/(Q + [\mathrm{Z}])$$

If $[\mathrm{Z}] \ll Q$, we obtain $r = P[\mathrm{Z}]/Q$ and the apparent order $n = 1$. On the other hand, if $Q \ll [\mathrm{Z}]$, we have $r = P$ and the apparent order $n = 0$. In many cases, a reaction is discussed in terms of *elementary reactions* such as A + BC ↔ AB + C. More complex reaction schemes, involving two or more elementary reactions, are accordingly denoted as *composite reactions*. Finally, with *molecularity* the number of molecules participating in the reaction is indicated. Clearly, for the reaction as given above the molecularity is 2. Molecularity and reaction order generally have different values, but for elementary reactions, with certain exceptions,[10] they have the same value.

10 Participation of a solvent is not easily detected since its molarity changes very little, while for participation of a catalyst after completing the reaction its concentration is unchanged.

The total reaction rate r is the difference between r_{for} for the reaction $\nu_A A + \nu_B B \to \nu_X X + \nu_Y Y$ and r_{rev} for the reaction $\nu_A A + \nu_B B \leftarrow \nu_X X + \nu_Y Y$, or

$$r = r_{\text{for}} - r_{\text{rev}} = -k_{\text{for}}[A]^{\nu_A}[B]^{\nu_B} + k_{\text{rev}}[X]^{\nu_X}[Y]^{\nu_Y}$$

At equilibrium $r = 0$ or $k_{\text{for}}/k_{\text{rev}} = [X]^{\nu_X}[Y]^{\nu_Y}/[A]^{\nu_A}[B]^{\nu_B}$ with K_c the *equilibrium constant* in terms of molarities.[11] For gas-phase reactions often the equilibrium constant K_P in terms of pressure P is used. The partial pressure is $P_j = (N_j/V)kT = c_j kT$, where k is Boltzmann's constant and c_j is the molarity of component j. Hence, we have

$$(kT)^{\Sigma_j \nu_j} K_c = (kT)^{\Sigma_P \nu_P - \Sigma_R \nu_R} K_c \equiv K_P$$

The rate equations can be integrated, and this yields the concentration as a function of time explicitly. We have, defining $c_{j,0} \equiv c_j(t = 0)$ for component j,

zeroth order: $-\mathrm{d}c_A/\mathrm{d}t = k_{\text{rea}} \quad \to \quad k_{\text{rea}}t = c_{A,0} - c_A$
first order: $-\mathrm{d}c_A/\mathrm{d}t = k_{\text{rea}}c_A \quad \to \quad k_{\text{rea}}t = \ln(c_{A,0}/c_A)$
second order: $-\mathrm{d}c_A/\mathrm{d}t = k_{\text{rea}}c_A^2 \quad \to \quad k_{\text{rea}}t = (1/c_A) - (1/c_{A,0})$
third order: $-\mathrm{d}c_A/\mathrm{d}t = k_{\text{rea}}c_A c_B \quad \to \quad k_{\text{rea}}t = (c_{B,0} - c_{A,0})^{-1}\ln(c_{A,0}c_B/c_{B,0}c_A)$

Various methods exist to determine the reaction order from the experimental data [75–77].

The rate of a reaction is influenced by several variables. Apart from the molarity [Z] of component Z, the temperature T is important. The change of equilibrium constants K_P or K_c with temperature is given by the *van't Hoff equation*

$$\partial \ln K_P/\partial T = \Delta H^\circ/RT^2 \quad \text{or} \quad \partial \ln K_c/\partial T = \Delta U^\circ/RT^2$$

where ΔH° and ΔU° are the enthalpy and energy under standard conditions, respectively. Empirically, the temperature dependence of the rate constant k_{rea} can often be described by an *Arrhenius equation*

$$k_{\text{rea}} = A\exp(-E_{\text{act}}/RT) \quad \text{or} \quad k_{\text{rea}} = A'T^m\exp(-E_{\text{act}}/RT) \tag{8.13}$$

where Eq. (8.13a) is used for limited temperature range with A the *pre-exponential factor*, dependent on the temperature T but often taken as constant, and E_{act} is the (apparent) *activation energy*. For a wider temperature range, one often uses Eq. (8.13b), where A' is a temperature-independent factor. The range of E_{act} for reactions is typically 40–120 kJ mol^{-1}, which implies an increase in the rate by a factor of 2–3 for every 10 K. The value of the parameter m can be positive as well as negative, but typically $|m| < 2$.

For reactions in general, the type of solvent, ionic strength, radiation, and the presence of catalysts are important, for which we refer to, for example, [75, 78, 79].

8.5.2 The Equilibrium Constant

Associating the macroscopic internal energy U with the statistical mechanics expression for the average energy $\langle H\rangle$, the macroscopic number of molecules N with the average number of molecules $\langle N\rangle$, and recalling the grand canonical partition function $kT\ln\Xi = PV$ (Section 7.1.2), the grand canonical entropy becomes

$$TS = -kT\sum_i p_i \ln p_i = -\mu\langle N\rangle + \langle H\rangle + kT\ln\Xi$$

11 This argument is based on the power law for the rate, but the result is generally valid.

The equilibrium condition at constant U and V for a system of several components is $\mathrm{d}S = \sum_j(\partial S_j/\partial N_j)\mathrm{d}N_j = -\sum_j(\mu_j/T)\mathrm{d}N_j = 0$, where the index j denotes the various reactants and products. If we have a reaction

$$\nu_{\mathrm{A}}\mathrm{A} + \nu_{\mathrm{B}}\mathrm{B} \leftrightarrow \nu_{\mathrm{X}}\mathrm{X} + \nu_{\mathrm{Y}}\mathrm{Y} \quad \text{or equivalently} \quad \sum\nolimits_{\mathrm{R}} \nu_{\mathrm{R}}\mathrm{R} \leftrightarrow \sum\nolimits_{\mathrm{P}} \nu_{\mathrm{P}}\mathrm{P}$$

where R and P denote reactants and products, respectively, we obtain

$$T^{-1}\sum_{\mathrm{R}} \nu_{\mathrm{R}}\mu_{\mathrm{R}} = T^{-1}\sum\nolimits_{\mathrm{P}} \nu_{\mathrm{P}}\mu_{\mathrm{P}} \quad \text{equivalent to} \quad \Delta G = 0$$

Using the relation $\mu_j = kT\ln(N_j/z_j)$, obtained from $\mu_j = \partial F/\partial N_j$ with z_j denoting the partition function of component j, the result is the *law of mass action*

$$\sum\nolimits_{\mathrm{R}} \nu_{\mathrm{R}} \ln(N_{\mathrm{R}}/z_{\mathrm{R}}) = \sum\nolimits_{\mathrm{P}} \nu_{\mathrm{P}} \ln(N_{\mathrm{P}}/z_{\mathrm{P}}) \quad \text{or, using} \quad N_{\mathrm{Z}}/V = c_{\mathrm{Z}} = [\mathrm{Z}]$$

$$\prod\nolimits_{\mathrm{P}} c_{\mathrm{P}}^{\nu_{\mathrm{P}}} \Big/ \prod\nolimits_{\mathrm{R}} c_{\mathrm{R}}^{\nu_{\mathrm{R}}} = K_c(T) \quad \text{with} \quad K_c(T) = \prod\nolimits_{\mathrm{P}} (z_{\mathrm{P}}/V)^{\nu_{\mathrm{P}}} \Big/ \prod\nolimits_{\mathrm{R}} (z_{\mathrm{R}}/V)^{\nu_{\mathrm{R}}} \tag{8.14}$$

where $K_c(T)$ is the *equilibrium constant*. The energy reference level for each factor in Eq. (8.14) is the same, but for the evaluation of the partition functions z it is more convenient to use the ground state for each species as reference. Let us take the chemical reaction AB + C ↔ A + BC with $K(T) = z_{\mathrm{A}}z_{\mathrm{BC}}/z_{\mathrm{AB}}z_{\mathrm{C}}$ as an example. Shifting the reference level of all species to an arbitrary level and denoting z with respect to the ground state for each species by z', we obtain, employing as usual $\beta = 1/kT$,

$$K(T) = \frac{\left(z_{\mathrm{A}}'\mathrm{e}^{-\beta E_{\mathrm{A}}}/V\right)\left(z_{\mathrm{BC}}'\mathrm{e}^{-\beta E_{\mathrm{BC}}}/V\right)}{\left(z_{\mathrm{AB}}'\mathrm{e}^{-\beta E_{\mathrm{AB}}}/V\right)\left(z_{\mathrm{C}}'\mathrm{e}^{-\beta E_{\mathrm{C}}}/V\right)} \equiv \frac{z_{\mathrm{A}}z_{\mathrm{BC}}}{z_{\mathrm{AB}}z_{\mathrm{C}}}\exp(-\beta\Delta E) \tag{8.15}$$

In the last step, the primes have been removed (using, from now on, as reference level for each of the species on the ground state) and $\Delta E \equiv (E_{\mathrm{A}} + E_{\mathrm{BC}}) - (E_{\mathrm{AB}} + E_{\mathrm{C}})$ represents the difference in ground state energies of the reactants and products.

8.5.3 Potential Energy Surfaces

Generally, a pictorial representation of the potential energy hypersurface – the *potential energy landscape* – is complex. The idea can be grasped from the simple example of a generic reaction between reactants R and products P. In Figure 8.5, a contour map for this reaction is given with as axes the generic structural coordinates X and Y. The map shows two valleys separated by a barrier and thermal fluctuations continuously attempt to move species from one valley to another. To calculate the rate $r(T)$ for the chemical reaction P ↔ R, we must take into account all trajectories on the potential energy surface that do pass a *dividing surface*, defined as a surface which cannot be passed without passing the barrier. In Figure 8.5a, for example, one of the dividing surfaces is given by $X = Y$. An upper limit to the rate $r(T)$ is then given by the *Wigner variational theorem* $r(T) \le j(T)$, where $j(T)$ represents the *flux*, the total amount of reactant systems crossing a dividing surface per unit volume and per unit time at a temperature T. All statistical methods for computing reaction rates are based on this principle but calculating all trajectories to estimate the flux $j(T)$ is an enormous task if done in a rigorous fashion. Therefore, approximate methods are normally introduced. Using a surface that depends on some parameters, the variation theorem provides the opportunity to select the lowest flux $j(T)$ by choosing the parameters in such a way that a minimum $j(T)$ is obtained.

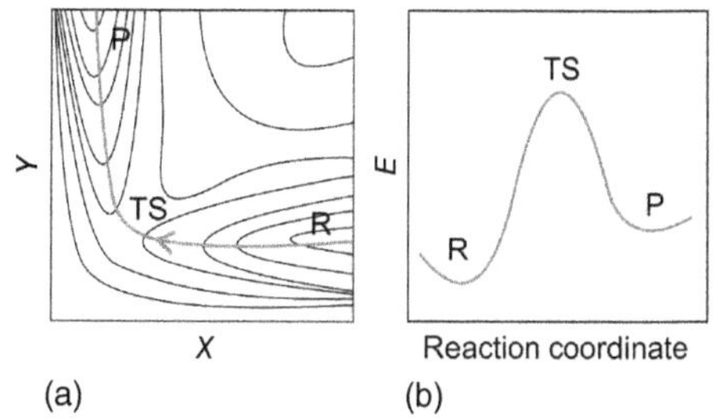

Figure 8.5 (a) Contour map of the potential energy surface using structural coordinates X and Y for the reaction R ↔ P. The transition state TS is at the top of the barrier between two valleys. (b) The potential energy along the trajectory from R to P as function of the reaction coordinate. Source: Van Speybroeck et al. [80]/Royal Society of Chemistry.

8.5.4 The Activated Complex

The simplest approach is to consider only the trajectory following the minimum energy path from one valley to the other valley. For the chemical reaction P ↔ R depicted in Figure 8.5a, this means the path from the configuration represented by point R to the configuration represented by point P. The coordinate along this path is called the *reaction coordinate*. The saddle point region between the two valleys is known as the *transition state*, while a system in the transition state is called an *activated complex*.

TST is based on a few assumptions. The first assumption is that the optimum dividing surface passes through the transition state and is perpendicular to the reaction coordinate. The second is that activated complexes are always in equilibrium with both the reactants and the products. This implies for the reaction A + BC ↔ AB + C that the forward reaction A + BC → $(ABC)^‡$ and the reverse reaction AB + C → $(ABC)^‡$, with $(ABC)^‡$ the activated complex, k_{for} and k_{rev} are given by

$$k_{for} = [(ABC)^‡]_{for}/[A][BC] \quad \text{and} \quad k_{rev} = [(ABC)^‡]_{rev}/[AB][C] \tag{8.16}$$

Since in Eqs. (8.16a,b) $[(ABC)^‡]_{for}$ and $[(ABC)^‡]_{rev}$ are equal, we have $[(ABC)^‡]_{for} = [(ABC)^‡]_{rev} = ½[(ABC)^‡]$. The combination of these two conditions together represents the *quasi-equilibrium assumption*. Finally, it is assumed that once the transition state is reached, the reactants do not return to the nonreacted state.

Equation (8.15) shows that k_{for} for the reaction A + BC ↔ $(ABC)^‡$ can be written as

$$k_{for} = \left(z_{(ABC)^‡}/z_A z_{BC}\right) \exp(-\beta\Delta E) \tag{8.17}$$

We assume that the partition function z for each species can be factorized as $z = z_{ele}z_{tra}z_{vib}z_{rot}$ with z_{ele}, z_{tra}, z_{vib}, and z_{rot} the electronic, translation, vibration, and rotation partition functions. Symmetry should be taken into account in z_{rot} by using the appropriate symmetry numbers. The contribution of z_{ele} reduces to the degeneracy number of the ground state, usually 1, since we assume the reaction proceeds on a potential energy surface, which exists only by virtue of the adiabatic assumption.

If we know the structure and relevant force constants for all species, we can construct the partition functions. For the activated complex z_{tra} and z_{rot} do not provide a problem because they can be constructed as for normal molecules, but z_{vib} requires some care. The normal coordinate analysis of an N-atom molecule yields $3N - 6$ normal coordinates for a nonlinear molecule (or $3N - 5$ normal coordinates for a linear molecule). For the activated complex all but one of these coordinates have a positive force constant. The last, negative coefficient corresponds to an imaginary frequency and represents the reaction coordinate. This implies that, in the transition state, a small fluctuation in the reaction coordinate leads to an unstable configuration with respect to this coordinate. It is customary[12,13] to treat this coordinate as a translation coordinate over a small length δ at the top of the potential energy barrier with a partition function $z = (\delta/h)(2\pi m^{\ddagger}kT)^{1/2}$. The complete partition function $z_{(ABC)}{}^{\ddagger}$ for the activated complex is then

$$z_{(ABC)}{}^{\ddagger} = (\delta/h)(2\pi m^{\ddagger}kT)^{1/2}z^{\ddagger} \tag{8.18}$$

where $z^{\ddagger}$ represents the partition function for the remaining coordinates, that is, the true vibration coordinates, the translation and rotation coordinates. Further, $m^{\ddagger}$ denotes the mass of the activated complex and k, T, and h have their usual meanings.

The concentration of the activated complex due to the forward reaction is given by Eq. (8.16a), where the forward equilibrium constant is given by Eq. (8.17). Since there are $[(ABC)^{\ddagger}]_{for}$ complexes per unit volume that populate the length δ and are moving forward, $r_{for} = [(ABC)^{\ddagger}]_{for}/\tau$, where τ is the average time to traverse the length δ, given by $\tau = \delta/\langle v\rangle$. Thus, we need the average velocity $\langle v\rangle$ in one direction over the length δ. Using the same approximation of free translatory motion again, we borrow from kinetic gas theory (Scction 8.1)

$$\langle v\rangle = \int_0^{\infty} v\exp(-\tfrac{1}{2}\beta m^{\ddagger}v^2)\,dv \Big/ \int_0^{\infty} \exp(-\tfrac{1}{2}\beta m^{\ddagger}v^2)\,dv = (2kT/\pi m^{\ddagger})^{1/2} \tag{8.19}$$

Combining Eqs. (8.16a)–(8.19) with $r_{for} = [(ABC)^{\ddagger}]_{for}/\tau = [(ABC)^{\ddagger}]v_{ave}/2\delta$, we obtain

$$r_{for} = k_{for}[A][BC] \quad \text{with} \quad k_{for} \equiv \frac{kT}{h}\frac{z^{\ddagger}}{z_A z_{BC}}\exp(-\beta\Delta E) \equiv \frac{kT}{h}K^{\ddagger} \tag{8.20}$$

A similar expression is obtained for the r_{rev}, and it can be verified that in equilibrium r_{for} and r_{rev} have the same value. Therefore, the overall rate can be calculated given the relevant information,[14] but a first-principle calculation requires the energy barrier for the reaction, and the structure and the force constants associated with the dynamics of the reactants, products, and activated complex. Typically, this information is incompletely available. This is also true for other reactions than gas reactions, although of course different arguments apply for details. The exponential dependence on the barrier energy is generally valid though and rationalizes the generally observed (modified) *Arrhenius-type behavior*.

Let us consider an even more simple configuration, namely $A \leftrightarrow (AB)^{\ddagger} \leftrightarrow B$, but now allow return and a significant deviation from equilibrium. This also implies a significant driving force so that $-\Delta G^{\ddagger}$ becomes $-\Delta G^{\ddagger} \pm W$, where W is the work done in the process $A \to B$. The forward and reverse rate constants become, respectively,

$$k_{for} \equiv (kT/h)\,e^{-\beta\Delta G^{\ddagger}}e^{\beta W} \quad \text{and} \quad k_{rev} \equiv (kT/h)\,e^{-\beta\Delta G^{\ddagger}}e^{-\beta W}$$

12 This approximation can be rationalized, see, for example, [81, 82].

13 Alternatively, we consider the reaction coordinate as the coordinate for which $\omega \to 0$. In that limit the vibrational contribution to z reads $kT/\hbar\omega$. For an ordinary vibration of a bond, the circular frequency ω would be the reciprocal of the time τ to make a complete vibration, that is, traversing the length δ twice. Here, we need the frequency for traversing δ only once, that is, 2ω. Hence, the rate $r = \tfrac{1}{2}[(ABC)^{\ddagger}]\,2\omega$, leading to the same final result.

14 Two other factors should be considered. First, a transmission coefficient κ to account for that not every activated complex reaching the top of the barrier and moving along the reaction coordinate leads to a reaction. Obviously, $\kappa \le 1$, but probably close to 1. Second, a tunneling factor $\zeta = 1-(\beta h v^{\ddagger})^2/24$, where $v^{\ddagger}$ is the imaginary frequency of vibration along the reaction coordinate. As $(v^{\ddagger})^2 < 0$, $\zeta \ge 1$, but usually also close to 1. In the absence of information, both factors are often omitted.

The net rate $k_{\text{rea}} = k_{\text{for}} - k_{\text{rev}}$ therefore reads, using $K^{\ddagger} = (kT/h)\exp(-\beta\Delta G^{\ddagger})$,

$$k_{\text{rea}} = \frac{kT}{h}\mathrm{e}^{-\beta\Delta G^{\ddagger}}(\mathrm{e}^{\beta W} - \mathrm{e}^{-\beta W}) = 2K^{\ddagger}\sinh(\beta W)$$

Since W is often small as compared kT, we can approximate by expanding $\exp(\pm W)$ to $1 \pm \beta W + \cdots$, yielding $k_{\text{rea}} = 2\beta K^{\ddagger} W$. Both the full sinh expression as well as its approximation are often used, in particular in the discussion of transport processes.

8.5.5 The Link to Experiment

We now need to connect these results to the experimental data. Experimentally, it is often observed that, if the logarithm of the rate constant is plotted against the reciprocal temperature, a straight line is obtained. The gradient of this line is used to define the (*empirical*) *activation energy* E_{act} by

$$E_{\text{act}} \equiv -R\,\partial \ln k_{\text{for}}/\partial(1/T) = RT^2 \partial \ln k_{\text{for}}/\partial T = RT^2(\partial \ln K^{\ddagger}/\partial T + T^{-1})$$

where in the rhs the van 't Hoff relation $\partial \ln K_c^{\ddagger}/\partial T = \Delta U^{\ddagger}/RT^2$, with $\Delta U^{\ddagger}$ the change in energy, can be used. Hence, we obtain

$$E_{\text{act}} = \Delta U^{\ddagger} + RT$$

and this equation provides a link between the experimentally observed E_{act} and $\Delta U^{\ddagger}$. Note that neither E_{act} nor $\Delta U^{\ddagger}$ is identical to ΔE, which appears in Eq. (8.17) because of the temperature dependence of the various partition functions. For the various partition functions, the temperature dependencies read

$$z_{\text{tra}} \sim T^{3/2}, \quad z_{\text{rot,lin}} \sim T, \quad z_{\text{rot,nonlin}} \sim T^{3/2} \quad \text{and} \quad z_{\text{ele}} \sim T^0$$

For the vibrations, however, the behavior is less simple. For $kT \ll \hbar\omega$, $Z_{\text{vib}} \sim T^0$, while for $kT \gg \hbar\omega$, $Z_{\text{vib}} \sim T^n$, where n is the number of vibrational modes of the molecule. At intermediate temperature, $Z_{\text{vib}} \sim T^a$ with $0 \le a \le n$. For a restricted (intermediate) temperature range, a is approximately constant and in total this yields

$$k_{\text{for}} \sim T^m \exp(-\Delta E/RT)$$

explaining the modified Arrhenius equation. As noted before, typically $|m| < 2$.

Finally, we note that the formalism can also be interpreted thermodynamically, and to this purpose we rewrite Eq. (8.20), $k_{\text{for}} \equiv kTK^{\ddagger}/h$, in terms of molar quantities

$$k_{\text{for}} = (kT/h)\,\mathrm{e}^{-\Delta G^{\ddagger}/RT} = (kT/h)\,\mathrm{e}^{\Delta S^{\ddagger}/R}\mathrm{e}^{-\Delta H^{\ddagger}/RT} = (kT/h)\,\mathrm{e}^{\Delta S^{\ddagger}/R}\mathrm{e}^{-\Delta U^{\ddagger}/RT}\mathrm{e}^{P\Delta V^{\ddagger}/RT}$$

with $\Delta S^{\ddagger}$, $\Delta H^{\ddagger}$, $\Delta U^{\ddagger}$, and $\Delta V^{\ddagger}$ the molar entropy, enthalpy, energy, and volume of activation, respectively. This expression is to be compared with the empirical expression $k_{\text{rea}} = A\exp(-E_{\text{act}}/RT)$. In a gas reaction, the term $P\Delta V^{\ddagger}$ may be put equal to $RT\Sigma_j \nu_j$, where $\Sigma_j \nu_j$ is the stoichiometric sum for activated complex formation. For a unimolecular reaction, the stoichiometric sum $\Sigma_j \nu_j = 0$ and $\Delta H^{\ddagger} = \Delta U^{\ddagger} = E_{\text{act}} - RT$, or $E_{\text{act}} = \Delta H^{\ddagger} + RT$. Consequently, $A = \mathrm{e}(kT/h)\exp(\Delta S^{\ddagger}/R)$. For a bimolecular reaction, $\Sigma_j \nu_j = -1$ and $\Delta H^{\ddagger} = \Delta U^{\ddagger} - RT = E_{\text{act}} - 2RT$, or $E_{\text{act}} = \Delta H^{\ddagger} + 2RT$. Consequently, $A = \mathrm{e}^2(kT/h)\exp(\Delta S^{\ddagger}/R)$. In solution, the volume term is nearly always negligible and therefore $\Delta H^{\ddagger} \cong \Delta U^{\ddagger}$. The pre-exponential factor A is thus related to the entropy.

Here, our brief overview on gas phase TST ends. It must be emphasized again that for a detailed calculation, a considerable amount of information is required.

Problem 8.9 ***Simple Dimers*** Dissociation of dimers (d) to molecules (m) in the gas phase can be described by $x_d \to 2x_m$ with x the mole fraction and $K = x_m^2/x_d$ the (mole fraction) equilibrium constant. Assume that initially N dimers are present, each with bonding energy $-\varepsilon$ with respect to molecules, and neglect vibration and rotation of the dimers. Show that $K = 4\exp(-\beta\varepsilon)/[2+\exp(\beta\varepsilon)]$ and $x_d = 0.576$ and $x_m = 0.424$ at $T = \varepsilon/k$.

Problem 8.10 ***The Dissociation of I_2*** Consider the reaction $I_2 \leftrightarrow 2I$ at 300 K. Note that the ground state for the I atom is $^2P_{2/3}$ (hence fourfold degenerate), while that for the I_2 molecule is $^1\Sigma_g{}^+$ (hence nondegenerate). For I_2, the rotational temperature $\theta_{rot} = 0.054$ K, the vibrational temperature $\theta_{vib} = 308$ K, the dissociation energy from the ground state $D = 1.5417$ eV, and the molar mass $\mathcal{M}_I = 127\ \text{g mol}^{-1}$.

(a) Show that the translational and electronic partition function for the I atom are $z_{tra} = \Lambda^{-3}V$ and $z_{ele} = 4\exp[-½(D/kT + \theta_{vib}/2T)]$, respectively.
(b) Show that the translational, vibrational, rotational, and electronic partition function for the I_2 molecule are $z_{tra} = 2^{-3/2}\Lambda^{-3}V$, $z_{vib} = [\exp(\theta_{vib}/2T) - \exp(-\theta_{vib}/2T)]^{-1}$, $z_{rot} = T/2\theta_{rot}$, and $z_{ele} = 1$, respectively.
(c) Show that $K_c = 32(\pi mkT/h^2)^{3/2}(\theta_{rot}/T)[1 - \exp(-\theta_{vib}/T)]e^{-D/kT}$, and that $K_P = kTK_c$.
(d) Calculate the numerical value for K_P at 300 K.

8.6 Dielectric Behavior

To describe dielectric behavior, we first introduce some basic aspects, next discuss the Debye–Langevin equation and finalize with some estimates for the dipole moment μ and polarizability α. Although discussed in terms of gases, many arguments apply to liquids as well, for which we also provide examples.

8.6.1 Basic Aspects

Let us first briefly describe macroscopic, (linear) dielectric behavior. In macroscopic terms, if we put in vacuum on a parallel plate capacitor,[15] with electrode area A and electrode distance d, an electrical voltage difference ϕ on the electrodes, a charge

$$q_0 = \int_0^t I(t)\,dt$$

will develop, where $I(t)$ is the current during the period of charging. The capacitance

$$C_0 = \varepsilon_0 A/d$$

is the proportionality constant[16] between the voltage applied and the charge q_0 developed at $t = \infty$, that is, $q_0 = C_0\phi$. The surface charge density σ_0 is defined by

$$\sigma_0 = q_0/A \quad \text{and thus} \quad \sigma_0 = \varepsilon_0\phi/d = -\varepsilon_0 E$$

---

15 We assume that edge effects as well as charge leakage through the capacitor can be neglected.
16 For a spherical capacitor with inner sphere radius a and inner radius of the outer spherical shell b, $C_0 = 4\pi\varepsilon_0/(a^{-1} - b^{-1})$, so that for solid sphere with radius R, $C_0 = 4\pi\varepsilon_0 R$.

where $E = |\boldsymbol{E}| = -\phi/d$ is the *electric field strength* and ε_0 is the *permittivity of vacuum*, which is a universal constant. The response to the applied electric field $\boldsymbol{E}$, the surface charge density σ_0, can be interpreted as a dipole moment per unit volume since its dimension C m^{-2} can be written as (C m) m^{-3}. In that case, the response is denoted as the *(di)electric displacement* $\boldsymbol{D}$, and so in vacuum we have $\boldsymbol{D} = \varepsilon_0\boldsymbol{E}$.

If we put a homogeneous dielectric material in the capacitor, the charge changes to $q = \varepsilon_r q_0$, with ε_r the *relative permittivity*, and thus the capacitance becomes $C = \varepsilon_r\varepsilon_0 A/d = \varepsilon A/d$ with $\varepsilon \equiv \varepsilon_r\varepsilon_0$ the *permittivity*. The relative permittivity ε_r is a material-specific function, dependent on temperature and pressure, and possibly also on the electric field. Equivalently, we say that the surface charge density changes from $\sigma_0 = q_0/A$ to

$$\sigma = q/A = \sigma_0 + \sigma_{\text{pol}}$$

where σ_{pol} is the extra surface charge density at the electrodes due to the polarization of the material in the capacitor. In terms of the field $\boldsymbol{E}$, the displacement $\boldsymbol{D}$ becomes

$$\boldsymbol{D} = \varepsilon\boldsymbol{E} = \varepsilon_r\varepsilon_0\boldsymbol{E} = \varepsilon_0\boldsymbol{E} + \boldsymbol{P}_E$$

where $\boldsymbol{P}_E$ is the *polarization*. Solving for $\boldsymbol{P}_E$, we obtain

$$\boldsymbol{P}_E = (\varepsilon_r - 1)\varepsilon_0\boldsymbol{E} \equiv \chi\varepsilon_0\boldsymbol{E}$$

with χ the (electric) *susceptibility*, for which it always holds that $\chi \geq 0$. Clearly, the vectors $\boldsymbol{E}$ and $\boldsymbol{D}$ are in principle related by a tensor $\boldsymbol{\varepsilon}$, but for isotropic and cubic materials $\boldsymbol{\varepsilon}$ reduces to the scalar ε, to which we restrict our discussion largely.

The variables $\boldsymbol{E}$ and $\boldsymbol{D}$ are also conjugated variables. To show this, consider that $\phi = -Ed$ and $\text{d}q = -A\text{d}(\varepsilon E)$. Therefore, the electrical work required to transfer an element of positive charge $\text{d}q$ from the negative to the positive electrode $đW = \phi\text{d}q = V_{\text{cap}}E\text{d}D$ with volume of the capacitor $V_{\text{cap}} = Ad$. Generalizing, the Gibbs equation (for a one-component system at temperature T and pressure P) thus becomes[17]

$$\text{d}U = T\text{d}S - P\text{d}V + \mu\text{d}n + V_{\text{cap}}\boldsymbol{E}\cdot\text{d}\boldsymbol{D}$$

Applying a Legendre transform using $G = U - TS + PV - V_{\text{cap}}\boldsymbol{E}\cdot\boldsymbol{D}$, we obtain

$$\text{d}G = -S\text{d}T + V\text{d}P + \mu\text{d}n - V_{\text{cap}}\boldsymbol{D}\cdot\text{d}\boldsymbol{E} \tag{8.21}$$

For normal dielectrics at low field, the permittivity is not a function of the field $\boldsymbol{E}$ and integration of Eq. (8.21) results in $G(T,P) = G^{(0)}(T,P) - \frac{1}{2}\varepsilon(T,P)E^2V_{\text{cap}}$, where $G^{(0)}(T,P)$ represents the value of $G(T,P)$ for zero field. From this expression, the entropy S, the volume V, and the energy U follow in the usual manner.

So, an applied field E introduces in a material a polarization[18] $P = (\varepsilon_r - 1)\varepsilon_0 E$, and our task is now to interpret P in microscopic terms. In general, we have several molecular contributions to the polarization. First, if the molecules are polar and have a dipole moment μ, we

---

17 More formally, starting from the expression for electrical work $đW = \int \boldsymbol{E}\cdot\text{d}\boldsymbol{D}\,\text{d}V$ [83] and the electric potential $\boldsymbol{E} = -\nabla\phi$, we have $đW = -\int \nabla\phi\cdot\boldsymbol{D}\,\text{d}V = -\int [\nabla\cdot\phi\text{d}\boldsymbol{D} - \phi\nabla\cdot\text{d}\boldsymbol{D}]\,\text{d}V = \int \phi\nabla\cdot\text{d}\boldsymbol{D}\,\text{d}V$ as by the Gauss theorem $\int \nabla\cdot\phi\text{d}\boldsymbol{D}\,\text{d}V = \int \phi\text{d}\boldsymbol{D}\cdot\boldsymbol{n}\,\text{d}A = 0$ because $\phi = 0$ at infinity. The remaining integral transforms as $\int \phi\text{d}\nabla\cdot\boldsymbol{D}\,\text{d}V = \int \phi\text{d}\rho\,\text{d}V$ using $\nabla\cdot\boldsymbol{D} = \rho$ with ρ the charge density (Appendix D), and reducing for a homogeneous configuration to $\phi\int \text{d}\rho\,\text{d}V = \phi V\text{d}q$ with $\text{d}q$ the charge.

18 In the remainder of this section we just use P for P_E.

have the alignment of these permanent dipoles due to the applied field. This process results in a polarization contribution labeled P_μ. Second, even if we have nonpolar molecules, all molecules have an electronic polarizability α_{ele} and molecular polarizability[19] α_{mol}, collectively denoted by[20] α, which results in a dipole moment induced by the applied field and is labeled P_α. The field experienced by the molecules is *not equal* to the applied field E as the surrounding molecules shield the reference molecule from the applied field, and a polar reference molecule will influence the surroundings, leading to a different shielding than for a nonpolar molecule. If we have N molecules in a volume V, we may write $P_\mu \equiv (N/V)\mu_\mu$ and $P_\alpha \equiv (N/V)\mu_\alpha$. Here μ_μ represents the (rotationally) averaged dipole moment $\langle\mu\rangle$, which is dependent on the *directional field* E_{dir} experienced by the molecules. For the induced dipole moment μ_α we have $\mu_\alpha = \alpha E_{\text{loc}}$, where E_{loc} is the *local field*. Although both fields E_{dir} and E_{loc} are proportional to the applied field E, they are not the same because a proportional increase in all three components of E will affect the magnitude of the induced dipole moment but not the directional effect for the permanent dipole. Hence, our task is to calculate E_{loc}, E_{dir}, and $\langle\mu\rangle$ so that P_μ and P_α can be determined.

Altogether, in microscopic terms the polarization can be interpreted as the number of molecules N with *effective dipole moment* μ_{eff} in a volume V, that is,

$$P = (\varepsilon_{\text{r}} - 1)\varepsilon_0 E = \rho\mu_{\text{eff}} = \rho(\mu_\mu + \mu_\alpha) \qquad \text{with} \qquad \rho = N/V = N_{\text{A}}/V_{\text{m}}$$

Here $V_{\text{m}} = \mathcal{M}/\rho$ is the molar volume, $\mathcal{M}$ is the molar mass, and ρ is the number density.

8.6.2 The Debye–Langevin Equation

The simplest case is to consider is a dipole moment μ_α induced by a local field E_{loc} in a nonpolar molecule, given by $\mu_\alpha = \alpha E_{\text{loc}}$, where α is the polarizability. Lorentz showed that under certain conditions, $E_{\text{loc}} = E(\varepsilon_{\text{r}} + 2)/3$ (Justification 8.1). So, for nonpolar molecules we obtain, using $P = \varepsilon_0(\varepsilon_{\text{r}} - 1)E = \rho\mu_\alpha$ and $\mu_\alpha = \alpha E_{\text{loc}} = \alpha E(\varepsilon_{\text{r}} + 2)/3$, the *Clausius–Mossotti equation*

$$\rho\alpha/3\varepsilon_0 = \alpha N_{\text{A}}/3\varepsilon_0 V_{\text{m}} \equiv P_{\text{m}}/V_{\text{m}} = (\varepsilon_{\text{r}} - 1)/(\varepsilon_{\text{r}} + 2) \tag{8.22}$$

linking microscopic terms, the polarizability α and number density $\rho = N_{\text{A}}/V_{\text{m}}$, to a macroscopic parameter, the relative permittivity ε_{r}. Traditionally (although unfortunately) $P_{\text{m}} \equiv \alpha N_{\text{A}}/3\varepsilon_0$ is called the *molar polarization*. Equation (8.22) is inadequate for polar molecules and we need to add a contribution due to the orientation of the dipoles, which we discuss in classical terms only.

From statistical mechanics we learned how to calculate the average dipole moment: first, determine the Hamilton function, that is, the energy in terms of coordinates and associated momenta; thereafter, evaluate the partition function; and finally, calculate the expectation value of μ. Here, decoupling of the kinetic and potential energy is relevant, which leads to a factor $\sin\theta$ in the configurational partition function, as briefly indicated below.[21] We discuss

---

19 While α_{ele} represents the displacement of the electrons with respect to the nuclei, α_{mol} represents the displacement of atoms (or atom groups) of the molecule relative to each other. Usually, $\alpha_{\text{mol}} \ll \alpha_{\text{ele}}$.
20 Often, the *polarizability volume* $\alpha' = \alpha/4\pi\varepsilon_0$ is used.
21 For a detailed general treatment, see [84]. In brief, the Hamilton function $\mathcal{H}$ can be written as $\mathcal{H} = f(\boldsymbol{P},\boldsymbol{p}) + g(\boldsymbol{Q},\boldsymbol{q},E)$, where $\boldsymbol{P}$ represent the "momentoids" (pseudo-momenta) associated with the orientation angles $\boldsymbol{Q}$ used as a coordinates, while $\boldsymbol{p}$ represent the momenta associated with the other

the behavior for a single molecule, because for non- (or weakly) interacting molecules their behavior is additive.

The Hamilton function $\mathscr{H}$ is the sum of the kinetic energy ϑ and potential energy $\boldsymbol{\Phi}$. If we take for a rigid molecule with mass M and moments of inertia A, B, and C as generalized coordinates for their position, the Cartesian coordinates $\boldsymbol{x}$ of the center of mass, and for their orientation the Eulerian angles (ξ,θ,ψ),[22] the kinetic energy is

$$\vartheta = ½M\left(\dot{x}_1^{\,2} + \dot{x}_2^{\,2} + \dot{x}_3^{\,2}\right) + ½\left(A\Omega_1^{\,2} + B\Omega_2^{\,2} + C\Omega_3^{\,2}\right)$$

Here, the components of the angular velocity about the A, B, and C axes are given by

$$\Omega_1 = -\dot{\xi}\sin\theta\cos\psi + \dot{\theta}\sin\psi, \qquad \Omega_2 = -\dot{\xi}\sin\theta\,\sin\psi + \dot{\theta}\cos\psi, \qquad \Omega_3 = \dot{\psi} + \dot{\xi}\cos\theta$$

Substitution in ϑ results after a somewhat lengthy calculation in the Jacobian determinant $MABC\sin^2\theta$. The potential energy $\phi = -\boldsymbol{\mu}\cdot\boldsymbol{E}$ is codetermined by the dipole moment $\boldsymbol{\mu}$ and the electric field $\boldsymbol{E}$. Therefore, using the element of solid angle $\mathrm{d}\boldsymbol{\omega} = \sin\theta\mathrm{d}\theta\mathrm{d}\psi$, the partition function becomes

$$Z = \int_0^{\infty} \exp(-\beta\vartheta)\,\mathrm{d}\boldsymbol{x}\mathrm{d}\boldsymbol{\omega}\mathrm{d}\xi \times \int_0^{2\pi}\int_0^{2\pi}\int_0^{\pi} \exp(-\beta\phi)\,\mathrm{d}\boldsymbol{\omega}\mathrm{d}\xi \equiv Z_\vartheta Z_\Phi$$

where the angle ranges are $0 \le \xi$, $\psi \le 2\pi$, and $0 \le \theta \le \pi$. Hence, Z separates in two factors with Z_ϑ representing the free translation and rotation and Z_Φ the influence of the electric field. Calculating $\langle\mu\rangle$, the kinetic parts and the integration over ψ cancel and $\langle\mu\rangle$ becomes

$$\langle\mu\rangle = Z_\Phi^{-1}\int_0^{\pi} \mu E\cos\theta\exp(\beta\mu E\cos\theta)\sin\theta\,\mathrm{d}\theta$$

Using $\sin\theta\mathrm{d}\theta = \mathrm{d}(\cos\theta)$ and $a \equiv \beta\mu E$ this average evaluates to

$$\begin{aligned}\langle\mu\rangle &= \mu L(a) \equiv \mu\{[\exp(a) + \exp(-a)]/[\exp(a) - \exp(-a)] - 1/a\}\\ &= \mu(a/3 - a^3/45 + \cdots) \cong \mu(a/3) = \mu^2E/3kT \ \text{ for } \ a \ll 1\end{aligned} \tag{8.23}$$

The function $L(a)$ is usually referred to as the *Langevin function*, for which the approximation for small a is nearly always valid. Recalling that we need the directional field E_{dir} for the field for a dipole, the total average dipole moment is

$$\langle\mu\rangle = (\mu^2/3kT)E_{\mathrm{dir}}$$

Debye took for the directional field E_{dir} the Lorentz field $E_{\mathrm{L}} = E(\varepsilon_{\mathrm{r}} + 2)/3$. Making the same assumption and the assumption that $\mu_\alpha = \alpha_{\mathrm{ele}}E_{\mathrm{loc}}$ adds to $\langle\mu\rangle$, μ_{eff} becomes

$$\mu_{\mathrm{eff}} = \mu_\alpha + \langle\mu\rangle = [\alpha_{\mathrm{ele}} + (\mu^2/3kT)]E_{\mathrm{L}} \equiv \alpha_{\mathrm{eff}}E_{\mathrm{L}}$$

referred to as the *Debye–Langevin equation* with the *effective polarizability* $\alpha_{\mathrm{eff}} = \alpha_{\mathrm{ele}} + \mu^2/3kT$. In terms of $P = \rho\mu_{\mathrm{eff}} = \varepsilon_0(\varepsilon_{\mathrm{r}} - 1)E$ we obtain the *Debye equation*, formally equivalent to the Clausius–Mossotti equation but with $\mu_{\mathrm{eff}} = \alpha_{\mathrm{eff}}E_{\mathrm{L}}$,

$$P_{\mathrm{m}}/V_{\mathrm{m}} \equiv [\alpha_{\mathrm{eff}}N_{\mathrm{A}}/3\varepsilon_0]/V_{\mathrm{m}} = (\varepsilon_{\mathrm{r}} - 1)/(\varepsilon_{\mathrm{r}} + 2) \tag{8.24}$$

---

coordinates $\boldsymbol{q}$. This procedure separates the rigid body rotational motion, which is independent of E and therefore can be omitted. Since $\boldsymbol{P}$ do not represent the real momenta associated with $\boldsymbol{Q}$, this procedure introduces a Jacobian determinant $\sin\theta$ for $g(\boldsymbol{Q},\boldsymbol{q},E)$. This is essentially the procedure outlined here.

22 Here, I use (ξ,θ,ψ) instead of the usual (ϕ,θ,ψ) [85], as ϕ is used for potential energy.

Thus, the effective polarizability increases with respect to α_{ele} due to the dipoles orienting in the field. However, as there is also within the molecule a charge shift contribution to α_{eff}, it is not clear that the simple additivity assumption is valid.

Hence, we consider the situation for the polar molecules anew, which, for the moment, we still consider as rigid with electronic polarizability α_{ele} and dipole moment $\boldsymbol{\mu}$. The molecular coordinates are referred to the principal axes $\lambda = (1,2,3)$ of the principal components α_λ of the polarizability tensor $\boldsymbol{\alpha}_{\text{ele}}$ of the molecule and we consider the polarization in the direction of the field, taken as z, with the direction cosines between λ and z given by l_λ. Some averages of l_λ over the unit sphere we need, denoted by an overbar, are $\overline{l_\lambda} = \overline{l_\lambda l_{\lambda'}} = 0$ and $\overline{l_\lambda^{\,2}} = 1/3$. The potential energy ϕ is given by

$$\phi = -\boldsymbol{\mu}\cdot\boldsymbol{E} - \tfrac{1}{2}\boldsymbol{E}\cdot\boldsymbol{\alpha}_{\text{ele}}\cdot\boldsymbol{E} = -E\sum\nolimits_\lambda \mu_\lambda l_\lambda - \tfrac{1}{2}E^2\sum_\lambda \alpha_\lambda l_\lambda^{\,2}$$

Because ϕ is typically small as compared to kT, we expand the partition function $Z_\Phi = Z = \int \exp(-\beta\phi)\mathrm{d}\omega$ to second order resulting in $Z = \int (1 - \beta\phi + \tfrac{1}{2}\beta^2\phi^2)\mathrm{d}\omega$. The first term just contributes $\int 1\,\mathrm{d}\omega = 4\pi$, and the second and third terms evaluate to

$$-\beta\int\phi\,\mathrm{d}\omega = 4\pi\left[\beta E\sum_\lambda\int\mu_\lambda l_\lambda\,\mathrm{d}\omega + \tfrac{1}{2}\beta E^2\sum_\lambda\int\alpha_\lambda l_\lambda^{\,2}\,\mathrm{d}\omega\right] = 4\pi\left[\tfrac{1}{2}\beta E^2\sum\nolimits_\lambda \alpha_\lambda/3\right]$$

$$\begin{aligned}\tfrac{1}{2}\beta^2\int\phi^2\,\mathrm{d}\omega &= 4\pi\left[\tfrac{1}{2}\beta^2E^2\sum\nolimits_{\lambda,\lambda'}\int\mu_\lambda\mu_{\lambda'}l_\lambda l_{\lambda'}\,\mathrm{d}\omega + \mathcal{O}(E^3)\right]\\ &= 4\pi\left[\tfrac{1}{2}\beta^2E^2\sum\nolimits_\lambda\int\mu_\lambda^{\,2}l_\lambda^{\,2}\,\mathrm{d}\omega + \tfrac{1}{2}\beta^2E^2\sum\nolimits_{\lambda\neq\lambda'}\int\mu_\lambda\mu_{\lambda'}l_\lambda l_{\lambda'}\,\mathrm{d}\omega\right]\\ &= 4\pi\left[\tfrac{1}{2}\beta^2E^2\sum\nolimits_\lambda\mu_\lambda^{\,2}/3\right]\end{aligned}$$

In this last contribution, the third-order terms in E are neglected. In total, Z becomes

$$Z = 4\pi[1 + \tfrac{1}{2}E^2(\beta\alpha + \mu^2\beta^2/3)] \quad \text{with} \quad \alpha = \sum\nolimits_\lambda \alpha_\lambda/3 \quad \text{and} \quad \mu^2 = \sum\nolimits_\lambda \mu_\lambda^{\,2}$$

The effective polarizability α_{eff} is again given by the Debye–Langevin expression

$$\alpha_{\text{eff}} = \frac{1}{E}\frac{kT}{Z}\frac{\partial Z}{\partial E} \cong \frac{kT}{4\pi}4\pi(\beta\alpha + \mu^2\beta^2/3) = \alpha + \frac{\mu^2}{3kT}$$

However, as we used a rigid molecule, we still need to consider the dynamics. We assume that the electronic motion is so fast that the electronic polarizability α_{ele} can follow the nuclear motion instantaneously, which is a plausible assumption. Clearly this does not apply to the molecular polarization, due to atomic shifts in the molecule, and we need the normal coordinates. In the presence of an electric field, the transformation to normal coordinates proceeds slightly different. With the Hamilton function, in component notation given by $\mathscr{H} = \vartheta + \Phi = \tfrac{1}{2}\sum_i m_i\dot{x}_i^{\,2} + (\tfrac{1}{2}\sum_{i,j}x_i\kappa_{ij}x_j - E\sum_i e_i l_i x_i)$, the equations of motion become $m_i\ddot{x}_i + \sum_j\kappa_{ij}x_j = e_i l_i E$. Diagonalizing with the transformation $x_i = \sum_j c_{ij}q_j$, we obtain $\vartheta = \tfrac{1}{2}\sum_j\dot{q}_j^{\,2}$ and $\Phi = \tfrac{1}{2}\sum_j\omega_j^{\,2}q_j^{\,2} - E\sum_j\pi_j q_j$ with $\pi_j = \tfrac{1}{2}\sum_k e_k l_k c_{kj}$ and the associated equations of motion $\ddot{q}_j + \omega_j^{\,2}q_j = E\pi_j$. The solutions ω_j are obtained from the determinantal equation $|\kappa_{ij} - m_i\omega^2\delta_{ij}| = 0$ and the moment in the direction of $\boldsymbol{E}$ is $\Sigma_j\pi_j q_j$. The polarizability $\alpha(\omega)$ as a function of frequency ω can be obtained by inserting $E = E_0\exp(-\mathrm{i}\omega t)$, solving for q_j and summing over j after multiplying with e_j. The final expression is of the form αE with $\alpha = \Sigma_j\pi_j^{\,2}/(\omega_j^{\,2} - \omega^2)$. For a single oscillator, this expression becomes $\alpha = e^2/[m(\omega_0{}^2 - \omega^2)]$, with $\omega_0{}^2 = \kappa/m$, and for the static case $\alpha = e^2/m\omega_0{}^2 = e^2/\kappa$, consistent with previous results (Section 4.1.4). Now we assume that for small vibrational amplitudes, the principal dipole moments depend linearly on the generalized coordinates, and we write $\mu_\lambda = \mu_{\lambda 0} + \Sigma_j b_{\lambda j} q_j$ or $\boldsymbol{\mu} = \boldsymbol{\mu}_0 + \boldsymbol{b}\cdot\boldsymbol{q}$. Hence, evaluating μ^2, we obtain

$$\mu^2 = \left\langle \sum\nolimits_\lambda \mu_\lambda{}^2 \right\rangle = \left\langle \sum\nolimits_\lambda \left(\mu_{\lambda 0} + \sum\nolimits_j b_{\lambda j} q_j\right)^2 \right\rangle$$

$$= \sum\nolimits_\lambda \left\langle \mu_{\lambda 0}{}^2 \right\rangle + \sum\nolimits_{\lambda,j} \langle b_{\lambda j} \mu_{\lambda 0} q_j \rangle + \sum\nolimits_{\lambda,\lambda',j,k} \langle b_{\lambda j} b_{\lambda' k} q_j q_k \rangle$$

$$= \mu_0^2 + \sum\nolimits_{\lambda,j} b_{\lambda j}{}^2 \left\langle q_j{}^2 \right\rangle /3 \quad \text{with} \quad \mu_0^2 = \sum\nolimits_\lambda \mu_{\lambda 0}{}^2$$

To calculate $\langle q^2 \rangle$, we invoke the virial theorem (Section 7.6.3), stating that for $\phi = ½\omega^2 q^2$, $\langle \phi \rangle = ½\omega^2 \langle q^2 \rangle = ½kT$ or $\langle q^2 \rangle = kT/\omega^2$. Hence, μ^2 becomes $\mu_0{}^2 + kT\Sigma_{j\lambda j} b_{\lambda j}{}^2/3\omega_j{}^2$ and

$$\alpha_{\text{eff}} = \left(\alpha_{\text{ele}} + \sum\nolimits_{\lambda,j} \frac{b_{\lambda j}{}^2}{3\omega_j{}^2}\right) + \frac{\mu_0{}^2}{3kT} \tag{8.25}$$

So, we regain the Deby–Langevin equation, but conclude that the dynamics result in a contribution to the polarizability that can be interpreted as the molecular polarizability α_{mol}. It appears that for molecules α_{mol} is usually small and Böttcher [87] estimates the contribution as 5% or $\varepsilon_\infty = 1.05n^2$. Further, the analysis shows that the dipole moment obtained is the dipole moment of the vibrational ground state, as there is no contribution of the dynamics to the dipole moment. Obviously, for really high field strength, higher-order terms in E for the expansions used contribute, but for (almost) all situations the above is applicable in view of that typically $\phi \ll kT$ (Problem 8.11).

The abovementioned model is applicable if the internal response is of the harmonic oscillator type with the coupling constants independent of the external coordinates. For other conditions, quantum-mechanical solutions must be used, which are, however, generally rather complex and for which we refer to the literature [84, 88].

The Debye equation links the effective polarizability α_{eff} and number density $\rho = N_\text{A}/V_\text{m}$ to the relative permittivity ε_r. A plot of $P_\text{m} = V_\text{m}(\varepsilon_\text{r} - 1)/(\varepsilon_\text{r} + 2)$ versus $1/T$ yields a slope $\mu^2/3k$, while the intercept provides α. For gases, the values so obtained for the dipole moment compare rather well with independently determined values, and the method can be considered as accurate. Figure 8.6a shows the graphs and Table 8.8 the early numerical $\varepsilon(P)$ data for HCl, HBr, and HI. Figure 8.6b shows similar data for the series CH_4, CH_3Cl, CH_2Cl_2, $CHCl_3$, and CCl_4, revealing clearly which molecules are polar and the magnitude of dipole moment and polarizability. Because for gases ε_r deviates only slightly from 1, $\varepsilon_\text{r} + 2$ can be approximated by 3, and P_m becomes linear in density ρ. For high density, deviations occur, see, for example, [87].

8.6.3 Frequency Dependence

So far, we have considered the effect of a static (or low-frequency) electric field. With increasing frequency ω, polar molecules will not be able to move fast enough to comply with the field, so that the dipole orientation remains random, and we are left with the molecular and electronic polarization (Figure 8.7a). In this case, we can use the Clausius–Mossotti equation if we take for ε_r the permittivity ε_∞ at a frequency where the

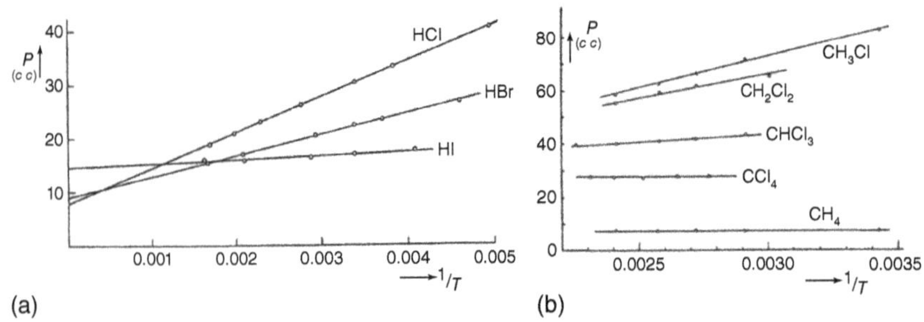

Figure 8.6 Molar polarization P of (a) HCl, HBr, and HI and (b) CH_4, CH_3Cl, CH_2Cl_2, $CHCl_3$, and CCl_4 as a function of $1/T$ as determined by Zahn (a) and Sänger (b). Source: Debye [86]/John Wiley and Sons.

Table 8.8 Dipole moment μ and polarizability volume $\alpha/(4\pi\varepsilon_0)$ for HF, HCl, HBr, and HI.

	r (pm)	μ_{lit} (D)	$\alpha_{lit}/(4\pi\varepsilon_0)$ (Å^3)	μ (D)	$\alpha/(4\pi\varepsilon_0)$ (Å^3)
HF	91.7	1.91	0.80	—	—
HCl	127.4	1.03	2.63	1.052	3.01
HBr	141.4	0.79	3.61	0.807	3.54
HI	160.9	0.38	5.45	0.388	5.58

Data for μ and α adapted from Debye [86]. Data for μ_{lit} and α_{lit} representing best literature values.

orientation polarization cannot longer follow the changing field but the molecular and electronic still can. For still higher, optical frequencies, we are left with the electronic polarization only. Such transitions show up in the ε–ω relationship.

The propagation of electromagnetic waves is described by the complex refractive index $n = n_1 + in_2 = (\varepsilon_r\mu_r)^{1/2}$ with n_1 the real part (refractive index), n_2 the imaginary part (absorption coefficient), and ε_r and μ_r the relative electric and magnetic permittivity, respectively. Since for a dielectric $\mu_r = 1$, $\varepsilon_r = n^2$ resulting in $\varepsilon_r = \varepsilon_1 + \varepsilon_2$ with $\varepsilon_1 = n_1{}^2 - n_2{}^2$ and $\varepsilon_2 = 2n_1n_2$. The transition from one regime to another not only shows up as a decrease in ε_1 above the transition frequency but also as a peak at this frequency, accompanied by a peak in ε_2 that serves to damp the transition, which otherwise would be a singularity. Using $n \equiv n_1$, we define the *molar refraction* R_m via

$$R_m \equiv V_m(n^2 - 1)/(n^2 + 2) = \alpha N_A/3\varepsilon_0 \tag{8.26}$$

conventionally addressed as the *Lorentz–Lorenz equation*. Empirically R_m appears to increase only slightly with temperature [89], typically by 10^{-4} K^{-1}, and to be largely independent on pressure and state of aggregation. In fact, an exact compensation of n by V_m cannot be achieved as V_m is temperature dependent. Böttcher [87] discussed a correction based on a fixed size for a molecular cavity (Justification 8.1). Nevertheless, the dependence on temperature is usually within the experimental error. For polar molecules at optical frequencies, we thus have $R_m = P_m$, a relation which should also hold for nonpolar liquids at all frequencies. Values for ε_1 and n for some liquids are shown in Table 8.9. Clearly for the left half of the table, the relationship $\varepsilon_1 = n^2$ is reasonably well obeyed, but for the right

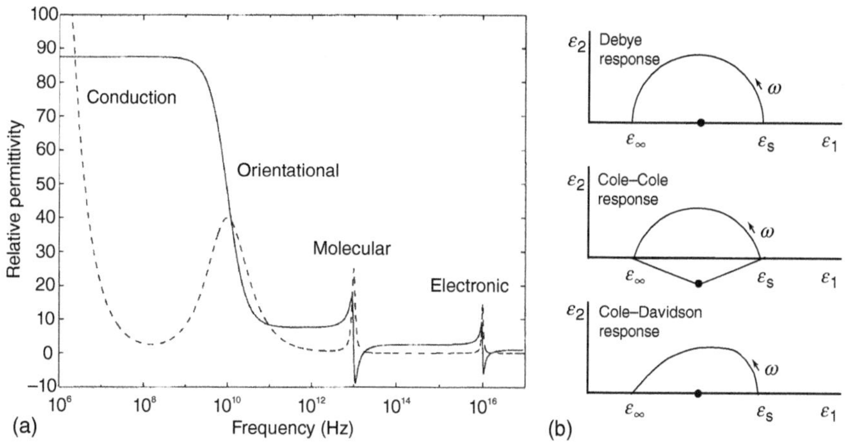

Figure 8.7 (a) The relative permittivity ε_1 (—) and ε_2 (- -) as a function of frequency ω, indicating the orientational, molecular, and electronic contribution to polarization. (b) Cole–Cole plots.

Table 8.9 Permittivity ε_1 and refractive index n_1 of several molecules.

	ε_1	n_1 (at 589 nm)		ε_1	n_1 (at 589 nm)
C_6H_6	2.284 (20 °C) 2.274 (25 °C)	1.498	H_2O (l)	80.37 (20 °C) 78.54 (25 °C)	1.333
$C_6H_5CH_3$	2.387 (20 °C) 2.379 (25 °C)	1.494	CH_3OH	33.62 (20 °C) 32.63 (25 °C)	1.326
c-C_6H_{12}	2.023 (20 °C) 2.015 (25 °C)	1.424	C_2H_5OH	25.3 (20 °C) 24.30 (25 °C)	1.359
n-C_6H_{14}	1.874 (20 °C) 1.890 (25 °C)	1.372	$C_6H_5NO_2$	35.74 (20 °C) 34.82 (25 °C)	1.550
CCl_4	2.238 (20 °C) 2.228 (25 °C)	1.468	C_6H_5Cl	5.708 (20 °C) 5.621 (25 °C)	1.523
CS_2 (l)	2.641 (20 °C)	1.629	NH_3	16.9 (25 °C)	—

Source: Weast [90]/Taylor & Francis.

half not at all. In general, this is due to cooperative effects, such as dimer formation and other structural effects, contributing to the polarization only at lower frequencies.

To describe the frequency dependence due to dipoles, we note that the polarization P_μ does not occur instantaneous (*dielectric relaxation*). Upon sudden application or removal of an external field E and starting at $P = P_0$, after sufficiently long time P tends to P_∞. We write $P(t) = P_0 + P_\mu(t)$ and assume that the rate $\mathrm{d}P_\mu/\mathrm{d}t = [P_\infty - P_0 - P_\mu(t)]/\tau$ with τ a (macroscopic) relaxation time. For sudden application of an external field E, $P_0 = P_\alpha$ and $P_\infty = P_\alpha + P_\mu(\infty)$, while for sudden removal we have $P_0 = P_\alpha + P_\mu(\infty)$ and $P_\infty = P_\alpha$. The solutions are $P = P_\alpha + P_\mu(\infty)[1 - \exp(-t/\tau)]$ and $P = P_\alpha + P_\mu(\infty)\exp(-t/\tau)$, respectively. For an alternating field, we have $E = E_0\exp(-\mathrm{i}\omega t)$, and, similarly, $P_\mu = A\exp(-\mathrm{i}\omega t)$. If we use for the static response $P = \varepsilon_0(\varepsilon_s - 1)\varepsilon_0 E$, with ε_s the static permittivity, and the high frequency response $P_\alpha = \varepsilon_0(\varepsilon_\infty - 1)E$, with $\varepsilon_\infty = n^2$ and n the refractive index, we obtain from dP_μ/dt the solution $P_\mu = [(\varepsilon_s - \varepsilon_\infty)E]/(1 - i\omega\tau)$. The permittivity becomes $\varepsilon(\omega) = \varepsilon_\infty + (\varepsilon_s - \varepsilon_\infty)/(1 - i\omega\tau)$ or

$$\varepsilon_1 = \varepsilon_\infty + (\varepsilon_s - \varepsilon_\infty)/(1 + \omega^2\tau^2) \quad \text{and} \quad \varepsilon_2 = (\varepsilon_s - \varepsilon_\infty)\omega\tau/(1 + \omega^2\tau^2)$$

where ε_1 and ε_2 are the real and imaginary parts of the complex relative permittivity $\varepsilon_r = \varepsilon_1 + i\varepsilon_2$ and known as the *Debye–(Drude) equations*, consistent with the oscillator model as discussed before. The imaginary part ε_2 shows a maximum $\varepsilon_2{}^{max} = ½(\varepsilon_s - \varepsilon_\infty)$ at $\omega\tau = 1$, where $\varepsilon_1 = ½(\varepsilon_s + \varepsilon_\infty)$. The *loss factor* δ is defined as $\tan\delta = \varepsilon_2/\varepsilon_1$ and shows a maximum $(\varepsilon_s - \varepsilon_\infty)/(\varepsilon_s\varepsilon_\infty)^{1/2}$ at $\omega\tau = (\varepsilon_s/\varepsilon_\infty)^{1/2}$. The Debye equation can be represented in the complex $\varepsilon_2 - \varepsilon_1$ plane as a semicircle from $(\varepsilon_1,\varepsilon_2) = (\varepsilon_s,0)$ at low ω to $(\varepsilon_1,\varepsilon_2) = (\varepsilon_\infty,0)$ at high ω, a so-called Cole–Cole plot (Figure 8.7b).

If the data neither conform to a single Debye equation, nor to a sum of two or three Debye terms, the *Cole–Cole equation* $\varepsilon(\omega) = \varepsilon_\infty + (\varepsilon_s - \varepsilon_\infty)/[1 - (i\omega\tau)^{1-a}]$ with $0 \le a \le 1$ can be used, which is represented in the ε_2–ε_1 plane also as a semicircle. However, its center is depressed downward from $[\varepsilon_s + \varepsilon_\infty/2,0]$ to $[\varepsilon_s + \varepsilon_\infty/2,-(\varepsilon_s - \varepsilon_\infty)\tan\pi a/2]$. It represents a symmetric relaxation time distribution with width a instead of a single relaxation time τ. Similarly, the *Cole–Davidson equation* $\varepsilon(\omega) = \varepsilon_\infty + (\varepsilon_s - \varepsilon_\infty)/(1 - i\omega\tau)^b$ with $0 \le b \le 1$ can be used, which represents a skew relaxation time distribution. Both equations are specializations of the *Havriliak–Negami equation* $\varepsilon(\omega) = \varepsilon_\infty + (\varepsilon_s - \varepsilon_\infty)/[1 - (i\omega\tau)^{1-a}]^b$. One might also use the *stretched exponential* or *Kohlrausch–Williams–Watts equation* $\phi = \exp[(at)^\alpha] = \int_0^\infty \exp(-t/\tau)g(\tau)d\tau$ with relaxation time distribution $g(\tau)$. For $\alpha \le 2/3$, $g(\tau)$ is monotonically decaying, while for $\alpha > 2/3$ a unimodal distribution arises. The physical significance of these relaxation time distributions is difficult to understand. For an attempt, see [91].

In general, the transition in ε_1 corresponds to a transition between energy levels and the peak in ε_2 to damping. The microscopic interpretation of τ depends on the type of transition. For the orientation relaxation, τ depends on the rotational inertia of the molecules, while for the molecular and electronic relaxation τ is concerned with translational inertia of atoms and electrons. For detailed reviews, see [92, 93].

Justification 8.1 ***The Local, Reaction, and Directional Fields*** To obtain the local, reaction, and directional fields, we first quote some results from electrostatics and thereafter describe the Clausius–Mossotti, Debye, and Onsager equations. Throughout we use $\mathscr{C} = (4\pi\varepsilon_0)^{-1}$. For further details, we refer to [87, 94].

We start with the *cavity field* $\boldsymbol{E}_{cav}$ (Appendix D) inside a spherical cavity with radius a embedded in a dielectric with relative permittivity ε_r (where here $\varepsilon_r \equiv \varepsilon_1$ as loss is ignored). For an applied external field $\boldsymbol{E}$, the cavity field is given by

$$\boldsymbol{E}_{cav} = 3\varepsilon_r\boldsymbol{E}/(2\varepsilon_r + 1)$$

If a nonpolarizable spherical molecule with dipole moment $\boldsymbol{\mu}$ is put in the center of such a cavity without any applied field $\boldsymbol{E}$, the molecule polarizes the dielectric and this polarization on its turn creates a *reaction field* $\boldsymbol{R}$ on the molecule. The reaction field $\boldsymbol{R}$ appears to be uniform, proportional to the dipole moment $\boldsymbol{\mu}$, and given by $\boldsymbol{R} = f\boldsymbol{\mu}$ with $f = 2\mathscr{C}(\varepsilon_r - 1)/(2\varepsilon_r + 1)a^3$. If the molecule is polarizable with polarizability α, the reaction field is due to the permanent dipole $\boldsymbol{\mu}$ and induced dipole $\alpha\boldsymbol{R}$, so that we have $\boldsymbol{R} = (\boldsymbol{\mu} + \alpha\boldsymbol{R})f$. Solving for $\boldsymbol{R}$, we obtain

$$\boldsymbol{R} = f\boldsymbol{\mu}/(1 - f\alpha)$$

Now, recall that in case a field $\boldsymbol{E}$ is applied, the dielectric displacement $\boldsymbol{D} = \varepsilon_0\varepsilon_r\boldsymbol{E} = \varepsilon_0\boldsymbol{E} + \boldsymbol{P}$ with polarization $\boldsymbol{P} = \varepsilon_0(\varepsilon_r - 1)\boldsymbol{E}$. The total polarization is thought to be the sum of $\boldsymbol{P}_\mu$ due to the dipole $\boldsymbol{\mu}$ of the molecules and $\boldsymbol{P}_\alpha$ due to the polarizability α of the molecules. We assume that $\boldsymbol{P}_\alpha$ equals the product $\rho\alpha\boldsymbol{E}_{\rm loc}$ with $\boldsymbol{E}_{\rm loc} = \boldsymbol{E}_{\rm cav} + \boldsymbol{R}$ the *local field* and $\rho = N/V$ the number density of the polarizable molecules. For $\boldsymbol{P}_\mu$ we have $\rho\langle\mu\rangle$ with $\langle\mu\rangle$ the rotational average of $\boldsymbol{\mu}$. Onsager argued that for $\boldsymbol{P}_\mu$ only the *directional field* $\boldsymbol{E}_{\rm dir}$, defined as $\boldsymbol{E}_{\rm dir} = \boldsymbol{E}_{\rm loc} - \langle\boldsymbol{R}\rangle$ with $\langle\boldsymbol{R}\rangle$ the rotational average of $\boldsymbol{R}$, is active. So, generally, using the Langevin estimate for $\langle\mu\rangle$,

$$\varepsilon_0(\varepsilon_r - 1)\boldsymbol{E} = \boldsymbol{P}_\alpha + \boldsymbol{P}_\mu = \rho\alpha\boldsymbol{E}_{\rm loc} + \rho\langle\boldsymbol{\mu}\rangle = \rho\alpha\boldsymbol{E}_{\rm loc} + \rho(\mu^2/3kT)\boldsymbol{E}_{\rm dir} \tag{8.27}$$

Let us focus first on nonpolar but polarizable molecules. The conventional approach to estimate the local field is due to Lorentz, who assumed that the reference molecule is surrounded by an imaginary sphere of such a size that the material outside the sphere can be treated as a continuum. If the reference molecule is removed, the total field in the sphere will be due to the applied field $\boldsymbol{E}$, the field $\boldsymbol{E}_{\rm sphere}$ due to free ends of the surrounding dipoles covering the sphere, and the field $\boldsymbol{E}_{\rm mol}$ due to molecules in the near vicinity of the reference molecule. The latter field averages to zero for sufficiently high symmetry [95]. The component of $\boldsymbol{E}_{\rm sphere}$ in the direction of $\boldsymbol{E}$ is calculated by integrating the field of infinitesimal rings perpendicular to the field direction. The apparent surface charge density on these rings is $P\cos\theta$, while their surface is $2\pi r^2\sin\theta{\rm d}\theta$, leading to a total charge on each ring of ${\rm d}q = 2\pi r^2P\sin\theta\cos\theta{\rm d}\theta$. According to Coulomb's law, each charge element contributes to the field by ${\rm d}E = \mathscr{C}r^{-2}\cos\theta{\rm d}q$. Combining yields $E_{\rm sphere} = \mathscr{C}P\int 2\pi\sin\theta\cos^2\theta\,{\rm d}\theta = P/3\varepsilon_0 = (\varepsilon_r - 1)E/3$. Due to symmetry, the other contributions vanish so that $\boldsymbol{E}_{\rm loc}$ becomes $\boldsymbol{E}_{\rm loc} = \boldsymbol{E} + \boldsymbol{P}/3\varepsilon_0 = \boldsymbol{E}(\varepsilon_r + 2)/3$. Introducing $\boldsymbol{E}_{\rm loc}$ in $\boldsymbol{P}_\alpha = \rho\alpha\boldsymbol{E}_{\rm loc} = \varepsilon_0(\varepsilon_r - 1)\boldsymbol{E}$ yields

$$\boldsymbol{P}_\alpha = \frac{3(\varepsilon_r - 1)}{(\varepsilon_r + 2)} \quad \text{or} \quad \frac{\varepsilon_r - 1}{\varepsilon_r + 2} = \frac{\rho\alpha}{3\varepsilon_0} \tag{8.28}$$

known as the *Clausius–Mossotti equation*. An alternative derivation, due to Onsager, uses the cavity field $\boldsymbol{E}_{\rm cav} = 3\varepsilon_r\boldsymbol{E}/(2\varepsilon_r + 1)$ and the reaction field $\boldsymbol{R}$. The latter is in this case $\boldsymbol{R} = (\alpha\boldsymbol{E}_{\rm cav} + \alpha\boldsymbol{R})f$. Solving for $\boldsymbol{R}$ leads to

$$\boldsymbol{R} = f\alpha\boldsymbol{E}_{\rm cav}/(1 - f\alpha)$$

Because $\boldsymbol{E}_{\rm loc} = \boldsymbol{E}_{\rm cav} + \boldsymbol{R}$, the result is

$$\boldsymbol{E}_{\rm loc} = \boldsymbol{E}_{\rm cav} + \frac{f\alpha}{1 - f\alpha}\boldsymbol{E}_{\rm cav} = \frac{1}{1 - f\alpha}\boldsymbol{E}_{\rm cav} = \frac{1}{1 - f\alpha}\frac{3\varepsilon_r}{2\varepsilon_r + 1}\boldsymbol{E}$$

Onsager further assumed $4\pi\rho a^3/3 = 1$. Putting this together with the expression for f, one obtains from Eq. (8.27) with $\boldsymbol{P}_\mu = 0$ (after some manipulation) again Eq. (8.28). This alternative derivation makes Onsager's (somewhat odd) assumption on the cavity size clear. Böttcher [87] assumed, more naturally, a fixed radius a and obtained as a correction factor h for P given by $h = 9\varepsilon_r/(\varepsilon_r + 2)[2\varepsilon_r + 1 - 2u(\varepsilon_r - 1)]$ with $u = \alpha'/a^3$. For CCl_4 $\alpha' = 10.8\,\text{Å}^3$ and $a = 3.05$ Å was obtained, in reasonable agreement with $\alpha' = 10.5\,\text{Å}^3$ as independently determined and $a = 2.85$ Å as estimated from $V_{\rm m}$.

We now turn to polar polarizable molecules. Debye used the Lorentz local field $\boldsymbol{E}_\mathrm{L} = \boldsymbol{E}(\varepsilon_\mathrm{r} + 2)/3$ for both $\boldsymbol{P}_\mu$ and $\boldsymbol{P}_\alpha$ in Eq. (8.27) and this leads to the *Debye equation*

$$\frac{\varepsilon_\mathrm{r}-1}{\varepsilon_\mathrm{r}+2} = \frac{\rho}{3\varepsilon_0}\left(\alpha + \frac{\mu^2}{3kT}\right) \tag{8.29}$$

Onsager [96] used $\boldsymbol{E}_\mathrm{dir}$ and $\langle \boldsymbol{R} \rangle$ to calculate $\boldsymbol{E}_\mathrm{loc}$ and obtained

$$\langle \boldsymbol{R} \rangle = \frac{f}{1-f\alpha}\langle \boldsymbol{\mu} \rangle = \frac{f}{1-f\alpha}\frac{\mu^2}{3kT}\boldsymbol{E}_\mathrm{dir} \quad \text{and} \quad \boldsymbol{E}_\mathrm{dir} = \boldsymbol{E}_\mathrm{cav} + f\alpha\boldsymbol{E}_\mathrm{dir} \quad \text{or}$$

$$\boldsymbol{E}_\mathrm{dir} = \frac{1}{1-f\alpha}\boldsymbol{E}_\mathrm{cav} = \frac{1}{1-f\alpha}\frac{3\varepsilon_\mathrm{r}}{2\varepsilon_\mathrm{r}+1}\boldsymbol{E} \tag{8.30}$$

The local field $\boldsymbol{E}_\mathrm{loc}$ is obtained from $\boldsymbol{E}_\mathrm{loc} = \boldsymbol{E}_\mathrm{dir} + \langle \boldsymbol{R} \rangle$ resulting in

$$\boldsymbol{E}_\mathrm{loc} = \boldsymbol{E}_\mathrm{dir} + \frac{f}{1-f\alpha}\frac{\mu^2}{3kT}\boldsymbol{E}_\mathrm{dir} = \left(1 + \frac{1}{1-f\alpha}\frac{\mu^2}{3kT}\right)\frac{1}{1-f\alpha}\frac{3\varepsilon_\mathrm{r}}{2\varepsilon_\mathrm{r}+1}\boldsymbol{E}$$

Finally, preserving as far as possible resemblance with Debye's equation, substituting $\boldsymbol{E}_\mathrm{dir}$ and $\boldsymbol{E}_\mathrm{loc}$ in Eq. (8.27) leads after considerable manipulation to

$$\frac{(\varepsilon_\mathrm{r}-1)(2\varepsilon_\mathrm{r}+1)}{3\varepsilon_\mathrm{r}} = \frac{\rho}{\varepsilon_0}\left[\alpha^* + \frac{(\mu^*)^2}{3kT}\right] \quad \text{with} \quad \frac{\mu^*}{\mu} = \frac{\alpha^*}{\alpha} = \frac{1}{1-f\alpha} \tag{8.31}$$

Further substitution of the expression for f, the condition $4\pi\rho a^3/3 = 1$ and the estimate $\mathscr{C}\,\alpha/a^3 = (\varepsilon_\infty - 1)/(\varepsilon_\infty^2 + 2)$ and explicitly solving for μ leads to the *Onsager equation*

$$\mu^2 = \frac{9\varepsilon_0 kTV_\mathrm{m}}{N_\mathrm{A}}\frac{(\varepsilon_\mathrm{r}-\varepsilon_\infty)(2\varepsilon_\mathrm{r}+\varepsilon_\infty)}{\varepsilon_\mathrm{r}(\varepsilon_\infty+2)^2} \tag{8.32}$$

Further refinements include the use of an ellipsoidal instead of a spherical cavity, the use of an ellipsoidal rather than a spherical molecule, and the use of a fixed cavity size for the Clausius–Mossotti equation [87]. These refinements generally all improve the agreement of the value for μ with results of independent experiments.

8.6.4 Estimating μ and α

One way to describe the dipole moment μ of a molecule is by group contributions for the individual bonds, the so-called *bond moments*. As an example, consider water, the total dipole moment of which is 1.85 D, due to the vector sum of two O—H bond moments. From the value for μ and the geometry of the molecule, one can derive that the bond moment of an O—H bond is 1.52 D. Using this process with a C—H bond moment of 0.4 D using toluene as reference [97], the bond moments of various bonds can be deduced (Table 8.10). Be aware that different authors use different values as reference; for example, Petro obtained C(sp^3)–H = 0.31 D, C(sp^2)–H = 0.63 D, and C(sp)–H = 1.05 D from infrared measurements [99]. The estimate for μ is not very accurate, the problem being that bond contributions are considered as independent, whereas a molecule is one entity of which the properties should be calculated with quantum methods. A slightly more reliable method uses groups of atoms in molecules. For example, in the case of nitro groups in aromatic compounds the dipole moment of the complete nitro group is used. For details, see Minkin et al. [97].

The molar refraction can also be estimated as a sum of contributions. Initially these contributions were taken as atomic contributions, but Denbigh [100] and Vogel et al. [101]

Table 8.10 Dipole moments for several chemical bonds using $C^{-}H^{+} \equiv 0.4$ D as reference.

Bond	μ (D)	Bond	μ (D)	Bond	μ (D)
C—H	0.4 (0.4)	C—S	0.9 (0.9)	H—O	1.51 (1.51)
C—F	1.39 (1.41)	C—Se	0.7 (0.8)	H—N	1.31 (1.31)
C—Cl	1.47 (1.46)	C=N	1.4	H—S	0.7 (0.68)
C—Br	1.42 (1.38)	C=O	2.4 (2.4)	Si—C	1.2
C—I	1.25 (1.19)	C=S	2.0	Si—H	1.0
C—N	0.45 (0.22)	C≡N	3.1 (3.6)	Si—N	1.55
C—O[a)]	0.7 (0.74)	N—O	(0.3)		

Data calculated from methyl derivatives (CH_3X) dissolved in benzene at 25 °C.
a) Ethers and alcohols. Adapted from Minkin et al. [97], data in parentheses from Smyth [98].

Table 8.11 Bond contributions to the molar refraction R_m (cm^3 mol^{-1}).

Bond	R_m (D)	R_m (V)	R_∞ (L)	Bond	R_m (D)	R_m (V)	R_∞ (L)
C—H	1.69	1.676	1.644	C—C[a)]	2.73	—	—
N—H	1.81	1.76	1.76	C=N	—	3.76	3.509
O—H	1.73	1.66[b)]	1.628	C=O	3.38	3.32[c)]	3.242[d)]
S—H	—	4.80	4.654	C≡C	6.4	5.87[e)]	5.670[f)]
C—C	1.25	1.296	1.254	C—F	1.72	—	—
C—N	1.54	1.57	1.490	C—Cl	6.53	6.51	6.361
C—O	1.51	1.54[g)]	1.490[h)]	C—Br	9.37	9.39	9.064
C—S	—	4.61	4.419	C—I	14.55	14.61	13.92
C=C	4.16	4.17	3.937				

D = Denbigh, V = Vogel, L = Le Fèvre.
a) Benzene derivatives and naphthalene derivatives 2.78.
b) Alcohols and acids 1.80.
c) Me-ketones 3.49.
d) Me-ketones 3.383.
e) Terminal and nonterminal 6.24.
f) Terminal.
g) Acetals 1.46.
h) Ethers and acetals 1.429.

showed that using bond contributions leads to a more consistent scheme. They used a set of the experimental data to determine bond refractions; the values so obtained are shown in Table 8.11. Denbigh used the experimental n_D-data as obtained at 589 nm (Na D line), leading to an average error between the experimental and additive molar refractions for 50 liquids of 0.7% and seldom larger than 1.5%. For 196 normal and branched alkanes, an average deviation of 0.43% was obtained by using R_m(C—H) = 1.674 cm^3 mol^{-1} and R_m(C—C) = 1.296 cm^3 mol^{-1}, respectively. This deviation can be reduced to 0.20% by applying a few branching correction rules. Vogel et al. analyzed a wide range of molecules and

Table 8.12 Experimental and calculated values of R_m ($cm^3\ mol^{-1}$) for several molecules.

Molecule	R_m (exp)	R_m (calc)	Molecule	R_m (exp)	R_m (calc)
CH_3OH	8.04	8.05	$C_6H_{13}OH$	30.76	30.72
C_2H_5OH	12.62	12.59	C_6H_6	25.05	25.14
C_3H_7OH	17.16	17.13	$C_6H_5CH_3$	29.82	29.67
C_4H_9OH	21.67	21.68	C_6H_5Cl	29.87	29.86
$C_5H_{11}OH$	26.26	26.22	C_6H_5Br	32.54	32.5

R_m (exp) denotes the value for the Le Fèvre fit to the Vogel data using Eq. (8.33).
Source: Adapted from Le Fèvre [89].

several frequencies yielding average bond refractions, for which Le Fèvre [89] estimated the accuracy as 1% or better. Thus, both schemes yield similar values. For example, for chloroform $CHCl_3$ ($R_m = 21.25\ cm^3\ mol^{-1}$), the Denbigh and Vogel data lead to 21.28 and $21.21\ cm^3\ mol^{-1}$, respectively. In the Lorenz-Lorentz expression the value n_∞, the value at infinite frequency, should be used. Sugden [102] estimated $R_\infty = 1.05R_D$, but a more consistent estimate is obtained from the wavelength dependence of R_m, described by classical dispersion theory as

$$R_\lambda = R_\infty + A/\left(\lambda^2 - \lambda_A{}^2\right) + B/\left(\lambda^2 - \lambda_B{}^2\right) + \cdots \quad \text{or} \quad R_\lambda \cong R_\infty + A/\left(\lambda^2 - \lambda_A{}^2\right) \tag{8.33}$$

with A, B, …, λ_A, λ_B, … constants. The approximation (A,λ_A) typically shows deviations in the third decimal place, so that this approximation is adequate for calculating α and μ. From R_m data at four wavelengths obtained by Vogel [101], Le Fèvre [89] determined bond R_∞ values (Table 8.11). In the absence of data for R_∞, these bond refractions lead to agreement with experiment with an average error of $0.05\ cm^3\ mol^{-1}$ (Table 8.12), only slightly larger than the experimental accuracy $\approx 0.001\ cm^3\ mol^{-1}$. Example 8.2 provides the details for estimating α and μ for methanol. Finally, we note that, since the polarization volume $\alpha' = \alpha/4\pi\varepsilon_0$ equals $(\sigma_{pol}/2)^3$ for a conducting sphere with radius $\sigma_{pol}/2$ (footnote 4, Chapter 4), α' should roughly parallel the molecular volume. Estimating the vdW radius as $\sigma_{vdW}/2 = (3b/16\pi N_A)^{1/2}$, with b the vdW constant, the correlation $\sigma_{vdW} \cong 0.84\sigma_{pol}$ can be obtained for typical small molecules.

Example 8.2 ***Methanol*** The dipole moment μ of methanol can be estimated using the contributions of the CH_3 group (equivalent to a C—H bond), the C—O bond and the O—H bond. These contributions are, respectively, 0.4, 0.7, and −1.51 D. Using $\theta = 109°$ as the C—O—H bond angle, and assuming a fixed orientation of the O—H group with respect to the molecular axis, the cosine rule $\mu^2 = (\mu_{CH} + \mu_{CO})^2 + \mu_{OH}{}^2 + 2(\mu_{CH} + \mu_{CO})\mu_{OH}\cos\theta$ provides $\mu_{OH} = 1.55$ D, to be compared with the experimental value $\mu = 1.69$ D. The agreement is good, but dependent on the assumption of fixed orientation. For free internal rotation of the O—H group, we obtain $\mu = \mu_{CH} + \mu_{CO} + \mu_{OH}\cos(180° - \theta) = 0.61$ D, deviating considerably from the experimental value. Actually, the O—H group exhibits restricted internal rotation with a rotation barrier of $372\ cm^{-1} = 1.80kT$ at room temperature. Although exact expressions for internal rotation are available (Section 7.5.2), these are cumbersome, and we assume a simple two-state model – a high-energy state with energy $1.80kT$ with respect to a low-energy state with energy $0kT$ – occupied with weights according to their Boltzmann factor. This leads to $e^0/(e^0+e^{1.8}) = 0.14$ and $e^{1.8}/(e^0+e^{1.8}) = 0.86$, so that $\mu = 0.86\times1.55+0.14\times0.61 = 1.42$ D, in reasonable agreement with the experiment. The polarizability α can be estimated by summing the contributions for the three C—H bonds, the C—O bond and the O—H bond to the molar refraction R_m. Using the Vogel scheme leads to $R_m = 8.23\,\text{cm}^3\,\text{mol}^{-1}$. From $R_m = N_A\alpha/3\varepsilon_0$, we have $\alpha = 3.63\times10^{-40}\,\text{C}^2\,\text{m}^2\,\text{J}^{-1}$ or $\alpha' = \alpha/4\pi\varepsilon_0 = 3.26\,\text{Å}^3$, in excellent agreement with the experimental value.

Problem 8.11 ***The Langevin Equation*** Derive Eq. (8.23). Verify that $L(a)\cong a/3$ for $a\ll1$. Show that this approximation is nearly always satisfied. Typical values are $\mu = 1$ D, $E = 10\,\text{kV}\,\text{cm}^{-1}$, and $T = 300$ K.

References

1 Moelwyn-Hughes, E.A. (1961). *Physical Chemistry*, 2e. Oxford: Pergamon Press.
2 Hirschfelder, Curtiss and Bird, 1964.
3 Chapman, S. and Cowling, T.G. (1970). *The Mathematical Theory of Non-uniform Gases*, 3e. Cambridge: Cambridge University Press.
4 Jeans, J.H. (1925). *The Dynamical Theory of Gases*, 4e. Cambridge: Cambridge University Press; see also (1940). *An Introduction to the Kinetic Theory of Gases*. Cambridge: Cambridge University Press.
5 Kennard, E.H. (1938). *Kinetic Theory of Gases*. New York: McGraw-Hill Book Comp.
6 Present, R.D. (1958). *Kinetic Theory of Gases*. New York: McGraw-Hill Book Comp.
7 Mayer, J.E. and Mayer, M.G. (1940). *Statistical Mechanics*. New York: Wiley.
8 Lucas, K. (2007). *Molecular Models of Fluids*. Cambridge: Cambridge University Press.
9 Swendsen, R.H. (2012). *Introduction to Statistical Mechanics and Thermodynamics*. Oxford: Oxford University Press.
10 Gill, E.K. and Morrison, J.A. (1966). *J. Chem. Phys.* 45: 1585.
11 Kozliak, E. and Lambert, F.L. (2008). *Entropy* 10: 274.
12 Ono, S. (1951). *J. Chem. Phys.* 19: 504.
13 Yang, C.N. and Lee, T.D. (1952). *Phys. Rev.* 87: 404 and 410.
14 Friedman, H.L. (1985). *A Course in Statistical Mechanics*. London: Prentice-Hall.
15 Hansen, J.-P. and McDonald, I.R. (2006). *Theory of Simple Liquids*, 3e. London: Academic Press.
16 Clisby, N. and McCoy, B.M. (2006). *J. Stat. Phys.* 122: 15.
17 Bannerman, M.N., Lue, L., and Woodcock, L.V. (2010). *J. Chem. Phys.* 132: 084507.
18 Hoover, W.G. and Ree, F.H. (1968). *J. Chem. Phys.* 49: 3609.
19 Tian, J., Jiang, H., and Mulero, A. (2019). *Phys. Chem. Chem. Phys.* 21: 13070.
20 Carnahan, N.F. and Starling, K.E. (1969). *J. Chem. Phys.* 51: 635.
21 Pieprzyk, S., Bannerman, M.N., Branka, A.C. et al. (2019). *Phys. Chem. Chem. Phys.* 21: 6886.
22 Sun, J.-X., Cai, L.-C., Wu, Q. et al. (2013). *Phys. Scr.* 88: 035005.
23 Ralston, A. (1965). *Introduction to Numerical Analysis*. New York: McGraw-Hill Book Comp.

24 Ree, F.H. and Hoover, W.G. (1964). *J. Chem. Phys.* 40: 939.

25 Mulero, A., Galán, C., and Cuadros, F. (2001). *Phys. Chem. Chem. Phys.* 3: 4991.

26 Wu, G.-W. and Sadus, R.J. (2005). *AIChE J.* 51: 309.

27 McQuarrie, D.A. (1973). *Statistical Mechanics.* New York: Harper and Row.

28 Solana, J.R. (2013). *Perturbation Theories for the Thermodynamic Properties of Fluids and Solids.* Boca Roton, FL: CRC Press.

29 Mulero, A. (2008). *Theory and Simulation of Hard-Sphere Fluids and Related Systems.* Berlin: Springer Verlag.

30 Putnam, W.E. and Kilpatrick, J.E. (1953). *J. Chem. Phys.* 21: 951.

31 Guggenheim, E.A. (1967). *Thermodynamics*, 5e. Amsterdam: North-Holland.

32 van der Waals, J.D. (1873). *Over de Continuïteit van den Gas- en Vloeistoftoestand* (On the Continuity of the Gaseous and Liquid State). PhD thesis. Leiden University.

33 Clausius, R.J.E. (1857). *Ann. Phys.* 100: 353; Hirn, G.A. (1863). *Moigno Cosmos* 22: 283.

34 Brush, S.G. (1961). *Am. J. Phys.* 29: 593.

35 Rowlinson, J.S. (1973). *Nature* 244: 415.

36 Kontogeorgis, G., Privat, R., and Jaubert, J.-N. (2019). *J. Chem. Eng. Data* 64: 4619.

37 Partington, J.R. (1949). *An Advanced Treatise on Physical Chemistry*, Vol. 1: *Fundamental Principles – The Properties of Gases.* London: Longmans.

38 Mathias, P.M. (2007). *Ind. Eng. Chem. Res.* 46: 6376.

39 Saturno, A.F. (1962). *J. Chem. Ed.* 39: 464.

40 Bevan Ott, J., Rex Goates, J.R., and Hall, H.T. (1971). *J. Chem. Ed.* 48: 515.

41 Redlich, O. and Kwong, J.N.S. (1949). *Chem. Rev.* 44: 233.

42 Kaye, G.W.C. and Laby, T.H. (1911). *Tables of Physical and Chemical Constants.* https://web.archive.org/web/20190506031327 and http://www.kayelaby.npl.co.uk/ (accessed 30 April 2021).

43 Carnahan, N.F. and Starling, K.E. (1972). *Am. Inst. Chem. Eng. J.* 18: 1184.

44 Román, F.L., Mulero, A., and Cuadros, F. (2004). *Phys. Chem. Chem. Phys.* 6: 5402.

45 Soave, G. (1972). *Chem. Eng. Sci.* 27: 1197.

46 Peng, D.-Y. and Robinson, D.B. (1967). *Ind. Eng. Chem. Fundam.* 15: 59.

47 De Santis, R., Gironi, F., and Narrelli, L. (1976). *Ind. Eng. Chem. Fundam.* 15: 183.

48 Beattie, J.A. and Bridgeman, O.C. (1928). *Proc. Am. Acad. Arts Sci.* 63: 229.

49 Benedict, M., Webb, G.B., and Rubin, L.C. (1940). *J. Chem. Phys.* 8: 334; (1942). *Ibid.* 10: 747.

50 Valderrama, J.O. (2003). *Ind. Eng. Chem. Res.* 42: 1603.

51 Abbott, M.M. (1973). *AIChE J.* 19: 596; (1979). *Cubic Equations of State: An Interpretative Review*, p. 47 in *Equations of State in Engineering and Research* (ed. K.C. Chao and R.C. Robinson Jr.), *Adv. Chem.*, Vol. 182, *Am. Chem. Soc.*, Washington DC.

52 Guggenheim, E.A. and McGlashan, M.L. (1951). *Proc. R. Soc. A* 206: 448.

53 Hamann, S.D., Lambert, J.A., and Thomas, R.B. (1955). *Aust. J. Chem.* 8: 149.

54 McGlashan, M.L. (1979). *Chemical Thermodynamics.* London: Academic Press.

55 Sengers, Kayser, Peters et al., 2000.

56 Woodcock, L.V. (2018). *Entropy* 20: 22.

57 Wei, Y.S. and Sadus, R.J. (2000). *Am. Inst. Chem. Eng. J.* 46: 169.

58 Smith, J.M. and van Ness, H.C. (1987). *Introduction to Chemical Engineering Thermodynamics*, 4e. New York: McGraw-Hill Book Comp.

59 Pitzer, K.S. (1939). *J. Chem. Phys.* 7: 583.

60 Guggenheim, E.A. (1945). *J. Chem. Phys.* 13: 253.

61 Scott, R.L. (1971). *Introduction*, Ch. 1 in *Physical Chemistry*, Vol. VIIIA: *Liquid State* (ed. D. Henderson), p. 1. New York: Academic Press.

62 Sen, G.-J. (1946). *Ind. Eng. Chem.* 38: 803.

63 Paynter, H.M. (1985). *J. Dyn. Syst. Meas. Control* 107: 233.

64 von Jüptner, H. (1913). *Z. Phys. Chem.* 85: 1.

65 Young, S. (1918). *Stoichiometry*. London: Longmans.

66 van Meurs, N. (1995). *Fluid Phase Equilib.* 106: 47.

67 Riedel, L. (1954). *Chem.-Ing. Tech. Z.* 26: 83, 259, 679; (1955). 27: 209 and 475; (1956). 28: 557.

68 Pitzer, K.S., Lippmann, D.Z., Curl, R.F. Jr., et al. (1955). *J. Am. Chem. Soc.* 77: 2427 and 3433; (1957). 79: 2369; and (1958) *Ind. Eng. Chem.* 50: 265.

69 Poling, Prausnitz and O'Connell, 2001.

70 Schreiber, D.R. and Pitzer, K.S. (1989). *Fluid Phase Equilib.* 46: 113.

71 Abbott, M.M. (1987). Personal communication on p. 92 in Smith and van Ness.

72 Lee, B.I. and Kessler, M.G. (1975). *AIChE J.* 21: 510.

73 Leland, T.W. and Chappelear, P.S. (1968). *Ind. Eng. Chem.* 60: 15.

74 Xiang, H.W. (2005). *The Corresponding-States Principle and Its Practice*. Amsterdam: Elsevier.

75 Arnaut, L., Formosinho, S., and Burrows, H. (2007). *Chemical Kinetics*. Amsterdam: Elsevier.

76 Connors, K.A. (1990). *Chemical Kinetics*. Weinheim: VCH.

77 Laidler, K.J. (1987). *Chemical Kinetics*, 3e. London: Harper and Row.

78 Buncel, E., Stairs, R., and Wilson, H. (2003). *The Role of Solvents in Chemical Reactions*. Oxford: Oxford University Press.

79 Moelwyn-Hughes, E.A. (1971). *The Chemical Statics and Kinetics of Solutions*. London: Academic Press.

80 Van Speybroeck, V., De Wispelaere, K., Van der Mynsbrugge, J. et al. (2014). *Chem. Soc. Rev.* 43: 7326.

81 Kreevoy, M.M. and Truhlar, D.G. (1986). *Transition State Theory*, Ch. 1 in *Investigation of Rates and Mechanisms of Reactions*, 4e (ed. C.F. Bernasconi), Vol. 1. New York: John Wiley & Sons.

82 Steinfeld, J.I., Francisco, J.S., and Hase, W.L. (1998). *Chemical Kinetics and Dynamics*, 2e. Upper Saddle River, NJ: Prentice-Hall.

83 Stratton, J.A. (1941). *Electromagnetic Theory*. New York: McGraw-Hill Book Comp.

84 van Vleck, 1932, reprint 1966.

85 Goldstein, H., Poole, C., and Safko, J. (2014). *Classical Mechanics*, 3e. San Francisco: Addison-Wesley.

86 Debye, P. (1929). *Polar Molecules*. New York: Chemical Catalogue Company.

87 Böttcher, C.J.F. (1973). *Theory of Electric Polarization*, 2e, Vol. 1. Amsterdam: Elsevier.

88 Fuller Brown, W. (1956). *Dielectrics*, Ch. 1 in *Handbuch der Physik*, Vol. XVII (ed. S. Flügge), p. 1. Berlin: Springer.

89 Le Fèvre, R.W.J. (1965). *Molecular Refractivity and Polarizability*, Ch. 1 in *Advances in Physical Organic Chemistry*, Vol. 3 (ed. V. Gold), p. 1. London: Academic Press.

90 Weast, R.C. (ed.) (1979). *Handbook of Chemistry and Physics*, 60e. Boca Raton, FL: CRC Press.

91 Iglesias, T.P., Vilão, G., and Reis, J.C.R. (2017). *J. Appl. Phys.* 122: 074102.
92 Kneser, H.O. (1965). *Relaxation Processes in Gases*, Ch. 3 in *Physical Acoustics*, Vol. IIA (ed. W.P. Mason), p. 133. New York: Academic Press.
93 Chernyi, G.G., Losev, S.A., Macheret, S.O. et al. (2004). *Physical and Chemical Processes in Gas Dynamics: Physical and Chemical Kinetics and Thermodynamics*, Vol. II. Reston, VA: American Institute Aeronautics Astronautics.
94 Fröhlich, H. (1958). *Theory of Dielectrics*, 2e. Oxford: Oxford University Press.
95 Aspnes, D.E. (1982). *Am. J. Phys.* 50: 704.
96 Onsager, L. (1936). *J. Am. Chem. Soc.* 58: 1486.
97 Minkin, V.I., Osipov, O.A., and Zhadanov, Y.A. (1970). *Dipole Moments in Organic Chemistry*. New York: Plenum Press.
98 Smyth, C.P. (1955). *Dielectric Behavior and Structure*. New York: McGraw-Hill Book Comp.
99 Petro, A.J. (1958). *J. Am. Chem. Soc.* 80: 4230.
100 Denbigh, K.G. (1940). *Trans. Faraday Soc.* 36: 936; Vickery, B.C. and Denbigh, K.G. (1949). *Trans. Faraday Soc.* 45: 62.
101 Vogel, A.I., Cresswell, W.T., Jeffery, G.H. et al. (1952). *J. Chem. Soc.* 98: 514.
102 Sugden, S. (1934). *Trans. Faraday Soc.* 30: 734.

Further Reading

Hirschfelder, J.O., Curtiss, C.F., and Bird, R.B. (1954). *Molecular Theory of Gases and Liquids*. New York: John Wiley & Sons; corrected ed. (1964). An about 1250-page classic containing, in spite of its age, still a wealth of information on theory and data.

Poling, B.E., Prausnitz, J.M., and O'Connell, J.P. (2001). *The Properties of Gases and Liquids*, 5e. McGraw-Hill Comp. An extensive review of estimation methods for many properties of gases and liquids with a short introduction for each topic.

Sengers, J.V., Kayser, R.F., Peters, C.J., and White, H.J. (2000). *Equations of State for Fluids and Fluid Mixtures*. Amsterdam: Elsevier. A broad discussion of equations of state, providing many details.

van Vleck, J.H. (1932, reprint 1966). *The Theory of Electric and Magnetic Susceptibilities*. London: Oxford University Press. Still a standard reference on what the title promises with many details.

9

Liquids

The second type of "bricks" are liquids. We start with an overview of the various approaches that exist, including a discussion of structural, some will say semiempirical, approaches. Thereafter, we briefly discuss some macroscopic characteristics, such as viscosity, and conclude with discussing dielectric properties.

9.1 Approaches to Liquids

As outlined in Section 1.5, liquids are more difficult to deal with than gases and solids. Attempts have been made to provide a broad picture of liquids based on a few simple facts and ideas [1]. One could try to extend the *virial approach* to liquids, but a few hardly surmountable problems exist (Section 8.2.1). Moreover, the coordination number of a molecule – the number of nearest-neighbor molecules around a reference molecule – is generally about 10 in the liquid state, and recalling that the nth virial coefficient deals with n-body interactions, a description of liquids in terms of the virial equation thus requires virial coefficients at least up to order 10, which are generally unknown.

Another route for (near-)spherical molecules is the *integral equation approach*, employing systematically statistical–mechanical concepts. Because sphericity is generally assumed, properties can be given in terms of their distance $r = |\mathbf{r}|$ only, although the approach can be extended to nonspherical molecules at the cost of considerable complexity. Moreover, the relevant equations must usually be solved numerically, even for spherical molecules. General considerations (Section 9.2) precede an outline (Section 9.3), followed by some results (Section 9.4). As an offspring, *scaled-particle theory* is discussed (Section 9.5). The intermediate position of liquids inevitably leads to the use of *structural models* based on lattices (Section 9.6). We also discuss the *generalized van der Waals theory* (Section 9.7), the *phonon theory* (Section 9.8), and the *quantum cluster equilibrium* model (Section 9.9). Nowadays full numerical methods, notably *Monte Carlo* (MC) and *molecular dynamics* (MD) simulations, are often used, for which we refer to [2–4].

Before discussing the various approaches, it might be useful to elaborate a bit on the meaning of structure for liquids [5, 6]. In a liquid, molecules vibrate and rotate, and with this intramolecular motion, a characteristic time τ_{int} is connected. They also move around as a whole (diffuse) with a characteristic time τ_{dif}. Now, suppose we take an instantaneous

Phases of Matter and their Transitions: Concepts and Principles for Chemists, Physicists, Engineers, and Materials Scientists, First Edition. Gijsbertus de With.

snapshot of the molecular whereabouts and their configuration and use this to describe the structure. This is the *instantaneous* or *I-structure*. In this picture, the intramolecular distances will show relatively large differences from their equilibrium values, say up to 10% or 15%, while the intermolecular distances vary widely. To get rid of these wild fluctuations, one might consider taking an average over a time interval $\tau_{\text{int}} < \Delta t < \tau_{\text{dif}}$. This represents the *vibrationally averaged* or *V-structure*. Finally, an average picture with a time interval $\Delta t \approx \tau_{\text{dif}}$ represent the *diffusionally averaged* or *D-structure* and, when taken consecutively in time, shows the path of molecules in time. While the I-structure can only be assessed via MD simulations, the V-structure is what is normally obtained from (inelastic neutron) scattering and infrared (IR) and Raman spectroscopic studies. X-ray diffraction (XRD), nuclear magnetic resonance (NMR), light scattering, and thermal properties provide an image of the D-structure. Clearly, the distinction between V- and D-structures hinges on the separation between τ_{int} and τ_{dif} and becomes increasingly less clear with increasing temperature.

9.2 Distribution Functions, Structure, and Energetics

Molecules in liquids are positioned neither regularly nor completely randomly, and we are forced to use a statistical description of their whereabouts. The *specific distribution function* for the distribution of N-specific molecules over N-specific positions in the NVT description is given by $n^{(N)}(\mathbf{r}_1, \mathbf{r}_2, \dots, \mathbf{r}_N)$ with $\mathbf{r}_j$ the Cartesian coordinates of particle j. For $n^{(M)}$ with $M < N$, we have (Section E.3)

$$n^{(M)} = \int \dots \int n^{(N)}(\mathbf{r}_1, \dots, \mathbf{r}_N)\, \mathrm{d}\mathbf{r}_{M+1} \dots \mathrm{d}\mathbf{r}_N$$

However, molecules are indistinguishable and we should use the *generic distribution function* $\rho^{(N)}$, describing any molecule at specific positions [7, 8]. Permutation arguments show that

$$\rho^{(M)}(\mathbf{r}_1, \dots, \mathbf{r}_M) = \frac{N!}{(N-M)!} \int \cdots \int n^{(N)}(\mathbf{r})\, \mathrm{d}\mathbf{r}_{M+1} \dots \mathrm{d}\mathbf{r}_N$$

with $\mathbf{r} = (\mathbf{r}_1, \mathbf{r}_2, \dots, \mathbf{r}_N)$. From this expression, we can also derive

$$\rho^{(N)} = N!n^{(N)}, \qquad (N-M)\rho^{(M)} = \int \dots \int \rho^{(M+1)}(\mathbf{r}_1, \dots, \mathbf{r}_{M+1})\, \mathrm{d}\mathbf{r}_{M+1} \qquad \text{and}$$

$$(N-M)!\rho^{(M)} = \int \dots \int \rho^{(N)}(\mathbf{r}_1, \dots, \mathbf{r}_N)\, \mathrm{d}\mathbf{r}_{M+1}..\mathrm{d}\mathbf{r}_N \tag{9.1}$$

The function $\rho^{(M)}$ thus describes the probability of M out of N molecules to be at positions $\mathbf{r}$. An extension for the momenta $\mathbf{p}$ and the angular coordinates $\boldsymbol{\omega}$ in case of nonspherical molecules is straightforward.

The simplest distribution function is the *singlet distribution function* $\rho^{(1)}$ that describes the likelihood to find any molecule in $\mathrm{d}\mathbf{r}_1$ at position $\mathbf{r}_1$. For a crystal, only specific positions $\mathbf{r}_1$ are allowed (neglecting defects and diffusion), but for a fluid all values of $\mathbf{r}_1$ are possible and hence $\rho^{(1)}$ should be independent of $\mathbf{r}_1$. So, we have

$$\rho^{(1)}(\mathbf{r}_1)\, \mathrm{d}\mathbf{r}_1 = \rho\, \mathrm{d}\mathbf{r}_1 \qquad \text{and} \qquad \int \rho^{(1)}(\mathbf{r}_1)\, \mathrm{d}\mathbf{r}_1 = N$$

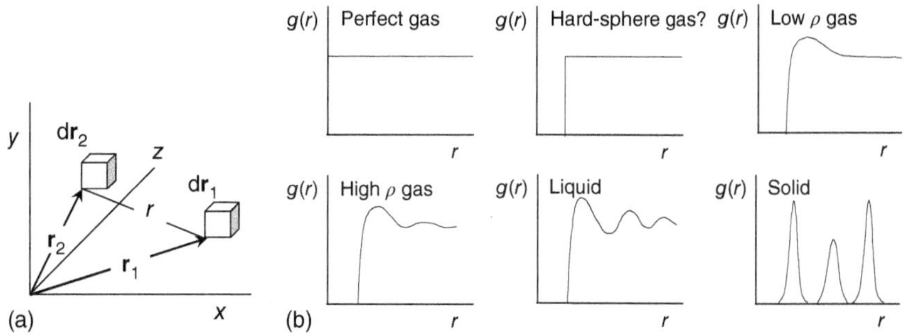

Figure 9.1 (a) The location of volume element $\mathbf{dr}_1$ located at $\mathbf{r}_1$ and the volume element $\mathbf{dr}_2$ located at $\mathbf{r}_2$ with their distance r. (b) Schematic of $g(r)$ for gases, liquids, and solids.

where $\rho = N/V$ is the number density.[1] The *doublet distribution function* $\rho^{(2)}$ describes the likelihood to find one molecule in $\mathbf{dr}_1$ at position $\mathbf{r}_1$ and another in $\mathbf{dr}_2$ at position $\mathbf{r}_2$ (Figure 9.1a). We have now

$$\rho^{(2)}(\mathbf{r}_1, \mathbf{r}_2)\,\mathbf{dr}_1\mathbf{dr}_2 \quad \text{and} \quad \int \rho^{(2)}\,\mathbf{dr}_1\mathbf{dr}_2 = (N-1)\int \rho^{(1)}\,\mathbf{dr}_1 = N(N-1) \tag{9.2}$$

Assuming that the potential energy between two molecules effectively has a relatively small range, we have for two particles at large distance $\rho^{(2)}(\mathbf{r}_1, \mathbf{r}_2) \cong \rho^{(1)}(\mathbf{r}_1)\rho^{(1)}(\mathbf{r}_2) = \rho^2$. For more close-by distances, we need the original expression $\rho^{(2)}(\mathbf{r}_1, \mathbf{r}_2)$, which we write as $\rho^{(2)}(\mathbf{r}_1, \mathbf{r}_2) = \rho^2 g^{(2)}(\mathbf{r}_1, \mathbf{r}_2)$ with $g^{(2)}(\mathbf{r}_1, \mathbf{r}_2)$ the *(pair) correlation function.*

For a microscopic interpretation we note that for any microscopic property $X = X_N(\mathbf{r}) = X_N(\mathbf{r}_1, \ldots, \mathbf{r}_N)$ depending on N (ordered) coordinates the expression for the canonical average $\langle X_N \rangle$, using Eqs. (7.22) and (9.1c), reads [7, 8]

$$\langle X_N \rangle = Q_N^{-1}\int X_N(\mathbf{r})\mathrm{e}^{-\beta\Phi(\mathbf{r})}\,\mathbf{dr} = \int X_N n^{(N)}\,\mathbf{dr} = (N!)^{-1}\int X_N \rho^{(N)}\mathbf{dr}$$

If $X_1 = \Sigma_i\, x_1(\mathbf{r}_i)$ and $X_2 = \Sigma_{i<j}\, x_2(\mathbf{r}_i, \mathbf{r}_j) = ½\Sigma_{i\neq j}\, x_2(\mathbf{r}_i, \mathbf{r}_j)$, we have

$$\langle X_1 \rangle = \int \rho^{(1)}(\mathbf{r}_1)x_1(\mathbf{r}_1)\,\mathbf{dr}_1 \quad \text{and} \quad \langle X_2 \rangle = ½\int \rho^{(2)}(\mathbf{r}_1, \mathbf{r}_2)x_2(\mathbf{r}_1, \mathbf{r}_2)\,\mathbf{dr}_1\mathbf{dr}_2 \tag{9.3}$$

If we are only interested in the correlation of distances, we can condense and write $\rho^{(2)}(\mathbf{r}_1, \mathbf{r}_2) = \rho^2 g^{(2)}(\mathbf{r}_1, \mathbf{r}_2) = \rho^2 g(r)$ with $g(r) \equiv g^{(2)}(r)$ and $r = |\mathbf{r}_1 - \mathbf{r}_2|$. From this expression, we can calculate the total number of atoms $N(r)$ in a spherical volume of radius r by integrating over both coordinates. Changing the coordinate system from $[\mathbf{r}_1, \mathbf{r}_2]$ to $[(\mathbf{r}_1 + \mathbf{r}_2)/2, \mathbf{r}_1 - \mathbf{r}_2] \equiv [\mathbf{R}, \mathbf{r}]$, integration over $\mathbf{R}$ yields the volume. The other integration over $r = |\mathbf{r}|$ (in a polar coordinate system) remains and we obtain

$$N(r) = \rho\int_0^r g(r)\,4\pi r^2\,\mathrm{d}r \tag{9.4}$$

for which the integrand $4\pi r^2 g(r)$ is often addressed as the *radial distribution function* (RDF). The function $g(r)$ is experimentally accessible by XRD and neutron diffraction.

1 Formally we require *translational invariance*, that is, $\rho^{(1)}(\mathbf{r}_1) = \rho^{(1)}(\mathbf{r}_1 + \mathbf{r}_1')$ for any $\mathbf{r}_1'$ (obviously not too close to the wall). This is only possible if $\rho^{(1)} = C$ with C a constant. Since we have on the one hand, $\int \rho^{(1)}(\mathbf{r})\,\mathbf{dr}_1 = N$, and on the other hand, $\int \rho^{(1)}(\mathbf{r})\,\mathbf{dr}_1 = C\int \mathbf{dr}_1 = CV$, we obtain $CV = N$ or $\rho^{(1)} = N/V$.

If we apply this to the perfect gas for which $\Phi = 0$ and consequently $\exp(-\beta\Phi) = 1$ and $Q_N = V^N$, we have for the pair distribution function

$$\rho^{(2)}(\mathbf{r}_1, \mathbf{r}_2)\,\mathrm{d}\mathbf{r}_1\mathrm{d}\mathbf{r}_2 = N^2\left(1 - \frac{1}{N}\right)\frac{\mathrm{d}\mathbf{r}_1}{V}\frac{\mathrm{d}\mathbf{r}_2}{V} = \rho^2\left(1 - \frac{1}{N}\right)\mathrm{d}\mathbf{r}_1\mathrm{d}\mathbf{r}_2 \cong \rho^2\mathrm{d}\mathbf{r}_1\mathrm{d}\mathbf{r}_2$$

This result matches basic probability considerations for N particles, each in an element $\mathrm{d}\mathbf{r}$, distributed randomly over a volume V, and shows that $g(r)$ is normalized to $1 - N^{-1}$, practically $\cong 1$ for large N.

If we could obtain a complete set of distribution functions, the description of liquid structure would be complete, and in that case, we would have a situation comparable to that of solids. However, such a set is unavailable, and we must usually be content with just the distribution function $\rho^{(2)}(r) = \rho^2 g(r)$. To summarize, the correlation function $g(r)$ essentially describes the probability to find two molecules at a distance r. For independent point molecules, $g(r) = 1$, and thus $\rho^{(2)}(r) = \rho^2$, but for correlated real molecules, given that a reference molecule is at the origin (with "probability" ρ), the probability to find a molecule at distance r from the origin is $\rho g(r)$. Alternatively, $\rho g(r)$ describes the local density around a reference molecule.

9.2.1 Structure

We now turn to some structural considerations for simple and normal liquids in terms of the correlation function $g(r)$. We first compare qualitatively the behavior of $g(r)$ for gases, liquids, and solids (Figure 9.1b). For a perfect gas, the density is everywhere the same, and hence no correlation is present. For a hard-sphere (HS) gas, one might expect a cut-off at the HS value with no correlation for larger values of r (we will come later to whether this is true or not). In a low-density real gas, there is some attraction, and hence a small peak at about twice the molecular radius is expected, whereas for a high-density gas, some further structure is anticipated. For a liquid, one expects even more structure as its density ρ is comparable to that of a solid. The first peak corresponds to the first shell of atoms around the reference atom or first *coordination shell*, while the second peak represents the second coordination shell. Finally, in the case of a solid, where the atoms remain largely at their equilibrium positions, one expects clear peaks due to the largely static coordination shells.

To calculate the *coordination number* (CN), the number of molecules z around a reference molecule, we use Eq. (9.4) and obtain for the first and second shell

$$z \equiv z^{(1)} = \rho\int_0^{M_1} g(r)\,4\pi r^2\,\mathrm{d}r \quad \text{and} \quad z^{(2)} = \rho\int_{M_1}^{M_2} g(r)\,4\pi r^2\,\mathrm{d}r$$

where M_1 and M_2 denote the first and second minimum in the correlation function.

A rough estimate for $z^{(1)}$ can be obtained as follows. For a dense liquid, we have approximately $\rho\sigma^3 \cong 0.64/0.74$ and we estimate that $M_1 \cong 2^{1/2}\sigma$. Then, by using $g(r) = 0$ for $r < \sigma$ and, say, $g(r) = 1.5$ for $\sigma < r < 2^{1/2}\sigma$, we obtain

$$z \cong 4\pi\rho\int_\sigma^{\sqrt{2}\sigma} g(r)\,r^2\,\mathrm{d}r \cong 4\pi\rho 1.5\left[\left(\sqrt{2}\sigma\right)^3 - \sigma^3\right]/3 \cong 10$$

confirming once more the relatively large CNs for liquids. Although the CN is a physically attractive concept, one should realize that its value shows considerable fluctuation.

Fisher [9] estimated using early Ar data [10] that $\sigma_z/z \cong 0.06$ and $\cong 0.26$ near T_{tri} and T_{cri}, respectively, where $\sigma_z^2 \equiv \langle z^2\rangle - \langle z\rangle^2$. For the second shell, this ratio is typically ≈ 0.47 near T_{tri}. Although more accurate data are nowadays available, the order of magnitude for σ_z/z is unlikely to change using these newer data. Similar fluctuations based on MC simulations are indicated by Woodcock [11].

So, close the triple point T_{tri}, the density of a liquid resembles that of a random close packing of spheres with $\eta \cong 0.64$ and $z \cong 10$, compared with $\eta = 0.74$ and $z = 12$ for the face-centered cubic (FCC) lattice, and implying an expansion of $\approx 15\%$ upon melting. It appears that near the, critical point T_{cri}, the intermolecular spacing for a wide range of liquids is given by $l = (V_{\text{cri}}/N)^{1/3} = (1.50 \pm 0.16)\sigma$. This implies an expansion from σ near T_{tri} to 1.5σ near T_{cri} or a volume expansion by a factor of 3.4. Such an amount of expansion cannot occur by a small rearrangement of the local coordination and introduces free volume. In hole theory, one takes the free volume units to be of molecular size, but this is not necessarily so [12]. Anyway, the expansion leads to a lower CN, although the nearest-neighbor distance changes only slightly.

In the experimental $g(r)$ data for Ar (Figure 9.2a), the structure anticipated for a monatomic liquid is visible with a clear first and second coordination shell. In this case

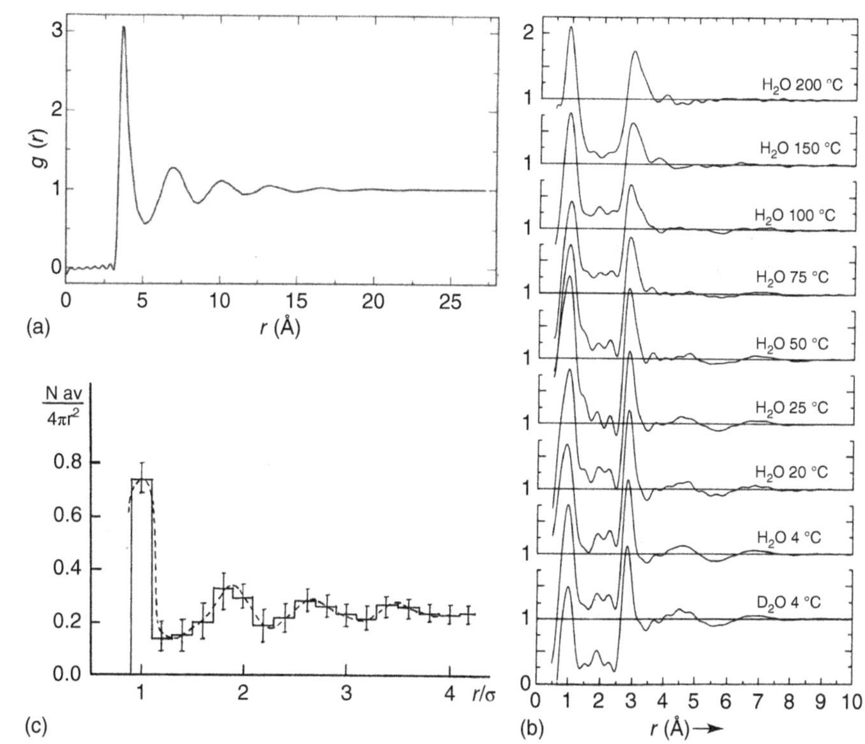

Figure 9.2 Correlation functions. (a) For Ar as determined using neutron-ray diffraction (NRD) at 85 K by Yarnell et al. [13]. Source: Yarnell et al. [13]/American Physical society (b) For H_2O as determined using X-ray diffraction (XRD) by Narten and Levy (page 311 in Vol. 1, [14]). Source: Franks [14]/Springer Nature; (c) For hard-spheres as determined experimentally by Scott [15] from a packing of close-packed steel balls. Source: Scott [15]/with permission of Elsevier.

also a third shell can be distinguished. Thereafter, the structure becomes "fuzzy," indicating the essentially random nature of liquids and the rapid fading out of the structure. The structure fades out more quickly at a higher temperature. For Ar the CN ranges from ≈ 10 at T_{tri} to ≈ 6 at T_{cri} and is mainly determined by the density and to a much lesser extent by temperature [16, 17]. Analysis of Kr data [18] showed a similar behavior for the CN (≈ 8.5 at 117 K and ≈ 4.0 at 210 K), with the nearest-neighbor distance changing only from 4.02 Å to 4.20 Å.

In Figure 9.2b, the correlation function for a molecular liquid, H_2O, is shown. The O—H and H—H distances are clearly visible, but the coordination shell structure is open to interpretation. Generally, the interpretation of the pair correlation function of molecular liquids requires insight into their chemical structure before conclusions can be drawn about the liquid structure.

In the past, several attempts were made to simulate the structure of a monatomic liquid by *analogous models*, providing a static – though qualitatively correct – picture of the structure of liquids. In particular, packings of steel balls (as used in ball bearings) were used. From well-defined random closed-packed assemblies of these balls, Bernal (actually his assistant Mason) was able to determine the reduced density [19–21] as $\eta_{RCP} = 0.6366 \pm 0.0004$. Moreover, the pair correlation function $g(r)$ was determined by counting the number of balls with increasing distance from a reference ball and averaging over several reference balls. The thus-obtained $g(r)$ closely resembles the experimentally determined $g(r)$ for monoatomic liquids, such as argon and several metals. Similar results obtained by Scott [15] are shown in Figure 9.2c. The CN was determined using intervals of 0.2σ, with σ the sphere diameter, which led to 9.3 ± 0.8. This determination of the CN was relatively straightforward (though tedious!) and subsequent computer simulations [12] essentially confirmed these findings. It should be borne in mind, however, that analogous models provide only a static picture of liquids, while the nature of fluids is essentially dynamic. For a brief history of this approach, see a paper by Finney [22].

Problem 9.1 *Boltzmann's Distribution Revisited* We expect that a relation exists between the energy $E = \mathcal{H}(\mathbf{p}, \mathbf{q})$ of N molecules with $\mathcal{H}$ the Hamilton function, and the probability of a configuration described by $n^{(N)}$. Show, by writing $n^{(N)}\mathbf{drdp} = f(\mathcal{H})\mathbf{drdp}$ and combining sets of M and N–M molecules, that $f(\mathcal{H})\cdot f(\mathcal{H}') = f(\mathcal{H} + \mathcal{H}')$. Further show that $f(\mathcal{H}) = \exp(-\beta\mathcal{H})$, with β a positive constant, is the only solution. Normalize the result, apply it to the perfect gas (Section 7.3) to show that $\beta = 1/kT$ and add the indistinguishability argument to obtain

$$\rho^{(N)}(\mathbf{r}, \mathbf{p}) = Z_N^{-1}\mathrm{e}^{-\beta\mathcal{H}} \quad \text{with} \quad Z_N = (N!h^{3N})^{-1}\int \dots \int \mathrm{e}^{-\beta\mathcal{H}}\,\mathbf{dr}_1 \dots \mathbf{dr}_N\,\mathbf{dp}_1 \dots \mathbf{dp}_N$$

Problem 9.2 *Liouville's Theorem* For systems in which the number of molecules is conserved, the total number of degrees of freedom (DoFs) and hence the number of phase points is also conserved, or $\mathrm{d}n^{(N)}/\mathrm{d}t = 0$ with $n^{(N)}$ the N-particle distribution function. Prove Liouville's theorem by showing that the distribution functions are only implicitly dependent on time, that is, $\partial n^{(N)}/\partial t = 0$, making use of $n^{(N)} = \exp(-\beta\mathcal{H})/Z_N$ and Hamilton's equations (Section 2.5).

9.2.2 Energetics

If the correlation function $g(r)$ is known, we can calculate the internal energy $U = \langle \mathscr{H} \rangle$ using Eq. (9.3b). Inserting $\Phi = ½\Sigma_{ij}\phi_{ij}(r_{ij})$ leads to $N(N-1)/2$ identical terms that can be reduced with $\rho^{(2)} = N(N-1)Q_N^{-1} \int \mathrm{e}^{-\beta\Phi}\,\mathrm{d}\mathbf{r}_3 \ldots \mathrm{d}\mathbf{r}_N = \rho^2 g(r)$ to

$$\langle \Phi \rangle = ½N(N-1)Q_N^{-1} \int \phi(r)\mathrm{e}^{-\beta\Phi}\,\mathrm{d}\mathbf{r} = ½N\rho \int \phi(r)g(r)\,4\pi r^2\mathrm{d}r$$

where $r \equiv r_{12}$. After adding the kinetic term $3kT/2$, the result is the *energy equation*

$$\langle \mathscr{H} \rangle/N = 3kT/2 + ½\rho \int \phi(r)\,g(r)\,4\pi r^2\,\mathrm{d}r \tag{9.5}$$

Similarly, though in a somewhat more complex fashion, an expression for the pressure P can be obtained. We start with the expression (Section 7.2)

$$P = -\left(\frac{\partial F}{\partial V}\right)_{N,T} = kT\left(\frac{\partial \ln Z}{\partial V}\right)_{N,T} = kT\left(\frac{\partial \ln[(N!\Lambda^{3N})^{-1}Q_N]}{\partial V}\right)_{N,T}$$

with the thermal wavelength $\Lambda \equiv (h^2/2\pi mkT)^{1/2}$. Before differentiating, we consider that P is independent of the shape of the container, so that we can take a cube and employ the coordinate transformation $\mathbf{r}_i = V^{1/3}\mathbf{r}_i'$. The expression for Q_N then becomes

$$Q_N = \int_0^{V^{1/3}} \ldots \int_0^{V^{1/3}} \mathrm{e}^{-\beta\Phi}\,\mathrm{d}\mathbf{r}_1 \ldots \mathrm{d}\mathbf{r}_N = V^N \int_0^1 \ldots \int_0^1 \mathrm{e}^{-\beta\Phi}\,\mathrm{d}\mathbf{r}_1{}' \ldots \mathrm{d}\mathbf{r}_N{}'$$

while for the intermolecular distance $r_{ij} = [(x_i - x_j)^2 + (y_i - y_j)^2 + (z_i - z_j)^2]^{1/2}$ we obtain the expression $r_{ij} = V^{1/3}[(x_i' - x_j')^2 + (y_i' - y_j')^2 + (z_i' - z_j')^2]^{1/2}$. Therefore, $\partial Q_N/\partial V$ becomes

$$\frac{\partial Q_N}{\partial V} = NV^{N-1} \int_0^1 \ldots \int_0^1 \mathrm{e}^{-\beta\Phi}\,\mathrm{d}\mathbf{r}_1{}' \ldots \mathrm{d}\mathbf{r}_N{}' - \beta V^N \int_0^1 \ldots \int_0^1 \frac{\partial \Phi}{\partial V}\mathrm{e}^{-\beta\Phi}\,\mathrm{d}\mathbf{r}_1{}' \ldots \mathrm{d}\mathbf{r}_N{}'$$

$$\text{with} \quad \frac{\partial \Phi}{\partial V} = \sum_{1 \le i < j \le N} \frac{\partial \phi}{\partial r_{ij}}\frac{\partial r_{ij}}{\partial V} = \sum_{1 \le i < j \le N} \frac{\partial \phi}{\partial r_{ij}}\frac{r_{ij}}{3V}$$

Transforming back, again noting that we have $N(N-1)/2$ identical terms and using the expression for $\rho^{(2)}$, one obtains the *pressure equation* (Section 8.2)

$$P_P = \rho kT - \frac{\rho^2}{6}\int r\frac{\mathrm{d}\phi(r)}{\mathrm{d}r}\,g(r)\,4\pi r^2\,\mathrm{d}r \tag{9.6}$$

In Eqs. (9.5) and (9.6), the first terms are due to the momenta (i.e. the kinetics), while the second term is due to the coordinates (i.e. the configuration). Thus, both equations relate a thermodynamic quantity to the structure, as represented by $g(r)$.

For future reference, we note that the compressibility κ_T can be calculated using the grand canonical ensemble (Justification 9.1). The final result

$$\kappa_T = (kT\rho)^{-1}\left[1 + \rho\int h(r)\,4\pi r^2\,\mathrm{d}r\right] \quad \text{where} \quad h(r) = g(r) - 1 \tag{9.7}$$

is known as the *compressibility equation*; this expression is derived without assuming pairwise additivity for the potential energy of the system and is thus generally valid.

Justification 9.1 ***The Compressibility Equation*** The derivation of the compressibility equation requires the grand canonical ensemble. We first recall the definition of the isothermal compressibility

$$\kappa_T = -V^{-1}(\partial V/\partial P)_T = \rho^{-1}(\partial \rho/\partial P)_{N,T} \quad \text{with } \rho = N/V$$

From $PV = kT \ln \Xi$ we have, using $\Xi = \sum_{N\geq 0}\frac{\lambda^N}{N!}\int \dots \int \exp(-\beta\Phi)\,\mathrm{d}\mathbf{r}_1 \dots \mathrm{d}\mathbf{r}_N$,

$$\frac{V}{kT}\left(\frac{\partial P}{\partial \rho}\right)_{N,T} = \left(\frac{\partial \ln \Xi}{\partial \rho}\right)_{N,T} = \frac{1}{\Xi}\frac{\partial}{\partial \rho}\sum_{N\geq 0}\frac{\lambda^N}{N!}\int \dots \int \exp(-\beta\Phi)\,\mathrm{d}\mathbf{r}_1 \dots \mathrm{d}\mathbf{r}_N$$

From the absolute activity $\lambda = \exp(\beta\mu)$ and $\partial \ln \Xi/\partial\rho = (\partial \ln \Xi/\partial\lambda)/(\partial\rho/\partial\lambda)$, we obtain for the numerator $\partial \ln \Xi/\partial\lambda$, using $Q_N \equiv \int \dots \int \exp(-\beta\Phi)\,\mathrm{d}\mathbf{r}_1 \dots \mathrm{d}\mathbf{r}_N$,

$$\left(\frac{\partial \ln \Xi}{\partial \lambda}\right)_{V,T} = \frac{1}{\Xi}\frac{\partial}{\partial\lambda}\sum_{N\geq 0}\frac{\lambda^N}{N!}Q_N = \frac{1}{\Xi}\sum_{N\geq 0}\frac{\lambda^{N-1}}{(N-1)!}Q_N = \frac{1}{\lambda}\int \rho\,\mathrm{d}\mathbf{r}_1 = \frac{V\rho}{\lambda}$$

where the distribution as defined for the grand canonical ensemble

$$\rho^{(s)}(\mathbf{r}_1, \dots, \mathbf{r}_s) \equiv \Xi^{-1}\sum_{N\geq 0}\frac{\lambda^N}{(N-s)!}\int \dots \int \exp(-\beta\Phi)\,\mathrm{d}\mathbf{r}_{s+1} \dots \mathrm{d}\mathbf{r}_N$$

is used. The denominator $\partial\rho/\partial\lambda$ is given by

$$\frac{\partial\rho}{\partial\lambda} = \sum_{N\geq 1}\frac{\partial}{\partial\lambda}\frac{\lambda^{N-1}}{(N-1)!}\frac{\lambda}{\Xi}\int \dots \int \exp(-\beta\Phi)\,\mathrm{d}\mathbf{r}_2 \dots \mathrm{d}\mathbf{r}_N$$

$$= \begin{pmatrix} \sum_{N\geq 2}\frac{\lambda^{N-2}}{(N-2)!}\frac{\lambda}{\Xi} + \sum_{N\geq 1}\frac{\lambda^{N-1}}{(N-1)!}\frac{1}{\Xi} \\ -\frac{1}{\Xi^2}\frac{\partial\Xi}{\partial\lambda}\sum_{N\geq 1}\frac{\lambda^N}{(N-1)!} \end{pmatrix}\int \dots \int \exp(-\beta\Phi)\,\mathrm{d}\mathbf{r}_2 \dots \mathrm{d}\mathbf{r}_N \quad \text{or}$$

$$\frac{\partial\rho}{\partial\lambda} = \frac{\rho^2}{\lambda}\int \mathrm{g}(r)\,\mathrm{d}\mathbf{r} + \frac{\rho}{\lambda} - \frac{\rho^2}{\lambda}\int \mathrm{d}\mathbf{r} = \frac{\rho}{\lambda}\left\{1 + \rho\int [\mathrm{g}(r) - 1]\,\mathrm{d}\mathbf{r}\right\}$$

where, for the last step, the definition $\mathrm{g}(\mathbf{r}_1, \mathbf{r}_2) = \rho^{(2)}(\mathbf{r}_1, \mathbf{r}_2)/\rho^{(1)}(\mathbf{r}_1)\rho^{(2)}(\mathbf{r}_2) = \rho^{(2)}(\mathbf{r}_1, \mathbf{r}_2)/\rho^2$ is used. Combining the expressions for $\partial\ln\Xi/\partial\lambda$ and $\partial\rho/\partial\lambda$ yields

$$\frac{V}{kT}\left(\frac{\partial P}{\partial V}\right)_{V,T} = \frac{V}{1 + \rho\int h(r)\,\mathrm{d}\mathbf{r}} \qquad \text{where } h(r) = \mathrm{g}(r) - 1$$

The isothermal compressibility is then given by

$$\kappa_T = \rho^{-1}(\partial\rho/\partial P)_{N,T} = (\rho kT)^{-1}\left[1 + \rho\int_0^\infty h(r)\,4\pi r^2\,\mathrm{d}r\right]$$

Equation (7.14), $\sigma_N{}^2 = \langle N^2\rangle - \langle N\rangle^2 = NkT\rho\kappa_T$, provides the number fluctuation σ_N.

One might wonder whether the compressibility equation and the pressure equation yield the same result for the pressure. Integrating Eq. (9.7), one obtains

$$P_\kappa = \int \frac{\partial P}{\partial V}\,\mathrm{d}V = \int \frac{\partial P}{\partial\rho}\,\mathrm{d}\rho = \int \frac{1}{\rho\kappa_T}\,\mathrm{d}\rho = kT\int\left[1 + \rho\int h(r)\,4\pi r^2\,\mathrm{d}r\right]^{-1}\mathrm{d}\rho \quad (9.8)$$

For the perfect gas $h(r) = 0$, and both P_κ and P_P, as obtained from Eq. (9.6), lead to the perfect gas EoS $P = \rho kT$. This is no longer true for a nonideal fluid for which $h(r)$ is unknown. Using approximations for $h(r)$ usually leads to different values for P_P and P_κ, indicating loss of *thermodynamic consistency* [23].

To obtain a complete description of the thermodynamic state, extra information is required, as is clear from the Gibbs equation $TdS = dU + PdV - \mu dn$, which shows that one needs information about the energy U, the pressure P, and the chemical potential μ to obtain a complete description of the thermodynamic state of the system. If the correlation function $g(r)$ would be known over a wide temperature range, we could use the energy (Eq. (9.5)), integrate the Gibbs–Helmholtz equation $\partial(F/T)/\partial(1/T) = U = \langle\mathscr{H}\rangle$ and so obtain the required information, since the chemical potential μ can be calculated from the Helmholtz energy F. Similarly, if $g(r)$ were to be known over a wide density range, we could use P_P and integrate $\partial(F/N)/\partial(1/\rho) = \rho^2 P$. Alas, neither the temperature nor the density dependence of $g(r)$ is usually available.

In the *coupling parameter approach*, a *coupling parameter* ξ is used, varying from 0 to 1, which controls the interaction between a reference molecule 1 and another molecule j by replacing ϕ_{1j} by $\xi\phi_{1j}$. The potential energy then becomes

$$\Phi(\xi) = \Phi(\mathbf{r}_1, \ldots \mathbf{r}_N, \xi) = \xi \sum\nolimits_{j=2}^{N} \phi(\mathbf{r}_{1j}) + \sum\nolimits_{2\leq i<j\leq N}^{N} \phi(\mathbf{r}_{ij}) \tag{9.9}$$

and a molecule is added to the system when ξ varies from 0 to 1. The partition function $Z_N(\rho,T)$ now reads $Z_N(\rho,T,\xi)$, while the correlation function $g(r;\rho,T)$ becomes $g(r;\rho,T,\xi)$. Because the number of particles N is very large, we can write

$$\mu = (\partial F/\partial N)_{V,T} \cong \Delta F/\Delta N = F(N,V,T) - F(N-1,V,T)$$

From $F = -kT \ln Z$, we obtain $-\beta F = \ln Q_N - \ln N! - 3N \ln \Lambda$, and hence we get

$$-\beta\mu = -\ln\lambda = \ln(Q_N/Q_{N-1}) - \ln N - \ln\Lambda^3 \tag{9.10}$$

Further, we have $Q_N(\xi = 1) = Q_N$ and $Q_N(\xi = 0) = VQ_{N-1}$. Therefore,

$$\ln\left(\frac{Q_N}{Q_{N-1}}\right) = \ln\left(\frac{Q_N(\xi=1)}{Q_N(\xi=0)}\right) + \ln V = \int_0^1 \frac{\partial \ln Q_N(\xi)}{\partial\xi}\,d\xi + \ln V \quad \text{with}$$

$$Q_N(\xi) = \int\cdots\int \exp[-\beta\Phi(\xi)]\,d\mathbf{r}_1 \ldots d\mathbf{r}_N \quad \text{and} \quad \lambda = \exp(\beta\mu)$$

the (absolute) activity. Calculating $\partial Q_N(\xi)/\partial\xi$ using Eq. (9.9) results in

$$\partial Q_N/\partial\xi = -\beta\int\cdots\int \exp[-\beta\Phi(\xi)]\left[\sum\nolimits_{j=2}^{N}\phi(r_{1j})\right] d\mathbf{r}_1 \ldots d\mathbf{r}_N$$

Finally, employing $\rho^{(N)}(\xi) = N!n^{(N)}(\xi)$ with $n^{(N)}(\xi) = Q_N(\xi)^{-1}\exp[-\beta\Phi(\xi)]$, and recognizing that we have $N-1$ identical integrals, we get

$$\frac{\partial \ln Q_N}{\partial\xi} = -\frac{\beta}{N}\int\cdots\int \phi(r_{12})\rho^{(2)}(\mathbf{r}_1,\mathbf{r}_2;\xi)\,d\mathbf{r}_1 d\mathbf{r}_2 = -\beta\rho\int_0^\infty \phi(r)g(r;\xi)\,4\pi r^2\,dr$$

where in the last step $\rho^{(2)}(\xi) = \rho^2 g(\xi)$ is employed. Substitution in Eq. (9.10) yields

$$\frac{\mu}{kT} = \ln\rho\Lambda^3 + \frac{\rho}{kT}\int_0^1\int_0^\infty \phi(r)g(r;\xi)\,4\pi r^2 dr\,d\xi \tag{9.11}$$

which is the required result. If $g(r;\xi)$ is known for a range of ξ-values, we can calculate all properties from $U = \langle\mathscr{H}\rangle$, P, and μ by integrating over ξ numerically.

The same procedure can also be used directly. If we have two states A and B with potential energy Φ_A and $\Phi_{B,}$ respectively, the transition from A to B can be described as $\Phi(\xi) = \Phi_A + \xi(\Phi_B - \Phi_A)$. For the partition function $Q(N, V, T, \xi)$, we obtain $Q(\xi) = \Sigma_j \exp[-\beta\Phi_j(\xi)]$ and thus $F(\xi) = -kT \ln Q(\xi)$. Therefore, we obtain

$$\Delta_{A\to B}F = \int_0^1 \frac{\partial F(\xi)}{\partial \xi}\,d\xi = -\int_0^1 \frac{kT}{Q(\xi)}\frac{\partial Q(\xi)}{\partial \xi}\,d\xi$$
$$= \int_0^1 \frac{kT}{Q(\xi)}\sum\nolimits_j \beta e^{-\beta\Phi_j(\xi)}\frac{\partial \Phi_j(\xi)}{\partial \xi}\,d\xi = \int_0^1 \left\langle \frac{\partial \Phi(\xi)}{\partial \xi} \right\rangle d\xi$$

Hence, if one can define for a process a potential energy $\Phi(\xi)$, a series of $\langle \partial\Phi(\xi)/\partial\xi \rangle$ values can be generated and $\Delta_{A\to B}F$ evaluated thereafter numerically. In this connection, the approach is generally addressed as *thermodynamic integration*, which is obviously a generally applicable method for which the only requirement is that a clearly defined process $A \to B$ as a function of some ξ can be discerned.

9.3 The Integral Equation Approach

We have seen that the energy U, Eq. (9.5), the pressure P, Eq. (9.6), and the chemical potential μ, Eq. (9.11), each contain two terms. The first term is due to the momenta, while the second term is due to the coordinates. Thus, all three equations in principle relate a thermodynamic quantity to the potential energy $\phi(r)$ and the structure as represented by the correlation function $g(r)$. In principle, if we determine $g(r)$ experimentally, we can calculate the thermodynamic properties for a liquid, given $\phi(r)$. However, despite the significant progress in experimental accuracy for determining $g(r)$, at present these methods are insufficiently accurate to calculate reliable thermodynamic properties, mainly because $\phi(r)$ has a minimum where $g(r)$ has a maximum. An alternative is to use integral equation methods. A review dealing with integral equations, the associated perturbation approach, and several other topics has been given by Barker and Henderson [24]. Other reviews are given in [25–27].

So, let us discuss how a theoretical expression for $g(r)$ can be obtained. We first introduce some new concepts leading to the Ornstein–Zernike equation. Thereafter, we deal with the Yvon–Born–Green equation and briefly introduce the Kirkwood, the hyper-netted chain, Percus–Yevick, and mean spherical approximation approaches. Readers not interested in the background of these approaches should consider them simply as labels for approximate solutions when comparing the results (Section 9.4).

9.3.1 The Ornstein–Zernike Equation

For the following, it is convenient to introduce some auxiliary functions. We recall that $\rho g(r_{12}) \equiv \rho g(\mathbf{r}_1, \mathbf{r}_2)$ describes the probability of finding a particle at distance $r_{12} = |\mathbf{r}_2 - \mathbf{r}_1|$ from the origin given that another particle is located at the origin. No correlation would intuitively imply $g \to 0$ for $r_{12} \to \infty$, but for $r_{12} \to \infty$, we actually have $g \to 1$. Therefore, we use the so-called *total correlation function* by $h(r_{12}) \equiv g(r_{12}) - 1$, for which obviously holds: $r_{12} \to \infty$, $h \to 0$. The total correlation between particle 1 and particle 2 is supposed to be composed of two contributions:

- the direct correlation between particle 1 and 2, represented by the *direct correlation function* $c(r_{12})$ and
- the indirect correlation of particle 1 and 2 via the influence of particle 1 on particle 3 and the influence of particle 3 on particle 2 summed over all such third particles (with density ρ). This results in the convolution $\rho \int c(r_{13})h(r_{32})\,\mathrm{d}\boldsymbol{r}_3$.

The total correlation is given by what is called the *Ornstein–Zernike* (OZ) *equation*

$$h(r_{12}) = c(r_{12}) + \rho \int c(r_{13})h(r_{32})\,\mathrm{d}\boldsymbol{r}_3 \tag{9.12}$$

The OZ equation is in fact one way of defining the direct correlation function $c(r_{12})$. Although the physical interpretation of $c(r_{12})$ is not directly as clear as that for $g(r)$ (Justification 9.2), it does have a simpler structure than $g(r)$, in particular it has a much shorter range. A formal solution for $h(r_{12})$ can be obtained by iteration yielding

$$h(r_{12}) = c(r_{12}) + \rho \int c(r_{13})c(r_{32})\,\mathrm{d}\boldsymbol{r}_3 + \rho^2 \int c(r_{13})c(r_{34})c(r_{42})\,\mathrm{d}\boldsymbol{r}_3\mathrm{d}\boldsymbol{r}_4 + \dots$$

This solution can be interpreted as a sum of "chains" of direct correlation functions between particles at $\boldsymbol{r}_1$ and $\boldsymbol{r}_2$. Since the OZ equation is a convolution, a practical solution for $h(r_{12})$ can be obtained from its Fourier transform (Section A.15). Using

$$\widetilde{h}(\boldsymbol{s}) = \int_0^\infty h(\boldsymbol{r}) \exp(\mathrm{i}\boldsymbol{s}\cdot\boldsymbol{r})\,\mathrm{d}\boldsymbol{r} \quad \text{and} \quad \widetilde{c}(\boldsymbol{s}) = \int_0^\infty c(\boldsymbol{r}) \exp(\mathrm{i}\boldsymbol{s}\cdot\boldsymbol{r})\,\mathrm{d}\boldsymbol{r}, \tag{9.13}$$

$$\widetilde{h}(s) = \widetilde{c}(s) + \rho\widetilde{c}(s)\widetilde{h}(s) \quad \text{or} \quad S(s) \equiv 1 + \rho\widetilde{h}(s) = [1 - \rho\widetilde{c}(s)]^{-1} \tag{9.14}$$

is obtained, where $s = |\boldsymbol{s}|$. The direct correlation function $c(r)$ is connected to the compressibility, as can be seen rewriting the compressibility equation, Eq. (9.7), to

$$\frac{\beta}{\rho\kappa_T} = \beta\left(\frac{\partial P}{\partial \rho}\right)_{N,T} = \frac{1}{1+\rho\int_0^\infty h(r)\,\mathrm{d}\boldsymbol{r}} = \frac{1}{1+\rho\widetilde{h}(0)} = 1 - \rho\widetilde{c}(0) \tag{9.15}$$

where Eq. (9.13) with $\boldsymbol{s} = \boldsymbol{0}$ is used. For both $c(r)$ and $g(r) = h(r) + 1$, expansions in ρ can be formulated, for which we refer to the literature, for example, [8, 25, 27].

So far, we have introduced several functions, which will be used again in the sequel, and for convenience we summarize them here. As before, ϕ denotes the pair potential in the total potential energy $\boldsymbol{\Phi} = ½\Sigma_{ij}\phi_{ij}$, $\beta = 1/kT$, and ρ the number density.

Boltzmann function	$e = e(r) = \exp[-\beta\phi(r)]$
Mayer function	$f = f(r) = e - 1$
Pair correlation function	$g = g(r)$
Potential of mean force	$W = W(r) = -kT\ln g$
Total correlation function	$h = h(r) = g - 1$
Direct correlation function	$c = c(r)$
Background correlation function	$y = y(r) = g/e$

To be concise, we omit in the following few paragraphs the arguments for each of these functions, except when it is required to be precise.

Justification 9.2 ***Linear Response and Direct Correlation*** Here, we show how $c(r)$ is related to the interaction potential of a fluid, considering only the linear response of the system. The treatment is brief, and we freely use some more advanced concepts. Linear response theory is discussed in detail by Pottier [28].

Suppose we have a perturbed system described by the Hamilton function $\mathscr{H} = \mathscr{H}_0 + \mathscr{H}_1(\mathbf{r},t) = \mathscr{H}_0 - b(\mathbf{r},t)B(\mathbf{r},t)$, where the external perturbation $\mathscr{H}_1$ is split in a time-dependent part representing the (displacement) field $b = b(\mathbf{r}, t)$ and the conjugate force $B = B(\mathbf{r}, t)$ of the system. Consider now a locally varying property $A(\mathbf{r}, t) = A_0 + \delta A(\mathbf{r}, t)$. Here, A_0 is the mean value of A and $\delta A(\mathbf{r}, t)$ its fluctuation, for which it is assumed that $\delta A(\mathbf{r}, t_0) = 0$ at time t_0, often taken as $t_0 = -\infty$. The *susceptibility* $\chi = \chi_{AB} = \partial\langle A(t)\rangle/\partial b$, describing the influence on A by b, is given by [28]

$$\delta A(t) = \int_{t_0}^{t} \chi(t-t')b(t')\,\mathrm{d}t' \qquad \text{with } \chi = -\beta\dot{C}_{AB}(\mathbf{r},t) = -\beta\mathrm{d}\left\langle A(t)^*B(0)\right\rangle/\mathrm{d}t \quad (9.16)$$

Here $C_{AB}(\mathbf{r},t)$ is a time-dependent correlation function obtained by generalizing C_{AB} (Section 7.1) with respect to time. This expression for $\delta A(t)$ is the most general one for linear response. This can be seen as follows. Generally, we have $\Delta A(t) = \chi(t)\Delta b$ for a small change Δb at time t. Hence, including the effect of changes at time $t - \theta$, we obtain $\Delta A(t) = \chi(t)b(t)\Delta t + \chi(t-\theta_1)b(\theta_1)\Delta t + \ldots = \Sigma_j\chi(t-\theta_j)b(\theta_j)\Delta t$. Taking infinitesimal steps so that $\Delta t \to \mathrm{d}t$ and $\Delta A \to \delta A$, we obtain Eq. (9.16a).

To derive Eq. (9.16b) for χ, we use the simplest possible perturbation, namely $b(t) = b\delta(t)$, where $\delta(\cdot)$ is the Dirac function. At time $t = 0^-$, that is, just before $t = 0$, the probability density in phase space $\rho = \rho(\mathbf{q},\mathbf{p},t)$ equals the equilibrium density $\rho_0 = \exp(-\beta\mathscr{H})/Z$ with Z the partition function. Time development of the system is described by the Liouville equation $\partial\rho/\partial t = \{\mathscr{H},\rho\}$ with $\{X,Y\} = (\partial X/\partial\mathbf{q})^{\mathrm{T}}(\partial Y/\partial\mathbf{p}) - (\partial X/\partial\mathbf{p})^{\mathrm{T}}(\partial Y/\partial\mathbf{q})$ the Poisson bracket (Section 2.5). Therefore, at $t = 0$, we have

$$\partial\rho/\partial t = \{\mathscr{H},\rho_0\} = \{\mathscr{H}_0,\rho_0\} - b\delta(t)\{B,\rho_0\}$$

Now $\partial\rho_0/\partial\mathbf{p} = -\beta\rho_0\partial\mathscr{H}/\partial\mathbf{p}$ and $\partial\rho_0/\partial\mathbf{q} = -\beta\rho_0\partial\mathscr{H}/\partial\mathbf{q}$ and thus $\{B, \rho_0\} = -\beta\rho_0\{B,\mathscr{H}\} = -\beta\rho_0\dot{B}(0)$ where the evolution equation $\dot{B} = \{B, \mathscr{H}\}$ is used. At time $t = 0^+$, that is, just after $t = 0$, we thus have

$$\partial\rho/\partial t = \partial\rho_0/\partial t + \beta b\delta(t)\rho_0\dot{B}(0)$$

Time integration yields

$$\rho(0^+) = \rho(0^-) + \beta b\rho_0\dot{B}(0) = \rho_0[1 + \beta b\dot{B}(0)]$$

We now calculate the expectation value

$$\langle A(t)\rangle = \int \rho_0 A(t)\,\mathrm{d}\mathbf{p}\mathrm{d}\mathbf{q} + \beta b\int \rho_0 A(t)\dot{B}(0)\,\mathrm{d}\mathbf{p}\mathrm{d}\mathbf{q} = \langle A\rangle + \beta b\left\langle A(t)\dot{B}(0)\right\rangle \qquad \text{or}$$

$$\chi_{AB}(t) = \partial\langle A\rangle/\partial b = \partial\langle\delta A\rangle/\partial b = \beta\left\langle A(t)\dot{B}(0)\right\rangle = \beta C_{A\dot{B}}$$

In full, we have $\left\langle A(t)\dot{B}(0)\right\rangle = \left\langle A(\mathbf{r},t)\dot{B}(\mathbf{0},0)\right\rangle = \left\langle A(\mathbf{r}-\mathbf{r}',t-t')\dot{B}(\mathbf{r}',t')\right\rangle$. It is relatively easy to prove that $\beta C_{A\dot{B}} = -\beta C_{\dot{A}B} = -\beta\dot{C}_{AB}$, so that $\chi_{AB}(t) = -\beta\dot{C}_{AB}$. Note that χ depends on the equilibrium properties of the system only.

Consider now a system perturbed with constant b switched on at $t = -\infty$, so that it is in equilibrium at $t = 0$, and for which the perturbation is switched off at $t = 0$. In that case, $\delta A(t) = b\int_{-\infty}^{0}\chi(t-t')\,\mathrm{d}t' = bH(t)\int_t^{\infty}\chi(\tau)\,\mathrm{d}\tau$ for $t \geq 0$, where $H(t)$ is the Heaviside function

(Section A.14), introduced as the effect of a change (the switch-off) can only be noticed afterwards (*causality*). Evaluating, we obtain

$$\delta A(t) = -\beta b H(t) \int_t^\infty \dot{C}_{AB}(t)\,\mathrm{d}t = -\beta b H(t)\; C_{AB}(t)\big|_t^\infty = -\beta b H(t)(\langle A\rangle\langle B\rangle - \langle AB\rangle)$$

$$= \beta b H(t) C_{\delta A\delta B} \quad \text{or} \quad \partial\delta A(t)/\partial b = \beta H(t) C_{\delta A\delta B} \tag{9.17}$$

The effect of this perturbation is thus fully described by the correlation function $C_{\delta A\delta B}$.

For further steps, it is profitable to work with the space–time Fourier transform, for a variable A reading

$$A(\mathbf{k},\omega) = \iint \exp[-\mathrm{i}(\omega t - \mathbf{k}\cdot\mathbf{r})]A(\mathbf{r},t)\,\mathrm{d}\mathbf{r}\mathrm{d}t$$

or, in short, $\widetilde{A} \equiv A(\mathbf{k},\omega) = F[A(\mathbf{r},t)]$ (Section A.15). In this notation, we have for Eq. (9.16a), using the convolution theorem, $\delta\widetilde{A} = \widetilde{\chi}\widetilde{b}$.

For an interacting system with interaction potential $\psi(\mathbf{r},\mathbf{r}')$ between the densities $\rho(\mathbf{r})$ and $\rho(\mathbf{r}')$, we have $\mathcal{H}_{\text{int}} = -\frac{1}{2}\int \rho(\mathbf{r})\psi(\mathbf{r},\mathbf{r}')\rho(\mathbf{r}')\,\mathrm{d}\mathbf{r}\mathrm{d}\mathbf{r}'$. Inserting $\rho(\mathbf{r}) = \rho_0 + \delta\rho(\mathbf{r})$, we obtain, neglecting the constant term with only ρ_0 representing the nonfluctuating system and the second-order contribution with $\delta\rho(\mathbf{r})$ and $\delta\rho(\mathbf{r}')$,

$$\mathcal{H}_{\text{int}} = \int \rho_0 \Psi(\mathbf{r})\,\mathrm{d}\mathbf{r} \quad \text{with } \Psi(\mathbf{r}) = -\int \psi(\mathbf{r},\mathbf{r}')\delta\rho(\mathbf{r}')\,\mathrm{d}\mathbf{r}'$$

where $\Psi(\mathbf{r})$ is the *mean-field potential*. In the Fourier space, Ψ becomes $\widetilde{\Psi} = -\widetilde{\psi}\delta\widetilde{\rho}$.

We now approximately calculate the time-independent response for a spatially homogeneous but fluctuating system by considering the interaction between particles as the effect of an external perturbation. To that purpose, we take for B and A the particle density ρ, so that using $b(\mathbf{r})B(\mathbf{r}) = a(\mathbf{r})A(\mathbf{r}) = a\rho$, we have $\delta\widetilde{\rho} = \widetilde{\chi}\widetilde{a}$ with $\widetilde{\chi} = \widetilde{\chi}_{\rho\rho}$, and where $\widetilde{y}$ now refers to the spatial Fourier transform of y. Using as total potential $\Phi = \Psi + a$, we also have $\delta\widetilde{\rho} = \widetilde{\chi}_0\widetilde{\Phi}$ with χ_0 the susceptibility of the noninteracting system (perfect gas). The fluctuation $\delta\widetilde{\rho}$ can thus be written as

$$\delta\widetilde{\rho} = \widetilde{\chi}_0(\widetilde{\Psi} + \widetilde{a}) = \widetilde{\chi}_0(-\widetilde{\psi}\widetilde{\chi} + 1)\widetilde{a} = \widetilde{\chi}\widetilde{a} \quad \text{or} \quad \widetilde{\chi} = \widetilde{\chi}_0/(1 + \widetilde{\chi}_0\widetilde{\psi})$$

Noting that $\langle\delta\rho(\mathbf{r})\delta\rho(\mathbf{0})\rangle$ is a convolution, we obtain $C_{\delta\widetilde{\rho}\delta\widetilde{\rho}} = \langle\delta\widetilde{\rho}^*(\mathbf{k})\delta\widetilde{\rho}(\mathbf{k})\rangle$ so that only fluctuations with the same wave vector correlate. Therefore, we have $\widetilde{\chi} = \beta\langle\delta\widetilde{\rho}^*\delta\widetilde{\rho}\rangle = \beta\rho_0(1 + \rho_0\widetilde{h}) = \beta\rho_0 S$. Here, $S = 1 + \rho_0\widetilde{h} = (1 - \rho_0\widetilde{c})^{-1}$ is the *structure factor*. For the noninteracting system $\widetilde{h} = 0$, so that $\widetilde{\chi}_0 = \beta\rho_0$. Finally, we thus obtain

$$\widetilde{\chi} = \widetilde{\chi}_0 S = \widetilde{\chi}_0/(1 - \rho_0\widetilde{c}) \quad \text{or} \quad \widetilde{c} = -\beta\widetilde{\psi} \tag{9.18}$$

Hence, we see that the direct correlation function $c(r)$ represents the interaction potential $\psi(r)$ ([29], for a slightly different result, see [30]). Several approximations are based on this result. For example, if we identify $\psi(r)$ with the pair potential $\phi(r)$, we have the *random-phase approximation* (RPA), while if we take the potential to be a Sutherland potential (Section 4.3), we employ the *mean-spherical approximation* (MSA).

9.3.2 The Yvon–Born–Green Equation

To obtain a useful integral equation, we start with the pair distribution function:

$$\rho^{(2)}(\mathbf{r}_1,\mathbf{r}_2) = \frac{N!}{Q_N(N-2)!}\int \exp[-\beta\Phi(\mathbf{r})]\,\mathrm{d}\mathbf{r}_3\ldots\mathrm{d}\mathbf{r}_N$$

For a pairwise additive potential $\Phi = ½\Sigma_{ij}\phi_{ij}$, differentiation with respect to $\mathbf{r}_1$ leads to

$$\frac{\partial \rho^{(2)}(\mathbf{r}_1,\mathbf{r}_2)}{\partial \mathbf{r}_1} = \frac{N!}{Q_N(N-2)!}\int -\beta \exp(-\beta\Phi)\frac{\partial \Phi}{\partial \mathbf{r}_1}\,\mathrm{d}\mathbf{r}_3 \ldots \mathrm{d}\mathbf{r}_N$$

For $\partial\Phi/\partial\mathbf{r}_1$, we obtain $\partial\Phi/\partial\mathbf{r}_1 = \partial\phi_{12}/\partial\mathbf{r}_1 + \sum_{k=3}^{k=N}(\partial\phi_{1k}/\partial\mathbf{r}_1)$ so that

$$\frac{\partial \rho^{(2)}(\mathbf{r}_1,\mathbf{r}_2)}{\partial \mathbf{r}_1} = -\beta\frac{\partial\phi(r_{12})}{\partial \mathbf{r}_1}\int \frac{N!}{Q_N(N-2)!}\exp(-\beta\Phi)\,\mathrm{d}\mathbf{r}_3 \ldots \mathrm{d}\mathbf{r}_N$$
$$-\beta\int \frac{\partial\phi(r_{13})}{\partial \mathbf{r}_1}\left[\int \frac{N!(N-2)}{Q_N(N-2)!}\exp(-\beta\Phi)\,\mathrm{d}\mathbf{r}_4 \ldots \mathrm{d}\mathbf{r}_N\right]\mathrm{d}\mathbf{r}_3$$

In the first integral we recognize $\rho^{(2)}$ and in the one in square brackets $\rho^{(3)}$ and, if we divide by ρ^2 and use $\rho^{(2)} = \rho^2 g^{(2)}$ and $\rho^{(3)} = \rho^3 g^{(3)}$, the overall result is

$$\frac{\partial g^{(2)}(\mathbf{r}_1,\mathbf{r}_2)}{\partial \mathbf{r}_1} = -\frac{g^{(2)}(\mathbf{r}_1,\mathbf{r}_2)}{kT}\frac{\partial\phi(r_{12})}{\partial \mathbf{r}_1} - \frac{\rho}{kT}\int g^{(3)}(\mathbf{r}_1,\mathbf{r}_2,\mathbf{r}_3)\frac{\partial\phi(r_{13})}{\partial \mathbf{r}_1}\,\mathrm{d}\mathbf{r}_3 \tag{9.19}$$

This is an infinitely nested set of coupled equations denoted as the *Yvon–Born–Green* (YBG) *hierarchy* [31, 32], which can be solved formally by iteration. However, for a concrete solution, a “closure” is required, and to that purpose we need to approximate. Suppose – at first sight not completely unrealistic – that $g^{(3)}(r_{12}, r_{23}, r_{31})$ is to a first approximation multiplicative or

$$g^{(3)}(r_{12}, r_{23}, r_{31}) \cong g^{(2)}(r_{12})\,g^{(2)}(r_{23})\,g^{(2)}(r_{31})$$

With this *superposition approximation* ([33], for a slightly grumpy review, see [34]), the integral equation can be solved numerically. The resulting *YBG equation* reads

$$\frac{\partial \ln g^{(2)}(\mathbf{r}_1,\mathbf{r}_2)}{\partial \mathbf{r}_1} = -\beta\frac{\partial\phi(r_{12})}{\partial \mathbf{r}_1} - \beta\rho\int g^{(2)}(\mathbf{r}_3,\mathbf{r}_1)g^{(2)}(\mathbf{r}_2,\mathbf{r}_3)\frac{\partial\phi(r_{13})}{\partial \mathbf{r}_1}\,\mathrm{d}\mathbf{r}_3$$

Using $W = -kT\ln g$ shows that: (i) W is composed of the direct interaction $\phi(r_{12})$ and the weighted interaction on particle 1 exerted by all particles 3 at arbitrary positions; and (ii) that W reduces to ϕ for $\rho \to 0$. Using a superposition for $g^{(4)}$ in terms of $g^{(3)}$, a better approximation can be made, although such an approach might suffer from a lack of energy conservation [35].

9.3.3 Other Integral Equations

Other integral equations can be derived in various ways. One approach uses thermodynamic integration combined with the pairwise additivity assumption $\Phi = ½\Sigma_{ij}\phi_{ij}$. The correlation function is taken to be dependent on the *coupling parameter* ξ and differentiated with respect to ξ, leading to an implicit equation for $\rho^{(2)}(\mathbf{r}_1, \mathbf{r}_2)$ in terms of $\rho^{(2)}(\mathbf{r}_1, \mathbf{r}_2)$ and $\rho^{(3)}(\mathbf{r}_1, \mathbf{r}_2)$. After quite some manipulation, the result is

$$-kT\ln g^{(2)}(\mathbf{r}_1,\mathbf{r}_2;\xi) = \xi\phi(\mathbf{r}_{12}) + \rho\int_0^{\xi}\int \phi(r_{13})\left[\frac{g^{(3)}(\mathbf{r}_1,\mathbf{r}_2,\mathbf{r}_3;\xi)}{g^{(2)}(\mathbf{r}_1,\mathbf{r}_2;\xi)} - g^{(2)}(\mathbf{r}_1,\mathbf{r}_3;\xi)\right]\mathrm{d}\mathbf{r}_3\mathrm{d}\xi$$

This expression gives $g^{(2)}$ in terms of $g^{(3)}$ and provides another example of a hierarchy of coupled equations, the so-called *Kirkwood hierarchy* [36]. Using the superposition approximation again results in the *Kirkwood equation*

$$-kT\ln g^{(2)}(\mathbf{r}_1,\mathbf{r}_2;\xi) = \xi\phi(\mathbf{r}_{12}) + \rho\int_0^{\xi}\int \phi(r_{13})\,g^{(2)}(\mathbf{r}_1,\mathbf{r}_3;\xi)\,[g^{(2)}(\mathbf{r}_2,\mathbf{r}_3) - 1]\,\mathrm{d}\mathbf{r}_3\mathrm{d}\xi$$

This integral equation also has to be solved numerically and shows again that for $\rho \to 0$, $W = -kT \ln g \to \phi$. It can be shown that the YBG and Kirkwood hierarchies are equivalent [37], but that their approximate solutions are not.

Another approach is the *Percus–Yevick* (PY) approximation obtained by systematic expansion of Q_N using a (graphical) cluster expansion or by functional differentiation [7, 27]. Here, we use another entrance as given by McQuarrie [8] using the OZ equation, Eq. (9.12). To solve the OZ equation for either $c(r)$ or $h(r)$ we need another relation, a "closure", between $c(r)$ and $h(r)$ for which $c(r)$ is considered to be the difference between $g(r) = \exp[-\beta W(r)]$ and an indirect correlation function $g_{\text{ind}}(r)$. We approximate $g_{\text{ind}}(r)$ by $g_{\text{ind}}(r) \cong \exp[-\beta(W - \phi)]$, that is, we assume that W minus the pair potential ϕ provides a good estimate for g_{ind}, or

$$c = g - g_{\text{ind}} \cong \exp(-\beta W) - \exp[-\beta(W - \phi)] = g - (g/e) = g - y$$

where the function $y = g/e$ is used. For c and h, we have the equivalent forms

$$c = g - y = ey - y = (e-1)y = fy \quad \text{and} \quad h = g - 1 = c + y - 1 = fy + y = -1 \tag{9.20}$$

Inserting $h = c + y - 1$ in the OZ equation, we obtain

$$y(r_{12}) = 1 + \rho \int f(r_{13})y(r_{13})[f(r_{32})y(r_{32}) + y(r_{32}) - 1]\,\mathrm{d}\boldsymbol{r}_3 \tag{9.21}$$

Since f is known, this integral equation in y, known as the *PY equation*, can be solved numerically for y and, once y is obtained, leads directly to $g = ye$.

The *hyper-netted chain* (HNC) approximation received its name from a type of diagrams included in the graphical cluster expansion analysis not discussed here. The HNC equation can be also obtained by expanding $y = \exp[-\beta(W - \phi)]$ to first order, so that $y = 1 - \beta(W - \phi)$. Hence, from $c = g - y$, we have the equivalent forms

$$c = g - y = g - [1 - \beta(W - \phi)] = h - \ln g - \beta\phi = h - \ln(g/e) = h - \ln y$$

Inserting $c = h - \ln y$ in the OZ equation results in one form of the *HNC equation*

$$\ln y(r_{12}) = \rho \int [h(r_{13}) - \ln y(r_{13})]h(r_{23})\,\mathrm{d}\boldsymbol{r}_3 \quad \text{or} \tag{9.22}$$

$$\ln[g(r_{12})/e(r_{12})] = \rho \int [g(r_{13}) - 1 - \ln g(r_{13}) - \beta\phi(r_{13})][g(r_{23}) - 1]\,\mathrm{d}\boldsymbol{r}_3 \tag{9.23}$$

This integral equation in g, since e and ϕ are known, can be solved numerically.

Although it appears that the HNC equation approximates the PY equation, the cluster expansion analysis shows that this is not true. While the HNC approximation neglects certain terms in the expansion because of their complexity, the PY approximation neglects even more terms. However, the original, physical but rather complex derivation of the PY equation is in terms of "collective coordinates," resembling the harmonic analysis of a crystalline solid [38]. While the HNC equation is thus a pure mathematical approximation, the PY equation has some physical basis.

The MSA consists of the core condition $g(r) = 0$ for $r < \sigma$ and the approximation $c(r) = -\beta\phi(r)$ for $r > \sigma$ in combination with the OZ equation. For HSs with $\phi(r) = \infty$ for $r < \sigma$, where σ is the hard-core diameter, this approach is identical to the PY approach, and an analytical solution is possible (Section 9.4). However, the importance of the MSA approximation is contained in the fact that it can also be solved analytically for other cases. In particular, if $\phi(r)$ represents a square-well potential, a Yukawa potential, the ion–ion, dipole–dipole, or quadrupole–quadrupole interaction, a solution can be found using Laplace methods. These solutions are used in the description of models for ionic solutions and polar fluids.

9.3.4 The Potential of Mean Force

We have seen that $W = -kT \ln g$ is a useful function. Generalizing, we define a quantity $W^{(s)}(\mathbf{r}_1,\ldots,\mathbf{r}_s)$ by

$$g^{(s)}(\mathbf{r}_1,\ldots,\mathbf{r}_s) \equiv \exp[-\beta W^{(s)}(\mathbf{r}_1,\ldots,\mathbf{r}_s)]$$

where the correlation function $g^{(s)}(\mathbf{r}_1,\ldots,\mathbf{r}_s)$ is given by

$$g^{(s)}(\mathbf{r}_1,\ldots,\mathbf{r}_s) = V^s \int \ldots \int \exp(-\beta\Phi)\,\mathrm{d}\mathbf{r}_{s+1}\ldots\mathrm{d}\mathbf{r}_N/Q_N$$

Taking logarithms on both sides and the gradient with respect to one of the particles, say j, one obtains

$$-\nabla_j W^{(s)}(\mathbf{r}_1,\ldots,\mathbf{r}_s) = \frac{\int\ldots\int(-\nabla_j\Phi)\exp(-\beta\Phi)\,\mathrm{d}\mathbf{r}_{s+1}\ldots\mathrm{d}\mathbf{r}_N}{\int\ldots\int\exp(-\beta\Phi)\,\mathrm{d}\mathbf{r}_{s+1}\ldots\mathrm{d}\mathbf{r}_N}$$

Since $F_j = -\nabla_j\Phi$ represents the force on molecule j for a fixed configuration of $\mathbf{r}_1,\ldots,\mathbf{r}_s$, the right-hand side is the mean force $\langle F_j^{(s)}(\mathbf{r}_1,\ldots,\mathbf{r}_s)\rangle$, the force averaged over the configurations of all molecules $s+1,\ldots,N$ not in the fixed set $1,\ldots,s$. Therefore,

$$-\partial W^{(s)}(\mathbf{r}_1,\ldots,\mathbf{r}_s)/\partial\mathbf{r}_j = \langle F_j(\mathbf{r}_1,\ldots,\mathbf{r}_s)\rangle$$

We see that for $s = 2$ we have $W(\mathbf{r}_1,\mathbf{r}_2) \equiv W^{(2)}(\mathbf{r}_1,\mathbf{r}_2)$, representing the *potential of mean force* between one particle held at position $\mathbf{r}_1$ and another held at position $\mathbf{r}_2$ in the field of all other molecules. As before, the functions g and W are related by

$$W(\mathbf{r}_1,\mathbf{r}_2) = -kT\ln g(\mathbf{r}_1,\mathbf{r}_2) \quad \text{or} \quad W(r) = -kT\ln g(r)$$

for isotropic systems. For $r = |\mathbf{r}_2 - \mathbf{r}_1| \to \infty$, $W(r) \to 0$ and $g \to 1$. From $W(r) = -kT \ln g(r)$, we see that, providing that $\beta W(r) \ll 1$, $g(r)$ permits the expansion

$$g(r) = 1 - \beta W(r) + \tfrac{1}{2}[\beta W(r)]^2 - \ldots$$

For low-density ρ, $W(r) \to \phi(r)$ because the two molecules considered are no longer affected by the other molecules. Hence, at low density, $g(r)$ reduces to

$$\lim_{\rho\to 0} g(r) = \exp[-\beta\phi(r)] \equiv e(r)$$

and there is a unique correspondence between $\phi(r)$ and $g(r)$. At higher density, one writes, using the *background correlation function* $y(r)$,[2]

$$g(r) = \exp[-\beta\phi(r)]y(r) \equiv e(r)y(r) \tag{9.24}$$

---

2 Also known as *cavity function*, as it describes the distribution of cavities in a HS fluid. Taking logarithms results in $\ln g(r) = -\beta\phi(r) + \ln y(r)$ or $\ln y(r) = -\beta[W(r) - \phi(r)] = \Delta W(r)$. Here, $\Delta W(r)$ represents the potential of mean force in excess over the interaction potential $\phi(r)$. This permits the expansion $y(r) = \exp[\Delta W(r)] = 1 + \Delta W(r) + \tfrac{1}{2}[\Delta W(r)]^2 + \ldots$.

Since $\mathrm{d}\exp(-\beta\phi)/\mathrm{d}r = -\beta\exp(-\beta\phi)\mathrm{d}\phi/\mathrm{d}r$, substituting Eq. (9.24) in the pressure Eq. (9.6) leads to the expression

$$\beta P/\rho = \tfrac{1}{6}\rho\int r(\mathrm{d}\,e(r)/\mathrm{d}r)\,y(r)\,4\pi r^2\,\mathrm{d}r \tag{9.25}$$

To obtain P as a function of ρ, we need the expansion for $y(r)$ in ρ reading

$$y(r) = 1 + A_{21}\rho + A_{22}\rho^2 + \ldots \tag{9.26}$$

so that g(r) becomes $g(r) = \exp[-\beta\phi(r)]y(r) \equiv \exp[-\beta\phi(r)](1 + A_{21}\rho + A_{22}\rho^2 + \ldots)$. In the same spirit, we write $g^{(3)}(r) = \exp[-\beta(\phi_{12} + \phi_{23} + \phi_{31})](1 + A_{31}\rho + A_{32}\rho^2 + \ldots)$. Substituting these expansions in the YBG hierarchy, Eq. (9.19), we obtain for both sides of the YBG equation a polynomial in the density ρ. The next step is equating equal order terms in ρ, and for the first-order terms in ρ, this leads to

$$\partial A_{21}(\mathbf{r}_1, \mathbf{r}_2)/\partial\mathbf{r}_1 = -\beta\int \exp[-\beta(\phi_{13} + \phi_{23})](\partial\phi(r_{13})/\partial\mathbf{r}_1)\,\mathrm{d}\mathbf{r}_3$$

Integration over $\mathbf{r}_1$ yields

$$A_{21}(\mathbf{r}_1, \mathbf{r}_2) = \int \exp[-\beta(\phi_{13} + \phi_{23})]\,\mathrm{d}\mathbf{r}_3 + C$$

where C is the integration constant. Since for $r_{12} \to \infty$, $g(r) \to 1$, we obtain from $g(r) = \exp(-\beta\phi)(1 + A_{21}\rho + \ldots)$ as the boundary condition $A_{21}(r_{12}) = 0$ for $r_{12} \to \infty$. Using $f(r) = \exp[-\beta\phi(r)] - 1 \equiv e(r) - 1$, this boundary condition is fulfilled if we write

$$A_{21}(\mathbf{r}_1, \mathbf{r}_2) = \int f(\mathbf{r}_1, \mathbf{r}_3)f(\mathbf{r}_2, \mathbf{r}_3)\,\mathrm{d}\mathbf{r}_3 = \int_0^\infty f(r_{13})f(r_{23})\,4\pi r_3^{\,2}\mathrm{d}r_3 \tag{9.27}$$

Inserting Eqs. (9.26) and (9.27) in Eq. (9.25), one obtains again the virial expression

$$P = \rho kT[1 + \rho B_2(\phi, T) + \rho^2 B_3(\phi, T) + \ldots] \quad \text{where} \tag{9.28}$$

$$B_2(\phi, T) = -\tfrac{1}{2}\int_0^\infty f(r)\,4\pi r^2\mathrm{d}r \quad \text{and} \quad B_3(\phi, T) = \tfrac{1}{3}\int_0^\infty f(r)A_{21}(r)\,4\pi r^2\mathrm{d}r \tag{9.29}$$

Example 9.1 ***The Hard-Sphere Fluid*** To illustrate these concepts, let us apply them to a HS fluid with a potential given by (Figure 9.3) $\phi(r) = \infty$ for $r \le \sigma$ and $\phi(r) = 0$ for $r > \sigma$. Hence, we have

$$f(r) = -1 \text{ for } r \le \sigma \quad \text{and} \quad f(r) = 0 \text{ for } r > \sigma$$
$$\exp(-\beta\phi_{\mathrm{HS}}) = 0 \text{ for } r \le \sigma \quad \text{and} \quad \exp(-\beta\phi_{\mathrm{HS}}) = 1 \text{ for } r > \sigma$$

The function $A_{21}(r)$ deviates from zero only if both $f(r_{12}) \ne 0$ and $f(r_{23}) \ne 0$ and for a HS fluid represents the volume of overlap of two spheres of radius $\sigma/2$ at distance r. Stereometry learns that the overlap function, labeled A_{HS}, reads (Figure 9.3)

$$A_{\mathrm{HS}}(r) = \frac{4\pi\sigma^3}{3}\left(1 - \frac{3r}{4\sigma} + \frac{r^3}{16\sigma^3}\right) \text{ for } r \le 2\sigma, \qquad A_{\mathrm{HS}}(r) = 0 \text{ for } r > 2\sigma$$

This implies that the correlation function $g_{\mathrm{HS}}(r)$ for HS fluid becomes

$$g_{\mathrm{HS}}(r) = h(r - \sigma)(1 + \rho A_{\mathrm{HS}}(r) + \ldots) \tag{9.30}$$

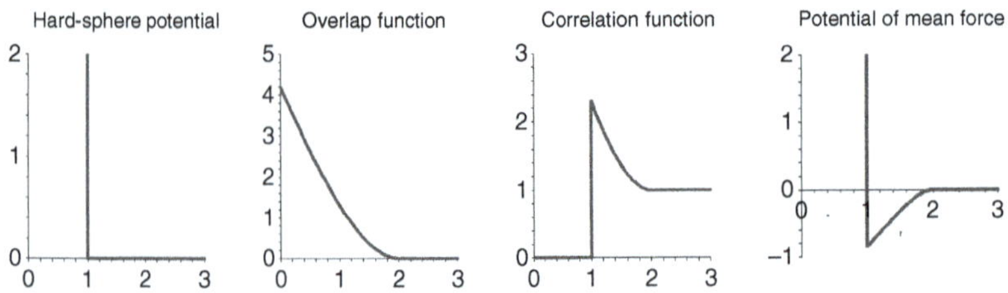

Figure 9.3 The hard-sphere potential ϕ_{HS}, the overlap function A_{HS}, the correlation function g_{HS}, and the potential of mean force W_{HS} as a function of r in units σ.

with the Heaviside function $h(x) = 0$ for $x < 0$ and $h(x) = 1$ for $x > 0$ representing $\exp(-\beta\phi_{HS})$. The second factor in Eq. (9.30) results in a peak in $g_{HS}(r)$ (Figure 9.3). The potential of mean force $W_{HS}(r) = -kT \ln g_{HS}(r)$ thus reads

$$\beta W_{HS}(r) = \beta\phi_{HS}(r) - \ln[1 + \rho A_{HS}(r) + \ldots]$$

The net result is an effective attractive force for $\sigma < r < 2\sigma$ (Figure 9.3), purely a result of geometrical restrictions. Since[3] $\mathrm{d}\exp(-\beta\phi_{HS})/\mathrm{d}r = -\beta \exp(-\beta\phi_{HS})\mathrm{d}\phi_{HS}/\mathrm{d}r = \beta\exp(-\beta\phi_{HS})\,\delta(r-\sigma)$, the pressure P, given by Eq. (9.25a), becomes[4]

$$P/\rho kT = 1 + {}^2\!/_3\pi\rho\sigma^3 y(\sigma) = 1 + 4\eta y(\sigma) = 1 + 4\eta g_{HS}(\sigma^+) \tag{9.31}$$

with $\eta = \pi\sigma^3\rho/6$ the packing fraction. Using Eq. (9.29), we obtain $B_2 = b \equiv 2\pi\sigma^3/3$ and $B_3 = 5b^2/8$; these results were already obtained in Section 8.2.

From Example 9.1 it is clear that the HS correlation function is not just a step function, as one might expect naively (Figure 9.1b), but rather shows an increased value near the reference molecule by the geometric restrictions imposed on the coordination shell, leading to an effective attraction. Briefly summarizing, the correlation function $g(r)$ on average should yield a value of 1. Since $g(r) = 0$ for $r < \sigma$, there should be a region where $g(r) > 1$. Because the potential of mean force $W(r) = -kT \ln g(r)$, the region where $g(r) > 1$ leads to $W(r) < 0$.

9.4 Comparison: Hard-Sphere and Lennard-Jones Results

Let us now discuss the results of some integral equation calculations. As before, we use the number density given by $\rho = N/V$ and the *packing fraction* $\eta = \rho b/4 = \pi\sigma^3\rho/6$ with $b = 2\pi\sigma^3/3$ (Section 8.2.1). The results of such calculations for the HS fluid at low ($\eta = 0.21$) and high density ($\eta = 0.47$) are compared with MC simulation results in Figure 9.4a. From this figure, it becomes clear that both the PY and HNC approximations result in similar pair correlations function as the MC simulation. For low density, $g(r)$ falls monotonically until about 1.5σ, shows a small second maximum at about 2.0σ, and thereafter $g(r) \cong 1$. The large first peak represents the nearest-neighbor distance, and for this density the structure

3 Because $\mathrm{d}h(r-r')/\mathrm{d}r = \delta(r-r')$, we have $\mathrm{d}\phi_{HS}(r-\sigma)/\mathrm{d}r = -\delta(r-\sigma)$. Direct integration of Eq. (9.6) cannot be done because $g_{HS}(r)$ is discontinuous at $r = \sigma$, but can be done using Eq. (9.25) since $y(r)$ is continuous at that point.

4 The notation $g(\sigma^+)$, often written as $g(\sigma)$, defines $g(\sigma^+) \equiv \lim_{r\to\sigma+0} g(r)$, implying that r approaches σ from the positive side. Similarly, $c(\sigma^-) \equiv \lim_{r\to\sigma-0} c(r)$, where r approaches σ from the negative side.

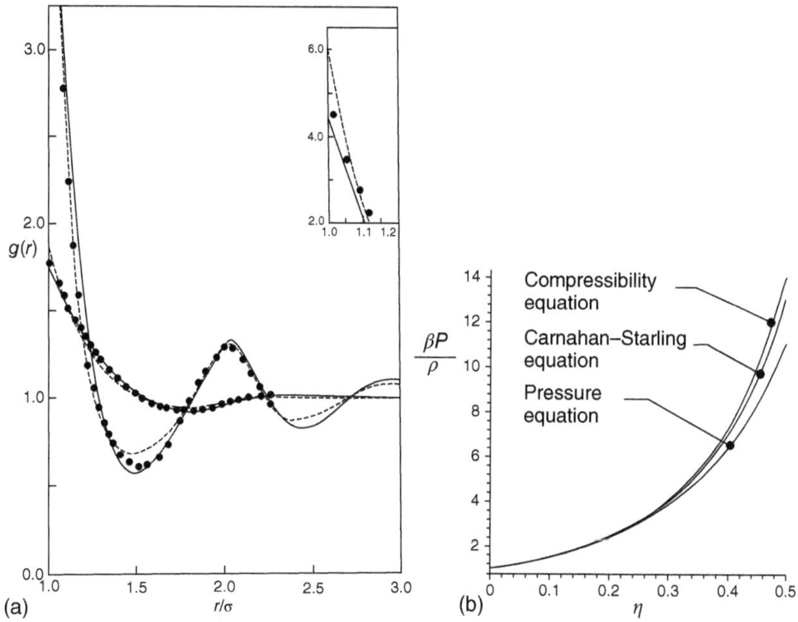

Figure 9.4 (a) The pair correlation function $g(r)$ and EoS for a hard-sphere fluid. (a) Results for $g(r)$ with - - - = HNC, —— = PY, and •••• = Monte Carlo simulation. The curves with the smaller intercept represent a state with $\eta = 0.21$. The other curves represent a state with $\eta = 0.47$ (intercept is shown in inset) [25]. (b) The EoS according to the pressure equation, the compressibility equation, and the Carnahan–Starling approximation for reduced density up to $\eta = 0.494$ ($V/V_0 = 1.50$) at which, according to simulations, the phase transition to a solid occurs.

becomes rapidly random. For high density there is more structure in $g(r)$ with clear second and third neighbor distances. Although the PY and HNC correlation functions are qualitatively quite satisfactory, they have two defects when compared with the MC simulation: the value at contact is too low for the PY solution and too high for the HNC solution, while the oscillations are slightly out of phase for both the PY and HNC solutions.

The results for a Lennard-Jones (LJ) fluid [39] at $\eta = 0.58$ and $T_{\text{red}} = kT/\varepsilon = 2.74$ using the PY, YBG, and HNC approximations are compared with the MC results in Figure 9.5a. Similar to the HS fluid, the overall behavior of the calculated results mimics the MC results reasonably well, although clear differences can be noticed. The YBG result mimics the MC results poorly for the second coordination shell, and all peaks in the calculated results are displaced with respect to the MC results. Figure 9.5b shows the comparison of PY results with experimental data for Ar. The qualitative features are well reproduced, but some differences exist, particularly at low densities. The influence of density is more pronounced than that of temperature, which is to be expected because the structure is largely determined by repulsive forces.

The next step is to calculate the equation of state (EoS). This can be done in two ways: (i) using $P = -\partial F/\partial V$, resulting in the *pressure equation* P_P, and (ii) using the *compressibility equation*, resulting in P_κ. We illustrate these expressions for the HS fluid, not to show the complexity for even a simple potential, but in view of their importance as a reference for more realistic potentials.

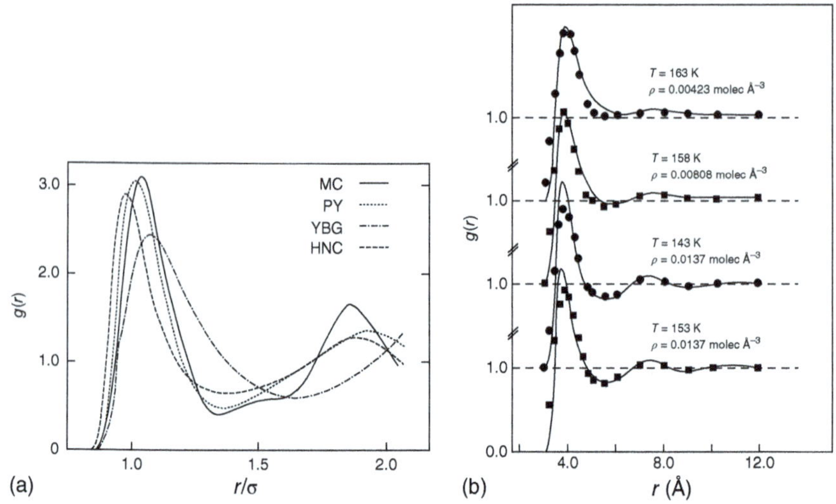

Figure 9.5 (a) The correlation function $g(r)$ for a Lennard-Jones fluid at $\eta = 0.58$ and $T_{red} = kT/\varepsilon = 2.74$, calculated using various methods [39]. (b) The correlation function $g(r)$ for Ar using the PY approximation [40] compared to the experimental data [41]. Source: Adapted from [40, 41].

Surprisingly, it appeared possible to solve the PY equation for the HS fluid using Laplace methods in terms of $c(r)$, which leads to [42–44]

$$c(r) = -\lambda_1 - \frac{\eta\lambda_1 r^3}{2\sigma^3} + \frac{6\eta\lambda_2 r}{\sigma}(r \leq \sigma) \quad \text{and} \quad c(r) = 0 \quad (r > \sigma) \quad \text{with}$$

$$\lambda_1 = (1+2\eta)^2/(1-\eta)^4 \quad \text{and} \quad \lambda_2 = (1+{}^1\!/\!_2\eta)^2/(1-\eta)^4$$

where $\eta = \rho\,[4\pi(\sigma/2)^3/3] = \pi\sigma^3\rho/6$ is used. A piece-wise, rather complex analytical solution for $g(r)$ is also available [45].

Because by definition $y = g/e$ is continuous at $r = \sigma$, we have for the HS fluid in the PY approximation $c(r) = -y(r)$ if $r \leq \sigma$ and $g(r) = y(r)$ if $r \geq \sigma$. Therefore,[3,4] $y(\sigma) = -c(\sigma^-) = g(\sigma^+)$, and we obtain, using Eq. (9.25),

$$\beta P/\rho = 1 + 4\eta y(\sigma) = 1 - 4\eta c(\sigma^-) = 1 + 4\eta g(\sigma^+)$$

Inserting the expression for $c(r)$, the result is

$$\beta P_P/P = (1 + 2\eta + 3\eta^2)/(1-\eta)^2$$

Using the compressibility, Eq. (9.15) can be shown to lead to

$$\beta P_\kappa/P = (1 + \eta + \eta^2)/(1-\eta)^3$$

The difference between P_P and P_κ shows the lack of thermodynamic consistency. An excellent description of the HS EoS (which means matching the simulation results with an average absolute deviation [AAD] of 0.33% over the whole range $0 \leq \eta \leq 0.49$) is a weighted average of the above two results, the *Carnahan–Starling* (CS) expression [46], Eq. (8.6), given by

$$\frac{\beta P_{CS}}{\rho} = \frac{1}{3}\frac{\beta P_P}{\rho} + \frac{2}{3}\frac{\beta P_\kappa}{\rho} = \frac{1+\eta+\eta^2-\eta^3}{(1-\eta)^3}$$

Since $\beta P/\rho = 1 + 4\eta y$, we also have

$$y_{\text{CS}} = (1 - \tfrac{1}{2}\eta)/(1 - \eta)^3$$

The reduced pressure $\beta P/\rho$ according to these three equations is shown in Figure 9.4b. A comparison of a variety of HS EoSs is made in [47, 48].

For realistic potentials, the integral equations can only be solved numerically and, consequently, the EoS can also be obtained only numerically. We will not pursue this topic further, but refer the reader to the literature, for example, [8, 27, 49]. Both the PY and HNC approximations are superior to the YBG approximation. Nowadays, the PY equation is often considered to be the best approximation for describing simple liquids with strong repulsions and short-range attractions, while for systems with long-range interactions, the HNC equation appears superior.

Problem 9.3 Derive Eqs. (9.14), (9.21), (9.22), and (9.23).

9.5 Scaled-Particle Theory

In Section 9.3.4, we indicated that using the potential of mean force for HS with diameter σ the pressure P can be described by Eq. (9.31), reading

$$\beta P/\rho = 1 + 4\eta y(\sigma) = 1 - 4\eta c(\sigma^-) = 1 + 4\eta g(\sigma^+)$$

with the various functions defined in Section 9.3. We also already mentioned that $y(r)$ is also denoted as the cavity function, and in this model we need only the value of $y(r)$ at $r = \sigma$ to obtain P. This observation led to *scaled-particle theory* (SPT) [50].

Consider a spherical cavity of radius r inside a fluid and let $p_0(r)$ be the probability that no molecular center lies inside this sphere. We take the value of the number density of HS just outside this cavity as $\rho G(r)$. Note that a cavity plays exactly the same role as an HS of diameter $2r - \sigma$. When $r = \sigma$, the cavity behaves as a typical molecule and $\rho G(\sigma)$ can be regarded as the number density at a distance σ from a specified particle. Thus, we take $G(\sigma) = g(\sigma)$, leading to

$$\beta P/\rho = 1 + 2\pi\rho\sigma^3 G(\sigma)/3 \tag{9.32}$$

Consequently, instead of calculating $g(r)$, we evaluate $G(r)$ at $r = \sigma$.

Now the probability of finding a particle in a spherical shell between r and $r + \mathrm{d}r$ is $4\pi r^2 \rho G(r)\mathrm{d}r$. Hence, the joint probability of finding an empty sphere of radius r and no particle in a shell between r and $r + \mathrm{d}r$ is given by

$$p_0(r + \mathrm{d}r) = p_0(r)[1 - 4\pi r^2 \rho G(r)\,\mathrm{d}r] \quad \text{or} \quad \mathrm{d}\ln p_0(r)/\mathrm{d}r = -4\pi r^2 \rho G(r) \tag{9.33}$$

where in the last step the limit $\mathrm{d}r \to 0$ is taken. For $r < \tfrac{1}{2}\sigma$, at most one sphere can lie in a cavity of radius r, implying that the probability $1 - p_0(r)$ is $4\pi r^3\rho/3$. The solution of Eq. (9.33) is thus

$$G(r) = [1 - 4\pi r^3 \rho/3]^{-1} \quad \text{for } 0 < r < \tfrac{1}{2}\sigma$$

One can show that $G(r)$ and its first derivative are continuous at $r = \tfrac{1}{2}\sigma$, but that the second derivative has a finite discontinuity due to the possible introduction of a second sphere. At

$r = \sigma/3^{1/2}$, two spheres can be included, and a discontinuity occurs in the fourth derivative. With increasing r, more and more spheres can be accommodated, and it appears that at the transition from accommodating L spheres to $L+1$ spheres, a discontinuity occurs in the $2L$th derivative of $G(r)$.

Considering next the behavior for large r, we recall first that

$$p_0(r) = \exp[-\beta W(r)] \tag{9.34}$$

where $W(r)$ is the work to create a cavity of radius r or the potential of mean force for this process. From Eqs. (9.33) and (9.34), we have

$$\mathrm{d}W(r) = 4\pi kTr^2 \rho G(r)\,\mathrm{d}r \quad \text{and} \quad \mathrm{d}W(r) = P\mathrm{d}V + \gamma[1-(2\delta/r)]\,\mathrm{d}A \tag{9.35}$$

as obtained from thermodynamics with the volume increment $\mathrm{d}V = 4\pi r^2\mathrm{d}r$, the surface increment $\mathrm{d}A = 8\pi r\mathrm{d}r$, the pressure P, the surface tension γ, and the Tolman length δ for a cavity of radius r (Section 11.2). Combining Eqs. (9.35a) and (9.35b) leads to

$$\rho kTG(r) = P + 2\gamma[1-(2\delta/r)]/r \quad \text{for} \quad r \gg \sigma$$

Although $G(r)$ is not an analytic function, it is still continuous because the discontinuities occur in the high-order derivatives. Hence, we use $G(r)$ as an interpolation function for the complete range $\tfrac{1}{2}\sigma < r < \infty$ and determine the quantities P, γ, and δ from Eq. (9.32) and the conditions that $G(r)$ and its first derivative should be continuous at $r = \tfrac{1}{2}\sigma$. This results, using $\eta = \pi\rho\sigma^3/6$, in

$$\frac{\beta P}{\rho} = \frac{1+\eta+\eta^2}{(1-\eta)^3}, \quad \frac{\beta\gamma}{\sigma\rho} = \frac{-3(\eta+\eta^2)}{4(1-\eta)^3} \quad \text{and} \quad \frac{\delta}{\sigma} = \frac{\eta}{4(1-\eta)}$$

Note that the expression for P is identical to the compressibility equation for HSs in the PY approximation and thus close to the CS approximation.

The above presents the basic model. Systematic relations [51] improving $G(r)$ can be generated and the single cavity approximation can be avoided. The model has been successfully applied to estimate interactions of, for example, hydrophobic solutes in water [52, 53].

9.6 Structural Models

In *structural models*, one uses structural and physical arguments to construct a lattice model that can handle normal and complex liquids. Clearly being approximate, such models nevertheless provide reasonable explanations and predictions.

In a lattice, atoms are located at fixed positions, while in a gas they can be everywhere within the enclosure of the gas. In a liquid, the molecules are much less confined than in a solid but much more so than in a gas. Consequently, their positions change more gradually, and it is tempting to use instead of the perfect gas as for the virial expansion a lattice as a reference configuration. In *(quasi-)lattice models*, one assumes a lattice-like structure on average for the liquid, at least for the first coordination shell. There is some, but only some, experimental justification for this. In Figure 9.6, the self-correlation function for Ar at 84.5 K is shown [54] describing the probability of finding a molecule, that was at time $t = 0$ at position $r = 0$, at a time t at position r. This function indicates how fast a molecule moves on average from its reference position at time $t = 0$. Typically, within 3 to 10×10^{-13} s

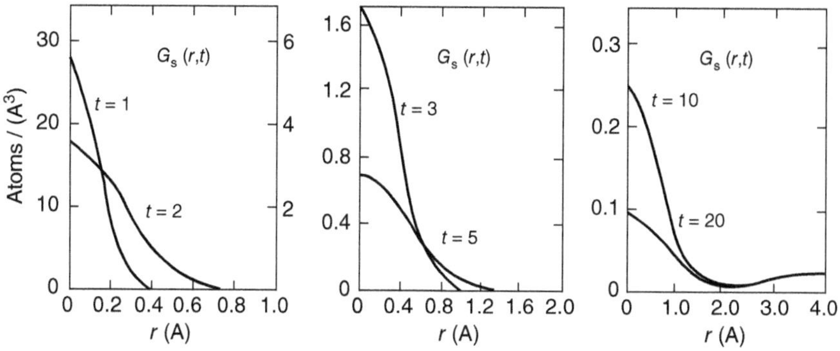

Figure 9.6 The self-correlation function for Ar at 84.5 K showing the probability to find a particle at position r at time t that was at position $r = 0$ at time $t = 0$ [54]. The time t is in units of 10^{-13} seconds. Source: Dasannacharya and Rao [54]/American Physical Society.

the original structure fades away, corresponding with about 3 to 10 vibrations, estimating the typical vibration frequency as 10^{13} s^{-1}. Moreover, XRD measurements on liquids show that intermolecular distances in liquids are very similar to those in the corresponding solid, although solids expand upon melting; for example, for Ar $\Delta_{mel}V \cong 14\%$. There is also some justification from simulations. So, a marginal resemblance with a lattice exists.

It appears that the structure of a liquid is largely determined by the repulsive and attractive interactions in the first coordination shell [55], while the remaining attractions mainly serve to provide cohesion, rendering the quasi-lattice concept useful. Critics of the quasi-lattice concept have often taken the lattice concept too literally. Actually, all that is required is a well-defined, average *coordination number* z and *nearest-neighbor* or *contact energy* u [56]. Hence, instead of a lattice, a random dense packing will do as well. If explicit spatial dependence of u is neglected, u will be T and P dependent. Separation of the internal DoFs from the translation DoFs is almost inevitably involved, which might not be always warranted.

For solutions, it is assumed that the molecules are sufficiently alike in size and shape so that they can pack in the same manner in the mixture as in the pure liquid, implying a constant value[5] for z. Such a temporary local structure around a reference molecule is denoted as a *cell* or a *cage*. The molecule we have in focus, the reference molecule, is called the *wanderer*, and its coordinating molecules constitute the *wall* of the cell. In this picture, each molecule thus "rattles" in a cell whose wall consists of the neighboring molecules. Since any molecule can be considered as a wanderer or as a wall molecule, this constitutes a basic inconsistency, which we must accept for the moment to make progress. Lattice models in which the number of cells equals the number of molecules are addressed as *cell models*, while models in which the number of cells is larger than the number of molecules are called *hole models*. In lattice models, one uses the concept of *free volume*,[6] that is, the volume swept

5 Actually, z being a nonconstant is the most important, but not often voiced, criticism that can be made for solutions. However, recent re-examination of quasi-lattice concepts rigorously showed that z is a constant for a solute belonging to the same "class" of molecules as the solvent [57].

6 The label "free volume" can easily mean different things to different people. Bondi [58] distinguished between (i) the *empty volume* $V_e = V - V_{vdW}$, with V_{vdW} the vdW volume, (ii) the *expansion volume* $V_{exp} = V - V_0$, with V_0 the (crystalline) volume at 0 K, and (iii) the *fluctuation volume* V_f, denoted here as the free volume. Actually, in many discussions, V_f is approximated by V_e, although it has been argued that $V_e = V - aV_{vdW}$ with $a \cong 1.3$ is a better choice [59].

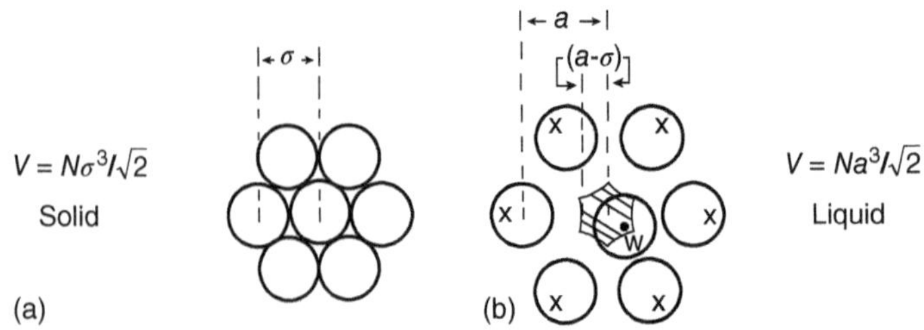

Figure 9.7 (a) A two-dimensional (2D) representation of the FCC lattice for the solid state. (b) Similar for the liquid state but with the distance between the atoms increased from σ to a so that the "smeared" diameter of the free volume, indicated by the hatched hexagon, becomes $a-\sigma$.

out by the center of gravity of the molecules due to thermal motion. A brief introduction to lattice models is [60] while [61, 62] review many details.

9.6.1 Cell Models

Let us use a cell model with a lattice-like structure in which we introduce some free volume. As the density of a liquid deviates not too much from that of the solid, we use for the lattice a dense, close-packed lattice such as the FCC lattice structure (Figure 9.7) with coordination number $z = 12$. For an FCC lattice with N atoms in volume V_0, the volume per molecule is $v_0 = V_0/N = \sigma^3/\gamma$, where $\gamma = 2^{1/2} = 1.414$ and σ is the molecular diameter.[7] Expanding this lattice slightly, we have $a^3/\gamma = v = V/N$. The configuration integral for the wanderer (molecule 1) represents the volume available to the wanderer weighted by a Boltzmann factor with the potential energy ϕ. This effective volume is called the *free volume* v_f and we have, using $\beta = 1/kT$,

$$Q(1) = v_\text{f} = \int \exp[-\beta\phi(\boldsymbol{r}_1)]\,\mathrm{d}\boldsymbol{r}_1 = \int \exp[-\beta\phi(r)]\,4\pi r^2 \mathrm{d}r$$

where the integration is over all of space, but as the wanderer cannot leave the cell may be taken over the cell. If we consider for the moment molecules as HSs, with for the average attraction $\overline{\phi}$ the van der Waals (vdW) expression [63], we have

$$\overline{\phi} \cong -a(T)/V \quad \text{inside the cell and} \quad \overline{\phi} = \infty \text{ outside the cell}$$

with $a(T) \geq 0$ a parameter that is dependent on temperature only. This leads to

$$Q(1) = \int \exp[-\beta\phi(\boldsymbol{r}_1)]\,\mathrm{d}\boldsymbol{r}_1 = \exp[-\beta\overline{\phi}]\,v_\text{f} \tag{9.36}$$

and results via $F_\text{con}(N) = -kT \ln Q(1)^N$ and $P = -\partial F/\partial V$ in the *Eyring EoS*

$$[P + Na(T)/V^2]\left(V - V_0^{1/3}\,V^{2/3}\right) = NkT \tag{9.37}$$

resembling the vdW equation and providing a useful semiempirical description of liquids. The factor $a(T)/V^2$ is sometimes called the *internal pressure*.

In a more sophisticated cell model by Lennard-Jones and Devonshire (LJD) [64], the FCC lattice is used in combination with the Lennard-Jones (LJ) potential $\phi_\text{LJ}(r)$. In the original LJD model, the interaction of the wanderer with its $z = 12$ nearest neighbors is calculated

7 T_ideal: $z = 10$, $\gamma = 4/3$; BCC: $z = 8$, $\gamma = 3^{3/2}/4 = 1.299$; SC: $z = 6$, $\gamma = 1$; D: $z = 4$, $\gamma = 3^{3/2}/8 = 0.650$.

by supposing that the neighbors are distributed with equal probability, that is, they are "smeared", over a sphere with a radius $a = (\gamma V/N)^{1/3} \equiv (\gamma/\rho)^{1/3}$, using a cell radius $a^* \geq 0.5a$. For this spherical cell, the pair potential ϕ is a function of the distance r from the center of the cell only. At the center, we have for each pair

$$\phi(0) = \tfrac{1}{2}\int_0^{\pi} \phi_{\mathrm{LJ}}(a)\,\sin\theta\,\mathrm{d}\theta$$

while for a position off-center, an extra contribution $\psi(r)$ must be added so that $\phi(r) = \phi(0) + \psi(r)$. The authors showed that, using $v_0/r_0^{\,3} = v/a^3$,

$$\phi(0) = \varepsilon\left[(r_0/a)^{12} - 2(r_0/a)^6\right] = \varepsilon\left[(v_0/v)^4 - 2(v_0/v)^2\right] \quad \text{and}$$
$$\psi(r) = \varepsilon\left[(r_0/a)^{12} l(r/a_1) - 2(r_0/a)^6 m(r/a_1)\right] \quad \text{where}$$
$$l(x) = (1 + 12x^2 + 25.2\,x^4 + 12x^6 + x^8)(1 - x^2)^{-10} - 1 \quad \text{and}$$
$$m(x) = (1 + x^2)(1 - x^2)^{-4} - 1$$

For a wanderer with z nearest neighbors, the partition function $Q = Q(1)^N$ becomes

$$Q = \exp[-\tfrac{1}{2}\beta z N\phi(0)]\,v_{\mathrm{f}}^N \quad \text{with} \quad v_{\mathrm{f}} = \int_0^{a^*} \exp[-\beta\psi(r)]\,4\pi r^2\mathrm{d}r$$

The choice of a^* depends on the "smearing" assumptions. It appears that $\psi(r)$ becomes large and positive for $r \to 0.5a$. Lennard-Jones and Devonshire therefore took $a^* = 0.5a$. Because the introduction of the LJ potential complicates the matter (see, e.g. [60, 64–66]), the necessary integrations must be performed numerically.

For the pressure sinuous curves are predicted below and monotonic curves above a certain temperature. Identifying this temperature with T_{cri}, the model predicts

$$P_{\mathrm{cri}} = 0.434\,\varepsilon/\sigma^3 \qquad V_{\mathrm{cri}} = 1.77\,N_{\mathrm{A}}\sigma^3 \qquad T_{\mathrm{cri}} = 1.30\,\varepsilon/k$$

to be compared with the "best" experimental corresponding states estimates

$$P_{\mathrm{cri}} = 0.116\,\varepsilon/\sigma^3 \qquad V_{\mathrm{cri}} = 3.14\,N_{\mathrm{A}}\sigma^3 \qquad T_{\mathrm{cri}} = 1.25\,\varepsilon/k$$

We see that the value of T_{cri} is about right, the value of V_{cri} is too small by a factor of about 2, and P_{cri} is too large by a factor of about 4.

It is not very realistic to expect that the wanderer will be permanently confined to the cell at or near T_{cri}. We expect that at or near the triple point T_{tri}, this assumption is more realistic and agreement with the experiment is better. For Ar, $T_{\mathrm{tri}} = 0.701\varepsilon/k$, and at this temperature the pressure is so low that the molar volume may be obtained by setting $P = 0$ in the EoS. Table 9.1 provides some data for Ar at T_{tri}. Overall, the values of the properties calculated with LJD theory are much closer to the experimental values for solids than for liquids, as might be expected.

Later non–nearest-neighbor effects were included for $\phi(0)$ and v_{f} via a lattice sum calculation (Section 10.6) [67], but this did not lead to significant improvement. Also, the volume of the cell was chosen as equal to the volume per molecule or $4\pi a^{*3}/3 = a^3/\gamma$, so that $a^* = 0.5527a$. The precise choice appears to be noncritical. Moreover, the "smearing" approximation for HSs appeared to have limited influence [65]. Allowing double occupancy of cells leads to somewhat improved results [68], but the properties appear to be insensitive to the precise potential used [69]. We further note that cell model calculations have been made using a random, close-packed structure and employing the LJ potential [70].

Table 9.1 Data for Ar at the triple point T_{tri}.

	kT_{tri}/ε	$V/N\sigma^3$	$U/N\varepsilon$	S/Nk	C_V/Nk
LJD theory	0.701	1.037	−7.32	−5.51	1.11
SLS theory	0.711	1.159	−6.19	−3.89	—
Solid Ar experimental	0.701	1.035	−7.14	−5.53	1.41
Liquid Ar experimental	0.701	1.178	−5.96	−3.64	0.85

SLS = significant liquid structure.
Experimental data are from [62].

The energy obtained compares favorably with MC and MD results, although the pressure appears to be rather sensitive to the correlation function used (conform Section 9.3). Extending the model by using a non-zero probability for being off-center for all molecules removes one of the basic inconsistencies (Section 9.6) and leads to slightly improved results [70, 71]. Overall, these conceptual improvements do not significantly improve the agreement with the experiment.

The most important deficit of cell models is the *communal entropy* problem, which is due to the assumption that the wanderer cannot leave the cell. To illustrate this problem, consider a perfect gas. The correct partition function Z and the Helmholtz energy F for the perfect N-atom gas are given by, using $\Lambda = (h^2/2\pi mkT)^{1/2}$,

$$Z = Z(N) = (\Lambda^{3N}N!)^{-1}V^N \quad \text{and} \quad F(N)_{\text{correct}} = -kT \ln Z(N)$$

respectively. For the cell model, Z for the wanderer and F for N atoms are given by

$$Z = Z(1) = \Lambda^{-3}(V/N) = \Lambda^{-3}v_{\text{f}} \quad \text{and} \quad F(N)_{\text{cell model}} = -NkT \ln Z(1)$$

respectively. The communal Helmholtz energy is thus given by

$$\Delta_{\text{com}}F = F(N)_{\text{correct}} - F(N)_{\text{cell model}}{}^{N} = -kT \ln(N^N/N!) = -NkT$$

Hence, for one mole the cell model estimates the molar entropy too low by $\Delta_{\text{com}}S = R$. Obviously, the above argumentation also applies to liquids because we know that also in liquids the atoms can move to any position, albeit somewhat restrictedly. For solids, where atoms essentially remain at their positions, the approach would be correct, and the cell model is thus describing a solid-like situation rather than a liquid-like situation. To obtain the correct limiting behavior for gases, Eyring [66] made the arbitrary assumption that $\Delta_{\text{com}}S$ is completely released during melting, which leads to an extra factor e^N in the partition function Z_N. The issue was resolved by Hoover and Ree [72] by using MC simulations for HSs. They essentially calculated $\Delta_{\text{com}}S$ by subtracting from the entropy of the unconstrained, normal HS fluid that of a constrained fluid. In the latter model, each sphere can move only within its own cell, in a similar fashion as in the cell model and consistent with the definition of $\Delta_{\text{com}}S$ (Justification 9.3). Data for varying system sizes were extrapolated to those for an infinite system, and from these calculations it appeared that the HS (disordered) fluid and (ordered) solid phases are in equilibrium with densities $\rho_{\text{L}} = (0.667 \pm 0.003)\rho_0$ and $\rho_{\text{S}} = (0.736 \pm 0.003)\rho_0$, respectively. Here $\rho_0 = v_0{}^{-1} = (\sigma^3/\gamma)^{-1}$ is the close-packed density with $\gamma = 2^{1/2}$. The melting pressure and temperature are related by $P_{\text{mel}} = (8.27 \pm 0.13)\rho_0 kT_{\text{mel}}$ (Figure 9.8). Hence the HS solid expands by about

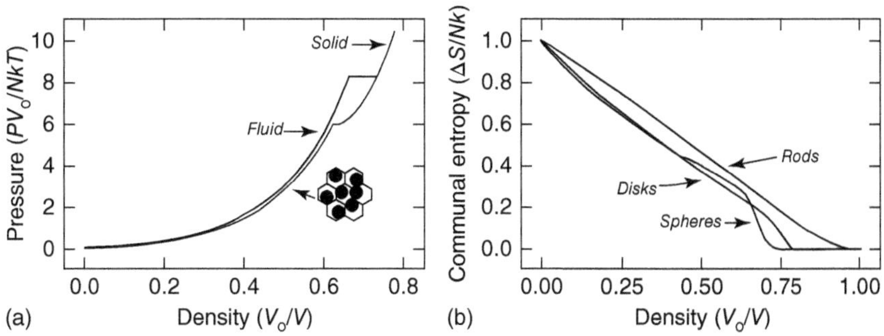

Figure 9.8 (a) The equation of state for three-dimensional (3D) hard spheres for the fluid and solid state, where the solid and fluid isotherms are connected by a tie line at equal chemical potential at $PV_0/NkT = 8.27$. (b) The communal entropy for 1D, 2D, and 3D hard spheres. Source: Hoover and Ree [72]/AIP Publishing.

10% upon melting. The density of the solid shows a cusp at $0.637\rho_0$ where the cell walls begin to become important. At this point, the crystal becomes mechanically unstable and without the walls it would rapidly disintegrate. It also appeared that $\Delta_{\text{com}}S$ decreases approximately linearly from $\Delta_{\text{com}}S = 1.0Nk$ at $\rho = 0$, corresponding to the factor e^N in Q, to $\Delta_{\text{com}}S \cong 0.25Nk$ at $\rho \cong 0.65\rho_0$ (Figure 9.8b). There is no simple way to either solve or circumvent the communal entropy problem for cell models, although some other approaches do so (Section 9.6.2).

To determine v_f one can use, for example, the thermal expansivity α_P combined with compressibility κ_T or the vapor pressure P_{vap} (Problem 9.5). Estimates based on α_P and κ_T data seem to be the most reliable if accurate data are available. Finally, we note that a single NVT MD simulation can provide a "calibration" of the cell model that takes into account the calculation of the cell properties, includes the particle correlations and accounts for $\Delta_{\text{com}}S$ [73]. The Helmholtz energies obtained give good agreement with the experiment over a wide T and P range.

Justification 9.3 ***Kirkwood's Analysis*** Consider a system of N interacting identical particles with partition function $Z_N \equiv (\Lambda^{3N}N!)^{-1}Q_N$ [36]. Divide coordinate space into N cells $\Delta_1, \Delta_2, \ldots, \Delta_N$, each of volume Δ, in an arbitrary manner, so that Q_N becomes a sum of terms in which the coordinates of the molecules are confined to a particular cell. Thus, Q_N will contain N^N integrals, corresponding to the N^N ways of positioning N molecules in N cells. Hence

$$Q_N = \sum\nolimits_{l_1=1}^{N} \sum\nolimits_{l_2=1}^{N} \cdots \sum\nolimits_{l_N=1}^{N} \int_{\Delta_{l_1}} \int_{\Delta_{l_2}} \cdots \int_{\Delta_{l_N}} \exp[-\beta\Phi(\mathbf{r})]\,\mathrm{d}\mathbf{r} \tag{9.38}$$

with $\Phi = \Sigma_{i<j}\phi_{ij}$. Now denote by $Q_N(m_1, m_2, \ldots, m_N)$ the various terms in Eq. (9.38) corresponding to m_1 molecules in cell 1, m_2 molecules in cell 2, …, and m_N molecules in cell N. Because the molecules are identical, there are $(N!/\Pi_i\, m_i!)$ ways of obtaining a particular set $\{m\} = m_1, m_2, \ldots, m_N$. Therefore, Q_N becomes

$$Q_N = \sum\nolimits_{\{m\}} \frac{N!}{\prod_i m_i!} Q_N(m_1, m_2, \ldots, m_N) \quad \text{or} \quad Q_N/N! = \overline{\sigma}^N Q_N^{(1)} \quad \text{in which}$$

$$Q_N^{(1)} \equiv Q_N(1,1,\ldots,1) \quad \text{and} \quad \overline{\sigma}^N \equiv \sum\nolimits_{\{m\}} Q_N(m_1, m_2, \ldots, m_N)/Q_N^{(1)} \prod\nolimits_i m_i!$$

The summation is over all sets $\{m\}$ subject to the constraint $\Sigma_i m_i = N$. At sufficiently high densities, the repulsive interaction prevents multiple occupancy, so that all $Q_N(m_1, m_2, \ldots, m_N)$ are zero except $Q_N^{(1)}$ and $\overline{\sigma} = 1$. On the other hand, for highly dilated gases there is effectively no interaction between the molecules and all $Q_N(m_1, m_2, \ldots, m_N)$ are equal. Hence, $\overline{\sigma} = \Sigma_{\{m\}}(\Pi_i\, m_i!)^{-1}$ so that $\overline{\sigma}^N = N^N/N! = \mathrm{e}^N$, using Stirling's approximation, or $\overline{\sigma} = \mathrm{e}$. Hence, generally $1 \leq \overline{\sigma} \leq \mathrm{e}$ and its change with density represents the *communal entropy*. The derivative of $\overline{\sigma}$ with respect to V contributes to P. To make the next step, we need the probability density

$$P_N^{(1)}(\mathbf{r}) = \left(Q_N^{(1)}\right)^{-1} \exp[-\beta\Phi(\mathbf{r})]$$

where $Q_N^{(1)}$ is the configurational partition function for a system of N interacting molecules, each of which is required to remain in its particular cell. We approximate $P_N^{(1)}(\mathbf{r})$ by a product of functions $s(\mathbf{r}_i)$, each dependent only on the coordinates of one molecule $\mathbf{r}_i$ with respect to the origin in the cell, that is,

$$P_N^{(1)} = \prod\nolimits_i s(\mathbf{r}_i) \qquad \text{with normalization condition} \quad \int_\Delta s(\mathbf{r}_i)\,\mathrm{d}\mathbf{r}_i = 1$$

The optimum choice for $s(\mathbf{r}_i)$ is obtained by minimizing $F = U - TS$ at constant T and Δ subject to the normalization constraint of $s(\mathbf{r}_i)$. Here,

$$F = -kT\ln(\Lambda^{3N}N!)^{-1}Q_N = -kT\ln \Lambda^{-3N}\overline{\sigma}^N\, Q_N^{(1)} = -NkT\ln \Lambda^{-3}\overline{\sigma} + F^{(1)}$$

where $F^{(1)} = -kT\ln Q_N^{(1)}$ is the configuration Helmholtz energy for N interacting molecules restrained to their cells. The configuration energy and entropy are given by

$$U_N^{(1)} \equiv \int_\Delta P_N^{(1)}(\mathbf{r})\Phi(\mathbf{r})\,\mathrm{d}\mathbf{r} \quad \text{and} \quad S_N^{(1)} \equiv -k\int_\Delta P_N^{(1)}(\mathbf{r})\ln P_N^{(1)}(\mathbf{r})\,\mathrm{d}\mathbf{r}$$

respectively. Therefore, $F^{(1)}$ becomes, after some manipulation,

$$F^{(1)}/NkT = \int_\Delta s(\mathbf{r})\ln s(\mathbf{r})\,\mathrm{d}\mathbf{r} + {}^1\!/\!_2\beta\int_{\Delta'}\int_\Delta Y(\mathbf{r}-\mathbf{r}')s(\mathbf{r})s(\mathbf{r}')\,\mathrm{d}\mathbf{r}\mathrm{d}\mathbf{r}'$$

with $Y(\mathbf{r}) = \Sigma_{j=2}\phi\,(\mathbf{R}_{1j} - \mathbf{r})$ as the contribution to the potential energy due to molecule 1 at position $\mathbf{r}$ with respect to the origin of cell 1 with all other particles located at their respective origins, and $\mathbf{R}_{1j}$ contains the origins of all j cells with respect to the origin of cell 1. To find the optimum, we need to solve

$$\delta F = \int_\Delta \left[kT\ln s(\mathbf{r}) + \int_{\Delta'} Y(\mathbf{r}-\mathbf{r}')s(\mathbf{r})s(\mathbf{r}')\,\mathrm{d}\mathbf{r}'\right]\delta s(\mathbf{r})\,\mathrm{d}\mathbf{r} = 0$$

$$\text{subject to the normalization condition} \int_\Delta \delta\, s(\mathbf{r})\,\mathrm{d}\mathbf{r} = 1 \tag{9.39}$$

Using Lagrange multipliers leads, again after some manipulation, to

$$s(\mathbf{r}) = \exp[\beta(\alpha-\psi)] \quad \text{with} \quad \exp(-\beta\alpha) = \int_\Delta \exp[-\beta\psi(\mathbf{r})]\,\mathrm{d}\mathbf{r}$$

$$\text{where} \quad \psi(\mathbf{r}) = \int_\Delta w(\mathbf{r}-\mathbf{r}')\exp\{\beta[\alpha-\psi(\mathbf{r}')]\}\,\mathrm{d}\mathbf{r}' \quad \text{and}$$

$$w(\mathrm{r}) = Y(\mathrm{r}) - Y_0 \quad \text{with} \quad Y_0 = \int_{\Delta'}\int_\Delta Y(\mathbf{r}-\mathbf{r}')\,s(\mathbf{r})s(\mathbf{r}')\,\mathrm{d}\mathbf{r}\mathrm{d}\mathbf{r}'$$

This solution provides the best approximation to $P_N^{(1)}(\mathbf{r})$ in terms of $s(\mathbf{r})$. The total partition function Z_N and the *free volume* v_f then become

$$Z_N = (\Lambda^{-3}\overline{\sigma}\, v_f)^N \exp(-\tfrac{1}{2}\beta N Y_0) \quad \text{and} \quad v_f \equiv \int_\Delta \exp[-\beta\psi(\mathbf{r})]\,\mathrm{d}\mathbf{r} \tag{9.40}$$

From these expressions, the thermodynamic properties can be obtained in the usual way after a choice for $\overline{\sigma}$ and $\psi(\mathbf{r})$ is made. As zeroth approximation, we assume that the molecules are located at the center of their cells and, using the Dirac delta function, we have $s(\mathbf{r}) = \delta(\mathbf{r})$, leading to

$$\psi(\mathbf{r}) = \sum\nolimits_{l=2}^{N}[\phi(\mathbf{R}_{1l}+\mathbf{r}) - \phi(\mathbf{R}_{1l})] \quad \text{and} \quad Y_0 = \sum\nolimits_{l=2}^{N}\phi(\mathbf{R}_{1l})$$

Restricting the sums to nearest neighbors, replacing this sum by an integral over a sphere of radius equal to the nearest-neighbor distance, and taking $\overline{\sigma} = \mathrm{e}$ yield the LJD model. We have $Y_0 = z\phi(0)$ and $\psi(r) = z[\phi(r) - \phi(0)]$, and the results become

$$Z_N = (\Lambda^{-3}\mathrm{e}\, v_f)^N \exp(-\tfrac{1}{2}\beta z N\phi(0)) \quad \text{and} \quad v_f = \int_\Delta \exp[-\beta\psi(\mathbf{r})]\,\mathrm{d}\mathbf{r} \tag{9.41}$$

For HSs the solution of Eq. (9.41) is $PV/NkT = [1-(v_0/v)^{1/3}]^{-1}$ [74]. This analysis represents a significant step in theory, rendering it clear that the problem is evaluating $\overline{\sigma}$. An expression for $\overline{\sigma}$ for HSs may be based on the MC results of Hoover and Ree [72], say, by a simple linear relation such as $\overline{\sigma} = Nk(1-1.15v_0/v)$ for the range $0 \le v_0/v \le 0.65$, but such an approach is ad hoc. Similarly, for LJ particles Kuramoto [75] proposed an expression for $\overline{\sigma}$ describing MD results for the EoS and melting curve well, but, as stated by the author, having no theoretical ground.

Problem 9.4 ***Free Volume via α_P and κ_T***

(a) Using F with the potential $\overline{\phi}$ and the appropriate Gibbs–Helmholtz expression, show that for the cell model the internal energy is $U/N_A = -\overline{\phi} + 3kT/2$, the pressure $P = \partial\overline{\phi}/\partial V + kT\,\partial\ln v_f/\mathrm{d}V$ and, noting that $\overline{\phi}$ and v_f are supposed to be functions of V only, that $(\partial P/\partial T)_{V,N} = k\,\partial\ln v_f/\mathrm{d}V$ with $v_f = 4\pi\gamma(v^{1/3} - v_0^{1/3})^3/3$.
(b) Show that, using expansivity α_P and compressibility κ_T, that $(\partial P/\partial T)_{V,N} = \alpha_P/\kappa_T$.
(c) Show that $\partial\ln v_f/\mathrm{d}V = (4\pi\gamma/3)^{1/3}/V^{2/3}v_f^{1/3}$ and thus that $v_f = (4\pi\gamma/3V^2)(k\kappa_T/\alpha_P)^3$.
(d) Taking a typical value for normal liquids, $\alpha_P \cong 1\times10^{-3}\ \mathrm{K}^{-1}$ and $\kappa_T \cong 1\times10^{-4}\ \mathrm{bar}^{-1}$, show that the typical value $V_m = 80\ \mathrm{cm}^3\ \mathrm{mol}^{-1}$, $V_f = N_A v_f \cong 0.44\ \mathrm{cm}^3\ \mathrm{mol}^{-1}$.

Problem 9.5 ***Free Volume via P_{vap}*** Consider a cell model with an average attraction potential $\overline{\phi}$ so that the total attraction $N\overline{\phi}$ represents the energy of evaporation $\Delta_{\mathrm{vap}}U$. Using $Q(1)$ given by Eq. (9.36), calculate the chemical potential μ of the cell model. Show, by equating the chemical potential for the liquid and the (perfect) gas, that the vapor pressure $P_{\mathrm{vap}} = (kT/v_f)\exp(-\Delta_{\mathrm{vap}}H/kT)$, where $\Delta_{\mathrm{vap}}H = \Delta_{\mathrm{vap}}U + PV_L$ is the enthalpy of vaporization. Compare its behavior with that of the Clapeyron–Clausius equation, Eq. (12.7). Also show that for chloroform, using the same data as in Problem 9.4, $V_f \cong 1.0\ \mathrm{cm}^3\ \mathrm{mol}^{-1}$.

Problem 9.6 ***Hildebrand's Rule and Trouton's Constant*** The result of Problem 9.5, assuming $\overline{\sigma} = \mathrm{e}$ for the liquid phase and $\Delta_{\mathrm{vap}}U = \Delta_{\mathrm{vap}}H - RT = a/V$, can be used to (approximately) validate Hildebrand's rule and to estimate the value of Trouton's constant (Section 13.1).

(a) Use Eq. (9.37) in combination with v_f while neglecting P with respect to a/V^2 to show that $\zeta PV/RT = [\Delta_{vap}H/RT - 1]^{-1} \exp(-\Delta_{vap}H/RT)$ with $\zeta = (4\pi/3)^{1/3}$. Accepting this result appears as general, show that Hildebrand's rule follows.
(b) Use that V varies only slightly in comparison with $\exp(-\Delta_{vap}H/RT)$ to show that the number density ρ is a universal function of $\Delta_{vap}H/RT$.
(c) Take $R/\zeta V$ constant and use the typical value $V_m = 80\,\text{cm}^3\,\text{mol}^{-1}$ at $T = 300$ K to show that one obtains Trouton's constant $\Delta_{vap}H = RT_n \cong 10.4$, with T_n the boiling temperature at $P = 1$ atm, and that $\Delta_{vap}H$ is insensitive to the precise value of ζ.

9.6.2 Hole Models

One way to improve the cell model is to remove the requirement that each cell is always occupied by only one molecule. If we take the number of cells M somewhat larger than the number of molecules N, the volume per cell V/M and the volume per molecule V/N become different. Accepting maximally one molecule per cell, we have *hole models* [76], with the empty cells addressed as holes. Instead of holes, molecules of another type are used, and this approach is also used for mixtures.

We recall that the interparticle pair potential contains a repulsive and an attractive part. The structure of a simple liquid appears to be largely determined by the repulsive part, while the attractive part mainly provides the "internal pressure" or the *cohesion*. In order to be able to calculate the Helmholtz energy in the thermodynamic limit, the pair potential must be more repulsive than $1/r^{3+\delta}$ for $r \to 0$ and decay more rapidly than $1/r^{3+\delta'}$ for $r \to \infty$, where δ and δ' are positive constants. An often-used model that takes both these interactions and conditions into account is the *lattice gas model*. The basic concept is again that the volume available to the fluid is divided into cells of molecular size. Usually, for simplicity, the cells are arranged in a regular lattice with coordination number z. Although not required in principle, normally these cells are occupied by one particle at most. This represents repulsion. Further, only nearest-neighbor cell attractive interactions are considered. The precise position of a particle in a cell is considered unimportant, so the particle and cell positions are the same.

If we position N_1 molecules of type 1 and N_2 molecules of type 2 on a lattice, there will be N_{11} (N_{22}) nearest-neighbor pairs 1-1 (2-2) of molecules of type 1 (2) with contact energy u_{11} (u_{22}). Similarly, we will have N_{12} pairs of the type in which molecules of type 1 and type 2 combine to a pair 1-2 with the contact energy u_{12}. Note that u corresponds to $\phi(0)$ of Section 9.6.1 and that $u < 0$. For any such a lattice some general relations exist. Drawing a line from each site occupied with a molecule of type 1 to the z neighboring sites, we will obtain zN_1 lines. If the pair is of type 1-1, there will be two lines between the sites, whereas if it is of type 1-2, there is only one line. A similar consideration applies for type 2-2. Therefore,

$$zN_1 = 2N_{11} + N_{12} \quad \text{and} \quad zN_2 = 2N_{22} + N_{12}$$

Consequently,

$$N_{11} = \tfrac{1}{2}zN_1 - \tfrac{1}{2}N_{12} \quad \text{and} \quad N_{22} = \tfrac{1}{2}zN_2 - \tfrac{1}{2}N_{12}$$

If we mix N_1 molecules of type 1 with N_2 molecules of type 2, the energy is

$$E = u_{11}N_{11} + u_{22}N_{22} + u_{12}N_{12} = wN_{12} + \tfrac{1}{2}zu_{11}N_1 + \tfrac{1}{2}zu_{22}N_2$$

where $w \equiv u_{12} - \tfrac{1}{2}u_{11} - \tfrac{1}{2}u_{22}$ is the *interchange energy*. Now we must distinguish between a mixture of molecules of types 1 and 2 assuming that no holes are present in the mixture, and a pure liquid with molecules (type 1) and holes (type 2). In the former case, the contribution $\tfrac{1}{2}z(u_{11}N_1 + u_{22}N_2)$ is a constant given by the composition, but in the latter case, it is variable since the number of holes depends on the temperature. Hence, formally, for this model, we must use the canonical ensemble for describing solutions and the grand canonical ensemble for describing pure fluids.

To obtain a basic hole model for a pure liquid, we take $u_{11} = u$ and $u_{12} = u_{22} = 0$, so that $w = -\tfrac{1}{2}u_{11} = -\tfrac{1}{2}u$, and $N = N_1$. The energy E for lattices with M sites occupied by N molecules, creating $N_{12} \equiv zX$ type 1-2 pairs, becomes

$$E = -\tfrac{1}{2}u_{11}N_{12} + \tfrac{1}{2}zu_{11}N_1 = uN_{11} \quad \text{or} \quad E = zXw + \tfrac{1}{2}zNu = \tfrac{1}{2}uz(N-X)$$

The last form is somewhat more convenient in calculations. For the configurational partition function Q_N, using the abbreviation $x = \exp(-z\beta w) = \exp(\tfrac{1}{2}\beta zu)$, we have

$$Q_N = q^N \sum\nolimits_X g(M,N,X)\mathrm{e}^{-\frac{1}{2}\beta z(N-X)u} = q^N x^{-N} \sum\nolimits_X g(M,N,X)x^X$$

Here, q is the internal partition function of the molecules, assumed to be separable, and $g(M,N,X)$ denotes the number of configurations with N molecules over M sites having zX pairs of type 1-2 all having the same energy (degeneracy factor). The summation Σ_X is over all possible configurations having zX pairs. Recalling the absolute activity $\lambda = \exp(\beta\mu)$ with μ the chemical potential, the grand partition function Ξ, with as usual P the pressure and V the volume, reads (Section 7.1.2)

$$\Xi = \exp(\beta PV) = \sum\nolimits_N \lambda^N Q_N$$

We will use here $PV = (Pv_0)\cdot(V/v_0) \equiv \chi \cdot M$, where v_0 is the volume of a cell and M is the number of cells, so that χ is the pressure in energy units. Hence,

$$\Xi = \sum\nolimits_N \lambda^N q^N x^{-N} \sum\nolimits_X g(M,N,X)x^X = \sum\nolimits_N a^N x^{-N} \sum\nolimits_X g(M,N,X)x^X$$

with $a \equiv q\lambda$. This is as far as we can go without approximation. An exact solution is only known for 2D lattices, and therefore we must approximate for the 3D case. The *zeroth approximation* (Section C.2) assumes that all molecules are distributed randomly over all sites in spite of their non-zero interaction. This implies the approximation

$$\sum\nolimits_X g(M,N,X)x^X \cong x^X \sum\nolimits_X g(M,N,X) = x^X M!/(M-N)!N!$$

where the last step can be made because the summation of $g(M, N, X)$ over all possible configurations X equals the number of configurations of N molecules over M sites. In this approximation we also have $zX = N\cdot z(M-N)/M = zN - zN^2/M$, leading together to

$$\Xi = \sum\nolimits_N a^N x^{-N^2/M} M!/(M-N)!N! \tag{9.42}$$

This sum is difficult to evaluate exactly, and we use the maximum-term method (Section 7.6.1)[8]. Denoting the terms in Eq. (9.42) by t_N, we thus approximate the sum by

8 This is equivalent to using the canonical partition function right away.

its maximum term $t_{\max}$, obtained by solving $\partial t_N/\partial N = 0$ or, equivalently, $\partial \ln(t_N)/\partial N = 0$. After some algebra, we obtain, using the density $\rho = N/M$,

$$\lambda = x^{2\rho}[\rho/(1-\rho)] \quad \text{or} \quad \beta\mu = \ln[\rho \exp(\beta\rho z u)/(1-\rho)]$$

Substituting the solution in $\beta\chi M = \ln\Xi = \ln t_{\max}$, the result is the EoS

$$\beta\chi = \ln[x^{\rho^2}/(1-\rho)] \quad \text{or} \quad \beta\chi = \tfrac{1}{2}\beta z u\rho^2 - \ln(1-\rho) \tag{9.43}$$

Depending on the temperature T, the pressure χ shows either continuously increasing curves with increasing density ρ, or sinuous curves indicating the presence of a discontinuous phase transition. One phase is the more condensed and considered as the liquid with density ρ_L, while the other is the less dense and considered as the vapor with density ρ_V. The critical point T_{cri} can be found by realizing the symmetry between vacant and non-vacant cells. In the two-phase region, $\rho_V + \rho_L = \rho$, and therefore by symmetry T_{cri} must be at $\rho_L = \rho_V = \tfrac{1}{2}$. The value of T_{cri} is obtained from $\partial(\beta\Phi)/\partial\rho = 0$ evaluated at $\rho = \tfrac{1}{2}$, leading to

$$z\beta_{\text{cri}}u = -4.00$$

Although qualitatively the basic physics of the liquid–gas transition is described correctly, the numerical accuracy is, as we will see shortly, not astonishing. In Section C.3, the *first approximation* solution based on the distribution of independent pairs of molecules (instead of independent single molecules) is discussed. It is also called the *quasi-chemical approximation* [77], as one can consider the interchange 1–1 + 2–2 → 1–2 + 2–1 as a quasi-chemical reaction described by the chemical equilibrium condition $X^2/(N_1 - X)(N_2 - X) = \eta^{-2}$ with $\eta = \exp(\beta w)$. In this approximation, $zu/kT_{\text{cri}} = 2z\ln[(z-2)/z]$, so that the result for a square 2D lattice ($z = 4$) is $zu/kT_{\text{cri}} = -5.54$, while the exact result is $zu/kT_{\text{cri}} = -7.05$.

For a pure liquid, we estimate, using as before the close-packed lattice, $z = 12$. Furthermore, to mimic results of other models using the LJ potential, we employ this potential here also, and take its minimum value $-\varepsilon$ for u and the position of the minimum $r_0 = 2^{1/6}\sigma$ for size. We also need an expression for the partition function q and take it conform the cell model as $q = v_f/\Lambda^3$. In principle, v_f depends on the *local configuration*, the number and arrangement of nearest neighbors. However, in the basic model we use here, we assume that we can use the value for $z = 12$ and discuss an improved model later. For the EoS $\chi = PV/M = Pv_0 = Pr_0{}^3/2^{1/2} = P\sigma^3$, we can use either the result of the zeroth or the first approximation. For the latter approximation [76], $z\varepsilon/kT_{\text{cri}} = -2z\ln[(z-2)/z]$ and $T_{\text{cri}} = 2.74\varepsilon/k$, to be compared with $T_{\text{cri}} = 3.0\varepsilon/k$ for the former approximation. From the associated EoS and the critical volume $v_{\text{cri}} = 2\sigma^3$, one can also calculate that $P_{\text{cri}}\sigma^3/\varepsilon = 0.469$, and therefore $P_{\text{cri}}v_{\text{cri}}/kT_{\text{cri}} = 0.342$. Table 9.2 compares these results with those of the LJD and the significant liquid structure (SLS) model.

An obvious shortcoming of the basic hole model is that v_f does not depend on the local configuration, although it should. To improve upon this, we write the partition function conform Eq. (9.41), using nearest-neighbor interactions only, so that we have with $\Phi_{\text{lat}} = \tfrac{1}{2}zN\phi(0) = \tfrac{1}{2}zNu$ and $\psi(\mathbf{r}) = z[\phi(\mathbf{r}) - \phi(0)] = z[\phi(\mathbf{r}) - u]$

$$Z_N = (\Lambda^{-3}v_f)^N \exp[-\tfrac{1}{2}\beta N z u] \quad \text{and} \quad v_f = \int_\Delta \exp[-\beta\psi(\mathbf{r})]\,d\mathbf{r}$$

Choosing again the FCC lattice $z_0 = 12$ and cell volume $a^3/\gamma = a^3/2^{1/2}$, we assume that the coordination number becomes $z = yz_0$, where y is a new parameter with range $0 \leq y \leq 1$.

Table 9.2 Critical data comparison for several theories.

	T_{cri}^*	V_{cri}^*	P_{cri}^*	$Z_{cri}^* = P_{cri}^* V_{cri}^* / RT_{cri}^*$	y_{cri}
LJD theory	1.30	1.77	0.434	0.591	1.00
Cernuschi and Eyring [76]	2.75	2.00	0.470	0.342	0.544
Henderson [78]	1.42	3.39	0.139	0.333	0.321
SLS theory	1.31	3.36	0.141	0.362	–
"Best" experimental[a)]	1.25	3.14	0.116	0.291	≈0.5

a) Different authors provide slightly different "best" values, even providing different values in the same text (e.g. [79]: $P_{cri} = 0.119$ and $0.141\ \varepsilon/\sigma^3$, $V_{cri} = 3.15$ and $3.36\ N\sigma^3$, $T_{cri} = 1.28$ and $1.31\ \varepsilon/k$).

Next, we define ω_k to be the fraction of vacant nearest-neighbor sites of molecule k. If all the molecules are at the origin of their cells, we have for the potential energy of one molecule $z(1-\omega_k)u$, and for the total potential energy

$$\Phi(0) = {}^1\!/\!_2 z(N-\Omega)u \quad \text{where} \quad \Omega = \Sigma_k\, \omega_k$$

When the molecules are not at their origin, we calculate, as in the cell model, the interaction of the wanderer with the nearest-neighbor molecule located at its origin. When the wanderer is at position $\mathbf{r}$ in the cell, its potential energy is $\psi(\mathbf{r}) = z[\phi(\mathbf{r}) - u]$. Using the "smearing" approximation with $z(1-\omega_j)$ nearest neighbors, we assume that the potential energy of the wanderer depends only on the number of nearest-neighbor molecules, and not on their precise configuration; this is obviously an oversimplification. Accepting this, we obtain for the potential energy

$$\Phi(\mathbf{r}) = {}^1\!/\!_2 z(N-\Omega)u + \sum\nolimits_{k=1}^{N} z(1-\omega_k)\overline{\psi}_k(\mathbf{r}) \quad \text{with} \quad \overline{\psi}_k = \overline{\phi(\mathbf{r}_k)} - u \tag{9.44}$$

where an overbar is used to indicate the smearing approximation. This results in

$$Q_N = \exp[-{}^1\!/\!_2 \beta z(N-\Omega)u] \prod\nolimits_k j(\omega_k)$$

with the *generalized free volume*

$$j(\omega_k) = \int_\Delta \exp[-\beta(1-\omega_k)z\overline{\psi}_k(r)]\, 4\pi r^2\, \mathrm{d}r$$

When no nearest-neighbor molecules are missing, $\omega_k = 0$, and $j(\omega_k)$ becomes the same as for the cell model in the "smearing" approximation. When all nearest-neighbor molecules are missing, $\omega_k = 1$, and $j(\omega_k)$ becomes the cell size. At high and low density, we therefore obtain, respectively,

$$Q_N = [j(0)]^N \exp\left[\frac{-zNu}{2kT}\right] \quad \text{and} \quad Q_N = [j(1)]^N \frac{M!}{N!(M-N)!} \cong \mathrm{e}^N \left(\frac{V}{N}\right)^N$$

While the first expression is the Einstein approximation for the partition function of a crystal, the second expression contains the factor e^N, necessary to obtain the perfect gas limit. In principle, the communal problem is thus avoided.

To calculate Q_N, it is necessary to use a simple relationship for $j(\omega)$. From Eq. (9.44) we see $j(\omega)$ at T yields the same value as v_f at $T^* = T/(1-\omega)$. We thus could use $v_f = v_f(T^*)$

to calculate $j(\omega)$. From Q_N the thermodynamic properties can then be calculated in the usual way. However, since v_f depends on the number and arrangement of holes, the relation between v_f and y is *not simple* [80].

A reasonably successful approach is based on the idea that a fluid contains solid-like and gas-like molecules. If v_f of a gas-like molecule with all its neighboring sites empty is given by $j(1)$ and v_f of a solid-like molecule with all neighboring sites occupied by $j(0)$, we take, following Henderson [78],

$$v_f = y_i j(0) + (1 - y_i) j(1)$$

where the probability to have solid-like and gas-like properties is y_i and $1 - y_i$, respectively. Using the LJ potential ϕ with parameters ε and σ, we obtain

$$u = \phi(0) = z\varepsilon\left[\left(1/v_*^4\right) - \left(2/v_*^2\right)\right]$$

where $v_* = v/\sigma^3$ for the FCC lattice with intermolecular distance a and volume $v = a^3/2^{1/2}$. If $\omega = V/M$ and $v = V/N$, we have $y_i = N/M = \omega/v$, and the Helmholtz energy $F = -kT\ln Z_N$, evaluated using the zeroth approximation, becomes

$$F = -NkT\ln \Lambda^{-3} v_f + \tfrac{1}{2}\frac{\omega}{v} N\psi(0) + NkT\ln\left(\frac{\omega}{v}\right) + NkT\left(\frac{v}{\omega} - 1\right)\ln\left(1 - \frac{\omega}{v}\right)$$

The volume per cell is determined by $(\partial F/\partial\omega)_{V,T} = 0$, while the EoS can be calculated from $PV = -v(\partial F/\partial v)_{\omega,T}$, which after some algebra, yields

$$\frac{PV}{NkT} = \frac{\omega - j(0)}{v - \omega + j(0)} + \frac{\omega}{v}\frac{Nu}{2kT} - NkT\frac{v}{\omega}\ln\left(1 - \frac{\omega}{v}\right)$$

The values for $j(0)$ and its temperature derivatives, needed for a numerical evaluation, can be calculated from tabulated data as given by Wentorff et al. [81]. Figure 9.9 shows the vapor pressure and the EoS for this extended hole theory in comparison with LJD theory results, and a clear improvement can be noted. Table 9.2 presents a comparison of the critical data with the results from some other theories. From these data, it is clear that for the

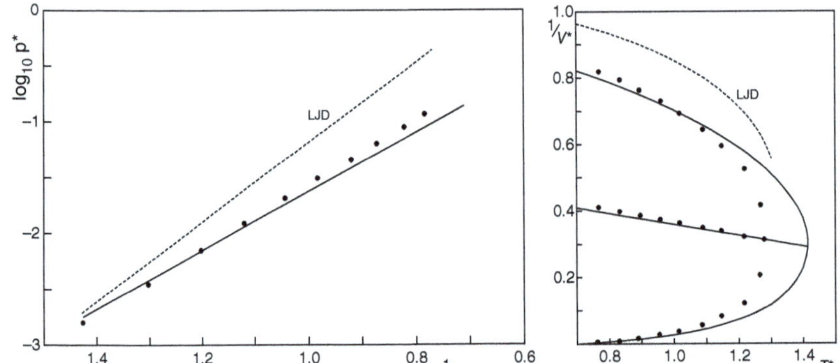

Figure 9.9 Hole theory results. (a) The logarithm of the reduced vapor pressure P^* for simple liquids versus reciprocal reduced temperature $1/T^*$. (b) The reduced density versus $1/V^*$ versus reduced temperature T^* for the hole theory ($T = (\varepsilon/k)T^*$, $V = (N/\sigma^3)V^*$, and $P = (\varepsilon/\sigma^3)P^*$). The solid line represents the hole theory results, the dots represent points from the empirical Guggenheim expressions, Eqs. (8.9), and the dotted line represents LJD results.

prediction of critical constants, the simple hole theory is inferior to the LJD theory, although the extended hole theory actually represents the experimental data quite well. Unfortunately, it appears [82] that the linear approximation for v_f compensates for the "smearing" approximation and the good agreement with experiment is somewhat fortuitous. Henderson [78] also calculated U, S and C_P, but the results obtained were inferior to those for the EoS. Neglecting the correlation between neighboring molecules and holes and the linear increase of v_f with the number of neighboring holes and neglecting the overlapping of neighboring collision spheres, seem to be the most important reasons, as these assumptions are valid only for high temperatures and low densities.

Summarizing, hole models can catch the qualitative aspects of liquids correctly, and the communal problem is in principle avoided. Although the first approximation is a clear improvement over the zeroth approximation, probably the most important thing to learn is that taking the hard core seriously (no two particles on the same site) already gives in the crudest approximation most of the realistic features of the liquid–vapor system, as indicated in the introduction to this section.

Problem 9.7 ***Hole Theory Critical Constants*** Calculate the critical constants using basic hole theory.

9.6.3 Some Other Implementations of Hole Theory

Hole theory resulted in several follow-up approaches, of which we indicate here briefly just two. Still, other hole models are available (e.g. [83, 84]).

The first approach we indicate is Eyring's *SLS theory* for which it is also argued that a liquid behaves in many cases as an intermediate between a gas and a solid [79]. The liquid structure is assumed to show some heterogeneity, containing a fraction of solid-like molecules and another fraction of holes. Holes of molecular size are favored because smaller ones cannot provide easy access to entering molecules, which limits the increase of entropy (but see [22]). In contrast, larger holes will require excessive energy without a compensating increase in entropy. These holes change one or more DoFs of a molecule surrounding the hole from vibration to translation, and these molecules thus act like vapor molecules. The concept does not imply that a liquid is a mixture of a solid and a gas. Rather, a molecule has solid-like properties for the time it vibrates around an equilibrium position but transforms instantaneously to gas-like behavior for one or more DoFs when it jumps into a neighboring hole.

In the simplest form, the gas-like regions behave like a perfect gas, whereas for the solid-like elements, the Einstein model (Section 10.8) is used. The crux is clearly to determine the fractions of gas-like (G) and solid-like (S) regions. The partition function of the liquids Z_L, therefore, can be written as

$$Z_L = (Z_S)^{N_S}(Z_G)^{N_G}/N_G! \tag{9.45}$$

Assuming holes of molecular size, in one mole of liquid $(V_L - V_S)/V_S$ moles of holes are present, where V is the molar volume. The number of accessible sites for a solid-like molecule is the number of holes around that molecule $n_h = n(V_L - V_S)/V_S$ with n a proportionality constant, multiplied by the probability that a molecule has the required energy to be able to jump into the holes, given by $\varepsilon_h = aE_SV_S/(V_L - V_S)N_A$ with a a proportionality constant and E_S the sublimation energy. The partition function Z_S becomes, using Z_S' as the partition function for the pure solid-like molecule,

$$Z_S = \left(1 + n_h e^{-\beta\varepsilon_h}\right) Z_S' \equiv \left(1 + n_h e^{-\beta\varepsilon_h}\right) e^{-\beta\phi/2}\{e^{-\theta/2T}/(1 - e^{-\theta/T})\}^3$$

As said, for Z_S' the simplest choice, the Einstein model, is used. The factor $\exp(-\beta\phi/2)$ takes into account that in a nearest-neighbor model $-\phi/2$ is the potential energy of a solid-like, vibrating molecule, and we interpret $-½N_A(\phi + 3k\theta)$ as the sublimation energy E_S. For Z_G we use the perfect gas expression with excess volume $V_L - V_S$ so that

$$Z_G = (V_L - V_S)/\Lambda^3 \quad \text{with} \quad \Lambda = (h^2/2\pi mkT)^{1/2}$$

Hence, according to Eq. (9.45), we have for the partition function Z_L of the liquid

$$Z_L = \left[e^{E_S/RT}(1 - e^{-\theta/T})^{-3}\left(1 + n_h e^{-\varepsilon_h/kT}\right)\right]^{N_S} [\Lambda^{-3}(V_L - V_S)]^{N_G}(N_G!)^{-1} \tag{9.46}$$

The parameters E_S, θ, and V_S have to be taken from theoretical or experimental solid state data, but the expression for ε_h and the parameters a and n can be evaluated theoretically [79]. Briefly, using the molar volume V_L of the liquid close to the melting point T_{mel} and the lattice coordination number z, n_h is given by

$$n_h = z\left(V_L - V_S\right)/V_L = \left(zV_S/V_L\right)\left(V_L - V_S\right)/V_S \quad \text{or} \quad n = zV_S/V_L$$

after comparison with $n_h = n(V_L - V_S)$. Further, if a molecule is to pre-empt a neighboring position in addition to its original position, it must have an additional kinetic energy above $3kT/2$ equal to or more than the kinetic energy ε_{kin} that the other $(n-1)$ neighboring molecules would otherwise introduce into the hole. If the average molecule divides its time equally between two neighboring sites, its energy density will be halved and as the molecule will be moving $(1/z)$th of its time in the direction of any neighbor, the average kinetic energy of the $(n-1)$ ordinary molecules will provide a hole with $\varepsilon_{kin} = ½(3kT/2)(n-1)/z$. This is also the value $\varepsilon_h = aE_SV_S/(V_L-V_S)N_A$ must have at T_{mel}, that is, $\varepsilon_h = \varepsilon_{kin}$ resulting in $a = N_A(V_L-V_S)\varepsilon_h/E_SV_S$. Since at T_{mel} the entropy of melting S_{mel} for a simple liquid is about $3N_Ak/2$ and $T_{mel}S_{mel} \cong E_{mel}$, we obtain for the energy of melting $E_{mel} = 3N_AkT_{mel}/2$ and thus $\epsilon_h = ½E_{mel}(n-1)/zN_A$. Moreover, since the holes introduced in the solid during melting are essentially accompanied by a potential energy change only, we also have $E_{mel} = (V_L-V_S)E_S/V_L$ and combining leads to

$$a = (n-1)\left(V_L - V_S\right)^2/2zV_LV_S.$$

These expressions result in $a = 0.00537$ and $n = 10.7$ for Ar, using $z = 12$ and the data given in Table 9.3. The best empirical fit [85] yields $a = 0.00534$ and $n = 10.8$, so that there is good agreement.

Using the same parameters as for the LJD theory, the triple point for Ar was predicted (Table 9.1), its critical constants (Table 9.2) as well as some P, V, and S data for Ar, Kr, and Xe (Table 9.3). They all show reasonable agreement with the experiment.

Although SLS theory yields fair agreement with the experimental data for noble gas liquids, the agreement is slightly less for V for molecules like CH_4 and N_2 that freely rotate in the liquid state. For molecules that do not rotate freely in the liquid phase, such as Cl_2, changes in Z_S must be made by treating the rotation in the solid-like part as a vibration. Doing so good agreement with experimental data can be reached. For example, for Cl_2 for the temperature range of 180–240 K, C_P as a calculated from $C_P = C_V + TV\alpha^2/\kappa_T$ with all

Table 9.3 Vapor pressure P, molar volume V, and entropy of vaporization $\Delta_{vap}S$ for Ar, Kr, and Xe as calculated from SLS theory, and the solid-state parameters used.

	T (K)	P_{calc} (atm)	P_{obs} (atm)	V_{calc} ($cm^3\ mol^{-1}$)	V_{obs} ($cm^3\ mol^{-1}$)	$\Delta_{vap}S_{calc}$ ($cal\ mol^{-1}\ K^{-1}$)	$\Delta_{vap}S_{obs}$ ($cal\ mol^{-1}\ K^{-1}$)
Ar (mp)	83.96	0.6874	0.6739	24.84	28.03	20.07	19.43
Ar (bp)	87.49	1.040	1.000	29.36	28.69	18.90	18.65
Kr (mp)	116.0	0.7605	0.7220	34.31	34.13	20.14	—
Kr (bp)	119.3	1.0660	1.000	34.90	—	19.15	17.99
Xe (mp)	161.3	0.8372	0.804	42.24	42.68	20.25	—
Xe (bp)	165.1	1.0623	1.000	42.84	—	19.50	18.29

	Ar	Kr	Xe
E_S ($cal\ mol^{-1}$)	1888.6	2740	3897.7
V_S ($cm^3\ mol^{-1}$)	24.98	29.6	36.5
θ (K)	60.0	45.0	39.2

data calculated from Z_N, differ by only ≈ 2% from the experimental values. The method has been applied to molten metals and salts, and quantum liquids, like H_2 and Ne, with fairly good results. The application to many liquids and phenomena has been reviewed by Eyring and Jhon [79, 86, 87].

Overall, we conclude that for describing liquids, SLS theory is much better than LJD theory, which resembles more a superheated solid than a liquid, and that it is rather successful in estimating thermodynamic properties, but also that its basis is not really clear as it has not been derived from a partition function by any mathematically well-defined approximation. Attempts have been made to clarify that basis [88, 89] and to replace the Einstein-type contribution based on empirical data by a LJD-type contribution based on intermolecular interactions. The basic feature of liquids – the pair correlation function $g(r)$ – is largely absent, although in a later stage attempts were made [90] to introduce $g(r)$, but that procedure must be characterized as artificial. Hence, despite its pragmatic success, the approach is largely abandoned.

The second model we indicate is the *empirical hole model* by Luck [91], introducing further semiempirical considerations. The starting point is the *ideal liquid*, defined by the *law of rectilinear diameters* (Section 8.4)

$$\rho_L + \rho_V = \rho_0 - AT \tag{9.47}$$

The deviation of the density of the real liquid ρ_L from the ideal density $\rho_L + \rho_V$ is given by ρ_V, so that the hole density $x = \rho_V/(\rho_L + \rho_V)$. At $T = T_{cri}$, $\rho_L = \rho_V$, and $x(T_{cri}) = ½$. It appears that the behavior of x as a function of reduced temperature $T_{red} = T/T_{cri}$ for many liquids is rather similar (Figure 9.10a).[9] As at T_{cri}, $\rho_{cri} = \rho_L = \rho_V$, we obtain from Eq. (9.47), with ρ_0 the density at 0 K, the experimentally well-obeyed relation

$$(\rho_L + \rho_V)/\rho_{cri} = \rho_0/\rho_{cri} - [(\rho_0/\rho_{cri}) - 2]T/T_{cri}$$

9 Note that this is another representation of Guggenheim's description of ρ_L and ρ_V (Section 8.4).

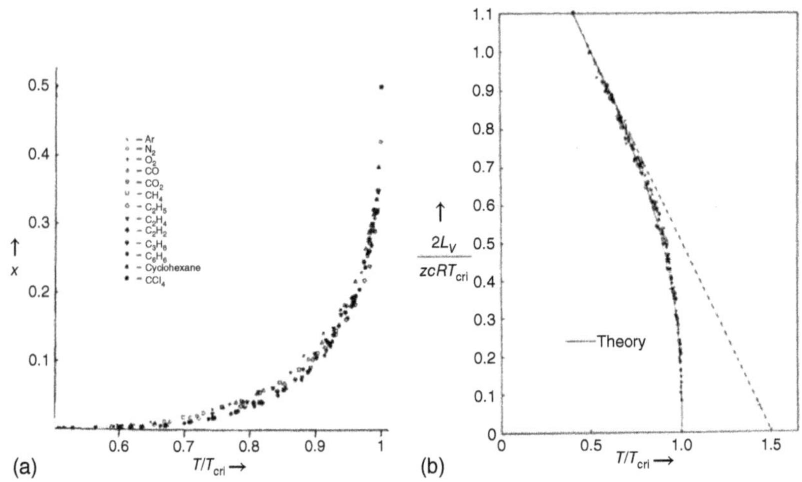

Figure 9.10 Typical properties of several liquids. (a) The fraction of holes x as a function of T/T_{cri}; (b) Experimental energies of vaporization L_V multiplied by $2/czRT_{cri}$ for the liquids Ar, Kr, Xe, N_2, CO, O_2, CO_2, CH_4, C_2H_6, C_6H_{14}, C_7H_{16}, C_6H_6, C_6H_5Cl, $C_{10}H_{10}$, SF_6, $(C_2H_5)_2O$, HCl, and NH_3. Source: Luck [92]/Science and Education Publishing.

deviating from the principle of corresponding states (PoCS) as ρ_0/ρ_{cri} is slowly increasing with T_{cri}, approximately described by $\rho_0/\rho_{cri} \cong 0.53 + 0.86 \log T_{cri}$ [93].

Intermolecular interactions are taken into account via $N_A\phi \sim 3RT_{cri}/2$. The proportionality constant is estimated from the collision diameter $\sigma_T^2 = \sigma_0^2(1+S/T)$ (Section 8.1) where the *Sutherland constant* S is related to the average potential energy ϕ via $S = -2\phi/3k$ [94]. Empirically, it appears that $S \cong cT_{cri}$ with the constant[10] $c \cong 1.0$ and hence $N_A\phi = -3RS/2 = 3cRT_{cri}/2$. Now the heat of evaporation at constant volume $L_V = H - P\Delta V \cong H - RT$. As at arbitrary temperature T, the liquid contains a fraction of holes x, the coordination number is taken as $z' = z(1-x)$ with z the coordination number at $T = 0$ K. Hence, $-\frac{1}{2}z'N_A\phi = -\frac{1}{2}z(1-x)N_A\phi = \frac{1}{2}z(1-x)\cdot 3cRT_{cri}/2 \equiv L_0(1-x)$.

However, there is a contribution for the liquid from the intermolecular DoFs. Assuming 1 DoF for each pair of molecules, consistent with available experimental data, an estimate is $\frac{1}{2}cz(1-x)RT_{cri}$. As this contribution becomes available upon evaporation, it must be subtracted. Moreover, at high temperature, ρ_V is relatively high and intermolecular interactions in the vapor phase must be subtracted. Estimating this contribution similar as for the liquid gives $\frac{1}{2}czx\cdot 3RT_{cri}/2$. Similarly, we must subtract the contribution of all the intermolecular DoFs, but that of the vapor $\frac{1}{2}zx\cdot cRT$ should remain and therefore must be added again. Taken all together, the total becomes

$$L_V = \tfrac{1}{2}czR(1-2x)\cdot[(3T_{cri}/2) - T]$$

At low T, L_V as a function of T/T_{cri} is represented by a straight line, for which the intercept divided by the slope yields $3RT_{cri}/2$ (Figure 9.10b). To include next nearest-neighbor

10 Initially Luck used $c \cong 1.2$ [91], clearly incorrect, and later changed without explanation to $c \cong 1/1.2 \cong 0.833$ [92] (just an error?). From a digitized image [91], I calculated $c = 0.783$ ($R^2 = 0.977$), and from Chapman and Cowling's data [95], $c = 0.969$ ($R^2 = 0.975$). Later, Luck also used $c = 1.0$ [96].

interactions for the FCC lattice with $z = 12$, Luck used the lattice sum for $n = 6$, so that $z = 14.5 \cong 1.2 \cdot 12$ (Section 10.6). Moreover, there is a 20% volume increase upon melting, resulting in $1.2cz/1.2 = cz \cong 12$. Experimentally, Luck quotes $(C_2H_5)_2O$ 14.5, $C_{10}H_{10}$ 14.4, C_6H_6 12.8, C_6H_5Cl 14.4, C_6H_5Br 12.9, C_6H_5I 13.8, NH_3 13.8, CCl_4 12.4, and HCl 10.1, with as average 13.2, so that only approximate agreement is obtained. However, the variability of cz explains the observed variation in Trouton's constant.

Using the Clapeyron–Clausius relation and approximating $\rho_L\rho_V/(\rho_L - \rho_V) \cong \rho_V = M/V_V = MP/RT$ leads for the vapor pressure P to

$$\ln P = -(L_0/RT) - (\tfrac{1}{2}cz - 1)\ln T + const.$$

As the $\ln T$ term is only slightly dependent on T and L_0 is constant, this explains nicely the experimentally frequently observed constancy of L_V as a function of T and naturally yields a logarithmic term, also occurring in the Kirchhoff equation (Section 13.1).

The enthalpy of vapor H_V and liquid under saturation conditions H_L can also be calculated using the same considerations. This leads to

$$H_V = \frac{cz}{2}\frac{3RT_{cri}}{2}(1-x) + \frac{cz}{2}RTx + \int_0^T C_V \mathrm{d}T - \frac{3RT}{2}x + PV_V \quad \text{and} \tag{9.48}$$

$$H_L = \frac{cz}{2}\frac{3RT_{cri}}{2}x + \frac{cz}{2}RT(1-x) + \int_0^T C_V \mathrm{d}T - \frac{3RT}{2}(1-x) + PV_L \tag{9.49}$$

In Eq. (9.48), the first term represents L_0, the evaporation enthalpy at 0 K, multiplied by the finite T correction $(1 - x)$, the second the enthalpy of the intermolecular DoFs, the third the enthalpy of the intramolecular DoFs, the fourth the decrease in translational DoFs due to association in the vapor phase, and the fifth the PV correction for the internal energy. A similar description applies to Eq. (9.49). Slightly better agreement is obtained when keeping 1 DoF of translation in the liquid state, an effect which possibly can be ascribed to motion of clusters of molecules. Overall, it appears possible to describe both H_V and H_L with a single cz value.

Differentiating Eq. (9.49) and adding $3R(1-x)/2$ to keep the translation DoF in the liquid, we obtain for the heat capacity C_σ on the saturation line

$$C_\sigma = R\frac{\mathrm{d}x}{\mathrm{d}T}\left[\frac{cz}{2}\frac{3T_{cri}}{2} - \left(\frac{cz}{2} - \frac{3}{2}\right)T\right] + R\left(\frac{cz}{2} - \frac{3}{2}\right)(1-x) + C_V^{idg}$$

where $C_V{}^{idg}$ represents the ideal gas contribution as calculated from the molecular characteristics. For Ar and C_6H_6, C_σ is in good agreement with experiment, in particular the rapid increase when approaching T_{cri} is captured well.

Water molecules are largely H-bonded with a tetrahedral configuration, albeit with considerable distortion and large fluctuations due to the shallow OH well depth. This leads to a certain fraction p of "free" OH bonds and suggests effectively two-state behavior between the free OH bonds (OH_{free}), the oxygen lone-pair electrons (O_{LP}) and the bonded OH bond (OH_{HB}) described by $OH_{free} + O_{LP} \leftrightarrow OH_{HB}$ with equilibrium constant $K = [OH_{HB}]/[OH_{free}][O_{LP}] = (1-p)/p^2c_{OH}$, where c_{OH} is the concentration of OH bonds [91, 97]. Further support for this approach is the fact that $\Delta_{HB}H = 15.5\,\mathrm{kJ\,mol^{-1}} \cong 6.2RT$, the experimentally observed single dielectric relaxation time and the presence of a sharp isosbestic point in the vibrational spectra. Using this two-state model the fraction p was determined as $\cong 0.1$ at 25 °C ([98], Figure 9.11a) and used to estimate other properties.

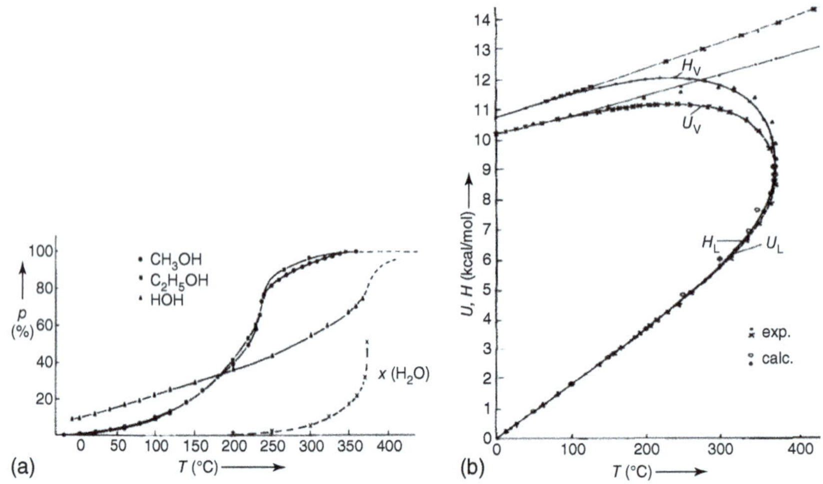

Figure 9.11 (a) The fraction p of non H-bonded OH groups in liquids H_2O, CH_3OH, and C_2H_5OH under saturation conditions, and the fraction holes x in water. Source: Adapted from Luck [98]. (b) The energy U and enthalpy H for the liquid phase (L) and for the vapor phase (V) at the saturation line for water. Source: Adapted from Luck [99].

For the heat of fusion $\Delta_{\text{fus}}H(\text{H}_2\text{O})$ it is assumed that the vdW contribution is similar to $\Delta_{\text{fus}}H(\text{H}_2\text{S}) = 2.4\,\text{kJ mol}^{-1}$. Hence, $\Delta_{\text{fus}}H(\text{H}_2\text{O}) = 2.4\,\text{kJ mol}^{-1} + 0.1$ (fraction NHBs) $\times\,2$ (H-bonds per molecule) $\times\,15.5\,\text{kJ mol}^{-1}$ (per H-bond) $= 5.5\,\text{kJ mol}^{-1}$, in good agreement with the experimental value $5.9\,\text{kJ mol}^{-1}$.

The internal energy for water under saturation conditions U_L with respect to the melting point T_mel contains four contributions, given by

$$U_\text{L} = \int_{T_\text{mel}}^{T} C_V^\circ \,\text{d}T + 2\Delta_\text{HB}H[p(T) - p(T_\text{mel})] + ZR(T - T_\text{mel}) - 3R(T - T_\text{mel})/2 \quad (9.50)$$

The first term describes the intramolecular contribution and the second term the energy for breaking the H-bonds. The third term is the contribution of the intermolecular DoFs. Attributing 1 DoF to each of the nearest neighbors, we have $Z = (1-p)Z'/2 + pZ''/2$ with $Z' = 4$ for the ordered part and $Z'' = 6$ (3 for the bonded neighbors +3 for the free OH bonds), resulting in $Z = 2 + p$. The fourth term is a correction for the loss of translation in the liquid. The total expression, containing only experimental determined parameters, describes U_L surprisingly well (Figure 9.11b). Similarly, for the energy of the vapor the expression

$$U_\text{V} = \int_{T_\text{mel}}^{T} C_V^\circ \,\text{d}T + \Delta_\text{sub}U - \Delta_\text{fus}H - 6Wx + (12-3)RTx/2 \quad (9.51)$$

is derived. Here, the first term represents again the intramolecular contribution, while the sublimation energy $\Delta_\text{sub}U$ minus the enthalpy of fusion $\Delta_\text{fus}H$ represents the enthalpy of evaporation $\Delta_\text{vap}H$. The third term is a correction for association in the gas phase estimated as half the pair energy W, taken as $8.8\,\text{kJ mol}^{-1}$, times the coordination $12x$. In the last term $12RTx/2$ takes in to account the intermolecular component for the enthalpy in the gas phase while $3RTx/2$ corrects for association. Again, the agreement is good (Figure 9.11b).

More briefly, the difference between Eqs. (9.50) and (9.51) represents $\Delta_\text{vap}H$, in good agreement with the experimental data. Density and C_V are less sensitive to the details of the model but are reasonably well described including the divergence of C_V toward T_cri.

About 50% of the C_V value is due to the H-bonds. Finally, there is a good description of the permittivity (Section 9.11).

To summarize, the model can explain many aspects and yield relatively accurate predictions for simple, normal liquids and complex liquids such as water [98, 99]. As the originator remarked, "these astonishing results could delude us into believing that the hole model is more than a mere empirical contrivance."

Problem 9.8 ***SLS Partition Function*** Show that the partition function of the SLS theory reduces to the perfect gas and solid-state partition functions if $V \to \infty$ and $V \to 0$, respectively. Derive the expression for P using the Einstein partition function for the solid-like phase.

9.7 The Generalized van der Waals Model

In this section, we describe a thermodynamic model as developed by Sandler [100, 101]. We recall that the configurational Helmholtz energy for N molecules reads $F_{\text{con}} = -kT\ln Q$ with Q the configuration partition function. This leads to

$$\mathrm{d}F_{\text{con}} = -kT\mathrm{d}(\ln Q) = \int \mathrm{e}^{-\beta\Phi}\mathrm{d}\Phi\mathrm{d}\mathbf{r} \Big/ \int \mathrm{e}^{-\beta\Phi}\mathrm{d}\mathbf{r} \quad \text{or} \quad \Delta F_{\text{con}} = \int \Delta\Phi\mathrm{e}^{-\beta\Phi}\mathrm{d}\mathbf{r} \Big/ \int \mathrm{e}^{-\beta\Phi}\mathrm{d}\mathbf{r}$$

neglecting higher-order terms in the last step. Equivalently, $\Delta F_{\text{con}} = \langle\Delta\Phi\rangle$. If we can write the total potential energy Φ as the sum of a repulsive and attractive part, that is, $\Phi = \Phi_{\text{rep}} + \Phi_{\text{att}} \equiv \Phi_{\text{rep}} + \Delta\Phi$, we have $F_{\text{con}} \cong F_{\text{rep}} + \langle\Phi_{\text{att}}\rangle_0$ using averaging with Φ_{rep} only. Defining $F_{\text{att}} \equiv \langle\Phi_{\text{att}}\rangle_0$, we have $Q = Q_{\text{rep}}Q_{\text{att}}$ and $U = U_{\text{rep}} + U_{\text{att}}$. Within the pair potential approximation (using the pair potential u, pair correlation function g, and density $\rho = N/V$) the potential part of the internal energy U is given by

$$U = ½N\rho \int_0^\infty u(r)g(r)\,\mathrm{d}\mathbf{r} \quad \text{but also by} \quad U = -\partial\ln Q/\partial\beta = kT^2\partial\ln Q/\partial T$$

As for $T = \infty$ only repulsions are relevant, we have

$$\ln Q - \ln Q_{\text{rep}} = \ln Q_{\text{att}} = \int_{T=\infty}^{T} (U_{\text{att}}/kT^2)\,\mathrm{d}T$$

Introducing a mean potential per molecule χ, corresponding to $½Y_0$ in Eq. (9.40), and the coordination number $z(r) = \rho\int_0^r g(r)\,\mathrm{d}\mathbf{r}$, the number of molecules present around a central molecule within a certain volume with radius r, we have from Q_{att}

$$-\beta N\chi \equiv \int_{T=\infty}^{T} (U_{\text{att}}/kT^2)\,\mathrm{d}T = ½N\rho \int_{T=\infty}^{T} (kT^2)^{-1} \int_0^r u_{\text{att}}(r)g(r)\,\mathrm{d}\mathbf{r}\,\mathrm{d}T$$

$$= ½N \int_{T=\infty}^{T} \left(\langle u_{\text{att}}\rangle/kT^2\right) z(r)\,\mathrm{d}T \ \text{ with } \ \langle u_{\text{att}}\rangle = \int_0^r u_{\text{att}}(r)g(r)\,\mathrm{d}\mathbf{r} \Big/ \int_0^r g(r)\,\mathrm{d}\mathbf{r}$$

In total, we have for the partition function

$$Z_{\text{GvdW}} = q^N(\Lambda^{3N}N!)^{-1}V_{\text{f}}^N\exp(-\beta N\chi) \equiv Q_{\text{kin}}Q_{\text{rep}}Q_{\text{att}} \tag{9.52}$$

In the last step, $Q_{\text{kin}} = q^N(\Lambda^{3N}N!)^{-1}$ is introduced with, as before, q the internal partition function and Λ the thermal wavelength. By introducing the free volume $V_{\text{f}} = Nv_{\text{f}}$, we consider the repulsion to be described by (effective) HS molecules. Equation (9.52) is referred to as the *Generalized van der Waals* (GvdW) *partition function* [102]. In order to make progress, choices for V_{f} and χ have to be made.

To calculate the free volume V_f, we use HS molecules with diameter σ. Using $P_\text{rep} = -\partial F_\text{rep}/\partial V = kT\ \partial\ln Q_\text{rep}/\partial V = kT\ \partial\ln V_\text{f}^{\,N}/\partial V$, we obtain from the vdW EoS for HS molecules, reading $\beta P = \rho/(1-b\rho)$ (Problem 9.9) for the free volume $V_\text{f} = V(1-\rho b)$, where $b\rho = 4\eta$ and $\eta = 4\pi(\sigma/2)^3\rho/3$. Similarly, for the CS model (Problem 9.9), the result is $V_\text{f} = V\exp[\eta(3\eta-4)/(1-\eta)^2]$.

To keep the expression for the mean potential per molecule χ simple, we use the square-well (SW) potential, given by $u(r) = \infty$ for $r \le \sigma$, $u(r) = -\varepsilon$ for $\sigma < r < R_\text{SW}\sigma$ and $u(r) = 0$ for $r \ge R_\text{SW}\sigma$ with $R_\text{SW} \cong 1.5-2.0$. For this potential, we have, using $r^* = R_\text{SW}\sigma$,

$$U_\text{att} = \tfrac{1}{2}N\rho\int_0^r u_\text{att}(r)g(r)\,\mathrm{d}\mathbf{r} = -\tfrac{1}{2}N\rho\varepsilon\int_0^{r^*} g(r)\,\mathrm{d}\mathbf{r} = -\tfrac{1}{2}N\varepsilon z(r^*) \quad \text{and}$$

$$-\beta N\chi = \tfrac{1}{2}N\int_{T=\infty}^{T}\left[\langle u_\text{att}\rangle z(r^*)/kT^2\right]\mathrm{d}T = -\tfrac{1}{2}\varepsilon N\int_{T=\infty}^{T}\left[z(r^*)/kT^2\right]\mathrm{d}T$$

Assuming as a first approximation $g(r) = 1$ and using $C \equiv 4\pi\sigma^3(R_\text{SW}{}^3-1)/3$, we obtain

$$z = C\rho, \quad U^\text{E} = -\tfrac{1}{2}Nz\varepsilon \quad \text{and} \quad \chi = -\tfrac{1}{2}C\varepsilon\rho = -\tfrac{1}{2}z\varepsilon$$

Calculating the total pressure using the vdW expression for the free volume V_f, the result is $P = NkT/V(1-\rho b) - a/V^2$ with $a = \tfrac{1}{2}CN^2\varepsilon$, which we recognize as the vdW EoS. As we know that actually at low density $g(r) = \exp[-\beta u(r)]$, we obtain as the second approximation

$$z = C\rho\exp(\beta\varepsilon), \quad U^\text{E} = -\tfrac{1}{2}Nz\varepsilon \quad \text{and} \quad \beta\chi = -\tfrac{1}{2}C\rho[\exp(\beta\varepsilon)-1] \tag{9.53}$$

The vdW EoS is regained, but now with $a(T) = CN^2[\exp(\beta\varepsilon)-1]/2\beta$. In general, we will have for the SW potential, the partition function

$$Z_\text{GvdW-SW} = q^N(\Lambda^{3N}N!)^{-1}V_\text{f}^N\exp\left\{-\tfrac{1}{2}N\varepsilon\int_{T=\infty}^{T}[z(r^*)/kT^2]\mathrm{d}T\right\}$$

A simple approximation for z for the SW potential can be obtained [103]. We write the expressions for a binary mixture, so that component 1 represents a molecule and component 2 represents a hole. The starting point is the first approximation $(N_1-X)(N_2-X) = X^2\eta^2$ with $\eta = \exp(\beta w)$ and $zX = M_{12}$, where M_{ij} is the total number[11] of molecules i around molecule j (Section C.3). Using the condition[12] $zN_1 = M_{11}+M_{21}$ leads to

$$M_{21} = (M_{22}M_{11})^{1/2}/\eta \quad \text{or} \quad M_{21}/M_{11} = (M_{22}/M_{11})^{1/2}/\eta$$

To obtain an approximate solution, we insert the zeroth approximation for M_{22} reading $M_{22} = zN_2x_2$. Similarly, we have $M_{11} = zN_1x_1$, so that $(M_{22}/M_{11})^{1/2} = (N_2{}^2/N_1{}^2)^{1/2}$, with as final result $M_{21}/M_{11} = (N_2/N_1)/\eta$. Combining with $zN_1 = M_{11}+M_{21}$ yields

$$M_{11} = zN_1^2/(N_1+N_2\eta^{-1})$$

We now transcribe to a pure liquid with volume V. In that case, in Eq. (9.53), $z = z_\text{max}$ with z_max the coordination number at close-packed density ($z_\text{max} = 18$ for $R_\text{SW} = 1.5$). Further using $N_2 = N_1(V-V_0)/V_0$ with $V_0 = N\sigma^3/2^{1/2}$ and $w = -\tfrac{1}{2}u = \tfrac{1}{2}\varepsilon$ leads for the coordination number $z(V,T) \equiv M_{11}/N_1$ of the pure liquid to

$$z \equiv M_{11}/N_1 = z_\text{max}V_0\exp(\tfrac{1}{2}\beta\varepsilon)/\left[V+V_0(\exp(\tfrac{1}{2}\beta\varepsilon)-1\right] \tag{9.54}$$

11 Note that in Section 9.6.2, N_{ij} was defined as the number of all i,j pairs, while M_{ij} represents the total number of molecules i around all molecules j. Hence for $i \ne j$, $M_{ij} = N_{ij}$, but for $i = j$, $M_{jj} = 2N_{jj}$.

12 This condition assumes that z is independent of x_j, which is not really true [57].

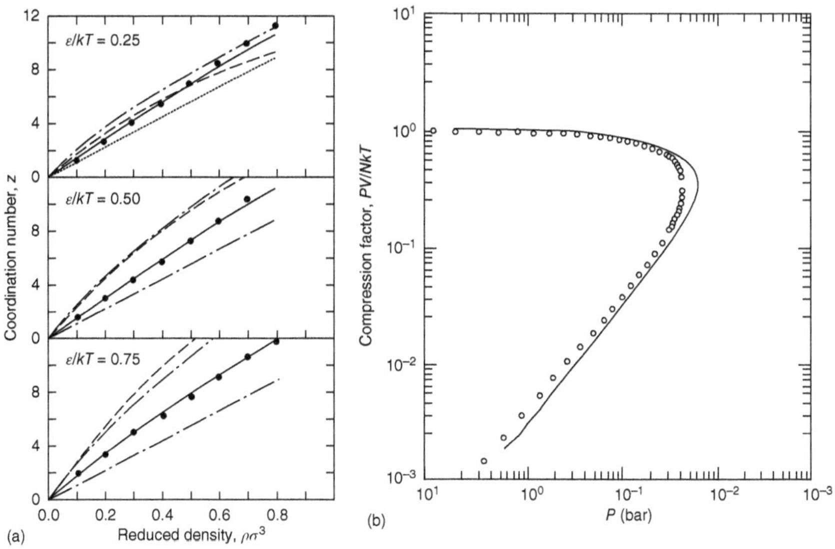

Figure 9.12 Generalized vdW model using the SW potential. (a) Coordination number z for three values of $\beta\varepsilon$. • MC simulations, ··· vdW EoS, -- Redlich-Kwong EoS, —·— Peng–Robinson EoS, — Eq. (9.54). (b) Compression factor PV/NkT for CH_4 using $P_{GvdW\text{-}SW}$ with $\varepsilon/k = 88.8$ K, $\sigma = 3.40$ Å and $R_{SW} = 1.85$ as calculated from the second virial coefficient, o experimental data.

The values agree closely with MC results [103] (Figure 9.12a). Combining this result with the CS pressure expression $\beta P_{CS}/\rho$ for the hard-core repulsion, the total pressure is $\beta P_{GvdW\text{-}SW}/\rho = \beta P_{CS}/\rho - z[1-\exp(-½\beta\varepsilon)]$. The compression factor $Z = PV/NkT$ agrees also favorably with MC results for the SW potential [104]. In fact, the absolute average deviation in Z for the GvdW-SW EoS in the one-phase region as compared to the MC results is 0.24, to be compared with 8.38, 2.16, and 1.33 for the vdW, Peng-Robinson, and Redlich-Kwong EoSs, respectively. For the critical data, the model provides $\varepsilon/kT_{cri} = 0.798$ (0.794), $Z_{cri} = 0.3344$ (0.2868) and $V_{cri}/V_0 = 6.11$ (4.25) with in parenthesis data from MD simulations [101]. For example, the Z-data for CH_4 (Figure 9.12b) using as only parameter z_{max} ($z_{max} = 42$ for $R_{SW} = 1.85$, fitted $z_{max} = 33.4$) are described well, the data obtained being $\varepsilon/kT_{cri} = 0.4466$ (0.4660), $Z_{cri} = 0.341$ (0.290) and $V_{cri}/V_0 = 5.83$ (5.91) with in parenthesis the experimental data.

Equation (9.54) does not satisfy the condition that for $\rho \to 0$, $z \sim \exp(\beta\varepsilon)$. Writing $M_{12}M_{21}/M_{11}M_{22} = f(\rho\sigma^3, \beta\varepsilon)$, a simple function for f satisfying this condition is $f = \exp(\alpha\beta\varepsilon)$ with $\alpha = (V-V_0)/V = (\rho_{max} - \rho)/\rho_{max}$. It shows proper behavior for $\rho \to 0$ and approximates $z \sim \exp(½\beta\varepsilon)$ for $\rho \to ½\rho_{max}$, using $½\rho_{max}$ as the approximate density for a liquid. However, in this way the density argument is used twice: for coordination and energy. Fortunately, the $\rho \to 0$ condition is not that important for liquids.

For other potentials than the SW potential, ε should be replaced by the average $-\langle u_{att}\rangle$, complicating matters considerably, and leading to different EoSs, depending on the choice for χ and V_f. The frequently used Redlich–Kwong, Soave–Redlich–Kwong, and Peng–Robinson EoSs (Section 4.2) all fit in this scheme [100].

Problem 9.9 ***The van der Waals and Carnahan–Starling Free Volume*** Show for the vdW and CS equations (Section 7.2), using $P = -\partial F/\partial V = kT\partial \ln V_f^N/\partial V$, that $V_f = V(1 - \rho b)$ and $V_f = V \exp[\eta(3\eta - 4)/(1 - \eta)^2]$, respectively.

9.8 Phonon Theory of Liquids

Although the heat capacity C_V of liquids is an important property, we have given it so far limited attention. Experimental data for many simple and complex liquids show that C_V decreases with temperature, from slightly larger than $3.0R$ near the melting point T_{mel} to $\cong 2.0R$ at high temperature [105]. In fact, a generally accepted theory to explain this decrease is absent and this presents a scientific as well as educational challenge [106]. Using Eq. (9.5) to calculate the internal energy U, we obtain $C_V = (\partial U/\partial T)_V$, but for that the potential $\phi(r)$ and the correlation function $g(r,T)$ must be known. These data must be rather precise as U, and thus C_V even more, is sensitive to details of $\phi(r)$ and $g(r)$. It is not clear whether this procedure has been shown to reproduce the C_V behavior, as interactions in liquids are strong and strongly dependent on the type.[13] As stated in Chapter 1, liquids behave intermediately between gases and solids. On the one hand, liquids flow and share this property with gases. Following this lead, liquids were mostly viewed as dense gases. On the other hand, the density of liquids is only slightly different from that of solids, but vastly different from gas densities. This suggests again using solid-state ideas.

The behavior of C_V and of quite a number of other properties is what the phonon theory of liquids, basically a relatively recent revival of old ideas by Frenkel [108], seeks to explain. The basic assumption is that the collective modes in a liquid can be described as vibrations of molecules in their cell or cage at a certain position for some time, followed by jumping to a new position accompanied by a *local relaxation event* (LRE), that is, a large-scale rearrangement of the molecules constituting the wall of the cell. Accordingly, the energy is divided into a vibration and a diffusion part, both having a kinetic and potential contribution. For the vibrations, Frenkel [108] argued that, like in solids, two types exist. The first type is the longitudinal waves. For the second type, the transverse waves, one normally assumes that they are absent in a liquid as the shear modulus $G \cong 0$, but Frenkel argued that at frequencies above $\omega_F = 2\pi/\tau$ the liquid *can* sustain transverse waves. Here, τ is the relaxation time, that is, the average time between LREs. The frequency ω_F, whose existence is experimentally well established [109], is called the *Frenkel frequency*. This approach [110], as it is based on solid state concepts, takes into account from the outset the strong interactions in liquids.

Accordingly, the total internal energy is given by

$$U = \vartheta_{lon} + \Phi_{lon} + \vartheta_{tra}(\omega > \omega_F) + \Phi_{tra}(\omega > \omega_F) + \vartheta_{dif} + \Phi_{dif}$$

with ϑ and Φ the kinetic and potential energy. We assume that $\Phi_{dif} \cong 0$, because the intermolecular interactions for both vibration and diffusion are the same and if Φ_{dif} would be significant and comparable to $\Phi_{tra}(\omega > \omega_F)$, strong restoring forces at low frequency would

13 This led to the phrase that "liquids do not allow calculations in a general form of the thermodynamic properties or even their dependence on temperature" [107], repeated in section 76 of [30].

result, leading to the existence of low-frequency vibrations instead of diffusion. Further, generally, $\vartheta_{\text{lon}} + \vartheta_{\text{tra}}(\omega > \omega_F) + \vartheta_{\text{dif}} = 3NkT/2$, which equals $(U_{\text{lon}} + U_{\text{tra}})/2$ according to the equipartition theorem. In the same spirit, $\Phi_{\text{lon}} = U_{\text{lon}}/2$ and $\Phi_{\text{tra}}(\omega > \omega_F) = U_{\text{tra}}(\omega > \omega_F)/2$. In total, using $U_{\text{tra}} = U_{\text{tra}}(\omega > \omega_F) + U_{\text{tra}}(\omega < \omega_F)$, we have

$$U = ½(U_{\text{lon}} + U_{\text{tra}}) + ½U_{\text{lon}} + U_{\text{tra}}(\omega > \omega_F) = U_{\text{lon}} + U_{\text{tra}}(\omega > \omega_F) + ½U_{\text{tra}}(\omega < \omega_F)$$

To estimate the number of active modes M, we consider that for N atoms we have $N = N\int_0^{\omega_D} g(\omega)\,d\omega$ longitudinal vibrations. For the N_1 transverse vibrations with $\omega > \omega_F$ and $g(\omega) = 3\omega^2/\omega_D{}^3$, we obtain from $2N\int_{\omega_F}^{\omega_D} g(\omega)\,d\omega$ the result $N_1 = 2N[1-(\omega_F{}^3/\omega_D)^3]$. As the number of transverse DoFs is conserved, the number of modes with $\omega < \omega_F$ is the difference $N_2 = 2N - N_1 = 2N(\omega_F/\omega_D)^3$. Hence, the number of active modes, $M = N + N_1 + ½N_2 = N[3-(\omega_F/\omega_D)^3]$. In the high-temperature approximation (HTA), the internal energy becomes $U = kTM = kTN[3-(\omega_F/\omega_D)^3]$. Neglecting for the moment, the temperature dependence of ω_F and ω_D (anharmonic contributions), we see that the decrease of the number of active DoFs as compared to solids is the major reason for the decrease in C_V.

To include anharmonicity, we use $f(\omega, T) = ½\hbar\omega + kT\ln\{1-\exp[\beta\hbar\omega(T)]\}$ for each vibrational mode $j = (p, \mathbf{q})$ and assume the ω_js to be temperature dependent. The vibrational Helmholtz energy $F_{\text{vib}} = \Sigma_j f_j$ yields the vibrational internal energy

$$U_{\text{vib}} = F_{\text{vib}} - T(\partial F_{\text{vib}}/\partial T) = U_0 + \Sigma_j[\exp(\beta\hbar\omega_j) - 1]^{-1}(\hbar\omega_j - T\partial\hbar\omega_j/\partial T) \qquad (9.55)$$

where U_0 is the zero-point energy. So far, everything is (almost) exact.

To progress, we need a model for ω_j and $\partial\omega_j/\partial T$ and for that, the quasi-harmonic Debye model (Section 10.8) with $3\,M$ DoFs and associated density of states (DoS) per DoF $g(\omega) = 3\omega^2/\omega_D{}^3$ is used. The geometric mean frequency ϖ is directly related to the Debye frequency ω_D via $\varpi = \omega_D e^{-1/3} = 0.717\omega_D$ [111], where $\hbar\omega_D = k\theta_D$ with θ_D the *Debye temperature*. We use the HTA $T > \theta_D$, so that the vibrational partition function reduces to $Z_{\text{vib}}(\omega,T) = (\beta\hbar\varpi)^{-3M}$ and the vibrational Helmholtz energy, neglecting zero-point energy, is given by $F_{\text{vib}}(\omega,T) = 3MkT\ln\beta\hbar\varpi$. Therefore, the entropy becomes

$$S_{\text{vib}} = -(\partial F_{\text{vib}}/\partial T)_V = 3Mk[1 - \ln\beta\hbar\varpi - (T/\varpi)(\partial\varpi/\partial T)]$$

Calculating the heat capacity, the result is

$$C_V = T\left(\frac{\partial S_{\text{vib}}}{\partial T}\right)_V = 3Mk\left[1 - \frac{2T}{\varpi}\left(\frac{\partial\varpi}{\partial T}\right)_V + \frac{T^2}{\varpi^2}\left(\frac{\partial\varpi}{\partial T}\right)_V^2 - \frac{T^2}{\varpi}\left(\frac{\partial^2\varpi}{\partial T^2}\right)_V\right] \quad \text{or}$$
$$\cong 3Mk[1 - (2T/\varpi)(\partial\varpi/\partial T)_V] \qquad (9.56)$$

neglecting second-order quantities. In the harmonic approximation, $\partial\varpi/\partial T = 0$ and $C_V = 3Mk$, consistent with the estimate above using U.

To assess the anharmonic contribution due to $(\partial\varpi/\partial T)_V$, we calculate first

$$P_{\text{vib}} = -(\partial F_{\text{vib}}/\partial V)_T = -(3MkT/\varpi)(\partial\varpi/\partial V)_T = 3MkT\gamma/V$$

where the (mode average) *Grüneisen parameter* $\gamma \equiv -(V/\varpi)(\partial\varpi/\partial V)_T$ is assumed to be a function of V only. Hence, the bulk modulus and its temperature derivative become

$$K_{\text{vib}} = -V(\partial P_{\text{vib}}/\partial V)_T = 3MkT\gamma(1-q)/V, \qquad (\partial K_{\text{vib}}/\partial T)_V = 3Mk\gamma(1-q)/V$$

where $q \equiv (\partial\ln\gamma/\partial\ln V)_T$ with typical value $q \cong 2$. For γ we use now the thermodynamic expression $\gamma = V\alpha K_T/C_V$ with the expansivity $\alpha = (1/V)(\partial V/\partial T)_P$, so that the derivative becomes $(\partial K_{\text{vib}}/\partial T)_V = 3Mk\alpha K_T(1-q)/C_V \equiv -3\,Mk\lambda K_T/C_V \cong -\lambda K_T$, and where the last step only can be made if we neglect second-order quantities as in that case $C_V = 3Mk$ will suffice. Note that, since $\lambda \equiv \alpha(q-1)$ and $q \cong 2$, the parameter λ can vary easily by a factor 2. Further assuming $\varpi^2 = cK$, where c is considered to be a constant, we obtain $-\gamma/V = (1/\varpi)(\partial\varpi/\partial T)_V = (1/2\,K)(\partial K/\partial T)_V$.[14] Since $K_T = K_0 + K_{\text{vib}}$, with K_0 the static energy contribution, we have $(1/\varpi)(\partial\varpi/\partial T)_V = (1/2K_T)(\partial K_T/\partial T)_V = -\tfrac{1}{2}\lambda$. By the way, the second-order terms in Eq. (9.56) vanish identically if $\varpi \sim \partial\varpi/\partial T$ (as is the case when $\varpi^2 = cK$ and $\partial\alpha/\partial T = 0$ are true). Moreover, $(\partial K_T/\partial T)_V = -\lambda K_T$ leads to $K_T = K_0\exp(-\lambda T)$ or $\kappa_T = 1/K_T = \kappa_0 \exp(\lambda T)$, as experimentally observed (Section 9.10).

Finally, to estimate $\omega_F = 2\pi/\tau$ we use the Maxwell model (Section 6.2.1), describing in the simplest possible way phenomenologically the viscoelastic behavior of materials. In this model, the shear strain γ is given by the sum of an elastic and viscous contribution so that the strain rate becomes $\text{d}\gamma/\text{d}t = G_\infty{}^{-1}\text{d}\sigma/\text{d}t + \eta^{-1}\sigma$, where G_∞, σ, and η are the (instantaneous or high-frequency) shear modulus, shear stress, and viscosity, respectively. When $\text{d}\gamma/\text{d}t$ becomes zero, the stress relaxes as $\sigma = \sigma_0 \exp(-t/\tau)$ with the *relaxation time* $\tau = \eta/G_\infty$. In reality, of course, τ has a distribution so that τ represents an average value. Experimentally, η can be described by the Vogel–Fulcher–Tammann (VFT) equation, $\eta = \eta_0\exp[A/(T - T_0)]$ with η_0, A, and T_0 parameters. The value of G_∞ can be taken at the maximum frequency of shear waves in the liquid. It is expected to have a similar value as the shear modulus G of the solid and as the T-dependence of G is much smaller than that of η, G_∞ is taken as a constant. While values for $\tau_D = 2\pi/\omega_D$ range from 0.1 to 1.0 ps, τ varies widely, typically from $\approx 10\tau_D$, even for a simple liquid like Ar, to $\approx 10^{15}\tau_D$ at the glass transition temperature, where $\eta \cong 10^{12}$ Pa s. Therefore, generally, $\tau > \tau_D$ or $\omega_F < \omega_D$.

The total expression for C_V in the HTA, neglecting terms in $\partial\omega_F/\partial T$, $\partial\omega_D/\partial T$, and $\partial\lambda/\partial T$, thus becomes

$$C_V = Nk(1+\lambda T)\left[3 - (\omega_F/\omega_D)^3\right]$$

For $0 < \omega_F \leq \omega_D$, C_V drops from $\cong 3.0Nk$ to $\cong 2.0Nk$. The factor $(1+\lambda T)$ explains why C_V exceeds $3Nk$ slightly just above T_{mel}, but the larger part of the temperature dependence for C_V is incorporated via $\omega_F = 2\pi G_\infty/\eta(T)$. Using for $3Nk$ the full (Debye) Z_{vib} or, alternatively, using $C_V = (\partial U/\partial T)_V$ with U given by Eq. (9.55), leads to applicability over a wider temperature range. Using data for τ_D and G_∞ from the solid state and $\eta(T)$ from the liquid state, in Figure 9.13 results for the C_V of Ar and CH_4 are given, showing that a good agreement with the experiment can be reached with physically quite reasonable values for $\lambda = \alpha(q-1)$. Similar agreement has been shown for several other simple liquids, liquid metals, and the complex liquids H_2S and H_2O, for the latter after correcting for rotation and configuration contributions.

It may seem a bit odd that a solid state model, the Debye model, is used for $g(\omega)$. Note though that, as long as collective modes are present, the existence of $g(\omega)$ is the only thing that counts, quite independent of whether a lattice structure is present or not, and in this sense the use of the quasi-harmonic Debye DoS is a pragmatic choice. In fact, the relation $\omega \sim |\boldsymbol{q}|$ with $\boldsymbol{q}$ the wave vector, as assumed for the Debye model, is experimentally shown to be (much) more applicable for liquids as for solids and therefore a not-to-far-off estimate for C_V is obtained. In a strict quasi-harmonic model, ϖ depends on $V(T)$ only, so

14 For a solid, we have $\varpi^2 = (K + 4G/3)/\rho$ and $G = 3\,K(1-2\nu)/2(1+\nu)$ with ρ the density and ν Poisson's ratio. If $\partial\nu/\partial T = 0$ (as experimentally it approximately is) and $\partial\rho/\partial T = 0$, we have $\varpi^2 = cK$. Neglecting the difference between K_T and the proper value K_S, I write here K.

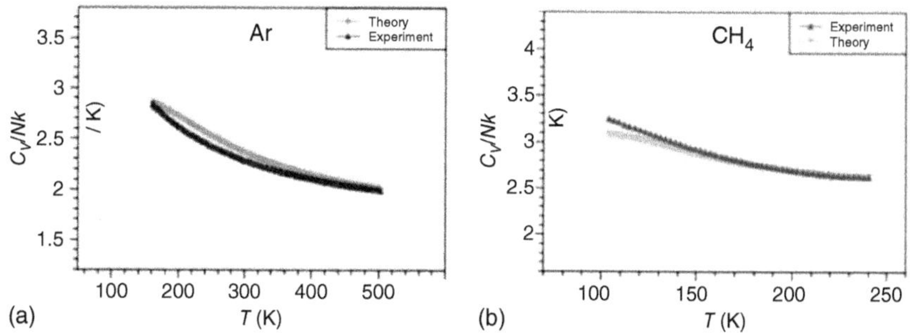

Figure 9.13 The heat capacity C_V according to phonon theory. (a) For Ar in the harmonic approximation ($\alpha = 0\,K^{-1}$) at 378 MPa using $\tau_{dif} = 0.5$ ps and $G_\infty = 0.18$ GPa. (b) For CH_4 in the anharmonic approximation at 50 MPa using $\tau_{dif} = 0.44$ ps, $G_\infty = 0.1$ GPa, and $\lambda = 1.4 \times 10^{-3}\,K^{-1}$ ($\alpha_{exp} = 3.1 \times 10^{-3}\,K^{-1}$), and with the rotational contribution $3R/2$ subtracted. Note that λ is used as fit parameter. Data for $\eta(T)$ and α are from the NIST Database (http://webbook.nist.gov/chemistry/fluid).

that explicit temperature dependence of ϖ is absent. This description is insufficient and in the quasi-harmonic phonon model, $\omega(V,T)$ has explicit temperature dependence, leading to the factor $(1 + \lambda T)$. For high η, $\tau \gg \tau_D$ and the term $(\omega_F/\omega_D)^3$ can be neglected. Hence, $C_V \cong 3Nk(1 + \lambda T)$ and C_V increases with temperature. For low η, the factor $(1 + \lambda T)$ is counteracted by the factor $[3 - (\omega_F/\omega_D)^3]$ leading to less temperature dependence, or even to a maximum in C_V. Due to experimental difficulties, such a maximum is not clearly observed, the experimental situation being that for low-η liquids, C_V decreases with T and for high-η liquids C_V increases with T. The more frequently employed C_P can be calculated from $C_P = C_V(1 + \alpha\gamma T)$, in which the factors $(1 + \lambda T)(1 + \alpha\gamma T)$ and $[3 - (\omega_F/\omega_D)^3]$ explain nicely the generally observed slight minimum in C_P.

It may be useful to indicate more fully why diffusive motion can be neglected. Recall that there are two approximately independent populations of molecules: one vibrating and one diffusing. This leads to $Z = Z_{vib}Z_{dif}$. If τ_{vib} represents the time spent for vibrations and a jump leading to an LRE takes about τ_{dif}, the jump probability is τ_{dif}/τ_{vib}. In equilibrium, this probability is equal to the ratio of diffusing molecules N_{dif} and the total number N, that is, $N_{dif}/N = \tau_{dif}/\tau_{vib}$. If the energy associated with diffusion $E_{dif} \sim N_{dif}$, this implies $E_{dif}/E_{vib} = \tau_{dif}/\tau_{vib} < 1$. As this is valid at any time, this relation applies to the average energies, and thus to C_V. However, the entropy depends on both Z_{vib} and Z_{dif}, which leads, given a diffusion model, to a communal contribution. The model has also been used to describe the glass transition behavior for glass-forming liquids [112] and (super)-critical behavior of liquids [113]. A review dealing with this (and some more) is available [114]. A rather rigorous analysis [115] shows that C_V is extremely sensitive to the values used for τ_F, as might have been expected. Nevertheless, the model can describe the thermal behavior of liquids satisfactorily.

Apart from the phonon theory, other theories deal with C_V, such as Eyring's SLS theory and Luck's empirical hole theory. Still other models exist, for example, Granato's interstitialcy approach based on an explicit model for molecular motion [106], the Rosenfeld–Tarazona expressions $U(\rho,T) = \alpha(\rho)T^{3/5} + \beta(\rho)$ and $C_V{}^E \sim T^{-2/5}$ based on HS perturbation theory [116], and the two-state model for the C_P of associating liquids [117]. Nevertheless, a generally accepted line of thought is still absent because these models are all based on different arguments. Finally, we mention that an empirical approach based on molecular volume [118] describes C_P of 1303 liquids at 298 K within 28 J mol^{-1} K^{-1}. Distinguishing between hydrogen bonding and non-hydrogen bonding liquids reduces this to about 20 J mol^{-1} K^{-1}.

Problem 9.10 ***The Debye Model*** Show that for the Debye model, $\varpi = \omega_D e^{-1/3}$.

Problem 9.11 ***Constant Volume Grüneisen Parameter*** Show that for any mode ω, $(\partial \ln\omega/\partial T)_V = \alpha(\gamma_T - \gamma_P)$ with $\gamma_T = (\partial \ln\omega/\partial V)_T$, $\gamma_P = (\partial \ln\omega/\partial V)_P$, and $\alpha = (\partial \ln V/\partial V)_P$. Use $\mathrm{d}\ln\omega(V,T) = (\partial \ln\omega/\partial V)_T \mathrm{d}V + (\partial \ln\omega/\partial T)_V \mathrm{d}T$ and the -1-rule for P, V, and T.

Problem 9.12 ***Entropy and Specific Heat*** Show that, if $Z = Z_{vib}Z_{dif}$ and $\varpi_{dif} \ll \varpi_{vib}$, $S \cong k[\ln Z_{vib} + (\partial \ln Z_{vib}/\partial T)_V + \ln Z_{dif}]$ and thus has a considerable diffusion component. Show that C_V depends mainly on Z_{vib}.

9.9 The Quantum Cluster Equilibrium Model

A relatively new approach, made feasible by present-day numerical calculations, is the *quantum cluster equilibrium* (QCE) model. The basic idea, due to Weinhold [119], is that, like in gases, in liquids clusters of molecules are present, and the liquid can be described as the collective of the weakly interacting clusters. For this idea to be applicable, the interactions between clusters should be less than those within clusters, and therefore, it is limited to molecules having secondary interactions like hydrogen bonding and other donor–acceptor bonding. In that case, the potential energy (surface) of these clusters can be calculated using modern quantum-chemical methods, whereafter their thermodynamics can be calculated by statistical mechanics, taking into account the internal DoFs for each type of cluster, the excluded volume, and the influence of all other clusters by means of a mean-field model.

The liquid is thus described by clusters W_j, considered to be an aggregate of j monomers W in thermodynamic equilibrium with all other clusters. Hence,

$$W \leftrightarrow \tfrac{1}{2}W_2 \leftrightarrow \tfrac{1}{3}W_3 \leftrightarrow \dots \leftrightarrow c^{-1}W_c$$

Each cluster type has its own partition function Q_j, containing a translational, rotational, vibrational, and electronic component. The total partition function is

$$Q = \prod\nolimits_j^c Q_j = \prod\nolimits_j^c q_j^{n_j}/n_j! \tag{9.57}$$

The equilibrium condition for N_A monomers, using the chemical potential μ_j, is by

$$\mu_1 \leftrightarrow \tfrac{1}{2}\mu_2 \leftrightarrow \tfrac{1}{3}\mu_3 \leftrightarrow \dots \leftrightarrow c^{-1}\mu_c \quad \text{with} \quad n_1 + 2n_2 + 3n_3 + \dots + cn_c = N_A \tag{9.58}$$

as constraint and n_j the number of clusters of size W_j. Combining these expressions, using the cluster W_r as a reference compound, yields

$$n_j = q_j(n_r/q_r)^{j/r}$$

Often, one chooses for W_r the monomer W_1. While for the perfect gas interactions between molecules could be neglected, in this case it is essential to take them into account, but as each cluster influences all other clusters, this results in a highly nonlinear problem. In the simplest approach, the excluded volume for each cluster depends on its translational partition function $q_j^{(\mathrm{tra})}$ and a correction for its electronic partition function $q_j^{(\mathrm{ele})}$ are included in a vdW-type way. Effects on the rotational and vibrational partition functions can be incorporated but are neglected here.

The excluded volume, weighted over all populations, is taken to be

$$V^{(\mathrm{exc})} = b\sum_j^c n_j V_j^{(\mathrm{exc})}$$

where a single "excluded-volume parameter" b is used for all clusters. Clearly, $V_j^{(\mathrm{exc})}$ is more complicated than for a spherical molecule, but can be calculated using various approximate schemes, such as in [120]. Hence, we have $q_j^{(\mathrm{tra})} = q_j^{(\mathrm{tra})}(b, \mathbf{n})$, where $\mathbf{n}$ denotes the set of all n_js. The effect on $q_j^{(\mathrm{ele})}$ is taken into account by a single mean-field parameter a by assuming $\Delta E(W_j) = a(j/V)$, where j/V is the number density of clusters of size j. Hence, similar as for the excluded volume effect, we have $q_j^{(\mathrm{ele})} = q_j^{(\mathrm{ele})}(a, \mathbf{n})$. In total, we have $q_j = q_j(a, b, \mathbf{n})$, which depends on parameters a and b as well as on all population numbers $\mathbf{n}$.

Combining Eqs. (9.58a) and (9.58b) results in a polynomial of order c, the solution of which determines $\mathbf{n}$ for given temperature T_0 and pressure P_0. Substitution in Eq. (9.57) provides the Helmholtz energy and pressure given by

$$F = -kT\ln Q \quad \text{and} \quad P = -\partial F/\partial V = kT\partial \ln Q/\partial V \tag{9.59}$$

In practice, values for V and $\mathbf{n}$ are searched that are consistent with P_0 through $P = -\partial F/\partial V$. This leads to a functional equation $P_0 = P[V^{(\alpha)}]$ whose roots $(V^{(\alpha)}, \mathbf{n}^{(\alpha)})$ satisfy Eqs. (9.58a) and (9.59b). Each root α corresponds to a Gibbs energy $G^{(\alpha)} = F^{(\alpha)} + P_0 V^{(\alpha)}$ and the root $V^{(\alpha)}$ with lowest $G^{(\alpha)}$ represents the stable phase for T_0 and P_0.

Solutions to the equations must be obtained by an iterative procedure by first assuming values for a, b, and $\mathbf{n}$ for a given pressure P_0 and temperature T_0, repeated until consistency is reached. From the final data, the most stable phase, G, L, or S, with the lowest $G^{(\alpha)}$ is obtained. The procedure has been described in detail in [119].

It appears that only a small number of cluster types need to be incorporated, which results (again!) from the balance of increased energy gain for larger clusters, due to the increased cooperativity within a cluster, and increased entropy gain for smaller clusters, due to their increased number. This effect has been shown clearly for water for which even a simple two-cluster model including monomers and one type of cluster, say six-membered rings, can describe its properties reasonably [121], thereby providing support for two-state models [122]. Figure 9.14 shows results for water using 3-21G basis set restricted Hartree–Fock (RHF) calculations for 18 types of clusters [123], which were corrected for superposition basis set [124] effects and by the standard factor 0.89 for vibration frequencies [125]. The dominating clusters appeared to be W_1, W_5, W_6, W_8, and W_{24}. Clearly, the final numerical accuracy is insufficient, but the main features of the phase diagram emerge clearly. In principle, there is no difficulty in calculating α_P, κ_T, and C_P, but the higher derivatives of Q required are subject to considerable error due to both model restrictions and numerical accuracy. Finally, we note that extension to multiple component systems is feasible.

9.10 Some Continuum Aspects

Microscopic theory provides useful insight but cannot always provide solutions for technical problems. Consequently, macroscopic considerations remain important. Here,

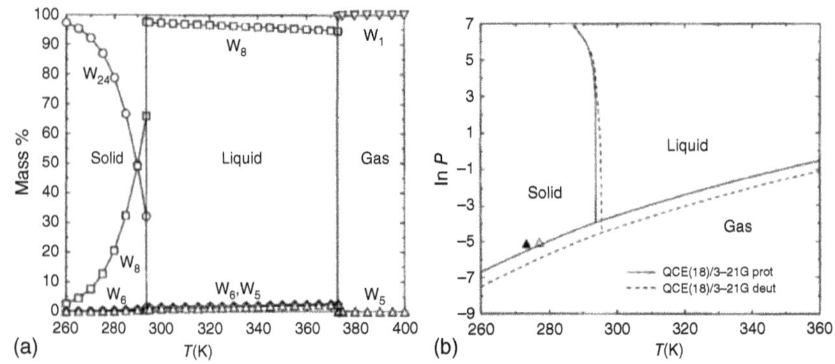

Figure 9.14 QCE results for water in the 18-cluster 3-21G model. (a) Dominating cluster populations (mass%) of normal water. (b) Phase diagram of normal (solid line) and heavy (dashed line) water. The calculated triple points are T_{tri} = 293.74 K, P_{tri} = 0.01915 atm for normal and T_{cri} = 295.78 K, P_{tri} = 0.01108 atm for heavy water. The experimental data are shown as triangles. Source: Ludwig and Weinhold [123]/Royal Society of Chemistry.

we briefly address only the typical values of some properties, the EoS and a few continuum aspects, as this is an extensive field in its own right.

Characteristic physical properties of liquids are the compressibility κ_T, the expansivity α_P, the heat capacity C_P, and the surface tension γ. For the latter, we refer to Section 11.2. Limiting ourselves to normal liquids, the typical range of κ_T is $(0.2-2.0)\times10^{-4}$ bar^{-1}. Generally, κ_T decreases with increasing pressure, with a smaller rate at higher pressure, and increases with increasing temperature, with a larger rate at lower pressure. The temperature dependence of κ_T can be described by

$$\kappa_T = \kappa_0 \exp(\beta T) \tag{9.60}$$

with κ_0 and β empirical constants (Table 9.4). The typical range of α_P for normal liquids is $(0.7-1.5)\times10^{-3}$ K^{-1} and increases with increasing temperature. Some data for κ_T and α_P are given in Section F.6. Values for C_P are generally somewhat higher than those for the corresponding solids at T just above T_{mel} and often described by a second- or third-order polynomial in T. Water behaves anomalously, however, and κ_T passes through a minimum at ≈45 °C, which Eq. (9.60) obviously cannot describe. An overview of the subtleties to consider for proper measurement of physical properties of liquids is given in [127]. Marcus [128] reviewed κ_T values (0.01 – 0.1 GPa^{-1}) and α_P values ($0.1-0.2\times10^{-3}$ K^{-1}) for liquid metals, the alkali metals somewhat deviating (κ_T: 0.2 – 0.6 GPa^{-1}, α_P: $0.1-0.2\times10^{-3}$ K^{-1}), and κ_T values for molten salts (0.1 – 0.7 GPa^{-1}).

Table 9.4 Compressibility parameters for a few compounds.

Molecule	T range	κ_0 (10^{-4} atm)	β (10^{-3} K^{-1})	ξ (–)
CS_2	168–353	12.55	6.73	7.5 (20 °C)
CH_3OH	175–331	18.6	6.30	8.8 (0 °C)
nC_5H_{12}	162–273	19.6	7.97	9.2 (10 °C)

Source: Moelwyn-Hughes [126]/Elsevier.

Table 9.5 Parameters for the Tait–Tammann isotherm for H_2O and CCl_4.

Molecule	unit	A (–)	B (25 °C)	B (45 °C)	B (65 °C)	B (85 °C)
H_2O	bar	0.1368	2996	3081	3052	2939
CCl_4	atm	0.0924	869	738	622	—

Source: Adapted from Moelwyn-Hughes [126].

For liquids, an expression equivalent to the perfect gas EoS does not exist. A simple EoS for liquids doing remarkably well is the *Tait equation* (TE) [129–132], given by

$$\frac{V - V_0}{PV_0} = -\frac{A}{B + P}$$

Here, V_0 represents the molar volume at reference pressure P_0 (usually 1 bar) with A and B are (positive) material specific parameters. Apparently, Tammann rewrote (or misquoted?) this expression for the first time [130] as

$$\frac{1}{V_0}\frac{dV}{dP} = -\frac{A}{B + P} \quad \text{or as its integrated form} \quad V_0 - V = AV_0 \ln\left(\frac{B + P}{B + P_0}\right)$$

Although there is little theoretical justification, both the Tait equation and the Tait–Tammann equation (TTE) represent the behavior of many types of liquids quite well up to pressures of about 1.5 kbar. Some values for TTE parameters are given in Table 9.5. It appears that generally A is only weakly dependent on temperature and covers actually a relatively narrow range, say 0.07–0.13, but that B decreases with increasing temperature. The anomalous behavior of water is reflected in its B-value by showing a maximum as a function of T. Using $p = P - P_0$, $\nu = V_0/A$, $\kappa_0 = [\nu(B+P_0)]^{-1}$, the TTE can be written [133] as $V/V_0 = 1 - \nu^{-1}\ln(1 + \nu\kappa_0 p)$, and from this expression one derives $\overline{K} \equiv (V/V_0)K = V_0(\partial P/\partial V) = K_0 + \nu p$, while experimentally, $K \equiv V(\partial P/\partial V) = K_0 + \gamma p$, with γ another constant, holds much more accurately. The latter leads to the modified equation [133] reading $V/V_0 = (1 + \gamma\kappa_0 p)^{-1/\gamma}$ and to $\kappa_T = \kappa_0(V/V_0)^\gamma$.

For the molar volume V_{sat} as a function of T and P of a *saturated liquid*, a liquid in equilibrium with its vapor, many empirical EoSs exist. An EoS [134], describing the behavior between T_{mel} and T_{n} and up to 300 MPa well, is $V = (a + bP)/[c - d(1 - \zeta)T + P]$ with all constants (a–d) and $\zeta > 0$. Both α_P and κ_T are predicted to increase with T and to decrease with P. It accounts for the experimentally observed intersection of α_P versus P isotherms and the T- and V-dependence of $(\partial P/\partial T)_V$. The purely empirical modification $V = (a + bP + eP^{1/2})/(c - dT + P + fP^{1/2})$ with (a–f) constants given for a wide range of data yields in an AAD of ~0.2%. For an overview of EoSs, see [135].

Before discussing an example of a continuum description, it might be useful to recall how a material in continuum theory is described. Generally the continuum behavior as a function T can be described by the Helmholtz energy $f(\boldsymbol{a},\boldsymbol{b},T)$ and the dissipation function $\phi(\boldsymbol{a}, \boldsymbol{b}, \dot{\boldsymbol{a}}, \dot{\boldsymbol{b}}, T)$ with $\boldsymbol{a}$ the external variables and $\boldsymbol{b}$ the internal variables. The conservative stress is given by $\boldsymbol{\sigma}^{(q)} = \rho\partial f/\partial \boldsymbol{a}$ and the dissipative stress $\boldsymbol{\sigma}^{(d)} = \xi\rho\partial\phi/\partial\dot{\boldsymbol{a}}$ with $\xi = \phi/[(\partial\phi/\partial\dot{\boldsymbol{a}})\cdot\dot{\boldsymbol{a}}]$. A *solid* is a material that can resist a deviatoric stress $\boldsymbol{\sigma}^{(q)\prime}$. The limiting case is an *elastic solid* defined by $\boldsymbol{b} = \boldsymbol{0}$ and $\boldsymbol{\sigma}^{(d)} = \boldsymbol{0}$ so that in this case the stress is quasi-conservative, depending only on $\boldsymbol{a}$ and T, as do all other state functions. A *fluid* cannot withstand a deviatoric stress. Here, the limiting case is a *viscous fluid*, defined by $\boldsymbol{a} = \boldsymbol{0}$ and $\boldsymbol{\sigma}^{(q)} = \boldsymbol{0}$, so that the stress is purely dissipative, depending on the deformation rate $\dot{\boldsymbol{a}}$ and T. Generally, both $\boldsymbol{\sigma}^{(q)}$ and $\boldsymbol{\sigma}^{(d)}$ differ from zero and internal variables $\boldsymbol{b}$ are required to describe flow behavior, for example, the viscoelastic behavior, and in these cases, $\boldsymbol{\sigma} = \boldsymbol{\sigma}^{(q)} + \boldsymbol{\sigma}^{(d)}$. Because the division of $\boldsymbol{b}$ in a conservative and dissipative part is to some extend arbitrary and $\boldsymbol{b}$ also appears in f, the resulting description depends on their precise choice.

With that in mind, we note that for isotropic materials we indicated in Section 5.7 that, for a stress s_{ij} and a strain t_{ij}, the specific Helmholtz energy f depends only the invariants of its arguments, which led in first order to

$$s_{ij} = \lambda\, t_{(1)}\delta_{ij} + 2\mu\, t_{ij} = \lambda\delta_{ij}\sum\nolimits_k t_{kk} + 2\mu\, t_{ij} = (\lambda + 2\mu/3)\sum\nolimits_k t_{kk} + 2\mu\, t_{ij}'$$

Here, $t_{(1)}$ is the first basic invariants of the tensor t_{ij} and t_{ij}' its deviator. A *Newton liquid* is an isotropic liquid for which the deviatoric stress is purely dissipative while the spherical stress is purely conservative. If the liquid is taken as incompressible, one obtains what is called a *Navier–Stokes liquid*. Using higher-order expressions for ϕ leads to *Reiner–Rivlin liquids*. For concise introductions, see [136] or [137]. In Example 9.2, we provide some details for a Newton liquid.

Example 9.2 ***The Newton Liquid*** From the Helmholtz energy f for isotropic materials,

$$\rho f = {}^1\!/\!_2\lambda\sum\nolimits_{i,j}\varepsilon_{ii}\varepsilon_{jj} + \mu\sum\nolimits_{i,j}\varepsilon_{ij}\varepsilon_{ij} \qquad \text{with } \lambda \text{ and } \mu \text{ elastic constants}$$

we obtain, using the bulk modulus K, the quasi-conservative pressure

$$P = \sum\nolimits_k \sigma_{kk}^{(q)} = \sum\nolimits_k \frac{\partial \rho f}{\partial \varepsilon_{kk}} = \left(\lambda + \frac{2}{3}\mu\right)\sum\nolimits_k \varepsilon_{kk} \equiv K\sum\nolimits_k \varepsilon_{kk} = K\Delta V/V$$

with $\Delta V/V$ the relative volume change. Similarly, we have the dissipation function $\rho\phi = \zeta\sum_{i,j}\dot{\varepsilon}_{ii}\dot{\varepsilon}_{jj} + 2\eta\sum_{i,j}\dot{\varepsilon}_{ij}\dot{\varepsilon}_{ij}$ with ζ and η viscosity coefficients leading to the dissipative stress

$$\sigma_{ij}^{(d)} = \xi\rho(\partial\phi/\partial\dot{\varepsilon}_{ij}) = \zeta\delta_{ij}\sum\nolimits_k \dot{\varepsilon}_{kk} + 2\eta\dot{\varepsilon}_{ij}$$

and the total stress $\sigma_{ij} = \sigma_{ij}^{(q)} + \sigma_{ij}^{(d)}$. If the *bulk viscosity* $\eta_V \equiv \zeta + 2\eta/3 = 0$, the liquid is incompressible and the expression for $\sigma_{ij}{}^{(d)}$ reduces to *Newton's law of viscosity*

$$\sigma_{ij}^{(d)} = 2\eta\left(\dot{\varepsilon}_{ij} - \delta_{ij}\sum\nolimits_k \dot{\varepsilon}_{kk}\right) = 2\eta\dot{\varepsilon}_{ij}'$$

with (*shear*) *viscosity* η and deviator $\dot{\varepsilon}_{ij}'$. The energy dissipated is $\phi = \sum_{i,j}\sigma_{ij}^{(d)}\dot{\varepsilon}_{ij} = \zeta\sum_{i,j}\dot{\varepsilon}_{ii}\dot{\varepsilon}_{jj} + 2\eta\sum_{i,j}\dot{\varepsilon}_{ij}\dot{\varepsilon}_{ij}$, reducing to $\phi = 2\eta\sum_{i,j}\dot{\varepsilon}_{ij}'\dot{\varepsilon}_{ij}'$ for a Newton liquid.

Problem 9.13 ***Compressibility*** Show that the dependence on volume for the Gibbs energy, estimated by integrating $\mathrm{d}G = V\mathrm{d}P - S\mathrm{d}T$ at constant T using for the compressibility $\kappa_T = \kappa_0(V/V_0)^\gamma$ meanwhile realizing that $V\mathrm{d}P = -\kappa_T{}^{-1}\mathrm{d}V$, leads to $\Delta G = [V/\kappa_T(\gamma - 1)][1 - (V/V_0)^{\gamma-1}]$.

9.11 Dielectric Behavior

For the discussion of the dielectric behavior, we use in this paragraph the relative permittivity $\varepsilon_r \equiv \varepsilon_1$. Phenomenologically, ε_r of many liquids can be described by $\varepsilon_r(T) = \varepsilon_r{}^\circ\ e^{-LT}$ with $\varepsilon_r{}^\circ$ and L parameters. Table 9.6 indicates that polar liquids typically have a (much) larger value for $\varepsilon_r{}^\circ$ and L than nonpolar liquids.

The question is now how well the approach as described in Section 8.6 dealing with the dielectric behavior of gases can be applied to liquids. Let us first consider (dilute) mixtures of polar molecules in a nonpolar solvent. Figure 9.15a shows an example: nitrobenzene in hexane. We consider that the molar polarization of a mixture P_m is given by the rule of mixtures

$$P_m = \Sigma_j P_{m,j}\, x_j$$

where x_j is the mole fraction of the component j. The molar polarization $P_{m,j}$ of each component j is, as before, given by

$$P_{m,j} = N_A\left(\alpha_j + \mu_j{}^2/3kT\right) = N_A \alpha_{\text{eff},j}$$

If the value of α_j of the nonpolar solvent is known, for example, from the refractive index, the value of μ_j of the solute can be calculated, given either temperature-dependent polarization data or independent data for α_j for the polar component. For the example mentioned

Table 9.6 Parameters $\varepsilon_r{}^\circ$ and L for various liquids.

Liquid	$\varepsilon_r{}^\circ$ (–)	L (10^{-3} K^{-1})	Liquid	$\varepsilon_r{}^\circ$ (–)	L (10^{-3} K^{-1})	Liquid	$\varepsilon_r{}^\circ$ (–)	L (10^{-3} K^{-1})
H_2O	311.2	4.63	C_6H_6	2.95	0.876	CCl_4	2.88	0.843
CH_3OH	157.6	5.39	$C_6H_5CH_3$	2.86	0.673	$CHCl_3$	12.0	3.33
C_2H_5OH	146.0	6.02	$C_6H_5NO_2$	164.7	5.21	$(CH_3)_2CO$	120.4	4.63

Source: Adapted from Moelwyn-Hughes [126].

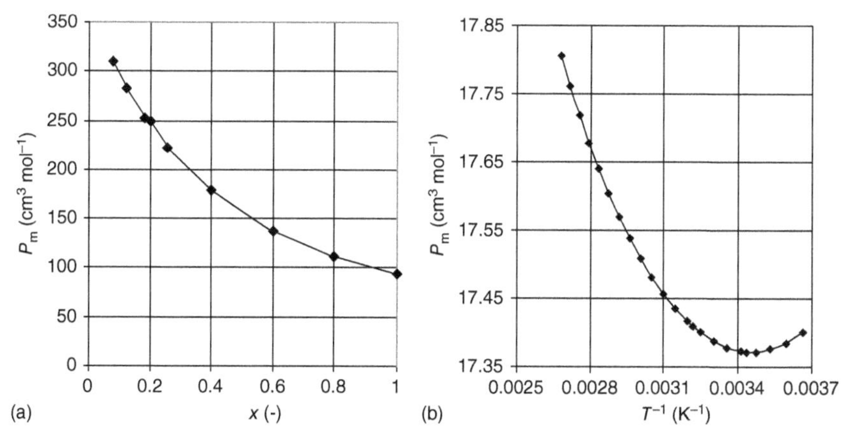

Figure 9.15 (a) The molar polarization P_m of the polar molecule $C_6H_5NO_2$ in the nonpolar solvent C_6H_{14} as a function of mole fraction x (the small discontinuity at $x \cong 0.2$ is due to merging of two data sets). (b) The anomalous behavior of P_m for water as function of inverse temperature.

$\mu_j = 3.9$ D ($\alpha_j \cong 0$ C^2 m^2 J^{-1}) was obtained as compared with $\mu_j = 4.2$ D for the molecule in the gas phase. The agreement is fair, which is true in general for molecules dissolved in nonpolar solvents. Because $P_{m,j}$ is only approximately described by a rule of mixtures using volume fractions for each component (Figure 9.15a), extrapolated values of P_m with $x_j \to 0$ are often used.

The situation is quite different for polar liquids, however. For water the dipole moment, $\mu = 1.85$ D, and the polarizability volume, $\alpha' = 1.48$ Å^3. If we calculate the molar polarization P_m and from that result the value of P_m/V_m, we obtain the value 4.2. Since $P_m/V_m = (\varepsilon_r - 1)/(\varepsilon_r + 2)$ should be always smaller than 1, this result is physically impossible. Such behavior is, therefore, occasionally called the "polarization catastrophe". For nonpolar compounds, however, with $\alpha_{eff} = \alpha_{mol} \cong \alpha_{ele}$ we note from $P_m/V_m \leq 1$ that $\alpha_{ele} = 4\pi\varepsilon_0\alpha' \leq V_m/N_A$ should hold. This is generally the case, as we have seen (Section 8.6.4), so that for nonpolar compounds the critical state is never reached. We return to the "catastrophe" in Chapter 17. Still, the dielectric behavior of water[15] (and other pure, polar liquids) is very unlike Debye behavior (Figure 9.15b), which is due to shielding and correlation.

With regard to shielding, a molecule experiences a field not equal to the applied field because the surrounding molecules shield the applied field. In the Clausius–Mossotti and Debye models, this effect was taken into account by using for both the local field E_{loc} and directional field E_{dir}, the Lorentz field. However, Onsager [139] argued that E_{dir} is different from E_{int}, and derived an expression for E_{dir}. Justification 8.1 presents an outline of that derivation for a spherical molecule in a spherical cavity with radius a, from which we quote,

$$\frac{(\varepsilon_r - 1)(2\varepsilon_r + 1)}{3\varepsilon_r} = \frac{N_A}{\varepsilon_0 V_m}\left(\alpha^* + \frac{\mu^{*2}}{3kT}\right) \tag{8.29}$$

Here, $\mu^*/\mu = \alpha^*/\alpha = 1/(1 - f\alpha)$ with $f = 2\mathscr{C}(\varepsilon_r - 1)/(2\varepsilon_r + 1)a^3$ and $\mathscr{C} = (4\pi\varepsilon_0)^{-1}$. In the derivation, Onsager made use of the approximation $4\pi N_A a^3/3V_m = 1$. This result implies, first, that the functional dependence on ε_r is changed as compared with the Debye result and, second, that both the dipole moment μ^* and polarizability α^* in the condensed state are typically enhanced by 20–50% as compared to the gas state values μ and α by the near presence of other molecules. For example, for $CHCl_3$ with $\varepsilon_r = 4.8$ and $n = 1.446$ we obtain $\mu^*/\mu = 1.24$, using for α the estimate $\mathscr{C}\alpha/a^3 = (\varepsilon_\infty - 1)/(\varepsilon_\infty + 2)$ with $\varepsilon_\infty \cong n^2$. Introducing the expressions for f, a, and α in Eq. (8.31), the explicit result for μ reads

$$\mu^2 = \frac{9\varepsilon_0 kTV_m}{N_A}\frac{(\varepsilon_r - \varepsilon_\infty)(2\varepsilon_r + \varepsilon_\infty)}{\varepsilon_r(\varepsilon_\infty + 2)^2} \tag{8.32}$$

usually denoted as the *Onsager equation*. The value of ε_∞, which sometimes is estimated as $\varepsilon_\infty \cong n^2$ (related to α_{ele}) and sometimes as $\varepsilon_\infty \cong 1.05n^2$ (related to $\alpha_{ele} + \alpha_{mol}$) [141], is critical. Nevertheless, from this equation estimates of μ for non-associating molecules can be obtained, generally within 10% as compared with results from independent gas-phase experiments (Table 9.7).

Turning to correlation, to calculate the average dipole moment we assumed that the alignment of the molecules was only due to the applied field. However, there is a correlation between neighboring molecules due to their mutual interaction. Molecules are polarizable and the dipole moment of the molecule surrounded by its neighbors $\boldsymbol{m} = \boldsymbol{\mu} + \alpha\boldsymbol{L}$, with $\boldsymbol{L}$

15 The anomalous behavior of H_2O is also reflected in its refractive index n. The relation $(n^2 - 1)/(n^2 + 2) = A\rho^B\exp(-CT)$ describes n to within a few digits of the seventh decimal place for optical wavelengths [138]. For the Na D-line, $A = 0.2064709$, $B = 0.88538$, and $C = 6.2037 \times 10^{-5}$ K^{-1}.

Table 9.7 Onsager dipole moment μ_0 and Kirkwood factor g for several polar molecules.

	T (°C)	V_m (cm^3 mol^{-1})	ρ (g cm^{-3})	ε_r (–)	ε_∞ (–)	μ (D)	μ_0 (D)	g (–)
C_6H_5Cl	25	102.2	1.107	5.61	2.55	1.73	1.40	0.61
$C_6H_5NO_2$	25	102.7	1.198	34.9	2.63	4.24	4.06	0.87
$(CH_3)_2CO$	25	73.86	0.785	19.11	2.0	2.89	3.03	0.97
CH_3Cl	25	57.47	1.005	9.68	1.93	1.86	1.74	0.9
CH_2Cl_2	25	64.47	1.385	8.93	2.35	1.57	1.37	1.01
$CHCl_3$	25	80.72	1.489	4.72	2.37	1.01	1.18	1.19
$N(CH_3)_3$	25	94.3	0.658	2.44	1.8	0.61	0.66	1.5
NH_3	25	28.3	—	16.9	1.53	1.47	—	1.65
H_2O	0	18.02	—	88.2	1.79	1.84	—	2.89
	25	18.07	—	78.5	1.79	1.84	—	2.81
	83	18.58	—	60.4	1.76	1.84	—	2.69

Data for V_m, ε_r, ε_∞, μ (gas phase dipole moment), and g are from [140]. Data for ρ and μ_0 from [141].

the local electric field due to the neighboring molecules, should be used. Modeling a polar fluid as a set of molecules embedded in a spherical cavity in a continuum with permittivity ε_r, as first devised by Kirkwood [142] and extended by Fröhlich [143], one obtains instead of the Onsager expression for μ^2 the same expression for $g\mu^2$ (Justification 9.4), where the *correlation factor* g is defined by

$$g = 1 + \sum\nolimits_{j=1}^{j=\infty} g_j \equiv 1 + \sum\nolimits_{j=1}^{j=\infty} N_j \left\langle \cos\gamma_j \right\rangle$$

with N_j the number of molecules in coordination shell j and $\langle \cos\gamma_j \rangle$ the average cosine of the angle between the dipole of a (fixed) reference molecule and the molecules in shell j. The molecular dipoles tend to line up but are counteracted by geometric packing effects. If the dipole moment is along the long axis of the molecule, packing effects forces more dipole moments to have an antiparallel configuration and $g < 1$. In case the dipole moment is along the short axis of the molecule, packing effects result in having more parallel dipole moments and $g > 1$. The correlation factor cannot be determined experimentally in an independent way, and accurate information on structure and intermolecular interactions is required for its calculation. Table 9.8 shows that there is neither correlation between g and ε, nor between g and μ. One would expect, however, that in a series of similar molecules using independent values for the dipole moments similar values for g are obtained. Table 9.8 shows that for aliphatic ketones this is indeed the case.

Hydrogen bonding constitutes a specific form of correlation between molecules and is particularly relevant for water. Assuming for water that every molecule is hydrogen-bonded to four neighbors and taking into account interactions up to the fourth shell with $N_1 = 4$, $N_2 = 11$, $N_3 = 22$, Pople [145] calculated $g_1 = 1.20$, $g_2 = 0.33$, and $g_3 = 0.07$, resulting in $g \cong 2.60$ at 0 °C (2.46 at 83 °C). It appears therefore that correlation with shells outside the first makes an appreciable contribution to ε_r. The permittivity obtained is $\varepsilon_r = 71.9$ at 0 °C (exp $\varepsilon_r = 88.0$) and $\varepsilon_r = 47.8$ at 83 °C (exp $\varepsilon_r = 59.9$) [145]. The dipole moment was estimated as $m = 2.15$ D, some 17% larger than for the gas state, $\mu = 1.85$ D. This estimate is too

Table 9.8 Values of the Kirkwood factor g for some aliphatic ketones.

Molecule	ε_r at 25 °C	μ (D)	$g\mu^2$ (D^2)	g (–)
Propanone	21.45	2.85	9.82	1.21
Methyl ethyl ketone	18.51	2.77	9.97	1.30
Methyl propyl ketone	15.45	2.72	9.69	1.29
Diethyl ketone	17.00	2.71	10.40	1.42
Methyl isobutyl ketone	13.11	2.72	9.20	1.24
Methyl amyl ketone	11.95	2.70	9.10	1.25
Dipropyl ketone	12.60	2.72	9.60	1.30
Methyl hexyl ketone	10.39	2.70	8.70	1.19

Source: Adapted from Minkin et al. [144].

small as Pople included for this calculation only the effect of the first coordination shell. For ice I_h at 0 °C, Coulson and Eisenberg [146] calculated, taking the first three shells into account, $m = 2.60$ D, some 40% larger than $\mu = 1.85$ D. Scaling the m data by 1.14, a factor obtained as the ratio between a nearest neighbor only and a three-shell calculation, yields $m = 2.45$ D and brings this result in good agreement with the experiment [5]. In the above calculations, $\varepsilon_\infty = n^2 \cong 1.77$ was taken, but the real value of ε_∞ should be used. Hill [147] showed that using $\varepsilon_\infty = 4.5$ (an estimate based on various experimental dispersion data), one finds $g \cong 1$ and concluded that, at least for water, association has no effect. For alcohols and some other associating liquids, $g \neq 1$ though (see, e.g. [141, 143]). A rather complex rigorous derivation [148] yields Eq. (9.63), and results for water, using $\varepsilon_r \cong 80$, $\varepsilon_\infty = 1.8$, and $\mu = 1.85$ D ($m = 2.95$ D), in $g \cong 2.8$. The debate continues [149] and seems, at least partly, to be related to whether or $\boldsymbol{\mu}$ or $\boldsymbol{m}$ is used, about which there is often confusion [150].

A somewhat more elaborate model to calculate ε_r [151] assumes tetrahedral configurations with a probability p for (single) H-bonds to occur and takes all shells into account, but neglects ring formation, cooperativity and volume changes. By straightforward geometrical considerations, the expression $g = 1 + 4p/(3 - p)$ is derived by summing over all neighboring shells. For N water molecules with n H-bonds, each with Helmholtz energy f, the total Helmholtz energy F is assumed to be

$$F = nf - kT\ln\Omega \quad \text{with} \quad \Omega = \frac{(2N)!}{n!(2N-n)!}\frac{(2N)!}{n!(2N-n)!}n!\left(\frac{\rho^\circ}{N}\right)^n$$

Here Ω takes into account all possible configurations that $2N$ donor and $2N$ acceptor sites can form (the first two factors), the way that n donor and acceptors can combine (the third factor) and the proximity criterion for bond formation ($\rho^\circ = \rho/\rho^*$, with ρ^* a close-packed reference volume taken at 4 °C, and approximated by $\rho^\circ = 1$). The usual optimization via $\partial F/\partial n = 0$ leads to $p/(1-p)^2 = 2K\rho$ or $p = (X-1)/(X+1)$ with $X = (1+8K\rho)^{1/2}$, where $K = \exp[-(\Delta H - T\Delta S)/kT]$ is the equilibrium constant for bond formation. Hence, the g-factor becomes $g = 1 + 2(X-1)/(X+2)$. For ice I_h, assuming $p = 1$, this yields $g = 3$ and via the Kirkwood–Fröhlich expression without any further parameter to a good agreement with the experiment ($\varepsilon_r = 91.3$, exp. $\varepsilon_r = 92.0$). For liquid water, the dielectric relaxation time τ is assumed to be controlled by the formation of 2-bonded molecules from 3-bonded molecules, as reorientation can only occur via 2-bonded molecules. Assuming as parameters ΔH, ΔS and ε_∞, experimental values for ε_r (over the T-range 270–650 K) and τ (over the T-range 280–380 K) for water were fitted yielding $\Delta H = -14.8\,\text{kJ mol}^{-1}$, $\Delta S = -19.2\,\text{J mol}^{-1}\,\text{K}^{-1}$, and $\varepsilon_\infty = 1.85$, the latter value being only slightly larger than the experimental value $n^2 = 1.77$. The values of ΔH and ΔS are sensitive because the dielectric properties in this model are related to the H-bond stoichiometry through p, which is a weak function of K. The ΔH-value obtained is in the range of estimates obtained otherwise (−6 upto $-20\,\text{kJ mol}^{-1}$), ΔS matches with $\Delta S = -19.7\,\text{J mol}^{-1}\,\text{K}^{-1}$ obtained from simulations [152], while the average number of H-bonds over the temperature range 270–650 K follows other estimates closely. In particular, at 300 K, the model yields $p \cong 10\%$, consistent with the spectroscopic data [97, 98]. Altogether, this two-state model for water provides an accurate description of its dielectric (and thermodynamic, Section 9.6.3) behavior.

Justification 9.4 ***The Kirkwood–Fröhlich Equation and beyond*** Here, we first deal with Kirkwood's equation for polar, but nonpolarizable molecules. Thereafter, we introduce an approximate modification given by Fröhlich [143] to deal with induction and finally indicate the exact equation. For full details, see [141].

We consider N identical molecules with dipole moment $\boldsymbol{\mu}$ and polarizability α. A molecule k at location $\boldsymbol{r}_k$ has an instantaneous dipole moment $\boldsymbol{m}_k = \boldsymbol{\mu}_k + \alpha\boldsymbol{L}_k$ with $\boldsymbol{L}_k$ the local field. The latter is given by $\boldsymbol{L}_k = \boldsymbol{E} - \sum_{k\neq l}\mathbb{T}_{kl}\cdot\boldsymbol{m}_k$ with the dipole field tensor $\mathbb{T}_{kl} = -\nabla\nabla\boldsymbol{r}_{kl}{}^{-1}$, the external field $\boldsymbol{E}$ and $\boldsymbol{r}_{kl} = \boldsymbol{r}_k - \boldsymbol{r}_l$. Employing the usual matrix convention for $\boldsymbol{m}$, $\boldsymbol{\mu}$ and $\mathbb{T}$, we can represent all these expressions by $\mathbf{m} = \boldsymbol{\mu} + \alpha(\mathbf{E} - \mathbb{T}\mathbf{m})$ or $\mathbf{m} = (\mathbb{I} + \alpha\mathbb{T})^{-1}(\boldsymbol{\mu} + \alpha\mathbf{E}) = \mathbf{m}_0 + \mathbb{A}\mathbf{E}$ with $\mathbf{m}_0 = (\mathbb{I} + \alpha\mathbb{T})^{-1}\boldsymbol{\mu}$ and $\mathbb{A} = \alpha(\mathbb{I} + \alpha\mathbb{T})^{-1}$. Further decomposing the matrix representation $\mathbb{I}$ of the unit tensor as $\mathbb{I} = (\mathbb{I} + \alpha\mathbb{T})(\mathbb{I} + \alpha\mathbb{T})^{-1} = (\mathbb{I} + \alpha\mathbb{T})^{-1} + \alpha\mathbb{T}(\mathbb{I} + \alpha\mathbb{T})^{-1}$, we write $(\mathbb{I} + \alpha\mathbb{T})^{-1} = \mathbb{I} - \alpha\mathbb{T}(\mathbb{I} + \alpha\mathbb{T})^{-1} = \mathbb{I} - \mathbb{A}\mathbb{T}$, so that $\mathbf{m}_0 = \boldsymbol{\mu} - \mathbb{A}\mathbb{T}\alpha$. The local field is

$$\mathbf{L} = \mathbf{E} - \mathbb{T}\mathbf{m} = \mathbf{E} - \mathbb{T}\mathbf{m}_0 - \mathbb{T}\mathbb{A}\mathbf{E} = (\mathbb{I} + \alpha\mathbb{T})^{-1}\mathbf{E} - \mathbb{T}\mathbf{m}_0 = \alpha^{-1}\mathbb{A}\mathbf{E} - \mathbb{T}\mathbf{m}_0$$

The total electrostatic energy of the system Φ_{ele} contains the potential energy of the molecules in the external field Φ_1, the potential energy of the molecules in other fields Φ_2, and the work of polarization Φ_3. Conform Eqs. (4.1) and (4.7), $\Phi_1 = -\mathbf{m}^{\mathrm{T}}\mathbf{E}$ and $\Phi_2 = ½\mathbf{m}^{\mathrm{T}}\mathbb{T}\mathbf{m}$, while Φ_3 is given by $\Phi_3 = \sum_j \int \mathbf{L}^{\mathrm{T}}\,\mathrm{d}\mathbf{m} = ½\alpha\mathbf{L}^{\mathrm{T}}\mathbf{L}$. Using these expressions, one can obtain $\Phi_2 + \Phi_3 = ½\boldsymbol{\mu}^{\mathrm{T}}\mathbb{T}\mathbf{m}_0 + ½\mathbf{E}^{\mathrm{T}}\mathbb{A}\mathbf{E}$. Using $\mathbf{m} = \mathbf{m}_0 + \mathbb{A}\mathbf{E}$ and adding the (short range) potential energy Φ_0, the total potential energy becomes

$$\Phi = \Phi_{\text{ele}} + \Phi_0 = ½\boldsymbol{\mu}^{\mathrm{T}}\mathbb{T}\mathbf{m}_0 - \mathbf{m}_0{}^{\mathrm{T}}\mathbf{E} - ½\mathbf{E}^{\mathrm{T}}\mathbb{A}\mathbf{E} + \Phi_0$$

Now consider the average dipole moment $\langle\mathbf{M}\rangle$ due to polarization, where, as usual, $\langle X\rangle = Q^{-1}\int X\exp[-\beta\Phi(\mathbf{x})]\mathrm{d}\mathbf{x}$ denotes the statistical mechanical average with Q the partition function and $\mathbf{x} = \mathbf{r}, \boldsymbol{\omega}$, describing the position coordinates $\mathbf{r}$ and orientation coordinates $\boldsymbol{\omega}$ (Section 4.1.1). Since we deal with an isotropic medium, we need to calculate $\langle\mathbf{M}^{\mathrm{T}}\mathbf{e}\rangle$, where $\mathbf{e}$ is the unit vector in the direction of the field $\mathbf{L}$. Using $\mathbf{P} = (\varepsilon_{\mathrm{r}} - 1)\varepsilon_0\mathbf{L} = \langle\mathbf{M}^{\mathrm{T}}\mathbf{e}\rangle/V$, we have $(\varepsilon_{\mathrm{r}} - 1)\mathbf{L} = \langle\mathbf{M}^{\mathrm{T}}\mathbf{e}\rangle/\varepsilon_0 V$, or, limiting to linear effects,

$$(\varepsilon_{\mathrm{r}} - 1) = (\varepsilon_0 V)^{-1}(\partial\left\langle\mathbf{M}^{\mathrm{T}}\mathbf{e}\right\rangle/\partial\mathbf{L})_{\mathbf{L}=0} = (\varepsilon_0 V)^{-1}(\partial\mathbf{E}/\partial\mathbf{L})_{\mathbf{L}=0}(\partial\left\langle\mathbf{M}^{\mathrm{T}}\mathbf{e}\right\rangle/\partial\mathbf{E})_{\mathbf{L}=0}$$

For $(\partial\langle\mathbf{M}^{\mathrm{T}}\mathbf{e}\rangle/\partial\mathbf{E})_{\mathbf{L}=0}$, we note that $\mathbf{M} = \Sigma_j\mathbf{m}_j$ with $\mathbf{m}_j$ the instantaneous dipole moment of molecule j. To evaluate $\mathbf{M}$, we consider a small (but macroscopic) sphere containing N molecules, and by taking N sufficiently large, the calculation can be as accurate as required. Carrying out the differentiation we find

$$\varepsilon_\mathrm{r} - 1 = \frac{1}{\varepsilon_0 V}\left(\frac{\partial \mathbf{E}}{\partial \mathbf{L}}\right)_{\mathrm{L}=0} Q^{-1}\left[\begin{array}{l}\int \frac{\partial \mathbf{M}^\mathrm{T}\mathbf{e}}{\partial \mathbf{E}} \mathrm{e}^{-\beta\Phi}\mathrm{d}\mathbf{r} - \beta\int (\mathbf{M}^\mathrm{T}\mathbf{e})\frac{\partial \Phi}{\partial \mathbf{E}}\mathrm{e}^{-\beta\Phi}\mathrm{d}\mathbf{r} \\ +\beta Q^{-1}\int (\mathbf{M}^\mathrm{T}\mathbf{e})\mathrm{e}^{-\beta\Phi}\mathrm{d}\mathbf{r}\int \frac{\partial \Phi}{\partial \mathbf{E}}\mathrm{e}^{-\beta\Phi}\mathrm{d}\mathbf{r}\end{array}\right]_{\mathrm{E}=0}$$

$$= \frac{1}{\varepsilon_0 V}\left(\frac{\partial \mathbf{E}}{\partial \mathbf{L}}\right)_{\mathrm{L}=0}\left[\left\langle \frac{\partial \mathbf{M}^\mathrm{T}\mathbf{e}}{\partial \mathbf{E}}\right\rangle - \beta\left\langle (\mathbf{M}^\mathrm{T}\mathbf{e})\frac{\partial \Phi}{\partial \mathbf{E}}\right\rangle\right]_{\mathrm{E}=0}$$

where the averages $\langle\ldots\rangle$ are taken at $\mathbf{E} = 0$ and the third term in the square brackets is zero because $\langle \mathbf{M}^\mathrm{T}\mathbf{e}\rangle = 0$ when $\mathbf{E} = 0$. Using $\mathbf{E} = E\mathbf{e}$, we obtain $-\partial\Phi/\partial E = \partial(E\mathbf{m}_0{}^\mathrm{T}\mathbf{e} + \frac{1}{2}E^2\mathbf{e}^\mathrm{T}\mathbb{A}\mathbf{e})/\partial E = \mathbf{m}_0{}^\mathrm{T}\mathbf{e} - E\mathbf{e}^\mathrm{T}\mathbb{A}\mathbf{e} = \mathbf{m}^\mathrm{T}\mathbf{e} = \mathbf{M}^\mathrm{T}\mathbf{e}$, where $\mathbf{m} = \mathbf{m}_0 + \mathbb{A}\mathbf{E}$ is used. Hence $\langle(\mathbf{M}^\mathrm{T}\mathbf{e})\,\partial\Phi/\partial E\rangle$ becomes $\langle(\mathbf{M}^\mathrm{T}\mathbf{e})^2\rangle = -\langle M^2\cos^2\theta\rangle = \langle M^2\rangle\langle\cos^2\theta\rangle = \langle M^2\rangle/3$. Similarly, $\partial\mathbf{M}^\mathrm{T}\mathbf{e}/\partial E = \partial\mathbf{m}^\mathrm{T}\mathbf{e}/\partial E = \mathbf{e}^\mathrm{T}\mathbb{A}\mathbf{e}$. Altogether we find the general expression

$$\varepsilon_\mathrm{r} - 1 = \frac{1}{\varepsilon_0 V}\left(\frac{\partial E}{\partial L}\right)_{L=0}\left[\left\langle \mathbf{e}^\mathrm{T}\mathbb{A}\mathbf{e}\right\rangle - \frac{1}{3kT}\left\langle M^2\right\rangle\right]_{E=0} \tag{9.61}$$

from which the Clausius–Mossotti and other equations can be derived [141].

Consider now N polar but nonpolarizable molecules enclosed by a sphere embedded in a continuum with permittivity ε_r. The field on that sphere is $E_\mathrm{cav} = 3\varepsilon_\mathrm{r}L/(2\varepsilon_\mathrm{r}+1)$ (Section D.2). Because we assume $\alpha = 0$, $\langle\mathbf{e}^\mathrm{T}\mathbb{A}\mathbf{e}\rangle = 0$ and the susceptibility $\varepsilon_\mathrm{r} - 1 = (\varepsilon_0 V)^{-1}\cdot[3\varepsilon_\mathrm{r}/(2\varepsilon_\mathrm{r}+1)]\cdot[\langle M^2\rangle/3kT]$, so that we need to evaluate $\langle M^2\rangle$ with $\mathbf{M} = \Sigma_j\boldsymbol{\mu}_j$. Now integration for $\langle M^2\rangle$ is done by first taking the average with the orientation of molecule j fixed and thereafter the average over the orientations of molecule j. To do so, we define $\mathrm{d}\mathbf{x} = \mathrm{d}\mathbf{x}^{(j)}\mathrm{d}\mathbf{x}_j$, where $\mathrm{d}\mathbf{x}^{(j)} = \mathrm{d}\mathbf{x}_1\ldots\mathrm{d}\mathbf{x}_{j-1}\mathrm{d}\mathbf{x}_{j+1}\ldots\mathrm{d}\mathbf{x}_N$ and the average moment $M_j^* = Q_j^{-1}\int\mathbf{M}\mathrm{e}^{-\beta\Phi}\mathrm{d}\mathbf{x}^{(j)}$ with $Q_j = \int\mathrm{e}^{-\beta\Phi}\,\mathrm{d}\mathbf{x}^{(j)}$. Further, we use $p(\mathbf{x}_j) = Q_j/Q^{-1}$. Hence, $\langle M^2\rangle = \sum_j\int p(\mathbf{x}_j)\mu_j^\mathrm{T}\mathbf{M}_j^*\mathrm{d}\mathbf{x}_j = N\int p(\mathbf{x}_j)\mu_j^\mathrm{T}\mathbf{M}_j^*\mathrm{d}\mathbf{x}_j$. Denoting the angle between dipoles j and k by θ_{jk}, we have $\boldsymbol{\mu}_j{}^\mathrm{T}\mathbf{M}_j^* = Q_j^{-1}\mu^2\sum_k^N\int\cos\theta_{jk}\mathrm{e}^{-\beta\Phi}\mathrm{d}\mathbf{x}^{(j)}$. Hence, $\langle M^2\rangle \equiv N\mu^2\Sigma_k\langle\cos\theta_{jk}\rangle = N\mu^2 Q_j^{-1}\Sigma_k\sum_k\int p(\mathbf{x}_j)\int\cos\theta_{jk}\mathrm{e}^{-\beta\Phi}\,\mathrm{d}\mathbf{x}^{(j)}$. Substituting this result in Eq. (9.61), we find the *Kirkwood equation*

$$\frac{(\varepsilon_\mathrm{r}-1)(2\varepsilon_\mathrm{r}+1)}{3\varepsilon_\mathrm{r}} = \frac{\langle M^2\rangle}{3\varepsilon_0 kTV} = \frac{1}{3\varepsilon_0 kTV}\mu^2\sum_{k=1}^N\left\langle\cos\theta_{jk}\right\rangle \equiv \frac{N_\mathrm{A}}{3kT\varepsilon_0 V_\mathrm{m}}\mu^2\mathrm{g}$$

identical to the Onsager equation for polar but nonpolarizable molecules (i.e. $\varepsilon_\infty = 1$), except for the extra factor g. It is well verified by MC simulations [153]. Restricting the content of the sphere to the z nearest neighbors only, $\mathrm{g} = 1 + z\langle\cos\theta_{jk}\rangle$.

The separation of the coordinates of molecule j and the rest becomes impossible when $\alpha \neq 0$. Although integration is possible, the complex expression resulting generally needs further approximations. A way out proposed by Fröhlich [143] uses that one can write $\varepsilon_\mathrm{r} = \varepsilon/\varepsilon_0\varepsilon_\infty$ as relative permittivity with respect to a medium with relative permittivity ε_∞ instead of $\varepsilon_\mathrm{r} = \varepsilon/\varepsilon_0$ with respect to vacuum with relative permittivity 1. The average moment squared becomes $\langle M^2\rangle/\varepsilon_\infty$, while the energy is given by Φ/ε_∞. The same algebra now yields

$$\varepsilon_\mathrm{r} - \varepsilon_\infty = \frac{1}{\varepsilon_0 V}\frac{3\varepsilon_\mathrm{r}}{2\varepsilon_\mathrm{r}+\varepsilon_\infty}\frac{\langle M^2\rangle}{3kT} \quad \text{or} \quad \frac{(\varepsilon_\mathrm{r}-\varepsilon_\infty)(2\varepsilon_\mathrm{r}+\varepsilon_\infty)}{3\varepsilon_\mathrm{r}} = \frac{N}{3\varepsilon_0 kTV}gm^2 \tag{9.62}$$

where $\mathbf{m}$ is the dipole moment of a molecule in a medium with permittivity ε_∞, which is given by $\mathbf{m} = (2\varepsilon_\infty + 2)\boldsymbol{\mu}/3$ (Section D.3). Some further algebra yields the *Kirkwood-Fröhlich equation*, equal to the Onsager result Eq. (8.32) but with μ^2 replaced by $g\mu^2$. Conventionally, $\varepsilon_\infty = n^2$ is used, but this is not always valid.

Later it appeared that the correct analogue of the Kirkwood equation for polarizable molecules, that is, for a sphere embedded in its own medium, is given by [150, 154]

$$\frac{(\varepsilon_{\mathrm{r}}-1)(2\varepsilon_{\mathrm{r}}+1)}{3\varepsilon_{\mathrm{r}}}-\frac{(\varepsilon_{\infty}-1)(2\varepsilon_{\infty}+1)}{3\varepsilon_{\infty}}=\frac{\langle M^2\rangle}{3\varepsilon_0 kTV}=\frac{gm^2}{3\varepsilon_0 kTV} \tag{9.63}$$

which, in hindsight, seems intuitively reasonable, but required the effort of many to render its (complex) derivation rigorous.

References

1 Weisskopf, V.F. (1977). *Trans. New York Acad. Sci.* 38: 202; Bernstein, H.J. and Weisskopf, V.F. (1987). *Am. J. Phys.* 55: 974.
2 Allen, M.P. and Tildesley, D.J. (1987). *Computer Simulation of Liquids*. Oxford: Clarendon Press.
3 Frenkel, D. and Smit, B. (2002). *Understanding Molecular Simulation: From Algorithms to Applications*, 2e. London: Academic Press.
4 Berendsen, H.J.C. (2007). *Simulating the Physical World*. Cambridge: Cambridge University Press.
5 Eisenberg, D. and Kauzmann, W. (1969). *The Structure and Properties of Water*. Oxford: Clarendon Press.
6 Malenkov, G.G., Teplukhin, A.V., and Poltev, V.I. (1989). *J. Struct. Chem.* 30: 605.
7 Friedman, H.L. (1985). *A Course on Statistical Mechanics*. Englewood Cliffs, NJ: Prentice-Hall.
8 McQuarrie, D.A. (1973). *Statistical Mechanics*. New York: Harper and Row.
9 Fisher, I.Z. (1964). *Statistical Theory of Liquids*. Chicago: University of Chicago Press.
10 Eisenstein, A. and Gingrich, N.S. (1942). *Phys. Rev.* 62: 261.
11 Woodcock, L.V. (1971). *Z. Naturforsch. A* 26 (2): 287.
12 Finney, J.L. (1970). *Proc. Roy. Soc. A* 319: 279.
13 Yarnell, J.L., Katz, M.J., Wenzel, R.G., and Koenig, S.H. (1973). *Phys. Rev. A* 7: 2130.
14 Franks, F. (ed.) (1972). *Water, a Comprehensive Treatise*. New York: Plenum Press.
15 Scott, G.D. (1960). *Nature* 188: 908; (1962). *Nature* 194: 956.
16 Brostow, W. and Sicotte, Y. (1973). *J. Stat. Phys.* 9: 339; (1975). *Physica A* 80: 513.
17 Pings, C.J. (1968). *Structure of Simple Liquids*, Ch. 10 in *Physics of Simple Liquids*, (ed. H.N.V. Temperley, J.S. Rowlinson, and G.S. Rushbrooke), p. 389. North-Holland, Amsterdam.
18 Clayton, G.T. and Heaton, L. (1961). *Phys. Rev.* 121: 649.
19 Bernal, J.D. and Mason, J. (1960). *Nature* 188: 910.
20 Bernal, J.D., Mason, J., and Knight, K.R. (1962). *Nature* 194: 957.
21 Bernal, J.D. (1964). *Proc. Roy. Soc. A* 280: 299.
22 Finney, J.L. (2013). *Phil. Mag.* 93: 31.
23 Raveché, H.J. and Green, M.S. (1969). *J. Chem. Phys.* 50: 5334.
24 Barker, J.A. and Henderson, D. (1976). *Rev. Mod. Phys.* 48: 587.
25 Watts, R.O. and McGee, I.J. (1976). *Liquid State Chemical Physics*. New York: John Wiley & Sons.
26 Boublik, T., Nezheda, I., and Hlavatý, K. (1980). *Statistical Mechanics of Simple Liquids and their Mixtures*. Amsterdam: Elsevier.

27 Hansen, J.-P. and McDonald, I.R. (2006). *Theory of Simple Liquids*, 3e. London: Academic Press; (1976). 1e; (1986). 2e.

28 Pottier, N. (2010). *Nonequilibrium Statistical Physics*. New York: Oxford University Press.

29 Schirmacher, W. (2015). *Theory of Liquids and Other Disordered Media*. Heidelberg: Springer.

30 Landau, L.D. and Lifshitz, E.M. (1980). *Statistical Physics*, 3e, Vol. 1. Oxford: Pergamon Press.

31 Yvon, J. (1935). *Actualités Scientifiques et Industrielles*, Vol. nr. 203. Paris: Herman and Cie.

32 Born, M. and Green, H.S. (1946). *Proc. Roy. Soc. London. A* 188: 10.

33 Kirkwood, J.G. and Boggs, E.M. (1942). *J. Chem. Phys.* 10: 394.

34 Grouba, V.D., Zorin, A.V., and Sevastianov, L.A. (2004). *Int. J. Mod. Phys. B* 18: 1.

35 Singer, A. (2004). *J. Chem. Phys.* 121: 3657; Chari, S.S.N., Inguva, R. and Murthy, K.P.N. (2016). arXiv:1608.02338.

36 Kirkwood, J.G. (1950). *J. Chem. Phys.* 18: 380.

37 Bellemans, A. *J. Chem. Phys.* 50: 2784.

38 Percus, J.K. and Yevick, G.J. (1958). *Phys. Rev.* 110: 1.

39 Broyles, J. (1962). *J. Chem. Phys.* 37: 2462.

40 Watts, R.O. (1969). *J. Chem. Phys.* 50: 984.

41 Mikolaj, P.G. and Pings, C.J. (1967). *J. Chem. Phys.* 46: 1401.

42 Baxter, R.J. (1968). *Phys. Rev.* 154: 170.

43 Thiele, E. (1963). *J. Chem. Phys.* 39: 474.

44 Wertheim, M.S. (1963). *Phys. Rev. Lett.* 10: 321.

45 Smith, W.R. and Henderson, D. (1970). *Mol. Phys.* 19: 411.

46 Carnahan, N.F. and Starling, K.E. (1965). *J. Chem. Phys.* 51: 635.

47 Mulero, A., Galán, C., and Cuadros, F. (2001). *Phys. Chem. Chem. Phys.* 3: 4991.

48 Wu, G.-W. and Sadus, R.J. (2005). *AIChE J.* 51: 309.

49 Solana, J.R. (2013). *Perturbation Theories for the Thermodynamic Properties of Fluids and Solids*. Boca Raton, FL: CRC Press.

50 Reiss, H. (1965). *Adv. Chem. Phys.* 9: 1.

51 Stillinger, F.H., Debenedetti, P.G., and Chatterjee, S. (2006). *J. Chem. Phys.* 125: 204504–204505.

52 Chandler, D. (2005). *Nature* 437: 650.

53 Ashbaugh, H.S. and Pratt, L.S. (2006). *Rev. Mod. Phys.* 78: 159.

54 Dasannacharya, B.A. and Rao, K.R. (1965). *Phys. Rev.* 137: A417.

55 Toxvaerd, S. and Dyre, J. (2011). *J. Chem. Phys.* 135: 134501.

56 Lucas, 2007.

57 Krooshof, G.J.P., Tuinier, R., and de With, G. (2018). *Fluid Phase Eq.* 465: 10.

58 Bondi, A. (1968). *Physical Properties of Molecular Crystals, Liquids and Glasses*. London: John Wiley & Sons.

59 Horn, N.R. (2016). *J. Membr. Sci.* 518: 289.

60 Pryde, J.A. (1966). *The Liquid State*. London: Hutchinson University Library.

61 Hirschfelder, J.O., Curtiss, C.F., and Bird, R.B. (1954). *Molecular Theory of Gases and Liquids*. New York: John Wiley & Sons; corrected ed. (1964).

62 Barker, J.A. (1963). *Lattice Theories of the Liquid State*. London: Pergamon Press.

63 Eyring, H. and Hirschfelder, J.O. (1937). *J. Phys. Chem.* 41: 249.

64 Lennard-Jones, J.E. and Devonshire, A.F. (1937). *Proc. Roy. Soc. A* 163: 53; (1938). 165: 1.

65 Buehler, R.J., Wentorf, R.H., Hirschfelder, J.O. et al. (1951). *J. Chem. Phys.* 19: 61.

66 Hirschfelder, J.O., Stevenson, D.P., and Eyring, H. (1937). *J. Chem. Phys.* 5: 896.

67 Wentorf, R.H., Buehler, R.J., Hirschfelder, J.O. et al. (1950). *J. Chem. Phys.* 18: 1484.

68 Pople, J. (1951). *Phil. Mag.* 12: 459.

69 Prigogine, I. and Raulier, S. (1942). *Physica* 9: 396.

70 Adams, D.J. and Matheson, A.J. (1972). *J. Chem. Soc. Far. Trans. II.* 68: 1536.

71 Taylor, W.J. (1956). *J. Chem. Phys.* 24: 454.

72 Hoover, W.G. and Ree, F.H. (1968). *J. Chem. Phys.* 49: 3609.

73 Henchman, R.H. (2003). *J. Chem. Phys.* 119: 400.

74 Wood, N.W. (1952). *J. Chem. Phys.* 20: 1334.

75 Kuramoto, Y. (1971). *Progr. Theor. Phys.* 45: 1724.

76 Cernuschi, F. and Eyring, H. (1939). *J. Chem. Phys.* 7: 547.

77 Guggenheim, E.A. (1935). *Proc. Roy. Soc. A* 148: 304; Rushbrooke, G.S. (1938). *Proc. Roy. Soc. A* 166: 296.

78 Henderson, D. (1962). *J. Chem. Phys.* 37: 631; (1963). *J. Chem. Phys.* 39: 54.

79 Eyring, H. and Jhon, M.S. (1969). *Significant Liquid Structures*. London: John Wiley & Sons; Jhon, M.S. and Eyring, H. (1978). *A Model of the Liquid State. Three Phase Partition Functions*, Ch. 3 in *Theoretical Chemistry*, Vol. 3 (ed. H. Eyring and D. Henderson), p. 55. This reference shows a large overlap with the first.

80 Rowlinson, J.S. and Curtiss, C.F. (1951). *J. Chem. Phys.* 19: 1519.

81 Wentorff, R.H., Buehler, R.J., Hirschfelder, J.O. et al. (1964). *J. Chem. Phys.* 40: 175.

82 Weissmann, M. (1964). *J. Chem. Phys.* 40: 175.

83 Weymann, H.D. (1961). *Kolloid-Z. und Z. Polymere* 181: 131; (1954). Kolloid-Z. 138, 41.

84 Smirnova, N.A. and Victorov, A.I. (1987). *Fluid Phase Eq.* 34: 285.

85 Fuller, E.J., Ree, T., and Eyring, H. (1959). *Proc. Natl. Acad. Sci.* 45: 1594.

86 Eyring, H. and Marchi, R.P. (1963). *J. Chem. Ed.* 40: 562.

87 Jhon, M.S. and Eyring, H. (1971). *The Significant Structure Theory of Liquids*, Ch. 5 in *Physical Chemistry*, Vol. VIIIA (ed. D. Henderson), p. 335. Academic Press, New York.

88 Henderson, D. (1963). *J. Chem. Phys.* 39: 1857.

89 Jhon, M.S. (1967). *J. Kor. Chem. Soc.* 11: 60.

90 Yoon, B.J., Jhon, M.S., and Eyring, H. (1981). *Proc. Nat. Acad. Sci.* 78: 6588.

91 Luck, W.A.P. and Ditter, W. (1971). *Tetrahedron* 27: 201; Luck, W.A.P. (1979). *Angew. Chem. Int. Ed.* 18: 350.

92 Luck, W.A.P. (1991). *How to Understand Liquids?* Ch. 3 in *Intermolecular Forces* (ed. P.L. Huyskens, W.A.P. Luck, and T. Zeegers-Huyskens), p. 55. Berlin: Springer.

93 Luck, W.A.P. (1972). *Ullmanns Encyklopädie Techn. Chemie*, 4e, Vol. I, p. 55, Verlag Chemie, Weinheim.

94 Sutherland, W. (1893). *Phil. Mag.* 36: 507.

95 Chapman, S. and Cowling, T.G. (1939, 1970). *The Mathematical Theory of Non-uniform Gases*, 3e. Cambridge: Cambridge University Press.

96 Luck, W.A.P. (2001). *Coll. Pol. Sci.* 279: 554.

97 Hare, D.E. and Sorensen, C.M. (1990). *J. Chem. Phys.* 93: 6954.

98 Luck, W.A.P. (1980). *Angew. Chem. Int. Ed.* 19: 28.

99 Luck, W.A.P. (1991). *Water-The most Anomalous Liquid*, Ch. 9 in *Intermolecular Forces* (ed. P.L. Huyskens, W.A.P. Luck, and T. Zeegers-Huyskens), p. 217. Berlin: Springer.
100 Sandler, S.I. (1985). *Fluid Phase Eq.* 19: 233.
101 Lee, K.H., Lombardo, M., and Sandler, S.I. (1985). *Fluid Phase Eq.* 21: 177.
102 Vera, J.H. and Prausnitz, J.M. (1972). *Chem. Eng. J.* 3: 1.
103 Sandler, S.I. (1990). Chem. Eng. Ed., Winter. 12.
104 Alder, B.J., Young, D.A., and Mark, M.A. (1972). *J. Chem. Phys.* 56: 3013.
105 Grimvall, G. (1975). *Phys. Scr.* 11: 381; Wallace, D.C. (1998). *Phys. Rev. E* 57: 1717; Dexter, A.R. and Matheson, A.J. (1968). *Trans. Far. Soc.* 64: 2632.
106 Granato, A.V. (2002). *J. Non-Cryst. Sol.* 307–310: 376.
107 Landau, L. and Lifshitz, E.M. (1969). *Statistical Physics*, 2e, Vol. 66, 74. Oxford: Pergamon Press.
108 Frenkel, J. (1946). *Kinetic Theory of Liquids*. Oxford: Oxford University Press (also Dover, 1953).
109 Pilgrim, W.C. and Morkel, C. (2006). *J. Phys.: Condens. Matter.* 18: R585.
110 Trachenko, K. (2008). *Phys. Rev. B* 78: 104201; Bolmatov, D. and Trachenko, K. (2011). *Phys. Rev. B* 84: 054106; Bolmatov, D., Brazhkin, V.V., and Trachenko, K. (2012). *Sci. Rep.* 2: 421.
111 Grimvall, G. (1999). *Thermophysical Properties of Materials*, 2e. Amsterdam: North-Holland; (1986). 1e.
112 Trachenko, K. and Brazhkin, V.V. (2011). *Phys. Rev. B* 83: 014201; Brazhkin, V.V. and Trachenko, K. (2014). *J. Phys. Chem. B* 118: 11417.
113 Bolmatov, D., Brazhkin, V.V., and Trachenko, K. (2013). *Nature Comm.* 4: 2331; Trachenko, K. and Brazhkin, V.V. (2009). *J. Phys.: Condens. Matter* 21: 425104.
114 Trachenko, K. and Brazhkin, V.V. (2016). *Rep. Prog. Phys.* 79: 016502.
115 Proctor, J.E. (2020). *Phys. Fluids* 32: 107105.
116 Rosenfeld, Y. and Tarazona, P. (1998). *Mol. Phys.* 95: 141.
117 Cerdeiriña, C.A., González-Salgado, D., Romani, L. et al. (2002). *J. Chem. Phys.* 120: 6648.
118 Naef, R. (2019). *Molecules.* 24: 1626.
119 Weinhold, F. (1998). *J. Chem. Phys.* 109: 367 and 373.
120 Silla, E., Villar, F., Nilson, O. et al. (1990). *J. Mol. Graphics.* 8: 168.
121 Weinhold, F. (2006). *Adv. Protein Chem.* 72: 121.
122 Jhon, M.S., Grosh, J., Ree, T. et al. (1965). *J. Chem. Phys.* 44: 1465; Davis, C.M. and Litovitz, T.A. (1965). *J. Chem. Phys.* 42: 2563; Biddle, J.W., Singh, R.K., Sparano, E.M. et al. (2017). *J. Chem. Phys.* 146: 034502.
123 Ludwig, R. and Weinhold, F. (2000). *Phys. Chem. Chem. Phys.* 2: 1613.
124 Boys, S.F. and Bernardi, F. (1970). *Mol. Phys.* 19: 553.
125 Hehre, W.J., Radom, L., Schleyer, P.v.R. et al. (1986). *Ab Initio Molecular Orbital Theory*. New York: John Wiley & Sons.
126 Moelwyn-Hughes, E.A. (1961). *Physical Chemistry*, 2e. Oxford: Pergamon Press.
127 Rowlinson and Swinton, 1982.
128 Marcus, Y. (2017). *J. Chem. Thermodyn.* 109: 11; (2013). *Ibid.* 61: 7.
129 Tait, P.G. (1889). *Voyages of HMS Challenger*, Vol. 2, part 4, p. 1. London: HMSO.
130 Tammann, G. (1896). *Z. phys. Chem.* 17: 620.
131 Hayward, A.T.J. (1967). *Brit. J. Appl. Phys.* 18: 969.
132 Dymond, J.H. and Malhotra, R. (1988). *Int. J. Thermophys.* 9: 941.

133 Ross MacDonald, J. (1966). *Rev. Mod. Phys.* 38: 669.
134 Toscani, S. and Szwarc, H. (2004). *J. Chem. Eng. Data.* 49: 163.
135 Poling, Prausnitz, and O'Connell, 2001.
136 Ziegler, H. (1983). *Introduction to Thermomechanics*, 2e. Amsterdam: North-Holland.
137 Hunter, S.C. (1983). *Mechanics of Continuous Media*, 2e. Chichester: Ellis Horwood Ltd.
138 Eisenberg, H. (1965). *J. Chem. Phys.* 43: 3887.
139 Onsager, L. (1936). *J. Am. Chem. Soc.* 58: 1486.
140 Buckingham, A.D. (1967). *Disc. Far. Soc.* 43: 205.
141 Böttcher, C.J.F. (1973). *Theory of Electric Polarization*, 2e, Vol. 1. Amsterdam: Elsevier.
142 Kirkwood, J.G. (1939). *J. Chem. Phys.* 7: 911.
143 Fröhlich, H. (1958). *Theory of Dielectrics*, 2e. Oxford: Oxford University Press.
144 Minkin, V.I., Osipov, O.A., and Zhadanov, Y.A. (1970). *Dipole Moments in Organic Chemistry*. New York: Plenum Press.
145 Pople, J.A. (1951). *Proc. Roy. Soc. A* 205: 163.
146 Coulson, C.A. and Eisenberg, D. (1966). *Proc. Roy. Soc. A* 291: 445.
147 Hill, N.E. (1963). *Trans. Faraday Soc.* 59: 344; (1970). *J. Phys. C* 3: 238.
148 Madden, P. and Kivelson, D. (2007). *A Consistent Molecular Treatment of Dielectric Phenomena*, Ch. 5 in *Advances in Chemical Physics*, Vol. 56 (ed. I. Prigogine and S.A. Rice), p. 467. New York: John Wiley & Sons.
149 Kaatze, U. (2018). *J. Mol. Liq.* 259: 304.
150 Felderhof, B.U. (1979). *J. Phys. C: Sol. State Phys.* 12: 2423; *Ibid*. 12: 3121.
151 Suresh, S.J. and Naik, V.M. (2000). *J. Chem. Phys.* 113: 9727; (2011). *Ibid*. 135: 086101.
152 Silverstein, K.A.T., Haymet, A.D.J., and Dill, K.A. (2000). *J. Am. Chem. Soc.* 122: 8037.
153 Valiskó, M. and Boda, D. (2009). *J. Chem. Phys.* 131: 164120.
154 Wertheim, M.S. (1978). *Mol. Phys.* 36: 1217.

Further Reading

Gallo, P. and Rovere, M. (2021). *Physics of Liquid Matter*. Cham, Switzerland: Springer Nature. Mainly dealing with microscopic, dynamic aspects but providing an introduction to mode-coupling theory and supercooled water.

Lucas, K. (2007). *Molecular Models of Fluids*. Cambridge: Cambridge University Press. Dealing mainly with activity coefficient and equation of state approach to fluids and containing a clear and detailed introduction to statistical mechanics.

Poling, B.E., Prausnitz, J.M., and O'Connell, J.P. (2001). *The Properties of Gases and Liquids*, 5e. McGraw-Hill Comp. A useful compilation of estimation methods, suggesting the most suitable method, and with a brief introduction for every property discussed.

Rowlinson, J.S. and Swinton, F.L. (1982). *Liquids and Liquid Mixtures*, 3e. London: Butterworth Scientific. Classic having a concise presentation but with many relevant details and remarks.

Sengers, J.V., Kayser, R.F., Peters, C.J., and White, H.J. (2000). *Equations of State for Fluids and Fluid Mixtures*. Amsterdam: Elsevier. A broad overview from an engineering point of view.

Wallace, D.C. (2002). *Statistical Physics of Crystals and Liquids: A Guide to Highly Accurate Equations of State*. Singapore: World Scientific. Clear presentation of basic, not necessarily simple, aspects.

10

Solids

In this chapter, we discuss the third type of "bricks": aspects of structure, bonding, and dynamics of solids. Contrary to our treatment of gases and liquids, the dielectric behavior of solids, which is strongly related to ferroelectricity, is not discussed in this chapter but in Chapter 17. We start with lattice concepts and deal subsequently with perfect structures. Thereafter, we continue with some dynamical aspects and an overview of defects in crystallographic structures. Finally, we discuss a macroscopic approach to solids. An overview of the various types of solids precedes this all.

10.1 Inorganics and Metals

Crystalline solids can be divided into single crystals and polycrystals and for both a regularly ordered structure exists at an atomic scale. This structure is largely maintained throughout the whole material in *single crystals*, while in *polycrystals* regions with different crystallographic orientation exist. These regions are referred to as *grains* and the boundaries between them as *grain boundaries*. X-ray diffraction indicates the presence of long-range order for crystalline solids (Figure 10.1a), while for *amorphous solids* and *glasses*, long-range order is absent (Figure 10.1b), although the local coordination may be similar to that in the corresponding crystalline solid.

Crystalline solids generally show a distinct melting point. Below the melting point the crystalline structure is present, while above the melting point an amorphous, liquid structure arises. Despite the long-range order, various defects may be present in single crystals (Section 10.10). They can be divided into *point defects* (interstitials, vacancies, substitutional atoms), *line defects* (dislocations), *planar defects* (stacking faults), and *volume defects* (inclusions, pores) (Figure 10.1c). Together with grain boundaries, the same defects play a role for polycrystals, for which their average grain size and grain size distribution are important for several properties.

Although high purity is often pursued, *impurities*, that is, foreign components, nevertheless are present. These impurities may dissolve in the lattice, if present below the solubility limit, or precipitate, if present in amounts above the solubility limit, sometimes forming new phases. In polycrystalline materials, impurities may segregate on the grain boundaries, leading to either amorphous or crystalline *grain boundary phases*. Obviously, these new phases, if they crystallize, can form new grains in the material. If the new phase remains

Phases of Matter and their Transitions: Concepts and Principles for Chemists, Physicists, Engineers, and Materials Scientists,
First Edition. Gijsbertus de With.

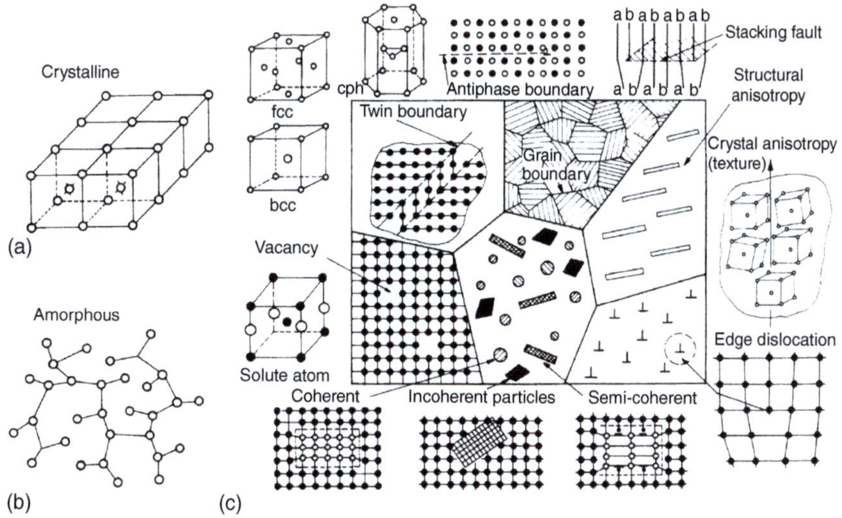

Figure 10.1 (a) Schematic of a crystalline solid. (b) Schematic of an amorphous solid. (c) Overview of the various microstructural features in inorganics and metals. Source: Zum-Gahr [1]/ Elsevier.

amorphous, it is generally referred to as a *glassy second phase*. If a foreign component is added on purpose, for example, to improve processing or properties, this component is called an *additive* or *dopant*. The type, structure, and number of phases, the number, geometric appearance (size, shape, etc.), and topological arrangement of the individual phase regions and their interfaces, and the type, structure, and geometry of lattice defects define what is called the *microstructure* (Figure 10.1c). The microstructure can be changed by *annealing*, that is, holding at high temperature below the melting point for some time.

In *molecular crystals* the building blocks are molecules, like rare gas atoms, alkanes, and diatomic elements, which are kept together by van der Waals (vdW) interactions. The cohesive energy ranges from 10 to 50 kJ mol^{-1} and these materials have typically a low electrical conductivity, melting point, stiffness, and strength.

Inorganics are materials that contain either ionically bonded atoms (Figure 10.2a) or covalently bonded atoms (Figure 10.2b).[1] In ionic bonding, positively and negatively charged ions attract each other nondirectionally with a cohesive energy ranging from 600 to 3500 kJ mol^{-1}. In covalent bonding, strong directional bonds are present between the atoms due to shared electrons with a cohesive energy range from 300 to 2000 kJ mol^{-1}. In both cases, generally a solid arises with a high melting point and corresponding high stiffness and low ductility. For these materials, the electrical conductivity is normally low, while the thermal conductivity can vary considerably. Halides (NaCl, KI, CsBr, etc.) and oxides (MgO, Al_2O_3, $BaTiO_3$, etc.) show mainly ionic bonding, while for carbides (SiC, WC, B_4C, etc.) and nitrides (AlN, TiN, Si_3N_4, etc.) the bonding is mainly covalent.

Metals are solids in which the bonding between the ions is collectively provided by the electrons (Figure 10.2c). The cohesive energy ranges from 100 to 800 kJ mol^{-1}. Depending

1 This classification is one of extremes and a gradual transition between categories exist. Nevertheless, this classification is useful and frequently used. For a more subtle classification, see Section 10.6.

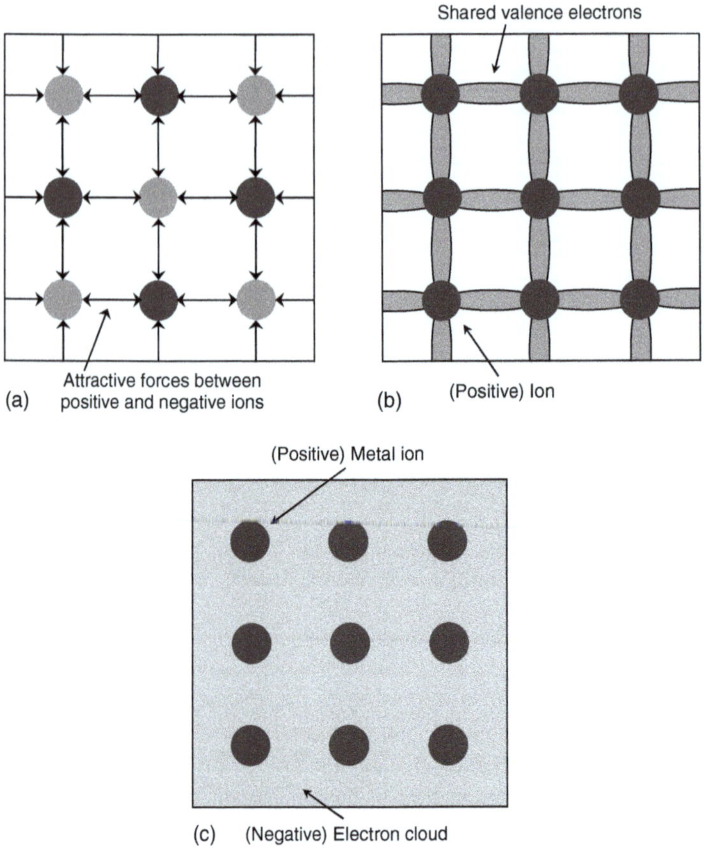

Figure 10.2 Schematics of bonding. (a) An ionically bonded material; (b) a covalently bonded material; and (c) a metallically bonded material. Adapted from [2].

on the type, metals can show a high melting point and corresponding high hardness and low ductility (e.g. Mo and W) or a relatively low melting point with associated lower hardness and higher ductility (e.g. Cu and Al). Due to the collective electrons, metals typically show a good electrical and thermal conductivity.

Adding other elements to a base metal is generally called *alloying*. Similar as for inorganics the additions can dissolve in the matrix, segregate at the grain boundaries, or lead to new phases. The solubility of additions is typically much higher as compared with crystalline inorganics. By alloying not only the chemical composition but also the microstructure can be influenced significantly, leading to widely varying properties for one and the same base metal. Particularly the strength can be increased, usually at the cost of lower ductility and toughness.

Apart from annealing, a typical feature of metals is the possibility to change their microstructure by mechanical *working* (rolling, forging, etc.). The change in microstructure is reflected in the properties of the metal, for example, by annealing and slow cooling the yield strength often becomes lower, while after annealing and quenching the yield strength increases. The fact that a wide range of microstructures can be obtained and that the final microstructure after annealing and/or working is largely independent of the original microstructure, is one of the most useful features of metals. A broad overview on materials is given in [3].

10.2 Polymers

Polymers consist of long chains of covalently bonded atoms, constructed from a set of repeating units, the *monomers*. Possibly the simplest example of a polymer is polyethylene, consisting of long chains of $-[CH_2-CH_2]-$ units using the monomer ethylene $CH_2{=}CH_2$. If the monomer is modified to $CH_2{=}CHX$, where X represents a certain chemical group, they are called *vinyl polymers*. If X is a methyl, phenyl, or chloride group, we have polypropylene, polystyrene, or polyvinylchloride, respectively. Other classes are *polyesters*, for example, polyethylene terephthalate, made from ethylene glycol and terephthalic acid, and *polyamides*, for example, nylon 66, made from adipic acid and hexamethylene diamine.

Nearly always a polymer material will consist of molecular chains of varying length and therefore of varying molecular weight. The distribution is characterized the *number average* M_n and *weight average* M_w, defined by

$$M_n = \sum\nolimits_i N_i M_i / \sum\nolimits_i N_i \quad \text{and} \quad M_w = \sum\nolimits_i (N_i M_i) M_i / \sum\nolimits_i N_i M_i$$

respectively, where N_i is the number of molecules with molecular weight M_i and the summation is over all molecular weights. The weight average is always larger than the number average, and in fact, the *dispersity* $Đ = (M_w/M_n) - 1$ represents the relative variance of the number distribution and therefore the width of that distribution.

Polymer chains show isomerism. *Chemical isomerism* in vinyl polymers is due to *head-to-head* ($-CH_2-CHX-CHX-CH_2-$) or *head-to-tail* ($-CH_2-CHX-CH_2-CHX-$) addition. *Steric isomerism* occurs when a side group is added to every alternate carbon atom. If the groups are all added in an identical way, we obtain *isotactic* polymers (Figure 10.3a). If there is an inversion for each monomer unit, we have *syndiotactic* polymers, while an irregular addition sequence leads to *atactic* polymers. The geometrical structure of the chains is largely determined by *rotational isomerism*. For units like –CHY–CHX–, the orientations with the lowest energy are *gauche* with the groups X and Y rotated through 60° along the C—C bond and trans with the groups X and Y rotated through 180° along the bond. The preferred orientation for units like –CHY=CHX– is cis with the groups X and Y on the same side of the double bond and trans with the groups X and Y on different sides of the double bond. For a C=C bond, the rotation barrier is considerably higher than that for a C—C bond.

Individual polymer chains, having a relatively large internal flexibility, form *coils*, the size of which depends on temperature and on whether the chain is in a melt or in solution. In the latter case, the size also depends on the solvent. Polymers in the molten state are amorphous, and the individual chains are entangled, that is, the chains get mixed up and are difficult to unravel since at various positions a kind of knots, *entanglements*, are formed.

For solid polymers, bonding between the molecules occurs via secondary interactions, such as vdW interactions and hydrogen bonds, or via cross-linking between the chains. The *vdW interaction* ranges from 10 to 40 kJ mol^{-1}. For polymers, *hydrogen bonding*, if

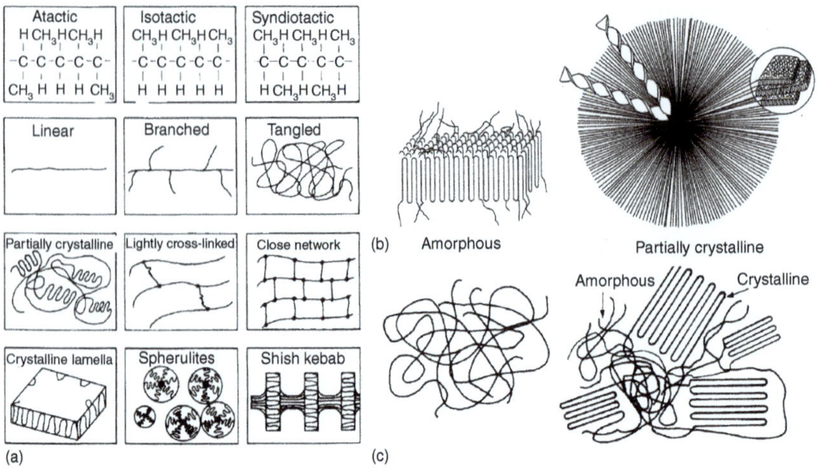

Figure 10.3 Microstructural features of polymers. (a) Network characteristics; (b) the structure of a lamella and spherulite; (c) amorphous structure, orientation, and crystallinity. Source: Zum-Gahr [1]/ with permission of Elsevier.

present, has a similar interaction strength. These bonding types result in rather soft materials with low melting temperature. By joining the chains at points along their length with a chemical bond a *cross-linked structure* (Figure 10.3a) arises, leading to materials typically showing no clear melting.[2] *Thermosets* are heavily cross-linked polymers. A less rigorous way to connect molecules to each other is via *chain branching*, where a secondary chain initiates from a point on the main chain. This leads to a more entangled structure as compared with linear polymers. *Thermoplastics* are linear and lightly branched polymers that are relatively easily deformed.

Many polymers, in particular atactic polymers, random copolymers, and highly branched polymers, when cooled down from a melt remain in a disordered or *amorphous* state (Figure 10.3c). Complete randomness is impossible in view of the covalent bonds between atoms in the chain, but generally, this cannot be detected by X-ray diffraction. If a polymer is stretched, the chains preferentially align along the stretch direction, so that the structure becomes somewhat oriented. This orientation may not be detectable by X-ray diffraction, but possibly can be detected by optical or other means. Such a structure is called *oriented amorphous*. Further stretching will lead to a strong *preferential orientation*, also detectable by X-ray diffraction. Overall, the amorphous state is characterized by the glass transition temperature T_g (Section 18.1).

Noncrystalline polymers above T_g behave like a viscous liquid. Unlimited flow can take place since the chains can slide along each other. Cross-linking can stop this process, and, in this way, one obtains *rubbers* (or *elastomers*). Lightly cross-linked rubber can be extended many times its own length and will return upon unloading rapidly to its original shape. For the prototype rubber, natural rubber, consisting of *cis*-isoprene, cross-linking was originally done with sulfur, often called *vulcanization*. Nowadays, cross-linking is usually done for natural rubber as well as for other polymers with chemicals like peroxides, or by electron or photon radiation.

2 With cross-linking in mind, a network oxide glass can be considered as an inorganic polymer.

Polymers, in particular those with a more regular chain structure, also can crystallize. Such *crystallized polymers* are macroscopically isotropic but microscopically nonhomogeneous. Generally, these polymers only partially crystallize, and alternating regions of order (crystallites) and disorder (amorphous regions) exist (Figure 10.3c). The crystallites typically have the shape of *lamellae* (Figure 10.3b), several tens of micrometers of lateral dimensions and about 10–20 nm in thickness. In crystallites, not only different chains align, but also a single chain participates in several lamellae. The crossover of chains, chain ends, and defects within the chain largely collect in the amorphous regions between the lamellae. In melt-grown lamellae, the end-to-end distance of the individual molecules remains largely preserved, so that the entanglement density is also largely preserved. In solution-grown lamellae, the end-to-end distance decreases considerably as compared to the value in solution leading to a much lower entanglement density. The lamellae orient typically more or less similarly, forming *stacks* with amorphous material in between, which in turn form superstructures. The *spherulite* (Figure 10.3a), in which stacks emanate radially from a certain nucleus, is the most important example. The collective of features is labeled *morphology*, a concept comparable to microstructure (although the noun morphology originally related to shape). Typically, the electrical and thermal conductivities of polymers are low. Overviews are given in [4–6].

10.3 Lattice Concepts

We now review some lattice concepts as required for the description of crystalline solids and for which two classic references are [7, 8]. To create a lattice, points in space must be arranged in such a way that every point has an identical neighborhood. This is realized by a regular stacking of *unit cells*, whose (not necessarily orthogonal) noncoplanar basis vectors are denoted by $\boldsymbol{a}_1$, $\boldsymbol{a}_2$, and $\boldsymbol{a}_3$. Such stacking can be done in only 14 ways in 7 crystal systems, resulting in a *Bravais lattice* (Figure 10.4a). If these points are actually occupied by molecules (or atoms or ions) and no other molecules are present in the cell, we have a *primitive lattice*, while if more molecules occupy the unit cell, we speak of a *lattice with a basis*. If we have N unit cells, each with r molecules, the molecules are labeled, according to the unit cell they occupy, by $\boldsymbol{n} = (n_1, n_2, n_3)$ and within the unit cell by $\alpha, \beta, \ldots$. The position of cell $\boldsymbol{n}$ is $\boldsymbol{r}_n = n_1\boldsymbol{a}_1 + n_2\boldsymbol{a}_2 + n_3\boldsymbol{a}_3$, while the position of molecule α in cell $\boldsymbol{n}$ is $\boldsymbol{r}_\alpha$. The position of a molecule with respect to the origin is $\boldsymbol{r}_{n\alpha} = \boldsymbol{r}_n + \boldsymbol{r}_\alpha$. In ideal crystalline solids, each unit cell contains the same amount of matter in the same configuration.

Instead of this conventional unit cell, it is often convenient to use the *Wigner–Seitz cell*, that is, the cell obtained by drawing the perpendicular bisectors of the translation vectors from a chosen center to the nearest equivalent lattice site (Figure 10.4b). Using a basis of unit vectors $\boldsymbol{a}_1$, $\boldsymbol{a}_2$, and $\boldsymbol{a}_3$, obviously different from the set indicated before, the volume of the Wigner–Seitz cell is given by $\Omega_{\text{WS}} = \boldsymbol{a}_1 \cdot (\boldsymbol{a}_2 \times \boldsymbol{a}_3)$. Each lattice point can be addressed by a *lattice vector* $\boldsymbol{l} = l_1\boldsymbol{a}_1 + l_2\boldsymbol{a}_2 + l_3\boldsymbol{a}_3$, where l_1, l_2, and l_3 are integers. If $\boldsymbol{r}$ denotes a position in the zeroth cell, periodicity requires that for any function $f(\boldsymbol{r})$, describing a property of the lattice, $f(\boldsymbol{r}+\boldsymbol{l}) = f(\boldsymbol{r})$. We write $f(\boldsymbol{r})$ as

$$f(\boldsymbol{r}) = \sum\nolimits_{\mathbf{g}} A_{\mathbf{g}} \exp(\mathrm{i}\mathbf{g}\cdot\boldsymbol{r}) \quad \text{with} \quad A_{\mathbf{g}} = \Omega_{\text{WS}}{}^{-1}\int_{\text{cell}} f(\boldsymbol{r}) \exp(-\mathrm{i}\mathbf{g}\cdot\boldsymbol{r})\,\mathrm{d}\boldsymbol{r}$$

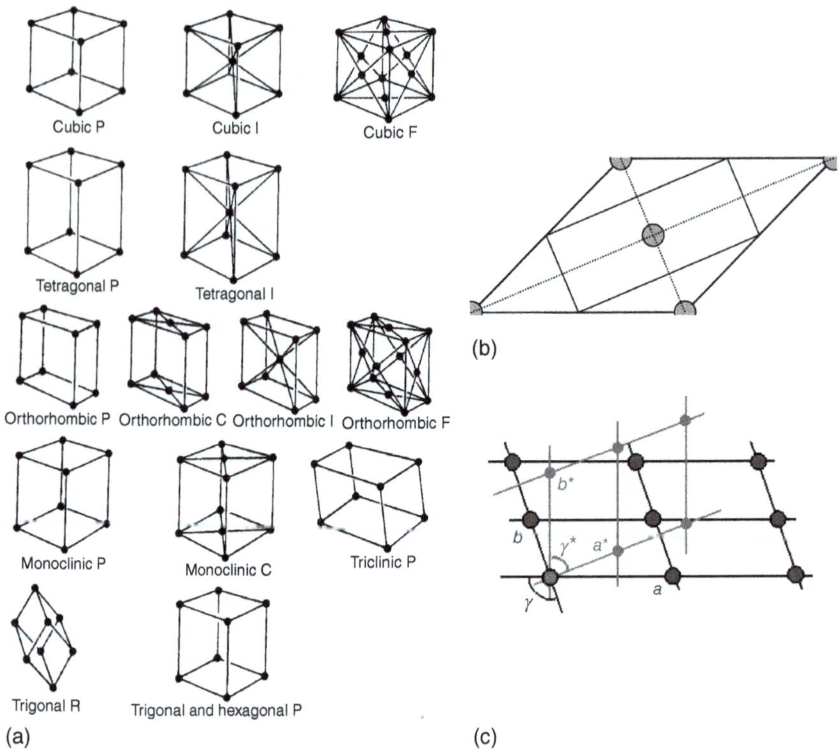

Figure 10.4 (a) The 14 Bravais lattices in 7 crystal systems. (b) The Wigner–Seitz cell in two dimensions (2D). (c) The 2D reciprocal lattice.

where $A_\mathbf{g}$ is the gth Fourier component of the function $f(\mathbf{r})$. The vector $\mathbf{g}$ is a *reciprocal lattice vector* (Figure 10.4c) for which the basis is given by

$$\mathbf{b}_1 = {\Omega_{\mathrm{WS}}}^{-1}(\mathbf{a}_2 \times \mathbf{a}_3) \quad \mathbf{b}_2 = {\Omega_{\mathrm{WS}}}^{-1}(\mathbf{a}_3 \times \mathbf{a}_1) \quad \text{and} \quad \mathbf{b}_3 = {\Omega_{\mathrm{WS}}}^{-1}(\mathbf{a}_1 \times \mathbf{a}_2)$$

One can show that $\mathbf{a}_i \cdot \mathbf{b}_j = \delta_{ij}$ so that the bases $\mathbf{a}_i$ and $\mathbf{b}_j$ form orthonormal sets. Any reciprocal lattice vector thus can be expressed as a linear combination of the basis vectors in reciprocal space, that is, $\mathbf{g} = g_1\mathbf{b}_1 + g_2\mathbf{b}_2 + g_3\mathbf{b}_3$. We note that:

- each vector of the reciprocal lattice $\mathbf{g}$ is normal to a set of lattice planes $\mathbf{l}$;
- the volume of the Wigner–Seitz cell in reciprocal space, called the *Brillouin zone* (BZ), is given by $\Omega_\mathrm{B} = (2\pi)^3/\Omega_{\mathrm{WS}}$; and
- the direct lattice is the reciprocal of its own reciprocal lattice.

Translational invariance requires that $H(\mathbf{l}) = H(\mathbf{0})$, where $H(\mathbf{0})$ and $H(\mathbf{l})$ are the Hamilton operators before and after a lattice translation $\mathbf{l}$ is applied. Similarly, $\Psi(\mathbf{r})$ and $\Psi(\mathbf{r}+\mathbf{l})$ denote the wave functions before and after the application of a lattice translation $\mathbf{l}$. Since $H(\mathbf{0})\Psi(\mathbf{r}) = \varepsilon\Psi(\mathbf{r})$ and $H(\mathbf{l})\Psi(\mathbf{r}+\mathbf{l}) = \varepsilon\Psi(\mathbf{r}+\mathbf{l})$, we also have $H(\mathbf{0})\Psi(\mathbf{r}+\mathbf{l}) = \varepsilon\Psi(\mathbf{r}+\mathbf{l})$, where ε is the eigenvalue of $\Psi(\mathbf{r})$. Moreover $\Psi(\mathbf{r}+\mathbf{a}_1)$ is indistinguishable from $\Psi(\mathbf{r})$ and thus $\Psi(\mathbf{r}+\mathbf{a}_1) = \lambda\Psi(\mathbf{r})$. From normalization we have $|\lambda|^2 = 1$ so that $\lambda = \exp(\mathrm{i}k_1)$ where k_1 is a constant. Similar results arise for $\mathbf{a}_2$ and $\mathbf{a}_3$. Since for a general translation it holds that $\Psi(\mathbf{r}+\mathbf{l}) = \Psi(\mathbf{r}+l_1\mathbf{a}_1+l_2\mathbf{a}_2+l_3\mathbf{a}_3)$, we obtain

$$\Psi(\mathbf{r}+\mathbf{l}) = \exp(\mathrm{i}\mathbf{k}\cdot\mathbf{l})\Psi(\mathbf{r}) \tag{10.1}$$

where the *wave vector* $\boldsymbol{k} = k_1\boldsymbol{b}_1 + k_2\boldsymbol{b}_2 + k_3\boldsymbol{b}_3$ is a vector in reciprocal space. Equation (10.1), *Bloch's theorem*, holds generally also for degenerate cases although here only indicated for the nondegenerate case. It states that for a perfect lattice any wave function satisfying the Schrödinger equation, a wave vector $\boldsymbol{k}$ exists such that a translation by a lattice vector $\boldsymbol{l}$ is equivalent to multiplying by a phase factor $\exp(\mathrm{i}\boldsymbol{k}\cdot\boldsymbol{l})$. Hence, we can label wave functions by their wave vector $\boldsymbol{k}$ and write

$$\Psi_{\boldsymbol{k}}(\boldsymbol{r}+\boldsymbol{l}) = \exp(\mathrm{i}\boldsymbol{k}\cdot\boldsymbol{l})\Psi_{\boldsymbol{k}}(\boldsymbol{r})$$

The wave vector $\boldsymbol{k}$ is only defined up to a reciprocal lattice vector and we take the smallest possible values as representative. In this so-called *reduced zone scheme*, we have $-\tfrac{1}{2}|\boldsymbol{b}_i| < |\boldsymbol{k}_i| \le \tfrac{1}{2}|\boldsymbol{b}_i|$ and the wave vector is restricted to the (first) BZ.

To count the number of allowed $\boldsymbol{k}$-vectors and avoid surface effects *periodic boundary* (or *Born–von Kármán*) *conditions* are invoked. These conditions imply that the last cell in all three lattice directions is connected to the first [9] so that

$$\Psi(\boldsymbol{r}+L_1\boldsymbol{a}_1) = \Psi(\boldsymbol{r}), \quad \Psi(\boldsymbol{r}+L_2\boldsymbol{a}_2) = \Psi(\boldsymbol{r}) \quad \text{and} \quad \Psi(\boldsymbol{r}+L_3\boldsymbol{a}_3) = \Psi(\boldsymbol{r})$$

where L_i is the number of cells in the i direction. Using Bloch's theorem, we have $\Psi_{\boldsymbol{k}}(\boldsymbol{r}+L_1\boldsymbol{a}_1) = \exp(\mathrm{i}\boldsymbol{k}\cdot L_1\boldsymbol{a}_1)\Psi_{\boldsymbol{k}}(\boldsymbol{r})$ and similar relations for $\boldsymbol{a}_2$ and $\boldsymbol{a}_3$. This implies that $\exp(\mathrm{i}\boldsymbol{k}\cdot L_1\boldsymbol{a}_1) = \exp(\mathrm{i}\boldsymbol{k}\cdot L_2\boldsymbol{a}_2) = \exp(\mathrm{i}\boldsymbol{k}\cdot L_3\boldsymbol{a}_3) = 1$, which can be obeyed only if

$$\boldsymbol{k} = k_1\boldsymbol{b}_1 + k_2\boldsymbol{b}_2 + k_3\boldsymbol{b}_3 = 2\pi\left(\frac{m_1}{L_1}\boldsymbol{b}_1 + \frac{m_2}{L_2}\boldsymbol{b}_2 + \frac{m_3}{L_3}\boldsymbol{b}_3\right)$$

where m_1, m_2, and m_3 are integers. For the (first) BZ $-\tfrac{1}{2}L_i < m_i \le \tfrac{1}{2}L_i$ and because the number of cells in the crystal is $L_1{\cdot}L_2{\cdot}L_3 = N$, the number of $\boldsymbol{k}$-vectors is N. Since $\Omega_{\mathrm{WS}} = V/N$, where V is the volume of the crystal, the volume per $\boldsymbol{k}$-vector is

$$\Omega_{\mathrm{B}}/N = (2\pi)^3/N\Omega_{\mathrm{WS}} = (2\pi)^3/V$$

In practice, N is large and for later reference we note that the sum over $\boldsymbol{k}$-vectors can be replaced by an integral according to

$$\textstyle\sum_{\boldsymbol{k}} \to \int \mathrm{d}\boldsymbol{k} = \dfrac{V}{(2\pi)^3}\displaystyle\iiint \mathrm{d}k_1\mathrm{d}k_2\mathrm{d}k_3$$

This result is independent of the shape of the crystal (Justification 19.2, [9]).

For any reciprocal lattice vector $\boldsymbol{g}$, Eq. (10.1) implies that $\boldsymbol{g}\cdot\boldsymbol{l} = 2\pi\cdot\text{integer}$, or defining $\boldsymbol{g} = 2\pi\boldsymbol{h}$, that

$$\boldsymbol{g}\cdot\boldsymbol{l} = (g_1\boldsymbol{b}_1 + g_2\boldsymbol{b}_2 + g_3\boldsymbol{b}_3)\cdot(l_1\boldsymbol{a}_1 + l_2\boldsymbol{a}_2 + l_3\boldsymbol{a}_3) = 2\pi(h_1l_1 + h_2l_2 + h_3l_3) = 2\pi\cdot\text{integer}$$

Hence h_1, h_2, and h_3 are integers. If one uses the conventional unit cell, they are normally labeled as *hkl* and called *Miller indices*. Specific planes in a crystal are then indicated by (*hkl*), while for the set of planes equivalent by lattice symmetry {*hkl*} is used. For directions in the crystal, [*hkl*] is used for a specific direction and <*hkl*> for the set of all directions equivalent to [*hkl*] by symmetry.

10.4 Crystalline Structures

The crystal structure of many compounds is based on a small number of simple packings, like the face-centered cubic (FCC) and hexagonal close packed (HCP) with packing density

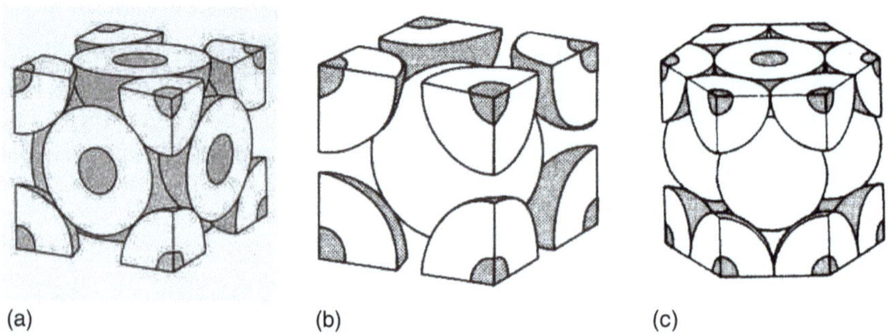

Figure 10.5 Three basic crystal structures: (a) FCC, (b) BCC, and (c) HCP.

$\eta_{FCC} = \eta_{HCP} = \pi/3\sqrt{2} \cong 0.741$, ideal tetragonal (T) with $\eta_T = 2\pi/9 \cong 0.698$, body-centered cubic (BCC) with $\eta_{BCC} = \pi\sqrt{3/8} \cong 0.680$, simple cubic (SC) with $\eta_{SC} = \pi/6 \cong 0.524$, and diamond (D) with $\eta_D = \pi\sqrt{3/16} \cong 0.340$ (Figure 10.5). Thus, a rather wide range in density exists. In a crystalline solid, a limited number of types of interstitial holes with a fixed ratio is present; for example, in the FCC lattice only tetrahedral holes are present, while for the BCC lattice only octahedral holes exist.

Of the 83 metal-like elements, 15 have the FCC, 25 have the HCP, 15 have the BCC, 11 have another, and 17 have an unknown structure. Only Po crystallizes in the SC structure. Small molecules, such as CO_2, and larger ones, such as naphthalene, usually also crystallize in a relatively simple lattice, for example, the orthorhombic lattice.

Inorganic materials usually crystallize in a more complex lattice with more than one atom per unit cell. We provide only two examples. The first is the CsCl structure (Figure 10.6a). In essence, this is a BCC lattice in which the alternate lattice points are occupied by positive and negative ions. The sublattice of each type of ion is the SC lattice. The second example is the NaCl structure (Figure 10.6b). This is a SC lattice whose lattice points are also alternately occupied by the positive and negative ions. The sublattice of each type of ion is the FCC lattice. Apart from metal halides, such as NaCl and KCl, some oxides also crystallize in this structure, for example, MgO. Since the radii r of the various ions vary a great deal, their coordination in AB crystals generally differs in different structures. The *ratio rules*

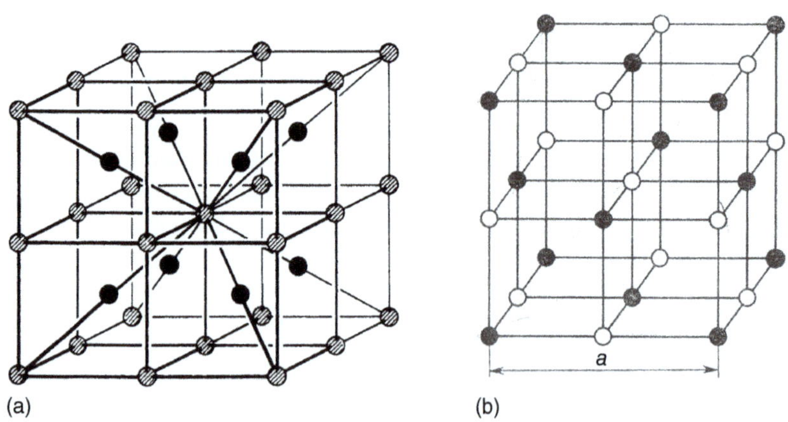

Figure 10.6 Two ionic crystal structures: (a) CsC and (b) NaCl.

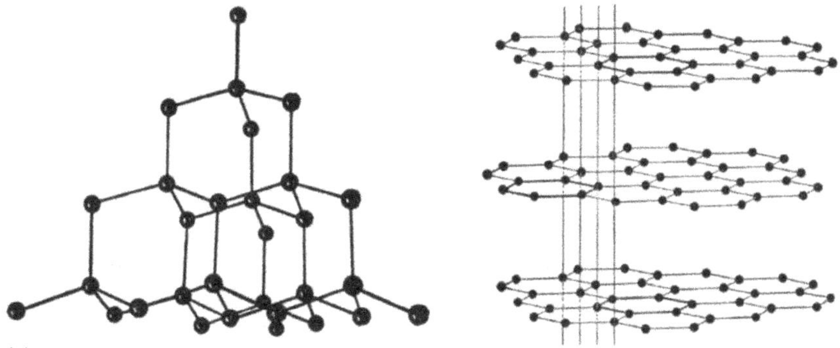

Figure 10.7 Covalent crystal structures. Source: Bragg and Bragg [12]/G. Bell: (a) diamond, (b) graphite.

state that for $r^+/r^- > 0.732$, typically the CsCl structure with an eightfold coordination is obtained. If $0.732 > r^+/r^- > 0.414$, the result is often the NaCl structure with a sixfold coordination, while for $0.414 > r^+/r^- > 0.225$ the ZnS structure with a fourfold coordination is preferred. In A_xB_y crystals, the coordination for each type of ion is usually different. For example, in the CaF_2 structure the Ca ions are surrounded by eight F ions, while each F ion is coordinated by four Ca ions. Together with the *principle of charge neutrality* for a WS cell, required as otherwise a high energy-consuming macroscopic charge would result, the ratio rules provide a guideline for which structure is formed. To estimate the size of the WS cell, the *sum rule* for the radii of the constituting ions is required (Section 10.6). Table 10.10 provides lattice constants for several cubic compounds. For a more complete description of lattices, see [10, 11].

The structure of covalently bonded materials is generally complex. An important basic type is the diamond lattice (D) in which each atom is bonded in a tetrahedral configuration with another atom (Figure 10.7a). As is well known, using solely C atoms, this leads to the hardest and stiffest existing material, diamond. This tetrahedral configuration can be realized by covalent bonding of sp^3 hybridized C atoms (Section 10.6.1). The structure can also be described as a connected network of puckered six rings of carbon atoms with a distance of 0.154 nm in the so-called chair configuration. The structure of many compounds is a variation on this basic structure, and IV–IV compounds (e.g. β-SiC), III–V compounds (e.g. GaAs), and II–VI compounds (e.g. CuS) often crystallize in the ZnS structure, which is diamond like but with alternating ions along the {111} direction. Consequently, when viewed along the {111} direction alternating layers of ions are present, which leads to two different types of surfaces after cleaving along these planes.

The vdW interaction is often important in layered structures, the most well-known example of which is graphite (Figure 10.7b). In this compound, sp^2 hybridized C atoms bond covalently to form layers, while the bonding between the layers is provided by the vdW interaction. The layers can be described by a connected network of flat six rings of C atoms with a distance of 0.142 nm. The distance between the layers is about 0.335 nm. Due to the relatively weak vdW interaction, graphite is easily deformed. Moreover, the thermal as well as electrical conductivity is high parallel to the layers, but low perpendicular to the layers.

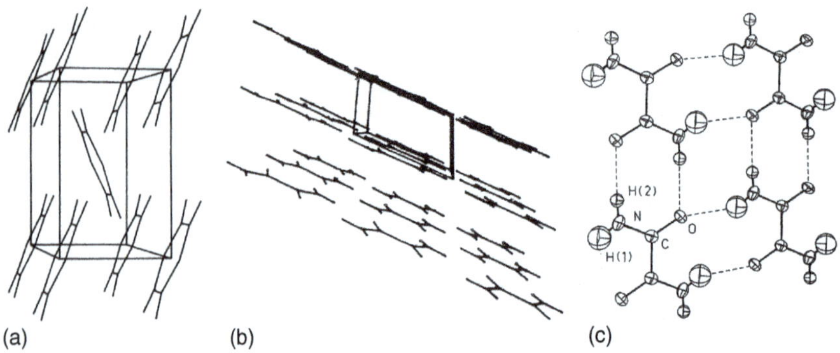

Figure 10.8 The crystallographic structure of (a) pyrazine [13] and (b, c) oxamide. Source: Adapted from de With and Harkema [13, 14].

Crystallization can also result solely due to vdW interactions in *molecular crystals*. The most well-known example is provided by the noble gas crystals. They generally crystallize in the FCC lattice. As another example we mention pyrazine, $C_4H_4N_2$, which crystallizes in an orthorhombic, body-centered unit cell [13] (space group P_{mnn}, Figure 10.8a). Owing to the weak vdW interactions, molecular crystals usually are relatively soft, and show a relatively low melting point, for pyrazine about 33 °C. Generally, the unit cell for molecular crystals has a low symmetry, the most common space group for organic molecules being $P2_1/c$.

Hydrogen bonding plays its role, an interesting example being provided by oxamide $C_2H_4N_2O_2$ [14] (space group $P\bar{1}$, Figure 10.8b,c). All molecules are planar within experimental accuracy and this plane deviates only about 0.65° from the *b*–*c* crystallographic plane. The molecular "sheets" are hydrogen bonded, while vdW interaction is present between the sheets. It will be no great surprise that the optical anisotropy is quite large, which is reflected in a refractive index difference $\Delta n \cong 0.3$, and that the thermal expansion is mainly perpendicular to the sheets. Hydrogen bonding occurs also in three dimensions, the most well-known example being ice (solid H_2O). The structure resembles the diamond structure in the sense that each H atom is linked with the lone-pair electrons of an O atom of a neighboring molecule.

10.5 Bonding: The Quantum-mechanical Approach

Having reviewed structure, we now turn to bonding. In Chapter 3, we presented an outline of Hartree–Fock (HF) theory. Here, we specialize to crystalline solids and discuss three approaches. First, the nearly free electron approximation, in which the lattice potential is supposed to be weak and electrons are nearly free to move and is typically used for metals. Second, the tight-binding approximation, in which the lattice potential is strong, and electrons remain most of the time near an ionic core, and which is more applicable for inorganics. Third, we discuss density functional theory.

10.5.1 The Nearly Free Electron Approximation

Metallic bonding is, metaphorically speaking, provided by the "sea" of electrons in which ions are submerged (Figure 12.2c). The valence electrons are not bound to specific atoms and the bonding is nondirectional. We discuss only simple metals in terms of the *Nearly Free Electron* (NFE) model. For non-simple metals, such as the transition and noble metals, more complex theories are required.

We first recall that the Schrödinger equation for a free electron of mass m, and the associated wave functions (plane waves) and eigenvalues are given by [15–17]

$$-\frac{\hbar^2}{2m}\nabla^2\psi(\boldsymbol{r}) = E\psi(\boldsymbol{r}), \quad \psi_k{}^0(\mathbf{r}) = L^{-3/2}\exp(\mathrm{i}\boldsymbol{k}\cdot\boldsymbol{r}) \quad \text{and} \quad E_k{}^0 = \frac{\hbar^2}{2m}k^2$$

where $\hbar = h/2\pi$ with h Planck's constant, E the energy, $\boldsymbol{r} = (x,y,z)$ the position vector, and $\boldsymbol{k} = (2\pi/L)(n_1,n_2,n_3)$ the wave vector with absolute value $k = |\boldsymbol{k}|$. The numbers n_1, n_2, and n_3 are the quantum numbers and periodic boundary conditions over a length L are used. Each state corresponding to a given $\boldsymbol{k}$ can contain two electrons of opposite spin according to Pauli's principle. At 0 K the lowest energy for N electrons is obtained by filling with electrons a sphere of volume $V = L^3$, the *Fermi sphere* with radius k_F. This results in $2(4/3)\pi k_F{}^3(L/2\pi)^3 = N$, since a unit volume of $\boldsymbol{k}$-space contains $V/(2\pi)^3$ states. Hence, $k_F = (3\pi^2\rho)^{1/3}$ with $\rho = N/V$ the electron density and the corresponding *Fermi energy* $E_F{}^\circ = \hbar^2 k_F{}^2/2m$. Considering two concentric spheres of radii k and $k + \mathrm{d}k$, the volume of the shell with thickness $\mathrm{d}k$ is $4\pi k^2\mathrm{d}k$. The *density of states* is therefore $n(k)\mathrm{d}k = (L/2\pi)^3 4\pi k^2\mathrm{d}k$ or in energy terms $n(E)\mathrm{d}E = ½4\pi(L/2\pi)^3(2mE/\hbar^2)^{1/2}(2m/\hbar^2)\mathrm{d}E = (L^3/2\pi^2)(2m/\hbar^2)^{3/2}E^{1/2}\mathrm{d}E$. The kinetic energy is given by $U_{\text{kin}} = \int_0^{k_F} En(E)\,\mathrm{d}E = 3E_F{}^\circ/5$. When $T \neq 0$ K, $n(E)$ is modified by the Fermi distribution [8] to $E_F \cong E_F{}^\circ\{1 - \frac{1}{6}(\pi kT)^2[n'(E_F)/n(E_F)]\}$ with $n'(E) = \partial n(E)/\partial E$. Because $n(E) \sim E^{1/2}$, at 300 K the correction is small as $kT \cong 0.025$ eV $\ll E_F \cong 5$ eV.

For nearly free electrons one approach is to expand the wave function, taking into account the Bloch condition, in a Fourier series [15, 16] given by

$$\Psi_k = \mathrm{e}^{\mathrm{i}\boldsymbol{k}\cdot\boldsymbol{r}}u(\boldsymbol{r}) = \mathrm{e}^{\mathrm{i}\boldsymbol{k}\cdot\boldsymbol{r}}\sum\nolimits_{k-g} c_{k-g}\mathrm{e}^{-\mathrm{i}\boldsymbol{g}\cdot\boldsymbol{r}} = \sum\nolimits_{k-g} c_{k-g}\mathrm{e}^{\mathrm{i}(\boldsymbol{k}-\boldsymbol{g})\cdot\boldsymbol{r}}$$

where $\boldsymbol{g}$ is a reciprocal lattice vector. Substituting these expressions in the Schrödinger equation, multiplying by $\exp[-\mathrm{i}(\boldsymbol{k} - \boldsymbol{g}')\cdot\boldsymbol{r}]$, integrating and realizing that the resulting integrals equal $\delta_{g'g}$, we obtain an infinite set of coupled, linear equations

$$\left[\frac{\hbar^2|\boldsymbol{k}-\boldsymbol{g}|^2}{2m} - E\right]c_{k-g} + \sum\nolimits_{g'} V_{g'-g}c_{k-g'} = 0$$

with $V_k = \int \exp(-\mathrm{i}\boldsymbol{k}\cdot\boldsymbol{r})V(r)\,\mathrm{d}\boldsymbol{r}$ the $\boldsymbol{k}$-th Fourier transform coefficient of the potential energy V. To solve this infinite set analytically, we must approximate. As a first approximation, we assume that all terms $V_{g'-g}$ are zero except those for which $\boldsymbol{g}' - \boldsymbol{g} = \boldsymbol{g}$ and $\boldsymbol{g}' - \boldsymbol{g} = -\boldsymbol{g}$. These two contributions are the dominant ones when Bragg reflection of an electron with wave vector $\boldsymbol{k}$ to an electron with wave vector $\boldsymbol{k} - \boldsymbol{g}$ at the zone boundary occurs (Figure 10.9a). Due to the lattice periodicity, these terms are equivalent, and both must be included. Omitting the constant term V_0, we obtain

$$\left[\frac{\hbar^2|\boldsymbol{k}|^2}{2m} - E\right]c_k + V_g c_{k-g} = 0 \quad \text{and} \quad V_{-g}c_k + \left[\frac{\hbar^2|\boldsymbol{k}-\boldsymbol{g}|^2}{2m} - E\right]c_{k-g} = 0$$

Solving for E as a function of $\boldsymbol{k}$, meanwhile using $V_{-g} = V_g{}^*$, the result is

$$E^\pm = ½\left(E_k{}^0 + E_{k-g}{}^0\right) \pm \left[\left(E_k{}^0 - E_{k-g}{}^0\right)^2 + |2V_g|^2\right]^{1/2} \tag{10.2}$$

For $\boldsymbol{k} \ll \boldsymbol{g}/2$, expanding the square root according to $(1+x)^{1/2} \cong 1+x/2$, we have $E^- \cong E_k{}^0 + |V_g|^2/(E_k{}^0 - E_{k-g}{}^0)^2$, reflecting a small correction to the free electron solution.

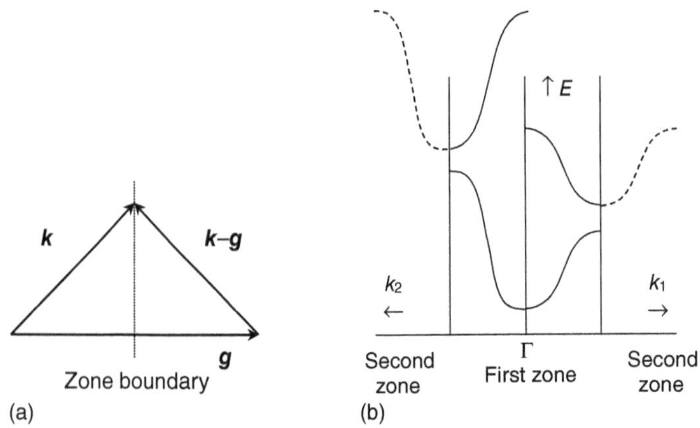

Figure 10.9 (a) Bragg reflection at the zone boundary. (b) Energy bands in the NFE approximation along k_1, showing the gap in the reduced (—) and extended (- - -) zone scheme, and along k_2, showing a gap which has overlap with a higher lying band in the k_1 direction.

For $\mathbf{k} = \mathbf{g}/2$, $E^- \cong E_{g/2}{}^0 - |V_g|$, and, similarly, $E^+ \cong E_{g/2}{}^0 + |V_g|$. Therefore, at the zone boundary the energy has a discontinuity and the allowed energy levels are clustered in *energy bands*, separated by a *band gap* $E_{\text{gap}} = 2|V_g|$ in this approximation. If more than two electron waves contribute significantly, the energy expression becomes correspondingly more complex, but the discontinuities at the zone boundary remain. Plotting the energy as a function of the component k_1, Figure 10.9b is obtained in which the allowed energy bands are indicated. Although in 1D this always implies the presence of an energy gap, in 3D the bands may overlap in the interior of the zone and generally they will do so. Therefore, in principle in the NFE approximation the electrons have access to a quasi-continuous energy spectrum and since the bands are typically only partially filled, materials in this description are electrically conducting.

The symmetry of the lattice determines which set of reciprocal lattice vectors is present and therefore also determines the shape of the BZ. For a SC lattice with lattice constant a, the shape of the first BZ is also a cube with lattice constant $2\pi/a$. The usual lattices FCC and BCC can be described as a superposition of several SC lattices. For example, the BCC lattice can be considered as two interpenetrating SC lattices of lattice constant a displaced by $(\frac{1}{2}a, \frac{1}{2}a, \frac{1}{2}a)$ relative to each other. The primitive lattice vectors are $a[11\bar{1}]/2$, $a[1\bar{1}1]/2$, and $a[\bar{1}11]/2$, resulting in reciprocal vectors $2\pi[110]/a$, $2\pi[101]/a$, and $2\pi[011]/a$, representing an FCC lattice in reciprocal space with lattice constant $4\pi/a$. The first BZ becomes a dodecahedron with 12 faces for $2\pi\langle 110\rangle/a$. Similarly, the FCC lattice can be considered as four interpenetrating SC lattices. The primitive lattice vectors are $a[110]/2$, $a[101]/2$, and $a[011]/2$, resulting in reciprocal vectors $2\pi[11\bar{1}]/a$, $2\pi[1\bar{1}1]/a$, and $2\pi[\bar{1}11]/a$, representing a BCC lattice in reciprocal space with lattice constant $4\pi/a$. The first BZ becomes a cubo-octahedron with eight hexagonal faces for $2\pi\langle 111\rangle/a$ and six cubic faces for $4\pi\langle 100\rangle/a$. The shapes of the BZ for the BCC and FCC lattice are shown in Figure 10.10.

To describe observed band structures the elements V_g should be small and positive, while, when calculated explicitly, they are large and negative due to the Coulomb potential of the core. However, the valence electron bands should be orthogonal to the core states, which has not been taken into account. This can be realized by using orthogonalized

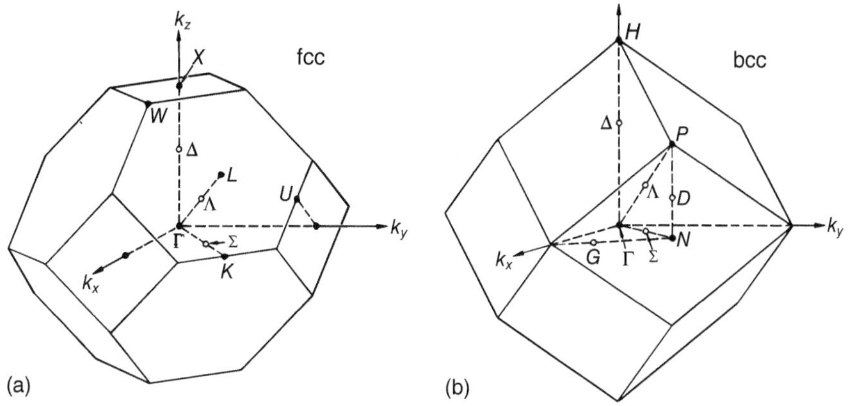

Figure 10.10 Brillouin zones with symmetry points indicated. (a) BCC lattice. (b) FCC lattice.

plane waves (OPWs) [18], given by $\Psi_{\boldsymbol{k}} = \Psi_{\boldsymbol{k}}{}^0 - \Sigma_{\rm c} S_{\boldsymbol{k}\rm c}\Psi_{\rm c}$, where the summation runs over all cores states $\Psi_{\rm c}$ and the coefficients $S_{\boldsymbol{k}\rm c}$ are the overlap integrals between plane waves and the core states. The latter are described by $H\Psi_{\rm c} = E_{\rm c}\Psi_{\rm c}$, where $E_{\rm c}$ is the energy of the core state. Repeating the above exercise with these OPWs and approximating E with $E_{\rm X}$ for the BZ boundary results in $E_{\rm gap} = 2|V_{\boldsymbol{g}} - \Sigma_{\rm c} S_{\boldsymbol{k}\rm c}{}^2(E_{\rm X} - E_{\rm c})|$. Because $E_{\rm X} - E_{\rm c} > 0$, the core contribution is repulsive, and the large negative term $V_{\boldsymbol{g}}$ is counteracted by a large positive core-orthogonality term. This implies that we can retain the NFE approximation if we replace the strong Coulomb potential (in atomic units) $V_{\rm C} = -2Z/r$ (with Z the valence) by a weak effective or pseudo-potential $V_{\rm pse}$. Assuming cancellation of the core-orthogonality and Coulomb terms to be exact over the core with radius $r_{\rm c}$, we obtain the *empty core pseudo-potential*, described $V_{\rm pse} = 0$ for $r < r_{\rm c}$ and $V_{\rm pse} = -2Z/r$ for $r > r_{\rm c}$ [19]. We now denote the *atomic volume* by $\Omega = V/N$, where N represents the number of atoms in a volume V, so that the size of the atomic volume may be expressed by the *Wigner–Seitz radius* $r_{\rm WS}$, which is the radius of a sphere with volume $\Omega = 4\pi r_{\rm WS}{}^3/3$. The electrostatic energy Φ_0 for the pseudo-potential is calculated from $\Phi_0 = \int_{r_{\rm c}}^{r_{\rm WS}} \rho(r)V_{\rm pse}(r){\rm d}\boldsymbol{r} = -3Z^2[(1/r_{\rm WS}) - (r_{\rm c}{}^2/r_{\rm WS}{}^3)]$ using $\rho = Z/\Omega$. Clearly, some criterion must be used to estimate $r_{\rm c}$. Generally, this pseudo-potential approach works quite well. We refer for further details to [20], while [21] provides a short review.

Neglecting the details of the band structure altogether, Girifalco [22] used a model in which the Z valence electrons are distributed spatially uniformly (Figure 10.2c) and the ions are represented by pseudo-potentials.

The *kinetic energy* $U_{\rm kin}$ for the electron gas[3] is repulsive and given by $3E_{\rm F}{}^\circ/5$ or

$$U_{\rm kin} = 2.210Z^{5/3}r_{\rm WS}{}^{-2}\ ({\rm Ry})$$

The *Coulomb energy* $U_{\rm Cou}$ results from the electron–core $U_{\rm pse}$ and electron–electron interactions $U_{\rm ee}$ and reads

$$U_{\rm Cou} = U_{\rm pse} + U_{\rm ee} = Z^2\left[\left(3r_{\rm c}{}^2/r_{\rm WS}{}^3\right) - (3/r_{\rm WS})\right] + (6/5)Z^2/r_{\rm WS}\ ({\rm Ry})$$

3 For derivations we refer to, for example, [23]. All results given are in Ry, as usual in the relevant literature.

Denoting $3Z^2/r_{WS} - 1.2Z^2/r_{WS}$ by $\alpha_C Z^2/r_{WS}$, the parameter $\alpha_C = 1.8$ resembles the Madelung constant for ionic solids. However, these contributions were calculated by integrating over one WS cell, although it should be over all cells. Because the WS cell is neutral, only a small contribution is expected from the neighboring cells. Doing such a calculation leads to $\alpha_C = 1.760$ for SC, $\alpha_C = 1.671$ for D, and $\alpha_C = 1.792$ for BCC, FCC, and HCP lattices [24], indeed deviating only slightly from $\alpha_C = 1.8$ for a single-point ion. Hence, the latter value is chosen.

We will need two further contributions. First, the *exchange energy*

$$U_{exc} = -0.9163\, Z^{4/3}/r_{WS}\ (\text{Ry})$$

which is attractive since parallel spins are kept apart by Pauli's exclusion principle, leading to a weaker mutual Coulomb repulsion. Second, the *correlation energy* [25]

$$U_{cor} = 0.0313Z\ \ln(r_{WS}/Z^{1/3}) - 0.115Z\ (\text{Ry})$$

which is also attractive due to the dynamic correlation between the electrons and is, even in this simplified model, only an approximate interpolation formula.

The total binding energy U thus becomes

$$U = U_{kin} + U_{Cou} + U_{exc} + U_{cor}$$

Table 10.1 Data for some metals in the Girifalco approach.

Metal	r_{WS}/a_0 exp.	r_c/a_0 from r_{WS}	U (Ry) theory	U (Ry) exp.	K/K_{non} theory	K/K_{non} exp.
Li	3.266	1.318	0.553	0.5205	0.63	0.50
Na	3.992	1.752	0.469	0.4620	0.83	0.80
K	4.860	2.267	0.396	0.3889	1.03	1.10
Rb	5.313	2.536	0.374	0.3716	1.14	1.55
Cs	5.699	2.765	0.345	0.3473	1.29	1.43
Be	2.356	0.751	2.439	2.278	0.45	0.27
Mg	3.346	1.366	1.834	1.747	0.73	0.54
Ca	4.122	1.829	1.535	1.4583	0.95	0.66
Sr	4.494	2.050	1.424	1.356	1.05	0.78
Ba	4.670	2.155	1.377	1.300	1.11	0.84
Zn	2.905	1.098	2.062	2.113	0.60	0.45
Cd	3.264	1.317	1.872	1.914	0.71	0.63
Hg	3.353	1.370	1.831	2.196	0.73	0.59
Al	2.99	1.11	—	—	0.69	0.32
Ga	3.16	1.20	—	—	0.74	0.33
In	3.48	1.37	—	—	0.83	0.39
Cu	2.67	0.91	—	—	0.45	2.16
Ag	3.02	1.37	—	—	0.71	2.94
Au	3.01	1.35	—	—	0.69	4.96

Source: Adapted from [15, 22].

and minimizing this equation with respect to r_{WS} yields the relation

$$\left(\frac{r_c}{r_{WS}}\right)^2 = \frac{1}{5} + \frac{0.1018}{Z^{2/3}} + \frac{0.00347 r_{WS}}{Z} - \frac{0.4911}{Z^{1/3} r_{WS}} \tag{10.3}$$

Choosing the parameter r_c so that the experimental Wigner–Seitz radius r_{WS} is matched [22] results in the values given in Table 10.1. We note that, typically, $2a_0 < r_{WS} < 6a_0$ and $r_c \approx 1 - 2a_0$. It appears that the contribution of the exchange-correlation energy cancels largely with the electron–electron Coulomb term. The bulk modulus $K = V\mathrm{d}^2U/\mathrm{d}V^2$, normalized by the bulk modulus of noninteracting electrons $K_{non} = 0.586r_s^{-5}$, is given by $K/K_{non} = 0.200Z + 0.815(r_c^2/r_s)$. Here, the contribution of U_{cor} is neglected since it contributes just a few percent. For simple (sp) metals, this theory results in reasonable estimates for U and K (Table 10.1), in which the contribution of the core is crucial. For the trivalent metals Al, Ga, and In and the (sd) metals Cu, Ag, and Au, unsatisfactory results are obtained.

Problem 10.1 ***Girifalco Bonding Energy 1*** Use the approximation $U = U_{kin} + U_{Cou}$ to show that its minimum $r_{min} = (4.42/\alpha_C)Z^{-2/3}$ the binding energy $U_{min} = 0.113\alpha_C{}^2Z^{7/3}$.

Problem 10.2 ***Girifalco Bonding Energy 2*** Derive Eq. (10.3). Plot the various contributions to U as a function of r_{WS}.

10.5.2 The Tight Binding Approximation

In the case of strong interaction of the electrons with the lattice, it is more appropriate to consider the atomic states as the zero approximation (Figure 12.2a,b). Therefore, in this case, we expand the wave function for lattice with lattice constant a as

$$\Psi_{\boldsymbol{k}} = \exp(\mathrm{i}\boldsymbol{k}\cdot\boldsymbol{r})u(\boldsymbol{r}) = \exp(\mathrm{i}\boldsymbol{k}\cdot\boldsymbol{r})\sum\nolimits_m \exp[-\mathrm{i}\boldsymbol{k}\cdot(\boldsymbol{r}+\boldsymbol{r}_m)]\phi(\boldsymbol{r}-\boldsymbol{r}_m)$$

where the labeling $\phi(\boldsymbol{r}-\boldsymbol{r}_m)$ or ϕ_m is used to identify the atomic orbital localized at $\boldsymbol{r}_m$, supposing for simplicity a Bravais lattice. We also suppose that the Schrödinger equation for the individual atoms

$$(\hbar^2/2m)\nabla^2\phi_n + (E_0 - V_n)\phi_n = 0 \tag{10.4}$$

is solved, assuming that the potential V_n is large near the lattice points and decreases rapidly with distance. As a zero-order approximation, we suppose that each electron is close to a particular lattice point, and we neglect the effect of other atoms. This state is highly degenerate since positioning the electrons in a similar orbit around each of the ions at the various lattice point results in the same energy for each atom and, assuming one valence electron per ion, the degeneracy is equal to the number of atoms in the crystal. Allowing now for interaction the degeneracy is removed and each level splits into a band. We explicitly assume that the presence of the other atoms introduces an energy change small as compared to the distance of the levels associated with Eq. (10.4). Calculating in this *tight binding* (TB) approximation, the expectation value of the wave function results in

$$\langle E_{\boldsymbol{k}}\rangle = \frac{\int \sum_{m'} \exp\left(-\mathrm{i}\boldsymbol{k}\cdot\boldsymbol{r}_{m'}\right)\phi_{m'}{}^* H \sum_m \exp(\mathrm{i}\boldsymbol{k}\cdot\boldsymbol{r}_m)\phi_m \,\mathrm{d}\boldsymbol{r}}{\int \sum_{m'} \exp\left(-\mathrm{i}\boldsymbol{k}\cdot\boldsymbol{r}_{m'}\right)\phi_{m'}{}^* \sum_m \exp(\mathrm{i}\boldsymbol{k}\cdot\boldsymbol{r}_m)\phi_m \,\mathrm{d}\boldsymbol{k}} = \frac{\sum_{m''} \exp\left(\mathrm{i}\boldsymbol{k}\cdot\boldsymbol{r}_{m''}\right)H_{m''}}{\sum_{m''} \exp\left(\mathrm{i}\boldsymbol{k}\cdot\boldsymbol{r}_{m''}\right)S_{m''}}$$

with $H_{m''} = \langle \phi_{m'}|H|\phi_m \rangle$ and $S_{m''} = \langle \phi_{m'}|\phi_m \rangle$. With $\sum_{m'''\neq m'} \langle \varepsilon_{m'}|V_{m'''}|\phi_{m'} \rangle = V_0$, we have

$$E_k = E_0 + \frac{V_0 + \sum_{m\neq m'} \sum_{m'''\neq m'} \exp\left(\mathrm{i}\boldsymbol{k}\cdot\boldsymbol{r}_{m''}\right) \left\langle \phi_{m'}|V_{m''}|\phi_{m} \right\rangle}{1 + \sum_{m''} \exp\left(\mathrm{i}\boldsymbol{k}\cdot\boldsymbol{r}_{m''}\right) S_{m''}} \tag{10.5}$$

where $\boldsymbol{r}_{m''} = \boldsymbol{r}_m - \boldsymbol{r}_{m'}$ is a non-zero lattice vector. Evaluating all integrals to obtain a self-consistent solution requires substantial calculation. Therefore, to have a simple model the *Hückel approximations* are often made. First, since in the TB method one assumes small overlap between different sites because the basis function ϕ_n decreases rapidly with distance from $\boldsymbol{r}_n$, one takes $\langle \phi_{m'}|V_{m'''}|\phi_m \rangle = 0$, so that only the simpler two-center integrals remain. Second, one even assumes that only the nearest-neighbor two-center integrals are non-zero. Finally, in the same spirit, one assumes that $\langle \phi_n|\phi_m \rangle = \int \phi_n{}^* \phi_m \mathrm{d}\boldsymbol{r} = \delta_{nm}$. In this approximation, the energy expression becomes

$$E_k = E_0 + V_0 + \sum\nolimits_{m'=m-1,m,m+1} \exp\left(\mathrm{i}\boldsymbol{k}\cdot\boldsymbol{r}_{m'}\right) \left\langle \phi_{m'}|V_m|\phi_m \right\rangle$$

Moreover, we write $\alpha_n = -\langle \phi_n|V_0 - V_n|\phi_n \rangle$ and $\beta_{n,m} = -\langle \phi_n|V_0 - V_n|\phi_m \rangle$, where $\beta_{n,m}$ depends on $\boldsymbol{n}$ and $\boldsymbol{m}$ only through their difference $|\boldsymbol{n} - \boldsymbol{m}|$. The values of α and β are generally positive, provided ϕ_n has only nodes inside the atomic cores.

To illustrate the ideas, let us consider the SC lattice, with six nearest neighbors at a distance a. Using only s-orbitals (Figure 10.11a) with energy E_1 and only the integrals $\beta(100) = \beta(010) = \beta(001) = \beta_{\mathrm{ss}}$, the energy expression becomes

$$E = E_{\mathrm{s}} - \alpha_{\mathrm{s}} - 2\beta_{\mathrm{ss}}[\cos ak_1 + \cos ak_2 + \cos ak_3]$$

The degenerated levels, all with energy E_{s}, thus split into a band centered at $E_{\mathrm{s}} - \alpha_{\mathrm{s}}$ and of width $12\beta_{\mathrm{ss}}$. For p-states on the SC lattice, the levels are triply degenerate, and we represent these states by $xf(r)$, $yf(r)$, and $zf(r)$. Considering the p_x-states, $\phi_n = x_n f(r_g)$ is the wave function with energy E_{p}. Since the ϕ_n is not spherically symmetric, we use for π overlap $\beta_{\mathrm{p}\pi} = -\langle \phi_{n_1,n_2+1,n_3}|V_0 - V_n|\phi_n \rangle$ and for σ overlap $\beta_{\mathrm{p}\sigma} = -\langle \phi_{n_1,n_2,n_3+1}|V_0 - V_n|\phi_n \rangle$, which leads to

$$E = E_{\mathrm{p}} - \alpha_{\mathrm{p}} - 2\beta_{\mathrm{p}\sigma} \cos ak_1 - 2\beta_{\mathrm{p}\pi}[\cos ak_2 + \cos ak_3]$$

Similar expressions result with permutation of k_1, k_2, and k_3 to account for $yf(\boldsymbol{r})$ and $zf(\boldsymbol{r})$. Thus, bands arise due to the interaction of the atoms with each other.

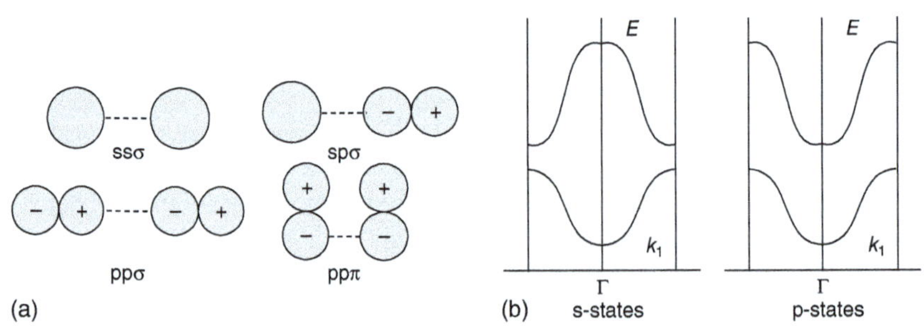

Figure 10.11 (a) The symmetry of the four basic types of interactions between s- and p-states. (b) Schematic picture of the energy bands in the tight-binding approximation for s-states and p-states.

Overall, for the TB approximation to be valid, the value of the βs must be small and therefore it is necessary that the levels E_s, E_p, … should be far apart. In the approximation discussed, the maximum of the s-states is obtained at $k_1 = k_2 = k_3 = \pi/a$ and $E = E_s - \alpha_s + 6\beta_{ss}$. Similarly, for the p-states the minimum is obtained at $k_1 = \pi/a$, $k_2 = k_3 = 0$, and $E = E_p - \alpha_p + 2\beta_{p\sigma} - 4\beta_{p\pi}$. We must therefore have $E_p - \alpha_p + 2\beta_{p\sigma} - 4\beta_{p\pi} > E_s - \alpha_s + 6\beta_{ss}$. All the energies of the second band lie above those of the first band and there are forbidden energy ranges, in contrast to the NFE case (Figure 10.11b). In conclusion, in the TB approximation the atomic levels spread to bands, which do not overlap. Therefore, the materials are electrically nonconducting if the gap is wide, or semiconducting if the gap is narrow as compared to kT.

The matrix elements, containing integrals over the basis functions, are dependent on the distance between neighboring atoms. Thus, the energy as a function of lattice constant can be obtained by using the variation principle. In *ab initio* calculations these integrals are calculated exactly, while in semiempirical calculations they are obtained by proper parameterization of the various occurring integrals on properties. The semiempirical approach, used extensively by Harrison [26], is capable to reproduce many features of the electronic structure and the resulting properties of materials and the TB approximation is extensively used in semiquantitative and qualitative discussions. In Example 10.1, the important example of Si is discussed.

Example 10.1 ***The Minimal Basis Set Band Structure of Si*** Silicon is an important covalently bonded material for which, within a simple TB model, the band structure can be calculated taking only 2s and 2p orbitals as basis functions. The nearest-neighbor and on-site matrix elements are estimated semiempirically as

$$\beta_{ss\sigma} = -1.94, \quad \beta_{sp\sigma} = 1.75, \quad \beta_{pp\sigma} = 3.10, \quad \beta_{pp\pi} = -1.08, \quad \alpha_s = -5.25, \quad \alpha_p = 1.20$$

with all values in eV and where σ and π denote their symmetry (Figure 10.11). The diamond lattice consists of two interpenetrating FCC lattices, separated by $\frac{1}{4}a[1,1,1]$ where a is the lattice constant. The WS cell can be defined by the vectors $\frac{1}{2}a[1,1,0]$, $\frac{1}{2}a[1,0,1]$, and $\frac{1}{2}a[0,1,1]$ and contains two atoms, one at (0,0,0) and the other at $\frac{1}{4}a(1,1,1)$. Since we have a basis set of four orbitals at each atomic site, each WS cell contains eight basis functions, and therefore we have eight bands.

Let $|\boldsymbol{m}j\alpha\rangle$ denote an atomic state at $\boldsymbol{r}_{\boldsymbol{m}j}$ and α the label for s, p_x, p_y, or p_z. Further, we denote with $|n\boldsymbol{k}\rangle$ an eigenstate with band index n ($1 \le n \le 8$) and wave vector $\boldsymbol{k}$. Using Bloch's theorem, we expand $|n\boldsymbol{k}\rangle$ as

$$|n\boldsymbol{k}\rangle = \sum\nolimits_{\boldsymbol{m}j,\alpha} \exp(\mathrm{i}\boldsymbol{k} \cdot \boldsymbol{r}_{\boldsymbol{m}j})\, c_{j,\alpha}{}^{(n)} |\boldsymbol{m}j\alpha\rangle$$

The expansion coefficients are obtained in the usual way by inserting the expansion in the Schrödinger equation $H|n\boldsymbol{k}\rangle = E^{(n)}|n\boldsymbol{k}\rangle$ and multiplying by $\langle \boldsymbol{0}j'\alpha'|$. The result is

$$\sum\nolimits_{j,\alpha} H_{j\alpha j'\alpha'}(\boldsymbol{k}) c_{j\alpha}{}^{(n)}(\boldsymbol{k}) = E^{(n)}(\boldsymbol{k}) c_{j'\alpha'}{}^{(n)}(\boldsymbol{k}) \quad \text{where}$$

$$H_{j\alpha j'\alpha'}(\boldsymbol{k}) = \sum\nolimits_{\boldsymbol{m}} \exp\left[\mathrm{i}\boldsymbol{k} \cdot \left(\boldsymbol{r}_{\boldsymbol{m}} + \boldsymbol{r}_j - \boldsymbol{r}_{j'}\right)\right] \langle \boldsymbol{0}j'\alpha'|H|\boldsymbol{m}j\alpha\rangle$$

For each $\boldsymbol{k}$-value an 8 by 8 matrix eigenvalue equation has to be solved. In this approximation, when $j = j'$, the matrix element $\langle 0j'\alpha'|H|\boldsymbol{m}j\alpha\rangle$ is zero unless $\boldsymbol{m} = 0$ and $\alpha = \alpha'$. Hence, the diagonal elements are either α_s or α_p depending on whether α denotes an s or p orbital.

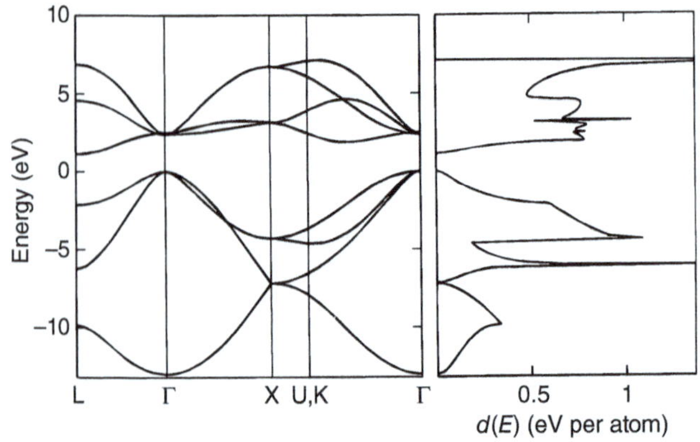

Figure 10.12 The band structure (left) and density of states $d(E)$ (right) for Si in the TB approximation using empirical matrix elements. Source: Sutton [16]/Oxford University Press.

For the off-diagonal terms with $j \neq j'$, the expression above reduces to just four terms, one for each neighbor. Figure 10.12 shows the resulting band structure along a few directions in the BZ and the density of states. In the latter, the lower part contains the four valence bands and the upper part the conduction bands, separated by a band gap of about 1.1 eV, indicating that Si is a semiconductor. The total energy can be calculated but since the matrix elements are constants, the energy as a function of lattice constant cannot be obtained in this approximation.

10.5.3 Density Functional Theory

The solution of the Schrödinger equation for a particular system, even in an approximate way, is generally an enormous task. For an N-electron system the wave function contains $3N$ variables (or $4N$ including spin). Therefore, directly after the discovery of quantum mechanics attempts were made to employ the (observable) electron density ρ directly. In fact, there are good arguments to do so since

- the electron density $\rho(\mathbf{r})$ integrates to the number of electrons, that is, $\int \rho(\mathbf{r})\mathrm{d}\mathbf{r} = N$;
- $\rho(\mathbf{r})$ has maxima, actually cusps, at the positions of the nuclei R_A; and
- the density at the position of the nucleus contains information about the nuclear charge Z via $\lim_{r_{iA}\to 0}[(\partial\overline{\rho}(r)/\partial r) + 2Z_\mathrm{A}\overline{\rho}(r)] = 0$, where $\overline{\rho}(r)$ denotes the spherically averaged electron density at the nuclear position of atom A.

Thus $\rho(\mathbf{r})$ provides all information necessary for specifying the Hamilton operator H for a specific structure.

The first step was the Thomas–Fermi–Dirac model (∼1928) in which for the densities of kinetic energy T, Coulomb energy E_C, and exchange energy E_xc the homogeneous electron gas expressions were used, so that the energies become a functional of $\rho(\mathbf{r})$. The model was applied primarily to atoms, but also to molecules and solids [27, 28]. For atoms a general solution can be obtained in nondimensional form. The E_xc expression is local in this approximation, contrary to the Hartree–Fock E_xc term, which led Slater [29] in 1951 to his $X\alpha$ method in which the latter is replaced by the former, thereby rendering solutions easier, while obtaining reasonable results.

The next important step was made only in 1964 by the discovery of two important theorems by Hohenberg and Kohn [30] and which has led to *density functional theory* (DFT). The first theorem states that 'the external potential $V_{ext}(\boldsymbol{r})$ is (within a constant) a unique *functional* of $\rho(\boldsymbol{r})$'. Since on its turn $V_{ext}(\boldsymbol{r})$ fixes H, we see that the full many-particle ground state is a unique functional of $\rho(\boldsymbol{r})$. Symbolically, we have

$$\rho_0 \rightarrow (n, Z_A, R_A) \rightarrow H \rightarrow \Psi_0 \rightarrow E_0$$

where ρ_0 and Ψ_0 denote the exact electron density and wave function, respectively. Since the ground-state energy $E_0[\rho(\boldsymbol{r})]$ is a functional of $\rho(\boldsymbol{r})$ (Section A.8), so must be its individual components and we write $E_0[\rho_0] = T[\rho_0] + E_{ee}[\rho_0] + E_{ext}[\rho_0] \equiv F_{HK}[\rho_0] + \int V_{ext}[\rho_0(\boldsymbol{r})]\,d\boldsymbol{r}$, using square brackets to indicate a functional. The last term contains the system-specific potential energy $V_{ext} = -\sum_A Z_A/r_{1A}$ due to nuclear–electron attraction. The kinetic energy T and electron–electron interaction E_{ee} are independent of N, Z_A, and R_A, and define $F_{HK}[\rho]$, the *Hohenberg–Kohn functional*. This implies that, if $F_{HK}[\rho]$ is calculated with ρ a n-representable density [31], it yields

$$F_{HK}[\rho] = T[\rho] + E_{ee}[\rho] = \langle\Psi|T + V_{ee}|\Psi\rangle$$

the expectation value of $T + V_{ee}$. Unfortunately, the form of F_{HK} is unclear. Only the classical Coulomb part $E_C[\rho]$ is evident so that we have

$$E_{ee}[\rho] = {}^1\!/\!_2\int \rho(\boldsymbol{r}_1)V_C\,d\boldsymbol{r}_1 + E_{ncl}[\rho] \equiv E_C[\rho] + E_{ncl}[\rho]$$

where the Coulomb potential $V_C = \int[\rho(\boldsymbol{r}_2/r_{12})]\,d\boldsymbol{r}_2$ is used. The term $E_{ncl}[\rho]$ is a non-classical contribution containing all effects of self-interaction, exchange, and correlation. The major challenge of DFT is to find explicit expressions for $T[\rho]$ and $E_{ncl}[\rho]$ because the ground-state electron density $\rho(\boldsymbol{r})$ determines all properties.

Suppose that a trial density $\rho(\boldsymbol{r})$ is employed. This density defines its own operator H_ρ and therefore Ψ_ρ. This wave function can be taken as a trial wave function in the variational principle using the true operator H generated by V_{ext}. Hence, we have

$$\langle\Psi_\rho|H_\rho|\Psi_\rho\rangle = T[\rho] + E_{ee}[\rho] + \int \rho(\boldsymbol{r})V_{ext}\,d\boldsymbol{r} = E[\rho] > E_0[\rho] = \langle\Psi_0|H|\Psi_0\rangle$$

This is the second theorem which states that the functional $F_{HK}[\rho(\boldsymbol{r})]$ that delivers the ground-state energy of the system provides the lowest energy if (and only if) $\rho(\boldsymbol{r}) = \rho_0(\boldsymbol{r})$. This theorem thus provides a variational principle for $\rho(\boldsymbol{r})$. However, this variational principle applies only to the exact functional with approximate densities. For approximate functionals the variational principle is not valid.

Another major step was made by Kohn and Sham [32]. For a system of noninteracting electrons, a Slater determinant of spin orbitals is an exact solution of the Schrödinger equation. The Hamilton operator H_S is then given by

$$H_S = -{}^1\!/\!_2\sum_i \left[\nabla_i^2 + V_S(\boldsymbol{r}_i)\right] \equiv \sum_i f_{KS}(i) \tag{10.6}$$

where V_S is a one-electron potential to be identified later and the one-electron operator f_{KS} is referred to as the *Kohn–Sham operator*. The Slater determinant, indicated by Θ to distinguish it from the conventional Ψ, reads

$$\Theta = N^{-1/2}\,|\,\theta_1(\boldsymbol{r}_1)\theta_2(\boldsymbol{r}_2)\ldots\theta_N(\boldsymbol{r}_N)\,|$$

and the spin orbitals $\theta_j(\boldsymbol{r}_j)$, addressed as *Kohn–Sham orbitals* to distinguish them from their HF counterparts, are determined from

$$f_{\mathrm{KS}}(1)\,\theta_j(\boldsymbol{r}_1) = \varepsilon_j\theta_j(\boldsymbol{r}_1)$$

Since we are interested in a real system with interacting electrons the potential V_S is chosen in such a way that the electron density $\rho(\boldsymbol{r})$ equals the exact density $\rho_0(\boldsymbol{r})$ or

$$\rho(\boldsymbol{r}) = \sum_j \theta_j{}^2(\boldsymbol{r}) = \rho_0(\boldsymbol{r})$$

Next Kohn and Sham suggested to calculate the kinetic energy T_S of the noninteracting, reference system with the same density as the real, interacting system

$$T_\mathrm{S} = -½\sum_j \langle\theta_j|\nabla^2|\theta_j\rangle$$

It can be shown that $T_\mathrm{S} \leq T$, where T is the kinetic energy of the real system. The functional $F_{\mathrm{HK}}[\rho(\boldsymbol{r})]$ is separated according to

$$F_{\mathrm{HK}}[\rho] = T_\mathrm{S}[\rho] + E_\mathrm{C}[\rho] + E_{\mathrm{xc}}[\rho]$$

defining the *exchange-correlation energy* as

$$E_{\mathrm{xc}}[\rho] \equiv \{T[\rho] - T_\mathrm{S}[\rho]\} + \{E_{\mathrm{ee}}[\rho] - E_\mathrm{C}[\rho]\} = T_\mathrm{c}[\rho] + E_{\mathrm{ncl}}[\rho]$$

The residual part T_c not covered by T_S is thus added to the nonclassical electrostatic contributions. Therefore, E_{xc} is a functional that contains everything that is unknown, that is, the self-interaction, exchange, correlation, and the remainder of the kinetic energy. Since the total energy is a functional of $\rho(\boldsymbol{r})$, T_S is also a functional of $\rho(\boldsymbol{r})$.

The total energy of the real system is thus

$$E[\rho] = T_\mathrm{S}[\rho] + E_\mathrm{C}[\rho] + E_{\mathrm{xc}}[\rho] + E_{\mathrm{ext}}[\rho]$$

where the only term for which no explicit form is present is E_{xc}. Now the variational theorem is applied in order to minimize this energy expression under the usual constraint of $\langle\theta_i|\theta_j\rangle = \delta_{ij}$. It can be shown that this procedure results in

$$\{-½\nabla^2 + [V_\mathrm{C} + V_{\mathrm{xc}} + V_{\mathrm{ext}}]\}\,\theta_i(\boldsymbol{r}_1) = [-½\nabla^2 + V_\mathrm{S}]\theta_i(\boldsymbol{r}_1) = \varepsilon_i\theta_i(\boldsymbol{r}_1)$$

identifying the potential V_S introduced in Eq. (10.6). The term V_{xc} is due to the exchange-correlation energy E_{xc}, given by the functional derivative $V_{\mathrm{xc}} = \delta E_{\mathrm{xc}}/\delta\rho$.

For DFT to be successful, the functional E_{xc} must be approximated as best as possible. Unfortunately, in contrast to the HF approach where systematic improvement can be realized by extending the basis set or using CI, no such systematic improvement seems possible for the DFT functionals, although a possible option is provided [33]. We discuss briefly two generally accepted approaches.

In the *local density approximation* (LDA), homogeneous electron gas expressions are used. The E_C and E_{xc} contributions are well known. For the correlation contribution, an explicit expression is not available but accurate Padé approximant interpolations to highly accurate calculations are available. The LDA overestimates the correlation energies typically by 100% and therefore overestimates bonding. Properties such as bond angles, bond lengths, and vibrational frequencies tend to agree surprisingly well with experiments for normal bonds, that is, hydrogen and vdW bonds excepted.

To improve upon this approximation, it is assumed that not only $\rho(\mathbf{r})$ but also $\nabla\rho(\mathbf{r})$ is important. Because in this approximation the integrals of the Fermi and Coulomb hole do not yield the required values of −1 and 0 electrons anymore, these requirements are used as constraints, leading to the *generalized gradient approximation* (GGA). The accuracy of the GGA is considerably better than that for the LDA. Many other approximations exist [34, 35]. Conceptual interpretation [36] and application to new fields [37] are made, but challenges remain to be met [38].

For actual calculation of the Kohn–Sham orbitals, typically an expansion in basis functions is used, although a full numerical solution is feasible. For molecules similar basis sets as used in the HF method are employed. For solids plane waves are often employed. Finally, we note that many complexities have not been addressed, for example, degenerate systems, open shell, excited state, and finite-temperature calculations [31, 39, 40]. Example 10.2 illustrates one particular application.

Example 10.2 ***Spinel $MgAl_2O_4$ Phase Transition* Using *DFT*** AB_2O_4 compounds form important crystals with the spinel structure. In this structure, the oxygen atoms form an FCC lattice. The A atoms are situated at the tetrahedral and the B atoms at the octahedral interstitial sites of the FCC network. However, for certain spinel compounds the A atoms exchange partially with the B atoms to form $(A_{1-x}B_x)[A_{x/2}B_{1-x/2}]_2O_4$, where () indicate the tetrahedral and [] the octahedral sites. For natural $MgAl_2O_4$ the inversion parameter $x \cong 0$, while for synthetic $MgAl_2O_4$ x depends on heat treatment. As another example, $TiMg_2O_4$ is actually $(Mg)[Ti_{x/2}Mg_{x/2}]_2O_4$. A high-pressure (HP) form of $MgAl_2O_4$ exists which might be important in the mantle of the earth.

For MgO, Al_2O_3, $MgAl_2O_4$, and the HP-form of $MgAl_2O_4$, DFT calculations [41] within the LDA were done with full relaxation of the unit cell size and the atomic parameters. The results in Figure 10.13 show that $MgAl_2O_4$ is stable with respect to $MgO + Al_2O_3$ and has a larger unit cell volume than the sum of $MgO + Al_2O_3$. The HP form is unstable at normal pressure and has a smaller unit cell volume with respect to the forming compounds. The calculations indicate that with increasing pressure $MgAl_2O_4$ first decomposes at pressure P_1 into $MgO + Al_2O_3$, while upon further increase at pressure P_2 these compounds

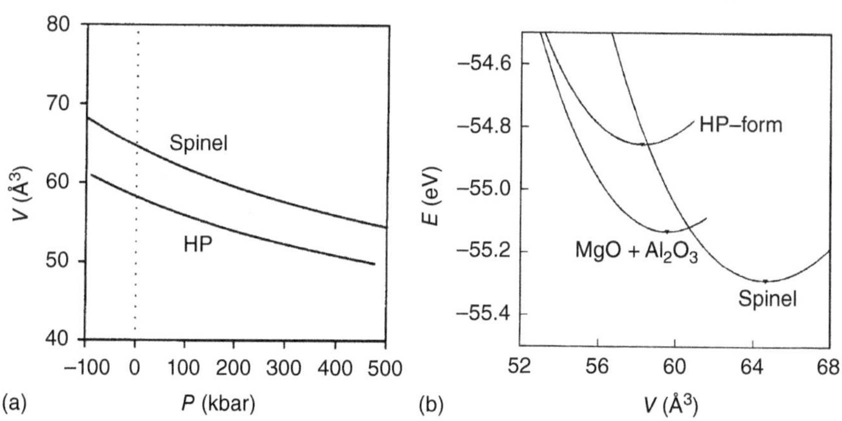

Figure 10.13 (a) The volume–pressure relation for $MgAl_2O_4$ and the HP form of $MgAl_2O_4$. (b) The energy–volume relation for $MgO + Al_2O_3$, $MgAl_2O_4$, and the HP form of $MgAl_2O_4$. Source: Adapted from Fang and de With [41].

combine again to the HP form. Taking into account the zero-point vibrational energy and the inversion contribution to the entropy, it is estimated that at 1800 K, $P_1 \cong 4$ GPa and $P_2 \cong 33$ GPa. The ranges for the experimental data are $8 < P_1 < 15$ GPa and $25 < P_2 < 40$ GPa. The theoretical estimates are rather sensitive to the precise values of the various contributions involved and the difference with the experimental result is probably mainly due to the relatively simple estimate for the inversion contribution. However, the unit cell volumes for both forms are in good agreement with experimental data, as are the estimates for the bulk modulus at 0 K.

In summary, the NFE and TB approaches represent extreme views on bonding in solids, but both result in a band structure with similar features. The success of the NFE method is primarily due the orthogonalization of the valence electron waves on the core states leading to a more or less flat pseudo-potential. While the NFE and TB schemes can be used for semiquantitative discussions, a full self-consistent calculation must be used for reliable predictions. In the DFT approach all properties are determined by the energy functional of the electron density. Because the exact electron density is not known, an approximate functional must be used, which using the variation principle, yields an approximate minimum energy. While for orbital methods a systematic improved accuracy can be realized, a systematic way for improving the functionals used in DFT seems to be absent. Further introductions are [15, 16, 42], while [20, 31, 39, 40, 43] provide detailed treatments.

10.6 Bonding: The Empirical Approach

For bonding and the resulting structures, historically also other approaches were developed. In view of their usefulness, they remain to be widely used. We first deal with some concepts such as atoms, ions, potentials, (partial) charges, and electronegativity. Thereafter, we discuss covalent, molecular, polar covalent, and metallic compounds. For a systematic treatment of effective pair potentials, see [44].

10.6.1 Atoms, Ions, and Electronegativity

For atoms, only for hydrogen an exact solution of the Schrödinger equation $(\hbar^2/2m)\nabla^2\Psi + e^2(4\pi\varepsilon_0 r)^{-1}\Psi = E\Psi$ can be obtained. The detailed procedure is elaborate (see, e.g. [45]). It appears that the wavefunction Ψ is separable due to the spherical symmetry in radial parts $R_{nl}(r)$ and *spherical harmonics* $Y_{lm}(\theta,\phi)$, describing the angular part. The solutions are labeled by four quantum numbers:

- The *principal quantum number* $n = 1, 2, 3, \ldots, \infty$. The energy of a level (n, l, m) is $E_n = -me^4[(4\pi\varepsilon)^2 2\hbar^2]^{-1}n^{-2} = -\frac{1}{2}n^{-2}E_\text{h} \cong 13.61n^{-2}$ eV, independent of l and m.
- The *azimuthal quantum number* $l = 0, 1, 2, \ldots, n-1$. Levels with $l = 0, 1, 2, \ldots$ are labeled by s, p, d, f, … . The overall angular momentum is $|\boldsymbol{l}| = [l(l+1)]^{1/2}\hbar$.
- The *magnetic quantum number* $m_l = -l, -l+1, \ldots, l-1, l$, representing the component of $\boldsymbol{l}$ in a specified direction, say z, so that $\boldsymbol{l}_z = m\hbar$ with m indicating all $2l+1$ degenerate levels m_l. In the presence of a magnetic field along the z direction, degeneracy is lifted and $\boldsymbol{l}$ precesses about its axis with its projection on the z-axis equal to one of the $2l+1$ possible values of m.
- Finally, there is the *spin quantum number* $m_s = \frac{1}{2}$ or $m_s = -\frac{1}{2}$.

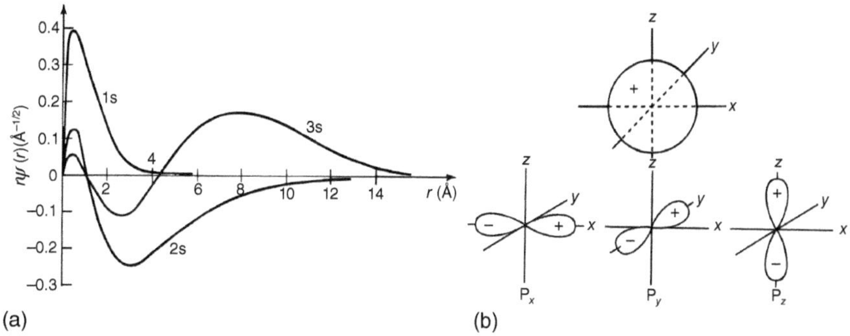

Figure 10.14 (a) The nodes in 1s, 2s, and 3s wavefunctions. (b) The shape of s and p orbitals. Source: Sutton [16]/Oxford University Press.

The *wavefunctions* $\Psi_{nlm} = R_{nl}(r)Y_{lm}(\theta,\phi)$ are orthonormal or $\int \Psi_{nlm}{}^{*}\Psi_{n',l',m'}\ \mathrm{d}\mathbf{r} = \delta_{n,n'}\delta_{l,l'}\delta_{m,m'}$. Since the $Y_{lm}(\theta,\phi)$ themselves are orthonormal, orthonormality for Ψ_{nlm} is guaranteed if $l \neq l'$ and/or $m \neq m'$. However, if $l = l'$ and $m = m'$ the radial components have to satisfy $\int R_{nl}R_{n'l'}r^2\mathrm{d}r = 0$. This leads to nodes in the radial functions, as illustrated in Figure 10.14a. To obtain real functions one takes linear combinations of the spherical harmonics. If done so, the first few wavefunctions are

$$\Psi_{1s} = \pi^{-1/2}a_0{}^{-3/2}\exp(-r/a_0), \quad \Psi_{2s} = (2\pi)^{-1/2}a_0{}^{-3/2}(2 - r/a_0)\exp(-r/2a_0)$$

$$\Psi_{2p_x} = F\sin\theta\cos\phi, \quad \Psi_{2p_y} = F\sin\theta\sin\phi \text{ and } \Psi_{2p_z} = F\cos\theta \quad \text{with}$$

$$F = 1/4(2\pi)^{-1/2}a_0{}^{-3/2}(r/a_0)\exp(-r/2a_0)$$

They are addressed as *orbitals* and their shape is shown in Figure 10.14b.

For *many-electron atoms*, the solution of the Schrödinger equation still separates in radial and angular parts and thus orbital labeling remains the same, apart from that one uses upper case symbols for $\boldsymbol{l}$ and $\boldsymbol{s}$. The electrostatic interaction is different for each orbital as the inner electrons shield partially the full nuclear charge for the outer electrons. The energy becomes a function of n and l, if magnetic interactions are present also of m_l, and if a magnetic field is applied also of m_s. It also leads to different radial parts, which can be determined by, for example, the Hartree–Fock method (Section 3.3.3). Often Slater's rules [46] are used to estimate the effective nuclear charge Z_{eff}. A particular electron configuration (e.g. $1s^22s^22p^2$) can be described for $Z \lesssim 40$ by $\boldsymbol{LS}$ coupling, where $\boldsymbol{L} = \Sigma_j \boldsymbol{m}_L$ is the total angular momentum and $\boldsymbol{S} = \Sigma_j \boldsymbol{m}_S$ is the total spin, where $\boldsymbol{m}_L$ and $\boldsymbol{m}_S$ are the angular momentum and spin of the individual electrons j. A configuration usually contains several *terms*, indicated by the *term symbol* ${}^{2S+1}L_J$, where $2S+1$ indicates the *multiplicity*. The *levels* $L = 0$, 1, 2, 3 … are conventionally labeled by S, P, D, F, … and $\boldsymbol{J} = \boldsymbol{L} + \boldsymbol{S}$. The latter is primarily relevant in the presence of magnetic interactions, and when neglected, often omitted. The filling of the various levels with electrons is controlled by the *Aufbau* principle, stating that an electron occupies orbitals in order from the lowest energy to the highest, considering Pauli's principle and Hund's rules. The Pauli principle is already alluded to, while *Hund's rules* say that (i) terms with the largest S are the lowest, (ii) for given S, terms with the largest

L values are the lowest, and (iii) for given L and S, if the subshell is less (more) than half full, the state with smallest (highest) value of J is the lowest. The procedure has been described many times [45, 47, 48], to which we refer for details, as are its basis and exceptions [49]. As example, a carbon atom with as *ground-state* configuration $1s^22s^22p^2$ has terms $^3P(L = 1, S = 1, m_S = \pm1, 0, m_L = \pm1, 0, E_{3P} = 29.6\ cm^{-1}$ with $^3P_0 = 0)$, $^1D\ (L = 2, S = 0, m_S = 0, m_L = \pm2, \pm1, 0, E_{1D} = 10.2 \times 10^3\ cm^{-1})$, and $^1S\ (L = 0, S = 0, m_S = 0, m_L = 0, E_{1S} = 21.6 \times 10^3\ cm^{-1})$. Its *valence state* $1s^22s2p^3$, in which sp, sp^2, or sp^3 hybrid orbitals can be formed, contains the terms $^5S\ (33.7 \times 10^3\ cm^{-1})$, $^3D\ (64.0 \times 10^3\ cm^{-1})$, and $^3P\ (75.3 \times 10^3\ cm^{-1})$. Hybridization thus costs energy, which is compensated by a larger gain in energy upon bond formation. For a clear discussion on hybridization and valence state, see [50].

Atoms have a certain size, of which the value depends on how it is determined. The *vdW radius* is the radius when two atoms are in proximity in a nonbonding configuration. By covalently bonding two atoms of the same kind, say A, an apolar binary compound is formed of which half of the bond length l_{AA} is taken as the *covalent radius* r_A (generally r_{cov}). If the molecule AA does not exist, data from various solids and/or small molecules are used. By removing or adding an electron from a neutral atom it becomes an ion. For the former process the *ionization potential* (IP) is required, while the latter requires the *electron affinity* (EA). For a negative ion the radius r_A^- is larger than r_A, while for a positive ion the radius r_A^+ is smaller than r_A. Two ions A^+ and B^- forming an ionic bond lead to the polar compound AB.

Pauling [51] stated that there is a continuous transition from apolar to polar bonding, and therefore bonds are partially ionic and partially covalent. To describe this, he defined the *electronegativity* (EN) X representing "the attraction of a neutral atom in a stable molecule for electrons." The value for X was estimated as the difference between the average bond energy of two homopolar molecules with E_{AA} and E_{BB} and that of the corresponding heteropolar compound E_{AB}. He defined $E_{AB} = ½(E_{AA} + E_{BB}) + k(X_A - X_B)^2$, based on carbon $X_C = 2.50$ and fluorine $X_F = 4.00$ with k a unit-dependent constant. So, the *ionic energy* $E_{ion} = k(X_A - X_B)^2$, dependent on the EN difference, was added to the *covalent energy* $E_{cov} = ½(E_{AA} + E_{BB})$. Using experimental data, he so obtained the EN for various atoms. At present, the actual values mostly used are those refined by Allred [52]. Actually, Pauling preferred the definition $E_{cov} = ½(E_{AA}E_{BB})^{1/2}$, but still used the arithmetic mean as the geometric mean was (is) often more difficult to apply. Mulliken [53] introduced another scale, by arguing that the IP represents the average electron attraction to the positive ion and EA represents the average electron attraction to the negative ion, and that their average $X = (IP + EA)/2$ should be a measure for the electron attraction for a neutral atom in its *valence state* and thus for the EN. For data, see Section F.15.

Allred and Rochow [54] defined EN as the force $X = 3.59 \times 10^{-3} Z_{eff}/r_{cov}{}^2 + 0.744$ experienced by an outer electron with the nucleus. Here Z_{eff} is the effective nuclear charge experienced by the valence electrons, often determined by Slater's rules [46], while r_{cov} is given in Å. The constants were added to mimic Pauling's scale as best as possible. Several other scales have been proposed and are compared in [55].

Sanderson [56] argued that X is related to the compactness of an atom and proposed $X = D/D_{inert}$, where D is the electron density at r_{cov} and D_{inert} the electron density at r_{cov} as interpolated linearly from rare gas atom data. Initially the numerical values so obtained were used, but later he rescaled the values so that $X_F = 4.00$ in order to be consistent with Pauling's scale, meanwhile revising slightly some values [57, 58] (Section F.15). Further he argued that, because an atom in a molecule with a high X-value attracts electrons, it becomes negatively charged, leading to a lower Z_{eff} and larger repulsion, and consequently to a lower X-value for that atom in the molecule. Similarly, an atom with a low X-value increases its X-value, and equilibrium is reached when the X-values are equal. This is the *principle of electronegativity equalization*, postulated to occur when $X = (X_A X_B)^{1/2}$. In equilibrium therefore, atoms in a molecule have a partial charge δ. As two electrons for a bond cannot be at the same time equally distributed over two atoms and partially accumulating at one of them, he argued that the ionic bond contribution E_{ion} and the covalent contribution E_{cov} to the bond energy E are complementary and given by $E = t_{\text{ion}}E_{\text{ion}} + t_{\text{cov}}E_{\text{cov}}$, where $t_{\text{ion}} + t_{\text{cov}} = 1$. Thus, contrary to Pauling's view where E_{ion} is added to E_{cov}, in Sanderson's view E_{ion} replaces part of E_{cov}. It is often stated (by Sanderson and others, quoting Parr [31]) that using DFT it can be shown that $X \cong -\mu$ with μ the chemical potential, so that the principle of EN equalization is analogous to the equalization of chemical potential in macroscopic thermodynamics. Hinze [59, 60] disproves that $X \cong -\mu$, but still accepts the principle of EN equalization.

A convenient representation of the character of a specific bond is given by its position in a bond triangle [61] in which the various corners indicate pure ionic, pure covalent, and pure metallic bonding (Figure 10.15). To render this representation quantitative [63], the *covalency* $\mathscr{C} = (X_A + X_B)/2$ and *ionicity* $\mathscr{I} = (X_B - X_A)$ are used, where $\mathscr{C}$ is a measure of the localization of an A—B bond and $\mathscr{I}$ a measure for its asymmetry. By delineating areas of compounds with similar properties, one can identify domains with a predominant metallic, ionic, or covalent character.

After this summary of concepts, we now turn to empirical bonding theories.

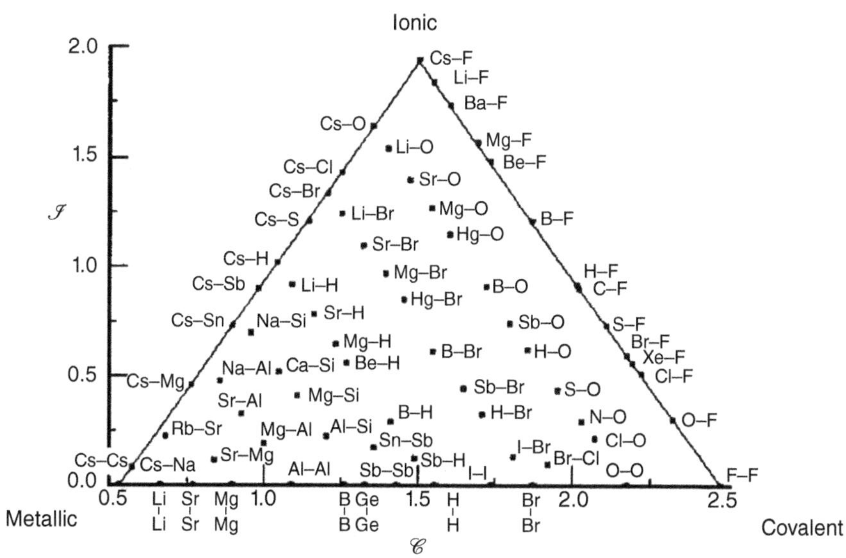

Figure 10.15 A bond diagram [61] showing the character of an A–B bond by its position in the triangle, as made quantitative by the covalency $\mathscr{C} = (X_A + X_B)/2$ and ionicity $\mathscr{I} = (X_B - X_A)$ using the scale of Martynov and Batsanov [62]. Other scales show pretty much the same pattern. Source: Adapted from [61, 62].

10.6.2 Covalent and Molecular Crystals

For *covalent crystals*, one approach to estimate the *bonding* or *cohesive energy* E_{coh} is to use the concept of *bond energy*, normally taken as the energy E_{AB} to break the bond in a binary compound AB. For example, $E_{\mathrm{CC}} \cong 83.2\,\mathrm{kcal\,mol^{-1}}$ [56], so that for diamond one estimates $E_{\mathrm{coh}} \cong \frac{1}{2}zE_{\mathrm{CC}}$ with $z = 4$ the coordination number and the ½ is used to avoid double counting. This results in $E_{\mathrm{coh}} = 166.4\,\mathrm{kcal\,mol^{-1}}$, while the experimental value is $169.4\,\mathrm{kcal\,mol^{-1}}$ [64]. For bond energy data, see Section F.16.

For molecular compounds the bond energy approach is only a very rough approximation. For example, the dissociation energy $E_{\mathrm{dis}}(\mathrm{H_2O}) = 222\,\mathrm{kcal\,mol^{-1}}$, so that $E_{\mathrm{OH}} = 111\,\mathrm{kcal\,mol^{-1}}$. The energy for the reaction $\mathrm{H_2O} \rightarrow \mathrm{OH} + \mathrm{H}$ is $120\,\mathrm{kcal\,mol^{-1}}$, so that only $102\,\mathrm{kcal\,mol^{-1}}$ is left to break the second OH bond. In fact, when one bond is broken, the remaining fragment reorganizes itself with a *reorganization energy* E_{reorg}, in this case $9\,\mathrm{kcal\,mol^{-1}}$. One of the few cases, if not the only one, where the effect can be tested without invoking charged species is CO_2. From CO_2 we have $E_{\mathrm{CO}} = 192\,\mathrm{kcal\,mol^{-1}}$, while from the reaction $\mathrm{CO_2} \rightarrow \mathrm{CO} + \mathrm{O}$ we have $127\,\mathrm{kcal\,mol^{-1}}$. The difference $E_{\mathrm{reorg}} = 65\,\mathrm{kcal\,mol^{-1}}$ should thus be added to E_{CO}, so that $E_{\mathrm{dis}}(\mathrm{CO}) = 257\,\mathrm{kcal\,mol^{-1}}$, which is indeed the experimental value.

For *molecular crystals* a relatively successful approach is to use atomic potentials, usually neglecting the directional dependence. An often-used expression for atomic potentials is the *Mie potential* (Section 4.3) given by $\phi_{\mathrm{M}} = \phi_{\mathrm{rep}} - \phi_{\mathrm{att}} = b/r^n - c/r^m$ where b, c, n, and m are parameters and $n > m$. A power law with $m = 6$ for the attractive part ϕ_{att} of the potential can be rationalized based on vdW interactions, while for the repulsive part ϕ_{rep} an often-made choice is $n = 12$, mainly for mathematical convenience, leading to the *Lennard-Jones* (LJ) *potential* (for data, see Sections F.1 and F.2). The repulsion is due to orbital overlap and on quantum-mechanical grounds expected to be better described by an exponential, which leads to the *exp-6 potential* $\phi = b\exp(-r/\rho) - c/r^m$.

For crystal lattices, the interactions must be summed over all atom pairs, and for power law interactions, we have

$$\Phi_{\mathrm{pow}} = \tfrac{1}{2}\sum\nolimits_{n,\alpha} \phi(\boldsymbol{r}_{n,\alpha}) = \tfrac{1}{2}(b/r^n)\sum\nolimits_{n,\alpha} (r/r_{n,\alpha})^n \equiv \tfrac{1}{2}bS_n/r^n = \tfrac{1}{2}B_n/r^n$$

where r is here the nearest-neighbor distance and $\boldsymbol{r}_{n,\alpha}$ is the distance to atom α in cell $\boldsymbol{n}$. The *lattice sums* S_n for the SC, BCC, FCC, and HCP lattices are tabulated [65, 66] for $n = 4$ to $n = 30$ (selected are values given in Table 10.2). Using these lattice sums, we have

$$\Phi_{\mathrm{M}} = \tfrac{1}{2}\sum\nolimits_{n,\alpha} \phi_{\mathrm{M}}(\boldsymbol{r}_{n,\alpha}) = \tfrac{1}{2}(bS_n r^{-n} - cS_m r^{-m}) = \tfrac{1}{2}(B_n r^{-n} - C_m r^{-m}) \tag{10.7}$$

The simplest applications are rare gas crystals with $m = 6$. Originally the parameters b, c, and n were fitted so as to reproduce the second virial coefficients. It appeared that the values for b and c can be reasonably determined for a given value of n, but that it is impossible to choose n so that both the equilibrium distance r_0 and lattice energy U_0 agree with the experimental values. Moreover, there is considerable latitude for the choice of n, indicating some self-compensating effects [9]. Alternatively, the parameters can be determined by fitting on macroscopic quantities such as equilibrium distance, lattice energy, and compressibility. In Section 4.4.3, we already indicated that such a simple potential even for rare gases is inadequate if one potential is to be used for many properties. Nevertheless, the LJ potential, extended, for example, by charges and polarizabilities, is frequently used for molecular and

Table 10.2 Lattice sums S_n for various lattices.

n	SC	BCC	FCC	HCP
4	16.532	22.639	25.338	—
6	8.402	12.253	14.454	14.555
8	6.946	10.355	12.802	12.802
10	6.426	9.564	12.311	12.312
12	6.202	9.114	12.132	12.132
14	6.098	8.817	12.059	12.059

polymeric crystals with a reasonable degree of success. General reviews are in [67, 68] while [69, 70] provide a broader perspective.

10.6.3 Ionic Crystals: The Classical Approach

In the classical approach to describe ionic crystals, one assumes that ions have their full charge, are essentially spherical with a constant radius, and contain a core in which most of the electron density is concentrated (Figure 10.2a). From X-ray diffraction experiments on solids, it appeared that there is also an outer shell containing a rather flat and small electron density over some range, rendering the assignment of *ionic radii* from these data difficult. Still, reference values for radii are needed and Pauling radii [51] are one such choice. Nowadays, the Shannon–Prewitt radii [71] are preferred (Section F.14). Several other additive sets are in use, typically self-consistent within about 5–10%.

The basic ionic bond is based on a balance between the electrostatic forces, described by the *Coulomb energy* Φ_C, and the short-range repulsion forces described by Φ_{rep}, so that again $\Phi = \Phi_C + \Phi_{rep}$. The Coulomb potential energy ϕ_C between two charges $q_A e$ and $q_B e$ separated by a distance r is given by [9, 72]

$$\phi_C = q_A q_B e^2/(4\pi\varepsilon_0 r)^{-1} \equiv q_A q_B a/r, \quad a \equiv e^2\mathscr{C} \quad \text{and} \quad \mathscr{C} \equiv (4\pi\varepsilon_0)^{-1}$$

where e is the unit charge and ε_0 is the permittivity of the vacuum. Writing the sum of the contributions for all ions pairs in terms of the nearest-neighbor distance r, it appears to be attractive, and we have

$$\Phi_C = -\tfrac{1}{2}\sum\nolimits'_{n,\alpha} \phi_C(\boldsymbol{r}_{n,\alpha}) = -\tfrac{1}{2}\frac{a}{r}\sum\nolimits'_{n,\alpha} q_0 q_{n,\alpha}(r/r_{n,\alpha}) \equiv -q_A q_B M\frac{a}{r}$$

where the prime over the summation sign excludes the atom at the origin. The lattice sum M, usually called the *Madelung constant*, has a value characteristic for the crystal structure. In view of the long-range nature of electrostatic interactions, special techniques are necessary to calculate the value of M accurately (Table 10.3). For A_xB_y compounds, such as CaF_2, a reduced Madelung constant $M' = 2M/(x+y)$ can be defined [73]. Clearly, $M' = M$ for $x = y = 1$, but, contrary to M, M' is nearly constant (Table 10.3), indicating that the electrostatic energies per atom differ only slightly.

The repulsion is for ionic compounds also described by either

$$\phi_{rep} = b\,\exp(-r/\rho) \quad \text{(Born–Mayer)} \quad \text{or} \quad \phi_{rep} = b/r^n \quad \text{(Born–Landé)}$$

Table 10.3 Madelung constants M for nearest-neighbor distance and unit charge for some crystals.

Structure	M	M'	Structure	M	M'
Cesium chloride (CsCl)	1.7627	1.7627	Fluorite (CaF_2)	2.5194	1.6796
Sodium chloride (NaCl)	1.7476	1.7476	Rutile (TiO_2)	2.4080	1.6053
Sphalerite (ZnS)	1.6381	1.6381	Anatase (TiO_2)	2.400	1.600
Wurtzite (ZnS)	1.6413	1.6413	Corundum (Al_2O_3)	4.1719	1.6688

Table 10.4 Born exponents n for various electron configurations.

Type	He	Ne	Ar	Kr	Xe
Example	Li^+	Na^+, Mg^{2+}, F^-, O^{2-}	K^+, Ca^{2+}, Cl^-, S^{2-}	Rb^+, Ag^+, Br^-	Cs^+, I^-
n	5	7	9	10	12

where (b,ρ) or (b,n) are parameters. The value of n varies for different electron configurations (Table 10.4, often used values with the average taken for a compound), while ρ is approximately constant at about 0.33. Most of the lattice energy is provided by attraction, and because the range of the repulsion is relatively short, often only nearest-neighbor repulsion is considered. Example 10.1 shows how to determine (b,ρ) from the bulk modulus K and equilibrium distance r_0. For example, for LiCl $r_0 = 2.572$ Å and $K = 29.3$ GPa and we obtain $\rho = 0.332$ Å and $b = 0.782 \times 10^{-16}$ J, resulting in a lattice energy $u_{\text{calc}} = 822\ \text{kJ mol}^{-1}$, to be compared with $u_{\text{exp}} = 844\ \text{kJ mol}^{-1}$.

Example 10.3 ***The Born–Mayer Model*** In binary salts each ion is coordinated with z ions of the opposite charge. Using $b\exp(-r/\rho)$ and neglecting all but the nearest-neighbor contributions for repulsion, the energy per unit cell $u(r)$ is

$$\phi(r) = -Ar^{-1} + Be^{-r/\rho}, \qquad A = e^2\mathscr{C}M \quad \text{and} \quad B = zb$$

where the symbols have the meaning defined previously. Since $\text{d}\ln V = 3\,\text{d}\ln r$ or $(1/V)\text{d}V = (3/r)\text{d}r$, one finds for the pressure P and the bulk modulus K

$$P = -\text{d}\phi/\text{d}V = -[-Ar^{-1} + B(r/\rho)e^{-r/\rho}]/3V \quad \text{and} \tag{10.8}$$

$$K = -V(\text{d}P/\text{d}V) = [-A/r^{-1} - B(r/\rho)e^{-r/\rho} + B(r/\rho)^2e^{-r/\rho}]/9V + P$$

For static equilibrium at $P = 0$ we have from Eq. (10.8)

$$A/r_0 = B(r_0/\rho)e^{-r_0/\rho} \tag{10.9}$$

Taking $r = r_0$, $P = 0$ and eliminating B using Eq. (10.9), we obtain

$$\phi(r_0) = -(A/r_0)(1 - \rho/r_0) \quad \text{and} \quad \rho/r_0 = (A/r_0)/(9V_0K + 2A/r_0)$$

Using $\phi_{\text{rep}} = b/r^n$ (Problem 10.3) yields $\phi(r_0) = -(A/r_0)(1 - 1/n)$. Both the factors $(1 - \rho/r_0)$ and $(1 - 1/n)$ are called *repulsion factor*. This basic ionic model provides reasonable estimates for simple lattices but neglects various terms. If we extend the potential $\Phi = B\exp(-r/\rho) - A/r$ with a vdW attraction term $-C/r^6$, a similar exercise as in Example 10.3 leads to an extra term to $\phi(r_0)$ given by

$$-\left(C/r_0{}^6\right)(1-6\rho/r_0) \quad \text{and} \quad \rho/r_0 = \left(A/r_0 + 6C/r_0{}^6\right) / \left(9V_0K + 2A/r_0 + 42C/r_0{}^6\right)$$

Because typically $\rho/r_0 \cong 0.13$, the contribution to the lattice energy is only about $0.2C/r_0{}^6$, much smaller than the actual vdW energy $C/r_0{}^6$ and due to a compensation via the $B\exp(-r/\rho)$ term. A similar effect occurs if we include next-nearest neighbors in the repulsion and zero-point energy. This illustrates the self-compensating effect, which for ionic crystals is even more clearly present than for molecular crystals. The analysis also shows that ionic radii can be constant only if the contribution C/r^m is negligible (which it is not) and if the charges are constant (which they are not, see Section 10.6.4). Using (b,n) the value of n varies considerably within a family (or group) of the periodic system, while the value of ρ is rather constant for atoms in a family but changes slightly from family to family [74]. The (b,ρ) set is capable to describe ionic crystals and molecules in the gas phase reasonably well so that it is the preferred set. Lattice energies can be predicted within a few percent with the basic model, but Baughan claims that the extended model gives lattice energies to about ½% with constant family values of ρ [74]. For further reviews, including polarizability, vdW interactions, and thermal effects, see [75–78].

In summary, although a successful model, partially due to self-compensating effects, assuming integer charges and non-variable radii appear to be untenable.

A successful offspring is the *bond valence model* [79]. While for ionic compounds one uses electrostatic bond strengths with integer valence numbers so that the anion and cation bond strengths are equal to the formal charge, in the bond valence approach non-integer valences are introduced. This leads to the valence sum rule $V = \Sigma_k v_k$ relating the valence V of an atom, equal to the number of electrons in its valence shell used for bonding, to the bond valence v_k between this atom and atom k. It appears that for an isoelectronic series this short-range interaction v can be described by $v = \exp[(r_0 - r)/b)]$, where r is the bond length and r_0 and b are parameters. For many bond types but not all, b is close to 0.37 Å. For unstrained structures, the valences of the bonds determined are also found to obey $\Sigma_{\text{loop}} v_j = 0$, where the summation indicates any set of bonds that form a closed loop in the bond network. For the ionic interaction, formal charges are used, which as Brown shows, is not as serious a restriction as it sounds because any differences in charge and covalent effects are included in the fitted values for b and r_0. All bonds are between an electron donor and an electron acceptor, and this effectively restricts the model to inorganic compounds. The equations indicated correspond to the statement that each atom shares its valence as equally as possible between the bonds that it forms. Given the parameters b and r_0, the model is mainly applied for complex solids to predict their structure or to rationalize their X-ray-determined structure. For a concise introduction, see [80].

Problem 10.3 *Power-law Repulsion* Using $\phi_{\text{rep}} = b/r^n$, show that the corresponding equations to Example 10.3 are:

$$A/r_0 = nB/r_0{}^n, \quad 1/n = (A/r_0)/(A/r_0 + 9V_0K) \quad \text{and} \quad \phi(r_0) = -(A/r_0)(1-1/n)$$

10.6.4 Ionic Crystals: Electronegativity Approaches

In the Born–Mayer approach formal charges were used. However, EN equalization requires that atoms in a molecule or lattice have a partial charge. First, we discuss the outright empirical but original *polar covalence model* of Sanderson, based on partial charges in combination with EN equalization, covalent bonding, and ionic bonding, and applied to molecules and solids. Second, we deal with the more transparent approach by Hinze, which we label the *orbital electronegativity model.* It uses different EN values for different valence states but is applied to molecules only. Both models provide partial charges, bond length, and bond energy.

The starting point of the *polar covalence model* is that experimentally it appears that bond length decreases with increasing polarity. With increasing partial positive charge, electrons feel a stronger attraction and decrease in radius, while the radius increases with increasing partial negative charge. Sanderson argued that the simplest way to represent this changing bond length $l_{AB} = r_A{}^\delta + r_B{}^\delta$ is to use the ionic radius $r^\delta = r - B\delta$, with δ the partial charge and B an atom-specific constant [56, 58]. Now to estimate δ and B, some reference compound is needed for which NaF was chosen by Sanderson. It was assumed[4] that $\delta_{Na} = 0.80$ and that the EN X changes linearly with δ. To obtain ionic radii $r_A{}^\delta$ and $r_B{}^\delta$, the experimental bond length l_{NaF} was divided in parts each inversely proportional to Z_{eff}. The latter is given by $Z_{eff} = Z - S$, using Pauling's screening constant $S = 4.52$ for both the Na and F atom [51], so that $Z_{eff} = 6.48$ for Na and $Z_{eff} = 4.48$ for F. By the way, Slater's rules yield $S = 4.15$. Taking into account that $\delta_{Na} = 0.80$ results in $Z_{eff} = 6.41$ and $Z_{eff} = 4.55$, respectively. This leads for NaF to $l_{NaF} = r_{Na} + r_F = x/Z_{Na} + x/Z_F$, or $x = Z_{Na}Z_F l_{NaF}/(Z_{Na} + Z_F)$, so that with $l_{NaF} = 1.93$ Å, $r_{Na}{}^{+0.80} = 0.80$ Å and $r_F{}^{-0.80} = 1.13$ Å. For specific data, see Section F.15–F.17.

Sanderson defined the *partial charge* δ as the ratio $\delta = X/X_1$, where X is the EN value for a particular atom in a molecule and X_1 the value this atom would have for a charge of unity (Section F.17). Taking $X_{NaF} = (X_{Na}X_F)^{1/2} = (0.56 \times 4.00)^{1/2} = 1.50$, the shift in X for Na in NaF is $\Delta X_{Na} = X_{NaF} - X_{Na} = 1.50 - 0.56 = 0.94$. Similarly, $\Delta X_F = 1.50 - 4.00 = -2.50$. As $\delta_{ion} = 0.80$ is assumed, the shifts for unit charge become $\Delta X_{Na,1} = 0.94/0.80 = 1.18$ and $\Delta X_{F,1} = 2.50/0.80 = 3.14$. For Na $r_{cov} = 1.54$ Å, while the ionic radius $r_{ion} = 0.81$ Å, so that $B\delta = r_{cov} - r_{ion} = 0.74$ Å or $B_{Na} = 0.74/0.80 = 0.92$ Å. Similarly, for F $r_{cov} = 0.68$ Å and $r_{ion} = 1.13$, so that $B_F = -0.56$ Å. Further values are derived from binary compounds containing either Na or F using experimental bond lengths, and by combination for other elements. Writing ΔX_{Na} as $(X_F{}^{1/2} - X_{Na}{}^{1/2})X_{Na}{}^{1/2} = (4.00^{1/2} - 0.56^{1/2})X_{Na}{}^{1/2} = 1.25X_{Na}{}^{1/2}$, we have $\Delta X_{Na,1} = \Delta X_{Na}/0.80 = 1.57X_{Na}{}^{1/2}$. This appears to apply to any atom, so that generally $\Delta X_1 = 1.57X^{1/2}$.

---

4 In the first paper $\delta_{Na} = 0.90$ was taken, quickly changed to 0.75 because $D_F{}^-$ should be smaller than D_{Ne} as $Z_{eff,F} < Z_{eff,Ne}$ and thus $X_F{}^- < 1.00$ [56, 81]. Thereafter, Sanderson rescaled the X-values taking as reference $X_F = 4.00$, "to invite more willing acceptance" [57, 58], meanwhile making some changes, in particular to $\delta_{Na} = 0.80$ (actually 0.797) for NaF. These data are used here. In his last review [82], $\delta_{Na} = 0.69$ for NaF and no longer reference was made to ionic radii and NaF as reference compound. Clarity in detail was, unfortunately, not Sanderson's forte, for example, the expression $\Delta X_{Na,1} = 1.57X_{Na}{}^{1/2}$ (see later) was originally introduced as "has been found to be expressible" [81] and later as "found to be equivalent to" [58] (1.57 was 2.08 before rescaling). Further, inconsistencies and numerical errors occur with some regularity (e.g. Slater's $S_{Na} = S_F = 4.15$ is given as 4.50 and B_F calculates as 0.562, while given as 0.520) and referencing is meagre. All this together, I guess, contributed to the limited acceptance of his approach in spite of its originality and practical success, clearly to Sanderson's regret [58]. Personally, I think the original presentation [56] is the clearest.

We now turn to E_{cov} and E_{ion}. For heteropolar compounds, because the experimental bond length r_{obs} differs from r_{cov}, Sanderson added a correction factor (r_{cov}/r_{obs}) to $E_{cov} = (X_A X_B)^{1/2}$, which thus became $E_{cov} = (r_{cov}/r_{obs})(X_A X_B)^{1/2}$. For the polar contribution the electrostatic expression $E_{ion} = c_{ion}/r_{obs}$ is used, where $c_{ion} = 332$ kcal Å mol^{-1}. The bond energy is then $E = t_{ion}E_{ion} + t_{cov}E_{cov}$ with $t_{ion} = (\delta_+ - \delta_-)/2$ and $t_{cov} = 1 - t_{ion}$. Alternatively, t_{ion} is calculated without relying on partial charges for binary compounds from $t_{ion} = (E_{exp} - E_{cov})/(E_{ion} - E_{cov})$. The values so obtained agree for 191 compounds within an average of 0.007 with those calculated from partial charges. Molecules containing N, O, F, P, S, and Cl atoms need special attention (Section F.16).

Before we illustrate the approach, we must make some modifications for nonmolecular solids. The ionic component becomes essentially the Born–Mayer expression $E_{ion} = c_{ion} f_{cor} M k_{rep} z^2/r_{obs}$ kcal mol^{-1}, with M the Madelung constant and $k_{rep} = 1 - 1/n$ the repulsion factor. The correction factor $f_{cor} = 1$ for univalent anions, but to obtain good agreement with the experimental data, one must use $f_{cor} = 0.63$ for oxides, $f_{cor} = 0.57$ for sulfides, and $f_{cor} = 0.54$ for selenides. The cause may partly be that the values of the repulsion factor k_{rep} used are obtained from the Born–Mayer model. The covalent component is changed by a factor n, representing the *effective number of bonds*. For example, in KCl there are eight valence electrons to be distributed, although the coordination number for the rock salt structure is 6. Hence, $n_{cov} = 4$. For most halides and oxides, $n_{cov} = 4$ applies but to obtain good agreement in some cases, it is necessary, in particular for small atoms, to use $n_{cov} = 3$ and occasionally $n_{cov} = 6$. For hydrides $E_{HH} = 104.2$ kcal mol^{-1} must be multiplied by the factor $1.00 - \delta_H$ when δ_H is positive. Note that, because r_{obs} for solids is different from r_{obs} for molecules, B-values for the solid state differ from those for the gaseous state. Finally, we note that in the Sanderson approach stability is judged by the *cohesive energy* E_{coh} for which the reference state is given by *atoms* in the gas state (i.e. the formation enthalpy of the compound plus atomization energy of the elements involved, Section F.15). In the Born–Mayer approach, the *lattice energy* E_{lat} is calculated for which the reference state is given by *ions* in the gas state (and to which the IP + EA must be added to compare it with E_{coh}).

In Figure 10.16 the correlation between calculated bond lengths and bond energies with the experimental values for the alkali halides is shown. The agreement with experimental values is generally quite good. The method has been applied with success to many halides, oxides, and sulfides, as well as organic molecules. Typical difference between calculated and experimental bond length values for many compounds is about 0.02 Å. For the bond energy the discrepancy between calculated and experimental values is typically between 2 and 5 kcal mol^{-1}.

The starting point of *orbital EN model* [59] is that the energy of an atom $E(q)$ is a function of charge q in the (hybrid) valence orbital expressed in a Taylor series. Thus, $E(q) = a + bq + cq^2$ with $0 \geq q \geq -2$, provided it is accepted that q can be a continuous variable. The orbital EN is defined by $X(q) = \partial E(q)/\partial q = b + 2c$. Using IP $= E(0) - E(-1)$ and EA $= E(-1) - E(-2)$ results in $b = (3\text{IP} - \text{EA})/2$ and $c = (\text{IP} - \text{EA})/2$, which in turn yields to $X(-1) = (\text{IP} + \text{EA})/2$. The "hardness" is defined by $\frac{1}{2}\partial^2 E(q)/\partial q^2 = c$, akin to the "hardness" η as defined by Parr and Pearson [83]. For data, see Section F.19.

The next step is accepting EN equalization, as suggested by Sanderson, but now between two orbitals. Thus, if for a binary molecule AB an orbital i on atom A joins with orbital j on atom B to form a bond, a transfer of charge Δq will occur from the less

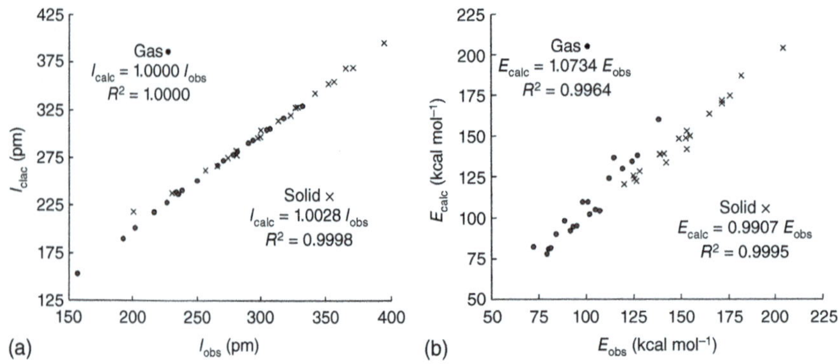

Figure 10.16 The polar covalence approach for gaseous and solid alkali halides. (a) Calculated bond length l_{calc} versus experimental length l_{obs}. (b) Calculated bond energy E_{calc} versus experimental energy E_{obs}. Graphs plotted according to the data in [58]. Source: Sanderson [58]/with permission of Elsevier.

electronegative orbital to the more electronegative one, thus lowering the energy of the system until the electronegativities of the bond forming orbitals are equalized. This leads to $X_i(q_i{}^\circ + \Delta q) = X_j(q_j{}^\circ - \Delta q)$ with q° the reference charge before bond formation. Using the definition of X directly provides $\Delta q = [X_j(q_j{}^\circ) - X_i(q_i{}^\circ)]/2(c_j + c_i)$, while the energy lowering $\Delta = [X_j(q_j{}^\circ) - X_i(q_i{}^\circ)]^2/4(c_j + c_i)$, which resembles the ionic energy expression in the original definition by Pauling.

For a general molecule, we must extend these concepts and consider an atom with m orbitals, forming m localized two electron bonds to its ligands. In each bond the charge Δq_k for $k = 1, \ldots, m$ will be transferred and thus the total net charge on the atom considered will be $Q = \Sigma_k \Delta q_k$. The EN of orbital i of this atom will not only depend on the charge Δq_i transferred to the bond it is engaged in, but also on the residual charge $r_i = Q - \Delta q_i = \Sigma_{k \neq i} \Delta q_k$, due to the charge transferred in all the other bonds the atom has formed. Consequently, the orbital EN becomes $X(q_i,r_i) = b_i(r_i) + 2c_i(r_i)q_i$, where b_i can be approximated by $b_i = b_i{}^0 + b_i{}^1 r_i$ as the values of r_i are usually small and c depends only weakly on r_i. With b^0, b^1, and c given as determined for the orbitals from atomic spectroscopic data, the orbital EN becomes $X(q_i,r_i) = b_i{}^0 + b_i{}^1 r_i + 2c_i q_i$. For a molecule with N bonds, each formed between two atomic orbitals, say i and j, of two adjacent atoms, we have the bond index (ij) for 1–N. The requirement of EN equalization within each bond $X_i(q_i{}^\circ + \Delta q_i, r_i) = X_j(q_j{}^\circ - \Delta q_j, r_j)$ results in N linear equations, which can be solved for $\Delta q_i = -\Delta q_j$. As the concept of EN equalization used is restricted to localized two electron bonds, two additional provisions are required: (i) in the case two atoms are joined by multiple (σ and π) bonds, these have to be treated together, otherwise an instability in the linear equations yields an unreasonable σ donation balanced by an almost equally large π back donation; (ii) in the case of conjugated systems, the corresponding dominant Lewis structures with localized bonds need to be treated separately and the results averaged appropriately. The final result obtained is not very sensitive to the weights used in averaging.

Using these concepts, it is possible to predict ESCA chemical shifts, proton affinities, NMR chemical shifts, force constants, group electronegativities, and hardness parameters for molecular fragments. Here, we deal only with bond lengths and energies. With as bond

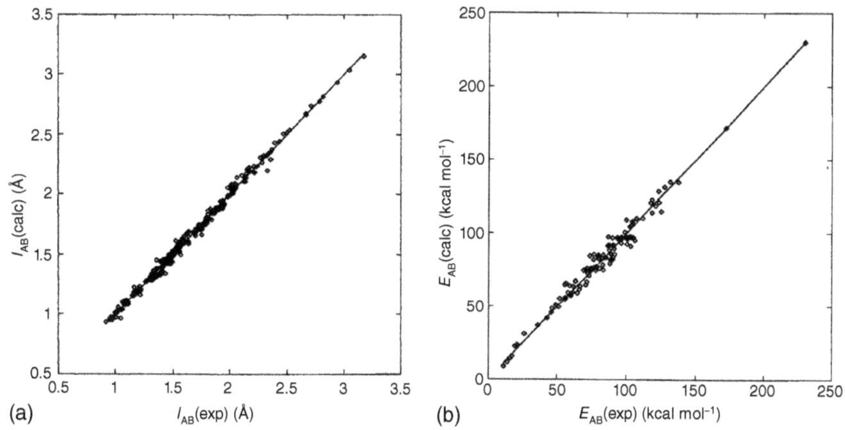

Figure 10.17 The orbital electronegativity approach. Source: Hinze [59]/with permission of Elsevier. (a) Calculated bond length l_{AB} versus experimental length. (b) Calculated bond energy E_{AB} versus experimental energy.

length $l_{AB} = r_A + r_B$, it appeared that in the expression $r = r° + dQ$ the parameter d can be written as $d = (\partial r/\partial Q)_{Q=0} = c_0 + c_1/c$ with c_0 and c_1 global parameters. This leads to

$$l_{AB} = r_A{}^° + r_B{}^° + c_0(Q_A + Q_B) + c_1(Q_A/c_A + Q_B/c_B) + c_2 Q_A Q_B/(c_A + c_B) l_{AB}$$

with c_2 another global parameter to account for the Coulomb interaction due to direct neighbors. Because the quadratic term is small, it is often sufficient to approximate in the last term l_{AB} by $l_{AB}{}^° = r_A{}^° + r_B{}^°$. The parameters, 51 covalent radii and the three global parameters c_0, c_1, and c_2 (Section F.19), were determined by fitting the expression to bond length data for 543 bonds, derived from spectroscopic structure determinations of the corresponding molecules in the gas phase. Regression (Figure 10.17a) leads to a correlation coefficient $r = 0.997$, a standard error $\sigma = 0.028$ Å, and a relative standard error of 1.3%. Using $l_{AB}{}^°$ yields an error of ≈5%, while using l_{AB} results in an error of ≈1.5%.

Inserting $\Delta = [X_j(q_j{}^°) - X_i(q_i{}^°)]^2/4(c_j + c_i)$ in $E_{AB} = ½(E_{AA} + E_{BB}) + \Delta_{AB}$ and introducing global parameters c_3 and c_4 to fit the data, results in

$$E_{AB} = \frac{1}{1 + |Q_A Q_B|}\left[½(E_{AA} + E_{BB}) + c_3 \frac{(X_A - X_B)^2}{4(c_A + c_B)} + c_4 \frac{Q_A Q_B}{l_{AB}}\right]$$

Bond weakening occurs if partial charges on atoms become significant, as reflected in the prefactor, and Coulomb interaction between partially charged neighbors is given by the last term. Regression (Figure 10.17b) yields $r = 0.987$ and $\sigma = 5\ \text{kcal mol}^{-1}$.

In summary, both the polar covalence and orbital EN model are empirical. The former model suffers from ununderstood aspects, like the need for deviating values for f_{cor} and n_{cov}, as admitted frankly by Sanderson. Although bond length values are calculated, the system is not self-consistent because experimental l_{AB} values are used for the bond energy. Nevertheless, the model is rather successful in predicting partial charges and energies for molecules and solids. The latter model provides a (much) more transparent procedure, is also quite successful, but is limited to molecules. As the system is linear, the solution of the N equation for the Δq_js provides self-consistent solutions. For both, the ionic radii calculated are generally different from other "established" radii, such as the Pauling or Shannon–Prewitt radii.

10.6.5 Metallic Crystals

For metals a successful empirical approach exists in which a scaled binding energy curve is used. Consider the binding energy U described by $U(r-r_0) \cong U_0 + ½U_0''(r-r_0)^2 + \cdots$ where $U_0 = U(r_0) = -|U_0|$ at the equilibrium distance r_0, $U_0'' = (\partial^2 U/\partial r^2)_{r=r_0}$ and, as usual in equilibrium, the first derivative $U_0' = 0$. As a scaled relation, we use $\phi(a) \cong -1 + ½\phi_0''(a)^2 + \cdots$ with $a = (r-r_0)/l$ and l as scaling parameter. Thus, $\phi(a) = U(r)/|U_0|$. One could scale further by using $l = r_0$, but the surprising move, yielding better results, is to fix the value of l by taking $\phi_0'' = 1$, which leads to $a^2 = U_0''(r-r_0)^2$ or $l = (|U_0|/U_0'')^{1/2}$. Evaluating U_0'' from the equilibrium bulk modulus at $T = 0\,\text{K}$, $K_0 = -V(\partial P/\partial V)_T$, we have $U_0'' = 12\pi r_0 K_0$, so that $l = (|U_0|/12\pi r_0 K_0)^{1/2}$, implying altogether that the $U(r)$ curve is scaled as

$$U(r) = U_0\phi(a) \equiv U_0\phi[(r-r_0)/l]$$

The scaled curves $\phi(a)$ of various metals as calculated from first principles calculations appear to lie closely on top of each other (Figure 10.18a), as do those from bimetallic adhesion, molecular binding, cohesion, and chemisorption of gas molecules on metals [84]. The scaled curve can be described empirically by

$$\phi(a) = -(1 + a + 0.05a^3)\exp(-a)$$

where the coefficient of a^3 has been obtained by matching the thermal expansivity of Cu (Section 10.9.2). Hence, the complete energy–density relationship is known whenever the equilibrium radius $r_0 = (3V_0/4\pi)^{1/3}$, the equilibrium binding energy U_0, and the equilibrium bulk modulus K_0 are known. The scaling length l is evidently related to a harmonic crystal and therefore to the linear elastic region, while the parameter $\eta_0 = r_0/l$ is used as a measure for the degree of anharmonicity. For the data, see Table 10.5.

Given the expressions above, the equation of state can be calculated as[5]

$$P = -(\partial U/\partial V)_T = -K_0 \frac{3[(V/V_0)^{1/3} - 1]}{(V/V_0)^{2/3}} \phi'$$

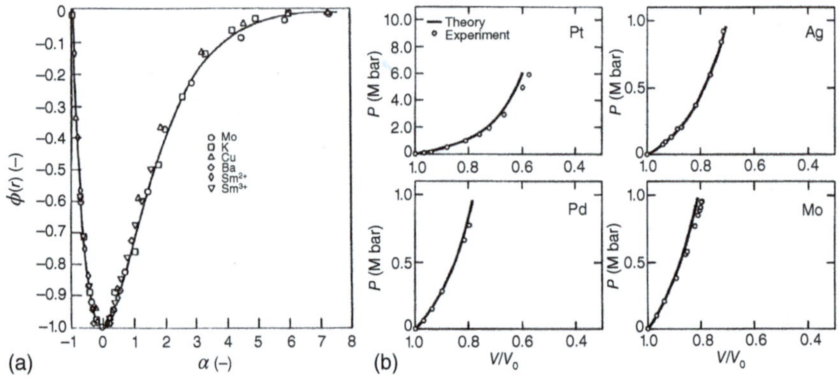

Figure 10.18 (a) The universal bonding (UB) potential $\phi(r) = U(a)/|U_0|$ for several metals as a function of scaled separation $a = (r-r_0)/l$. (b) The EoS for Pt, Pd, Ag, and Mo (o experimental data, – UB fit). Source: Rose et al. [84]/American Physical society.

5 For convenience: $\phi'' = e^{-a}(1-1.3a + 0.3a^2 - 0.05a^3)$ and $\phi''' = e^{-a}(-2.3 + 1.9a - 0.45a^2 + 0.05a^3)$.

Table 10.5 The universal bonding (UB) potential parameters l (Å) and r_0 (Å) for various metals.

Li	Be										B	C
0.553	0.312											
1.719	1.25											
Na	Mg										Al	Si
0.562	0.316										0.336	0.344
2.080	1.77										1.58	1.68
K	Ca	Sc	Ti	V	Cr	Fe	Co	Ni	Cu	Zn	Ga	Ge
0.651	0.483	—	0.340	0.310	0.254	0.274	0.262	0.270	0.272	0.215	—	0.348
2.573	2. 18		1.62	1.49	1.42	1.41	1.39	1.38	1.41	1.54		1.76
Rb	Sr	Y 0.47	Zr	Nb	Mo	Ru	Rh	Pd	Ag	Cd 0.214	In	Sn
0.658	—	1.99	0.395	0.336	0.265	0.245	—	0.237	0.269	1.73	0.360	—
2.750			1.77	1.63	1.55	1 0.48		1.52	1.60		1.84	
Cs	Ba	La	Hf	Ta	W	Os	Ir	Pt	Au	Hg	Tl	Pb
0.714	0.558	—	0.373	0.330	0.274	—	0.230	0.237	0.236	—	0.331	0.303
2.977	2.46		1.74	1.62	1.56		1.50	1.53	1 0.59		1.90	1.93

Below each element the values of l and r_0 are given in two consecutive lines. Note the systematics for the parameters.
Source: Rose et al. [84]/American Physical society.

Figure 10.18b shows the results for four metals. Similar agreement is obtained for several other metals. In summary, using experimental values for r_0, U_0, and K, and the empirical potential $\phi(a)$, sometimes exaggeratedly referred to as the "universal bonding" (UB) potential, the EoS is well described. In Section 10.9.2, we show that this model provides some further useful results.

Several non-pair potential methods exist, of which we only mention two. The atomistic *Miedema model* [85] is based on the idea that 'an atom is the smallest particle of a substance that still contains all the properties of the substance' and has been applied to pure metals and alloys. The enthalpy H is described as a function of the surface tension γ, which on its turn depends on the electron density. Upon alloying the electron density adjusts itself, resulting in an ΔH responsible for stability of alloys. In the *embedded atom method* [86], the energy is given as a sum of pair potentials, typically Morse potentials $\phi(r_{ij})$ with parameters p_1, p_2, and p_3, plus an embedding function with as arguments the total electronic charge density q_j at the atomic sites. The q_js are obtained from the sum of the q_{ij}s, described by $r_{ij}{}^{-6}\Sigma_i p_i \exp(-p_i r_{ij})$. A cutoff distance for the q_js and for $\phi(r_{ij})$ is used, which is also adjusted for smoothness at cutoff. The embedding parameters are determined by fitting to various experimental data, such as the sublimation energy, lattice constants, and vacancy formation energy. Both models provide useful predictions in the absence of experimental data.

10.7 Lattice Dynamics

So far, in sketching the structure and bonding of crystalline solids we used a static image, taking only into account the potential energy of molecules. However, molecules vibrate around their equilibrium positions, and we must include their kinetic energy as well. We focus here on some basic aspects of the coupled dynamics of atoms in crystals, usually denoted as *lattice dynamics* to estimate the thermal expansivity α, the heat capacity C, and the bulk modulus K [7–9, 87, 88].

To describe waves in a lattice, we need again wave vectors, here labeled $\boldsymbol{q}$, and the periodic boundary conditions. The potential energy $\Phi = \Phi(\boldsymbol{r}_{n\alpha})$, given as a function of the coordinates of all atoms as indicated by $\boldsymbol{r}_{n\alpha}$ and expanded for small displacements $\boldsymbol{u}_{n\alpha}$ in a Taylor series around the equilibrium position $\boldsymbol{r}_{n\alpha}$, yields

$$\Phi(\boldsymbol{r}_{n\alpha} + \boldsymbol{u}_{n\alpha}) = \Phi(\boldsymbol{r}_{n\alpha}) + ½\sum\nolimits_{m\beta} \boldsymbol{u}_{n\alpha} \cdot \boldsymbol{\Phi}_{n\alpha}^{m\beta} \cdot \boldsymbol{u}_{m\beta} + \cdots \quad \text{with} \quad \boldsymbol{\Phi}_{n\alpha}^{m\beta} = \frac{\partial^2 \Phi(\boldsymbol{r}_{n\alpha})}{\partial \boldsymbol{u}_{n\alpha} \partial \boldsymbol{u}_{m\beta}} \tag{10.10}$$

The linear terms are absent since the expansion is around the equilibrium minimum energy position, and higher-order terms are neglected for the time being. Equation (10.10) indicates the potential energy for a "generalized" harmonic oscillator, and hence the approximation is called the *harmonic approximation* (HA). The second derivatives of the potential energy $\boldsymbol{\Phi}_{n\alpha}^{m\beta}$, the *coupling constants*, are the generalized spring constants.

The next step is to assume that Newton's laws are valid, and this leads to

$$m_\alpha \ddot{\boldsymbol{u}}_{n\alpha} + \sum\nolimits_{m\beta} \boldsymbol{\Phi}_{n\alpha}^{m\beta} \boldsymbol{u}_{m\beta} = 0 \quad \text{or} \quad m_\alpha \ddot{u}_{n\alpha i} + \sum\nolimits_{m\beta j} \Phi_{n\alpha i}^{m\beta j} u_{m\beta j} = 0 \tag{10.11}$$

where m_α denotes the mass of the atoms associated with the label α, and the components $u_{n\alpha i}$ have been introduced. For rN atoms in total, this leads to $3rN$ coupled differential equations. For the coupling constants several relations apply. The most important one is due to translational invariance, which implies that the 3×3 coupling constant matrix depends only on the difference $\boldsymbol{m} - \boldsymbol{n}$, and therefore,

$$\Phi_{n\alpha i}^{m\beta j} = \Phi_{0\alpha i}^{(m-n)\beta j}$$

If we assume periodic boundary conditions, the displacements $\boldsymbol{u}_{n\alpha}$ can be written as plane waves $h_{n\alpha}(\boldsymbol{q})$ with wave vector $\boldsymbol{q}$, frequency $\omega(\boldsymbol{q})$, and amplitude $\boldsymbol{U}_\alpha(\boldsymbol{q})$, given by

$$\boldsymbol{u}_{n\alpha}(\boldsymbol{q}) = \boldsymbol{U}_\alpha(\boldsymbol{q})h_{n\alpha}(\boldsymbol{q}) \equiv \boldsymbol{U}_\alpha(\boldsymbol{q})(m_\alpha)^{-1/2}\exp\{\mathrm{i}[\boldsymbol{q}\cdot\boldsymbol{r}_{n\alpha} - \omega(\boldsymbol{q})\,t]\}$$

Substitution in Eq. (10.11) results in the eigenvalue equation

$$-\omega^2(\boldsymbol{q})U_{\alpha i}(\boldsymbol{q}) + \sum\nolimits_{\beta j}\left\{\sum_{\boldsymbol{m}}(m_\alpha m_\beta)^{-1/2}\Phi_{0\alpha i}^{m\beta j}\exp[\mathrm{i}\boldsymbol{q}\cdot(\boldsymbol{r}_{m\beta} - \boldsymbol{r}_{0\alpha})]\right\}U_{\beta j}(\boldsymbol{q}) = 0$$

$$-\omega^2(\boldsymbol{q})\boldsymbol{U}_\alpha(\boldsymbol{q}) + \sum\nolimits_\beta \boldsymbol{D}_{\alpha\beta}(\boldsymbol{q})\boldsymbol{U}_\beta(\boldsymbol{q}) = 0 \quad \text{or} \quad [\mathbf{D}(\boldsymbol{q}) - \omega^2(\boldsymbol{q})\mathbf{1}]\mathbf{U}(\boldsymbol{q}) = 0 \qquad (10.12)$$

with the amplitude column $\mathbf{U}(\boldsymbol{q})^\mathrm{T} = [\mathbf{U}_1(\boldsymbol{q})^\mathrm{T}, \ldots, \mathbf{U}_\alpha(\boldsymbol{q})^\mathrm{T}, \ldots, \mathbf{U}_r(\boldsymbol{q})^\mathrm{T}]$ and the *dynamical matrix* $\mathbf{D} = \boldsymbol{D}_{\alpha\beta} = D_{\alpha i,\beta j}$. Equation (10.12) is a linear homogeneous equation of order $3r$. In general, such a system has only solutions if the determinant vanishes, and thus if

$$\det[D_{\alpha i,\beta j}(\boldsymbol{q}) - \omega^2(\boldsymbol{q})\delta_{\alpha i,\beta j}] = 0$$

In case we have a primitive unit cell, $r = 1$, and for every wave vector $\boldsymbol{q}$ we must solve a system of three equations. For $r \neq 1$, for every wave vector $\boldsymbol{q}$, Eq. (10.12) has $3r$ solutions $\omega_p(\boldsymbol{q})$, denoted as *branches* indicated by the branch label p. Since $\mathbf{D}$ is Hermitian, the eigenvalues $\omega_p(\boldsymbol{q})$, known as *dispersion relations*, are real. The branches for which $\omega_p(\boldsymbol{q}) = 0$ at $\boldsymbol{q} = 0$ are called the *acoustic(al) branches*, while for $\omega_p(\boldsymbol{q}) \neq 0$ at $\boldsymbol{q} = 0$ the term *optic(al) branches* is used. In crystals there are always three acoustic branches, while the number of optic branches depends on the number of atoms per unit cell. The eigenvectors, collectively denoted by $\mathbf{e}_p(\boldsymbol{q})^\mathrm{T} = [\mathbf{e}_{p,1}(\boldsymbol{q})^\mathrm{T}, \ldots, \mathbf{e}_{p,\alpha}(\boldsymbol{q})^\mathrm{T}, \ldots, \mathbf{e}_{p,r}(\boldsymbol{q})^\mathrm{T}]$, describe the direction of motion of the atoms for *mode* $j = (p,\boldsymbol{q})$, that is, for branch p at wave vector $\boldsymbol{q}$. They are orthonormalized via $\Sigma_\alpha \boldsymbol{e}_{p,\alpha}(\boldsymbol{q})^* \cdot \boldsymbol{e}_{l,\alpha}(\boldsymbol{q}) = \delta_{pl}$ with $\boldsymbol{e}_{p,\alpha}(\boldsymbol{q})$ the *polarization vector* of atom α and δ_{pl} the Kronecker delta. Using the *amplitude vectors* $\boldsymbol{U}_{p,\alpha}(\boldsymbol{q}) = N^{-1/2}A_p(\boldsymbol{q})\mathbf{e}_{p,\alpha}(\boldsymbol{q})$ with the number of cells N and amplitude $A_p(\boldsymbol{q})$, the total can be written as $\mathbf{U}_p(\boldsymbol{q})^\mathrm{T} = [\mathbf{U}_{p,1}(\boldsymbol{q})^\mathrm{T}, \ldots, \mathbf{U}_{p,\alpha}(\boldsymbol{q})^\mathrm{T}, \ldots, \mathbf{U}_{p,r}(\boldsymbol{q})^\mathrm{T}]$. Each mode $(p,\boldsymbol{q})$ can be considered as a quasi-particle, addressed as a *phonon*. The general solution becomes $\boldsymbol{u}_{n\alpha} = \Sigma_{p\boldsymbol{q}}\boldsymbol{U}_{p,\alpha}(\boldsymbol{q})h_{n\alpha}(\boldsymbol{q}) \equiv \boldsymbol{u}_\alpha$, where the last step can be made because the atoms in all cells behave identically.

From Bloch's theorem and the periodic boundary conditions, the number of allowed $\boldsymbol{q}$-values or *states* was shown to be N. For large N, the states are densely packed and form a quasi-continuous distribution. The *density of states* $g(\omega)$ is $\Sigma_{p\boldsymbol{q}}\delta[\omega - \omega_p(\boldsymbol{q})]$ with $\delta(x)$ the Dirac delta function, or replacing the sum by an integral

$$g(\omega)\mathrm{d}\omega = (2\pi)^{-3}V\sum\nolimits_p \int \delta[\omega - \omega_p(\boldsymbol{q})]\,\mathrm{d}\boldsymbol{q} \quad \text{where} \quad \int_0^\infty g(\omega)\,\mathrm{d}\omega = 3rN$$

is the normalization condition as we have rN atoms, each with three degrees of freedom. If we separate the wave vector increment $\mathrm{d}\boldsymbol{q}$ in a length element $\mathrm{d}q$ perpendicular to the surface $\omega_p(\boldsymbol{q}) = \omega$, and an element of surface area $\mathrm{d}S_{\boldsymbol{q}}$ on that surface, we can write $\mathrm{d}\boldsymbol{q} = \mathrm{d}S_{\boldsymbol{q}}\mathrm{d}q$. With $\mathrm{d}\omega_p = |\nabla_{\boldsymbol{q}}\omega_p(\boldsymbol{q})|\mathrm{d}q$, one obtains

$$g(\omega) = (2\pi)^{-3}V \sum\nolimits_p \int_{S_q} |\nabla_{\boldsymbol{q}}\omega_p(\boldsymbol{q})|^{-1}\mathrm{d}S_{\boldsymbol{q}} \tag{10.13}$$

from which it follows that a flat region in the dispersion relation corresponds to a singularity in the density of states (*van Hove singularity*).

The properties are calculated as usual. Because all modes $j = (p,\boldsymbol{q})$ contribute independent of each other, the vibrational partition function is $Z_{\mathrm{vib}} = \Pi_j\, z_j$ with

$$z_j = z(\omega_j) = \sum\nolimits_n \exp[-\beta\hbar\omega_j(n + ½)] = \exp(-y_j/2)/[1 - \exp(-y_j)]$$

where $\beta = 1/kT$, k is Boltzmann's constant, and $y_j = \beta\hbar\omega_j$. Therefore, the vibrational Helmholtz energy becomes $F_{\mathrm{vib}} = -kT\ln Z_{\mathrm{vib}} = \Sigma_j f_j$ with

$$f_j = f(\omega_j) = -kT\,\ln z(\omega_j) = ½\hbar\omega_j + kT\ln[1 - \exp(-\beta\hbar\omega_j)]$$

For $T \gg 0$, expanding the exponentials, we have $z_j = (\beta\hbar\omega_j)^{-1}$ and $Z = (\beta\hbar\varpi)^{3rN}$ with ϖ the *geometric mean frequency* defined by $3rN\ln\varpi = \Sigma_j\ln\omega_j$. In the HA, F_{vib} is thus the sum of the contributions of the various modes $j = (p,\boldsymbol{q})$, due to the fact that the lattice potential $\boldsymbol{\Phi}$ was limited to second-order terms only. To incorporate effectively higher-order terms, the *quasi-harmonic approximation* (QHA) can be used, in which the frequency ω is assumed to be a function of volume V only. Using the bulk modulus $K_T = -V(\partial P/\partial V)_T$, $\boldsymbol{\Phi}$ is written as

$$\boldsymbol{\Phi} = \boldsymbol{\Phi}_0 + ½K_T(V - V_0)^2/V$$

with V_0 the volume at the potential energy minimum. The total Helmholtz energy $F = F_{\mathrm{vib}} + \boldsymbol{\Phi}$ should be minimal and thus $\partial F/\partial V = 0$. Evaluating results in

$$\frac{(V - V_0)}{V} = \frac{1}{VK_T}\sum\nolimits_j \gamma_j\varepsilon(\omega_j) \equiv -\frac{1}{VK_T}\sum\nolimits_{p,\boldsymbol{q}} \frac{\partial \ln \omega_p(\boldsymbol{q})}{\partial \ln V}\varepsilon\,[\omega_p(\boldsymbol{q}), T]$$

and where the mean energy of an oscillator reads (Section 7.5.1)

$$\varepsilon(\omega, T) = \hbar\omega\,(\overline{n} + ½) \quad \text{with} \quad \overline{n} = \hbar\omega\,[\exp(\beta\hbar\omega) - 1]^{-1} \tag{10.14}$$

The parameters $\gamma_j = \gamma_p(\boldsymbol{q}) \equiv -\partial\ln\omega_p(\boldsymbol{q})/\partial\ln V$ are addressed as *mode Grüneisen parameters*, which characterize the anharmonicity and may vary considerably with $(p,\boldsymbol{q})$.

Let us finally calculate the mean square displacement tensor $\langle \boldsymbol{u}_\alpha \boldsymbol{u}_\alpha^{\,*}\rangle$. We first need to establish the value of the amplitude $A_p(\boldsymbol{q})$ and to that purpose we calculate the kinetic energy $\vartheta = ½\sum_{n\alpha i} m_\alpha \dot{u}_{n\alpha i}^{\,2}$ and the potential energy $\boldsymbol{\Phi} = ½\sum_{n\alpha i,m\beta j} u_{n\alpha i}\boldsymbol{\Phi}_{n\alpha i}^{m\beta j} u_{m\beta j}$. Both yield after some calculation $½N\Sigma_p A_p^{\,2}\omega_p^{\,2}$, in consonance with the virial theorem, so that the total energy $E = \vartheta + \boldsymbol{\Phi} = N\Sigma_p A_p^{\,2}\omega_p^{\,2} \equiv \Sigma_p\varepsilon_p$. Hence, each mode $(p,\boldsymbol{q})$ contributes $\varepsilon_p = NA_p^{\,2}\omega_p^{\,2}$ and $A_p^{\,2} = \varepsilon_p/N\omega_p^{\,2}$, where ε_p is given by Eq. (10.14). Substituting A_p in $\boldsymbol{U}_{p,\alpha}(\boldsymbol{q})$, we obtain

$$\langle \boldsymbol{u}_\alpha \boldsymbol{u}_\alpha\rangle = (Nm_\alpha)^{-1}\Sigma_{p,\boldsymbol{q}}\varepsilon_p(\boldsymbol{q})\omega_p(\boldsymbol{q})^{-2}\left(\boldsymbol{e}_{p,\alpha}\boldsymbol{e}_{p,\alpha}^{\,*}\right)$$

The mean square displacement in any direction is given by $\mathrm{tr}\langle \boldsymbol{u}_\alpha \boldsymbol{u}_\alpha\rangle = \Sigma_i\langle u_{\alpha i}^{\,2}\rangle$, so that for a primitive cubic lattice ($r = 1$, $u_{\alpha i} = u$, $m_\alpha = m$) we have

$$\langle u^2\rangle = (3Nm)^{-1}\sum\nolimits_{pq}\varepsilon_p(\boldsymbol{q})\omega_p(\boldsymbol{q})^{-2} = (3Nm)^{-1}\int \varepsilon(\omega)\omega^{-2}g(\omega)\,\mathrm{d}\omega$$

Here the orthonormality relation $\Sigma_\alpha\, \boldsymbol{e}_{p,\alpha}(\boldsymbol{q})^* \cdot \boldsymbol{e}_{l,\alpha}(\boldsymbol{q}) = \boldsymbol{e}_p(\boldsymbol{q})^* \cdot \boldsymbol{e}_l(\boldsymbol{q}) = \delta_{pl}$ is used and the sum is replaced by an integral. Hence, $\langle u^2 \rangle$ can be determined without knowledge of the polarization vectors. This is, however, not true for lattices with a basis.

For further evaluation, one needs the density of states $g(\omega)$ which can be calculated, given $\Phi(\boldsymbol{r}_{n\alpha})$, by solving the dynamical equations for a sufficient number of $\boldsymbol{q}$-values within the (irreducible part of the first) BZ. As results so obtained are material specific, we pay attention to two simple, general models.

10.8 Two Simple Models

The simplest vibrational model one can take, proposed by Einstein (1907), assumes independent vibrations with frequency $\omega_{\rm E}$ for all atoms (Section 7.5.1). This implies a density of states $g(\omega) = 3rN\delta(\omega - \omega_{\rm E})$, with $\omega_{\rm E}$ the *Einstein frequency* for this *Einstein model*. The result $F_{\rm vib} = -kT\ln Z_{\rm vib} = \Sigma_{p\boldsymbol{q}} f_{p,\boldsymbol{q}} = 3rNf(\omega_{\rm E})$ applies.

Another simple but frequently used model, first proposed by Debye in 1912, assumes that the crystal is an isotropic elastic continuum with sound velocity $c_{\rm L}$ for the longitudinal waves and $c_{\rm T}$ for the two (degenerated) transverse waves, given by

$$c_{\rm L} = \left(K_S/\rho\right)^{1/2} \left[\frac{3(1-\nu)}{(1+\nu)}\right]^{1/2} \quad \text{and} \quad c_{\rm T} = \left(K_S/\rho\right)^{1/2} \left[\frac{3(1-2\nu)}{2(1+\nu)}\right]^{1/2}$$

where K_S is the isentropic (conventionally called adiabatic) bulk modulus, ν Poisson's ratio and ρ the mass density. For a continuum, $\omega = cq$ and the gradient $|\nabla_{\boldsymbol{q}}\omega|$ becomes the sound velocity c_p for branch p. Because the behavior is isotropic, c_p is independent of the wave vector $\boldsymbol{q}$, and for each branch p, the surface $S_{\boldsymbol{q}}$ represents a sphere with radius q on which ω is constant. The surface in Eq. (10.13) is thus $S_{\boldsymbol{q}} = 4\pi q^2$. Consequently, the density of states $g(\omega)$ for each branch p becomes

$$g_p(\omega) = \frac{V}{(2\pi)^3} \int_{S_{\boldsymbol{q}}} |\nabla_{\boldsymbol{q}}\omega|^{-1} \mathrm{d}S_{\boldsymbol{q}} = \frac{V}{(2\pi)^3} \frac{4\pi q^2}{c_p} = \frac{V}{2\pi^2} \frac{\omega^2}{c_p{}^3}$$

Replacing $c_p{}^{-3}$ by its average value $c_{\rm m}{}^{-3}$, we have

$$g(\omega)\mathrm{d}\omega = \frac{V}{2\pi^2} \frac{\omega^2}{c_{\rm m}{}^3} \mathrm{d}\omega \quad \text{with} \quad \frac{1}{c_{\rm m}{}^3} = \frac{1}{3}\left(\frac{1}{c_{\rm L}{}^3} + \frac{2}{c_{\rm T}{}^3}\right)$$

This $g(\omega)$ is assumed to apply to anisotropic real materials at all frequencies, although actually valid only at low frequencies for isotropic materials. Since the total number of states is $3rN$, we normalize $g(\omega)$ by taking a maximum frequency $\omega_{\rm D}$ given by

$$3rN = \frac{V}{2\pi^2} \frac{1}{c_{\rm m}{}^3} \int_0^{\omega_{\rm D}} \omega^2 \,\mathrm{d}\omega = \frac{V}{2\pi^2} \frac{1}{c_{\rm m}{}^3} \frac{1}{3} \omega_{\rm D}{}^3 \quad \text{or} \quad \omega_{\rm D} = \left(\frac{6\pi^2 \cdot 3rN}{V}\right)^{1/3} c_{\rm m}$$

The density of states thus becomes $g(\omega) = 3rN(3\omega^2/\omega_{\rm D}{}^3)$. The cut-off frequency $\omega_{\rm D}$ is denoted as the *Debye frequency*, and the model as such as the *Debye model*.

Introducing a *characteristic temperature* θ via $\hbar\omega \equiv k\theta$, with θ denoting either $\theta_{\rm E}$ or $\theta_{\rm D}$, for both models the behavior of $F_{\rm vib}(T,V)$ scales as $a(T,V) = Tb(\xi)$, where $b(\xi)$ is the same function of the variable $\xi = T/\theta$ for all materials, and the material is characterized by characteristic temperature $\theta(V)$, dependent on V only [9]. Using

$$\left(\frac{\partial b}{\partial \ln \theta}\right)_T = -\frac{T}{\theta} \frac{\mathrm{d}b(\xi)}{\mathrm{d}(\xi)} = -\left(\frac{\partial b}{\partial \ln T}\right)_V$$

one obtains the useful expression

$$\left(\frac{\partial a}{\partial V}\right)_T = -\frac{\gamma}{V}\left(\frac{\partial a}{\partial \ln \theta}\right)_T = \frac{\gamma T}{V}\left(\frac{\partial b}{\partial \ln T}\right)_V = \frac{\gamma T}{V}\left(\frac{\partial (a/T)}{\partial \ln T}\right)_V = \frac{\gamma}{V}\left[T\left(\frac{\partial a}{\partial T}\right)_V - a\right] \tag{10.15}$$

where $\gamma = -\text{d}\ln\theta/\text{d}\ln V$ is used, supposed to be a material characteristic parameter. For the Debye model we obtain $\gamma = ⅓ + \text{d}\ln c_\text{m}/\text{d}\ln V$. Assuming $\nu = const.$, only approximately true, one obtains $\gamma = a(\text{d}K/\text{d}P) - b$ with $a = ½$ and $b = ⅙$. Various other formulations exist and, generally, $a = ½$ is well-obeyed but typically $b \cong 0.9$ [89].

10.9 Properties

In our approach to the thermal behavior of solids, we have to take two contributions to the Helmholtz energy F into account. The first contribution is due to the lattice potential energy Φ, while the second stems from the thermal vibrations, F_vib. For C_V we need only F_vib. We introduce the Φ contribution to F when considering α and K_T.

10.9.1 Heat Capacity

In order to estimate the energy involved in the dynamics, we need the mean energy of an oscillator $\varepsilon(\omega,T)$, as given in Eq. (10.14). Using the density of states $g(\omega)$, the internal energy contribution $U(T)$, due to the dynamics, may now be written as

$$U(T) = \int_0^\infty g(\omega)\,\varepsilon(\omega, T)\,\text{d}\omega$$

while the specific heat capacity at constant volume C_V is given by

$$C_V = \partial U(T)/\partial T = \int_0^\infty g(\omega)\,\frac{\partial \varepsilon(\omega, T)}{\partial T}\,\text{d}\omega$$

We now need to insert $g(\omega)$. Taking the Einstein model, using $y = \beta\hbar\omega$, we obtain

$$\frac{U}{3rN} = \int_0^\infty \delta(\omega - \omega_\text{E})\,\varepsilon(\omega, T)\,\text{d}\omega = \varepsilon(\omega_\text{E}, T) = kT\left(\frac{y}{\exp(y) - 1} + ½y\right) \tag{10.16}$$

$$\frac{C_V}{3rN} = \int_0^\infty \delta(\omega - \omega_\text{E})\,\frac{\partial \varepsilon(\omega, T)}{\partial T}\,\text{d}\omega = \frac{\partial \varepsilon(\omega_\text{E}, T)}{\partial T} = k\frac{y^2 \exp(y)}{[\exp(y) - 1]^2}$$

In Eq. (10.16), $½\hbar\omega_\text{E}$ is the *zero-point energy*. If we take the Debye model, we have

$$\frac{U}{3rN} = \frac{3}{\omega_\text{D}{}^3}\int_0^{\omega_\text{D}} \omega^2 \varepsilon(\omega, T)\,\text{d}\omega = kT\left[\frac{3}{x^3}\int_0^{\theta_\text{D}/T} \frac{y^3}{\text{e}^y - 1}\,\text{d}y\right] + \frac{3k\theta_\text{D}}{8}$$

where the *Debye temperature* θ_D, defined by $\hbar\omega_\text{D} = k\theta_\text{D}$, $x = \theta_\text{D}/T$, and $y = \beta\hbar\omega$ are used, and $3k\theta_\text{D}/8$ represents the *zero-point energy*. Similarly, we have

$$\frac{C_V}{3rN} = \frac{3}{\omega_\text{D}^3}\int_0^{\omega_\text{D}} \omega^2 \frac{\partial \varepsilon(\omega, T)}{\partial T}\,\text{d}\omega = k\left[\frac{3}{x^3}\int_0^{\theta_\text{D}/T} \frac{y^4 \text{e}^y}{(\text{e}^y - 1)^2}\,\text{d}y\right]$$

For $kT \gg \hbar\omega_\text{D}$, y is small, the integrand may be expanded to give for the integral $⅓(\theta_\text{D}/T)^3[1 - (\theta_\text{D}/T)^2/20 + \cdots]$, and C_V reduces to

$$C_V = 3rNk\left[1 - (\theta_D/T)^2/20 + \cdots\right] \cong 3rNk \quad (T \gg \theta_D)$$

the well-known *law of Dulong and Petit*, which can also be obtained from the Einstein model. For $kT \ll \hbar\omega_D$, one obtains, by extending the upper limit of the integration from θ_D/T to ∞, in which case the integral tends to the constant $4\pi^4/15$,

$$C_V = 3rNk\left[3(T/\theta_D)^3(4\pi^4/15)\right] \quad (T \ll \theta_D)$$

which is also experimentally well established. The C_V for the Einstein model at low T, though, behaves like $k(\beta\hbar\omega)^2\exp(-\beta\,\hbar\omega)$, and therefore decreases much faster with T.

Table 10.6 provides a selected set of Debye temperatures,[6] while Figure 10.19a shows $g(\omega)$ for both models as well as a realistic $g(\omega)$ for Si. From this figure it is clear that the

Table 10.6 Debye temperature θ_D (K) for various compounds from low-temperature C_V data.

Pb	88	BaF_2	177	CO	80	C_3H_8	100
Hg	97	CaF_2	474	NO	119	F_2	100
Au	173	$CaTiO_3$	750	N_2O	127	Cl_2	115
Pt	237	Al_2O_3	1030	H_2O	192	Br_2	86
Cu	313	SiO_2	1217	D_2O	185	I_2	76
Al	389	Ar	84	H_2S	130	HF	177
Si	645	Ne	63	CO_2	140	HCl	128
Be	1000	Kr	63	CS_2	104	HBr	92
C	1910	Xe	55	COS	95	HI	73
NaCl	281	H_2	91	CH_4	78	HCN	160
KCl	230	N_2	68	CCl_4	85	NH_3	210
KBr	177	O_2	89	C_2H_6	131	$N(CH_3)_3$	143

Source: Moelwyn-Hughes [90]/Elsevier.

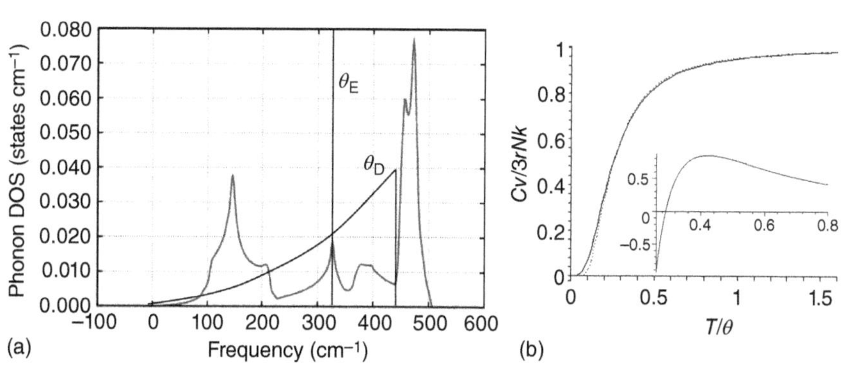

Figure 10.19 (a) Density of states for Si (T_{mel} = 1687 K) with the corresponding Debye temperature θ_D = 645 K and Einstein temperature $\theta_E = 0.75.\cdot 645 = 484$ K. (b) The heat capacity $C_V/3rNk$ as a function of T/θ for the Debye (–) and Einstein model (…) for $\theta_E = 0.75\theta_D$. Inset: $[C_{V,E}(\theta_E) - C_{V,D}(\theta_D)]/C_{V,D}(\theta_D)$ in %.

6 Various definitions of θ_D (or γ for that matter) are possible. For an overview, see [91].

Debye density of states is only well obeyed at low $\boldsymbol{q}$. In spite of the fact that the Debye density of states is far from realistic, the predictions and experimental values for C_V match closely, if the material contains only one kind of atoms (metals, noble gases) or atoms of similar mass and bonding characteristics. This is because C_V, being a property averaged over the density of states, is rather insensitive to the precise distribution of states. If the bond type, coordination and mass of the various atoms in the unit cell become increasingly different, the Debye approximation becomes increasingly more inadequate. However, it is still frequently used for more complex solids in view of its simplicity. Finally, we note that using the Einstein model with θ_E corresponding to the arithmetic mean frequency of the Debye spectrum $\theta_E = 3\theta_D/4$, the temperature behavior of C_V above $0.26\theta_D$ deviates less than 1% from the Debye theory result (Figure 10.19b).

10.9.2 Thermal Expansivity

For an estimate of the thermal expansion coefficient $\alpha = V^{-1}(\partial V/\partial T)$, we need the contribution duc to the lattice potential energy $\Phi(V)$ to the Helmholtz energy F. We assume that we may expand Φ in terms of $V - V_0$, where V_0 is the volume at 0 K, and that, using $\Phi_0' \equiv \mathrm{d}\Phi(V_0)/\mathrm{d}V$, $\Phi_0'' \equiv \mathrm{d}^2\Phi(V_0)/\mathrm{d}V^2, \ldots$, we may write,

$$\Phi = \Phi_0 + \frac{1}{2!}\Phi_0''(V - V_0)^2 + \frac{1}{3!}\Phi_0'''(V - V_0)^3 + \cdots$$

since $\Phi_0' = 0$ at equilibrium. Omitting all terms after the one in Φ_0'', the lattice energy is called *harmonic*. The term in Φ_0''' and higher ones denote *anharmonic terms*. At $T = 0$ K, $\Phi'' \equiv \mathrm{d}^2\Phi/\mathrm{d}V^2 = \Phi_0''$, while at any T we have $\Phi' = \Phi_0''(V - V_0)$ and $\Phi'' = \Phi_0'' + \Phi_0'''(V - V_0)$, neglecting higher-order terms.

The second contribution stems from the thermal vibrations. Using either the Einstein or the Debye model, we can write for the vibrational contribution $F_{\text{vib}}(T,V)$ to the Helmholtz energy, according to Eq. (10.15),

$$\frac{\partial F_{\text{vib}}}{\partial V} = \frac{\gamma}{V}\left(T\frac{\partial F_{\text{vib}}}{\partial T} - F_{\text{vib}}\right) = \frac{\gamma}{V}(-TS_{\text{vib}} - F_{\text{vib}}) = -\frac{\gamma}{V}U_{\text{vib}}$$

Using $F = \Phi + F_{\text{vib}}$, we calculate from $P = -\partial F/\partial V$ the equation of state, and obtain

$$P + \frac{\mathrm{d}\Phi}{\mathrm{d}V} = -\frac{\partial F_{\text{vib}}}{\partial V} = \frac{\gamma}{V}U_{\text{vib}} \tag{10.17}$$

This is the *Mie–Grüneisen equation of state* [92], with γ the (*thermodynamic*) *Grüneisen parameter*. We now consider the product αK_T and rewrite it as

$$\alpha K_T = -\frac{1}{V}\left(\frac{\partial V}{\partial T}\right)_P V\left(\frac{\partial P}{\partial V}\right)_T = -\left(\frac{\partial V}{\partial T}\right)_P\left(\frac{\partial P}{\partial V}\right)_T = \left(\frac{\partial T}{\partial P}\right)_V^{-1} = \left(\frac{\partial P}{\partial T}\right)_V$$

Here, in the third step the −1-rule for P, V, and T or $(\partial V/\partial T)_P(\partial P/\partial V)_T(\partial T/\partial P)_V = -1$ is used. From Eq. (10.17), we find that $(\partial P/\partial T)_V = \gamma C_V/V$ (remember that Φ is a function of V only), and therefore we obtain the *Grüneisen relation*

$$\alpha = \frac{\gamma C_V}{K_T V} \quad \text{or using} \quad \frac{K_S}{K_T} = \frac{C_P}{C_V}, \quad \alpha = \frac{\gamma C_P}{K_S V} \tag{10.18}$$

In this approximation, α is thus proportional to C_V. Since for $T \to 0$, $C_V \to 0$, and therefore $\alpha \to 0$, conform to the third law. Clearly, there is only one thermodynamic γ and in the QHA we obtain $\gamma_{\text{QHA}} = \Sigma_{pq}\gamma_p(\boldsymbol{q})c_{V,p}(\boldsymbol{q})/\Sigma_{pq}c_{V,p}(\boldsymbol{q})$ with $c_{V,p}(\boldsymbol{q})$ the heat capacity of mode $p,\boldsymbol{q}$. In absence of information, one often uses $\gamma_p(\boldsymbol{q}) = \gamma$.

Thermal expansion and vibrations are thus linked via γ. Noting that the characteristic frequency $\omega \sim (\Phi'')^{1/2}$, and using $\Phi'' = \Phi_0'' + \Phi_0'''(V - V_0) + \cdots$, we have

$$\frac{\mathrm{d}\omega}{\mathrm{d}V} = \omega\frac{\Phi_0'''}{2\Phi''} \cong \omega\frac{\Phi_0'''}{2\Phi_0''} \quad \text{and} \quad \gamma = -\frac{\mathrm{d}\ln\theta}{\mathrm{d}\ln V} = -\frac{V}{\omega}\frac{\mathrm{d}\omega}{\mathrm{d}V} \cong -\frac{V}{2}\frac{\Phi_0'''}{\Phi_0''}$$

corresponding to $\omega/\omega_0 = (V/V_0)^{-\gamma}$, if $\gamma = const.$ is assumed. Hence, if the vibrations are harmonic, $\Phi_0''' = 0$, therefore $\gamma = 0$ and α vanishes. While the calculation of Φ_0'' is already somewhat involved, the calculation of Φ_0''' and higher-order terms is complex and complicates the solution of the dynamical equation tremendously. In fact, the principal axes transformation to obtain the dispersion relations cannot be applied any longer successfully. The usual procedure is to use the harmonic solutions as a basis for perturbation theory to obtain more accurate answers, but the approach is complex and we refer to the literature [93–95].

Given the above, the heat capacity at constant pressure C_P can be expressed explicitly in terms of C_V. To that purpose, we note that from thermodynamics we have

$$C_P = T\left(\frac{\partial S}{\partial T}\right)_P = T\left(\frac{\partial S}{\partial T}\right)_V + T\left(\frac{\partial S}{\partial V}\right)_T\left(\frac{\partial V}{\partial T}\right)_P = C_V + T\left(\frac{\partial S}{\partial V}\right)_T \alpha V$$

Using the Maxwell relation $(\partial S/\partial V)_T = (\partial P/\partial T)_V$, the relation $(\partial P/\partial T)_V = \alpha K_T$ and the Grüneisen relation $\alpha = \gamma C_V/VK_T$, the expression for C_P becomes

$$C_P = C_V(1 + \alpha\gamma T) \tag{10.19}$$

In Section 10.6, the UB potential $\phi(a)$ was introduced, which, using the present results, can yield some further useful estimates. Accepting again $\omega \sim (\phi'')^{1/2}$, the Grüneisen parameter is given by $\gamma = -(r/6l)(\phi'''/\phi'')$, or at zero pressure by $\gamma_0 = 0.383\eta_0$, agreeing with an average absolute difference (AAD) of 20% with 28 experimental data points [96]. The bulk modulus is given by $K = U''/12\pi r|U_0|$, or at $P = 0$, by $K_0 = U_0''/12\pi r_0|U_0| = U_0/12\pi r_0 l^2$, while the variation of K with pressure P is given by $(\partial K/\partial P)_T = 1 + r\phi'''/3l\phi''$, or at $P = 0$, $(\partial K/\partial P)_T = 1 + 2.3\eta_0/3$. The thermal expansivity α can be calculated from Eq. (10.18), which results, using $C_V \cong 3k$, in $\alpha = 3.45k/\eta_0|U_0|$, agreeing with an AAD of 12% with 18 experimental data points [97]. This confirms the usefulness of the UB potential approach as not only the EoS, but also γ (or α or $(\partial K/\partial P)_T$ for that matter) are reasonably described.

10.9.3 Bulk Modulus

To obtain the isothermal bulk modulus K_T, we multiply the equation of state, Eq. (10.17), with V, and differentiate with respect to V, resulting in

$$P + V\frac{\partial P}{\partial V} + \frac{\partial \Phi}{\partial V} + V\frac{\partial^2\Phi}{\partial V^2} = \gamma\frac{\partial U_{\text{vib}}}{\partial V} = \frac{\gamma^2}{V}(TC_V - U_{\text{vib}})$$

In the last step, Eq. (10.15) has been used again. Using $K_T = -V\partial P/\partial V$, we find

$$K_T = P + \Phi' + V\Phi'' - \gamma^2(TC_V - U_{\text{vib}})/V$$

At $T = 0\,\text{K}$, $C_V = 0$, while in the Debye approximation the zero-point energy is $U_{\text{vib}} = 3k\theta_{\text{D}}/8$, so that $P + \Phi' = \gamma U_{\text{vib}}/V = 3\gamma k\theta_{\text{D}}/8V$, resulting in $K_T = K_0 + 3(\gamma + \gamma^2)k\theta_{\text{D}}/8V$ with $K_0 \equiv V\Phi_0''$. Hence, even at $T = 0\,\text{K}$ there is in principle a vibrational contribution to the bulk modulus, albeit a small one. For any temperature there are three contributions. The term $V\Phi''$ contains the lattice contribution and a thermal expansion contribution. The term Φ' is also a thermal expansion contribution, while the last term is due to vibrations. To make this clearer, the expansions $\Phi' = \Phi_0''(V-V_0)+\cdots$ and $\Phi'' = \Phi_0'' + \Phi_0'''(V-V_0)+\cdots$ are used, yielding to first order in $V-V_0$, and neglecting P because at atmospheric pressure, P is small as compared to Φ',

$$K_T = K_0 + K_0\left(2+\frac{V_0\Phi_0'''}{\Phi_0''}\right)\frac{V-V_0}{V_0} - \frac{\gamma^2}{V}(TC_V - U_{\text{vib}})$$

From $\Phi' = \gamma U_{\text{vib}}/V$, Eq. (10.17), and $\Phi' = \Phi_0''(V-V_0)+\ldots$, we find using $K_0 = V_0\Phi_0''$,

$$(V-V_0)/V_0 = \gamma U_{\text{vib}}/VK_0$$

Hence the final result is

$$K_T = K_0 + \left(2+\frac{V_0\Phi_0'''}{\Phi_0''}\right)\frac{\gamma U_{\text{vib}}}{V} - \frac{\gamma^2}{V}(TC_V - U_{\text{vib}})$$

Here, we recognize in the first term a pure lattice term, the second represents the thermal expansion contribution, and the third is a pure vibrational contribution. If the vibrations are harmonic or $\Phi_0''' = 0$, thus $\gamma = 0$ and $\partial K_T/\partial T$ vanishes. In the above derivations, $\gamma = const.$ is assumed, but if $\gamma = \gamma(V)$ is used we obtain identical results, except that for K_T the term $-(\mathrm{d}\gamma/\mathrm{d}V)[U_{\text{vib}} + (3rNk\theta_D/8V)(\mathrm{d}\gamma/\mathrm{d}V)]$ must be added [77]. Obviously, values for the actual V-value must be used.

Summarizing, in this approach the thermal expansion coefficient α is proportional to the heat capacity C_V and the bulk modulus K_T contains a lattice, an expansion, and a vibrational contribution. Given values for θ, γ, V_0, and K_0 (or V and K_T for that matter), one can estimate α and $\partial K/\partial T$ using the proper harmonic oscillator expressions.

Typical values for θ range from $\approx$ 100 K for soft metals (like Pb) via 400–500 K for engineering metals (like Al and Fe) and 1000 K for ceramic materials (like Al_2O_3) to $\approx$2000 K for diamond. For vdW crystals the approach has been used [67] as well, with θ ranging from about 50–70 K for the noble gas crystals and cyclohexane via $\approx$100 K for benzene and C_{60} to $\approx$140 K for naphthalene and anthracene. The value for γ is typically 1–2 for many materials.

This concludes our discussion on the structure and dynamics of perfect crystals. It will be clear that even in this approximate treatment, the argument is somewhat complex, but we note that the overall behavior can be explained qualitatively very well, if not semiquantitatively. Further progress along these lines appears difficult and for specific materials detailed and complex, atomistic calculations are required.

10.10 Defects

Real solids are not perfect and contain defects. For crystalline solids one distinguishes between zero-, one-, two-, and three-dimensional defects. While for a 0D defect only a single atom deviates from the perfect crystallographic order, for a 1D, 2D, and 3D defect, this is a (connected) line, plane, or volume of atoms, respectively [3]. We limit the discussion mainly to 0D and 1D defects. Although by introducing defects in solids the order is disturbed, the structure remains overwhelmingly regular. For example, near the melting point, though they are thermodynamically required, in a metal typically only about 3–4‰ of vacancies are present. Early overviews are in [98, 99].

10.10.1 Zero-dimensional Defects

Several types of zero-dimensional defects are known (Figure 10.20). The basic types for unary crystals are *vacancies* (missing atoms at a lattice point), interstitials (atoms not positioned at lattice points), *substitutional impurities* (foreign atoms at a lattice point), and *interstitial impurities* (foreign atoms not positioned at lattice points). In all cases, locally a deformation of the lattice arises with an associated energy term that contributes, in addition to electrostatic terms, to the formation energy of the defect. Obviously, a larger energy value is associated with interstitial than with substitutional atoms. In binary crystals, one can distinguish between Frenkel defects and Schottky defects. For a *Frenkel defect*, an ion is displaced to an interstitial position. This is usually a (positive ion or) cation as (negative ions or) anions typically have a larger size. For a *Schottky defect* a pair of ions is moved to the surface, leaving behind a vacant cation and anion site. During diffusion, all these defects will migrate, for example, an atom at a regular lattice position migrates to a vacancy leaving behind another vacancy, so that the original vacancy has effectively moved. Similarly, an interstitial molecule may migrate to a neighboring hole in the lattice. A somewhat more complex point defect mechanism is the *interstitialcy* mechanism, for

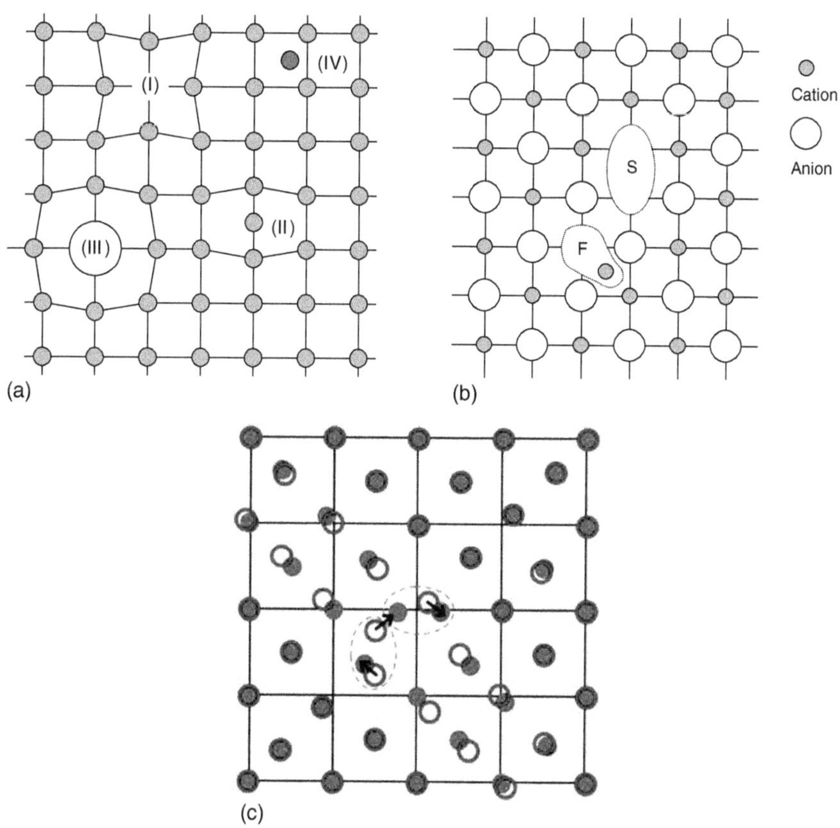

Figure 10.20 (a) Schematic of a vacancy (I), interstitial (II), substitutional impurity, (III) and interstitial impurity (IV); (b) Frenkel (F) and Schottky (S) defects; (c) elementary jump of an interstitialcy in an FCC lattice. Open circles, before jump; closed circles, after jump.

which, during migration, an interstitial atom displaces an atom at a regular lattice position to an interstitial position, thereby occupying itself in the regular position.

To estimate the number of vacancies, consider a unary crystal with N molecules, n vacancies and coordination number z. If $\varepsilon_L(V)$ denotes the lattice energy per molecule at volume V and $\varepsilon_H(V)$ the energy to displace a molecule to the surface, the energy to create a vacancy is given by $\varepsilon(V) = \varepsilon_H(V) - \varepsilon_L(V)$, where the value for $\varepsilon(V)$ has to be taken at the volume V for the actual temperature considered. If the vacancies are independent, the number of possible configurations in which n vacancies can be indistinguishably distributed over $N+n$ sites is $W = (N+n)!/N!n!$, leading to the mixing entropy $S_{\text{mix}} = k\ln W$. For the Helmholtz energy F, we thus obtain

$$F = U - TS = N\varepsilon_L + n\varepsilon - kT\ln W = N\varepsilon_L + n\varepsilon - kT\ln[(N+n)!/N!n!]$$

At equilibrium, $\partial F/\partial n = 0$ and, using Stirling's approximation and $\beta = 1/kT$, we obtain

$$n = (N+n)\exp(-\beta\varepsilon) \cong N\exp(-\beta\varepsilon)$$

The last step can be made since usually $n \ll N$. An exponential increase with T thus results, and only at $T = 0$ K point defects are thermodynamically not required.

The above essentially considers only S_{mix}. Let us discuss the influence of vibrations on this description [100] using the Einstein model with frequency $\omega \equiv \omega_E$. For this model the vibrational Helmholtz energy at sufficiently high temperature is $3NkT\ln(\beta\hbar\omega)$, so that the total Helmholtz energy of a perfect crystal becomes

$$F = N[\varepsilon_L + 3kT\ln(\beta\hbar\omega)]$$

For a defect lattice, we assume that the vibration frequency of the z neighbors of the vacancy decreases in the direction of the vacancy to ω' and that it remains constant at ω in the two other directions. In this case, F is given by

$$F = N\varepsilon_L + n\varepsilon + kT(3N - nz)\ln\beta\hbar\omega + nzkT\ln\beta\hbar\omega' - kT\ln[(N+n)!/N!n!]$$

From the equilibrium condition $\partial F/\partial n = 0$, we obtain in the usual way

$$n = (N+n)A\exp[-\beta\varepsilon(V)] \cong NA\exp[-\beta\varepsilon(V)] \tag{10.20}$$

with $A \equiv (\omega/\omega')^z$. Normally, experiments are done at constant pressure instead of volume. Thus, we consider defect formation enthalpies and assuming that the change in defect enthalpy h by a change in T is given by

$$\mathrm{d}h = (\partial h/\partial T)\mathrm{d}T = h'(\partial V/\partial T)\mathrm{d}T = \alpha Vh'\mathrm{d}T \quad \text{with} \quad h' = \partial h/\partial V$$

and where the thermal expansivity $\alpha = V^{-1}(\partial V/\partial T)$ is used, we obtain

$$h(T) = h(0) + \int_0^T \mathrm{d}h = h(0) + \alpha VTh'$$

if α and h' are temperature independent. Inserting in Eq. (10.20) yields

$$n = NAB\exp(-h(0)/kT) \quad \text{with} \quad B \equiv \exp(-\alpha Vh'/k) \tag{10.21}$$

and where $h(0)$ is the enthalpy to form a vacancy at absolute zero. Since $h' < 0$, $B > 1$, as is A. The exponential relation is thus maintained. The factor $AB = \exp(s/k)$ represents the entropy per defect s, which is, as will be clear, difficult to estimate accurately with this model, but experimentally metals typically has the value $0.5k \leq s \leq 1.5k$ with Si being an exception [101]. The formation entropies reported for Si range between about 6–15k at

Table 10.7 Defect formation enthalpy h and entropy s for some crystals.

Vacancies			Schottky defects			Frenkel defects[a)]		
	h_V (eV)	s_V (k)		h_S (eV)	s_S (k)		h_F (eV)	s_F (k)
Al	0.75	2.4	NaCl	2.44	9.8	AgCl	1.45–1.55	5.4–12.2
Cu	1.18	1.6–3.0	NaI	2.00	7.6	AgBr	1.13–1.28	6.6–12.2
Ag	1.09	~1.5	KCL	2.54	9.0			
Au	0.94	1	KBr	2.53	10.3			

a) h_F and s_F may vary considerably with temperature.
Source: Allnatt and Lidiard [104]/Cambridge University Press.

high temperatures. Seeger and Chik (1968) proposed that in Si the self-interstitial is the dominating point defect and not the vacancy as is normally the case for elemental crystals [102]. The matter is still under debate [103].

For MX crystals the expressions for Schottky and Frenkel defects are

$$n_S = NAB\ \exp(-\tfrac{1}{2}\beta h_S) \quad \text{and} \quad n_F = (NN^*)^{1/2}AB\ \exp(-\tfrac{1}{2}\beta h_F)$$

where h_S and h_F are the corresponding formation enthalpies (Table 10.7) and N^* is the number of interstitial positions. The extra factor two in the exponent, due to the generation of pairs, is also present in the factor B. The factor A for Schottky defects is still $A = (\omega/\omega')^{z'}$ and one expects that ω' for both vacancies are lower than ω thereby reinforcing their effects. For Frenkel defects $A^2 = (1/\omega')(\omega/\omega'')^{z''}(\omega/\omega''')^{z'''}$ where ω' for the interstitial and ω'' for its z'' neighbors are likely increased, but ω''' for the z''' vacancy neighbors decreased, thereby tending to cancel their effects.

As an example, consider NaCl having Schottky defects and for which $z = 6$, $\alpha \cong 1.2\times10^{-4}\ \text{K}^{-1}$ and $\varepsilon \cong 56\ \text{kcal mol}^{-1}$. Estimating $\omega_E/\omega_E' \cong 2$ and $\partial\ln\varepsilon/\partial V \cong 2$, we have $A \cong 64$ and $B \cong \exp(6.7)$, so that $AB \cong 3.1\times10^4$, corresponding to $s \cong 10.30k$, somewhat fortuitously agreeing with the experimental value (Table 10.7).

As the energies involved for point defects in ionic solids are relatively high, thermal equilibration takes some time. Point defects also have a *migration enthalpy* h_{mig}, which determine their mobility, typically with a value of about 50–400 kJ mol^{-1}. The typical diffusion time t over a length L is given by $t = L^2/D$ with $D = D_0\exp(-h_{mig}/kT)$ the diffusion coefficient. Taking $D \sim 10^{-12}\ \text{m}^2\ \text{s}^{-1}$ and $L \sim 0.1$ mm, we arrive at several hours before equilibration is reached.

For a vacancy in a vdW crystal, an estimate for the defect formation energy ε can be made using $\varepsilon = \varepsilon_H - \varepsilon_L$. In a nearest-neighbor approximation $\varepsilon_L = \tfrac{1}{2}z\phi$ with z the coordination number and ϕ the bond energy. In the same spirit $\varepsilon_H = z\phi$, and it follows that $\varepsilon_H = 2\varepsilon_L$ or $\varepsilon = \varepsilon_L$. This estimate neglects relaxation and typically yields a too high value by a factor of ⅔ to ¾. For other crystals, simple estimates are unreliable but empirical relations exist. For example, for heavy and alkali metal halides h_S (eV) $= 2.14\times10^{-3}T_{mel}$ with T_{mel} the melting point [105]. For metal halides also the relation $h_S \cong 1.1E_{ato}$ is obtained, while for metal oxides $h_S \cong 0.63E_{ato}$ with E_{ato} the atomization energy [106].[7] Typical vacancy

7 For oxides, Me_aO_b, the atomization energy $E_{ato} = \Delta H + aL_{Me} + (b/2)D_{O_2}$ with ΔH the heat of formation, L_{Me} the sublimation energy of the metal, and $D_{O2} = 498.2\ \text{kJ mol}^{-1}$ the dissociation energy of O_2.

energies are ≈10–50 kJ mol^{-1} for molecular crystals, ≈50–150 kJ mol^{-1} for metals, and ≈150–250 kJ mol^{-1} for ionic crystals (Table 10.7). An extensive overview is [107] and for a modern treatment, see [5].

10.10.2 One-dimensional Defects

The most important one-dimensional defect is the dislocation. There are two basic types: the *edge dislocation*, essentially an extra half-plane of atoms inserted in the lattice and the *screw dislocation*, essentially a "staircase" in the lattice (Figure 10.21a). The line representing the end of the half-plane and the center of the staircase is the *dislocation line*, and is characterized by the unit tangent vector $\boldsymbol{l}$. An important characteristic of dislocations is their *Burgers vector* $\boldsymbol{b}$, which can be found by making a closed loop around a dislocation line and counting the mismatch in lattice vectors (Figure 10.21b). In fact, the Burgers vector $\boldsymbol{b}$ is an invariant characteristic.

For a pure edge dislocation, $\boldsymbol{b}$ and $\boldsymbol{l}$ are perpendicular, that is, $\boldsymbol{b}\cdot\boldsymbol{l} = 0$. The slip plane in this case is spanned by $\boldsymbol{b}$ and $\boldsymbol{l}$ and characterized by its normal $\mathbf{g} = \boldsymbol{b}\times\boldsymbol{l}$. For a screw dislocation, $\boldsymbol{b}$ is parallel to $\boldsymbol{l}$, that is, $\boldsymbol{b}\cdot\boldsymbol{l} = |\boldsymbol{b}| = b$. In this case, the slip plane also contains both $\boldsymbol{b}$ and $\boldsymbol{l}$, but, since $\boldsymbol{b}$ and $\boldsymbol{l}$ are parallel for a screw dislocation ($\mathbf{g} = \boldsymbol{b}\times\boldsymbol{l} = \boldsymbol{0}$), $\boldsymbol{b}$ and $\boldsymbol{l}$ do not determine a plane and a screw dislocation is not restricted to a specific plane. Edge and screw dislocations are the extremes. Usually, a dislocation line is curved, and a so-called *mixed nature* is present (Figure 10.21a). Generally, the number of dislocations is characterized by the *dislocation density* $\rho = L/V$, where L denotes the total length of dislocation lines in a volume V.

Dislocations have an excess energy that can be considered as a sum of the energy of the core and the outer region. The core is the inner part of a dislocation, the energy of which has to be estimated by atomic models. For the outer region, elasticity theory can be used because the displacements of the atoms from their equilibrium configuration in the ideal crystal are small. As the stress field for a dislocation is given by $\alpha G^2/2\pi r$ [108], the associated elastic

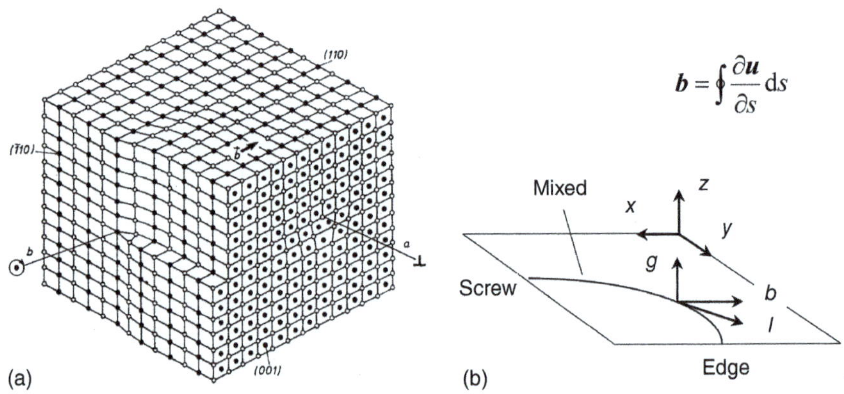

Figure 10.21 (a) Binary cubic lattice containing a mixed dislocation. At point "a" the dislocation has a pure edge character while at point "b" the character is purely screw; (b) Schematic of a dislocation indicating the dislocation line $\boldsymbol{l}$ (the line from "a" to "b" in (a)), the slip plane $\boldsymbol{g}$ (the plane containing the line from "a" to "b" in (a)) and Burgers vector $\boldsymbol{b}$. The Burgers vector convention used is indicated with the contour integral around a dislocation line $\boldsymbol{l}$, where $\boldsymbol{u}$ is the displacement and ds an element of the contour.

energy for isotropic materials can be estimated – taking right away a few steps elaborated on below – as

$$U = \int dU_{\text{ela}} = \tfrac{1}{2}\int \sigma b \mathrm{d}r = \int_{r_0}^{R} \alpha \frac{Gb^2h}{4\pi}\frac{\mathrm{d}r}{r} = \alpha \frac{Gb^2h}{4\pi} \ln r \Big|_{r=r_0=b}^{r=R=500b} \cong \alpha \frac{Gb^2h}{2} \tag{10.22}$$

where G is the shear modulus, b is the length of the Burgers vector, h is the length of the dislocation, α is a factor, 1 for a screw dislocation and $(1-\nu)^{-1}$ an edge dislocation, and ν is Poisson's ratio. Furthermore, r_0 denotes the dislocation core radius and R the range of influence of a dislocation. Let us make first an estimate for lower limit r_0. The domain for which the deformation can be described by continuum elasticity theory starts at about 4–5 times the Burgers vector. A logical choice is thus $r_0 = 4b$ or $r_0 = 5b$. The interior of the dislocation, the *core*, cannot be described by continuum elasticity. Using detailed models, an estimate of the core energy can be made and this results in $U_{\text{core}} \cong 0.07-0.15U_{\text{ela}}$. A reasonable way to incorporate this contribution is to use as lower limit $r_0 = b$. For an estimate of the upper limit, we need the dislocation density ρ, which can vary widely for different materials and conditions, a typical value being $\rho \cong 10^{10}\,\text{cm}^{-2}$. Since mobile dislocations will take positions with net resolved shear stress close to zero and the stress of a dislocation decreases rapidly with increasing distance, the average distance between dislocations $\rho^{-1/2}$ is often used for an estimate of the upper limit R. For a typical Burgers vector of length $b = 0.2$ nm and the typical value $\rho^{-1/2} = 10^{-5}$ cm, the upper limit is thus estimated as $R \cong 500b$.

For a mixed dislocation, representing a mix between a screw and an edge dislocation with Burgers vectors $b_{\text{screw}} = b\cos\gamma$ and $b_{\text{edge}} = b\sin\gamma$ superimposed on each other, the α factor becomes $\cos^2\gamma + \sin^2\gamma/(1-\nu) = (1-\nu\cos^2\gamma)/(1-\nu)$, where γ is the angle between $\mathbf{b}$ and $\mathbf{l}$. As $\langle\cos^2\gamma\rangle = \tfrac{1}{2}$, $\langle\alpha\rangle = \tfrac{1}{2}(1-\tfrac{1}{2}\nu)/(1-\nu) \cong 0.63$ for the typical value $\nu = \tfrac{1}{3}$.

Because the logarithmic dependence of the energy U on R and r_0 is weak, U is only weakly dependent on the presence of other dislocations, and we can simplify the energy expression to $U \cong \tfrac{1}{2}\alpha Gb^2h$. Therefore, one often uses a constant, and specific energy $u = U/h \cong \tfrac{1}{2}\alpha Gb^2$. Note that the energy is by no means concentrated at the center of the dislocation; for $R/r_0 = 500b/b$, half of the energy is in the region between about $22b$ and $500b$. To estimate the order of magnitude of the energy per atoms along the dislocation, we take $G = 50$ GPa and $b = 0.25$ nm, which leads to $U \cong 150\,\text{kJ mol}^{-1}$, comparable to the energy of a point defect in metals and inorganics.

During the formation of point defects, the loss in energy is compensated by the gain in configurational entropy, rendering the Helmholtz energy of a point defect negative. That is why point defects are a thermodynamic necessity. For flexible dislocations, Cottrell [108] estimated as upper limit the configurational entropy change per atom along the dislocation as $s_{\text{con}} = -(a/h)kT\ln(A/a^2) - kT\ln z$ with a the lattice spacing, h the length of the dislocation line, A the cross-section area of the crystal perpendicular to the dislocation line, and $z \cong 6$ the number of neighboring atomic sites into which the path of the dislocation line may deviate in each atom plane through which it passes. The first term is small, about $10^{-6}kT$, and the second is about $-1.8kT$ for $z \cong 6$. For the vibrational entropy change per atom s_{vib}, using the Einstein–Grüneisen approximation at high temperatures, to first order $3kT\ln(\beta\hbar\omega) - 3kT\ln(\beta\hbar\omega_0) \cong -3kT\gamma\mathrm{d}V/V$ with $\gamma \cong 2$. While for screw dislocations $\mathrm{d}V/V = 0$, for edge dislocations the contributions of the atoms above the glide plane cancel with those of below that plane. Hence, for both cases only the atoms in the core contribute. Estimating for these atoms $\gamma\mathrm{d}V/V \approx 1$, $s_{\text{vib}} \cong -3kT$. Overall though, the Helmholtz energy F of a dislocation equals nearly the energy U, and as $U > 0$, dislocations are not thermodynamically stable [108]. Nevertheless, they are nearly always present as grown-in defects. An extensive literature on dislocations exists, the standard treatises being Nabarro [109] and Hirth and Lothe [110].

10.10.3 Other Defects

In solids several other defects can be present of which we discuss briefly stacking faults and grain boundaries. So-called *extended defects* are discussed in [10, 111].

To describe *stacking faults* [11], consider a crystal as a stacking of lattice planes, for example, the FCC lattice as the sequence ...ABC|ABC... and the HCP lattice as ...AB|AB.... A stacking error can occur, for example, in the FCC lattice ...ABC|ABC... → ...ABC|BCA... or ...ABC|ABC... → ...ABC|ACBCA... In the first example, the total number of lattice planes of each type is conserved but not in the second. These two types are generally indicated as *intrinsic* and *extrinsic stacking faults*. When a stacking sequence is permanently changed to ...ABC|ACBACB..., we have a *twin*. With these stacking faults, energy is associated, for metals typically 0.05–0.2 J m^{-2}. These defects occur in all three material classes but are of limited importance in polymers. In well-annealed metals, the number of stacking faults is small since these faults generally cost relatively a large amount of energy. In some inorganic materials, the required energy is small and stacking faults can be abundantly present [10, 111].

In polycrystalline materials, *grain boundaries* delineate areas of a different crystallographic orientation. With these grain boundaries a grain boundary energy is associated, ranging from about 0.1 to about 1 J m^{-2}. For *small-angle grain boundaries*, the crystallographic orientation difference between two grains is small and can be characterized by the difference in angle θ between the normal vectors associated with both grains. This can occur by a simple *tilt* (no rotation of grain boundary plane, only tilting the lattice, corresponding to a set of pure edge dislocations) or by a *twist* (no tilt, only rotation in the grain boundary plane, corresponding to a set of pure screw dislocations). A small-angle grain boundary can be considered as an array of interacting dislocations and an estimate for the energy of small-angle grain boundaries can be made based on the behavior of arrays of dislocations (Figure 10.22). For a simple tilt boundary, this leads to

$$U_{\text{gb}} = \frac{Gb}{4\pi(1-\nu)}\theta(A - \ln\theta) \quad \text{with} \quad \theta = \frac{b}{h} \quad \text{and} \quad A = 1 + \ln\frac{b}{2\pi r_0}$$

where h is the dislocation spacing along the grain boundary and the other symbols have the same meaning as before. This expression shows an asymptotic behavior to a constant value, typically $U_{\text{gb}} \cong 0.5\ \text{J m}^{-2}$, above $\theta \cong 15°$. This expression appears to agree with the experiment up to rather high values of θ, in fact much better and extending to a much larger tilt angle than could be anticipated from the model.

Such a description cannot be made for *wide-angle grain boundaries*, which are generally considered as nonmatching areas between the lattices of two grains (Figure 10.22). For particular orientations though, the lattices can match, each with a different repetitive unit. The resulting superlattices are addressed as *coincident site lattices* (CSLs). The structure

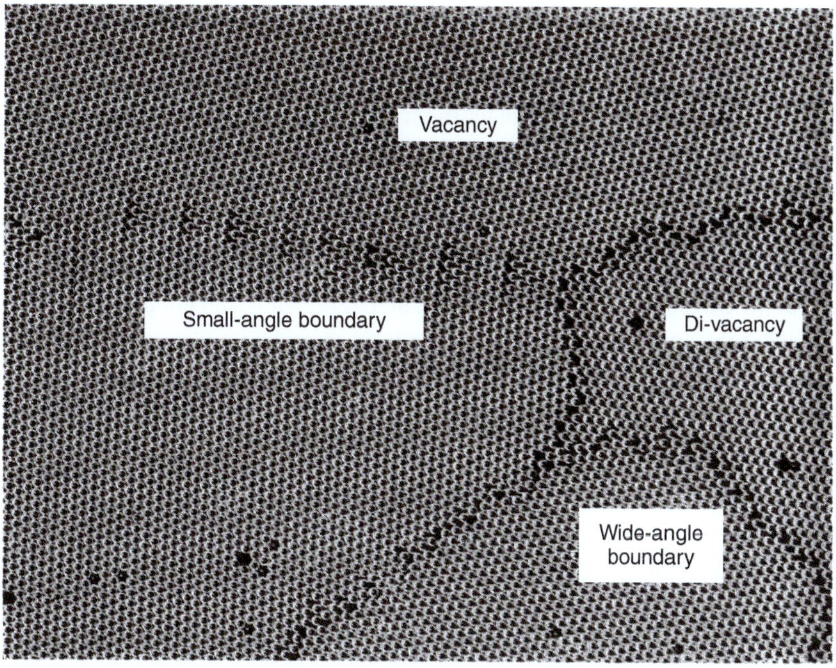

Figure 10.22 Image of a bubble raft, providing a 2D analogue of crystalline materials [112], and showing a wide-angle grain boundary, essentially a narrow random region, and a small-angle grain boundary, constructed from a line of dislocations. Also visible are vacancies and a di-vacancy.

Table 10.8 Intrinsic stacking fault energy γ_{sf}, twin-boundary energy γ_{twin}, and grain boundary energy γ_{gb} (data in parenthesis are from [115]) for some metals in comparison with their surface energies γ [110].

Metal	γ_{sf} (mJ m^{-2})	γ_{twin} (mJ m^{-2})	γ_{gb} (mJ m^{-2})	γ_{gb} (mJ m^{-2}), T (°C)	γ (mJ m^{-2})
Ag	16	8	790	(375, 950)	1140
Al	166	75	325	(324, 450)	980
Au	32	15	364	(378, 1000)	1485
Cu	45	24	625	(625, 950)	1725
Ni	125	43	866	(866, 1060)	2280
Pt	322	161	1000	(660, 1300)	3000

Source: Adapted from [110, 115].

of real grain boundaries is frequently based on CSLs and CSLs play an important role in twinning phenomena and for coherent inclusions. Estimating U_{gb} for wide-angle boundaries tends to be complex. For a tutorial, see Ref. [113].

Impurity elements can segregate to grain boundaries and surfaces. In crystalline materials, the driving force is related to the difference in size and/or charge. In inorganics, small amounts of impurities may segregate dramatically to the grain boundaries, influencing the material properties to a large extent. For example, the presence of just a few ppm of CaO in polycrystalline Al_2O_3 [114] with a segregation factor of about 600 leads to several atomic percent of Ca at the grain boundaries. This in turn leads, apart from a small degradation in mechanical properties, to a large decrease in sodium corrosion resistance.

To conclude, Table 10.8 provides the data for the stacking fault, twin boundary, and grain boundary energies for some metals in comparison to their surface energies.

10.11 Thermo-elasticity

We now turn to the continuum approach of solids, and although we do not discuss lattice symmetry in detail,[8] we need to say first some words on its effect on physical properties. Further, we limit the discussion to thermo-elastic behavior.

For all physical properties represented by tensors, the influence of the lattice symmetry on the physical properties is given by *Neumann's principle*, stating that *the symmetry elements of any physical property of a crystal must include at least all the symmetry elements of the point group of the crystal.* Physical properties thus may, and often do, possess more symmetry than the point group. For example, cubic crystals are optically isotropic and therefore exhibit more symmetry (isotropy) than required (cubic). Neumann's principle implies that the numerical values of a property before and after applying a symmetry operation to the crystal are identical. Since a symmetry operation can be considered as a coordinate axis transformation, an orthogonal matrix R_{ij} can describe the symmetry operation. A property represented by a tensor such as the fourth-order tensor C_{ijkl} changes by applying a coordinate axis transformation in $C_{pqrs}{}^*$ according to $C_{pqrs}{}^* = \Sigma_{ijkl} R_{ip} R_{jq} R_{kr} R_{ls} C_{ijkl}$. Hence Neumann's principle requires

$$C_{pqrs}{}^* = \sum\nolimits_{pqrs} R_{ip} R_{jq} R_{kr} R_{ls} C_{ijkl} \equiv C_{pqrs}$$

or, in words, the tensor transforms to itself for every symmetry element.

10.11.1 Elastic Behavior

The constitutive behavior of materials can be described, as emphasized in Chapter 5, by the Helmholtz function f and the dissipation function ϕ. For (thermo-)elastic materials, internal variables and dissipative stresses are absent, so that $\phi = 0$ [121], which implies the *absence of hysteresis*. Therefore, we deal with f only. Furthermore, we noted that by taking $f = f(\varepsilon_{ij}, T)$ with the SDG strains ε_{ij} and temperature T as kinematical variables, the stresses $\sigma_{ij}{}^{(q)}$ and specific entropy s (Section 5.2) are given by

$$\sigma_{ij}^{(q)} = \rho(\partial f/\partial \varepsilon_{ij}) \quad \text{and} \quad s = -\partial f/\partial T \tag{10.23}$$

where f acts as a potential for σ_{ij} and s for a material with mass density ρ. From now on, we omit the superscript (q) in this section as internal variables are excluded.

Let us first consider only isothermal changes and use $f = f(\varepsilon_{ij})$. We consider as *reference state* an equilibrium state having no stress at the reference temperature. For small deviations from equilibrium f can be developed in a Taylor series resulting in

$$f(\varepsilon_{ij}) = f_0 + \sum\nolimits_{ij} f_{ij}\varepsilon_{ij} + \tfrac{1}{2}\sum\nolimits_{ijkl} f_{ijkl}\varepsilon_{ij}\varepsilon_{kl} + \cdots$$

8 A brief summary of the symmetry group theory is given by Nye [116], Appendix B. For a complete group theoretical discussion, see [117] or for physical discussion, see [11] and [118–120].

with $f_{ij} = (\partial f/\partial \varepsilon_{ij})_0$, $f_{ijkl} = (\partial^2 f/\partial \varepsilon_{ij}\partial \varepsilon_{kl})_0$, and the subscript 0 denoting the reference state. In equilibrium the terms f_{ij} are zero, and therefore $f - f_0$ is of second order. Consequently, the density ρ can be considered as constant and Eq. (10.23) reduces to

$$\sigma_{ij} = \partial w/\partial \varepsilon_{ij} \quad \text{where} \quad w = \rho(f - f_0)$$

is interpreted as the *strain energy density*. For adiabatic conditions a similar treatment yields the same expressions, but with f replaced by the specific internal energy u.

The general expression for the stress in an elastic solid, with a stress-free reference configuration, is thus

$$\sigma_{ij} = \sum\nolimits_{kl} C_{ijkl}\varepsilon_{kl} \quad \text{or} \quad \boldsymbol{\sigma} = \mathbb{C} : \boldsymbol{\varepsilon} \quad \text{with} \quad \mathbb{C} = C_{ijkl} = \rho f_{ijkl} \tag{10.24}$$

denoting the *elastic (stiffness) constants*. This expression is often labeled as *Hooke's law* but in this case, law really means constitutive relation. The total number of elements of the elastic stiffness tensor is $3^4 = 81$. However, the number of independent elements is reduced due to the symmetry of both $\boldsymbol{\sigma}$ and $\boldsymbol{\varepsilon}$. This allows for the index exchanges $ij \leftrightarrow ji$ and $kl \leftrightarrow lk$, reducing the number to $6^2 = 36$. Since the order of differentiation in Eq. (10.24) is immaterial, we also have the index exchange $ij \leftrightarrow kl$. This reduces the number of independent elements to $(6 \cdot 5)/2 + 6 = 21$. Crystal symmetry will reduce the number of required elastic constants further.

As discussed in Section 5.7, if a constitutive relation connects two symmetric second-order tensors, here the stress σ_{ij} and the strain ε_{ij}, its general form representing *anisotropic materials* is according to the Hamilton–Cayley theorem (Section A.10), given by

$$\sigma_{ij} = C_{ij} + \sum\nolimits_{kl} C_{ijkl}\varepsilon_{kl} + \sum\nolimits_{klmn} C_{ijklmn}\varepsilon_{kl}\varepsilon_{mn}$$

with as material properties C_{ij}, C_{ijkl}, and C_{ijklmn}. For *isotropic materials* the relation itself must be isotropic and of the form

$$\sigma_{ij} = a(\varepsilon_{(1)}, \varepsilon_{(2)}, \varepsilon_{(3)})\delta_{ij} + b(\varepsilon_{(1)}, \varepsilon_{(2)}, \varepsilon_{(3)})\varepsilon_{ij} + c(\varepsilon_{(1)}, \varepsilon_{(2)}, \varepsilon_{(3)})\sum\nolimits_k \varepsilon_{ik}\varepsilon_{kj}$$

with a, b, and c functions of the basic invariants $\varepsilon_{(1)}$, $\varepsilon_{(2)}$, and $\varepsilon_{(3)}$ of the tensor ε_{ij}. For a linear theory with a stress-free reference state, we obtain (see also [116])

$$\sigma_{ij} = \lambda\varepsilon_{(1)}\delta_{ij} + 2\mu\varepsilon_{ij} = \lambda\delta_{ij}\sum\nolimits_k \varepsilon_{kk} + 2\mu\varepsilon_{ij}$$

where the temperature-dependent scalars λ and μ are *Lamé's constants*.

Decomposing this expression in isotropic and deviatoric parts, the latter indicated by a prime, yields $\Sigma_k\sigma_{kk} = (3\lambda + 2\mu)\Sigma_k\varepsilon_{kk}$, $\sigma_{ij}' = 2\mu\varepsilon_{ij}'$, $\Sigma_k\varepsilon_{kk} = \Sigma_k\sigma_{kk}/(3\lambda + 2\mu)$, and $\varepsilon_{ij}' = \sigma_{ij}'/2\mu$, from which we obtain the inversion

$$\varepsilon_{ij} = \varepsilon_{ij}' + \sum\nolimits_k \varepsilon_{kk}\delta_{ij} = \frac{1}{2\mu}\left(\sigma_{ij} - \frac{\lambda}{3\lambda + 2\mu}\sum\nolimits_k \sigma_{kk}\delta_{ij}\right) \tag{10.25}$$

10.11.2 Stress States and the Associated Elastic Constants

Let us consider for isotropic materials a *uniaxial stress* state, where, say, only $\sigma_{11} \neq 0$, as occurs, for example, for a rod in tension. In that case, we obtain from Eq. (10.25)

$$\varepsilon_{11} = \frac{\lambda + \mu}{\mu\,(3\lambda + 2\mu)}\sigma_{11} \equiv \frac{1}{E}\sigma_{11}, \quad \varepsilon_{22} = \varepsilon_{33} = -\frac{\lambda}{2\mu\,(3\lambda + 2\mu)}\sigma_{11} \equiv -\frac{\nu}{E}\sigma_{11}$$

and $\varepsilon_{23} = \cdots = 0$. The constant E is *Young's modulus*, while ν is *Poisson's ratio*.

Table 10.9 Relations between the various elastic constants for isotropic materials.

	E*, *G	***E*, *ν***	***G*, *ν***	***λ*, *μ***	***K*, *G***	***K*, *E***
E	E	E	$2(1+\nu)G$	$\dfrac{\mu(3\lambda+2\mu)}{(\lambda+\mu)}$	$\dfrac{9KG}{3K+G}$	E
ν	$\dfrac{E-2G}{2G}$	ν	ν	$\dfrac{\lambda}{2(\lambda+\mu)}$	$\dfrac{3K-2G}{6K+2G}$	$\dfrac{3K-E}{6K}$
K	$\dfrac{EG}{9G-3E}$	$\dfrac{E}{3(1-2\nu)}$	$\dfrac{2G(1+\nu)}{3(1-2\nu)}$	$\lambda+2\mu/3$	K	K
$G=\mu$	G	$\dfrac{E}{2(1+\nu)}$	G	G	G	$\dfrac{3KE}{9K-E}$
λ	$\dfrac{G(E-2G)}{3G-E}$	$\dfrac{E\nu}{(1+\nu)(1-2\nu)}$	$\dfrac{2G\nu}{(1-2\nu)}$	λ	$K-2G/3$	$\dfrac{3K(3K-E)}{9K-E}$

Two other useful relations are $\mu/(\lambda+\mu)=1-2\nu$ and $\mu/(\lambda+2\mu)=\nu/(1-\nu)$.

Similarly, a *hydrostatic stress*, $\sigma_{ij}=-p\delta_{ij}=\Sigma_k\sigma_{kk}\delta_{ij}/3$, which occurs, for example, in bodies during immersion in a liquid, yields

$$\varepsilon_{kk}=-p/K \quad \text{where} \quad K=\lambda+2\mu/3$$

is the *bulk modulus* with as inverse the *compressibility* $\kappa=1/K$.

A *simple shear stress*, for which, for example, only $\sigma_{23}=\sigma_{32}\neq 0$, results in

$$\varepsilon_{11}=\cdots=0, \quad \varepsilon_{31}=\varepsilon_{12}=0, \quad \gamma_{23}=\varepsilon_{23}+\varepsilon_{32}=G^{-1}\sigma_{23} \quad \text{where} \quad G=\mu$$

is the *shear modulus* and $\gamma_{23}=\varepsilon_{23}+\varepsilon_{32}$ is called a *pseudo-vector component*. Useful relations between the various elastic constants are given in Table 10.9.

The expression for Poisson's ratio $\nu=(3K-2G)/(6K+2G)$ shows that for $K\to\infty$, $\nu\to ½$, a property called *incompressibility*. For rubbers high values of ν are obtained as for a typical rubber $K=1$ GPa and $G=1$ MPa, leading to $\nu=0.4995$. While loosely speaking rubbers are called incompressible, they actually have a low ratio G/K.

10.11.3 Elastic Energy

Let us consider the amount of energy associated with the deformation of a material element. If we accept the various strains as independent coordinates, this energy is usually called the *strain energy density*. In fact, we have seen that it is the Helmholtz energy density $\rho f=\rho f(\varepsilon_{ij},T)$ for isothermal conditions. From the strain energy density definition, we obtain, using $\sigma_{ij}=\Sigma_{kl}C_{ijkl}\varepsilon_{kl}$,

$$w=\int\sigma_{ij}(\varepsilon_{kl})\mathrm{d}\varepsilon_{ij}=\sum\nolimits_{ijkl}\int C_{ijkl}\varepsilon_{kl}\,\mathrm{d}\varepsilon_{ij}=½\sum\nolimits_{ijkl}C_{ijkl}\varepsilon_{ij}\varepsilon_{kl}=½\sum\nolimits_{ij}\sigma_{ij}\varepsilon_{ij} \tag{10.26}$$

The expression $w=½\Sigma_{ij}\sigma_{ij}\varepsilon_{ij}$ is sometimes referred to as *Clapeyron's equation*.

For isotropic materials inserting Eq. (10.24) in Eq. (10.26) results in

$$w=½\lambda\sum\nolimits_{ij}\varepsilon_{ii}\varepsilon_{jj}+\mu\sum\nolimits_{ij}\varepsilon_{ij}\varepsilon_{ij}=½\lambda\varepsilon_{(1)}{}^2+\mu\varepsilon_{(2)}=½K\varepsilon_{(1)}{}^2+G\varepsilon_{(2)}{}'$$

where, as before, $\varepsilon_{(1)} = \Sigma_k \varepsilon_{kk}$ and $\varepsilon_{(2)} = \Sigma_{ij}\varepsilon_{ij}\varepsilon_{ij}$ denote the basic invariants of the strain tensor ε_{ij} and the prime indicates the deviatoric nature. It is thus possible to express the strain energy density of an isotropic body as the sum of a term $w_{\text{sha}} = G\varepsilon_{(2)}{'}$ only dependent on the shape change, and a term $w_{\text{vol}} = \frac{1}{2}K\varepsilon_{(1)}{}^2$ only dependent on the volume change. This energetic decoupling is characteristic for isotropic materials and does not occur for anisotropic materials.

Considering the various stresses as independent coordinates, we obtain the *complementary strain energy density* or *stress energy density*

$$w^{(c)} = \int \varepsilon_{ij}(\sigma_{kl})\,\mathrm{d}\sigma_{ij} = \tfrac{1}{2}\sum\nolimits_{ijkl} S_{ijkl}\sigma_{ij}\sigma_{kl}$$

where the tensor S_{ijkl} is the inverse of C_{ijkl} (or, equivalently, $\mathbb{S} = \mathbb{C}^{-1}$). Because $w^{(c)} = \sigma_{ij}\varepsilon_{ij} - w$ and recognizing that the Gibbs energy density ρg is given by $\rho g = \rho f - \Sigma_{ij}\sigma_{ij}\varepsilon_{ij}$ (similar to $G = F + PV$), we observe that $w^{(c)} = -\rho g(\sigma_{ij};T)$ for isothermal conditions. For a linear elastic material, $w^{(c)}$ is numerically equal to w because $w^{(c)} = \Sigma_{ij}\sigma_{ij}\varepsilon_{ij} - w = \Sigma_{ij}(\sigma_{ij}\varepsilon_{ij} - \frac{1}{2}\sigma_{ij}\varepsilon_{ij}) = \frac{1}{2}\Sigma_{ij}\sigma_{ij}\varepsilon_{ij}$.

Finally, we note that for stability reasons the strain energy w is positive definite ($w > 0$). Therefore, it holds for isotropic materials that $K \geq 0$ and $G \geq 0$. From $E = 9KG/(3K + G)$, we also have $E \geq 0$. Since $G = E/2(1 + \nu)$ and $K = E/3(1 - 2\nu)$, it follows that $-1 < \nu < \frac{1}{2}$. For most materials, though, $\nu > 0$ with a value of about 0.2–0.4.

10.11.4 A Matter of Notation

To be able to represent stresses and strains, and elastic constants as column matrices and square matrices, respectively, we introduce here a frequently used notation.

For the strains ε_{ij} the following abbreviations are used:

$$\varepsilon_1 = \varepsilon_{11} \quad \varepsilon_2 = \varepsilon_{22} \quad \varepsilon_3 = \varepsilon_{33} \quad \varepsilon_4 = \varepsilon_{23} + \varepsilon_{32} \quad \varepsilon_5 = \varepsilon_{13} + \varepsilon_{31} \quad \varepsilon_6 = \varepsilon_{12} + \varepsilon_{21}$$

This allows us to write the strain tensor as a 6×1 column matrix given by

$$\boldsymbol{\varepsilon} = (\varepsilon_1, \varepsilon_2, \varepsilon_3, \varepsilon_4, \varepsilon_5, \varepsilon_6)^{\mathrm{T}}$$

The components ε_i are called *pseudo-vector components*, while the components ε_{ij} are addressed as *tensor components*. The components ε_4, ε_5, and ε_6 were indicated before by γ_{23} ($=\gamma_{32}$), γ_{13} ($=\gamma_{31}$), and γ_{12} ($=\gamma_{21}$). While the tensorial components relate to half the change in angle between two unit vectors $\boldsymbol{n}$ and $\boldsymbol{m}$, the pseudo-vector components, being twice as large, relate directly to that change. A similar notation can be used for the gradient ∇ and displacement $\boldsymbol{u}$. Writing ∇ and $\boldsymbol{u}$ as

$$\mathbf{G} = G_{ij} = \begin{pmatrix} \partial/\partial x & 0 & 0 \\ 0 & \partial/\partial y & 0 \\ 0 & 0 & \partial/\partial z \\ 0 & \partial/\partial z & \partial/\partial y \\ \partial/\partial z & 0 & \partial/\partial x \\ \partial/\partial y & \partial/\partial x & 0 \end{pmatrix} \quad \text{and} \quad \mathbf{u} = u_j = \begin{pmatrix} u_x \\ u_y \\ u_z \end{pmatrix}$$

the relation $\boldsymbol{\varepsilon} = \nabla\boldsymbol{u}$ can also be written as $\varepsilon_i = \Sigma_j G_{ij} u_j$ or, equivalently, $\boldsymbol{\varepsilon} = \mathbf{G}\mathbf{u}$.

Similar to the strain case, the pseudo-vector notation for stresses σ_{ij} is given by

$$\sigma_1 = \sigma_{11} \quad \sigma_2 = \sigma_{22} \quad \sigma_3 = \sigma_{33} \quad \sigma_4 = \sigma_{23} = \sigma_{32} \quad \sigma_5 = \sigma_{31} = \sigma_{13} \quad \sigma_6 = \sigma_{12} = \sigma_{21}$$

which allows us to write the stress tensor as a 6×1 column matrix

$$\boldsymbol{\sigma} = (\sigma_1, \sigma_2, \sigma_3, \sigma_4, \sigma_5, \sigma_6)^{\mathrm{T}}$$

Note the difference in definition, as compared with the strain, for elements 4, 5, and 6.

Finally, to make full use of the advantages of matrix calculation and for easy representation, the fourth-order elasticity tensor C_{ijkl} is often written as a 6×6 square matrix C_{mn}, whereby the following index changes are applied:

$$11 \to 1 \quad 22 \to 2 \quad 33 \to 3 \quad 23 = 32 \to 4 \quad 31 = 13 \to 5 \quad 12 = 21 \to 6$$

To get consistent results, the compliances should transform as $S_{ijkl} \to S_{mn}$ if both m and n belong to the set (1,2,3), $2S_{ijkl} \to S_{mn}$ if only one of m or n belong to the set (1, 2, 3), and $4S_{ijkl} \to S_{mn}$ if both m and n belong to the set (4,5,6). Although the benefits of this notation from a computational point of view are clear, it relies on a strict adherence to the ordering of the elements, not always done in the same way by different authors.

Summarizing, for a homogeneous elastic material the stress–strain relation in index, direct, pseudo-vector, and matrix notation reads, respectively, as

$$\sigma_{ij} = \sum\nolimits_{kl} C_{ijkl}\varepsilon_{kl}, \quad \sigma = \mathbb{C} : \varepsilon, \quad \sigma_i = \sum\nolimits_j C_{ij}\varepsilon_j \quad \text{and} \quad \boldsymbol{\sigma} = \mathrm{C}\boldsymbol{\varepsilon}$$

The corresponding inverse relation are

$$\varepsilon_{ij} = \sum\nolimits_{kl} (C^{-1})_{ijkl}\sigma_{kl} = \sum\nolimits_{kl} S_{ijkl}\sigma_{kl}, \quad \varepsilon = \mathbb{S} : \boldsymbol{\sigma}, \quad \varepsilon_i = \sum\nolimits_j S_{ij}\sigma_j \quad \text{and} \quad \boldsymbol{\varepsilon} = \mathbb{S}\boldsymbol{\sigma}$$

where $\mathbb{C}$ denote the elastic stiffnesses and $\mathbb{S}$ their inverses, the elastic compliances. For isotropic materials the matrices $\mathbb{C}$ and $\mathbb{S}$ in terms of E and ν read in full

$$C_{11} = C_{22} = C_{33} = (1 - \nu)E/(1 + \nu)(1 - 2\nu), \quad C_{44} = C_{55} = C_{66} = E/2(1 + \nu)$$

$$C_{12} = C_{13} = C_{23} = \nu E/(1 + \nu)(1 - 2\nu) \qquad C_{45} = C_{46} = C_{56} = 0$$

$$S_{11} = S_{22} = S_{33} = 1/E \qquad S_{44} = S_{55} = S_{66} = 2(1 + \nu)/E$$

$$S_{12} = S_{13} = S_{23} = -\nu/E \qquad S_{45} = S_{46} = S_{56} = 0$$

10.11.5 Anisotropic Materials

So far, we have discussed isotropic materials, for which only two elastic constants remain, but materials are often anisotropic. Single-crystal materials are never isotropic, only accidentally nearly isotropic, and with decreasing symmetry of the crystal an increasing number of elastic constants is required. A full discussion of the influence of symmetry upon the number of elastic constants is given from a group theoretical point of view in [117] or from a physical point of view in [116, 119], where also a useful table providing the full symmetry relations for all crystal lattices is given. We here just illustrate two important symmetries in Justification 10.1.

Justification 10.1 ***Reduction by Crystal Symmetry*** The reduction of the number of elastic constants due to crystal symmetry is based on Neumann's principle. Since each symmetry operation can be represented by an orthogonal matrix R_{ij}, we have according to Neumann's principle, $C_{ijkl}^* \equiv C_{ijkl}$. For reflection with respect to the $x_1 - x_2$ mirror normal to x_3, an inversion, and a n-fold axis around x_3 ($n = 2, 3, 4$ and 6), we have, respectively,

$$R_{ij} = \begin{pmatrix} 1 & 0 & 0 \\ 0 & 1 & 0 \\ 0 & 0 & -1 \end{pmatrix}, \quad R_{ij} = \begin{pmatrix} -1 & 0 & 0 \\ 0 & -1 & 0 \\ 0 & 0 & -1 \end{pmatrix} \quad \text{and} \quad R_{ij} = \begin{pmatrix} \cos 2\pi/n & \sin 2\pi/n & 0 \\ -\sin 2\pi/n & \cos 2\pi/n & 0 \\ 0 & 0 & 1 \end{pmatrix}$$

For cubic crystals, the axes $x_1 - x_2 - x_3$ have fourfold symmetry, but in fact cubic symmetry is defined by four threefold axes along the $\langle 111 \rangle$ directions. After some straightforward but tedious algebra (as such a calculation is often called in the literature) we obtain [116]

$$C_{ijkl} = C_{12}\delta_{ij}\delta_{kl} + C_{44}(\delta_{ik}\delta_{jl} + \delta_{il}\delta_{jk}) + (C_{11} - C_{12} - 2C_{44})I_{ijkl}$$

where $I_{ijkl} = 1$ if all indices are equal and zero otherwise. Therefore, cubic crystals have three independent constants, namely C_{11}, C_{12}, and C_{44}. For isotropic materials a similar procedure leads to

$$C_{ijkl} = C_{12}\delta_{ij}\delta_{kl} + C_{44}(\delta_{ik}\delta_{jl} + \delta_{il}\delta_{jk}) \quad \text{with} \quad 2C_{44} = C_{11} - C_{12}$$

The number of independent constants is therefore reduced to two.

The positive definiteness of the strain energy $w = ½\Sigma_{ijkl}C_{ijkl}\varepsilon_{ij}\varepsilon_{kl} = ½\Sigma_{ij}C_{ij}\varepsilon_i\varepsilon_j$ requires that $w > 0$ for all strains, implying that the eigenvectors of C_{ij} are all positive. For cubic symmetry this leads to

$$C_{44} > 0, \quad C_{11} > | C_{12} | \quad \text{and} \quad C_{11} + 2C_{12} > 0$$

Similar relations can also be derived for other crystal symmetries.

For cubic crystals, the lattice constants are all equal, that is, $a = b = c$, while for the lattice angles it holds that $\alpha = \beta = \gamma = 90°$, while the three necessary, independent elastic constants are $C_{11} = C_{22} = C_{33}$, $C_{12} = C_{13} = C_{23}$, and $C_{44} = C_{55} = C_{66}$, with the remaining elements zero. Values for some cubic crystals are given in Table 10.10. The degree of anisotropy is indicated by the *anisotropy parameter* $A = 2C_{44}/(C_{11} - C_{12})$, also given in Table 10.10. Evidently, anisotropy varies to a large extent.

We can also calculate Young's modulus in any direction in the crystal. For a cubic crystal, Young's modulus E in the direction $\boldsymbol{n} = (n_1,n_2,n_3)$ is given by

$$E^{-1}(\boldsymbol{n}) = S_{11} - 2S\left(n_1^{\,2}n_2^{\,2} + n_2^{\,2}n_3^{\,2} + n_3^{\,2}n_1^{\,2}\right) \tag{10.27}$$

where n_1, n_2, and n_3 denote the direction cosines and $S = (S_{11} - S_{12} - ½S_{44})$. The elastic compliance constants S_{ij} are given explicitly by

$$S_{11} = (C_{11} + C_{12})/C, \quad S_{12} = -C_{12}/C \quad \text{and} \quad S_{44} = C_{44}^{\,-1} \tag{10.28}$$

where $C = (C_{11} - C_{12})/(C_{11} + 2C_{12})$. If we calculate Young's modulus in [111, 100] directions, it appears that the anisotropy can result in a Young's modulus for the two directions differing more than a factor 2.

Table 10.10 Density, lattice constant, and elastic constants for some cubic materials at 298 K.

Material	ρ (g cm^{-3})	a (Å)	C_{11} (GPa)	C_{12} (GPa)	C_{44} (GPa)	A (–)
Ar[a)]	1.771	5.256	5.29	1.35	1.59	0.81
HMT[b)]	1.331	6.927	16.4	4.3	5.2	0.86
Adamantane[c)]	1.0730	—	6.15	3.1	3.0	1.97
Cyclohexane[d)]	0.8329	—	2.8	2.4	0.35	1.75
NaCl	2.163	2.814	49.47	12.88	12.87	0.70
KCl	1.984	3.14	40.69	7.11	6.31	0.38
CsCl	3.988	3.56	36.44	8.82	8.04	0.58
Al	2.697	4.049	106.8	60.41	28.34	1.22
Fe	7.867	2.866	226	140	116	2.70
Ni	8.908	3.524	248.1	154.9	124.2	2.67
MgO	3.597	3.010	297.1	95.4	156.1	1.55
Fe_3O_4	5.175	8.397	273.0	106.0	97.1	1.16
$Y_3Al_5O_{12}$	4.553	11.954	334	112	115	1.04
(Mn,Zn)Fe_2O_4[e)]	5.133	8.840	234	142	88.5	1.92

a) At 4.2 K.
b) HMT = hexamethylenetetramine [122].
c) At 293 K [123].
d) At 279 K [124] (only about 0.2% of all organic crystals have a cubic space group [125]).
e) $Mn_{0.45}Zn_{0.50}Fe_{2.05}O_4$ [126].
Source: Rumble [127]/Taylor & Francis.

A similar relation can be derived for Poisson's ratio ν for transverse strain in the direction $\boldsymbol{m}$ and longitudinal strain in the direction $\boldsymbol{n}$, and reads

$$\nu(\boldsymbol{m},\boldsymbol{n}) = -\frac{S_{12} + S\left(n_1{}^2m_1{}^2 + n_2{}^2m_2{}^2 + n_3{}^2m_3{}^2\right)}{S_{11} - 2S\left(n_1{}^2n_2{}^2 + n_2{}^2n_3{}^2 + n_3{}^2n_1{}^2\right)}$$

As before, n_1, n_2, n_3, m_1, m_2, and m_3 denote direction cosines. A rather complete discussion on orientational dependence of moduli in cubic crystals is given in [128].

Another important example is orthorhombic symmetry. For an orthorhombic crystal the lattice constants $a \neq b \neq c$, while the lattice angles $\alpha = \beta = \gamma = 90°$. The required nine independent elastic constants are $C_{11}, C_{22}, C_{33}, C_{12}, C_{13}, C_{23}, C_{44}, C_{55}$ and C_{66} with the remaining elements zero. This symmetry is not only important for inorganic crystals but also for composites and polymer crystals. An example of the latter is the polyethylene crystal with the chains fully ordered along the z-axis and for which the theoretically calculated elastic compliance constants are [129]

$$S_{11} = 14.5, \quad S_{22} = 11.7, \quad S_{33} = 0.317, \quad S_{44} = 31.4, \quad S_{55} = 61.7, \quad S_{66} = 27.6$$

$$S_{12} = -4.78, \quad S_{13} = -0.019, \quad S_{23} = -0.062, \quad \text{all in GPa}^{-1}/100$$

In the direction of the chains, we find $E_3 = S_{33}{}^{-1} = 312.5$ GPa, while the shear modulus G_6 is equal to $G_6 = S_{66}{}^{-1} = 3.6$ GPa. This polymer thus shows a high stiffness in the chain direction, consistent with its use as reinforcement fiber in composites, but perpendicular to the chain the material is very compliant. For comparison, Young's and shear moduli of steel are about 200 and 80 GPa, respectively.

Not only single crystals are anisotropic but polycrystalline and natural materials can also show anisotropy in their elastic (and other) properties. During casting of a metal, an approximately random distribution of grain orientation arises. In many cases the metal thereafter is deformed during which a preferential orientation, often labeled *texture*, may develop. Natural materials, such as wood, are also anisotropic. While the trunk of a tree has approximately an axial symmetry, beams cut from a trunk exhibit approximately orthorhombic symmetry.

Problem 10.4 *Modulus Anisotropy* Plot Eq. (10.27) for $S > 0$ and $S < 0$. Discuss qualitatively the difference.

Problem 10.5 *Compliances and Constants for Cubic Crystals* Derive Eq. (10.28).

10.11.6 The Effect of Temperature

In many cases the conditions are neither isothermal, nor adiabatic [121]. In that case thermal response cannot be neglected and we have to assume that the Helmholtz energy f is dependent on the strains ε_{ij} and the temperature T, that is, $f = f(\varepsilon_{ij},T)$. If we measure the strain from a stress-free reference state at a reference temperature T_0 and restrict ourselves to small temperature deviations $\Delta T = T - T_0$, we can expand $f(\varepsilon_{ij},T)$ in a power series in ε_{ij} and ΔT and truncate the series after the quadratic terms. In the reference state $\sigma_{ij} = \rho\partial f/\partial\varepsilon_{ij} = 0$ and $T = T_0$, the linear terms in ε_{ij} are absent and thus

$$\rho f = \rho f_0 - \rho s_0\Delta T + {}^1\!/\!_2\sum\nolimits_{ijkl} C_{ijkl}\varepsilon_{ij}\varepsilon_{kl} + \sum\nolimits_{ij} c_{ij}\varepsilon_{ij}\Delta T - \rho c\Delta T^2/2T_0 \tag{10.29}$$

where the coefficients f_0, s_0, C_{ijkl}, c_{ij}, and c have to be determined and the factors ρ and ρ/T_0 have been added for convenience. Taking $\varepsilon_{ij} = 0$ and $T = T_0$ shows that f_0 is the reference Helmholtz energy. Equation (10.29) describes anisotropic materials. Restricting ourselves to isotropic materials, the third term on the right-hand side can be written as

$${}^1\!/\!_2\lambda\sum\nolimits_{ij}\varepsilon_{ii}\varepsilon_{jj} + \mu\sum\nolimits_{ij}\varepsilon_{ij}\varepsilon_{ij} = {}^1\!/\!_2\lambda{\varepsilon_{(1)}}^2 + \mu\varepsilon_{(2)}$$

In this case also c_{ij} must be isotropic and the appropriate form is $a\delta_{ij}$. It is convenient to use $a = -(3\lambda + 2\mu)\alpha$ and, if we do so, the expression for f reduces to

$$\rho f = \rho f_0 - \rho s_0\Delta T + {}^1\!/\!_2\lambda\sum\nolimits_{ij}\varepsilon_{ii}\varepsilon_{jj} + \mu\sum\nolimits_{ij}\varepsilon_{ij}\varepsilon_{ij} - (3\lambda + 2\mu)\alpha\varepsilon_{kk}\Delta T - \rho c\Delta T^2/2T_0$$

The stress is calculated as before from $\sigma_{ij} = \rho\partial f/\partial\varepsilon_{ij}$, resulting in

$$\sigma_{ij} = \lambda\sum\nolimits_k\varepsilon_{kk}\delta_{ij} + 2\mu\varepsilon_{ij} - (3\lambda + 2\mu)\alpha\delta_{ij}\Delta T = [\lambda\varepsilon_{kk} - (3\lambda + 2\mu)\alpha\Delta T]\,\delta_{ij} + 2\mu\varepsilon_{ij} \tag{10.30}$$

Its inversion can be obtained in the same way as before and reads

$$\varepsilon_{ij} = \frac{1}{2\mu}\left\{\left(\sigma_{ij} + \left[2\mu\alpha\Delta T - \frac{\lambda}{3\lambda + 2\mu}\sum\nolimits_k\sigma_{kk}\right]\delta_{ij}\right)\right\} \tag{10.31}$$

The strains are similar as before but for an additional term $\alpha(T-T_0)\delta_{ij}$, which results from the change in temperature and leads to a uniform volume change $3\alpha(T-T_0)$. The parameter α may thus be interpreted as the *linear thermal expansion coefficient*.

Similarly using $s = -\partial f/\partial T$, we obtain the entropy:

$$\rho s = -\rho\,\partial f/\partial T = \rho s_0 + (3\lambda + 2\mu)\alpha \sum_k \varepsilon_{kk} + \rho c\Delta T/T_0 \tag{10.32}$$

so that s_0 is the entropy in the reference state. Calculating the *heat capacity*, given by the derivative $T\partial s/\partial T$, the result is cT/T_0, so that c can be interpreted as the heat capacity c_V at the reference temperature. This completes the interpretation of the constants in the Helmholtz expression.

Further considerations of the thermoelastic equations show that heat conduction and deformation become coupled. Generally, however, the interaction coefficient that occurs is small so that this coupling is usually neglected. We refer to the literature for details, for example, Fung [130] or Boley and Weiner [131].

Problem 10.6 ***Thermo-elastic Strain*** Prove Eq. (10.31).

Problem 10.7 ***Internal Energy*** Show that the internal energy $u = f + Ts$, taken as a function of ε_{ij} and T, is given by

$$\rho u = \rho u_0 + {}^1\!/_2\lambda\varepsilon_{ii}\varepsilon_{jj} + \mu\varepsilon_{ij}\varepsilon_{ij} + (3\lambda + 2\mu)\alpha T_0\varepsilon_{kk} + (\rho c/2T_0)\left(T^2 - T_0{}^2\right)$$

where $u_0 = f_0 + T_0 s_0$ is the internal energy in the reference state. Transform $u = u(\varepsilon_{ij},T)$ to $u = u(\varepsilon_{ij},s)$ with as natural variables ε_{ij} and s.

10.12 Final Remarks

The present chapter forms a broad overview of solids and many more things could be said. For this we refer to treatments of solid state physics and chemistry, for example, Elliott [132] and Gersten and Smith [3], and for many details to, for example, [5, 8, 133]. We also hardly touched molecular solids, for which an accessible entrance is Saito [134].

References

1 Zum-Gahr, K.-H. (1987). *Microstructure and Wear of Materials*. Amsterdam: Elsevier.
2 Guy, A.G. (1960). *Elements of Physical Metallurgy*, 2e. Reading, MA: Addison-Wesley Publ. Comp.
3 Gersten and Smith, 2001.
4 Gedde, U.W. (1995). *Polymer Physics*. London: Chapman and Hall.
5 Phillips, 2001.
6 Young, R.J. and Lovell, P.A. (2011). *Introduction to Polymers*, 3e. Boca Raton, FL: CRC Press.
7 Seitz, F. (1940). *Modern Theory of Solids*. New York: McGraw-Hill Book Comp.
8 Ziman, J.M. (1972). *Principles of the Theory of Solids*, 2e. London: Cambridge University Press.

9 Born and Huang, 1954.

10 Greenwood, N.N. (1968). *Ionic Crystals, Lattice Defects and Non-stoichiometry*. London: Butterworth.

11 Kelly, A.A. and Knowles, K.M. (2012). *Crystallography and Crystal Defects*, 2e. Chichester: John Wiley & Sons.

12 Bragg, W.H. and Bragg, W.L. (1933). *The Crystalline State*, Vol. I, *A General Survey*. London: G. Bell and Sons Ltd.

13 de With, G., Harkema, S., and Feil, D. (1976). *Acta Cryst. B* 32: 3178; de With, G. (1976). *J. Appl. Cryst.* 9: 502.

14 de With, G. and Harkema, S. (1977). *Acta Cryst. B* 33: 2367; de With, G. (1977). *J. Appl. Cryst.* 10: 353.

15 Pettifor, D.G. (1995). *Bonding and Structure of Molecules and Crystals*. Oxford: Clarendon Press.

16 Sutton, A. (1993). *Electronic Structure of Materials*. Oxford: Clarendon Press.

17 Mott, N.F. and Jones, H. (1936). *The Theory of Metals*. Oxford: Clarendon Press (also Dover, 1958).

18 Herring, C. (1940). *Phys. Rev.* 57: 1169.

19 Ashcroft, N.W. (1966). *Phys. Lett.* 23: 48.

20 Martin, R.M. (2020). *Electronic Structure*, 2e. Cambridge: Cambridge University Press.

21 Schwerdtfeger, P. (2011). *ChemPhysChem*. 12: 3143.

22 Girifalco, L.A. (1976). *Acta Metall.* 24: 759.

23 Kittel, C. (1963). *Quantum Theory of Solids*. New York: John Wiley & Sons.

24 Carr, W.J. (1961). *Phys. Rev.* 122: 1437.

25 Nozieres, P. and Pines, D. (1958). *Phys. Rev.* 111: 442.

26 Harrison, 1980.

27 Gombás, P. (1949). *Die Statistische Theorie des Atoms und ihre Anwendungen*. Vienna: Springer-Verlag.

28 March, N.H. (1957). *Adv. Phys.* 6: 1.

29 Slater, J.C. (1951). *Phys. Rev.* 81: 385.

30 Hohenberg, P. and Kohn, W. (1964). *Phys. Rev. B* 864: 136.

31 Parr, R.G. and Yang, W. (1989). *Density-functional Theory of Atoms and Molecules*. Oxford: Oxford University Press.

32 Kohn, W. and Sham, L.S. (1965). *Phys. Rev.* 140: A113.

33 Kanungo, B., Zimmerman, P.M., and Gavini, V. (2019). *Nat. Commun.* 10: 4497.

34 Toulouse, J. (2023). *Review of Approximations for the Exchange-Correlation Energy in Density-Functional Theory*, Ch. 1 in *Density Functional Theory* (ed. E. Cancès and G. Friesecke), p. 1. Cham: Springer.

35 Tran, F., Doumont, J., Kalantari, L. et al. (2019). *J. Appl. Phys.* 126: 110902.

36 Geerlings, P., De Proft, F., and Langenaeker, W. (2003). *Chem. Rev.* 103: 1793.

37 Makkar, P. and Ghosh, N.N. (2021). *RSC Adv.* 11: 27897.

38 Cohen, A.J., Mori-Sánchez, P., and Yang, W. (2012). *Chem. Rev.* 112: 289.

39 Dreizler, R.M. and Gross, E.K.U. (1990). *Density Functional Theory*. Berlin: Springer.

40 Koch, W. and Holthausen, M.C. (2001). *A Chemist's Guide to Density Functional Theory*. Weinheim: Wiley-VCH.

41 Fang, C.M. and de With, G. (2002). *Philos. Mag. A* 82: 2885.

42 Singleton, J. (2001). *Band Theory and the Electronic Structure of Solids*. Oxford: Oxford University Press.
43 Prasad, R. (2014). *Electronic Structure of Materials*. Boca Raton: CRC Press.
44 Stillinger, F.H., Sakai, H., and Torquato, S. (1970). *J. Chem. Phys.* 24: 3677.
45 Pauling, L. and Wilson, E.B. (1935). *Introduction to Quantum Mechanics*. New York: McGraw-Hill Book Comp. (also Dover, 1985).
46 Slater, J.C. (1930). *Phys. Rev.* 36: 57.
47 Slater, J.C. (1951). *Quantum Theory of Matter*. New York: McGraw-Hill Book Comp.
48 Pilar, F.L. (1968). *Elementary Quantum Chemistry*. London: McGraw-Hill Book Comp.
49 Kutzelnigg, W. and Morgan, J.D. III. (1996). *Z. Phys. D* 36: 197.
50 Coulson, C.A. (1961). *Valence*, 2e. Oxford: Oxford University Press.
51 Pauling, L. (1960). *The Nature of the Chemical Bond*, 3e. Ithaca, NY: Cornell University Press.
52 Allred, A.L. (1961). *J. Inorg. Nucl. Chem.* 17: 215.
53 Mulliken, R.S. (1934). *J. Chem. Phys.* 2: 782.
54 Allred, A.L. and Rochow, E.G. (1958). *J. Inorg. Nucl. Chem.* 5: 264.
55 Sproul, G.D. (2020). *ACS Omega* 5: 11585.
56 Sanderson, R.T. (1976). *Chemical Bonds and Bond Energy*, 2e. New York: Academic Press.
57 Sanderson, R.T. (1983). *J. Am. Chem. Soc.* 105: 2259.
58 Sanderson, R.T. (1983). *Polar Covalence*. New York: Academic Press.
59 Hinze, J. (1999). *The Concept of Electronegativity of Atoms in Molecules*, Ch. 7 in *Pauling's Legacy: Modern Modelling of the Chemical Bond* (ed. Z.B. Maksić and W.J. Orville-Thomas), p. 189. Amsterdam: Elsevier and references therein.
60 Bergmann, D. and Hinze, J. (1987). *Electronegativity and Charge Distribution*, Ch. 7 in *Electronegativity*, Vol. 66 (ed. K.D. Sen and C.K. Jørgensen), p. 145. Berlin: Springer-Verlag.
61 Jensen, W.B. (1998). *J. Chem. Educ.* 75: 817; This is paper II of a series of three, worthwhile reading (I: *Ibid.* 679 and III: *Ibid.* 961).
62 Martynov, A.I. and Batsanov, S.S. (1980). *Russ. J. Inorg. Chem.* 25: 1737.
63 Jensen, W.B. (1995). *J. Chem. Educ.* 72: 395.
64 Brewer, L. (1977). Lawrence Berkeley National Laboratory. LBNL Report #: LBL-3720; Although unpublished, this reference has been quoted many times, https://escholarship.org/uc/item/08p2578m.
65 Lennard-Jones, J.E. and Ingham, A.E. (1925). *Proc. R. Soc A* 107: 636.
66 Kihara, T. and Kuba, S. (1952). *J. Phys. Soc. Jpn.* 7: 348.
67 Kitaigorodsky, A.I. (1973). *Molecular Crystals and Molecules*. New York: Academic Press.
68 Pertsin, A.J. and Kitaigorodsky, A.I. (1987). *The Atom-Atom Potential*. Berlin: Springer-Verlag.
69 Finnis, M. (2003). *Interatomic Forces in Condensed Matter*. Oxford: Oxford University Press.
70 Fyta, M. (2016). *Computational Approaches in Physics*. San Rafael, CA: Morgan & Claypool Publishers.
71 Shannon, R.D. and Prewitt, C.T. (1969). *Acta Cryst. B* 25: 925; Shannon, R.D. (1976). *Acta Cryst. A* 32: 751.

72 Slater, J.C. (1939). *Introduction to Chemical Physics*. New York: McGraw-Hill Book Comp. (also Dover, 1970).
73 Templeton, D.H. (1953). *J. Chem. Phys.* 21: 2097.
74 Baughan, E.C. (1959). *Trans. Faraday Soc.* 55: 736.
75 Jenkins, H.D.B. and Waddington, T.C. (1971). *Nat. Phys. Sci.* 237: 5.
76 Ladd, M.F.C. and Lee, W.H. (1959). *J. Inorg. Nucl. Chem.* 11: 264; (1964). *Prog. Solid State Chem.* 1: 37. New York: Pergamon Press.
77 Tosi, M.P. (1964). *Cohesion of Ionic Solids in the Born Model*, Ch. 1 in *Solid State Physics*, Vol. 16 (ed. F. Seitz and D. Turnbull), p. 1. New York: Academic Press.
78 Waddington, T.C. (1959). *Adv. Inorg. Chem. Radiochem.* 1: 157.
79 Brown, I.D. (2017). *The Chemical Bond in Inorganic Chemistry – The Bond Valence Model*, 2e. Oxford: Oxford University Press; (2002). 1e.
80 Brown, I.D. (2000). *J. Chem. Educ.* 77: 1070.
81 Sanderson, R.T. (1955). *J. Chem. Educ.* 32: 140.
82 Sanderson, R.T. (1988). *J. Chem. Educ.* 65: 112, 227.
83 Parr, R.G. and Pearson, R.G. (1983). *J. Am. Chem. Soc.* 105: 7512.
84 Rose, J.H., Smith, J.R., Guinea, F. et al. (1984). *Phys. Rev. B* 29: 2963.
85 de Boer, F.R., Mattens, W.C.M., Boom, R. et al. (1988). *Cohesion in Metals*. Amsterdam: North-Holland.
86 Daw, M.S., Foiles, S.M., and Baskes, M.I. (1993). *Mater. Sci. Rep.* 9: 251; Voter, A.F. and Chen, S.P. (1987). *Mater. Res. Soc. Symp.* 82: 175.
87 Donovan, B. and Angress, J.F. (1971). *Lattice Vibrations*. London: Chapman and Hall.
88 Böttger, H. (1983). *Principles of the Theory of Lattice Dynamics*. Weinheim: Physik Verlag.
89 Vočadlo, N.L. and Price, G.D. (1984). *Phys. Earth Planet. Inter.* 82: 261.
90 Moelwyn-Hughes, E.A. (1961). *Physical Chemistry*, 2e. Oxford: Pergamon Press.
91 Grimvall, G. (1999). *Thermophysical Properties of Materials*, 2e. Amsterdam: Elsevier; (1989). 1e. Amsterdam: North-Holland.
92 Grüneisen, E. (1926). *Geiger-Scheel's Handbuch der Physik*, Vol. 10, p. 1. Berlin: Springer.
93 Barron, T.H.K. and Klein, M.L. (1974). *Perturbation Theory of Anharmonic Crystals*, Ch. 7 in *Dynamical Properties of Solids*, Vol. 1 (ed. G.K. Horton and A.A. Maradudin), p. 391. Amsterdam: North-Holland.
94 Leibfried, G. (1961). *Theory of Anharmonic Effects in Crystals*, Ch. 5 in *Solid State Physics*, Vol. 12 (ed. F. Seitz and D. Turnbull), p. 276. New York: Academic Press.
95 Wallace, D.C. (1972). *Thermodynamics of Crystals*. New York: John Wiley & Sons, Inc. (also Dover, 1998).
96 Girifalco, L.A. and Kniaz, K. (1997). *J. Mater. Res.* 12: 311.
97 Guinea, F., Rose, J.H., Smith, J.R. et al. (1984). *Appl. Phys. Lett.* 44: 53.
98 Flynn, C.P. (1972). *Point Defects and Diffusion*. Oxford: Clarendon Press.
99 Stoneham, A.M. (1975). *Theory of Defects in Solids*. Oxford: Oxford University Press.
100 Mott, N.F. and Gurney, R. (1940). *Electronic Processes in Ionic Crystals*. Oxford: Clarendon Press.

101 von Ammon, W., Sattler, A., and Kissinger, G. (2017). *Defects in Monocrystalline Silicon*, Ch. 5 in *Springer Handbook of Electronic and Photonic Materials* (ed. S. Kasap and P. Capper), p. 111. Berlin: Springer Verlag.

102 Seeger, A. and Chik, K.P. (1968). *Phys. Stat. Sol.* 29: 455.

103 Suezawa, M. (2017). *Jpn. J. Appl. Phys.* 56: 048005.

104 Allnatt, A.R. and Lidiard, A.B. (1993). *Atomic Transport in Solids*. Cambridge: Cambridge University Press.

105 Barr, L.W. and Dawson, D.K. (1971). *Proc. Br. Ceram. Soc.* 19: 151.

106 Kofstad, P. (1972). *Non-stoichiometry, Diffusion and Electrical Conductivity in Binary Metal Oxides*. New York: Wiley-Interscience.

107 Varotsos, P.A. and Alexopoulos, K.D. (1986). *Thermodynamics of Point Defects and Their Relation with Bulk Properties*. Amsterdam: North-Holland.

108 Cottrell, A.H. (1953). *Dislocations and Plastic Flow in Crystals*. Oxford: Oxford University Press.

109 Nabarro, F.R.N. (1976). *Theory of Crystal Dislocations*. Oxford: Oxford University Press (also Dover, 1987).

110 Hirth, J.P. and Lothe, J. (1982). *Theory of Dislocations*, 2e. New York: McGraw-Hill Book Comp.

111 Mrowec, S. (1980). *Defects and Diffusion in Solids*. Amsterdam: Elsevier.

112 Smith, C.S. (1964). *Rev. Mod. Phys.* 36: 524; The bubble raft analogue was invented by Bragg, L. and Nye, J.F. (1947). *Proc. R. Soc. London* 190: 474.

113 Fischmeister, H.J. (1985). *J. Phys. Colloids* 46: C4–C3.

114 de With, G., Vrugt, P.J., and van de Ven, A.J.C. (1985). *J. Mater. Sci.* 20: 1215.

115 Murr, L.E. (1975). *Interfacial Phenomena in Metals and Alloys*. Reading, MA: Addison-Wesley Publishing Company.

116 Nye, J.F. (1957). *Physical Properties of Crystals*. London: Oxford University Press.

117 Bhagavantam, S. (1966). *Crystal Symmetry and Physical Properties*. London: Academic Press.

118 Wooster, W.A. (1973). *Tensors and Group Theory for the Physical Properties of Crystals*. Oxford: Clarendon Press.

119 Newnham, R.E. (2005). *Properties of Materials*. Oxford: Oxford University Press.

120 Haussühl, S. (2007). *Physical Properties of Crystals*. Weinheim: Wiley-VCH.

121 Ziegler, H. (1983). *Introduction to Thermomechanics*, 2e. Amsterdam: North-Holland.

122 Haussühl, S. (1958). *Acta Cryst.* 11: 58.

123 Damien, J.C. (1975). *Solid State Commun.* 16: 1271.

124 Ahmad, S.F., Kiefte, H., and Clouter, M.J. (1978). *J. Chem. Phys.* 69: 5468.

125 Belsky, V.K., Zorkaya, O.N., and Zorky, P.M. (1995). *Acta Cryst. A* 51: 651.

126 de With, G. and Damen, J.P.M. (1981). *J. Mater. Sci. Lett.* 16: 838.

127 Rumble, J.R. (ed.) (2021). *CRC Handbook of Chemistry and Physics (Internet Version)*, 102e. Boca Raton, FL: CRC Press/Taylor & Francis; accessed 15 May 2022.

128 Knowles, K.M. and Howie, P.R. (2015). *J. Elast.* 120: 87.

129 Tashiro, K., Kobayashi, M., and Tadakoro, H. (1978). *Macromolecules* 11: 914.

130 Fung, Y.C. (1965). *Foundations of Solid Mechanics*. Englewood Cliffs, NJ: Prentice-Hall, Inc.

131 Boley, B.A. and Weiner, J.H. (1960). *Theory of Thermal Stresses*. New York: John Wiley & Sons.

132 Elliott, 1998.
133 Ashcroft, N.W. and Mermin, N.D. (1976). *Solid State Physics*. Orlando, FL: Harcourt College Publishers Inc.
134 Saito, K. (2020). *Chemical Physics of Molecular Condensed Matter*. Singapore: Springer.

Further Reading

Born, M. and Huang, K. (1954). *Dynamical Theory of Crystal Lattices*. Oxford: Oxford University Press. A classic and now somewhat outdated, but still a highly readable and illuminating book.

Elliott, S.R. (1998). *The Physics and Chemistry of Solids*. Chichester: John Wiley & Sons. A well-written text, largely focusing on the physics rather than the chemistry.

Gersten, J.I. and Smith, F.W. (2001). *The Physics and Chemistry of Materials*. New York: John Wiley & Sons. An intermediate, modern text on materials science dealing with many topics.

Girifalco, L.A. (2000). *Statistical Mechanics of Solids*. Oxford: Oxford University Press. An intermediate level overview on many topics for solids but does not deal with bonding.

Phillips, R. (2001). *Crystals, Defects and Microstructures*. Cambridge: Cambridge University Press. An illuminating overview on exactly what the titles promises.

Boyd, R.H. and Phillips, P.J. (1993). *The Science of Polymer Molecules*. Cambridge: Cambridge University Press. A highly useful and broad overview on polymer architecture, synthesis and solutions.

Harrison, W.A. (1980). *Electronic Structure and the Properties of Solids*. San Francisco: W.H. Freeman and Comp. (also Dover, 1989). An inspired and detailed exposition on the use of semi-empirical methods in solid state problems.

11

Interfaces

We now turn to the "cement" between the "bricks," that is, to interfaces. They influence the various transitions, but they also themselves show a (roughening) transformation, and segregation of a component to an interface can be considered a transformation as well. We limit the discussion mainly to the thermodynamics and structure of planar (non-electrified) surfaces of one- and two-component systems.

11.1 Thermodynamics of Interfaces

Indicating first our notation, in this chapter, components are labeled with a subscript, say j. The reference compound, usually the matrix, is labeled as 1. In considering two phases, we must indicate the phase, and we do so by a superscript (α) and (β) for the two bulk phases and (σ) for the interface. Generally, a *surface* is the border between a gas and a liquid or solid, while an *interface* is the border between condensed phases.

There are two approaches to describe thermodynamically the interfacial behavior of, say, a liquid and a vapor. The first approach, by van der Waals, Bakker, and Guggenheim, takes the interface as a real phase with volume, entropy, and so on (Figure 11.1a). Hence, there are three phases to consider: the liquid, the vapor, and the interfacial phase, also called *interphase*. For each phase, the normal thermodynamic relations apply. There is only an additional work term for the interphase, to which we will come shortly. The second approach, by Gibbs, associates the interface between two phases with a *dividing* (or *geometrical*) *plane*, positioned in such a way that the total amount of a particular component is the same as if both bulk phases remained homogeneous up to that dividing plane (Figure 11.1a). Although the dividing plane has volume zero, it nevertheless has excess properties such as entropy and energy, which consequently can be positive or negative. The amount of component j, in excess over the amount when both phases remain homogeneous up to the dividing plane, referred to as component 1, is indicated by $n_j^{(1)}$ or per unit area as the *surface excess* $\Gamma_j^{(1)} = n_j^{(1)}/A$, where A is the surface area. Clearly, $\Gamma_1^{(1)} = 0$.

There is a significant difference in surroundings of the molecules at the surface of, say, a liquid, as compared with the bulk, resulting in a lesser number of intermolecular interactions, even if the average structure on the surface apart from the cut remains the same as in the bulk. After surface creation, relaxation takes place, resulting in an average structure in the surface different from the one in the bulk, but still with excess energy (Figure 11.1b). Generally, the thickness τ of the interphase is small, say, a few molecular diameters or about

Phases of Matter and their Transitions: Concepts and Principles for Chemists, Physicists, Engineers, and Materials Scientists, First Edition. Gijsbertus de With.

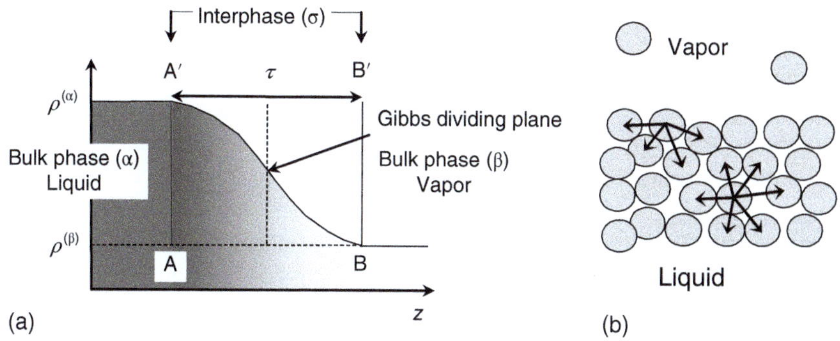

Figure 11.1 (a) Schematic representation of the density profile over a liquid surface. (b) Schematic representation of the liquid–vapor interface, showing the difference in environment for a molecule at the surface and in the bulk.

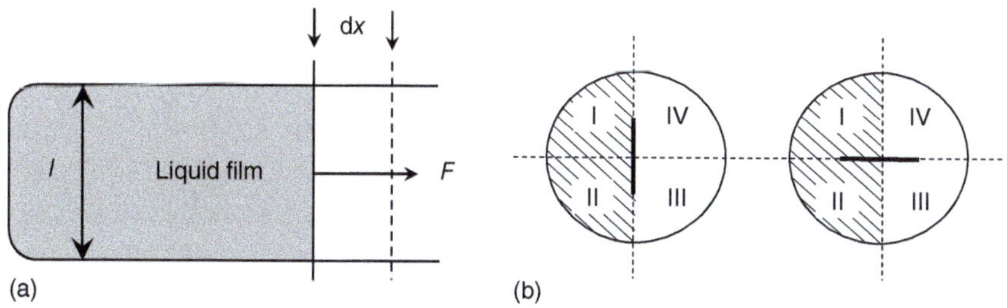

Figure 11.2 (a) Schematic representation of a sliding-wire experiment to extend the area A of a liquid film at a constant temperature T by $dA = 2l dx$, requiring a force F. Because the interface region of the film has a smaller density than the bulk of the film, during the process the volume changes and the process occurs at a constant pressure P. Since the interfacial thickness τ is typically less than about 1 nm, the interphase volume $V^{(\sigma)}$ is small as compared to the bulk volume $V^{(L)}$, and to a high degree of precision the process still occurs at a constant volume V. For a single-component liquid, because the work done is $F dx = \gamma dA$, the surface tension γ represents both the surface stress per unit length and the Helmholtz (Gibbs) energy per unit area. (b) The difference in attractive interactions between a test plane parallel and perpendicular to a surface. The hatched area represents the bulk phase α, while the non-hatched area represents the vapor phase β. The range of interaction, as felt by a molecule in the center, is indicated by the circles. The decrease in attraction in quadrants I + II and III + IV for an orientation of the test plane parallel to the surface will be larger than the decrease in attraction between quadrants I + IV and II + III for a perpendicular orientation because the density is low in both quadrants III and IV and high in both quadrants I and II. In the bulk, the difference vanishes.

1 nm for normal liquids and molecular solutions. Although for interfaces the dissimilarity between both phases is less, they nevertheless have an interface energy.

So far, we have treated materials as homogeneous and obeying Euler's condition, so that internal energy U and entropy S are additive. As surfaces have a surface energy, the work involved if we create additional surfaces adds to the various thermodynamic potentials. A well-known elementary, though not so easy to carry out, experiment (Figure 11.2a) shows that for the extension of a surface by dA an amount of work $\delta W = \gamma dA$ is needed, where γ is often called the *surface* or *interfacial tension* for interfaces between two phases α and β (although it represents surface work). Stability requires that $\gamma > 0$, consistent with the missing interactions in the molecular picture.

To consider the work $\gamma\,dA$ a bit more carefully, we use a system (Figure 11.1a), where the planes AA′ and BB′ are positioned in the bulk phases α and β parallel to the surface, so that their distance is τ and the volume of the interphase is $V^{(\sigma)} = A\tau$. The pressure P is the perpendicular force per unit area on an infinitesimal virtual test plane in the bulk phase, resulting from the collisions of all molecules on one side of this plane. Consider now for a static liquid an infinitesimal cube with the x–y plane parallel to the surface. The pressure on the $yz-$, $xz-$, and $xy-$plane is P_{xx}, P_{yy}, and P_{zz}, respectively, and equilibrium of forces requires that $\partial P_{xx}/\partial x = \partial P_{yy}/\partial y = \partial P_{zz}/\partial z = 0$ (Section 5.4) From $\partial P_{zz}/\partial z = 0$, we have $P_{zz} \equiv P_{\text{n}}$ with P_{n} the *normal pressure*, constant everywhere in the fluid. Because the bulk is isotropic with $P_{xx} = P_{yy} = P_{zz} = P$ where P is the *hydrostatic pressure*, we have $P_{\text{n}} = P$. For the interphase, only the xy-plane is isotropic with $P_{xx} = P_{yy} \equiv P_{\text{t}}$, where P_{t} is the *tangential pressure* [1]. This implies that in the interphase P_{xx}, P_{yy}, and P_{zz} are functions of z only. The difference between $P = P_{\text{n}}$ and $P_{\text{t}}(z)$ integrated over the thickness represents γ, or

$$\gamma = \int_{-\infty}^{+\infty} [P - P_{\text{t}}(z)]\,\text{d}z \cong \int_{-1/2\tau}^{+1/2\tau} [P - P_{\text{t}}(z)]\,\text{d}z \tag{11.1}$$

where the last step can be made as for $|z| > {}^1/_2\tau$ we have $P \cong P_{\text{t}}$. Summarizing, the surface energy corresponds to a surface stress parallel to the surface (Figure 11.2b).

Consider now a change in area dA, associated with a change in thickness dτ and a change in volume to $V^{(\sigma)} + \text{d}V^{(\sigma)}$ for a constant amount of material. For the work by the force in the direction perpendicular to the interface, we have $-PA\text{d}\tau$. The work by the force in the direction parallel to the interface is $-(P\tau - \gamma)\,\text{d}A$. The total work done is

$$\delta W = -PA\text{d}\tau - (P\tau - \gamma)\,\text{d}A = -P(A\text{d}\tau + \tau\text{d}A) + \gamma\text{d}A = -P\text{d}V^{(\sigma)} + \gamma\text{d}A$$

and this expression replaces $-P\text{d}V^{(\alpha)}$ as used for the bulk phase.

In the Guggenheim picture, the Helmholtz energy F of the system is defined by

$$F \equiv F^{(\alpha)} + F^{(\beta)} + F^{(\sigma)} \tag{11.2}$$

where $F^{(\alpha)}$, $F^{(\beta)}$, and $F^{(\sigma)}$ denote the Helmholtz energies of the two homogeneous phases and the interphase, respectively. A similar expression applies for U, V, S, and n_i. For the bulk phase, we have

$$\text{d}F^{(\alpha)} = -S^{(\alpha)}\text{d}T - P\text{d}V^{(\alpha)} + \sum\nolimits_j \mu_j \text{d}n_j^{(\alpha)} \quad \text{and} \quad F^{(\alpha)} = -PV^{(\alpha)} + \sum\nolimits_j \mu_j n_j^{(\alpha)} \tag{11.3}$$

Because Eq. (11.3a) is homogeneous of the first degree in the extensive properties $V^{(\alpha)}$ and $n_j^{(\alpha)}$, Eq. (11.3b) follows from Euler's theorem (Section A.3). Clearly, the intensive properties T, P, and μ_i do not require a phase superscript, since in equilibrium they are constant throughout the phases. For the interphase, we similarly obtain

$$\text{d}F^{(\sigma)} = -S^{(\sigma)}\text{d}T - P\text{d}V^{(\sigma)} + \sum\nolimits_j \mu_j \text{d}n_j^{(\sigma)} + \gamma\text{d}A, \quad F^{(\sigma)} = -PV^{(\sigma)} + \sum\nolimits_j \mu_j n_j^{(\sigma)} + \gamma A \tag{11.4}$$

From Eqs. (11.4a) and (11.2), we obtain

$$S^{(\sigma)} = -\left(\frac{\partial F^{(\sigma)}}{\partial T}\right)_{V^{(\sigma)},n_j^{(\sigma)},A} \quad \text{and} \quad \gamma = \left(\frac{\partial F^{(\sigma)}}{\partial A}\right)_{T,V^{(\sigma)},n_j^{(\sigma)}} = \left(\frac{\partial F}{\partial A}\right)_{T,V,n_j}$$

For the Gibbs energy,[1] we use

$$G^{(\alpha)} \equiv F^{(\alpha)} + PV^{(\alpha)} \quad \text{and} \quad G^{(\sigma)} \equiv F^{(\sigma)} + PV^{(\sigma)} - \gamma A \quad \text{so that} \tag{11.5}$$

$$\mathrm{d}G^{(\sigma)} = -S^{(\sigma)}\mathrm{d}T + V^{(\sigma)}\mathrm{d}P + \sum\nolimits_j \mu_j \mathrm{d}n_j^{(\sigma)} - A\,\mathrm{d}\gamma \quad \text{and} \quad G^{(\sigma)} = \sum\nolimits_j \mu_j n_j^{(\sigma)} \tag{11.6}$$

Equation (11.6b) corresponds to a similar expression for the bulk phase. Because $F = F^{(\alpha)} + F^{(\beta)} + F^{(\sigma)} = -PV + \Sigma_j \mu_j n_j + \gamma A$, this requires that $G = F + PV$ is taken as $G = G^{(\alpha)} + G^{(\beta)} + G^{(\sigma)} + \gamma A$, deviating from the form of Eq. (11.2). Calculating dG, we find

$$\mathrm{d}G = -S\mathrm{d}T + V\mathrm{d}P + \sum\nolimits_j \mu_j \mathrm{d}n_j + \gamma \mathrm{d}A \quad \text{and} \quad \gamma = (\partial G/\partial A)_{T,P,n_j}$$

consistent with the sliding-wire experiment (Figure 11.2a).

Differentiating Eq. (11.4b) and subtracting from the result Eq. (11.4a), we obtain

$$\mathrm{d}\gamma = -s^{(\sigma)}\mathrm{d}T + \tau\,\mathrm{d}P - \sum\nolimits_j \Gamma_j \mathrm{d}\mu_j \quad \text{with} \quad s^{(\sigma)} = \frac{S^{(\sigma)}}{A}, \quad \Gamma_j = \frac{n_j^{(\sigma)}}{A}, \quad \tau = \frac{V^{(\sigma)}}{A} \tag{11.7}$$

Equation (11.7) is the surface analogue of the Gibbs–Duhem equation, known as the *Gibbs adsorption equation* (GAE). Using $\mu_j = \mu_j^* + RT \ln a_j$ with a_j the activity, we obtain

$$\mathrm{d}\mu_j = RT\,\mathrm{d}\ln a_j \quad \text{and} \quad \mathrm{d}\gamma = -s^{(\sigma)}\mathrm{d}T + \tau \mathrm{d}P - RT\sum\nolimits_j \Gamma_j\,\mathrm{d}\ln a_j$$

It can be shown that dγ is independent of the precise choice for the position of the planes AA′ and BB′ [2, 3]. Taking the limit $\tau \to 0$ and referring to the surface excess $\Gamma_j^{(1)}$ with respect to the resulting dividing plane determined by component 1 is equivalent to using the Gibbs approach with $V^{(\sigma)} = 0$. Approximating the activity a_j by the concentration c_j, the result is the GAE as usually employed

$$\mathrm{d}\gamma \cong -s^{(\sigma)}\mathrm{d}T - RT\sum\nolimits_j \Gamma_j^{(1)}\mathrm{d}\ln c_j = -s^{(\sigma)}\mathrm{d}T - RT\sum\nolimits_j \left(\Gamma_j^{(1)}/c_j\right)\mathrm{d}c_j$$

For a two-component system at constant T and P, we have $\Gamma_2^{(1)} = -(RT)^{-1}(\partial\gamma/\partial \ln c_2)_T = -(c_2/RT)(\partial\gamma/\partial c_2)_T$. Evidently, $\Gamma_2^{(1)}$ can be obtained by measuring the change in surface tension γ with concentration c_2.

According to the phase rule for a single-component system, γ is a function of one variable only and, if we take T, we can write dγ/dT. In that case, Eq. (11.7) reduces to

$$\mathrm{d}\gamma = -s^{(\sigma)}\mathrm{d}T + \tau\mathrm{d}P - \Gamma\mathrm{d}\mu$$

Employing $\mathrm{d}\mu = -S_\mathrm{m}{}^{(\alpha)}\mathrm{d}T + V_\mathrm{m}{}^{(\alpha)}\mathrm{d}P = -S_\mathrm{m}{}^{(\beta)}\mathrm{d}T + V_\mathrm{m}{}^{(\beta)}\mathrm{d}P$, where $S_\mathrm{m}{}^{(\alpha)}$ indicates the molar entropy and $V_\mathrm{m}{}^{(\alpha)}$ indicates the molar volume, to eliminate dμ and dP, we obtain

$$-\frac{\mathrm{d}\gamma}{\mathrm{d}T} = \left(s^{(\sigma)} - \Gamma S_\mathrm{m}^{(\alpha)}\right) - \left[\left(\tau - \Gamma V_\mathrm{m}^{(\alpha)}\right)\frac{S_\mathrm{m}^{(\beta)} - S_\mathrm{m}^{(\alpha)}}{V_\mathrm{m}^{(\beta)} - V_\mathrm{m}^{(\alpha)}}\right] \cong s^{(\sigma)} - \Gamma S_\mathrm{m}^{(\alpha)} \tag{11.8}$$

The last step can be made only for surfaces between a condensed phase α and a vapor phase β. In the interphase, the density is comparable to that in phase α, so that $\tau/\Gamma \cong V_\mathrm{m}{}^{(\alpha)}$,

1 The Gibbs energy of the interphase can be defined also by transforming with respect to the volume work only, that is, using $G'^{(\sigma)} \equiv F^{(\sigma)} + PV^{(\sigma)}$ instead of with respect to all work terms as in Eq. (11.7b). This choice results in $G'^{(\sigma)} = \Sigma_j \mu_i n_j^{(\sigma)} + \gamma A$ (deviating from the form for the bulk) and $\mathrm{d}G'^{(\sigma)} = -S^{(\sigma)}dT - V^{(\sigma)}dP + \Sigma_j \mu_j \mathrm{d}n_j^{(\sigma)} + \gamma \mathrm{d}A$, so that $\gamma = (\partial G'/\partial A)_{T,P,n}$ and $(\partial S^{(\sigma)}/\partial A)_{T,P,nj} = (\partial\gamma/\partial T)_{A,P,nj}$. In this case, we have $G = G^{(\alpha)} + G^{(\beta)} + G'^{(\sigma)}$ (consistent with Eq. (11.3)) and leading again to $\gamma = (\partial G/\partial A)_{T,P,n}$, but (at constant T and P) $dG = -(w_\mathrm{max} - PdV)$ is *not* the net work $-(w_\mathrm{max} - PdV + \gamma dA)$, as it is for Eq. (11.6).

but $\tau/\Gamma \ll V_m{}^{(\beta)}$. Therefore, remote from the critical temperature, $(\tau/\Gamma) - V_m{}^{(\alpha)} \ll V_m{}^{(\beta)} - V_m{}^{(\alpha)}$. Similarly, employing $\mu = G_m{}^{(\alpha)} = U_m{}^{(\alpha)} - TS_m{}^{(\alpha)} + PV_m{}^{(\alpha)} = G_m{}^{(\beta)} = U_m{}^{(\beta)} - TS_m{}^{(\beta)} + PV_m{}^{(\beta)}$ and $g^{(\sigma)} = \Gamma\mu = u^{(\sigma)} - Ts^{(\sigma)} + P\tau - \gamma$ with $u^{(\sigma)} = U^{(\sigma)}/A$ and $g^{(\sigma)} = G^{(\sigma)}/A$, and eliminating $S_m{}^{(\alpha)}$, $S_m{}^{(\beta)}$, and $s^{(\sigma)}$, the result is

$$\gamma - T\frac{d\gamma}{dT} = \left(u^{(\sigma)} - \Gamma U_m^{(\alpha)}\right) - \left[\left(\tau - \Gamma V_m^{(\alpha)}\right)\frac{U_m^{(\beta)} - U_m^{(\alpha)}}{V_m^{(\beta)} - V_m^{(\alpha)}}\right] \cong u^{(\sigma)} - \Gamma U_m^{(\alpha)} \qquad (11.9)$$

where the last step can be made for the same reason as for Eq. (11.8). For a created surface of unit area, we may thus interpret γ as the work done, $-T(d\gamma/dT)$ as the heat absorbed (the difference in entropy of surface and bulk for the same amount of matter, multiplied by T), and $\gamma - T(d\gamma/dT)$ as the increase in internal energy (the difference in internal energy of surface and bulk for the same amount of matter).

Using the Gibbs approach for this system so that $V^{(\sigma)} = n_1{}^{(1)} = 0$, we obtain

$$dF^{(\sigma)} = -S^{(\sigma)}\,dT + \gamma dA \quad \text{and} \quad F^{(\sigma)} = \gamma A \quad \text{or} \quad f^{(\sigma)} = F^{(\sigma)}/A = \gamma$$

and we see again that the surface tension is equal to the surface Helmholtz energy per unit area.[2] The (*specific*) *surface internal energy* is

$$u^{(\sigma)} = U^{(\sigma)}/A = (F^{(\sigma)} + TS^{(\sigma)})/A = \gamma - T(d\gamma/dT)$$

as expected and consistent with Eqs. (11.8) and (11.9).

Finally, consider the equilibrium of an infinitesimal nonplanar interfacial area with pressure $P^{(\beta)}$ on the concave side and $P^{(\alpha)}$ on the other side at constant T and n_j. In the Gibbs approach, we have from Eqs. (11.2), (11.3a), and (11.4a),

$$dF = dF^{(\alpha)} + dF^{(\beta)} + dF^{(\sigma)} = -P^{(\alpha)}dV^{(\alpha)} - P^{(\beta)}dV^{(\beta)} + \gamma dA = 0$$

Since $V = V^{(\alpha)} + V^{(\beta)}$ is constant, we have $dV^{(\beta)} = -dV^{(\alpha)}$, resulting in

$$\Delta P = (P^{(\beta)} - P^{(\alpha)}) = \gamma(dA/dV^{(\beta)}) \qquad (11.10)$$

---

2 The terms "surface tension," "surface stress," and "surface energy" are often used indiscriminately but making the distinction is useful. Since all work is given by the product of a "generalized displacement," for surface work the increase in area dA, and a "generalized force," for surface work called *surface stress* Ψ, we have for a single-component, single-phase system, $\delta W = \Psi dA$. At constant T and V, ΨdA contributes to the Helmholtz energy F. Upon increasing A, for liquid surfaces, the average distance between molecules does not change (conventionally, but confusingly, called a *plastic increase*). For solid surfaces, this average distance *does* change (labeled as *elastic increase*). Assuming for the *surface Helmholtz energy* $F^{(\sigma)} = A\gamma$, we obtain for Ψ the *Shuttleworth equation*

$$\Psi = dF^{(\sigma)}/dA = \gamma + A(d\gamma/dA)$$

with $d\gamma/dA$ taken at constant T, μ, and c [4]. This equation has been heavily debated (see, e.g. [5]). A large part of the confusion is due to the use of different reference configurations [6, 7]. Note that in the GAE, both γ and dA refer to coordinates in the deformed state (Euler coordinates, Section 5.4, [8]). For liquids $d\gamma/dA = 0$, therefore $\delta W = \gamma dA$ so that the specific *surface work* γ equals Ψ and, consequently, is often called *surface tension*. For solids, in principle, the full expression is required. Measuring Ψ for solids is a difficult task since different methods lead to different values and solid surfaces are often not in equilibrium, as their relaxation might be kinetically hampered. Logically, surface energy should refer to surface internal energy $u^{(\sigma)}$, similarly as energy refers to internal energy, but this rule is seldom obeyed. The quantities γ and $f^{(\sigma)}$ have the same dimension and are numerically equal as long as relaxation is fast. Note that $u^{(\sigma)}$ has also the same dimension as γ but is *not* numerically equal.

From differential geometry, we learn that $dA/dV = r_1^{-1} + r_2^{-1}$, where r_1 and r_2 are the radii of curvature in two orthogonal directions in the infinitesimal plane dA. It can be shown that $r_1^{-1} + r_2^{-1} = R_1^{-1} + R_2^{-1}$ where R_1 and R_2 are the minimum and maximum radius of curvature in two orthogonal directions or the *principal curvatures*. Using $dA/dV = R_1^{-1} + R_2^{-1}$ in Eq. (11.10), one usually refers to it as the *Laplace equation*, relating γ to ΔP over a curved surface. For a sphere, $R_1 = R_2 = R$ and

$$P^{(\beta)} - P^{(\alpha)} = 2\gamma/R \tag{11.11}$$

The above result can also be obtained using $V = 4\pi R^3/3$ and $A = 4\pi R^2$. It will be clear that for a planar surface $R_1 = R_2 = \infty$, resulting in $P^{(\beta)} - P^{(\alpha)} = 0$.

Problem 11.1 ***Spreading Coefficient*** A liquid L will spread spontaneously on a substrate S (either a solid or another liquid) when spreading decreases the Gibbs energy and, hence, when $\gamma_{LV} + \gamma_{SL} < \gamma_{SV}$, or if the *spreading coefficient* $S = \gamma_{SV} - (\gamma_{LV} + \gamma_{SL}) > 0$, where γ_{XY} is the interfacial tension between X and Y with the label V denoting vapor. Table 11.1 shows the various γ-values for pure water (WV), benzene (BV), and interface WB as well the values for mutually saturated liquids. Based on these data, assess the behavior of a drop of pure benzene put on top of the surface of pure water.

Table 11.1 Surface tension data for water (W), benzene (B), and their interface.

Surface tension	γ_{WV} (mJ m^{-2})	γ_{BV} (mJ m^{-2})	γ_{WB} (mJ m^{-2})
Pure liquid	72.8	28.8	35.0
Mutually saturated liquid	62.2	28.9	35.0

Problem 11.2 ***Constant μ, P, and T*** Show for a planar system with bulk phases α and β and an interphase σ that the chemical potential μ, pressure P, and temperature T are constant throughout the system.

Problem 11.3 ***Dividing Plane Independent $\Gamma_j^{(1)}$*** Show for a planar system containing bulk phases α and β and an interface σ that the surface excess $\Gamma_j^{(1)}$ is independent of the position of the dividing plane. To do so, express $n_j^{(\sigma)}$ in terms of n_j, $c_j^{(\alpha)}$, $c_j^{(\beta)}$ and $V^{(\beta)}$, subtract $n_1^{(\sigma)}(c_j^{(\alpha)} - c_i^{(\beta)})/(c_1^{(\alpha)} - c_1^{(\beta)})$ from the result and note that one side is independent of the position of the dividing plane.

11.2 One-Component Surfaces: Semiempirical Considerations

We now discuss the surface energy of one-component systems from a semiempirical point of view. The focus is on liquids, but many arguments apply to solids as well. To start the discussion, we note that the values for γ of metals is typically 1000–2000 mJ m^{-2} [9], while for salts γ typically ranges from 500 to 1000 mJ m^{-2} [10, 11]. For molecular liquids, the range is typically 10–100 mJ m^{-2}. For further introductions including experimental procedures, see [12–14].

Table 11.2 Surface tension $\gamma = a - bt$ as a function of temperature t (°C) for various liquids.

Liquid	Range (°C)	a (mJ m^{-2})	b (mJ m^{-2} K^{-1})	Liquid	Range (°C)	a (mJ m^{-2})	b (mJ m^{-2} K^{-1})
CS_2	10–50	35.29	0.1484	$(CH_3)_2CO$	20–90	22.26	0.122
CCl_4	15–105	29.49	0.1224	$(C_2H_5)_2O$	15–30	18.92	0.0908
CCl_3H	15–75	29.91	0.1295	CH_3OH	10–60	24.00	0.0773
C_6H_6	10–80	31.50	0.1287	C_2H_5OH	10–70	24.05	0.0832
$C_6H_5CH_3$	10–100	30.90	0.1889	C_3H_7OH	10–90	25.26	0.0777
C_6H_5OH	40–140	43.54	0.1068	C_4H_9OH	10–100	27.18	0.0898

Source: Adapted from Jasper [11].

Table 11.3 Eötvös constant K and Stefan's ratio j for various liquids.

Liquid	K (erg K^{-1} mol$^{2/3}$)	T_{cri} (K)	$U^{(\sigma)}$ (cal mol^{-1})	L_V (cal mol^{-1})	j (–)
Ar	2.02	150.7	745	2278	0.431
N_2	2.00	126.0	612	1711	0.358
CO	2.00	132.9	647	1981	0.326
CS_2	2.022	546.2	2784 (2210)	8444 (6610)	0.330 (0.325)
CCl_4	2.105	556.3	2951 (2342)	11 220 (8000)	0.263 (0.287)
C_6H_6	2.104	556.6	2952 (2343)	11 434 (8180)	0.258 (0.301)

K Eötvös constant (1 erg = 10^{-7}J), T_{cri} critical temperature, $U^{(\sigma)}$ surface internal energy, L_V heat of evaporation. Note that $U^{(\sigma)} = \gamma A$ with molar area $A = \kappa N_A^{1/3} V_m^{2/3}$ and $\kappa = 2^{1/3}$ for FCC (001). In parenthesis, values using $\kappa = 1$. Data for L_V in parenthesis are from [18]. Note that, although values for L_V differ, values for Stefan's ratio j calculated from the data in parenthesis are rather similar.
Source: Moelwyn-Hughes [19]/with permission of Elsevier.

Experimentally, γ decreases with the temperature T (Table 11.2) and vanishes at the critical temperature T_{cri}. One of the first empirical expressions for $\gamma(T)$, applicable to simple and normal liquids, is *Eötvös' rule* [15] stating that, if V_m is the molar volume,

$$\gamma V_m^{2/3} = K(T_{cri} - T) \quad \text{with} \quad K = 2.1 \text{ erg } K^{-1} \text{ mol}^{2/3} \quad \text{for } T < 0.9T_{cri} \tag{11.12}$$

The constant[3] K is supposed to be a universal constant valid for all liquids (Table 11.3), which appears to be approximately true for many nonpolar and non-hydrogen bonding liquids. The value for hydrogen-bonding liquids varies with the type of compound, typically $0.95 - 1.5$ erg K^{-1} mol$^{2/3}$ for alcohols and $0.9 - 1.7$ erg K^{-1} mol$^{2/3}$ for organic acids. For H_2O, $K \cong 0.87$ erg K^{-1} mol$^{2/3}$ at 0 °C and $K \cong 1.21$ erg K^{-1} mol$^{2/3}$ at 140 °C [20]. These low values indicate molecular association for these compounds.

To compare properly one should use data for a mole of molecules in the surface. Estimating the molecular area as $a \equiv A/N_A = \kappa(V_m/N_A)^{2/3}$, the *molar surface (Helmholtz) energy*

3 Eötvös used $K = 2.27$ erg K^{-1}mol$^{2/3}$ but the present value represents the state of affairs somewhat better. Slightly more precise estimates are obtained with the *Eötvös–Ramsey–Shields rule* $\gamma V_m^{2/3} = K'(T_{cri} - T - \delta)$, where $K' \cong 2.12$ erg K^{-1}mol$^{2/3}$ and $\delta \cong 6$. There is no physical argument whatsoever for this modification [16], although [17] claims there is a mathematical argument.

$F_m^{(\sigma)} = \gamma A = \kappa N_A^{1/3}\gamma V_m^{2/3}$, where κ is a numerical factor close to 1. For example, using a face-centered cubic (FCC) lattice model, one has $\kappa = 2^{1/3} = 1.26$ for the (100) plane, $2^{5/6} = 1.78$ for the (110) plane, and $2^{-2/3}3^{1/2} = 1.09$ for the (111) plane. Often one takes $\kappa = 1$, corresponding to the (100) plane for simple cubic packing. To obtain the molar surface internal energy $U_m^{(\sigma)}$, we use the Gibbs–Helmholtz relation, resulting in

$$U_m^{(\sigma)} = \partial(F^{(\sigma)}/T)/\partial(1/T) = \kappa N_A^{1/3} V_m^{2/3}[\gamma - T(\mathrm{d}\gamma/\mathrm{d}T) - 2\alpha\gamma T/3] \tag{11.13}$$

From Eq. (11.12) one obtains $U_m^{(\sigma)} = \kappa N_A^{1/3} K T_{cri}$ and $S_m^{(\sigma)} = \kappa N_A^{1/3} K$, indicating the approximations made as compared to Eq. (11.13).

For crystalline solids and in the quasi-lattice picture for liquids, the N molecules in the bulk have coordination number z and nearest-neighbor interaction u. The potential energy Φ and heat of evaporation at constant volume L_V at $T = 0$ K are

$$\Phi = \tfrac{1}{2}Nzu \quad \text{and} \quad L_V = -\Phi$$

If we cleave the system into two parts, we have molecules at the surface with surface coordination number s and the work expended $W^{(\sigma)}$ is

$$W^{(\sigma)} = \tfrac{1}{2}N(s - z)u$$

because the number of molecules N, which in the bulk had a potential energy $\tfrac{1}{2}zu$, have now a potential energy $\tfrac{1}{2}su$. At $T = 0$ K, $U^{(\sigma)} = W^{(\sigma)}$ [18, 21]. Because $u < 0$ and $s < z$, we regain $U^{(\sigma)} > 0$. Dividing $U^{(\sigma)}$ by L_V, we obtain *Stefan's ratio*:

$$j \equiv U^{(\sigma)}/L_V = 1 - (s/z)$$

Using FCC packing for the bulk and a (111) or (011) plane for the surface, we have $z = 12$ and $s = 9$ or 8, resulting in $j = \tfrac{1}{4}$ or $\tfrac{1}{3}$, respectively. While experimentally for simple liquids $j \cong 0.35$ is observed, for normal liquids $j \cong 0.25$ is adequate (Table 11.3). The experimental values are temperature dependent though.

One can estimate γ somewhat better by invoking the Hamaker–Israelachvili expression $W = H/12\pi d^2$ [22] for the work required to separate two materials having a planar interface to a distance d. The *Hamaker constant H* is obtained by summing pair interactions over the surface of the two materials and is a characteristic for the material combination. Lifshitz theory ([23], see also [24]), employing the (complex) permittivity thereby avoiding pairwise addition of interactions, provides the same expression for W, but both approaches do not take surface relaxation into account. Estimating γ, one takes $2\gamma = W = H/12\pi d_0^2$, where d_0 is now the effective distance between two planes in the material, typically having a smaller value than the distance of closest approach for molecules. Using the "universal value" $d_0 = 0.165$ nm provides an estimate accurate to about 15% even for liquids like ethanol and acetone. Only for highly polar hydrogen-bonding liquids, such as H_2O, CH_3OH, and $(CH_3)_2NCHO$, the approach seriously underestimates the value for γ. Surface relaxation is implicitly taken into account via the value for d_0.

Another estimate for j results from Luck's empirical hole model (Section 9.6.3). We use

$$L_V = \tfrac{1}{2} c_L z R(1 - 2x) \cdot [(3T_{cri}/2) - T]$$

with x the fraction of holes in the fluid. The constant $c_L \cong 1.2 \cdot (1.2)^{-1} \cdot c = c$ takes care of next-nearest-neighbor interactions (1.2), density decrease upon melting (1/1.2), and the proportionality constant $c \cong 1.0$ between the potential energy $N_A u$ and the kinetic

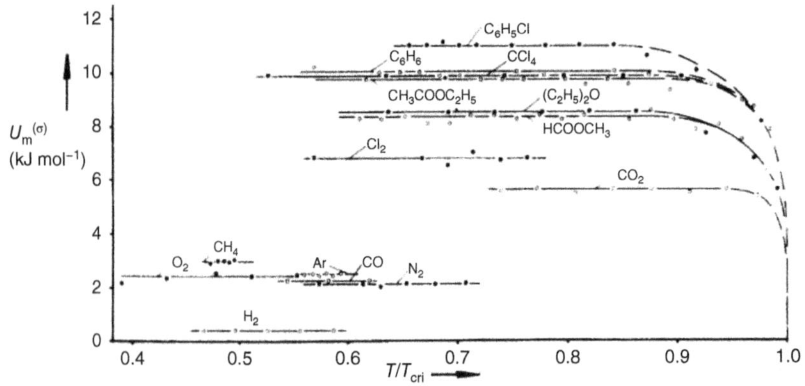

Figure 11.3 Surface energy $U_m^{(\sigma)}$ as a function of T/T_{cri} for several liquids. Source: Luck [25]/John Wiley & Sons.

energy at T_{cri}. In this model, $U_m^{(\alpha)} = L_V - RT = \frac{1}{2}cz(1-2x)\cdot 3RT_{cri}/2$. A similar equation with z replaced by $z-s$ holds for $U_m^{(\sigma)}$, which indeed appears experimentally to be proportional to T_{cri}. Using $z = 12$ and $s = 9$, the expected value for the Eötvös constant $K = 3(z-s)c/4 \cong 2.25\,\text{erg K}^{-1}\,\text{mol}^{2/3}$. The values of $U_m^{(\sigma)}$ for simple and normal liquids appear to be constant for $T/T_{cri} < 0.9$ (Figure 11.3) with $(z-s)c$ in the range of 2.6–2.9 with as average 2.85 (excepting H_2 and He), fairly close to the model estimate $(z-s)c = 3$. For $T > 0.9T_{cri}$, $U_m^{(\sigma)}$ is no longer temperature independent because of the decreasing factor $1-2x$. However, this factor is still too large and better represented by $1-4x$ [25–27], possibly because of more holes being present in the interphase than in the bulk. For $j = U_m^{(\sigma)}/L_V$, we thus have

$$j = (z-s)(1-4x)/\left[z(1-2x)\left(1-\tfrac{2}{3}T_{red}\right)\right] \quad \text{with} \quad T_{red} = T/T_{cri}$$

For $T \to 0$ K, $j \to \frac{1}{4}$ and j behaves approximately linear in the range from $0.4T/T_{cri}$ to $0.9T/T_{cri}$, as has been observed for several liquids [26]. For polar liquids $U_m^{(\sigma)}$ is constant up to $\approx 0.5T/T_{cri}$ and thereafter shows an upswing before it approaches zero at T_{cri}. Both the plateau level and the maximum are fairly well estimated [28].

Empirically extending the T-dependence of γ, but still using the fact that γ vanishes at T_{cri}, van der Waals [29] and Ferguson [30, 31] suggested to use for liquids

$$\gamma = \gamma_0 X^{1+r} \quad \text{where } X = 1 - T/T_{cri} \tag{11.14}$$

with γ_0 and r constants. For simple liquids, excellent agreement was obtained with $r = 2/9 \cong 0.222$, while for a range of organic compounds, the average value $r = 0.210 \pm 0.015$ describes the experimental data well [30, 31]. A compilation for 85 normal liquids [32] yields $r = 0.249$ with a sample standard deviation of 0.054. Katayama [33] argued that γ is determined by the difference in interactions between the liquid and vapor phases and therefore is related to the density difference. Defining $y = \Delta\rho/\mathcal{M}$ or $yV_{cri} = \Delta\rho/\rho_{cri}$, with $\mathcal{M}$ the molar mass and $\Delta\rho = \rho_L - \rho_V$ with ρ_L and ρ_V the density of the liquid and vapor, respectively, he proposed to use, based on six liquids,

$$\gamma y^{-2/3} = c(T_{cri} - T) \quad \text{with} \quad c = 2.04\,\text{erg K}^{-1}\text{mol}^{2/3} \tag{11.15}$$

Guggenheim [34] showed that for simple liquids the density difference can be accurately represented by $\Delta\rho/\rho_{cri} = 7X^{1/3}/2$ (Section 8.4). Consequently, we can write

$$y = (7\rho_{cri}/2\mathcal{M})X^{1/3} \quad \text{or} \quad \gamma = cT_{cri}(7\rho_{cri}/2\mathcal{M})^{2/3}X^{11/9} \equiv \gamma_0 X^{11/9}$$

after eliminating y using Eq. (11.15), thereby explaining nicely the empirically observed relation with $r = 2/9$. Eliminating T/T_{cri} between Eq. (11.15) and $\Delta\rho/\rho_{cri} = 7X^{1/3}/2$, results in

$$\gamma = C(\Delta\rho/\rho_{cri})^{11/3} \quad \text{with} \quad C = cT_{cri}(2/7)^3(\rho_{cri}/\mathcal{M})^{2/3}$$

In 1923, McLeod [35], apparently unaware of the results of Katayama and Ferguson, suggested a relation between γ and the densities ρ_L and ρ_V given by

$$\gamma = C'(\Delta\rho)^n \quad \text{with} \quad n = 4 \quad \text{and} \quad C' = (cT_{cri})^6/\gamma_0^5\mathcal{M}^4$$

The constant C', not given by McLeod, can be obtained by eliminating X from Eqs. (11.14) and (11.15) and taking $r = 1/5$. In fact, the data of Ramsay and Shields [16] used by McLeod show that n is somewhat smaller than 4. Sugden [36, 37] rewrote the McLeod expression to

$$\mathcal{P} = \gamma^{1/4}\mathcal{M}/\Delta\rho = (cT_{cri})^{3/2}/\gamma_0{}^{5/4}$$

where the parameter $\mathcal{P}$, denoted as *parachor*, is supposed to be a constant, different for different liquids. Experimentally $\mathcal{P}$ appears to be indeed independent of temperature over a reasonable range. Sugden pointed out that $\mathcal{P}$ can be interpreted as the molar volume corrected for thermal motion via the surface tension and related $\mathcal{P}$ to the molecular structure. A group contribution method, with contributions fitted to surface tension data, is discussed by Quayle [38]. His data were modified (Section F.13) by Poling, Prausnitz, and O'Connell [39], who state that their data are often as accurate as those from Quayle, except for compounds like benzonitrile, carbon disulfide, carbon tetrachloride and pyridine where the errors are unacceptably high. However, there are some serious flaws in this correlator, as demonstrated clearly by Exner [40], but unfortunately his critique has been largely neglected.

Because the expression $\gamma = \gamma_0 X^{1+r}$ with $X = 1 - T/T_{cri}$ is in principle only applicable close to T_{cri} (but nevertheless describes $\gamma(T)$ well [32]), Somayajulu [41] proposed to use $\gamma = aX^{5/4} + bX^{9/4} + cX^{13/4}$. He showed that for molecular liquids this expression provides a pretty good fit, generally within 1%, from T_{tri} to T_{cri}. Larger deviations appeared to be due largely to inconsistent experimental data [42].

The surface energy also obeys the principle of corresponding states (PoCS). Guggenheim [2] noticed from dimensional considerations that $\gamma_0 V_{cri}{}^{2/3}/T_{cri}$ should be a common function of T/T_{cri}. For the set Ne, Ar, Xe, N_2, O_2, and CH_4, the average value is (4.30 ± 0.16) erg K^{-1} mol$^{-2/3}$, where $\pm$ indicates the sample standard deviation (Table 8.6). An improved correlation is obtained using the acentric factor ω (Section 8.4). As a first step, the reduced density $\rho_{red} = \rho/\rho_{cri}$, taken as a function of $T_{red} = T/T_{cri}$ and ω, is used. Following Riedel [43, 44], we can use the correlation[4]

$$\rho/\rho_{cri} = 1 + 0.85(1 - T_{red}) + (1 - T_{red})^{1/3}(1.89 + 0.91\omega)$$

Inserting $T_{red} = 0$, the (hypothetical) density ρ_0 at 0 K is obtained. As an aside, we note that for $0.3 < T_{red} < 0.8$ it appears that the value of ρ/ρ_0 is independent of ω and described

4 In fact, Riedel used a parameter $\alpha = 5.808 + 4.93\omega$, which is obviously fully equivalent to ω.

well ($R^2 = 0.9999$) by $\rho/\rho_0 = 0.9720 - 0.2541T_{\text{red}} - 0.2473T_{\text{red}}{}^2$, obtained from fitting the data given in [44]. Using ρ/ρ_0 and ω, a single density measurement is sufficient to estimate $\rho(T)$ and thus the thermal expansivity α. The second step is using $\gamma = \gamma_0 X^{1+r}$, where γ_0 is interpreted as the (hypothetical) surface tension at 0 K. A correlation for γ_0 as a function of ω and the (hypothetical) volume at $T = 0$ K, $V_0 = (\mathscr{M}/\rho)(\rho/\rho_0)$ is given by Pitzer[5] [44, 45]

$$\gamma_0 V_0^{2/3}/RT_{\text{cri}} = 2.24 \times 10^{-8} + 1.42 \times 10^{-8}\omega \quad \text{or} \quad \gamma_0 V_0^{2/3}/T_{\text{cri}} = 1.86 + 1.18\omega \quad (11.16)$$

The fit of Eq. (11.16) is rather good – typically the deviations for normal fluids are less than 5% – and this led Pitzer [44, 45] to suggest that γ_0 could be used to test whether liquids behave as "normal" (Section 1.5). Now given γ_0, $\gamma = \gamma_0 X^{1+r}$ provides γ.

Finally, we note that γ is taken as size independent, but becomes size dependent for drops with very small radius r, as expressed by $\gamma = \gamma_0(1 - 2\delta/r)$ [46], where γ_0 is the surface tension for a flat surface and δ is the *Tolman length* with a magnitude in the order of a few tenths of the molecular diameter. For details, we refer to the literature.

Problem 11.4 Verify the various κ-values as given for the FCC lattice.

11.3 One-Component Surfaces: Theoretical Considerations

We now turn to a more theoretical approach. Experimentally, it appears that the thickness of the interphase is only a few molecular diameters. The question thus arises, in particular for liquids: is the surface intrinsically somewhat diffuse or intrinsically sharp, but blurred by motion of the surfaces? The first point of view, initiated by van der Waals [29], led to what is called density functional theory, while the second, due to Mandelstam [47], led to capillary wave theory.

11.3.1 Density Functional Theory

The classical *density functional theory* (DFT) is akin to QM-DFT (Section 10.5.3) based on the theorem that the electron density is a unique functional[6] of the Hamilton operator in the presence of the "external" field due to the nuclei. This theorem was extended by Mermin [48] to non-zero temperatures. In classical DFT, one proceeds likewise: the density of molecules ρ is a unique functional of the Hamilton function $\mathscr{H}$ in the presence of the external field due to the surface. The theory is involved and our presentation is necessarily concise, omitting mathematical proofs [49–51].

We use a Hamilton function $\mathscr{H} = \vartheta + \Phi + \mathscr{V}$ for N molecules, each with mass m, kinetic energy ϑ, potential energy Φ, and external potential $\mathscr{V}$. As the interphase can exchange molecules with the bulk, we use the grand partition function $\Xi = \text{tr}\exp\{\beta(\mu N - \mathscr{H}\} \equiv (N!h^{3N})^{-1}\Sigma_N \int \exp\{\beta(\mu N - \mathscr{H}\}\,\text{d}\mathbf{p}\text{d}\mathbf{q}$ with the Gibbs probability density given by $p^* = \Xi^{-1}\exp\{\beta(\mu N - \mathscr{H})\}$, associated grand potential $\Omega = -kT\ln\Xi$, dependent temperature T, volume V, and chemical potential μ (Section 7.1.2), and where the abbreviation

---

5 Note that Guggenheim used V_{cri} and Pitzer used V_0.

6 In this section, I use $[\cdot]$ to indicate a functional and use $(\cdot)$ and $\{\cdot\}$ for grouping variables.

tr (trace) is introduced. Mermin considered a functional $\Omega[p] = \mathrm{tr}\, p(\mathscr{H} - \mu N + kT \ln p)$, which for $p = p^*$ reduces to the equilibrium Ω. Using the Gibbs inequality $x \ln x - x + 1 \geq 0$ for any $x \geq 0$, one can show that the functional $\Omega\,[p]$ obeys the inequality $\Omega[p] \geq \Omega[p^*]$ for $p \neq p^*$. The equilibrium density $\rho_0(\boldsymbol{r}) = \langle \rho(\boldsymbol{r}) \rangle$, with $\rho(\boldsymbol{r}) = \Sigma_j \delta(\boldsymbol{r} - \boldsymbol{r}_j)$ and where for any X, $\langle X \rangle = \mathrm{tr}\, p^* X$. Because p^* is a functional of $\mathscr{V}$, ρ_0 is also a functional of $\mathscr{V}$ and, one can show, also a functional of $\rho_0(\boldsymbol{r})$ [49]. This implies that for a given $\boldsymbol{\Phi}$, the Helmholtz energy $\mathscr{F}\,[\rho] = \mathrm{tr}\, p^*(\vartheta + \boldsymbol{\Phi} + kT \ln p^*)$ is a unique functional of $\rho(\boldsymbol{r})$. This expression is also valid for external potentials. Using

$$\Omega[\rho] = \int \rho(\mathbf{r}) \mathscr{V}(\mathbf{r})\, \mathrm{d}\mathbf{r} + \mathscr{F}\,[\rho] - \mu \int \rho(\mathbf{r}) \mathscr{V}(\mathbf{r})\, \mathrm{d}r$$

one can show that for $\rho(\boldsymbol{r}) = \rho_0(\boldsymbol{r})$, $\Omega[\rho] = \Omega[\rho_0] = \Omega$. It is also the minimum for $\Omega[\rho]$, so that $\Omega[\rho] \geq \Omega[\rho_0]$. The proper ρ is then given by the functional derivative $\delta\Omega[\rho]/\delta\rho(\mathbf{r})|_{\rho 0} = 0$. Because $\mathscr{F}\,[\rho_0]$ is the intrinsic Helmholtz energy, we obtain for the total Helmholtz energy, $F = \mathscr{F}\,[\rho_0] + \int \rho(\mathbf{r}) \mathscr{V}(\mathbf{r})\, \mathrm{d}\boldsymbol{r}$. Defining the *intrinsic chemical potential* by $\mu_{\mathrm{int}} = \delta\mathscr{F}\,[\rho]/\delta\rho(\boldsymbol{r})$, we also have $\mathscr{V}(\boldsymbol{r}) + \mu_{\mathrm{int}}[\rho_0; \boldsymbol{r}] = \mu$. If $\mathscr{F}\,[\rho]$ can be determined, this expression determines the equilibrium density ρ_{equ}. For example, if $\mathscr{V}(\boldsymbol{r}) = 0$, $\mathscr{F}\,[\rho]$ reduces to

$$\mathscr{F}_{\mathrm{idl}}[\rho] = kT \int \rho(\mathbf{r}) \{\ln \Lambda^3 \rho(\mathbf{r}) - 1\}\, \mathrm{d}\mathbf{r}, \quad \Lambda = (h^2/2\pi mkT)^{1/2} \quad \text{and} \quad \mu_{\mathrm{int}} = kT \ln\{\Lambda^3 \rho(\boldsymbol{r})\}$$

yielding the barometric expression $\rho(\boldsymbol{r}) = z \exp\{-\beta\mathscr{V}(\boldsymbol{r})\}$ with activity $z = \Lambda^{-3} \exp(\beta\mu)$.

Interactions are incorporated by defining $\mathscr{F}\,[\rho] \equiv \mathscr{F}_{\mathrm{idl}}[\rho] - X[\rho]$. For a given $\boldsymbol{\Phi}$, $X[\rho]$ is a unique functional of ρ. The intrinsic chemical potential can be written as

$$\beta\mu_{\mathrm{int}}[\rho_0; \boldsymbol{r}] = \ln\{\Lambda^3 \rho(\boldsymbol{r})\} - c[\rho; \boldsymbol{r}]$$

where $c[\rho; \boldsymbol{r}] = \beta\delta\boldsymbol{\Phi}[\rho]/\delta\rho(\boldsymbol{r})$ results from the interactions. The equilibrium density distribution becomes $\rho_0(\boldsymbol{r}) = z \exp\{-\beta\mathscr{V}(\boldsymbol{r}) + c[\rho_0; \boldsymbol{r}]\}$, so that $c[\rho_0; \boldsymbol{r}]$ acts as an effective one-body potential. Higher-order derivatives can be obtained by further differentiation, for example, $c[\rho; \boldsymbol{r}_1, \boldsymbol{r}_2] = \delta c[\rho; \boldsymbol{r}_1]/\delta\rho(\boldsymbol{r}_2)$, which is the OZ direct correlation function for the inhomogeneous fluid.

If we take $u(\boldsymbol{r}) = \mu - \mathscr{V}(\boldsymbol{r})$, $\Omega[\rho] = -\int \rho(\mathbf{r}) u(\mathbf{r})\, \mathrm{d}\mathbf{r} + \mathscr{F}\,[\rho]$, differentiation yields

$$\delta\boldsymbol{\Phi}[\rho]/\delta u(\boldsymbol{r}) = -\rho(\boldsymbol{r}) + \int \{\delta\rho(\boldsymbol{r}')/\delta u(\boldsymbol{r})\}\{\delta\mathscr{F}\,(\boldsymbol{r}')/\delta\rho(\boldsymbol{r}) - u(\boldsymbol{r}')\}\, \mathrm{d}\boldsymbol{r}'$$

with as result that for $\rho = \rho_0$, $\delta\boldsymbol{\Phi}[\rho]/\delta u(\boldsymbol{r}) = -\rho_0(\boldsymbol{r})$, which can also be directly derived from Ξ. Applying a second differentiation one can show [49] that the *density–density correlation function* $\mathscr{G}(\boldsymbol{r}_1, \boldsymbol{r}_2)$ is given by

$$\begin{aligned}\mathscr{G}(\mathbf{r}_1, \boldsymbol{r}_2) &\equiv kT\delta\rho(\boldsymbol{r}_1)/\delta u(\boldsymbol{r}_2) \\ &= \langle\{\rho(\boldsymbol{r}_1) - \rho_0(\boldsymbol{r}_1)\}\{\rho(\boldsymbol{r}_2) - \rho_0(\boldsymbol{r}_2)\}\rangle = \rho^{(2)}(\boldsymbol{r}_1, \boldsymbol{r}_2) + \rho_0\delta(\boldsymbol{r}_1 - \boldsymbol{r}_2) - \rho_0(\boldsymbol{r}_1)\rho_0(\boldsymbol{r}_2)\end{aligned}$$

with $\rho^{(2)}(\boldsymbol{r}_1, \boldsymbol{r}_2)$ the usual two-particle correlation function (Section 9.2). Rewriting $\rho_0(\boldsymbol{r}) = z \exp\{-\beta\mathscr{V}(\boldsymbol{r}) + c[\rho_0; \boldsymbol{r}]\}$ as $c[\rho_0; \boldsymbol{r}] = \ln \Lambda^3 \rho_0(\boldsymbol{r}) - \beta u(\boldsymbol{r}_1)$ and differentiating with respect to $\rho_0(\boldsymbol{r}_2)$, we obtain

$$c^{(2)}(\boldsymbol{r}_1, \boldsymbol{r}_2) \equiv c^{(2)}(\rho_0; \boldsymbol{r}_1, \boldsymbol{r}_2) = \delta[(\boldsymbol{r}_1 - \boldsymbol{r}_2)/\rho_0(\boldsymbol{r}_1) - \beta\delta u(\boldsymbol{r}_2)]/\delta\rho_0(\boldsymbol{r}_2)$$

The second term is the inverse of $\mathscr{G}(\boldsymbol{r}_1,\boldsymbol{r}_2)$ or $-\mathscr{G}^{-1}(\boldsymbol{r}_1,\boldsymbol{r}_2)$, given by the definition $\int \mathscr{G}(\boldsymbol{r}_1,\boldsymbol{r}_3)\mathscr{G}^{-1}(\boldsymbol{r}_3,\boldsymbol{r}_2)\mathrm{d}\boldsymbol{r}_3 = \delta(\boldsymbol{r}_1-\boldsymbol{r}_2)$. Substituting $\mathscr{G}(\boldsymbol{r}_1,\boldsymbol{r}_2)$ and $c^{(2)}(\boldsymbol{r}_1,\boldsymbol{r}_2)$ in this definition, we obtain

$$\begin{aligned}\rho^{(2)}(\boldsymbol{r}_1,\boldsymbol{r}_2) - \rho_0(\boldsymbol{r}_1)\rho_0(\boldsymbol{r}_2) &= \rho_0(\boldsymbol{r}_1)\rho_0(\boldsymbol{r}_2)c^{(2)}(\boldsymbol{r}_1,\boldsymbol{r}_2)\\ &\quad + \rho_0(\boldsymbol{r}_2)\int \{\rho^{(2)}(\boldsymbol{r}_1,\boldsymbol{r}_3) - \rho_0(\boldsymbol{r}_1)\rho_0(\boldsymbol{r}_3)c^{(2)}(\boldsymbol{r}_1,\boldsymbol{r}_2)\}\,\mathrm{d}\boldsymbol{r}_3\end{aligned}$$

For a uniform fluid having constant ρ_0, $c^{(2)}(\boldsymbol{r}_1,\boldsymbol{r}_2) = c^{(2)}(r)$, where $r = |\boldsymbol{r}_1 - \boldsymbol{r}_2|$ and $\rho^{(2)}(\boldsymbol{r}_1,\boldsymbol{r}_2) = \rho_0^2 g(r)$ with $g(r)$ the usual pair correlation function, and this expression reduces to the usual OZ equation $g(r) - 1 = c^{(2)}(r) + \rho_0 \int \{g(r) - 1\}c^{(2)}(r')\}\mathrm{d}\boldsymbol{r}'$.

If we assume that $\Phi = \frac{1}{2}\Sigma_{i\neq j}\phi(r_i, r_j)$, one can prove [49] that the correlation function $\rho^{(2)}(\boldsymbol{r}_1,\boldsymbol{r}_2) = 2\delta\Omega[\rho_0]/\delta\phi(\boldsymbol{r}_1,\boldsymbol{r}_2)$ for fixed $u(\boldsymbol{r})$. It also follows that

$$\rho^{(2)}(\boldsymbol{r}_1,\boldsymbol{r}_2) = 2\delta\mathscr{F}[\rho_0]/\delta\phi(\boldsymbol{r}_1,\boldsymbol{r}_2) = -2\delta\mathscr{F}[\rho_0]/\delta\Phi[\rho_0]$$

Functionally integrating, using as parameter $0 \le \alpha \le 1$, results in

$$\mathscr{F}[\rho_0] = \mathscr{F}_{\mathrm{ref}}[\rho_0] + \tfrac{1}{2}\iint \{\rho^{(2)}(\phi_\alpha;\boldsymbol{r}_1,\boldsymbol{r}_2)\psi(\boldsymbol{r}_1,\boldsymbol{r}_2)\}\,\mathrm{d}\boldsymbol{r}_1\mathrm{d}\alpha$$

where $\phi_\alpha(\boldsymbol{r}_1,\boldsymbol{r}_2) \equiv \phi_{\mathrm{ref}}(\boldsymbol{r}_1,\boldsymbol{r}_2) + \alpha\{\phi(\boldsymbol{r}_1,\boldsymbol{r}_2) - \phi_{\mathrm{ref}}(\boldsymbol{r}_1,\boldsymbol{r}_2)\}$ and $\mathscr{F}_{\mathrm{ref}}[\rho_0]$ corresponds to an initial nonequilibrium reference system with $\phi_{\mathrm{ref}}(\boldsymbol{r}_1,\boldsymbol{r}_2)$ and ρ_0. If $\phi_{\mathrm{ref}}(\boldsymbol{r}_1,\boldsymbol{r}_2) = 0$,

$$\mathscr{F}[\rho_0] = \mathscr{F}_{\mathrm{idl}}[\rho_0] + \tfrac{1}{2}\iint \{\rho^{(2)}(\phi_\alpha;\boldsymbol{r}_1,\boldsymbol{r}_2)\phi(\boldsymbol{r}_1,\boldsymbol{r}_2)\}\,\mathrm{d}\boldsymbol{r}_1\mathrm{d}\alpha$$

For uniform fluids, this expression represents thermodynamic integration (Section 9.4).

Now, we are ready to deal with the interface, which we assume to be planar (stabilized by a weak gravitational field) and taken as the x–y plane. Using the Helmholtz density $f[\rho]$, we write

$$\mathscr{F}[\rho] = \int f[\rho]\,\mathrm{d}\boldsymbol{r} \qquad \text{with} \quad f[\rho] = f_0\{\rho(\boldsymbol{r})\} + \boldsymbol{f}\cdot\nabla\rho(\boldsymbol{r}) + \nabla\rho(\boldsymbol{r})\cdot\mathbb{F}\cdot\nabla\rho(\boldsymbol{r}) + \ldots$$

and where f_0, $\boldsymbol{f}$, and $\mathbb{F}$ are functions of $\rho(\boldsymbol{r})$. Because $f[\rho]$ is a unique functional, this expression must be invariant under rotations and reduces to

$$f[\rho] = f_0\{\rho(\boldsymbol{r})\} + f_2^{(\mathrm{a})}\nabla^2\rho + f_2^{(\mathrm{b})}\nabla\rho\cdot\nabla\rho + \ldots$$

Writing $f_2^{(\mathrm{a})}\nabla^2\rho$ as $\nabla\cdot\{f_2^{(\mathrm{a})}\nabla\rho\} - (\mathrm{d}f_2^{(\mathrm{a})}(\rho)/\mathrm{d}\rho)\nabla\rho\cdot\nabla\rho$, the divergence term vanishes upon integration over $\boldsymbol{r}$, so that we can write $f[\rho] = f_0\{\rho(\boldsymbol{r})\} + f_2(\rho)\nabla\rho\cdot\nabla\rho + \ldots$. The intrinsic chemical potential becomes

$$\mu_{\mathrm{int}} = \delta\mathscr{F}[\rho]/\delta\rho(\boldsymbol{r}) = f_0'(\rho) + f_2'(\rho)|\nabla\rho|^2 + 2f_2(\rho)\nabla^2\rho + \ldots$$

where $f'(\rho) = \partial f/\partial\rho$. From $u(\boldsymbol{r}) = \mu - \mathscr{V}(\boldsymbol{r})$, we also obtain

$$\mu = \mathscr{V}(\boldsymbol{r}) + \mu\{\rho_0(\boldsymbol{r})\} - f_2'(\rho_0)|\nabla\rho_0(\boldsymbol{r})|^2 - 2f_2(\rho_0)\nabla^2\rho_0(\boldsymbol{r}) + \ldots$$

so that if $f_2(\rho_0)$ can be determined in some way, all thermodynamic properties follow.

For slowly varying density distributions, we have $\rho^* = \rho - \rho_{\mathrm{uni}}$ with ρ_{uni} the density of the uniform fluid, and we can use the functional expansion

$$\begin{aligned}\mathscr{F}[\rho] &= \mathscr{F}[\rho_{\mathrm{uni}}] + \int \{\delta\mathscr{F}[\rho]/\delta\rho(\boldsymbol{r})\}\rho^*(\boldsymbol{r})\,\mathrm{d}\boldsymbol{r}\Big|_{\rho_{\mathrm{uni}}}\\ &\quad + \tfrac{1}{2}\int \{\delta^2\mathscr{F}[\rho]/\delta\rho(\boldsymbol{r})\delta\rho(\boldsymbol{r}')\}\rho^*(\boldsymbol{r})\}\rho^*(\boldsymbol{r}')\,\mathrm{d}\boldsymbol{r}\mathrm{d}\boldsymbol{r}'\Big|_{\rho_{\mathrm{uni}}} + \ldots\end{aligned}$$

where the densities must be set to the constant density ρ_{uni} after differentiation. The second derivative relates to the OZ equation via

$$\delta^2\mathscr{F}[\rho]/\delta\rho(\mathbf{r})\delta\rho(\mathbf{r}') = \rho^{-1}(\mathbf{r})\delta(\mathbf{r}_1 - \mathbf{r}_2) - c[\rho; \mathbf{r}_1, \mathbf{r}_2]$$

which for a uniform system can only depend on r. Therefore, $\mathscr{F}[\rho]$ can be written as

$$\mathscr{F}[\rho] = \mathscr{F}[\rho_{\text{uni}}] + \int \mu(\rho_{\text{uni}})\rho^*(\mathbf{r})\,\mathrm{d}\mathbf{r}$$
$$+ \tfrac{1}{2}kT \int \{\rho^{-1}(\mathbf{r})\delta(\mathbf{r}_1 - \mathbf{r}_2) - c[\rho; \mathbf{r}_1, \mathbf{r}_2]\}\rho^*(\mathbf{r})\rho^*(\mathbf{r}')\,\mathrm{d}\mathbf{r}\mathrm{d}\mathbf{r}' + \ldots$$

For the second term, we take the Fourier transform, resulting in $\tfrac{1}{2}kTV^{-1}\Sigma_{\mathbf{q}}\{c[\rho_{\text{uni}}, q] - \rho_{\text{uni}}{}^{-1}\rho^*(\mathbf{q})\rho^*(-\mathbf{q})\}$ where $c[\rho_{\text{uni}}, q]$ is the Fourier transform of $c(\mathbf{r})$ given by $c[\rho_{\text{uni}}, q] - \rho_{\text{uni}}{}^{-1} = 1/kT\chi$ with χ the linear density response function (Section 9.5). This expression is valid as long as $|\rho^*(\mathbf{r})/\rho_{\text{uni}}| \ll 1$. If $\rho(\mathbf{r})$ is also slowly varying, only low-order Fourier coefficients will be significant and we can expand $c[\rho_{\text{uni}}, q]$ as

$$c[\rho_{\text{uni}}, q] = a(\rho_{\text{uni}}) + b(\rho_{\text{uni}})q^2 + d(\rho_{\text{uni}})q^4 + \ldots$$

Substituting and back transforming, the result is

$$\mathscr{F}[\rho] = \mathscr{F}[\rho_{\text{uni}}] + \int \mu(\rho_{\text{uni}})\rho^*(\mathbf{r})\,\mathrm{d}\mathbf{r}$$
$$+ \tfrac{1}{2}kT\{a(\rho_{\text{uni}}) - \rho^{-1}\} \int \{\rho^*(\mathbf{r})\}^2\,\mathrm{d}\mathbf{r} + b(\rho_{\text{uni}}) \int \{\nabla\rho^*(\mathbf{r})\}^2\,\mathrm{d}\mathbf{r} + \ldots$$

Comparison with the original expansion yields $f_0'(\rho_{\text{uni}}) = \mu(\rho_{\text{uni}})$, $f_0''(\rho_{\text{uni}}) = -kT\{a(\rho_{\text{uni}}) - \rho_{\text{uni}}{}^{-1}\}$ and $f_2(\rho_{\text{uni}}) = -\tfrac{1}{2}kTb(\rho_{\text{uni}})$. The first two expressions have been obtained before but the third provides the new relation, $f_2(\rho) = \tfrac{1}{2}m = \tfrac{1}{2}\cdot\tfrac{1}{6}\int r^2c(\rho, r)\,\mathrm{d}\mathbf{r}$ with m the *influence parameter*[7].

Using the expansion to second order results in the *square gradient approximation*:

$$\mathscr{F}[\rho] = \mathscr{F}[\rho_{\text{uni}}] + \int f_0\{\rho(r)\} + f_2\{\rho(r)\{\nabla\rho^*(\mathbf{r})\}\}^2\,\mathrm{d}\mathbf{r}$$

The stress tensor is given by $\sigma = -\omega[\rho_0]I + 2f_2(\rho_0)\nabla\rho_0\nabla\rho_0$ with internal energy density (or local pressure) $\omega[\rho_0] = f[\rho_0] - \mu\rho_0(r)$, from which the pressures $P_{\text{t}} = -\omega[\rho_0]$ and $P_{\text{n}} = -\omega[\rho_0] + 2f_2(\rho_0(z)|\rho_0'(z)|^2$ follow. The surface energy can now be obtained from Eq. (11.1), $\gamma = \int_{-\infty}^{+\infty}[P_{\text{n}}(z) - P_{\text{t}}(z)]\,\mathrm{d}z$, and becomes

$$\gamma = 2\int_{-\infty}^{+\infty} f_2(\rho_0)|\rho_0'|^2\,\mathrm{d}z = 2\int_{\rho_{\text{L}}}^{\rho_{\text{V}}} f_2(\rho_0)\rho_0'\,\mathrm{d}\rho = 2\int_{\rho_{\text{L}}}^{\rho_{\text{V}}} \{2f_2(\rho_0)f(\rho)\}^{1/2}\,\mathrm{d}\rho \quad (11.17)$$

so that γ can be evaluated from $f(\rho)$ without knowing $\rho(z)$. Because $P_{\text{n}}(z) = P$, this also follows from $\Omega = -PV + \gamma A$. According to Evans, the so-called square gradient approximation

7 An interesting analogy exists between gradient theory and classical mechanics (Chapter 5). Denoting "associates with" by $\leftrightarrow$, we have $m \leftrightarrow$ mass m, $z \leftrightarrow$ time t and $\rho(z) \leftrightarrow$ position x. As $-F(\rho) \leftrightarrow$ potential energy Φ, $F'(\rho) = f(\rho) - \mu_0 \leftrightarrow$ force F, so that $f'(\rho) - \rho''(z) = 0$ corresponds with Newton's second law. The kinetic energy $\vartheta \leftrightarrow \tfrac{1}{2}m\rho'(z)^2$ and thus the Lagrange function $L = \vartheta - V \leftrightarrow \tfrac{1}{2}m\rho'(z)^2 + F(\rho)$, while the Hamilton function $H = \vartheta + V \leftrightarrow \tfrac{1}{2}m\rho'(z)^2 + f(\rho) = 0$, as $f(\rho_{\text{V}}) = f(\rho_{\text{L}}) = 0$. Minimizing γ as an integral over $\tfrac{1}{2}m\rho'(z)^2 + F(\rho)$ corresponds with Hamilton's principle, which minimizes the action, that is, the integral over L.

(SGA) is probably the simplest description of a liquid–vapor surface, which however is only valid near T_{cri} as $\rho(r)$ does change too fast near T_{tri}.

If a gradient in ρ in the interphase would not exist, the Helmholtz energy of the system would be minimized by an infinitely sharp transition between a region with density ρ_{L} and density ρ_{V}. However, a gradient $\rho'(z)$ does exist and stabilizes the liquid at the densities in the interphase, since keeping the liquid at a density different from ρ_{L} and ρ_{V} will cost a significant amount of Helmholtz energy.

Integrating $\rho'(z) = \{2f(\rho/m\}^{1/2}$ yields a parametric representation of the density profile $z = (m/2)^{1/2} \int_{\rho(0)}^{\rho(z)} \{f(\rho)\}^{-1/2}\,\text{d}\rho$. At liquid–gas coexistence, $f(\rho)$ has two minima of equal depth. One is located at $\rho = \rho_{\text{L}}$ and another at $\rho = \rho_{\text{V}}$. A simple parameterization, valid near the critical point, is

$$f(\rho) = C_\rho(\rho_{\text{L}} - \rho)^2(\rho - \rho_{\text{V}})^2 \quad \text{with} \quad C_\rho = C_\rho(T) \tag{11.18}$$

a parameter. Substituting Eq. (11.18) in the expression for z, we obtain

$$z = (m/2C_\rho)^{1/2} \int_{\rho(0)}^{\rho(z)} [(\rho_{\text{L}} - \rho)(\rho - \rho_{\text{V}})]^{-1}\,\text{d}\rho = \xi \ln\{(\rho - \rho_{\text{V}})/(\rho_{\text{L}} - \rho)\}$$

Here, $\xi = (m/2C_\rho)^{1/2}/(\rho_{\text{L}} - \rho_{\text{V}})$ is the *correlation length*, a characteristic length for the interfacial width. Solving for ρ we find

$$\rho(z) = ½(\rho_{\text{L}} + \rho_{\text{V}}) - ½(\rho_{\text{L}} - \rho_{\text{V}})\tanh(z/2\xi)$$

For this profile, the density ρ changes continuously from ρ_{L} to ρ_{V}, is antisymmetric with respect to the mid-point z_0 as a consequence of the assumed form of Eq. (11.18), and diverges at the critical point. Within the mean-field theory (Section 15.2), it appears that $\rho_{\text{L}} - \rho_{\text{V}} \sim (T_{\text{cri}} - T)^{1/2}$, so that ξ diverges as $(T_{\text{cri}} - T)^{-1/2}$.

Substituting Eq. (11.18) in Eq. (11.17c), we find for γ close to T_{cri}

$$\gamma = (2mC_\rho)^{1/2} \int_{\rho_{\text{V}}}^{\rho_{\text{L}}} (\rho_{\text{L}} - \rho)(\rho - \rho_{\text{V}})\,\text{d}\rho = 6^{-1}(2mC_\rho)^{1/2}(\rho_{\text{L}} - \rho_{\text{V}})^3$$

Hence, in this formulation, $\gamma \sim (\rho_{\text{L}} - \rho_{\text{V}})^3 \sim (T_{\text{cri}} - T)^{3/2}$. In fact, in the mean-field theory, the critical exponents[8] for the correlation length $\xi \sim (T_{\text{cri}} - T)^{-\nu}$, the compressibility $\kappa \sim (T_{\text{cri}} - T)^{-\gamma}$, and density difference $\rho_{\text{L}} - \rho_{\text{V}} \sim (T_{\text{cri}} - T)^{\beta}$ are ν = ½, γ = 1, and β = ½. Hence, for $\gamma \sim (T_{\text{cri}} - T)^{-\mu}$, we have $\mu = -\nu + \gamma + 2\beta = 3/2$. In the renormalization theory, Eq. (11.18) is replaced by a different but qualitatively similar function and the exponents become $\nu \cong 0.63$, $\gamma \cong 1.24$, and $\beta \cong 0.33$. Hence, $\mu \cong 1.26$, in agreement with the experiment and close to $\mu = 11/9 \cong 1.22$ in Guggenheim's formula.

It was shown [52] for a variety of liquids that $\gamma\kappa_T = l$ is rather constant, where l is intrinsic length scale l. Similarly, for 24 organic liquids near 298 K [53] the empirical relation $\gamma\kappa_T = 27.4 \pm 3.4$ pm was obtained, where ± indicates the sample standard deviation. This relation can be rationalized for monatomic liquids by using DFT arguments [54]. Sanchez [55] provided $\gamma(\kappa_T/\rho)^{1/2} = C_\gamma^{\ 1/2}$, with $C_\gamma^{\ 1/2} = 2.78 \pm 0.13$ (erg cm^2/g)$^{1/2}$ as a temperature-independent constant, and where he attributed about half of the variation to experimental errors. Although $C_\gamma^{\ 1/2}$ was shown to be proportional to the second moment of $c(r)$ and empirically related to the Lennard-Jones energy parameter ε and size parameter

8 See Chapter 15. The use of symbols like γ, β and μ for the exponents as used here is largely standardized, but should not be confused with surface tension γ, $\beta = 1/kT$ and chemical potential μ.

σ via $C_\gamma^{1/2} = 0.26(\varepsilon\sigma^2/\mathscr{M})^{1/2}$ with $\mathscr{M}$ the molecular mass, the physical implication of this correlation is unclear.

One of the nice aspects of DFT is that, in principle, any EoS can be used, whether experimental or theoretical. Moreover, although, in principle, derived for conditions near T_{cri}, it appears that its validity covers a much wider range [56]. Despite being rather successful, some conceptual problems do exist: the density over the interphase is considered to be a well-defined function of the height in the interphase and gravity is taken to localize the interface and to stabilize it against capillary waves [57]. Finally, we repeat Evans' warning that it is easy to forget that for an arbitrary EoS the underlying Hamilton function is probably unknown or does not exist [51]. For a treatment of all these aspects, we refer to the literature. More sophisticated treatments are given in [58–62] as well as in the reviews [49, 50, 63]. While [51, 64] provide high-level introductions, [65] is easy going.

11.3.2 Capillary Wave Theory

An interface can sustain various types of waves, but here we only discuss capillary waves. A *capillary wave* is a 2D wave propagating on the surface of condensed matter for which the restoring force is provided by the surface tension. Capillary wave theory (CWT) provides a mesoscopic view of the interface, which is considered as a mathematical 2D surface that can fluctuate. This interface can have a certain intrinsic width and is not limited to an infinitely sharp interface. Thermal fluctuations create waves on the interface, covering, in principle, the whole spectrum of wavelengths, from atomic dimensions to macroscopic lengths. A fluctuation is described by the locations of the interface points $z = z(\boldsymbol{r})$ with $\boldsymbol{r} = (x, y)$. The deviation of the interface from a flat configuration requires the Helmholtz energy

$$\Delta F_\gamma = \gamma\,\Delta A \cong \gamma \int_A [(1 + |\nabla z|^2)^{1/2} - 1]\,\mathrm{d}\boldsymbol{r} \cong \tfrac{1}{2}\gamma \int_A |\nabla z|^2\,\mathrm{d}\boldsymbol{r} \tag{11.19}$$

where ΔA is the excess area when a small-amplitude fluctuation $z(\boldsymbol{r})$ is introduced and γ the *bare (Helmholtz) interface energy* or the interface energy without any capillary waves present. The probability of a given fluctuation is proportional to $\exp(-\beta\Delta F_\gamma)$.

To obtain an explicit expression, we use Fourier amplitudes. Assuming periodic boundary conditions in the x–y plane and a square area of linear size $L = A^{1/2}$, we write

$$z(\boldsymbol{r}) = \sum\nolimits_{\boldsymbol{q}} a_{\boldsymbol{q}} \exp(\mathrm{i}\boldsymbol{q}\cdot\boldsymbol{r}) \quad \text{with} \quad \boldsymbol{q} = (2\pi/L)\boldsymbol{n} \quad \text{and} \quad \boldsymbol{n} = 0, \pm1, \pm2, \ldots \tag{11.20}$$

and where $\boldsymbol{q}$ is a 2D wave vector. Substituting in Eq. (11.19), we have

$$\Delta F_\gamma = \tfrac{1}{2}\gamma \int_A \sum\nolimits_{\boldsymbol{q},\boldsymbol{q}'} a_{\boldsymbol{q}} a_{\boldsymbol{q}'} (\boldsymbol{q}\cdot\boldsymbol{q}') \exp[\mathrm{i}(\boldsymbol{q}-\boldsymbol{q}')\cdot\boldsymbol{r}]\,\mathrm{d}\boldsymbol{r} = \tfrac{1}{2}\gamma A \sum\nolimits_{\boldsymbol{q}} a_{\boldsymbol{q}}^2 q^2$$

To evaluate $a_{\boldsymbol{q}}^2 = a_{\boldsymbol{q}} a_{-\boldsymbol{q}}$, we write $a_{\boldsymbol{q}} = (\alpha_{\boldsymbol{q}} + \mathrm{i}\beta_{\boldsymbol{q}})/2^{1/2}$, so that $a_{\boldsymbol{q}}^2 = \tfrac{1}{2}(\alpha_{\boldsymbol{q}}^2 + \beta_{\boldsymbol{q}}^2)$, $\mathrm{d}a_{\boldsymbol{q}}\mathrm{d}a_{-\boldsymbol{q}} = \mathrm{d}\alpha_{\boldsymbol{q}}\mathrm{d}\beta_{\boldsymbol{q}}$ and $\langle a_{\boldsymbol{q}} a_{-\boldsymbol{q}}\rangle = \tfrac{1}{2}(\langle\alpha_{\boldsymbol{q}}^2\rangle + \langle\beta_{\boldsymbol{q}}^2\rangle)$. Using the equipartition theorem, we have $\Delta F_{\boldsymbol{q}} \equiv \tfrac{1}{2}A\gamma\cdot\tfrac{1}{2}(\langle\alpha_{\boldsymbol{q}}^2\rangle + \langle\beta_{\boldsymbol{q}}^2\rangle)q^2 = \tfrac{1}{2}kT$ per mode $\boldsymbol{q}$, so that $\langle a_{\boldsymbol{q}}^2\rangle = kT/\gamma A q^2$. Using the inverse Fourier transform, we obtain the height correlation function [57, 66]

$$H(\boldsymbol{r}) \equiv \lim_{L\to\infty} \frac{1}{L^2} \int_{-L/2}^{L/2} z^*(\boldsymbol{r}) z(0)\,\mathrm{d}\boldsymbol{r} = \sum\nolimits_{\boldsymbol{q}} \langle a_{\boldsymbol{q}}^2\rangle \mathrm{e}^{-\mathrm{i}\boldsymbol{q}\cdot\boldsymbol{r}}$$

The root mean-square amplitude or *(extrinsic) width* $h = [H(0)]^{1/2}$ is thus given by

$$\langle h^2\rangle \equiv A^{-1} \int_A \langle z^2(\boldsymbol{r})\rangle\,\mathrm{d}r = \sum\nolimits_{\boldsymbol{q}} \langle a_{\boldsymbol{q}}^2\rangle = (kT/\gamma A) \sum\nolimits_{\boldsymbol{q}} q^{-2}$$

In the thermodynamic limit, where $\sum_{\boldsymbol{q}} \ldots = (2\pi)^{-2} A \int \ldots \, \mathrm{d}\boldsymbol{q}$, we have

$$\langle h^2 \rangle = \frac{kT}{4\pi^2\gamma} \int_{q_{\min}}^{q_{\max}} \frac{\mathrm{d}\boldsymbol{q}}{q^2} = \frac{kT}{4\pi^2\gamma} \int_{q_{\min}}^{q_{\max}} \frac{2\pi q \,\mathrm{d}q}{q^2} = \frac{kT}{2\pi\gamma} \ln \frac{q_{\max}^2}{q_{\min}^2}$$

where the integration is now taken over a circle with radius q. The amplitudes of long-wavelength fluctuations ($\boldsymbol{q}_{\max} \to 0$) diverge, and these fluctuations disrupt the interface completely. Thus, in this picture, fluid interfaces must be stabilized by external fields, gravity being the main one. If gravitational energy is taken into account, a term $mg(\rho_L - \rho_V)z(\boldsymbol{r}) \equiv mg\Delta\rho z(\boldsymbol{r})$ has to be added, so that

$$\Delta F_g = mg\Delta\rho \int_A \int_{-\infty}^{+\infty} z(\boldsymbol{r}) \,\mathrm{d}z\mathrm{d}\boldsymbol{r} = \tfrac{1}{2} mg\Delta\rho \int_A h(\boldsymbol{r})^2 \mathrm{d}\boldsymbol{r}$$

with m the mass of the particles and g the acceleration of gravity. This changes the Helmholtz energy to $\Delta F = \Delta F_\gamma + \Delta F_g$, expressed in terms of Fourier amplitudes as

$$\Delta F = \tfrac{1}{2}\gamma A \sum_{\boldsymbol{q}} a_{\boldsymbol{q}}{}^2 (q^2 + l^{-2}) \qquad \text{with the } \textit{capillary length } l = [\gamma/mg\Delta\rho]^{1/2}$$

From equipartition, we now have $\langle a_{\boldsymbol{q}}{}^2 \rangle = kT/A\gamma(q^2 + l^{-2})$, while $H(\boldsymbol{r})$ and h become

$$H(\boldsymbol{r}) = \frac{kT}{4\pi^2\gamma} \int \frac{\mathrm{e}^{-i\boldsymbol{q}\cdot\boldsymbol{r}}}{q^2 + l^{-2}} \mathrm{d}\boldsymbol{q} \quad \text{and} \quad h^2 = \frac{kT}{4\pi^2\gamma} \int \frac{\mathrm{d}\boldsymbol{q}}{q^2 + l^{-2}} = \frac{kT}{2\pi\gamma} \ln \frac{q_{\max}^2 + l^{-2}}{q_{\min}^2 + l^{-2}}$$

The capillary length l introduces a minimum wave vector l^{-1} for the fluctuations and prevents any divergence of the fluctuation amplitudes. The maximum allowed wave vector results from the fact that the interface is not a true continuum system and therefore $q_{\max} \cong \xi^{-1}$, where ξ is the bulk *correlation length*. If we take $L \to \infty$ but g finite, $q_{\min} = 2\pi/L \to 0$ and

$$h^2 \cong (kT/2\pi\gamma) \ln \left(q_{\max}^2 l^2 + 1 \right) \cong (kT/\pi\gamma) \ln(l/\xi)$$

where we used the fact that ξ is typically a few Ångströms and l a few hundred micrometers. Since $h \to \infty$ if g $\to 0$, gravity localizes the interface. If we take g $\to 0$ but L finite, then

$$h^2 \cong (kT/\pi\gamma) \ln(q_{\max}/q_{\min}) = (kT/\pi\gamma) \ln(L/\xi)$$

so now it is the finite area of the system that localizes the interface.

Although h diverges, this divergence is a relatively slow logarithmic one, which practically means that, even though the effect of capillary waves can easily be observed, very large samples or extremely weak external fields are necessary for the divergence to become problematic. This approach uses γ as a parameter, which should be calculated by, for example, full DFT or its SGA form, or statistical mechanics (see, e.g. [67]). However, it provides a description of the widening of the interface due to the thermally created capillary waves and yields a contribution to γ, whether the interface is infinitely sharp or not. The contribution of the capillary waves to γ is $\Delta\gamma_{\mathrm{CW}} = \partial\Delta F/\partial A = \partial\Sigma_{\boldsymbol{q}}\Delta F_{\boldsymbol{q}}/\partial A = \tfrac{1}{2}kT \cdot \partial\Sigma_{\boldsymbol{q}} 1/\partial A = \tfrac{1}{2}kT \cdot (A/4\pi^2) \int \mathrm{d}\boldsymbol{q} = (kT/8\pi)(q_{\max}{}^2 - q_{\min}{}^2)$. This results for liquids typically in a few mJ m^{-2}, so although small, not altogether negligible. The prefactor depends on the details of the precise model used. A detailed discussion is [68], while [58] provides a brief review.

To conclude this part, we note that Stillinger [69] attempted to reconcile CWT and DFT. However, a generally accepted theory seems still lacking (but see [70]). While the basic theory [49, 71] is well developed, experimental developments, for example, using optical and X-ray scattering techniques [72], continuously provide new insights.

Problem 11.5 ***Capillary Waves*** Verify the expression for $\Delta\gamma_{\text{CW}}$.

11.4 Solid Surface Structure

The thermodynamic arguments in Section 11.1 apply to liquids and solids. However, while liquid surfaces are isotropic and flat, solid surfaces are anisotropic and have structure.

The simplest surface is an *ideal* or *bulk-like surface*, the structure of which is like that of the corresponding lattice plane in the bulk. Such surfaces occur primarily in metals. For example, the {111}, {110}, and {100} planes in the FCC structure, the {111} and {110} planes in the body-centered cubic (BCC) structure, and the {0001} and $\{10\bar{1}0\}$ planes in the hexagonal close-packed (HCP) structure. They also occur in oxides, for example, {100} NiO and {110} in III-V compounds (e.g. GaAs) or II-VI compounds (e.g. ZnS). For an ideal surface γ can be estimated using the bond pair model (Section 11.2) with bulk coordination number z, nearest-neighbor, temperature-independent bond energy u, and heat of sublimation $\Delta_{\text{sub}}H = \frac{1}{2}N_{\text{A}}zu$. For the (111) plane in an FCC lattice with $z = 12$, the surface coordination number $s = 9$. Hence, taking $\kappa_{\text{SC}}(100) = 1$, that is, for the number of atoms per unit surface $N = (N_{\text{A}}/V_{\text{m}})^{2/3}$ with V_{m} the molar volume, instead of $\kappa_{\text{FCC}}(111) = 1.09$, we obtain

$$U^{(\sigma)} = \tfrac{1}{2}(z - s)uN = \Delta_{\text{sub}}H/4V_{\text{m}}^{\;2/3}N_{\text{A}}^{\;1/3}$$

More generally, for FCC lattices (and some others) Friedel et al. [73] derived that $u^{(\sigma)} = 2(2h + l)(h^2 + k^2 + l^2)^{-1/2}u/a^2$, with Miller indices $h \geq k \geq l$ and a the lattice constant (Table 11.4). Experimentally, one often obtains an average over many crystal planes, resulting in a value of Stefan's ratio $j \cong \frac{1}{3} - \frac{1}{4}$. For metals, this estimate yields systematically too high values. Miedema's model (Section 10.6) rationalizes the same expression explicitly for metals and his (essentially fitted) value for $j \cong \frac{1}{6}$ describes the data well. For example, for Ag 1.20 (1.14) J m^{-2}, Au 1.54 (1.41) J m^{-2}, and Cu 1.80 (1.67) J m^{-2}, with the experimental values given in parenthesis.

Generally, low-index solid surfaces (for crystals often called *facets*) are only nominally planar and have *steps* or *ledges*. For FCC and BCC metals, the step height typically is monatomic, while for HCP metals it is generally doubled. In the steps *kinks* (Figure 11.4a) are present representing a deviation from the overall step direction. The planar area between the steps is called a *terrace*. On both terraces and ledges atoms can adsorb (*ad-atoms*) and vacancies can arise. Together, these features represent the *terrace–ledge–kink* (TLK) *model*. A convenient way to display the anisotropy of γ (and $u^{(\sigma)}$) is the *Wulff plot*, in which $\gamma(\boldsymbol{n})$ is plotted as a function of surface orientation indicated by a normal unit vector $\boldsymbol{n}(\theta,\phi)$ in a polar diagram (Figure 11.4b). At certain low-index planes (*singular surfaces*), minima in γ occur which are usually called cusps (although a cusp mathematically is a minimum with an infinite value for the derivative). Essentially because moving away from the low-index orientation steps and kinks adds energy to γ, for nearby surfaces (*vicinal surfaces*) with a small angle θ to the singular surface with surface energy γ_{sin} and containing noninteracting ledges with height h, we can write $\gamma = \gamma_{\text{sin}} \cos\theta + (\varepsilon_{\text{led}}/h) \sin|\theta|$, where ε_{led} is the excess Helmholtz energy for a unit length of

Table 11.4 Surface energy (mJ m^{-2}) for Ag, Au, and Cu.[a]

Metal	ϕ (fJ bond^{-1})	a^2 (Å^2)	(100)	(110)	(111)	Exp. solid	Exp. liquid
Ag	7.92	16.60	1905	1350	1650	1140 (907)	879 (1100)
Au	9.8	16.58	2425	1715	2100	1410 (1027)	1140 (1338)
Cu	9.4	13.02	2890	2080	2500	1670 (1047)	1300 (1535)

[a] According to [73], experimental data are from [10], T in °C in parenthesis.
Source: Adapted from Friedel et al. [73].

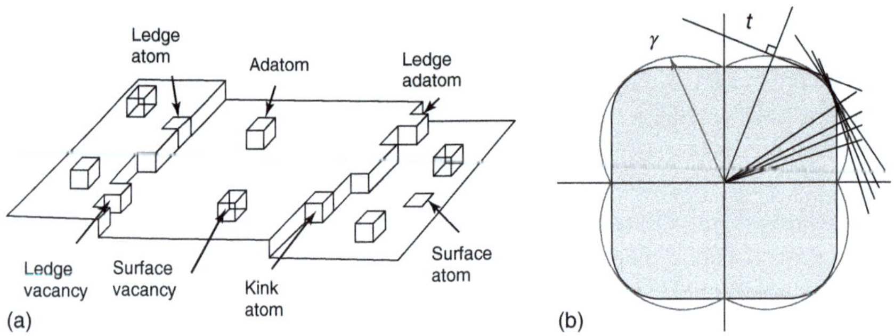

Figure 11.4 (a) The TLK model. (b) A 2D Wulff plot showing several orientations with associated γ and tangent line t. The equilibrium shape is geometrically similar to a body formed by all tangent lines.

ledge. This expression can provide an experimental estimate for ε, for example, for ledges on a (100) Cu plane, $\varepsilon_{\text{led}} \cong 0.13$ eV per ledge atom. A similar relation provides an estimate for the excess Helmholtz energy of kinks $\varepsilon_{\text{kink}}$ as a function of ϕ [74]. At fixed volume the shape of the crystal (*habit*) is determined by the minimum of the surface integral $\int \gamma(\boldsymbol{n})\text{d}A$ [75]. Although the shape of the crystal can be determined from a Wulff plot, the reverse procedure returns only the ratios of the γs for various surfaces. Contrary to simple structures, in complex oxides several terminations of a crystallographic plane can exist, which complicates the matter considerably. An overview of surface structure of single crystals is [76], while [10, 77] focus on their thermodynamics.

To lower their surface energy, atoms in the surface region exhibit in general a *relaxation* from the ideal lattice positions, leading in several cases to *reconstruction*. For a relaxed surface the overall structure is still like an ideal surface while for a reconstructed surface a clear symmetry break occurs.

For relaxed surfaces of metals typically a bond contraction of a few percent occurs between the first and second layers. The contraction is larger for the more open surfaces. For example, generally a contraction of ~0% occurs for the {111} FCC and {100} FCC (or even a widening as for Pt and Pd) and a contraction of ~10% for {110} FCC, to as large as 16% for Pb. Deeper layers generally alternate in contraction and widening. For II–IV and III–V compounds, both the bond angle and bond length vary. Contractions as well as elongations of ~5% can occur in bond length while the bond angle ranges from 90° to 120°, deviating considerably from the ideal 109.5°. The low value $j \cong 1/6$ is probably due

Table 11.5 Surface energy ($J\ m^{-2}$) of spinel $MgAl_2O_4$ [78].

Surface	γ (non-relaxed)	γ (relaxed)	γ (hydrated)
(100)	4.0	2.5	0.6 (5)
(110)	5.6	2.7	0.2 (8)
(111)	8.4	3.1	0.1 (7)

Only the most stable orientations are given. The number of adsorbed water molecules are given in parenthesis.

to relaxation, although also obtained for a (100) SC lattice plane. For oxides the situation is rather mixed. While for {100} NiO, MgO, CaO, and CoO there is little contraction, for {111} CoO a contraction of about 17% is present and for {100} $SrTiO_3$ even buckling of the surface occurs.

Reconstruction of a surface leads to a unit cell with a symmetry different from the corresponding lattice plane. Frequently, the size of the surface unit cell increases, for example, for Si the {111} 7×7 reconstruction. For a *displacive* reconstruction, the displacements are small, but breaking of symmetry occurs, for example, in {100} 1×1 Mo and W. In a *missing row* surface, a row of atoms is missing from the surface, leading to a 2×1 unit cell, as in {110} Ir, Pt, and Au. Since decreasing coordination often leads to a lower bond length, top layers can be *closer packed* than the bulk, resulting in an approximately hexagonal structure, as in {100} Ir, Pt, and Au.

The surface energy generally can be lowered further by external agents, the most prominent one being H_2O, present nearly everywhere. For metals, the influence of gases, such as CO and NO, has been extensively studied. For complex oxides, H_2O is often the most important. For example, pair potential calculations including polarization effects on spinel $MgAl_2O_4$ [78] indicate displacements of ions to a depth of two unit cells upon H_2O adsorption. The surface energies of non-relaxed, relaxed, and hydrated surfaces of (100), (110), and (111) orientations (Table 11.5) show, apart from a large change in γ upon relaxation, a reversed order of preferred planes.

11.4.1 Surface Roughening

It appears that solid surfaces often show a roughening transformation at a certain T_R below which they remain nominally smooth but above they become rough. To model this phenomenon, both continuum and atomistic models have been used.

The continuum model for solids uses CWT but with the stabilization provided by a periodic lattice potential $V(z) = V[1 - \cos(2\pi z/a)]$, preferentially locating the surface on planes $z = na$, with a the lattice constant and n integer. Using Eq. (11.20), the Fourier transform of the energy becomes

$$\Delta E(a_{\boldsymbol{q}}) = \gamma_0 A + \Delta E_{\boldsymbol{q}} + V(a_{\boldsymbol{q}}) = \gamma_0 A + \tfrac{1}{2}\gamma_0 A \sum\nolimits_{\boldsymbol{q}} a_{\boldsymbol{q}}^2 \boldsymbol{q}^2 + V(a_{\boldsymbol{q}})$$

calculated similarly as $\Delta F_{\boldsymbol{q}}$ (Section 11.3.2) but with γ replaced by γ_0, where γ_0 is the surface *internal* energy. The lattice potential leads to a coupling between the modes, complicating the calculation of the partition function $Z = \int \exp(-\beta\Delta E(a_{\boldsymbol{q}}) \mathrm{d}a_{\boldsymbol{q}}$. To calculate Z,

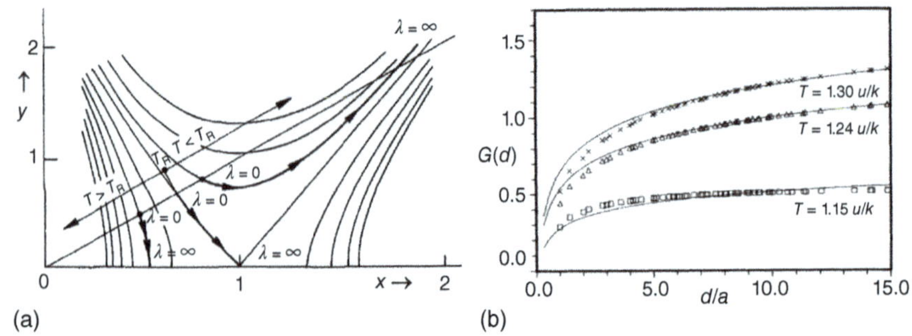

Figure 11.5 Kosterlitz–Thouless trajectories y as a function of x. Source: Lapujoulade [79]/ American Physical Society. (a) All starting points for iteration lie on the straight line $y = x$ corresponding γ_0 and V. (b) The height correlation function $G(d)$ as a function of d/a (–) fitted to MC simulation results at various temperatures, where T_R corresponds with $T = 1.24u/k$. Source: Shugard et al. [80]/with permission of Elsevier.

one normally applies a renormalization procedure in real space (Section 15.4).[9] In brief, by defining a new cut-off $\boldsymbol{q}_{cut}$, we divide the variables in a group having a large wavevector $\boldsymbol{q}_{lrg}(\boldsymbol{q}_{cut} < \boldsymbol{q}_{lrg} < \boldsymbol{q}_{max} = \pi/L)$ and one having a small wavevector $\boldsymbol{q}_{sml}(\boldsymbol{q}_{min} = \pi/\xi < \boldsymbol{q}_{sml} < \boldsymbol{q}_{cut})$. Integration of Z over the variables with $\boldsymbol{q}_{lrg}$ yields the partition function $Z = \int \exp(-\beta\Delta E(a_{lrg}, a_{sml})\text{d}a_{\boldsymbol{q}} \equiv \exp[-\beta E^{(1)}(a_{sml})]$, where $E^{(1)}(a_{sml})$ is the effective energy for the small wavevector components. By expanding $V(a_{\boldsymbol{q}})$ to second order, one can show that $\Delta E^{(1)}(a_{sml})$ has the same form as $\Delta E(a_{\boldsymbol{q}})$ but with renormalized variables $\gamma^{(1)}(a_{cut})$ and $E^{(1)}(a_{cut})$. This procedure is iterated by defining a new cut-off $\boldsymbol{q}_{cut} + \text{d}\boldsymbol{q}_{cut}$, and calculating $\gamma^{(2)}(a_{cut} + \text{d}a_{cut})$ and $E^{(2)}(a_{cut} + \text{d}a_{cut})$ until convergence with solutions γ^* and V^* is reached. In terms of the dimensionless variables $\lambda = \ln(\boldsymbol{q}_{min}/\boldsymbol{q}_{cut})$, $x = 2\beta\gamma(\lambda)a/\pi$ and $y = 4\pi\beta V(\lambda)/\boldsymbol{q}_{cut}{}^2$, one obtains after some algebra in the continuum limit

$$\text{d}y/\text{d}\lambda = 2y(x-1)/x \quad \text{and} \quad \text{d}x/\text{d}\lambda = A(2/x)\,y^2/2x$$

with $A(2/x)$ a slowly varying function [81]. These equations define the *Kosterlitz–Thouless class* [82] of phase transitions, the solutions of which (Figure 11.5a) show two types of behavior, dependent on whether $T < T_R$ or $T > T_R$. For $T < T_R$ there is always a $\boldsymbol{q}_{cut}$ above which the renormalized potential V^* goes to infinity, fluctuations are pinned, and the surface is called *smooth*. For $T > T_R$, V^* always vanishes on a large scale and the surface is denoted as *rough*. Hence, $T_R = 2\gamma^* a^2/\pi k$ is identified as the critical temperature given by the fixed point at $x = 1$ and $y = 0$. Assuming an isotropic surface (which {100} SC is not, but {111} FCC is), the height correlation function becomes

$$\begin{aligned} G(d) &= \langle [H(\mathbf{r}+\mathbf{d}) - H(\mathbf{r})]^2 \rangle = 2\sum\nolimits_{\boldsymbol{q}} \langle h^2 \rangle (1 - \cos \boldsymbol{q}\cdot\mathbf{d}) \\ &= (kT/2\pi^2\gamma) \int_{\boldsymbol{q}_{min}}^{\boldsymbol{q}_{cut}} q^{-1} \int_0^{2\pi} [1 - \cos(qd\,\cos\theta)]\,\text{d}\theta\text{d}q \\ &= (kT/\pi\gamma) \int_{\boldsymbol{q}_{min}}^{\boldsymbol{q}_{cut}} q^{-1}[1 - J_0(qd)]\,\text{d}q \\ &\cong (kT/\pi\gamma)\,\ln(\pi d/\xi) \end{aligned}$$

9 It seems inevitable that in dealing with many topics, some new topics require concepts not discussed before. I tried to minimize this, but I preferred here to discuss interfaces before continuous transitions.

The last step is made by approximating the zeroth-order Bessel function J_0 with a Heaviside step function at $qr = 1$ and is only valid for large r, as we must have $G(0) = 0$. The extrinsic width is still given by $\langle h^2 \rangle = (kT/\pi\gamma_\infty^*)\ln(L/\xi)$ with $\gamma_\infty^* = \gamma^*(\lambda = \infty)$. As an expansion of V is used, V must be small and fails for $T < T_R$ when $V^* \to \infty$. Although this renormalization procedure clearly shows the universal nature of the roughening transition, atomistic models are evidently needed.

The classic atomistic model is by Burton, Cabrera, and Frank [83]. Here we use an instructive model by Mullins [84]. He used the order–disorder transformation in 2D. for {100} SC (and FCC) crystal surfaces using a three-level nearest-neighbor bond model with bond energy u, and which includes surface atoms, surface adatoms, and surface vacancies. Equilibrium is supposed as removing or adding a kink atom regenerates another similar kink but otherwise leaves the crystal unaltered.

For the {100} SC surface, creating a vacancy on a terrace requires $5u$ but adding that atom (to conserve matter) to a kink adds $3u$. As at a high(er) temperature, the vacancy fraction becomes large, we must correct for divacancies for which it is assumed that there is no binding energy with vacancies and no binding energy between divacancies. For $n_{\text{vac}}^{(t)}$ vacancies per unit surface area terrace, the probability that an adjacent site is also a vacancy is $4n_{\text{vac}}^{(t)}/n^{(t)}$, where $n^{(t)}$ is the surface density of atoms for the {100} planes. Hence, their density is $\frac{1}{2} \cdot 4(n_{\text{vac}}^{(t)})^2/n^{(t)}$, where one must multiply by $\frac{1}{2}$ to avoid counting divacancies twice. The energy of a divacancy is the difference between $4u$ for the eight free half bonds of a divacancy and $5u$ for the five free half bonds of two vacancies, so that in total, this results in $-2u(n_{\text{vac}}^{(t)})^2/n^{(t)}$. The vacancy formation energy thus becomes $U_{\text{vac}}^{(t)} = 2un_{\text{vac}}^{(t)} - 4u(n_{\text{vac}}^{(t)})^2/n^{(t)}$.

If we have $n_{\text{ad}}^{(t)}$ adatoms per unit area and assume no overhanging atoms on vacancies, the number of complexions for the $n_{\text{sur}} = n^{(t)} - n_{\text{ad}}^{(t)}$ available surface atoms is $W = \exp(S/k) = n_{\text{sur}}!/(n_{\text{sur}} - n_{\text{vac}}^{(t)})!\,(n_{\text{vac}}^{(t)})!$. The Helmholtz energy thus becomes

$$F = U = -TS = 2un_{\text{vac}}^{(t)} - \frac{4u\left(n_{\text{vac}}^{(t)}\right)^2}{n^{(t)}} - kT\ln\frac{n_{\text{sur}}!}{\left(n_{\text{sur}} - n_{\text{vac}}^{(t)}\right)!\,n_{\text{vac}}^{(t)}!}$$

Using the condition $\partial F/\partial n_{\text{vac}}^{(t)} = 0$, Stirling's approximation and $\beta = 1/kT$ results in

$$\frac{n_{\text{vac}}^{(t)}}{n^{(t)} - n_{\text{vac}}^{(t)} - n_{\text{ad}}^{(t)}} = \exp-\left\{2\beta u\left[1 - 2\left(n_{\text{vac}}^{(t)}/n^{(t)}\right)\right]\right\} \tag{11.21}$$

A similar expression results for adatoms with $n_{\text{vac}}^{(t)}$ replaced by $n_{\text{ad}}^{(t)}$. Using that $n_{\text{ad}}^{(t)} = n_{\text{vac}}^{(t)}$, the mole fraction $x \equiv x_{\text{vac}} = n_{\text{vac}}^{(t)}/n^{(t)} = x_{\text{ad}} = n_{\text{ad}}^{(t)}/n^{(t)}$, Eq. (11.21) becomes

$$x/(1 - 2x) = \exp-[2\beta u(1 - 2x)] \tag{11.22}$$

The surface roughness R is defined as the number of free bonds parallel to the surface per site, which for {100} SC yields

$$R = U_{\text{vac}}/(\tfrac{1}{2}u)n_{\text{sur}} = 8x(1 - x) \tag{11.23}$$

Simultaneously solving Eqs. (11.22) and (11.23) results in Figure 11.6a, which shows that below $T \cong 0.5u/k$, the surface remains essentially smooth. Estimating u from the sublimation enthalpy, which in this bond model is given by $\Delta_{\text{sub}}H = \frac{1}{2} \cdot zN_{\text{A}}u$ or from the surface energy in the same approximation given by $\gamma = \frac{1}{2} \cdot (z - s)n_{\text{sur}\{100\}}u = \frac{1}{6} \cdot \Delta_{\text{sub}}H/V_{\text{m}}^{2/3}N_{\text{A}}^{1/3}$, it unfortunately appears that $T_{\text{R}} \gg T_{\text{mel}}$.

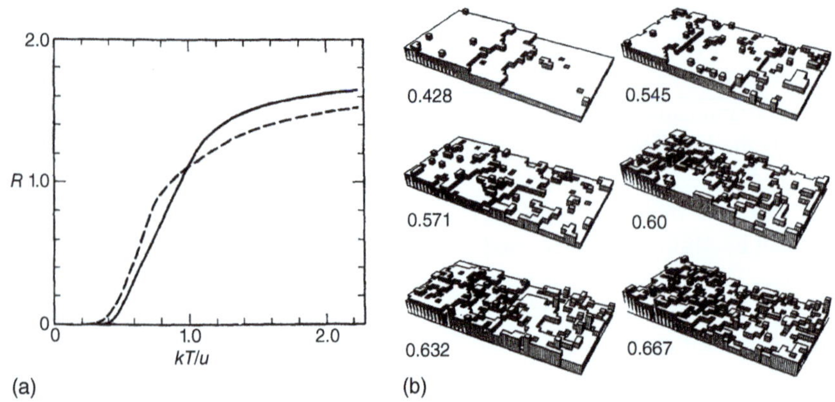

Figure 11.6 (a) Surface roughness R for {100} SC surfaces as a function of $1/\beta u$ according to Mullins [84] (–) and Burton et al. [83] (- -). (b) Perspective images of representative (20,1,0) surface configurations at various values of $2kT/u$ with $2kT_R/u \cong 0.64$. Source: Leamy [85]/with permission of Elsevier.

As experimental roughening does occur, we clearly miss something. Similar to the roughening of terraces, ledges can also be roughened. If we use the same procedure as before, for an [100] ledge on an {100} surface, the formation energy of $n_{vac}^{(1)}$ ledge vacancies equals $U_{vac}^{(1)} = un_{vac}^{(1)} - u(n_{vac}^{(1)})^2/n^{(1)}$. Counting complexions for the number of available ledge sites $n^{(1)}$, the equilibrium expression for ledge vacancies becomes

$$\frac{n_{vac}^{(1)}}{n^{(1)} - n_{vac}^{(1)} - n_{ad}^{(1)}} = \exp - \left[\beta u\left(1 - 2x_{vac}^{(1)}\right)\right]$$

A similar expression is obtained for $n_{ad}^{(1)}$ ledge adatoms, where $n_{vac}^{(1)}$ in the numerator on the lhs is replaced by $n_{ad}^{(1)}$ and $x_{vac}^{(1)}$ by $x_{ad}^{(1)}$ on the rhs. Adding ledge roughness brings T_R down to about $\frac{1}{2}T_{mel}$. A ledge vacancy, a divacancy, and so on, are all equivalent to two opposite kinks. The number of kinks $n^{(k)}$ is given by $n^{(k)}/n^{(1)} = \exp(-\beta U^{(k)}) = \exp(-\frac{1}{2}\beta u)$. This expression can be used to estimate ledge and kink formation energies if microscopic images as a function of T are available.

Here, the zeroth-order solution of the order–disorder problem was used (Section C.2) for which the critical temperature $T_R = zu/2k = 3u/k$. Burton et al. [83] used the first-order solution as well as the exact Onsager solution. The latter for a square lattice yields $T_R = u/k\ln(1+2^{1/2}) \cong 1.13u/k$, but even that is still a poor estimate: when a large cluster is formed, another cluster can easily build on top of its surface.

Clearly, the three-level model is insufficient, and the three-level approximation can be avoided by using the *solid on solid* (SoS) model. Considering the same SC lattice with forbidden overhangs, we have columns of height z relative to the flat surface at $T = 0$ K, or $z_j = z(x_j, y_j)$ where (x_j, y_j) are the coordinates of column j. The excess energy $U(z_j)$ is related to the number of free vertical faces of the cubes in the column and given by $U(z_j) = \frac{1}{2}u\Sigma_{j,\delta} f(z_j - z_{j-\delta})$ where δ runs over the neighboring columns. Using the bond model, the function $f(z_j - z_{j-\delta}) = |h_j - h_{j-\delta}|$. This defines the *absolute SoS* (ASoS) *model* [86], from which we recover the three-level model if we restrict of z_j to -1, 0, and 1 (In the discrete gaussian SoS [DGSoS] model [87], one uses $f(z_j - z_{j-\delta}) = (z_j - z_{j-\delta})^2$, but we will not discuss that variant.). This model must be solved numerically which can be done by the

MC method in the (μ,T,V) ensemble (Section 7.6) using as probability $\exp[-\beta(\Delta U - \mu)]$. From the results one can obtain for a distance d between columns, the height correlation function $G(d) = \langle\overline{(z_i - z_j)^2}\rangle$. The results of such an ASoS calculation for the SC lattice (Figure 11.5b) shows that for $T < T_R$, $G(d)$ tends to a constant value for a large value of d, while for $T > T_R$, $G(d)$ diverges weakly. Before we indicated that $G(d) = (kT/\pi\gamma_0^*)(\ln d + c)$ where c is a constant. By fitting the $G(d)$ curves with this expression, it was found that $T_R \cong 1.24u/k$, while the DGSoS model yields $T_R \cong 1.46u/k$. Images from MC simulations on (20,1,0) SC surfaces at various temperatures [85] show that considerable roughness is already present below T_R. A body-centered SoS (BCSoS) model [88] describing the {100} and {111} FCC surfaces can be solved analytically and the condition that $h_i - h_j = 0$ or $\pm$ ½, resulting in $T_R = u/k \ln 2 \cong 1.44u/k$.

To conclude, we note that the SoS roughening transitions also belong to the Kosterlitz–Thouless class [82]. The non-singular part of the Helmholtz energy as a function of T when approaching T_R follows $F \sim \exp[-c'/(|T - T_R|^{1/2})]$ with c' another constant and vanishes at and above T_R. All thermodynamic quantities and their derivatives are continuous to any order, thereby explaining why the roughening transition is rather smooth as compared to the order–disorder continuous transition model. The disappearance of crystal facets is a characteristic of roughening and has been observed experimentally. For example, for Pb below 323 K the bounding facets are (111), (110), (100), and (112). For 323 K < T < 393 K, the (112) facets are absent, while above 393 K up to T_{mel}, the only remaining facets are the (111) and (110). The latter planes are the most close packed and for them $T_R > T_{mel}$.

11.5 Adsorption at Interfaces

To lower the interface energy for two-component systems, one of the components may enrich at the interface. It is called *segregation* (the usual word for enrichment at any interface of a solid) or *adsorption* (the usual word for enrichment at a fluid interface). These phenomena are quite general; their thermodynamics are identical apart from the choice of the reference state (solid for segregation and gas for adsorption), but their kinetics generally have a rather different characteristic time. Emphasizing this generic behavior, we just call it *adsorption*. As adsorption requires a second component, we first discuss solutions, deal thereafter with the thermodynamics and statistics of adsorption, and finalize with a discussion on adsorption isotherms.

11.5.1 Solutions

The simplest model of a solute 2 in a solvent 1 is the *perfect solution* model for which the mixing entropy $\Delta_{mix}S_{perfect} = -R(n_1 \ln x_1 + n_2 \ln x_2)$ and the mixing enthalpy $\Delta_{mix}H_{perfect} = 0$. If one assumes that $\Delta V = 0$, $\Delta_{mix}U = \Delta_{mix}H$. As perfect behavior is often used as a reference, for any quantity Z, the *excess quantity* $Z^E = \Delta Z - \Delta Z_{perfect}$ is defined.

The next simplest model is the *regular solution* (RS) model with $\Delta_{mix}S$ as for the perfect solution but with $\Delta_{mix}H$ either $\Delta_{mix}H > 0$, which occurs if the 1–2 interaction is unfavorable, or $\Delta_{mix}H < 0$, which is the case when the 1–2 interaction is favorable. In Section 9.13, we derived, using a lattice model with coordination number z, the random distribution of molecules over the lattice having 1–1, 2–2, and 1–2 nearest-neighbor interactions with a

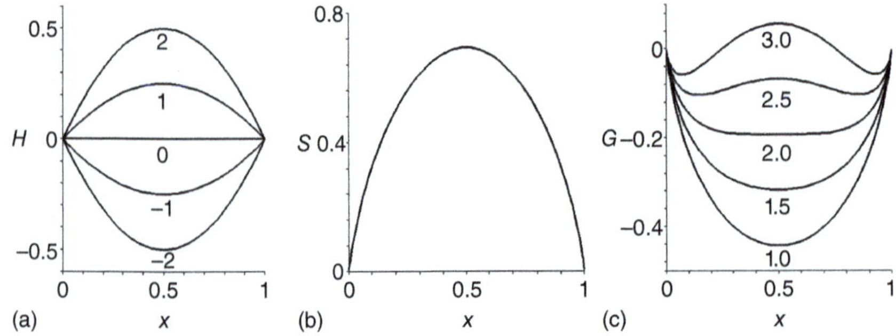

Figure 11.7 The regular solution model. (a) The enthalpy of mixing $H = \Delta_{mix}H/RT$, (b) the entropy of mixing $S = \Delta_{mix}S/R$, and (c) the Gibbs energy of mixing $G = \Delta_{mix}G/RT$ for various values of the interaction parameter zw/RT, all as a function of mole fraction x.

bond energy $u_{ij}(= -|u_{ij}| < 0)$, that the *interchange energy* w and mixing energy $\Delta_{mix}U$, equal to $\Delta_{mix}H$ if $\Delta V = 0$ as normal for lattice models,[10] are

$$w \equiv u_{12} - \tfrac{1}{2}u_{11} - \tfrac{1}{2}u_{22} \quad \text{and} \quad \Delta_{mix}H = \Delta_{mix}U = nN_A x_1 x_2 zw$$

Here, zw represents the interchange energy for replacing one molecule 1 (2) by one molecule 2 (1) in the matrix of molecules 1 (2). For now, we interpret U as internal energy but revisit this interpretation later. We have for the excess Gibbs energy $G^E = H^E - TS^E = U^E = Nzwx_1x_2$, sometimes called *Porter's equation*. The molar Gibbs energy of mixing becomes

$$\Delta_{mix}G_m = H^E + \Delta_{mix}S_{perfect} = N_A[zwx_1x_2 + kT(x_1 \ln x_1 + x_2 \ln x_2)] \tag{11.24}$$

Figure 11.7 shows the behavior of $\Delta_{mix}H_m$, $\Delta_{mix}S_m$, and $\Delta_{mix}G_m$ for various values of the parameter zw/RT. Dependent on the value of zw, $\Delta_{mix}G_m$ shows a single or a double minimum; the latter situation leads to phase separation.

To calculate the activity coefficient f_j, we use the chemical potential $\mu_j = \mu_j^\circ + RT \ln f_j x_j$, so that we have the molar excess Gibbs energy[11] $G_m^{\ E}$ and f_j

$$G_m^{\ E} = RT\Sigma_j n_j \ln f_j \quad \text{and} \quad \partial G_m^{\ E}/\partial n_j = RT \ln f_j$$

Using $G^E = Nzwx_1x_2 = N_A zwn_1n_2/(n_1 + n_2)$ and $\beta = 1/kT$, the result is

$$\ln f_1 = \beta zwx_2^{\ 2} \quad \text{and} \quad \ln f_2 = \beta zwx_1^{\ 2}$$

Note that for a dilute solution ($x_2 \ll 1$), $\mu_1 \cong \mu_1^\circ + RT \ln x_1$ but that $\mu_2 \cong \mu_2^\circ + RT \ln x_2 + N_A zw$, still resulting in a non-perfect solution.

To assess phase behavior, we calculate the critical point for phase separation from

$$\frac{\partial \mu_1}{\partial x_1} = \frac{\partial \mu_2}{\partial x_1} = \frac{\partial^2 \mu_1}{\partial x_1^2} = \frac{\partial^2 \mu_2}{\partial x_2^2} = 0$$

10 $H^E - U_V^{\ E} = T(\partial P/\partial T)_V \Delta V + \ldots$ is known if $(\partial P/\partial T)_V = -(\partial V/\partial T)_P(\partial P/\partial V)_T = -\alpha\kappa$ and ΔV are known.

11 Alternatively, using $G^E = ng^E$, $RT \ln \gamma_1 = g^E + x_2 \partial g^E/\partial x_1$ and $RT \ln \gamma_2 = g^E - x_1 \partial g^E/\partial x_1$, or using the auxiliary function $Q = g^E/x_1x_2$, $RT \ln \gamma_1 = x_2^{\ 2}(Q + x_1 \partial Q/\partial x_1)$ and $RT \ln \gamma_2 = x_1^{\ 2}(Q - x_2 \partial Q/\partial x_1)$.

Inserting μ_j and $\ln f_j$, we obtain

$$\frac{\partial \mu_1}{\partial x_1} = \frac{\partial \ln a_1}{\partial x_1} = \frac{1}{1-x_2} - 2x_2\beta zw = 0 \qquad \frac{\partial^2 \mu_1}{\partial x_1^2} = -\frac{1}{(1-x_2)^2} + 2\beta zw = 0$$

$$\frac{\partial \mu_2}{\partial x_2} = \frac{\partial \ln a_2}{\partial x_1} = \frac{1}{x_2} - 2(1-x_2)\beta zw = 0 \qquad \frac{\partial^2 \mu_2}{\partial x_2^2} = -\frac{1}{x_2^2} + 2\beta zw = 0$$

These equations are satisfied for $x_1 = x_2 = 0.5$ (as is to be expected on symmetry grounds) and $(\beta zw)_{\text{cri}} = 2$. The critical temperature T_{cri} for the phase separation is thus

$$T_{\text{cri}} = zw/2k \tag{11.25}$$

At this point, $\ln f = 0.5$ or $f = 1.649$, while the activity $a = fx = 0.824$. Using the differential $\partial\Delta_{\text{mix}}G/\partial x_2 = 0$, the complete solution of the phase separation problem is

$$\ln(x_2/x_1) = \beta zw(2x_2 - 1)$$

Using the transformation $s = 2x_2 - 1$, we obtain

$$\beta zw = s^{-1}\ln[(1+s)/(1-s)] = 2s^{-1}\tanh^{-1}s \quad \text{or} \quad s = \tanh(½\beta zws)$$

This is identical to the *zeroth-order* approximation for the order–disorder problem (Section C.2).

The total pressure P and at T_{cri} in the RS model become

$$P = P_1^* f_1 x_1 + P_2^* f_2 x_2 \quad \text{and} \quad P_{\text{cri}} = 0.824\left(P_1^* + P_2^*\right) \tag{11.26}$$

to be compared with $P = (P_1^* + P_2^*)$ for completely insoluble liquids and $P_{\text{cri}} = 0.5(P_1^* + P_2^*)$ for completely soluble liquids. To estimate zw, we use the experimental vapor pressure P_{exp} for equimolar solutions and Eq. (11.26), resulting in $\beta zw = 4\ln[2P_{\text{exp}}/(P_1^* + P_2^*)]$. It appears that normally the value of zw decreases with increasing temperature, approximately according to

$$zw = a + bT + cT\ln T \cong a + bT$$

For example, for the system $CS_2 - CHCl_3$, $zw(\text{J mol}^{-1}) = 5305 - 13.2T$, and for the system $(C_2H_5)_2O - CHCl_3$, $zw(\text{J mol}^{-1}) = -7368 + 15.5T$, making it clear that zw cannot be related to the values for u_{ij} only, since the u_{ij} are assumed to be temperature independent. Experimental values for some systems are given in Table 11.6. Problem 11.6 reveals why the approximate description $zw = a + bT$ is in principle insufficient.

There remains to be discussed the physical interpretation of the parameter w and how we can obtain a value for w. So far, we deliberately often addressed the parameters u_{ij} as interactions and $2w$ was interpreted as an energy (enthalpy). Guggenheim [90] realized that w represents a value averaged over all accessible states of the remainder of the system, each weighted with their relevant Boltzmann factor. Hence, w represents, if taken at constant V, the Helmholtz energy $F(T, V, x_1)$, or, if taken at constant P, the Gibbs energy $G(T, P, x_1)$. Of course, the mixing entropy is still to be added. If we interpret $N_A zwx_1x_2$ as $G_m^{\,E}$, with $w_m = N_A w(T, P)$, we have

$$S_m^E = -\frac{\partial G_m^E}{\partial T} = -x_1x_2\frac{\partial zw_m}{\partial T}, \quad H_m^E = G_m^E - T\frac{\partial G_m^E}{\partial T} = x_1x_2\left(zw_m - T\frac{\partial zw_m}{\partial T}\right)_P$$

Table 11.6 Parameters $H^{E^*} = H^E/Rx_1x_2$, $S^{E^*} = S^E/Rx_1x_2$ and $C^{E^*} = C^E/Rx_1x_2$ for some systems.

System	T (K)	H^{E^*} (–)	S^{E^*} (K^{-1})	T (K)	H^{E^*} (–)	S^{E^*} (K^{-1})	C^{E^*} (–)[a]
CCl_4–C_6H_6	298	52.4	0.044	343	86.0	0.15	0.76
C_6H_6-*cyclo*-C_6H_{12}	293	400	0.84	343	324	0.60	−1.6
C_6H_6-1,2-$C_2H_4Cl_2$	298	32	0.060	343	28	0.060	0.009
System	**T (K)**	**H^{E^*} (–)**	**S^{E^*} (K^{-1})**	**System**	**T (K)**	**H^{E^*} (–)**	**S^{E^*} (K^{-1})**
CCl_4-*cyclo*-C_6H_{12}	313	68	0.11	C_6H_6–$C_6H_6CH_3$	353	24	0.076
CS_2–$(CH_3)_2CO$	308	700	0.64	*n*-C_5F_{12}–*n*-C_5H_{12}	277	744	0.56

a) The heat capacity C^{E^*} is assumed a constant, applicable to both temperatures.
Source: Adapted from Pitzer [89].

A nice illustration is given by the system CCl_4–C_6H_{12} for which G^E and H^E data are available [91]. These data are described well by the single expression [2]

$$zw_m\ (\text{J mol}^{-1}) = 1176 + 1.96\,T\ln T - 14.18T \qquad \text{so that}$$

$$(zw_m - T\partial zw_m/\partial T)\ (\text{J mol}^{-1}) = 1176 - 1.96T$$

Longuet-Higgins [92] in his theory of *conformal* solutions provided a more rigorous way to show that w contributes to $F(T, V, x_1)$, meanwhile avoiding a lattice-like model.

The RS model represents a *symmetrical mixture*, meaning it shows symmetrical behavior with respect to the components 1 and 2. Experimentally this is not so in many cases. However, although H^E, S^E, and V^E may exhibit complex behavior, G^E is often still rather symmetric and reasonably well described by $G^E = Nzwx_1x_2$. There are a few deficiencies that could be remedied. First, we assumed that the coordination is random while there are preferred interactions. This assumption is thus not correct, but the correction for this effect appears to be small. Second, the enthalpy estimate is based on nearest-neighbors on a lattice but as indicated the RS model can be freed of the lattice concept as well as the nearest-neighbor approximation. The effect of this is also small. The quasi-chemical (and higher-order) approximation provides a better estimate for T_{cri} but the overall behavior remains symmetric.

Another improvement, to accommodate for the unequal size and shape of molecules, is to extend the entropy expression. We derive in Appendix C the *Guggenheim–Staverman entropy* expression, which for a binary solution is given by

$$\Delta_{mix}S/k = -N_1\ln\phi_1 - N_2\ln\phi_2 + \tfrac{1}{2}z[q_1N_1\ln(\theta_1/\phi_1) + q_2N_2\ln(\theta_2/\phi_2]$$

with mole fraction x_j, volume fraction $\phi_j = r_jx_j/\Sigma_jr_jx_j$, and surface fraction $\theta_j = q_jx_j/\Sigma_jr_jx_j$. Here, $r_j = V_j/V_{ref}$ and $q_j = A_j/A_{ref}$ are nondimensional size and area parameters, respectively, in practice typically calculated using the vdW values for CH_2 unit with $V_{ref} = 15.17\ \text{cm}^3\ \text{mol}^{-1}$ and $A_{ref} = 2.503 \times 10^9\ \text{cm}^3\ \text{mol}^{-1}$ in combination with (the arbitrary choice) $z = 10$. Neglecting internal contacts, that is, taking $q_j \sim r_j$, this expression reduces to the *Flory–Huggins* (FH) *entropy* expression

$$\Delta_{mix}S/k = -N_1\ln\phi_1 - N_2\ln\phi_2$$

Surprisingly, a recent reconsideration [93] of Guggenheim's results [94, 95] showed that the coordination number z is not a free parameter. In his derivation, Guggenheim introduced the z-r-q-rule reading $\frac{1}{2}z = (r_j - 1)/(r_j - q_j)$, valid for open-chain molecules (i.e. for linear or branched molecules without rings). Essentially, one can derive, without using the z-r-q-rule, that in the quasi-lattice model on the one hand the entropy difference $\Delta S/k = \Sigma_j N_j \ln \phi_j + \frac{1}{2}z\Sigma_j N_j q_j \ln(\theta_j/\phi_j)$, and on the other hand that $\Delta S/k = \Sigma_j N_j \ln(\theta_j/x_j) - (1 - \frac{1}{2}z)\Sigma_j N_j r_j \ln(\theta_j/\phi_j)$. The activity coefficients can be calculated from these expressions via $\ln f_j = \partial(\Delta S/k)/\partial N_j$, and since both formulae for $\ln f_j$ must yield the same result, it appears that generally $zq_j/2 = [1-(\phi_j/x_j)]/[1-(\phi_j/\theta_j)]$. Hence, the value for z is implicit in the model, rendering an independent choice for z obsolete. This expression results in a z-value of an alkane or 1-alcohol in a mixture of alkanes or 1-alcohols, which stays constant over the entire concentration range for each mixture, but is dependent on the number of carbon atoms [96]. Because $zq_j/2$ can be estimated using either a lattice model or an off-lattice (fluid) model [93], this frees again the quasi-chemical approximation from the quasi-lattice concept (see [95, 97]).

The excess Gibbs energy as used in the RS model can be systematically but largely empirically extended. As a first approximation, we wrote for the RS model $G_m{}^E = zw_m x_1 x_2$, where w_m is a temperature-dependent parameter. Corrections may be obtained by expanding in $x_1 - x_2$. Hence, we write the *Redlich–Kister expansion* as[12]

$$G^E/x_1x_2 = A + B(x_1 - x_2) + C(x_1 - x_2)^2 + \cdots$$

The inclusion of these terms generates systematic improvements in the description.

When $A = B = C = \cdots = 0$, $G^E = 0$, $RT \ln f_1 = RT \ln f_2 = 0$, and $\gamma_1 = \gamma_2 = 1$. Accordingly, we regain the perfect solution.

When $B = C = \cdots = 0$, $G^E = Ax_1x_2$ and $RT \ln f_1 = Ax_2{}^2$ and $RT \ln f_2 = Ax_1{}^2$. Obviously, these results represent the RS model, and the expressions are symmetric in nature. For infinite dilution, one obtains $RT \ln f_1{}^\infty = RT \ln f_2{}^\infty = A$.

When $C = \cdots = 0$, $G^E/x_1x_2 = A + B(x_1 - x_2)$, implying a deviation from the RS model. Equivalently, since $x_1 + x_2 = 1$, we may write $G^E/x_1x_2 = A(x_1 + x_2) + B(x_1 - x_2)$ or defining $A_{21} = A + B$ and $A_{12} = A - B$,

$$G^E/x_1x_2 = A_{21}x_1 + A_{12}x_2$$

a result known as the *Margules equation* [100]. This leads to

$$RT \ \ln f_1 = x_2^2[A_{12} + 2x_1(A_{21} - A_{12})] \quad \text{and}$$

$$RT \ \ln f_2 = x_1^2[A_{21} + 2x_2(A_{12} - A_{21})]$$

For $x_1 \to 0$, $RTf_1{}^\infty = A_{12}$ and for $x_2 \to 0$, $RT \ln f_2{}^\infty = A_{21}$. Obviously, this approach can be extended, although, an expansion further than third order is hardly ever required in view of the experimental accuracy [95]. We note that if the coefficients B, D, … vanish, we have a *symmetrical mixture*. This approach is flexible and can describe rather different types of behavior, that is, continuously increasing G^E, continuously decreasing G^E, and a G^E showing a maximum or minimum.

---

12 See [98]. However, the expansion was already used by Guggenheim [99].

One may also expand the reciprocal x_1x_2/G^E in $x_1 - x_2$ to obtain the (generalized) *van Laar expansion* [101]:

$$x_1x_2/G^E = A' + B'(x_1 - x_2) + \dots$$

Equivalently, we may write for a two-parameter truncation, since $x_1 + x_2 = 1$,

$$x_1x_2/G^E = A'(x_1 + x_2) + B'(x_1 - x_2) = (A' + B')x_1 + (A' - B')x_2$$

Defining $1/A_{21}' = A' + B'$ and $1/A_{12}' = A' - B'$, we obtain

$$\frac{x_1x_2}{G^E} = \frac{x_1}{A_{21}'} + \frac{x_2}{A_{12}'} \quad \text{or} \quad \frac{G^E}{x_1x_2} = \frac{A_{12}'A_{21}'}{A_{12}'x_1 + A_{21}'x_2} \tag{11.27}$$

known as the *van Laar equation*. The expressions for the activity coefficients are

$$RT\ \ln f_1 = A_{12}'\left(1 + \frac{A_{12}'x_1}{A_{21}'x_2}\right)^{-2} \quad \text{and} \quad RT\ \ln f_2 = A_{21}'\left(1 + \frac{A_{21}'x_2}{A_{12}'x_1}\right)^{-2}$$

In this case, for $x_1 \to 0$, $RT \ln f_1^\infty = A_{12}'$ and for $x_2 \to 0$, $RT \ln f_2^\infty = A_{21}'$.

Some algebraic manipulation, meanwhile identifying $A_{21}' = CV_1$ and $A_{12}' = CV_2$ where V_j is the partial volume of component j and C a constant, shows that Eq. (11.27) is equivalent to an expansion of G^E in the volume fraction $\phi_1 = V_1x_1/V_m$, where $V_m \equiv V_1x_1 + V_2x_2$ is the molar volume of the mixture, that is, $G_m^E = C\phi_1\phi_2V_m$. This relation fits data well for many binary systems of normal liquids, where C, as rationalized by Scatchard [102], is given by

$$C = \left[(-\Delta U_{11}/V_1)^{1/2} - (-\Delta U_{22}/V_2)^{1/2}\right]^2 \equiv (\delta_1 - \delta_2)^2$$

with the molar energy of vaporization $-\Delta U_{jj} = \Delta_{\text{vap}}U_j = \Delta_{\text{vap}}H_j - RT$ and the *solubility parameter* $\delta_j = (\Delta_{\text{vap}}U_j/V_j)^{1/2}$. For normal liquids, the ratio $A_{12}'/A_{21}' \cong V_1/V_2$; for example, for the mixture benzene and iso-octane [103] at 45 °C $A_{12}'/A_{21}' = 0.562$, while $V_1/V_2 = 0.536$. For complex liquids, significant deviations occur.

The above expressions show a relatively large flexibility in fitting the experimental data and a semiempirical theoretical basis can be given. However, the equations are somewhat complex to extent to multicomponent systems and the temperature dependence of the parameters involved is not explicitly included.

One alternative for describing $G^E = ng^E$ is using (Legendre) *orthogonal polynomials* $L_j(x_2)$ with property $\int L_iL_j\mathrm{d}x_2 = (2j+1)^{-1}\delta_{ij}$ or $L_j(y)$ with $y = x_1 - x_2$ and $\int L_iL_j\mathrm{d}y = 2(2j+1)^{-1}\delta_{ij}$ [104]. The first few polynomials for the expansion $g^E = \Sigma_{j=0}c_jx_1x_2L_j$, with as parameter c_j, are

$$L_0 = 1, \quad L_2 = 1 - 2x_2 = y \quad \text{and} \quad L_3 = 1 - 6x_2 + 6x_2^2 = ½(3y^2 - 1)$$

Additional terms can be generated from $L_n(y) = [(2n-1)yL_{n-1}(y) - (n-1)L_{n-2}(y)]/n$. Highly asymmetric excess functions can be fitted, adding higher-order terms should not affect the lower-order coefficients, and the same coefficients can be used for $L_j(x_2)$ and $L_j(y)$. For partial quantities, the same coefficients apply as for the polynomials[11] $g_j^E = g^E + (1-x_j)\partial g^E/\partial x_j$. Conversion of the coefficients for various expansions is also discussed in [104]. A slightly different approach is presented in [105].

An example of the use of all these concepts is given by a description of the eutectic melting temperature T_{eut} of *deep eutectic solutions* (or solvents, DESs). DESs are binary mixtures of usually solid components which exhibit a remarkably large melting point depression compared to ideal behavior, resulting often in liquids at a much lower temperature than expected. This extended liquid range in combination with the choice of constituents offers prospects for designing new solvents. Using the FH entropy expression in combination with the Redlich–Kister expansion, employing orthogonal polynomials, the phase diagrams of a number of solutions consisting of polyols or dicarboxylic acids with quaternary ammonium salts could be described rather well than using a temperature-independent enthalpy alone [106]. Using $\Delta_{\text{mix}}S_{\text{perfect}}$ one requires quadratic terms in $y = x_1 - x_2$, while using $\Delta_{\text{mix}}S_{\text{FH}}$ only linear terms are required. The corresponding change $\Delta_{\text{mix}}H \cong 0.5 - 1.0RT$, illustrating the effect of the choice of the entropy expression upon the resulting enthalpy value. It is worth noting that the choice of the reference volume can influence $\Delta_{\text{mix}}H$ by another 0.2 RT.

Problem 11.6 Show that, if $zw = a + bT$, the parameters a and b relate to the enthalpy and entropy, respectively. Indicate why, considering data from Table 11.6, this description cannot be sufficient. Using $zw = a + bT + cT \ln T$, interpret a, b, and c.

11.5.2 Thermodynamics of Adsorption

For two-component solutions, we neglect the contributions due to the interphase thickness from the start, as their effect is negligible and the resulting equations are cumbersome [2]. As before we denote the mole fraction of the solvent by x_1 and of the solute by x_2. The adsorption behavior, described by the GAE, Eq. (11.7), becomes

$$d\gamma = -s^{(\sigma)}dT + \tau dP - \Gamma_1 d\mu_1 - \Gamma_2 d\mu_2 \cong -s^{(\sigma)}dT - \Gamma_1 d\mu_1 - \Gamma_2 d\mu_2$$

Substituting the expression for $d\mu_1$, reading

$$d\mu_1 = -S_1 dT + V_1 dP + \frac{\partial \mu_1}{\partial x_2}dx_2 \cong -S_1 dT + \frac{\partial \mu_1}{\partial x_2}dx_2$$

and the corresponding expression for $d\mu_2$, we have

$$-d\gamma = (s^{(\sigma)} - \Gamma_1 S_1 - \Gamma_2 S_2)\,dT + \left(\Gamma_1 \frac{\partial \mu_1}{\partial x_2} + \Gamma_2 \frac{\partial \mu_2}{\partial x_2}\right)dx_2$$

Inserting the Gibbs–Duhem equation, $x_1(\partial\mu_1/\partial x_2) + x_2(\partial\mu_2/\partial x_2) = 0$, the result is

$$-d\gamma = (s^{(\sigma)} - \Gamma_1 S_1 - \Gamma_2 S_2)\,dT + \Gamma_2^{(1)}\frac{\partial \mu_2}{\partial x_2}dx_2 \quad \text{with} \quad \Gamma_2^{(1)} = \Gamma_2 - \frac{x_2\Gamma_1}{x_1} \tag{11.28}$$

From this expression, the temperature derivative is given as

$$-(\partial\gamma/\partial T)_x = s^{(\sigma)} - \Gamma_1 S_1 - \Gamma_2 S_2 \tag{11.29}$$

From $\mu_j = G_j = F_j + PV_j \cong F_j = U_j - TS_j$ and $G^{(\sigma)} = F^{(\sigma)} + PV^{(\sigma)} - \gamma A \cong F^{(\sigma)} - \gamma A = n_1^{(\sigma)}\mu_1 + n_2^{(\sigma)}\mu_2$, we have $u^{(\sigma)} - Ts^{(\sigma)} - \gamma = \Gamma_1\mu_1 + \Gamma_2\mu_2$, so that

$$\gamma - T(\partial\gamma/\partial T)_x = u^{(\sigma)} - \Gamma_1 U_1 - \Gamma_2 U_2$$

Note that $u^{(\sigma)} - \Gamma_1 U_1 - \Gamma_2 U_2$ can be interpreted as the energy per unit area of interface minus the energy of the same amount of material in the bulk liquid state. A similar interpretation holds for $s^{(\sigma)} - \Gamma_1 S_1 - \Gamma_2 S_2$. From Eq. (11.28), we can calculate similarly the composition derivative, as

$$-\left(\frac{\partial \gamma}{\partial x_2}\right)_T = \Gamma_2^{(1)} \frac{\partial \mu_2}{\partial x_2} = RT\Gamma_2^{(1)} \frac{\partial \ln a_2}{\partial x_2} = RT\Gamma_2^{(1)} \frac{\partial \ln x_2}{\partial x_2} = RT \frac{\Gamma_2^{(1)}}{x_2} \tag{11.30}$$

where $a_j = f_j x_j$ denotes the activity with activity coefficient f_j for the mole fraction x_j. The last two steps can be made for a perfect mixture with $f_j = 1$. For solutes with a measurable vapor pressure, we can also replace $\ln a_2$ with $\ln P_2$. One can show that both partial derivatives are invariant to a shift in position of the Gibbs dividing plane. The theory discussed applies to all types of solutes.

To obtain an adsorption isotherm, that is, an equation relating γ with x_2, we first need some definitions [107–110]. The chemical potential μ_j in the bulk and surface phase of the solution are indicated by $\mu_j^{(\alpha)}$ and $\mu_j^{(\sigma)}$, respectively, while superscripts $^{(*\alpha)}$ and $^{(*\sigma)}$ apply for the pure components. A similar convention is used for the activity $a_j = f_j x_j$, while the partial molar surface areas are indicated by A_j and A_j^*. Now, we consider a process consisting of [109]:

- Removal of a portion of liquid from the bulk solution and formation of a pure liquid, described by $\mu_j^{(*\alpha)} - \mu_j^{(\alpha)} = -RT \ln a_j^{(\alpha)}$,
- Formation of pure surface phases for these portions, given by $\mu_j^{(*\sigma)} - \mu_j^{(*\alpha)} = \gamma_j A_j^*$,
- Mixing of the pure liquid surface phases to form an equilibrium surface layer, for which $\mu_j^{(\sigma)} - \mu_j^{(*\sigma)} = -RT \ln a_j^{(\sigma)}$, and
- Adding the newly formed layer to the surface of the bulk solution, described by $\mu_j^{(\sigma)} - \mu_j^{(\sigma)} = 0$.

This process is equivalent to the creation of new solution surface as characterized by $\mu_j^{(\sigma)} - \mu_j^{(\alpha)} = A_j \gamma$. Therefore,

$$A_j \gamma = -RT \ln a_j^{(\alpha)} + \gamma_j A_j^* + RT \ln a_j^{(\sigma)}$$

Introducing the activity coefficient f_j, one has for a binary mixture [108]

$$\gamma = \frac{A_1^*}{A_1}\gamma_1 + \frac{RT}{A_1} \ln \frac{f_1^{(\sigma)} x_1^{(\sigma)}}{f_1^{(\alpha)} x_1^{(\alpha)}} = \frac{A_2^*}{A_2}\gamma_2 + \frac{RT}{A_2} \ln \frac{f_2^{(\sigma)} x_2^{(\sigma)}}{f_2^{(\alpha)} x_2^{(\alpha)}}$$

Using $A_1 x_1^{(\sigma)} + A_2 x_2^{(\sigma)} = A$, we obtain

$$A\gamma = x_1^{(\sigma)} A_1^* \gamma_1 + x_1^{(\sigma)} RT \ln \frac{f_1^{(\sigma)} x_1^{(\sigma)}}{f_1^{(\alpha)} x_1^{(\alpha)}} + x_2^{(\sigma)} A_2^* \gamma_2 + x_2^{(\sigma)} RT \ln \frac{f_2^{(\sigma)} x_2^{(\sigma)}}{f_2^{(\alpha)} x_2^{(\alpha)}}$$

In the Guggenheim picture, the partial molar area A_j has to be interpreted as $A_j = V_j/\tau$ with V_j the molar partial volume and τ the thickness of the interphase, so that A_j represents a size average for the solute over the thickness τ. Similarly, $A_j^* = V_j^*/\tau^*$. To calculate γ often insufficient data are available and further assumptions must be made.[13] To summarize, the above approach is in principle a general description of adsorption, the general nature of which has been emphasized by Kaptay [112] as well.

One example of this approach is the work of Randles and Behr [110], who discuss solutions of C_6H_6 and $CHCl_3$ with $(CH_3)_2CO$, and CS_2 with C_6H_{12}, C_6H_6, $(CH_3)_2CO$

13 Butler's paper [107] already contains all the essential results. See [111] for an illuminating discussion.

and $(C_2H_5)_2O$, as well as H_2O with CH_3OH, C_2H_5OH, C_4H_9OH, and dioxane (1,4-dioxacyclohexane, $C_4H_8O_2$). They calculated the activity coefficients from a Redlich–Kister description and used the interphase thickness τ and a factor K, representing the change in interchange energy w for the interphase, as parameters. For the organic liquids used, K ranges from 0.8 to 1.0, indicating that nonideality in the surface layer is similar to that of the bulk (except for CS_2–C_6H_6 with $K = 0.4$ for unclear reasons), while the τ-values range from 5.0 to 5.6 Å. For the H_2O–alcohol solutions, $K \cong 0.5$ (except for CH_3OH and $C_4H_8O_2$ with $K = 0.9$, possibly related to orientation at the surface). The value for τ systematically increases with the chain length from 4.1 to 5.3 Å ($C_4H_8O_2$, $\tau = 5.2$ Å), nicely comparing to the value for butanol.

11.5.3 Statistics of Adsorption

To augment the thermodynamic description of surfaces, we now "change gear" to a statistical mechanical approach. To that purpose consider a binary system with bulk composition $A_{1-x}B_x$ and surface composition $A_{1-y}B_y$ (solvent A = component 1, solute B = component 2). We use again a lattice typc model, assume that only the first molecular layer has a composition different from the bulk and that for both the bulk and surface the components are distributed randomly. We use s_t for the coordination number (CN) for "bonds" in the surface layer, and s_n for the CN for "bonds" between the surface layer and the first bulk layer, so that the total surface coordination number reads $s = s_t + s_n$. For the bulk CN, we use $z = s_t + 2s_n$ (e.g. for a FCC (111) plane $s_t = 6$ and $s_n = 3$, so that for FCC bulk $z = s + s_n = s_t + 2s_n = 12$). As before, u_{ij} refers to the pair interaction between components, indicated by the subscripts i and j, while the phases are indicated by the superscripts (σ) and (α). We start with a perfect solution and thereafter consider the influence of nonideality, closely following [95, 113].

Because we keep μ_j, V, and T constant, it is proper to use the grand canonical partition function, for the bulk given by $\Xi = PV$ (Section 7.1). For a surface A, $kT \ln \Xi^{(\sigma)} = -\gamma A$. It will be expedient to use the activity $\lambda_j = \exp(\beta\mu_j)$ with $\beta = 1/kT$. To avoid many superscripts, we use N_j and M_j for the number of molecules in the bulk and in the surface, respectively. Similarly, we use q_j and r_j for the internal partition function of the component j in the bulk and in the surface, respectively. While q_j contains a factor $\exp(-½\beta z u_{jj})$, r_j will contain a factor $\exp\left(-½\beta s u_{jj}\right)$. We will also use the chemical potential of the component j in phases α and σ, given by

$$\mu_j^{(\alpha)} = \mu_j^{(*\alpha)} + kT \ln x_j \quad \text{and} \quad \mu_j^{(\sigma)} = \mu_j^{(*\sigma)} + kT \ln y_j \tag{11.31}$$

with $\mu_j^{(*\alpha)}$ and $\mu_j^{(*\sigma)}$ the standard chemical potentials of the pure components in the bulk and the surface, respectively, and $x_j \equiv x_j^{(\alpha)}$ and $y_j \equiv x_j^{(\sigma)}$ their mole fractions.

Let us first consider a pure one-component system j. The total Helmholtz energy is $F = F^{(\alpha)} + F^{(\sigma)} = -N_j kT \ln q_j - M_j kT \ln r_j$ and we have $\mu_j^{(*\alpha)} = -kT \ln q_j$, equivalent to $\lambda_j^{(*\alpha)} q_j = 1$. Similarly, $\lambda_j^{(*\sigma)} r_j = 1$. For surface area A, the surface tension γ_j becomes

$$\begin{aligned} A\gamma_j &= F - (N_j + M_j)\mu_j^{(*\alpha)} = F - (N_j + M_j)kT \ln \lambda_j^{(*\alpha)} = F^{(*\sigma)} - M_j kT \ln \lambda_j^{(*\alpha)} \\ &= -M_j kT \ln \lambda_j^{(*\alpha)} r_j = M_j kT \ln(q_j/r_j) \quad \text{or} \quad a_j^* \gamma_j = kT \ln(q_j/r_j) \end{aligned}$$

with the partial molecular area $a_j^* = \partial A/\partial M_j = A/M_j$ for the one-component system. Equivalently, for a one-component system at equilibrium we have $\mu_j^{(\alpha)} = \mu_j^{(\sigma)}$, while the surface tension is given by $A\gamma_j = F_j^{(\sigma)} - F_j^{(\alpha)}$ (Section 11.1), so that we again obtain

$$\mu_j^{(*\sigma)} - \mu_j^{(*\alpha)} = a_j^* \gamma_j = kT \ln(q_j/r_j) \tag{11.32}$$

Turning now to solutions, the partition function $Q^{(\sigma)}$ of a surface layer is given by

$$Q^{(\sigma)} = \frac{M!}{M_1!M_2!} r_1^{M_1} r_2^{M_2} \exp\left(-\beta U^{(\sigma)}\right) \quad \text{where } M_j = y_j M$$

Summing over all values of M_1 and M_2 subject to $M_1 + M_2 = M$, $\Xi^{(\sigma)}$ becomes

$$\Xi^{(\sigma)} = \sum\nolimits_{M_1} \sum\nolimits_{M_2} \exp[\beta(M_1\mu_1 + M_2\mu_2)]\, Q^{(\sigma)}$$

Alternatively summing y_2 over all values between 0 and 1, and using $\lambda_j = \exp(\beta\mu_j)$, we obtain

$$\begin{aligned}\Xi^{(\sigma)} &= \sum\nolimits_{y_2} \frac{M!}{(y_1M)!(y_2M)!} (\lambda_1 r_1)^{y_1M} (\lambda_2 r_2)^{y_2M} \exp(-\beta u^{(\sigma)}) \\ &= \sum\nolimits_{y_2} (\lambda_1 r_1/y_1)^{y_1M} (\lambda_2 r_2/y_2)^{y_2M} \exp(-\beta u^{(\sigma)})\end{aligned} \tag{11.33}$$

where the last step is made by using Stirling's approximation.

The simplest case is the perfect solution for which $u^{(\sigma)} = 0$ and thus

$$\Xi^{(\sigma)} = \sum\nolimits_{y_2} (\lambda_1 r_1/y_1)^{y_1M} (\lambda_2 r_2/y_2)^{y_2M}$$

Using the maximum-term method (Justification 7.4) by replacing the sum by its largest term and using the condition $\partial \ln \Xi^{(\sigma)}/\partial y_2 = 0$, we obtain

$$\Xi^{(\sigma)} = \left(\frac{\lambda_1 r_1}{y_1}\right)^{y_1M} \left(\frac{\lambda_2 r_2}{y_2}\right)^{y_2M} \quad \text{and} \quad \frac{\lambda_1 r_1}{y_1} = \frac{\lambda_2 r_2}{y_2} = \lambda_1 r_1 + \lambda_2 r_2 \tag{11.34}$$

From $\mu_j = \mu_j^{(*\alpha)} + kT \ln x_j$ we have $\ln\lambda_j = \ln(x_j/q_j)$ and substituting in $\lambda_1 r_1/y_1 = \lambda_2 r_2/y_2$, meanwhile using $a_j^*\gamma_j = kT \ln q_j/r_j$, results in

$$kT \ln(y_2/y_1) = kT \ln(x_2/x_1) - \left(a_2^*\gamma_2 - a_1^*\gamma_1\right) \tag{11.35}$$

This is the adsorption equation for perfect solutions. For the second term on the rhs, we can write $a_j^*\gamma_j = ½(z-s)u_{jj}$. Equivalently, from Eqs. (11.31) and (11.32), we obtain

$$\mu_1^{(*\sigma)} = \mu_1^{(\sigma)} + kT \ln(x_1/y_1) \quad \text{and} \quad \mu_2^{(*\sigma)} = \mu_2^{(\sigma)} + kT \ln(x_2/y_2) \tag{11.36}$$

If we subtract Eq. (11.36b) from Eq. (11.36a) and insert Eq. (11.32) for both components, we regain Eq. (11.35).

To obtain an expression for γ, we substitute Eq. (11.34b) in Eq. (11.34a). Hence,

$$\Xi^{(\sigma)} = (\lambda_1 r_1 + \lambda_2 r_2)^M = \exp(-\beta A\gamma) = \exp(-\beta Ma\gamma)$$

where $A \equiv aM \equiv (y_1 a_1^* + y_2 a_2^*)M$. Consequently,

$$\begin{aligned}\exp(-\beta a\gamma) &= \lambda_1 r_1 + \lambda_2 r_2 = \lambda_1^* x_1 r_1 + \lambda_2^* x_2 r_2 = x_1 r_1/q_1 + x_2 r_2/q_2 \quad \text{or} \\ \exp(-\beta a\gamma) &= x_1 \exp\left(-\beta a_1^*\gamma_1\right) + x_2 \exp\left(-\beta a_2^*\gamma_2\right)\end{aligned} \tag{11.37}$$

For a perfect solution therefore, γ is given by an additive rule for $\exp(-\beta a_j^*\gamma_j)$.

The next simplest case is probably a RS. The energy $u^{(\sigma)}$ is evaluated as for a RS in the bulk in the zeroth approximation. Creating a surface $s_t M y_1 y_2$ "1-2 bonds" are realized in the first layer, $s_n M(y_1 x_2 + x_1 y_2)$ "bonds" are realized between the first and second layers, while $\frac{1}{2}s_n M(x_1x_2+x_2x_1)$ "bonds" are deleted. Using as before the interchange energy $2w = 2u_{12} - u_{11} - u_{22}$, we have

$$U^{(\sigma)} = Mw[s_t y_1 y_2 + s_n(y_1x_2 + x_1y_2 - x_1x_2)]$$

$$= Mw\left[y_1\left(s_t y_2^2 + s_n x_2^2\right) + y_2\left(s_t y_1^2 + s_n x_1^2\right)\right] \equiv Mwy_1X + Mwy_2Y$$

to be inserted in Eq. (11.33). The second step can be made using $y_1 + y_2 = x_1 + x_2 = 1$. The grand partition function becomes

$$\Xi^{(\sigma)} = \sum\nolimits_{y_2} (\lambda_1 r_1\, e^{\beta wX}/y_1)^{y_1 M} (\lambda_2 r_2\, e^{\beta wY}/y_2)^{y_2 M}$$

Using the maximum-term method again, the condition $\partial \ln \Xi^{(\sigma)}/\partial y_2 = 0$ leads to

$$\ln(\lambda_1 r_1/y_1) - \beta w\left(s_t y_2^2 + s_n x_2^2\right) = \ln(\lambda_2 r_2/y_2) - \beta w\left(s_t y_1^2 + s_n x_1^2\right) \tag{11.38}$$

For a RS, we obtained $\mu_j^{(\alpha)} = \mu_j^{(*\alpha)} + kT\ln x_j + zw(1-x_j)^2$, equivalent to $\ln\lambda_j = \ln(x_j/q_j) + \beta zw(1-x_j)^2$. Substituting λ_j, we obtain after some calculation

$$kT\ln\frac{y_2}{y_1} = kT\ln\frac{x_2}{x_1} + \tfrac{1}{2}(z-s)(u_{22}-u_{11}) - w[s_t(1-2y_2) - s(1-2x_2)] \tag{11.39}$$

where $s = z - s_n$ and $\mu_j^{(*\alpha)} = \frac{1}{2}zu_{jj}$ are used. Comparing to Eq. (11.35), we see that the surface tension term $-(a_2^*\gamma_2 - a_2^*\gamma_2)$ reappears as $\frac{1}{2}(z-s)(u_{22}-u_{11})$, while the interaction term containing w is obviously absent for the perfect solution.

To obtain the surface tension γ for a RS, the grand partition function $\Xi^{(\sigma)} = \exp(-\beta\gamma A)$ can be rewritten, using Eq. (11.38), as

$$\Xi^{(\sigma)} = \left\{\frac{\lambda_1 r_1}{y_1}\exp\left[-\beta w\left(s_t y_2^2 + s_n x_2^2\right)\right]\right\}^M = \left\{\frac{\lambda_2 r_2}{y_2}\exp\left[-\beta\, w\left(s_t {y_1}^2 + s_n {x_1}^2\right)\right]\right\}^M$$

which results, using $\ln\lambda_j = \ln(x_j/q_j) + \beta zw(1-x_j)^2$ and $\langle a\rangle = A/M$, in

$$\langle a\rangle\gamma = kT\ln\frac{y_1 q_1}{x_1 r_1} + s_t w\left(y_2^2 - x_2^2\right) - s_n w x_2^2 = kT\ln\frac{y_2 q_2}{x_2 r_2} + s_t w\left(y_1^2 - x_1^2\right) - s_n w x_1^2$$

Finally, recalling $a_j^*\gamma_j = kT\ln(q_j/r_j)$, we obtain

$$\begin{aligned}\langle a\rangle\gamma &= a_1^*\gamma_1 + kT\ln(y_1/x_1) + s_t w y_2^2 - (s_t + s_n)wx_2^2 \\ &= a_2^*\gamma_2 + kT\ln(y_2/x_2) + s_t w {y_1}^2 - (s_t + s_n)w{x_1}^2\end{aligned} \tag{11.40}$$

Given $\gamma_1, a_1^*, \gamma_2, a_2^*$, and x_2, solving the second equation numerically for y_2, γ can be calculated from the first, altogether a somewhat complex procedure.

Estimating u_{jj} from the heat of evaporation in the nearest-neighbor approximation given by $L_j = -\frac{1}{2}zN_A u_{jj}$ and w from the heat of solution L_{12} of component 1 in component 2 (or vice versa from L_{21}), we have

$$\Delta L_{12} = -N_A zw = \Delta L_{21} \quad \text{with} \quad 2w = 2u_{12} - u_{11} - u_{22}$$

Note that the heat of vaporization is counted positive, while the bond energy is counted negative. In the zeroth approximation of RS theory, $\Delta L_{12} = \Delta L_{21}$. Experimentally, this condition is only exceptionally fulfilled, which indicates the need for further improvement. Further, the expression for γ contain only the partial areas a_j^* of the pure components, as the lattice model does not allow a volume change for the solution. This also explains the presence of $\langle a\rangle = A/M$ instead of $a = y_1 a_1 + y_2 a_2$.

Adding y_1 times Eq. (11.40a) and y_2 times Eq. (11.40b), differentiating and meanwhile using $\ln\lambda_j = \ln(x_j/q_j) + \beta zw(1-x_j)^2$, one obtains

$$-\langle a\rangle\,\mathrm{d}\gamma = y_1kT\mathrm{d}\ln\lambda_1 + y_2kT\mathrm{d}\ln\lambda_2 + 2s_\mathrm{t}w(y_2 - x_2)\mathrm{d}x$$

Hence, since we have the term $2s_tw(y_2 - x_2)\mathrm{d}x$, we see that the model does not satisfy the GAE, $-a\mathrm{d}\gamma = y_1kT\,\mathrm{d}\ln\lambda_1 + y_2kT\,\mathrm{d}\ln\lambda_2$, while actually it should. Two possible solutions have been forwarded. The first solution is to assume that the second layer is also affected by adsorption [114]. It appears that solving the relevant equations, the second layer indeed has a composition different from the bulk (the surface excess changes maximally 10%), but also that the surface tension is hardly affected. A general model for an arbitrary number of layers affected is available [115, 116]. The second solution claims that assuming the total number of 1–1 and 2–2 bonds of the second layer to be equal to that in the bulk [108] is inconsistent with the monolayer model. Taking the Gibbs dividing plane just below the surface layer, it is argued that the second layer already belongs to the bulk phase. Therefore, $(s_\mathrm{t}+s_\mathrm{n})x_2^{\,2}$ and $(s_\mathrm{t}+s_\mathrm{n})x_1^{\,2}$ in Eq. (11.40) should read $zx_2^{\,2}$ and $zx_1^{\,2}$, respectively. In that case, the GAE is satisfied. Whatever solution is taken, the quasi-lattice model is approximate anyway and for a more realistic prediction, a more realistic model should be adopted.

Despite these shortcomings, reasonable results are obtained. We provide just one example for the fully miscible system, ether-acetone [117]. For this system, γ ranges from 19 mJ m^{-2} to 25 mJ m^{-2}. Assuming regular behavior with the same partial areas for bulk and surface, the average deviation $\Delta = |\gamma_\mathrm{calc} - \gamma_\mathrm{exp}|$ for both 288 and 303 K over the full range is $\cong 0.08$ mJ m^{-2} with as largest deviation 0.25 mJ m^{-2} or $\cong 1\%$.

11.5.4 Adsorption Isotherms

The pioneer of the field was Langmuir [118], who in 1917 studied adsorption of solutions of amphiphilic molecules with different chain lengths. Adsorption leads to a *surface pressure* $\pi = \gamma_0 - \gamma$, with γ_0 the surface tension of the pure liquid and γ the surface tension of the solution, which is usually measured with a *surface pressure balance* or *Langmuir–Blodgett trough* (Figure 11.8a). On one side of a movable barrier, surfactant is added to the surface, while the other side has a pure water surface. A film of amphiphilic molecules as a function of available area A shows structural changes akin to gas compression (Figure 11.8b). In the gas-like region G the tails of the adsorbed molecules still make a considerable amount of contact with the liquid surface and can be described by equations like $\pi A = NkT$ or $\pi(A - A_0) = NkT$. In the liquid-expanded region $\mathrm{L_e}$, the surface contains still a disordered, homogeneous distribution of adsorbed molecules, but the tails lift off the surface. In this region, a vdW-like description can be applied. There is an intermediate region $\mathrm{L_e}$/G where gas-like and liquid-like areas coexist. Occasionally, a separate phase I, intermediate between the liquid-condensed phase $\mathrm{L_c}$ and $\mathrm{L_e}$ phase, is observed. The $\mathrm{L_c}$ phase is more solid-like because the tails are already aligned, probably non-perpendicular to the surface. The adsorbed molecules are relatively close packed at the surface but are still rather mobile. In the solid-like region $\mathrm{L_c}$, adsorbed molecules are really close-packed and compressed to each other with a contact area per molecule corresponding to the crystallographic cross-sectional area and thickness agreeing with that as calculated from the bulk density. Grazing incidence X-ray diffraction (XRD) typically indicate a centered orthorhombic unit cell, consistent with the chains being perpendicular to the surface. The transition

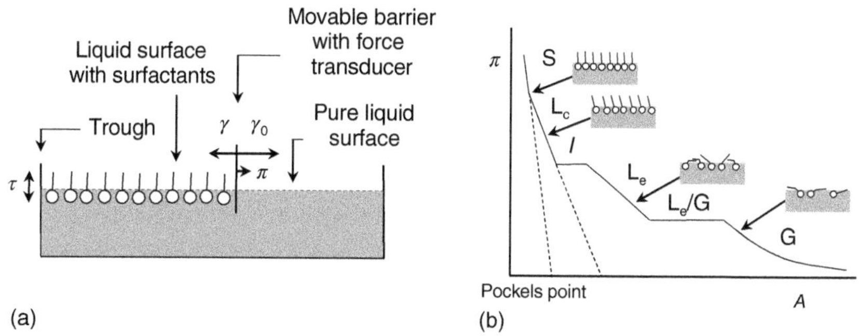

Figure 11.8 (a) Schematic representation of a Langmuir–Blodgett trough. (b) Schematic representation of the surface pressure $\pi = \gamma_0 - \gamma$ versus surface area A and the associated structure, showing the regimes of gas-like (G), liquid-expanded (L_e), liquid-condensed (L_c), and solid (S).

point between the S and L_c phases, the *Pockels point*, is a measure for the area a_{mol} of the adsorbed molecules, since at that point they are just close packed.

For the verification of the GAE, we indicate only one rather convincing experiment using a solution of tritiated sodium dodecyl sulfate (SDS) [119]. The surface tension γ and the adsorption Γ were independently determined as a function of the bulk concentration c. These measurements (Figure 11.9) indicate that γ starts to decrease at about $1\,\text{mmol}\,\text{l}^{-1}$, Γ saturates at about $3\,\text{mmol}\,\text{l}^{-1}$, corresponding to about $3\times10^{-10}\,\text{mol}\,\text{cm}^2$, while the critical micelle concentration (CMC) is about $7\,\text{mmol}\,\text{l}^{-1}$. From the $\gamma - c$ data using a Debye–Hückel activity coefficient, the calculated Γ agrees very well with the experimental Γ, confirming the validity of the GAE. The cross-section of SDS as determined from π is estimated as $a_{mol} \cong 50\,\text{Å}^2\,\text{mol}^{-1}$ (Figure 11.9c).

Although the interpretation of the abovementioned behavior of the adsorption isotherm is straightforward, generally the relevant experiments are loaded with intricacies, for example, the effects of impurities and polydispersity, the ratio between solubility and micelle formation equilibrium constants, and establishing equilibrium itself. Hence, the debate continues [120, 121], and a straightforward interpretation is not always possible without additional data. It appears that for nonionic surfactants, γ-data and neutron reflection data agree up to the CMC [122] as adsorption shows either a clear plateau or something very close to a plateau just below or at the CMC. For anionic surfactants, γ-data are often seriously compromised by traces of divalent ions and adsorption does not generally reach saturation at the CMC, making it difficult to obtain the limiting Gibbs slope. Moreover, a significant width of the micelle size distribution may make it impossible to apply the Gibbs equation over a sufficiently large concentration range below the CMC [123]. For cationic surfactants, the analysis of γ-data is again limited by the effects of the width of the micelle distribution, just as was found for the anionic surfactants. This width may, however, be a more serious problem for cationic surfactants because many of them have low aggregation numbers for which the width is expected to be larger. Extrapolation into the micellization region to obtain the limiting surface excess is not a reliable option because adsorption increases through and above the CMC. Neutron reflection data show that there is no plateau in the adsorption at the CMC, at least for the three surfactants measured. Thus γ-data cannot be easily modeled in a concentration region, where γ is

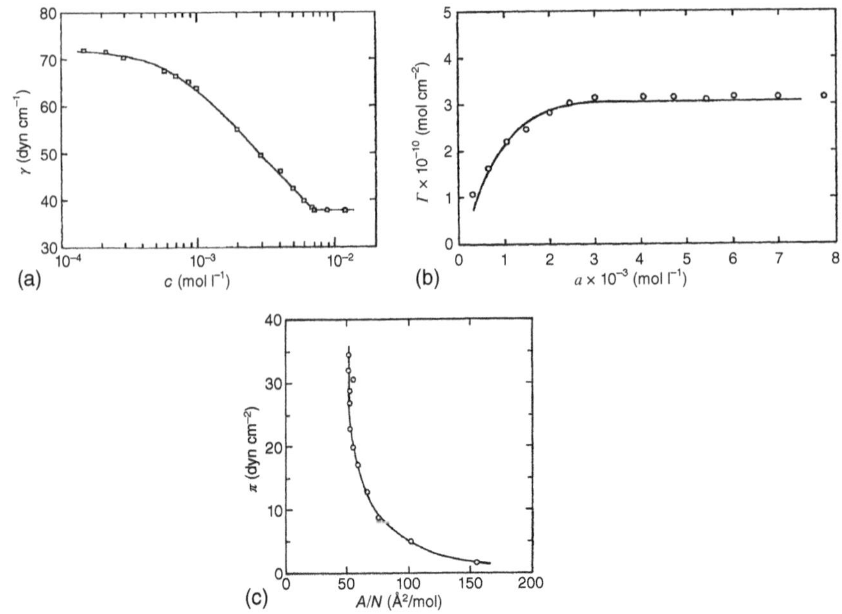

Figure 11.9 SDS adsorption. Source: Tajima et al. [119]/The Chemical Society of Japan. (a) Surface tension γ as a function of SDS concentration c using the Wilhelmy method. (b) Adsorption Γ as a function of SDS activity $a = f_{\mathrm{DH}}c$ with c measured by radiotracer methods and f_{DH} the Debye–Hückel activity coefficient (o), and as calculated (–). (c) Surface pressure π as a function of area per molecule A/N.

changing significantly with coverage. The combination of these two problems presents serious obstacles for the determination of Γ_{CMC} using the GAE [124]. The authors emphasize though that their conclusions do not in any way challenge the validity of the GAE; it is purely the difficult task of making the proper measurements of the necessary accuracy.

We now turn to some adsorption isotherms and discuss a kinetic approach and a statistical-mechanical approach. For the kinetic approach, we assume that the surface is a plane with a certain density of surface sites, that adsorption occurs essentially only in the first surface layer, and that adsorbed species do not interact. The process in the surface layer can be represented as a chemical reaction where the solute dissolved in the bulk (B) "reacts" with an empty surface site (S) to an occupied adsorbed site[14] (Ad), that is, $\mathrm{B} + \mathrm{S} \leftrightarrow \mathrm{Ad}$. We also assume that the molecules at the surface do not interact. Since the concentration of the solute is supposed to be low, we approximate the activity a by the concentration $c = N_{\mathrm{B}}/V$, where N_{B} is the number of dissolved molecules in the volume V. The fraction of occupied surface sites $\theta = N_{\mathrm{Ad}}/N_{\mathrm{max}}$, where N_{Ad} is the number of occupied sites and N_{max} is the number of sites at monolayer coverage. Hence, $N_{\mathrm{S}}/N_{\mathrm{max}} = 1 - \theta$. The equilibrium constant K for this "reaction" is

$$K = \theta/(1-\theta)c \quad \text{or} \quad \theta = Kc/(1+Kc) \tag{11.41}$$

14 If we replace the concentration in the condensed phase $c = N/V$ by $P/kT = N/V$, where P is the pressure in the gas phase, we model the adsorption of a gas at a surface.

For liquids and gases, this expression is usually denoted as the *Langmuir isotherm*, but for solids sometimes as the *McLean isotherm*. The adsorption $\Gamma \equiv \Gamma_2^{(1)}$ is, with $\Gamma_{\max}$ a proportionality constant representing monolayer coverage,

$$\Gamma = \Gamma_{\max}\theta \quad \text{or} \quad \Gamma^{-1} = \Gamma_{\max}^{-1} + (\Gamma_{\max}Kc)^{-1}$$

Before we discuss the behavior of the Langmuir isotherm, we first derive it via statistical mechanics. In this approach, we equate the chemical potential of the solute in the bulk phase $\mu^{(\alpha)}$ with that of the surface phase $\mu^{(\sigma)}$. With a small concentration of solute, we may assume ideal behavior with the N solute molecules freely moving through the matrix with the volume V with the potential energy $\phi^{(\alpha)}$. Hence, we use for the partition function $z^{(\alpha)}$ of a single molecule (Section 7.2)

$$z^{(\alpha)} = z_{\text{int}}^{(\alpha)} z_{\text{ext}}^{(\alpha)} \exp(-\beta\phi^{(\alpha)})$$

where $z_{\text{int}}^{(\alpha)}$ is the internal partition function and $z_{\text{ext}}^{(\alpha)} = V/\Lambda^3$ is the external partition function, representing the translation motion of the molecule as a whole, with $\Lambda \equiv (h^2/2\pi mkT)^{1/2}$ the thermal length. As usual, $\beta = 1/kT$. For the total partition function $Z^{(\alpha)}$ and Helmholtz energy $F^{(\alpha)}$ in the bulk we have, respectively,

$$Z^{(\alpha)} = (N!)^{-1}(z^{(\alpha)})^N \quad \text{and} \quad F^{(\alpha)} = -kT \ln Z^{(\alpha)} \quad \text{so that}$$

$$\mu^{(\alpha)} = (\partial F^{(\alpha)}/\partial N)_{V,T} = -kT \ \ln\left[Vz_{\text{int}}^{(\alpha)} \exp(-\beta\phi^{(\alpha)})/\Lambda^3 N\right]$$

For the surface we suppose that we have M (localized) sites of which O are occupied. Hence, the fraction of adsorbed molecules $\theta = O/M$ and the fraction of non-occupied sites is $1-\theta = (M-O)/M$. The partition function for a single molecule becomes

$$z^{(\sigma)} = z_{\text{int}}^{(\sigma)} z_{\text{ext}}^{(\sigma)} \exp(-\beta\phi^{(\sigma)})$$

where similar labels are used as for the bulk. The total partition function $Z^{(\sigma)}$ and Helmholtz energy $F^{(\sigma)}$ for the surface are, respectively,

$$Z^{(\sigma)} = \frac{M!}{O!(M-O)!}(z^{(\sigma)})^O \quad \text{and} \quad F^{(\sigma)} = -kT \ln Z^{(\sigma)}$$

where the factor $M!/O!(M-O)!$ is introduced to account for indistinguishability. Employing Stirling's approximation for the factorials, we have the surface partition function $F^{(\sigma)} = -kT \ln[M^M(z_{\text{int}}^{(\sigma)} z_{\text{ext}}^{(\sigma)})^O \exp(-\beta\phi^{(\sigma)})/O^O(M-O)^{M-O}]$ and the chemical potential becomes

$$\mu^{(\sigma)} = \left(\frac{\partial F^{(\sigma)}}{\partial O}\right)_{M,T} = -kT \ \ln\left[\frac{1-\theta}{\theta} z_{\text{int}}^{(\sigma)} z_{\text{ext}}^{(\sigma)} \exp(-\beta\phi^{(\sigma)})\right]$$

Equating $\mu^{(\alpha)}$ with $\mu^{(\sigma)}$ we obtain, using for the bulk concentration $c_{\text{B}} = N/V$,

$$\frac{\theta}{1-\theta} = b(T)c \quad \text{with} \quad b(T) = \left(\frac{z_{\text{int}}^{(\sigma)} z_{\text{ext}}^{(\sigma)}}{z_{\text{int}}^{(\alpha)}}\right)\Lambda^3 \exp[-\beta(\phi^{(\sigma)} - \phi^{(\alpha)})] \tag{11.42}$$

equivalent to Eq. (11.41) if we take[15] $K = b(T)$.

---

15 Often $z_{\text{ext}}^{(\sigma)} = 1$ is taken on the argument that the sites are localized but the molecules still vibrate with respect to the surface leading to $z_{\text{ext}}^{(\sigma)} \neq 1$.

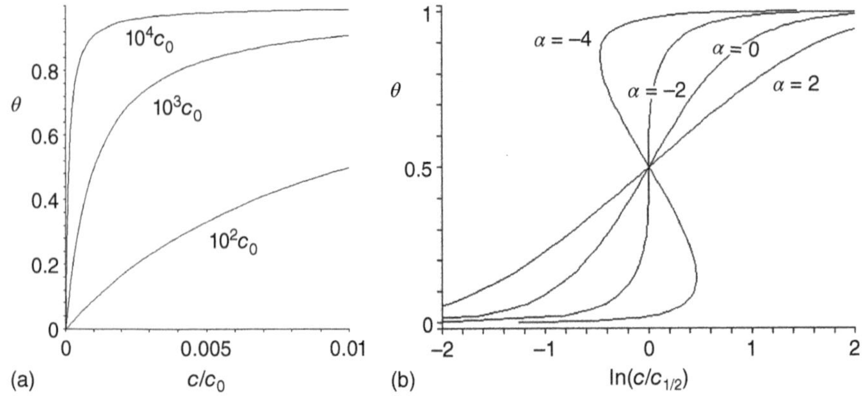

Figure 11.10 (a) The Langmuir isotherm $\theta(c) = K(c/c_0)/[1 + K(c/c_0)]$ for $K = 10^2c_0$, 10^3c_0, and 10^4c_0; (b) The Fowler–Guggenheim adsorption isotherm for various values of $\alpha = sw/kT$. The critical value is $\alpha = -2$ while $\alpha = 0$ corresponds to the Langmuir isotherm.

Let us now evaluate the Langmuir behavior. For small c, the increase in $\theta \cong Kc$, while for large c we have $\theta \to 1$. This behavior is displayed in Figure 11.10a for various values of K. Recalling that the equilibrium constant for adsorption K is related to the Gibbs energy of adsorption G_{ads} by $K = \exp(-G_{ads}/kT)$ (compare with Eq. (11.42), having determined K as a function of T, all other thermodynamic properties, like the heat of adsorption, can be determined in the usual way.

Obviously, other assumptions lead to different adsorption isotherms. In fact, if we assume that the molecules at the surface do interact, so that the energy changes by $2sw\theta$, with s the surface coordination number and $2w$ the energy, when a new pair of neighbors is formed, but that the surface entropy is not affected,[16] we have

$$\mu^{(\sigma)} = \left(\frac{\partial F^{(\sigma)}}{\partial O}\right)_{M,T} = -kT \ln\left(\frac{1-\theta}{\theta} z_{int}^{(\sigma)} z_{ext}^{(\sigma)} \exp[-\beta(\phi^{(\sigma)} + 2sw\theta)]\right)$$

This leads to the *Fowler–Guggenheim adsorption isotherm*

$$\frac{\theta}{1-\theta} = b'(T)c \quad \text{with} \quad b'(T) = \left(\frac{z_{int}^{(\sigma)} z_{ext}^{(\sigma)}}{z_{int}^{(\alpha)}}\right) \Lambda^3 e^{-\beta(\phi^{(\sigma)}+2sw\theta-\phi^{(\alpha)})} = b(T)e^{-2\beta sw\theta} \tag{11.43}$$

Plotting θ as a function of $\ln(c/c_{1/2})$, where $c_{1/2}$ is the concentration at $\theta = 1/2$, this isotherm for $\alpha = sw/kT < -2$ produces a van der Waals shape with two loops, characteristic of a phase transformation. The critical temperature is $T_{cri}{}^{(\sigma)} = -sw/2k$, occurring at $\theta_{cri} = 1/2$ (Figure 11.10b). In view of the fact that the solution is obtained via the zeroth approximation, more sophisticated solutions lead to different values for $T_{cri}{}^{(\sigma)}$. For example, the first-order approximation leads for a square lattice ($s = 4$) to $kT_{cri}{}^{(\sigma)}/w = -2.773$ while the exact solution yields $kT_{cri}{}^{(\sigma)}/w = -3.526$ (Sections C.2 and C.3).

Another extreme is adsorbed molecules moving freely, like a 2D gas, as suggested by the surface pressure π, for low c experimentally given by $\pi = \lambda c$ with λ a constant. Using the

16 That is, we use the zeroth-order approximation for a regular solution model for the surface (Section 11.5.1). Note that $w < 0$, as otherwise no attraction between the adsorbed molecules occurs.

GAE, $-\,d\gamma = d\pi = RT\Gamma\,d\ln c$, in conjunction with $\pi = \lambda c$, we obtain

$$\pi = \Gamma NkT \quad \text{or} \quad \pi A = NkT \quad \text{with} \quad A = 1/\Gamma$$

resembling the expression for the perfect gas law and suggesting similar improvements for the surface as have been made for bulk gases. The simplest extension is *Volmer's equation* $\pi(A - A_0) = NkT$, where A_0 denotes the excluded area, similar to the Clausius equation in 3D. Employing the vdW approximation, in analogy with the 3D expression $(P + a\rho^2)(1 - b\rho) = \rho RT$ (Section 8.3), we have for 2D

$$(\pi + a\theta^2)(1 - b\theta) = \theta\, kT \quad \text{or} \quad (\pi + a\sigma^{-2})(\sigma - b) = kT$$

with a and b parameters. Here θ (or $\sigma = 1/\theta$) in 2D corresponds to ρ (or $V_m = 1/\rho$) in 3D, and the *van der Waals adsorption isotherm* becomes, with the constant $\varepsilon = 2a/b$,

$$\frac{\theta}{1-\theta} = b''(T)c \quad \text{with} \quad b''(T) = \left(\frac{z_{\text{int}}^{(\sigma)}}{z_{\text{int}}^{(\alpha)}}\right)\Lambda \mathrm{e}^{-\left[\frac{\theta}{1-\theta} + \frac{(\phi^{(\sigma)} - \phi^{(\alpha)}) - \varepsilon\theta}{kT}\right]} \tag{11.44}$$

This isotherm predicts a phase change, similarly as in the case of localized adsorption with attractive interaction. Characteristics are $\theta_{\text{cri}} = 1/3$ and $kT_{\text{cri}}^{(\sigma)}/\varepsilon = -\,4/27 = -\,0.148$. This isotherm is not symmetric around $\theta = 1/2$. While for $\theta > 1/2$, the vdW isotherm resembles the Langmuir isotherm, for $\theta < 1/2$, it decreases much more rapidly and resembles the Fowler–Guggenheim isotherm.

Using the GAE, $\Gamma = -\,(RT)^{-1}\partial\gamma/\partial \ln c$, adsorption isotherms can be used to estimate $\Delta\gamma$ upon adsorption. Substituting the Langmuir isotherm $\Gamma = \Gamma_{\max}\theta = \Gamma_{\max}Kc/(1 + Kc)$ in the GAE and solving for γ, one obtains the *Szyszkowski equation* [125, 126]

$$\gamma = \gamma_0 - RT\Gamma_{\max} \ln(1 + Kc)$$

describing the change in γ at low concentration often rather well.

Feinerman [127] proposed a model combining the GAE, written as $d\gamma/d\mu_1 = -\,\Gamma_1 + (x_1/x_2)\Gamma_2$, with the expression $d\gamma/d\mu_1 = X\gamma + Y$, where X and Y are constants. For $\gamma_1 > \gamma_2$ and boundary conditions $x_1 \to 0$, $\gamma = \gamma_2$ and $d\gamma/d\mu_1 = 0$, and $x_1 \to 1$, $\gamma = \gamma_1$ and $d\gamma/d\mu_1 = K/RT$ with the constant $K = \lim_{x_1 \to 1} d\gamma/d\mu_1$, the equation to solve is $d\gamma/dx_1 = K(\gamma - \gamma_2)/x_1(\gamma - \gamma_2)$, where $\mu_1 = \mu_1{}^* + RT \ln x_1$ is used. Solving leads to

$$\gamma = (\gamma_1 - \gamma_2)x_1^B + \gamma_2 = (\gamma_1 - \gamma_2)(1 - x_2)^B + \gamma_2$$

where $B = K/(\gamma_1 - \gamma_2) = RTX$. The solution also holds for nonideal mixtures with the activity coefficient f_1, in which case x_1 must be replaced by f_1x_1. Plotting B (Figure 11.11) show that for any value of B above, say 5, the surface for $x_2 > 0.2$ is already mainly filled with the solute. In fact, H_2O–C_2H_5OH mixtures show such behavior [2].

Since Langmuir, this field for liquid and solids developed tremendously. For the many interesting phenomena, detailed discussions are available elsewhere [128, 129].

Problem 11.7 ***Van der Waals Isotherm*** Using the 2D vdW equation $(\pi + a\sigma^{-2})(\sigma - b) = kT$ and the Gibbs adsorption equation $\Gamma = -\,(RT)^{-1}\partial\gamma/\partial \ln c$, derive the vdW adsorption expression Eq. (11.44). Do so by writing $d\ln c$ as a function of $d\pi$ and integrating the resulting expression term by term.

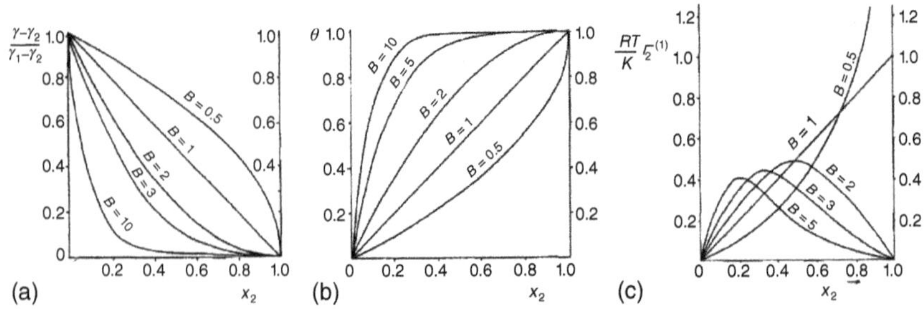

Figure 11.11 Results for the Feinerman model. Source: Feinerman [127]/Springer Nature. (a) The ratio $(\gamma - \gamma_2)/(\gamma_1 - \gamma_2)$ as a function of x_2 for various values of B. (b) The coverage $\theta = 1 - (1 - x_2{}^B)$ as a function of x_2 for various values of B. (c) The adsorption $RT\Gamma_2{}^{(1)}/K = x_2(1 - x_2)^B$ as a function of x_2 for several values of B.

11.6 Final Remarks

The molecular structure of interfaces is still intensively investigated. The literature is extended and many – often opposing – views have been expressed. Important experimental methods, such as sum-frequency spectroscopy [130] and neutron reflection [131, 132], as well as theoretical [66, 133] and experimental methods for pure liquids [72], solutions [134], and solids [135] have been reviewed.

It seems that the eminent physicist Wolfgang Pauli used to say [136], "God made the bulk; surfaces were invented by the devil," which is a succinct way of saying that surfaces are bestowed with many subtleties. We only touched on a few aspects, and even for these, more considerations can be made, indicating the richness of the field. In particular, the effects of curved and electrified surfaces were omitted, despite their importance, but further discussion is outside our scope, and we refer to the literature.

References

1 Orowan, E. (1970). *Proc. Roy. Soc. A* 316: 473; Goodman, F.O. (1990). *Surf. Sci. Lett.* 232: 224.

2 Guggenheim, E.A. (1967). *Thermodynamics*, 5e. Amsterdam: North-Holland.

3 Butt, Graf and Kappl, 2006.

4 Müller, P. and Saúl, A. (2004). *Surf. Sci. Rep.* 54: 157.

5 Kramer, D. and Weismüller, J. (2007). *Surf. Sci.* 601: 3042.

6 Cahn, J.W. (1979). *Thermodynamics of Solid and Fluid Surfaces*, Ch. 1 in *Interfacial Segregation* (eds. W.C. Johnson and J.M. Blakely) *Am. Soc. Metals*, p. 3. Metals Park, Ohio.

7 Nozières, P. and Wolf, D.E. (1987). *Z. Phys. B* 70: 399.

8 de With, G. (2006). *Structure, Deformation, and Integrity of Materials*. Weinheim: Wiley-VCH.

9 Howe, 1997.

10 Alonso, J.A. and March, N.H. (1989). *Electrons in Metals and Alloys*. London: Academic Press.

11 Jasper, J.J. (1972). *J. Phys. Chem. Ref. Data* 1: 841.
12 Brown, R.C. (1947). *Proc. Phys. Soc.* 59: 429–711; (1974). *Contemp. Phys.* 15: 301.
13 Yarnold, G.D. (1960). *Contemp. Phys.* 1: 267.
14 Lindford, R.G. (1973). *Surface Thermodynamics of Solids*, Ch. 1 in *Solid State Surface Science*, Vol. 2 (ed. M. Green), p. 1. New York: Marcel Dekker; (1978). *Chem. Rev.* 78, 81.
15 Eötvös, R. (1886). *Ann. Phys.* 263: 448.
16 Ramsay, W. and Shields, J. (1893). *Phil. Trans. A* 184: 647.
17 Pászli, I. and László, K. (2007). *Coll. Pol. Sci.* 285: 1505.
18 Dunken, H., Klapproth, H., and Wolf, K.L. (1940). *Koll. Z.* 91: 232.
19 Moelwyn-Hughes, E.A. (1961). *Physical Chemistry*, 2e. Oxford: Pergamon Press.
20 Adam, N.K. (1941). *The Physics and Chemistry of Surfaces*. London: Oxford University Press.
21 Skapski, J. (1948). *J. Chem. Phys.* 16: 386.
22 Israelachvili, J.N. (2006). *Intermolecular and Surface Forces*, reprinted, 2e. Amsterdam: Academic Press.
23 Mahanty, J. and Ninham, B.W. (1976). *Dispersion Forces*. New York: Academic Press.
24 Hough, D.B. and White, L.R. (1980). *Adv. Coll. Interface Sci.* 14: 3.
25 Luck, W.A.P. (1979). *Angew. Chem. Int. Ed.* 18: 350.
26 Luck, W.A.P. and Ditter, W. (1971). *Tetrahedron* 27: 210.
27 Luck, W.A.P. (2001). *Coll. Pol. Sci.* 279: 554.
28 Luck, W.A.P. (1980). *Angew. Chem. Int. Ed.* 19: 28.
29 van der Waals, J.D. (1894). *Z. phys. Chem.* 13: 716, for an English translation, see (1979). *J. Stat. Phys.* 20: 197.
30 Ferguson, A. (1916). *Phil. Mag.* 31: 37.
31 Ferguson, A. and Kennedy, J.S. (1936). *Trans. Far. Soc.* 32: 1474.
32 Lin, H., Duan, Y.-Y., and Min, Q. (2007). *Fluid Phase Eq.* 254: 75.
33 Katayama, M. (1916). *Rep. Tôhoku Univ.* 4: 373.
34 Guggenheim, E.A. (1945). *J. Chem. Phys.* 13: 253.
35 McLeod, D.B. (1923). *Trans. Far. Soc.* 19: 38.
36 Sugden, S. (1924). *J. Chem. Soc.* 125: 32–1177.
37 Sugden, S. (1930). *The Parachor and Valency*. London: Routledge.
38 Quayle, O.R. (1953). *Chem. Rev.* 53: 439.
39 Poling, B.E., Prausnitz, J.M., and O'Connell, J.P. (2001). *The Properties of Gases and Liquids*, 5e. London: McGraw-Hill Comp.
40 Exner, O. (1962). *Nature* 196: 890. and (1997). *J. Phys. Org. Chem.* 10: 797, with the inspiring title: *How to get wrong results from good experimental data*.
41 Somayajulu, G.R. (1988). *Int. J. Thermophys.* 9: 559.
42 Mulero, A., Parra, M.I., and Cachadina, I. (2013). *Fluid Phase Eq.* 339: 81.
43 Riedel, L. (1954). *Chem.-Ing. Tech.* 26: 679.
44 Curl, R.F. and Pitzer, K.S. (1958). *Ind. Eng. Chem.* 50: 265.
45 Pitzer, K.S. (1995). *Thermodynamics*, 3e. New York: McGraw-Hill Book Comp.
46 Blokhuis, E.M. and Kuipers, J. (2006). *J. Chem. Phys.* 124: 074701.
47 Mandelstam, L. (1913). *Ann. Phys.* 41: 609.
48 Mermin, N.D. (1965). *Phys. Rev. A* 137: 1441.
49 Evans, R. (1979). *Adv. Phys.* 28: 143.

50 Evans, R. (1992). *Density Functionals in the Theory of Nonuniform Fluids*, Ch. 3 in *Fundamentals of Inhomogeneous Fluids* (ed. D. Henderson), p. 85. New York: Dekker.
51 Evans, R. (2009). *Density Functional Theory for Inhomogeneous Fluids I: Simple Fluids in Equilibrium*, Lecture notes 3rd Warsaw School of Statistical Physics.
52 Egelstaff, P.A. and Widom, B. (1970). *J. Chem. Phys.* 53: 2667.
53 Freeman, G.R. and March, N.H. (1998). *J. Chem. Phys.* 109: 10521.
54 Bhatia, A.B. and March, N.H. (1978). *J. Chem. Phys.* 68: 19999.
55 Sanchez, I.C. (1983). *J. Chem. Phys.* 79: 405.
56 Ted Davis, H. and Scriven, L.E. (1982). *Adv. Chem. Phys.* 49: 357.
57 Buff, F.P., Lovett, R.A., and Stillinger, F.H.J. (1965). *Phys. Rev. Lett.* 15: 621.
58 Rowlinson and Widom, 1982.
59 Ted Davis, H. (1995). *Statistical Mechanics of Phases, Interfaces and Thin Films*. New York: Wiley-VCH.
60 Safran, S.A. (1994). *Statistical Thermodynamics of Surfaces, Interfaces and Membranes*. Boulder, CO: Westview Press.
61 Hansen, J.-P. and McDonald, I.R. (2006). *Theory of Simple Liquids*, 3e. London: Academic Press (1e 1976, 2e 1986).
62 Yang, A.M., Fleming, P.D., and Gibbs, J.H. (1976). *J. Chem. Phys.* 64: 3732.
63 Lutsko, J.F. (2010). *Adv. Chem. Phys.* 144: 1.
64 Löwen, H. (2006). *Density Functional Theory for Inhomogeneous Fluids II: Freezing, Dynamics, Liquid crystals*, Lecture notes 3rd Warsaw School of Statistical Physics.
65 Widom, B. (1996). *J. Phys. Chem.* 100: 13190.
66 Weeks, J.D. (1977). *J. Chem. Phys.* 67: 3106.
67 Wertheim, M.S. (1976). *J. Chem. Phys.* 65: 2377.
68 Kayser, R.F. (1986). *Phys. Rev. Lett. A* 33: 1948.
69 Stillinger, F.H. (2008). *J. Chem. Phys.* 128: 204705.
70 Blokhuis, E.M. and Bedeaux, D. (1990). *Physica A* 164: 518.
71 Toxvaerd, S. (1972). *Progr. Surf. Sci.* 3: 189.
72 Penfold, J. (2001). *Rep. Prog. Phys.* 64: 777.
73 Friedel, J., Cullity, B.D., and Crossard, C. (1953). *Acta Met.* 1: 79.
74 Giesen, M. (2001). *Progr. Surf. Sci.* 68: 1.
75 Herring, C. (1951). *Phys. Rev.* 82: 87; (1953). *The Use of Classical Macroscopic Concepts in Surface-Energy Problems*, Ch. 2 in *Structure and Properties of Solid Surfaces* (eds. R. Gomer and C.S. Smith), p. 5. Chicago: University of Chicago Press.
76 Van Hove, M.A. (1993). *Crystal Surfaces*, Ch. 8 in *Structure of Solids*, Vol. I (eds. R.W. Cahn, P. Haasen, and E.J. Kramer), p. 481. Weinheim: VCH Verlag.
77 Meier, G.H. (2014). *Thermodynamics of Surfaces and Interfaces*. Cambridge: Cambridge University Press.
78 Fang, C.M., de With, G., and Parker, S.C. (2001). *J. Am. Ceram. Soc.* 84: 1553.
79 Lapujoulade, J. (1994). *Surf. Sci. Rep.* 20: 191.
80 Shugard, W.J., Weeks, J.D., and Gilmer, G.H. (1978). *Phys. Rev. Lett.* 41: 1399.
81 Nozières, P. and Gallet, F. (1987). *J. de Phys.* 48: 353.
82 Kosterlitz, J.M. and Thouless, D.J. (1973). *J. Phys.* C6: 1181; Kosterlitz, J.M. (1974). *Ibid.* C7: 1046.
83 Burton, W.K., Cabrera, N., and Frank, F.C. (1950). *Phil. Mag.* 243A: 299.
84 Mullins, W.W. (1959). *Acta Met.* 7: 746.

85 Leamy, H.J. and Gilmer, G.H. (1974). *J. Cryst. Growth* 24–25: 499.
86 Shugard, W.J., Weeks, J.D., and Gilmer, G.H. (1978). *Phys. Rev. Lett.* 41: 4978.
87 Chui, S.T. and Weeks, J.D. (1976). *Phys. Rev. Lett. B.* 14: 4978.
88 van Beijeren, H. (1977). *Phys. Rev. Lett.* 38: 993.
89 Pitzer, K.S. *Thermodynamics*, 3e. New York: McGraw-Hill Book Comp.
90 Guggenheim, E.A. (1937). *Trans. Far. Soc.* 33: 151.
91 Adcock, D.S. and McGlashan, M.L. (1954). *Proc. Roy. Soc. A* 226: 266.
92 Longuet-Higgins, H.C. (1951). *Proc. Roy. Soc. A* 205: 247; Brown, W.B. (1957). *Proc. Roy. Soc. A* 240: 561.
93 Krooshof, G.J.P., Tuinier, R., and de With, G. (2018). *J. Mol. Liq.* 266: 467.
94 Guggenheim, E.A. (1948). *Trans. Faraday Soc.* 44: 1007.
95 Guggenheim, E.A. (1952). *Mixtures*. Oxford: Clarendon Press.
96 Krooshof, G.J.P., Tuinier, R., and de With, G. (2018). *Fluid Phase Eq.* 465: 10.
97 Lucas, 2007.
98 Redlich, O. and Kister, A.T. (1948). *Ind. Eng. Chem.* 40: 341; Redlich, O., Kister, A.T., and Turnquist, C.E. (1952). *Chem. Eng. Progr. Symp. Ser.* No. 2, Vol. 48: 49.
99 Guggenheim, E.A. (1935). *Proc. Roy. Soc. A* 148: 304.
100 Margules, M. (1895). *Sitzungsber. Akad. Wiss. Wien* 104: 1243.
101 van Laar, J.J. (1906). *Sechs Vorträge über das thermodynamische Potential*. Brunswick, Germany: Vieweg Verlag; (1910). *Z. Phys. Chem.* 72: 723; (1913). *Ibid.* 89: 599.
102 Scatchard, G. (1931). *Chem. Rev.* 8: 321; (1949). *Ibid.* 44: 7.
103 Prausnitz, J.M., Lichtenthaler, R.N., and Azeredo, E.G. (1999). *Molecular Thermodynamics of Fluid-Phase Equilibria*, 3e. Englewood Cliffs, NJ: Prentice-Hall.
104 Pelton, A.D. and Bale, C.W. (1986). *Metall. Trans. A* 17: 1057.
105 Klaus, R.L. and van Ness, H.C. (1967). *Chem. Eng. Progr. Symp. Ser.* 63: 88.
106 Kollau, L.B.J.M., Vis, M., van den Bruinhorst, A. et al. (2020). *J. Mol. Liq.* 302: 112155.
107 Butler, J.A.V. (1932). *Proc. Roy. Soc. A* 135: 348.
108 Hoar, T.P. and Melford, B.A. (1957). *Trans. Far. Soc.* 53: 315.
109 Sprow, F.B. and Prausnitz, J.M. (1965). *Trans. Far. Soc.* 62: 1105.
110 Randles, J.E.B. and Behr, B. (1972). *J. Electroanal. Chem.* 35: 389.
111 Kaptay, G. (2015). *Langmuir* 31: 5796; Korosz, J. and Kaptay, G. (2017). *Coll. Surf. A* 533: 296.
112 Kaptay, G. (2020). *Adv. Coll. Interface Sci.* 283: 102212.
113 Sparnaaij, M.J. (1984). *Surf. Sci. Rep.* 4: 101.
114 Defay, R. and Prigogine, I. (1950). *Trans. Far. Soc.* 46: 199.
115 Ono, S. (1947). *Mem. Fac. Eng. Kyushu Univ.* 10: 195; 10: 195 and (1950). *Ibid.* 12: 1.
116 Murakami, T., Ono, S., Tamura, M. et al. (1951). *J. Phys. Soc. Japan* 6: 309.
117 Prigogine, I. (1948). *Trans. Far. Soc.* 44: 626.
118 Langmuir, I. (1917). *J. Am. Chem. Soc.* 39: 1848. This is paper 2 of a series of 3 pioneering surface chemistry papers, the other two being (1916). *Ibid.* 38: 1848 and (1918). *Ibid.* 40: 1361.
119 Tajima, K., Muramatsu, M., and Sasaki, T. (1970). *Bull. Chem. Soc. Japan* 43: 1991.
120 Menger, F.M., Shi, L., and Rizvi, S.A.A. (2009). *J. Am. Chem. Soc.* 131: 10380; (2010). Langmuir 26, 1588.
121 Laven, J. and de With, G. (2012). *Langmuir* 27: 7958 with as title the rhetorical question: *Should the Gibbs analysis be revised*?

122 Li, P.X., Li, Z.X., Shen, H.-H. et al. (2013). *Langmuir* 29: 9324.
123 Li, P.X., Li, Z.X., Ma, K. et al. (2013). *Langmuir* 29: 9335.
124 Li, P.X., Thomas, R.K., and Penfold, J. (2014). *Langmuir* 30: 6739.
125 von Szyszkowski, B. (1908). *Z. Phys. Chem.* 64: 385.
126 Meissner, H.P. and Michaels, A.S. (1949). *Ind. Eng. Chem.* 41: 2782.
127 Feinerman, A.E. (1974). *Coll. Pol. Sci.* 152: 582.
128 Hiemenz and Rajagopalan, 1997; Adamson and Gast, 1997.
129 Rosen, M.J. and Kunjappu, J.T. (2012). *Surfactants and Interfacial Phenomena*, 4e. Hoboken, NJ: John Wiley & Sons.
130 Richmond, G.L. (2001). *Ann. Rev. Phys. Chem.* 52: 357.
131 Penfold, J. and Thomas, R.K. (1990). *J. Phys. Condens. Matter* 2: 1369.
132 Thomas, R.K. (2004). *Ann. Rev. Phys. Chem.* 55: 391.
133 Tarazona, P., Chaćon, E., and Bresme, F. (2012). *J. Phys. Condens. Matter* 24: 284123.
134 Shukla, R.K., Kumar, A., Srivastava, U. et al. (2013). *Can. J. Phys.* 91: 211.
135 Mamonova, M.V., Prudnikov, V.V., and Prudnikova, I.A. (2013). *Surface Physics*. Boca Raton, FL: CRC Press.
136 Jamtveit, B. and Meakin, P. (1999). *Growth, Dissolution, and Pattern Formation in Geosystems*, p. 291. Dordrecht: Kluwer Academic Publishers.

Further Reading

Adamson, A.W. and Gast, A.P. (1997). *Physical Chemistry of Surfaces*, 6e. New York: John Wiley & Sons. After several revisions, this book has become a classic dealing with various surfaces, providing a broad overview on all aspects of surfaces.

Butt, H.-J., Graf, K., and Kappl, M. (2006). *Physics and Chemistry of Interfaces*, 2e. Weinheim: Wiley-VCH. Nice introductory book, about equally divided between principles and applications.

Butt, H.-J. and Kappl, M. (2010). *Surface and Interfacial Forces*. Weinheim: Wiley-VCH. A modern overview with many examples and many details.

Hiemenz, P.C. and Rajagopalan, R. (1997). *Principles of Colloid and Interface Chemistry*, 3e. New York: Marcel Dekker. A didactic overview, focusing on the interfacial aspects of colloid chemistry.

Rowlinson, J.S. and Widom, B. (1982). *Molecular Theory of Capillarity*. Oxford: Oxford University Press (see also Dover, 1988). Detailed exposition of capillary theory.

Howe, J.M. (1997). *Interfaces in Materials*. New York: John Wiley & Sons. A broad overview providing many experimental illustrations of the concepts treated.

Desjonquères, M.C. and Spanjaard, D. (1996). *Concepts in Surface Physics*, 2e. Berlin: Springer-Verlag. A theoretical treatment from the point of view of solid-state physics.

Zangwill, A. (1988). *Physics at Surfaces*. Cambridge: Cambridge University Press. A broad overview with a focus on solids from a physical point of view.

12

Phase Transitions: General Aspects

In Chapters 2–7, we discussed the "tools" and in Chapters 8–11, the "bricks" and "cement." Now we turn to the "house," that is, the various transitions between gases, liquids, and solids. In this chapter, we deal with aspects related to all types of transitions. We start with some general considerations and stability arguments and thereafter deal with some equations generally valid for discontinuous and continuous transitions.

12.1 Some General Considerations

By changing the conditions, such as P or T, for many materials, a transition from one phase to another can be induced and, under certain conditions, two phases of the same material coexist. Since each of the two phases has its own Gibbs energy expression, in equilibrium the chemical potentials of these phases are equal. The Gibbs energy G is always continuous over the transition, but the partial derivatives $\partial G/\partial T$ and $\partial G/\partial P$ may be discontinuous (Figure 12.1). In that case, the phase transition is denoted as *discontinuous*[1] (or *first order*), while for the situation where the first derivatives are continuous, but the higher derivatives are either zero or infinite, one speaks of a *continuous* (or *second-order*) phase transition.

The angle of intersection of the G_1 and G_2 curves for phases 1 and 2, respectively, determines the entropy and volume change associated with the phase transformations and, hence, the type of phase transition. Experimentally, it appears that by moving along the liquid–vapor (L–V) coexistence line over the critical point (CP), the differences in properties, in particular the density, between the liquid and gas phases vanish in a continuous way, and the transition is called continuous (Figure 12.2). Moving across the L–V curve from one phase to the other leads to a discontinuous change of properties and the transition is denoted as discontinuous.

Following the coexistence (vapor pressure) line between liquid and vapor in the P–T diagram, starting at the triple point T_{tri} and passing the normal boiling point T_n, we end at the

1 In the past, the transitions were often labeled as first and second order, according to the discontinuity of their first- or second-order derivatives of the Gibbs energy. However, the second-order "class" appeared to be more complex than anticipated and therefore these transitions are nowadays often labeled as continuous, due to the fact that in all cases a continuous transition from a one-phase state to a two-phase state occurs with a continuous change in order parameter over the transition. Although the label "first order" stuck, I refer to this transition for consistency as discontinuous, the more so since the density behavior for materials is discontinuous in this case.

Phases of Matter and their Transitions: Concepts and Principles for Chemists, Physicists, Engineers, and Materials Scientists, First Edition. Gijsbertus de With.

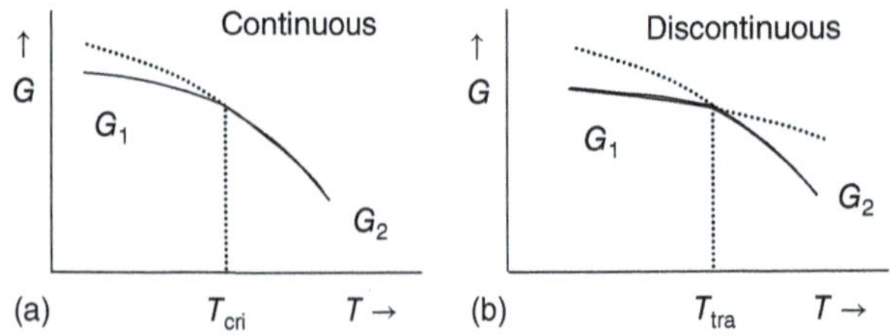

Figure 12.1 Schematic of the behavior of the Gibbs energy G for two phases around (a) a continuous phase transition and (b) a discontinuous phase transformation. In both cases, the stable states below the transition temperature have a Gibbs energy G_1, while above the transition temperature the Gibbs energy is G_2. The continuous transition occurs at the critical temperature T_{cri} with a continuous change in G, that is, $\partial\Delta G/\partial T = 0$, where $\Delta G = G_2 - G_1$. The dotted line indicates the metastable continuation of the high-temperature G below T_{cri}. The discontinuous transition occurs at a certain transition temperature T_{tra} with a discontinuous change in G ($\partial\Delta G/\partial T \neq 0$).

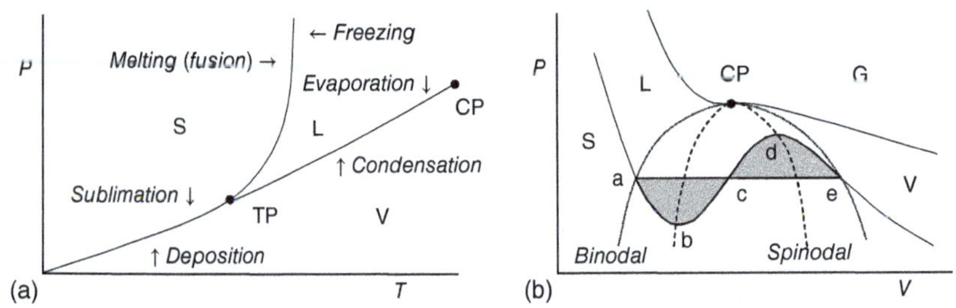

Figure 12.2 (a) Schematic of the phase equilibrium between the solid (S), liquid (L), and vapor (V) phase in the P–T plane, showing the triple point (TP) and the critical point (CP). These are natural reference points since the melting temperature T_{mel} and the boiling temperature T_n depend on the environment, in particular the pressure P. For water, for example, $P_{cri} = 218.3$ atm, $T_{cri} = 374.15\,°C$, $\rho_{cri} = 320\,kg\,m^{-3}$, and $T_{tri} = 0.01\,°C$. While the transition *across* a coexistence line relates to a discontinuous phase transition, the transition over the critical point *along* the coexistence line relates to a continuous phase transition. (b) Schematic of the phase equilibrium in the P–V plane. The horizontal line indicates the equal area Maxwell construction. Above the CP only gases (G) can exist. The binodal line indicates the demarcation of global stability, while the spinodal line indicates the limits of local stability.

CP with temperature T_{cri}. In this process, the density of the liquid ρ_L decreases, while the density of the vapor ρ_V increases. At T_{cri}, $\rho_V = \rho_L$. Moreover, for $T < T_{cri}$ a meniscus, that is, a sharp transition region between liquid and vapor, is present except for temperatures close to T_{cri} (say, within 1°), where this meniscus widens and suddenly disappears at T_{cri} (Figure 12.3).[2]

Before discussing some ideas on how to describe transitions, let us first discuss some stability considerations [1]. For this we need the concept of *convexity*. A curve is *convex* (or convex up) if the chord is above the curve, whereas a curve is *concave* (or convex down) if the chord is below the curve (Figure 12.4a). Thermodynamics requires that for changes in entropy we always have $(\Delta S)_U \geq 0$, which implies that entropy S is concave over its entire domain. If we obtain from a model or from experimental data a curve like abcdefg in Figure 12.4b, the above implies that the envelope abhfg has to be considered as the acting

2 This transition process is strongly influenced by gravity, and the density profile over the meniscus near T_{cri} can be described by an expression akin to the barometric formula (Problem 12.1).

Figure 12.3 The disappearing of the meniscus of benzene along the coexistence curve from far below (a) to close to (b) and just below (c) to just above T_{cri} (d).

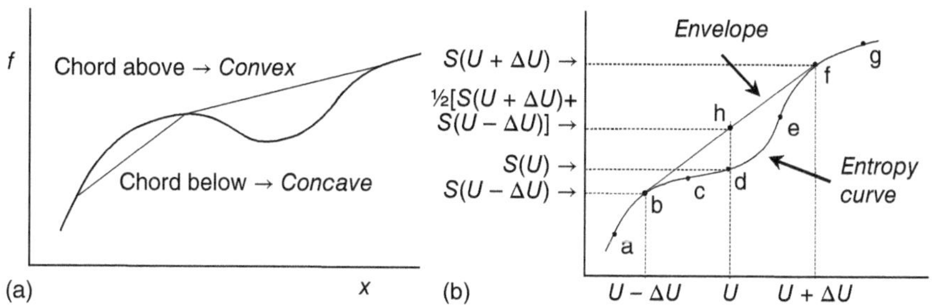

Figure 12.4 (a) Concave and convex intervals of a function; (b) The relation to stability. If the curve abcdefg represents the entropy S, say, as obtained from a model, as a function of energy U, the envelope abhfg represents the entropy to be used since the entropy curve should be always concave. The range bcdef represents global instability (binodal), while the range cde (with c and e the inflection points) represents local instability (spinodal).

entropy function. Considering for the moment S as a function energy U and volume V, concavity of S results in

$$S(U+\Delta U, V) + S(U-\Delta U, V) \le 2S(U,V) \quad \text{or} \quad (\partial^2 S/\partial U^2)_V \le 0 \tag{12.1}$$

where the reduction to the differential form only follows if $\Delta U \to 0$. Similarly,

$$S(U, V+\Delta V) + S(U, V-\Delta V) \le 2S(U,V) \quad \text{or} \quad (\partial^2 S/\partial V^2)_U \le 0 \tag{12.2}$$

where again the differential form only follows if $\Delta V \to 0$. In fact, these considerations also apply for a combined change of U and V, so that

$$S(U+\Delta U, V+\Delta V) + S(U-\Delta U, V-\Delta V) \le 2S(U,V) \tag{12.3}$$

leading again to Eqs. (12.1) and (12.2) as well as to (Problem 12.3)

$$\frac{\partial^2 S}{\partial U^2}\frac{\partial^2 S}{\partial V^2} - \left(\frac{\partial^2 S}{\partial V \partial U}\right)^2 \ge 0 \tag{12.4}$$

So, concavity leads to conditions for global stability while the differential forms lead to conditions for local stability. For example, Eq. (12.1) for n moles leads to

$$\left(\frac{\partial^2 S}{\partial U^2}\right)_V = -\frac{1}{T^2}\left(\frac{\partial T}{\partial U}\right)_V = -\frac{1}{nT^2 C_V} \le 0 \quad \text{or} \quad C_V \ge 0$$

indicating that the molar heat capacity C_V must be positive for a system to be thermally stable. Although Eqs. (12.2) and (12.4) can be used to establish further stability requirements, it is easier to employ thermodynamic potentials instead of the entropy. To that purpose, we recall that the principle $(\Delta S)_U \geq 0$ is always equivalent to $(\Delta U)_S \leq 0$, stating that the energy U should be always convex. This leads to

$$U(S + \Delta S, V + \Delta V) + U(S - \Delta S, V - \Delta V) \geq 2U(S, V)$$

$$\frac{\partial^2 U}{\partial S^2} \geq 0, \qquad \frac{\partial^2 U}{\partial V^2} \geq 0 \quad \text{and} \quad \frac{\partial^2 U}{\partial S^2}\frac{\partial^2 U}{\partial V^2} - \left(\frac{\partial^2 U}{\partial V \partial S}\right)^2 \geq 0$$

which is fully analogous to the condition for the entropy. It is convenient to also consider the Helmholtz energy F and Gibbs energy G. We first note that, if we have $U(S,V)$ and $F(T,V)$, $T = \partial U/\partial S$ and $S = -\partial F/\partial T$, respectively. Therefore,

$$\frac{\partial T}{\partial S} = \frac{\partial^2 U}{\partial S^2} \quad \text{and} \quad \frac{\partial S}{\partial T} = -\frac{\partial^2 F}{\partial T^2} \quad \text{or} \quad \frac{\partial^2 F}{\partial T^2} = -\left(\frac{\partial^2 U}{\partial S^2}\right)^{-1}$$

Hence, the sign of $\partial^2 F/\partial T^2$ is the negative of the sign of $\partial^2 U/\partial S^2$, implying that if U is a convex function of S, F is a concave function of T. This type of result holds for all transforms, so that we obtain for F and G, respectively,

$$\left(\frac{\partial^2 F}{\partial T^2}\right)_V \leq 0, \quad \left(\frac{\partial^2 F}{\partial V^2}\right)_T \geq 0, \quad \left(\frac{\partial^2 G}{\partial T^2}\right)_P \leq 0 \quad \text{and} \quad \left(\frac{\partial^2 G}{\partial P^2}\right)_T \leq 0 \tag{12.5}$$

Summarizing, the thermodynamic potentials, that is, the energy and its Legendre transforms, are convex functions of their extensive variables and concave functions of their intensive variables.[3]

From Eq. (12.5), we easily obtain

$$\left(\frac{\partial^2 F}{\partial V^2}\right)_T = -\left(\frac{\partial P}{\partial V}\right)_T = \frac{1}{V\kappa_T} \geq 0 \quad \text{or} \quad \kappa_T \geq 0$$

indicating that the isothermal compressibility κ_T must be positive for a system to be mechanically stable. From combining $C_V \geq 0$ and $\kappa_T \geq 0$ with $\kappa_T - \kappa_S = TV\alpha^2/C_P$ and $C_P - C_V = TV\alpha^2/\kappa_T$ (Eq. (2.23)), one can further infer that

$$C_P \geq C_V \geq 0 \quad \text{and} \quad \kappa_T \geq \kappa_S \geq 0$$

The first of these equations states that, upon the addition of heat to a system, the temperature rises, but more at constant volume than at constant pressure. The second-state equation states that, upon decreasing the volume of the system, the pressure increases, but more at constant temperature than at constant entropy.

Problem 12.1 ***Near Critical Point Density Profile*** Near the CP, the density profile of molecules with molar mass $\mathcal{M}$ over the meniscus is significantly influenced by gravity. If the gravitational potential is (approximately) given by $\phi = gh$ with g the acceleration of gravity and h the height, the equilibrium condition reads $\mathrm{d}(\mu + \mathcal{M}\phi) = 0$.

(a) Calculate the expression for the molar volume V_m, the pressure P, and the density $\rho = \mathcal{M}/V_\mathrm{m}$ as a function of height h with respect to the meniscus level h_0.

(b) In many cases, properties are scaled with respect to their value at critical point, for example, $T_\mathrm{red} = T/T_\mathrm{cri}$, $P_\mathrm{red} = P/P_\mathrm{cri}$, and $V_\mathrm{red} = V/V_\mathrm{cri}$. What is the appropriate expression for μ_red in terms of P_cri, V_cri, and T_cri?

3 A similar exercise for the Massieu functions, that is, the entropy and its Legendre transforms, leads to concave functions of their extensive variables and convex functions of their intensive variables.

(c) The scaled equilibrium condition reads $\mathrm{d}\mu_{\mathrm{red}} = -\mathrm{d}h_{\mathrm{red}}$. Calculate h_{cri} in $h_{\mathrm{red}} = h/h_{\mathrm{cri}}$.
(d) Consider Ne and H_2O. For which compound is the width h_{cri} of the transition zone between liquid and gas near the CP larger? What does the order of magnitude of h_{cri} indicate to you? Use data as given in Appendix F.

Problem 12.2 ***Mechanical Stability*** Using the relation $\partial^2 U/\partial V^2 \geq 0$, show that $\kappa_S \geq 0$.

Problem 12.3 ***Thermodynamic Stability*** Show by expanding the left-hand side of Eq. (12.3) in a Taylor series to second order that $S_{UU}(\Delta U)^2 + 2S_{UV}\Delta U\Delta V + S_{VV}(\Delta V)^2 \leq 0$, where $S_{UU} \equiv \partial^2 S/\partial U^2$, $S_{UV} \equiv \partial^2 S/\partial U\partial V$, and $S_{VV} \equiv \partial^2 S/\partial V^2$. Recalling that $S_{UU} \leq 0$, show that this expression can be written as $[S_{UU}\Delta U + S_{UV}\Delta V]^2 + [S_{UU}S_{VV} - S_{UV}{}^2](\Delta V)^2 \geq 0$ and that this subsequently leads to Eq. (12.4).

12.2 The Clapeyron and Clapeyron–Clausius Equation

For discontinuous transitions, we have $\mu_2 = \mu_1$, but $(\partial\mu_2/\partial T)_P \neq (\partial\mu_1/\partial T)_P$ and $(\partial\mu_2/\partial P)_T \neq (\partial\mu_1/\partial P)_T$. This implies that $s_2 \neq s_1$ and $v_2 \neq v_1$. For equilibrium between phases, their chemical potential μ should therefore be equal, as otherwise transfer of matter occurs until $\mathrm{d}G = (\mu_2 - \mu_1)\mathrm{d}n = 0$ is fulfilled. If we consider that each phase has its own Gibbs energy function,[4] the cross-over temperatures between solid and liquid and liquid and vapor determine the melting and boiling temperatures, respectively. The effect of pressure P is illustrated in Figure 12.5, showing that, upon increasing P, the melting point rises if the molar volume of the solid $V_{\mathrm{m}}(\mathrm{S}) <$ the molar volume of the liquid $V_{\mathrm{m}}(\mathrm{L})$, while the melting point decreases if $V_{\mathrm{m}}(\mathrm{S}) > V_{\mathrm{m}}(\mathrm{L})$.

The coexistence curves for two phases, say, L and V, in the P–T plane (Figure 12.2a) can be obtained from the *Clapeyron equation*, which, using $\Delta G = \Delta H - T\Delta S = 0$ or, equivalently, $\mu_{\mathrm{L}} - \mu_{\mathrm{V}} = 0$, with the Maxwell relation $\mathrm{d}P/\mathrm{d}T = \mathrm{d}S/\mathrm{d}V$, reads

$$\frac{\mathrm{d}P}{\mathrm{d}T} = \frac{\Delta_{\mathrm{vap}}S}{\Delta V} = \frac{\Delta_{\mathrm{vap}}H}{T\Delta V} \tag{12.6}$$

When liquid and vapor are both present in equilibrium, we have $\Delta G = G_{\mathrm{V}} - G_{\mathrm{L}} = 0$. Hence, we have (Figure 12.2b), since $G = F + PV$, the relation $F_{\mathrm{a}} - F_{\mathrm{e}} = -P(V_{\mathrm{a}} - V_{\mathrm{e}})$. But the work required to go from vapor to liquid is also $F_{\mathrm{a}} - F_{\mathrm{e}} = -\int_{\mathrm{e}}^{\mathrm{a}} P\mathrm{d}V$. We conclude that the (gray) area (Figure 12.2b) described by the curve abc must equal the (gray) area described by the curve cde. Hence, phase equilibrium is determined by the horizontal line for which these two areas are equal.[5] This is *Maxwell's equal area rule. Metastable states* can occur for the range ab (*supersaturation*) and de (*superheating*), while the range bcd represents *unstable states*. For the van der Waals equation of state (vdW EoS), a parametric solution linking ρ_{L} and ρ_{V} exists [3]. The Clapeyron equation evidently can be applied to solid–solid and solid–liquid transitions as well.

4 In principle, phase transitions should emerge from a single partition function. As that partition function is generally unknown, one resides to a partition function for each of the phases and thus to a Gibbs function for each of the phases and describes the transition by equating the Gibbs functions at the transition temperature. The reason why this is allowed has been explained clearly in [2].
5 Note, though that in practice, doing reversible work along the curve bcd is impossible.

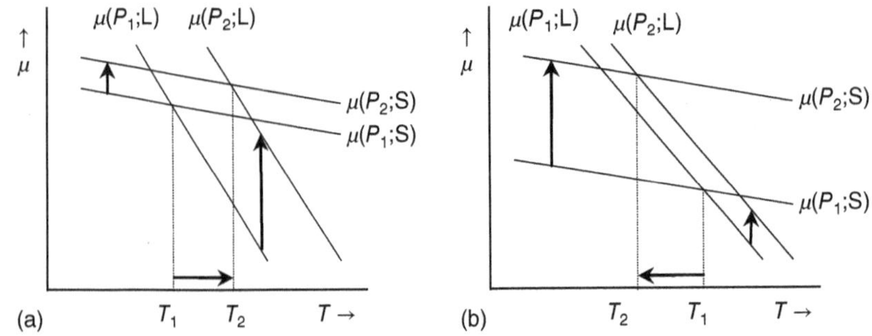

Figure 12.5 The effect of pressure on the melting point. (a) $V_m(S) < V_m(L)$ leading to rising of melting point ($T_2 > T_1$); (b) $V_m(S) > V_m(L)$ leading to lowering of melting point ($T_2 < T_1$).

When the phase transition of a substance is between the gas phase and a condensed phase (C) and occurs at $T \ll T_{cri}$ of that substance, the specific volume of the gas phase V_C is much larger than that of the condensed phase V_C. Therefore, one may approximate $\Delta V = V_{GC} = V_G(1 - V_C/V_G)$ by V_G. For sufficiently low pressure the perfect gas law can be invoked so that $V_G = RT/P$ with R, as usual, the gas constant. Inserting in the Clapeyron equation leads to the *Clapeyron–Clausius equation*:

$$\frac{dP}{dT} = \frac{\Delta_{vap}H}{T\Delta V} \cong \frac{P\Delta_{vap}H}{RT^2} \quad \text{or} \quad \frac{d\ln P}{d(1/T)} = -\frac{\Delta_{vap}H}{R} \tag{12.7}$$

We deal with the Clapeyron–Clausius equation further in Section 13.1.

12.3 The Mosselman Solution for the Clapeyron Equation

The Clapeyron equation is not an exact differential equation, and in order to arrive at an explicit relation between the pressure P and temperature T, one has to integrate the equation. Because the enthalpy and volume changes involved are both functions of T and P, this cannot be done directly. It is generally believed that the Clapeyron equation is unsolvable without preliminary substitution of a simple expression for $\Delta H(T)$ and $\Delta V(T)$. However, Mosselman et al. [4] found the integration factor $\Delta V/T$ that transforms the Clapeyron equation into an exact differential equation. Using this integrating factor, one can write

$$T^{-1}\Delta V(T,P)\,dP + \Delta_{vap}H(T,P)\,dT^{-1} = 0$$

satisfying Euler's condition for integrability

$$\partial\Delta_{vap}H(T,P)/\partial P = \partial[\Delta V(T,P)/T]/\partial T^{-1}$$

The solution is

$$T^{-1}\int_{P_0}^{P} \Delta V(T,P)\,dP + \int_{T_0^{-1}}^{T^{-1}} \Delta H(T,P_0)\,dT^{-1} = 0$$

$$T_0^{-1}\int_{P_0}^{P} \Delta V(T_0,P)\,dP + \int_{T_0^{-1}}^{T^{-1}} \Delta H(T,P)\,dT^{-1} = 0 \tag{12.8}$$

Because $\partial\Delta_{\text{vap}}H/\partial T = \Delta C_P$, integration of Eq. (12.8a) results in

$$T^{-1}\int_{P_0}^{P} \Delta V(T,P)\,\mathrm{d}P + \int_{T_0^{-1}}^{T^{-1}} \left(\int_{T_0}^{T} \Delta C_P(T,P_0)\,\mathrm{d}T\right)\mathrm{d}T^{-1} = \Delta_{\text{vap}}H(T_0,P_0)\,\left(T_0^{-1} - T^{-1}\right) \tag{12.9}$$

Equation (12.9) permits the calculation of ΔH at any coexistence situation (T_0,P_0) from the (T,P) data if the EoSs and ΔC_P at one pressure are known. This pressure is not necessarily P_0 since conversion is possible by means of the EoSs. A similar result for ΔV is obtained from Eq. (12.8b) but with the heat capacity C_P replaced by the isothermal compressibility κ_T. If one is interested in ΔH at some other pressure P_1, a somewhat involved procedure yields

$$\begin{aligned} &T^{-1}\int_{P_1}^{P} \Delta V(T,P)\,\mathrm{d}P - T_0^{-1}\int_{P_1}^{P_0} \Delta V(T_0,P)\,\mathrm{d}P + \int_{T_0^{-1}}^{T^{-1}} \left(\int_{T_0}^{T} \Delta C_P(T,P_1)\,\mathrm{d}T\right)\mathrm{d}T^{-1} \\ &\quad = \Delta_{\text{vap}}H(T_0,P_1)\,\left(T_0^{-1} - T^{-1}\right) \end{aligned} \tag{12.10}$$

If the gas phase has the EoS $V_G(T,P) = RT/P + \Sigma_j c_j(T)P^j$ and the condensed state a similar EoS with the RT/P term omitted, the difference in molar volume is $\Delta V(T,P) = RT/P + \Sigma_j \Delta c_j(T)P^j$. Substitution in Eq. (12.9) results for the gas phase in

$$\begin{aligned} &R\ln P + \int_{T_0^{-1}}^{T^{-1}} \left(\int_{T_0}^{T} \Delta C_P(T,P_0)\,\mathrm{d}T\right)\,\mathrm{d}T^{-1} + T^{-1}\sum\nolimits_j (j+1)^{-1}\Delta c_j(T)\left(P^{j+1} - P_0^{\,j+1}\right) \\ &\quad = \Delta_{\text{vap}}H(T_0,P_0)\,\left(T_0^{-1} - T^{-1}\right) + R\ln P_0 \end{aligned}$$

Without the ln term this result applies to the condensed state. Plotting the values of the left hand side of this equation versus ($T_0^{-1} - T^{-1}$), $\Delta_{\text{vap}}H(T_0,P_0)$ is obtained as the slope, for which iteration of P_0 is required. For arbitrary states $\Delta_{\text{vap}}H$ is obtained via substitution of $\Delta V(T,P) = RT/P + \Sigma_j \Delta c_j(T)P^j$ in Eq. (12.10), resulting in

$$\begin{aligned} &R\ln P + \int_{T_0^{-1}}^{T^{-1}} \left(\int_{T_0}^{T} \Delta C_P(T,P_1)\,\mathrm{d}T\right)\,\mathrm{d}T^{-1} + T^{-1}\sum\nolimits_j (j+1)^{-1}\Delta c_j(T)\left(P^{j+1} - P_1^{\,j+1}\right) \\ &\quad = \Delta_{\text{vap}}H(T_0,P_1)\,\left(T_0^{-1} - T^{-1}\right) + R\ln P_0 + T_0^{-1}\sum\nolimits_j (j+1)^{-1}\Delta c_j(T_0)\left(P_0^{\,j+1} - P_1^{\,j+1}\right) \end{aligned} \tag{12.11}$$

Using this expression, $\Delta_{\text{vap}}H$ can be directly calculated for arbitrary states without an intermediate calculation for the coexistence line. However, the accuracy in the (T,P) data must be one order of magnitude or more better for this differential method than for the conventional integral method.

Figure 12.6a displays the results of this analysis for ethanol, from which the temperature dependence of $\Delta_{\text{vap}}H$ is clearly demonstrated. Figure 12.6b shows the conventional ln P versus $1/T$ plot, resulting in an average $\Delta_{\text{vap}}H = 41.6$ kJ mol^{-1}.

Equation (12.11) can also be used to rationalize several empirical expressions. Putting $P_1 = 0$ shows that ΔC_P is independent of pressure and that the C_P terms of the gas phase are separated from its volume expansion coefficients c_j, so that only the latter account for the intermolecular interactions. As examples, Mosselman et al. [4] used the expressions $\Delta C_P = \Delta a + \Delta bT + \Delta cT^2$ and $V = RT/P + B$, where for gases $B = b + KT\exp(J/T)$ is the

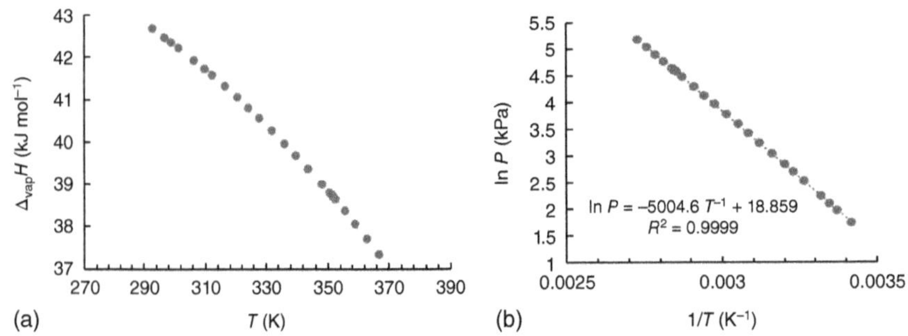

Figure 12.6 (a) The vaporization enthalpy $\Delta_{vap}H$ (kJ mol^{-1}) of ethanol versus temperature T (K). (b) The conventional ln P versus 1/T plot. Source: Both plots are after data adapted from Mosselman et al. [4].

second virial coefficient valid for dimerization and the square-well potential, and for condensed phases $B = b$. This leads to

$$\ln P = A + B/T + C\ln T + DT + ET^2 + F\exp(J/T) \tag{12.12}$$

$$A = R^{-1}\left[-\Delta a - \Delta a\ln T_0 - bT_0 - {}^1\!/_2\Delta cT_0^{\ 2} + \Delta H(p_1 = 0, T_0)T_0^{\ -1} + R\ln P_0 + KP_0\mathrm{e}^{J/T_0}\right]$$

$$B = R^{-1}\left[\Delta aT_0 + {}^1\!/_2\Delta bT_0^{\ 2} + {}^1\!/_3\Delta bT_0^{\ 3} - \Delta H(p_1 = 0, T_0)\right]$$

$$C = R^{-1}\Delta a, \qquad D = {}^1\!/_2R^{-1}\Delta b, \qquad E = R^{-1}\Delta c/6, \qquad F = -R^{-1}K$$

Equation (12.12) is similar to the empirical expression by Scott and Osborn [5], which accurately reproduces the experimental vapor pressures for both normal and abnormal fluids, like water. By applying various sets of constraints on the coefficients, a whole series of empirical vapor pressure equations can be obtained [4, 6].

Although Mosselman et al. applied their results to gas-condensed phase equilibria only, it can be applied generally for first-order transitions of pure compounds. As Lobo and Ferreira [6] pointed out, no specific conditions are applied, and the analysis depends only on a reference point (P_0,T_0) and on suitable analytical expressions for ΔV and C_P. These authors applied the approach to sublimation, vaporization, fusion, and solid–solid equilibria, and the equations they presented provide a comprehensive and systematic means to describe two-phase transition lines over a large temperature range within a few percent of the measured equilibrium pressures.

12.4 The Ehrenfest–Prigogine–Defay Equations

Before it was clear that the first derivative of μ could not be only zero but also infinite, Ehrenfest [7] classified phase transitions according to their derivatives of the Gibbs energy. Because $\mathrm{d}G(T,P,n) = -S\mathrm{d}T + V\mathrm{d}P + \mu\mathrm{d}n$, we have

$$S = -\partial G/\partial T \equiv -G_T, \qquad V = \partial G/\partial P \equiv G_P \qquad \text{and} \qquad \mu = \partial G/\partial n \equiv G_n$$

Here and in the remainder of this section, the subscript X for G indicates the derivative of G with respect to X (for other variables it denotes, as usual, that the subscript variable is kept constant). Transitions between phases 1 and 2 with $\Delta\mu = \mu_2 - \mu_1 = 0$ were labeled as first order and transitions with $\partial\Delta\mu/\partial T = \partial\Delta\mu/\partial P = 0$ as second order. For the latter (continuous) transitions, we thus have

$$(\partial^2\Delta\mu/\partial T^2)_P \neq 0, (\partial^2\Delta\mu/\partial P^2)_T \neq 0 \quad \text{and} \quad (\partial^2\Delta\mu/\partial T\partial P) \neq 0 \tag{12.13}$$

Often experimental thermodynamical data are summarized as data for the heat capacity $c_P = T(\partial S/\partial T)_P = -T(\partial^2 G/\partial T^2)_P \equiv TG_{T,T}$, expansivity $\alpha_P = V^{-1}(\partial V/\partial T)_P = V^{-1}(\partial^2 G/\partial T\partial P)_P \equiv V^{-1}G_{T,P}$ and compressibility $\kappa_T = V^{-1}(\partial V/\partial P)_T = V^{-1}(\partial^2 G/\partial P^2)_T \equiv V^{-1}G_{P,P}$. For a continuous transition, we thus have $\Delta c_P \neq 0$, $\Delta\alpha_P \neq 0$ and $\Delta\kappa_T \neq 0$. We recall that generally for a thermodynamical potential Z, we have $Z_{X,Y} = Z_{Y,X}$.

A relation analogous to the Clapeyron equation can be derived for continuous transitions. Taking the temperature derivative from the expressions $(\partial\Delta\mu/\partial T)_P = -\Delta s = 0$ and $(\partial\Delta\mu/\partial P)_T = \Delta v = 0$,[6] we obtain the *Ehrenfest relations*:

$$\mathrm{d}\Delta s = -\frac{\partial\Delta\mu_T}{\partial T}\mathrm{d}T - \frac{\partial\Delta\mu_T}{\partial P}\mathrm{d}P = \frac{\Delta c_P}{T}\mathrm{d}T - V\Delta\alpha_P\mathrm{d}P = 0 \quad \text{or} \quad \frac{\mathrm{d}P}{\mathrm{d}T} = \frac{1}{VT}\frac{\Delta c_P}{\Delta\alpha_P}$$

$$\mathrm{d}\Delta v = \frac{\partial\Delta\mu_P}{\partial T}\mathrm{d}T + \frac{\partial\Delta\mu_P}{\partial P}\mathrm{d}P = V\Delta\alpha_P\mathrm{d}T - V\Delta\kappa_T\mathrm{d}P = 0 \quad \text{or} \quad \frac{\mathrm{d}P}{\mathrm{d}T} = \frac{\Delta\alpha_P}{\Delta\kappa_T}$$

Combing leads to the *Ehrenfest ratio*

$$\Pi = \Delta c_P\Delta\kappa_T/VT(\Delta\alpha_P)^2 = 1 \tag{12.14}$$

In many cases, though, next to the external variables **a**, internal variables **b** play a role [8]. Therefore, we rederive Eq. (12.14) introducing internal variables **b** for a closed system but restrict their number initially to 1. From $\mathrm{d}G(T,P,b) = -S\mathrm{d}T + V\mathrm{d}P - B\mathrm{d}b$, we now have $B = -\partial G/\partial b$ or $\partial B/\partial b = -\partial^2 G/\partial b^2 = -G_{b,b}$. We also have $\partial B/\partial P = -\partial V/\partial b = -G_{P,b}$ and $\partial B/\partial T = \partial S/\partial b = -G_{T,b}$. As an aside we note, inserting $S = (H - G)/T$ in $\partial S/\partial b$, that this expression can be written as $\partial B/\partial T = \partial S/\partial b = (\partial H/\partial b - \partial G/\partial b)/T = (h_{T,P} + B)/T$ with $h_{T,P} \equiv \partial H/\partial b$. Expressing $\mathrm{d}b(T,P,B)$ in terms of B, we obtain

$$\mathrm{d}b = \frac{\partial b}{\partial B}\frac{\partial B}{\partial T}\mathrm{d}T + \frac{\partial b}{\partial B}\frac{\partial B}{\partial P}\mathrm{d}P + \frac{\partial b}{\partial B}\mathrm{d}B = \frac{\partial b}{\partial B}\left(\frac{\partial B}{\partial T}\mathrm{d}T + \frac{\partial B}{\partial P}\mathrm{d}P + \mathrm{d}B\right) \quad \text{or}$$

$$\mathrm{d}b = -G_{b,b}{}^{-1}(G_{T,b}\mathrm{d}T + G_{P,b}\mathrm{d}P + \mathrm{d}B)$$

The next step is to write c_P, α_P, and κ_T as functions of b. For c_P we have

$$c_P = T(\partial S/\partial T)_P = T(\partial S/\partial T)_{P,b} + T(\partial S/\partial b)_{T,P}(\partial b/\partial T)_P = c_{P,b} - T(G_{T,b})^2/G_{b,b}$$

as long as b is not at its equilibrium value. When equilibrium is reached, $T(\partial S/\partial b)_{T,P} = 0$ and $c_{P,b}$ becomes $c_P \equiv T(\partial S/\partial T)_{P,b=b(\mathrm{equ})}$. Similarly, we may write

$$(\partial V/\partial P)_T = (\partial V/\partial P)_{T,b} + (\partial V/\partial b)_{T,P}(\partial b/\partial P)_T$$

$$-V^{-1}(\partial V/\partial P)_T = -V^{-1}(\partial V/\partial P)_{T,b} - (G_{P,b})^2/VG_{b,b} \quad \text{or} \quad \kappa_T = \kappa_b - (G_{P,b})^2/VG_{b,b}$$

In a fully analogous way, we also obtain

$$(\partial V/\partial T)_P = (\partial V/\partial T)_{P,b} + h_{T,P}G_{P,b}/VG_{b,b} \quad \text{or} \quad \alpha_P = \alpha_b - h_{T,P}G_{P,b}/VG_{b,b}$$

6 This is equivalent to using l'Hôpital's rule, which states that if the limit (a/b) is of the form $0/0$ or ∞/∞, one can use limit $(a/b) =$ limit (a'/b'), provided limit (a'/b') exists and that the derivative $b' \neq 0$.

This leads to

$$\Delta c_P = -TG_{T,b}{}^2/G_{b,b}, \Delta\kappa_T = -G_{P,b}{}^2/VG_{b,b} \quad \text{and} \quad \Delta\alpha_P = h_{T,P}G_{P,b}/VTG_{b,b}$$

Combining these expressions, we regain the Ehrenfest ratio, which is thus also valid when a single internal variable is relevant. In the connection with internal variables, it is known as the *Prigogine–Defay equation.*

When more than one internal variable is present, the Ehrenfest–Prigogine–Defay equality is no longer valid. The simplest way to see that is replacing $B\mathrm{d}b$ (or $b\mathrm{d}B$) everywhere by $\mathbf{B}^\mathrm{T}\mathbf{db}$ (or $\mathbf{b}^\mathrm{T}\mathbf{dB}$) [9, 10]. Introducing $\mathbf{v} = (\partial V/\partial\mathbf{b})_{T,P}$, $\mathbf{h} = (\partial H/\partial\mathbf{b})_{T,P}$ and $\mathbb{X} = (\partial\mathbf{B}/\partial\mathbf{b})_{T,P}$, we obtain analogously

$$\Delta c_P = \mathbf{h}^\mathrm{T}\mathbb{X}\mathbf{h}/T, \qquad \Delta\kappa_T = \mathbf{v}^\mathrm{T}\mathbb{X}\mathbf{v}/V \quad \text{and} \quad \Delta\alpha_P = \mathbf{v}^\mathrm{T}\mathbb{X}\mathbf{h}/TV \tag{12.15}$$

Because the matrix $\mathbb{X}$ is symmetrical, it can be diagonalized by an orthogonal transformation for which the dissipation $\mathbf{B}^\mathrm{T}\mathbf{db}$ does not change. We now invoke the Schwartz inequality stating that $(\mathbf{x}^\mathrm{T}\mathbf{y})^2 \leq (\mathbf{x}^\mathrm{T}\mathbf{x})(\mathbf{y}^\mathrm{T}\mathbf{y})$ for any $\mathbf{x}$ and $\mathbf{y}$, and where the equality only holds if $\mathbf{x}$ is proportional to $\mathbf{y}$. Applying this inequality to Eq. (12.15) with $\mathbb{X}$ diagonalized, we obtain

$$TV(\Delta\alpha_P)^2 \leq \Delta c_P \Delta\kappa_T \tag{12.16}$$

Therefore, the Ehrenfest–Prigogine–Defay equality becomes an inequality. The equality Eq. (12.14) can be recovered therefore not only when there is one single internal variable, but also if $\mathbf{h} = \lambda\mathbf{v}$ where λ is a constant. A priori there is no reason to expect such a proportionality.

The considerations of the previous paragraph, however, depend on the existence of a single transition temperature. This is not the case for glasses with a transition temperature range [11, 12], which we discuss in Chapter 18.

12.5 Landau and Landau-like Theory

One rather general approach to phase transitions and transformations is a theory originally by Landau [13] for continuous transitions, see also [14]. Remember that in thermodynamics the equilibrium situation is conveniently obtained by minimizing the "generalized" Helmholtz energy $F(\eta;T,V)$, dependent on temperature T, volume V, with respect to the internal variable, here labeled η. We use the label "generalized" because the "generalized" F becomes the thermodynamic F only after minimization with respect to η. In general, η varies with position, that is, it is a field quantity, and we must use the energy *density* f. For a uniform distribution of η we can use the energy itself, but we will still use the energy density. The internal variable η can be a local average of an underlying microscopic parameter describing the physics of the system. For example, for a ferromagnetic material η is the magnetization m, which in the lattice model (Appendix C) results as the average of the magnetic spins σ with value ± 1. Above T_cri, $m = 0$, while below T_cri the magnetization is $\pm m$. Similarly, for ferroelectric materials η is the polarization. The internal variable η can also be a purely macroscopic quantity. For example, for a fluid the internal variable is conveniently taken as the density difference $\Delta\rho$ between liquid and gas.

In the Landau approach, we use $\mathscr{L}(\eta;T,V) = f(\eta) - h\eta$, typically representing the grand potential density (Section 7.1). We assume that the Helmholtz part $f(\eta)$ of $\mathscr{L}(\eta)$ near T_{cri} can be expanded in a power series in η, where for generality a gradient term for inhomogeneous systems is included:

$$\mathscr{L}(\eta) = a_0 + a_1\eta + {}^1\!/_2 a_2\eta^2 - {}^1\!/_3 a_3\eta^3 + {}^1\!/_4 a_4\eta^4 + \cdots + \lambda(\nabla\eta)^2 - h\eta$$

The driving force h is the conjugate variable to the internal variable η, so that $h = \partial f(\eta)/\partial\eta$. For a fluid the driving force is the chemical potential μ (with $\eta = \Delta\rho$), while for a ferromagnetic material mentioned above, it is the external magnetic field H (with $\eta = m$). In principle, all coefficients a_n depend on temperature T. The parameter a_0 is a reference energy and can be omitted. Equilibrium requires that $\partial\mathscr{L}/\partial\eta = 0$, and in the case $h = 0$, this requires that a_1 is absent. Further note that $a_4 > 0$ should hold if the system is to be stable. In cases the internal variable can have only the two values $\pm\eta$, as for the magnetization and polarization cases, the third-order terms must also be absent because $\mathscr{L}(\eta)$ should equal $\mathscr{L}(-\eta)$. For fluids this is obviously not the case if we use $\eta = \Delta\rho = \rho - \rho_{\text{cri}}$, but we assume this condition to be true for the moment. We also consider homogeneous systems with $\nabla\eta = 0$. So, we are left with $\mathscr{L}(\eta) = f(\eta) = {}^1\!/_2 a_2\eta^2 + {}^1\!/_4 a_4\eta^4$, or to ease the notation slightly, with

$$f(\eta) = {}^1\!/_2 a\eta^2 + {}^1\!/_4 c\eta^4 \tag{12.17}$$

As usual, equilibrium is reached when $\partial f(\eta)/\partial\eta = 0$, and the solutions are found to be

$$\eta_0 = 0 \quad \text{and} \quad \eta_\pm = \pm(-a/c)^{1/2}$$

For $a(T) < 0$, we have a critical temperature T_{cri} where above only one solution exists. If we expand a with respect to temperature, we have $a = \alpha + a'\Delta T + \mathcal{O}(\Delta T^2)$, where $a' = \partial^3\mathscr{L}/\partial\eta^2\partial T$ and $\Delta T = T - T_{\text{cri}}$. For $T < T_{\text{cri}}$, η should be non-zero arbitrarily close to T_{cri}, so that we must have $\alpha = 0$ and hence $a = a'\Delta T$. A similar expansion for c will lead to second-order temperature effects and, hence, we take c as constant. This implies that we have parameters $a' > 0$ and $c > 0$. Figure 12.7 displays the $f(\eta;T,V)$ curves as a function of temperature. For $T < T_{\text{cri}}$ we have two coexistent states, while for $T > T_{\text{cri}}$ only one state survives. We thus describe a continuous transition.

From $f(\eta)$ the various properties can now be calculated. For example, if we write for the entropy $s = -\partial f/\partial T = -(\partial f/\partial\eta)(\partial\eta/\partial T)$, the two factors are

$$\partial f/\partial\eta = (a + c\eta^2)\eta \quad \text{and} \quad \partial\eta/\partial T = {}^1\!/_2(-a/c)^{-1/2}(-a'/c) = -a'/2c\eta$$

so that for $T < T_{\text{cri}}$ the entropy s and specific heat c_V become

$$s = -(a + c\eta^2)a'/2c \quad \text{and} \quad c_V = T\partial s/\partial T = a'^2\,T/2c$$

For $T > T_{\text{cri}}$ the internal variable $\eta = 0$ and we have $s = c_V = 0$. Thus, the specific heat due to the transition thus increases linearly for $T < T_{\text{cri}}$ and jumps to zero at $T = T_{\text{cri}}$.

The relevant response function is the *isothermal susceptibility* χ_T, given by

$$\chi_T(h) \equiv (\partial\eta/\partial h)_T = [\partial^2 f(\eta)/\partial\eta^2]^{-1} = [a + 3c\eta(h)^2]^{-1}$$

where $\eta(h)$ is a solution of $h = a\eta + c\eta^3$. For $T > T_{\text{cri}}$ we have $\eta = 0$ and $\chi_T = a^{-1}$. Similarly, for $T < T_{\text{cri}}$ at $h = 0$, we have $\eta^2 = -a'\Delta T/c$ and $\chi_T = (-2a)^{-1}$. Note that $\chi_T(T < T_{\text{cri}}) = -{}^1\!/_2\chi_T(T > T_{\text{cri}})$ is always true for this model.

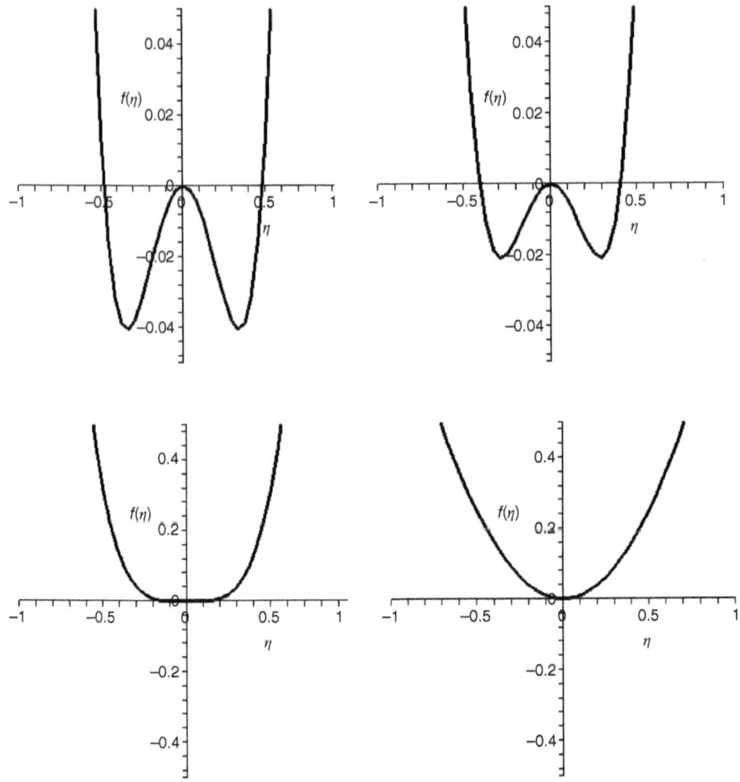

Figure 12.7 The Helmholtz energy curve f for a continuous transition along the coexistence curve for T changing from below T_{cri} (upper) to T_{cri} (lower left) to above T_{cri} (lower right).

Let us now consider whether a Landau-like approach also can be applied to discontinuous transitions by admitting a cubic term:

$$\mathscr{L}(\eta) \cong {}^1\!/\!_2 a\eta^2 - {}^1\!/\!_3 b\eta^3 + {}^1\!/\!_4 c\eta^4 - h\eta \tag{12.18}$$

We may assume that $c > 0$, because $\mathscr{L}(\eta)$ has the symmetry $(|c|,\eta) \leftrightarrow (-|c|,-\eta)$. For $h = 0$, we obtain from the equilibrium condition $\partial f(\eta)/\partial\eta = 0$ the solutions

$$\eta_0 = 0 \quad \text{and} \quad \eta_\pm = (b/2c) \pm \sqrt{(b/2c)^2 - (a/c)}$$

There are three real solutions if $b^2 > 4a/c$.

To illustrate the behavior possibly most clearly, we scale η and a as $\eta = (b/c)y$ and $a = (b^2/c)x$. The Helmholtz energy $f(\eta)$ becomes $\varphi(y) = {}^1\!/\!_2 xy^2 - {}^1\!/\!_3 y^3 + {}^1\!/\!_4 y^4$, and the nondimensional solutions become $y_0 = 0$ and $y_\pm = {}^1\!/\!_2 \pm ({}^1\!/\!_4 - x)^{1/2}$. Figure 12.8a shows a plot of y versus x. For decreasing temperature, the transformation takes place at $x^* = 0$ with a jump from $y = 0$ to $y = 1$, while for increasing temperature it occurs at $x^{**} = 1/4$ with a jump from $y = 1/2$ to $y = 0$. For $x^* = 0$ with $y = 1$, $\varphi(y) = -1/12$, while for $x^{**} = {}^1\!/\!_4$ with $y = {}^1\!/\!_2$, $\varphi(y) = 1/192$. The equilibrium transformation occurs at x_0 when $\varphi(y) = \partial\varphi(y)/\partial y = 0$. Solving y^2 from $\partial\varphi(y)/\partial y = 0$ yields $y^2 = y - x$ and substituting in $\varphi(y) = 0$ results in $y = 3x$. Substituting this back in $y^2 = y - x$, it follows that $x_0 = 2/9$ and y jumps from $y = 0$ to $y = 2/3$.

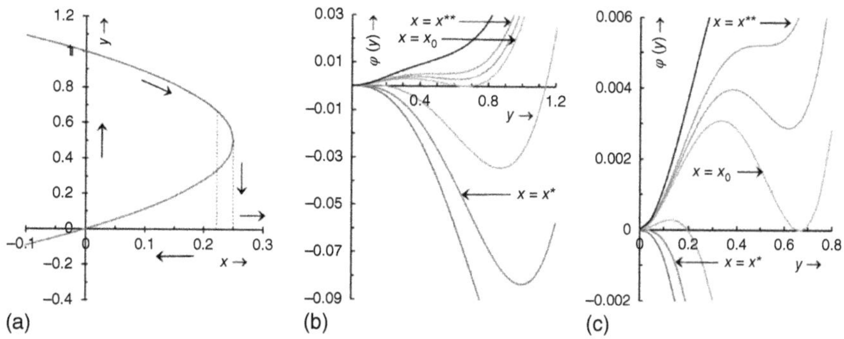

Figure 12.8 The discontinuous transition: (a) the nondimensional internal variable $y = (c/b)\eta$ as a function of the nondimensional temperature $x = (c/b^2)a$. The arrows indicate the behavior with increasing and decreasing x, illustrating the hysteresis loop; (b, c) the shape of the nondimensional Helmholtz energy curve $\varphi(y)$ across the coexistence curve for various values of x.

The $\varphi(y)$ curves as a function of $y(x)$ (Figure 12.8b) show for $x > x^{**}$ only one minimum. At $x = x^{**}$ an inflection point appears developing in a secondary minimum, which becomes at $x = x_0$ equal to the one at $\eta = 0$. For $x = x^*$ the first minimum becomes an inflection point, and the new minimum becomes the global minimum. So, transforming back, for decreasing temperature η jumps from $\eta = 0$ to $\eta = 2b/3c$ at $T^* = T_0 - 2b^2/9a'c$, while with increasing temperature η jumps from $\eta = b^2/2c$ to $\eta = 0$ at $T^{**} = T_0 + b^2/36a'c$, so that we have a *hysteresis* loop. Clearly, the cubic term in η gives rise to a discontinuous transition so that Landau-like theory can describe both types of transition. From $a = a'(T - T_0)$ we note that η at the transitions is not arbitrarily small, as required by Landau theory. Strictly speaking, T_0 is thus not a critical point, and therefore we use T_0 instead of T_{cri}.

To conclude our general discussion on these types of theories, we remark (i) that some authors use $a = a'(T - T^*)$ instead of $a = a'(T - T_0)$ because T^* is often directly accessible, and (ii) that discontinuous transitions can also be described by extending f with $\frac{1}{6}\eta^6$ instead of by $\frac{1}{3}\eta^3$ as done initially by Devonshire [15], but that this leads to the same conclusions. An extensive reference for these theories is [16]. Landau and Landau-like theories will be used in several of the chapters that follow.

Problem 12.4 ***Helmholtz Energy Differential*** Using $G = F - PV$, show that $\mathrm{d}f = \mu \mathrm{d}\rho - s\mathrm{d}T$ with $f = F/V$, $s = S/V$, and $\rho = n/V$.

References

1 Callen, H. (1985). *Thermodynamics and an Introduction to Thermostatistics*. New York: John Wiley & Sons.

2 Slater, J.C. (1939). *Introduction to Chemical Physics*. New York: McGraw-Hill Book Comp. (also Dover, 1970).

3 Lekner, J. (1982). *Am. J. Phys* 50: 164.

4 Mosselman, C., van Vugt, W.H., and Vos, H.J. (1982). *J. Chem. Eng. Data* 27: 246.

5 Scott, D.W. and Osborn, A.G. (1979). *J. Phys. Chem.* 83: 2714.
6 Lobo, L.Q. and Ferreira, A.G.M. (2001). *J. Chem. Thermodyn.* 33: 1597.
7 Ehrenfest, P. (1933). *Proc. Kon. Akad. Wetensch.* 36: 153.
8 Priogine, I. and Defay, R. (1954). *Chemical Thermodynamics*. London: Longmans, Green and Co.
9 Davies, R.O. and Jones, G.O. (1953). *Adv. Phys.* 2: 370.
10 Davies, R.O. and Jones, G.O. (1953). *Proc. R. Soc. A* 217: 26.
11 Schmelzer, J. (2006). *J. Chem. Phys.* 125: 184511.
12 Tropin, T.V., Schmelzer, J.W.P., Gutzow, I. et al. (2012). *J. Chem. Phys.* 136: 124502.
13 Landau, L.D. (1937). *Phys. Z. Sowjet* 11: 26; Translated in *Collected Papers of L.D. Landau* (ed. D. ter Haar), p. 193. New York: Gordon and Breach.
14 Landau, L.D. and Lifshitz, E.M. (1980). *Statistical Physics*, 3e, Vol. 1. Oxford: Pergamon Press.
15 Devonshire, A.F. (1949). *Philos. Mag.* 40: 1040; (1951). 42: 1065.
16 Tolédano, J.-C. and Tolédano, P. (1987). *The Landau Theory of Phase Transitions*. Singapore: World Scientific Publishing Company.

Further Reading

Fultz, B. (2020). *Phase Transitions in Materials*, 2e. Cambridge: Cambridge University Press. A textbook on phase transformations and thermodynamics of materials emphasizing the level of atoms and electrons, and covering topics from classical metallurgy to nanoscience and magnetic phase transitions (1e, 2014).

Hillert, M. (2012). *Phase Equilibria, Phase Diagrams and Phase Transformations*, 2e. Cambridge: Cambridge University Press. A somewhat idiosyncratic book, full of ideas and concepts on phase diagrams and thermodynamic models, mainly applied to inorganic materials (1e, 1998).

Papon, P., Leblond, J., and Meijer, P.H.E. (2006). *The Physics of Phase Transitions: Concepts and Applications*, 2e. Berlin: Springer-Verlag. An extensive, relatively high-level level presentation of transitions and transformations including their dynamics for a variety of materials and experimental methods.

Skripov, V.P. and Faizullin, M.Z. (2006). *Crystal-Liquid-Gas Phase Transitions and Thermodynamic Similarity*. Weinheim: Wiley-VCH. Booklet dealing both with crystal-liquid and liquid-gas transitions from the thermodynamic similarity point of view.

Smirnov, B.M. and Berry, R.S. (2008). *Phase Transitions of Simple Systems*. Berlin: Springer. Theoretical approach based on classical physics, mainly applied to solid-fluid transitions in small and large systems of hard spheres, Lennard-Jones molecules and inert gases.

13

Discontinuous Phase Transitions: Liquids ↔ Gases

After having discussed some general aspects of phase transitions, we now focus on the discontinuous transitions between liquids and gases. We deal first with the thermodynamics and second with the kinetics of evaporation. Thereafter, we close with a discussion on the reverse transition, namely condensation.

13.1 Thermodynamics of Evaporation

Evaporation is not to be confused with boiling. Evaporation is a continuous process with negligible or no energy barrier at the surface of a liquid. In boiling, gas bubbles are created in the bulk of the liquid by a nucleation process, for which the presence of a nucleation barrier is characteristic (Section 13.3), and which thereafter escape in view of their much lower density compared to the liquid. In a transition from the liquid to the gas state, the volume changes largely and the evaporation enthalpy[1] L involved is mainly determining the vapor pressure P. Typically, P changes are from about 1 mbar at T_{tri} via 1 bar at T_{n} to about 50 bar at T_{cri}. Obviously, L changes with T, and denoting the derivative along the phase boundary curve by $\mathrm{d}L/\mathrm{d}T$, we have

$$\frac{\mathrm{d}L}{\mathrm{d}T} = \left(\frac{\partial L}{\partial T}\right)_P + \left(\frac{\partial L}{\partial P}\right)_T \frac{\mathrm{d}P}{\mathrm{d}T}$$

Using $(\partial L/\partial T)_P = \partial(T\Delta S)/\partial T = \Delta S + T(\partial \Delta S/\partial T) = L/T + \Delta C_P$, $(\partial L/\partial P)_T = T(\partial \Delta S/\partial P)_T = -T(\partial \Delta V/\partial T)_P$, and substituting $L/T\Delta V$ for $\mathrm{d}P/\mathrm{d}T$, we obtain *Planck's equation* [1]:

$$\frac{\mathrm{d}L}{\mathrm{d}T} = \frac{L}{T} + \Delta C_P - \frac{L}{\Delta V}\left(\frac{\partial V}{\partial T}\right)_P = \Delta C_P + \frac{L}{T}\left[1 - \left(\frac{\partial \ln \Delta V}{\partial \ln T}\right)_P\right]$$

This expression can be used to calculate one of the quantities involved if the others are known. For example,[2] for water at 100 °C and 1 atm, $L = 539\,\text{cal g}^{-1}$, $\mathrm{d}L/\mathrm{d}T = -0.64\,\text{cal g}^{-1}\,\text{K}^{-1}$, $V_{\text{V}} = 1674\,\text{cm}^3\,\text{g}^{-1}$, $V_{\text{L}} = 1\,\text{cm}^3\,\text{g}^{-1}$, $\partial V_{\text{V}}/\partial T = 4.813\,\text{cm}^3\,\text{g}^{-1}\,\text{K}^{-1}$, $\partial V_{\text{L}}/\partial T = 0.001\,\text{cm}^3\,\text{g}^{-1}\,\text{K}^{-1}$, and $C_{P,\text{L}} = 1.01\,\text{cal g}^{-1}\,\text{K}^{-1}$. From these data we obtain $C_{P,\text{V}} = 0.47\,\text{cal g}^{-1}\,\text{K}^{-1}$, as was also obtained by direct measurement. For evaporation one

1 Here I denote the evaporation enthalpy, normally indicated by $\Delta_{\text{vap}}H$, for brevity by L.
2 In this chapter, I use the subscripts L and V for liquid and vapor. Note that V indicates volume.

Phases of Matter and their Transitions: Concepts and Principles for Chemists, Physicists, Engineers, and Materials Scientists, First Edition. Gijsbertus de With.

often takes $\Delta V = V_V - V_L \cong V_V = RTZ/P$, with the compression factor $Z = PV/RT = 1$ for a perfect gas, so that $dL/dT = \Delta C_P$, consistent with the *Kirchhoff equation*

$$L = L_0 + \int_0^T [C_P(\text{vapor}) - C_P(\text{liquid})]dT \equiv L_0 + \int_0^T \Delta C_P dT \tag{13.1}$$

simply obtained from integrating $dH = C_P dT$ for both phases. For the same example, $C_{P,V} = 0.37\,\text{cal g}^{-1}\,\text{K}^{-1}$, yielding a difference of some 20%. Making the same approximations for the Clapeyron equation, $dP/dT = L/T\Delta V$ (Section 12.2), results in the *Clapeyron–Clausius equation*:

$$d\ln P/dT = L/RT^2 \quad \text{or} \quad d\ln P/d(1/T) = -L/R \tag{13.2}$$

This simplified equation is often used to estimate $L(T_n)$ at the normal boiling point T_n at 1 atm. However, the approximations $Z = 1$ and $\Delta V = V_V$ can lead to significant errors, as shown above (Problem 13.1). Still accepting these approximations and integrating this simplified equation, we obtain, using Eq. (13.1),

$$\ln P = -L_0/RT + \int_0^T (RT^2)^{-1} \left(\int_0^T \Delta C_P dT \right) dT + const.$$

Assuming that $\Delta C_P = \Delta C_{P,0}[1 - (T/T_{cri})]$, the result is

$$\ln P = -\frac{L_0}{RT} + \frac{\Delta C_{P,0}}{R}\left(\ln T - \frac{T}{2T_{cri}}\right) + const. \quad \text{or}$$

$$\ln P = a - b\ln T + (bT/2T_{cri}) - c/T \cong a - b\ln T - c/T \tag{13.3}$$

where a, b, and c are positive constants. The last step can be made if the temperature dependence of ΔC_P can be neglected so that $\Delta C_P = \Delta C_{P,0}$ [2]. In this approximation, the evaporation enthalpy is thus $L = L_0 + \Delta C_{P,0}T$. The experimental vapor pressure indeed can be described between T_{mel} and T_n, generally within less than 2%, by

$$\log P = a' - b'\log T - c'/T$$

with $a' = a/2.303$, $b' = b$, and $c' = c/2.303$ constants.[3] Evaluating L yields

$$L = RT^2(d\ln P/dT) = R(c - bT) \quad \text{and therefore}$$

$$L_0 = Rc \quad \text{and} \quad \Delta C_{P,0} = -Rb$$

One step further is to assume that $\Delta C_{P,0} = 0$ or L is constant, resulting in $\log P = a' - c'/T$. As this expression can describe P over only a limited temperature range, one often uses the empirical *Antoine equation*: [3]

$$\ln P(T) = A - [B/(T + C)] \tag{13.4}$$

Tables F.3 and F.12 provide some data for (a', b', c') and (A, B, C). As both Eqs. (13.3) and (13.4) contain three parameters, Eq. (13.3) describes the empirical data rather well, and has a clear theoretical foundation, the "raison d'être" for the purely empirical Eq. (13.4) is unclear. A similar approximate expression is $\ln P = A + B/T^\alpha$, for which Korsten [4] showed that, with $\alpha = 1.30$, data for alkanes, alkylbenzenes, and various other compounds can be reproduced rather well.

---

3 Often the experimental data are given as in terms of $\log P$ instead of $\ln P$.

For a series of molecules, the values for L_0 and b vary systematically with the structure. Analyzing the data for aliphatic hydrocarbons, one obtains [5]

$$L_0(C_nH_{2n+2}) = 1.1 + 1.7n \quad \text{and} \quad b(C_nH_{2n+2}) = 0.5 + 0.94n$$

where n denotes the number of carbon atoms, and the value is in kcal mol^{-1}. For aliphatic alcohols, similarly,

$$L_0(C_nH_{2n+1}OH) = 10.0 + 2.2n \quad \text{and} \quad b(C_nH_{2n+1}OH) = 2.5 + 1.8n$$

These types of correlations can be used to estimate unknown data. For example, estimating the data for $C_{12}H_{26}$ at 400 K and $C_{18}H_{38}$ at 500 K, we obtain 12.0 and 15.3 kcal mol^{-1}, to be compared with the experimental data 12.3 ± 1.3 and 15.3 ± 0.7 kcal mol^{-1}, respectively.

Values of L for some compounds are given in Table 14.1. Empirically, the evaporation enthalpy L at T_n is often described by *Trouton's rule*:

$$L/RT_n = C \quad \text{with} \quad C \cong 10.2 \tag{13.5}$$

The correlation is not particularly accurate in general and C ranges from about 8–13 (Table F.4). Compounds that form dimers in the liquid state have a large Trouton constant, for example, NO with C = 13.7. Furthermore, compounds that form dimers in the gaseous state have a small Trouton constant, for example, CH_3COOH with C = 7.5 and HF with C = 3.1. Other differences are less easily rationalized. A much more accurate rule is *Hildebrand's rule*, which states that L divided by the vaporization temperature T_{vap} for all liquids is the same when vaporization is achieved at a temperature where the vapor density is the same. The value of L/T_{vap} = 93 J K^{-1} mol^{-1}, when the vapor density is taken as Rc = 0.1 or $c = 2.02 \times 10^{-2}$ mol l^{-1} [6].

13.1.1 Evaporation in the Presence of an Inert Gas

For the derivation of Clapeyron–Clausius equation, Eq. (13.2), we considered the liquid–vapor equilibrium. In many cases, though, another gas or mixture of gases, in particular air (A), is also present. Let us consider the isothermal change of the vapor pressure of a liquid in the presence of air. At equilibrium, $\mu_L(P) = \mu_V(P_V)$, with the total pressure $P = P_A + P_V$, where P_A and P_V are the pressures of air and vapor, respectively. The equilibrium condition, thus

$$[\partial\mu_L(P)/\partial P]\mathrm{d}P = [\partial\mu_V(P_V)/\partial P_V]\mathrm{d}P_V \quad \text{or} \quad V_L\mathrm{d}P = V_L[\mathrm{d}(P_A + P_V)] = V_V\mathrm{d}P_V$$

using $\partial\mu/\partial P = V$. Solving leads to

$$\mathrm{d}P_V/\mathrm{d}P_A = V_L/(V_V - V_L) \cong V_L/V_V$$

sometimes referred to as *Poynting's equation*. For water, more or less at standard temperature and pressure (STP) conditions, the perfect gas approximation is applicable, leading to $P_V^{-1}\mathrm{d}P_V/\mathrm{d}P_A = V_L/RT$ with as solution $\ln(P_{V1}/P_{V2}) = (P_{A2} - P_{A1})V_L/RT$. We conclude that for water at 20 °C with $V_L \cong 18$ cm^3 mol^{-1} and $\Delta P = P_{A2} - P_{A1} = 1$ atm, the vapor pressure increase is only ≅0.1% and, hence, negligible.

One could object that dissolution of air in the liquid (water) has been neglected. True, however, Raoult's law shows that dissolution leads to $P_V = P_V^* x_w$ with P_V^* the vapor pressure of pure water and x_w the mole fraction of water in the air–water solution. At 0 °C about 29 cm^3 of air dissolves in 1 l of water in equilibrium with air, which corresponds at 1013 mbar (1 atm) to $N = PV/kT = 7.8 \times 10^{20}$. Using the mass density of water ρ_w,

Avogadro's constant N_A, and the molar mass of water $\mathscr{M}_W$, the number density of water is calculated as $\rho_W N_A/\mathscr{M}_W = 3.4\times10^{28}\,\text{m}^{-3}$. So, 1 l of water contains 3.4×10^{25} water molecules and the mole fraction of air in water is 2.3×10^{-5} leading to a pressure decrease of $\cong 1.4\times10^{-4}$ mbar. This decrease is about 35 times smaller than the increase due to the overall pressure increase.

Physically, the situation is thus clear. On the one hand, the increase in pressure leads to an increase in density, which increases the repulsion or, equivalently, decreases the attraction. Hence, with increased pressure, it is easier for molecules to escape from the liquid, which leads to a higher P_V. On the other hand, the dissolution of a gas in the liquid leads to a lower density. The resulting reduced density leads to a reduced attraction, or, equivalently, increased repulsion. Hence, this effect leads to a lower P_V. The example discussed shows that for water the increase in vapor pressure due to the increased overall pressure, although small, is much larger than the decrease in vapor pressure due to the dissolution of air.

Problem 13.1 ***Clapeyron versus Clapeyron–Clausius*** For benzene [7] at the normal boiling point $T_n = 353$ K and at pressure $P = 1$ atm, $\partial P/\partial T = 3.121\times10^{-3}\,\text{cm}^3\,\text{K}^{-1}\,\text{mol}^{-1}$. Estimate $L(T_n)$ using the Clapeyron–Clausius equation and the Clapeyron equation. Calculate ΔV using for V_G the virial expansion $PV/RT = 1 + B_2/V \cong 1 + B_2P/RT$ (Section 8.2) with $B_2 = -960\,\text{cm}^3\,\text{mol}^{-1}$ and $V_L = 96\,\text{cm}^3\,\text{mol}^{-1}$. Comment on the difference.

13.2 Kinetics of Evaporation

So far, we have discussed the thermodynamic aspects of evaporation. However, the kinetic aspects are at least as important, but despite their importance the situation, as we will see, is not completely clear, which is largely due to the complexity of doing proper experiments. The classical approach is kinetic theory (CKT), but some others exist. In the sequel, we discuss mainly CKT, inspired by Cammenga's review [8], and thereafter deal briefly with some other approaches.

13.2.1 Classical Kinetic Theory

Ever since the pioneering contributions of Hertz and Knudsen [9], the kinetic theory of gases has been employed to understand the evaporation of liquids. In this approach, in spite of evaporation occurring, the overall system is assumed to be in equilibrium and all nonequilibrium effects are supposed to take place in a thin layer at the liquid with a thickness approximately equal to the molecular mean free path, the so-called *Knudsen layer*. Using a half-range Maxwell distribution function for z, we obtain for the number flux, the number of molecules N striking unit liquid surface per unit time,

$$-\frac{1}{A}\frac{\partial N}{\partial t} = n_V\chi_V^3\int_{-\infty}^{+\infty}\int_{-\infty}^{+\infty}\int_{-\infty}^{0}\dot{z}\exp[-m(\dot{x}^2+\dot{y}^2+\dot{z}^2)/2kT]\,\mathrm{d}\dot{x}\mathrm{d}\dot{y}\mathrm{d}\dot{z}$$
$$= n_V\chi_V^3\frac{1}{\chi_V}\frac{1}{\chi_V}\frac{1}{2\pi\chi_V^2} = \frac{n_V}{2\pi\chi_V} = \frac{n_V\langle v_V\rangle}{4}$$

Here A denotes the liquid surface area, $\dot{x}, \dot{y}$, and $\dot{z}$ denote the components of the velocity of the molecules with mass m, $n = N/V$ their number density, $\langle v\rangle = (8kT/\pi m)^{1/2}$ their average speed, and $\chi = (m/2\pi kT)^{1/2} = m/l$ (Section 8.1). If the vapor obeys the perfect gas law $P = \rho kT$ and using the mass density $\rho = n\mathcal{M}/N_A$ with molar mass $\mathcal{M}$, the number flux and mass flux become, respectively,

$$-\frac{1}{A}\frac{\partial N}{\partial t} = \frac{P}{(2\pi mkT)^{1/2}} \quad \text{and} \quad -\frac{\mathcal{M}}{N_A}\frac{1}{A}\frac{\partial N}{\partial t} \equiv J_{LV} = P_V\left(\frac{\mathcal{M}}{2\pi RT_V}\right)^{1/2} = P_V\chi_V$$

This expression applies for evaporation into vacuum. At equilibrium between a liquid and a vapor, the rate of condensation must equal the rate of evaporation. These rates must be considered as independent if molecular scattering in the gas phase is negligible. In that case evaporation is maximal and the rate reads

$$-\frac{\mathcal{M}}{N_A}\frac{1}{A}\frac{\partial N}{\partial t} \equiv J_{LV} = {}^1\!/\!_4\rho_V v_L = P_L(\mathcal{M}/2\pi RT_L)^{1/2} = P_L\chi_L \tag{13.6}$$

where P_L is the saturation vapor pressure corresponding to the surface temperature of the liquid T_L. Experimentally, it was found that the evaporation rate varies considerably, and for nominally the same conditions, it is often several decades smaller than predicted. That is why Marcellin [10] and later Knudsen [9] introduced the evaporation coefficient ζ_L to account for this deviation. After including ζ_L and a corresponding condensation coefficient ζ_V, the net mass flux J can be expressed as $J = \zeta_L J_{LV} - \zeta_V J_{VL}$, which results in what is called the *Hertz–Knudsen equation*

$$J = \left(\frac{\mathcal{M}}{2\pi R}\right)^{1/2}\left(\frac{\zeta_L P_L}{T_L^{1/2}} - \frac{\zeta_V P_V}{T_V^{1/2}}\right) \cong \zeta(P_L - P_V)\left(\frac{\mathcal{M}}{2\pi RT}\right)^{1/2} = \zeta(P_L - P_V)\chi \tag{13.7}$$

where the last step can be made if $T_L \cong T_V$ and one takes $\zeta = \zeta_L = \zeta_V$, as is often done.

In this analysis, several simplifications are made:

- The evaporation and condensation rates are equal. This cannot be true if net evaporation or condensation occurs, implying a finite interface velocity.
- There is no temperature jump between liquid and vapor. We will come to this assumption later.
- The equilibrium velocity distribution is used. This also cannot be true if net evaporation or condensation occurs, implying, as said, a finite interface velocity.
- The vapor behaves as a perfect gas. This is not a bad approximation since for water at 298 K and saturation pressure, one calculates only $\cong 0.2\%$ deviation from perfect behavior.
- No backscattering occurs near the liquid surface. The kinetic theory of gases provides for the mean free path $\delta = (n_V\pi\sigma^2 2^{1/2})^{-1} = kT/P_V\pi\sigma^2 2^{1/2}$, where σ is the molecular collision diameter. In the absence of data, one assumes molecular close packing using the density ρ_L, that is, one uses $\sigma = (2^{1/6}/N_A{}^{1/3})(\mathcal{M}/\rho_L)^{1/3}$. At 1 bar and 300 K, $\delta/\sigma \approx 200$, so that the assumption of independent rates is warranted.

We have thus to correct for the use of the equilibrium (half-range) Maxwell distribution and the movement of the interface [11]. Indicating the drift velocity of the vapor molecules due the velocity distribution asymmetry by u and using the factor $f = 1 - \rho_V/\rho_L$ for the interface motion, we have to add to Eq. (13.6)

$$\begin{aligned}
&-\zeta_V n_V \chi_V^3 \int_{-\infty}^{+\infty}\int_{-\infty}^{+\infty}\int_{-\infty}^{0} \dot{z}\exp\left\{-m\left[\dot{x}^2 + \dot{y}^2 + (\dot{z} - uf)^2\right]/2kT\right\}\mathrm{d}\dot{x}\mathrm{d}\dot{y}\mathrm{d}\dot{z} \\
&= -\zeta_V n_V \chi_V^3 \chi_V^{-1}\chi_V^{-1}\int_{-\infty}^{0} \dot{z}\exp\left\{-m\left[(\dot{z} - uf)^2\right]/2kT\right\}\mathrm{d}\dot{z}
\end{aligned}$$

This expression can be reduced using $s = \pi^{1/2}\chi_V uf$ (with some complexity) to

$$-\zeta_V n_V \chi_V \Gamma \quad \text{with} \quad \Gamma \equiv \exp(-s^2) - s\pi^{1/2}[1 - \text{erf}(s)]$$

where erf denotes the error function. Using $\chi = (m/2\pi kT)^{1/2}$ and $\langle v\rangle = (8kT/\pi m)^{1/2}$ the rate expression, known as the *Hertz–Knudsen–Schrage equation*, becomes

$$-A^{-1}(\partial N/\partial t) = \tfrac{1}{4}\zeta_L n_L v_L - \tfrac{1}{4}\zeta_V n_V v_V \Gamma \tag{13.8}$$

Expanding the exponential and error function into series, we have

$$\exp(-s^2) = \sum\nolimits_{n=0}(-s^2)^n/n! = 1 - s^2 + s^4/2 - \cdots \quad \text{and}$$

$$\text{erf}(s) = 2\pi^{-1/2}\sum\nolimits_{n=0}\frac{(-1)^n s^{2n+1}}{n!(2n+1)} = 2\pi^{-1/2}\left(s - \frac{s^3}{3} + \frac{s^5}{10} - \cdots\right) \quad \text{so that} \tag{13.9}$$

$$\Gamma \cong 1 - \pi^{1/2}s + s^2 - \frac{s^4}{6}\cdots \quad \text{or} \quad \Gamma \cong 1 - \frac{2uf}{v_V} + \frac{4}{\pi}\left(\frac{uf}{v_V}\right)^2 - \frac{8}{3\pi^2}\left(\frac{uf}{v_V}\right)^4 \cdots$$

using $s = \pi^{1/2}\chi_V uf = uf(\mathcal{M}/2RT_V)^{1/2} = 2uf/\pi^{1/2}v_V$. Taking only the first two terms, that is linearizing the equation, and inserting $v_V = (8kT_V/\pi m)^{1/2}$, we obtain $\Gamma \cong 1 - fu(\pi\mathcal{M}/2RT_V)^{1/2}$. Multiplying Eq. (13.8) by $\mathcal{M}/N_A$, inserting the approximate expression for Γ, and using $\zeta = \zeta_L = \zeta_V$, the result is

$$J = \frac{2\zeta}{2 - f\zeta}\left(\frac{\mathcal{M}}{2\pi R}\right)^{1/2}\left(\frac{P_L}{T_L^{1/2}} - \frac{P_V}{T_V^{1/2}}\right) \tag{13.10}$$

so that in this approximation the Hertz–Knudsen equation is altered by the factor $2\zeta/(2 - f\zeta)$. If $\zeta \cong 1$, this implies a factor 2 as f is usually close to 1. It has been pointed out that the Hertz–Knudsen–Schrage equation does not conserve energy and momentum [12] because of reflection of molecules from the surface. Taking these conditions into account, still using the Maxwell distribution, leads for diffuse reflection to a factor $16\zeta/(16 - 7f\zeta)$.[4] A generalization of the classical Hertz–Knudsen–Schrage laws for the evaporation mass and energy fluxes at a liquid–vapor interface has been derived from the kinetic theory with a simple model for a velocity-dependent condensation coefficient [14] and which can, when combined with an independently obtained heat flux, describe the experiments examined quite well.

One way to obtain the heat flux $\dot{q}$ runs as follows. For condensation, we can write

$$\dot{q} = h_V(T_V - T_L)$$

where h_V is the condensation coefficient of heat transfer. If we multiply J with $\Delta_{vap}H/\mathcal{M}$, where $\Delta_{vap}H$ denotes the enthalpy of vaporization, we obtain[5]

$$\dot{q} = \frac{\Delta_{vap}H}{\mathcal{M}}J = \frac{\Delta_{vap}H}{\mathcal{M}}\zeta_V(P_V - P_L)\left(\frac{\mathcal{M}}{2\pi RT_V}\right)^{1/2}$$

Using the Clapeyron–Clausius equation, Eq. (13.2), we may write

$$P_V - P_L \cong \frac{dP}{dT}(T_V - T_L) \cong \frac{\Delta_{vap}H}{V_V T_V}(T_V - T_L) \quad \text{and} \quad h_V \cong \frac{\zeta_V(\Delta_{vap}H)^2}{V_V\left(2\pi\mathcal{M}RT_V^3\right)^{1/2}} \tag{13.11}$$

4 An approximate solution of the Boltzmann transport equation for large condensation rates leads to a limiting factor 1.67 instead of $16/9 \cong 1.78$ [13].

5 Actually, we have to add Δc_p as $T_L \neq T_V$, but as $\Delta_{vap}H/\mathcal{M} \gg c_p(T_V - T_L)$, Δc_p can often be neglected.

If we lower the vapor pressure from its saturated value P_V to P_V', the heat required is taken from the liquid adjacent to the surface. The heat conduction involved is governed by Fourier's law $\partial T/\partial t = a\partial^2 T/\partial x^2$, where $a = \lambda/\rho C_P$ is the thermal diffusivity with thermal conductivity λ, mass density ρ, and heat capacity C_P. The boundary conditions are, with heat transfer coefficient h and surface temperature T_{sur},

$$(\partial T/\partial x)_{x=0} = h(T_{sur} - T_V) \quad \text{for } x = 0,$$

$$T_{sur} - T_V = T_{sur,0} - T_V \equiv \Delta T \quad \text{and} \quad (\partial T/\partial x)_{x=0} = 0 \quad \text{for} \quad t = 0$$

The coefficient h can be related to h_V by using the expression $\dot{q} = -\lambda(\partial T/\partial x)$, so that $h = h_V/\lambda = \zeta(\Delta_{vap}H)^2 P/\lambda R(2\pi\mathcal{M}RT_V{}^5)^{1/2}$. Solving Fourier's equation, using $z = h(at)^{1/2}$, leads to

$$T(x,t) = T_V + \Delta T\{\text{erf}(xh/2z) + \exp(z^2 + hx)\,\text{erfc}\,[(xh/2z) + z]\} \tag{13.12}$$

so that the surface temperature T_s at $x = 0$ becomes

$$T_{sur}(0,t) = T_V + \Delta T[\exp(h^2 at)\,\text{erfc}\,(h\sqrt{at})]$$

Expanding the complementary error function erfc(z) into a series for large z leads to

$$\text{erfc}\,(z) = \pi^{-1/2}\exp(-z^2)(1 - 1/2z^2 + 3/4z^4 - 15/8z^6 + \cdots)$$

and we obtain for long times $T_{sur}(t) = T_V + \Delta T/h(\pi at)^{1/2}$. Expanding the error function erf(x) and exponential function exp(x) using Eq. (13.9), we obtain for short times $T_{sur}(t) = T_{sur,0} - \Delta T2\,h(at/\pi)^{1/2}$. For $\partial T/\partial x$ one can calculate from Eq. (13.12)

$$\frac{\partial T}{\partial x} = \Delta T\left[\frac{e^{-xh/2z} - e^{x^2/4at}}{(\pi at)^{1/2}} + h\,e^{z^2}e^{hx}\,\text{erfc}\left(\frac{xh}{2z} + z\right)\right]$$

which leads for the surface to

$$(\partial T/\partial x)_{x=0} = \Delta T\left[h\exp(h^2 at)\,\text{erfc}\,(h\sqrt{at})\right] \tag{13.13}$$

Let us estimate some numbers for water. To avoid convection (remember the anomalous density of water as a function of T), we consider a situation at $T_L = 4\,°C$ with a saturated vapor pressure $P_V = 813$ Pa, which we reduce to $P_V' = ½P_V$. From Eq. (13.11a) we calculate $\Delta T \cong 7.1$ K. Using $\lambda = 0.562\,\text{W m K}^{-1}$, $a = 1.34\times10^{-7}\,\text{m}^2\,\text{s}^{-1}$, and $\zeta = 1$, we obtain $h = 2.84\times10^5\,\text{m}^{-1}$. Figure 13.1a shows the surface temperature as a function of time as well as the approximate solutions for short and long times and we conclude that the temperature drops extremely fast: from 4 to −3 °C in about 10^{-2} seconds. From Eq. (13.13) the temperature gradient is calculated to be about 13.9 K cm^{-1}. Experimentally, for these conditions, Rudolph [15] obtained $\Delta T \cong 6.8$ K and $(\partial T/\partial x)_{x=0} \cong 12.3\,\text{K cm}^{-1}$, so that the above estimates are in good agreement with experiment. Here, the heat flux was treated as being directly proportional to the mass flux although, according to irreversible thermodynamics, they are more intricately coupled. However, ΔT appears to be the most important factor [14], accounting for the good agreement.

As stated before, widely varying and often rather small values for ζ have been obtained experimentally, in particular for polar liquids. Several reviews discuss the values obtained for water. While Eames et al. [16] conclude that the true evaporation constant ζ is unity under all conditions, Marek and Straub [17] emphasize the effects of impurities as an important reason for obtaining low values. At any rate, various structural reasons have been advocated to explain the differences. One suggestion that has been advanced is an

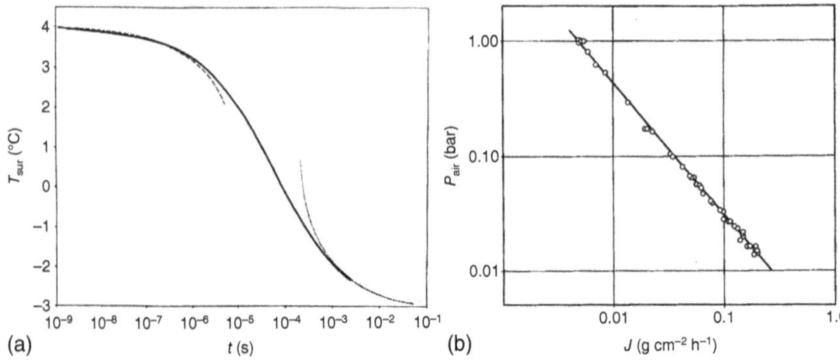

Figure 13.1 (a) Surface temperature T_{sur} of water evaporating at an initial value of $P_V' = ½P_V$ as a function of time t. The dotted lines indicate the approximate solutions. (b) Evaporation rate J of water at 22 °C and the air partial pressure P_{air} at which the evaporation rate was measured.

entropy effect due to rotational restriction in the liquid state. Such an effect is indeed present but, using Eq. (13.7) or Eq. (13.10), this effect is already incorporated via the vapor pressure. It has also been suggested that a proper orientation of the molecules approaching the interface is required. The average residence time τ_{res} of a molecule in the surface can be estimated as the ratio of the number density of molecules in the surface $n_A \cong n_V$ and the rate of hitting the surface $A^{-1}\partial N/\partial t$, that is, $\tau_{res} = n_A/(A^{-1}\partial N/\partial t) = 4\mathcal{M}^{1/3}\rho_L{}^{2/3}/vN_A{}^{1/3}\rho_V$. For water at 20 °C, $\tau_{res} \cong 4\times10^{-7}$ seconds. The orientational relaxation time can be estimated using Debye's theory yielding $\tau_{ori} = 4\pi b^3\eta/3kT$, where $b \cong \sigma/2$ is a size measure and η the viscosity. For water, $\sigma/2 = 1.35\times10^{-10}$ m and $\eta = 1.00\times10^{-2}$ g cm^{-1} s^{-1}, leading to $\tau_{ori} \cong 2.5\times10^{-12}$ seconds. Consequently, the molecules have ample time to orient. In the residence time the molecules can diffuse about $\langle r^2\rangle^{1/2} \cong (6D\tau_{res})^{1/2} \cong 7.4\times10^{-6}$ cm (using for the self-diffusion coefficient $D = 2.02\times10^{-9}$ m^2 s^{-1}) or about 200 molecular diameters.

Actually, several workers [8, 16, 18, 19] express that the evaporation–condensation coefficient $\zeta = 1$, unless secondary effects are present, such as contamination, limited heat transfer, diffusion in the vapor phase, the presence of non-condensable entities, restriction to flow in the vapor phase, or inaccurate data of physical quantities required to evaluate ζ. This is supported by an estimate of ζ from transition state theory (Section 8.5), which we outline now. Evaporation is seen as a first-order reaction, L ↔ V, that is, $-A^{-1}\partial N/\partial t = k_{eva}n_A = k_{eva}(N_A\rho_L/\mathcal{M})^{2/3}$, so that the rate of mass loss reads $-mA^{-1}\partial N/\partial t = k_{eva}m(N_A\rho_L/M)^{2/3}$. Here, the rate constant is given by

$$k_{eva} = (kT/h)(Q^{\ddagger}/Q)\exp(-E_{act}/RT)$$

where $Q^{\ddagger}$ is the partition function of the activated complex, Q is the partition function of the liquid, and E_{act} is the activation energy for evaporation. For $Q^{\ddagger}$ we have $Q^{\ddagger} = [2\pi mkT/h^2]v^{2/3}Q_{int}{}^{\ddagger}$ with $Q_{int}{}^{\ddagger}$ the partition function for the (remaining) internal modes. Using the cell theory for liquids (Section 9.7), we have $Q = (2\pi mkT/h^2)^{3/2}v_fQ_{int}$, where v_f denotes the free volume per molecule and Q_{int} is the partition function due to the internal modes. If we further assume that the internal modes are not affected, we obtain, meanwhile using $(N_A\rho_L/M)^{2/3}v^{2/3} = 1$ and $N_Am = \mathcal{M}$,

$$-\frac{\mathcal{M}}{N_A}\frac{1}{A}\frac{\partial N}{\partial t} = \left(\frac{\mathcal{M}}{2\pi RT}\right)^{1/2}\frac{RT}{N_Av_f}\exp(-E_{act}/RT) \tag{13.14}$$

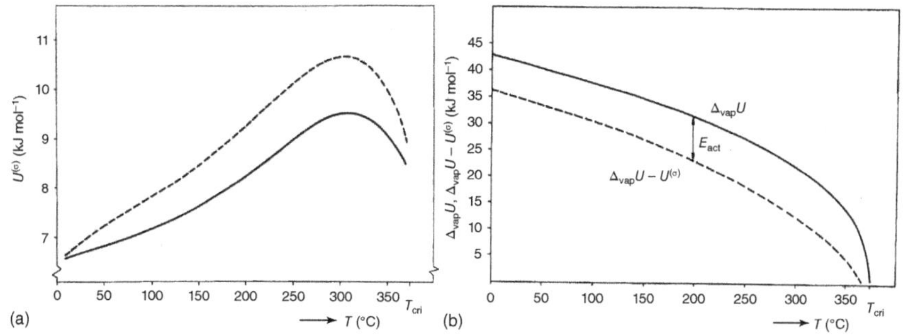

Figure 13.2 (a) Molar energy of surface formation $U^{(\sigma)}$ for water, calculated without (dashed curve) and with (solid curve) thermal expansion taken into account (T_{cri} is the critical temperature). (b) Molar internal energy of evaporation $\Delta_{vap}U$ and the difference $\Delta_{vap}U - U^{(\sigma)}$ for water.

Cammenga [8] has emphasized that E_{act} should not be identified with the energy of vaporization $\Delta_{vap}U = \Delta_{vap}H - P(V_V - V_L)$. The latter can be split into the energy $U^{(\sigma)}$ to bring the molecule from the interior of the liquid to the surface and the energy to remove it from the surface into the vapor phase. Only the last contribution represents the activation energy for evaporation. From Eq. (11.13), we have:

$$U^{(\sigma)} = \frac{\partial(F^{(\sigma)}/T)}{\partial(1/T)} = N_A^{1/3}(\mathcal{M}/\rho_L)^{2/3}[\gamma - T(\partial\gamma/\partial T) - 2\alpha\gamma T/3] \tag{13.15}$$

Substituting $E_{act} = \Delta_{vap}U - U^{(\sigma)}$ in Eq. (13.14), meanwhile using $PV_V = RT$ and reading Eq. (13.15) for $U^{(\sigma)}$, comparison with Eq. (13.6) yields:

$$\zeta = (V_V/N_A v_f)\exp\left[-\left(\Delta_{vap}U - U^{(\sigma)}\right)/RT\right]$$

Data for water for $U^{(\sigma)}$, $\Delta_{vap}U$ and $\Delta_{vap}U - U^{(\sigma)}$ as a function of T are given in Figure 13.2a,b, and show that $E_{act} \approx \Delta_{vap}U/6$, except near T_{cri}. Estimates using various methods for the free volume of water[6] yield $v_f \cong 0.5\,\text{cm}^3\,\text{mol}^{-1}$. Substituting all numerical values results in $\zeta = 1.0 \pm 0.2$, where the relatively large error is due to the simplicity of the model used and the (in)accuracy of the data used. Consequently, in this case (approximate) theory and experimental analysis agree that $\zeta \approx 1.0$. However, we note that other authors [17, 20] state that CKT theory does not provide consistent results when applied to evaporation experiments and that reasons for this suggested in the literature are not satisfactory.

13.2.2 Secondary Effects

The evaporation rate is influenced by many factors, of which we address only two, namely the effect of an inert gas and the effect of surfactants.

In Section 13.1.1, we briefly discussed the effect of an inert gas on the vapor pressure. A similar effect is present for the evaporation rate J. We consider the 1D stationary case for flat liquid surfaces. At the surface, we have the saturation pressure P_V corresponding to the

6 Estimating v_f for water is tricky as the molecules are not in van der Waals contact, resulting in a value strongly dependent on the assumptions made. In complex liquids, such as water, the tetrahedrally biased directional bonding renders packing efficiency a criterion of limited relevance.

surface temperature T_L, while at a distance z (of order δ) from the surface, we have P_z. From here diffusion occurs through the gas to the position l of the absorber for the vapor with pressure P_A. Fick's first law $J = -D\partial c/\partial z \cong [D/(l-z)](\mathcal{M}/RT)(P_z - P_A)$, with D the interdiffusion coefficient, applies. However, the rate is also equal to $J = \zeta\,(\mathcal{M}/2\pi RT)^{1/2}\,(P_V - P_z)$. Eliminating P_z, we obtain

$$J = \left\{\left(\frac{\mathcal{M}}{2\pi RT}\right)^{1/2} \Big/ \left[\frac{1}{\zeta} + \left(\frac{RT}{2\pi\mathcal{M}}\right)^{1/2}\frac{l-x}{D}\right]\right\}(P_V - P_A) \tag{13.16}$$

This expression has two limiting forms. First, for $l - x \ll 1$, one regains Eq. (13.7) for pure evaporation. Second, if the process is controlled by diffusion, $1/\zeta$ can be neglected, resulting in $J \cong [D/(l-z)]\,(P_V - P_A)$. From kinetic gas theory we know that D is inversely proportional to the pressure. Figure 13.1b for water at 22 °C shows that J is indeed proportional to the air pressure P_{air}.

Surfactants, besides having effect on the surface tension γ, have also an influence on the evaporation rate J. This is usually described by adding a resistance factor $R_{sur}(RT/2\pi\mathcal{M})^{1/2}$ to the denominator of Eq. (13.16). Experiments show that R_{sur} normally increases with the surface concentration of the surfactant, until a complete monolayer is obtained, after which R_{sur} remains approximately constant. For a homologous series, such as n-alcohols, R_{sur} increases exponentially with the chain length. A surface coverage as low as 0.001 can cause a decrease of 0.01–0.03. The effect on j for other liquids than water seems to be small.

Three possible explanations for evaporation retardation have been given. The first explanation uses *energy barrier theory*, which attributes the retardation to the work to be done to create a hole in the surfactant layer. However, while as indicated, a rather low surface coverage for one surfactant can retard evaporation significantly, the effect of another surfactant at monolayer coverage is negligible. This renders the energy barrier theory at least incomplete. It has become clear that a distinction must be made between soluble and insoluble surfactants. While soluble surfactants hardly influence the evaporation rate, insoluble surfactants decrease the rate substantially [21]. The second explanation uses *permeation theory*, in which it is assumed that J is determined by the number of water molecules penetrating the fluctuating holes in the monolayer. Although partially successful, for example, for the effect of n-alcohols on the surface pressure–surface coverage relation, it cannot explain the effect of chain length and of temperature, while very low values for ζ for pure water have to be taken to fit the experimental data. Hence, this theory is also insufficient. The third explanation uses *microconvection theory*, in which it is argued that the heat flow required to evaporate gives rise to temperature differences, which in turn induce density and surface tension differences. The surface flow due to the latter is then retarded by the surface pressure of the surfactant, possibly reinforced by surface viscosity and elasticity effects. Again, the effect at low surface coverage is difficult to explain. In summary, none of the theories can explain the effect of surfactants on J.

13.2.3 Other Approaches

Of the other approaches to evaporation, we only briefly mention statistical rate theory (SRT) [22] and linear irreversible thermodynamics, in this connection often called near-equilibrium thermodynamics (NET). Badam et al. [20] reviewed CKT, SRT, and NET, in particular with respect to the temperature gradient at the liquid–gas interface. SRT is based on probability considerations for the number of possible states of a vapor at T_V in equilibrium with a liquid at T_L, representing the steady-state conditions during evaporation, and employs Boltzmann's entropy to obtain rate expressions. SRT provides explicit expressions for ζ and provides an equilibrium molecular exchange rate between liquid and vapor but does not take the temperature jump at the interface into account. Bond and Struchtrup [14] show that for the experimental conditions of [22], the evaporation and condensation mass flux is mostly driven by energy flow, while the mass flow essentially determines the liquid interface temperature. The latter is almost independent of the coefficients that appear in the interface mass flow expression, implying that these can have rather arbitrary values and still give results in good agreement with experiment. This is the most likely reason why a wide variety of values for these coefficients can be found in the literature.

Bedeaux and Kjelstrup [23] derived expressions for evaporation and heat fluxes based on the entropy production equation from linear irreversible thermodynamics (NET). The local equilibrium assumption is evidently used and for this case validated [24]. Further, the gas phase was assumed to be ideal, and by using appropriate thermodynamic relations, the authors expressed the final flux equations without neglecting the cross coefficients effects, which were determined from the experimental results of Fang and Ward [25]. The experimental results confirms that the vapor phase heat flux is linear and a strong function of the direct (diagonal) force, that is, the temperature jump, and a weak function of the cross (non-diagonal) component, that is, the pressure jump force across the free surface. Badam et al. [20] concluded that the NET expression for vapor phase heat flux as in given [5] seems to predict the behavior of the evaporation process, although they also mention that in some cases unphysical results were obtained. Interestingly, it was pointed out by Bond and Struchtrup [14] that the evaporation mass flux expression of NET is similar to the linearized form of the SRT expression.

Summarizing, most experiments have been done for water and the temperature jump across the liquid–vapor interface during the evaporation process is well established. The vapor phase temperature is always higher than the liquid phase temperature close to liquid–vapor interface at a slow rate of evaporation, but the jump is a strong function of vapor side heat flux. The jump also depends on the vapor pressure of the system. As the vapor pressure decreases, the temperature jump across the liquid–vapor interface increases.

13.3 The Reverse Transition: Condensation

In Sections 13.1 and 13.2, we addressed evaporation. In this section, the reverse transition, condensation, is discussed. Here, we need to distinguish between film-like and drop-like condensations. In film-like condensation, the same arguments apply as for evaporation, albeit with as net result condensation. In drop-like condensation, however, as well as in boiling, nucleation processes are involved. First, we discuss the vapor pressure of drops and bubbles and second, nucleation of drops and bubbles in the gas and liquid phases. Finally, we briefly deal with molecular simulations for film evaporation/condensation and drop nucleation.

13.3.1 Drops and Bubbles

The vapor pressure of liquid drops and in gas bubbles is different from the vapor pressure above a flat surface [26]. We first discuss the matter in general and thereafter become specific. Suppose a spherical particle of phase β with radius r and interface energy γ is formed in phase α. For the mechanical equilibrium we have, according to the Laplace equation, Eq. (11.11),

$$P^{(\beta)} - P^{(\alpha)} = 2\gamma/r \quad \text{or} \quad \mathrm{d}P^{(\beta)} - \mathrm{d}P^{(\alpha)} = \mathrm{d}(2\gamma/r) \tag{13.17}$$

where we use the convention that the β-phase is always on the concave side (chord inside) of the particle. Moreover, we have the chemical equilibrium, determined by

$$\mu^{(\beta)} = \mu^{(\alpha)} \quad \text{or} \quad \mathrm{d}\mu^{(\beta)} = \mathrm{d}\mu^{(\alpha)} \tag{13.18}$$

The simplest way to arrive at the governing equation is to use the Gibbs–Duhem relation for each of the two phases, which are, using the molar entropy S and molar volume V,

$$\mathrm{d}\mu^{(\alpha)} = -S^{(\alpha)}\mathrm{d}T + V^{(\alpha)}\mathrm{d}P^{(\alpha)} \quad \text{and} \quad \mathrm{d}\mu^{(\beta)} = -S^{(\beta)}\mathrm{d}T + V^{(\beta)}\mathrm{d}P^{(\beta)} \tag{13.19}$$

Using Eqs. (13.18) and (13.19) at constant temperature leads (again) to the *Poynting equation* $V^{(\alpha)}\mathrm{d}P^{(\alpha)} = V^{(\beta)}\mathrm{d}P^{(\beta)}$, which on its turn combined with Eq. (13.17) leads to

$$(V^{(\alpha)} - V^{(\beta)})\mathrm{d}P^{(\alpha)}/V^{(\beta)} = (V^{(\alpha)} - V^{(\beta)})\mathrm{d}P^{(\beta)}/V^{(\alpha)} = \mathrm{d}(2\gamma/r) \tag{13.20}$$

This is the generic equation for spherical particles, fixating the pressure if the radius is given and vice versa.

Now let us specify that the particle is a liquid drop in a gas, that is, the α-phase is vapor (V)[7] and the β-phase is liquid (L). Assuming that the volume of the drop $V^{(\beta)} = V^{(\mathrm{L})}$ is negligible compared to $V^{(\alpha)} = V^{(\mathrm{V})}$, that is, we take $V^{(\mathrm{V})} - V^{(\mathrm{L})} \cong V^{(\mathrm{V})}$, and that the vapor phase obeys the perfect gas law $V^{(\mathrm{V})} = RT/P^{(\mathrm{V})}$, we have

$$\frac{V^{(\mathrm{V})} - V^{(\mathrm{L})}}{V^{(\mathrm{L})}}\mathrm{d}P^{(\mathrm{G})} \cong \frac{V^{(\mathrm{V})}}{V^{(\mathrm{L})}}\mathrm{d}P^{(\mathrm{V})} = \frac{RT}{V^{(\mathrm{L})}}\frac{\mathrm{d}P^{(\mathrm{V})}}{P^{(\mathrm{V})}} = \mathrm{d}\left(\frac{2\gamma}{r}\right)$$

which, upon integration from $P = P^{\square}$ for $r = \infty$ to $P^{(\mathrm{V})}$ for a drop of radius r and assuming that the liquid is incompressible, that is, $V^{(\mathrm{L})}$ is constant, leads to the *Kelvin equation*

$$\ln\frac{P^{(\mathrm{V})}}{P^{\square}} = \frac{V^{(\mathrm{L})}}{RT}\frac{2\gamma}{r} \tag{13.21}$$

It shows that the vapor pressure increases as the drop size decreases and thus that larger drops will grow at the expense of smaller ones. Moreover, it shows that in an infinite volume of vapor at constant T and P, a drop satisfying Eq. (13.21) is unstable. If some liquid evaporates (condenses), the vapor pressure of the drop increases (decreases) and thus the drop continue to evaporate (condense). Obviously, in a sufficiently small finite volume such a drop can be stable.

Assuming that $V^{(\beta)} = V^{(\mathrm{L})}$ cannot be neglected but that the liquid is still incompressible, a similar integration leads to

$$\ln\frac{P^{(\mathrm{V})}}{P^{\square}} = \frac{V^{(\mathrm{L})}}{RT}\left[\frac{2\gamma}{r} + (P^{(\mathrm{V})} - P^{\square})\right]$$

Assuming again that $V^{(\mathrm{V})} - V^{(\mathrm{L})} \cong V^{(\mathrm{V})}$, but assuming that the liquid is compressible with compressibility κ, leads to $\ln(P^{(\mathrm{V})}/P^{\square}) = (V^{(\mathrm{L})}/RT)(2\gamma/r)(1 - \kappa\gamma/r)$, but the effect of this correction is generally rather small [26].

---

7 In Chapters 1 and 12, a (non-condensable) gas is labeled G, while a (condensable) vapor is labeled V. Note, therefore, that V (vapor) should not be confused with V (volume).

Specifying that the nucleus is a vapor bubble in a liquid (α = L and β = V), a similar calculation can be made. Approximating $V^{(\alpha)} - V^{(\beta)}$ by $-V^{(\beta)} = -V^{(V)}$ or not, and assuming an incompressible liquid in addition to $V^{(V)} = RT/P^{(\beta)} = RT/P^{(V)}$, leads to

$$\ln\frac{P^{(V)}}{P^{\square}} = -\frac{V^{(L)}}{RT}\frac{2\gamma}{r} \quad \text{and} \quad \ln\frac{P^{(V)}}{P^{\square}} = \frac{V^{(L)}}{RT}\left[-\frac{2\gamma}{r} + (P^{(V)} - P^{\square})\right]$$

Hence, to first order, the vapor pressure difference for inside a bubble is the negative of that outside a droplet and decreases with increasing bubble size. This symmetry is destroyed upon using a better approximation, but the correction terms are usually small. Estimating the order of magnitude using water at 18 °C with $\gamma = 73$ mJ m^{-2} as an example, we obtain for $r = 10^{-6}$ m, $P(\text{drop})/P^{\square} = 1.001$, $P(\text{bubble})/P^{\square} = 0.999$, and a pressure difference between either drop or bubble and surrounding fluid of 1.46 bar. For $r = 10^{-8}$ m, $P(\text{drop})/P^{\square} = 1.115$, $P(\text{bubble})/P^{\square} = 0.897$, and a pressure difference of 146 bar. Clearly, considerable pressure differences can occur in small drops and bubbles. Finally, we note that we kept γ/r as one variable before integration so that these relations remain valid if γ varies with r. Further, the ratio $S = P^{(V)}/P^{\square} = P/P^{\square}$ is often called the *supersaturation* and we have in the above approximations, $\Delta\mu = kT \ln S$.

13.3.2 Classical Nucleation Theory

In a nucleation process, a barrier for the formation of a cluster of molecules is involved, which can be described in terms of the Gibbs energy difference ΔG between two phases (here the liquid phase and vapor phase) and the interfacial tension associated with the interface between these phases. These clusters, here labeled as *embryos*, form by fluctuations and normally disaggregate again, which makes it a thermally activated process. However, once a cluster has reached a critical size, called a *nucleus*, it remains stable and grows. Nucleation can take place homogeneously, that is, freely in the volume of the original phase, or heterogeneously, that is, on surfaces of the container or foreign, impurity particles. We deal with here *homogeneous classical nucleation theory* (CNT), providing somewhat general expressions to be able to use them for several cases. Let us again suppose that a spherical cluster of phase β is formed in phase α in a system at constant temperature T and constant external pressure $P^{(\alpha)}$. Here and in the sequel, we use several expressions based on assuming a spherical cluster with radius R_N and containing N molecules with molecular volume $\Omega = V^{(\beta)}/N$. The number of molecules $N = (4\pi/3\Omega)R_N{}^3$, or equivalently, $R_N = (3\Omega/4\pi)^{1/3}N^{1/3}$. Hence, the surface area of a cluster $A_N = 4\pi R_N{}^2 = (36\pi)^{1/3}\Omega^{2/3}N^{2/3} = a_1N^{2/3}$ so that $\mathrm{d}A_N/\mathrm{d}N = 2\Omega/R_N = 2A_N/3$. In the *capillarity approximation*, which implies using thermodynamic quantities for few-molecule systems, the Gibbs energy of formation of a cluster ΔG may be written as

$$\Delta G = \Delta G_{\text{vol}} + \Delta G_{\text{sur}} \equiv (4/3)\pi R_N^3 g + 4\pi R_N^2\gamma = \Omega N g + (36\pi^{1/3})\Omega^{2/3}N^{2/3}\gamma$$

where γ is the surface Gibbs energy between α and β. The difference in Gibbs energy for α and β, given by $\Delta G_{\text{vol}} = G^{(\beta)} - G^{(\alpha)}$, is for a single-component system per unit volume equal to $g = (\mu^{(\beta)} - \mu^{(\alpha)})/V^{(V)}$ with $V^{(V)}$ the (assumed constant) molar volume for phase β and $(\mu^{(\beta)} - \mu^{(\alpha)})$ the difference in chemical potential. Assuming that β is the more stable phase, $g < 0$ and ΔG has the form shown in Figure 13.3a. The maximum value, labeled

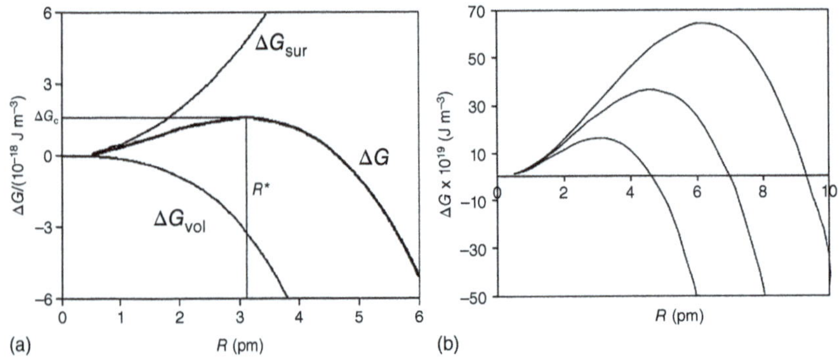

Figure 13.3 Schematic of nucleation. Source: Diogo and Moura Ramos [27]. (a) Volume and surface contribution to the total ΔG as a function of R. (b) ΔG as a function of R for different degrees of undercooling of a melt. The curves represent, from the bottom to the top, ΔG at $\Delta T = 15$, 20, and 30 K, respectively, using $\Delta_{\text{mel}}S = 8.6 \times 10^5\,\text{J K}^{-1}\,\text{m}^{-3}$ and $\gamma = 4 \times 10^{-2}\,\text{N m}^{-1}$.

ΔG^*, occurs at the critical radius R^*, equivalent to a critical number of molecules N^* in the cluster and represents the *Gibbs energy of formation* of the *nucleus*. Setting, as usual, $\partial\Delta G/\partial R = \partial\Delta G/\partial N = 0$, one obtains

$$R^* = -2\gamma/g, \quad N^* = (32\pi/3\Omega)(\gamma/g)^3 \quad \text{and} \quad \Delta G^* = 16\pi\gamma^3/3g^2$$

The expression for R^* is often called the *Thomson–Gibbs equation*. For $R < R^*$, the cluster is unstable and disappears with time, while for $R > R^*$, the cluster is stable and grows with time (Figure 13.3a).

For the vapor to liquid transition, we use the identification $\alpha \to$ V (bulk vapor) and $\beta \to$ L (liquid drop). We consider at constant T and constant total pressure Π, for example, due to the abundant presence of an inert gas, a drop with vapor pressure $P(V)$ and bulk liquid with saturation pressure $P^{\square}$. Because the difference in μ in the drop and bulk liquid equals the difference of the μs of the vapor with which they are in equilibrium, from $\mu = \mu^\circ + RT\ln P$ we have

$$\Delta\mu = RT\ln(P/P^{\square}) = RT\ln S \tag{13.22}$$

where for the last equality the supersaturation S is used. Combining with the Kelvin equation, Eq. (13.21), the result is

$$\Delta\mu = 2\gamma V^{(\text{L})}/R_N \quad \text{or} \quad \Delta\mu = 2\gamma\Omega/R_N \text{ per molecule} \tag{13.23}$$

Hence, for the transfer of dN molecules at constant total pressure Π from the bulk vapor phase to a liquid drop containing N molecules with radius R_N, we can write $\Delta\mu\text{d}N = 2\gamma\Omega\text{d}N/R_N$. Because $\text{d}A_N/\text{d}N = 2\Omega/R_N$, this leads to a drop in the *Gibbs energy of formation*

$$\Delta G_N = \int_0^N \Delta\mu\text{d}N = \int_0^N (2\gamma\Omega/R_N)\text{d}N = \gamma\int_0^{A_N}\text{d}A_N = \gamma A_N$$

where we assumed that γ is independent of R_N. For an assessment thereof, see [28].

If no inert gas is present, so that bulk and drop experience only their own vapor pressure, a similar calculation yields $\Delta\mu = 2\gamma\Omega/R_N + (P^{(\text{V})} - P^{\square})\Omega$. The *Helmholtz energy of formation* is thus $\Delta F_N = \int_0^N(\Delta\mu - \Delta P\Omega)\,\text{d}N = \gamma A_N$. In fact, $\Delta F_N = \gamma A_N$ is always true whatever the pressure and only at equal pressure for drop and bulk, $\Delta G_N = \Delta F_N$. Anyway, numerical values of ΔG_N and ΔF_N differ insignificantly for condensed phases.

13.3.3 Nucleation Kinetics

Now the question arises how the sizes of clusters are distributed [29]. To address this, we use first a thermodynamic approach and consider equilibrium between clusters, containing N molecules (N-clusters) with concentration c_N and chemical potential μ_N. Here $\mu_N = (\partial G/\partial c_N)_{T,P}$ with G the total Gibbs energy. For equilibrium, represented by $NX_1 \leftrightarrow X_N$, where X stands for the chemical symbol of the molecule, we require $\mu_N = N\mu_1$. To determine μ_N, we consider a reversible isothermal cycle with total concentration of molecules (including clusters) c and the total pressure for all steps kept constant:

(1) Extract an infinitesimal number of N-clusters at c_N and convert them to single molecules. Because $\mu_N = N\mu_1$, the change in Gibbs energy is $\Delta G(1) = 0$.
(2) Change the concentration of these single molecules from c_1 to the concentration $c^{\square}$ of the saturated vapor. The change in Gibbs energy is $\Delta G(2) = NkT\ln(c^{\square}/c_1)$.
(3) Convert these molecules to liquid at pressure $P^{\square}$ with $\Delta G(3) = 0$.
(4) Convert the resulting liquid to isolated N-clusters with $\Delta G(4) = \gamma A_N$.
(5) Dissolve these clusters in a gas phase consisting of N-clusters at concentration c. This step has $\Delta G(5) = \xi_N$, of which the value has to be discussed.
(6) Change the concentration of the clusters from c to c_N and reintroduce them in the initial gas phase with $\Delta G(6) = kT\ln(c_N/c_1)$.

The total change $\Delta G = \Sigma_j \Delta G(j) = 0$ and thus

$$NkT\ln(c^{\square}/c_1) + \gamma A_N + \xi_N + kT\ln(c_N/c) = 0 \quad \text{or}$$
$$c_N = c\exp\{-\beta[\gamma A_N - NkT\ln(c_1/c^{\square}) + \xi_N]\} \quad \text{with} \quad \beta = 1/kT \tag{13.24}$$

Here $NkT\ln(c_1/c^{\square}) = NKT\ln S = kT\ln S^N$ and the constant ξ_N is possibly a function of N. Hence, N-clusters are distributed as $c_N \sim \exp(-\beta\gamma A_N)$.

A rather similar result can be obtained from kinetic arguments for which we use $c_N = c_1\exp(-\beta\Delta G_N) = c_1\exp(-\beta N\Delta\mu)$ with for the moment c_1 a constant [30, 31]. To describe the change of c_N with time, we consider an N-cluster with a forward rate f_N and a backward rate b_N. The overall rate of change for c_N is than given by

$$\frac{\partial c_N}{\partial N} = f_{N-1}c_{N-1} - b_N c_N - f_N c_N + b_{N+1}c_{N+1} \equiv J_{N-1} - J_N \quad \text{with} \quad J_N = f_N c_N - b_{N+1}c_{N+1}$$

For the vapor to liquid transition, kinetic gas theory provides $f_N = P^{(V)}A_N/(2\pi mkT)^{1/2}$. In equilibrium, $J_N = 0$ so that we obtain $b_{N+1} = f_N^{\square}c_N^{\square}/c_{N+1}^{\square}$. Further, $\Delta G_N = \gamma A_N$ implies that $c_N^{\square} = c_1^{\square}\exp(-\beta\gamma A_N)$ and if we divide the expression for J_N by $f_N c_N^{\square} S^N$, where $S = f_N/f_N^{\square} = P/P^{\square}$, we find from inserting b_{N+1} that

$$\frac{J_N}{f_N c_N^{\square} S^N} = \frac{1}{S^N}\frac{c_N}{c_N^{\square}} - \frac{1}{S^{N+1}}\frac{c_{N+1}}{c_{N+1}^{\square}}$$

Summing over $N = 1$ to a large value for $N - 1$ yields

$$\sum_1^{N-1} \frac{J_N}{f_N c_N^{\square} S^N} = 1 - \frac{c_N}{S^N c_N^{\square}} \cong 1 \tag{13.25}$$

where the last step can be made because $c_N/S^N c_N{}^{\square}$ approaches zero asymptotically for large N. In equilibrium, J_N neither depends on time, nor on N. Extending the sum to ∞, the final result is

$$J_{\text{kin}} = \left(\sum\nolimits_1^{\infty} \frac{1}{f_N c_N^{\square} S^N}\right)^{-1} = \left(\int_0^{\infty} \frac{\mathrm{d}N}{f_N c_N^{\square} S^N}\right)^{-1} \tag{13.26}$$

where in the last step we replaced the sum by an integral, valid if N is large enough.

In the past, an artificial slightly out-of-equilibrium state was used as a reference configuration, for which $J^{\text{ref}} = 0$ was assumed [30]. In this version, labeled CNT and largely due to Frenkel [32], Becker and Döring [33], and Zeldovich [34], $S > 0$ and the factor $c_N{}^{\square}$ is replaced by $c_N{}^{\text{ref}}$. Transforming, we obtain

$$c_N{}^{\text{ref}} = c_1{}^{\square} \exp(-\beta N \Delta\mu^{\text{ref}}) = c_1{}^{\square} \exp[-\beta N(\Delta\mu^{\square} - kT \ln S)] = c_N{}^{\square} S$$

so that $J_{\text{kin}} = S^{-1} J_{\text{CNT}}$, where S^{-1} is called the *Courtney factor*. The distribution following $c_N{}^{\text{ref}}$ is called the *Frenkel distribution* [32], while the distribution $c_N{}^{\square}$ is called the *Courtney distribution* [35]. In view of the discrepancies often observed between the experimental data and CNT predictions (Figure 13.4), this difference, also related to the value for the constant ξ_N, led to an extensive discussion [29, 38]. We will come back to this point briefly later.

Returning to J_{kin}, we have for the Gibbs energy at equilibrium $\Delta G = \gamma a_1 N^{2/3}$ with $a_1 = (36\pi)^{1/3} \Omega^{2/3}$. The largest contributions are from clusters with N close to N^* and therefore we develop ΔG in a Taylor series to second order to obtain

$$\Delta G \cong \Delta G^* + \tfrac{1}{2}\Delta G''(N - N^*)^2 + \cdots \qquad \text{and} \qquad c_N \cong c_{N^*} \exp[-\tfrac{1}{2}\beta \Delta G''(\Delta N)^2]$$

where $\Delta N = N^* - N$ and $\Delta G''$ (<0) is evaluated at N^*. The distribution for c_N becomes a Gauss distribution having a standard deviation $\sigma_{\Delta N} = (-\beta \Delta G'')^{-1/2}$. The height of the nucleation barrier is clearly ΔG^*, but if we consider thermal fluctuations, configurations

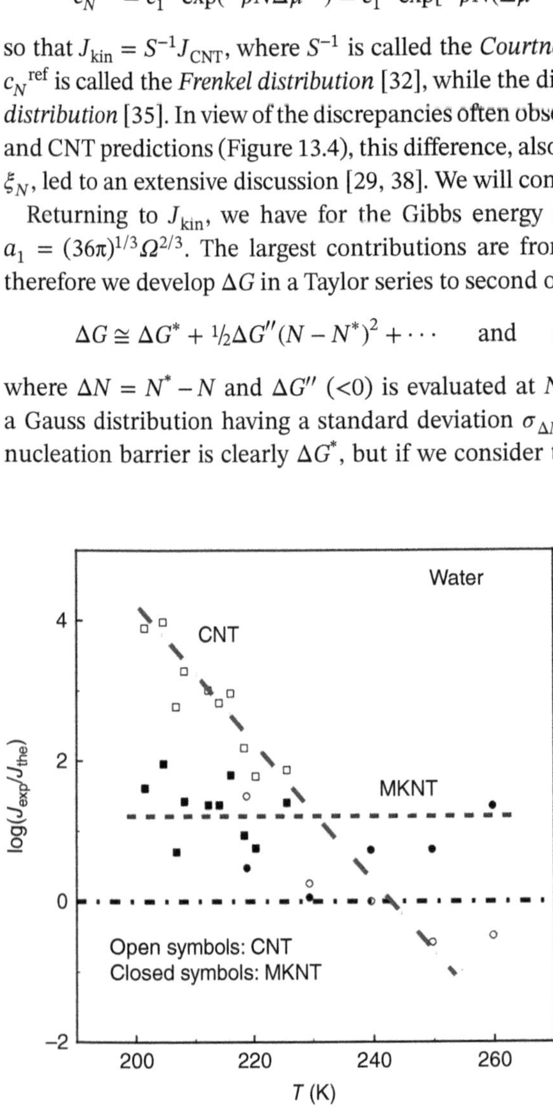

Figure 13.4 Relative nucleation rate $J_{\text{rel}} = J_{\text{exp}}/J_{\text{the}}$ for water. The circles represent the experimental results of Wölk et al. [36], while the squares represent those of Labetski et al. [37]. The dashed lines indicate the temperature dependence of J_{rel}. The horizontal dot-dash line represents $J_{\text{exp}} = J_{\text{the}}$. Source: After Kalikmanov [30].

having a ΔG less by about kT can obviously also pass. This corresponds to the range $\Delta N_{\text{fluc}} = (-\Delta G''/2kT)^{-1/2} = 2^{1/2}\sigma_{\Delta N}$. After integration, the net rate J_{kin} becomes

$$J_{\text{kin}} = \mathcal{Z} f_{N^*} c^{\square}_{N_*} \quad \text{with}$$
$$\mathcal{Z} = (-\beta\Delta G''/2\pi)^{1/2} = \tfrac{1}{3}(\beta\gamma a_1/\pi)^{1/2}(N^*)^{-2/3}$$

Here $\mathcal{Z}$ is the *Zeldovich factor*, clearly related to the thermal fluctuations. The second step follows from differentiating $\Delta G = \gamma a_1 N^{2/3}$. One can show [30] that the *consistency condition* of being in agreement with the law of mass action, Eq. (8.14), is satisfied by the Courtney distribution but not by the Frenkel distribution. Both distributions do not satisfy the *limiting condition* $c_N = c_1$ for $N = 1$, but this aspect is of lesser importance as N is supposed to be large.

To compare predictions with experiments, we use water as an example. Figure 13.4 shows the relative rates for water using two types of experiments over a temperature range of 200–260 K. Clearly, the experimental data are consistent, but CNT does not describe the overall behavior and shows a value several orders of magnitude. This is often the case for CNT, whether the Courtney factor is used or not. One factor advanced for the discrepancy is that the enthalpy for small clusters may be highly temperature dependent. However, introducing this dependence does not seem to solve the matter [29].

13.3.4 Modifications

There are some aspects to consider that can improve the classical result. First, it is often assumed that the degree of supercooling or supersaturation is small so that g can be taken as constant or linearly dependent on ΔT or ΔP. This is not necessarily true and can be remedied if an estimate for the properties of the metastable phase can be made. Second, the basic theory assumes a spherical cluster with properties identical to those of a macroscopic volume of the new phase and with a sharp interface. While asphericity can be taken into account by assuming another shape, both the assumption of identical properties and the sharp interface are doubtful. This can be easily appreciated by calculating N^*, usually resulting in a rather limited number of molecules, for example, for water at about 210 K, 15–20 molecules. For such a small-number-of-molecules particles, the total interaction per unit volume will be different from that in the bulk, leading to a difference in both the Gibbs energy G and surface Gibbs energy γ compared to bulk values.

The latter aspect has been taken approximately into account by Kalikmanov [30] in what he called *Mean-field kinetic nucleation theory* (MKNT), of which we provide a summary here. For small clusters the concept of interfacial tension is doubtful and to address this, the number of molecules for a cluster is divided according to $N = N_{\text{vol}} + N_{\text{sur}}$, where the N_{vol} molecules in the bulk of the cluster behave like a bulk liquid, but the N_{sur} at the interface are treated differently. This represents a core–shell model. We write for the internal potential energy $\Phi_N = -E_0 + E_N = -Ne_0 + e_1 N_{\text{sur}}$. Here, $-|E_0|$ is the binding energy, E_N is the surface energy of the N-cluster, and e_1 is the *surface energy per surface particle*. Both e_0 and e_1 are considered as material constants. Introducing Φ_N in the partition function Q_N of the N-cluster, we have $Q_N = V\exp(\beta N e_0)X_N$ with $X_N = (N!)^{-1}\int \exp(-\beta e_1 N_{\text{sur}})d\mathbf{r}$. From the Helmholtz energy $F_N = -kT\ln Q_N$, we have for the entropy $S_N = -\partial F_N/\partial T = k[\ln X_N + \beta e_1\langle N_{\text{sur}}\rangle + \ln V]$, where $X_N{}^{-1}\partial X_N/\partial\beta = -e_1\langle N_{\text{sur}}\rangle$ is used. The volume entropy per molecule is then given by $s_0 = \lim_{N\to\infty} S_N/N$. Introducing a temperature-dependent volume parameter v_0, representing the average volume per molecule in the cluster, and scaling X_N with v_0, we obtain

$$X_N = v_0^{N-1}(N!)^{-1} \int \exp(-\beta e_1 N_{\text{sur}})\,\mathrm{d}\boldsymbol{x} = v_0^{N-1} \sum\nolimits_1^N g(N, N_{\text{sur}}) \exp(-\beta e_1 N_{\text{sur}})$$

with $\boldsymbol{x} = \boldsymbol{r}/{v_0}^{1/3}$. In the last step, the integral is replaced by a sum over the possible N_{sur} values, and the degeneracy factor g indicates the number of complexions in the N-cluster with the same number N_{sur} of the surface molecules. Evaluating X_N using the maximum term method (Justification 7.4), the result is

$$X_N = v_0^{N-1} g(N, \langle N_{\text{sur}}\rangle) \exp(-\beta e_1 \langle N_{\text{sur}}\rangle)$$

If we write $S_N = S_{\text{vol}} + S_{\text{sur}} = Ns_0 + s_1\langle N_{\text{sur}}\rangle$ and combine the expressions for s_0 and X_N, the result, omitting arguments, is

$$s_1\langle N_{\text{sur}}\rangle = kN \ln v_0 + k \ln g - Ns_0 \quad \text{or} \quad g = v_0^{-N} \exp[k^{-1}(Ns_0 + s_1\langle N_{\text{sur}}\rangle)]$$

where s_1 is the configurational *surface entropy per particle*. Combining the expressions for Q_N, X_N, and g, the result is

$$Q_N/V = C\{\exp[\beta(e_0 + s_0 T)]\}^N \{\exp[-\beta(e_1 - s_1)T]\}^{\langle N_{\text{sur}}\rangle} \quad \text{where} \quad C = {v_0}^{-1}$$

Substituting Q_N in the number density of N-clusters, $\rho_N = (Q_N/V)\zeta^N$, yields

$$\rho_N = Cy^N x^{\langle N_{\text{sur}}\rangle} \quad \text{with fugacity} \quad \zeta = \exp(\beta\mu^{(V)})\Lambda^{-3}$$

$$y \equiv \zeta\{\exp[\beta(e_0 + s_0 T)]\}^N \quad \text{and} \quad x \equiv \{\exp[-\beta(e_1 - s_1)T]\}^{\langle N_{\text{sur}}\rangle}$$

Here y characterizes the chemical potential of the vapor and x the temperature. While ρ_N decreases continuously for $y < 1$ and $y = 1$, for y slightly larger than 1, ρ_N shows a minimum and thereafter diverges, indicating the onset of condensation. Identifying saturation with $y = 1$, we find that $\exp[\beta(e_0 + Ts_0)] \equiv 1/\zeta^{\square}$ with $\zeta^{\square}$ the fugacity at saturation, so that the number density at saturation becomes

$$\rho_N^{\square} = C\{\exp[-\beta(e_1 - Ts_1)]\}^{\langle N_{\text{sur}}\rangle} = C[\exp(-\beta f)]^{\langle N_{\text{sur}}\rangle}$$

Here $f \equiv e_1 - Ts_1$ is the *Helmholtz energy per surface molecule*.

Now one needs an estimate for f and $\langle N_{\text{sur}}\rangle$. From the compression factor at saturation, $Z_{\text{sat}} = P_{\text{sat}}V_{\text{sat}}/RT_{\text{sat}}$, where the various terms are evaluated using the expressions for F_N, ρ_N, and ${\rho_N}^{\square}$, one finds that $Z_{\text{sat}} = 1 + {Z_{\text{sat}}}^{\text{E}}$, where ${Z_{\text{sat}}}^{\text{E}}$ is the excess compression factor. Therefore, the microscopic surface tension originates from the nonideality of the vapor and can be determined from a suitable equation of state [30]. The simplest option is to use the virial expansion to second order.

For an estimate of $\langle N_{\text{sur}}\rangle$, the core–shell model of a cluster is again used. The density of the shell is given by $\rho_{\text{sur}} = \xi\rho_{\text{vol}}$, with $\xi < 1$ and the intermolecular distance by $r_{\text{sur}} = \lambda r_{\text{vol}}$. For this model, the λ-range is $1 \le \lambda < 2$, as for $\lambda = 1$ there is no shell and for $\lambda = 2$ there is a complete double layer of molecules on the initial core, rendering the cluster a core without a shell again. From this model, we have

$$N = \langle N_{\text{vol}}\rangle + \langle N_{\text{sur}}\rangle = (4\pi/3)\left\{\rho_{\text{vol}}R_{\text{vol}}^3 + \xi\rho_{\text{vol}}\left[(R_{\text{vol}} + \lambda r_{\text{vol}})^3 - R_{\text{vol}}^3\right]\right\} \tag{13.27}$$

Using instead of λ and ξ, the parameters λ and $\omega = \xi\lambda$, Eq. (13.27) becomes

$$X = -3\omega X^2 - 3\omega\lambda X + (N - \omega\lambda^2) \quad \text{where} \quad X = R_{\text{vol}}/r_{\text{sur}} = \langle N_{\text{vol}}\rangle^{1/3} \tag{13.28}$$

having a unique real positive root if $N - \omega\lambda^2 > 0$. At the limit $N_{vol} = 1$, we have $N = 1 + N_{sur}$, which leads to $\omega = N_{sur}/(3 + 3\lambda + \lambda^2)$ and thus to $\lambda = \sqrt{(N_{sur}/\omega) - (3/4)} - 3/2$. As $\lambda > 1$, the limits for ω are $0 < \omega < N_{sur}/7$. From $\langle N_{sur}\rangle = N - \langle N_{vol}\rangle = N - X^3$, with X the solution of Eq. (13.28), we obtain $X = N^{1/3}(1-\alpha)^{1/3}$ with $\alpha = 1 - X^3/N$. From the limit for small clusters, it follows that to first order $\alpha = 3\omega N^{-1/3}$, while from the limit for large clusters it follows that $\langle N_{sur}\rangle = 3\omega N^{2/3}$. Comparing with $\rho_{sat}(N)$ results in $\rho_{sat}(N) \sim \exp(-3\omega f N^{2/3})$, while CNT yields $\rho_{sat}(N) \sim \exp(-\gamma N^{2/3})$. Hence, $\omega = \gamma/3f$.

The next step is introducing the coordination number z_{vol} of the bulk. The basic idea is that in an amorphous solid or liquid, z_{vol} has the same dependency of density as that for crystalline solids but allowing non-integer numbers. Using the BCC ($z_{vol} = 8$, $\eta = 0.6802$), SC ($z_{vol} = 6$, $\eta = 0.5236$), and T ($z_{vol} = 4$, $\eta = 0.3401$) lattices, a fit provides $z_{vol} = 5.5116\eta^2 + 6.1383\eta + 1.275$, where the *packing fraction* $\eta = \pi\sigma^3/6\Omega$ (Section 8.2.1) with molecular volume Ω is used.[8] For liquids, σ should be replaced by the value for the first peak in pair correlation function $g(r)$ and Ω by $1/\rho$ with ρ the number density of the liquid. If structural data are not available, an HS estimate can be used.

Having all the results required, we now insert them in the nucleation rate J, which we write as

$$J = J_0'/\sum\nolimits_N \exp(\beta\Delta G_N) \quad \text{with} \quad J_0' = \rho_{sat}^V f_1^\square S,\ f_1^\square = a_1 P^\square(2\pi mkT)^{-1/2} \quad \text{and}$$

$$\beta\Delta G_N = \ln N^{-2/3} - \ln S^N + \beta\gamma a_1(\langle N_{sur}\rangle - 1) \cong \ln S^{-N} + \beta f(\langle N_{sur}\rangle - 1)$$

Here $\langle N_{sur}\rangle = N$ for $N \le z_{vol}$ and $\langle N_{sur}\rangle = N - X^3$ for $N \ge z_{vol} + 1$, with z_{vol} given as above. In fact, $\beta\Delta G_N$ increases for small N linearly and (neglecting a spurious maximum considered as an artefact) reaches a maximum when $d\Delta G_N/dN = -\ln S + f d\langle N_{sur}\rangle/dN = 0$. Expanding ΔG_N to second order, we have $\Delta G_N = \Delta G^* + ½\, G''(\Delta N)^2$, where $\Delta G''$ is evaluated at N^*, and leading to

$$\sum\nolimits_{N=1}^{N=\infty} e^{\beta\Delta G_N} \cong e^{\beta\Delta G^*}\int_{-\infty}^{+\infty} \exp\left[½\beta\Delta G''(\Delta N)^2\right] d\Delta N = e^{\beta\Delta G^*}/\mathcal{Z}$$

where $\mathcal{Z} = (-\beta\Delta G''/2\pi)^{1/2}$ is the *Zeldovich factor*. The nucleation rate therefore is

$$J = J_0 \exp\left(-\beta\Delta G_{N^*}\right) \quad \text{with} \quad J_0 = \mathcal{Z}\rho_{sat}^V f_{N^*}, \quad f_{N^*} = f_1^\square(N^*)^{2/3}S \quad \text{and}$$

$$\beta\Delta G^* = -N^*\ln S + \beta f[\langle N_{sur}\rangle - 1]$$

Making again the comparison for water (Figure 13.4), we note that the temperature dependence is now predicted correctly, although the absolute value is still off by an order of magnitude. For other compounds similar improvements can be noticed.

Two further aspects deserve some attention. The first is, as mentioned before, that in the classical literature $\xi_N = 0$ was mostly used, but it was pointed out by Becker [39] that this is incorrect. A statistical mechanical estimate for ξ_N [38] results in $\xi_N = -kT\ln(Z_{tra}Z_{rot}/N)_N$, where Z_{tra} and Z_{rot} are the translational and rotational partition function for an N-cluster. Generally, a large discrepancy between experimental data and theoretical estimates resulted. Extensive discussions have been held about this issue, but it seems that the matter is solved by Reiss et al. [40] using rather involved and subtle arguments, not reproduced here. Essentially it appears that the Helmholtz energy associated with the translational

8 Whether this is the optimal expression for z can be questioned as only data up to $z_{vol} = 8$ are used, while for $z > 8$, z_{vol} increases much less than the fit for $z_{vol} \le 8$. For example, for FCC ($z = 12$, $\eta = 0.7405$), $z_{vol} = 8.84$.

degrees of freedom of the clusters that serve as embryos for the formation of drops is responsible. The authors showed, unless extreme care is exercised, that the evaluation of this contribution can involve a redundant counting of molecular configurations that can lead to a predicted nucleation rate in error by many orders of magnitude. This error arises from the fact that the drop used to model the cluster cannot be located more accurately than within the fluctuation of its volume. In a lattice model, as often used to count the number of complexions for the evaluation of Z_{tra}, the lattice parameter may not be less than the cube root of the volume of fluctuation.

The second deals with *heterogeneous nucleation* for which another already existing surface, be it that of a container or impurity particle, must be taken into account. In this case, the model configuration is that of a spherical cap of radius R on that surface with contact angle θ. From an in-plane force balance, we obtain $W = \gamma^{(\beta)} - \gamma^{(\alpha)} = -\gamma\cos\theta$, normally addressed as *Young's equation*, where γ is the interface Gibbs energy between phases α and β and $\gamma^{(x)}$ is the interface Gibbs energy between phase x and air. This leads to

$$\Delta G = \pi R^2(1 - \cos^2\theta)W + 2\pi R^2(1 - \cos\theta)\gamma - \pi R^3(2 + \cos\theta)(1 - \cos\theta)^2 g/3$$

Solving as before results in

$$R^* = -2\gamma/g \quad \text{and} \quad \Delta_{\text{het}}G^* = \Delta_{\text{hom}}G^* f(\theta) \quad \text{with } f(\theta) = (2 + \cos\theta)(1 - \cos\theta)^2/4\mathfrak{X} \tag{13.29}$$

the *heterogeneity function*. Since $0 \le \theta \le \pi/2$, $0 \le f(\theta) \le 1$ and the presence of another surface (nearly) always renders nucleation more easy. In case $\theta \to 0$, $\Delta_{\text{het}}G^* \to 0$. Consequently, heterogeneous nucleation is the rule in the presence of already existing surfaces.

Boiling and drop-like condensation possibly present the simplest examples of nucleation processes for which the above descriptions can be used, but in principle they apply to all nucleation processes, albeit with different degrees of additional complications. For example, liquid–solid nucleation (freezing) also occurs in principle via a similar nucleation process as for boiling or condensation (Section 14.6). In this section, we dealt only with CNT and a few modifications, but in fact, nucleation theory is a field in its own right with a large literature. The books by Abraham [41] and Kashchiev [42] deal mainly with CNT. However, alternative theories have been presented. Itkin and Kolesnichenko [43] presented the "microscopic condensation theory," while Kalikmanov [30] reviews CNT, the use of density functional theory, and some other approaches.

Problem 13.2 ***Heterogeneous Nucleation*** Verify Eq. (13.29) as well the expression for $f(\theta)$.

13.3.5 Molecular Aspects

In our treatment of evaporation, we dealt with a kinetic theory in which several assumptions are made that were likely in the days of their genesis, but nowadays can be assessed using molecular dynamics (MD) simulations. Of the various simulation studies, we discuss mainly the results of an insightful study made by Tabe et al. [44].

Their basic approach was to use to MD simulations for 12 000 Ar atoms at 85 K, using a LJ potential with $\sigma_{\text{LJ}} = 0.3405$ nm, $\varepsilon_{\text{LJ}} = 1.635 \times 10^{-21}$ J, a cut-off of 1.5 nm, and the leapfrog method with time step $\Delta t = 5$ fs for solving the equations of motion. The simulation box was a symmetric set-up of 8.0, 8.0, and 17.5 nm with at the center of the box a liquid layer surrounded by gas molecules, so that periodic boundary conditions could be applied. The liquid–gas interface was well described by a hyperbolic tangent function located at $z_m = 4.24$ nm from the center of the box in the z-direction with the 10–90% thickness $\delta = 0.765$ nm (corresponding to the tanh normalizing factor 0.455δ). All calculations were done with reduced variables defined by $\bar{z} = (z - z_m)/\delta$. The vapor phase was considered to start at $\bar{z}_V = 3\delta$ and for molecules with $\bar{z} < \bar{z}_V$, the reaching position z_R, defined as the minimum value of $\bar{z}$ that a molecule has reached, and the staying time t_S, defined as the time that a molecule had stayed on the liquid-phase side of the vapor boundary, were obtained. Further, the incident velocity v_{in}, the normal velocity when a molecule passed through the vapor boundary toward the liquid phase ($v_{in} < 0$), and outgoing velocity v_{out}, the normal velocity toward the vapor phase ($v_{out} > 0$), were determined. Using the k-means[9] method of data clustering, molecules were classified into two types of clusters based on the similarity of the molecular variables z_R and t_S: molecules with shorter t_S and larger z_R and considered as reflecting (labeled reflection molecules), and molecules with longer t_S and smaller z_R considered as condensed and thereafter evaporated (labeled condensation–evaporation molecules).

Figure 13.5a shows that the molecules that did not pass through the vapor–liquid interface (the molecules with $0.0 < z_R < 3.0$) returned with a short t_S. There was a large increase in t_S of molecules whose z_R values were within $z_R < 0.0$. This means that t_S tends to increase when molecules pass through the vapor–liquid interface and reach the liquid phase, which is considered as condensation. The average t_S values of molecules with $z_R \geq 0.0$ and $z_R < 0.0$

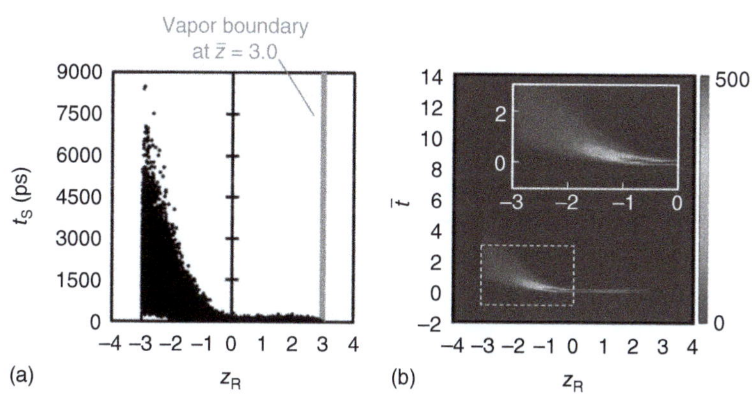

Figure 13.5 Relationship between z_R and t_S of molecules in $-3.0 < z_R < 3.0$. (a) Dataset representing relationship between z_R and t_S of molecules for the $\cong 7.1 \times 10^4$ trials in the region $-3 < zR < +3$ out of 10^5 trials. (b) Bivariate histogram of data points in $z_R - \bar{t}$ coordinate plane, where the insert shows the dotted area in more detail. The bar at the left side indicates the increasing values.

9 In the k-means method, a set of particles is divided into several clusters with initially their centroids placed at random positions and assigning particles to the closer centroid as a cluster. Recalculating the positions of centroids of each cluster, iterating the assignment of particles to clusters, and recalculating the positions of centroids until convergency is obtained results in a division for which the mean square distance of each particle to its associated centroid is minimal [45].

are approximately 25.8 and 7.40×10^2 ps, respectively. To make the distribution of the number of data points clearer, after binning their number was counted and normalized as $\bar{t} = t_S/\sigma$ to approximately unify the scales in the t_S and z_R directions. Here, $\sigma = 6.99 \times 10^2$ ps is the standard deviation of t_S derived from the t_S of molecules whose z_R values were within $-3.0 < z_R < 3.0$. Figure 13.5b shows higher values for $0.0 < z_R < 3.0$, which is the region between the vapor–liquid interface and the vapor boundary, and it exhibits quite small values for $z_R < -2.0$, which is the region inside the bulk liquid phase. In addition, the values gradually change in the region $-2.0 < z_R < 0.0$, indicating the transition region.

Further, the data were classified using z_R and t_S by the k-means method into a cluster with reflection molecules ($\cong 4.5 \times 10^4$, average $t_S = 52.0$ ps) and a cluster with condensation/evaporation molecules ($\cong 2.7 \times 10^4$, average $t_S = 1.04 \times 10^3$ ps), which appeared to be separated at approximately $z_R = -1.0$ (Figure 13.6a). To simplify, the clusters were distinguished the line $\bar{z} = -1$ (Figure 13.6b) for which about 5.8% of the reflection molecules and about 0.79% of the condensation/evaporation molecules had to be excluded. From this result it was not deemed necessary to include t_S in the classification criteria, as most of the molecules with short t_S are classified as reflection molecules. From the distribution of t_S of the molecules (Figure 13.6b) it appeared that many of the molecules with $-1.0 < z_R \leq 0.0$ had t_S in the range $t_S < 200$ ps, whereas the molecules with $-2.0 \leq z_R \leq -1.0$ had a widely distributed t_S with $t_S > 100$ ps.

There is a small peak in t_S around $t_S < 60$ ps (Figure 13.6b). The order of this time coincides with about 33 ps for a molecule to reciprocate the range $-1 \leq \bar{z} \leq 3$ at the most probable speed of the Maxwell distribution. Hence, molecules with t_S less than 100 ps should be classified as reflection molecules because of their early return to the vapor phase. Molecules with $-1.0 < z_R \leq 0.0$ occupy a large part of the range $t_S < 100$ ps (Figure 13.6b). In fact, approximately 95% of molecules with $t_S < 100$ ps have the z_R values within $-1.0 < z_R \leq 0.0$. This indicates that most of the molecules with a short t_S are classified as reflection molecules for the liquid boundary at $\bar{z} = -1$.

The distributions for the incident velocity v_{in} and the outgoing velocity v_{out} of reflection molecules are rather similar, but both have lower v_{in} and v_{out} than the most probable speed

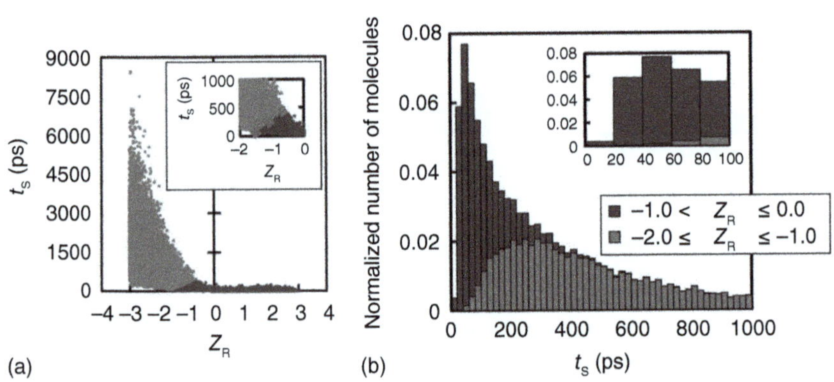

Figure 13.6 (a) Staying time distribution t_S for reflection molecules (dark gray) and condensation–evaporation molecules (light gray). Inset shows that the two clusters are divided by a dotted line at approximately $z_R = -1.0$. (b) Number of molecules normalized by the total number of molecules with $-2.0 \leq z_R \leq 0.0$ in the two z_R ranges for $-2.0 \leq z_R \leq 0.0$ with time interval $\Delta t_S = 20$ ps and the inset showing the distribution for $t_S \leq 100$ ps.

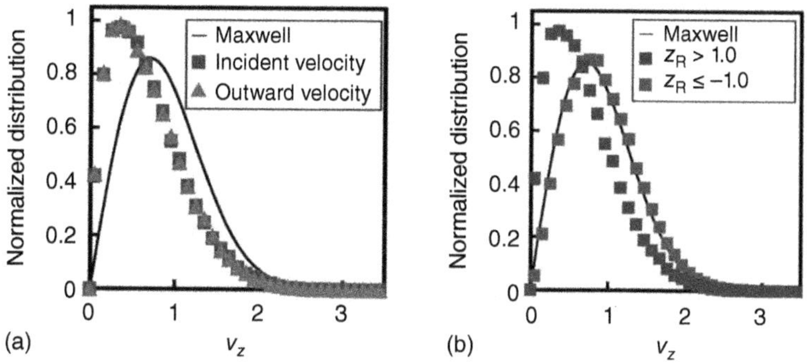

Figure 13.7 Molecular distribution functions. (a) Molecular velocity distribution functions of reflection molecules for the normalized incident velocity v_{in} and normalized outgoing velocity v_{out}. (b) Molecular velocity distribution functions of reflection molecules whose z_R values lie in $z_R > -1.0$ and condensation–evaporation molecules whose z_R values lie in $z_R \leq -1.0$.

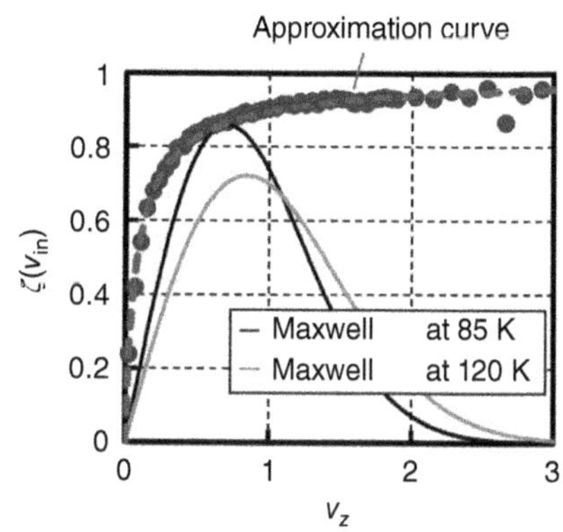

Figure 13.8 Condensation coefficient ζ_V as a function of incoming velocity v_{in}.

of the Maxwell distribution (Figure 13.7a). Thus, molecules with a low incident velocity are easily reflected. The velocity distribution function of the condensation/evaporation molecules is very similar to the Maxwell distribution, unlike that of the reflection molecules (Figure 13.7b). The condensation coefficient ζ_V was found to be a function of v_{in} (Figure 13.8). The value of ζ_V increases strongly when $v_{in} < 1.0$, but becomes almost constant when $v_{in} > 1.0$ although ζ_V fluctuates due to the small number of sample molecules for $v_{in} > 2.0$. This means that molecules with $v_{in} < 1$ are less likely to condense into the liquid phase than molecules with a high ξ_{in}.

Using for Ar a truncated and shifted LJ potential at 2.5σ, Heinen et al. [46] examined evaporation into vacuum at $T_{red} = T/T_{cri} = 0.74$. The authors varied the non-thermostated transition region between the bulk liquid and the interface having width L. By increasing L from 5.2σ to 208σ, the interface temperature T_{int} dropped by 17% and the evaporation flux decreased by a factor of 4.4. From a series of simulations for increasing values of L, an asymptotic value of T_{int} for $L \to \infty$ can be estimated which is 21% lower than the bulk liquid temperature T_L. For three values of L (5.2σ, 10.4σ, and 15.6σ), the evaporation flux calculated. This was done by adjusting T_L for 10.4σ and 15.6σ to the same T_{int} as for 5.2σ. In this way it was found that the evaporation flux was solely determined by the interface temperature T_{int}, independent on T_L or L.

Although in the above study the velocity distribution is approximately a Maxwell distribution, clear differences can be observed. In an earlier study, Ishiyama et al. [47] also employed MD simulations between 70 and 130 K for LJ particles with similar characteristics as used by Tabe et al. for Ar. In this case, the velocity distribution of spontaneously evaporating molecules was found to be nearly a half-Maxwell distribution at low temperatures. Moreover, the (overall) condensation coefficient was close to unity below the triple-point temperature, decreasing gradually with increasing temperature. The density profile for liquid–vapor equilibrium was also well described by a tanh function with the same normalized width. For evaporation into vacuum only at low temperatures the profile could be described by a tanh function, while at higher temperatures there is clear difference, having a small hump near $\overline{z} = 0$.

Most simulations are for simple LJ particles, often mimicking Ar. In a study for water, using the TIP3 potential, and ethanol, using the OPLS potential [48], Ishiyama et al. noted that, despite the formation of molecular clusters in the vapor phase and the presence of the preferential orientation of molecules at the interface, essentially the same results as for Ar were obtained. The evaporation coefficients of water and methanol were found to be decreasing functions of temperature, and to approach unity with decreasing temperature.

In summary, these studies reveal, first, that some aspects are well assessed by CNT, but others, as expected, less so, and second, that there is still no complete agreement between the various simulation studies.

References

1 Planck, M. (1945). *Treatise on Thermodynamics*, 3e, after the 7th German ed. Dover (originally Longmans, Green & Co., 1926).

2 Romps, D.M. (2021). *Q. J. R. Meteorolog. Soc.* 147: 3493.

3 Antoine, C. (1888). *Compt. Rend.* 107: 681 and 728; Easier: Thomas, G.W. (1940). *Chem. Rev.* 38: 1.

4 Korsten, H. (2000). *Ind. Eng. Chem. Res.* 39: 813.

5 Moelwyn-Hughes, E.A. (1961). *Physical Chemistry*, 2e. Oxford: Pergamon Press.

6 Hildebrand, J.H. and Scott, R.L. (1950). *The Solubility of Non-electrolytes*, 3e. New York: Reinhold Publishing Corporation.

7 Adapted from Rowlinson, J.S. and Swinton, F.L. (1982). *Liquids and Liquid Mixtures*, 3e. London: Butterworth.

8 Cammenga, H.K. (1980). *Evaporation Mechanisms of Liquids*, Ch. 4 in *Current Topics in Materials Science*, Vol. 5 (ed. E. Kaldis), p. 335. Amsterdam: North-Holland.

9 Hertz, H. (1882). *Ann. Phys. Chem.* 17: 177; Knudsen, M. (1915). *Ann. Phys. Chem.* 47: 697.

10 Marcellin, R. (1914). *Compt. Rend.* 158: 1419–1674.

11 Schrage, R. (1953). *A Theoretical Study of Interphase Mass Transfer*. New York: Columbia University Press.

12 Barrett, J. and Clement, C. (1992). *J. Colloid Interface Sci.* 150: 352.

13 Labuntsov, D.A. and Kryukov, A.P. (1979). *Int. J. Heat Mass Transfer* 22: 989.

14 Bond, M. and Struchtrup, H. (2004). *Phys. Rev. E* 70: 061605.

15 Cammenga, H.K. and Rudolph, B.-E. (1976). *Proc. 5th Int. Symp. Fresh Water from the Sea* 1: 231; Rudolp, B.-E. (1975). PhD thesis. Technical University of Braunschweig.

16 Eames, I.W., Marr, N.J., and Sabir, H. (1997). *Int. J. Heat Mass Transfer* 40: 2963.

17 Marek, R. and Straub, J. (2001). *Int. J. Heat Mass Transfer* 44: 39.

18 Hickman, K. (1954). *Ind. Eng. Chem.* 46: 13.

19 Wylie, K.F. and Brodkey, R.S. (1972). *Prog. Heat Mass Trans.* 6: 195.

20 Badam, V.K., Kumar, V., Durst, F. et al. (2007). *Exp. Therm. Fluid Sci.* 32: 276.

21 Moroi, Y. (2007). *Croat. Chem. Acta* 80: 331.

22 Rahimi, P. and Ward, C.A. (2005). *Int. J. Thermodyn.* 8: 1 and references therein.

23 Bedeaux, D. and Kjelstrup, S. (1999). *Physica A* 270: 413.

24 Johannessen, E. and Bedeaux, D. (2003). *Physica A* 330: 354.

25 Fang, G. and Ward, C. (1999). *Phys. Rev. E* 59: 417.

26 Defay, R. and Prigogine, I. (1966). *Surface Tension and Adsorption*. London: Longmans.

27 Diogo, H.P. and Moura Ramos, J.J. (2006). *J. Chem. Ed.* 83: 1389.

28 Schmelzer, J.W.P., Abyzov, A.S., and Baidakov, V.G. (2019). *Entropy* 21: 670.

29 Strickland-Constablc, R.F. (1968). *Kinetics and Mechanism of Crystallization*. London: Academic Press.

30 Kalikmanov, V.I. (2013). *Nucleation Theory*. Dordrecht: Springer.

31 Katz, J.L. and Wiedersich, H. (1977). *J. Colloid Interface Sci.* 61: 351; Katz, J.L. and Donohue, M.D. (1979). *Adv. Chem. Phys.* 40: 137.

32 Frenkel, J. (1946). *Kinetic Theory of Liquids*. Oxford: Oxford University Press.

33 Becker, R. and Döring, W. (1935). *Ann. Phys.* 24: 719.

34 Zeldovich, J. (1942). *J. Exp. Theor. Phys. USSR* 12: 525.

35 Courtney, J. (1961). *J. Chem. Phys.* 35: 2249.

36 Wölk, J. and Strey, R. (2001). *J. Phys. Chem. B* 105: 11683.

37 Labetski, D.G., Holten, V., and van Dongen, M.E.H. (2004). *J. Chem. Phys.* 120: 6314.

38 Lothe, J. and Pound, G.M. (1962, 2080). *J. Chem. Phys.* 36; Hirth, J.P. and Pound, G.M. (1963). *Condensation and Evaporation*, Vol. XI, *Progress in Materials Science*. Oxford: Pergamon Press.

39 Becker, R. (1949). *Trans. Faraday Soc.* 5: 56.

40 Reiss, H., Kegel, W.K., and Katz, J.L. (1997). *Phys. Rev. Lett.* 78: 4506; (1998). *J. Phys Chem. A* 102: 8548.

41 Abraham, F.F. (1974). *Homogeneous Nucleation Theory*. New York: Academic Press.

42 Kashchiev, D. (2000). *Nucleation, Basic Theory with Applications*. Oxford: Butterworth-Heinemann.

43 Itkin, A.L. and Kolesnichenko, E.G. (1997). *Microscopic Theory of Condensation in Gases and Plasma*. Singapore: World Scientific Publishing Company.

44 Tabe, H., Kobayashi, K., Fujii, H. et al. (2021). *PLoS One* 16: e024866.

45 Jain, A.K. (2010). *Pattern Recognit. Lett.* 31: 651.

46 Heinen, M., Vrabec, J., and Fischer, F. (2016). *J. Chem. Phys.* 145: 081101.

47 Ishiyama, T., Yano, T., and Fujikawa, S. (2004). *Phys. Fluids* 16: 2899.

48 Ishiyama, T., Yano, T., and Fujikawa, S. (2004). *Phys. Fluids* 16: 4713.

Further Reading

Faust, J.A. and House, J.E. (ed.) (2018). *Physical Chemistry of Gas–Liquid Interfaces*. Amsterdam: Elsevier. After an introduction to simulations, this book focuses on experimental methods to study the liquid-gas interface and the various aspects of aerosols.

Fujikawa, S., Yano, T., and Watanabe, M. (2011). *Vapor–Liquid Interfaces, Bubbles and Droplets – Fundamentals and Applications*. Berlin: Springer-Verlag. This book is concerned in detail with the basics of evaporation and condensation at the vapor-liquid interface using both a microscopic and macroscopic approach.

Kalikmanov, V.I. (2013). *Nucleation Theory*. Heidelberg: Springer. A modern treatment focusing on vapor-gas systems and includes classical nucleation theory, simulations and density functional theory for unary and binary systems.

14

Discontinuous Phase Transitions: Solids ↔ Liquids

In this chapter, we continue with our discussion of discontinuous phase transitions, but now from the solid to the liquid state – melting – and vice versa – freezing. Although melting is ubiquitous, understanding of the "why," "how," and "what" has not been established completely, while for freezing the description of the nucleation process will be seen to be loaded with complications. A brief follow-up on nucleation concludes this chapter.

14.1 Melting or Fusion

Upon melting a solid, the most eye-catching macroscopic change is that the solid becomes a fluid, implying a tremendous change in viscosity and loss of shear modulus, as indicated in Chapter 1. Other macroscopic properties change far less dramatically, although some microscopic properties change to a large extend as well. Let us first discuss a few phenomenological changes. We note upfront that a significant effort has been paid to metals and less to other materials, so that more examples from metals are used as is normally done in this book.

Typically, solids expand upon melting. To illustrate the molar volume increase, Table 14.1 provides some typical data. Typically, $\Delta V/V_S \cong 10\%$ for molecular liquids, while for metals and ionic compounds $\Delta V/V_S \cong 4\%$ and 20%, respectively. A simple argument [3] indicates the reason for this increase. Consider an arbitrary plane in the liquid. For this plane the arrangement of molecules must be such as to allow them to pass to another plane while still being in contact with (a number of) neighbors. If we take a close-packed configuration for this plane, the molecule is sixfold coordinated in this plane and the molecular area is $o_L = \frac{1}{2}3^{1/2}\sigma^2$, where σ is the diameter of the molecule. It is then not unreasonable to suppose that the volume available to the molecule in the liquid is $v_L = {o_L}^{3/2} = 2^{-3/2}3^{3/4}\sigma^3$. If we compare v_L with the volume available in the body-centered cubic (BCC) lattice, $v_{BCC} = 2^2 3^{-3/2}\sigma^3$, we obtain $v_L/v_{BCC} = 1.05$, providing an order of magnitude estimate. Similarly, for the face-centered cubic (FCC) structure, $v_{FCC} = 2^{-1/2}\sigma^3$, and we obtain $v_L/v_{FCC} = 1.14$, in good correspondence with the value for inert gases, indeed crystallizing in the FCC or hexagonal close packed (HCP) structure. This estimate indicates clearly the general trend but, as many other factors play a role, it should not be taken too seriously.

Phases of Matter and their Transitions: Concepts and Principles for Chemists, Physicists, Engineers, and Materials Scientists, First Edition. Gijsbertus de With.

Table 14.1 Volumes V and enthalpies ΔH at the melting point T_{mel} and boiling point T_n.

Liquid	T_{mel} or T_{tri} (K)	V_S ($cm^3\,mol^{-1}$)	V_L ($cm^3\,mol^{-1}$)	$\Delta V/V_S$ (%)	$\Delta_{mel}H$ ($kcal\,mol^{-1}$)	$\Delta_{vap}H$ ($kcal\,mol^{-1}$)	T_n (K)
Ne	24.6	14.03	16.18	15.3	0.080	0.415	27.17
Ar	83.8	24.61	28.14	14.4	0.281	1.500	87.44
Kr	1165.0	29.65	34.13	15.2	0.391	2.315	121.4
Xe	161.4	37.09	42.69	15.1	0.549	3.210	164.4
H_2O	273.1	19.82	18.18	−8.3	1.436	9.705	373.2
CH_4	90.7	30.94	33.63	8.6	0.226	2.228	111.7
CD_4	89.8	29.2	31.7	8.6	—	—	—
CCl_4	250.4	87.9	91.87	4.5	0.577	7.262	348.4
C_2H_4	104.0	39.06	43.63	11.7	0.801	—	—
C_6H_6	278.5	77.28	88.28	11.4	2.348	7.366	353.3
$C_{10}H_8$	353.2	112.2	130.9	11.7	4.550	9.700	491.1
NaF	1265	16.4	21.5	31.1	7.81	213	—
NaCl	1073	29.6	37.7	27.4	7.22	183	—
KF	1133	23.4	30.4	29.8	6.28	190	—
KCl	1043	40.5	48.8	20.5	6.41	165	—
KBr	611	45.0	56.0	24.4	2.84	159	—
Bi	544	21.6	20.8	−4.8	2.51	47.8	—
Cd	594	13.5	14.1	4.7	1.46	27.0	—
Hg	234	14.1	14.7	3.6	0.58	15.5	630.5
Sn	505	16.5	17.0	2.8	1.72	68.0	—
Pb	601	18.5	19.4	4.8	1.22	46.7	1887

Source: Adapted from [1, 2].

Table 14.1 also provides the enthalpies of fusion $\Delta_{mel}H$ and vaporization $\Delta_{vap}H$ for most compounds. These data show that for molecular liquids the ratio $\Delta_{mel}H/\Delta_{vap}H$ typically has a value of 0.10–0.25. For metals this ratio is about 0.025–0.045, while for ionic compounds it is 0.020–0.040. Molecular compounds are in the solid state well packed but in the liquid state they may rotate, which requires some extra space, and this leads to an increased average distance between the molecules. The structure of such a liquid is more open, much less ordered than the corresponding solid, and the work necessary to pull the molecules apart results in a higher value for $\Delta_{mel}H$.

Possibly the simplest fluid–solid transition occurs in hard-sphere systems and we follow the description as given by Lekkerkerker and Tuinier [4]. On the one hand, we know that the fluid state equation of state (EoS) can be described accurately by the Carnahan–Starling EoS (using $\eta = \rho[4\pi(\sigma/2)^3/3]$, Section 8.2.1)

$$\beta P/\rho = (1+\eta+\eta^2-\eta^3)/(1-\eta)^3$$

The corresponding chemical potential (Problem 8.7) is given by

$$\beta\mu = \ln(\rho\Lambda^3) + (8 - 9\eta + 3\eta^2)\eta/(1 - \eta)^3 \quad \text{or, equivalently, by}$$

$$\beta\mu = \ln(\rho\Lambda^3) + (3 - \eta)/(1 - \eta^3) - 3$$

On the other hand, we know that the solid state EoS can be reasonably well described by the Lennard-Jones–Devonshire (LJD) model (Section 9.6.1). For this model, the Helmholtz energy $F = -kT\ln Z$, using the free volume[1] $v_{\mathrm{f}} = 8(v^{1/3} - v_*^{1/3})^3$ with $v_* = 4\pi(\sigma/2)^3/3$ (footnote 3, Section 8.1.1), is given by

$$F = NkT\{\ln(27\Lambda^3/8v_{\mathrm{f}}) - 3\ln[(\eta_{\mathrm{CP}}/\eta) - 1]\}$$

so that the pressure becomes

$$\beta P/\rho = \left[1 - (\eta/\eta_{\mathrm{CP}})^{1/3}\right]^{-1} \cong 3\eta/(1 - \eta/\eta_{\mathrm{CP}})$$

where $\eta_{\mathrm{CP}} = \rho\pi\sigma^3/6 \cong 0.741$ for a close-packed crystal. The chemical potential is then

$$\beta\mu = \ln(\Lambda^3/v^*) + 27/8\eta_{\mathrm{CP}}{}^3 + 3\ln\eta/(1 - \eta/\eta_{\mathrm{CP}}) + 3/(1 - \eta/\eta_{\mathrm{CP}})$$

The coexistence criteria, $P_{\mathrm{F}}(\eta_{\mathrm{F}}) = P_{\mathrm{S}}(\eta_{\mathrm{S}})$ and $\mu_{\mathrm{F}}(\eta_{\mathrm{F}}) = \mu_{\mathrm{S}}(\eta_{\mathrm{S}})$, yield $\eta_{\mathrm{F}} = 0.491$ (0.494) and $\eta_{\mathrm{S}} = 0.541$ (0.545) at a pressure of $\beta Pv^* = 6.01$ (6.12) with in parenthesis the Monte Carlo results of Hoover and Ree [5]. A good agreement is thus observed. The thermodynamic stability is governed by maximum entropy, and the equilibrium configuration of a hard-sphere system is the one that maximizes the entropy. At low density the disordered (fluid) state corresponds to maximum entropy, while at higher density crystalline configurations are optimum in this respect. It is satisfying to observe that this transition is not only observed "in silico," but that, with proper precaution to realize a hard-sphere system, the transition is also observed experimentally [4].

According to Dash [6], the question of how a solid melts can be addressed using three approaches. The first states simply that melting is a discontinuous transition, in which solid and liquid can coexist at a certain pressure for each temperature, and that the process has no intermediate states. The second approach assumes that melting is a nearly continuous transition, as suggested by the small increase in volume, small enthalpy of fusion, and small change in specific heat. According to this approach, if studied with sufficient resolution, it should be possible to discern intermediate stages. The third approach acknowledges that continuous melting is typical for almost all crystals but accepts that interfaces play an important role. Limiting ourselves to approaches 2 and 3, the second is often addressed as *mechanical* or *bulk melting*, while the third is often denoted as *thermodynamic* or *surface-mediated melting*.

Clearly, in principle, any treatment dealing with melting should consider both the solid and the liquid phase. Hoover and Ross [7] considered that for purely inverse power law (pair) potentials $\phi(r) = \varepsilon(\sigma/r)^n$, all properties can be scaled with the characteristic length $(V/N)^{1/3}$ and the relative importance of any configuration in scaled space is always the same at fixed $\rho(\varepsilon/kT)^{3/n}$. This applies to both solids and liquids. In reality attractive interactions are present, and this scaling is not obeyed. However, over a limited temperature–density interval and considering that repulsive interactions dominate, this

1 Note that for liquids, generally, $v_{\mathrm{f}} = 4\pi(v^{1/3} - v^{*1/3})^3/3$ is used.

scaling might still be approximately valid, thereby explaining why one-phase approaches might still yield reasonable results.

14.2 Mechanical or Bulk Melting

Many approaches have been forwarded to rationalize and predict melting temperatures. They were initially based on the most important aspects to consider, that is, the vibrational and lattice instabilities, often using only one of the two. Other contributing terms include: (i) an electronic term, mostly for metals; (ii) an orientational term, mostly for polar molecules; (iii) a configuration term due to the distortion in shape of the molecules, mostly for nonspherical flexible molecules; and (iv) an association term due to the local microstructure of the melt, mostly for polyatomic anions. The sometimes observed melting over a certain temperature interval has been advanced to support the continuous melting hypothesis. However, impurities and polycrystallinity are often present and both broaden the melting transition. Nevertheless, one should realize that *disorder* is present in solids, either intrinsically (point defects) or because of processing or thermal excitations (point defects, dislocations, and orientational disorder). The characteristic energy, associated with each of these defects, governs their increase with temperature and this increase may accelerate the approach to the melting point, although the structure remains crystalline until the transition itself. In fact, most types of these defects have been proposed as being responsible for melting. From this list of aspects, it is probably clear that a thorough discussion of all these aspects requires a book by itself. Hence, we limit ourselves to general considerations. Since the melting curve is thermodynamically defined as the locus of the *PVT* points where the Gibbs energy G of the solid (S) and the liquid (L) are equal, and since the differences in the G_L and G_S are typically small, rather accurate and consistent approximations are required. This has limited the applicability of first principle approaches to only relatively simple solids so far. Attempts to model more complex solids thus require other approaches. In the sequel, we address a few of these approaches.

14.2.1 Vibrational Instability

Probably the most well-known vibrational instability approach is based on a paper by Lindemann [8], although Lindemann himself nowhere suggested that his approach could be used to estimate the melting point T_{mel}. Actually, he attempted to estimate the Einstein frequency of solids ω_E – Einstein published his theory just three years earlier – by assuming that at melting, due to their thermal motion, the molecules would just make contact.[2] Writing for the hard-sphere atom diameter $\sigma = d(1-\delta)$, where d is the atom–atom distance and δ the gap between the atoms expressed as a fraction of d, each atom has to be displaced on average by $\sigma\delta/2$ for contact with another atom. Equating at T_{mel} the kinetic energy to $U_{kin} = \int_0^{\sigma\delta/2} ax\,dx = a\sigma^2\delta^2/8 = ½m\omega_E{}^2u^2$ (a = force constant, $\omega_E = (a/m)^{1/2}$), with the high-temperature limit of the Einstein (harmonic oscillator)

2 Sutherland [9] explicitly introduced the idea that T_{mel} is related to the displacements of the atoms in 1890, but it seems that Lindemann was not aware of this paper.

function $\varepsilon(\omega;T_{\text{mel}}) = \hbar\omega_{\text{E}}\{[\exp(\beta\hbar\omega_{\text{E}})-1]^{-1}+½\} \cong kT_{\text{mel}}$, he obtained the relation $\theta_{\text{E}} = \hbar\omega_{\text{E}}/k = (\hbar/k)\,(8T_{\text{mel}}/m\delta^2\sigma^2)^{1/2}$. Here, m is the mass of the molecule and θ_{E} is the Einstein temperature. For CaF_2 Lindemann estimated $\delta \cong 0.05$ from the dielectric permittivity using the Clausius–Mossotti relation. Using $\nu_{\text{E}} = \omega_{\text{E}}/2\pi = c(T_{\text{mel}}/\mathcal{M}V_{\text{m}}^{2/3})^{1/2}$, with $c \cong 2.06 \times 10^{12}$ an empirical proportionality constant in cgs units, $\mathcal{M}$ the molar mass and V_{m} the molar volume, he further obtained not unreasonable agreement with the experimental "Reststrahlen" frequencies, that is, the optic lattice frequencies at wave vector $\boldsymbol{q} = 0$, for several metals and salts. Later the argument was reversed by others to estimate from T_{mel} from θ_{E} (or the Debye temperature θ_{D} for that matter).

"The concept of direct contact of neighboring atoms at fusion is factitious," as Gilvarry [10] called it, and he was the first, it seems, to assume that the root mean square displacement $\langle u^2\rangle^{1/2}$ of an atom from its equilibrium position reaches a critical fraction ξ of the nearest-neighbor distance $r_0 = (V_{\text{m}}/N_{\text{A}})^{1/3} \equiv \Omega^{1/3}$ of atoms at fusion,[3] that is, $\langle u^2\rangle = \xi^2 r_0{}^2$, with ξ the *Gilvarry ratio*. The mean square displacement for a primitive cubic lattice is given by $\langle u^2\rangle = (3mN)^{-1}\int_0^{\omega_{\max}} \varepsilon(\omega;T)\omega^{-2}g(\omega)\,d\omega$ and using the Debye approximation for the vibrational DoS $g(\omega) = 9N\omega^2/\omega_{\text{D}}{}^3$ and the high-temperature limit of $\varepsilon(\omega;T) \cong kT$, one obtains $\langle u^2\rangle = 3kT/m\omega_{\text{D}}{}^2 = 3\hbar^2T/mk\theta_{\text{D}}{}^2$, where ω_{D} and θ_{D} are the Debye frequency and temperature, respectively. Hence, $T_{\text{mel}} = \xi^2(k/3\hbar^2)m\theta_{\text{D}}{}^2r_0{}^2$. The values for ξ as obtained by Gilvarry for 10 metals are indeed approximately constant and are given by $\xi_{\text{FCC}} \cong 0.11$, $\xi_{\text{BCC}} \cong 0.13$ and $\xi_{\text{HCP}} \cong 0.09$. Cho [11] refined these values by considering 54 metals, leading to[4] $\xi_{\text{FCC}} \cong 0.096$, $\xi_{\text{BCC}} \cong 0.121$, and $\xi_{\text{HCP}} \cong 0.069$.

The reformulation of the Lindemann rule by Gilvarry employs only harmonic terms in the potential energy (while anharmonic terms, and more importantly, bond breaking are involved). It also neglects the possible effect of lattice defects near the melting point and, in the usual treatment, restricts the discussion to monatomic solids. Moreover, melting is linked to individual atomic properties (while it is a cooperative process) and the solid alone (while it should be linked to both liquid and solid).

Attempting to put Lindeman's rule on a firm statistical-mechanical basis, Ross [7, 12] argued that, when the melting transition is viewed at an atomistic level, we always see the same picture when properly scaled. Postulating that for all points along the melting curve the solid always occupies the same fraction of configuration space, which implies scaling the configurational partition function Q by using the reduced coordinates $\mathbf{x} = \mathbf{r}/V^{1/3}$ with V the volume of the system and $\mathbf{r}$ the coordinates of the N molecules so that $Q = V^NQ^*$, we have $Q^*(T_{\text{mel}},V_{\text{mel}}) = const.$ As a general theory of liquids was (is) lacking, the LJD model (Section 9.6.1) was used, in which $Q = v_{\text{f}}{}^N\exp[-\beta\Phi(\mathbf{0})]$ with free volume $v_{\text{f}} = \int\exp(-\beta\Delta\Phi)d\mathbf{r}$, $\Delta\Phi = \Phi(\mathbf{r})-\Phi(\mathbf{0})$ and $\beta = 1/kT$, leading to $v_{\text{f}}{}^*(T_{\text{mel}},V_{\text{mel}}) = const.$ Using for $\Phi(\mathbf{r})$ the LJ potential in good agreement with the thermodynamic data is shown for Ar. Reducing $\Delta\Phi$ to a harmonic oscillator potential, $\Delta\Phi = ½ar^2 = ½aV^{2/3}x^2 = ½m\omega^2V^{2/3}x^2$, the requirement $v_{\text{f}}{}^* = const.$ leads right away to $\beta\Delta\Phi = const.$ or at melting to $T_{\text{mel}} \sim m\omega^2V_{\text{mel}}{}^{2/3}/k$, which is Lindemann's rule. Obviously, this approach does not incorporate two-phase considerations and, in the case of the harmonic oscillator potential, anharmonicity.

3 In this chapter, we use Ω instead of the previously used v_0 to avoid the use of double subscripts.
4 Be aware that a direct comparison for various treatments for ξ is complicated by the use of $\langle u^2\rangle$ or $\langle \boldsymbol{uu}^{\text{T}}\rangle$ (Section 10.7) and $\Omega = v_0 = r_0{}^3$ or $v^* = v_0/\gamma$ (footnote 4, Chapter 8).

In an approach, taking both solid and liquid states as well as anharmonicity into account, Lawson [13], examining 74 distinct elements,[5] also rationalized the Lindemann rule. The starting point is the entropy difference $\Delta S = S_L - S_S$ between liquid (L) and solid (S). For the liquid with molecules with mass m, a Sackur–Tetrode-like expression S_{ST} (Eq. 7.19) as derived from $F = -kT\ln Z$ with $Z = z^N$ instead of $Z = z^N/N!$ and the volume taken as the atomic volume $\Omega = V_m/N_A$, was used together with a correlation term S_{cor} (for the perfect gas $S_{cor} = -k\ln(1/N!)$. Hence, the entropy S_L per molecule in the liquid phase is given by

$$S_L = S_{ST} + S_{cor} = k\ \ln(e\Omega/\Lambda^3) + S_{cor} \quad \text{with} \quad \Lambda \equiv (h^2/2\pi mkT)^{1/2}$$

For the solid, the Wallace expression [14] $S = 3k\ln(eT/\theta_0)$ with θ_0 the characteristic entropy temperature [15] was used. The parameter θ_0 corresponds with the zeroth moment or, equivalently, the geometric mean of the phonon frequencies, $k\theta_0 = \hbar\varpi$. Since θ_0 values are hard to get, Lawson inserted, based on Debye theory and backed-up by a strong correlation, the expression $\theta_0 = e^{-1/3}\theta_D$ with θ_D the Debye temperature as obtained from either low temperature heat capacity or elastic behavior data [15]. For convenience of getting data, an empirical relation, namely $1.3k\theta_D = k\theta_{ela} = \hbar(K_T\Omega/m)^{1/2}(6\pi^2/\Omega)^{1/3}$ with θ_{ela} the Debye temperature as derived from room temperature elastic data, was introduced. Note that using the bulk modulus K_T the expression for θ_{ela} ignores the shear modes. They were left out since data for these modes are more difficult to obtain, particularly at high temperatures (for the same reason, the temperature dependence of K_T and Ω was neglected). Hence, the entropy per molecule for the solid is given by

$$S_S = 3k\ \ln(1.3e^{4/3}T/\theta_{ela})$$

Finally, introducing the effect of anharmonicity via the thermodynamic relation $C_P - C_V = \alpha^2 K_T\Omega T$, the corresponding high-temperature entropy term $9\gamma^2k^2T/K_T\Omega$ was added, where $\gamma = \alpha K_T\Omega/C_V$ is the thermodynamic Grüneisen parameter and α is the volume expansivity. All together this leads, via $\Delta S - S_{cor} = S_{ST} - S_S - S_{anh}$, to

$$\Delta S - S_{cor} = k\ \ln\frac{e\Omega}{\Lambda^3} - 3k\ \ln\frac{e^{4/3}T}{\theta_{ela}/1.3} - \frac{9\gamma^2k^2T}{K_T\Omega} \tag{14.1}$$

Lawson divided the structures examined in FCC, BCC, HCP, and in more complex, "open" structures. Fitting $-S_{cor} = S_{ST} - S_S - S_{anh} - \Delta S$ by $-S_{cor} = a + b\gamma$, given the experimental data for ΔS, shows that a strong correlation between $(\Delta S - S_{cor})$ and γ is obtained. As also a strong correlation between ΔS and γ is observed, S_{cor} can be estimated. Using these fits T_{mel} can be calculated from Eq. (14.1). Figure 14.1a shows the comparison with the experimental data, indicating good agreement in view of the approximations made. As for many elements $S_{anh} \ll S_{cor}$, generally T_{mel} is determined by a balance between S_S, the Debye entropy for the solid, and S_{cor}, the correlation entropy of the liquid. The dimensionless ratio $kT_{mel}/K_T\Omega$ plotted versus γ also shows a good correlation (Figure 14.1b).

The Gilvarry ratio ξ was calculated as $\xi = \langle u^2\rangle^{1/2}/2r^*$ (note the two) with $\langle u^2\rangle = 3\hbar^2T_{mel}/mk\theta_{ela}{}^2$ and $r^* = (3\Omega/4\pi)^{1/3}$, leading to $\xi = (0.513kT_{mel}/K_T\Omega)^{1/2}$. Inserting numbers, one obtains approximately $\xi \cong 0.1 - 0.01\gamma$ or $\xi \cong 0.08$ using $\gamma \cong 2$, typical for FCC, BCC, and HCP structures. However, this estimate seems to fail for the open structures. As the considerations are essentially based on Debye's model, this might have been expected. In conclusion, Lawson rationalized Lindemann's rule and showed that it is well obeyed

5 The paper by Lawson also provides a useful tabulation of all the data used.

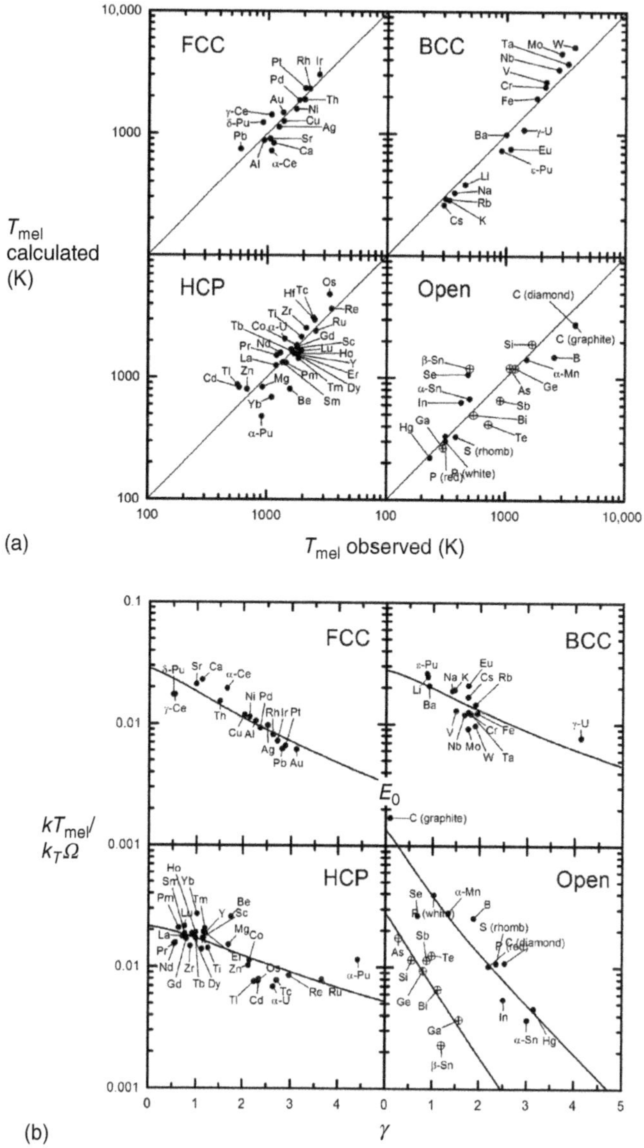

Figure 14.1 Melting characteristics for elements. (a) Melting temperature T_{mel} calculated versus T_{mel} observed. (b) The ratio $kT_{mel}/K_T\Omega$ versus Grüneisen parameter γ. Actually, Lawson divided the open structures into two categories. Hence, the panel labeled "open" shows two correlations. Source: Lawson [13]/Taylor & Francis.

for simple structures but much less well for less symmetric structures. More generally, since all versions of Lindemann's rule contains many approximations, one can hope at best only for approximate agreement with the experiment, and this has been amply shown by various authors.[6] Nevertheless, in the absence of any experimental data, Lindemann's rule may be the best bet.

6 Note, however, that some authors state "good" agreement for complex solids [16] when appropriate parameters are taken, while others [17] deny such agreement. Obviously, considering the Lindemann rule

14.2.2 Lattice Instability

Another approach, now largely abandoned, is based on the lattice (in)stability of solids. Increasing the temperature of a solid will reduce the elastic constants and it was postulated by Born [18] that at melting the shear elastic constant $G = C_{44}$ would vanish. He calculated an explicit expression for the Gibbs energy of a BCC crystal using (approximate) quasi-harmonic lattice dynamics and determined the variation of the elastic moduli with T/θ_{D}. Moreover, he derived the stability conditions for cubic crystals at zero pressure. By adding external loading [19], in the hydrostatic case by adding $-P\mathbb{I}$ with P the pressure and $\mathbb{I}$ the unit tensor, the stability conditions are

$$C_{11} + 2C_{12} + P > 0, \quad C_{11} - C_{12} - 2P > 0, \quad C_{44} - P > 0$$

With $P = 0$ Born's theory also leads to an expression similar to Lindemann's expression $\nu_{\mathrm{D}} \cong c(T_{\mathrm{mel}}/MV_{\mathrm{m}}{}^{2/3})^{1/2}$ with $c \cong 1.62 \times 10^{12}$ in cgs units. An obvious drawback is that the theory is a single-phase theory that contains no distinct description of the melt, while both phases must be involved, and thus fails to account for the discontinuous, first-order character of melting. Because it is an essentially homogeneous theory, it does not explain the occurrence of superheating, the metastability of the melt, and the heterogeneous nucleation and growth features of the melting process. Observations show that none of the shear moduli is zero or near zero in the solid state at the melting point [20]. In an attempt to remedy the situation, Tallon et al. [21] used the observation that the shear moduli of a solid seem to extrapolate to zero at V_{L}. Hence, they proposed that melting occurs when the solid can transform isothermally to a state of zero shear modulus. Obviously, although this interpretation introduces the required discontinuous nature, it does not apply to crystals with $\Delta V_{\mathrm{mel}} < 0$. Wang et al. [22] have shown with molecular dynamics (MD) simulations using LJ potentials for Au that indeed the elastic constants vanish upon increasing the volume, but that it is the constant $G' = C_{11} - C_{12}$ that becomes zero first. An extensive discussion of instability theory has been provided by Boyer [23].

14.2.3 Vacancies

Frenkel [24] recognized that for S–L equilibrium, crystals must contain thermally accessible defects. The only thermally accessible defects are vacancies, interstitials, and interstitialcies. From an exposition on fusion [2, 24], it became clear though that the normal increase in the amount of vacancies in a crystal with temperature cannot account for the increase in volume upon melting. If we have N atoms and $N\alpha$ holes with $\alpha \ll 1$ on a lattice with $N(1 + \alpha)$ sites, the number of possible configurations is

$$W = [N(1 + \alpha)]!/N!(N\alpha)! \quad \text{or} \quad W = [(1 + \alpha)^{1+\alpha}/\alpha^{\alpha}]^N$$

using Stirling's approximation $N! = (N/\mathrm{e})^N$. Hence, for the entropy we have

$$\Delta S = k \ln W = Nk[(1 + \alpha)\ln(1 + \alpha) - \alpha \ln \alpha]$$

The ratio $\Delta_{\mathrm{mel}}H/\Delta_{\mathrm{vap}}H$ for metals is typically 0.04 and this suggest that $\alpha \approx 0.04$, so that $\Delta S \approx 0.17Nk = 0.33\,\mathrm{cal\,K^{-1}\,mol^{-1}}$. A typical experimental value for entropy of fusion is

as a scaling rule (or as an example of the principle of corresponding states [PoCS]), this depends on which of the aspects not dealt with remain constant over a series of compounds considered.

$\Delta_{mel}S \approx 2\,\text{cal K}^{-1}\,\text{mol}^{-1}$, so that a vacancy fraction $\alpha \approx 0.04$ cannot account for this. To account for such a value of $\Delta_{mel}S$, a value of $\alpha \approx 1$ would be required, leading to a liquid metal density of about half that of solid metal. Similarly, for molecular compounds, $\Delta_{mel}H/\Delta_{vap}H \approx 0.1$, and hence $\alpha \approx 0.1$, corresponding $\Delta_{mel}S \approx 0.34Nk = 0.66\,\text{cal K}^{-1}\,\text{mol}^{-1}$, while experimentally $\Delta_{mel}S \approx 5\,\text{cal K}^{-1}\,\text{mol}^{-1}$ is observed. This leads again to rather large changes in density. Since these changes are not observed, other effects, to which we come later, must come into play. A more fundamental objection is that the concept of vacancies in a lattice as a model for a liquid implies long-range order in a liquid, which is, however, not observed.

Nevertheless, the vacancy theory was elaborated by Gorecki [25], who, focusing on metals and ignoring the entropy argument, assumed that the mechanism of melting is connected to the introduction of vacancies in the solid phase. It was shown that for a range of FCC, HCP, and BCC metals the vacancy concentration c, defined by $c = N\alpha/(N + N\alpha)$, is given by

$$c = \exp[(S_{vac}/k) - (E_{vac}/kT)]$$

where $S_{vac}/k = 4.1$ and $E_{vac}/T_{mel} = 80.4\,\text{J K}^{-1}$ are the empirically determined entropy and enthalpy (energy) of vacancy formation, respectively. Hence, one easily calculates that $c_{vac}(T_{mel}) = 0.0037$, in good agreement with the experimental data for all the metals studied. Such a low bulk concentration of vacancies will not lead to lattice instability. A further correlation of the enthalpy of melting $\Delta_{mel}H$ with E_{vac} yielded $\Delta_{mel}H/E_{vac} = 0.127$, sufficient for a vacancy concentration increase $\Delta c = 0.13$ upon melting. A similar correlation using the electrical resistivity yielded $\Delta c = 0.07$. Experimentally, it appears that the volume increase upon melting is $\Delta V/V_S \cong 0.05$ for FCC and HCP metals and $\Delta V/V_S \cong 0.025$ for BCC metals. As it is also known that for FFC and HCP metals the mean atomic volume for a defect lattice $\Omega_{vac} \cong \frac{1}{2}\Omega$ [26], the model implies $\Delta c \cong 2\Delta V/Vs \cong 0.10$. Overall, Δc thus appears to be $\Delta c \cong 0.10$.

Using a factor ζ due to lattice relaxation around a vacancy, $\Omega_{vac} = \Omega + \zeta c\Omega$. Considering that $\Omega = 4\pi r^3/3$ and $\Omega_{vac} = 4\pi(r - \Delta r)^3/3$, one obtains the relation $c = \zeta^{-1}[(\Omega_{vac}/\Omega) - 1] = \zeta^{-1}\{[(r - \Delta r)/r]^3 - 1\}$. Furthermore, the empirical correlation $x \equiv \Delta r/r = 3.0E_{vac}/EV_m$ with E Young's modulus, and $x_{FCC} = 0.24$, $x_{HCP} = 0.21$, and $x_{BCC} = 0.37$ was used. Hence, $\Omega_{vac} = (1 - 3x + 3x^2 - x^3)\Omega$, which leads to $\zeta_{FCC} = -0.56$, $\zeta_{HCP} = -0.51$ and $\zeta_{BCC} = -0.75$, for FCC and HCP consistent with the experimental estimate $\Delta V/V_S \cong 0.05$. Interpreting for the liquid $r - \Delta r$ as the radius of the first coordination shell leads to, using the available experimental data, $\Delta c \cong 0.09$, in good agreement with the aforementioned estimates. Note that, although one might expect upon melting an increase in radius of the first coordination sphere, actually a small decrease occurs. The model was further used to calculate dT_{mel}/dP, with results also in fair agreement with the experiment. In conclusion, the approach is quite successful describing melting (of metals), but the reason why $\Delta c \cong 0.1$ has not been made clear.

14.2.4 Interstitials

Another early attempt [27], essentially employing the interstitial holes in a lattice, uses in its simplest form for both the liquid and the solid state the LJD theory (Section 9.6.1). In that theory the FCC lattice was used for the liquid using the Einstein approximation for the vibrations and we derived an approximate partition function Z reading, using $\beta = 1/kT$ and $\Lambda = (h^2/2\pi mkT)^{1/2}$,

$$Z = f^N \equiv [\Lambda^{-3} v_\mathrm{f} \exp(-\beta\Phi/N)]^N \quad \text{with} \quad v_\mathrm{f} = \int \exp[-\beta\Delta\phi(r)]\, 4\pi r^2 \mathrm{d}r$$

Here, f is the single-particle partition function,[7] $\Phi = \frac{1}{2}Nz\phi_0$ with ϕ_0 the potential at the center $r = 0$ of the cell, z the coordination number, $\Delta\phi(r) = \phi(r) - \phi_0$ the potential at position r within the cell with respect to its center, and v_f the free volume. Further, recall that the FCC lattice contains an equal number of lattice sites (here labeled as α-sites) and interstitial, or hole, sites (here labeled as β-sites), the latter also being arranged in an FCC manner. It is now assumed that in the solid state the molecules occupy the α-sites only, while during melting the β-sites become available. For the fusion process, we then have an order–disorder description at hand, for which we can use the conventional zeroth or first approximation (Sections C.2 and C.3). As the zeroth approximation yields essentially the same results as the first and is much simpler, it is used here.

If the relative occupancy of α-sites is given by $X = N_\alpha/N$, $X = 1$ represents the ordered state, while $X = \frac{1}{2}$ denotes complete disorder. In the zeroth approximation in a disordered state there will be XN molecules on α-sites and the α–α interaction becomes $\frac{1}{2}Xz\phi_0 \cdot XN$ (instead of $\frac{1}{2}z\phi_0 \cdot N$ for the ordered, i.e. solid state). Since the α- and β-lattices are similar, the β–β interaction becomes $\frac{1}{2}(1 - X)z\phi_0 \cdot (1 - X)N$, while for the α–β interaction the result is $(1 - X)z\chi_0 \cdot XN$ with χ_0 the difference in energy between α- and β-sites. The total energy therefore becomes, using $\psi = \chi_0 - \phi_0$,

$$\tfrac{1}{2}zN\phi_0(1 - 2X + 2X^2) + zN\chi_0 X(1 - X) \quad \text{or} \quad \Phi + zN\psi X(1 - X)$$

Next, we need the number of ways $W(X)$ of choosing N_α molecules over the α-sites and N_β molecules over the β-sites, which is given by

$$W(X) = \frac{N!}{(N - N_\alpha)!N_\alpha!}\frac{N!}{(N - N_\beta)!N_\beta!} = \left\{\frac{N!}{(NX)![N(1 - X)]!}\right\}^2$$

Finally, we assume that the motion of a molecule on an α-site is the same as on a β-site. In that case, the partition function becomes

$$Z = f^N W(X) \exp[-\beta zN\psi X(1 - X)]$$

Minimizing the Helmholtz energy $F = -kT\ln Z$ with respect to X, one obtains, after some calculation (Appendix C),

$$2X_\mathrm{m} - 1 = \tanh[\tfrac{1}{4}\beta z\psi(2X_\mathrm{m} - 1)] \tag{14.2}$$

This equation is always satisfied when $X_\mathrm{m} = \frac{1}{2}$, but when $\beta z\psi/4 > 1$, there may be another root larger than $\frac{1}{2}$, which, if it exists, minimizes F. Otherwise $X_\mathrm{m} = \frac{1}{2}$ is the solution. For $\beta z\psi/4 \gg 1$, $X_\mathrm{m} \cong 1$ and the order is nearly perfect; in fact, it describes the solid with a few interstitials. When $\beta z\psi/4$ decreases, the disorder increases and when $\beta z\psi/4 \le 1$, $X_\mathrm{m} = \frac{1}{2}$ and there is complete disorder. Substituting the value of X_m in Z, we obtain

$$Z = Z_1 \cdot Z_2 \equiv f^N \cdot W(X) \ \exp[-\beta zN\psi X_\mathrm{m}(1 - X_\mathrm{m})]$$

7 Here f is used (instead of the conventional z) to avoid confusion with the coordination number z.

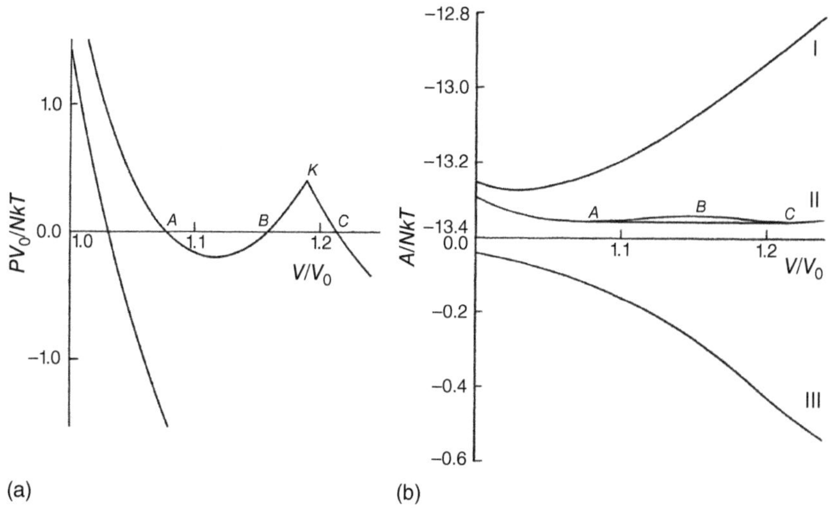

Figure 14.2 The LJD fusion model for Ar at $P = 0$. (a) The pressure as a function of volume, where the lower curve represents the pressure for the ordered state and the upper curve the total pressure. (b) The Helmholtz energy F for the ordered state (I), the disorder contribution (III), and the total (II). The points A, B, and C correspond to the same points as in (a). Source: Lennard-Jone and Devonshire [27]/The Royal Society.

so that the Helmholtz energy becomes $F = F_1 + F_2$ where F_1 is the Helmholtz energy for the ordered state and F_2 is the additional Helmholtz energy due to disorder (Figure 14.2b). For the pressure $P = -\partial F/\partial V$, we obtain $P = P_1 + P_2$, with

$$P_1 = NkT(\partial \ln f/\partial V) \quad \text{and} \quad P_2 = -zNX_\text{m}(1 - X_\text{m})(\partial\psi/\partial V)$$

If we further assume that only repulsion contributes, we have $\psi = \psi_0(r_0/r)^{12} = \psi_0(V_0/V)^4$, and P can be calculated as function of T, since for a given volume V, Φ and ψ are known, while X_m can be calculated from Eq. (14.2). This results in a curve as shown in Figure 14.2a (the peak K is sharp due to the use of the zeroth approximation). The pressure P is zero for three values of V/V_0. The smallest one at $V/V_0 = 1.078$ (A) corresponds to the volume of the solid at melting, while the largest one at $V/V_0 = 1.214$ (C) corresponds to the volume of the liquid. Using the equal-area rule, ψ_0 can be determined in such a way that the melting pressure for a particular compound is zero. This has been done for Ar ($T_\text{mel} = 83.8$ K) and this leads to $\Delta V = V_\text{L} - V_\text{S} = 1.214V_0 - 1.078V_0 = 0.136V_0$. For the entropy the result is $S = 1.70\,k$ and both values are in reasonable agreement with experiment. It was found that for this state, $kT/\varepsilon \cong 0.7$, where ε is the depth of the LJ potential used. The melting point T_mel is thus $T_\text{mel} \cong 0.7\varepsilon/k$. Using this rule, reasonably good agreement for T_mel was obtained for Ne, N_2, CO, CH_4, and H_2. Using a slightly different formulation [28], agreement between theory and experiment for the melting curves is very good for the rare gases, and fair for NH_3, N_2, CO_2, and CH_4.

Since molecules generally have not only translational but also orientational degrees of freedom (DoFs), any theory of melting of molecular crystals should take these DoFs into account, as done by People and Karasz [29] in the framework of the interstitial approach.

This approach takes properly into account that both solid and liquid should be involved and results in quite reasonable values for the various properties. The model also provides an explanation for Lindemann's rule. However, it predicts a critical point for the solid–liquid transition, similar as for the liquid–vapor transition, which has never been observed experimentally, but using a compressible lattice model this can be avoided [30]. The isobars $S(T)$ display a monotonic rise instead of being S shaped, as required for melting being a discontinuous transition. Finally, as discussed in Chapter 9, the liquid state is poorly described, as insufficient disorder is introduced.

14.2.5 Dislocations

Several melting theories are based on dislocations, for which the basic idea is that a solid becomes a liquid when the solid is saturated with dislocations. Such dislocation theories of melting (DTMs) [31] have appealing aspects. First, although dislocations have an excess energy over the ideal lattice, the Gibbs energy of a solid containing a dense network of dislocations and that of a liquid remain comparable. This leads to a low enthalpy of fusion, comparable with the core energy of dislocations. Second, the fluidity can be ascribed to the mobility of a dense network of dislocations [32]. Pairs of opposite sign dislocations (loops) can be created thermally and, most importantly, the presence of dislocations reduces the formation energy of additional dislocations, so that a cooperative effect creating an avalanche of dislocations can be envisaged to occur. When the Gibbs energy difference $\Delta G = G_L - G_S = 0$, a discontinuous transition occurs. As both G_L and G_S are involved, one can consider DTMs as two-phase theories. There is some evidence for dislocation-like structures in liquids [33].

The birth of DTMs can be considered to be a paper by Mott, rapidly followed by others [31, 34]. Here, we consider mainly a variant originated by Ninomiya [35] and in slightly modified form applied by Poirier [36]. In this model, the energy U of a lattice with dislocations contains essentially three terms. First, the energy of the core of the dislocations ΔU_{cor}. This is usually considered to be a constant per unit dislocation length for which the enthalpy of melting appears to be a rather good estimate [37]. Measured in units $\mu b^2/4\pi$ with μ the shear modulus and b the Burgers vector length, the core energy per unit length can be expressed as $u_{cor} = c\mu b^2/4\pi$ with $c \cong 1$. The total core energy ΔU_{cor} is thus $\rho V_m u_{cor}$ with the dislocation density $\rho = L/V$, where L is the length of dislocations per unit volume V of crystal and V_m is the molar volume.

Second, the elastic energy of the dislocations ΔU_{ela}. Here, we use the logarithmic energy expression $u_{ela} = (\alpha\mu b^2/4\pi)\ln(R/r_0)$ for the energy unit length (Section 10.10.2), as we need an explicit dependence on ρ. The constant $\alpha = 1$ for screw dislocations and $\alpha = 1/(1-\nu)$ for edge dislocations, ν is Poisson's ratio, R is the average distance between dislocations, which is estimated as[8] $(\pi\rho)^{-1/2}$, and r_0 is the core radius. The total elastic energy is thus $\Delta U_{ela} = \rho V_m u_{ela}$. However, the introduction of dislocations leads to a volume strain $\varepsilon(\rho)$, so that $V = V_0(1+\varepsilon)$, thereby reducing μ and increasing b. Using $\mathrm{d}\ln\mu/\mathrm{d}\ln V = -(2\gamma + ⅓)$ as given by Slater [2], $\mu = \mu_0[1-(2\gamma + ⅓)\varepsilon]$, where γ is the Grüneisen parameter, while $b = b_0(1+\varepsilon)^{1/3}$.

Third, the energy due to the expansion of the material as a whole is given by $\Delta U_{exp} = ½N_A\Omega K\varepsilon^2$, where K is the bulk modulus and $\Omega = V_m/N_A$ the atomic volume.

8 The average stressed cross-section along a dislocation is ρ^{-1}. As the core is considered to be circular with area $\pi r_0^{\,2}$, and u_{ela} should become zero when $R = r_0$, the proper estimate is $R = (\pi\rho)^{-1/2}$.

The total energy change is thus $\Delta U = \Delta U_{cor} + \Delta U_{ela} + \Delta U_{exp}$ or

$$\Delta U = \rho V_m(4\pi)^{-1}c\mu b^2 + \rho V_m(4\pi)^{-1}\alpha\mu b^2 \ln(\rho^{-1/2}/\pi^{1/2}r_0) + \tfrac{1}{2}\rho\Omega K\varepsilon^2 \tag{14.3}$$

The value of ε_{dis}, corresponding to the dislocation density ρ, is obtained by minimizing ΔU with respect to ε. The result is, taking care of the strain dependence of V, b, and μ, and neglecting small terms [35],

$$\varepsilon_{dis} \cong c_{dis}\mu_0 b_0^2(\gamma - \tfrac{1}{3})\rho/2\pi K \cong \mu_0 b_0^3(\gamma - \tfrac{1}{3})/6\pi K\Omega \tag{14.4}$$

where $c_{dis} = c + \alpha\ln(\rho^{-1/2}/\pi^{1/2}r_0)$. The last step can be made as saturation of dislocations is given by the condition $\pi^{1/2}r_0 \cong \rho^{-1/2}$ and $c \cong 1$, so that $c_{dis} \cong 1$, while the maximum dislocation density is estimated as $\rho_{max} \cong b/3\Omega$ [38]. The factor $b_0{}^3/\Omega$ depends only on the lattice type and is $\sqrt{2}$, $3\sqrt{3}/4$, and $4/\sqrt{3}$ for FCC, BCC, and (ideal) HCP lattices, respectively (footnote 7, Chapter 9). The volume change upon melting becomes $\Delta V = N_A\Omega\varepsilon_{dis} = V_m\varepsilon_{dis}$, while the internal energy change ΔU, obtained by substituting Eq. (14.4) in Eq. (14.3), reads

$$\Delta U = V_m K\varepsilon_{dis}/2(\gamma - \tfrac{1}{3}) - \tfrac{1}{2}V_m K\varepsilon_{dis}^2$$

The entropy of the dislocations ΔS contains a configurational and vibrational contribution. The first is generally negligible (Section 10.10.2), while the vibrational entropy contains a term due to the decrease in lattice frequencies brought about by the dilatation ε_{dis} and a term due to the vibration of the dislocation lines. When a crystal, with average (or Einstein) vibration frequency ω contains dislocations, the vibrations associated with the dislocations change frequency to ω'. Hence, in the high-temperature approximation to vibrations, we have, using $f = \Omega\rho/b$,

$$\Delta S_{vib} = kN(3-f)\ln\beta\hbar\omega + kNf\ln\beta\hbar\omega' = 3Nk\ln\beta\hbar\omega + Nkf\ln\lambda$$

where $\lambda = \omega'/\omega$, estimated by Ninomiya [35] to be $\cong$0.13, so that $\ln\lambda \cong -2.0$. Moreover, due to the dilatation, the lattice frequency itself changes and using the Grüneisen relation, $\mathrm{d}\ln\omega/\mathrm{d}\ln V = -\gamma$ with γ the Grüneisen parameter, we obtain $\omega = \omega_0(1-\gamma\varepsilon)$. From the Helmholtz energy $\Delta F = \Delta U - T\Delta S$, the total dilatation is given by $\partial\Delta F/\partial\varepsilon = 0$ and this results in $\varepsilon = (3k\gamma/K\Omega)T + \varepsilon_{dis}$. Inserting ε in the expression for ΔF, the entropy $\Delta S = -\partial\Delta F/\partial T$ becomes

$$\Delta S = N_A k(3\varepsilon_{dis}\gamma - f_{max}\ln\lambda) = N_A k(3\varepsilon_{dis}\gamma + 2f_{max})$$

The melting temperature T_{mel} is obtained by setting $\rho = \rho_{max}$ and taking the Helmholtz energy $\Delta F = \Delta U - T\Delta S = 0$ (Figure 14.3a). The final expression becomes

$$T_{mel} = \frac{\Omega K}{2k}\frac{\varepsilon_{max}[1 - \varepsilon_{max}(\gamma - \tfrac{1}{3})]}{(\gamma - \tfrac{1}{3})(3\varepsilon_{max}\gamma + 2f_{max})}$$

From the Clausius–Clapeyron equation $\mathrm{d}T/\mathrm{d}P = \Delta V/\Delta S$, we also have

$$\frac{\mathrm{d}T_{mel}}{\mathrm{d}P} = \frac{\Delta V}{\Delta S} = \frac{\Omega}{2k}\frac{\varepsilon_{max}}{3\varepsilon_{max}\gamma + 2f_{max}} \quad \text{or} \quad \frac{\mathrm{d}\ln T_{mel}}{\mathrm{d}P} = \frac{2(\gamma - \tfrac{1}{3})}{K[1 - \varepsilon_{max}(\gamma - \tfrac{1}{3})]} \tag{14.5}$$

Equation (14.5) differs from the differential Lindemann expression $\mathrm{d}\ln T_{mel}/\mathrm{d}P = 2(\gamma - \tfrac{1}{3})/K$ only by the factor $[1 - \varepsilon_{max}(\gamma - \tfrac{1}{3})]^{-1}$, which is somewhat larger than 1 for $\gamma > \tfrac{1}{3}$. The latter condition is always true and thus the DTM also provides some justification for Lindemann's rule. Estimates for c, ρ_{max}, and λ already being made, we need

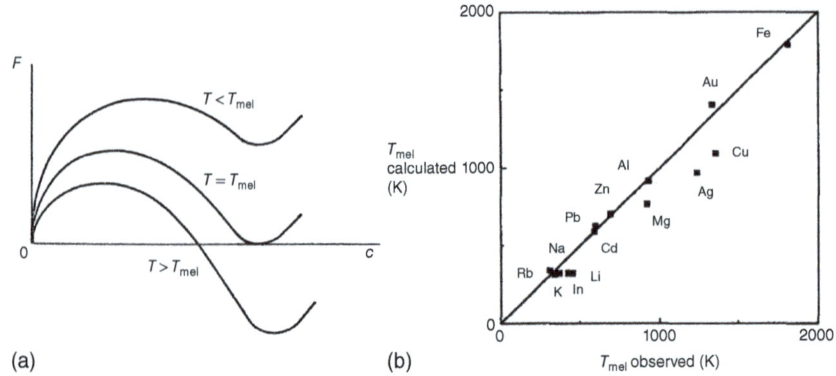

Figure 14.3 DTM illustrated. (a) Helmholtz energy F as a function of dislocation concentration c as used in [35, 36]. Source: Adapted from [35, 36]. (b) The correlation between calculated and experimental T_{mel} according to [36]. Source: Adapted from Poirier [36].

only estimates for the material properties μ, K, and γ, and knowing the crystal structure, T_{mel} can be calculated. For Fe Poirier [36] obtained $T_{mel} = 1786$ K, while experimentally $T_{mel} = 1808$ K is obtained. A good agreement with experiment is thus observed.

It appeared, however, that the agreement for Fe was a somewhat lucky shot as the results for other metals differ significantly from the experimental data, but taking the experimental values for ΔV and ΔS, Poirier [36] obtained good agreement for the 14 metals studied (Figure 14.3b). A recent revival of the DTM by Burakovsky et al. [39], using a slightly different approach, the melting temperatures and latent heats are estimated for about 70 elements with an accuracy of about 20%.

14.2.6 Interstitialcies

A relatively recent "defect" approach is based on the interstitialcy (or a dumb-bell interstitial in which two atoms try to occupy the same lattice site), as studied primarily for FCC metals, in particular Cu by Granato [40]. It is argued that dislocations have a too high energy per atomic length to be generated thermally. As indicated in Section 14.2.3, thermal accessibility of defects involved in melting is deemed necessary. While vacancies are thermally accessible, they result in a too low value for the melting entropy S_{mel}. Interstitialcies are also thermally accessible but have a strong coupling to external shear stress, low-frequency resonance modes, and an extended linear string-like character (Figure 14.4a), leading to large entropy effects. Based on these considerations, a liquid is considered to be a crystal containing a certain percentage of interstitialcies in thermal equilibrium, for which an interstitialcy concentration-dependent Gibbs energy expression was derived.

It is supposed that the volume V_0, the bulk modulus K_0, shear modulus μ_0, evaluated for the static lattice, and their pressure derivatives K' and μ' are given, and that the Helmholtz energy $F_0(V,\varepsilon)$ with ε the shear strain can be expressed in terms of these parameters. Accepting the Einstein model for vibrations at high temperatures with frequency ω_E, the Helmholtz energy F at finite temperature for N atoms is given by $F = F_0 + 3NkT\ln(\beta\hbar\omega_E)$ where k, β, $\hbar$, and ω_E have their usual meaning. Introducing n defects, we must add the

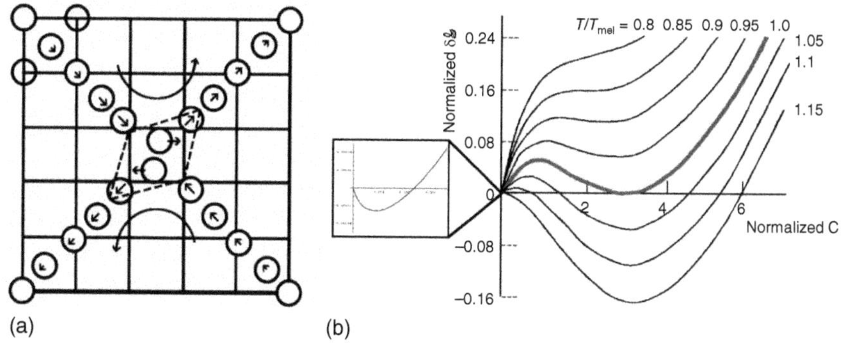

Figure 14.4 Interstitialcy theory illustrated. (a) Interstitialcy configuration in a [100] direction in an FCC lattice. Arrows indicate displacements of atoms along close-packed ⟨110⟩ directions on a {100} plane for an applied shear stress and the so-called E_g resonance mode. (b) Normalized Gibbs energy as function of normalized interstitialcy concentration. Source: Adapted from Granato [40].

work for creating these defects ΔF_{for}, the change in vibrational behavior ΔF_{vib}, and a configurational term ΔF_{con}.

The first contribution is assumed to be $\text{d}f_{\text{for}} \equiv \text{d}F_{\text{for}}/N = (\alpha_\mu \mu\Omega + \alpha_K K\Omega)\text{d}c$ with Ω the atomic volume and $c = n/N$ the concentration of interstitialcies, or, equivalently, $f_{\text{for}} = \int(\alpha_\mu \mu\Omega + \alpha_\mu K\Omega)\text{d}c$. It is expected, based on simulations, that for the constants α_μ and α_K we have $\alpha_K/\alpha_\mu \ll 1$, so that the work of formation is mainly due to shear deformation (a fit for Cu yields $\alpha_\mu \cong 0.9$ and $\alpha_K \cong 0.03$). The second contribution is due to the five low-frequency resonance modes with frequency ω_R and the six high-frequency local modes with frequency ω_L for each interstitialcy, as described by Dederichs et al. [41]. This leads to $\text{d}f_{\text{vib}} = -kTc[5\ln(\omega_E/\omega_R) + 6\ln(\omega_E/\omega_L)]$. The third contribution is the configurational entropy and reads $-TS_{\text{con}} = -kTc[1 + \ln(z/c)]$ with the degeneracy factor z as an interstitialcy can be oriented along any of z directions ($z = 3$ for FCC lattices).

The dependence of K on shear strain ε is neglected and μ is described by $\mu(V,\varepsilon,c) = \mu_0(V,\varepsilon) + \alpha_\mu \int (\partial^2\mu/\partial\varepsilon^2)(\Omega/\Omega_0)\text{d}c$ or, equivalently by $\partial\mu/\partial c = \alpha_\mu(\partial^2\mu/\partial\varepsilon^2)(\Omega/\Omega_0)$. As μ must be periodic in displacement x with lattice distance b for lattice planes separated by a distance d, the simplest periodic even function, $\mu = \mu(V,c)\cos(2\pi d\varepsilon/b)$ with $\varepsilon = x/d$, is chosen. Hence, $\partial^2\mu/\partial c^2 = -\xi\mu$ with $\xi = 4\pi^2 d^2/b^2$ leads to $\mu = \mu_0(V,\varepsilon)\exp(-\alpha_\mu \xi c)$, since $\Omega/\Omega_0 \cong 1$. For $\alpha_\mu \cong 1$ and $d \cong b$, $\xi \cong 4\pi^2 \cong 40$, so that a concentration of, say, 3%, should reduce μ to ≅0.3 of its original value. For Cu the dependency of μ on c is known and indeed shows a rapid decrease of the elastic constant C_{44} with $\text{d}\ln C_{44}/\text{d}c \cong 31$ (for C_{ij}s, Section 10.11.4).

For the Gibbs energy $G = F + PV$, we must add a term $3NkT\ln[\omega_E(V)/\omega_E(V_0)]$. Also, the expectation is that ω_R is reduced under constant pressure conditions. As a precise analysis is complex, the behavior is approximated by $\omega_R = \omega_{R0}(1 + \lambda\alpha_\mu\xi c)$, where one expects $\lambda < 1$. Using $r = (\omega_E/\omega_R)/(1 + \lambda\xi c)$, $\ln\omega_E/\omega_R = 0.87r - 1.05$ is a good approximation according to Granato. With $V = \partial G/\partial P$ and $\omega_E(V)/\omega_E(V_0) = (V/V_0)^\gamma$, the (normalized) Gibbs energy change y (Figure 14.4b) can be obtained as a function of the (normalized) defect concentration x and (normalized) temperature t (after some manipulation) as

$$y = xX - tx\left\{6\eta\gamma\,(\gamma - 1/3)\,\frac{\mu_0}{K}X + \frac{4.35(\omega_E/\omega_R)}{1 + ax} + \ln\,\frac{b}{x} - 4.25\right\} \quad \text{with}$$

$$X = (1 - q)(1 - e^{-x})/x + q$$

Here $y = [\Delta G(1-q)\xi]/NG_0V_0$, $x = \alpha_\mu \xi c$, $t = kT/\eta G\Omega$, $t_{rel} = T/T_{mel}$, $\eta\mu V = \alpha_\mu \mu V + \alpha_K KV$, $1 - q = \alpha_\mu/\eta$, $a = \lambda/\eta(1-q)$ and $b = z\eta(1-q)\xi(\omega_E/\omega_L)^6$. Moreover, $\mu' = (2\gamma + 1/3)\mu/K$ where the Grüneisen parameter γ [2] is used. We show here the full expression for ΔG just to indicate that an analytical theory is obtained. For $\omega_L/\omega_E = 1.73$ was taken from theory [41]. The other parameters have been fitted on data for Cu [41], imposing $\omega_E/\omega_R > 1$, which allows taking $\eta = 1$, and, using $2\mu = C_{44} + (C_{11} - C_{12})/2$, yields $\alpha\mu_0\Omega_0 = 3.94$ eV, in agreement with experiment. As for $\xi < 30$ no solution could be obtained, $\xi = 35$ was fixed. Using $S_{mel} = 1.15\,k$ at T_{mel} with $\omega_E/\omega_R = 5.3 \pm 7\%$ led to $\lambda = 0.206 \pm 10\%$, $q = 0.075 \pm 22\%$, and $x = 3.0 \pm 17\%$, so that $c(T_{mel}) = 0.093 \pm 17\%$.

For this choice of parameters, Figure 14.4b displaying the behavior of y ($\sim\Delta G$) versus x ($\sim c$), shows very similar overall behavior to that of DTM. The shallow minimum (shown in the inset) at low c is due to the fact that interstitialcies are thermally accessible. The figure also shows that undercooling to $t \cong 0.85$ is possible. For $t > 1$, that is, a temperature above T_{mel}, there are three solutions of which the one with the highest x is interpreted as equilibrium melting. Melting actually occurs for $y = dy/dx = 0$, providing the relation $h_{for} = \eta\mu_0\Omega_0 \simeq \xi\, kT_{mel}$ between T_{mel} and the enthalpy of formation of an isolated interstitialcy h_{for}. Assuming $\eta = 1$, the data for Cu yield $\xi = 33.7$. Using the Grüneisen relation $c_V = \alpha K\Omega/\gamma$ with α the thermal expansivity and γ Grüneisen's parameter together with the Dulong and Petit high-temperature value $c_V \cong 3k$, results in $\alpha T_{mel} \cong \gamma\mu/\xi K$. From Poisson's ratio $\nu = (3K - 2\mu)/2(3K + \mu)$, one obtains $\mu/K = 3(1 - 2\nu)/2(1 + \nu) \cong$ *const.*, so that $\alpha T_{mel} \cong C$ with $C = 3(1 - 2\nu)\gamma/2(1 + \nu)\xi$ a constant. For Cu, with $\nu = 0.35$ and $\gamma = 3.0$, $C_{cal} \cong 0.030$, to be compared with $C_{exp} \cong 0.024$. This approach thus interprets melting as reaching an interstitialcy-driven shear instability with no relation to a critical vibration amplitude [42]. Because Lindemann's rule $\langle u^2\rangle^{1/2} = aC$ with a the lattice constant can be approximated by $\Delta a = aC$, the correlation $(1/a)(\Delta a/T_{mel})T_{mel} \cong \alpha T_{mel} = C$ results and therefore his rule can be rationalized by the interstitialcy theory, but, as said, without invoking a critical vibration amplitude.

The relation $T_{mel} = \mu_0\Omega_0/k\xi$ can be tested independently and yields with $\mu_0 = 48$ GPa and $\Omega_0 = 1.26 \times 10^{-29}$ m^3 for Cu $T_{mel} = 1303$ K, to be compared with the experimental value of 1358 K, and equivalent to $dT_{mel}/d(\mu_0\Omega_0) = 386$ K eV^{-1}. A fit on 62 elements lead to $dT_{mel}/d(\mu_0\Omega_0) = 228$ K eV^{-1} and Granato et al. [42] suggest that the actual datapoints represent approximate formation enthalpies of the interstitialcies. In the absence of adequate data, this correlation can be easily off by a factor of 1.5–2.0, though. As ξ is substance dependent and several metals show a phase transformation below T_{mel}, thereby changing their values for μ, K, ν, and γ, a strict linear relationship $\alpha T_{mel} \cong$ *const.* is not expected anyway.

Finally, as an aside, we recall the intriguing specific heat C_P behavior for liquids [43]. The value of C_P generally decreases with T near T_{mel} (before increasing again near T_{cri} due to the continuous transition at T_{cri}). From $C_V = -T\partial^2 f/\partial T^2$ with $f = f_{for} + TS(c)$ one obtains, using $\partial f/\partial c = 0$ which is equivalent to $T\partial S/\partial c = \partial f_{for}/\partial c$, $\Delta C_V = (\partial f_{for}/\partial c)(\partial c/\partial T)$. For the crystalline state $\partial f_{for}/\partial c$ is the formation energy of a single defect and $\partial c/\partial T$ increases exponentially with T. For the liquid state $\partial f_{for}/\partial c = \alpha_\mu \mu(c)\Omega + \alpha_K K\Omega$ and $\partial c/\partial T$ is approximately constant, and C_V follows the c and T dependence of μ. It appears that at T_{mel}, $\mu(T_{mel})/\mu_0 \cong \exp(-\alpha_\mu \xi c) \cong \exp(-3) \cong 0.05$, so that the two factors in ΔC_V are comparable near T_{mel}. With increasing T, c increases, μ decreases and, hence, C_V decreases. Upon undercooling, C_V increases until no further defects are formed and then returns to the classical value 3*Nk*. The prediction $C_P = C_V(1 + \alpha\gamma T)$ describes the experimental data for several metals well.

14.2.7 Simulations

Nowadays MD and other simulation techniques are illuminating tools resulting in detail that otherwise is hard to obtain. In simulations for bulk materials, generally full periodic boundary conditions are used. In order to be able to do the same in the presence of a surface, the simulation box is usually filled with a liquid layer surrounded by vapor and where either the vapor or liquid layer is periodically connected. Using such MD simulations with a LJ potential for Ar, Han [44] clearly showed the difference between bulk and thermodynamical melting (Figure 14.5). The author mainly analyzed the volume and energy per atom, both in the bulk and at the surface, from which this difference is clearly visible.

Focusing on bulk melting and using the energy landscape for Cu and Al, Samanta et al. [45] employed advanced simulations to indicate that in these cases melting occurs via multiple, competing pathways involving the formation and migration of point defects and dislocations. Each path is characterized by multiple barrier crossings arising from multiple metastable states for the solid. At temperatures approaching superheating, melting becomes a single-barrier process, while at the limit of superheating, the melting mechanism is driven by a vibrational instability. Comparing their results with nucleation theory, the authors suggest that classical nucleation theory (CNT) for melting should be revised.

We further mention three other approaches. First, Holian [46] used the LJD model as an ingredient in a hybrid approach. The LJD model, in either the angular or smeared form and when corrected by the classical harmonic correlational entropy, gives a satisfactory model of a classical solid. The Helmholtz energy of the fluid was obtained from the Hansen–Ree analytic fit to MC EoS calculations. It appeared that at high densities along the melting curve, the anharmonic correction to the correlational Helmholtz energy approaches a small constant compared to the harmonic contribution and the resulting predictions of the solid–fluid coexistence curves are in excellent agreement with computer experiments. This hybrid model demonstrates that both anharmonicity and long-wavelength-correlated motion must be properly incorporated. Second, a comparable route was followed by Bhattacharya et al. [47] using the cell model and employing

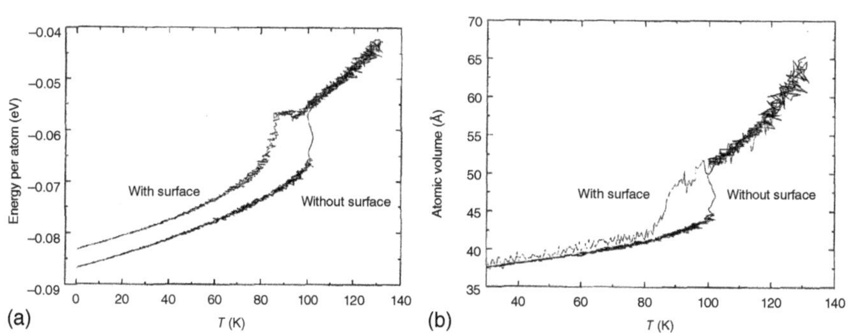

Figure 14.5 MD Argon simulations. (a) The energy per atom Ar for a simulation box with and without surface; (b) the volume per atom Ar for a simulation box with and without surface. Source: Han [44]/with permission of the Korean Ceramic Society.

embedded atom method (EAM) potentials to account for many body interaction effects. The Helmholtz energy so obtained was used to determine melting curves of FCC metals. For this purpose, the liquid-phase Helmholtz energy was calculated using the corrected rigid spheres model of Kerley [48]. In this modified perturbation theory, the energy of a fluid molecule is defined by a function that depends on the local configuration of its neighbors and does not need explicit knowledge of interaction potential. For Al, Cu, Ni, and Pt, the results match well with the available experimental/theoretical data. Third, we recall the rather different quantum cluster equilibrium (QCE) simulation approach of Weinhold (Section 9.9), based on weakly interacting clusters of molecules that do interact strongly within a cluster. The model focuses on water and predicts the phase diagram quite well.

14.3 Thermodynamic or Surface-Mediated Melting

In 1935, Peierls [24, 49] discussed disorder introduced in 1D and 3D lattices by thermal vibrations, followed by discussions on 2D lattices by others [24, 50]. It appears that 1D lattices can have only long-range positional order at $T = 0\,\text{K}$ and that in the thermodynamic limit at any temperature above absolute zero long-range is destroyed by the thermal vibrations. In 3D models, long-range order persists at any temperature, as long as the material is solid. In 2D, possibly expected, long-range order is destroyed by thermal motion at a finite temperature and the mean-square displacement grows logarithmically with distance between the molecules, although in view of the weak divergence quasi-long-range order remains. This prompted extensive modeling research on phase transitions in surface films [51], which on its turn stimulated both experimental tests and simulations. From all this a 3D melting picture emerged in which surfaces play an important role [52, 53].

The importance of surfaces, already indicated by Tammann and Stranski [54] as well as by Frenkel [24], is discussed in several reviews [55, 56]. As summarized by Dash [6], surface melting is a consequence of wetting of a solid by its melt, occurring when the surface energy of the combined solid–liquid–vapor system is lower than that of the "dry" solid, that is, when $\gamma_{SV} > \gamma_{SL} + \gamma_{LV}$. Although this is usually the case, there are also liquids that do not wet their own crystal, for example, Ga, Hg, *p*-methylaniline, and phenyl salicylate [57]. The (possibly local) reduction of the surface energy by the liquid renders premelting of the solid energetically favorable. As premelting proceeds, the short-range order of the interface gradually evolves from crystalline order to liquid-like disorder. Approaching the melting temperature T_{mel}, the disordered or *quasi-liquid* layer grows to a film with thickness δ which in its upper layers is indistinguishable from the liquid but retains some solid-like order over a few molecular distances of the solid interface. Premelting does not occur for all surfaces though. For example, for FCC crystals generally the relatively open {110} surfaces do show complete premelting (i.e. δ diverges), while the close-packed {111} surfaces do not show premelting. The {100} surfaces with intermediate packing density show incomplete premelting (i.e. δ remains finite when $T \to T_{mel}$).

14.3.1 Melting of Nanoparticles

Surface melting can be explained in a thermodynamic way as follows. We consider a crystal with bulk melting temperature T_{mel}, melting enthalpy per unit volume L,[9] and chemical

9 We denote in this section the melting enthalpy, normally indicated by $\Delta_{mel}H$, for brevity by L.

potential μ_S. For such a crystal with surface area A and containing N atoms (State 1), the Gibbs energy is given by $G_1 = N\mu_S + A\gamma_{SV}$. For a crystal covered with a quasi-liquid layer containing N' atoms (state 2), $G_2 = (N–N')\mu_S + N'\mu_L + A\Gamma$, where Γ depends on the thickness δ of the liquid layer. For $\delta = 0$, $\Gamma = \gamma_{SV}$ while for $\delta = \infty$, $\Gamma = \gamma_{SL} + \gamma_{LV}$, and Γ should take an intermediate value accounting for interactions between the two interfaces for a thin layer. Using the *spreading coefficient* $S = \gamma_{SV} - \gamma_{SL} - \gamma_{LV}$ (Problem 11.1), one can write $\Gamma = \gamma_{SL} + \gamma_{LV} + S\exp(-\delta/\xi)$. Thus, with $\Delta\mu = \mu_L - \mu_S$,

$$\Delta G = G_2 - G_1 = N'\Delta\mu - AS[1 - \exp(-\delta/\xi)] \equiv N'\Delta\mu - ASf(\delta) \tag{14.6}$$

Here, $f(\delta) = 1 - \exp(-\delta/\xi)$ is chosen for short-range interactions (as in metals) with ξ the correlation length. Minimizing ΔG with respect to N', meanwhile using $N' = A\rho_L\delta$, yields $\delta = \xi\ \ln(S/\xi\rho_L\Delta\mu) = \xi\ \ln(\lambda_{sho}/t)$ with $t = (T_{mel} - T)/T_{mel} \equiv \Delta T/T_{mel}$ and $\lambda_{sho} = St/\xi\rho_L\Delta\mu$. For ΔT small, $\Delta\mu \cong (\partial\Delta\mu/\partial T)(T - T_{mel})$. Moreover, in equilibrium $\Delta\mu = L - T_{mel}(-\partial\Delta\mu/\partial T) = 0$, so that we also have $\Delta\mu \cong Lt$ and $\lambda_{sho} \cong S/\xi\rho_L L$. We can also write $\delta = \xi\ \ln[(T_{mel} - T_{sur})/(T_{mel} - T)]$, where $T_{sur} = T_{mel}(1 - \lambda_{sho})$ signals the onset of surface melting at $\delta(T_{sur}) = 0$. So, surface melting only occurs when $S > 0$ and the solid is stable without a liquid layer below T_{sur}. When $T_{sur} < T < T_{mel}$, a liquid surface layer forms with thickness δ, which when $T \to T_{mel}$, diverges, that is, $\delta \to \infty$. The logarithmic divergence is consistent with the data for metals with $\xi \cong 1.5$ monolayer distance, ≅5–6 Å for metals like Al and Pb [52], typically assuming $\rho_L = \rho_S = \rho$. For materials with long-range interaction, such as vdW crystals, the approximation $f(\delta) = 1 - (\xi/\delta)^2$ is more appropriate [58] and leads similarly to $\delta = \lambda_{lon}/t^{1/3}$ with $\lambda_{lon}{}^3 = 2\xi^2 S/T_{mel}\rho(\partial\Delta\mu/\partial T) \cong 2\xi^2 S/\rho L$ [59].

A clear indication for the importance of surface melting is the decrease in T_{mel} with particle size, predicted in 1909 and first observed in 1954 [60], and for which the example of Au shown in Figure 14.6a [61]. An extension of the model described above provides a size-dependent T_{mel} [62]. The system is supposed to change from a homogeneous solid sphere of radius r containing N atoms (state 1) to a solid particle of radius r' covered with a surface molten layer with thickness δ containing N' atoms (state 2), so that $r' = r - \delta$ (Figure 14.6b). The Gibbs energy for state 1 reads $G_1 = N\mu_S + 4\pi r^2\gamma_{SV}$, while for state 2 we now have $G_2 = (N–N')\mu_S + N'\mu_L + 4\pi r^2[\gamma_{SL}(r'/r)^2 + \gamma_{LV} + S'\exp(-\delta/\xi)]$ with $S' = \gamma_{SV} - [\gamma_{LV} + \gamma_{SL}(r'/r)^2]$. With $\Delta G = G_2 - G_1$, we obtain

$$\Delta G = N'\Delta\mu - 4\pi r^2 S' f(\delta) = N'(\mu_L - \mu_S) - 4\pi r^2 S'[1 - \exp(-\delta/\xi)] \tag{14.7}$$

This expression generally shows a maximum and minimum, which are interpreted as corresponding to a critical liquid layer thickness δ_{cri} and equilibrium liquid layer thickness δ_{equ}, respectively. With increasing T, δ_{equ} increases and δ_{cri} decreases, until complete melting occurs when $\delta_{equ} = \delta_{cri}$. Calculating the minimum of ΔG, assuming L constant, leads to

$$\Delta T/T_{mel} = (2\gamma_{SL}/\rho_L L r')[1 - \exp(-\delta/\xi)] + (S'r^2/\rho_L L\xi r'^2)\exp(-\delta/\xi) \tag{14.8}$$

For very small particle size, ΔG does not show a minimum, but only a maximum which disappears at a certain temperature T_0, suggesting that a particle solid at $T < T_0$ becomes entirely liquid for $T > T_0$ as the barrier height is zero. In other words, such a particle is liquid always. From this theory, one can estimate an upper bound using $\delta \to 0$ (onset instability) as $T_{mel}{}^{(upp)}/T_{mel} = 1 - (2\gamma_{SL}/\rho_L rL)$, as well as a lower bound using $\Delta G \to 0$ (no driving force) as $T_{mel}{}^{(low)}/T_{mel} = [1 - (3/\rho_S rL)][\gamma_{SV} - \gamma_{LV}(\rho_S/\rho_L)^{2/3}]$. It appears that experimental

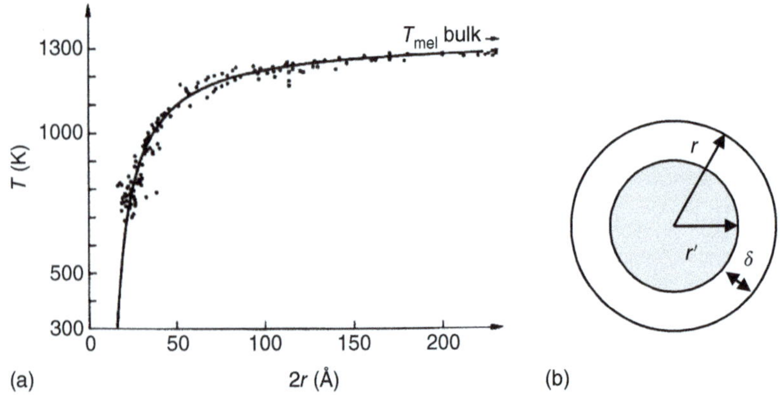

Figure 14.6 The size dependence of the melting point. (a) T_{mel} for Au as a function of the diameter of particles $2r$. Source: Buffat and Borel [61]/American Physical Society. (b) Schematic of a solid particle covered with a liquid shell.

data for Sn, In, Bi, and Pb are limited by these bounds (data not shown) [63, 64]. For Pb particles, using $\xi \cong 20$ Å, as compared to $\xi \cong 6$ Å for a flat Pb surface, the model describes the data well.

One might expect that L varies with particle size [64]. Assuming again the two-state model as described above, the volume fraction of molten material is given by $x = [r^3 - (r-\delta)^3]/r^3 \cong 3\delta/r$ for $\delta \ll r$. For a particle with radius r, $L = (1-x)L_S + xL_L = L_S + (3\delta/r)(L_L - L_S)$, where L_L and L_S are the melting enthalpy of the solid and quasi-liquid, respectively. It follows that there is a critical size $r_{cri} = 3\delta(1 - L_L/L_S)$ for which L vanishes. Calorimetric measurements show that (for particles embedded in a matrix) the intercept of a plot of L/L_S versus $1/r$ is $\cong$1 (as it should). From the slope (assuming a fixed value $\delta \cong 1$ nm) the value for L_L can be estimated. The negative values obtained for L_L indicate that L_L cannot be interpreted right away as the pure liquid enthalpy, but that other (structural) effects play a role. Finally, L varies not only with r but also with T. For this dependency, an estimate for $\mathrm{d}L/\mathrm{d}T$ is needed, but a combined analysis seems not to be available, while the required data are likely unknown. The assumption $\rho_L = \rho_S = \rho$ is easy to avoid, but ρ_L for the quasi-liquid is also typically unknown.

Nevertheless, overall agreement with experiments is quite good, considering all the experimental difficulties. Discussions of other aspects of nanoparticles, such as superheating, either when embedded in a matrix or covered with an (insoluble) layer, and the effects of shape, size distribution, and applied stress are available [65, 66].

14.3.2 Vacancies Revisited

A relatively simple, but attractive model [67] for thermodynamic melting of metals assumes that vacancies are the most probable defects responsible for melting, uses the correlations as given by Gorecki [25], and tries to clarify why there is a critical vacancy concentration $c_{cri} \cong 0.1$, why and how the lattice becomes unstable at c_{cri}, and how surface (pre-)melting is related to vacancy concentration and migration.

Recalling from Section 14.2.3 that the vacancy concentration in the bulk of the lattice c_{lat} is given by $c_{lat} = \exp(4.1 - E_{lat}/kT)$, and assuming a similar relation for the vacancy

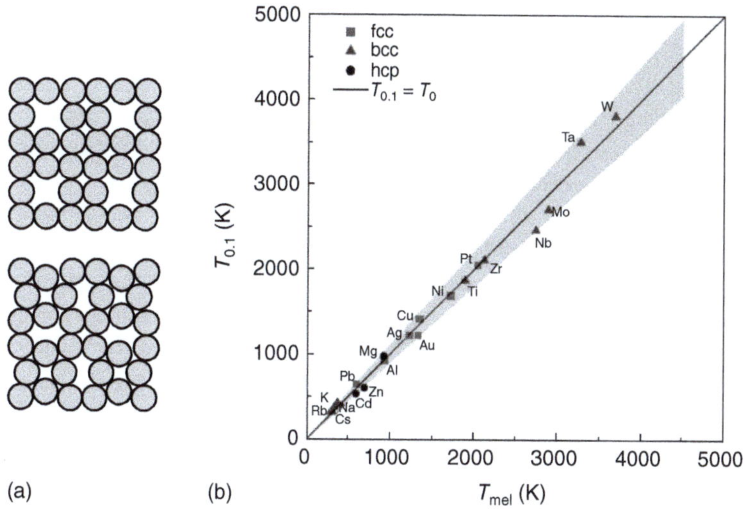

Figure 14.7 (a) Schematic of a lattice with about 10% vacancies, showing the original lattice (upper part) and the restructured lattice (lower part). (b) Correlation of $T_{0.1}$ with T_{mel}. Source: Adapted from Mei and Lu [67].

concentration at the surface of the lattice c_{sur}, we need to estimate E_{lat} and E_{sur}.[10] To estimate E_{sur}, we consider that the vacancy formation energy is related to the bonding energy of the atoms. While for FCC and HCP metals we have 12 nearest neighbors in the bulk, at the surface there are only 8.[11] Hence, it is reasonable to estimate that $E_{sur} = 2E_{lat}/3$. We know already that $E_{lat}/T_{mel} = 80.4\,\text{J K}^{-1}$. Therefore, taking Al ($T_{mel} = 1235$ K) as an example, at T_{mel} we have $c_{lat} \cong 0.0033$ and $c_{sur} \cong 0.086$. In fact, it appears that $c_{sur} \cong 0.1$ considering all metals for which reliable data are available. The doubt about the "universal" value of the bulk concentration $c_{lat} \cong 0.0037$ (Section 14.2.3) being too small to induce a lattice instability may thus be answered by considering the surface concentration $c_{sur} \cong 0.1$, which is comparable to the concentration increase upon melting Δc. Defining $T_{0.1}$ as the temperature at which c_{sur} reaches the critical value $c^* = 0.1$, it appears empirically that $T_{0.1} = T_{mel}$, valid for the FCC, HCP, and BCC metals considered (Figure 14.7b). Hence, close to T_{mel}, although the bulk is still a crystal, the surface is already molten. Assuming that the vacancy formation energy at the solid–liquid interface is comparable to E_{sur},[12] once surface melting has occurred, additional vacancies will be formed at this interface until the whole crystal is molten. This answers the question as to how and why c increases from $c \cong 0.0037$ in the solid state to $c \cong 0.1$ in the liquid state.

To understand why the lattice becomes unstable at $c \cong 0.1$, we note that the crystal can be viewed as an aggregation of clusters of atoms, each of which has about 10 nearest neighbors around a vacancy (Figure 14.7a). However, such a configuration of clusters is unstable

10 Note that we denote the *vacancy* energy (E_{vac} in Section 14.2.3) here by E_{lat} as we need values for the energy of vacancy formation in the bulk (E_{lat}) and at the surface (E_{sur}).

11 One might expect that the FCC (111) with 9 nearest-neighbors leading to a factor ¾ would be used, but Gorecki used 8 (possibly with FCC (100) in mind) and this was taken over by Mei and Lu [67]. The estimate ¾ instead of ⅔ will decrease the correlation somewhat.

12 The authors [67] state that the vacancy formation energy at the solid–liquid interface and solid–air interface are similar, but this is arguable in view of the similar solid and liquid densities.

because local disordering can take place if sufficiently high mobility exists and leads to a configuration akin to an interstitialcy, as first suggested by soap bubble raft simulations [68]. This local disordering will eventually lead to the collapse of the whole crystal. It was found that the local disordering around a single vacancy can occur when the concentration of migrating atoms c_{mig} reaches a critical value $c_{mig}{}^* \cong 0.17$ for FCC and HCP lattices. As $c_{mig}{}^* \cong 2c^*$, this indicates that one needs at least two mobile atoms simultaneously. Considering the mobility of atoms, the concentration of mobile atoms at the surface is estimated as $c_{mig} = \exp(4.1 - E_{mig}/kT)$, where the migration energy of a vacancy E_{mig} at the surface is taken, similar as for their formation, by 2/3 of the experimental bulk values [69]. Using the Al example again, one calculates that at T_{mel}, $c_{mig} \cong 0.34$, so that restructuring indeed can take place and that at $T = 820$ K, $c_{mig} \cong c_{mig}{}^* \cong 0.17$ and $c_{sur} \cong 0.04$, which implies that at this temperature surface disordering (roughening) occurs. This disordered layer becomes a liquid when $c_{sur} \cong 0.1$ at T_{mel}. Considering several metals, this prediction appeared to be in good agreement with experiment.

In summary, vacancies are more easily formed at the surface than in the bulk and the mobility of atoms at the surface is much larger than in the bulk. Premelting (roughening) of the surface occurs if the concentration of mobile atoms c_{mig} reaches a critical value, estimated as $c_{mig}{}^* \cong 0.17$. If the concentration of surface vacancies c reaches a critical value, estimated as $c^* \cong 0.10$, the lattice becomes locally unstable and restructures around a vacancy. This disordered domain acts as a liquid nucleus. Once both these critical values are reached, disordering proceeds through the whole crystal, eventually leading to a molten crystal. A similar analysis using a defective lattice and employing equality of lattice and liquid entropy as the melting criterion has been given by Fecht [70] and led to $c^* \cong 0.08$.

14.3.3 Dislocations Revisited

The DTM as discussed in Section 14.2.5 refrained from incorporating surfaces. In a somewhat different form of DTM, Kristensen et al. [71] did take the liquid–solid interface into account and estimated the rate of dislocation formation as $\dot{c}_+ = A\exp[-\beta(E_{for} + E_{mig})]$ with A a constant, E_{for} and E_{mig} the formation and migration energies of a dislocation segment, and $\beta = 1/kT$, as usual. For the annihilation rate of dislocations, similarly $\dot{c}_- = B\exp(-\beta E_{mig})\rho^a$, with B another constant, ρ the dislocation density and a a constant that would be 2 if segments can be considered as small compared to their distance. The energy of a dislocation segment of unit length may be expressed, similarly as before but more condensed, as $U = (\alpha\mu b^2/4\pi)\ln(\zeta R/b)$, where $\zeta = (b/r_0)\exp(4\pi U_{cor}/\alpha\mu b^2)$ with U_{cor} and r_0 the core energy and radius, respectively. Introducing the dimensionless dislocation density $c = \rho b^2$, the excess energy of crystal with volume V may be expressed as $U_V = (\alpha\mu Vc/4\pi)\ln(\zeta/2c^{1/2})$. Here, $\rho^{-1/2} = 2r_0$ is used instead of $\rho^{-1/2} = \pi^{1/2}r_0$ as used before, but the effect of this difference is negligible. Ignoring lattice relaxations, the formation energy $U_{for}(c)$ as a function of density c may thus be estimated as $U_{for}(c) = (\alpha\mu V_{dis}/4\pi)\ln(\zeta/2c^{1/2})$ with $V_{dis} = Lb^2$, where L is the dislocation length. It is assumed that this elastic energy expression is valid up to the maximum concentration of dislocations $c_0 = \zeta^2/4e$ at which U_{for} will attain its maximum. In equilibrium $\dot{c}_+ = \dot{c}_-$ and this leads to

$$c = (A/B)^{1/a}(\zeta/2c^{1/2})^{-\beta\alpha\mu V_{dis}/4\pi a} \tag{14.9}$$

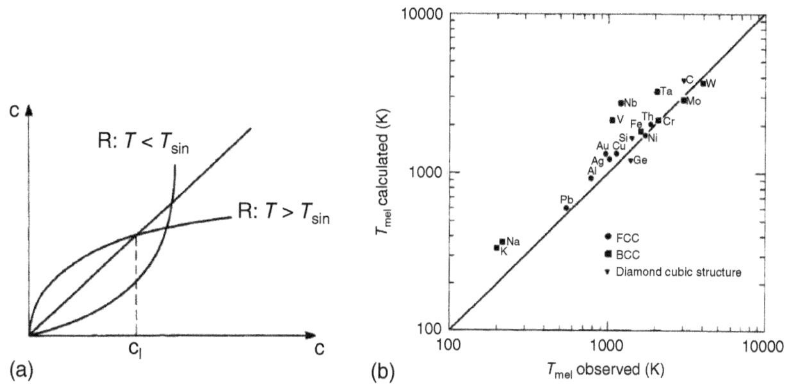

Figure 14.8 DTM according to [71]. (a) The lhs (L) and rhs (R) for $T < T_{sin}$ and $T > T_{sin}$ of Eq. (14.10) with $T_{sin} = \alpha G V_d / 8\pi k a$. (b) The correlation between calculated and experimental T_{mel}. Source: Damgaard Kristensen et al. [71]/IOP Publishing.

This expression has the form $z = Zz^{\eta}$ with as solutions $z = 0$ and $z = Z^{1/(1-\eta)}$. Hence

$$c = 0 \quad \text{and} \quad c = \left[(2/\zeta)(A/B)^{4\pi/\beta\alpha\mu V_{dis}}\right]^{2\beta\alpha\mu V_{dis}/(8\pi a - \alpha\mu V_{dis})} \tag{14.10}$$

and the latter expression has a singularity for the temperature $T_{sin} = \alpha\mu V_{dis}/8\pi k a$. The behavior of Eq. (14.10) is shown in Figure 14.8a. For $T < T_{sin}$, the inevitably present fluctuations will lead ultimately to the first solution, describing the solid without any dislocations. For $T > T_{sin}$ they will lead to the second solution, describing the solid saturated with dislocations. Hence, this temperature is interpreted as an instability temperature, actually the melting temperature in the absence of an interface.

The presence of an interface will affect the balance between creation and annihilation of dislocations as a surface can act as a sink or source of dislocations. Assuming a planar interface at $x = 0$ with the liquid phase being described by a dislocation density c_0, for local equilibrium to exist one must have

$$[\mathrm{d}c(x_0)/\mathrm{d}t]\mathrm{d}x = \dot{c}_+\mathrm{d}x - \dot{c}_-\mathrm{d}x + j(x_0) - j(x_0 + \mathrm{d}x) = 0$$

where $j(x)$ is the flux at x, assumed to be given Fick's first law $j = -D\mathrm{d}c/\mathrm{d}x$ with $D = D_0\exp(-\beta E_{mig})$ the diffusion constant for dislocations. Hence, it follows that

$$A/D_0 \exp(-\beta E_{for}) - Bc^a/D_0 + \mathrm{d}^2c/\mathrm{d}x^2 = 0$$

or, using $E_{for}(c) = (\alpha\mu V_{dis}/4\pi)\ln(\zeta/2c^{1/2})$ and $T_{sin} = \alpha\mu V_{dis}/8\pi k a$, that

$$\mathrm{d}^2c/\mathrm{d}x^2 - k_B c^a + k_A(4c/\zeta^2)^{a\tau}$$

where $k_A = A/D_0$, $k_B = B/D_0$ and $\tau = T_{sin}/T$. This expression may be integrated to

$$\frac{1}{2}\left(\frac{\mathrm{d}c}{\mathrm{d}x}\right)^2 - \frac{k_B}{1+a}c^{a+1} + \frac{k_A}{1+a\tau}\left(\frac{4}{\zeta^2}\right)^{a\tau}c^{a\tau+1} = C \quad \text{with as solution}$$

$$x = \pm\int_{c_0}^{c}\left(\frac{2k_B}{1+a}c^{a+1} - \frac{2k_A}{1+a\tau}\left(\frac{4}{\zeta^2}\right)^{a\tau}c^{a\tau+1} + 2C\right)^{-1/2}\mathrm{d}c$$

The integration constant C must be chosen as $C = 0$ to ensure that $\mathrm{d}c/\mathrm{d}x = 0$ for $c = 0$, while the negative root has to be chosen to avoid a physically unacceptable negative dislocation concentration. Some analysis of these expressions shows that equilibrium at the interface can only be attained if, for the relative temperature $\tau = \tau_0$,

$$\frac{k_B}{1+a}c_0^{a+1} - \frac{k_A}{1+a\tau_0}\left(\frac{4}{\zeta^2}\right)^{a\tau_0} c_0^{a\tau_0+1} = 0 \quad \text{or} \quad c_0 = \left[\left(\frac{B}{A}\right)\left(\frac{1+a\tau_0}{1+a}\right)\right]^{1/a(\tau_0-1)} \left(\frac{\zeta}{2}\right)^{2\tau_0/(\tau_0-1)}$$

Consequently, τ_0 is interpreted to represent the melting temperature via $\tau_0 = T_{\text{sin}}/T_{\text{mel}}$. Estimating values for A, B, a, and ζ, meanwhile using $c_0 = \zeta^{\,2}/4\text{e}$ yields an implicit expression for τ_0. It appeared that for diamond-type (D) lattices with $\zeta \cong 4$, $a\tau_0$ is close to 3.8, while for FCC, HCP, and BCC lattices with $\zeta \cong 2$, $a\tau_0$ ranges from 4.9 to 6.1, so that the average 5.5 was used. To estimate V_{dis}, the details of the dislocation formation process should be considered. The energetically most favorable process is the formation of a dislocation dipole [72], and assuming this process occurs with a Burgers vector length of the nearest-neighbor distance d_{nn} and a dislocation line length of $\cong 2d_{\text{nn}}$, we have $V_{\text{dis}} = 2d_{\text{nn}}^{\,3}$. Hence, the expression for T_{mel} becomes

$$T_{\text{mel}} = \mu d_{\text{nn}}^3 / f\pi(1-\nu)k$$

with $f = 22.0$ for FCC, HCP, and BCC and $f = 15.2$ for D structures. Figure 14.8b shows that a reasonable correlation between experimental and calculated values of T_{mel} is obtained for 19 metals. Apart from that, the above approach takes in to account the interface, thereby differentiating between the instability and melting temperature.

Finally, we note that also an approach that employs both dislocations and point defects to explain melting has been proposed [73]. The treatment of dislocations is 3D, while the treatment of the vacancies is 2D. Using the grand potential Ω, the transition between Ω_{dis} and Ω_{vac} is taken as T_{mel}, which for the series Li, Na, K, Rb, and Cs, shows good agreement with experiment.

Problem 14.1 ***The LJD Entropy*** Show that the entropy corresponding to F_2 reads $S_2 = -2(1 - X_{\text{m}})\ln(1 - X_{\text{m}}) - 2X_{\text{m}}\ln X_{\text{m}}$ and that this function increases continuously with X_{m}.

Problem 14.2 ***Nanoparticle Melting*** Verify Eqs. (14.6), (14.7), and (14.8).

14.4 Polymer Melting

So far, we discussed largely melting of small molecules or monomeric compounds. Polymers generally do not completely crystallize, and their melting behavior is more complicated than that of low-molar-mass compounds. As usual the equilibrium melting temperature is determined by the balance of enthalpy and entropy. For example, for polyethylene $\Delta_{\text{mel}}H = 4.142\ \text{kJ mol}^{-1}$, $\Delta_{\text{mel}}S = 9.9\ \text{J K}^{-1}\text{mol}^{-1}$ resulting in $T_{\text{mel}} = 145.5\ °\text{C}$. Similarly, for poly(1,4-cis-isoprene), $\Delta_{\text{mel}}H = 4.393\ \text{kJ mol}^{-1}$, $\Delta_{\text{mel}}S = 14.2\ \text{J K}^{-1}\text{mol}^{-1}$, and $T_{\text{mel}} = 35.5\ °\text{C}$ [74]. Peculiar to polymers is that their melting and crystallization temperatures are not the same (we return to crystallization mechanisms later). The melting point T_{mel} observed is always larger than the crystallization temperature T_{cry} and a plot of T_{mel} versus T_{cry} is usually rather straight. However, generally a melting trajectory is present and the melting

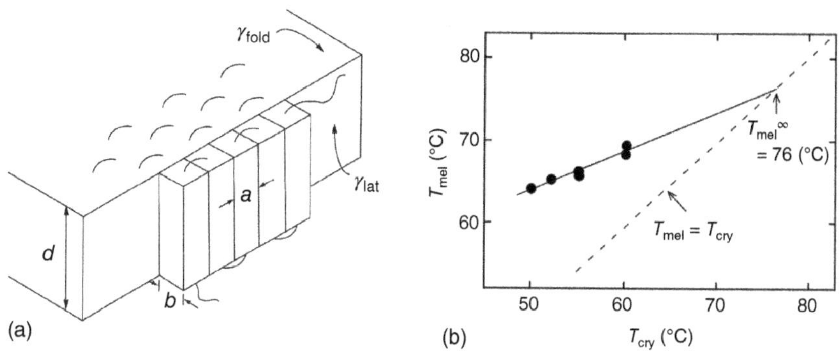

Figure 14.9 Polymer melting. (a) The Hoffman–Lauritzen model. (b) Melting temperature T_{mel} versus crystallization temperature T_{cry} for poly(dl-propylene oxide) with the equilibrium melting temperature $T_{mel}{}^{\infty}$. Source: Young and Lovell [76]/Taylor & Francis.

temperature depends on the crystallization temperature used before and on the heating rate q^*. Nevertheless, the concept of equilibrium temperature $T_{mel}{}^{\infty}$ is introduced representing the melting temperature of an infinite crystal.

Although devised for crystallization, the most often used model is that of Hoffman and Lauritzen [75], which builds on the nucleation model for small molecules. It covers several aspects of melting and therefore we discuss it briefly. An embryo is supposed to have a cylinder-like shape, which grows to a platelet shape with thickness d and to which monomers with cross-section ab add (Figure 14.9a). They are supposed to do that with an area ad parallel to the lateral surface of the platelet and fold sequentially over the lateral surface of the platelet having a surface energy γ_{lat}. The surface energy of the planar surface of the platelet, the *fold plane*, is γ_{fold}. The change in surface Gibbs energy is given by $\Delta_{sur}G_n = 2bd\gamma_{lat} + 2nab\gamma_{fold}$, where n is the number of strands. The change in bulk Gibbs energy is $\Delta_{cry}G_n = -nabd\Delta_{fus}G$, where $\Delta_{fus}G$ is estimated from $\Delta_{fus}G = \Delta_{fus}H - T_{mel}{}^{\infty}\Delta_{fus}S$. If $\Delta_{fus}S$ is not very temperature dependent, $\Delta_{fus}G = \Delta_{fus}H - T\Delta_{fus}H/T_{mel}{}^{\infty}$ and introducing the undercooling $\Delta T = T_{mel}{}^{\infty} - T$, we obtain $\Delta_{fus}G = \Delta T\Delta_{fus}H/T_{mel}{}^{\infty}$. At given n, the expression for $\Delta G_n = \Delta_{cry}G_n + \Delta_{sur}G_n$ shows that it is maximal when d is small and decreases as d increases. The equilibrium value $d°$ is obtained at $\Delta G_n = 0$. Normally n is large, the term $2bd\gamma_{lat}$ in $\Delta_{sur}G_n$ can be neglected and combining ΔG_n with $\Delta_{fus}G$ yields $d° \cong 2\gamma_{fold}T_{mel}{}^{\infty}/\Delta T\Delta_{fus}H$. This is, although a simplified model, indeed the experimentally observed proportionality.

We take the intercept between the melting and crystallization curve as $T_{mel}{}^{\infty}$ (Figure 14.9b). Upon forming a platelet with width x and length y, the change in surface energy is $\Delta_{sur}G = 2(x+y)d\gamma_{lat} + 2xy\gamma_{fold}$, while the change in bulk energy is $\Delta_{vol}G = 2xyd\Delta_{fus}G$ and thus the overall change $\Delta G = 2xyd\Delta_{fus}G - 2(x+y)d\gamma_{lat} - 2xy\gamma_{fold} \cong 2xyd\Delta_{fus}G - 2xy\gamma_{fold}$. In equilibrium $\Delta G = 0$, therefore $T_{mel} = T_{mel}{}^{\infty} - 2\gamma_{fold}T_{mel}{}^{\infty}/d\Delta_{fus}H$ and the melting temperature of a finite size crystal is always less that for an infinite crystal $T_{mel}{}^{\infty}$. If the relation between T_{mel} and d can be determined experimentally, a fit provides $T_{mel}{}^{\infty}$ and γ_{fold} if $\Delta_{fus}H$ is known from, for example, calorimetry. Upon annealing just below T_{mel} so that sufficient mobility is present, the polymers relax due a decrease in $\Delta_{sur}G$, resulting in an increase in thickness d and decrease in area xy. Upon melting, solidification, and remelting, T_{mel} thus increases. It also explains the dependence of T_{mel} on the heating rate q^* as q^* codetermines the amount of relaxation that can take place.

Various factors determine T_{mel} of a polymer. Most important is the chemical structure. We mention a few factors. First, the stiffness of the chains. Groups such –O–, –O–O– and –CO–O– increase the flexibility and lead to a lower T_{mel}, while phenyl groups in the main chain decrease the flexibility and lead to a higher T_{mel}. Second, the presence of polar groups such as amide groups –CONH– allows intermolecular hydrogen bonding thereby increasing T_{mel}. Third, the presence of side groups. For example, polypropylene, which can be considered as ethylene with regular CH_3 side groups, has a reduced chain flexibility as compared to polyethylene and has a higher T_{mel}. Bulky side groups will do the same, but long and flexible side groups generally reduce T_{mel}. Also the amount of branching is important as more side chains and chain ends reduce the packing density, thereby lowering T_{mel}. The melting point thus can be manipulated significantly but these changes also influence the glass transition temperature, which on its turn might influence crystallization.

For polymers single-chain single crystals do exist and such crystals melt by simple consecutive detachment of chain segments from the crystalline substrate and their diffusion into the melt. However, complications in the melting process occur for a semicrystalline polymer where chains are shared between different crystals. The distribution of entanglements is highly heterogeneous as the entanglements are mostly confined to the amorphous regions, whereas the crystalline regions are void of them. This will influence the process of detachment from the surface. Experimentally, a clear distinction in different melting processes can be observed, by the differences in the activation energies required for the consecutive detachment of chain segments or of segments having topological constraints. The consecutive detachment of free chain segments starts at the melting temperature predicted from the Gibb–Thomson equation, whereas higher temperature or time is required if the chain has to overcome the constraints [77]. Usually, the heterogeneous distribution of entanglements is lost on melting, and the entanglements are uniformly distributed along the chain, characterized by the molar mass between entanglements. With increasing molar mass $\mathcal{M}$, the number of entanglements increases, and the melt viscosity follows the relationship $\eta_0 \sim \mathcal{M}^{3.4}$. However, for polymers with a low number of entanglements per unit volume, the melting kinetics strongly influence the resulting melt state [78]. On slow heating a long-lived heterogeneous melt state can be generated that shows long-lasting low plateau modulus, high crystallization rate, and enhanced solid-state drawability.

14.5 The Influence of Pressure

Melting is influenced by pressure, as indicated by the Clapeyron–Clausius equation. For example, for water at 0 °C, $\Delta_{mel}H$ = 80 cal g^{-1}, V_L = 1 cm^3 g^{-1} and V_V = 1.09 cm^3 g^{-1}, resulting in dT_{mel}/dP = −0.0074 K atm^{-1}, to be compared with the experimental value −0.0075 K atm^{-1}. Consequently, the triple point is T_{tri} = 0.0075 °C.[13]

The effect of pressure on the melting point over a wider pressure range is almost universally described by the empirical *Simon–Glatzel equation* [79], reading $\Delta P/a = (T/T_0)^b - 1$, where a and b are parameters, $\Delta P = P - P_0$ with P_0 and T_0 reference pressure and temperature, often taken as the triple point. In many cases P_0 is also neglected, as normally $P_0 \ll P$.

13 We can calculate similarly the gradient of the sublimation and evaporation curve. For H_2O, given that at the triple point (P_{tri} = 4.58 mmHg, T_{tri} = 0.0075 °C), $\Delta_{sub}H$ = 684 cal g^{-1}, $\Delta_{vap}H$ = 604 cal g^{-1}, V_V = 205 000 cm^3 g^{-1}, dP_{SV}/dT = 0.384 mmHg K^{-1}, and dP_{LV}/dT = 0.339 mmHg K^{-1}.

Table 14.2 Simon–Glatzel parameters for various liquids.

Molecule	*P* and *T* ranges	Δ (K)[a]	T_0 (K)	P_0 (kPa)	*a* (MPa)	*b* (–)
Ar FCC [81]	1.3–6.3 GPa 296–740 K	—	82.9[e]	—	210	1.556
Ar FCC [82]	0–6.3 GPa 83–840 K	2[b)e)]	83.80	68.90	224.5 (3.2)	1.5354 (0.0044)
Ar FCC [83]	1.3–4.2 GPa 294–495 K	21[c]	83.80	69	244	1.476
CH_4 [83]	2.4–3.2 GPa 410–558 K	20[c]	90.69	117	208	1.698
Benzene 99.8% [84]	1.3–174 MPa 280–323 K	0.044[d]	278.24	—	347.9	2.7111
Cyclohexane 99.5% [84]	0.1–85 MPa 280–323 K	0.040[d]	278.88	—	289.8	1.7838
Cyclohexane 99.95% [84]	0.1–85 MPa 280–323 K	0.040[d]	279.55	—	280	1.8262
Water (ice VII) [81]	2–13 GPa 350–750 K	50 MPa[c]	354.8[f]	2.17 GPa[f]	1253	3.0
Methanol 99.5% (water 0.01%) [85]	9.1–265 MPa 176–208 K	0.12[d]	175.17	—	358.6	3.2443
Ethanol 99.7% (water 0.2%) [85]	8.5–199 MPa 160–183 K	0.12[d]	158.37	—	436.9	2.6432

a) Δ = Deviation.
b) average deviation.
c) rms deviation.
d) maximum deviation.
e) calculated from the data given.
f) triple-point ice VI-Ice VII-liquid.

Equivalently, in differential form, we can write $d[dP/d(\ln T_{mel})]/dP = b$. The Simon–Glatzel expression fits experimental fusion curves reasonably well by a proper choice of the two parameters a and b for molecular solids but overestimates T_{mel} for metals and ionic compounds [80]. A few examples for molecular compounds are given in Table 14.2. From the data sets given for Ar, it will be clear that the precise values depend on the pressure range used.[14] The data for cyclohexane indicate that a small amount of impurity can lead to significantly different parameters. Many data, mainly for molecular compounds, have been collected by Babb [86]. However, the Simon–Glatzel equation can only describe increasing melting points with increasing pressure, while experimentally decreasing melting points with increasing pressure do occur.

The Ross approach to rationalize the Lindemann rule (Section 14.2.1) can also be used to rationalize the Simon–Glatzel equation. As at high pressures repulsion is dominant, using

14 The results depend also on the experimental techniques used, as high-pressure experiments are loaded with pitfalls. For a discussion of these problems, we refer to the literature, for example [81].

a potential with repulsion only, $\phi = \varepsilon(r_0/r)^n$ with ε and n constants, may suffice for a lattice with coordination number z. Neglecting thermal energy, the reduced potential in the harmonic approximation using $x = r/V^{1/3}$ becomes

$$\phi = ax^2 \equiv \varepsilon(v^*/v)^{n/3}c_n x^2$$

with $v^* = r_0{}^3/\gamma$ ($\gamma = 2^{1/2}$ for FCC) and c_n a number dependent on the crystal structure and the repulsion exponent n. Hence, here the force constant a is specified in terms of volume. Again assuming $v_f{}^* = const.$, leads via $\beta\phi = const.$ at melting to $(z\varepsilon/kT_{\text{mel}})(v^*/v_{\text{mel}})^{n/3}c_n = const.$ Writing $V_{\text{mel}} = V_0 - \Delta V$ with $\Delta V = V_0 - V_{\text{mel}}$ and V_0 a reference volume at reference temperature T_0, we obtain

$$T/T_0 = (V_0/V_{\text{mel}})^{n/3} = (1 - \Delta V/V_0)^{n/3} \cong 1 + \tfrac{1}{3}n\Delta V/V_0 + \tfrac{1}{3}n(n-1)\tfrac{1}{2}(\Delta V/V_0)^2 + \cdots \tag{14.11}$$

Using Eq. (14.11) and the reduced Helmholtz expression $F = \tfrac{1}{2}N\Phi(0) - NkT(\ln Vv_f{}^*)$ one can obtain $P(T,V) = P(0,V) + 3\gamma kT/V$ with $\gamma = (n/6 + \tfrac{1}{3})$ the Grüneisen constant for this model. Substituting Eq. (14.11) in the pressure expression leads to

$$P_{\text{mel}} = P(0, V_{\text{mel}}) + \frac{3\gamma kT_0}{V_0}\left(\frac{T_{\text{mel}}}{T_0}\right)^{1+3/n} \quad \text{or} \quad P_{\text{mel}} = P(0, V_{\text{mel}}) + a\left(\frac{T_{\text{mel}}}{T_0}\right)^{b}$$

with $a = 3\gamma kT_0/V$ and $b = 1 + 3/n$, in which we recognize the Simon–Glatzel equation. Referring to his calculations on Ar, Ross indicated that neither $P(0,V_{\text{mel}})$, nor γ are really constants, with as conclusion that the Simon–Glatzel equation cannot be extrapolated with any confidence.

The rationalization given above improves on the one given by Salter [87] in which the Lindemann rule and the Grüneisen EoS were used to derive the Simon–Glatzel expression. The Simon–Glatzel relation can also be justified [88] by the Lennard-Jones–Devonshire theory [27]. Neglecting second- and higher-order terms in Eq. (14.11) leads to the so-called *Kraut–Kennedy relation*. This equation is not without dispute [89], as it has been defended as a non-approximated equation, although the various data and rationalizations, such as the one in Section 14.2.1, show otherwise.

One of the reasons that the Simon–Glatzel curve should not be used as an extrapolation equation is that repulsion as described by $\phi_{\text{LJ}} = \varepsilon(r_0/r)^n$ is too steep for small distances, that is, at high pressures, and is possibly better described by $\phi_{\text{M}} = A\exp(-\alpha r)$ [81]. As for power-law repulsion $n = -(r\, \text{d}\ln\phi_{\text{LJ}}/\text{d}\ln r)$, using the same expression for ϕ_{M} leads to an effective $n = \alpha r$. Approximating the lattice energy by $U = z\phi_{\text{M}} = \tfrac{1}{2}zA\exp(-\alpha r_1)$ with z the coordination number and r_1 the nearest-neighbor distance, the pressure $P = -\partial U/\partial V$ for a close-packed configuration becomes

$$P = (2^{1/2}/3)\left(\tfrac{1}{2}z\alpha A/r_1^2\right)\exp(-\alpha r_1)$$

As shown above, power-law repulsion leads to $P = aT^{1+3/n}$ so that $\text{d}P/P = (1 + 3/n)\text{d}T/T = (1 + 3/\alpha r_1)\text{d}T/T$. Integrating we obtain

$$\ln T = -\alpha r_1 + \ln(\alpha r_1 + 3) + C$$

The latter two equations together describe the melting curve parametrically as a function of r_1. For $P \to \infty$, $r_1 \to 0$ and the exponential repulsion predicts a limiting maximum temperature $T = \exp(\ln 3 + C)$, whereas the power-law repulsion leads to a divergence. Hence, the results deviate, to a degree depending on the pressure range used. However, this can only be part of the story, as $\ln T$ does not decrease for $P \to \infty$.

Kechin [90] showed that the Simon–Glatzel equation can be derived from the Clapeyron equation (Section 12.2), written as $d(\ln T_{mel})/dP = \Delta V/L$ with ΔV the volume change and L the enthalpy for melting, using a first-order Taylor series in ΔP. Later he extended the Simon–Glatzel equation by developing $d(\ln T_{mel})/dP$ in a Taylor series to second order. This leads to

$$d(\ln T_{mel})/dP \equiv y \cong y_0 + y_0{}'\Delta P + \tfrac{1}{2}y_0{}''(\Delta P)^2 + \ldots$$

Where the single and double primes indicate the first- and second-order derivatives, respectively. Employing a (1,1) Padé approximation,[15] the result can be written as

$$d(\ln T_{mel})/dP = (1 - \beta\Delta P)/\sigma(\alpha + \Delta P)$$

with the parameters σ, α, and β given by

$$\sigma = -y_0{}''/2y_0y_0{}', \quad \alpha = -2y_0{}'/y_0{}'' \quad \text{and} \quad \beta = (y_0{}''/2y_0{}') - (y_0{}'/y_0)$$

This expression can be integrated to

$$T_{mel} = T_0(1 + \Delta P/a)^b \exp(-c\Delta P) \tag{14.12}$$

with $a = \alpha$, $b = (1 + \alpha\beta)/\sigma = y_0{}'\alpha^2$ and $c = \beta/\sigma = -y_0{}'\alpha - y_0$. After some manipulation one obtains $\alpha = (L/\Delta V)_0/\sigma$, $\sigma = \gamma - \beta(L/\Delta V)$, where $\gamma = \Delta(L/\Delta V)/\Delta P$, and $\beta = \Delta\gamma/\Delta(L/\Delta V)$ so that all parameters have an identification in terms of thermodynamic quantities. The equation appears to describe the experimental data well (Figure 14.10). If $c = 0$, the *Kechin equation*, Eq. (14.12), reduces to the Simon–Glatzel equation with $1/\sigma$ equal to b. As $y_0{}' < 0$, the parameter $b > 0$, and $\Delta P = 1/\beta$ corresponds to a maximum in the

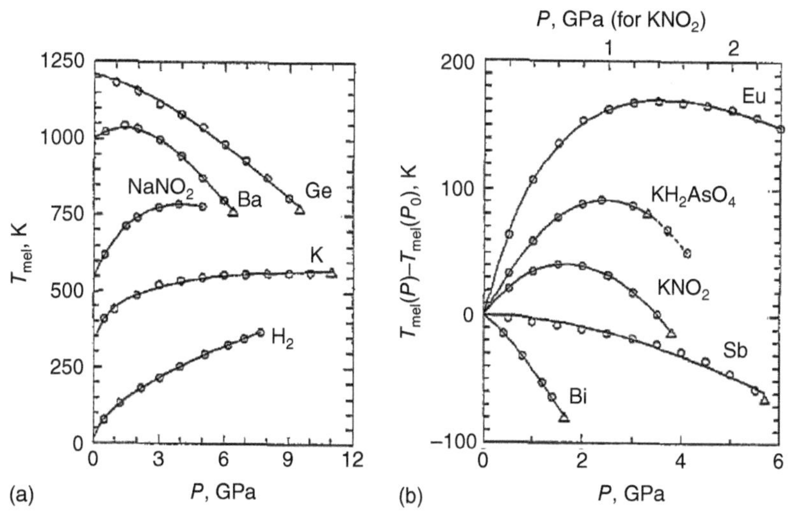

Figure 14.10 Melting curves for various compounds with ∘ experimental data points, Δ the high-pressure triple point, and the solid line, the fit to the Kechin equation. Source: Kechin [90]b/IOP Publishing.

15 See Chapter 8, footnote 4. For a second-order Taylor expansion $f(x) = c_0 + c_1x + c_2x^2$, the (1,1) Padé approximant reads $f(x) = \{c_0 + [c_1 - (c_0c_2/c_1)x]\}/[1 - (c_2/c_1)x]$.

melting curve. The hypothesis that all materials show a maximum in their melting curve was first suggested by Tammann and refined by Kawai and Inokuty [91] by postulating that the maximum could occur at positive as well as negative pressure. The latter case corresponds to a decreasing-with-pressure melting curve. As for the Kechin curve, the maximum corresponds to a negative pressure if $\beta < 0$, the equation can represent both increasing-with-pressure and decreasing-with-pressure melting curves (Figure 14.10).

Finally, it may be useful to explain the occurrence of a maximum in the melting curve by a simple model [92], in which it is supposed that the short-range structure in the liquid is similar to that of the solid near the melting line. If a second structure appears in the solid as a high-pressure phase, in the liquid phase a second structure with a similar short-range order as the high-pressure solid phase may also appear, though at a lower pressure. Because of the disorder, the transition to the higher-density phase occurs continuously in the liquid, while it can only occur with a jump in the solid phase. The liquid may thus become denser than the solid in a certain P–T range and the melting curve will show a maximum.

14.6 Other Aspects

Some materials show deviations from the general behavior. Here we discuss two such deviations. First, a puzzle for almost 130 years on two different melts from the same solid, and second, the odd–even effect for chain molecules.

Although most compounds do show a clear-cut melting point, there are exceptions. Acetaldehyde phenylhydrazone (APH, $C_8H_{10}N_2$), first prepared by Fischer in 1877, seemed to have two distinct forms with melting point 56 and 98 °C. The low-melting solid can be converted to the high-melting form by slurrying in weakly alkaline solution or by allowing ammonia vapor to permeate the solid for a few minutes. Conversely, the high-melting solid can be converted to the low-melting solid by slurrying in a weakly acid solution or by treatment with acidic vapors. Intrigued by the possible structural differences and reasons for the sensitivity to trace acid or base exposure, Bernades et al. [93] investigated this phenomenon with diffraction, infrared (IR), differential scanning calorimetry (DSC), nuclear magnetic resonance (NMR), calorimetry, microscopy, and simulations. All samples had identical IR, solid-state NMR spectra, and identical crystal structures, but exhibited sharp melting points varying from 56 to 101 °C. NMR studies of the melts provided the key to the understanding of this behavior: differently melting samples did so because they initially melted to liquids with different proportions of the Z and E isomers,[16] although given enough time they all tended to the same equilibrium proportion. The thermodynamic equilibrium state for the melt is one with an isomer ratio E/Z = 1.7, while for the solid state corresponds to a pure Z compound. For instantaneous isomerization in the melt, the melting temperature should be found at the thermodynamic melting temperature but for relatively slow isomerization it will result in a long(er) lifetime of the metastable melt. This isomerization process is catalyzed by traces of acid or base as usual. So, in polymorphism different structures melt to the same liquid, but in this case, the same structure melts to different metastable liquids.

16 E isomers have the substituents preferably on the opposite sides of a double bond, while Z isomers have them preferably on the same side. The letter E stems from “entgegen” (German: opposite) and the letter Z from “zusammen” (German: together). For relatively simple compounds, E means a trans isomer and Z a cis isomer, but for more complex compounds, the E–Z designation is more refined.

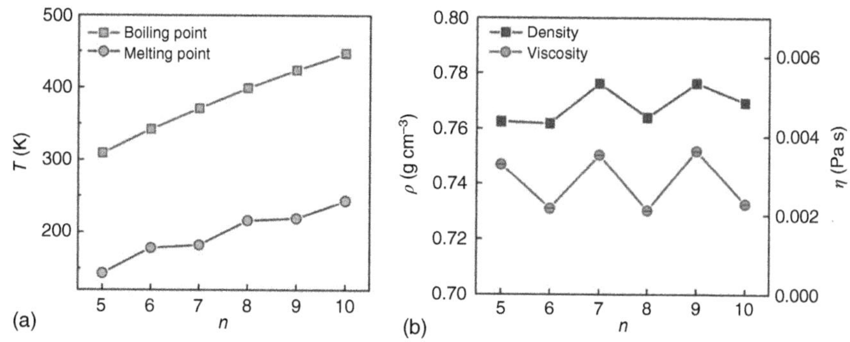

Figure 14.11 The odd–even effect. (a) Melting point and boiling of the first few alkanes. (b) Density and viscosity of the same first few alkanes. Source: Yang et al. [94]/John Wiley & Sons.

Several molecules do show the odd–even effect, n-alkanes (C_nH_{2n+2}) being a well-known example [95]. The melting points of n-alkanes show a zigzag pattern rather than a monotonic trend as a function of the number of carbon atoms (Figure 14.11a). This odd–even effect also holds for most of the α- and α,ω-substituents [96–98], the latter showing a larger effect [96]. The effect becomes more significant for short chains: the shorter the chain length, the larger the difference in melting points between two n-alkanes differing in length by one carbon atom. Other physical properties, such as sublimation enthalpy, solubility, and modulus, display a similar odd–even effect [96–98]. Although the odd–even effect was discovered about 150 years ago [95], its molecular origin was revealed only relatively recently [98] by using controlled crystal growth and single-crystal diffraction to solve the crystal structures of several n-alkanes. It appeared that the intermolecular distances between the CH_3 end groups are responsible for the alternation in the densities and melting points. The even-numbered n-alkanes have optimal intermolecular contacts at both ends, whereas the odd ones possess these only at one end, while at the other end their distances are larger. This leads to a less dense packing for the odd n-alkanes and their densities are lower with relatively lower melting points as a result. The effect diminishes with increasing chain length because the van der Waals attraction becomes dominating. Broadhurst [99] also studied long alkanes (14 paraffins, $n = 44–100$) and showed that the melting point can be described well by $T_{mel} = T_{mel}°[(n+a)/(n+b)]$, where $T_{mel}° = 141.1$ °C, $a = -1.5$ and $b = 5.0$. Associated thermodynamical data as well as some refinements are available as well [100]. Structure studies on n-alkyl carboxylic acids ($n = 6–16$) [101] showed that molecules form hydrogen-bonded dimers arranged into bilayers with a rectangular packing arrangement in the plane perpendicular to the dimer long axes. In all structures the carboxyl groups are identically disposed and the packing density within bilayers is comparable so that the alternating crystal density can be attributed solely to alternating packing density between bilayers. The odd–even effect has also been observed for adsorbed layers [102, 103].

In contrast, the boiling points of n-alkanes increase monotonically as a function of the molecular weight (Figure 14.11a), thus suggesting that the odd–even effect does not exist in the liquid state. Yang et al. [94] showed, possibly somewhat surprisingly, that the odd–even effect also exists in the liquid for the translational diffusivity. They used quasi-elastic neutron scattering experiments to measure the relaxation dynamics of the liquid n-alkanes. It appeared that odd-numbered n-alkanes exhibit up to 30 times slower dynamics than even-numbered n-alkanes near their respective melting points. Hence, an odd–even effect does exist in the liquid state (Figure 14.11b). The effect is more prominent in the dynamic quantities than in the thermodynamic quantities and the authors suggest that mechanisms other than periodic packing should be scrutinized to gain a more thorough understanding. Further, they speculate that because the symmetries of odd and even n-alkane molecules in an all-trans configuration are different, although this is not the favorable configuration in the liquid state, that an extra CH_3 group switches the structural symmetry of neighboring n-alkanes and thus affects the local packing structures of the n-alkane molecules. The effect thus should diminish with increasing temperature.

14.7 Melting in Perspective

In spite of that melting is a familiar and well-defined phenomenon, formulating a general theory of melting appears rather intractable, and none of the proposed models or explanations is generally accepted. It also appeared that many studies focused on a limited group of related compounds. The challenges associated with formulating a general theory are linked to the following aspects:

- Melting is conventionally considered as a physical process. Although many solids melt by redistributing relatively weak vdW forces, many other solids melt by redistributing relatively strong interactions, akin to chemical bonding.
- Although melting involves generally clear characteristics, such as a precise melting point, loss of long-range order, and reversibility, there is a large variety of interactions between the constituents.
- As we seen at several occasions before, the (structural) information available for liquids is much less than that for solids. This is reinforced by difficulties to obtain quantitative information for the interface between a melting solid and a liquid.
- In many cases, small amounts of impurities may obscure the nature of the melting process by exhibiting disproportional effects, in particular for the onset of fusion.

Based on these considerations, the "set/liq" model for melting was proposed by Galwey [104]. The starting point is again that the enthalpy and density changes that occur on melting are relatively small, leading to the assumption that at T_{mel} no significant changes in component sizes or shapes, intercomponent bonding, constituent packing control, or other physical properties (except viscosity) occur. In this model, the liquid is represented by dynamic equilibria between small domains composed of particles packed in alternative lattice arrays that rapidly and continually interconvert. These domains are supposed to be small, say, a few nm, so that they escape XRD detection. Each domain is bounded by less ordered interfaces that maintain coherent contact with all contiguous domains, analogous to a grain boundary, and provides lubrication for ready relative movements of the domains, thereby accounting for fluidity. The essential feature is thus the continual rapid interconversion of small zones, representing equilibrium between all possible, locally ordered domains.[17] With this qualitative model at hand, many aspects of melting for

17 In my view, the model resembles relatively strongly the SLS model (Section 9.6.3) and the original "hole" model of Frenkel [24] (not to be confused with hole model of Chapter 9), although no references

various types of materials were (obviously) qualitatively discussed. Han and Kim [105] also proposed a model containing rotationally free atomic clusters, resembling the "set/liq" model, but referring neither to Frenkel nor to Galwey.

A review on the thermodynamics of melting is given by Stishov [106], while Poirier [107] discusses inorganic materials without taking interfaces in to account and Nabarro [108] deals with dislocation mechanisms only. Bilgram [109] focuses on dynamics, Löwen [110] deals with superheating, dynamics, and a comparison with colloids, while Haymet [111] reviews melting from the density functional theory point of view. Both Kofman et al. [65] and Mei and Lu [66] primarily pay attention to surface effects and superheating, and Ram [112] discusses equilibrium theory in relation to freezing. Another view is presented by Garai [113], who describes melting as resonance between the energy of the vibrating atoms in the surface layer and supposes that if this" uniform" energy is higher than the energy corresponding to a metastable transition state, all surface atoms lose their position stability.

The scaling idea used by Ross [12] (Section 14.2.1) was revived by Dyre et al. [114, 115] and called "hidden scale invariance." The basic idea is that the repulsive part of the potential dominates so that scale invariance, being exactly true for homogeneous potential functions [116], is in reality approximately true. An application of the scaling idea is given in [117], where in essence one MD simulation carried out at a single coexistence state point is used to employ the scaling concepts for further predictions. The theory quantifies the deviations from the hard-sphere predicted melting-line invariants and is validated by simulations of the standard 12-6 LJ system. It is claimed that the theory applies for the sizable class of systems characterized by hidden-scale invariance, but detailed mechanisms are not discussed.

In spite of the ubiquitous presence of melting, it is fair to say that, although the thermodynamics are well understood, the mechanisms involved are much less so. Even brief browsing the literature on melting makes it clear that it is massive and contains many, sometimes diverging, views. Moreover, although generality often is claimed, some of the approaches are clearly dependent on the type of material. Nearly all types of defects have been proposed. Although in 3D melting dislocations provide only one of the routes, in 2D melting dislocations definitely do play an important role for which Joós [57] provided an extensive review. Apart from the general materials science problem that various types of materials may be controlled by different mechanisms, even for one type of material, several options exist to explain many of the phenomena involved. Indeed, several theories can describe the overall behavior of melting reasonably well and, possibly because of that, some authors claim universal applicability of their theory, but usually illustrate it by using data for one type of materials. Nevertheless, clearly for normal melting surfaces are an important factor for all types of materials. Alternative, attractive simple views still surface regularly (see, e.g. Section 14.3), so that a final verdict is hardly possible, although it may well be that a generally applicable mechanistic theory does not exist.

---

were made. With respect to SLS theory, the differences are that the ordered domains contain all the various lattice types a crystal can have instead of only the most stable one, and that the interfacial zones are also densely but randomly packed instead of being gas-like. The "hole" model describes a liquid, in the words of Frenkel, as particles "distributed in a non-uniform way forming more or less compact 'bunches,' separated from each other by fluctuating cracks, or, to be more exact – a relatively compact mass with a density but slightly below that of the corresponding crystal by fluctuating fissures." The only difference with the "set/liq" model to me seems to be the presence of only one type of ordered domain, based on the most stable lattice type.

14.8 The Reverse Transition: Freezing or Solidification

After having discussed melting (or fusion), we now turn to the reverse transition, freezing (or solidification), which is evidently also important. Sometimes freezing is called crystallization, but not all solidification processes are also crystallization processes. In these transitions, nucleation processes are involved, the treatment of which, as said, is a field in its own right [118–120]. Basic nucleation theory is dealt with in Section 13.3.2, which should be read in conjunction, and applies to all nucleation processes, albeit with different degrees of additional complications. After nucleation, growth of nuclei also occurs.

14.8.1 Nucleation and Growth

During freezing several molecules form a solid cluster. The barrier involved in the liquid–solid nucleation process can again be described in terms of the Gibbs energy difference ΔG between the phases, solid and liquid in this case, and their interfacial tension. We derived for a cluster of β in a matrix of α that the change in Gibbs energy is given, using the same notation as in Section 13.3.2, by

$$\Delta G = \Delta G_{\text{vol}} + \Delta G_{\text{sur}} \equiv (4/3)\pi r^3 g + 4\pi r^2 \gamma = N v g + (36\pi^{1/3}) N^{2/3} v^{2/3} \gamma$$

which lead for a *nucleus* to

$$r_N^* = -2\gamma/g, \qquad N^* = (32\pi/3v)(\gamma/g)^3 \qquad \text{and} \qquad \Delta G_{\text{hom}}^* = 16\pi\gamma^3/3g^2$$

For freezing we use the melting temperature T_{mel} as a reference. At $T = T_{\text{mel}}$, $\rho h = -\Delta_{\text{mel}}H$ and $\rho s = -\Delta_{\text{mel}}S = -\Delta H_{\text{mel}}/T_{\text{mel}}$, with $\Delta_{\text{mel}}H$ and $\Delta_{\text{mel}}S$ the melting enthalpy and melting entropy per unit volume, respectively. For $T \neq T_{\text{mel}}$, with $\Delta T = T_{\text{mel}} - T$ the *degree of supercooling*, we have

$$\Delta H_V = -\Delta_{\text{mel}}H - \int_{T_{\text{mel}}}^{T} C_P \, \mathrm{d}T \qquad \text{and} \qquad \Delta S_V = -\Delta_{\text{mel}}S/T_{\text{mel}} - \int_{T_{\text{mel}}}^{T} (C_P/T) \, \mathrm{d}T$$

with ΔC_P the difference between the heat capacity of the two phases. Hence

$$\Delta_{\text{mel}}G = -\Delta_{\text{mel}}H\Delta T/T_{\text{mel}} - \int_{T_{\text{mel}}}^{T} \Delta C_P \, \mathrm{d}T + T\int_{T_{\text{mel}}}^{T} (\Delta C_P/T) \, \mathrm{d}T \tag{14.13}$$

Approximating $\Delta C_P = 0$, we have $\Delta_{\text{mel}}G = -\Delta_{\text{mel}}H\Delta T/T_{\text{mel}}$, and consequently

$$\Delta G = -4\pi r^3 \Delta_{\text{mel}}H\Delta T/3T_{\text{mel}} + 4\pi r^2 \gamma \tag{14.14}$$

For the maximum we now obtain, using $\Delta_{\text{mel}}S = \Delta_{\text{mel}}H/T_{\text{mel}}$,

$$r^* = 2\gamma/\Delta_{\text{mel}}S\Delta T \qquad \text{and} \qquad \Delta_{\text{hom}}G^* = 16\pi\gamma^3/3(\Delta_{\text{mel}}S\Delta T)^2 \tag{14.15}$$

Hence, increasing ΔT induces a lower critical radius r^* and lower barrier $\Delta G_{\text{hom}}{}^*$. However, at lower temperature the kinetics become slower, and one expects a maximum in the nucleation rate as a function of T.

The derivation of the cluster size distribution (Section 13.3.2) yielded

$$p = n/n_0 = \exp(-\beta\Delta G) \qquad \text{or} \qquad p^* = n^*/n_0 = \exp(-\beta\Delta_{\text{hom}}G^*)$$

with p^* the probability for creating a critical nucleus of size n^*. If we write as pseudo-reaction $n\text{B} = \text{B}_n$, where B represents the aggregating molecules and B_n a nucleus with n B molecules, the corresponding equilibrium constant is $K = [B_n]/[B]^n$ with [B] the concentration of B. For the nucleus, with $-\Delta_{hom}G^* = -RT \ln K$, we obtain

$$[B^*] = [B^*]^{n^*} \exp(-\beta\Delta_{hom}G^*) \quad \text{and} \quad J^* = p^*[B^*]$$

where J^* is the *nucleation rate of nuclei*. The growth of supercritical clusters will occur via diffusion with as effective diffusion coefficient

$$D = (kT/h)\lambda^2 \exp(-\beta\Delta G_D) \quad \text{or} \quad \exp(-\beta\Delta G_D) = h/3\pi\eta\lambda^3$$

after inserting the Stokes–Einstein expression, $D = kT/3\pi\lambda\eta$. Here k, T, and h have their usual meaning, while λ represents the mean free path in the liquid and η the viscosity. Overall, the homogeneous nucleation rate thus becomes

$$J_{nucl} = J_0 \exp[-\beta(\Delta_{hom}G^* + \Delta G_D)] \quad \text{or} \quad J_{nucl} = (C_1/\eta)\exp(-\beta\Delta_{hom}G^*)$$

with J_0 and C_1 constants. The rate J is thus promoted by a lower temperature because $\Delta_{hom}G^*$ becomes smaller but is retarded at the same time because η becomes larger.

Similar arguments can be applied for the growth rate after nucleation. The frequency of attachment of molecules to the growing cluster w^+ is proportional to the number density of the liquid n_L, to the surface area of the cluster $4\pi r^2$, and is given by

$$w^+ = C_2 4\pi r^2 n_L \exp(-\beta\Delta U) \equiv C_3 \exp(-\beta\Delta U)$$

with C_2 a constant, C_3 a slightly temperature-dependent pre-exponential factor, and ΔU the activation energy barrier for rearrangement of the molecules in the liquid when crossing the cluster boundary to occupy their proper positions in the solid. The frequency of detachment of molecules from the growing cluster w^- is given by

$$w^- = C_4 \exp[-\beta(\Delta_{mel}H + \Delta U)]$$

so that for the growth rate we obtain

$$J_{growth} = D(w^+ - w^-) = D\exp(-\beta\Delta U)[C_3 - C_4 \exp(-\beta\Delta_{mel}H)]$$

At $T = T_{mel}$, the growth rate is zero and therefore $C_3 = C_4 \exp(-\beta\Delta_{mel}H)$. Overall,

$$J_{growth} = a\exp(-\Delta_{mel}S/k)\exp(-\beta\Delta U)[1 - \exp\left[-\beta\Delta_{mel}H\left(T^{-1} - T_{mel}^{-1}\right)\right] \quad (14.16)$$

with $a = C_4 D$ and $\Delta_{mel}S = \Delta_{mel}H/T_{mel}$. If for $\Delta_{mel}G$, Eq. (14.13), we may assume that $\Delta C_P = 0$, $g = n_L\Delta\mu = -\Delta_{mel}H\Delta T/T_{mel}$ and the term in square brackets in Eq. (14.16) can be replaced by $[1-\exp(-\beta\Delta G_V)]$. Both nucleation and growth take place below T_{mel}, but above the glass transition temperature T_g if the melt is glass-forming. Their optimum does not necessarily occur at the same temperature.

An essential difference for the nucleation of solids compared to liquids is the anisotropy of the solid phase nucleated. Not only the bulk properties of the solid are anisotropic, which leads to an intricate discussion on Gibbs energy for solids [121, 122], but also the surfaces are anisotropic, which leads to different surface energies for various surfaces. In principle, nuclei are no longer spherical and the effects of strain should be considered as well [122]. This renders freezing from a modeling point of view more complex than condensation [123, 124].

14.8.2 Some Further Remarks

Many other aspects remain. We cannot deal with all of them here, but only indicate a few of them that can improve the classical monomer addition–removal results, and thereafter we discuss heterogeneous nucleation in some detail.

First, we note that it is often assumed that the degree of supercooling or supersaturation is small so that g can be taken as linearly dependent on the ΔT or ΔP. This is not necessarily true and can be remedied if an estimate for the properties of the metastable phase can be made. Second, basic CNT theory assumes spherical nuclei with properties identical to those of a macroscopic volume of the new phase and with a sharp interface. While non-sphericity can be taken into account by assuming another shape, both the assumption of identical properties and the sharp interface are doubtful, as can be easily appreciated by calculating N^*, usually resulting in a rather limited number of molecules. For particles with such a small number of molecules, the total interaction per unit volume will be different from that in the bulk, leading to a difference in both the bulk Gibbs energy G and interface Gibbs energy γ, but actually indicating that a continuum treatment, the capillarity approximation, is in principle inappropriate. Third, in particular for solidification from solutions, several mechanisms have been proposed that deviate considerably from the classic monomer addition–removal nucleation picture (Figure 14.12a). They often involve the aggregation of small, amorphous or semicrystalline particles that are formed by a more or less classical nucleation process. These aggregates may, wholly or partially, rearrange to yield another stage, from which further growth takes place [127]. We just quote the example for silica [126] to produce monodisperse particles via the Stöber-Yokoi procedure and analyzed by high-resolution cryogenic TEM, pH monitoring, and ^{29}Si magic-angle-spinning NMR spectroscopy. It appeared that the formation proceeds through the synchronized association of uniform 2.3 nm primary particles with exceptional colloidal stability to larger units called assemblies. Growth of these assemblies occurs through the continuous addition of primary particles to their surface, concomitant with a collapse of the primary particles upon association, both in terms of morphology and chemistry (Figure 14.12b). This mechanism is consistent with previously reported observations. Such meso-sized clusters play a role for many systems [128]. In crystallization from solutions often surface-active agents are used, usually to promote growth, but also to retard it. For example, such small molecular agents play an important role in, what is called, biomineralization [129, 130], but polymers are also frequently used [131]. Overall, the crystallization process is even more complex than solidification [132]. For the active field of crystallization, see, for example, [133] for a critical discussion on classical and non-classical nucleation mechanisms [134] and for recent developments [135].

For heterogeneous nucleation an already existing surface, be it that of the container or a (deliberately added or impurity) particle, must be considered as briefly addressed in Section 13.3.4. The CNT barrier for homogeneous nucleation is given by $\Delta_{\text{hom}}G^* = 16\pi\gamma^3/3g^2$ where $g = (\mu_S - \mu_L)/\Omega$ with chemical potential μ and molecular volume Ω. With the use of the heterogeneity function f and replacing $(\mu_S - \mu_L)$ by $kT\ln S$ (Section 13.3.2), we obtained for the CNT rate for heterogeneous nucleation $J = J_0 \exp[-16\pi\gamma^3\Omega^2 f/(kT)^3(\ln S)^2]$.

A more detailed analysis by Liu [136], extending heterogenous nucleation to curved surfaces, shows that under normal conditions, nucleation includes a sequence of progressive processes, each characterized by different *heterogeneity functions* $f(m,x)$. The latter are a

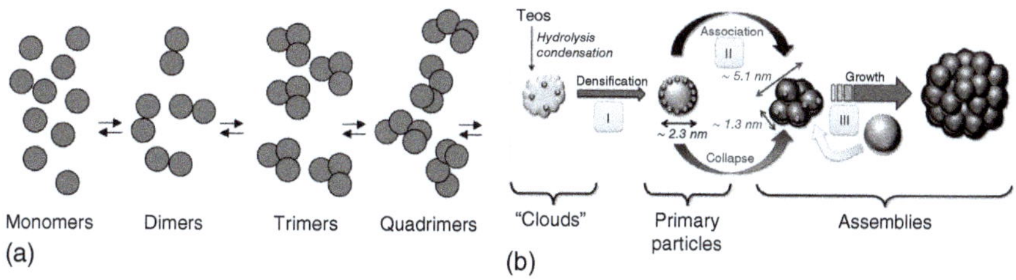

Figure 14.12 Schematics of nucleation. (a) According to classical nucleation theory (CNT) by monomer addition and removal. Source: Ford [125]/with permission of SAGE Publishing. (b) More complex as proposed, for example, for silica in [126]. Source: Adapted from Carcouët et al. [126].

generalization by Fletcher [137] of f to curved surfaces (particles). Here $m = (\gamma_{PL} - \gamma_{PS})/\gamma_{SL}$, where the subscripts refer to solid (S), liquid (L), and particle (P), and $x = R_P/R^*$ with particle radius R_P and critical nucleus radius R^*.

The analysis was done assuming, as usual, steady-state conditions with single monomer addition or removal, with density of foreign bodies N^0 and ΔG_g the energy barrier to form a g-mer. One then has $n_g = n_1 \exp(-\Delta G_g/kT)$, where n_g and n_1 denote the density of g-mer and of monomer, respectively. The rate of molecule addition is $\alpha = \beta_{st}\nu^{\neq}$, where $\nu^{\neq}$ is the collision rate of monomers with an embryo and β_{st} the sticking probability, defined as $\beta_{st} = (\nu a/\lambda)\exp(-\Delta G_{st}/kT)$ with ν the vibration frequency of structural units in the neighborhood of the surface, a the dimension of structural units in the direction parallel to the crystal surface, and λ the average distance between two kinks at the surface. Further, ΔG_{st} is the activation energy for addition at steps. Under steady-state conditions the rate α' at which embryos lose monomers should be equal to the rate of addition α and one has $\alpha_g n_g - \alpha_{g+1}' n_{g+1} = 0$. Further, if there is growth, with a surface concentration of g-mers f_g, the formation rate of critical nuclei per unit volume and time around a nucleating particle J' is equal to the steady-state growth of embryos at the surface. That is, we have $\alpha_g f_g - \alpha_g' f_{g+1} = const.$, subject to the boundary conditions $f_{gcri} = 0$ (because whenever a $(g_{cri}-1)$-mer becomes a g_{cri}-mer, it disappears from the $(g_{cri}-1)$-population) and $f_g/n_g \to 1$ when $g \to 1$ (because both n_g and f_g are large numbers when g is small). Given this, Liu derived [138]

$$J^{*\prime} = f'(m,x)[f(m,x)]^{1/2} J_0 \exp[-\Delta_{hom} G^* f(m,x)] \quad \text{where}$$

$$J_0 = 4n_1^2 \alpha \beta_{st} \Omega (\gamma_{SL}/kT)^{1/2} \quad \text{and} \quad f'(m,x) = ½ + (1 - xm)/2w$$

As the average nucleation rate J in the solution depends on the density and size of foreign particles in the system, it is given by $J = 4\pi a(R_P)^2 N^0 J^{*\prime}$. The influence of foreign bodies is thus not only present in the exponential nucleation barrier term but also the pre-exponential term. For homogeneous nucleation one has $f''(m,x) = f(m,x) = 1$ and $4\pi a(R_P)^2 N^0 \to 1$. The function $f(m,x)$ is shown in Figure 14.13a.

Assuming that the relation between the incubation time of nucleation τ is proportional to $1/J$, a linear relation between $\ln\tau$ and $[\ln(1+S)]^{-2}$ for a given $f(m,x)$ should be obtained (Figure 14.13b). At low supersaturation, nucleation will be controlled by the process with a small heterogeneity function $f(m,x)$, which results from a strong interaction and good structural match between the foreign bodies and the crystallizing phase. At high supersaturation, nucleation on foreign particles with a weak interaction and poor structural

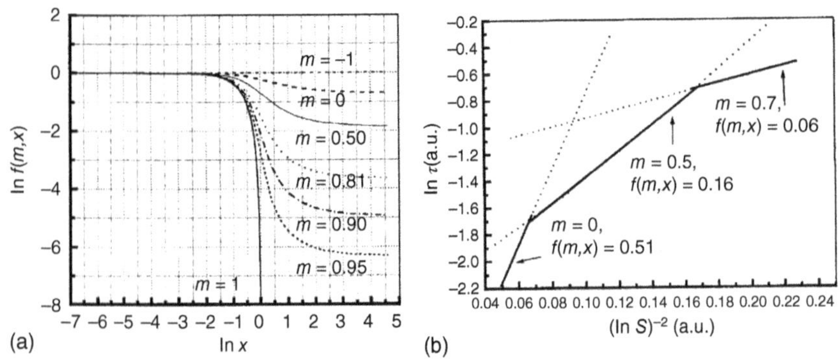

Figure 14.13 Heterogeneous nucleation. (a) The heterogeneity factor $f(m,x)$ as a function of relative particle size x for various m values. (b) The incubation time τ as a function of supersaturation S. Both quantities are in arbitrary units (a.u.). Source: Liu [136]/American Institute of Physics.

match with the crystallizing phase ($f(m,x) \to 1$) will govern the kinetics. The model was confronted with experiments for paracetamol nucleating from an aqueous solution, N-lauroyl-L-glutamic acid di-n-butylamide nucleating from an isostearyl alcohol solution and $CaCO_3$ from an aqueous solution. For the former two crystals, three rather linear parts were discerned in the $\ln\tau$ versus $[\ln(1+S)]^{-2}$ plot, indicating that the change of $f(m,x)$ at different supersaturations can be attributed to the change in m rather than to x. For $CaCO_3$ also three regimes were detected but at low supersaturation the relation is somewhat nonlinear. This was attributed to the relatively high interface energy of $CaCO_3$ compared to that of the organic crystals, leading to larger R^* values that render the nonlinear influence of x in the pre-exponential factor larger. However, at high supersaturation for all three crystals the limiting slope in the $\ln\tau$ versus $[\ln(1+S)]^{-2}$ plot for $f(m,x) \to 1$ was not obtained. Further, for $CaCO_3$ the slope under microgravity conditions is four times larger as observed by Liu and yields an interfacial energy close to that obtained from contact angle measurements. Genuine homogeneous nucleation, the upper limit of heterogeneous nucleation, may thus not be easily achievable. The balance between homogeneous and heterogeneous nucleation is also discussed in [139]. We further refer to an early review [140] and a recent one on atomistic mechanisms [141].

14.8.3 Polymers and Metals

Solidification of metals and polymers are both important fields and complex. We briefly mention some aspects of metal solidification, referring for details to the literature, and thereafter discus polymers somewhat more extensively.

The solidification of metals is important for castings, where a liquid metal of appropriate temperature and composition is introduced in a mold under controlled conditions of cooling, pouring, and possibly atmosphere. As ingots are usually large, temperature gradients are inevitable and enthalpy effects may decrease the local temperature in the ingot. Although homogeneous nucleation may occur, most likely heterogeneous nucleation dominates, also because the wall of the mold is the coolest. Deliberately introduced particles may act as nucleation sites as well. Finally, segregation of a component might occur. Together these effects render the solidification process inhomogeneous and resulting in different microstructures at different positions. The basic nucleation mechanism seems to be reasonably well caught by CNT based on monomer addition–removal kinetics. Broad and detailed overviews are given in [142, 143].

Polymer solidification in which chain connectivity and thus the multiple length and time scales involved see to it that it is an extensive and complex subject. The final properties of semicrystalline polymer materials are affected by the starting melt, the path of nucleation, the organization of lamellar crystals. and the evolution of the final crystalline structures. Numerous different, or even conflicting, models and theories have been proposed, but none of them is fully satisfactory [144]. The relevant literature is huge, and we limit the discussion to some basic aspects. Extensive treatises are [74, 145] while recent reviews are [146, 147].

In view of that polymers are generally flexible long chains of coupled monomeric units, one could expect that they do not form crystals at all. In principle, a periodic structure in three dimensions can be obtained if the chains form identical helices (including stereoregularity), orient the helix axes parallel to each other, and pack in a regular manner. Crystals of short-chain molecules such as the *n*-alkanes are indeed composed of stacks of layers with their interfaces occupied by the end groups that cannot be incorporated in the interior parts of the layer. In a melt (and in solution as well), polymers are coiled and penetrate each other so that the helical state is not obtained: disentangling the chains is entropically expensive and would require a too long a time. Nevertheless, they crystallize, and it appears that polymeric solids are built up as two-phase structures, composed of platelike crystallites that are separated by amorphous regions. Cooling a melt to a temperature at which a polymer crystallizes results in a semicrystalline state with this character. However, similarly as for alkanes, polymer crystallization requires that the chain entanglements are removed from the melt which cannot be done within the available time and the entanglements are just shifted into the amorphous intercrystalline regions. Since the crystal thickness is small compared to the chain length, a given chain returns into the same or the adjacent crystal after an excursion into the amorphous region (Figure 14.9a). The platelet thickness depends on the crystallization temperature and generally increases with rising temperature. In the final morphology *spherulites* occur, the structure of which results from repeated branching and splaying of the crystal lamellae. The consequence is that the radial growth rate of a spherulite is equal to the lateral growth rate of the constituent lamellar crystallites. In fact, polymer crystals grow in the two lateral directions only while growth in the chain direction (normal to the plate surface) is blocked by folds and loops. The growth rate decreases exponentially with increasing temperature. Gedde [148] provided a compact description of crystalline polymers.

Crystallization of polymers can be broadly divided into three groups: (i) during polymerization, (ii) induced by orientation, and (iii) under quiescent conditions. The mechanism of crystallization during polymerization can be the successive polymerization and crystallization or the simultaneous polymerization and crystallization [150]. In the latter case, the final polymer crystal is obtained due to chemical reactions at the liquid–solid interface (Figure 14.14a). Crystallization induced by orientation can be realized by, for example, stretching a sample for some time, whereby a distortion of the chains from their most probable conformation results, thereby decreasing the conformational entropy and easing the transformation to a crystalline state. Both natural rubber and polyisobutylene crystallize relatively easily under stretched conditions, whereas they crystallize slowly otherwise [74].

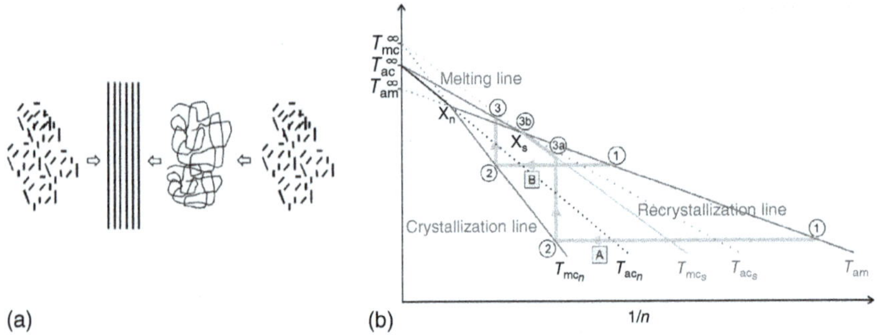

Figure 14.14 Polymer crystallization. (a) The left part represents crystallization during polymerization while the right part indicates crystallization after polymerization. (b) T-n^{-1} diagram for polymer platelet in a melt (label a) dealing with three phases: mesomorphic (m), native crystalline (c_n), and stabilized crystalline (c_s). Lines of size-dependent phase transitions: T_{mcn} between mesomorphic and native crystalline layers; T_{acn}, T_{mcs}, T_{acs}, and T_{am} with corresponding meanings. Two routes for an isothermal crystallization followed by heating: route A for low crystallization temperatures and route B for high crystallization temperatures. Triple points X_n (X_s) with coinciding Gibbs energies of the melt, a mesomorphic layer, and a native stabilized crystalline layer with the same thickness. Source: Strobl [149]/American Physical Society.

In the 1970s, focusing on crystallization under normal conditions, the conventional view for crystallization, due to Hoffmann and Lauritzen [75], arose. Their theory was derived originally for flexible polymers like polyethylene and is basically an extension of the regular monomeric addition theory for small molecules. Analysis indicated that the maximum growth rate is achieved by crystallites whose thickness is near the stability limit, that is, the melting point, of the lamellar crystallites. The latter is for a crystal with thickness d and heat of fusion $\Delta_{mel}H$ given by the Gibbs–Thomson equation $T_{mel}(d) = T_{mel}{}^{\infty} - (2\gamma^{E}T_{mel}/d\Delta_{mel}H)$ describing the suppression of T_{mel} below the equilibrium value for a macroscopic body $T_{mel}{}^{\infty}$. This suppression is caused by the excess surface energy γ^{E} of the fold surface. Hence, the lamella thickness becomes

$$d = (2\gamma^{E}T_{mel}{}^{\infty}/\Delta_{mel}H\left(T_{mel}{}^{\infty} - T\right) + \delta$$

where δ is a small excess driving force required to realize crystallization. The resulting growth rate is given by

$$u = u_0 \exp\left(-T_A{}^{*}/T\right) \exp\left(-T_G/\left(T_{mel}{}^{\infty} - T\right)\right]$$

Here the first exponential term expresses the temperature dependence of the segmental mobility in the melt; for temperatures of, say, $T_g + 100$, it obeys an Arrhenius law with some effective activation temperature $T_A{}^{*}$. The second exponential term refers to the energy of activation associated with the placement of a secondary nucleus on the growth face. It diverges together with d at $T_{mel}{}^{\infty}$. For the parameter T_G the theory yielded an expression of the form $T_G = K/T$, with K determined by $\Delta_{mel}H$, γ^{E} and the surface energy γ_{lat} of the growth face.

The Hoffman–Lauritzen treatment thus implies that the lamellae grow by direct attachment of chain sequences from the melt onto essentially smooth lateral faces, and the lamellar thickness is determined by the supercooling below the equilibrium melting point as given by the Gibbs–Thomson equation. A detailed discussion on the growth kinetics and mechanisms (up to about 1994) is given in [148].

In the 1990s new insights appeared. From crystallizing polyethylene just below the triple point it appeared that the formation of crystals occurred from a mesomorphic "hexagonal" phase and it was speculated that this may also happen under normal pressure [151]. Further, the scattering which arose before the appearance of the crystallites was interpreted as indicating the build-up of a precursor phase in the first step of polymer crystallization [152]. Finally, time- and temperature-dependent small-angle X-ray scattering experiments contradicted the basic assumption of control of the lamellar thickness by the supercooling below the equilibrium melting point [153]. The lamellar thickness turned out to be determined by the supercooling below another temperature, which is always located above the equilibrium melting point. In addition, the thickness is hardly affected by the presence of co-units.

While all these data arose from global measurements, more experimental options became available with time, such as modulated high-speed calorimetry, synchrotron radiation with resolution down to about a micrometer, time-dependent recording of IR during crystallization, and simultaneous small- and wide-angle X-ray diffraction (XRD). Finally, atomic force microscopy (AFM), capable of providing real-space images of local structures in time- and temperature-dependent in-situ studies with nanometer-range resolution can be used.

Based on this information, Strobl advocated that crystallization occurs generally via a mesomorphic phase,[18] as mentioned already for PE [151], the detailed arguments being provided in his reviews [149, 154]. If true, this implies varying melting temperatures, dependent on the conditions used (Figure 14.14b). It seems that not many attempts to use the theory have been published but for poly(ε-caprolactone) Sheth [155] did so. The lamellar crystal thickness as a function of T was consistent with both the Hoffman–Lauritzen and the Strobl models. However, in contrast to the predictions of Strobl's model, the value of the mesomorph-to-crystal equilibrium transition temperature was very close to the zero-growth temperature. Moreover, the lateral block sizes (obtained using wide-angle XRD) and the lamellar thicknesses were not found to be controlled by the mesomorph-to-crystal equilibrium transition temperature. Hence, the author concluded that the crystallization of poly(ε-caprolactone) is not mediated by a mesophase.

Finally, we note that as memory effects in polymer melts are pronounced, self-nucleation may occur from remaining crystallized entities in the melt [156].

14.8.4 Water

In view of its importance, ice nucleation from water is a highly researched topic [157] on which we elaborate a bit. For water CNT has been reviewed in detail [158]. The authors considered various methods to obtain the thermodynamic and kinetic parameters as well as the sensitivity of calculated nucleation rate for the parameters chosen. This sensitivity is controlled firstly by the interfacial tension (determining the barrier for the nucleation process) and led to a difference of 25 orders of magnitude depending on the choice of the parameters. Of second importance is the activation energy of the nucleation, which can lead to a variation of 16 orders of magnitude. This led to the conclusion that temperature behavior assumed for these parameters generally does not match with predicted nucleation rates. For water the authors recommend a particular description of the interfacial tension [159] combined with the activation energy from self-diffusion experiments [160].

18 Mesomorphic phases are phases or 'forms' that occur in between (or 'in the middle' between) the traditional phases of solid, liquid, or gas.

Frequently one wishes to avoid water solidification (ice formation). Because this is clearly a highly relevant topic, ice-phobic surfaces have drawn widespread attention because of their ability to repel incoming water droplets, suppress ice nucleation, and/or lowering ice adhesion strength. As time elapses and temperature lowers sufficiently, ice accretion becomes inevitable, and for various outdoor anti-icing applications the lowest ice adhesion strength possible is desirable. Coatings are the usual answer, for example [161].

Passive anti-icing strategies are the use of smooth surfaces, textured surfaces, slippery surfaces, and sub-surface textured surfaces, each having their own pros and cons [162]. Smooth surfaces usually need to be tuned by surface chemistry and elastic modulus to lower ice adhesion strength, possibly combinable with other anti-icing strategies to reduce ice adhesion strength. For textured surfaces to maintain surface ice-phobicity, the longevity of surface chemistry and hierarchical structures is a critical issue. To enhance surface ice-phobicity, textured surfaces can introduce elastic/flexible substrates and/or hollow sub-surface structures to lower the ice adhesion strength. For slippery surfaces, low ice adhesion strength can be realized by designing the proper lubricating mechanisms and combining them with other passive anti-icing strategies. The use of subsurface textured surfaces combined with other passive anti-icing strategies seems to be become realistic.

Obviously, sometimes ice formation is desirable. For this a set of physical descriptors for heterogeneous ice nucleation was recently identified via machine learning [163], apart from a preferable lattice match to a low-index face of ice as known from Vonnegut's work [164]. These factors are (i) local tetrahedral ordering induced in liquid water, (ii) density reduction of liquid water near the surface, and (iii) corrugation of the adsorption energy landscape felt by water with a smooth energy landscape preferred to a rugged one. Two of these descriptors are directly related to intrinsic surface properties (lattice match and adsorption energy landscape), while the others relate to interfacial water. The authors concluded that no single descriptor could predict the nucleation ability reasonably well. However, other potentially influencing factors such as macroscopic hydrophobicity and large-scale roughness are relevant as well. So, quantitative understanding of heterogeneous ice nucleation and the in-silico design of materials to control ice formation has not been reached yet.

Problem 14.3 ***Heterogeneous Nucleation*** Assess the errors made by using Eq. (14.14) in view of all other approximations.

References

1 Moelwyn-Hughes, E.A. (1961). *Physical Chemistry*, 2e. Oxford: Pergamon Press.
2 Slater, J.C. (1939). *Introduction to Chemical Physics*. New York: McGraw-Hill Book Comp. (also Dover, 1970).
3 Clusius, K. and Weigand, K. (1940). *Z. Phys. Chem. B* 46: 1.
4 Lekkerkerker, H.N.W. and Tuinier, R. (2011). *Colloids and the Depletion Interaction*. Dordrecht: Springer.
5 Hoover, W.G. and Ree, F.H. (1968). *J. Chem. Phys.* 49: 3609.
6 Dash, J.G. (1999). *Rev. Mod. Phys.* 71: 1737; (2002). *Contemp. Phys.* 43: 427.
7 Hoover, W.G. and Ross, M. (1971). *Contemp. Phys.* 12: 339.

8 Lindemann, F.A. (1910). *Physik. Z.* 11: 609.
9 Sutherland, W. (1890). *Phil. Mag.* 30: 318; (1891). *Phil. Mag.* 32: 31.
10 Gilvarry, J.J. (1956). *Phys. Rev.* 102: 308; *Ibid.* 103: 1700.
11 Cho, S.-A. (1982). *J. Phys. F: Met. Phys.* 12: 1069.
12 Ross, M. (1969). *Phys. Rev.* 184: 233; Kuramoto, Y. (1971). *Progr. Theor. Phys.* 45: 1724.
13 Lawson, A.C. (2009). *Phil. Mag.* 89: 1757.
14 Wallace, D.C. (1991). *Proc. Roy. Soc. A* 433: 615 and 631; see also Wallace, (2002).
15 Grimvall, G. (1986, 1999). *Thermophysical Properties of Materials*, 2e. Amsterdam: North-Holland.
16 Mulargia, F., Quareni, F., and Boschi, E. (1987). *Il Nuovo Cim. C* 10: 479.
17 Wolf, G.H. and Jeanloz, R. (1984). *J. Geophys. Res.* 89: 7821.
18 Born, M. (1939). *J. Chem. Phys.* 7: 591; (1940). *Proc. Cambridge Philos. Soc.* 36: 160.
19 Wang, J., Li, J., Yip, S. et al. (1995). *Phys. Rev. B* 52: 12627.
20 Slagle, O.D. and McKinstry, H.A. (1967). *J. Appl. Phys.* 38: 437.
21 Tallon, J.L., Robinson, W.H., and Smedley, S.I. (1977). *Nature* 266: 337; Tallon, J.L. and Robinson, W.H. (1977). *Phil. Mag.* 36: 741; Tallon, J.L. (1982). *Nature* 299: 188.
22 Wang, J., Li, J., Yip, S. et al. (1997). *Physica A* 240: 396.
23 Boyer, L.L. (1985). *Phase Transit.* 5: 1.
24 Frenkel, J. (1946). *Kinetic Theory of Liquids*. Oxford: Oxford University Press.
25 Gorecki, T. (1974). *Z. Metalkd.* 65: 426; *Ibid.* (1976). 67: 269; *Ibid.* (1977). 68: 231.
26 Schultze, G.E.R. (1967). *Metallphysik*. Berlin: Akademie-Verlag; Takamura, J.I. (1965). *Point Defects*, Ch. 14 in *Physical Metallurgy* (ed. R.W. Cahn), p. 681. Amsterdam: North-Holland Publ. Comp.
27 Lennard-Jones, J.E. and Devonshire, A.F. (1939). *Proc. Roy. Soc. A* 170: 464.
28 Robinson, L.B., Au-Yeung, V.S.T., and Yassen, H.A. (1980). *Phys. Rev.* 21: 2352.
29 Pople, J.A. and Karasz, F.E. (1961). *J. Phys. Chem. Solids* 18: 28; Karasz, F.E. and Pople, J.A. (1961). *J. Phys. Chem. Solids* 20: 294.
30 Domb, C. (1956). *J. Chem. Phys.* 25: 783; Mori, H., Okamoto, H., and Isa, S. (1972). *Progr. Theor. Phys.* 47: 1087.
31 Kuhlmann-Willsdorf, D. (1965). *Phys. Rev.* 140: 1599A; Cotterill, R.M.J. (1980). *J. Cryst. Growth* 48: 582.
32 Shockley, W. (1952). *Dislocation Models of Grain Boundaries*, Ch. 13 in *L'Etat Solide* (ed. R. Stoops), p. 431. Institut Int. de Physique Solvay. The discussion on a liquid as being a solid with a high density of dislocations is on page 477.
33 Cotterill, R.M.J. (1979). *Phys. Rev. Lett.* 42: 1541.
34 Mott, N.F. and Gurney, R.W. (1939). *Trans. Far. Soc.* 35: 364; Mizushima, S.J. (1960). *J. Phys. Soc. Jpn.* 15: 70; Ookawa, A. (1960). *J. Phys. Soc. Jpn.* 15: 2191; Siol, M. (1961). *Z. Phys.* 164: 93.
35 Ninomiya, T. (1978). *J. Phys. Soc. Jpn.* 44: 263.
36 Poirier, J.-P. (1986). *Geophys. J. R. Astr. Soc.* 85: 315; Poirier, J.-P. and Price, G.D. (1992). *Phys. Earth Planet. Int.* 69: 153; Poirier, J.-P. and Shankland, T.J. (1993). Geophys. *J. Int.* 115: 147.
37 Cotterill, R.M.J. and Doyama, M. (1966). *Phys. Rev.* 145: 465.
38 Suzuki, H. (1983). *Dislocation Model of Liquids. Basic Concepts and Phenomena Related to Atomic Motions*, Ch. 17 in *Topological Disorder in Condensed Matter*

(ed. F. Yonegawa and T. Ninomiya), p. 222. Berlin: Springer-Verlag; Jensen, E.J., Damgaard Kristensen, W., and Cotterill, R.M.J. (1973). *Phil. Mag.* 27: 623.

39 Burakovsky, L. and Preston, D.L. (2000). *Solid State Comm.* 115: 341; Burakovsky, L., Preston, D.L., and Silbar, R.R. (2000). *Phys. Rev. B* 61: 15011; Burakovsky, L., Preston, D.L., and Silbar, R.R. (2000). *J. Appl. Phys.* 88: 6294.

40 Granato, A.V. (1992). *Phys. Rev. Lett.* 68: 974; (2014). *Eur. Phys. J.* 87: 18 and references therein.

41 Dederichs, P.H., Lehmann, C., Schober, H.R. et al. (1978). *J. Nucl. Mater.* 69, 70: 176.

42 Granato, A.V., Joncich, D.M., and Khonik, V.A. (2010). *Appl. Phys. Lett.* 97: 171911.

43 Granato, A.V. (2002). *J. Non-Cryst. Sol.* 307–310: 376.

44 Han, J.-H. (2007). *J. Korean Ceram. Soc.* 47: 412.

45 Samanta, A., Tuckerman, M.E., Yu, T.-Q. et al. (2014). *Nature* 346: 729.

46 Holian, B.L. (1980). *Phys. Rev. B* 22: 1394.

47 Bhattacharya, C., Srivastava, M.K., and Menon, S.V.G. (2011). *Physica B* 406: 4035.

48 Kerley, G.I. (1980). *J. Chem. Phys.* 73: 469; *Ibid.* 73: 478 and 73: 487.

49 Peierls, R.E. (1935). *Ann. Inst. Henri Poincaré.* 5: 177.

50 Jancovici, B. (1967). *Phys. Rev. Lett.* 19: 20; Dash, J.G. and Bretz, M. (1972). *J. Low Temp. Phys.* 9: 291.

51 Gulacsi, G. and Gulacsi, M. (1998). *Adv. Phys.* 47: 1.

52 van der Veen, J.F., p. 289; Bienfait, M and Gay, J.-M., p. 307; Levi, A.C., p. 327; all in (1990). *Phase Transitions in Surface Films 2*, NATO ASI Series B Physics (ed. H. Taub, Torzo, G. Lauter, H.J. and Fain, S.C.), Vol. 267, Plenum Press, New York.

53 van der Veen, J.F. (1999). *Surf. Sci.* 433–435: 1.

54 Tammann, G. (1910). *Z. Phys. Chem.* 68: 257; Stranski, I.N. (1942). *Naturwissenschaften* 30: 425.

55 Dash, J.G., Fu, H.-Y., and Wettlaufer, J.S. (1995). *Rep. Progr. Phys.* 58: 115.

56 Wettlaufer, J.S. (1999). *Crystal Growth, Surface Phase Transitions and Thermomolecular Pressure*, Ch. 3 in *Ice Physics and the Natural Environment* (ed. J.S. Wettlaufer, J.G. Dash, and N. Untersteiner), Vol. NATO-ISI Series 1, p. 39. Berlin: Springer.

57 Joós, B. (1996). *The Role of Dislocations in Melting*, Ch. 55 in *Dislocations in Solids* (ed. F.R.N. Nabarro and M.S. Duesbery), p. 505. Amsterdam: Elsevier.

58 Nenow, D. and Trayanov, A. (1986). *J. Cryst. Growth.* 79: 801 and references therein.

59 Baker, M.B. and Dash, J.G. (1989). *J. Cryst. Growth.* 97: 770.

60 Pawlow, P. (1909). *Z. Phys. Chem.* 65: 1; Takagi, M. (1954). *J. Phys. Soc. Jpn.* 9: 359.

61 Buffat, P. and Borel, J.-P. (1976). *Phys. Rev. A* 13: 2287; Borel, J.-P. and Chatelain, A. (1985). *Helv. Phys. Acta* 58: 443.

62 Kofman, R., Cheyssac, P., Aouaj, A. et al. (1994). *Surf. Sci.* 303: 231.

63 Allen, G.L., Bayles, R.A., Gile, W.W. et al. (1986). *Thin Solid Films* 144: 297.

64 Sheng, H.W., Lu, K., and Ma, E. (1998). *Acta Mater.* 46: 5195.

65 Kofman, R., Cheyssac, P., and Garrigos, R. (1990). *Phase Transit.* 24–26: 283.

66 Mei, Q.S. and Lu, K. (2007). *Prog. Mater. Sci.* 52: 1175.

67 Mei, Q.S. and Lu, K. (2008). *Phil. Mag. Lett.* 88: 203 and references therein.

68 Fukushima, E. and Ookawa, A. (1955). *J. Phys. Soc. Jpn.* 10: 970.

69 Wollenberger, H.J. (1996). *Point Defects*, Ch. 18 in *Physical Metallurgy*, 4e (ed. R.W. Cahn and P. Haasen), p. 1621. Amsterdam: North-Holland.

70 Fecht, H.F. (1992). *Nature* 356: 133.

71 Damgaard Kristensen, W., Klæstrup Kristensen, J., and Præstgaard, E. (1984). *Phys. Scri.* 30: 421.

72 Cotterill, R.M.J. and Kristensen, J.K. (1974). *Phil. Mag.* 30: 245.

73 Liu, K. and Chen, H. (2008). *Phys. Lett. A* 372: 1109.

74 Mandelkern, L. (2002, 2004). *Crystallization of Polymers*, Vol. 1: *Equilibrium Concepts*, 2e, Vol. 2: *Kinetics and Mechanisms*, 2e. Cambridge: Cambridge University Press.

75 Hoffman, J., Davis, G., and Lauritzen, J. (1976). *The Rate of Crystallization of Linear Polymers with Chain Folding*, Ch. 7 in *Treatise on Solid State Chemistry* (ed. N.B. Hannay), Vol. 3, p. 497. New York: Plenum Press.

76 Young, R.J. and Lovell, P.A. *Introduction to Polymers*, 3e. Boca Raton, FL: CRC Press.

77 Lippits, D.R., Rastogi, S., and Höhne, G.W.H. (2006). *Phys. Rev. Lett.* 96: 218303.

78 Rastogi, S., Lippits, D.R., Peters, G.W.M. et al. (2005). *Nat. Mater.* 4: 635.

79 Simon, F.E. and Glatzel, G. (1929). *Z. anorg. allg. Chem.* 178: 309.

80 Kennedy, G.C. and Vaidya, S.N. (1970). *J. Geophys. Res.* 75: 1019.

81 Datchi, F., Loubeyre, P., and LeToullec, R. (2000). *Phys. Rev.* 61: 6535.

82 Ferreira, A.G.M. and Lobo, L.Q. (2008). *J. Chem. Therm.* 40: 618.

83 Abramson, E.H. (2011). *High Pressure Res.* 31: 549.

84 Sun, T.F., Kortbeek, P.J., Biswas, S.N. et al. (1987). *Ber. Bunsen Ges. Phys. Chem.* 91: 1013.

85 Sun, T.F., Schouten, J.A., Trappeniers, N.J. et al. (1988). *Ber. Bunsen Ges. Phys. Chem.* 92: 652.

86 Babb, S.E. (1963). *Rev. Mod. Phys.* 35: 400.

87 Salter, L. (1954). *Phil. Mag.* 45: 369.

88 de Boer, J. (1952). *Proc. Roy. Soc. A* 215: 5.

89 Kraut, E.A. and Kennedy, G.C. (1966). *Phys. Rev.* 151: 668; Gilvarry, J.J. (1966). *Phys. Rev. Lett.* 16: 1089; Kennedy, G.C. and Vaidya, S.N. (1970). *J. Geophys. Res.* 75: 1019.

90 Kechin, V.V. (2002). *Phys. Rev. B* 65: 052102; (1995). *J. Phys.: Condens. Matter* 7: 531; (1989). *High Temp. High Press.* 21: 297; (1994). *High Temp. High Press.* 26: 439.

91 Tammann, G. (1903). *Kristallisieren und Schmelzen*. Leipzig: J.A. von Barth Verlag; Kawai, N. and Inokuty, Y. (1968). *Jpn. J. Appl. Phys.* 7: 989; (1970). *Ibid.* 9: 31.

92 Rapoport, E. (1967). *J. Chem. Phys.* 46: 2891.

93 Bernades, C., Carravetta, M., Coles, S.J. et al. (2019). *Cryst. Growth Des.* 19: 907.

94 Yang, K., Cai, Z., Jaiswal, A. et al. (2016). *Angew. Chem. Int. Ed.* 55: 14090.

95 Baeyer, A. (1877). *Ber. Dtsch. Chem. Ges.* 10: 1286.

96 Badea, E., Della Gatta, G., D'Angelo, D. et al. (2006). *J. Chem. Thermodyn.* 38: 1546.

97 Mishra, M.K., Varughese, S., Ramamurty, U. et al. (2013). *J. Am. Soc. Chem.* 135: 8121.

98 Boese, R., Weiss, H., and Blaser, D. (1999). *Angew. Chem. Int. Ed.* 38: 988.

99 Broadhurst, M.G. (1962). *J. Chem. Phys.* 36: 2578.

100 Broadhurst, M.G. (1963). *J. Res. Natl. Bur. Stand.* 67A: 233; (1966). *Ibid.* 70A: 481.

101 Bond, A.D. (2004). *New. J. Chem.* 28: 104.

102 Morishige, K. and Kato, T. (1999). *J. Chem. Phys.* 111: 7095.

103 Pradeilles, J.A., Zhong, S., Baglyas, M. et al. (2020). *Nature Chem.* 12: 475.

104 Galwey, A.K. (2005). *J. Therm. Anal. Calorim.* 79: 219; (2006). *Ibid.* 86: 561.

105 Han, J.-H. and Kim, D.-Y. (2003). *Acta Mater.* 51: 5439.

106 Stishov, S.M. (1975). *Sov. Phys.-Usp.* 17: 625.

107 Poirier, J.-P. (2000). *Introduction to the Physics of the Earth Interior*, 2e. Cambridge: Cambridge University Press.

108 Nabarro, F.R.N. (1967). *Theory of Crystal Dislocations*. Oxford: Oxford University Press (also Dover, 1987).

109 Bilgram, J.H. (1987). *Phys. Rep.* 153: 1.

110 Löwen, H. (1994). *Phys. Rep.* 237: 249.

111 Haymet, A.D.J. (1987). *Ann. Rev. Phys. Chem.* 38: 89.

112 Ram, J. (2014). *Phys. Rep.* 528: 121.

113 Garai, J. (2010). *Sol. State Comm.* 150: 1710.

114 Bacher, A.K., Schrøder, T.B., and Dyre, J.C. (2014). *Nature Comm.* 5: 5424.

115 Dyre, J.C. (2014). *J. Phys. Chem. B* 118: 10007.

116 Klein, O. (1919). *Medd. K. Vetenskapsakad. Nobelinst.* 5: 1.

117 Pedersen, U.R., Costigliola, L., Bailey, N.P. et al. (2016). *Nature Comm.* 7: 12386.

118 Abraham, F.F. (1974). *Homogeneous Nucleation Theory*. New York: Academic Press.

119 Kalikmanov, V.I. (2013). *Nucleation Theory*. Dordrecht: Springer.

120 Kashchiev, D. (2000). *Nucleation, Basic Theory with Applications*. Oxford: Butterworth-Heinemann.

121 Kamb, W.B. (1961). *J. Geophys. Res.* 66: 259.

122 McLellan, A.G. (1980). *The Classical Thermodynamics of Deformable Materials*. Cambridge: Cambridge University Press.

123 Diogo, H.P. and Moura Ramos, J.J. (2006). *J. Chem. Ed.* 83: 1389.

124 Balibar, S. and Caupin, F. (2002). *C. R. Phys.* 7: 988.

125 Ford, I.J. (2004). *Proc. Instn Mech. Engrs.* 218: 883.

126 Carcouët, C.C.M.C., van de Put, M.W.P., Mezari, B. et al. (2014). *Nano Lett.* 14: 1433.

127 De Yoreo, J.J., Gilbert, P.U.P.A., Sommerdijk, N.A.J.M. et al. (2015). *Science* 349: 498 and references therein.

128 Zong, S., Wang, J., Huang, X. et al. (2021). *IUCrJ* 9: 215.

129 De Yoreo, J.J. and Vekilov, P.G. (2003). *Rev. Mineral. Geochem.* 54: 57.

130 Oosterlaken, B.M., Vena, M.P., and de With, G. (2021). *Adv. Mater.*, 2004418 (1–27).

131 Lemanowicz, M., Mielańczyk, A., Walica, T. et al. (2021). *Polymers.* 13: 2695.

132 Kashchiev, D. and van Rosmalen, G.M. (2003). *Cryst. Res. Technol.* 38: 555.

133 Tiller, W.A. (2010). *The Science of Crystallization*. Cambridge: Cambridge University Press.

134 Gebauer, D., Raiteri, P., Gale, J.D. et al. (2018). *Am. J. Sci.* 318: 969.

135 Gebauer, D., Gale, J.D., and Cölfen, H. (2022). *Small* 2107735.

136 Liu, X.Y. (2000). *J. Chem. Phys.* 112: 9949.

137 Fletcher, N.H. (1958). *J. Chem. Phys.* 29: 572; Qian, M. and Ma, J. (2009). *J. Chem. Phys.* 130: 214709.

138 Liu, X.Y. (1999). *J. Chem. Phys.* 111: 1628.

139 Khaleghi, A., Sadrameli, S.M., and Manteghian, M. (2020). *Rev. Inorg. Chem.* 40: 167.

140 Binsbergen, F.L. (1973). *Prog. Solid State Chem.* 8.

141 Fan, Z. and Men, H. (2022). *Metals* 12: 1547.

142 Fredriksson, H. and Åkerlind, U. (2012). *Solidification and Crystallization Processing in Metals and Alloys*. Chichester, UK: Wiley.

143 Campbell, J. (2015). *Complete Casting Handbook*, 2e. Amsterdam: Butterworth-Heinemann.

144 Tang, X., Chen, W., and Li, L. (2019). *Macromolecules* 52: 3575.
145 Piorkowska, E. and Rutledge, G.C. (ed.) (2013). *Handbook of Polymer Crystallization.* Hoboken, NJ: John Wiley & Sons.
146 Zhang, M.C., Guo, B.-H., and Xu, J. (2017). *Crystals* 7: 4.
147 Xu, J., Reiter, G., and Alamo, R.G. (2021). *Crystals* 11: 304.
148 Gedde, U. (1995). *Polymer Physics*. London: Chapman and Hall.
149 Strobl, G. (2009). *Rev. Mod. Phys.* 81: 1287.
150 Wunderlich, B. (1976). *Macromolecular Physics*, Crystal Nucleation, Growth, Annealing, Vol. 2. New York: Academic Press.
151 Rastogi, S., Hikosaka, M., Kawabata, H. et al. (1991). *Macromolecules* 24: 6384; Keller, A., Hikosaka, M., Rastogi, S. et al. (1994). *J. Mater. Sci.* 29: 2579.
152 Imai, M., Kaji, K., Kanaya, T. et al. (1995). *Phys. Rev. B* 52: 12696; Olmsted, P., Poon, W., McLeish, T. et al. (1998). *Phys. Rev. Lett.* 81: 373.
153 Hauser, G., Schmidtke, J., and Strobl, G. (1998). *Macromolecules* 31: 6250.
154 Strobl, G. (2006). *Progr. Pol. Sci.* 31: 398.
155 Sheth, S.S. (2013). *Self-Consistency of the Lauritzen-Hoffman and Strobl Models of Polymer Crystallization Evaluated for Poly(ε-Caprolactone) Fractions and Effect of Composition on the Phenomenon of Concurrent Crystallization in Polyethylene Blends*, PhD diss., Virginia Polytechnic Institute and State University.
156 Sangroniz, L., Cavallo, D., and Müller, A.J. (2020). *Macromolecules* 53: 4581.
157 Sojoudi, H., Wang, M., Boscher, N.D. et al. (2016). *Soft Matter* 12: 1938.
158 Ickes, L., Welti, A., Hoose, C. et al. (2015). *Phys. Chem. Chem. Phys.* 17: 5514.
159 Reinhardt, A. and Doye, J.P.K. (2013). *J. Chem. Phys.* 139: 1.
160 Zobrist, B., Koop, T., Luo, B.P. et al. (2007). *J. Phys. Chem.* C111: 2149.
161 Shamshiri, M., Jafari, R., and Momen, G. (2021). *Surface & Coatings Techn.* 424: 127656.
162 He, Z., Zhuo, Y., Zhang, Z. et al. (2021). *Coatings* 11: 1343.
163 Fitzner, M., Pedevilla, P., and Michaelides, A. (2020). *Nature Comm.* 11 (1-9): 4777.
164 Turnbull, D. and Vonnegut, B. (1952). *Ind. Eng. Chem.* 44: 1292 and references therein.

Further Reading

Davey, R.J. and Garside, J. (2000). *From Molecules to Crystallizers: an Introduction to Crystallization*, Oxford Chemistry Primers n° 86. New York: Oxford University Press. This book aims to summarize and explain the links between the molecular recognition processes that govern crystallization and the practical aspects required to engineer a crystallization process to make a product with desirable properties.

Markov, I.V. (2017). *Crystal Growth for Beginners: Fundamentals of Nucleation, Crystal Growth and Epitaxy*, 3e. Singapore: World Scientific Publishing Co. This book focuses on crystallization in pure systems and contains a biographical description of the founders of modern crystal growth science and technology.

Ubbelohde, A.R. (1978). *The Molten State of Matter*. Chichester: John Wiley & Sons. An underrated classical review on early literature providing many details on a broad spectrum of materials.

15

Continuous Phase Transitions: Liquids ↔ Gases

In Chapters 13 and 14, we discussed discontinuous phase transitions. In this chapter, we discuss aspects of continuous phase transitions, mainly focusing on the liquid–gas transition. Continuous phase transitions provide a particularly rich part of chemical physics and considerable effort has been paid to describe and understand them. This has led to an image that captures the essentials well. In the sequel, we first discuss some experimental facts, thereafter mean-field theory, scaling, and renormalization.

15.1 Limiting Behavior

We recall that thermodynamic equilibrium is determined by minimization of the appropriate potential with respect to the (relevant) internal variable(s). For the liquid–gas transition, the density difference $\Delta\rho = \rho - \rho_{\text{cri}}$ is single valued above T_{cri}, has two values below T_{cri}, and $\Delta\rho$ is conventionally used as the internal variable. In connection with transitions, the internal variable used is often called the *order parameter*. For liquids, both single-compound liquids and liquid mixtures exhibit continuous transitions. As an example, in Figure 15.1 the *reduced density* $(\rho_{\text{L}} - \rho_{\text{V}})/2\rho_{\text{cri}}$ of CO_2 versus the *reduced temperature* $t \equiv (T - T_{\text{cri}})/T_{\text{cri}}$ is given, while Figure 15.2 shows the difference in volume fraction $\phi^{(1)} - \phi^{(2)}$ for CCl_4 in C_7F_{14} versus t. In both cases, the behavior close to T_{cri} is described by a power law [1, 2].

This type of experiment has led to the definition of various critical exponents. A critical exponent of a function, say a for a function $f(x)$ of x, is defined by

$$a \equiv \lim_{x\to 0} [\mathrm{d}\ln f(x)/\mathrm{d}\ln x]$$

and we write $f(x) \sim x^a$. This does not imply that $f(x) = Ax^a$ but rather that $f(x) = Ax^a(1 + Bx^b + \ldots)$ with $b > 0$, and where Ax^a is called the *singular part* of $f(x)$ and $1 + Bx^b + \ldots$ the *regular part* or *background function*. The exponent a defines the rate of approach of $f(x)$ to zero for $a > 0$ or to infinity for $a < 0$. If $a = 0$, the result is ambiguous. In this case, $f(x)$ shows either a discontinuity or else has a logarithmic singularity, that is, $f(x)$ behaves like $\ln(T - T_{\text{cri}})$.

For liquids, the density difference $\Delta\rho$ along the saturation curve for $T < T_{\text{cri}}$ is

$$[\rho_{\text{L}}(T) - \rho_{\text{V}}(T)]/2\rho_{\text{cri}}(T) = B(-t)^{\beta}(1 + \ldots)$$

Phases of Matter and their Transitions: Concepts and Principles for Chemists, Physicists, Engineers, and Materials Scientists, First Edition. Gijsbertus de With.

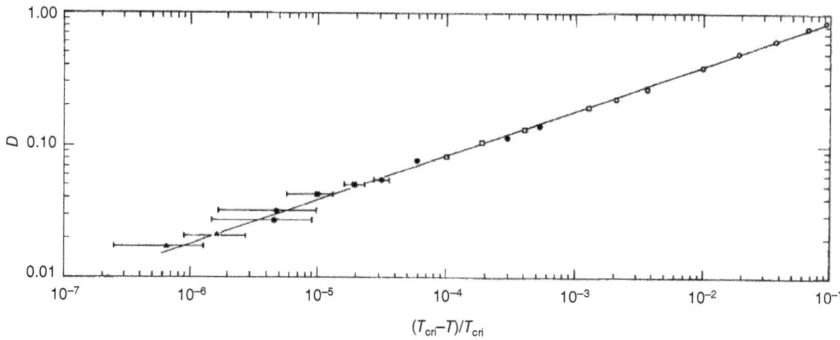

Figure 15.1 The reduced density $D = (\rho_L - \rho_V)/2\rho_{cri}$ of CO_2 versus relative temperature $-t = (T_{cri} - T)/T_{cri}$ with $T_{cri} = 304.18$ K, described by $(\rho_L - \rho_V)/2\rho_{cri} = B(-t)^\beta$ with $B = 1.85$ and $\beta = 0.340 \pm 0.015$. Source: Heller [1]/IOP Publishing.

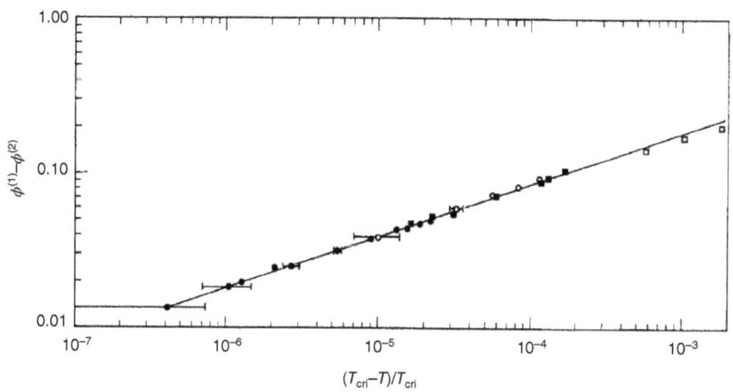

Figure 15.2 The difference in volume fraction $\phi^{(1)}$ in the upper phase and $\phi^{(2)}$ in the lower phase for CCl_4 in C_7F_{14} versus the relative temperature $-t = (T_{cri} - T)/T_{cri}$ with $T_{cri} = 301.786$ K, described by $\phi^{(1)} - \phi^{(2)} = B(-t)^\beta$ with $B = 1.81$ and $\beta = 0.335 \pm 0.020$. Source: Heller [1]/IOP Publishing.

where B and β are parameters[1] and $(1 + \ldots)$ is the background function to which the critical exponent β is insensitive. Experimentally, the range for β is $0.3 < \beta < 0.4$.

The response function with respect to pressure – the compressibility κ_T – behaves similarly but becomes infinite at $T = T_{cri}$. For $T < T_{cri}$ (for $\rho = \rho_L$ or $\rho = \rho_V$ on the saturation curve) and $T > T_{cri}$ (for $\rho = \rho_{cri}$), κ_T is described by, respectively,

$$\kappa_T/\kappa_T{}^\circ = C'(-t)^{-\gamma'}(1 + \ldots) \quad \text{and} \quad \kappa_T/\kappa_T{}^\circ = C(t)^{-\gamma}(1 + \ldots)$$

where C, C', γ, and γ' are parameters and $\kappa_T{}^\circ = 1/P_{cri}{}^\circ = m/kT\rho_{cri}$ is the compressibility of an ideal gas with molecules of mass m and density ρ_{cri} at T_{cri}. Experimentally, the range for γ (γ') is $1.2 < \gamma < 1.4$.

Such a power–law relation also holds for the heat capacity C_V given, in a two-phase system for $T < T_{cri}$ (for $\rho_L + \rho_V = 2\rho_{cri}$) and $T > T_{cri}$ (for $\rho = \rho_{cri}$), by, respectively,

$$C_V/Nk = A'(-t)^{-\alpha'}(1 + \ldots) \quad \text{and} \quad C_V/Nk = A(t)^{-\alpha}(1 + \ldots)$$

1 Here, I use a largely standardized labeling for the critical exponents. Note that these labels are also used for other properties.

where A, A', α, and α' are parameters. Experimentally, the range for α (α') is $-0.2 < \alpha < 0.2$.

For the pressure P along the critical isotherm $T = T_{\text{cri}}$, data are described by

$$\frac{P - P_{\text{cri}}}{P^{\circ}_{\text{cri}}} = D\left|\frac{\rho - \rho_{\text{cri}}}{\rho_{\text{cri}}}\right|^{\delta} \text{sgn}(\rho - \rho_{\text{cri}})$$

where $\text{sgn}(I) = 1$ if $I > 0$ and $\text{sgn}(I) = -1$ if $I < 0$. Experimentally, we have $4 < \delta < 5$.

Finally, for liquids the (total) pair correlation function $h(r) = g^{(2)}(r) - 1$ (Section 9.1) is important. For large r, $h \to 0$ and for $T \neq T_{\text{cri}}$, $r \gg \xi$ behaves like

$$h(r) = \pi^{(1-d)/2} kT \exp(-r/\xi)/2^{(1+d)/2} \lambda r^{(d-2)/2} \xi^{(d-3)/2} \quad \text{for} \quad T \neq T_{\text{cri}} \quad \text{and}$$

$$h(r) = f(r) \exp(-r/\xi)/r^{d-2} \quad \text{for} \quad T = T_{\text{cri}} \tag{15.1}$$

Here $f(r)$ is a weakly varying function of r and d is the dimension of space, and the *coherence* or *correlation length* ξ measures the range of correlation in the liquid. In essence, ξ is a measure of the greatest distance from a given particle at which this particle can still cause nonrandom effects on the particle density. Under normal conditions, the order of magnitude for ξ is a few molecular diameters, but for $T \to T_{\text{cri}}$ and $\rho \to \rho_{\text{cri}}$, the value of ξ diverges to macroscopic size. In this case, high- and low-density regions appear on an increasingly larger scale until macroscopic regions of different densities in the fluid lead to the two-phase regions for $T \leq T_{\text{cri}}$. However, the small(er)-scale fluctuations do not disappear and within high-density regions there will be smaller regions of lower density and vice versa. As soon as ξ approaches the wavelength of visible light, say 0.5–0.8 μm, these fluctuations lead to the phenomenon of *critical opalescence*. Using the experimental data, it appears that ξ also scales with T and for $T < T_{\text{cri}}$ (for $\rho = \rho_{\text{L}}$ or $\rho = \rho_{\text{V}}$ on the saturation curve) and $T > T_{\text{cri}}$ (for $\rho = \rho_{\text{cri}}$) is given by, respectively,

$$\xi = \xi_0'(-t)^{-\nu'} \quad \text{and} \quad \xi = \xi_0(t)^{-\nu}$$

with ξ_0 and ξ_0' the *bare coherence lengths*. Using Eq. (15.1), Fisher [3] showed that for $d = 2$, $h(r) \sim \ln r$, which is physically impossible. Therefore, he introduced the additional exponent η (obeying $0 < \eta < 2$) for $r \to \infty$ at ($\rho = \rho_{\text{cri}}$) and ($\rho = \rho_{\text{cri}}$ and $T = T_{\text{cri}}$), respectively, by

$$h(r) \to H_d \exp(-r/\xi)/r^{d-2+\eta} \quad \text{and} \quad h(r) \to H_d/r^{d-2+\eta}$$

It appears that in three dimensions (3D), the value of η is small, typically about 0.04. Moreover, η is related to ν and γ (Example 15.1).

Example 15.1 ***The Fisher Relation*** In Chapter 9, we indicated that the compressibility κ_T is given in terms of $h(r)$ by

$$\rho\kappa_T = \beta\left[1 + \rho\int_0^{\infty} h(r)\, 4\pi r^2\, \mathrm{d}r\right]$$

Using $h(r) \to H_d \exp(-r/\xi)/r^{d-2+\eta}$ we obtain

$$\rho\int_0^{\infty} h(r)\, 4\pi r^2\, \mathrm{d}r \to 4\pi\rho\int_0^{\infty} \frac{H_3 \mathrm{e}^{-r/\xi}}{r^{1+\eta}} r^2\, \mathrm{d}r = 4\pi\rho H_3 \xi^{2-\eta}\int_0^{\infty} x^{\eta-1} \mathrm{e}^{-x}\, \mathrm{d}x \sim \xi^{2-\eta}$$

with $x = r/\xi$. Because $\kappa_T \sim t^{-\gamma}$, $\kappa_T \sim 1 + \rho\int_0^{\infty} h(r)\, 4\pi r^2\, \mathrm{d}r \sim \xi^{2-\eta}$ and $\xi \sim t^{-\nu}$, we have

$$t^{-\gamma} = \xi^{2-\eta} = (t^{-\nu})^{2-\eta} \quad \text{or} \quad \gamma = (2 - \eta)\nu$$

Table 15.1 Critical points and their order parameters.

System	Order parameter	Example	T_{cri} (K)
Liquid–gas	Density difference	H_2O	647.05
Fluid mixture	Fraction of components	CCl_4–C_7F_{14}	301.78
Binary alloy	Fraction species on sublattice	Cu–Zn	739
Ferromagnetic	Magnetization	Fe	1044.0
Antiferromagnetic	Sublattice magnetization	FeF_2	78.26
Ferroelectric	Polarization	$(NH_2CH_2COOH)_3 \cdot H_2SO_4$	322.5

Source: Adapted from Ma [4].

There appear to be more relations between the critical exponents. They were originally derived as (thermodynamic) inequalities between the various critical exponents and usually named after their discoverer. We have, for example,

$$\text{Rushbrooke's inequality } \alpha' + 2\beta + \gamma' \geq 2, \tag{15.2}$$

$$\text{Griffiths' inequality } \alpha' + \beta(1+\delta) \geq 2 \text{ and} \tag{15.3}$$

$$\text{Widom's inequality } \gamma' \geq \beta(\delta - 1) \tag{15.4}$$

Later, we will see that actually they are equalities and moreover that $\alpha' = \alpha$ and $\gamma' = \gamma$. The derivation of some of them is easy, but for others it is quite involved. As an example, we derive the Rushbrooke inequality. To do so we need *Stanley's lemma* that states that for $f(x) \sim x^a$ and $g(x) \sim x^b$ with $f(x) \leq g(x)$ and for $0 < x < 1$, so that $\ln x < 0$, we have $a \geq b$. From thermodynamics, we know that $C_P - C_V = TV\alpha^2/\kappa_T$ (all quantities positive, Section 6.1.6), so that $C_P \geq TV\alpha^2/\kappa_T$. Because we have for the heat capacity $C_P \sim (-t)^{-\alpha'}$, for the compressibility $\kappa_T \sim (-t)^{-\gamma'}$, and for the thermal expansion coefficient $\alpha \sim (-t)^{\beta-1}$, we obtain at once $-\alpha' \leq \gamma' + 2(\beta - 1)$ or $\alpha' + 2\beta + \gamma' \geq 2$ (for $0 < \alpha' < 1$, while for $\alpha' \leq 0$, the most we can say is that $2\beta + \gamma' \geq 2$). Note the resemblance with the Ehrenfest ratio (Section 12.4).

Finally, it appears that the critical exponents for several phenomena are essentially the same (Table 15.1). This happens to be the case if for two phenomena both the spatial dimensionality, for example, 2D or 3D, and the dimensionality of the order parameter, for example, a scalar or vector, are the same. These phenomena are said to belong to the same *universality class*. For the phenomena listed in Table 15.1, the lattice gas model (Appendix C), with a scalar order parameter, forms the prototype model. The liquid–gas transition also belongs to this class (Problem 15.3). We limit ourselves entirely to this class, and some experimental results are given in Table 15.2.

15.2 Mean-Field Theory: Landau Theory

In Chapter 12, we introduced Landau theory as a general thermodynamic theory of phase transformations and transitions. In this theory, a "generalized" Helmholtz energy $F(\eta;T,V)$, dependent on temperature T and volume V, is minimized with respect to the internal variable (order parameter) for which we use here the symbol η. We noted that in general η varies

Table 15.2 Experimental values of critical exponents for various fluids.

Compound	α	β	γ	δ	ν	η
CO_2	0.10	0.321	1.24	4.85	0.57	~0
Xe	0.11	0.329	1.23	4.74	—	0.05
N_2	—	0.327 ± 0.002	1.23 ± 0.01	—		
Ne	—	0.327 ± 0.002	1.25 ± 0.01	—		

Source: Adapted from Hocken and Moldover [5].

with position and we should use the *energy density*. Although we can use the energy itself for a uniform distribution of η as employed here, we will still use the energy density. For the liquid–gas transition, the order parameter is conveniently taken as the density difference $\Delta\rho$ between liquid and gas.

In general, we use $\mathscr{L}(\eta)$ representing the grand potential density (Section 7.1) and assume that the Helmholtz part $f(\eta)$ of $\mathscr{L}(\eta)$ near T_{cri} can be expanded in a power series in η:

$$\mathscr{L}(\eta) = f(\eta) - h\eta = a_0 + a_1\eta + \tfrac{1}{2}a_2\eta^2 - \tfrac{1}{3}a_3\eta^3 + \tfrac{1}{4}a_4\eta^4 + \ldots - h\eta$$

The driving force h is the conjugate variable to the order parameter η, that is $h = \partial f(\eta)/\partial\eta$. For a homogeneous fluid with order parameter $\eta = \Delta\rho$, the driving force h is the chemical potential μ. For continuous transitions, equilibrium requires that a_1 is absent and that $a_4 > 0$ for the system to be stable. If the order parameter η can have only the two values $\pm\eta$, as for the magnetization and polarization cases, the third-order terms must also be absent because $\mathscr{L}(\eta)$ should equal $\mathscr{L}(-\eta)$. For fluids this is clearly not the case. If we assume this to be true, we essentially discuss the magnetic or electric problem. For $h = 0$, we are left with $\mathscr{L}(\eta) = a_2\eta^2 + a_4\eta^4$ and equilibrium is reached when $\partial\mathscr{L}(\eta)/\partial\eta = 0$, with as solutions

$$\eta_0 = 0 \ \text{and} \ \eta_\pm = \pm(-a_2/a_4)^{1/2}$$

The coefficient a_2 was expanded with respect to temperature as $a_2 = a'\Delta T$, but a_4 was taken as constant. For $T < T_{cri}$ two equivalent minima exist. At $T = T_{cri}$, these minima merge to a single minimum, representing the continuous transition. A single minimum persists for $T > T_{cri}$ (Figure 12.7).

We now calculate the critical exponents for Landau theory. For $T < T_{cri}$, the solution is $\eta = (-a_2/a_4)^{1/2} = (-a'\Delta T/a_4)^{1/2}$ and comparing with $\eta \sim (T - T_{cri})^\beta$ the results is $\beta = \tfrac{1}{2}$.

For $T > T_{cri}$, $\eta = 0$ and thus $\mathscr{L}(\eta) = 0$. For $T < T_{cri}$, $\eta = (-a_2/a_4)^{1/2}$ and thus $\mathscr{L}(\eta) = \tfrac{1}{2}a_2(-a_2/a_4) + \tfrac{1}{4}a_4(-a_2/a_4)^2 = -a_2{}^2/2a_4$. Because $C_V = -T\partial^2\mathscr{L}/\partial T^2$, we obtain

$$C_V = 0 \ (T > T_{cri}) \quad \text{and} \quad C_V = a'^2T/2a_4 \ (T < T_{cri})$$

The heat capacity thus changes discontinuously at T_{cri}, which can be represented by taking the exponent $\alpha = 0$. To compute the other exponents, we need to let $h \neq 0$. Equilibrium is obtained from $\partial\mathscr{L}/\partial\eta = 0$ and, using $\mathscr{L}(\eta) = \tfrac{1}{2}a_2\eta^2 + \tfrac{1}{4}a_4\eta^4 - h\eta$, we obtain $h = a_2\eta + a_4\eta^3$. On the critical isotherm, $a_2 = 0$ and thus $h \sim \eta^3$ so that $\delta = 3$.

The relevant response function, the *isothermal susceptibility* χ_T, is given by

$$\chi_T(h) \equiv (\partial\eta/\partial h)_T = [\partial^2 f(\eta)/\partial\eta^2]^{-1} = [a_2 + 3a_4\eta(h)^2]^{-1}$$

with $\eta(h)$ a solution of $h = a_2\eta + a_4\eta^3$. For $T > T_{cri}$ we have $\eta = 0$ and $\chi_T = a_2{}^{-1}$. Similarly, for $T < T_{cri}$ at $h = 0$, we have $\eta^2 = -a_2/a_4$, $\chi_T = (-2a_2)^{-1}$, and thus $\gamma = \gamma' = 1$.

15.2.1 Landau-Like Theory: Fluid Transitions

Let us now return to our case of the liquid–gas transition for which $a_1, a_3 \neq 0$. In this case, $\mathscr{L}(\eta)$ reads, using the chemical potential μ as the proper driving force,

$$\mathscr{L}(\eta) = f(\eta) - h\eta = a_0 + a_1\eta + {}^1\!/\!_2 a_2\eta^2 - {}^1\!/\!_3 a_3\eta^3 + {}^1\!/\!_4 a_4\eta^4 + \ldots - \mu\eta \tag{15.5}$$

The general solution of Eq. (15.5) is discussed in Section 12.5. For the fluid case, however, we apply a trick by writing $\eta = \eta_0 + \Delta\eta$. If we do so, we obtain

$$\begin{aligned}\mathscr{L}(\eta_0 + \Delta\eta) &= (a_1 - \mu)(\eta_0 + \Delta\eta) + {}^1\!/\!_2 a_2(\eta_0 + \Delta\eta)^2 \\ &\quad - {}^1\!/\!_3 a_3(\eta_0 + \Delta\eta)^3 + {}^1\!/\!_4 a_4(\eta_0 + \Delta\eta)^4 - \mu(\eta_0 + \Delta\eta) \quad \text{or} \\ \mathscr{L} &= A(\eta_0) + B(\eta_0)\Delta\eta + C(\eta_0)\Delta\eta^2 + D(\eta_0)\Delta\eta^3 + E(\eta_0)\Delta\eta^4 \quad \text{with} \\ &\quad A(\eta_0, \mu) = (a_1 - \mu)\eta_0 + {}^1\!/\!_2 a_2\eta_0^2 - {}^1\!/\!_3 a_3\eta_0^3 + {}^1\!/\!_4 a_4\eta_0^4, \\ &\quad B(\eta_0) = (a_1 - \mu) + a_2\eta_0 - a_3\eta_0^2 + a_4\eta_0^3, \\ &\quad C(\eta_0) = {}^1\!/\!_2 a_2 - a_3\eta_0 - 3a_4\eta_0^2/2, \qquad D(\eta_0) = -{}^1\!/\!_3 a_3 + a_4\eta_0, \qquad E(\eta_0) = {}^1\!/\!_4 a_4\end{aligned}$$

Each of the functions $B(\eta_0), \ldots, E(\eta_0)$ can be considered as a function of temperature, which we expand to first order in T. The function $A(\eta_0,\mu)$ depends on η_0 and μ so that a first-order expansion in T and μ with respect to T_{cri} and μ_{cri} becomes $A(\eta_0,\mu) = a_{1,1}\Delta T + \Delta\mu + \mathcal{O}(\Delta^2 T)$ with $\Delta T = T - T_{\text{cri}}$ and $\Delta\mu = \mu - \mu_{\text{cri}}$ using the notation $a_{i,j} = \partial^{i+j}\mathscr{L}/\partial\eta^i\partial T^j$. We can eliminate the linear and cubic terms in $\Delta\eta$ by setting them equal to zero. First solving the cubic term $D(\eta_0)$ for η_0 leads to $\eta_0 = a_3/3a_4 = -a_{3,1}\Delta T/3a_4$. Substitution in the linear term $B(\eta_0)$ results, to first order in ΔT, to

$$\begin{aligned}B(\eta_0) &\cong (a_1 - \mu) + a_2\eta_0 - \ldots \\ &= (a_{1,1}\Delta T - \Delta\mu) + a_{2,1}\Delta T\eta_0 \cong (a_{1,1}\Delta T - \Delta\mu)\end{aligned}$$

where the last step can be made because $\Delta T\eta_0$ is of second order in ΔT. Since $B(\eta_0) = 0$ requires $\Delta\mu = a_{1,1}\Delta T$, we see that this condition is already satisfied. Therefore, along the line determined by $\eta_0 = a_{3,1}\Delta T/3a_4$, the Landau function $\mathscr{L}(\eta)$ is

$$\mathscr{L} = A(\eta_0) + C(\eta_0)\Delta\eta^2 + E(\eta_0)\Delta\eta^4 = A(\eta_0) + {}^1\!/\!_2 a_{2,1}\Delta T\Delta\eta^2 + {}^1\!/\!_4 a_4\Delta\eta^4$$

which has the same form as $\mathscr{L}(\eta)$ for the magnetic problem. The solution accordingly is also the same and for $T < T_{\text{cri}}$ is given by

$$\Delta\eta = \pm(-a_{2,1}\Delta T/a_4)^{1/2}$$

All critical exponents are thus also the same as before. The complete solution is obtained by superposing the $\Delta\eta$ solution on the η_0 behavior. To interpret η_0 we recall that $\eta = \eta_0 + \Delta\eta = \rho - \rho_{\text{cri}}$. Using $\eta_{\text{L}} = \eta_0 + \Delta\eta$ and $\eta_{\text{G}} = \eta_0 - \Delta\eta$, we see that $\rho_{\text{L}} + \rho_{\text{G}} = 2\eta_0$, so that η_0 equals the average density and represents the *law of rectilinear diameters* (Section 8.4). Because $\eta_0 = -a_{3,1}\Delta T/4a_4$, the average density is a linear function of temperature, as approximately confirmed by experimental evidence. However, expanding all functions to second order leads to nonlinear behavior [6].

In conclusion to this part, we note that the order of magnitude of the critical exponents as obtained from the mean-field approach is correct, but that quantitative agreement is missing (Table 15.3). The Landau expression is approximate because fluctuations are neglected. The criterion whether we can neglect these or not was given by Levanyuk and

Table 15.3 Critical exponents according to various models.

Model	α	β	γ	δ	ν	η
Mean field	0	1/2	1	3	1/2	0
2D lattice gas[a)]	0	1/8	7/4	15	1	1/4
2D lattice gas[b)]	−0.27	0.72	0.83	2.2	—	—
2D lattice gas[c)]	—	—	—	—	1.07	—
2D lattice gas[d)]	—	—	—	—	0.901	—
3D lattice gas[e)]	0.110(1)	0.3265(3)	1.2372(5)	4.789(2)	0.6301(4)	0.0364(5)

a) Exact solution due to Onsager [7].
b) Triangular lattice with (3 sites → 1 cell) renormalization leading to $\lambda_1 = 1.623$ and $\lambda_1 = 2.121$ so that $p = \ln \lambda_1/2 \ln 3^{1/2} = 0.441$ and $q = \ln \lambda_2/2 \ln 3^{1/2} = 0.684$.
c) Square lattice with removing half the sites [8]).
d) Square lattice with (5 sites → 1 cell) renormalization [8].
e) Data from [9].

Ginzburg and rationalized by Amit [10] (see also [11]). The conditions for the validity are discussed carefully in [12] as well as in [11, 13]. In brief, we need to consider inhomogeneous systems and consider the term $\lambda(\nabla\eta)^2$. It appears that the value of λ depends on the range of the interactions and it follows that $\xi_0 = (\lambda/at)^{1/2}$ and $\xi_0' = (-\lambda/2at)^{1/2}$ for $T > T_{cri}$ and $T < T_{cri}$, respectively. Further, one can show that for $d > 4$ and $t \to 0$, Landau theory is self-consistent, while for $d = 4$ it requires a logarithmic correction. For $d < 4$, fluctuations become important at the *Ginzburg temperature* T_G, as expressed by $t_G = |T_G - T_{cri}|/T_{cri} = ½(k/4\pi\Delta C_V\xi_0^{\ 3})^2$ for $d = 3$. It thus depends on λ whether Landau theory close to T_{cri} can be considered as self-consistent or not. For some phenomena this is the case very close to T_{cri}. For the liquid–gas transformation (Figure 15.1) and a few others, this is not the case and further steps are required. These include scaling, using generalized homogeneous functions applied to thermodynamic potentials, and renormalization theory that states that phenomena behave the same at different length scales near the critical point.

Problem 15.1 ***Stanley's Lemma*** Show that for $f(x) \sim x^a \le g(x) \sim x^b$, for sufficiently small x, $a \ge b$.

Problem 15.2 ***Inequalities*** Verify that Eqs. (15.9)–(15.11) are satisfied as equalities for mean-field theory.

Problem 15.3 ***The van der Waals Critical Exponents*** The vdW expression, $[P + a/V^2](V - b) = NkT$, rewritten as polynomial in V, reads $V^3 - [b + (NkT/P)]V^2 + (a/P)V - (ab/P) = 0$. Show that the vdW expression leads to the mean-field theory values of the critical exponents via the following steps:

(a) Comparing with the expression $(V - V_{cri})^3 = V^3 - 3V_{cri}V^2 + 3V_{cri}^{\ 2}V - V_{cri}^{\ 3}$, show that $T_{cri} = 8a/27Nkb$, $P_{cri} = a/27b^2$ and $V_{cri} = 3b$ and thus that $P_{cri}V_{cri}/RT_{cri} = 3/8$.
(b) Show that $(P_{red} + 3/V_{red}^{\ 2})(3V_{red} - 1) = 8T_{red}$, where $Z_{red} \equiv Z/Z_{cri}$ with $Z = T, P$ or V.

(c) Defining $\omega \equiv (\rho - \rho_{\text{cri}})/\rho_{\text{cri}}$, $p \equiv (P - P_{\text{cri}})/P_{\text{cri}}$, and $t \equiv (T - T_{\text{cri}})/T_{\text{cri}}$, show that $p(2-\omega) = 8t(1+\omega)+3\omega^3$.
(d) To calculate the exponent β, consider the behavior near the critical point. Show that the vdW expression reduces to $\omega(3\omega^2+12t) = 0$ using that for $\omega = 0$, $p = 4t$.
(e) Show that this solution leads to $\omega = \pm 2(-t)^{1/2} = \pm(1-T_{\text{red}})^{1/2}$ or $\beta = \frac{1}{2}$.
(f) To calculate the exponent δ, consider the behavior at T_{cri}, that is, at $t = 0$. Show that the vdW expression becomes $p(2-\omega) = 3\omega^3$.
(g) Show that this solution leads to $p \cong (3\omega^3/2)[1+(\omega/2)+\ldots]$ or $\delta = 3$.
(h) To calculate the critical exponent γ, show that the compressibility $\kappa_T \sim (\partial p/\partial\omega)_T \sim (24t+18\omega^2-6\omega^3)/(2-\omega)^2$. Show that this solution for $t > 0$ leads to $\kappa_T \sim 6t$ or $\gamma = 1$.

15.3 Scaling

In the previous sections, we saw that the mean-field theory delivers values for the critical coefficients, but that the agreement with experiment is only approximate. In this section, we show that the thermodynamic inequalities are equalities using the concept of scaling. We first discuss homogeneous functions and thereafter consider thermodynamic potentials as homogeneous functions and lattice scaling.

15.3.1 Homogeneous Functions

We recall that a homogeneous function is defined by $f(\lambda r) = g(\lambda)f(r)$, in which $g(\lambda)$ is an unspecified function of the scale factor λ. For example, if $f(r) = Br^2$, we have $f(\lambda r) = \lambda^2 f(r)$ so that $g(\lambda) = \lambda^2$.

The general expression for $g(\lambda)$ appears to be λ^p, which can be derived as follows. Suppose we have the function $f[\lambda(\mu r)]$, depending on λ and μ as well as r. We write

$$f[\lambda(\mu r)] = g(\lambda)f(\mu r) = g(\lambda)g(\mu)f(r) \quad \text{and}$$

$$f[\lambda(\mu r)] = g(\lambda\mu)f(r) \quad\rightarrow\quad g(\lambda\mu) = g(\lambda)g(\mu)$$

Now, differentiating both sides with respect to μ, denoting $\text{d}g(x)/\text{d}x$ by g', we obtain

$$\frac{\partial}{\partial\mu}g(\lambda\mu) = \lambda g'(\lambda\mu) \quad \text{and} \quad \frac{\partial}{\partial\mu}g(\lambda)g(\mu) = g(\lambda)g'(\mu) \tag{15.6}$$

Let $\mu = 1$ and set $g'(\mu = 1) \equiv p$. Using Eq. (15.6a) leads to

$$\frac{g'(\lambda)}{g(\lambda)} = \left[\frac{\text{d}}{\text{d}\lambda}\ln g(\lambda)\right] = \frac{p}{\lambda} \quad \text{or} \quad \ln g(\lambda) = p\ln\lambda + c \quad \text{or} \quad g(\lambda) = \text{e}^c\lambda^p$$

Consequently,

$$g'(\lambda) = p\text{e}^c\lambda^{p-1} \quad \text{but} \quad g'(1) = p \;\rightarrow\; c = 0 \quad \text{and} \quad g(\lambda) = \lambda^p$$

If a homogeneous function is dependent on two variables, we have $f(\lambda x,\lambda y) = \lambda^p f(x,y)$. Let us take $\lambda = 1/y$, resulting in $f(\lambda x,\lambda y) = f(x/y,1) = y^{-p}f(x,y)$. If we denote $f(z,1)$ by $F(z)$, we have $f(x,y) = y^p F(x/y)$. So, a homogeneous function $f(x,y)$ may be written as y^p times a function of x/y. Similarly, $f(x,y) = x^p G(y/x)$ with $G(z) = f(1,z)$. This implies that, by suitable choice of variables, say $f(x,y)/x^p$ and y/x, all data collapse on a single curve. We now generalize the concept by writing $f(\lambda^a x,\lambda^b y) = \lambda f(x,y)$. One could think that writing $f(\lambda^a x,\lambda^b y) = \lambda^p f(x,y)$ provides an even more general form, but that is not the case because this expression is equivalent to $f(\lambda^{a/p}x,\lambda^{b/p}y) = \lambda f(x,y)$ and merely constitutes rescaling of a and b. For such a *generalized homogeneous function* of two variables, there are at most two undetermined parameters.

15.3.2 Scaling Potentials

One of the problems of the Landau(-like) mean-field approach is that the various thermodynamic potentials are not analytical and therefore cannot be expanded, strictly speaking, with respect to the critical temperature. It was Widom [14] who considered thermodynamic potentials as generalized homogeneous functions. So, let us write the Gibbs energy $G(t,P)$ where we use $t = (T - T_{\text{cri}})/T_{\text{cri}}$. We assume that the singular part of G is a generalized homogeneous function of t and P, and therefore we write

$$G(\lambda^p t,\lambda^q P) = \lambda G(t,P)$$

Differentiating with respect to P yields

$$\lambda^q \frac{\partial G(\lambda^p t,\lambda^q P)}{\partial \lambda^q P} = \lambda \frac{\partial G(t,P)}{\partial P} \quad \text{or} \quad \lambda^q V(\lambda^p t,\lambda^q P) = \lambda V(t,P)$$

Now for the moment we take $P = 0$ to obtain

$$V(t,0) = \lambda^{q-1} V(\lambda^p t,0)$$

In the spirit of the procedure used before for homogeneous functions, we use

$$\lambda = (-t)^{-1/p} \quad \text{leading to} \quad V(t,0) = (-t)^{(1-q)/p} V(-1,0)$$

Approaching T_{cri} via $t \to 0^-$, the latter notation implying that t approaches zero from the negative side, the result is

$$V(t,0) \sim (-t)^\beta \quad \text{with} \quad \beta = (1-q)/p$$

In a similar vein, we take $t = 0$ and let the pressure P go to zero. The first step yields

$$V(0,P) = \lambda^{q-1} V(0,\lambda^q P)$$

and taking

$$\lambda = P^{-1/q} \quad \text{leads to} \quad V(0,P) = P^{(1-q)/q} V(0,1)$$

From $P \to 0$ it follows that

$$V(0,P) \sim P^{1/\delta} \quad \text{with} \quad \delta = q/(1-q)$$

Solving the results $\beta = (1-q)/p$ and $\delta = q/(1-q)$ for p and q lead to

$$p = \frac{1}{\beta}\frac{1}{\delta+1} \quad \text{and} \quad q = \frac{\delta}{\delta+1} \tag{15.7}$$

The next step is to differentiate the Gibbs energy G twice with respect to P, leading to

$$\lambda^{2q}\kappa_T(\lambda^p t,\lambda^q P) = \lambda \kappa_T(t,P)$$

Using the same procedure as before, that is, taking again $P = 0$ and assuming

$$\lambda = (-t)^{-1/p} \quad \text{leads to} \quad \kappa_T(t,0) = (-t)^{-(2q-1)/p}\kappa_T(-1,0)$$

Approaching T_{cri} via $t \to 0^-$ the result is

$$\kappa_T(t,0) \sim (-t)^{-\gamma'} \quad \to \quad \gamma' = (2q-1)/p$$

Substituting Eq. (15.7), the final result is

$$\gamma' = \beta(2\delta - 1)$$

which we recognize as Widom's (in)equality. We conclude for now there are only two independent critical exponents, p and q, instead of the original three, β, δ, and γ'.

Repeating the last procedure but now with using $t \to 0^+$, we obtain

$$\lambda = t^{-1/p}, \quad \kappa_T(t,0) = t^{-(2q-1)/p}\kappa_T(-1,0) \quad \text{and}$$

$$\kappa_T(t,0) \sim t^{-\gamma} \quad \text{with} \quad \gamma = (2q-1)/p$$

Hence, $\gamma' = \gamma$ stating that the critical exponents below and above T_{cri} are equal.

Continuing with differentiating G twice with respect to T leads to

$$\lambda^{2p} C_P(\lambda^p t, \lambda^q P) = \lambda C_P(t,P)$$

Taking $P = 0$ and assuming

$$\lambda = (-t)^{-1/p} \quad \text{leads to} \quad \alpha' = 2 - (1/p)$$

From a similar calculation we obtain $\alpha = \alpha'$. From the last result and $p = [\beta(\delta-1)]^{-1}$, we conclude that $\alpha + \beta(\delta+1) = 2$, which we recognize as Griffiths' (in)equality. Finally, combining $\gamma = \beta(\delta - 1)$ with $\alpha + \beta(\delta+1) = 2$, we rediscover Rushbrooke's (in)equality, $\alpha + 2\beta + \gamma = 2$.

In conclusion to this part, we note that the thermodynamic inequalities have become equalities. Moreover, we showed that the values of the critical exponents are equal above and below the critical temperature, and that the number of independent critical exponents has been reduced to two. The remaining questions are (i) why is a power law describing the critical behavior? and (ii) what are the values of the critical exponents? The first question will be answered in the next section, while a sketch of an answer to the second question will be given in Section 15.4.

15.3.3 Scaling Lattices

The next step in the elucidation-of-transitions process was provided by Kadanoff [15]. Essentially, the idea of scaling is applied to a lattice. One can think of the lattice gas model for which a Hamilton function describes the equilibrium distribution of occupied and nonoccupied lattice sites. The correlation between the sites is given by the (total) correlation function $h(r) \equiv \langle \rho(r)\rho(0)\rangle - \langle \rho^2 \rangle$, and we use the *coherence length* ξ as a measure for the range of the correlation function (Figure 15.3).

Consider therefore a d-dimensional lattice with N sites and lattice constant a, and partition this lattice in cells (or blocks) of size La with $L \gg 1$ (Figure 15.4a). Hence, there are $n = N/L^d$ cells, each containing L^d sites. Further consider that we are close to T_{cri} so that the correlation length $\xi \gg La$. Next, we associate with each cell a parameter that states whether the cell as a whole is considered to be occupied, or not. For this process we use a

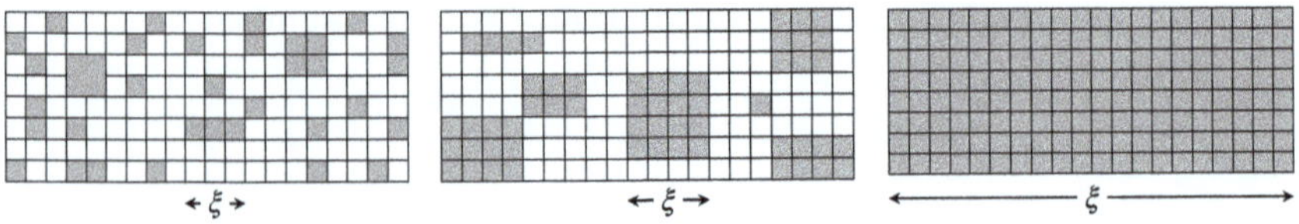

Figure 15.3 The coherence length ξ for the lattice gas model for $T \ll T_{\text{cri}}$, $T < T_{\text{cri}}$ and $T = T_{\text{cri}}$.

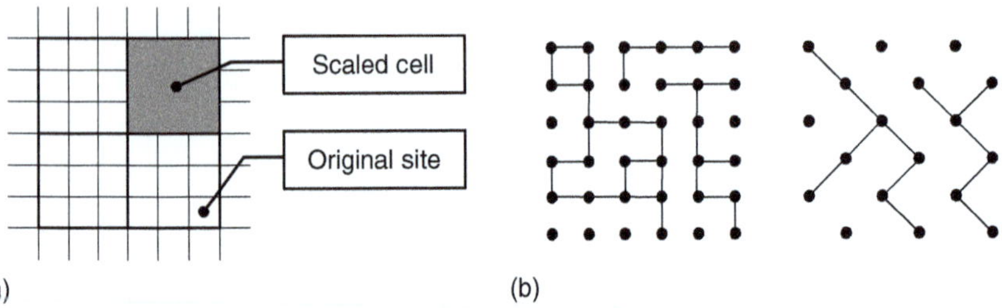

Figure 15.4 Lattice scaling. (a) Terminology used, illustrated for a 2D square lattice. (b) The decimation process for a 2D square lattice in which every other site of the site lattice is deleted and in the resulting cell lattice a bond between cells is present if on the original site lattice at least one connection via two bonds existed.

definite rule. For example, in the case of averaging over three lattice sites in a triangular lattice, we obtain a cell for which we apply the *majority rule*, that is, a value of 1 is taken if at least two of three original sites have also the value 1. Hence, 000 → 0 (occurrence 1), 001 → 0 (occurrence 3), 011 → 1 (occurrence 3), and 111 → 1 (occurrence 1). For a square lattice one can use the *decimation process* (Figure 15.4b). Finally, we assume that for the cells the same Hamilton function applies as for the sites. If that is the case, then one might expect that the thermodynamics of the site model would be the same as that for the cell model. Denoting the relative temperature difference with respect to T_{cri} as before by t and the pressure by P, we have for a cell (site) t_{cell} (t_{site}) and P_{cell} (P_{site}). The Gibbs energy g per site of the site and per cell for the cell model then relate as

$$g(t_{\text{cell}}, P_{\text{cell}}) = L^d g(t_{\text{site}}, P_{\text{site}})$$

Now we need relations between t_{cell} and t_{site} on the one hand, and P_{cell} and P_{site} on the other hand. We expect that the pressure of the cell model is proportional to the pressure of the site model, but that the proportionality constant B will be dependent on L. The same assumption can be made for t. So, we have

$$t_{\text{cell}} = A(L)t_{\text{site}} \quad \text{and} \quad P_{\text{cell}} = B(L)P_{\text{site}}$$

Although it is possible to continue without further assumptions [16], we assume that

$$A(L) = L^x \quad \text{and} \quad B(L) = L^y$$

with x and y arbitrary numbers (Justification 15.1). This leads right away to

$$g(L^x t_{\text{site}}, L^y P_{\text{site}}) = L^d g(t_{\text{site}}, P_{\text{site}})$$

This expression, stating that g is a generalized homogeneous function of t and P, leads essentially to scaling with exponents $p = x/d$ and $q = y/d$.

Justification 15.1 ***Power Laws as Scale-Free Functions*** One might ask why a power relation $(x/x_0)^b$ is more scale free than, say, an exponential relation $\exp(x/x_0)$, since for both the dimensionless number x/x_0 with x_0 scale length is involved. Suppose that we have a power relation reading $y/y_0 = a(x/x_0)^b$, and that we have data available for the range $x = x_0/2$ to $x = 2x_0$. This implies that the ratio between maximum and minimum measured values is 4^b. If we scale x_0 to cx_0 and use the same range of data, we still have a factor of 4^b between maximum and minimum measured values. This implies that a superposition of these data ranges can be realized by a simple linear change of scale, and the data appear the same no matter on what scale they are is probed. No other function can do this.

15.4 Renormalization

The plot thickens, as they say, and the next step is considerably more complex than the previous ones. That is why only a sketch of an answer to the remaining questions will be given. Full details can be found in the references given in the Section "Further Reading".

To continue, we need fluctuations for which in Figure 15.5 snapshots from an early Monte Carlo (MC) simulation of the 2D lattice gas model using 64×64 cells are shown. From these pictures and other evidence, it became clear that near T_{cri} *fluctuations occur on every length scale*, and that the structure of the pictures is rather independent of the density, suggesting *self-similar behavior*. The *coherence length* ξ, which measures the range of the correlation function $g(r)$, scales as $\xi/\xi_0 \sim |t|^{-\nu}$ above as well as below T_{cri}. In order to have self-consistency, fluctuations in $g(r)$ should be much smaller than $\langle \rho^2 \rangle$. We have $\langle \rho^2 \rangle \sim t^{2\beta}$ and $g(r) \sim \exp(-r/\xi)/r^{d-2+\eta}$. Hence, for $r \cong \xi$, we obtain $t^{2\beta} > (t^{-\nu})^{-(d-2+\eta)}$ or, using Stanley's lemma, $2\beta < \nu(d-2+\eta)$. Using scaling theory, this inequality also appears to be an equality. Using the mean-field values $\beta = ½$, $\nu = ½$ and $\eta = 0$, results in $d \geq 4$. This result states that for fluctuations to be unimportant, the dimensionality should be at least 4, which implies that in 3D, *fluctuations do matter*. One could think that adding fluctuations to the mean-field result would remedy the situation, but since the fluctuations are more divergent than the mean values, this approach is not fruitful.

The answers to the question, "What are the numerical values of the critical exponents?" can be found in the approach, normally called *renormalization group* (RG) theory [18]. In the simplest form of this theory, we choose a transformation that transforms the original Hamilton function $\mathscr{H}$ of the site lattice with lattice constant a to a new Hamilton function $\mathscr{H}_1$ of the cell lattice with lattice constant La, without changing the expression for the partition function, so that $Z(\mathscr{H},N) = Z(\mathscr{H}_1,N/L^d)$. Further, we use the usual lattice model approximation in which a site (cell) is occupied or not, representing repulsion, and only nearest-neighbor interactions with parameter J, representing attraction, are considered. In fact, we used the idea in the previous paragraph with respect to the Gibbs energy of the lattice, but here we employ the Hamilton function. The basic idea is that this transformation can be described by $r_1 = f(r)$ so that the parameters r_1 of the Hamilton function $\mathscr{H}_1$ of the cell lattice can be obtained from the parameters r of the Hamilton function $\mathscr{H}$ of the site

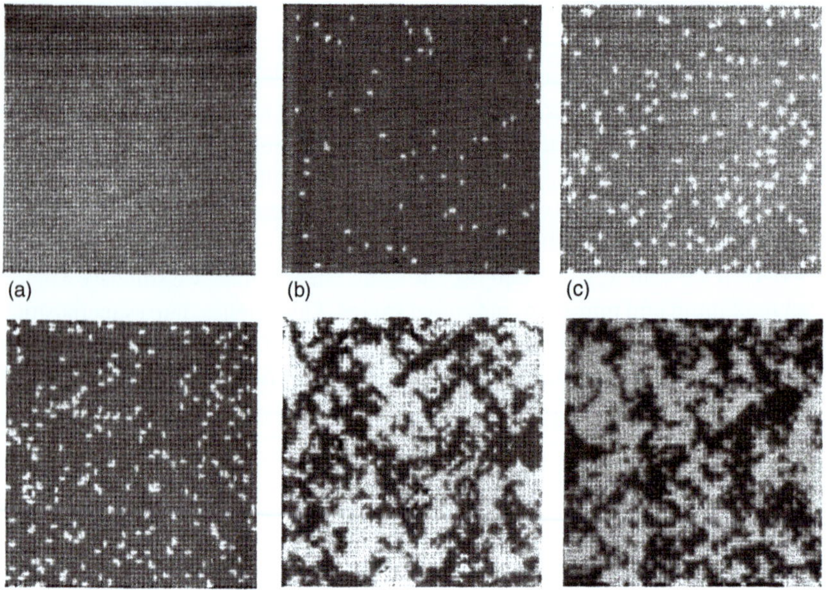

Figure 15.5 Images (cathode ray tube (CRT) screen shots) of lattice gas model results using a 64 × 64 2D lattice for (a) $T = 0$; (b) $T = T_{cri}/4$; (c) $T = T_{cri}/2$; (d) $T = 3T_{cri}/4$; (e) $T \cong T_{cri}$; (f) $T > T_{cri}$ [17].

lattice. This process can be iterated, that is, $r_2 = f(r_1) = f[f(r)]$ and so on, for increasingly larger blocks until La reaches ξ. We can describe this process also by $r_{i+1} = f(r_i)$. It may happen that for $i \to \infty$ we obtain $r^* = f(r^*)$, where r^* is called a *fixed point*. As an example, we use the function

$$r_{i+1} = cr_i(1 - r_i) \quad \text{with} \quad 0 < r < 1 \text{ and } c > 1 \tag{15.8}$$

The fixed points are then the solutions of $r^* = cr^*(1 - r^*)$, which are $r^* = 0$ and $r^* = 1 - 1/c$. Stability can be probed by adding a small perturbation ε_i to r^* so that $x_i = r^* + \varepsilon_i$. From substitution in $r_{i+1} = f(r_i)$ and expanding $f(r_i)$ in a Taylor series while keeping only the linear term, we obtain

$$\varepsilon_{i+1} = (\mathrm{d}f/\mathrm{d}r)_{r=r^*}\varepsilon_i$$

If $(\mathrm{d}f/\mathrm{d}r)_{r=r^*} < 1$, during each iteration the distance to r^* will decrease and the fixed point will be called *stable*. If on the other hand, $(\mathrm{d}f/\mathrm{d}r)_{r=r^*} > 1$, the distance will increase and the fixed point is called *unstable*. For our example, Eq. (15.8), $\mathrm{d}f/\mathrm{d}r = c(1 - 2r)$ and since $c > 1$, $r^* = 0$ is an unstable fixed point and $r^* = 1 - 1/c$ is a stable fixed point.

From the above procedure we can determine not only the fixed points but also the way how they are approached. Using $r_1 = f(r)$, one obtains

$$r_1 - r^* = f(r) - f(r^*) \cong (\mathrm{d}f/\mathrm{d}r)_{r=r^*}(r - r^*) + \dots \tag{15.9}$$

Now, during the transformation the Gibbs energy per particle $g(r)$ must be unaltered, that is, $g(a_1) = L^d g(a)$ or, equivalently, for the partition function $Z(\mathscr{H},N) = Z(\mathscr{H}_1,N/L^d)$. This scaling can only be done if the cell size is less than the coherence length ξ. Since for $T \to T_{cri}$,

$\xi \to \infty$, at T_{cri} the scaling can be done for any cell size. Hence, we identify the critical point of the phase transition T_{cri} with the fixed-point T^*. Taking $r = J/kT$, with J the interaction energy and k Boltzmann's constant, we can rewrite Eq. (15.9) for r not too far removed from r^* as

$$T_1 - T^* \cong (df/dr)_{r=r^*}(T - T^*) \quad \text{or} \quad t_1 \cong (df/dr)_{r=r^*}t$$

On comparison with the general expression $t_1 \equiv (T - T_{cri})/T_{cri} = L^x t$, we obtain

$$x = \ln\, (df/dr)_{r=r^*} / \ln L$$

If there are more parameters, say $r = r(t)$ as before and $s = s(P)$ with, for example, $s = P/kT$ for the pressure ensemble, we have $f = f(r,s)$ resulting in the following two equations:

$$r^* = f_1(r^*, s^*) \quad \text{and} \quad s^* = f_2(r^*, s^*)$$

Using the same procedure, we obtain

$$\begin{pmatrix} (r_1 - r^*) \\ (s_1 - s^*) \end{pmatrix} = \begin{pmatrix} \left(\dfrac{df_1}{dr}\right)_{r=r^*} & \left(\dfrac{df_1}{ds}\right)_{r=r^*} \\ \left(\dfrac{df_2}{dr}\right)_{s=s^*} & \left(\dfrac{df_2}{ds}\right)_{s=s^*} \end{pmatrix} \begin{pmatrix} (r - r^*) \\ (s - s^*) \end{pmatrix} \tag{15.10}$$

Because r_1 and s_1 are coupled, we transform to normal coordinates to obtain the scaling behavior, and for this we must find the eigenvalues of the derivative matrix. For convenience, we label the derivative matrix as $\mathbf{T}$ and denote the set $(r - r^*)$ and $(s - s^*)$ by the column $\delta\mathbf{k}$, so that Eq. (15.10) reads $\delta\mathbf{k}_1 = \mathbf{T}\delta\mathbf{k}$. However, as $\mathbf{T}$ is nonsymmetric,[2] generally its left and right eigenvectors will be different. Using $\boldsymbol{\varphi}$ as the column for the left eigenvectors, we write $(\boldsymbol{\varphi}^{(\alpha)})^{\mathrm{T}}\mathbf{T} = \lambda^{(\alpha)}(\boldsymbol{\varphi}^{(\alpha)})^{\mathrm{T}}$ or $\Sigma_j \varphi_j^{(\alpha)} T_{ji} = \lambda^{(\alpha)}\varphi_i^{(\alpha)}$, where the superscript α labels the eigenvalues λ. Despite $\mathbf{T}$ being a nonsymmetric matrix, the eigenvectors do not mix with one another under an RG transformation. This can be shown by introducing the normal coordinates $u^{(\alpha)} = (\boldsymbol{\varphi}^{(\alpha)})^{\mathrm{T}}\delta\mathbf{k}$, leading to

$$u_1^{(\alpha)} = (\boldsymbol{\varphi}^{(\alpha)})^{\mathrm{T}}\delta\mathbf{k}_1 = (\boldsymbol{\varphi}^{(\alpha)})^{\mathrm{T}}\mathbf{T}\delta\mathbf{k} = \lambda^{(\alpha)}(\boldsymbol{\varphi}^{(\alpha)})^{\mathrm{T}}\delta\mathbf{k} = \lambda^{(\alpha)}u^{(\alpha)}$$

If we assume that all eigenvalues are real, the local geometry of the Hamilton function surface around a fixed point is thus described by the normal coordinates $u^{(\alpha)}$, which are also scale invariant. This implies that the Hamilton function and therefore the Gibbs energy per particle $g(\mathbf{k})$ can be written in terms of the normal coordinates $u^{(\alpha)}$. We write $g(u^{(1)},u^{(2)},\ldots) = L^{-d}g(\lambda^{(1)}u^{(1)}, \lambda^{(2)}u^{(2)},\ldots)$ and we identify the coordinates $u^{(\alpha)}$ as the scaling parameters of the problem. Since in general at T_{cri} all $u^{(\alpha)}$ are zero, we can determine for a given coordinate, say number 1, its behavior close to T_{cri}, by setting $u^{(2)}, u^{(3)}, \ldots$ equal to zero so that we have $g(u^{(1)}, 0, \ldots) = L^{-d}g(\lambda^{(1)}u^{(1)}, 0, \ldots)$. If $r = r(t)$ and $s = s(P)$ are the only

---

2 A square matrix $\mathbf{A}$ of order m, operating on a column $\mathbf{u}$, yields a new column $\mathbf{u}'$. If this operation yields a multiple of $\mathbf{u}$, say $\lambda\mathbf{u}$, the resulting equation $(\mathbf{A} - \lambda\mathbf{I})\mathbf{u} = 0$ is an *eigenvalue equation* with λ the *eigenvalue*, where $\mathbf{I}$ is the unit matrix and $\mathbf{u}$ is the *eigenvector*. This equation can be solved by $\det(\mathbf{A} - \lambda\mathbf{I}) = 0$, yielding m eigenvalues and associated eigenvectors. Collecting the set of columns $\mathbf{u}$ in the matrix $\mathbf{U}$, the complete set of equations can be written as $\boldsymbol{\Lambda} = \mathbf{U}^{\mathrm{T}}\mathbf{A}\mathbf{U}$, where $\boldsymbol{\Lambda}$ is the diagonal matrix containing all eigenvalues. For a symmetric matrix $\mathbf{A}$, $\mathbf{U}$ is an orthogonal matrix, that is, $\mathbf{U}^{-1} = \mathbf{U}^{\mathrm{T}}$. For a nonsymmetric matrix $\mathbf{A}$, this is no longer the case, and we have to keep track of whether the eigenvector is positioned on the left or right side of the matrix $\mathbf{A}$, that is, whether we solve $\mathbf{A}\mathbf{u} = \lambda\mathbf{u}$ or $\mathbf{u}^{\mathrm{T}}\mathbf{A} = \lambda\mathbf{u}^{\mathrm{T}}$.

variables of interest, we have $g(r,s) = L^{-d}g(\lambda^{(1)}r, \lambda^{(2)}s)$ and comparing this expression with the Widom scaling function $g(\lambda^p t,\lambda^q P) = \lambda g(t,P)$, the result becomes

$$\lambda = L^d, \quad p = \frac{1}{d}\frac{\ln \lambda^{(1)}}{\ln L} \quad \text{and} \quad q = \frac{1}{d}\frac{\ln \lambda^{(2)}}{\ln L}$$

In Section 15.3, we have shown that all exponents follow from p and q. Hence, all critical exponents are determined by the scaling behavior of the eigenvalues of the linearized transformation matrix **T** near the critical point.

Applying this process, the critical exponents of the lattice model have been obtained to high accuracy (Table 15.3). As an illustration of the accuracy of some approximate results, the values for the 2D lattice gas using different lattices and blocks are also given. Note also the good agreement of the 3D values as calculated using renormalization theory with the experimental values (Table 15.2). To conclude this part, we note that the above outlined procedure is the simplest possible way of renormalizing real space. Moreover, we note that the inverses of these transformations do not exist so that the designation "renormalization group" actually should read "renormalization semi-group." However, the nomenclature is conventional.

This approach is apparently abstract, but we refrain from detailing one or another model. It is useful, however, to illustrate the methods by the much simpler example of percolation (Justification 15.2), which has its own intrinsic interest. If we mix electrically conducting and insulating particles at random, using an increasing fraction of conducting particles, the mixture will at first be insulating; however, at a certain fraction a transition to a conductive mixture occurs. This change occurs over a small volume fraction change of conductive particles, and thereafter the conductivity still increases, albeit slowly. This phenomenon is called *percolation*, and the point at which the transition occurs is termed the *percolation threshold*. The phenomenon can be considered as a geometrical phase transition where, at the percolation threshold infinite, connected clusters of conductive particles appear. For 2D (i.e. for disks) the transition occurs at 50 vol%, whereas for 3D (i.e. for spheres), it occurs at about 16 vol%. The development of these clusters in a percolation model is similar to that of clusters of molecules in phase transformations. Each realization of a packing at a certain volume fraction can be considered as a snapshot of a configuration of two types of molecules. Of course, this analogue, given by [19], is static in time, but the basic idea is the same.

The physical problem is now to derive when and how electrical conduction across a macroscopic region occurs in terms of the microdetails of the packing of the particles. As a concrete example, we use a 2D structure having a triangular lattice with renormalization by three particles (disks). For the original packing the microdescription is in terms of sites that are either occupied or unoccupied with a conductive particle. We denote the probability that any site is occupied by p, independent of whether any other site is occupied or not. The occupancy distribution of sites provides the exact microdescription of the system for which we only know the single site probability p.

In the spirit of the renormalization theory, we increase the length scale to b, that is, we increase the area by $b^2 = 3$. At the higher level of description, here thus a cell of three sites, and using the majority rule, a cell is occupied if and only if two or more of the sites are occupied (Figure 15.6). Let us call p_1 the probability that a cell of three sites is occupied. The probability p_1 is then the sum of probabilities of four mutually exclusive states. Using p for occupied sites and $1-p$ for unoccupied sites, we have once p^3 and three times $p^2(1-p)$.

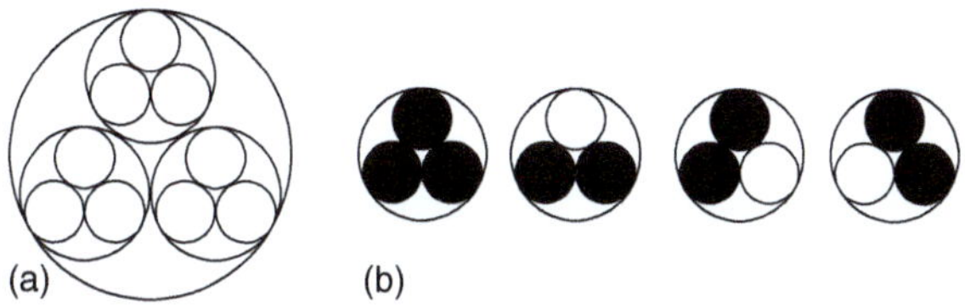

Figure 15.6 Percolation and renormalization by taking three spheres together to which the majority rule is applied. (a) Showing a schematic of the original packing; (b) showing the four possible "renormalized" sites.

The percolation threshold (critical point) occurs if the original probability p equals the renormalized probability p_1. We thus have

$$p_1 = p^3 + 3p^2(1-p)$$

The solution of this equation does not require iteration but yields directly as solutions $p^* = 0$ and $p^* = 1$, which represent the stable fixed points, and the solution $p^* = 0.5$ representing the unstable fixed point. The unstable fixed-point solution thus appears to be the exact solution for this 2D percolation problem, which however is fortuitous.

Let us now calculate the critical exponent ν, which describes the onset of the percolation phenomenon. We take the coherence length ξ for the original packing to be given by $\xi = C|p-p_{\text{cri}}|^{-\nu}$, while for the renormalized packing, we have $\xi_1 = C|p_1-p_{\text{cri}}|^{-\nu}$ with $\xi_1 = \xi/b$. Combining results in

$$b|p_1 - p_{\text{cri}}|^{-\nu} = |p - p_{\text{cri}}|^{-\nu} \qquad \text{or}$$

$$\frac{1}{\nu} = \frac{\ln[(p_1 - p_{\text{cri}})/(p - p_{\text{cri}})]}{\ln b} = \frac{\ln \lambda}{\ln b} \qquad \text{with} \qquad \lambda = \frac{p_1 - p_{\text{cri}}}{p - p_{\text{cri}}} \cong \frac{\text{d}p_1}{\text{d}p}$$

Expanding p_1 as $p_1 = p^* + \lambda^*(p-p^*) + \ldots$ with $\lambda^* = (\text{d}p_1/\text{d}p)_{p=p^*}$ and substitution in the original equation $p_1 = p^3 + 3p^2(1-p)$ yields

$$\text{d}p_1/\text{d}p = 3p^2 + [6p(1-p) - 3p^2] = 6p(1-p)$$

so that $\lambda^* = 3/2$. Because $b^2 = 3$, we obtain

$$\nu = \frac{\ln b}{\ln \lambda} = \frac{\ln 3^{1/2}}{\ln(3/2)} = 1.355$$

The exact value is $\nu = 4/3$. Approaching the threshold, the coherence length ξ increases until ξ diverges at the threshold itself. This implies that conductive clusters span the complete system, and the system becomes macroscopically conductive.

Justification 15.2 ***Percolation*** Percolation is a generic term applied to many phenomena and for which an often-used model system deals with an infinite lattice on which clusters of occupied sites are embedded. Each site is occupied randomly with *probability* p and a *cluster* is defined as a group of occupied lattice sites connected by nearest-neighbor distances. At low probability, that is, at low-volume fraction of occupied sites, only isolated clusters exist. At some higher-volume fraction, the so-called *percolation limit* p_{cri}, infinite clusters start to span the infinite lattice. This is *site percolation* in which the occupancy of sites is considered. Another type is *bond percolation* in which one considers the bonds between sites. Percolation using no lattice at all, as in the example of conductive spheres, is referred to as *continuum percolation*. We limit ourselves to site percolation. We denote with *cluster strength* P the probability that an occupied site belongs to an infinite cluster. Hence, for $p < p_{\text{cri}}$ we have $P = 0$, while for $p > p_{\text{cri}}$, P is finite. The *correlation function* $g(l)$ describes the probability that a particular site at distance l belonging to the same finite cluster is occupied, given that the origin site is occupied, and the *correlation length* ξ is the root-mean-square distance between two occupied sites in the same finite cluster and represents its typical radius.

For clusters containing s sites, we define the *cluster number* n_s as the number of such s clusters divided by the number of all lattice sites. The number of clusters per lattice site, irrespective of their size, becomes $N = \Sigma_s n_s$, while the probability per site that a site belongs to a cluster of size s is sn_s. As all occupied sites belong to a cluster, we have for $p < p_{\text{cri}}$ for the *probability* p and the *mean cluster size* S, the expressions

$$p = \sum\nolimits_{s=1}^{s=\infty} sn_s \quad \text{and} \quad S = \sum\nolimits_{s=1}^{s=\infty} s\left(sn_s/\sum\nolimits_{s=1}^{s=\infty} sn_s\right) = p^{-1}\sum\nolimits_{s=1}^{s=\infty} s^2 n_s$$

The sum $\Sigma_l g(l)$ represents the total probability of sites connected to an occupied site so that $S = 1 + \Sigma_l g(l)$. A site is either nonoccupied (probability $1 - p$) or occupied, and occupied sites either belong to a finite (probability sn_s) or an infinite cluster (probability pP). Therefore, denoting by Σ_s a summation over finite clusters,

$$1 = (1 - p) + \sum\nolimits_s sn_s + pP \quad \text{or} \quad P = 1 - p^{-1}\sum\nolimits_s sn_s$$

All these quantities are thus related to one or more of the kth moments of the $n_s(p)$ distribution, defined as $\Sigma_s\, s^k n_s(p)$.

Assuming that only one unique typical cluster size exists, one can show that near p_{cri}, $n_s(p) = s^{-\tau} f_\pm(|p - p_{\text{cri}}|s^\sigma)$, where $f_\pm$ is a homogeneous function of $|p - p_{\text{cri}}|s^\sigma$ and in the $\pm$ symbol the + indicates $p > p_{\text{cri}}$ and the – indicates $p < p_{\text{cri}}$. Replacing in $\Sigma_s\, s^k n_s(p)$ both s and n_s by their typical values s_ξ and $n_{s\xi}$, respectively, taking into account the additional factor $s^{1+k-\tau}$ due to the summation, and further that $s_\xi \sim |p - p_{\text{cri}}|^{-1/\sigma}$, we have $\Sigma_s\, s^k n_s(p) \sim |p - p_{\text{cri}}|^{(\tau-1-k)/\sigma}$. Hence, we have

$$\begin{aligned}
N &\sim \Sigma_s s^0 n_s(p) \sim |p - p_{cri}|^{2-\alpha} && \text{with} \quad 2 - \alpha = (\tau - 1)/\sigma, \\
P &\sim \Sigma_s s^1 n_s(p) \sim |p - p_{cri}|^{\beta} && \text{with} \quad \beta = (\tau - 2)/\sigma \ \text{and} \\
S &\sim \Sigma_s s^2 n_s(p) \sim |p - p_{cri}|^{-\gamma} && \text{with} \quad \gamma = (3 - \tau)/\sigma
\end{aligned}$$

The labels for the exponents, such as α and β, are conventional, similar as in Section 15.1. Moreover, in $\xi \sim |p - p_{\text{cri}}|^{-\nu}$, the exponent appears to be $\nu = (\tau - 1)/d\sigma$, where d is the dimensionality of the lattice. Therefore, all exponents are related to τ and σ. The exponents α, β, γ, and ν are universal, that is, do not depend on the type of lattice or on the type of percolation, but do depend on the dimensionality of the lattice. For two dimensions, we have $\alpha = -2/3$, $\beta = 5/36$, $\gamma = 43/18$, and $\nu = 4/3$, or equivalently, $\tau = 187/91$ and $\sigma = 36/91$. However, p_{cri} depends on the type of lattice and equals 0.5000 for a triangular and 0.5927 for a square lattice.

As a simple example, we take the *Bethe* (pseudo-)*lattice* (or infinite, regular *Cayley tree*), that is, a structure with at each site z connections and having no loops. For the first shell around the origin we have z sites, for the second shell $z(z - 1)$ sites, and therefore for the

l-th shell $z(z-1)^l$ sites. The probability that a site at distance l and all sites between this site and the origin are occupied is p^l, while the total number of sites in shell l is $z(z-1)^l$. Consequently, the correlation function $g(l)$ is given by

$$g(l) = z(z-1)^{l-1}p^l = \frac{z}{z-1}p(z-1)^l = \frac{z}{z-1}e^{l\ln[p(z-1)]} = \frac{z}{z-1}e^{-l/\xi}$$

Therefore, for $p(z-1) < 1$, we have an exponential decrease of the cluster size, while for $p(z-1) > 1$, we have an exponential increase. Hence, the critical value occurs at $p_{\text{cri}}(z-1) = 1$, resulting in $p_{\text{cri}} = (z-1)^{-1}$. The correlation length ξ in $\exp(-l/\xi)$ becomes

$$\xi \sim \frac{-1}{\ln p(z-1)} = \frac{-1}{\ln p/p_{\text{cri}}} = \frac{1}{\ln p_{\text{cri}} - \ln p} \sim \frac{1}{p_{\text{cri}} - p}$$

where $\ln p = \ln[1-(1-p)] \approx -(1-p)$ is used. Thus, in the neighborhood of $p_{\text{cri}}, \nu = 1$. Similarly, the mean cluster size $S = 1 + \Sigma_l g(l)$, so that we have, for $p < p_{\text{cri}}$,

$$S = 1 + \sum_{l=1}^{l=\infty} g(l) = 1 + \frac{z}{z-1}\sum_{l=1}^{l=\infty} p(z-1)^l = 1 + \frac{z}{z-1}\frac{(1+p)}{1-(z-1)p} = \frac{p_{\text{cri}}(1+p)}{p_{\text{cri}} - p}$$

where $p_{\text{cri}} = (z-1)^{-1}$ is used. Thus $\gamma = 1$. To calculate the cluster strength P, consider the product of the probability p with P. We also denote the probability that an arbitrary site is not connected to infinity with q. The latter is equal to the probability that the end of a branch is not occupied or $1-p$, plus the probability that this site is occupied, but none of the $z-1$ branches lead to infinity, that is, pq^{z-1}. In total therefore, $q = 1-p+pq^{z-1}$. If we take, as an example $z = 3$, we obtain the solutions $q = 1$ and $q = (1-p)/p$. Moreover, the probability that the origin is occupied but that no branch leads to infinity is given by either $p(1-P)$ or pq^z, leading for $z = 3$ to $P = 1-q^3$. For $q = 1$, we have $P = 0$, and this is the solution for $p < p_{\text{cri}}$. For $q = (1-p)/p$, the result is $P = 1-[(1-p)/p]^3$, providing the solution for $p > p_{\text{cri}}$. For $z = 3$ we have $p_{\text{cri}} = ½$, and, developing P in a Taylor series around p_{cri}, the strength of the infinite cluster can be shown to be described by $P \sim p - p_{\text{cri}}$, corresponding to $\beta = 1$.

In summary, for the Bethe lattice we have, $\alpha = -1$, $\beta = 1$, $\gamma = 1$, $\nu = 1$, corresponding to $\tau = 5/2$ and $\sigma = ½$, but, due to its peculiar nature, $\nu = (\tau-1)/d\sigma$ is not obeyed. For regular 2D and 3D lattices, solving the percolation problem is not so straightforward and we refer for general reviews to Stauffer and Aharony [20] and Christensen and Moloney [21].

Problem 15.4 Show that for the decimation process, as illustrated in (Figure 15.4b), $p_1 = p^4 + 4p^3(1-p) + 2p^2(1-p)^2 = 2p^2 - p^4$. Further show that the fixed points are $p^* = 0, 1$, and $½(5^{1/2}-1) \cong 0.618$ and that the critical exponent $\nu \cong 0.817$.

15.5 Final Remarks

We have come a long way: from a phenomenological description recognizing universality classes, via mean-field theory, generalized homogeneous functions, lattice gas, and scaling to renormalization. These seven pillars (of wisdom?) constitute the modern basis of continuous phase transition theory. The overall picture is somewhat complex, in particular the renormalization part, but capable of catching most of the essentials of continuous phase transformations quite well. Finally, we note that a link between RG theory and regular theory describing the behavior of liquids at far from critical conditions can be made. Such theories are called *crossover theories*. See, for example, for the vdW fluid [22].

References

1 Heller, P. (1967). *Rep. Progr. Phys.* 30: 731.

2 Sengers, J.V. and Levelt Sengers, J.M.H. (1978). *Critical Phenomena in Classical Fluids*, Ch. 4 in *Progress in Liquid Physics* (ed. C.A. Croxton), p. 103. Chichester: John Wiley & Sons.

3 Fisher, M.E. (1964). *J. Math. Phys.* 5: 944; (1967). *Rep. Prog. Phys.* 30: 615.

4 Ma, S.-K. (1976). *Modern Theory of Critical Phenomena*. Reading, MA: W.A. Benjamin.

5 Hocken, R. and Moldover, M.R. (1976). *Phys. Rev. Lett.* 37: 29.

6 Barieau, R. (1966). *J. Chem. Phys.* 45: 3175.

7 Onsager, L. (1944). *Phys. Rev.* 65: 117.

8 Gitterman and Halpern, 2004.

9 Pelissetto, A. and Vicari, E. (2002). *Phys. Rep.* 368: 549.

10 Amit, D.J. (1974). *J. Phys. C* 7: 3369.

11 Goldenfeld, 1993; Uzunov, 1993.

12 Landau, L.D. and Lifshitz, E.M. (1980). *Statistical Physics*, 3e, Vol. 1. Oxford: Pergamon Press.

13 Uzunov, 1993.

14 Widom, B. (1965). *J. Chem. Phys.* 43: 3892 and 3898.

15 Kadanoff, L.P. (1966). *Phys.* 2: 263; Kadanoff, L.P., Götze, W., Hamblen, D. et al. (1967). *Rev. Mod. Phys.* 39: 395.

16 Cooper, M.J. (1968). *Phys. Rev.* 168: 183.

17 Ogita, N., Ueda, A., Matsubara, T. et al. (1969). *J. Phys. Soc. Jpn. Suppl.* 26: 145.

18 Wilson, K.G. (1971). *Phys. Rev. B* 4: 3174 and 3184; See also Wilson, K.G. and Kogut, J. (1974). *Phys. Rep.* 12: 75.

19 Chowdhury, D. and Stauffer, D. (2000). *Principles of Equilibrium Statistical Mechanics*. Weinheim: Wiley-VCH.

20 Stauffer, D. and Aharony, A. (1994). *Introduction to Percolation Theory*, 2e. London: Taylor and Francis; See also Stauffer, D. (1979). *Phys. Rep.* 54: 1.

21 Christensen, K. and Moloney, N.R. (2005). *Complexity and Criticality*. London: Imperial College Press.

22 Kostrowicka Wyczalskowska, A., Sengers, J.V., and Animosov, M.A. (2004). *Physica A* 334: 482.

Further Reading

Binney, J.J., Dowrick, N.J., Fisher, A.J., and Newman, M.E.J. (1992). *The Theory of Critical Phenomena*. Oxford: Clarendon Press. A broad introductory overview and dealing, amongst others with numerical simulations, real space normalization and diagrammatic perturbation theory.

Gitterman, M. and Halpern, V.H. (2004). *Phase Transitions*. Singapore: World Scientific. Called by the authors "a brief account with modern applications," the latter comprising transitions in quantum systems, universality, random and small world systems and self-organized criticality.

Goldenfeld, N. (1992). *Lectures on Phase Transitions and the Renormalization Group*. Reading, MA: Taylor & Francis. A compact treatment which has become a classic.

Herbut, I. (2007). *A Modern Approach to Critical Phenomena*. New York: Cambridge University Press. A detailed graduate treatment with as aim not only to become familiar, but also to learn how to actually use the theory to explain and compute.

Stanley, H.E. (1971). *Introduction to Phase Transitions and Critical Phenomena*. New York: Oxford University Press. Despite its age, still worth reading and emphasizing the analogy between the ferromagnetic and liquid-gas transition.

Uzunov, D.I. (1993). *Introduction to the Theory of Critical Phenomena*. Singapore: World Scientific Publ. Comp. Another broad overview taking fairly small steps with many details.

16

The Liquid Crystal Transformation

We now turn our attention to two phase transformations: in this chapter to one in the liquid state – the liquid crystal transformation – and in next chapter to one in the solid state – the ferroelectric transformation. They serve as examples for other phase transformations in the liquid and solid states. Liquids are generally assumed to be isotropic (and also often as homogeneous). The term liquid crystal is an apt term to indicate that it is a liquid with a crystalline aspect, namely directionality.

16.1 Nature and Types

Crystalline solids have a long-range, three-dimensional (3D) order, while liquids show short-range order but are overall disordered. In between there are systems showing short-range order in some direction(s) and long-range in some other. Orientational order combined with positional disorder occurs in liquid crystals. There are a few prerequisites to form a liquid crystal. The molecules involved should

- have a highly anisotropic shape, or
- be a polymer with (at least partly) a high rigidity or
- be a polymer that forms rod-like structures in solution by secondary interactions.
- Another important class nowadays is that of polymers having side groups that are mesogenic (display liquid crystal properties), such as siloxanes and acrylates. The materials can be cross-linked and, in that case, are called *liquid crystal elastomers*.

Such molecules are often called *mesogens*. Liquid crystalline order is characterized by an *internal variable* ξ, in this context often called the *order parameter* (Section 16.2.1). For $\xi = 1$ the system is fully ordered, while for $\xi = 0$ the system is fully disordered. Two examples of molecules of the first type are shown in Figure 16.1.

In a *nematic* liquid crystal, rod-like molecules are preferentially oriented along a certain direction, the *director*, but they have no positional order. If the molecule is also *chiral*, that is, it has no mirror plane or in other words there is a left-hand and right-hand form of the molecule, there is a preference for a left-hand one to a right-hand one with as result that they form a helix with a well-defined pitch. Such a type is called *chiral nematic*, or more commonly, *cholesteric*. The pitch typically has a value about equal to the wavelength of light, which leads to observable effects in an optical microscope. There is also the *smectic*

Phases of Matter and their Transitions: Concepts and Principles for Chemists, Physicists, Engineers, and Materials Scientists, First Edition. Gijsbertus de With.

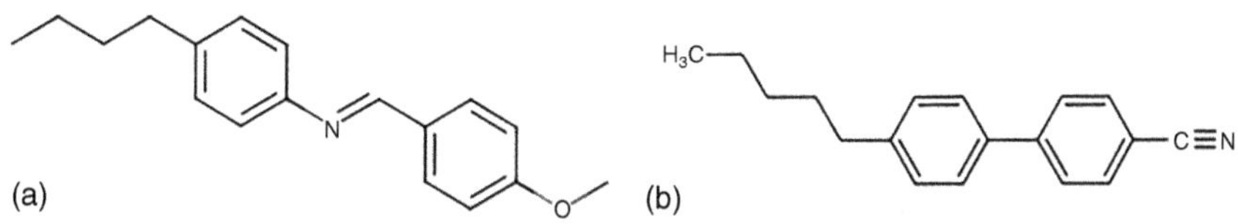

Figure 16.1 (a) Structure of *N*-(4-methoxybenzylidene)-4-butylaniline (MBBA) with a crystalline–nematic transition at 22.1 °C and a nematic–isotropic transformation at 44 °C. (b) Structure of 4′-pentyl[1,1′-biphenyl]-4-carbonitrile (5CB) with a crystalline–nematic transition at 22.5 °C and a nematic–isotropic transformation at 35.0 °C.

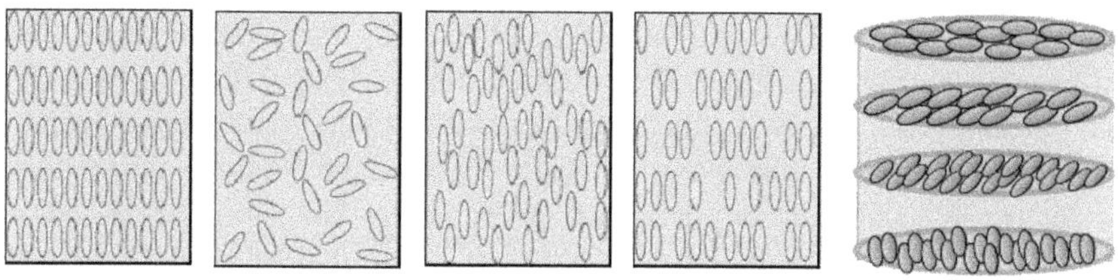

Figure 16.2 While solids have preferred orientation and periodicity and liquids have neither preferred orientation nor periodicity, the nematic phase shows some preferred orientation but no periodicity, and the smectic phase shows some orientation combined with some periodicity. Chiral molecules can have a chiral nematic or cholesteric phase.

form in which the molecules arrange in sheets or layers that can slide over one another. They are ordered along one direction. In fact, there are many different smectic phases, all characterized by different types and degrees of positional and orientational order [1]. For example, in the Smectic A phase, the molecules are oriented along the layer normal, while in the Smectic C phase, they are tilted away from it, both being liquid-like within the layers. Finally, molecules can also be disk-like, and in the *columnar* phase such molecules are stacked in columns in a disorderly fashion, while the columns as such are orderly packed. Hence, they show 2D positional order. Figure 16.2 shows a schematic of these phases.

The simplest and only change we focus on in the next section is probably from the nematic (N) phase at low temperatures to an isotropic (I) phase at high temperatures. In our terminology, this is the nematic–isotropic (N–I) transformation with transformation temperature $T_{\text{N–I}}$, although in the literature it is commonly called the N–I transition. It is a discontinuous transformation, for which there are two types. The first is the *thermotropic* one, for which the transformation is caused by a temperature change. They typically are comprised of a one-component type 1 mesogen or of, in practical applications, a mixture of type 1 mesogens, that is, of molecules having a highly anisotropic shape. For this type, ξ decreases with increasing temperature with a typical range $\xi = 0.3 - 0.5$ near $T_{\text{N–I}}$. The second is the *lyotropic* one, in which generally polymer molecules with a high rigidity (type 2) or forming rod-like structures in solution (type 3) transform by a change in concentration. However, low-molecular-weight liquid crystal systems do exist, for example, surfactants that form micelles, either regular or inverse. With increasing concentration, they can form a cubic, hexagonal, cubic bicontinuous, and lamellar structure, giving rise to a liquid crystal. In this case, ξ increases near $T_{\text{N–I}}$ with increasing concentration with a typical range $\xi = 0.6–0.8$. Finally, mixtures of liquid crystal molecules are used that show eutectic behavior and biphasic regions at the N–I transformation.

The field of liquid crystals is wide. Chandrasekhar [1] provided a broad overview, while de Gennes and Prost [2] focused on the theory, as did Vertogen and de Jeu [3]. Another overview is given by Blinov [4], a nice-to-read introduction from the molecular side is by van der Schoot [5], while simulations are discussed by Zanonni [6]. A compact introduction is given by Andrienko [7].

16.2 The Nematic–Isotropic Transformation

Generally, for a transformation to occur the balance between temperature-dependent energy and entropy contributions has to shift with a change in temperature. In the case of liquid crystal molecules, at low temperatures, an ordered phase is preferred (has a lower Helmholtz energy F or Gibbs energy G), while at higher temperatures the reverse is true. For smaller molecules the van der Waals (vdW) interactions are the most likely cause of a lower value for F or G and the dominating balance is that between the energy (enthalpy) change ΔH and orientational entropy ΔS_{ori}. For larger molecules, the geometrical considerations with respect to packing (thus, entropy considerations) are most likely determining. In this case, the relevant balance is between the orientational entropy change ΔS_{ori} and the positional entropy change ΔS_{pos}. These two situations largely but not exclusively correspond to the classification in thermotropic and lyotropic types. The energy–entropy balance is described by using an approach based on an internal variable, which for liquid crystal molecules is related to their orientation. The first step is thus to find the proper internal variable, which we thereafter can use in thermodynamics.

16.2.1 The Orientation as Internal Variable

To describe the orientation of molecules, one could think of a unit vector $\boldsymbol{\nu}^{(\kappa)}$ along the length of the molecule κ describing its direction [8]. Generally, a liquid crystal molecule has a center of inversion or, if not, the molecules have (almost) equal probability for being oriented parallel or antiparallel. In this case, we call the orientation *uniaxial* and the relevant internal variable ξ should be even in $\boldsymbol{\nu}^{(\kappa)}$. However, this is not necessarily so, and in these cases *ferroelectric nematic* structures may arise. If there is also order in the plane perpendicular to $\boldsymbol{\nu}^{(\kappa)}$, we call the orientation *biaxial*.

Restricting our discussion to uniaxial nematics, we note that the average of $\boldsymbol{\nu}^{(\kappa)}$ cannot be used for ξ because it yields zero by being even in $\boldsymbol{\nu}^{(\kappa)}$. Therefore, basically the invariant tensor $\mathbb{A}^{(\kappa)} = \langle \boldsymbol{\nu}^{(\kappa)}\boldsymbol{\nu}^{(\kappa)} \rangle$ constructed from $\boldsymbol{\nu}^{(\kappa)}$ (Sections A.11–A.12) is used. Using a spherical coordinate system (Section A.7), every molecule has its own unit vector

$$\boldsymbol{\nu}^{(\kappa)} = (\sin\theta\cos\phi, \sin\theta\sin\phi, \cos\theta)^{\text{T}}$$

with θ the polar angle and ϕ the azimuthal angle. For uniaxial nematics, the orientation probability function, generally given by $p(\theta,\phi)$, reduces to $p(\theta)$. Each molecule deviates with angle θ from the average unit vector in the z-direction $\boldsymbol{n} \equiv \langle \Sigma_\kappa \boldsymbol{\nu}^{(\kappa)} \rangle$, called the *director*, while ϕ is randomly distributed. In this coordinate system, $A_{11} = \langle \sin^2\theta \rangle \langle \cos^2\phi \rangle$, $A_{22} = \langle \sin^2\theta \rangle \langle \sin^2\phi \rangle$, and $A_{33} = \langle \cos^2\theta \rangle$. The former two reduce to $A_{11}\ A_{22} = ½\langle \sin^2\theta \rangle$ because for a random distribution of ϕ, $\langle \cos^2\phi \rangle = \langle \sin^2\phi \rangle = ½$. Clearly $\mathbb{A}$ is diagonal. For perfect order we would like to have $A_{33} = \langle \cos^2\theta \rangle = 1$, while all other components are zero.

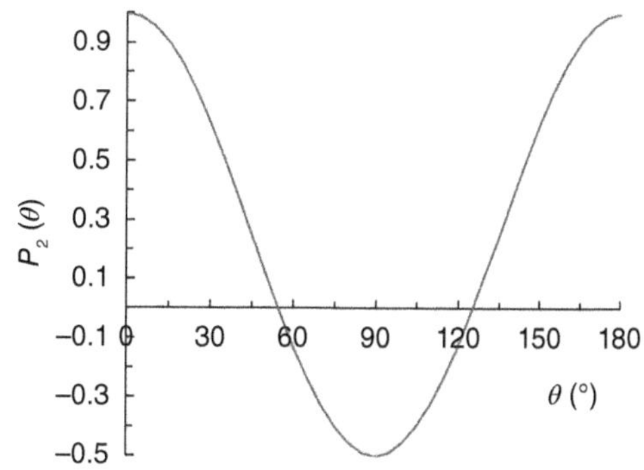

Figure 16.3 The function $P_2(\theta) = ½(3\cos^2\theta - 1)$ as a function of polar angle θ.

Moreover, because in the isotropic phase we would also like to have $\xi = 0$, we employ the traceless invariant form

$$\mathbb{Q} = \langle \mathbb{A} - ⅓\mathbb{I} \rangle \qquad \text{or rather} \qquad \mathbb{Q} = ½\langle 3\mathbb{A} - \mathbb{I} \rangle$$

so that for perfect order we obtain indeed $Q_{33} = 1$ and $Q_{11} = Q_{22} = -½$, while tr $\mathbb{Q} = 0$. For partial order we have, rewriting $\sin^2\theta$ in terms of $\cos^2\theta$, the expression

$$\mathbb{Q} = ½\langle 3\mathbb{A} - \mathbb{I} \rangle = \xi \begin{pmatrix} -½ & 0 & 0 \\ 0 & -½ & 0 \\ 0 & 0 & 1 \end{pmatrix} \qquad \text{with} \qquad \xi = ½\langle 3\cos^2\theta^{(\kappa)} - 1 \rangle$$

The function $P_2(\cos\theta) \equiv ½(3\cos^2\theta - 1)$ is the *second Legendre polynomial* and shown in Figure 16.3. We note, although tr $\mathbb{Q} = 0$, that tr $\mathbb{Q}^2 = 3\xi^2/2$ and consequently we also have $\xi = (2\ \text{tr}\ \mathbb{Q}^2/3)^{1/2}$. For a continuum description the average should be taken over representative volume element (Section 5.1) and this average is in general position dependent. So, a general representation of $\mathbb{Q}(\boldsymbol{x})$ is

$$\mathbb{Q}(\boldsymbol{x}) = (V/N) \sum\nolimits_{\kappa} ½(3\mathbb{A}^{(\kappa)} - \mathbb{I})\, \delta(\boldsymbol{r} - \boldsymbol{r}^{(\kappa)})$$

with $\delta(\cdot)$ the Dirac delta function for position $\boldsymbol{r}$ and the factor V/N is added to make $\mathbb{Q}$ dimensionless. The internal variable is in this case thus a tensor.

If the nematic order is not along the z-axis but along an arbitrarily oriented director $\boldsymbol{n}$, we have $\mathbb{Q} = ½\xi(3\boldsymbol{nn} - \mathbb{I})$ or $Q_{ij} = ½\xi(3n_in_j - \delta_{ij})$. In this description, ξ represents the magnitude of the order parameter, while the factor $½(3\boldsymbol{nn} - \mathbb{I})$ represents the direction of the order by means of the unit vector $\boldsymbol{n}$. Finally, we note that, because $\xi = ½\langle 3\cos^2\theta - 1 \rangle$, in theory we have $-½ \le \xi \le 1$. Generally, a typical nematic is a prolate molecule, ξ is between 0 and 1, and the system has some partial nematic order with the molecular orientations attracted toward $\boldsymbol{n}$. If $\xi = 0$, the system is in the isotropic state. When the nematic is an oblate molecule, we can have $-½ \le \xi < 0$ and the system is in a peculiar *negative nematic* state with the molecular orientations repelled from $\boldsymbol{n}$. This negative nematic state does not frequently occur.

16.2.2 The Discontinuous Transformation

Now that we have the proper internal variable, we can use the machinery of the Landau-like theory for discontinuous transformations (Section 12.5, this section should be read in conjunction). Because the Helmholtz energy density f is a scalar, f can only depend on the (basic) invariants of the orientation tensor $\mathbb{Q}$. These are tr $\mathbb{Q}$, tr $\mathbb{Q}^2$ and tr $\mathbb{Q}^3$ (Section A.10). As discussed, the term tr $\mathbb{Q} = 0$, but the terms tr $\mathbb{Q}^2$ and tr $\mathbb{Q}^3$ survive. Generally, therefore, there are two relevant characteristics, but the simplest approach is to use only tr $\mathbb{Q}^2$. We do so here and use $\xi = (2\ \mathrm{tr}\,\mathbb{Q}^2/3)^{1/2} = \langle P_2(\theta)\rangle$ as an internal variable. We recall that the Helmholtz energy density f for such a transformation, using the general notation of Section 12.5, is given by

$$f(\eta) = \tfrac{1}{2}a\xi^2 - \tfrac{1}{3}b\xi^3 + \tfrac{1}{4}c\xi^4, \qquad a = a'(T - T_0), \qquad a' > 0 \quad \text{and} \quad c > 0$$

Without loss of generality, we also assume that $b > 0$, because $f(\xi)$ has the symmetry $(|b|,\xi) \leftrightarrow (-|b|,-\xi)$. For equilibrium we required that $\partial f/\partial \xi = 0$, which led to

$$\xi_0 = 0 \quad \text{and} \quad \xi_\pm = b/2c \pm \sqrt{(b/2c)^2 - a/c}$$

It appears that at high temperature T with $a > 0$, f has one minimum at $\xi = 0$. Upon decreasing T, this minimum gradually decreases in depth, and an inflection point in the $f - \xi$ curve occurs at $T = T^{**}$. Upon further decreasing T, the inflection point becomes a second minimum which becomes equal in depth as that of the first minimum at $T = T_0$. Decreasing T still further, the second minimum deepens until at $T = T^*$ the first minimum becomes an inflection point in $f - \xi$ curve. Still decreasing T, only the second minimum at $\xi = \xi_+$ survives (Figure 12.8). Thus, in the current context $T^* < T < T_0$ is the metastability range for the isotropic phase for supercooling and $T_0 < T < T^{**}$ is the metastability range for the nematic phase for superheating. After determining the relevant parameters, thermodynamic properties can be calculated in the usual way. For example, for the compound 5CB the values for the coefficients in the Landau expansion, with the error in parenthesis, are $a' = 0.13(0.01) \times 10^6\ \mathrm{J\,m^{-3}\,K^{-1}}$, $b = 1.6(0.2) \times 10^6\ \mathrm{J\,m^{-3}}$, $c = 3.9(0.3) \times 10^6\ \mathrm{J\,m^{-3}}$ in combination with $T^* = 307.2\ \mathrm{K}$, leading to $T_0 = 308.3\ \mathrm{K}$ and $T^{**} = 308.4\ \mathrm{K}$ [9]. The critical value of internal variable at T_0 changes to only $\xi_0 = 2b/3c = 3a'(T_0 - T^*)/b = 0.27$, while the enthalpy $L = \tfrac{1}{2}a'T_0\xi_0^{\,2} = 1.47\ (0.15)\ \mathrm{J\,kg^{-1}}$, comparing favorably with the differential scanning calorimetry (DSC) value $L = 1.41\ (0.15)\ \mathrm{J\,kg^{-1}}$.

As indicated in Section 12.5, Landau theory is a mean-field theory that neglects fluctuations and basically the expansion of $\mathscr{L}$ as an analytical function in ξ cannot hold for too close to the transformation temperature T_0. Fortunately, for liquid crystals the domain where fluctuations do become important is not so large. Moreover, the difference between T^* and T^{**} is small, as indicated above for the example of 5CB.

Problem 16.1 ***Biaxial Nematics*** Show that for a biaxial nematic $Q_{ij} = \xi(3n_in_j - \delta_{ij}) + B(l_il_j - m_im_j)$, where the vectors $\boldsymbol{n}$, $\boldsymbol{l}$, and $\boldsymbol{m}$ form a local cartesian coordinate system.

16.3 Alternative Approaches

The pro of the thermodynamic theory discussed in Section 16.2, its wide applicability, is at the same time also its con, namely the absence of a molecular picture, and thus other approaches have been forwarded. In many of these approaches, the orientation distribution function $p(\theta)$, describing the distribution of the polar angles of the unit vectors $\boldsymbol{v}$ of the molecules with respect to the director $\boldsymbol{n}$, plays an important role. Using $p(\theta)$, we can calculate the (magnitude of the) internal variable ξ from

$$\xi = \int_0^{\pi} P_2(\theta)p(\theta)\sin\theta\,\mathrm{d}\theta$$

In the following, we deal with Maier–Saupe theory, largely corresponding with thermotropic systems, and Onsager theory, largely corresponding with lyotropic systems. For the latter often polymers are involved, which usually are in the coil state and exhibit an overall globular shape. Before they can form a liquid crystal system, they must transform to the helix state, which shows an overall rod-like shape, by sufficiently strong secondary interactions.

16.3.1 Maier–Saupe Theory

For a microscopic theory of the N–I transformation we need to express entropy and energy (enthalpy) in terms of the orientation function $p(\theta)$. The entropy S_{ori} in the nematic phase due to orientation is given, analogously to the discrete case, by

$$S_{\text{ori}} = -k\int_0^{2\pi}\int_0^{\pi} p(\theta)\ln p(\theta)\,\mathrm{d}\Omega \qquad \text{where} \qquad \mathrm{d}\Omega = \sin\theta\,\mathrm{d}\theta\mathrm{d}\phi$$

is an element of solid angle. For an isotropic state, we have $p(\theta) = 1/4\pi$, so that the entropy change ΔS for the transformation from isotropic to nematic state becomes

$$\Delta S = -k\int_0^{\pi} p(\theta)\ln[4\pi p(\theta)]\sin\theta\,\mathrm{d}\theta\mathrm{d}\phi \tag{16.1}$$

We further need the energy ΔU and for this Maier and Saupe [10] advanced that one could use a mean-field expression, assuming that the molecules experience an average orientation-dependent potential Φ while their translation is unaffected by Φ. They proposed to use

$$\Phi(\mathbf{v}, \mathbb{Q}) = -\alpha\,\mathrm{tr}\{\tfrac{2}{3}\mathbb{Q}\,[\tfrac{1}{2}(3\mathbf{vv} - \mathbb{I})]\}$$

where α is a parameter expressing the strength of the attraction between two neighboring molecules. The normalized parameter α/kT is sometimes called the *nematic potential*. Maier and Saupe originally ascribed the physical origin of α to the anisotropic part of the vdW interaction. It appeared that the values estimated from London theory are much smaller than the values obtained by fitting, and that the isotropic part alone already could induce alignment, so that α is now usually considered as an empirical parameter. From Φ we obtain, using the one-molecule partition configurational function $q = \int\exp[-\beta\Phi(\mathbf{v})]\mathrm{d}\mathbf{v}$ with as usual $\beta = 1/kT$, for the orientation function $p(\theta) = \exp[-\beta\Phi(\mathbf{v})]/q$. Self-consistency can be enforced by requiring that $\mathbb{Q} = \int[\tfrac{1}{2}(3\mathbf{vv} - \mathbb{I})]\,p(\mathbf{v})\,\mathrm{d}\mathbf{v}$ or $\xi = \int P_2(\cos\theta)\,p(\theta)\,\sin\theta\,\mathrm{d}\theta\,\mathrm{d}\phi$. If we use again a polar axis system with $\boldsymbol{n}$ aligned along the z-axis, we obtain for the zz component

$$Q_{zz} = \xi = (2\pi/q)\int_0^{\pi} P_2(\cos\theta)\exp[\beta\alpha\xi P_2(\cos\theta)]\sin\theta\,\mathrm{d}\theta$$

One of the solutions is $\xi = 0$, which corresponds to the isotropic state. The second solution, for the nematic state, can only be found numerically, although it can be written in a parametric form after integration by parts [7], given by

$$\xi = \frac{1}{2}\left[\frac{1}{xD(x)} - \frac{1}{x^2}\right] - \frac{1}{6} \qquad \text{using} \qquad x^2 = \beta\alpha\xi$$

where $D(x) = \exp(-x^2)\int_0^x \exp t^2\,dt$ is Dawson's integral. The energy now becomes

$$\Delta U = \langle \Phi(\mathbf{v}, \mathbb{Q}) \rangle = -\tfrac{1}{2}\alpha \mathrm{tr}\left[\tfrac{2}{3}\mathbb{Q} \int \tfrac{1}{2}(3\mathbf{vv} - \mathbb{I})p(\mathbf{v})\,d\mathbf{v}\right] = -\tfrac{1}{2}\alpha \mathrm{tr}\,\tfrac{2}{3}\mathbb{Q}\mathbb{Q} = -\tfrac{1}{2}\alpha\xi^2$$

Adding ΔU and $T\Delta S$, the Helmholtz energy ΔF becomes

$$\Delta F = \Delta U - T\Delta S = -\tfrac{1}{2}\alpha\xi^2 + kT\int_0^\pi p(\theta)\ln[4\pi p(\theta)]\sin\theta\,d\theta d\phi \tag{16.2}$$

Alternatively, as a simple 'Ansatz,' we use $\Delta U = -\tfrac{1}{2}\alpha\xi^2$ just as a phenomenological expression. In that case, to obtain $p(\theta)$, we proceed as follows. The first step is to obtain the most probable distribution for a given value of ξ. In other words, to find the stationary value of the entropy integral $\int p(\theta)\ln p(\theta)\,\sin\theta\,d\theta d\phi$ subject to the condition $\xi = \int \tfrac{1}{2}(3\cos^2\theta - 1)p(\theta)\,\sin\theta\,d\theta d\phi = \int P_2(\cos\theta)p(\theta)\,\sin\theta\,d\theta d\phi = const.$ Using λ as the Lagrange multiplier, the Euler condition (Section A.8), here just $\partial[p\ln p + \lambda pP_2]/\partial p = 0$, becomes after solving

$$\ln p(\theta) + 1 + \lambda P_2 = 0 \quad \text{or} \quad p(\theta) = \exp(3\lambda\cos^2\theta)$$

which leads to the same distribution function.

For a given value of ξ, ΔS can be calculated according to Eq. (16.1), the result of which is shown in Figure 16.4a. The Helmholtz energy, as calculated from Eq. (16.2), is given in Figure 16.4b and shows a minimum for $\alpha/kT > 4.55$, dependent on the value of the nematic potential α/kT. Another way to display the result is to plot ξ as a function $\sigma = \alpha/kT$ (Figure 16.4c), which clearly indicates the discontinuous character of the transformation. Finally, Figure 16.4d shows one example how well the theory works, even for a

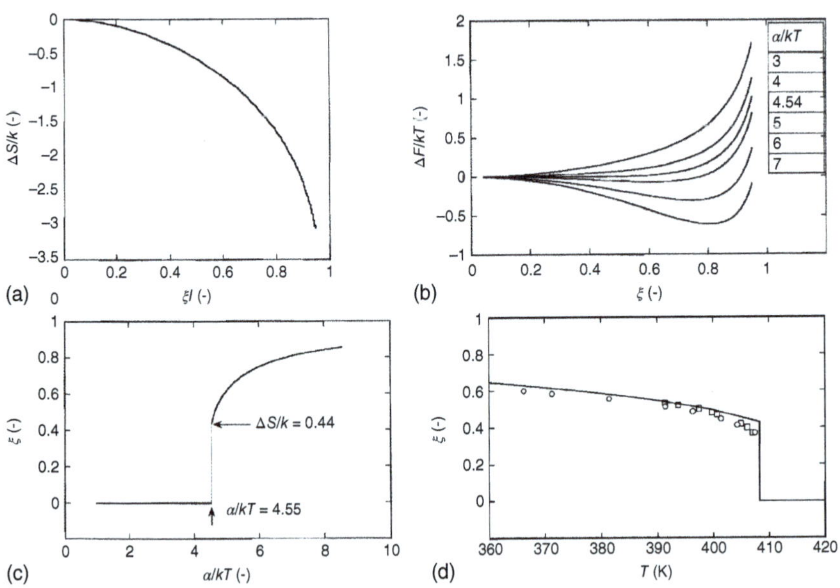

Figure 16.4 (a) The entropy $S(\xi)$ as a function of the internal variable ξ; (b) the Helmholtz energy $F(\xi)$; (c) the internal variable ξ as a function of the nematic potential α/kT; (d) the internal variable ξ as a function of temperature T for p-azoxyanisole (PAA). The solid line represents the Maier–Saupe theory result for $\alpha = const.$, as derived from the transformation temperature. Source: Jones [11]/Oxford University Press.

temperature-independent value of α. The basic reason behind its success is that, although both short- and long-range forces are important in determining the molecular orientation, the short-range part of the pair distribution function does not change appreciably at the nematic–isotropic transition. Consequently, it is the long-range contribution to the pair potential, on which the Maier–Saupe theory is built, which determines this transformation and provides a good description of nematic liquid crystals [12]. Another factor, already indicated and relevant for all mean-field theories, is the small domain where fluctuations are important. Obviously, the theory can be improved by including a temperature dependent value for α, like in Landau theory. A discussion of variations on the Maier–Saupe theory is given by Picken [13].

McMillan's model [14] is an extension of the Maier–Saupe theory to describe the phase transition of a liquid crystal from a nematic to a smectic A phase. It predicts that the phase transition can be either continuous or discontinuous depending on the strength of the short-range interactions between the molecules. As a result, it allows for a triple critical point where the nematic, isotropic, and smectic A phases meet. Although it predicts the existence of a triple critical point, it does not successfully predict its value. The model utilizes two order parameters that describe the orientational and positional order of the liquid crystal.

16.3.2 The Coil–Helix Transformation

As stated in Section 16.1, if molecules form rod-like structures by secondary interactions, such molecules may also exhibit liquid crystalline behavior. Polymer molecules normally have a coiled configuration, but a helix can be formed if relatively strong secondary interactions, normally hydrogen bonds, occur between near-by monomers. Overall, such a helix presents a rod-like shape. As usual with transformations, also the coil–helix transformation is dictated by a changing balance between energy and entropy with temperature. If we assume that the polymer molecule is constructed from a set of N segments that are partly in the coil state and partly in the helix state (Figure 16.5a), we have to calculate the energy change for the transformation between a completely coiled molecule to one containing, say, h helix segments and certain number, say, $2g$ junctions between helix and coil segments. Both the energy and entropy changes depend on these variables.

Starting with entropy, we use that the number of configurations (complexions) in the helix state Ω_h and coil state Ω_c are given by

$$\Omega_\text{h} = h!/g!(h-g)! \quad \text{and} \quad \Omega_\text{c} = (N-h)!/g!(N-h-g)!$$

The total number of complexions is $\Omega = \Omega_\text{h}\Omega_\text{c}$ so that the entropy change becomes

$$\Delta S_\text{c}(N,h,g) = k \ln \Omega_\text{h}\Omega_\text{c}$$

This expression can be easily simplified using Stirling's approximation, leading to

$$\Delta S_\text{c}(N,h,g) = h \ln h + N^* \ln N^* - 2g \ln g - h^* \ln h^* - N^{**} \ln N^{**}$$

where we defined $h^* \equiv h - g$, $N^* \equiv N - h$, and $N^{**} \equiv N - h - g$. The energy change is the sum of the energy for the helix state ΔU_h and that of the coil state ΔU_c, so that the Helmholtz energy change becomes

$$F(N,h,g) = h\Delta F_\text{c} + 2g\Delta F_\text{c} - T\Delta S_\text{c}(N,h,g)$$

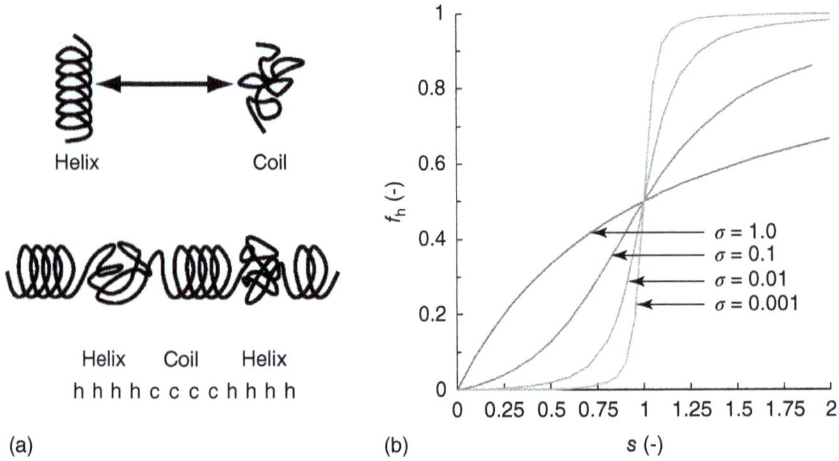

Figure 16.5 The helix–coil transformation. (a) Schematic of the configuration change. (b) The fraction f_h as a function of s. For $s < 1, f_h < 0.5$ always so that the coil-like state is preferred, while for $s > 1, f_h > 0.5$ and the helix-like state is preferred. Further, for large σ, the transformation range is wide, while for small σ it is narrow.

Minimizing $\Delta F(N,h,g)$ with respect to h and g leads to

$$\frac{N^*h^*}{hN^{**}} = \exp(-\beta\Delta U_c) \equiv s \text{ and } \frac{g^2}{N^*h^*} = \exp(-2\beta\Delta U_g) \equiv \sigma \tag{16.3}$$

where, as usual, $\beta = 1/kT$. The parameter s represents the preference of a segment for the helix state over the coil state. For $s < 1$ the coil state is preferred while for $s > 1$ the helix state is preferred. The parameter σ represents the cooperativity of the system. For $\sigma = 1$ there is no cooperativity, and the state of the segment is independent of its neighbors. With increasing value of ΔU_g, σ decreases and for $\sigma = 0$ there are no junctions between helix and coil segments.

Eliminating g from Eq. (16.3), meanwhile introducing $f_h = h/N$, we obtain

$$f_h = \tfrac{1}{2}\{1 + (s-1)/[(s-1)^2 + 4s\sigma]^{1/2}\}$$

Figure 16.5 shows the behavior of f_h as a function of s for several values of σ, illustrating that for $s < 1$ the coil state is always favorable, while for $s > 1$ the helix state is preferred. We also note that the nature of the transformation depends heavily on the value of σ. For a large value of σ, the transformation range is really wide. For $\sigma = 1$, the fraction f_h becomes $f_h = s/(s+1)$, the segments are independent of each other and f_h represents a Boltzmann factor.

It appears that, typically, the value of σ is about 10^{-4} to 10^{-3} and thus the transformation range is relatively small. However, the width remains finite for $N \to \infty$ and for a finite value of s, a certain fraction of coil-like segments will be always present. Hence, relatively long molecules have less chance to become fully rigid and normally contain a certain fraction of coil segments. Only relatively short molecules can become fully rigid.

16.3.3 Onsager Theory

In 1949, Onsager showed that the transformation of fully rigid rods from the isotropic to the nematic state is mainly due to their elongated shape and associated excluded volume, that is, due to their anisotropic hardcore repulsion [15]. Similar to hard spheres, the transformation is controlled by entropy changes only, which are due to the excluded volume effect. Recall that the entropy S for a random distribution of N point particles in a volume V is given by

$$S_{\text{ide}} = k \ln(aV/N)$$

where a is a constant. However, molecules have a volume b and taking that into account, the entropy becomes

$$S = k \ln[a(V - Nb)/N) = S_{\text{ide}} + k \ln(1 - bN/V) \cong S_{\text{ide}} + kbN/V$$

The approximation in the last step implies that we essentially truncate the virial expansion at the level of the second virial coefficient. Using the *density* $\rho = N/V$, the Helmholtz energy F consequently is

$$F = F_{\text{ide}} + kTbN/V = F_0 + kT \ln \rho + kT\rho b$$

In the Onsager model, molecules are represented by rigid sphero-cylinders, which for brevity we will continue to address as rods. The volume of a rod with length L and diameter D is given by $v_0 = {}^1\!/\!_4\pi LD^2 + {}^1\!/\!_6\pi D^3$ or $v_0 \cong {}^1\!/\!_4\pi LD^2$ for slender rods. Clearly, for rods the situation is more complex as compared with spherical particles because the degree of alignment is important. This affects the system in two ways. First, the alignment itself, which we describe by the orientation function $p(\theta)$, and the excluded volume becomes a function of θ. The associated entropy change, as indicated in Section 16.3, is given by $\Delta S = -k\int p(\theta) \ln[4\pi p(\theta)] \sin\theta \, \mathrm{d}\theta$. Second, the excluded volume becomes a function of θ. To deal with that, we consider two hard rods of length L, diameter D, and having an angle γ with respect to each other. To calculate the excluded volume for two such hard rods, we take the relative angle between them. This defines the base area $L^2|\sin\gamma|$ of the excluded volume, taken as a parallelogram, while the thickness D of the rods determines the thickness of the so-formed parallelepiped, and the excluded volume becomes $2L^2D|\sin\gamma|$. The average value for a random distribution of rods is $\langle|\sin\gamma|\rangle = \pi/4$. For a nonrandom distribution, this average becomes

$$\langle|\sin\gamma|\rangle = \iint p(\theta)p(\theta')\sin\gamma\sin\theta\sin\theta' \, \mathrm{d}\theta\mathrm{d}\theta'$$

so that $\langle|\sin\gamma|\rangle$ can be calculated if $p(\theta)$ is known. If we write for the *volume fraction* rods $\phi = \pi\rho LD^2/4$, we obtain $F = F_0 + F_{\text{ori}} + F_{\text{exc}}$, or

$$F = F_0 + kT[\ln L\phi/D + \int p(\theta) \ln[4\pi p(\theta)] \, \mathrm{d}\Omega + L^2D\rho\langle|\sin\gamma|\rangle \tag{16.4}$$

where all constant factors are included in F_0. We conclude that F is a function of the product $\phi L/D$ only, with as before ϕ the volume fraction and where $(L+D)/D \cong L/D$ is the *aspect ratio*. This functional cannot be solved analytically but can be solved numerically. An overview on how to do that is given in [16]. It appears that for $\phi L/D < 3.290$ the isotropic state ($\xi = 0$) is preferred, while for $\phi L/D > 4.191$ the nematic state has the lowest value for F. At the latter concentration, $\xi = 0.792$. A transformation range is observed for all values of ϕ. For the Onsager theory to be applicable, we need a low value of ϕ and consequently a rather large value of L/D.

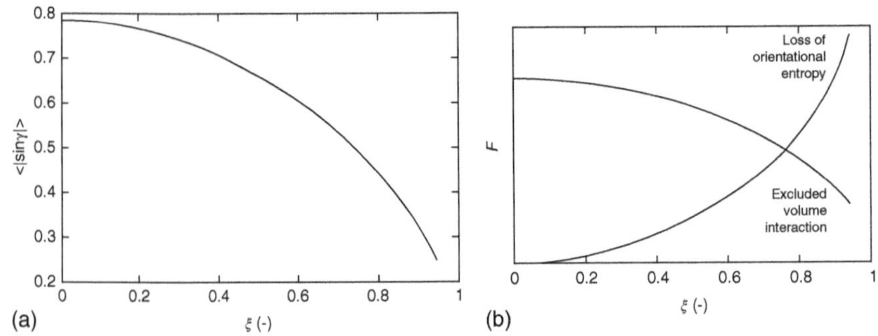

Figure 16.6 (a) The average ⟨|sinγ|⟩ as a function of ξ. (b) The Helmholtz energy F as a function of ξ. Source: Jones [11]/Oxford University Press.

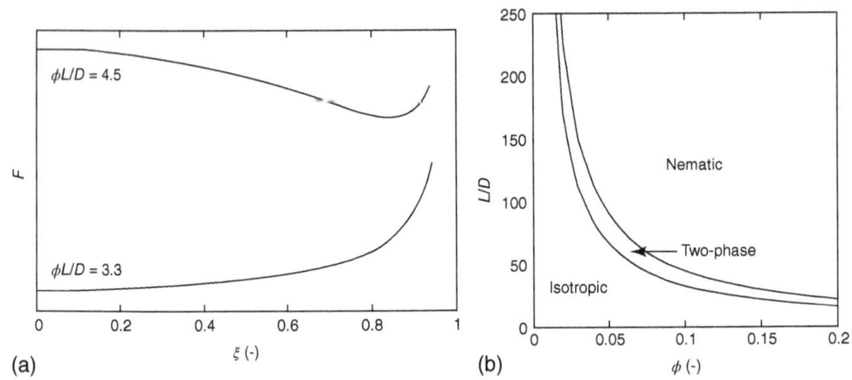

Figure 16.7 (a) The Helmholtz energy F as a function of ξ for two values of $\phi L/D$. (b) The phase diagram showing the isotropic, two-phase, and nematic regions. Source: Jones [11]/Oxford University Press.

We can also approximate by using a trial function for the minimization of F. The trial function with parameter α as proposed by Onsager is

$$p(\theta) = (\alpha/\pi)\cosh[\alpha\cos(\theta)]/\sinh\alpha$$

For $\alpha = 0$, the distribution is uniform, while with increasing value of α the function $p(\theta)$ develops peaks at $\theta = 0$ and $\theta = \pi/2$. Similarly as before, we us as internal variable $\xi = \frac{1}{2}\langle(3\cos^2\theta - 1)\rangle$. With this function $\langle|\sin\gamma|\rangle$ can be calculated, the result of which is shown in Figure 16.6, together with the associated Helmholtz energy. Clearly, with an increasing value of ξ, F_{ori} increases while F_{exc} decreases. Both depend on $\phi L/D$, which increases with increasing volume fraction ϕ and increasing aspect ratio L/D. This dependence is shown in Figure 16.7, together with the phase diagram. We note that for a small value of $\phi L/D$ the state with the lowest value of F is the isotropic state ($\xi = 0$), but that for a large value of $\phi L/D$ the nematic state has the lowest value. To be precise, for $\phi L/D < 3.340$ the isotropic state ($\xi = 0$) is preferred, while for $\phi L/D > 4.486$ the nematic state is the most favorable, in this case at $\xi \cong 0.848$. These results are not too far off from the exact numerical results.

The abovementioned analysis clearly shows that the phase behavior of rod-like molecules is largely determined by the strong dependence of the orientational entropy on the shape anisotropy of the molecules (rods) involved.

16.4 Some Extensions

Formally, the Onsager theory applies to infinitely thin hard rods, so that for real systems there are many additional aspects to be discussed. Among them are that rods are not infinitely thin, show size dispersity, may have attractive and repulsive interactions, and may be semiflexible. Onsager [15] already addressed particle repulsion and dispersity. A broad overview of these and other aspects is given in [17].

Here, we discuss only an approximate way to address the N–I transformation, based on an empirical correction in combination with scaled particle theory (SPT), as given by Tuinier [18]. For thin, rigid hard rods, the phase transformation occurs at a low volume fraction and the virial expansion could therefore be truncated at the level of the second virial coefficient, leading to an exact theory for infinitely thin particles [15, 17]. For rods with a small(er) aspect ratio, the theory can be extended but yields complex results. Another approach is to use a semiempirical approach to incorporate higher virial coefficients at least approximately.

Rendering the Onsager expression dimensionless, the Helmholtz energy for N rods with length L and diameter D in a volume V becomes [19]

$$f_{\mathrm{N}}/NkT = -1 + \ln c + \sigma + c\phi + const. \tag{16.5}$$

where $c[p] = b\rho$ is the dimensionless concentration with $b = \frac{1}{4}\pi L^2 D$ the second virial coefficient of rods in the isotropic state (or, equivalently, the excluded volume) and $\rho = N/V$. The dimensionless orientational entropy s_{ori} is given by $\sigma[p] = -s_{\mathrm{ori}}/k$ while $c\rho[p] = -s_{\mathrm{pac}}/k$ is the dimensionless packing entropy s_{pac}. We saw that $\sigma[p]$ is given by

$$\sigma[p] = \int p(\theta)\ln[4\pi p(\theta)]\,\mathrm{d}\Omega$$

while ϕ can be calculated using

$$\phi[p] = (4/\pi)\iint p(\theta)p\,(\theta')|\sin\gamma|\,\mathrm{d}\Omega\mathrm{d}\Omega'$$

where γ is the angle between two rods and $\mathrm{d}\Omega$ an element of solid angle. Minimizing the Helmholtz energy f under the constraint that $\int p(\theta)\sin\theta\,\mathrm{d}\theta$ yields

$$\ln[4\pi p(\theta)] = \lambda - (8c/\pi)\int p\,(\theta')|\sin\gamma(\theta,\theta')|\,\mathrm{d}\Omega'$$

where λ is the Lagrange multiplier. For the isotropic state, $\sigma_{\mathrm{I}} = 0$ and $\rho_{\mathrm{I}} = 1$, so that

$$f_{\mathrm{I}}/NkT = -1 + \ln c + c + const. \tag{16.6}$$

In the nematic state, we have the full expression for f. As said, for the nematic state an analytic solution for the Helmholtz energy is not available, but the numerical solution provides σ_{N} and ϕ_{N}. From Eqs. (16.5) and (16.6), the chemical potential $\mu = \partial f/\partial N$ and pressure $P = -\partial f/\partial V$ for each state can be determined. For the isotropic state, we obtain from Eq. (16.6)

$$P_{\mathrm{I}}b/kT = c_{\mathrm{I}} + c_{\mathrm{I}}^2 \quad \text{and} \quad \mu_{\mathrm{I}}/kT = \ln c_{\mathrm{I}} + 2c_{\mathrm{I}} + const.$$

while Eq. (16.5) for the nematic state yields

$$P_{\mathrm{N}}b/kT = c_{\mathrm{N}} + c_{\mathrm{N}}^2\phi[p] \quad \text{and} \quad \mu_{\mathrm{N}}/kT = \ln c_{\mathrm{N}} + \sigma[p] + 2c_{\mathrm{N}}\phi[p] + const.$$

For equilibrium we equate μ_I with μ_N and P_I with P_N and solving the Onsager equation numerically yields, as stated before, $c_I = 3.290$ and $c_N = 4.191$ with at the latter concentration, $\xi = 0.792$.

For the Onsager trial function, it appears that for large values of α, $p(\theta)$ can be approximated by $p(\theta) = N(\alpha)\exp(-\frac{1}{2}\alpha\theta^2)$ with $N(\alpha) \cong \alpha/4\pi$ for $0 \le \theta \le \pi/2$. For the range $\pi/2 \le \theta \le \pi$, θ must be replaced by $\pi - \theta$. In this case, σ and ϕ can be calculated [17] as $\sigma = \ln\alpha - 1$ and $\phi = 4/(\pi\alpha)^{1/2}$. The Helmholtz energy becomes

$$f_N/NkT = \ln c + \ln\alpha - 1 + 4c/\sqrt{\pi\alpha} + const.$$

and minimizing with respect to α, we obtain $\alpha = 4c^2/\pi$, so that the f_N reduces to

$$f_N/NkT = 3\ln c + \ln(4/\pi)\alpha + const.$$

In the same way as before, we obtain for the nematic state from Eq. (16.5)

$$P_N b/kT = 3c_N \quad \text{and} \quad \mu_N/kT = \ln(4/\pi) + 3 + 3\ln c_N + const.$$

The coexistence equations become

$$c_I + c_I^2 = 3c_N \quad \text{and} \quad \ln c_I + 2c_I = 3\ln c_N + \ln(4/\pi) + 3$$

which can be solved to yield $c_I = 3.451$ and $c_N = 5.122$, having a difference of 5%, respectively, 22% with the exact values. At the latter concentration $\alpha \cong 33.40$, leading to $\xi = 0.910$.

The approximation of $p(\theta)$ by a Gauss function is rather approximate and the simplest way to remedy this a bit is by introducing a constant correction Δs. In this way the Helmholtz energy becomes

$$f_N/NkT = \ln(4/\pi) + 3\ln c - \Delta s/k + const.$$

The value for Δs can be determined from the value for the exact numerical value for S_{ori} and that in the Gauss approximation, which leads to

$$\Delta s/k = \sigma_G - \sigma_{num} + c_N(\phi_G - \phi_{num})$$

When we use the value of $c_N = 4.191$, it follows from $\alpha = 4c^2/\pi$ that $\alpha = 22.36$. From $\sigma = \ln\alpha - 1$ and $\phi = 4/(\pi\alpha)^{1/2}$ we obtain $\sigma = 2.107$ and $\phi = 0.477$. Inserting the exact numerical values $\sigma = 1.601$ and $\phi = 0.565$, we obtain $\Delta s/k = 2.107 - 1.601 + 4.191(0.477 - 0.565) = 0.139$. With this correction $c_I = 3.26$ and $c_N = 4.62$, deviating only 1% and 10% from the numerical values, respectively.

Let us now apply this correction for rods with arbitrary aspect ratio in an attempt to go beyond the second virial approximation. For this we invoke SPT along the formulation of Cotter [20]. She derived for the isotropic state that

$$f_I/NkT = \ln(L/D) + \ln y + ay - \tfrac{1}{2}by^2 + const.$$

where the expressions for a and b are

$$a = 3 + \phi(r-1)^2/(3\xi - 1) \quad \text{and} \quad b = [12r(2r-1) + 12\phi r(r-1)^2]/(3r-1)^2$$

with $r = L/(L + D)$. For the nematic state the corresponding result is

$$f_N/NkT = \ln(L/D) + \ln y + \sigma + Py - \tfrac{1}{2}Qy^2 - 0.139 + const.$$

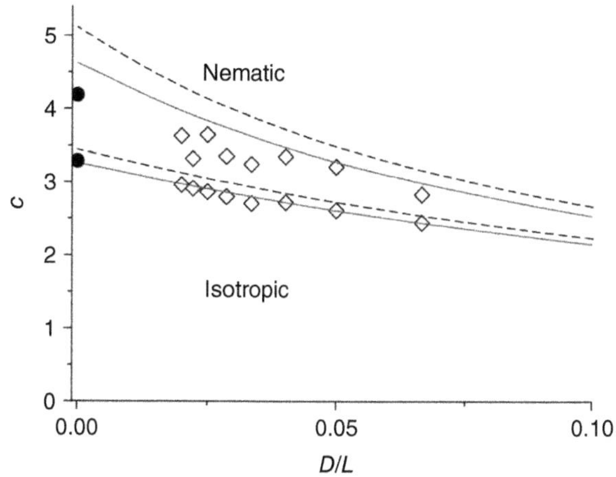

Figure 16.8 Isotropic–nematic phase coexistence for hard sphero-cylinders as a function of the inverse aspect ratio D/L [18]. ● = Onsager limit, ◊ = MC results [21], - - - = SPT Gauss approximation, – – = SPT corrected. Source: Adapted from [18, 21].

where we included the correction indicated before. If we minimize f with respect to α, the result is

$$a = \frac{36}{\pi}\frac{(r-1)^4}{(3r-1)^2}\left(y + \frac{2ry^2}{3r-1}\right)^2$$

Using these expressions, the values for c_{I} and c_{N} can be calculated. From f_{I} the chemical potential can be calculated as

$$\mu_{\text{N}}/kT = \ln(L/D) + \ln y + (1+2a)y + (a+3b/2)y^2 + by^3 + [\sigma - 0.139] + const.$$

where the term in square brackets is to be included for the nematic state only. Finally, the osmotic pressure in the isotropic and nematic state become

$$Pv_0/kT = y + a_{\text{X}}y^2 + b_{\text{X}}y^3$$

with v_0 the volume of the rod and where the subscript X denotes either I for the isotropic or N for the nematic state. In this way third-order and higher-order virial coefficients are included in a semi-empirical way. Implicitly, we assumed that the correction for $p(\theta)$ remains the same for various aspect ratios.

A comparison with Monte Carlo (MC) simulation results (Figure 16.8) using the corrected Helmholtz energy expressions via the rationalized ΔS correction shows that the N–I transformation is described rather well, in any case better than in the Gauss approximation. Finally, we note that in [3] both the Maier–Saupe and the Onsager theory are rationalized from the general expression for mean-field potentials and that in [5] both theories are compared, highlighting their common and noncommon aspects.

16.5 Elastic Energy and Defects

As some positional order is present for liquid crystals, one might expect that its elastic behavior is not isotropic. This is indeed the case, and the elastic behavior can be described with three elastic constants, rather similar as for cubic crystals. For a general deformation, the director $\boldsymbol{n}(\boldsymbol{r})$ as a function of position $\boldsymbol{r}$ plays the role of displacement as used in elasticity. There are three basic deformations, called *splay*, *twist*, and *bend* (Figure 16.9).

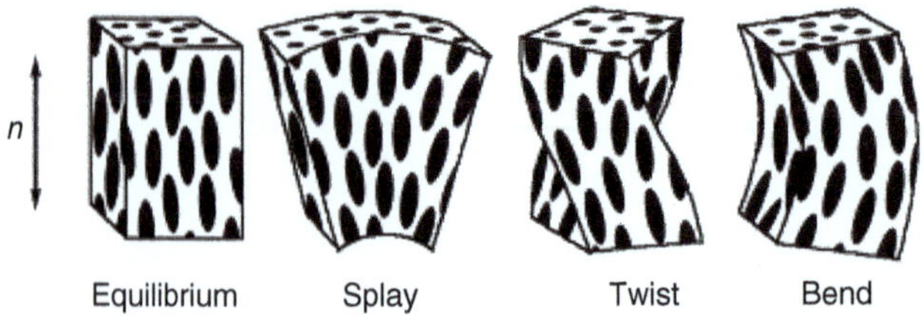

Figure 16.9 Deformations for a nematic liquid crystal showing equilibrium, splay, twist, and bend.

In Justification 16.1, we provide some details to show that the specific elastic Helmholtz energy f_{ela} for a nematic liquid crystal can be written as

$$f_{\text{ela}} = \tfrac{1}{2}K_{11}(\nabla \cdot \boldsymbol{n})^2 + \tfrac{1}{2}K_{22}(\boldsymbol{n} \cdot \nabla \times \boldsymbol{n})^2 + \tfrac{1}{2}K_{33}(\boldsymbol{n} \times \nabla \boldsymbol{n} \times \boldsymbol{n})^2 \tag{16.7}$$

Justification 16.1 ***Elastic Energy*** In a local Cartesian coordinate system (x,y,z), where $\boldsymbol{n}$ is taken parallel to the z-axis, we have in components for

$$\begin{aligned}
&\text{splay:}\quad s_1 = a_1 = \partial n_x/\partial x &&\text{and}\quad s_2 = a_5 = \partial n_y/\partial y,\\
&\text{twist:}\quad t_1 = -a_4 = -\partial n_y/\partial x &&\text{and}\quad t_2 = a_2 = \partial n_x/\partial y,\\
&\text{bend:}\quad b_1 = a_3 = \partial n_x/\partial z &&\text{and}\quad b_2 = a_6 = \partial n_y/\partial z
\end{aligned}$$

To first order (small deformations) the director is given by

$$n_x = s_1x + t_2y + b_1z + \mathcal{O}(r^2),\qquad n_x = -t_1x + s_2y + b_2z + \mathcal{O}(r^2),\qquad n_z = 1 + \mathcal{O}(r^2)$$

Because we take the deformations small, we can expand the Helmholtz (or Gibbs) elastic energy per unit volume as

$$f_{\text{ela}} = \sum\nolimits_k K_k a_k + \sum\nolimits_{k,l} K_{kl} a_k a_l$$

where the K_j and K_{jl} are elastic constants and the symbols a_j as given above are used. Symmetry considerations for the nematic state show that the constants are not all independent and, after some analysis, leads to two independent constants K_j and to five independent constants K_{jl}. If the nematic is (or can be considered as) uniaxial, we also have $K_1 = K_{12} = 0$, while if it is non-chiral, additionally $K_2 = 0$. For a nonpolar and non-chiral nematic the elastic energy reduces to

$$f_{\text{ela}} = \tfrac{1}{2}K_{11}(s_1 + s_2)^2 + \tfrac{1}{2}K_{22}(t_1 + t_2)^2 + \tfrac{1}{2}K_{33}\left(b_1{}^2 + b_2{}^2\right)$$

Realizing that $s_1 + s_2 = \partial n_x/\partial x + \partial n_y/\partial y = \nabla \cdot \boldsymbol{n}$, $t_1 + t_2 = -\partial n_y/\partial x + \partial n_x/\partial y = \nabla \times \boldsymbol{n}$ and $b_1{}^2 + b_2{}^2 = (\boldsymbol{n} \times \nabla \times \boldsymbol{n})^2$, the elastic energy per unit volume f_{ela} for a nematic system is [22]

$$f_{\text{ela}} = \tfrac{1}{2}K_{11}(\nabla \cdot \boldsymbol{n})^2 + \tfrac{1}{2}K_{22}(\boldsymbol{n} \cdot \nabla \times \boldsymbol{n})^2 + \tfrac{1}{2}K_{33}(\boldsymbol{n} \times \nabla \boldsymbol{n} \times \boldsymbol{n})^2$$

which is Eq. (16.7) and where K_{11}, K_{22}, and K_{33} are the *elastic* (or *Frank*) *constants*. Minimizing $\int f_{\text{ela}}\,\mathrm{d}V$ with as constraint $\boldsymbol{n}^2 = 1$, the Euler equation becomes

$$\begin{aligned}
&-K_{11}\nabla(\nabla \cdot \boldsymbol{n}) + K_{22}[A(\nabla \times \boldsymbol{n}) + (\nabla \times (A\boldsymbol{n}))]\\
&\quad + K_{33}\{[(\nabla \times \boldsymbol{n}) \times \boldsymbol{B}] + [\nabla \times (\boldsymbol{B} \times \boldsymbol{n})]\} + \mu\boldsymbol{n} = 0
\end{aligned}$$

where $A = \boldsymbol{n} \cdot (\nabla \times \boldsymbol{n})$, $\boldsymbol{B} = \boldsymbol{n} \times (\nabla \times \boldsymbol{n})$ and μ is the Lagrange multiplier. This equation determines the director $\boldsymbol{n}$. In case $K_{11} = K_{22} = K_{33} = K$, $f_{\text{ela}} = \tfrac{1}{2}K[(\nabla \cdot \boldsymbol{n})^2 + (\nabla \times \boldsymbol{n})^2]$.

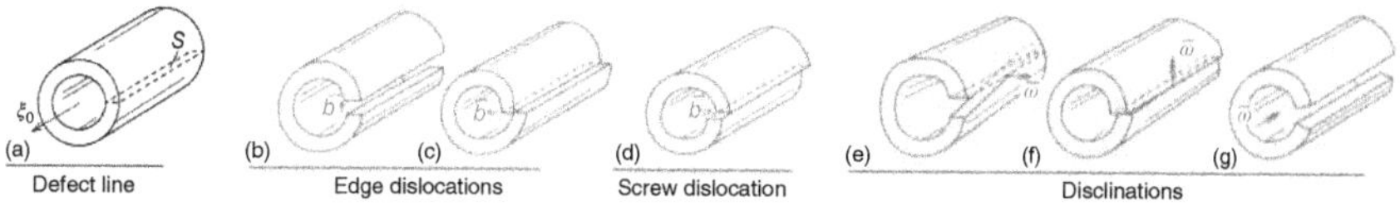

Figure 16.10 Volterra's distortions. (a) Reference cylinder with defect line ξ_0 and cut surface S; (b, c) edge dislocations with Burgers vector $\boldsymbol{b}$; (d) screw dislocation with Burgers vector $\boldsymbol{b}$; (e, f) twist disclinations with Frank vector $\boldsymbol{\omega}$; (g) wedge disclination with Frank vector $\boldsymbol{\omega}$. Source: Cordier et al. [23]/Springer Nature.

The elastic constants typically have a value of a few times 10^{-12} N. Because K_{22} is difficult to determine, sometimes the approximation $K_{33} = 2K_{22}$ is made. For example, for the molecule 5CB, at $T = T_0 - 10$ K, $K_{11} \cong 6.2\times10^{-12}$ N, $K_{22} \cong 3.9\times10^{-12}$ N and $K_{33} \cong 8.2\times10^{-12}$ N [24], indicating that $K_{33} = 2K_{22}$ is only approximately fulfilled.

The magnitude of the attraction between liquid crystal molecules is usually small. Again as an example, for the molecule 5CB the enthalpy change is ~0.8 kJ mol^{-1} [1]. Therefore, many surfaces with a preferred orientation (introduced by rubbing or by having grooves) induce a strong influence on the director of the neighboring liquid crystal molecules. If the preferred alignment is perpendicular to the surface, the boundary condition is called *homeotropic*, while when it is parallel to the surface it is called *homogeneous*. The latter is the more frequently occurring case.

The director does not always vary smoothly through space but changes at some points discontinuously (for a general discussion on topological defects, see [8]). In 1905, Volterra [25] described discontinuities in a continuum. One type of such discontinuities are the *dislocations* in crystals (Figure 16.10, Section 10.10.2). In a continuum theory they are described by a cut in a hollow cylinder along the axis, which is translated by the Burgers vector $\boldsymbol{b}$, either parallel or perpendicular to the axis. The axis of the cylinder corresponds with the dislocation line. Another type, the *disclinations*, are descriptions for defects in the director. They are described by a twist of such a cut, again either parallel or perpendicular to the axis. Disclinations, like dislocations, increase the elastic energy. Let us assume that $K_{11} = K_{22} = K_{33} = K$ and take $\boldsymbol{n}$ in the xy-plane making an angle $\theta(r) = s\phi(r) + const.$ with the x-axis, where $r^2 = x^2 + y^2$. The energy density becomes $f_{\text{ela}} = ½\,K(\nabla\theta)^2$ and using $\nabla\theta|_{r,z} = r^{-1}(\text{d}\theta/\text{d}\phi) = r^{-1}s$, we have $f_{\text{ela}} = ½\,K(s/r)^2$ [2]. The total energy is, therefore,

$$F_{\text{ela}}/L = \int_a^R f_{\text{ela}} 2\pi r\,\text{d}r = \pi K s^2 \ln(R/a) \tag{16.8}$$

with L the length of the disclination line, R the range of influence of a disclination (either the distance to other disclinations or to the walls), and a the disclination core radius, typically ≈ 5 nm. The *strength* $s = \text{d}\theta/\text{d}\phi$, describing changes in θ around the disclination line with angle ϕ, must be (half-)integer for the director to match again at some distance from the axis. In fact, $s = \pm½$ or $s = \pm1$, corresponding to screw and edge dislocations. The orientation pattern of the molecules viewed along the axis is shown in Figure 16.11. According to Eq. (16.8), similar as for dislocations, disclinations have a *line tension*, which typically amounts to ≈30 K. Because the associated entropy is negligible, they are, like dislocations, thermodynamically not required. The mathematics of disclinations is more complex as for dislocations and an extensive review series is [27]. Two other brief reviews are [28, 29].

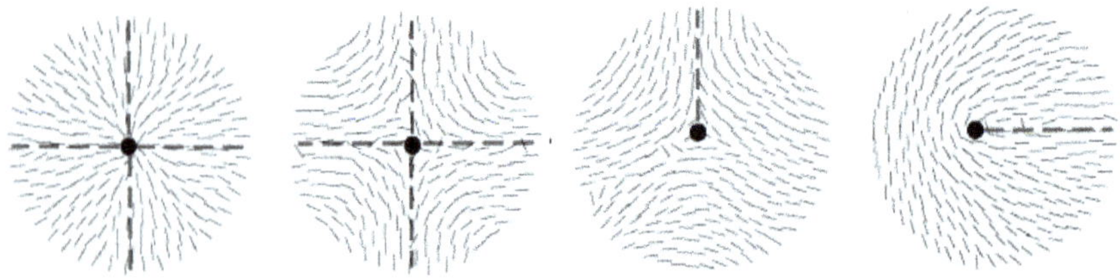

s = 1 s = −1 s = −1/2 s = 1/2

Figure 16.11 The orientation pattern of molecules for the various type of disclinations. Source: Libre Texts Engineering [26].

16.6 The Fréedericksz Transformation

Liquid crystals interact with electromagnetic fields, as all materials do. However, although many liquid crystal molecules do not have an inversion point and consequently have a dipole moment, they are typically oriented either parallel or antiparallel in equal amounts. Upon applying an electromagnetic field, the molecules align along the field but do not change their orientation. The latter occurs only in ferroelectric liquid crystals. The values of the permittivity parallel ($\varepsilon_{\parallel}$) and perpendicular ($\varepsilon_{\perp}$) to the director are different and the displacement $\boldsymbol{D}$ is given by

$$\boldsymbol{D} = \varepsilon_{\perp}\boldsymbol{E} + (\varepsilon_{\parallel} - \varepsilon_{\perp})(\boldsymbol{n}\cdot\boldsymbol{E})\boldsymbol{n}$$

The energy per unit volume u_{fie} due to the electric field $\boldsymbol{E}$ is

$$f_{\text{fie}} = -\frac{1}{4\pi}\int \boldsymbol{D}\cdot \mathrm{d}\boldsymbol{E} = -\frac{\varepsilon_{\perp}}{8\pi}E^2 - \frac{\Delta\varepsilon}{8\pi}(\boldsymbol{n}\cdot\boldsymbol{E})^2$$

where $\Delta\varepsilon = \varepsilon_{\parallel} - \varepsilon_{\perp}$. A similar description can be given for a magnetic field, as was done originally by Fréedericksz and Repiewa in 1927 [30], but we limit the discussion here to an electric field, in its modern form given by Saupe [31].

As said, the magnitude of the attraction between liquid crystal molecules is usually small and that is why the boundary conditions do play an important role for alignment. Similarly, an electric field can change the orientation. Suppose a thin film of nonconducting molecules is enclosed in a thin cell with two surfaces having homogeneous but perpendicular boundary conditions. Also suppose that at the surfaces the alignment anchoring with the boundary is complete. In that case, in the absence of a field, the liquid crystal system shows only a twist deformation along the, say, z-direction. Therefore, the director $\boldsymbol{n}$ depends only on z. Upon applying a field along the z-direction, the molecules try to align with the field, thereby introducing a splay deformation as their mutual interaction tries to keep them in plane. The competition between splay energy and field energy results in the critical field required to induce such a reorientation. If the deformation out of plane is small and characterized by a deviation $\delta\boldsymbol{n}(z)$ with respect to the unperturbed director $\boldsymbol{n}_0$, we have

$$f_{\text{tot}} = f_{\text{ela}} + f_{\text{fie}} = \tfrac{1}{2}K_1[\partial\delta\boldsymbol{n}(z)/\partial z]^2 - (\Delta\varepsilon E^2/8\pi)[\delta\boldsymbol{n}(z)]^2 \tag{16.9}$$

with K_1 and $\Delta\varepsilon$ as defined. At the surfaces, $z = 0$ and $z = d$, where d is the thickness of the film, $\delta\boldsymbol{n} = 0$ because the alignment is prescribed by the surfaces. If we use $\delta\boldsymbol{n}(z) = \delta n \sin(\pi z/d)$, substitute this expression in Eq. (16.9) and integrate, we obtain

$$\int_0^d f_{\text{tot}}\,\mathrm{d}z = \delta n^2[(K_1\pi^2/4d) - (\Delta\varepsilon E^2 d/16\pi)]$$

A small distortion will therefore lead to a decrease in Helmholtz energy if the field E exceeds a critical field E_{cri}, given by

$$E_{\text{cri}} = (2\pi/d)\sqrt{\pi K_1/\Delta\varepsilon}$$

Note that the critical voltage $E_{\text{cri}}d$ is independent of the thickness d. For $E \ll E_{\text{cri}}$ the director remains largely undisturbed with parallel orientations throughout the cell. However, because $\Delta\varepsilon = \varepsilon_{\parallel} - \varepsilon_{\perp}$ is not necessarily small compared to $\varepsilon_{\perp}$, at a relatively small field already an angle-dependent factor for the critical voltage is introduced given by $\cong [1 + 1/4(1 + \Delta\varepsilon/\varepsilon_{\perp})]$. For $E > E_{\text{cri}}$ the director in the bulk becomes parallel to the field and thus perpendicular to the surfaces.

In general, the situation is more complex. First, we assumed infinitely strong anchoring at the surfaces. This is not guaranteed, but an interface energy can be introduced in the analysis to deal with this effect. Second, because $\Delta\varepsilon$ is not necessarily small compared to the average permittivity, the electric field in a distorted nematic is not homogeneous in the z-direction. Moreover, we assumed a nonconductive fluid, which is also not necessarily the case. Finally, for more complex geometries, the field is no longer along one of the principal axes of the nematic, and all three deformations become active, resulting in a critical field given by [32]

$$E_{\text{cri}} = (2\pi/d)\sqrt{\pi/\Delta\varepsilon}[K_1 + 1/4(K_3 - 2K_2)]^{1/2}$$

The Fréedericksz transformation forms the basis for liquid crystal displays. The display surfaces are made from crossed polarizers and in the zero-field condition without liquid crystal molecules no light is transmitted. Adding the liquid crystal molecules, the surfaces of the display impose a parallel alignment on the molecules, but as the surfaces have a perpendicular orientation, the liquid crystal molecules are twisted by 90° between the top and bottom surfaces, thereby changing the polarization of the light so that some light is transmitted. By applying a field $E > E_{\text{cri}}$ perpendicular to the polarizers, the alignment becomes parallel to the field in the bulk of the system and the liquid crystal acquires a cylindrical symmetry showing no birefringence along the axis. Consequently, the light polarized by the first polarizer is blocked by the second one and the display is dark. Because at the surfaces generally no complete anchoring occurs, there is a splay deformation near these surfaces, but as both surfaces are oriented perpendicular this has no overall effect, although it helps to increase the switching speed via the locally stored elastic energy. Finally, we note that although we did not discuss mixtures of liquid crystal molecules, they are indispensable in practical displays.

Problem 16.2 ***From Rods to Spheres*** Show that the pressure expression for hard rods reduces to that for hard spheres if we take the aspect ratio as 1 and $p = 1/4\pi$.

References

1 Chandrasekhar, 1992.

2 de Gennes and Prost, 1993.

3 Vertogen and de Jeu, 1988.

4 Blinov, 2011.

5 van der Schoot, 2022.

6 Zannoni, C. (2022). *Liquid Crystals and their Computer Simulations*. Cambridge: Cambridge University Press.

7 Andrienko, D. (2018). *J. Mol. Liq.* 267: 520.

8 Chaikin and Lubensky, 1995.

9 Coles, H.J. (1978). *Mol. Cryst. Liq. Cryst.* 49: 67.

10 Maier, W. and Saupe, A. (1958). *Z. Naturforsch. A* 13: 564; (1959). *Ibid.* 14: 882; (1960). *Ibid.* 15: 287.

11 Jones, R.A.L. (2002). *Soft Condensed Matter*. Oxford: Oxford University Press.

12 Luckhurst, G.R. and Zannoni, C. (1977). *Nature* 267: 412.

13 Picken, S.J. (2010). *Liquid Crystals* 37: 977.

14 McMillan, W.L. (1971). *Phys. Rev. A* 4: 1238.

15 Onsager, L. (1949). *Ann. N.Y. Acad. Sci.* 51: 627.

16 van Roij, R. (2005). *Eur. J. Phys.* 26: S57.

17 Vroege, G.J. and Lekkerkerker, H.N.W. (1992). *Rep. Progr. Phys.* 55: 1241.

18 Tuinier, R. (2016). *Adv. Cond. Matter Phys.* 5871826: 1–6.

19 Lekkerkerker, H.N.W. and Tuinier, R. (2011). *Colloids and the Depletion Interaction*. Dordrecht: Springer.

20 Cotter, M.A. (1977). *J. Chem. Phys.* 66: 1098.

21 Bolhuis, P. and Frenkel, D. (1997). *J. Chem. Phys.* 106: 666.

22 Frank, F.C. (1958). *Disc. Faraday Soc.* 25: 19.

23 Cordier, P., Demouchy, S., Beausir, B. et al. (2014). *Nature* 507: 51.

24 Dunmur, D.A., Fukuda, A. and Luckhurst, G.R., eds. (2001). *Physical Properties of Liquid Crystals: Nematics*, EMIS Data Reviews Series No. 25, Inst. Elec. Eng. (INSPEC), London.

25 Volterra, V. (1905). *Nuovo Cim.* 10: 361; (1906). *Ibid.* 11: 5, 144, 205 and 338; (1906). *R. C. Accad. Lincei* 15: 519.

26 Libre Texts Engineering, https://eng.libretexts.org/Bookshelves/Materials_Science/TLP_Library_II (accessed 29 October 2022).

27 de Wit, R. (1970). *Fundamental Aspects of Disclination Theory*, Vol. 1 and 2, Nat. Bur. Stand. (U.S.) Spec. Publ. 317, p. 651; (1973). *J. Res. Natl. Bur. Stand.* 77A: 49; (1973). *Ibid.* 77A: 359; (1973). *Ibid.* 77A: 607.

28 Kleman, M. and Lavrentovich, O.D. (2006). *Phil. Mag.* 86: 4117.

29 Alexander, G.P., Chen, B.G., Matsumoto, E.A. et al. (2001). *Rev. Mod. Phys.* 84: 497.

30 Fréedericksz, V. and Repiewa, A. (1927). *Z. Phys.* 42: 532; Fréedericksz, V. and Zolina, V. (1933). *Trans. Faraday Soc.* 29 (140): 919.

31 Saupe, A. (1960). *Z. Naturforsch.* 15a: 815.

32 Yang, K.H. (1983). *Appl. Phys. Lett.* 43: 171.

Further Reading

Blinov, L.M. (2011). *Structure and Properties of Liquid Crystals*. Dordrecht: Springer. A detailed treatment dealing with structure, physical properties and electro-optics discussing the matter at a relatively slow pace.

Chaikin, P.M. and Lubensky, T.C. (1995). *Principles of Condensed Matter Physics*. Cambridge: Cambridge University Press. Contains many topics on hard and soft condensed matter including some, rather compact high-level discussion on liquid crystals.

Chandrasekhar, S. (1992). *Liquid Crystals*, 2e. Cambridge: Cambridge University Press. A broad and classic overview on basics, types and nature of various types of liquid crystals and devices.

de Gennes, P.G. and Prost, J. (1993). *The Physics of Liquid Crystals*, 2e. Oxford: Oxford University Press. The more technical classic on the topic at stake.

van der Schoot, P. (2022). *Molecular Theory of Nematic (and other) Liquid Crystals*. Cham, CH: Springer. A modern introduction with a final chapter comparing Maier-Saupe and Onsager theory.

Singh, S. (2002). *Liquid Crystals: Fundamentals*. River Edge, NJ: World Scientific. A comprehensive book with a strong theoretical emphasis, given on the level of de Gennes/Prost and Chandrasekhar, but in style of Vertogen and de Jeu. It also includes descriptive discussions when appropriate.

Vertogen, G. and de Jeu, W.H. (1988). *Thermotropic Liquid Crystals, Fundamentals*. Berlin: Springer-Verlag. Rather rigorous treatise with a thorough introduction of the Landau theory and an emphasis on statistical theories.

17

Dielectric Behavior and the Ferroelectric Transformation

After we dealt in Chapter 16 with the liquid crystal transformation, in this chapter we discuss an example of a phase transformation in the solid state, namely the ferroelectric transformation. As ferroelectric behavior of solids is closely linked with dielectric behavior, this topic is discussed first. Thereafter, we deal with the ferroelectric transformation in terms of macroscopic and microscopic considerations.

17.1 Preliminaries and Dielectric Materials

To discuss dielectric behavior in some detail, we first deal with a few of its general aspects. Thereafter, we deal with the types of dielectrics that we can anticipate, followed by a discussion of the various mechanisms. General references are [1, 2].

17.1.1 General Remarks

As discussed before (Section 8.6), the *polarization* $\boldsymbol{P} = (\varepsilon_r - 1)\varepsilon_0\boldsymbol{E} = \varepsilon_0\chi\boldsymbol{E}$ is key to dielectric behavior. Here, $\varepsilon \equiv \varepsilon_0\varepsilon_r$ is the *permittivity*, ε_0 is the *permittivity of the vacuum*, ε_r is the *relative permittivity* of the material involved, χ is the *susceptibility*, and $\boldsymbol{E}$ is the externally applied electric field. In microscopic terms, $\boldsymbol{P} = \rho_{dip}\boldsymbol{p} = \rho_{dip}\Sigma_j\boldsymbol{p}_j$ with ρ_{dip} the density of induced dipoles $\boldsymbol{p}_j = \alpha_{eff,j}\boldsymbol{E}_{loc}$ having an effective polarizability $\alpha_{eff,j}$ in the local field $\boldsymbol{E}_{loc}$ experienced, the latter often taken as the Lorentz field $\boldsymbol{E}_L = \boldsymbol{E}(\varepsilon_r + 2)/3$. The effective polarizability α_{eff} contains in general several contributions (Figure 17.1). At high frequency in the ultraviolet (UV) optical range, we have the *electronic contribution* α_{ele} due to the distortion of the bound electron clouds and a *molecular contribution* α_{mol} due to displacements of atoms in molecules. At lower frequency in the infrared (IR) range, there is the *vibrational contribution* α_{vib} due to the relative displacement of the ions or molecules as a whole. At still lower frequency in the microwave range, the *orientational contribution* α_{ori} due to the orienting dipole moments arises. Finally, at low frequency or static fields a *space–charge contribution* α_{sch} due to the accumulation of charge at boundaries may occur. Near-corresponding resonances α_{eff} varies rapidly but in between these resonances α_{eff} is approximately independent of the frequency (Figure 8.7), although with different values for different regimes. As usual for solids, the description of the behavior for various material classes is different and not all mechanisms contribute at low frequency to all types of materials.

Phases of Matter and their Transitions: Concepts and Principles for Chemists, Physicists, Engineers, and Materials Scientists, First Edition. Gijsbertus de With.

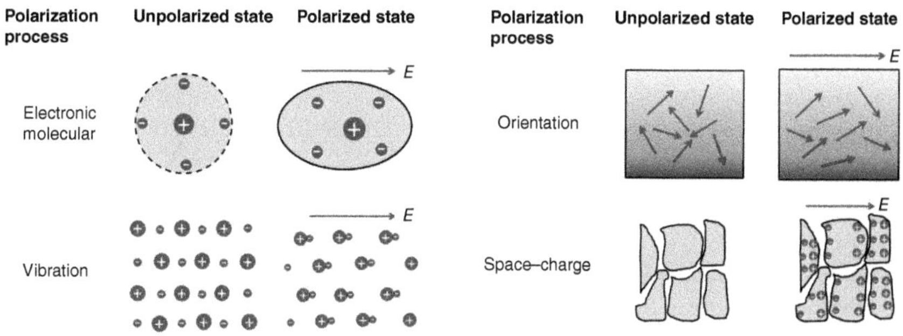

Figure 17.1 The various polarization mechanisms. Source: Liu et al. [3]/Royal Society of Chemistry.

Basically *Maxwell's equations* are involved, given for stationary media by [4–6]

$$\nabla \times \boldsymbol{B} = \mu_0\left(\boldsymbol{J} + \varepsilon_0\frac{\partial \boldsymbol{E}}{\partial t}\right), \qquad \nabla \times \boldsymbol{E} = -\frac{\partial \boldsymbol{B}}{\partial t}, \qquad \nabla \cdot \boldsymbol{B} = 0, \qquad \nabla \cdot \boldsymbol{E} = \frac{\rho}{\varepsilon_0}$$

with $\boldsymbol{B}$ the *magnetic field*,[1] $\boldsymbol{E}$ the *electric field*, $\boldsymbol{J}$ the *charge current*, ρ the *charge density*, and μ_0 the *permeability of free space*. Writing $\rho = \rho_b + \rho_f$ and $\boldsymbol{J} = \boldsymbol{J}_b + \boldsymbol{J}_f$, they become

$$\nabla \times \boldsymbol{H} = \boldsymbol{J}_f + \frac{\partial \boldsymbol{D}}{\partial t}, \qquad \nabla \times \boldsymbol{E} = -\frac{\partial \boldsymbol{B}}{\partial t}, \qquad \nabla \cdot \boldsymbol{B} = 0, \qquad \nabla \cdot \boldsymbol{D} = \rho_f \tag{17.1}$$

Here, ρ_f is the *free* (or true) *charge*, $\boldsymbol{J}_f$ is the *free current*, $\boldsymbol{H} = \boldsymbol{B}/\mu_0 - \boldsymbol{M}$ is the *magnetic intensity* with *magnetization* $\boldsymbol{M}$, and $\boldsymbol{D} = \varepsilon_0\boldsymbol{E} + \boldsymbol{P}$ is the *dielectric displacement*, with the *bound* (or polarization) *charge* $\rho_b = -\nabla \cdot \boldsymbol{P}$ and the *bound current* $\boldsymbol{J}_b = \nabla \times \boldsymbol{M} + \partial\boldsymbol{P}/\partial t$. In vacuum we thus have $\boldsymbol{D} = \varepsilon_0\boldsymbol{E}$, but in matter this expression is $\boldsymbol{D} = \varepsilon_0\boldsymbol{E} + \boldsymbol{P} \equiv \varepsilon_0\boldsymbol{E} + \varepsilon_0\chi\boldsymbol{E} = \varepsilon_0(1+\chi)\boldsymbol{E} \equiv \varepsilon_r\varepsilon_0\boldsymbol{E} \equiv \varepsilon\boldsymbol{E}$. Similarly, $\boldsymbol{H} = \boldsymbol{B}/\mu_0$ becomes $\boldsymbol{H} = \boldsymbol{B}/\mu$, where $\mu = \mu_r\mu_0$ with *relative permeability* μ_r. The solution of Maxwell's equations is only possible if the constitutive equations $\boldsymbol{D} = \varepsilon_r\varepsilon_0\boldsymbol{E}$ and $\boldsymbol{H} = \boldsymbol{B}/\mu_r\mu_0$ are given. Generally, ε_r and μ_r are material tensors, dependent on temperature, pressure, and frequency, but for isotropic and cubic materials they become scalars. For typical field strengths ε_r is independent of $E = |\boldsymbol{E}|$, but a similar approximation cannot be made for μ_r.

After writing $\boldsymbol{B} = \mu\boldsymbol{H}$ in Eq. (17.1b) and thereafter eliminating $\boldsymbol{H}$ by inserting the result in Eq. (17.1a) and evaluating, meanwhile using $\nabla \times \nabla \times \boldsymbol{E} = \nabla(\nabla \cdot \boldsymbol{E}) - \nabla^2\boldsymbol{E}$, $\nabla \cdot \boldsymbol{D} = 0$ for a charge-free region and $\boldsymbol{J}_f \equiv \sigma_0\boldsymbol{E}$ with σ_0 the *direct conductivity*, we obtain

$$\nabla^2\boldsymbol{E} = \mu\varepsilon\frac{\partial^2\boldsymbol{E}}{\partial t^2} + \mu\sigma_0\frac{\partial \boldsymbol{E}}{\partial t}$$

If we use as trial solution $\boldsymbol{E} = \boldsymbol{E}_0\exp[\mathrm{i}(\boldsymbol{k}\cdot\boldsymbol{r} - \omega t)]$, the result is $k^2 = |\boldsymbol{k}^2| = \mu\varepsilon\omega^2 + \mathrm{i}\mu\sigma_0\omega$ showing that the wave vector $\boldsymbol{k}$ is complex. Introducing $c^{-2} = \mu\varepsilon$, where c has the dimension of speed, we can write $k^2 = (\omega^2/c^2)(1 + \mathrm{i}\mu\sigma_0/\varepsilon_0\omega)$. In vacuum $\sigma_0 = 0$ and we have the *speed of light* $c = (\varepsilon_0\mu_0)^{-1/2}$, while in matter the speed of light is $v = (\varepsilon_0\varepsilon_r\mu_0\mu_r)^{-1/2} = c/n$ with $n = (\varepsilon_r\mu_r)^{1/2}$ the *refractive index*. Because $\boldsymbol{k}$ is a complex quantity, n is a complex quantity

1 While the labels for $\boldsymbol{E}$ and $\boldsymbol{D}$ are universally accepted, $\boldsymbol{B}$ is labeled here, omitting the magnetic, as field and $\boldsymbol{H}$ as intensity conform [4], the argument being that $\boldsymbol{E}$ and $\boldsymbol{B}$ are the analogues determining the electric and magnetic forces. Guggenheim [7] characterized the generally used names for $\boldsymbol{B}$ and $\boldsymbol{H}$, mostly induction and field [5, 6], as extremely confusing. Further, Maxwell's own exposition of his equations is a mess, and you will not find the present compact and elegant structure, largely due to Lorentz [8], in Maxwell's writings [9].

$n = n_1 + in_2$ with n_1 the real and n_2 the imaginary components. For dielectrics, $\mu_r = 1$ is usually an excellent approximation and for this case $\varepsilon_r = n^2$. Hence, we have a *complex permittivity* $\varepsilon_r = \varepsilon_1 + i\varepsilon_2$ with $\varepsilon_1 = n_1^2 - n_2^2$ and $\varepsilon_2 = 2n_1n_2$. To obtain n_1 and n_2 we note that $\varepsilon_1^2 + \varepsilon_2^2 = (n_1^2 + n_2^2)^2$. After taking the square root and either adding or subtracting $\varepsilon_1 = n_1^2 - n_2^2$, the result is $n_1 = \{½[(\varepsilon_1^2 + \varepsilon_2^2) + \varepsilon_1]\}^{1/2}$ and $n_2 = \{½[(\varepsilon_1^2 + \varepsilon_2^2) - \varepsilon_1]\}^{1/2}$.

In the absence of free currents, $\nabla \times \boldsymbol{H} = \partial \boldsymbol{D}/\partial t$, where the latter for a periodic electric field $\boldsymbol{E} = \boldsymbol{E}_0\exp(-i\omega t)$ is given by $\partial \boldsymbol{D}/\partial t = -i\omega\varepsilon_r\boldsymbol{E}$, and thus $\nabla \times \boldsymbol{H} = -i\omega\varepsilon_r\boldsymbol{E} = \omega\varepsilon_2 + i\omega\varepsilon_1$. Hence, adding $\boldsymbol{J}_f = \sigma_0\boldsymbol{E}$, we have for the total current $\nabla \times \boldsymbol{H} = \boldsymbol{J}_f + \boldsymbol{J}_b = \sigma\boldsymbol{E}$ with *complex conductivity*[2] $\sigma = \sigma_1 - i\sigma_2$, where $\sigma_1 = \sigma_0 + \omega\varepsilon_2$ and $\sigma_2 = \omega\varepsilon_1$. Using the *loss tangent* $\tan\delta = (\omega\varepsilon_2 + \sigma_0)/\omega\varepsilon_1$, an incident electric field $\boldsymbol{E} = \boldsymbol{E}_0\exp[i(k_x x - \omega t)]$ decays with the factor $\exp(-\delta k_x x)$. The power density $½\boldsymbol{E}\cdot\boldsymbol{D} = ½\varepsilon E^2$ decays with $\exp(-2\delta k_x x)$, describing the loss in power (conversion to heat) with distance x.

Generally, for linear response the real and imaginary components are not independent. They are connected via the *Kramers–Kronig transform* (Section A.3) and, given *full knowledge* of $x_1(\omega)$, where x is α, ε, n or σ, $x_2(\omega)$ can be calculated and vice versa.

17.1.2 Dielectric Materials

We now briefly review the type and nature of dielectric materials. There exist two ways to describe them. First, along the types of materials involved and, second along the types of polarizations involved. For the first, we just note that dielectrics are electrical insulators (or very poor) conductors and that within the category of inorganic materials oxides predominate, although nowadays polymers are also an important category. For the second, we have nonpolar, polar, and dipolar materials.

Nonpolar materials consist out of molecules without a permanent dipole moment. Examples of nonpolar materials are the noble gas liquids and solids, solids like Si, Ge, and C (diamond), hydrocarbons like paraffins, and polymers like polyethylene. Variations in polarizability occur mainly in the UV optical region. The main contribution to α_{eff} is α_{elc} so that $\varepsilon_r \cong n^2$ (Table 8.9) with n typically about 1.5, rutile (TiO_2, $n = 2.61$) and diamond ($n = 2.38$) being on the higher side. Using as model the homogenously conducting sphere or the Bohr atom (footnote 4 of Chapter 4), the polarizability $\alpha_{ele} = 4\pi\varepsilon_0 r^3$. Introducing the density $\rho = (2r)^{-3}$, the susceptibility $\chi_1 = P/\varepsilon_0 E = \rho\alpha_{ele}/\varepsilon_0 E = (2r)^{-3}\cdot 4\pi\varepsilon_0 r^3 \cdot E/\varepsilon_0 E = \pi/2$ so that $\varepsilon_1 \cong 1 + \chi_1 \cong 2.5$ or $n_1 \cong 1.6$, right away explaining the order of magnitude.

The class of polar materials contains compounds containing molecules without dipole moment but having locally relatively large charge density differences. The most important representatives are ionic solids, like alkali halides. These materials show changes in polarizability in the IR as well as the optical/UV region. Dipolar materials consist of molecules that do have a dipole moment, which largely determines their orientational polarizability. Both polar and dipolar compounds do have a somewhat larger relative permittivity, say, 5–15.

For a long time, the focus in dielectric behavior was on obtaining high permittivity values. However, for electronic circuitry insulating, low-permittivity materials are as least as important as they co-determine the electric losses and the more so with decreasing feature sizes. The dielectric behavior of polymers is diverse, and we indicate here only briefly a

2 Taking the electric field as $\boldsymbol{E} = \boldsymbol{E}_0\exp(-i\omega t)$, ε and n are defined with a + sign, while σ is defined with a − sign. Taking $\boldsymbol{E} = \boldsymbol{E}_0\exp(i\omega t)$, the signs of ε, n, and σ change, with the same final result.

Table 17.1 Group contributions P_j to the molar polarization $P_m = \Sigma_j P_j$ [10]; ϕ = benzene ring.

Group	P_j (cm^3 mol^{-1})	Group	P_j (cm^3 mol^{-1})	Group	P_j (cm^3 mol^{-1})
$-CH_3$	5.64	>C=O	(10)	$-CF_2-$	6.25
$-CH_2-$	4.65	–COO–	15	$-CCl_2-$	17.7
–CH<	3.62	–CONH–	30	–CHCl–	13.7
>C<	2.58	–O–COO–	22	–S–	8
–ϕ	25.5	–F	(1.8)	–OH (alcohol)	(6)
–ϕ–	25.0	–Cl	(9.5)	–OH (phenol)	≈20
–O–	5.2	–C≡N	11		

Source: van Krevelen and te Nijenhuis [10]/with permission of Elsevier.

group contribution approach [10] to estimate the value of their permittivity. The basis is the molar polarization $P_m/V_m = (\varepsilon_1 - 1)/(\varepsilon_1 + 2) = \alpha_{eff}N_A/3\varepsilon_0 V_m$ where $V_m = \mathcal{M}/\rho$ is the molar volume with molar mass $\mathcal{M}$ and mass density ρ, and $\alpha_{eff} = \alpha + \mu_0{}^2/3kT$ with polarizability α and (ground state) dipole moment μ_0 (Section 8.6.2). This quantity is supposed to be the sum of the contributions of chemical groups constituting the polymer molecule (Table 17.1). For slightly polar polymers one neglects the dipole contribution so that the molar polarization becomes the molar refractivity. Reasonable agreement for the refractive index n can be obtained as judged by a group of 18 polymers, for which the AAD is 0.07 or 4%. Clearly, dipole moments do affect the result, but estimates for the dipole moments of groups from the difference between the above estimate and the experimental permittivity are low compared with the dipole moments of the polar groups in liquids, typically by a factor of 3–5. Reference [11] deals mainly with polymers including loss, relaxation, and breakdown, while [12] presents many details on theory, experimental techniques, and materials, again with a focus on polymers.

In this classification electric order or *spontaneous polarization* is not included, and we turn to that in Section 17.6. For now, we just mention that all materials not showing this phenomenon are called *paraelectric*. Clearly, there are also materials that do have phase transformations where the permittivity changes insignificantly, for example, $LaAlO_3$, $Cu_2Fe_2O_4$ or K_2ReCl_6, and the term *orthoelectric* has been coined for them. In Sections 17.2–17.5, we turn to the various mechanisms for dielectric behavior. We start on the high frequency side, and in each next section, we lower the frequency.

17.2 Electronic Polarization

The first contribution to polarization is the electronic contribution α_{ele} contributing to all materials. In a classical description as first introduced by Lorentz [8], one assumes that the material can be described by a set of oscillators that represent bound particles, generally electrons. We take these oscillators as isotropic. Applying a uniform electric field $\boldsymbol{E}$ along the x-axis displaces the center of gravity of the negative charges $-|e|$ of a particle with mass m with respect to that of the positive charge by x, thereby introducing an induced dipole moment $\boldsymbol{p} = -ex$. For small displacements the restoring force with force constant a is elastic,

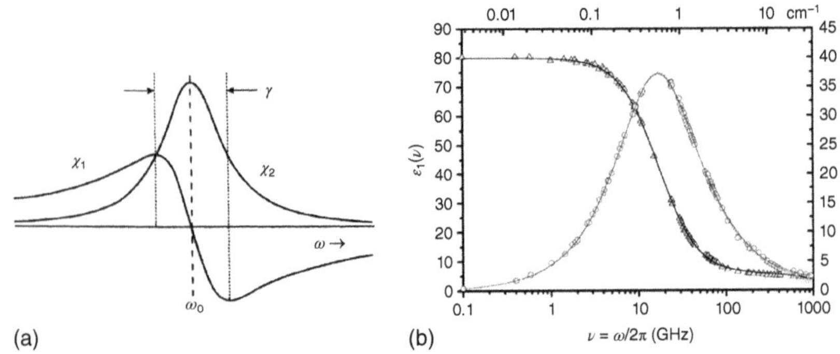

Figure 17.2 (a) Real and imaginary part of the electronic susceptibility χ according to the (Lorentz) oscillator model. Source: Peiponen et al. [13]/Springer Nature. (b) Real part $\varepsilon_1(\nu)$ and imaginary part $\varepsilon_2(\nu)$ of the Debye relaxation spectrum for water. Source: Beneduci [14]/with permission of Elsevier.

and we introduce a damping constant γ, in solids related to absorption in the molecules, and the fact that the oscillators representing the molecules are coupled. The equation of motion for such an oscillator is

$$m\ddot{x} + m\gamma\dot{x} + ax = -e\boldsymbol{E}$$

Using as trial solution $x = x_0\exp(-\mathrm{i}\omega t)$ in combination with $\boldsymbol{E} = \boldsymbol{E}_0\exp(-\mathrm{i}\omega t)$ we obtain for the displacement x and susceptibility $\chi = \varepsilon_\mathrm{r} - 1$

$$x = -\frac{e\boldsymbol{E}}{m\left[\left(\omega_0{}^2 - \omega^2\right) - \mathrm{i}\gamma\omega\right]} \quad \text{and} \quad \chi = \frac{\boldsymbol{P}}{\varepsilon_0\boldsymbol{E}} = \frac{e^2}{m\varepsilon_0\left[\left(\omega_0{}^2 - \omega^2\right) - \mathrm{i}\gamma\omega\right]}$$

respectively, with $\omega_0 = (a/m)^{1/2}$ the resonance frequency of the non-damped oscillator. For N identical oscillators in volume V, the number density $\rho = N/V$ and splitting $\chi = \Sigma_j\chi_j$ in a real and imaginary component, we obtain

$$\chi_1 = \frac{\rho e^2}{m\varepsilon_0}\frac{\omega_0{}^2 - \omega^2}{\left[\left(\omega_0{}^2 - \omega^2\right) + \gamma^2\omega^2\right]} \quad \text{and} \quad \chi_2 = \frac{\rho e^2}{m\varepsilon_0}\frac{\gamma\omega}{\left[\left(\omega_0{}^2 - \omega^2\right) + \gamma^2\omega^2\right]}$$

Abbreviating $\rho e^2/m\varepsilon_0$ by C, the real part has a maximum $\chi_1 \approx 2\gamma\omega_0 C$ at $\omega = \omega_0 - \gamma$, $\chi_1 = 0$ at $\omega = \omega_0$ and has a minimum $\approx -2\gamma\omega C$ at $\omega = \omega_0 + \gamma$, while the imaginary part has a maximum $\chi_2 = C/\omega_0\gamma$ at $\omega = \omega_0$ (Figure 17.2a). Using the refractive index $n = (\varepsilon_\mathrm{r})^{1/2} = (1 + \chi)^{1/2}$, we see that at low frequency ($\omega < \omega_0 - \gamma$) we have limited absorption and the material is largely transparent. This behavior is characteristic for ionic crystals (and polymers) in the visible range. For larger frequency ($\omega_0 - \gamma < \omega < \omega + \gamma$) the materials are absorbing and at the center of the absorption line and for $\sigma_0 = 0$ the conductivity also shows a maximum $\sigma_1 = \omega\varepsilon_2 = \omega_0\chi_2 = C/\gamma$. For $\omega_0 + \gamma < \omega < \omega_0$, we might have for the permittivity $\varepsilon_1 < 1$ and in that case, there is total reflection until $1 + \chi_1 = \chi_2$. For a still larger frequency, we have $\varepsilon_1 > 1$, rendering again a slightly absorbing transparent material. This type of behavior, often called *Debye relaxation*, is also shown by many other compounds, Figure 17.2b showing data for water. Many details on classical dispersion theory, outside our scope, can be found in [13, 15].

Quantum mechanically, one uses time-dependent perturbation theory (Section 3.3.4) for the Hamilton operator $H = H_0 + H_1 = H_0 - (e\Sigma_j x_j)\boldsymbol{E}$, where H_0 is the Hamilton operator

Table 17.2 Polarizability α of some atoms and alkali halide ions.

Ions	Free ion α (Å^3)	Solid state α (Å^3)	Atom	α (10^{-40} F m^2)	Ions	Free ion α (Å^3)	Solid state α (Å^3)
			He	0.22	Li^+	0.029	0.029 (0.029, 0.029)
F^-	1.04	0.644 (0.876, 0.876)	Ne	0.45	Na^+	0.179	0.408 (0.312, 0.285)
Cl^-	3.66	2.960 (3.063, 3.005)	Ar	1.84	K^+	0.83	1.334 (1.136, 1.149)
Br^-	4.77	4.158 (4.276, 4.168)	Kr	2.69	Rb^+	1.40	1.979 (1.758, 1.707)
I^-	7.1	6.431 (6.517, 6.294)	Xe	4.37	Cs^+	2.42	3.335 (3.015, 2.789)

Free ions: Source: Adapted from Pauling [22], solid state: [19], and data in parenthesis: [20] and [21].

in the absence of the field $\boldsymbol{E}$ and x_j is the displacement of the electrons [16]. We represent the ground state by $|0\rangle$, one of the allowed excited by $|j\rangle$, and abbreviate $\langle 0|\Sigma_j ex_j|j\rangle$ as $P_j = P_j^* = \langle j|\Sigma_j ex_j|0\rangle$. Introducing the *oscillator strength* $f_j = (2m/e^2\hbar)\omega_j P_j^2$, one obtains for the complex polarizability

$$\alpha = \alpha_1 + \alpha_2 = \frac{e^2}{m\varepsilon_0}\sum\nolimits_{j\neq 0}\left[\frac{f_j}{\omega_j^{\,2} - \omega^2} - \frac{\mathrm{i}\pi f_j\omega\delta(\omega_j - \omega)}{2\omega}\right]$$

which, apart from the line shape, is identical to the classical result. However, if one assumes that for the system in state j the coefficient is not given by $c_j(0) = 1$ (Section 3.3.4) but by $c_j(0) = \exp(-\gamma t)$, one can show that the line shape becomes again Lorentz-like with γ the *natural line width*, typically of order $10^8\ \text{s}^{-1}$. Experimentally a much larger width is observed, as the natural width is normally masked by interactions between molecules (and in gases by the Doppler effect). For solids as described in the tight-binding approximation, the ground state becomes a sum over the valence-band states, while j runs over conduction-band states. One expects that the narrower the band gap, the larger α_{ele}, which is indeed approximately the case [1].

Empirically, from experimental refractivity data the polarizability of ionic crystals can be reasonably well described as the sum of the polarizabilities of contributing ions using the Clausius–Mossotti equation [17, 18]. For alkali halides, Tessman et al. [19] provided an often-used set by minimizing the absolute deviations for α_{ele} for which the relative sample standard deviation is 7.3%. This set has been revised [20] by minimizing the relative deviations for α_{ele} and resulting in 2.8%, and further reducing it by using newer data [21] to 1.4% (Table 17.2). In all the above cases, α_{ele} for Li^+ was taken as 0.029 Å^3, its precise value being relatively unimportant in view of its low contribution. Shannon [23] provided a similar set using 129 oxides and 25 fluorides, with a typical deviation of 0.05 Å^3 for the smaller and 0.1 Å^3 for the larger cations. With such a set α_{ele} can thus be well described, but the α-values for the individual ions differ from the values for the free ions (Table 17.2).

17.3 Vibrational Polarization

Lowering the frequency, the vibrational contribution α_{vib} adds to the electronic contribution α_{ele}. While α_{ele} shows a resonance frequency ω_{UV} in the UV part of the optical spectrum, for α_{vib} the resonance frequency ω_{IR} is in the IR part of the optical spectrum. This contribution

Table 17.3 The permittivity ε_0 and ε_∞ for various crystals.

Crystal	ε_s (–)	ε_∞ (–)	Crystal	ε_s (–)	ε_∞ (–)	Crystal	ε_s (–)	ε_∞ (–)
LiF	9.27	1.92	NaBr	5.99	2.62	RbF	5.91	1.93
LiCl	11.05	2.75	NaI	6.60	2.91	RbCl	5.00	2.19
LiBr	12.10	3.16	KF	6.05	1.85	RbBr	5.00	2.33
LiI	11.3	3.80	KCl	4.68	2.13	RbI	5.00	2.63
NaF	6.00	1.74	KBr	4.78	2.33	CsCl	7.20	2.60
NaCl	5.62	2.25	KI	4.94	2.69	CsBr	6.51	2.78

Source: Bunget and Popescu [1]/with permission of Elsevier.

is most important for ionic compounds, and we focus on binary ionic compounds with a cubic unit cell, approximately following the outlines given in [24, 25], but differing in detail. In the low-frequency region $\omega \ll \omega_{\mathrm{IR}}$ we denote the permittivity by ε_{S}, while for the region $\omega_{\mathrm{IR}} \ll \omega \ll \omega_{\mathrm{UV}}$ it is labeled ε_∞. The difference $\Delta\varepsilon = \varepsilon_{\mathrm{S}} - \varepsilon_\infty$ is thus a measure for the vibrational contribution to ε_{r} and for ionic crystals it is about 2–9 units (Table 17.3).

The maximum frequency of the lattice vibrations is about $10^{13}\ \mathrm{s}^{-1}$, while the far IR frequency range is typically quoted as 2×10^{14} to $4\times10^{14}\ \mathrm{s}^{-1}$, corresponding to a wavelength of about 800–2500 nm and thus much larger than a unit cell. This means that a *long wavelength approximation* can be used [26]. Indicating the wave vector of phonons by $\boldsymbol{q}$ and of electromagnetic waves with $\boldsymbol{k}$, we must have $\hbar\omega(\boldsymbol{q}) = \hbar\omega(\boldsymbol{k})$ to conserve energy and $\hbar\boldsymbol{q} = \hbar\boldsymbol{k}$ to conserve momentum.

Semi-classically, the interaction energy is $-\boldsymbol{M}\cdot\boldsymbol{E}$, where $\boldsymbol{M}$ is the total dipole moment of the crystal with respect to its equilibrium configuration and $\boldsymbol{E}$ is the electric field. If we describe the ions as point charges, we have $\boldsymbol{M} = \Sigma_{\boldsymbol{n},\lambda} Z_\lambda e \boldsymbol{u}_{\boldsymbol{n},\lambda}$ with Z_λ the valency of ion λ, e the unit charge, and $\boldsymbol{u}_{\boldsymbol{n},\lambda}(\boldsymbol{q})$ the displacement of ion λ in unit cell $\boldsymbol{n}$ (Section 10.7). For the displacements we use the general solution $\boldsymbol{u}_{\boldsymbol{n},\lambda} = \Sigma_{p,\boldsymbol{q}} \boldsymbol{U}_{p,\lambda}(\boldsymbol{q}) h_{\boldsymbol{n},\lambda}(\boldsymbol{q})$ with amplitude vector $\boldsymbol{U}_{p,\lambda}(\boldsymbol{q}) = N^{-1/2} A_p(\boldsymbol{q}) \boldsymbol{e}_{p,\lambda}(\boldsymbol{q})$, having eigenvectors $\boldsymbol{e}_{p,\lambda}(\boldsymbol{q})$ for branch p, and space–time factor $h_{\boldsymbol{n},\lambda}(\boldsymbol{q}) = (m_\lambda)^{-1/2} \exp[\mathrm{i}(\boldsymbol{q}\cdot\boldsymbol{r}_{\boldsymbol{n},\lambda} - \omega t)]$. The polarization $\boldsymbol{P}$ for a binary ionic compound with $|Z_\lambda| = Z$ becomes, omitting the time factor,

$$\boldsymbol{P} = \frac{\boldsymbol{M}}{V} = \frac{Ze}{VN^{1/2}} \sum\nolimits_{\boldsymbol{n},p,\boldsymbol{q}} A_p(\boldsymbol{q}) \left[\frac{\boldsymbol{e}_{p,+}(\boldsymbol{q})}{m_+{}^{1/2}} - \frac{\boldsymbol{e}_{p,-}(\boldsymbol{q})}{m_-{}^{1/2}} \right] \exp(\mathrm{i}\boldsymbol{q}\cdot\boldsymbol{r}_{\boldsymbol{n}})$$

where we use + and – to label the ions, and V is the volume of the crystal. Summing over $\boldsymbol{n}$ shows that only modes with $\boldsymbol{q} = 0$ contribute to $\boldsymbol{P}$ so that $\boldsymbol{P}$ reduces to

$$\boldsymbol{P} = \frac{ZeN^{1/2}}{V} \sum\nolimits_p A_p(\boldsymbol{0}) \left[\frac{\boldsymbol{e}_{p,+}(\boldsymbol{0})}{m_+{}^{1/2}} - \frac{\boldsymbol{e}_{p,-}(\boldsymbol{0})}{m_-{}^{1/2}} \right]$$

For a binary compound, the orthonormalization condition $\Sigma_\alpha \boldsymbol{e}_{p,\alpha}(\boldsymbol{q})^* \cdot \boldsymbol{e}_{l,\alpha}(\boldsymbol{q}) = \delta_{pl}$ reduces for $\boldsymbol{q} = 0$ to $\boldsymbol{e}_{p,+}\cdot\boldsymbol{e}_{l,+} + \boldsymbol{e}_{p,-}\cdot\boldsymbol{e}_{l,-} = 0$, where p now refers to an acoustic branch and l to an optic branch. For an acoustic branch at $\boldsymbol{q} = 0$, we moreover have $\boldsymbol{e}_{p,-} = (m_-/m_+)^{1/2}\boldsymbol{e}_{p,+}$, because all atoms move in the same direction (as results from $\boldsymbol{u}_{\boldsymbol{n},\alpha}$ for $\boldsymbol{q} = \boldsymbol{0}$ and $\omega = 0$). Combining these expressions leads to $\boldsymbol{e}_{p,+}\cdot[\boldsymbol{e}_{l,+} + (m_-/m_+)^{1/2}\boldsymbol{e}_{l,-}]$ or, because $\boldsymbol{e}_{p,+} \neq 0$ and the square bracket term cannot be perpendicular to all acoustic branches, to $(m_+)^{1/2}\boldsymbol{e}_{l,+} = -(m_-)^{1/2}\boldsymbol{e}_{l,-}$

for the optic branches. Because at $\boldsymbol{q} = 0$, $\boldsymbol{u}_{n,\alpha,l} = (Nm_\alpha)^{-1/2}A_l\boldsymbol{e}_{l,\alpha}$, we have equivalently $m_+\boldsymbol{u}_{l,+} = -m_-\boldsymbol{u}_{l,-}$ so that for the optic branch all positive ions move oppositely to all negative ions. This produces an oscillating dipole with whom the electromagnetic radiation interacts. Using for the number density $\rho = N/V$ with N the number of unit cells and V the total volume, and omitting the label $\boldsymbol{q} = \boldsymbol{0}$, the polarization $\boldsymbol{P}$ becomes

$$\boldsymbol{P} = \rho Ze(\boldsymbol{u}_+ - \boldsymbol{u}_-) \equiv \rho Ze\boldsymbol{u} \quad \text{with} \quad \boldsymbol{u} = \boldsymbol{u}_+ - \boldsymbol{u}_-$$

Implicitly we assumed that the phonons and electromagnetic waves could be treated independently, which, however, is not the case.

Because the interaction between ions and the field propagates with finite velocity, we must solve these equations simultaneously with Maxwell's equations. The starting point for a coupled treatment is setting up equations in terms of the electric field $\boldsymbol{E}$ and the relative displacement $\boldsymbol{u}$ of the two ions in the unit cell. First, we recall that the field experienced by the ions is the local field $\boldsymbol{E}_{\text{loc}}$, which we take as the Lorentz field $\boldsymbol{E}_{\text{L}} = \boldsymbol{E} + \boldsymbol{P}/3\varepsilon_0$. The polarization $\boldsymbol{P}$ is due to the displacement $\boldsymbol{u}$ of the ions and their polarizability $\alpha = \alpha_+ + \alpha_-$. Adding both contributions, we have $\boldsymbol{P} = \rho(Ze\boldsymbol{u} + \alpha\boldsymbol{E}_{\text{L}})$. After inserting $\boldsymbol{E}_{\text{L}}$, we obtain $\boldsymbol{P} = \rho ZeC\boldsymbol{u} + 2\alpha C\boldsymbol{E}$, where we introduced the abbreviation $C^{-1} = 1 - \rho\alpha/3\varepsilon_0$. The ions move according to the equations of motion

$$m_+\ddot{\boldsymbol{u}}_+ = -a(\boldsymbol{u}_+ - \boldsymbol{u}_-) + Ze\boldsymbol{E}_{\text{L}} \quad \text{and} \quad m_-\ddot{\boldsymbol{u}}_- = -a(\boldsymbol{u}_+ - \boldsymbol{u}_-) - Ze\boldsymbol{E}_{\text{L}}$$

with a the coupling or force constant. By subtracting $\ddot{\boldsymbol{u}}_-$ from $\ddot{\boldsymbol{u}}_+$ we obtain

$$\mu_m\ddot{\boldsymbol{u}} = -a\boldsymbol{u} + Ze\boldsymbol{E}_{\text{L}} \quad \text{with} \quad 1/\mu_m = 1/m_+ + 1/m_-$$

the reduced mass. We now write these two expressions in a renormalized form by using $\boldsymbol{w} = (\rho\mu_m)^{1/2}\boldsymbol{u}$. After what is called some straightforward algebra, we obtain

$$\ddot{\boldsymbol{w}} = b_{11}\boldsymbol{w} + b_{12}\boldsymbol{E} \quad \text{and} \quad \boldsymbol{P} = b_{21}\boldsymbol{w} + b_{22}\boldsymbol{E}$$

where

$$b_{11} = -a/\mu_m, \quad b_{12} = b_{21} = (\rho/\mu_m)^{1/2}ZeC \quad \text{and} \quad b_{22} = 2\alpha C \tag{17.2}$$

The next step uses Maxwell's equations, Eq. (17.1), in the absence of free charges ($\rho_{\text{f}} = 0$ and $\boldsymbol{j}_{\text{f}} = \boldsymbol{0}$). If we assume that $\boldsymbol{w}$, $\boldsymbol{E}$, $\boldsymbol{B}$, and $\boldsymbol{P}$ vary as $\exp[\mathrm{i}(\boldsymbol{q}\cdot\boldsymbol{r} - \omega t)$, it follows that

$$-\omega^2\boldsymbol{w} = b_{11}\boldsymbol{w} + b_{12}\boldsymbol{E}, \quad \boldsymbol{q}\cdot\boldsymbol{D} = 0, \quad \boldsymbol{q}\cdot\boldsymbol{B} = 0, \quad \boldsymbol{q}\times\boldsymbol{E} = -\omega\boldsymbol{B}, \quad \boldsymbol{q}\times\boldsymbol{B} = -\omega\mu_0\boldsymbol{D} \tag{17.3}$$

Inserting Eq. (17.3a) in Eq. (17.2b) results in

$$\boldsymbol{P} = \left[b_{22} - {b_{12}}^2/(b_{11} + \omega^2)\right]\boldsymbol{E} \quad \text{or} \quad \varepsilon(\omega) = 1 + b_{22} - {b_{12}}^2/(b_{11} + \omega^2) \tag{17.4}$$

At low frequency ($\omega \to 0$), $\ddot{\boldsymbol{w}} = 0$ and therefore $\boldsymbol{P}(0) = (b_{22} - {b_{12}}^2/b_{11})\boldsymbol{E} = \varepsilon_0(\varepsilon_{\text{s}} - 1)\boldsymbol{E} = \varepsilon_0\chi_{\text{s}}\boldsymbol{E}$ with ε_{s} (χ_{s}) the *static permittivity* (*susceptibility*). At high frequency (but below the electronic region), $\boldsymbol{w} = 0$ and $\boldsymbol{P}(\infty) = b_{22}\boldsymbol{E} = \varepsilon_0(\varepsilon_\infty - 1)\boldsymbol{E} = \varepsilon_0\chi_\infty\boldsymbol{E}$ with ε_∞ the *high frequency permittivity*. This implies that

$$\varepsilon_{\text{r}}(\omega) = \varepsilon_1(\omega) = \varepsilon_\infty + (\varepsilon_{\text{s}} - \varepsilon_\infty){\omega_0}^2/\left({\omega_0}^2 - \omega^2\right) \tag{17.5}$$

where we used the free oscillator frequency $-b_{11} \equiv {\omega_0}^2 = a/\mu_m$. Note that in this model the permittivity is real. Inserting Eq. (17.4) in $\boldsymbol{q}\cdot\boldsymbol{D} = \boldsymbol{q}\cdot(\varepsilon_0\boldsymbol{E} + \boldsymbol{P}) = 0$, we find that $\varepsilon_{\text{r}}(\omega)(\boldsymbol{q}\cdot\boldsymbol{E}) = 0$. We now must distinguish between whether $\varepsilon_{\text{r}}(\omega) = 0$ or $\boldsymbol{q}\cdot\boldsymbol{E} = 0$.

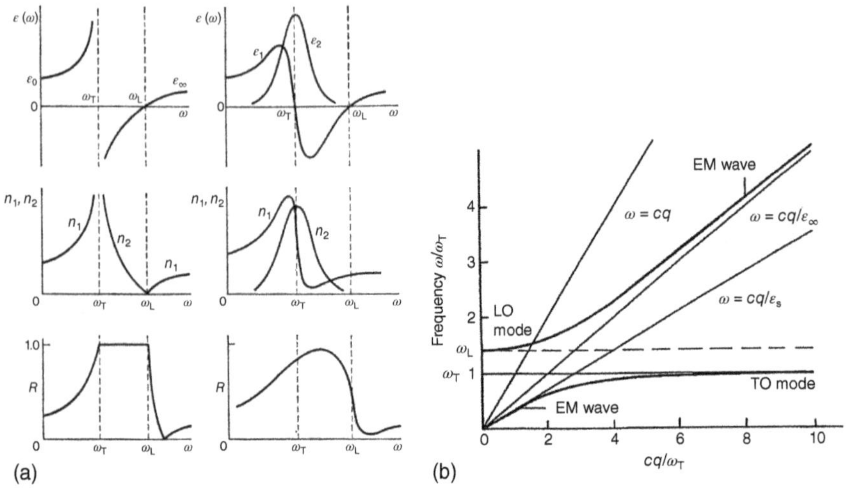

Figure 17.3 (a) Schematic of the behavior of the permittivity ε, refractive index $n = n_1 + in_2$, and reflectivity R for a typical alkali halide, showing in the left column ideal behavior with zero damping and in the right column more realistic behavior with constant damping. Source: Donovan and Angress [24]/Taylor & Francis. (b) Coupling of electromagnetic and transverse acoustic waves. Source: Elliott [25]/John Wiley & Sons.

When $\varepsilon_r(\omega) = 0$, $\varepsilon_0\boldsymbol{E} = -\boldsymbol{P}$ as follows from $\boldsymbol{q}\cdot(\varepsilon_0\boldsymbol{E} + \boldsymbol{P}) = 0$, and therefore $\boldsymbol{q}\times\boldsymbol{B} = 0$. Hence, either $\boldsymbol{B} = 0$ or $\boldsymbol{B}\parallel\boldsymbol{q}$, where $\parallel$ denotes "parallel to." From $\boldsymbol{q}\cdot\boldsymbol{B} = 0$ it follows that $\boldsymbol{B} = 0$ or $\boldsymbol{B}\perp\boldsymbol{q}$, where $\perp$ denotes "perpendicular to." Thus for $\boldsymbol{B} = 0$, from which it follows using $\boldsymbol{q}\times\boldsymbol{E} = -\omega\boldsymbol{B} = \boldsymbol{0}$, that $\boldsymbol{E}\parallel\boldsymbol{q}$, that is, we have *longitudinal waves*. Solving $\varepsilon_1(\omega) = 0$ leads to $\omega_L^{\ 2}/\omega_0^{\ 2} = \varepsilon_s/\varepsilon_\infty$ so that longitudinal waves show no dispersion. For the uncoupled electrostatic case the same solution is obtained and, hence, retardation does not influence the longitudinal optic (LO) waves.

If $\boldsymbol{q}\cdot\boldsymbol{E} = 0$, $\boldsymbol{E}\perp\boldsymbol{q}$, so that we now have *transverse waves*. From $\boldsymbol{q}\times\boldsymbol{E} = -\omega\boldsymbol{B}$ it follows that also $\boldsymbol{E}\perp\boldsymbol{B}$, so that we have $q = \omega B/E$ and $qB = \omega\mu_0(\varepsilon_0 E + P)$, where, as usual, $X = |\boldsymbol{X}|$. If we eliminate B, these two expressions result in

$$q^2E/\mu_0\omega^2 = \varepsilon_0 E + P \quad \text{or} \quad q^2c^2E/\omega^2 = \varepsilon_1(\omega)E$$

where $c = (\varepsilon_0\mu_0)^{-1/2}$ is the speed of light in vaccum. Because $\boldsymbol{E}\neq 0$, we have

$$\frac{q^2c^2}{\omega^2} = \varepsilon_\infty + \frac{(\varepsilon_s - \varepsilon_\infty)\omega_0^{\ 2}}{\left(\omega_0^{\ 2} - \omega^2\right)} \quad \text{or} \quad \omega^4 - \omega^2\left[\omega_L^{\ 2} + \frac{q^2c^2}{\varepsilon_\infty}\right] + \frac{q^2c^2\omega_0^{\ 2}}{\varepsilon_\infty} = 0 \tag{17.6}$$

showing that transverse waves are mixtures between lattice and electromagnetic waves (Figure 17.3b) [27]. This explains dispersion for the transverse optic (TO) waves and permittivity, and thus for the refractive index in the optical/UV regime.

For $q\to\infty$, Eq. (17.6) shows that $\omega\to\omega_0$ and thus we can identify ω_0 with ω_T and replace ω_0 in Eq. (17.5) by ω_T. Using $\varepsilon_1(\omega) = 0$ for Eq. (17.5) also provides a relation between ω_L and ω_T. Together, we thus have

$$\frac{\omega_L^{\ 2}}{\omega_T^{\ 2}} = \frac{\varepsilon_s}{\varepsilon_\infty} \quad \text{and} \quad \varepsilon_1(\omega) = \varepsilon_\infty + \frac{(\varepsilon_s - \varepsilon_\infty)\omega_T^{\ 2}}{\omega_T^{\ 2} - \omega^2} = \varepsilon_\infty\frac{\omega_L^{\ 2} - \omega^2}{\omega_T^{\ 2} - \omega^2} \tag{17.7}$$

From the former, known as the *Lyddane–Sachs–Teller* (LST) *relation* [28], the value for ω_L can be determined. For nonpolar materials such as C (diamond), $b_2 = 0$ and thus

$\varepsilon_s = \varepsilon_\infty$ and $\omega_L = \omega_T$, but generally, $\omega_L > \omega_T$. The latter describes the behavior of the ionic permittivity and shows an infinite discontinuity at $\omega = \omega_T$. From $\varepsilon_1 = n_1^2 - n_2^2$ and $\varepsilon_2 = 2n_1n_2 = 0$ we conclude that we have either $n_1 = 0$ and $n_2 \neq 0$ or $n_1 \neq 0$ and $n_2 = 0$. For $\omega < \omega_T$ and $\omega > \omega_L$, $\varepsilon_1 > 0$ and therefore $n_1 > 0$ and $n_2 = 0$ and the material is transparent. For $\omega_T < \omega < \omega_L$, $\varepsilon_1 < 0$ and $n_1 = 0$ with $n_2 \neq 0$. From the reflection coefficient (at normal incidence) $R = [(n_1 - 1)^2 + n_2^2]/[(n_1 + 1)^2 + n_2^2]$, it follows that $R = 1$ so that the material is 100% reflecting. The left column of Figure 17.3a shows the behavior of ε_1, n_1, n_2, and R. Experimentally, however, a different behavior is observed. Clearly, we "forgot" absorption and to include that we add a damping factor $-\gamma\dot{\omega}$, like in Section 17.2, to the equation of motion Eq. (17.2a). This leads to the same expression as before but with $b_{11} + i\gamma\omega$ instead of b_{11}. The permittivity becomes $\varepsilon_r(\omega) = \varepsilon_1 + i\varepsilon_2$, and if we use $(\varepsilon_s - \varepsilon_\infty)\omega_T^2 = \varepsilon_\infty\omega_P^2$, with ω_P the *ionic plasma frequency* where $\varepsilon_1 = \varepsilon_2$, we obtain

$$\varepsilon_r(\omega) = \varepsilon_\infty\left[1 + \frac{\omega_P^2}{\left(\omega_T^2 - \omega^2\right) - i\gamma\omega}\right] \quad \text{or}$$

$$\varepsilon_1(\omega) = \varepsilon_\infty\left[1 + \frac{\omega_P^2\left(\omega_T^2 - \omega^2\right)}{\left(\omega_T^2 - \omega^2\right) + \gamma^2\omega^2}\right] \quad \text{and} \quad \varepsilon_2(\omega) = \frac{\varepsilon_\infty\omega_P^2\gamma\omega}{\left(\omega_T^2 - \omega^2\right) + \gamma^2\omega^2}$$

Because γ/ω_T typically is ≈0.05, this behavior, displayed in right column of Figure 17.3a, shows that $\varepsilon_1 \cong 0$ for ω_T and ω_L. In the region $\omega_T < \omega < \omega_L$, $\varepsilon_1 < 0$, and this leads to strong reflection, a phenomenon of importance called the *Reststrahlen* effect (Section 14.2.2). The maximum reflectivity can be related to ω_T, but the analysis is involved [29]. The introduction of damping also modifies the LST relation to [18]

$$\frac{\varepsilon_s}{\varepsilon_\infty} = \frac{[\mathrm{Re}(S_L)]^2 + \gamma_L^2/4}{[\mathrm{Re}(S_T)]^2 + \gamma_T^2/4}$$

where

$$S_L = \left[\omega_0^2(\varepsilon_s/\varepsilon_\infty) - \tfrac{1}{4}\gamma_L^2\right]^{1/2} - \tfrac{1}{2}i\gamma_L \quad \text{and} \quad S_T = \left(\omega_0^2 - \tfrac{1}{4}\gamma_T^2\right)^{1/2} - \tfrac{1}{2}i\gamma_T$$

For other than binary cubic crystals, the LST relation can be generalized including the effect of damping [30]. Much more can be said about the optical properties of solids, for which we, however, must refer to the literature. Fox [31] provides a readable introduction, while Wooten [32] and Dressel and Grüner [33] provide an advanced treatment.

17.3.1 Three Models

To relate the behavior as described above to molecular parameters, several models exist of which we discuss three. In this paragraph we omit damping.

The *rigid-ion model* assumes that the ions are point masses with valency Z and neglects polarizability. Therefore, $b_{22} = 2\alpha C = 0$, and this model does not satisfy as it predicts $\varepsilon_\infty = 1 + b_{22} = 1$, due to the neglect of polarizability.

In the *polarizable-ion model*, the nuclei with charge Xe are coupled by a "spring" with force constant S to the *electron* cloud with mass m_e and charge Ye, where $X + Y = Z$. In an electric field $\boldsymbol{E} = \boldsymbol{E}_0\exp(-i\omega t)$, this leads to oscillating dipole moment $\boldsymbol{M} = (Ye)^2\boldsymbol{E}/m_e(\omega_{UV}{}^2 - \omega^2)$, where $\omega_{UV}{}^2 = S/m_e$. The polarizability of the ions is for $\omega \ll \omega_{UV}$ given by $\alpha = (Ye)^2/S$. The total polarization is thus given, as before, by $\boldsymbol{P} = \rho(Ze\boldsymbol{u} + \alpha\boldsymbol{E}_L)$, while the permittivity, also given before, is

$$\varepsilon_1(\omega) = \varepsilon_\infty + (\varepsilon_s - \varepsilon_\infty)\omega_0{}^2/\left(\omega_0{}^2 - \omega^2\right)$$

which reduces, as it must, to the rigid-ion model expression for $\alpha = 0$. From the expressions for b_{22} and $\varepsilon_\infty = 1 + b_{22}$, we obtain

$$(\varepsilon_\infty - 1)/(\varepsilon_\infty + 2) = \rho(\alpha_+ + \alpha_-)/3\varepsilon_0$$

in which we recognize the *Clausius–Mossotti equation*.

As shown by Szigeti [34], the parameters a and Z can be related to ε_s, ε_∞, and ε_T. Eliminating Ze between b_{11} and b_{12}, we obtain the *first Szigeti relation*

$$a = \mu_m\omega_T{}^2(\varepsilon_s + 2)/(\varepsilon_\infty + 2)$$

A simple model relates the force constant a to the bulk modulus K (Justification 17.1) and leads to $a = 3\Omega K/r_0{}^2$ where r_0 is the nearest neighbor distance and Ω the cell volume. Upon combining these two expressions, we have

$$K = \left(\mu_m\omega_T{}^2/6r_0\right)(\varepsilon_s + 2)/(\varepsilon_\infty + 2)$$

Data for a few crystals (Table 17.4) show that this expression is reasonably well obeyed.

If b_{11}, b_{22}, and $(\varepsilon_s - \varepsilon_\infty)$ are used to eliminate a, we obtain the *second Szigeti relation*

$$(Z^*e)^2 = \left(9\mu_m\varepsilon_s\Omega\omega_T{}^2\right)(\varepsilon_s - \varepsilon_\infty)/(\varepsilon_\infty + 2)^2$$

We labeled the solution Z^*e because it appears that Z^*e is *not equal* to the formal Ze (Table 17.4). For the ionic plasma frequency, we obtain $\omega_P{}^2 = (Z^{**}e)^2/\mu_m\Omega\varepsilon_0\varepsilon_\infty$, where $Z^{**}e$ is another effective charge given by $Z^{**}e = Z^*e(\varepsilon_\infty + 2)/3$. Using this notation, the static permittivity ε_s, the LST relation, and the permittivity ε_1 become, respectively,

$$\varepsilon_s = \varepsilon_\infty\left[1 + \left(\omega_P{}^2/\omega_T{}^2\right)\right], \quad \omega_P{}^2 = \omega_L{}^2 - \omega_T{}^2 \quad \text{and} \quad \varepsilon_1(\omega) = \varepsilon_\infty\left[1 + \omega_P{}^2/\left(\omega_T{}^2 - \omega^2\right)\right] \tag{17.8}$$

reducing to the rigid-ion equations by taking $\varepsilon_\infty = 1$. Generally, results from the polarizable ion model are in worse agreement with experiment than those from the rigid-ion model.

Table 17.4 Bulk modulus K (GPa) and effective valency Z^* (–) for alkali halide crystals.

Crystal	K_{exp}	K_{calc}	Z^*	Crystal	K_{exp}	K_{calc}	Z^*	Crystal	K_{exp}	K_{calc}	Z^*
LiF	65.4	65.4	0.87					RbBr	15.1	12.6	0.82
NaF	57.0	47.4	0.93	KCl	18.6	17.7	0.80	RbI	15.9	10.4	0.89
NaCl	23.7	23.5	0.74	KBr	15.7	14.9	0.76	CsCl	19.4	16.8	0.84
NaBr	17.4	19.7	0.69	KI	11.8	11.7	0.69	CsBr	16.3	14.2	0.79
NaI	13.4	14.1	0.71	RbCl	16.8	15.0	0.84	TlCl	40.2	20.4	1.08

Data from Szigeti [34] but converted to bulk modulus, as that seems to me more natural. Present-day experimental values differ but will not change the approximate agreement.
Source: Adapted from Szigeti [34].

According to Szigeti, the difference $Z^* - Z$, due to the distortion of the electron density in the region of overlap from that of spherical ions, is responsible.

Justification 17.1 ***Force Constant and Bulk Modulus*** Generalizing the Born(-Landé or -Mayer) potential for binary salts slightly to $\phi(r) = -A/r + z\phi_{\text{rep}}(r)$, we obtain from the equilibrium condition $\text{d}\phi/\text{d}r = 0$ (Example 10.3), $A/r_0^2 = -z\phi_{\text{rep}}'(r_0)$. Here, A is the attractive or Madelung term, z is the coordination number, while the prime denotes differentiation with respect to the variable. Using the Helmholtz energy F, we have using $\text{d}\Omega/\Omega = 3\text{d}r/r$ for the bulk modus $K = -V(\text{d}P/\text{d}V)_0 = \Omega(\text{d}^2F/\text{d}\Omega^2)_0$. Approximating F by ϕ, we obtain

$$K = \left[\frac{r}{3}\frac{\text{d}}{\text{d}r}\left(\frac{r}{3\Omega}\frac{\text{d}\phi}{\text{d}r}\right)\right]_0 = \frac{r_0^2}{9\Omega}\left(\frac{\text{d}^2\phi}{\text{d}r^2}\right)_0$$

which upon inserting ϕ results in $\left(r_0^2/9\Omega\right)\left[-\left(2A/r_0^3\right) + z\phi_{\text{rep}}''(r_0)\right]$. Eliminating the Madelung term results in $K = \left(zr_0^2/9\Omega\right)\left\{\phi_{\text{rep}}''(r_0) + \left[2\phi_{\text{rep}}'(r_0)/r_0\right]\right\}$. To obtain the force constant, we expand as usual ϕ_{rep} in a Taylor series leading to $a_{ij} = \text{d}^2\phi_{\text{rep}}/\text{d}x_i\text{d}x_j$ or $\boldsymbol{a} = \text{d}^2\phi_{\text{rep}}/\text{d}\boldsymbol{x}\text{d}\boldsymbol{x}'$, so that the displacements are $\boldsymbol{u} = \boldsymbol{a}\cdot\boldsymbol{x}$. However, we need only the isotropic part of $\boldsymbol{a}$, so that $a = \frac{1}{3}\nabla^2\phi = \frac{1}{3}(\text{d}^2\phi_{\text{rep}}/\text{d}r^2 + 2r^{-1}\text{d}\phi_{\text{rep}}/\text{d}r)_0$. Hence $a = 3\Omega K/zr_0^2$.

To account for the electronic polarizability, Dick and Overhauser [35] introduced the *shell model*. In its basic form each ion is again described by a core with charge X, coupled to a spherical electronic shell with charge Y, so that $Z = X + Y$, but the shells can move with respect to the cores. The force constants between the cores and shells are $S_\pm$, while the core displacements are $u_\pm$ and the shell displacements $v_\pm$. The resulting equations of motion of the shells and cores are

$$0 = -(S_+ + a)\boldsymbol{v}_+ + S_+\boldsymbol{u}_+ + a\boldsymbol{v}_- + Y_+e\boldsymbol{E}_\text{L}, \qquad 0 = -(S_- + a)\boldsymbol{v}_- + S_-\boldsymbol{u}_- + a\boldsymbol{v}_+ + Y_-e\boldsymbol{E}_\text{L}$$

$$m_+\ddot{\boldsymbol{u}}_+ = -S_+\boldsymbol{u}_+ + S_+\boldsymbol{v}_+ + X_+e\boldsymbol{E}_\text{L}, \qquad m_-\ddot{\boldsymbol{u}}_- = -S_-\boldsymbol{u}_- + S_-\boldsymbol{v}_- + X_-e\boldsymbol{E}_\text{L}$$

where the mass of the shells has been taken as zero, corresponding to the adiabatic approximation. Elimination of $\boldsymbol{v}_+$ and $\boldsymbol{v}_-$ leads after some algebra to

$$\mu\ddot{\boldsymbol{u}} = -a^*\boldsymbol{u} + Z^*e\boldsymbol{E}_\text{L}$$

with

$$a^* = \frac{a\,S_+S_-}{S_+S_- + a(S_+ - S_-)} \quad \text{and} \quad Z^* = Z + \frac{Y_-S_-^{-1} - Y_+S_+^{-1}}{a^{-1} + S_+^{-1} + S_-^{-1}} \tag{17.9}$$

The polarization becomes

$$\boldsymbol{P} = \rho e(X_+\boldsymbol{u}_+ + X_-\boldsymbol{u}_- + Y_+\boldsymbol{v}_+ + Y_-\boldsymbol{v}_-)$$

leading after eliminating again $\boldsymbol{v}_+$ and $\boldsymbol{v}_-$ to

$$\boldsymbol{P} = \rho Z^*e\boldsymbol{u} + \rho\alpha^*\boldsymbol{E}_\text{L} \quad \text{with} \quad e^{-2}\alpha^* = \frac{Y_+^2}{S_+} + \frac{Y_-^2}{S_-} - \frac{\left(Y_-S_-^{-1} - Y_+S_+^{-1}\right)^2}{a^{-1} + S_+^{-1} + S_-^{-1}} \tag{17.10}$$

From Eq. (17.10b) it follows that the polarizability deviates by the third term from the polarizability given by the sum of the ion polarizabilities $\alpha_+ + \alpha_- = e^2Y_+^2/S_+ + e^2Y_-^2/S_-$ in the polarizable ion model. The effective valency Z^* is usually smaller than the formal Z. This can be seen from the second term in Eq. (17.9) by writing its numerator $Y_-/S_- - Y_+/S_+$ as $(\alpha_-/Y_-e^2)-(\alpha_+/Y_+e^2)$, which is negative because normally $\alpha_+ \gg \alpha_-$. Furthermore, $Z^* = 0$ only when $Z = 0$ and $Y_+/S_+ = Y_-/S_-$, that is, for identical atoms, and thus $Z^* \neq 0$ is not characteristic for the bonding mechanism. Finally, if $S_\pm \to \infty$, $\alpha^* = 0$, $Z^* = Z$, and $a^* = a$, and the model reduces to the rigid-ion model.

Note that for all three models the basic form of the equations of motion and the polarization is the same and only the further assumptions differ. After the application of the shell model to dielectric properties [35], the shell model has been also been applied to calculate phonon dispersion curves, first it seems by [36], with much better results than the polarizable ion model. Further improvement can be obtained by the "breathing shell" model [37] in which the shell radii vary during their motion.

17.4 Orientational Polarization

Lowering again the frequency, we come to the orientational contribution. In many cases molecules have a dipole moment, and in gases and liquids they can orient more or less freely in an electric field, but for solids this is generally difficult. In some cases, though, some orientational freedom is available. A well-known example is HCl, which shows a sudden increase from $\varepsilon_1 \cong 4$ to $\varepsilon_1 \cong 16$ at $T_0 \cong 99$ K and thereafter decreases with temperature [38]. At the melting point, $T_{\rm mel} \cong 159$ K, there is only a small discontinuity but $\varepsilon_{\rm r}$ decreases further at more or less the same rate with increasing temperature. This behavior is attributed to rotational freedom in the solid state. Other examples showing the same phenomenon are disubstituted ethanes, such as $C_2H_4(C{\equiv}N)_2$ ($T_0 \cong -45$ °C and $T_{\rm mel} \cong 54.2$ °C, [39]), and small molecules like H_2S ($T_0 \cong 103.5$ K and $T_{\rm mel} \cong 187.7$ K, [40]).

To describe the rotational behavior in the solid state, we use the general theory (Justification 9.4) which, although derived for liquids, does not contain any reference to structure and it thus also applicable to solids. We use Eq. (9.62), rewritten as

$$(\varepsilon_1 - \varepsilon_\infty)(2\varepsilon_1 + \varepsilon_\infty)/3\varepsilon_1 = (\rho/3\varepsilon_0 kT)gm^2 \equiv gB(T)/T \tag{17.11}$$

Here, we take the correlation factor g = 1, while $\rho = N/V$ is the density of the dipoles, and $m^2 = \langle m^2 \rangle$ is the average dipole moment squared. Here, we need to be precise about whether $\varepsilon_\infty = n^2$ or $\varepsilon_\infty + \Delta\varepsilon = cn^2$, with $c \cong 1.05$ a constant, is used conform our discussion in Section 8.6.2. Generally, $\Delta\varepsilon \ll 1$ and $\varepsilon_\infty \cong n^2$, which on its turn, except for ionic solids, is about equal to $\varepsilon_1(T = 0)$, which we denote here by $\varepsilon_{\rm s}$. In principle, if m is no function of temperature, $B(T)$ is independent of temperature. If we approximate ε_∞ by $\varepsilon_{\rm s}$, we have

$$B(T) = (\rho/3k\varepsilon_0)m^2 = (\varepsilon_1 - \varepsilon_{\rm s})(2\varepsilon_1 + \varepsilon_{\rm s})T/3\varepsilon_1$$

deviating by a term $\Delta\varepsilon T(\varepsilon_1 + 2\varepsilon_{\rm s} - \Delta\varepsilon)/3\varepsilon_1$ from Eq. (17.11). Hence, from the experimental data of ε_1 as a function of T, $B(T)$ can be calculated. If $B(T)$ were really independent of T, an increasing value of $B(T)$ with decreasing T would indicate an increasing parallel orientation of the dipoles. Similarly, a decreasing value of $B(T)$ with decreasing T indicates an increasing antiparallel orientation of the dipoles. Since B is generally not independent of T, a decrease indicates now increasing antiparallel orientation or a decrease of dipole moment with T, and vice versa if $B(T)$ increases with T. If the order–disorder temperature is below T_{mel}, this model explains the transition, otherwise free orientation occurs at T_{mel}. As examples, we quote the experiments on dichloropropane $(CH_3)_2CCl_2$ showing an increasing value of B with T [41], on tertiary butyl chloride $(CH_3)_3CCl$ with a decreasing value of B with T [41], and on pentamethyl chloro-benzene $C_6(CH_3)_5Cl$ with $B(T) \cong const.$ [42].

Another way to describe this phenomenon is to use mean-field theory. If we write for the local field $\boldsymbol{E}_{loc} = \boldsymbol{E} + c\boldsymbol{P}$, where $c = c_L = 1/3\varepsilon_0$ if the local field is given by the Lorentz field, and recall that $\boldsymbol{P} = \rho\alpha_{eff}\boldsymbol{E}_{loc}$, the polarization $\boldsymbol{P}$ is given by $\rho\alpha_{eff}\boldsymbol{E}/(1 - c\rho\alpha_{eff})$ and thus the susceptibility by $\chi = \rho\alpha_{eff}/(1 - c\rho\alpha_{eff})$. Further, recall that $\alpha_{eff} = \alpha + \mu^2/3kT$. Hence,

$$\chi^{-1} = \frac{(1 - c\rho\alpha_{eff})}{\rho\alpha_{eff}} = \frac{T(1 - c\rho\alpha_{eff}) - T_0}{\rho\alpha_{eff}T + C} \quad \text{with} \quad T_0 = \frac{c\rho\mu^2}{3k} \text{ and } C = \frac{\rho\mu^2}{3k} \tag{17.12}$$

where C is the Curie constant and T_0 is the asymptotic temperature. For $\theta = T_0/(1 - c\rho\alpha_{eff})$, we have $1/\chi = 0$, so that θ could be considered as the transition temperature. Using reasonable numbers, $c\rho\alpha_{eff}$ ranges from 0.05 to 0.5, so that the difference between θ and T_0 can be large. Moreover, for $T \to \infty$, χ approaches the asymptotic value $\rho\alpha_{eff}/(1 - c\rho\alpha_{eff})$. If $\alpha \ll \mu^2/3kT$, Eq. (17.12) reduces to $\chi = C/(T - T_0)$ with T_0 the transition temperature, analogous to the well-known Curie–Weiss result for the magnetic case. For dielectrics α generally contributes significantly to α_{eff} and therefore the full expression must be used. Below T_0, χ is described as usual way in mean-field theory.

17.5 Space-Charge Polarization

Further lowering the frequency, we arrive in the space–charge region. Generally, at interfaces charges are less mobile and may accumulate there. These space charges also polarize the material, a phenomenon labeled as *space–charge* (or *interfacial*) *polarization*. This accumulation applies particularly to grain boundaries in polycrystalline materials, but also to the interfaces at the electrodes. The simplest model to analyze such behavior is probably the Maxwell–Wagner double layer model. An equivalent diagram for this model consists out of a series connection of two units containing a capacitor with capacitance C parallel to a resistor with resistance R. We denote the units by the subscript 1 (electrode interface) and 2 (grain boundaries). In the stationary state both units carry an equal current described by

$$C_1\frac{d\Psi_1}{dt} + \frac{\Psi_1}{R_1} = C_2\frac{d\Psi_2}{dt} + \frac{\Psi_2}{R_2}$$

while the total voltage $\Psi = \Psi_1 + \Psi_2$. Combining leads to

$$C\frac{d\Psi_1}{dt} + \frac{\Psi_1}{R} = \frac{\Psi}{R_2} \quad \text{and} \quad C\frac{d\Psi_2}{dt} + \frac{\Psi_2}{R} = \frac{\Psi}{R_1}$$

where $C = C_1 + C_2$ and $1/R = 1/R_1 + 1/R_2$. The boundary conditions at $t = 0$ are $\Psi_1 = C_2\Psi/C$ and $\Psi_2 = C_1\Psi/C$, so that the above expressions integrate to

$$\Psi_1 = \Psi\frac{R_1}{R_1 + R_2}\left[1 - \frac{\tau - \tau_2}{\tau}e^{-t/\tau}\right] \quad \text{and} \quad \Psi_2 = \Psi\frac{R_2}{R_1 + R_2}\left[1 - \frac{\tau - \tau_1}{\tau}e^{-t/\tau}\right]$$

with unit time constants $\tau_j = R_jC_j$ and overall time constant $\tau = RC = (R_1\tau_2 + R_2\tau_1)/(R_1 + R_2)$. For $t \to \infty$, the voltages become $\Psi_j = \Psi R_j/(R_1 + R_2)$. Although in principle two units contribute to τ, it depends on the difference between R_j and C_j and their temperature dependencies whether τ_1 and τ_2 can be experimentally distinguished. Taking reasonable values for all parameters, τ appears to be of the order of seconds.

Problem 17.1 ***Three Models*** Verify the details of the three models (Section 17.3.1) used to describe $\varepsilon_r(\omega)$.

Problem 17.2 ***Dielectric Relaxation*** Review the applicability of the relaxation expressions given in Section 8.6.3 for solids.

17.6 Ferroelectric Materials

In Section 10.1, we indicated that crystals can be divided into seven crystal systems. The systems further divide into 32 point groups (crystal classes) according to their symmetry with respect to a point. Of these, 11 are centrosymmetric. If a uniform mechanical stress is applied, the resulting small displacement of charge is symmetrically distributed about the center of symmetry. Applying an electric field provides a strain which upon reversal of the field remains the same. The resulting effect, called *electrostriction*, increases quadratically with the field and occurs in all materials.

The materials in the other 21 non-centrosymmetric crystal classes (except point group 432) produce a linear strain upon applying an electric field and, conversely, produce a voltage upon applying a strain and are called *piezoelectric*. Of the 21 classes, 10 have a unique polar axis and possess *spontaneous polarization*. This effect might be, however, difficult to detect directly as these charges may be compensated by external charges. Moreover, in polycrystalline materials the overall polarization is zero due to the random orientation of the various grains. As this polarization is temperature dependent, it can be detected by measuring the current upon suddenly changing the temperature. Accordingly, these materials are called *pyroelectric*.

A subgroup of pyroelectric compounds becomes *ferroelectric* when it has at least two stable and polarized equivalent states in the absence of an electric field and when such a state can be shifted to another by a suitable field. This effect occurs below a certain temperature, the *transformation temperature* T_0. The crystal is called *uniaxial* if it has only two equivalent states and *multiaxial* otherwise. Above T_0 the permittivity ε_1 often can be described to a good approximation by a Curie–Weiss form of expression $\varepsilon_1 = C/(T - T_0)$ with C the *Curie constant*. For completeness, we mention *antiferroelectrics* containing in the polar phase more than one type of polarizable unit of which the dipoles may align antiparallel [43].

To minimize depolarization effects[3] at the surface in the polarized state, different regions in the material, called *domains*, polarize in different directions. Domains are separated by

---

3 So far, depolarization was not mentioned. In the absence of electrodes, charges accumulate at the surfaces of a polarized material. This leads to a depolarizing electric field $\varepsilon_0\boldsymbol{E}_{\text{dep}} = -N_{\text{dep}}\boldsymbol{P}$ that acts against the local field $\boldsymbol{E}_{\text{loc}}$ with N_{dep} the depolarization factor depending only on the geometry of the sample. Only for certain geometries and orientations N_{dep} is zero. For example, for a thin plate with the applied field $\boldsymbol{E}$ parallel to the plate $N_{\text{dep}} = 0$ ($\boldsymbol{E}_{\text{loc}} = \boldsymbol{E} + \boldsymbol{P}/3\varepsilon_0$) or for a long thin cylinder with $\boldsymbol{E}$ parallel to the axis $N_{\text{dep}} = 0$ ($\boldsymbol{E}_{\text{loc}} = \boldsymbol{E} + \boldsymbol{P}/3\varepsilon_0$). For a thin plate with $\boldsymbol{E}$ perpendicular to the plate $N_{\text{dep}} = 1$ ($\boldsymbol{E}_{\text{loc}} = \boldsymbol{E} - 2\boldsymbol{P}/\varepsilon_0$), or for a thin long cylinder with $\boldsymbol{E}$ perpendicular to the axis $N_{\text{dep}} = ½$ ($\boldsymbol{E}_{\text{loc}} = \boldsymbol{E} + \boldsymbol{P}/6\varepsilon_0$), while for a sphere $N_{\text{dep}} = ⅓$ ($\boldsymbol{E}_{\text{loc}} = \boldsymbol{E}$). At domain boundaries depolarizing effects occur.

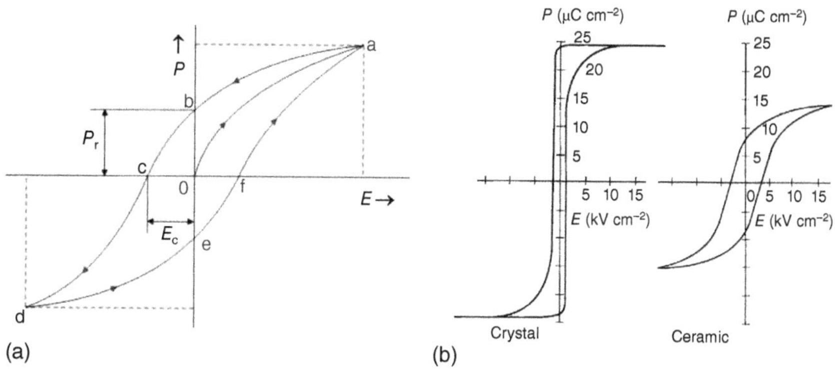

Figure 17.4 (a) Polarization P of a ferroelectric upon reversal of the electric field E. Indicated are the remanent polarization P_r and coercive field E_c. For a linear dielectric the response is a straight line through the origin given by $P = \varepsilon_0 \chi E$. (b) Hysteresis loop for a (100) crystal plate and a high-quality ceramic of $BaTiO_3$. Source: Jaffe et al. [44]/with permission of Elsevier.

domain walls who do not interrupt the lattice structure, but only change the polarization. They have typically a wall energy of tens of mJ m^{-2} and a thickness less than 5 nm. While a uniaxial ferroelectric shows only 180° domains, multiaxial ferroelectrics also typically show 90° domains. The presence of equivalent states implies that when an electric field is applied, the polarization changes nonlinearly and a *hysteresis loop* occurs upon reversal of the field, contrary to a regular dielectric where the polarization changes linearly with the field (Figure 17.4a). The domain structure is crucial for this behavior and domains can be (partially) aligned by an appropriate field, a process called *poling*. While for a single crystal often a "square" loop results, ceramics mostly show a more "slanted" shape (Figure 17.4b).

For spontaneous polarization to occur, a phase transformation from a nonpolar state is required. As a reference often a so-called *prototype phase* is used, generally nonpolar but not necessarily so, from which the ferroelectric state can be reached by small structural changes upon lowering the temperature. Ferroelectrics are often characterized by having either a *displacive transformation* or an *order–disorder transformation*. For the former, if an ion is slightly displaced from equilibrium, the force from the local electric field increases faster than the restoring force below the transformation temperature. This leads to an asymmetrical shift in the equilibrium ion positions (change of point group) and hence to a permanent dipole moment. For the latter, there is a dipole moment in each unit cell, but at high temperature they are pointing in random directions. Upon decreasing the temperature and passing the transformation temperature, the dipoles order and point in the same direction within a domain. These descriptions apply in the extremes of a well-defined single well and a deep double well with negligible tunneling, but often the ferroelectric transformation demonstrates elements of both.

Considering now the materials involved, several types can be distinguished for which Kao [2] provided a compact description. Broadly speaking, two main categories can be discerned: the double oxides and hydrogen-bonded (HB) crystals, by and large coinciding with a mainly displacive and order–disorder nature. For the latter, triglycine sulphate (TGS) and potassium dihydrogen phosphate (KDP) are characteristic. The structure of these compounds is complex (TGS in Figure 17.5a) but they show the usual $\varepsilon_1(T)$ behavior (Figure 17.5b). For KDP the transformation is discontinuous but near continuous [47], and

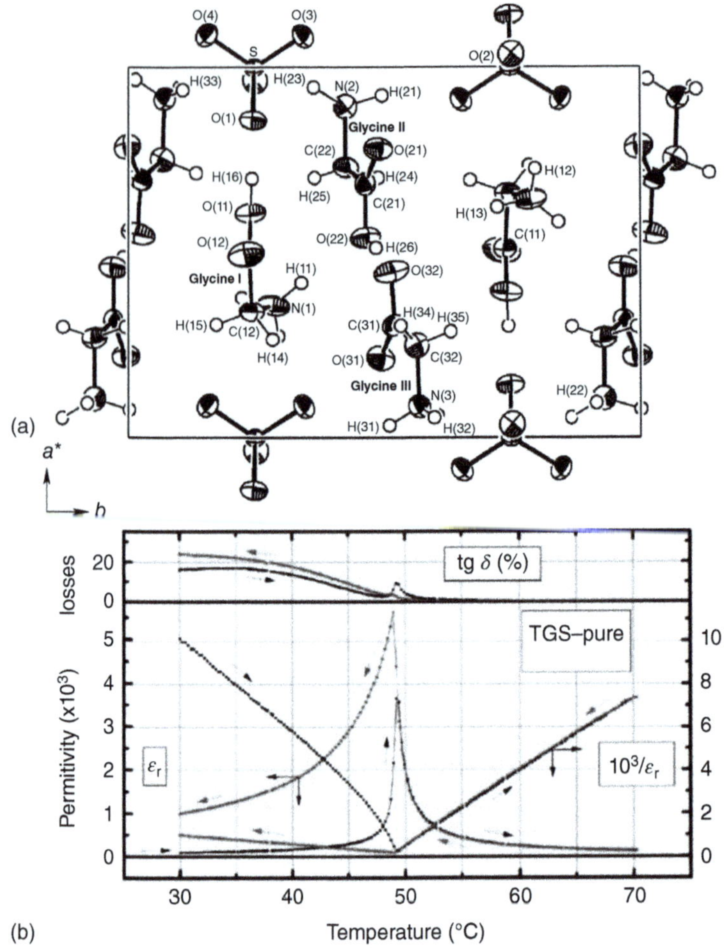

Figure 17.5 (a) [001] projection of the crystal structure of the unit cell of TGS. Source: Kikuta et al. [45]/with permission of Elsevier. (b) The permittivity ε_r and loss tangent δ of TGS above and below the transformation temperature. Source: Berbecaru et al. [46]/with permission of Elsevier.

for TGS it is continuous [48]. The value of the Curie constant is typically a few hundred degrees.

For the double oxides, perovskites with a relatively simple crystal structure form the most important class, changing from cubic to tetragonal. $BaTiO_3$ and $PbTiO_3$ are typical examples (Figure 17.6a). $BaTiO_3$ was considered for a long time as a model ferroelectric but appeared to be more complicated than originally thought [43]. First, this compound shows several transformations. Accordingly, the lattice constant changes several times abruptly with temperature (Figure 17.6c), as does its permittivity (Figure 17.6d). Second, its dynamical behavior, grossly anharmonic, is also far from typical. In this sense $PbTiO_3$ acts more as a model displacive ferroelectric. Both perovskites show a discontinuous transformation. The transformation temperature T_0 can be largely tuned by adding other ions. For $BaTiO_3$, for example, both $SrTiO_3$ and $PbTiO_3$ are fully miscible over the complete range. $SrTiO_3$ itself is not ferroelectric but has a virtual (negative) T_0 near 0 K and with increasing Sr content the

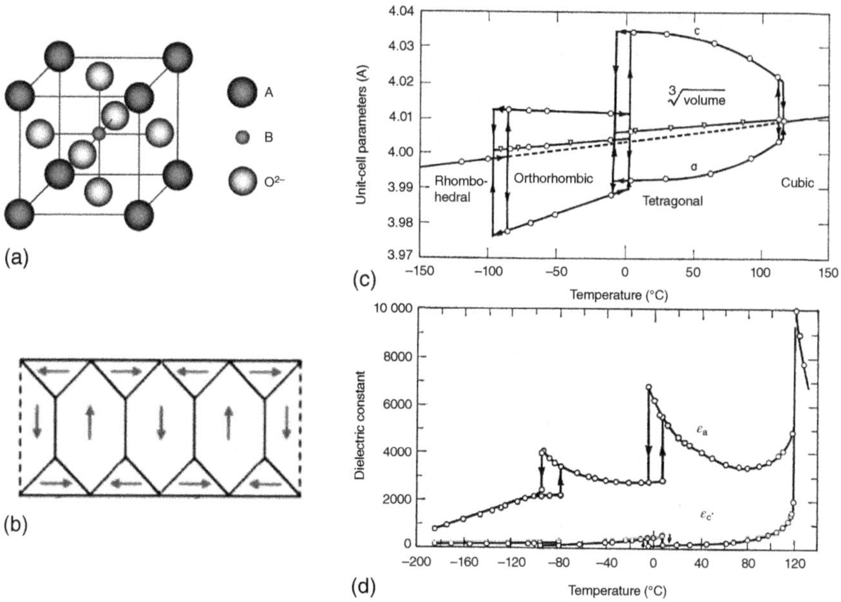

Figure 17.6 (a) The unit cell of perovskite ABO_3. (b) Schematic of 180° and wedge domains (angle exaggerated). (c) The lattice constants $a(T)$ and $c(T)$ of single crystal $BaTiO_3$. Source: Jona and Shirane [49]/with permission of Elsevier. (d) The permittivity $\varepsilon_r(T)$ of single crystal $BaTiO_3$. Source: Jona and Shirane [49].

T_0 of $BaTiO_3$ decreases linearly over the whole composition range. Adding $PbTiO_3$ increases T_0 of $BaTiO_3$ linearly over the whole composition range up to $T_0 \cong 490\,°C$ for $PbTiO_3$ itself. The Curie constant of perovskites is typically high, say 10^4 K or more.

For multiaxial compounds, like $BaTiO_3$, minimizing the electrostatic energy so that $\nabla \cdot \boldsymbol{P} \cong 0$ at the domain boundary can be satisfied by 90° and 180° walls in the interior of a crystal. However, this is not true for the also frequently appearing wedge domains (Figure 17.6b), but their wedge angles are usually so small that they do not contribute a significant depolarization energy. They are instrumental though in easy changing the polarization upon reversal of $\boldsymbol{E}$. Zhirnov [50] estimated for single 180° walls a thickness of about 5–10 Å and an energy of about 10 mJ m^{-2}, while for single 90° walls he obtained 50–100 Å and 2–4 mJ m^{-2}, respectively. Generally, both the domains themselves and the domain walls contribute to ε_1. For polycrystalline $BaTiO_3$ with grain size d larger than about 10 μm, for example, $\varepsilon_1 \cong 1700$ is constant with a volume contribution of about 1000 and a domain wall contribution of about 700. The domain wall contribution is normally determined by the length of the 90° walls, but in this range there are also domain walls that are mixed with a constant distance of about 10 μm and that act as an artificial grain boundary. Because grain boundaries form an almost insurmountable barrier for domain walls, this limits the length of the 90° walls. Below $d \cong 10$ μm, ε_1 increases with d, showing a maximum at ≅1 μm, and the 90° domain width w decreases as $w \sim d^{1/2}$ due to the shift in balance between depolarization and domain wall energy. The smaller the grains, the more ε_1 is determined by the 90° walls and ε_1 shows a maximum of about 5000 for $d \cong 0.9$ μm. With decreasing d still further, the tetragonal distortion decreases, and the structure becomes pseudo-cubic at $d \cong 0.3$ μm [51]. Another effect of domains is that the

Table 17.5 Characteristics of some typical ferroelectrics.

Material	T_0 (°C)	C (K)[a)]	Order	P_{sp} (μC cm^{-2})[b)]	$\Delta_{tra}H$ (cal mol^{-1})
$BaTiO_3$	120	1.3×10^4	1	20 (T_0), 26 (RT)	47
$PbTiO_3$	490	8.9×10^3	1	57–75 [53]	1150
KH_2PO_4 (KDP)	−150	259	1→2	4.8 (T_0−27)	87
$(NH_2CH_2COOH)_3H_2SO_4$ (TGS)	49	255	2	2.8 (RT)	150

a) $\varepsilon_1 = C/(T - T_0)$.
b) Spontaneous polarization at the temperature indicated.
Source: Burfoot [54]/with permission of Cengage Learning.

overall domain structure produces near-complete compensation for the polarization. The pyroelectric effect only turns up when the domains are sufficiently aligned by poling. $Pb(Zr_{0.415}Ti_{0.585})O_3$) is an example of a ferroelectric compound that after poling can have a high pyroelectric coefficient (at $T_0 \cong 170$ °C, 4.8×10^{-4} μC m^{-2} K^{-1}, [52]).

Table 17.5 summarizes some properties of the ferroelectrics indicated. The transformation temperature T_0 for polycrystalline ferroelectrics is essentially the same as for single crystals but changes with pressure. For $BaTiO_3$, for example, T_0 increases with 2.6×10^{-3} K bar^{-1} and decreases with 5.7×10^{-3} K bar^{-1} for TGS. The maximum polarization P_{max} that can be reached is, however, lower than that of single crystals. For example, for $BaTiO_3$ the maximum is 84% [55], but in practice due to the presence of pores and stresses typically 50% or less. Still, P_{max} obtained is large and the spontaneous polarization $P_{sp} = 0.26$ μC m^{-2} of single-crystalline $BaTiO_3$, for example, implies a charge $+e$ and $-e$ separated by ≅1 Å for every cell.

In summary, ferroelectrics are a subgroup of pyroelectrics, just as pyroelectrics are a subgroup of piezoelectrics, while the latter are a subgroup of dielectrics. While originally ferroelectric behavior was considered as exceptional, nowadays many ferroelectric compounds are known. Jona and Shirane [49] provide many experimental details on single crystals. Other references are the books by Fatuzzo and Merz [56] and Burfoot [54]. The classic is the book by Lines and Glass [43] to which Rabe et al. [57] is complementary. Grindlay [58] provided a detailed exposition of the macroscopic approach, as did Salje [59]. Domain structure is discussed in [60, 61].

17.7 Ferroelectric Behavior

There are, as usual, several approaches to deal with a phenomenon and ferroelectrics are no exception. The macroscopic approach uses Landau or Landau-like theory (Section 12.5), while the microscopic approach is often based on lattice dynamics and the concept of soft modes.

17.7.1 The Thermodynamic Approach

In Chapter 6, we introduced internal variables and made use of them in the several chapters. Here we employ them to deal with the ferroelectric phase transformation. Essentially, the displacement $\boldsymbol{D}$ should be used as an internal variable, but because for ferroelectrics $\varepsilon_1 \gg 1$, the polarization $\boldsymbol{P}$ can be used as well.

Before discussing the ferroelectric case, however, let us first review the coupled dielectric–elastic case in which thermo–electro–mechanical coupling occurs. The conjugated variables involved are electric field and displacement ($\boldsymbol{E}$,$\boldsymbol{D}$), mechanical stress and strain ($\boldsymbol{\sigma}$,$\boldsymbol{\varepsilon}$), and temperature and entropy (T,S). To avoid confusion, we will denote the couple ($\boldsymbol{\sigma}$,$\boldsymbol{\varepsilon}$) in this chapter by ($\mathbb{X}$,$\boldsymbol{x}$). We recall that the mechanical work, using pseudo-vector notation (Section 10.11.4), is $\mathrm{d}W_\mathrm{M} = \mathbb{X}\cdot\boldsymbol{x}$ and the electrical work $\mathrm{d}W_\mathrm{E} = \boldsymbol{E}\cdot\mathrm{d}\boldsymbol{D}$, so that the change in internal energy is $\mathrm{d}U = T\mathrm{d}S + \mathbb{X}\cdot\mathrm{d}\boldsymbol{x} + \boldsymbol{E}\cdot\mathrm{d}\boldsymbol{D}$. Of the various Legendre transforms that can be applied, the ones using the Helmholtz energy $F = U - TS$ and elastic Gibbs energy $G_\mathrm{ela} = U - TS - \mathbb{X}\cdot\boldsymbol{x}$ are the most relevant here. From the latter it follows that

$$\mathrm{d}G_\mathrm{ela} = -S\mathrm{d}T - \boldsymbol{x}\cdot\mathrm{d}\mathbb{X} + \boldsymbol{E}\cdot\mathrm{d}\boldsymbol{D} \tag{17.13}$$

with $S = -\partial G_\mathrm{ela}/\partial T$, $\boldsymbol{x} = -\partial G_\mathrm{ela}/\partial\mathbb{X}$ and $\boldsymbol{E} = \partial G_\mathrm{ela}/\partial\boldsymbol{D}$. The Maxwell relations

$$\frac{\partial S}{\partial\mathbb{X}} = \frac{\partial\boldsymbol{x}}{\partial T},\quad \frac{\partial S}{\partial\boldsymbol{D}} = -\frac{\partial\boldsymbol{E}}{\partial T}\quad\text{and}\quad \frac{\partial\boldsymbol{E}}{\partial\mathbb{X}} = -\frac{\partial\boldsymbol{x}}{\partial\boldsymbol{D}}$$

also result. Taking the linear differentials of S, $\boldsymbol{x}$, and $\boldsymbol{E}$ results in

$$\begin{aligned}
\mathrm{d}S &= \frac{\partial S}{\partial T}\mathrm{d}T + \frac{\partial S}{\partial\mathbb{X}}\cdot\mathrm{d}\mathbb{X} + \frac{\partial S}{\partial\boldsymbol{D}}\cdot\mathrm{d}\boldsymbol{D},\\
\mathrm{d}\boldsymbol{x} &= \frac{\partial\boldsymbol{x}}{\partial T}\mathrm{d}T + \frac{\partial\boldsymbol{x}}{\partial\mathbb{X}}\cdot\mathrm{d}\mathbb{X} + \frac{\partial\boldsymbol{x}}{\partial\boldsymbol{D}}\cdot\mathrm{d}\boldsymbol{D}\quad\text{and}\\
\mathrm{d}\boldsymbol{E} &= \frac{\partial\boldsymbol{E}}{\partial T}\mathrm{d}T + \frac{\partial\boldsymbol{E}}{\partial\mathbb{X}}\cdot\mathrm{d}\mathbb{X} + \frac{\partial\boldsymbol{E}}{\partial\boldsymbol{D}}\cdot\mathrm{d}\boldsymbol{D}
\end{aligned}$$

and leads to linear relations for all dependent variables. Similarly, for F we obtain

$$\mathrm{d}F = -S\mathrm{d}T + \mathbb{X}\cdot\mathrm{d}\boldsymbol{x} + \boldsymbol{E}\cdot\mathrm{d}\boldsymbol{D}$$

and thus $S = -\partial F/\partial T$, $\mathbb{X} = \partial F/\partial\boldsymbol{x}$ and $\boldsymbol{E} = \partial F/\partial\boldsymbol{D}$. Moreover,

$$\frac{\partial S}{\partial\boldsymbol{x}} = -\frac{\partial\mathbb{X}}{\partial T},\quad \frac{\partial S}{\partial\boldsymbol{D}} = -\frac{\partial\boldsymbol{E}}{\partial T}\quad\text{and}\quad \frac{\partial\boldsymbol{E}}{\partial\boldsymbol{x}} = \frac{\partial\mathbb{X}}{\partial\boldsymbol{D}}$$

If we consider the linear differentials of S, $\mathbb{X}$, and $\boldsymbol{E}$, we have

$$\begin{aligned}
\mathrm{d}S &= \frac{\partial S}{\partial T}\mathrm{d}T + \frac{\partial S}{\partial\boldsymbol{x}}\cdot\mathrm{d}\boldsymbol{x} + \frac{\partial S}{\partial\boldsymbol{D}}\cdot\mathrm{d}\boldsymbol{D},\\
\mathrm{d}\mathbb{X} &= \frac{\partial\mathbb{X}}{\partial T}\mathrm{d}T + \frac{\partial\mathbb{X}}{\partial\boldsymbol{x}}\cdot\mathrm{d}\boldsymbol{x} + \frac{\partial\mathbb{X}}{\partial\boldsymbol{D}}\cdot\mathrm{d}\boldsymbol{D}\quad\text{and}\\
\mathrm{d}\boldsymbol{E} &= \frac{\partial\boldsymbol{E}}{\partial T}\mathrm{d}T + \frac{\partial\boldsymbol{E}}{\partial\boldsymbol{x}}\cdot\mathrm{d}\boldsymbol{x} + \frac{\partial\boldsymbol{E}}{\partial\boldsymbol{D}}\cdot\mathrm{d}\boldsymbol{D}
\end{aligned} \tag{17.14}$$

so that also in this formulation S, $\mathbb{X}$, and $\boldsymbol{E}$ are all linearly dependent on T, $\boldsymbol{x}$, and $\boldsymbol{D}$. All derivatives being either vectors or tensors, many coefficients must be determined, and their number depends on the crystal system of the compound involved. Symmetry will reduce that number, as has been discussed in detail by Nye [62]. In the isotropic isothermal case, the remaining relations for ($\boldsymbol{x}$,$\boldsymbol{E}$) and ($\mathbb{X}$,$\boldsymbol{E}$) contain just three parameters as a consequence of symmetry and the Maxwell relations.

For nonlinear behavior the usual point of departure is in principle to extend the expansion (17.13) from a prototype phase with $\boldsymbol{D} = \boldsymbol{x} = 0$ to a higher order. However, this leads rapidly to a complex formulation if all the (phenomenological) coefficients are considered. Hence, further assumptions are made, and a common assumption is that the strains x_j are small and that only terms in x_jx_k and $x_j\Pi_kD_l$ need to be retained. Limiting ourselves to uniaxial crystals with $\boldsymbol{D}$ oriented along the uniaxial axis (and isotropic materials), we have in the ferroelectric phase $\boldsymbol{E} = \boldsymbol{x} = 0$. In the isothermal case, we can write $G_{\text{ela}} = \frac{1}{2}\alpha D^2 + \frac{1}{4}\gamma D^4 + \frac{1}{6}\delta D^6$, where the coefficients α, γ, and δ are temperature dependent. For a first-order theory, only the temperature dependence of α needs to be considered (Section 12.5). If we than assume that $\alpha = \beta(T - T^{\otimes})$ and that γ and δ are constants, we can apply the Landau-like theory (Section 12.5), which, by the way, in this connection is normally called *Devonshire theory* [63]. In this theory, the transformation occurs at $T = T^{\otimes}$, where $T^{\otimes} = T_0$ for discontinuous transformations and $T^{\otimes} = T_{\text{C}}$ for continuous transformations. Now denoting the reciprocal permittivity by κ, we have for a continuous transformation

$$\kappa_- = \beta(T_{\text{C}} - T) \quad (T < T_{\text{C}}) \qquad \text{and} \qquad \kappa_+ = 2\beta(T - T_{\text{C}}) \quad (T > T_{\text{C}})$$

with the usual relation $\kappa_-/\kappa_+ = -2$. For a discontinuous transformation, similarly

$$\kappa_- = (3\gamma^2/4\delta) + 8\beta(T_{\text{C}} - T) \quad (T < T_{\text{C}}) \qquad \text{and}$$
$$\kappa_+ = \beta(T - T_0) = (3\gamma^2/16\delta) + \beta(T - T_{\text{C}}) \quad (T > T_{\text{C}})$$

with $T_{\text{C}} - T_0 = (3\gamma^2/16\beta\delta)$ resulting in $[\partial\kappa_-/\partial T]/[\partial\kappa_+/\partial T] = -8$ in the regions near T_{C}. The constant term leads to thermal hysteresis for the change from below T_{C} to above T_{C} and vice versa. The associated *metastability range* $T_1 - T_0 = \gamma^2/4\beta\delta$, where T_1 is the highest temperature for which the zero-field ferroelectric phase can exist and T_0 the lowest temperature for which the nonpolar phase can exist. For $BaTiO_3$, for example, $T_1 - T_0 \cong 13$ K. Often T_{C} is just a few degrees below T_0 (Section 12.5) and, somewhat unfortunately, in that case the transformation is labeled as *near continuous*. In practice, it is often cumbersome to distinguish between both types. A characteristic for discontinuous transformation is the presence of a heat of transformation $\Delta_{\text{tra}}H$, but this quantity may be rather small rendering experimental detection difficult.

These descriptions clearly provide a macroscopic picture with a wide applicability range. Two examples will suffice. For single-crystal $BaTiO_3$, a full Devonshire description up to sixth order using temperature-dependent coefficients [64] can describe almost all electrical phenomena, including double and triple hysteresis loops, accurately. The authors also showed that the introduction of eighth-order coefficients, as sometimes is done although these coefficients have an unclear physical interpretation, is unnecessary. For polycrystalline $PbTiO_3$ [65], the excess specific heat, excess enthalpy, excess entropy, and spontaneous tetragonal deformation near the $P_{\text{m3m}} \leftrightarrow P_{\text{4mm}}$ transition were examined, leading to a Devonshire polynomial with temperature-independent higher-order dielectric stiffness coefficients, which provides a satisfactory and self-consistent description of the single-domain ferroelectric behavior of this compound. However, these descriptions provide no clue as to the microscopic picture to which we turn now.

17.7.2 The Microscopic Approach

From the microscopic considerations that follow, it will become clear that anharmonic effects and thereby nonlinear effects are involved. To make clear why this is so, we start with the local field $\boldsymbol{E}_{\text{loc}} = \boldsymbol{E} + c\boldsymbol{P}$, where the polarization $\boldsymbol{P}$ is linked to the permittivity

ε_r, the density ρ and effective polarizability α_{eff} of the polarizable units (molecules) via $\boldsymbol{P} = \varepsilon_0(\varepsilon_r - 1)\boldsymbol{E} = \rho\alpha_{eff}\boldsymbol{E}_{loc}$. This results in the general equations

$$\boldsymbol{E}_{loc} = (1 - c\rho\alpha_{eff})^{-1}\boldsymbol{E} \quad \text{and} \quad \chi = \varepsilon_r - 1 = (1 - c\rho\alpha_{eff})^{-1}\rho\alpha_{eff}$$

From $\boldsymbol{P} = \rho\alpha_{eff}\boldsymbol{E}_{loc}$ it follows that the susceptibility $\chi = \partial\boldsymbol{P}/\partial\varepsilon_0\boldsymbol{E}$ diverges at $c\rho\alpha_{eff} = 1$.

Debye used for $\boldsymbol{E}_{loc}$ the Lorentz field $\boldsymbol{E}_L = \boldsymbol{E} + c_L\boldsymbol{P} = \gamma_L\boldsymbol{E}$ with $c_L = 1/3\varepsilon_0$ and $\gamma_L = (\varepsilon_r + 2)/3$. Writing $y = \rho\alpha_{eff}/3\varepsilon_0$, he obtained

$$y = (\varepsilon_r - 1)/(\varepsilon_r + 2)$$

Further, for polar molecules (Section 8.6.1)

$$\alpha_{eff} = \alpha + \alpha_\mu \equiv (\alpha_{ele} + \alpha_{mol}) + \mu^2/3kT \tag{17.15}$$

As indicated, Lorentz considered a fictive cavity with a polarizable continuum replacing the material outside the cavity and a cubic array or a random array of equal point dipoles for material inside the cavity. As in the derivation for c_L all dipoles point in the same direction, c_L neglects all correlation between molecules and, in fact, is valid only if proper symmetry conditions for material inside the cavity are imposed [66]. In case $\alpha \ll \alpha_\mu = \mu^2/3kT$, we have $\boldsymbol{P} = \rho\mu^2\varepsilon_0\boldsymbol{E}/(1 - c_L\rho\alpha_\mu)3kT$. Introducing a critical temperature, defined by $T_D = \varepsilon_0 c_L\rho\mu^2/3k$, the susceptibility χ becomes $\chi = c_L^{-1}T_D/(T - T_D)$. Thus, if $T < T_D$, the susceptibility diverges. If one calculates in this model T_D for water, one obtains ≅1500 K. Clearly, we can reach the "polarization catastrophe," contrary to the case for nonpolar molecules (Section 9.19).

To avoid the catastrophe, Onsager, being aware of the presence of permanent dipole moments and therefore distinguishing between a local and a directional field, estimated the local field as (Justification 8.1)

$$\boldsymbol{E}_O = \left(1 + \frac{1}{1 - f\alpha}\frac{\mu^2}{3kT}\right)\frac{1}{1 - f\alpha}\frac{3\varepsilon_r}{2\varepsilon_r + 1}\boldsymbol{E}$$

with $f = 2C(\varepsilon_r - 1)/(2\varepsilon_r + 1)r_m^{\ 3}$, where $C = (4\pi\varepsilon_0)^{-1}$ and r_m is the radius of the molecule (or, more generally, the polarizable unit). In his model, he also took a molecular-sized cavity and replaced the outside by a polarizable continuum. But for the inside he used a polarizable point dipole, thereby avoiding the required canceling of the molecular entities within the cavity. This point dipole polarizes the continuum, which on its turn influences the point dipole. In case $\alpha \ll \alpha_\mu = \mu^2/3kT$, we have $\boldsymbol{E}_O = 3\varepsilon_r\alpha_\mu\boldsymbol{E}/(2\varepsilon_r + 1) = [9\varepsilon_r\alpha_\mu/(2\varepsilon_r + 1)(\varepsilon_r + 2)]\gamma_L\boldsymbol{E}$. The Onsager local field E_O is thus always considerably smaller than the Lorentz local field E_L as ε_r typically amounts to 5–10. Taking $\boldsymbol{E}_{loc} = \boldsymbol{E} + c_O\boldsymbol{P}$ with $c_O = \varepsilon_r/(2\varepsilon_r + 1) = 3c_L/(2\varepsilon_r + 1)$, we now have as a result $\boldsymbol{P} = N\mu^2(\varepsilon_0\boldsymbol{E} + c_O\boldsymbol{P})/3kT$. For y the result is

$$y = (\varepsilon_r - 1)(2\varepsilon_r + 1)/9\varepsilon_r$$

and in this model the catastrophe is thus avoided. While the Lorentz field neglects all correlations, the Onsager neglects local correlations, as incorporated by the Kirkwood factor g (Section 9.19), but is still fairly successful because typically g is close to 1. However, the linearized Langevin equation resulting $\alpha_\mu = \mu^2/3kT$ was used and Pirenne [67] showed that using the full Langevin expression one still obtains a critical temperature $T_O = 0.221T_D$. For water, therefore, $T_O \cong 340$ K, rendering water still ferroelectric in this model.

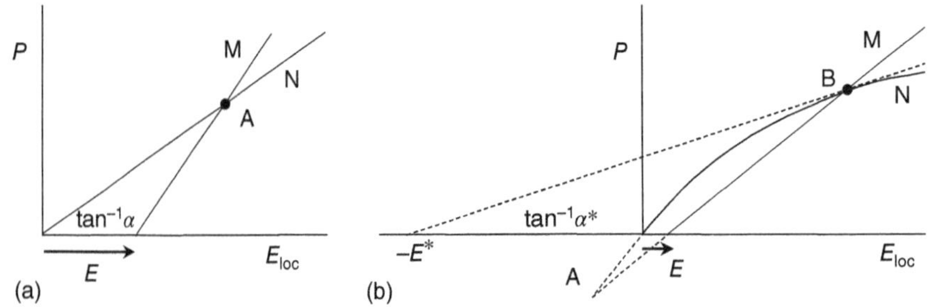

Figure 17.7 Dielectric behavior in linear and nonlinear approximations. (a) Linear behavior with $\gamma\rho\alpha < 1$, where line M representing $E_{\text{loc}}(P)$ has a larger slope than the line N representing $P(E_{\text{loc}})$. If a shift δP from point A occurs, the smaller shift δE_{loc} disfavors the fluctuation and the configuration is stable. (b) Nonlinear behavior with $\gamma\rho\alpha < 1$, where line M has a smaller slope than line N, showing the nonstable solution A and the stable (nondiverging) solution B.

Plotting $E_{\text{loc}} = E + cP$ and $P = \rho\alpha E/(1 - c\rho\alpha)$ in one graph, one can summarize linear behavior (Figure 17.7a) and nonlinear behavior (Figure 17.7b). In the former case, a stable solution is obtained when $c\rho\alpha < 1$, but an unstable one for $c\rho\alpha \geq 1$. In the latter case, a stable solution $P = \rho\alpha^*(E + E^*)/(1 - c\rho\alpha^*)$ still can be obtained when $c\rho\alpha \geq 1$ with α^* and E^* as defined in Figure 17.7b.

All this implies that for a microscopic approach the linear behavior and associated harmonic lattice dynamics used so far used are inadequate. We recall that for the displacements the general solution $\boldsymbol{u}_{n,\lambda} = \Sigma_{p,\boldsymbol{q}}\boldsymbol{U}_{p,\lambda}(\boldsymbol{q})h_{n,\lambda}(\boldsymbol{q})$ applies, with amplitude vector $\boldsymbol{U}_{p,\lambda}(\boldsymbol{q}) = N^{-1/2}A_p(\boldsymbol{q})\boldsymbol{e}_{p,\lambda}(\boldsymbol{q})$ having eigenvectors $\boldsymbol{e}_{p,\lambda}(\boldsymbol{q})$ for branch p and space–time factor $h_{n,\lambda}(\boldsymbol{q}) = (m_\lambda)^{-1/2}\exp[\mathrm{i}(\boldsymbol{q}\cdot\boldsymbol{r}_{n,\lambda} - \omega t)]$. Because the displacement of an atom in a particular mode $(p',\boldsymbol{q}')$ depends on $N^{-1/2}$, it is small. However, this ceases to be true when the frequency of this particular mode tends to zero. In that case, its energy tends to kT, its amplitude can become infinitely large, and its displacement is only limited by the anharmonicity of the potential. Such a mode is usually called a *soft mode*. If the displacement is to be the same for every unit cell (thus independent of $\boldsymbol{n}$), this requires that $\boldsymbol{q}' = 0$. When $\boldsymbol{q}'$ is at a special position in the Brillouin zone, several modes can be degenerate (have the same frequency but different eigenvectors) and become unstable simultaneously. Which of these modes develops depends on the details of the anharmonic potential [68].

For ionic crystals (Section 17.3), such a mechanism can be described somewhat more quantitively in relatively simple terms. Using the shell model, we obtain for the TO frequency after some manipulation

$$\omega_{\text{T}}^{\,2} = \frac{a^*}{\mu_m}\left[1 - \frac{(\varepsilon_\infty + 2)(Z^*e)^2}{9\varepsilon_{\text{s}}\Omega a^*}\right] \tag{17.16}$$

and, if ω_{T} is to be zero, for the LO frequency

$$\omega_{\text{L}}^{\,2} = \frac{(\varepsilon_\infty + 2)^2(Z^*e)^2}{9\varepsilon_{\text{s}}\varepsilon_\infty\mu_m\Omega} \tag{17.17}$$

As Eq. (17.17) cannot vanish and if ε_{s} has to show a singularity ($\varepsilon_{\text{s}} \to \infty$), it follows from the LST relation $\omega_{\text{L}}^{\,2}/\omega_{\text{T}}^{\,2} = \varepsilon_{\text{s}}/\varepsilon_\infty$ that $\omega_{\text{T}} \to 0$ (neglecting damping). Because the lattice frequencies are temperature independent in the harmonic approximation, we must look to anharmonicity for an explanation. We use the quasi-harmonic formulation (Section 10.9), which introduces a temperature dependence for the unit cell volume and, through the shell model, also for a^* and Z^*. Here the asterisk indicates that the shell model is used. For the situation where ω_T is small, we write $\Omega = \Omega_0(1+\alpha_P T)$, where α_P is the thermal expansivity, and do so similarly for a^* and Z^*. Upon substitution these relations in Eq. (17.15) and requiring that ω_T should vanish at $T = T_0$, we obtain

$$\omega_T^{\ 2} = \delta(T - T_0)$$

with δ a complicated function having the same order of magnitude as α_P. Alternatively, if ω_T should vanish at T_0 we just assume that in the neighborhood of T_0 we can approximate the behavior of $\omega_T^{\ 2}$ by a linear relation. Inserting the expression for $\omega_T^{\ 2}$ in Eq. (17.8) results in

$$\varepsilon_s = \varepsilon_\infty + \frac{\varepsilon_\infty \omega_P^{\ 2}}{\omega_T^{\ 2}} = \varepsilon_\infty + \frac{\varepsilon_\infty\left(\omega_L^{\ 2} - \omega_T^{\ 2}\right)}{\omega_T^{\ 2}} \cong \varepsilon_\infty + \frac{C}{T - T_0} \quad \text{with} \quad C = \frac{\varepsilon_\infty \omega_L^{\ 2}}{\delta}$$

in which we recognize the Curie–Weiss expression. If the transformation leads to a symmetry for which the point group is a subgroup of that of the paraelectric phase, a continuous transformation is possible. Otherwise, the transformation is discontinuous.

The behavior of soft optic modes, $\omega_T \to 0$ as $T \to T_0$ from above, has been amply verified experimentally for perovskites and other compounds. The soft modes lead to a finite shift in atomic positions, as discussed by Cochran [68]. For $BaTiO_3$, for example, the responsible mode results in an upward shift of 0.05 Å of the Ba^{2+} ions and to a downward shift of 0.1 and 0.04 Å for the Ti^{4+} and O^{2-} ions, respectively, for which Slater [69] provided a detailed model. The detailed microscopic theory of ferroelectrics is complicated for which Lines and Glass [43] presented an overview. For further details we refer to the literature [70–73].

Problem 17.3 ***Devonshire's Formulation*** Derive the relevant equations from $G_{ela} = \frac{1}{2}\alpha D^2 + \frac{1}{4}\gamma D^4 + \frac{1}{6}\delta D^6$ as has been done in Section 12.5 for the expression $G_{ela} = \frac{1}{2}\alpha D^2 - \frac{1}{3}D^3 + \frac{1}{4}\gamma D^4$.

References

1 Bunget and Popescu, 1984.

2 Kao, 2004.

3 Liu, S., Shen, B., Haoa, H. et al. (2019). *J. Mater. Chem. C* 7: 15118.

4 Robinson, F.N.H. (1973). *Electromagnetism*. Oxford: Oxford University Press.

5 Jackson, J.D. (2001). *Classical Electrodynamics*, 3e. Hoboken: John Wiley & Sons.

6 Panofsky, W.K.H. and Phillips, M. (1955). *Classical Electricity and Magnetism*. Reading, MA: Addison-Wesley Publishing Company; my favorite.

7 Guggenheim, E.A. (1967). *Thermodynamics*, 5e. Amsterdam: North-Holland.

8 Lorentz, H.A. (1916). *The Theory of Electrons and Its Applications to the Phenomena of Light and Radiant Heat*, 2e. Leipzig: B.G. Teubner and New York: G.E. Sechert & Co. (also Dover, 1952).

9 Wilczek, F. (2012). *Sci. Am.* 306: 24.

10 van Krevelen, D.W. and te Nijenhuis, K. (2009). *Properties of Polymers*, 4e. Amsterdam: Elsevier.
11 Raju, G.G. (2003). *Dielectrics in Electric Fields*. New York: Marcel Dekker.
12 Kremer, F. and Schönhals, A. (2003). *Broadband Dielectric Spectroscopy*. Berlin: Springer-Verlag.
13 Peiponen, K.-E., Vartiainen, E.M., and Asakura, T. (1999). *Dispersion, Complex Analysis and Optical Spectroscopy: Classical Theory*. Berlin: Springer-Verlag.
14 Beneduci, A. (2008). *J. Mol. Liq.* 138: 55.
15 Christy, R.W. (1972). *Am. J. Phys.* 40: 1403.
16 Seitz, F. (1940). *Modern Theory of Solids*. New York: McGraw-Hill Book Comp.
17 Mott, N.F. and Gurney, R. (1940). *Electronic Processes in Ionic Crystals*. Oxford: Clarendon Press.
18 Born, M. and Huang, K. (1954). *Dynamical Theory of Crystal Lattices*. Oxford: Oxford University Press.
19 Tessman, J.R., Kahn, A.H., and Shockley, W. (1953). *Phys. Rev.* 92: 890.
20 Pirenne, J. and Kartheuser, E. (1964). *Physica* 30: 2005.
21 Jaswal, S.S. and Sharma, T.P. (1973). *J. Phys. Chem. Solids* 34: 509.
22 Pauling, L. (1927). *Proc. R. Soc. A* 114: 191.
23 Shannon, R.D. (1993). *J. Appl. Phys.* 73: 348.
24 Donovan, B. and Angress, J.F. (1971). *Lattice Vibrations*. London: Chapman and Hall.
25 Elliott, S.R. (1998). *The Physics and Chemistry of Solids*. Chichester: John Wiley & Sons.
26 Huang, K. (1951). *Proc. R. Soc. A* 208: 352.
27 Kittel, C. (1987). *Quantum Theory of Solids*, 2e. New York: John Wiley & Sons.
28 Lyddane, R.H., Sachs, R.G., and Teller, E. (1941). *Phys. Rev.* 59: 673.
29 Havelock, T.H. (1924). *Proc. R. Soc. A* 105: 488.
30 Chang, I.F., Mitra, S.S., Plendl, J.N. et al. (1968). *Phys. Stat. Sol.* 23: 663.
31 Fox, M. (2010). *Optical Properties of Solids*, 2e. New York: Oxford University Press.
32 Wooten, F. (1972). *Optical Properties of Solids*. New York: Academic Press.
33 Dressel, M. and Grüner, G. (2003). *Electrodynamics of Solids – Optical Properties of Electrons in Matter*. Cambridge: Cambridge University Press.
34 Szigeti, B. (1949). *Trans. Faraday Soc.* 45: 155; (1950). *Proc. R. Soc A* 204: 51.
35 Dick, B.G. and Overhauser, A.W. (1958). *Phys. Rev.* 112: 90; see also Hanlon, J.E. and Lawson, A.W. (1959). *Phys. Rev.* 113: 472.
36 Woods, A.D.B., Cochran, W., and Brockhouse, B.N. (1960). *Phys. Rev.* 119: 980.
37 Schröder, U. (1966). *Solid State Commun.* 4: 347.
38 Swenson, R.W. and Cole, R.H. (1954). *J. Chem. Phys.* 22: 284.
39 White, H. and Morgan, S.O. (1937). *J. Chem. Phys.* 5: 655.
40 Smyth, C.P. and Hitchcock, C.S. (1934). *J. Am. Chem. Soc.* 56: 1084.
41 Turkevich, A. and Smyth, C.P. (1940). *J. Am. Chem. Soc.* 62: 2468.
42 White, H., Biggs, B.S., and Morgan, S.O. (1940). *J. Am. Chem. Soc.* 62: 16.
43 Lines and Glass, 1977.
44 Jaffe, H., Cook, W.R., and Jaffe, B. (1971). *Piezoelectric Ceramics*. London: Academic Press.
45 Kikuta, T., Hamatake, D., Yamazaki, T. et al. (2007). *Ferroelectrics* 347: 65.
46 Berbecaru, C., Alexandru, H.V., Pintilie, L. et al. (2005). *Mater. Sci. Eng. B* 118: 141.
47 Reese, W. (1969). *Phys. Rev.* 181: 905.

48 Alexandru, H.V. and Berbecaru, C. (2002). *Mater. Sci. Semicond. Process.* 5: 159.
49 Jona, F. and Shirane, G. (1962). *Ferroelectric Crystals*. Oxford: Pergamon Press (also Dover, 1993).
50 Zhirnov, V.A. *Sov. Phys. – JETP* 8: 822; Kholodenko, L.P. (1961). *Sov. Phys. – JETP* 3: 3142.
51 Arlt, G., Hennings, D., and de With, G. (1985). *J. Appl. Phys.* 58: 1619.
52 Kamel, T.M., Kools, F.X.N.M., and de With, G. (2007). *J. Eur. Ceram. Soc.* 27: 2471; Kamel, T.M. and de With, G. (2007). *J. Mater. Res.* 22: 3348.
53 Damjanovic, D. (1998). *Rep. Prog. Phys.* 61: 1267.
54 Burfoot, 1967.
55 Jaffe, H. (1958). *J. Am. Ceram. Soc.* 41: 494.
56 Fatuzzo, E. and Merz, W.J. (1967). *Ferroelectricity*. Amsterdam: North-Holland Publ. Comp.
57 Rabe, Ahn and Triscone, 2007.
58 Grindlay, J. (1970). *An Introduction to the Phenomenological Theory of Ferroelectricity*. Oxford: Pergamon Press.
59 Salje, 1990.
60 Sidorkin, A.D. (2006). *Domain Structure in Ferroelectrics and Related Materials*. Cambridge: Cambridge International Science Publishing.
61 Meier, D., Seidel, J., Gregg, M. et al. (2020). *Domain Walls: From Fundamental Properties to Nanotechnology Concepts*. Oxford: Oxford University Press.
62 Nye, J.F. (1957). *Physical Properties of Crystals*. London: Oxford University Press.
63 Levanyuk, A.P., Misirlioglu, I.B., and Okatan, M.B. (2020). *Ferroelectrics* 569: 310.
64 Lu, X., Li, H., and Cao, W. (2013). *J. Appl. Phys.* 114: 224106.
65 Rossetti, G.A. Jr., Cline, J.P., and Navrotsky, A. (1998). *J. Mater. Res.* 13: 3197.
66 Aspnes, D.E. (1982). *Am. J. Phys.* 50: 704.
67 Pirenne, J. (1949). *Helv. Phys. Acta* 22: 479.
68 Cochran, W. (1960). *Adv. Phys.* 9: 387; (1963). *Rep. Prog. Phys.* 26: 1.
69 Slater, J.C. (1950). *Phys. Rev.* 78: 748.
70 Thomas, H. (1969). *IEEE Trans. Magn.* 5: 874; (1971). p. 15 in *Structural Phase Transitions and Soft Modes* (ed. E.J. Samuelsen). Oslo: Universitetsvorleiget.
71 Kwok, P.C. and Miller, P.B. (1966). *Phys. Rev.* 151: 387.
72 Lines, M.E. (1969). *Phys. Rev.* 177: 797; *Ibid.* 177: 812; *Ibid.* 177: 819.
73 Pytte, E. (1972). *Phys. Rev. Lett.* 28: 985.

Further Reading

Bunget, I. and Popescu, M. (1984). *Physics of Solid Dielectrics*. Amsterdam: Elsevier. A somewhat old-fashioned but broad and interesting review of dielectric properties. Unfortunately, the many typographical errors do not ease the reading.

Burfoot, J.C. (1967). *Ferroelectrics – An Introduction to the Physical Principles*. Princeton, NJ: D. van Nostrand Company. Old but highly readable introduction using only elementary methods.

Debye, P. (1929). *Polar Molecules*. New York: Chemical Catalogue Company. Although old, still illustrative on basic thoughts.

Fröhlich, H. (1958). *Theory of Dielectrics*, 2e. Oxford: Oxford University Press. The classic on macroscopic dielectric behavior providing a wide overview.

Kao, K.C. (2004). *Dielectric Phenomena in Solids* – with Emphasis on Physical Concepts of Electronic Processes. Amsterdam: Elsevier. A broad overview, dealing approximately with the same topics as Bunget and Popescu.

Lines, M.E. and Glass, A.M. (1977). *Principles and Applications of Ferroelectrics and Related Materials*. Oxford: Oxford University Press. The classical reference in the field and delivering what the title indicates.

Rabe, K.M., Ahn, C.H., and Triscone, J.-M. (2007). *Physics of Ferroelectrics – A Modern Perspective*. Berlin: Springer-Verlag. A set of chapters supplementing Lines and Glass.

Salje, E.K.H. (1990). *Phase Transitions in Ferroelectric and Co-elastic Crystals*. Cambridge: Cambridge University Press. Introducing Landau theory and focusing on domains, specific heat, coupling between internal variables, gradient coupling and some kinetic aspects.

van Vleck, J.H. (1932, reprint 1966). *The Theory of Electric and Magnetic Susceptibilities*. London: Oxford University Press. Still a standard reference on what the title promises with many details.

18

The Glass Transition

So far, we have discussed transitions with a clear transition temperature, such as melting for crystalline solids. In this penultimate chapter, we deal with a transition for noncrystalline materials, characterized by a temperature range in which the properties gradually change from solid-like to liquid-like, the so-called glass transition.

18.1 What Is a Glass?

In noncrystalline solids, like in liquids and in contrast to crystalline solids, long-range order is absent, although the local coordination of a specific atom in the noncrystalline state may not be that different from the coordination of that atom in the corresponding crystalline state (Figure 18.1a). Noncrystalline materials can be essentially realized by vitrifying an undercooled liquid, by deposition of a gas on a cold substrate, or by disrupting the crystalline order by radiation or mechanical action, and can be divided into glasses and amorphous materials [1]. While *amorphous materials* upon heating (re)turn irreversibly to the crystalline state, *glasses* can pass progressively and reversibly from the so-called *vitreous state* to a more fluid state. For a glass, properties change over a certain temperature range, the *glass transition range*, more quickly as compared with their regular temperature dependence. This is most clearly seen from the rigidity. Although glasses do not show a clear melting point where rigidity disappears, they do show with increasing temperature, a gradual change from elastic to viscous behavior in the transition range. The *glass transition temperature* T_g, located approximately in the middle of the transition range, characterizes the transition, while the overall behavior is denoted as *viscoelastic*. Thus, unlike during melting/freezing, where several properties change abruptly at the melting temperature (in this chapter labeled) T_m, properties for a glass change gradually from one regime to another in the glass transition range, as illustrated for the specific volume in Figure 18.1b. Whether crystallization occurs or not, depends on the balance between nucleation and relaxation. The T_g is usually determined by extrapolation of the behavior of the undercooled liquid and vitreous regions. Characteristic is that T_g depends on the cooling rate q^* used: the lower q^*, the lower the T_g obtained.

Qualitatively, this behavior can be understood through relaxation. At high q^* only limited relaxation of the liquid structure can occur before the temperature has decreased so far that further relaxation is very slow. At lower q^* more relaxation can take place, thus continuing

Phases of Matter and their Transitions: Concepts and Principles for Chemists, Physicists, Engineers, and Materials Scientists, First Edition. Gijsbertus de With.

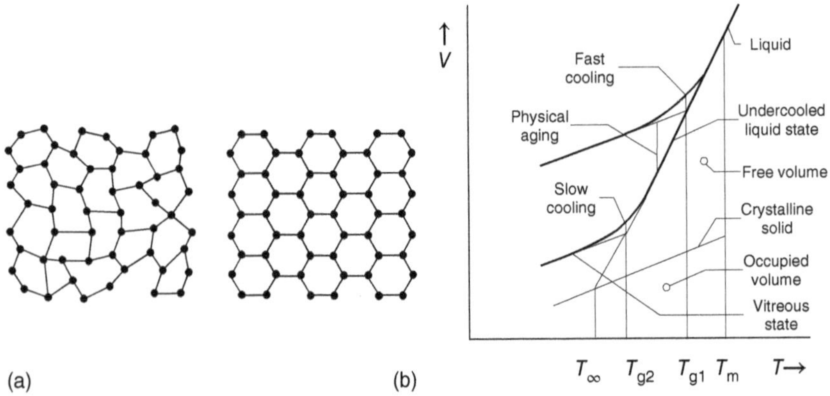

Figure 18.1 (a) Two-dimensional (2D) schematic of an amorphous structure with no long-range order and a crystalline structure with long-range order. (b) The (idealized) change in specific volume V for the glass transition at cooling rate 1 > cooling rate 2 leading to glass transition temperatures T_{g1} and T_{g2}. Also shown is the normal melting behavior at melting point T_m. The enthalpy shows similar behavior.

the liquid-like regime to lower temperature and hence leading to a lower T_g. A glass cooled at a high rate can, when kept for a sufficiently long time at a sufficiently high temperature $T < T_g$ in the transition regime, relax to its undercooled liquid state at T, a process often denoted as *physical aging* when it occurs undesirably, or as *annealing* when it occurs by design. Such a glass is called a *stabilized glass* with a *fictive temperature* $T_{fic} = T$, and during annealing at $T < T_{fic}$ it contracts, while at $T > T_{fic}$ it expands. For a heating rate close to the initial q^*, $T_{fic} \cong T_g$. Physical aging stands in contrast to *chemical aging* where a slow chemical reaction modifies the chemical constitution. The latter effect is relatively unimportant for inorganic glasses but may be considerable for polymers.

A common feature of disordered solids is the inability of a system to find a unique ground state, often labeled by (the misleading term) *frustration* [2]. This applies also for noncrystalline materials for which a model is a *random close-packed* (RCP) structure of single-sized spheres, as first presented by Bernal. This packing can be described by the polyhedral shape of its interstitial holes. The five basic types of holes that can be discerned (Figure 18.2a) are actually somewhat distorted, but an RCP packing contains a certain fraction of each type [3, 5] (Table 18.1); their frequency represents the frozen structure. Similarly, we have for the coordination polyhedra the number of faces, on average 14.2, and edges per face, on average 5.2 (Table 18.2). The reduced density for RCP spheres η_{RCP} is about 0.637, compared to $\eta_{FCC} = 0.741$ and $\eta_{BCC} = 0.680$. On average, the volume per molecule v_{mol} is thus about 15% larger than v_{mol} for an FCC crystal, corresponding to a lower average coordination number (CN). This excess volume for RCP spheres can be described by the distribution of v_{mol}. In a crystal, v_{mol} is the same for every molecule. In a noncrystalline material, there are molecules with a surrounding for which v_{mol} is less than the RCP average, while for others it is larger; this distribution has a considerable width (Figure 18.2b, [4]). The static (or average) structure of glasses can also be described, similar as for liquids, by the correlation function $g(r)$ (Section 9.1). The average CN as determined from $g(r)$ is about 9, compared to 12 for the FCC lattice. In Section 18.1.1 it will become clear though that many glasses are not formed by single-type, sphere-like structural units.

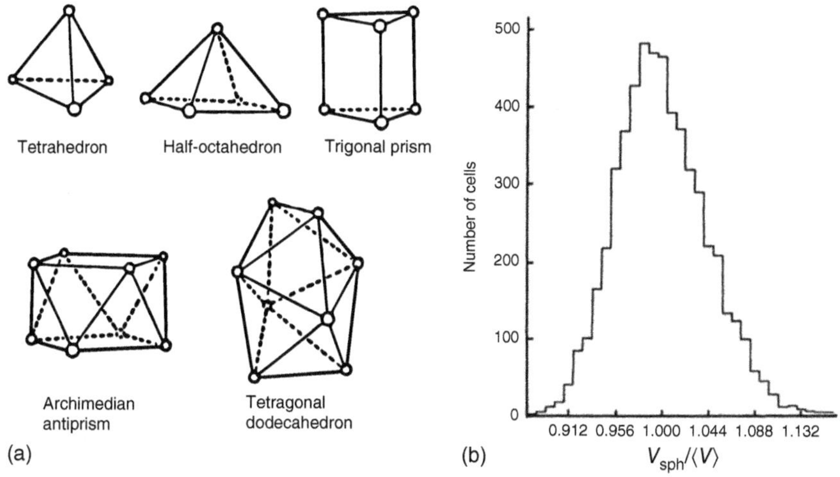

Figure 18.2 (a) The five canonical holes of Bernal. Source: Frost [3]/with permission of Elsevier. (b) The volume per sphere V_{sph} distribution for a random close packing of spheres with the maximum at $V_{sph}^* = V_{sph}/\langle V\rangle = 0.984$ with $\langle V\rangle = 0.637$ the random close-packed relative density. Source: Finney [4]/Royal Society of Chemistry.

Table 18.1 Frequency of canonical holes.

Hole	Number %	Volume %	Hole	Number %	Volume %
Tetrahedron	73.0	48.4	Tetragonal dodecahedron	3.1	14.8
Half-octahedron	20.3	26.9	Archimedean antiprism	0.4	2.1
Trigonal prism	3.2	7.8			

Source: Adapted from [3, 5].

Table 18.2 Frequency of the number of faces and edges per face of RCP coordination polyhedra.

Edges/face	%	Edges/face	%	Faces/cell	%	Faces/cell	%
3	4	6	29	12	5	15	27
4	19	7	6	13	20	16	11
5	41	8	1	14	34	17	3

Source: Adapted from Finney [4].

Illuminating as this picture is, it ignores the changes due to temperature change and the dynamics. We have seen that the structure of an undercooled liquid is frozen in, characterized by T_g. Therefore, we define a glass as a thermodynamically unstable, but kinetically stabilized undercooled melt with a structure for $T < T_g$ corresponding to that of T_g. In the simplest possible picture, the *Simon picture*, a glass below T_g is in the vitreous state and does not change with $T(t)$, while for $T_g < T < T_m$ it is an undercooled liquid in metastable equilibrium.

Problem 18.1 ***Glass versus Crystal*** Using Figure 18.2b, estimate the density of the glass relative to that of the crystal. Also estimate the local density variations that can be expected.

18.1.1 Glassy Materials

In this section, we briefly review materials that can form glasses. For most inorganic materials that form glasses, silica or SiO_2 is the basic component. Solid amorphous silica or *fused silica* has a low thermal expansivity and a highly homogeneous network structure in which the SiO_4 tetrahedra are connected to each other with all corners shared, thereby forming a network. Other *network-forming oxides* are, for example, GeO_2 and B_2O_3, while Al_2O_3 and Bi_2O_3 form glasses only in the presence of other network-forming oxides. By adding *network-modifying oxides* such as Na_2O, K_2O, CaO, and Al_2O_3, the properties can be varied over a wide range. The full connectivity of the SiO_4 tetrahedra is lost and charge compensation is provided by the other cations. These modifications affect all properties. For inorganic glasses T_g typically ranges from 500 to 1000 °C. Although macroscopically homogeneous, on a microscopic scale, glasses generally are somewhat inhomogeneous (microphase separation). General references for inorganic glasses are [1, 6, 7].

Depending on the degree of crystallinity, polymers may or may not show a clear melting point. Therefore, depending on the temperature, a polymer may exhibit elastic/brittle, viscoelastic, or viscous behavior, like inorganic glasses. Amorphous polymers do show a glass transition temperature T_g rather than a melting point, for which the typical range is from −120 to 220 °C. As for inorganic glasses, the lower the cooling rate, the lower the T_g obtained. Contrary to what is often maintained, it seems that the T_g as determined by thermal or mechanical means for polymers differs insignificantly [8]. Apart from cooling rate, T_g also depends on chemical structure, molecular weight, branching, and cross-linking. The more flexible the chains, the lower T_g while bulky and/or polar side groups tend to increase T_g. Because there is more free volume associated with a chain end than with a chain middle, T_g increases with increasing molecular weight M. The behavior is for large number-averaged molecular weight M_n approximately described by the *Fox-Flory equation* [9]

$$T_g = T_{g_\infty} - k_1/M_n$$

where $T_{g\infty}$ indicates the T_g for very high M_n and k_1 can be estimated in principle from molecular data, but usually is obtained by fitting to experimental data for a given polymer. If these data are not available, one can use the estimate $k_1 = 0.002\,715 T_{g\infty}{}^3$ [10]. Blends for which components 1 and 2 have weak or no specific intermolecular interaction can be reasonably well described by the *Fox equation* [11]

$$1/T_g = w_1/T_{g1} + w_2/T_{g2}$$

where w_1 and w_2 are the weight fractions of the components with T_{g1} and T_{g2}. Similarly, a small amount of branching reduces T_g while a large amount restricts mobility and therefore increases T_g. Cross-linking increases the density, implying that the free volume decreases. Hence, cross-linking increases T_g, and the behavior for a homologous series with large M_n and a small degree of cross-linking can be approximately described by the *Fox–Loshaek equation* [12]

$$T_g = T_{g_\infty} - k_1/M_n + k_2\rho$$

with ρ the number of cross-links per gram and k_2 a parameter that can be estimated from molecular data, but normally is obtained by fitting the experimental data. However, molecular weight is not necessarily a practical design parameter to control T_g because the range over which it can be changed, without altering the physical properties of the polymer, is small [13]. Roth [14] provided an extensive review.

If sufficiently rapidly cooled from the melt (splat cooling), some metals can solidify in the amorphous state. Typical cooling rates are, for example, for Al alloys 10^6 K s^{-1}. These amorphous metals, sometimes called *metal glasses*, can have quite different properties, in particular ductility and wear resistance, from their crystalline counterparts. It is also possible to produce compositions that cannot be produced as crystalline alloys. Although real metallic glasses can be realized, amorphous metals normally crystallize easily upon annealing. General introductions to metallic glasses are [15, 16], while [17] reviews their mechanical behavior.

Small organic molecules can also form glasses. Examples are 3-methylhexane, 2,3-dimethylpentane, 2-butanol, glycerol, sucrose, and glucose. As experimental conditions for these materials are generally easier than that for both inorganic glasses and polymer glasses, these materials are widely used in experimental investigations, revealing many aspects of the glass formation process. The dynamics of such glasses is dealt with in [18]. *Organic glasses*, polymers with an irregular sequence of monomers or polymers with complex molecules that cannot form a compact regular network under cooling, are also employed, mainly for (electro-)optical purposes.

To conclude this section, we note that glass-forming compounds generally can be considered as *n-mers* (or *polymers* in a generalized interpretation), constructed from *monomers* with a degree of association x dependent on T. For inorganic and polymer glasses generally the ratio $T_g/T_m \cong 2/3$, while for metallic glasses it is about 1/2.

18.1.2 Property Changes at T_g

The properties of glasses change in the glass transition region, but most significantly at about T_g. First and foremost, the viscosity η changes and, contrary for normal liquids, generally cannot be described by $\eta = \eta_0\exp(\Delta H/RT)$, as applicable to small-molecule liquids for a limited temperature range. Near exceptions are SiO_2 over a range 10^4–10^{13} P (or dPa s) [19] and GeO_2 over a range 10^3–10^{13} P [20]. Conventionally, one uses the *Vogel–Fulcher–Tammann* (VFT) *equation* given by

$$\eta = \eta_0 \exp[\Delta H/k(T - T_\infty)] \tag{18.1}$$

with η_0, ΔH, and T_∞ empirical constants, implying that the activation enthalpy $\Delta H = \Delta H_0 T/(T - T_\infty)$, where $T_\infty/T_m \cong 0.5$ for inorganic and polymer glasses and $T_\infty/T_m \cong 0.3$ for metal glasses. The viscosity at T_g is often quoted as 10^{13} P, but falls, dependent on q^*, typically in the range 10^{12}–10^{14} P [21]. The data for SiO_2 and glycerol (Figure 18.3a, [22]) show that η changes orders of magnitude over a small temperature range.

Most other properties do change largely in the transition region. An excess C_P as indicated by the area between the curves for the crystal and the undercooled melt (Figure 18.3b) leads to a residual entropy $S_0 = \int_{T_g}^{T_m}(C_P/T)\,dT \neq 0$. For glycerol, $S_0 = 5.6$ cal K^{-1} mol^{-1} [23]. Hence, glasses do not behave according to the third law. Schultz [24] studied many other

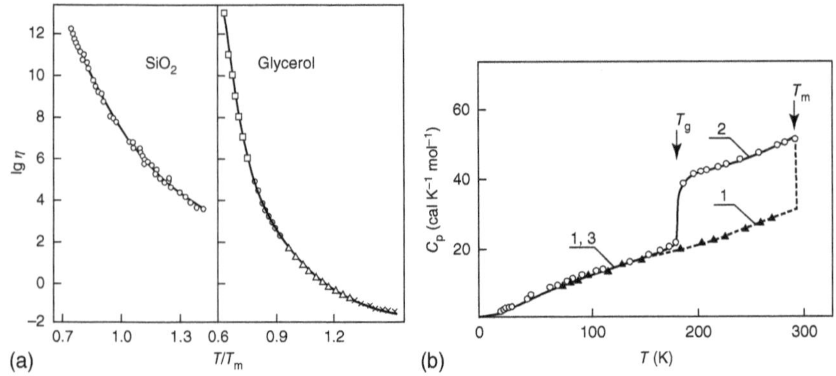

Figure 18.3 (a) The viscosity η (dPa s) of SiO_2 and glycerol as function of the relative temperature T/T_m [22]. (b) The heat capacity C_P of glycerol as function of the relative temperature T. Line 1 (▲) indicates the crystal, line 2 (o) indicates the undercooled melt, and line 3 (o) indicates the glass [22]. The area between lines 1 and 2 corresponds with the residual entropy S_0. Source: Adapted from Gutzow and Schmelzer [22].

properties for glycerol (Figure 18.4). His data clearly show that for all these properties significant changes occur at about the same temperature, the T_g, which for glycerol is about −80 °C. Other glasses do behave similarly. Detailed treatments of relaxation phenomena are given in [25, 26].

The glass transition is one of the deepest challenges in physics [27], for which there is no agreement as to what is its governing parameter [28]. Therefore, many ways to deal with glasses have been proposed, of which we discuss just a few. The thermodynamic approach discusses deviations from classical thermodynamic behavior, while the structural approach has structure in focus. The phonon theory of liquids, employing wave concepts, is applied to glasses as well. Finally, mode-coupling theory (MCT) is an approach focusing on the dynamics of a glass.

18.2 The Thermodynamic Approach

Before we discuss the thermodynamics of the glass transition, we must set the terminology. Using the superscript (g) for the glass state and the superscripts (f) and (c) for the liquid (fluid) and solid (crystalline) states, we can write

$$\Delta Z(T) = Z^{(f)}(T) - Z^{(c)}(T) \quad \text{and} \quad \Delta Z^{(g)}(T) = Z^{(g)}(T) - Z^{(c)}(T)$$

where Z is a general label for the enthalpy H, entropy S, or Gibbs energy G. As

$$\Delta H(T) = \Delta_m H - \int_T^{T_m} \Delta C_P \,dT \quad \text{and} \quad \Delta S(T) = \Delta_m S - \int_T^{T_m} (\Delta C_P/T)\,dT \quad (18.2)$$

we have for the change in Gibbs energy $\Delta G = \Delta H - T\Delta S$, the expression

$$\Delta_m G(T) = \Delta_m S(T_m - T) - \int_T^{T_m} \left[\int_T^{T_m} (\Delta C_P/T)\,dT\right] dT$$

where $\Delta_m G \equiv \Delta G(T_m)$, $\Delta_m H \equiv \Delta H(T_m)$ and $\Delta_m S \equiv \Delta S(T_m)$ denote the changes in G, H and S at T_m. At the melting temperature T_m, $\Delta_m G = 0$. We noted before that the structure is

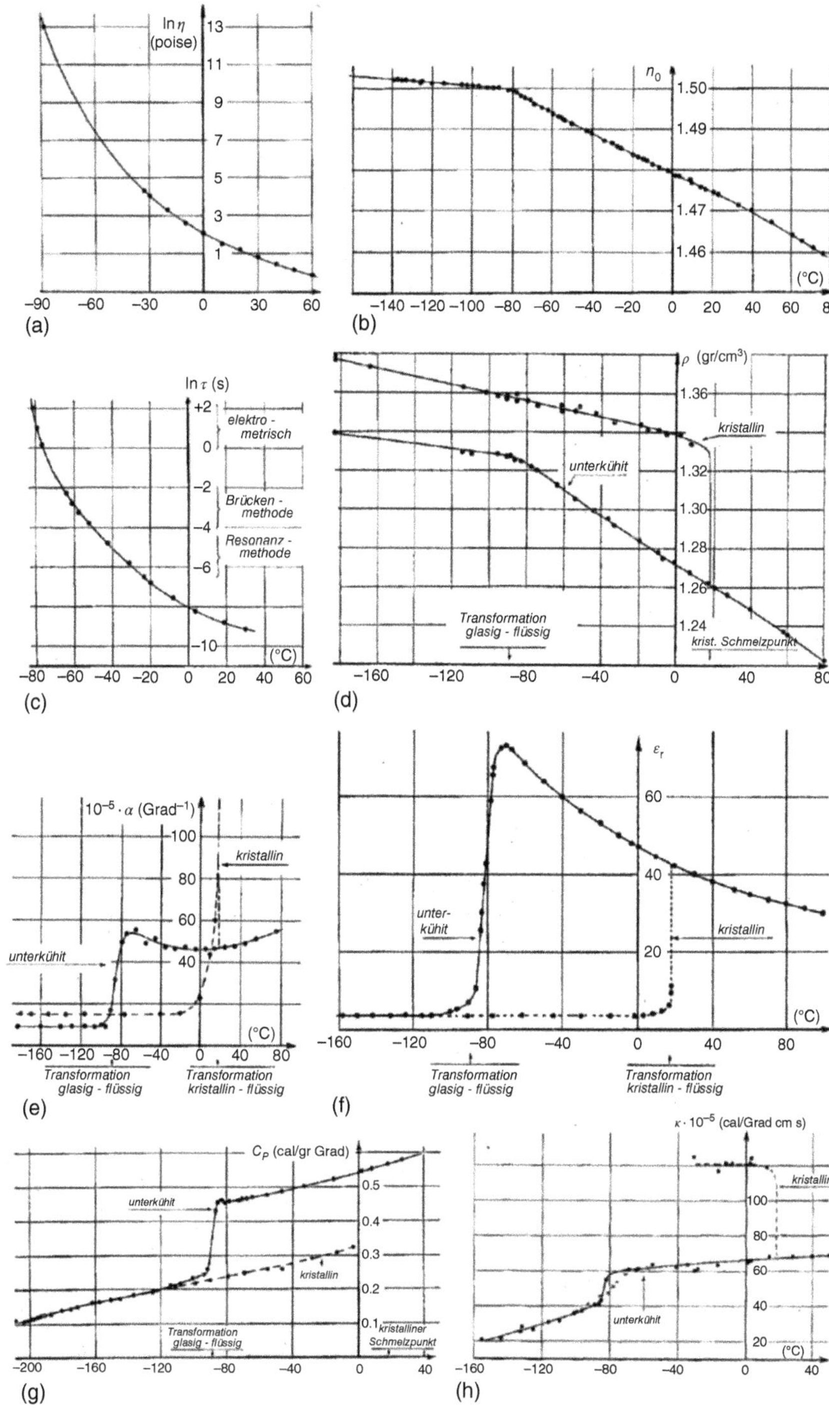

Figure 18.4 Change in properties for glycerol glass as a function of temperature displaying the original figures from Ref. [24]. Note that all large changes in the properties occur at about $T_g \cong -80\,°C$. (a) Viscosity η (data reproduced but not measured by Schultz and not covering the full range); (b) refractive index n; (c) dielectric relaxation time τ; (d) density ρ; (e) expansivity α; (f) static permittivity ε_r; (g) heat capacity C_p; (h) thermal conductivity κ.

frozen in at about T_g. Assuming that $\Delta C_P \cong 0$ for $T < T_g$, as experimentally approximately true, in these expressions the limit T can be taken as T_g. From Eq. (18.2b), we then obtain

$$\Delta S(T) = \Delta S(T_g) \equiv \Delta_g S = const.$$

Consequently, the entropy below T_g has an approximately constant value and does not approach $S = 0$ when $T \to 0$ K. Extrapolating ΔS from above T_g linearly to $T = 0$ K results in a negative entropy at $T = 0$ K, which is impossible according to the third law. This is usually addressed as *Kauzmann's paradox*, although Kauzmann was well aware that extrapolation only to $\Delta S(T_K) = 0$ is allowed. Nowadays, the temperature T_K where $\Delta S = \Delta S^{(f)} - \Delta S^{(c)}$ is often called the *Kauzmann temperature* [29]. This suggests that there could be some kind of transition for $T > 0$ K that leads to $\Delta S(T_K) \cong 0$.

However, we also noted the dependence of T_g on cooling rate q^* and from both phenomena it will be clear that, despite that most properties of a glass at room temperature are quite constant in time, a glass is not real in equilibrium because thermodynamic equilibrium states are path independent. Path-dependent behavior can be described with one or more internal variables. So, we suppose that we can describe its thermodynamic behavior using the Gibbs energy dependent on a set of internal variables, generally labeled by **b** or b_j, but here by $\boldsymbol{\xi}$ or ξ_j, in addition to the usual external variables, generally labeled by **a** or a_j, but here taken specifically as pressure P, temperature T, and mole numbers n_j. Upon approaching equilibrium, the values of the ξ_js change with $\partial G/\partial \boldsymbol{\xi} > 0$, while at equilibrium with $\boldsymbol{\xi} = \boldsymbol{\xi}^{(equ)}(T)$ we have $\partial G/\partial \boldsymbol{\xi} = 0$. Stability requires that $\partial^2 G/\partial \boldsymbol{\xi}^2 > 0$. In Section 18.1.1, we noted that glass-forming compounds generally can be considered as *n-mers* (or *polymers* using a generalized interpretation), constructed from *monomers* with a degree of association x dependent on T. Gutzow and Debrova [30], following Milchev [31], considered that for such compounds three internal variables are necessary to describe the thermodynamic state of glass-forming melts with sufficient accuracy: one for the volume accessible to the molecules (i.e. the free volume v_f), one for their size (or degree of association), and one for their possibility to appear in different conformations, as determined by their rigidity. Nevertheless, we use here first the simplest possible description with only one internal variable ξ, largely along the line of [22], but differing in details.

Now, to be concrete, consider the heat capacity. Using $S = S(T,\xi)$, we have

$$C_P^{(c)} = T\left(\frac{\partial S}{\partial T}\right)_{P,\xi=\xi^{(equ)}} \mathrm{d}T \quad \text{and} \quad C_P^{(g)} = T\left(\frac{\partial S}{\partial T}\right)_{P,\xi} \mathrm{d}T + T\left(\frac{\partial S}{\partial \xi}\right)_{P,T}\left(\frac{\partial \xi}{\partial T}\right)_P \tag{18.3}$$

for the solid and glass states. This implies that we have for $C_P^{(g)}$ two contributions. The first term represents the vibrational contribution of the undercooled melt, which we label $C_P^{(f,vib)}$, and the second term represents the configurational contribution, which we label $C_P^{(f,con)}$. In the Simon picture and approximating $C_P^{(f,vib)}$ with $C_P^{(c)}$, we get $\Delta C_P^{(g)} = C_P^{(f,con)} = const.$, as the undercooled liquid is in metastable equilibrium with $\partial \xi/\partial T = 0$ for $T_g \le T \le T_m$. For $T \le T_K$ we have $\Delta C_P^{(g)} = 0$, as otherwise $\Delta S(T = 0) \le 0$ would result. Hence, $\Delta C_P^{(g)} = T(\partial S/\partial \xi)_{T,P}(\mathrm{d}\xi/\mathrm{d}T)_P = C_P^{(f,con)}$.

Now we are ready to discuss the thermodynamics of undercooled melts, for which we use a second-order Taylor expansion of ΔG with respect to T_m. That is, we write

$$\Delta G(T) = G_{T=T_m}{}'(T - T_m) + \tfrac{1}{2} G_{T=T_m}{}''(T - T_m)^2 + \cdots$$

or

$$\Delta G(T) = T_m \Delta_m S(1 - \theta)[1 - a_0(1 - \theta)] \tag{18.4}$$

where $\theta = T/T_m$, $\Delta S = -\Delta G' = -\partial \Delta G/\partial T$ and $\Delta C_P = -T\Delta G'' = T\partial \Delta S/\partial T$ are used and $a_0 = \Delta_m C_P/\Delta_m S$. The ratio a_0 is clearly determining the overall behavior. If we accept again that $\Delta C_P(T) = const.$ for $T_K \le T \le T_g$ and $\Delta C_P(T) = 0$ for $0 \le T \le T_K$, we have

$$a_0 = const. \quad \text{for } \theta_K \le \theta \le 1 \quad \text{and} \quad a_0 = 0 \quad \text{for } 0 \le \theta \le \theta_K$$

Hence, we obtain

$$\Delta S(T)/\Delta_m S = 1 + a_0 \ln \theta \quad \text{or} \quad T_K = T_m \exp(-1/a_0)$$

From Eq. (18.4), we can also derive

$$\Delta H(T)/T_m \Delta_m S = 1 - a_0(1 - \theta_X)$$

where

$$\theta_X = \theta \quad \text{for} \quad \theta_K \le \theta \le 1 \quad \text{and} \quad \theta_X = \theta_K \quad \text{for} \quad 0 \le \theta \le \theta_K$$

If $T_g/T_m \cong 2/3$ holds, linear extrapolation of $\Delta S(T)$ typically results in $T_K/T_m \cong 1/2$ or $T_K/T_g \cong 3/4$. For glycerol, with $T_m = 291$ K and $T_g \cong 178$ K, $\Delta_m C_P = 88$ J K^{-1} mol^{-1} and $\Delta_m S = 62.9$ J K^{-1} mol^{-1}, so that $a_0 = 1.41$ and $T_K = 144$ K, in agreement with these rules of thumb.[1]

Let us deal now with vitrified melts. In the Simon picture, below T_{fic} both the entropy change $\Delta S(T) = \Delta S^{(g)}$ and the enthalpy change $\Delta H(T) = \Delta H^{(g)}$ are constant, but dependent on the cooling rate q^*. Therefore, from $\Delta G^{(g)}|_{T<Tfic} = \Delta H^{(g)} - T\Delta S^{(g)}$, we obtain that for $T \to 0$ K, $\Delta G^{(g)} \to \Delta H^{(g)}$, and we expect a linear increase in $\Delta G^{(g)}$ from $T = T_g$ to $\Delta H^{(g)}$ at $T = 0$. Figure 18.5 shows the behavior upon lowering the temperature for a regular solid and, in a normalized way, for a glass.

To assess the dependency of the vitrification process on the cooling rate $q^* = -\mathrm{d}T/\mathrm{d}t$, we recall that, in the Frenkel picture of a liquid (Section 9.15), molecules vibrate for a certain average time τ around a temporary position, whereafter they jump about a molecular diameter to new temporary position. Moreover, for a relaxation-type internal variable, we noted in Section 6.2 that such an internal variable ξ with relaxation time $\tau_\xi = (\xi - \xi^{(equ)})/(\mathrm{d}\xi/\mathrm{d}t)$ varies with the external variable a with applied rate $\tau_a = a/(\mathrm{d}a/\mathrm{d}t)$ noticeably only when $\tau_\xi \cong \tau_a$. In fact, this relation can be obtained from linear irreversible thermodynamics (Section 6.4) most easily if we calculate the affinity $\mathcal{A} = -(\partial G/\partial \xi)_{P,T}$, instead of from the entropy S, from the Gibbs energy $\mathrm{d}G = -S\mathrm{d}T + V\mathrm{d}P - \mathcal{A}\mathrm{d}\xi$. Expanding G similarly as S to second order in a Taylor series, we obtain $\partial \mathcal{A}/\partial \xi = -(\partial^2 G/\partial \xi^2)_{P,T=T_g}(\xi - \xi^{(equ)})$, so that $\mathrm{d}\xi/\mathrm{d}t = L\mathcal{A}$ with L the phenomenological coefficient. In the present case we have $\tau_a = \tau_T = T/q^*$. This leads immediately to $\tau_T{}^{-1}\tau_\xi = [T^{-1}q^*\tau_\xi]_{T=T_g} \cong 1$. Multiplying by T and taking the derivative with respect to T, one obtains $\mathrm{d}(q^*\tau_\xi)/\mathrm{d}T = q^*\mathrm{d}(\tau_\xi)/\mathrm{d}T \cong 1$. Now we assume, with Frenkel, that $\tau_\xi/\tau_{\xi,0} = \exp(\beta U)$, where $\beta = 1/kT$ and U an activation energy,

---

1 Gutzow and Schmelzer [22] provide $T_g = 193$ K in their main text and $T_g = 178$ K in their table. Further, they quote Wunderlich [32] for their ΔC_P data. However, Wunderlich states that $\Delta C_P = 0.212$ cal K^{-1} g^{-1}. With $\mathcal{M} = 92.09$ g mol^{-1} this leads to $\Delta C_P = 81.8$ J K^{-1} mol^{-1}. Moreover, Wunderlich used $T_g = 183$ K, so that from his data $a_0 = 1.30$ and $T_K = 84.8$ K, showing the sensitivity of the estimate.

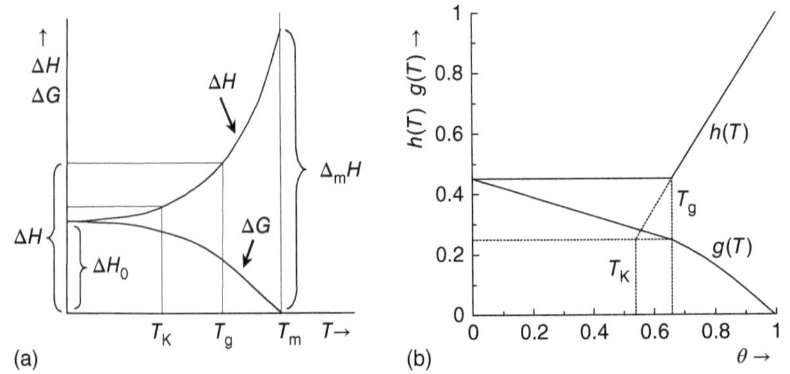

Figure 18.5 (a) Temperature dependence of the enthalpy ΔH and Gibbs energy ΔG between solid and liquid states for regular solids in consonance with the third law. (b) The normalized dependence $h(T) \equiv \Delta H(T)/T_m\Delta_m S$ and $g(T) \equiv \Delta G(T)/T_m\Delta_m S$ as a function $\theta = T/T_m$ for glasses.

which we take here independent of T. Taking the derivative with respect T and thereafter logarithms, we obtain $\ln q^* + \ln \tau_{\xi,0} + U/kT = 0$. Considering that $T = T_g$, this leads to

$$T_g^{-1} = (k/U)\ln(C/\tau_0) - (k/U)\ln q^* \quad \text{or} \quad T_g^{-1} = A - B\ln q^*$$

with $C = kT_g^2/U$, $A = B\ln(C/\tau_0)$, and $B = k/U$. This represents in fact the experimentally observed *Bartenev–Ritland relationship* [33]. It can be rewritten in terms of $\theta = T/T_m$, which leads to

$$\theta_g^{-1} = \varepsilon^{-1}[\ln(C/\tau_0) - \ln q^*] \quad \text{with} \quad \varepsilon = U/kT_m$$

It appears that for glass-forming oxides ε ranges from about 15 to 20. The parameter C has been found by Bartenev to be $C = 1$–5 K according to [22]. Using $\varepsilon = 17.5$, $C = 3$ K and taking for τ_0 the typical vibrational time $\tau_0 \cong 10^{-13}$ seconds in combination with $q^* = 100\ \text{K s}^{-1}$, we obtain $\theta_g \cong 0.73$, consistent with the empirical rule $T_g/T_m \cong 2/3$.

The separation procedure for any property $Z(\xi)$, as illustrated by Eq. (18.3), was in fact already introduced in Section 12.4, where we used the general label b for an internal variable instead of ξ as used here. It led to, the subscript for G denoting differentiation,

$$\Delta c_P = -TG_{T,b}{}^2/G_{b,b},\ \Delta\kappa_T = -G_{P,b}{}^2/VG_{b,b} \quad \text{and} \quad \Delta\alpha_P = h_{T,P}G_{P,b}/VTG_{b,b} \qquad (12.16)$$

and the Ehrenfest–Prigogine–Defay (EPD) ratio, Eq. (12.14),

$$\Pi = \Delta C_P \Delta\kappa_T/TV(\Delta\alpha_P)^2 = 1 \qquad (12.14)$$

However, for glasses experiments show that generally $\Pi > 1$. We indicated (Section 12.4) that this is to be expected if more than one internal variable is present, which is generally the case for glasses (Section 18.4). However, Schmelzer et al. [34, 35] pointed out, considering the width of the glass transition region, that one internal variable would still do. This result has been contested by Garden et al. [36]. Their critique boils down to the fact that the expression as derived in [35] is not an EPD ratio because the derivatives $(\partial A/\partial T)_P$ and $(\partial A/\partial P)_T$ are neglected, rendering the process considered an equilibrium process, which it is not. Schmelzer in his reply, argued that these derivatives are small, referring to a submitted paper[2] and therefore that a single internal variable will still do. More recently, a similar presentation [37] was given, which was contested by Zanotto et al. [38], arguing that relaxation was not included in the glass definition and glasses do ultimately crystallize. Schmelzer et al. replied that their arguments were largely ignored and, since everything flows at large time scales, crystallization should not be included in a definition. Moreover, they pointed out that several materials exist, notably polymer glasses, that do not crystallize at all because they have no plausible competing crystal structure, although normally glass substances are only kinetically prevented from crystallizing on human time scales. Furthermore, they argued, as is well established, that glasses do have a residual entropy.[3] A general reference on the thermodynamics of the glassy state, focusing on two-temperature thermodynamics and exactly solvable models, is [39].

18.3 The Structural Approach

As indicated, the mechanical behavior of glasses in the transition region is viscoelastic, which can be assessed using either quasi-static or frequency-dependent measurements. By using data collected at various temperatures and/or frequencies, a substantial part of the complete response curve can be determined. This is done by shifting the results for any temperature horizontally with a factor a_T to the response at the arbitrarily chosen reference temperature T_0. Moreover, a vertical shift with a factor $b_T = \rho T/\rho_0 T_0$ is applied to correct for density effects. At the reference temperature, clearly, no shift is applied. For a dynamic, frequency-dependent experiment, we have

$$E^*(T,\omega) = E^*(T_0,\omega_0) = b_T E^*(T_0, a_T\omega) \quad \text{or} \quad E^*(T,\ln\omega) = b_T E^*(T_0, \ln\omega + \ln a_T)$$

while for a quasi-static, time-dependent experiment, we similarly write

$$E(T,t) = E(T_0,t_0) = b_T E(T_0, t/a_T) \quad \text{or} \quad E(T,\ln t) = b_T E(T_0, \ln t - \ln a_T)$$

where $\omega = 2\pi/\tau$ is the frequency while E^* and E are moduli. These equations represent what is called *time–temperature equivalence*. The shift factor b_T is often close to 1, therefore neglected and we do so here. The shift factor ln a_T is often described by the *Williams–Landel–Ferry* (WLF) *equation*

$$\log a_T = \frac{-c_1{}^\circ (T-T_0)}{c_2{}^\circ + (T-T_0)} \quad \text{with} \quad T_0,\ c_1{}^\circ \text{ and } c_2{}^\circ \text{ constants}$$

Frequently one takes T_g for T_0. For this choice, $c_1{}^g$ and $c_2{}^g$ are in the range $c_1{}^g = 14$–18 K and $c_2{}^g = 30$–70 K. The constants $c_1{}^\circ$ and $c_2{}^\circ$ for any T_0 can be calculated from $c_1{}^\circ = c_1{}^g c_2{}^g/(c_2{}^g + T_0 - T_g)$ and $c_2{}^\circ = c_2{}^g + T_0 - T_g$. The temperature dependence of the real part of E^* and E is often much less than that of the imaginary part and the viscosity η essentially being the imaginary part of the modulus, the WLF equation applies also to η and the average relaxation time $\tau = \eta/E$. In this case, $a_T = \eta_T/\eta_{T_g}$ and the η-form of the WLF equation is approximately valid in the range $T_g < T < T_g + 100$, while at higher

2 The reference is: J.W.P. Schmelzer and T.V. Tropin, "Modeling the glass transition: Some comments", J. Chem. Phys. (submitted). However, I was unable to retrieve this reference, neither from J. Chem. Phys. nor from anywhere else.
3 The whole discussion seems to me to be largely a cat and dog fight.

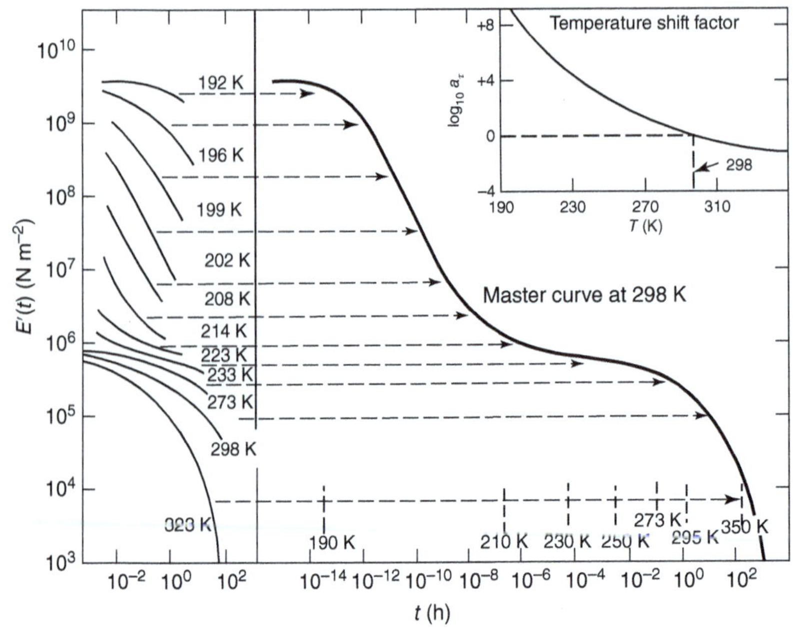

Figure 18.6 Example of time–temperature equivalence demonstrated on data for polyisobutylene with viscosity-average molecular weight 1.35×10^6 showing the storage modulus as a function of relaxation time for various temperatures, the master curve at 298 K, and the temperature dependence of the shift factor a_T. Source: Adapted from Tobolsky and Catsiff [42].

temperatures the viscosity behaves Arrhenius-like. For T_0 chosen as T_g, c_1 and c_2 are approximately, but only approximately, constants given by $c_1 = 17.44$ and $c_2 = 51.6$ K [40]. As for polymers $T_g - T_0 \cong 43$ K, a better choice according to van Krevelen [41] is to use $T_0 = T_g + 43$ K, leading to $c_1 = 9.51$ and $c_2 = 94.6$ K.

Figure 18.6 shows the classic example of polyisobutylene [42]. Time–temperature equivalence is often used for the analysis of viscoelastic properties of polymers and explicitly assumes that all viscoelastic processes have the same temperature dependence. However, in many cases the equivalence breaks down because it seems to apply to individual modes only and not the average behavior.

The WLF equation, equivalent to the VFT equation ($c_2 = T_0 - T_\infty$, $2.303c_1c_2 = \Delta H/k$ and $c_1T_0 = \log[\eta(T_0)/\eta_0]$), is for the viscosity of glasses a central issue and still a matter of debate. We discuss here *free volume theory* (FVT) leading to a discontinuous transition and, rather briefly, *continuous transformation theory* (CTT) leading to a continuous transition.

18.3.1 Free Volume Theory

A simple way to derive the WLF equation is using free volume arguments. The free volume v_f is defined by $v_f = v - v_0$, where v is the total macroscopic volume and v_0 the molecular, "occupied" volume, which includes the volume of molecules as represented by their van der Waals (vdW) volume and the volume associated with their vibrational motion (Section 9.11). We assume that v below T_g increases to first order linearly with temperature, while above T_g the volume v shows an extra expansion due to v_f, also linear to first order with T

and suggesting that the extra free volume above that of the RCP structure largely starts to be introduced at about T_g. Neglecting that T_g is dependent on q^*, we have for the fractional free volume $f = v_f/v$, the relation

$$f = f_g + \alpha_f(T - T_g) \tag{18.5}$$

with $f_g = f(T_g)$ and α_f the expansivity of v_f. We take for the shift factor $a_T = \eta_T/\eta_{T_g}$, which combined with the empirical *Doolittle equation* for monomeric liquids [43]

$$\eta = a \exp(bv/v_f) = a \exp(b/f)$$

where a and b are constants, with b having a typical value $b \cong 1$, results in

$$\ln a_T = b\left(f^{-1} - f_g^{-1}\right) \quad \text{or} \quad \ln a_T = -\frac{(b/f_g)(T - T_g)}{(f_g/\alpha_f) + T - T_g}$$

after substituting Eq. (18.5). This expression has the form of the WLF equation. For many amorphous polymers $f_g \cong 0.025$, while a reasonable value for $\alpha_f \cong 4.8 \times 10^{-4}\ \text{K}^{-1}$. Substituting these values, we obtain $c_1{}^g = b/f_g \cong 40$ and $c_2{}^g = f_g/\alpha_f \cong 52$, providing an interpretation for the constants $c_1{}^g$ and $c_2{}^g$. It also leads to a discussion whether a real transition temperature, given by $T_g - c_2{}^g$, exists at or near T_K.

In a somewhat more sophisticated treatment of FVT by Turnbull and Cohen [44], it is assumed that molecular transport can be based on hard spheres (HSs) with mass m and diameter σ having repulsive interactions only and the attractive interactions affecting only the average density. The material is supposed to be a Frenkel-like liquid having holes with a size distribution, for which a crystallization pathway is excluded. Each molecule is supposed to "rattle" in a cage formed by its nearest neighbors. Diffusion is treated as translation of a molecule across the void within its cage, and therefore occurs not because of the usual activated process but rather as the result of the redistribution of free volume within the liquid without energy change. The self-diffusion coefficient D for such a model can be expressed as

$$D = {}^1\!/\!_3\langle u\rangle \int_0^\infty p(v)c(v)a(v)\,\mathrm{d}v = {}^1\!/\!_3\langle u\rangle \int_{v^*}^\infty p(v)c(v)a(v)\,\mathrm{d}v \tag{18.6}$$

with $p(v)$ the probability that an element of free volume v lies between v and $v + \mathrm{d}v$ and $\langle u\rangle = (3kT/m)^{1/2}$ the average gas kinetic velocity. For the correlation factor $c(v)$ for displacements within a cage with free volume v, we expect that $c(v) \to 0$ as $v \to 0$ and $c(v) \to 1$ as $v \to \infty$. The function $a(v)$ describes the jump distance as a function of volume. The value of the self-diffusion coefficient D is zero unless v exceeds a critical volume v^* just large enough to permit another molecule to jump in after the displacement. Hence, the lower limit in Eq. (18.6) can be replaced by v^*.

We now need an expression for $p(v)$. If the total range of the free volume values for a molecule is divided into small regions j having average value v_j, we have

$$\gamma \sum_j N_j v_j = V_f \quad \text{and} \quad \sum_j N_j = N \tag{18.7}$$

with N_j the number of particles having v_j, N the total number of particles, $V_f = N\bar{v}$ the total free volume with $\bar{v}$ the average free volume, and γ a numerical factor to correct for overlap of free volume, likely to have a value between ½ and 1. The number of ways to redistribute the free volume without changing values of N_j is $W = N_j!/\Pi_j N_j!$. Maximization at constant N and V_f using the Lagrange multipliers λ and β leads to

$$N_j = \exp[-(\lambda + \beta v_j)] \quad \text{or} \quad p(v) = (\gamma/v_f)\exp(-\gamma v/\overline{v})] \tag{18.8}$$

after evaluating γ and β from Eq. (18.7) and passing to the continuum limit for v_j.

To calculate D, as a first approximation we use for the correlation function a step function (Section 9.14): $c(v) = 0$ for $v < v^*$ and $c(v) = 1$ for $v > v^*$. It corresponds to dividing the molecules sharply into two categories, one "gas-like" and the other "solid-like." Further, also as a first approximation, we use $a(v) = \alpha v$. Substituting these approximations and Eq. (18.8) in Eq. (18.6b), the result is

$$D = {}^1\!/\!_3\langle u\rangle\frac{\gamma}{\overline{v}}\int_{v^*}^{\infty}\alpha v\,\exp\left(-\frac{\gamma v}{\overline{v}}\right)\mathrm{d}v = {}^1\!/\!_3\langle u\rangle\alpha\left[v^* + \left(\frac{\overline{v}}{\gamma}\right)\right]\exp\left(-\frac{\gamma v^*}{\overline{v}}\right)$$

This expression reduces to $D = {}^1\!/\!_3\langle u\rangle\alpha v^*\exp(-\gamma v^*/\overline{v})$ when $v^* \gg \overline{v}/\gamma$ and to $D = {}^1\!/\!_3\langle u\rangle\alpha\overline{v}/\gamma$ when $v^* \ll \overline{v}/\gamma$. Dividing by ${}^1\!/\!_3\langle u\rangle\alpha\overline{v}/\gamma$ yields for the correlation factor:

$$\langle c\rangle = [1 + (\gamma v^*/\overline{v})]\exp(-\gamma v^*/\overline{v})$$

Both D and $\langle c\rangle$ extrapolate to the proper limit when $\overline{v} \to 0$ or $\overline{v} \to \infty$. From Monte Carlo (MC) simulations of HSs (Section 8.2) it became clear that an HS system shows a discontinuous transition from the liquid state at $v = 1.50v_0$ to an FCC solid state with $v = 1.36v_0$, where v_0 is the volume per particle for a close-packed lattice. Both values are higher than $v_{\mathrm{RCP}} = 1.16v_0$ resulting from random close-packed HSs, which is identified with the metastable "ideal" glass state of monoatomic fluids. Using the Stokes–Einstein equation $D = kT/3\pi\sigma\eta$, the above also leads to the Doolittle equation, although the Stokes–Einstein relation is not generally valid for supercooled liquids [45–47].

With the extra free volume appearing only above a certain T_2, we have $\overline{v} = \alpha v_{\mathrm{m}}\Delta T$, where α and v_{m} are the mean value of thermal expansivity and molecular volume over the range $\Delta T = T - T_2$. If the vdW volume of the particle is independent of pressure P, we may also introduce pressure dependence by writing $\alpha v_{\mathrm{m}}\Delta T - \kappa v_P\Delta P$, where α and v_{m} are evaluated at $P = 0$ and v_P is the average molecular volume over the range $\Delta P = P - P_0$. The reduced expression for D then becomes

$$D = {}^1\!/\!_3\langle u\rangle\alpha v^*\exp[-\gamma v^*/(v_{\mathrm{m}}\alpha\Delta T - v_P\kappa\Delta P)] \tag{18.9}$$

Clearly, the temperature dependence of the free volume is to be explained and for that we consider the free volume a bit closer. In Section 9.1, we indicated that v_f depends on the expansion factor with respect to a close-packed reference. For a cage radius r close to the value r_0 for the minimum in potential energy for a particle in its cage, the potential is approximately parabolic, the particle vibrates, and no extra free volume above that of the RCP structure arises. With increasing temperature $r > r_0$, the potential becomes relatively flat over the dimensions of the cage with respect to kT (except near the cage walls, where it rises steeply). At a certain T_2, a molecule within a cage can move more or less freely and a redistribution of v without energy change can take place. This transition temperature between the two regimes can be identified with $T_{\mathrm{g}} - c_2{}^{\mathrm{g}}$, although, dependent on the cooling rate, the structure is already jammed at T_{g}. This argument rationalizes the behavior of the volume as shown in Figure 18.1b.

For a few vdW fluids, such as carbon tetrachloride, 2,2- and 2,3-dimethyl butane, and *i*-butyl bromide, the reduced expression for D was fitted as a function of T and P using γv^* and T_0 as parameters with quite acceptable results. The value obtained for v^* is close to the vdW volume v_0 ($0.66v_0 < v^* < 0.86v_0$ for $\gamma = 1$ and $1.32v_0 < v^* < 1.7v_0$ for $\gamma = ½$), while T_0 is close to T_g. It was also shown for a somewhat wider series of molecules that the principle of corresponding states (PoCS) holds for the viscosity [44].

In a later stage, a "new", more solid FVT was constructed by Grest and Cohen, but that did not change the overall picture significantly. For the elaborate mathematical details, we refer to [48]. One of the criticisms on FVT is that, if $v_f \sim \Delta T$, which, if true at all temperatures, leads to $1/\eta \sim \Delta T$ and thus to $\eta \to \infty$ for $T \to T_\infty$. Such an essential singularity is difficult to imagine. The new FVT [49] leads to

$$v_f = (v_a/T_0)[\Delta T_0 + (\Delta T^2 + TT_1)^{1/2}]$$

with v_a, T_0, and T_1 constants. At high T, $v_f \sim \Delta T$ but v_f vanishes only when $T \to 0$, consistent with VFT behavior at high T and a change to Arrhenius behavior near T_g, as experimentally often observed for inorganic, ionic, and network glasses. For polymers this change occurs at an unobservable high viscosity because the pre-exponential factor η_0 increases with molecular weight M as M^ν with $\nu = 3.0$–3.4.

FVT has been applied to polymers as well, but in this case purely free volume explanations have met serious objections. Clearly, polymers do not have spherical structural units. Moreover, from the experimental data on β-relaxations it is inferred that v_0 is not a constant for constant v while varying T or P [50]. This implies that a significant reinterpretation of v_f for polymers, one that is likely not to be so clearly linked with the geometrical starting point of FVT, is required.

18.3.2 Continuous Transition Theory

The second approach, the CTT due to Gibbs, DiMarzio and Adam [51], was developed in its simplest form by taking as the basic unit a linear polymer. These authors used Flory–Huggins lattice statistics for the number of complexions to calculate the entropy ΔS. Their analysis required that V and H are functions of P and T only. Using the transition state theory, they derived that the frequency of molecular jumps ν for polymeric liquids is given by

$$\nu = A\exp(-n\Delta G/kT) \cong (kT/h)\exp(-n\Delta G/kT) \qquad (18.10)$$

with ΔG the Gibbs energy of a single segment of the n cooperating segments rearranging. The other symbols have their usual meaning. If s_n is the entropy of a unit of n segments, assumed to be independent of the temperature, the total entropy S for a mole of segments is $S = N_A s_n/n$. They calculated that at a temperature T_S the entropy ΔS becomes zero, thereby predicting a continuous transition at that temperature. This implies that n becomes infinite and that there are no configurations left for further rearrangement. The entropy ΔS can be calculated directly from difference in heat capacity ΔC_P between the glass and the supercooled liquid state at T_g according to

$$\Delta S(T) = \Delta C_P \ln(T/T_S) = \Delta C_P \ln(1 + \Delta T/T_S) \cong \Delta C_P \Delta T/T_S \quad \text{with} \quad \Delta T = T - T_S$$

where it is assumed that ΔC_P is constant over the temperature range considered. Substituting all these in Eq. (18.10) results in a relaxation time τ given by

$$\tau = \nu^{-1} = \tau_0 \exp[b/(T - T_S)]$$

with $\tau_0 = 1/A$ and $b = N_A s_n \Delta G/k\Delta C_p$. This is the WLF equation if we identify T_S with $T_g - c_2{}^g$. To calculate T_g, Gibbs and DiMarzio used $\Delta\varepsilon$, the energy difference between the trans and gauche conformations of a hydrocarbon-type polymer chain, and 2α, the hole formation energy. Using experimental data for $\Delta\varepsilon$ and 2α, they calculated T_g for monodisperse polystyrene T_g as a function of the degree of polymerization. The results agree quite well with the experimental data. We refrain from further discussion as further developments for CTT have been described by Donth [26].

Eisenberg and Saito [52] observed that when $\Delta\varepsilon/k$ and $2\alpha/k$ are plotted against each other, a straight line with a slope of 1 is observed within the experimental error. Because $\Delta\varepsilon/k$ and $2\alpha/k$ are both a function of v_f alone, this implies that v_f at T_g is apparently constant. The product $T_g\Delta\alpha$, with $\Delta\alpha$ the difference between the expansivities above and below T_g, is also a constant so that this quantity too is a function only of v_f. The "universal" values obtained are $v_f = 0.037$ and $T_g\Delta\alpha = 0.116$. The latter value is close to 0.113, obtained by Simha and Boyer [53] independent of CTT. Hence, for homopolymers the CTT and FVT approaches are practically equivalent. This equivalence was also discussed for alloys [54] and polymers [55].

Be that as it may, at about the same time Macedo and Litovitz [56] combined VFT and CTT. These authors tested their theory on B_2O_3, which is notoriously difficult to describe properly but for which data over a large η-range are available. In a quasi-lattice picture, the probability of a jump from one site to another is given by $p = p_E p_V$, where p_E is the probability of attaining sufficient energy to break bonds and p_V is the probability that there is sufficient local free volume for a jump to occur. They classified liquids as free volume limited when $p_E > p_V$ and energy limited when $p_E < p_V$. Above 800 °C, for B_2O_3 $p_E \cong p_V$, while at lower temperatures it becomes free volume limited, thereby explaining its resistance to description. As possibly expected, SiO_2 and GeO_2 are energy limited ($p_E < p_V$). Other materials tested were vdW liquids ($p_E \cong p_V$), alcohols ($p_E < p_V$), Ar and Hg ($p_E \cong p_V$), some silicate mixtures ($p_E \gg p_V$), and some polymers ($p_E > p_V$). Clearly, whether a liquid is free volume limited or not depends on the temperature range considered. A statistical–mechanical derivation of the Macedo–Litovitz expression is available [57].

However, Brummer [58] pointed out that all volume effects are incorporated in the P_V term while the activation energy E_{act} is taken as independent of density ρ. The latter is not true, as one might expect, and which experiments indeed show. For example, for formamide E_{act} doubles on a 10% decrease in molar volume. The success of the equation proposed probably lies in the similarity of the dependencies of E_{act} and P_V on V and must be regarded as somewhat fortuitous. Evaluation of E_{act} should not be made only as a function of T, but also as a function of ρ; in other words, one must obtain E_{act} directly from its temperature dependence at constant ρ.

This discussion shows the considerable confusion that exist(ed) over many years. Actually, the discussion whether a "real" transition is underlying the glass transition is still open to debate but seems to be somewhat academic as sufficiently low cooling rates to reach this transition are impossible. Further, in a long-lasting crusade, Mnyukh [59] contested on good grounds the existence of continuous transformations in condensed phases in general and provided strong arguments that such transformations are always discontinuous. There are other experimental arguments that no "real" transition occurs but only a freezing in of the random structure in one of its, almost equivalent from an energetically point of view, nonequilibrium structures above the crystalline state [60, 61]. This is, in fact, also the point of view of the lattice gas, phonon, and mode-coupling theories, to which we turn in Sections 18.4 and 18.5.

18.4 The Lattice Gas Approach

In Chapter 9, we discussed the lattice gas model for liquids. This approach can be extended to glasses by adding some further considerations about the allowable configurations of a connected chain, a *n-mer* or *polymer*, of building blocks, *monomers*, employing again a more liberal interpretation of the term polymer.

We use here the model initiated by Gutzow [62, 63] in which a three-dimensional (3D) lattice with coordination number z is filled with linear chains, monomers, and holes. For a given temperature it is assumed that of the $N + N_0$ lattice sites, N are occupied, either with n-mers or monomers, and N_0 are empty (holes). As an internal variable the fraction of holes $\xi = N_0/(N + N_0)$ is used. For n-mers N_j with x_j as their degree of association, the average $\overline{x} = \Sigma_j x_j N_j / \Sigma_j N_j$, so that $N = \overline{N}\overline{x}$ with $\overline{N}$ the total number of chains. Further it is assumed, following Flory, that for the $\overline{x} - 1$ bonds of the polymer a fraction f is in a completely flexible state, while a fraction $1 - f$ is in a completely rigid state.

After a somewhat complex counting of the number of complexions W, avoiding approximations made in CTT, the entropy change is given by [31, 62, 64]

$$\Delta S = k \ln W = \Delta_\xi S + \Delta_{\overline{x}} S + \Delta_f S + \Delta_{\text{conf}} S \quad \text{with} \tag{18.11}$$

$$\Delta_\xi S = -R\left[\frac{\eta}{1-\eta}\ln\eta + \frac{1}{\overline{x}}\ln(1-\eta)\right], \qquad \Delta_{\overline{x}} S = -\frac{R}{\overline{x}}[\overline{x}\ln\overline{x} - (\overline{x}-1)\ln(\overline{x}-1)]$$

$$\Delta_f S = -R\left(1 - \frac{1}{\overline{x}}\right)[f\ln f + (1-f)\ln(1-f)], \quad \Delta_{\text{conf}} S = -R\left(1 - \frac{1}{\overline{x}}\right) f \ln\left[\frac{(z-1)}{\text{e}}\right]$$

Here $\Delta_\xi S$ is due to free volume changes, $\Delta_{\overline{x}} S$ is due to mixing of chains of different length, $\Delta_f S$ is due to the interchange of flexible and rigid bonds, and $\Delta_{\text{conf}} S$ is due to the configurations of an average chain with degree of association $\overline{x}$ and flexibility f. The entropy is thus fully determined by ξ, f, and $\overline{x}$, and in this approximation independent of the type of size distribution. The term $\Delta_{\overline{x}} S$ vanishes for both $\overline{x} = 1$ and $\overline{x} \to \infty$ and has a maximum $-R\ln 2$ at $\overline{x} = 2$. This term is thus particularly important for small association numbers as do occur, for example, in some glass-forming alloys.

For the energy it is assumed that

$$U = U_0 N_0 + (U_1 f - U_2)\overline{N}(\overline{x} - 1) \tag{18.12}$$

where U_0 represents the molar energy to form a hole. The parameter U_1 represents the molar energy for the excitation of a rigid bond to a flexible bond and U_2 is a parameter representing that, upon formation of an associated chain with a degree of association $\overline{x}$, the molar energy of the system decreases by $U_2(\overline{x} - 1)$. After inserting all the relevant expressions in

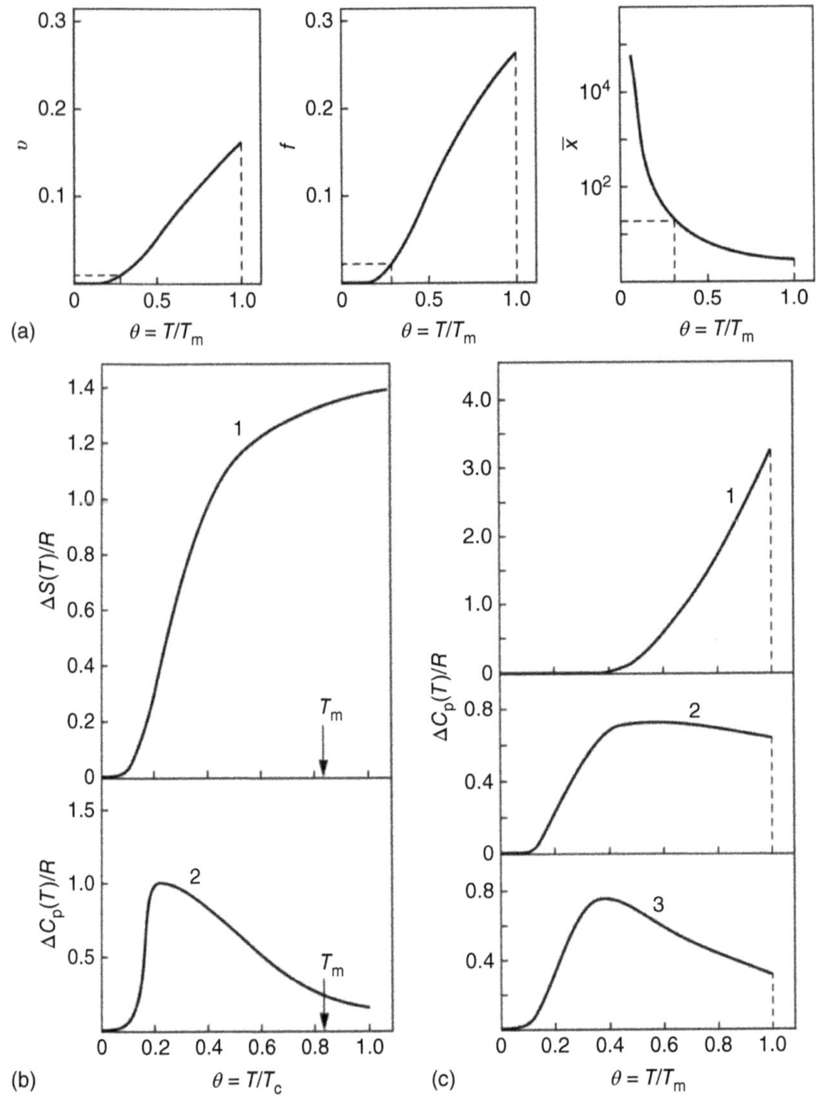

Figure 18.7 Results from the Gutzow–Milchev–Petroff analysis for glasses [22]. (a) Temperature dependencies of the relative free volume ξ, the flexibility f, and the mean degree of polymerization $\overline{x}$ for a model melt [65]. Dashed lines indicate the frozen-in values of the internal variables for $T/T_m = 0.3$. Source: Adapted from [22, 66]. (b) Temperature dependence of (1) the configurational entropy ΔS and (2) the specific heat ΔC_p of polyethylene (in reduced variables) with energy constants reflecting the real system. Source: Adapted from Milchev and Gutzow [66]. (c) Temperature dependence of the specific heat difference ΔC_p for three model melts with different degrees of complexity. Source: Adapted from Petroff et al. [67]: (1) Simple liquid without association, (2) liquid with flexible T-dependent linear n-mer aggregates, and (3) liquids with short aggregates as observed in some alloys.

the expression for the Helmholtz energy and minimizing with respect to ξ, f, and $\overline{x}$, one obtains after a somewhat lengthy calculation

$$\ln \xi = -\left[\frac{U_0}{RT} + (1-\xi)(1-\overline{x}^{-1})\right], \quad \ln f = -\frac{U_1}{RT}\left[\frac{z-1}{e} + \exp\left(-\frac{U_1}{RT}\right)\right]^{-1} \tag{18.13}$$

$$\overline{x} = 1 + \frac{1}{1-f}\left[(1-\xi)\exp\left(\frac{U_2}{RT}\right)\right]$$

Figure 18.7a shows the internal variables ξ, f, and $\overline{x}$ as a function of T for a model system representing an under-cooled polymer melt, calculated using representative U_0, U_1, and U_2 values. Inserting the equilibrium values of the internal variables in Eqs. (18.11)–(18.13), the temperature dependence of the entropy and heat capacity of the system can be obtained (Figure 18.7b).

A few general conclusions can be made. First, CTT [51] predicts that ΔS becomes zero at $T_S \neq 0$, and hence a continuous transition was suggested. It seems that this result is largely due to a too approximate counting of the number of complexions [31]. For the improved counting, ΔS and the other thermodynamic functions show a gradual decrease to zero entropy with decreasing T without any indication that a phase transition occurs (Figure 18.7a). Usually, the value $\Delta S = 0$ is approached for temperatures $T/T_m \sim 0.1$–0.2. A review of the third law aspects is given in [68].

Second, the temperature dependence of ΔC_P is determined largely by the structure and the complexity of the building blocks of this model system. For simple and normal liquids containing only monomers, monotonically increasing concave ΔC_P-curves are found, while for complex liquids, where association can occur, ΔC_P generally shows a maximum in the range $0 < T < T_m$. In the range $T_g < T < T_m$, $\Delta C_P \cong const.$ is typically a good approximation. The analysis shows that all types of temperature dependencies of ΔC_P which are observed experimentally can be modeled. In MC simulations of lattice models, employing n-mers with a temperature-dependent flexibility a λ-like peak in ΔC_P is observed, which, however, seems to be due to ordering of the n-mers.

Clearly, discrete lattices are used to describe continuous systems, and this may significantly affect the conformational possibilities of the arrangement of the n-mers as compared with continuous systems. Nevertheless, the model can describe the glass transition semi-quantitatively. It is related to *kinetically constrained models* for the glass transition, essentially cellular automatons with dynamical rules that try to approximate the cooperative nature of relaxation in the supercooled state [69].

18.5 Phonon Theory for Glasses

In Section 9.15, we introduced the phonon theory of liquids. This theory has been applied to glasses as well [70]. As stated in Section 18.1, the most eye-catching phenomenon for the glass transition is the loss of rigidity, so that elasticity is a relevant property [71]. The phonon theory attempts to understand the glass transition on the basis of thermoelastic properties, different for the liquid and glass states.

The basic premise of phonon theory, using ideas forwarded by Frenkel [72], is that each change in configuration requires a jump of a molecule from its surrounding *cage*, accompanied by large-scale rearrangement of the cage molecules, a *local relaxation event* (LRE) lasting on the order of Debye vibration period $\tau_0 \approx 0.1$ ps. We need to consider two basic aspects. First, the Maxwell liquid relaxation time $\tau = \eta/G_\infty$, where η is liquid viscosity and G_∞ the instantaneous shear modulus. Frenkel offered a microscopic interpretation for τ as the time between LREs at one point in space in a liquid [72]. Second, an LRE requires an increase of local volume. The activation barrier for an LRE at constant cage volume is large due to strong short-range interatomic repulsions [28, 72]. Hence, the molecules of the cage need to increase their volume v_c in order for the central molecule to be able to escape. In doing so, elastic work is performed to deform the surrounding liquid, equal to the activation barrier U for an LRE [28, 72]. As such a barrier is to be surmounted by temperature fluctuations, we have $\tau = \tau_0 \exp(U/kT)$ [72].

On the short timescale of an LRE (when $\tau_0 < \tau$), the surrounding liquid can be viewed as an elastic medium [72] and high-frequency waves with $\omega > 1/\tau$ are propagating in the liquid as in a solid [72], as has been experimentally confirmed (Section 9.15, [73]). Because an LRE elastically deforms the surrounding liquid rapidly, it induces such elastic waves. They deform the cages around other molecules in a liquid, thereby affecting their relaxation time because, as indicated above, jumping of a molecule depends on the configuration of its cage. The wavelengths being in the order of the intermolecular spacings, the frequency ω is in the order of Debye frequency $1/\tau_0$. This implies that in almost the entire range of τ relevant for the glass transition, $\omega > 1/\tau$. Therefore, LREs interact via the elastic waves they induce, and this is, in this model, the physical origin of cooperativity of relaxation in a liquid. With c being the speed of sound, the *liquid elasticity length* $d_{ela} = c\tau$ increases with decreasing temperature mainly because τ increases. Although the way in which these waves are generated by LREs may be system specific, the interaction of LREs via induced elastic waves is general and should apply to all liquids undergoing a glass transition.

In between LREs, the molecules vibrate harmonically with small amplitudes, like in solids. An LRE requires large-scale molecular displacements, say, $\Delta r \cong 1$–2 Å. For such large values of Δr, the anharmonicity of the potential is important. As a decrease by Δr results in a repulsive interaction that is always stronger than the attractive interaction for an increase by Δr, during an LRE a short-lived expansion of the local cage volume v_c around the LRE occurs. This local volume expansion Δv_c is propagated away from an LRE by the high-frequency elastic waves.

When a sphere expands in a static elastic medium, no compression takes place at any point. The strain components u from an expanding sphere (noting that $u \to 0$ as $r \to \infty$) are $u_{rr} = -2b/r^3$, $u_{\theta\theta} = u_{\phi\phi} = b/r^3$ [74], resulting in pure shear. Hence, the energy to quasi-statically expand a sphere of radius r by an amount Δr depends on the shear modulus G only: $E = 8\pi G r \Delta r^2$ [72]. However, there is compression at the front of an expanding wave. At time t, the wavefront causes an outward displacement of the molecules on a sphere of radius ct, taking place within τ_0 because the wave is due to a remote LRE that lasts approximately τ_0. This causes the displacement of a concentric spherical shell with radius of $ct + x$, where x is on the order of the intermolecular spacing. The result is a compressive stress in this layer of thickness x, which exists over a time approximately equal to the period of vibration of the molecules τ_0. This results in more work needed to expand the cage, thereby increasing the activation barrier U for the central LRE and slowing down the dynamics. As the wave propagates further, the *dynamic compressive stress* (DCS) between two layers thereafter disappears, and shear deformation is established as indicated above.

The activation energy U is equal to the total work required to expand the cage by an amount required for an LRE to occur [72]. We write $U = U_0 + U_1$. Here, U_0 is the high-temperature activation barrier that depends on liquid interatomic forces and structure, but not on the elastic interactions of LREs. In other words, it is noncooperative or intrinsic. The term U_1 represents the interaction between LREs, the cooperative term, and is equal to the additional work necessary to expand the cage for escape of the central molecule as delivered by the waves from other LREs. The associated increase in cage volume is, say $\Delta v_c \cong a^3$, with a the intermolecular spacing. If a DCS wave generated by an LRE j at distance r from the central molecule generates a stress given by $s_j(r)$, such a wave contributes the work $\Delta v_c s_j(r)$ to U_1, and thus the total work is $U_1 = \Sigma_j \Delta v_c s_j(r)$. For U, we thus have, from N_{rem} remote LREs,

$$U = U_0 + U_1 = U_0 + \sum_j^{N_{rem}} \Delta v_c s_j(r) \tag{18.14}$$

Because the central event relaxes during time τ, waves from all events located a distance $c\tau$ away from the central one, have sufficient time to propagate to the center. Further, a remote event also relaxes during time τ, and it contributes only one wave during the time of relaxation of the central event. Each LRE inside a sphere with radius $d_{ela} = c\tau$ thus contributes only once, thereby accounting for the dynamical nature of LREs as well as for wave dissipation. Next, we consider that the compressive stress $s(r)$, present in the spherical shell of thickness x only, decreases with r because it propagates like an elastic wave. As $s(r)$ is due to the strain u_{rr}, which decays like r^{-3}, $s(r)$ decays likewise. If s_0 is the value of $s(r)$ at the cage boundary at $d_0/2$ from its center, we have $s(r) = s_0(d_0/2r)^3$. Accordingly, inserting the above and taking the continuum limit, we can rewrite Eq. (18.14) as

$$U = U_0 + a^3 \int_{d_0/2}^{d_{ela}} 4\pi\rho r^2 s(r)\,\mathrm{d}r = U_0 + 3\Delta v_c s_0 \ln(2d_{ela}/d_0)$$

with local density $\rho = 6/\pi d_0{}^3$ and diameter of the LRE region under consideration d_0. Because U increases with d_{ela} and $d_{ela} = c\tau$ increases with $\tau = \tau_0 \exp(U/kT)$, U is defined in a self-consistent way. With $a \cong c\tau_0$, we obtain $d_{ela} = a\exp(U/kT)$, and thus

$$U = AT/(T - T_0) \quad \text{with} \quad T_0 = 3\Delta v_c s_0/k \quad \text{and} \quad A = U_0 - kT_0 \ln(d_0/2a)$$

corresponding to the activation energy expression for the VFT equation. In phonon theory, the physical origin of the VFT equation is thus the increase in the value of d_{ela} with decreasing temperature, resulting in an increase in the number of LREs that contribute to U.

Because s_0 is due to the local expansion, its value can be estimated as $s_0 = \alpha K T^*$, where α and K are the expansivity and bulk modulus, respectively, and T^* is the local temperature for an LRE, which should be in the order of T_m. If we take $a^3 \cong 1\,\text{Å}^3$, $\alpha \cong 5 \times 10^{-4}\,\text{K}^{-1}$, and $K \cong 10\,\text{GPa}$, the result is $T_0 = 0.5T_m$, in a rough agreement with the experimental data.

The effect of pressure can also be accounted for as well. With increasing external pressure P, the cage volume change increases as $\Delta v_c = \Delta v_c{}^\circ + \Delta v_c{}^\circ K^{-1} P$, where $\Delta v_c{}^\circ$ is the volume change at P_0, usually 1 bar. Moreover, a term $\Delta v_c P$ must be added to U. The value of $T_0(P) = T_0(P_0) + (\partial T_0/\partial P)_{P=P0} P = 3s_0 k^{-1}(\Delta v_c{}^\circ + \Delta v_c{}^\circ K^{-1} P)$ and for small values of P, $K = K_0 + (\partial K/\partial P)_{P=P0} P \equiv K_0 + K_0{}' P$. The overall result is $T_0(P) = T_0(P_0) + 3s_0 k^{-1} \Delta v_c{}^\circ (K_0 + K_0{}'P)^{-1} P \cong 3s_0 k^{-1} \Delta v_c{}^\circ [1 - K_0{}'(P/K_0)](P/K_0)$ and predicting an initial linear increase changing to nonlinear with increasing P, as experimentally observed.

Generally, two so-called crossovers occur for the glass transition. The first one is from Arrhenius at high T to VFT behavior at lower T. The phonon theory predicts that at a certain high temperature T_1, d_{ela} approaches a. At T_1, U reduces to U_0, thereby explaining the high-temperature Arrhenius behavior. At high temperatures, we have $d_{ela} < d_0$, the latter having a value of $\cong$10 Å for small molecules to $\cong$ 100 Å for large molecules. Hence, the waves due to LREs do not interact and each LRE shows an exponential relaxation. Lowering T leads to interactions, which can be shown to have stretched exponential behavior [75]. The corresponding relaxation time $\tau_1 = d_0/c = d_0\tau_0/a$, leading to τ_1 = 1–10 ps by taking $\tau_0 \cong 10^{-13}$ ps, corresponds with $\tau_1 \cong$ 1–30 ps as observed experimentally. The second crossover is from VFT behavior to more Arrhenius-like behavior at still lower T. At a certain T_2, d_{ela} equals the macroscopic size of the system L. At T_2, all LREs interact and lowering T still further will not lead to more interacting LREs. The corresponding relaxation time $\tau_2 = d_0/c = L\tau_0/a$ and, taking $L \cong$ 0.1–10 mm, results in $\tau_2 = 10^{-7}$–10^{-5} s, the range where other properties also show a qualitative change.

Whether T_0 represents a transition temperature or not, can also be addressed. From $d_{ela} = a\exp(U/kT)$ and $U = A/(T - T_0)$, we have $d_{ela} = a\exp[A/(T - T_0)]$. Upon approaching T_0, the temperature reaches $T_2 > T_0$ when $d_{ela} = L$ and the abovementioned crossover occurs. Hence, the singularity at T_0 for VFT behavior is shifted to $T = 0$ for Arrhenius behavior. The phonon theory thus predicts freezing in of the random structure and associated dynamics instead of a phase transition. Finally, we note that the behavior of C_V of glasses (and liquids) according to the phonon theory is given in [76].

18.6 Mode-Coupling Theory

Both the lattice gas and phonon theories of the glass transition are semi-empirical. They treat glasses in much the same way as liquids are treated in structural models. For liquids we started with what we called the integral equation approach, which is in principle ab-initio. MCT tries to do the same job for glasses. The glass transition can be seen as a transition from ergodic to non-ergodic behavior (essentially meaning not visiting all phase points, Section 19.3), or in other words, the system gets kinetically trapped. The theory is inevitably somewhat complex, and we only sketch its main line of approach. Detailed theory is given in [77] and a diagrammatic approach in [78]. Both Reichmann and Charbonneau [79], and Janssen [80] provided accessible introductions, here used as a guide for a compact treatment.

The starting point of MCT is the time-dependent density correlation function or intermediate scattering function $F(q,t)$ for wavenumber $q = |\boldsymbol{q}|$. It is the Fourier transform of the self-correlation function (Figure 9.6), probing correlations in particle density fluctuations over a certain wavenumber q and over a time interval t [81]. It thus characterizes the molecular configuration of a material at a time t with respect to $t = 0$ and changes in time-dependent structure with temperature are reflected in a changing $F(q,t)$. The function $F(q,t)$ is the generalization of the static structure factor $S(q) = F(q,0) = 1 + \int g(r)\exp(-\mathbf{i}\boldsymbol{q}\cdot\boldsymbol{r})\mathrm{d}\boldsymbol{r}$, as used in scattering experiments. This Fourier transform of the (pair) correlation function $g(r)$ (Section 9.6.1) probes in Fourier space the probability of finding a particle at a distance $r \cong 2\pi/q$ away from any other particle. The structure factor is also given by $S(q) = [1 - \rho c(q)]^{-1}$ with $c(q)$ the Fourier transform of the direct correlation function $c(r)$.

If we have N particles with mass m and position $\boldsymbol{r}_j(t)$ at time t, the number density is given by $\rho(\boldsymbol{r},t) = \Sigma_j\delta(\boldsymbol{r} - \boldsymbol{r}_j)$ or equivalently $\rho(\boldsymbol{q},t) = \int \rho(\boldsymbol{r},t)\exp[\mathrm{i}\boldsymbol{q}\cdot\boldsymbol{r}(t)]\mathrm{d}\boldsymbol{r} = \Sigma_j\exp[\mathrm{i}\boldsymbol{q}\cdot\boldsymbol{r}_j(t)]$, and $F(q,t) = N^{-1}\langle\rho(-\boldsymbol{q},0)\rho(\boldsymbol{q},t)\rangle$, with as usual $\langle\cdot\rangle$ the canonical average and $q = |\boldsymbol{q}|$. To obtain an equation of motion for $F(q,t)$, the Mori–Zwanzig formalism [82] is employed. The basic idea is to use in the description two mutually orthogonal subspaces, one containing the variables of interest and other containing the other variables, to describe how the dynamics of the relevant variables evolves over time in the presence of the others. One of the ways to solve such a multivariable problem is, if a separation of time scales in the dynamics is possible, to try for the system at hand (Section 6.1) to separate slow and fast variables and integrate the fast ones out (as done in Sections 19.4 and 19.5 for quantum systems). For details, we refer to [81, 83].

For the description of glasses, we take for the slow and relevant variables the density $\rho(\boldsymbol{q},t)$ and the associated current $j(\boldsymbol{q},t) = \mathrm{d}\rho(\boldsymbol{q},t)/\mathrm{d}t \equiv \dot{\rho}(\boldsymbol{q},t) = \mathrm{i}\Sigma_j(\boldsymbol{q}\cdot\boldsymbol{r}_j)\exp(\mathrm{i}\boldsymbol{q}\cdot\boldsymbol{r}_j)$. For large length scale, $j(\boldsymbol{q},t)$ vanishes so that overall the density is conserved. A convenient notation, combining $\rho(\boldsymbol{q},t)$ and $j(\boldsymbol{q},t)$ in a (pseudo-)vector, is

$$\boldsymbol{A}(t) = \begin{pmatrix} A_1(t) \\ A_2(t) \end{pmatrix} = \begin{pmatrix} \rho(\boldsymbol{q},t) \\ j(\boldsymbol{q},t) \end{pmatrix}$$

so that we can write

$$F(q,t) = N^{-1}\left\langle A_1{}^*(0)A_1(t)\right\rangle \quad \text{and} \quad \mathbb{C}_{\alpha\beta}(t) \equiv \left\langle A_\alpha{}^*(0)A_\beta(t)\right\rangle$$

or $NF(t) = \langle A_1{}^*(0)A_1(t)\rangle = C_{11}$ and $\mathbb{C}(t) = \langle\boldsymbol{A}^*(0)\cdot\boldsymbol{A}(t)\rangle$. In this way we can use the projection operator $P_A = \boldsymbol{A}^*\cdot\mathbb{C}(0)^{-1}\cdot\boldsymbol{A}$ or $P_A = \Sigma_{\alpha\beta}A_\alpha{}^*\mathbb{C}_{\alpha\beta}(0)^{-1}A_\beta$, extracting all the slow character P_AX defined through P_A for an arbitrary variable X, leaving the remaining part of X orthogonal to A. The procedure is analogous to that used for QM. We now invoke the Liouville operator $\mathbb{L}$ (Section 2.5) and write $\boldsymbol{A}(t) = \exp(\mathbb{L})\boldsymbol{A}(0)$. Separating the unit operator in $\mathbb{1} = P_A + \mathbb{1} - P_A$ and the Liouville operator similarly in a slow part and its orthogonal remainder, one can obtain after a somewhat lengthy procedure the *generalized Langevin equations*

$$\mathrm{d}\boldsymbol{A}(t)/\mathrm{d}t = \mathrm{i}\mathbb{W}\cdot\boldsymbol{A}(t) - \int_0^t \mathbb{K}(t')\cdot\boldsymbol{A}(t-t')\,\mathrm{d}t' + \boldsymbol{f}(t) \quad \text{and}$$

$$\mathrm{d}\mathbb{C}(t)/\mathrm{d}t = \mathrm{i}\mathbb{W}\cdot\mathbb{C}(t) - \int_0^t \mathbb{K}(t')\cdot\mathbb{C}(t-t')\,\mathrm{d}t' \tag{18.15}$$

in which $\mathbb{W}$ is the so-called *frequency operator*, a tensor describing the slow part of $\mathrm{d}\boldsymbol{A}(t)/\mathrm{d}t$, and $\boldsymbol{f}(t)$ a fast fluctuating vector function defined by $\boldsymbol{f}(t) = \exp[\mathrm{i}(\mathbb{1} - P_A)\mathbb{L}t]\times\mathrm{i}(\mathbb{1} - P_A)\mathbb{L}\boldsymbol{A}(0)$. The *memory function* $\mathbb{K}(t)$ is given by $\mathbb{K}(t) = \langle\boldsymbol{f}^*(0)\boldsymbol{f}(t)\rangle\cdot\mathbb{C}^{-1} = \langle\boldsymbol{f}^*(0)\mathrm{e}^{(1-P_A)\mathbb{L}t}\boldsymbol{f}(0)\rangle\cdot\mathbb{C}^{-1}$. This implies that first all the slow character from $\mathrm{d}\boldsymbol{A}(t)/\mathrm{d}(t)$ is removed using the complementary projection operator $\mathbb{1} - P_A$, which is subsequently propagated in time in the fast subspace orthogonal to $\boldsymbol{A}$. The time autocorrelation function of $\boldsymbol{f}(t)$ yields $\mathbb{K}(t)$, which represents the dissipative term that ultimately breaks the conservation of $\boldsymbol{A}$. That is to say, $\mathbb{K}(t)$ and $\boldsymbol{f}(t)$ describe how the slow variable $\boldsymbol{A}$, which at $t = 0$ can be completely described in the slow subspace, will evolve under the influence of the environment. So far, all what is said is exact.

Generally, there is no exact solution for $\mathbb{K}(t)$ and thus some approximation must be made to proceed. One way is using a superposition approximation for $\mathbb{K}(t)$, akin to the Yvon–Born–Green (YBG) approach for liquids. This is done by, using as basis $\rho(\boldsymbol{q}_1,t)\rho(\boldsymbol{q}_2,t)$, the description

$$f(t) = \sum\nolimits_{\boldsymbol{q}_1,\boldsymbol{q}_2} \rho(\boldsymbol{q}_1, t)\rho(\boldsymbol{q}_2, t)$$

so that

$$\mathbb{K}(t) = \sum\nolimits_{\boldsymbol{q}_1,\boldsymbol{q}_2,\boldsymbol{q}_3,\boldsymbol{q}_4} \langle \rho^*(\boldsymbol{q}_1, 0)\rho^*(\boldsymbol{q}_2, 0)\mathrm{e}^{(1-P_A)Lt}\rho(\boldsymbol{q}_3, t)\rho(\boldsymbol{q}_4, t)\rangle$$

The four-point correlation functions $\langle\rho^*(\boldsymbol{q}_1,0)\rho^*(\boldsymbol{q}_2,0)\rho(\boldsymbol{q}_3,t)\rho(\boldsymbol{q}_4,t)\rangle$ are now approximated by the product $\langle\rho^*(\boldsymbol{q}_1,0)\rho^*(\boldsymbol{q}_2,0)\rangle\langle\rho(\boldsymbol{q}_3,t)\rho(\boldsymbol{q}_4,t)\rangle$, meanwhile replacing $\exp[1 - P_A)Lt]$ by $\exp(Lt)$, necessary as otherwise zero would result. The next step is recalling that $\langle\rho^*(\boldsymbol{q}_1,0)\rho(\boldsymbol{q}_2,t)\rangle = \langle\rho(-\boldsymbol{q}_1,0)\rho(\boldsymbol{q}_2,0)\rangle = NF(q,t)$, which implies that $F(q,t)$ is governed by the same type of memory function as applies for $A_{11}(t)$, but for many different wavenumbers. After quite a bit of calculation [79], the result for C_{21} is

$$\frac{\mathrm{d}^2F(q,t)}{\mathrm{d}t^2} + \frac{kTq^2}{mS(q)}F(q,t) + \int_0^t K(q,t')\frac{\mathrm{d}F(q,t-t')}{\mathrm{d}t}\,\mathrm{d}t' = 0 \quad \text{where} \tag{18.16}$$

$$K(q,t) = \frac{\rho kT}{16\pi^3 m}\int \left|V_{\boldsymbol{q}',\boldsymbol{q}-\boldsymbol{q}'}\right|^2 F(q',t)F(|\boldsymbol{q}-\boldsymbol{q}'|,t)\,\mathrm{d}\boldsymbol{q}' \quad \text{and}$$

$$V_{\boldsymbol{q}',\boldsymbol{q}-\boldsymbol{q}'} = (\boldsymbol{q}\cdot\boldsymbol{q}')c(q') + \boldsymbol{q}\cdot(\boldsymbol{q}-\boldsymbol{q}')c(|\boldsymbol{q}-\boldsymbol{q}'|)$$

Here k is Boltzmann's constant and m and c are again the mass of the particles and the Fourier-transformed direct correlation function. Equation (18.16) is a closed, self-consistent equation, subject to the boundary conditions $F(q,0)/S(q) = 1$ and $\mathrm{d}F(q,0)/\mathrm{d}t = 0$.

Analytically solving the MCT equations seems impossible and we have to resort to numerical solutions, for which details are given in [84]. If for a given temperature T and density ρ, $S(q)$ is known, MCT can predict the dynamics of the system. To predict temperature and density dependence clearly requires the availability of a set $S(q;T,\rho)$. Looking at the results, first and foremost, MCT *can predict a glass transition*. This is due to the highly nonlinear nature of MCT which renders it sensitive to rather small changes in $S(q)$. On the downside, the transition occurs at far too high temperature as compared with experimental T_gs. This is due to missing ergodicity-restoring relaxation mechanisms, such as particles "hopping" out of their local cages to resist freezing, which keep the system longer in the liquid state. This results in overestimating caging. Nevertheless, caging is predicted, so that for a molecule temporarily only vibration and rotation in a cage remains. MCT is also capable of predicting the power laws associated with the dynamics, as indicated in Figure 18.8a, albeit it predicts constant exponents while experimentally they are often temperature dependent. Furthermore, scaling properties in the β-relaxation regime are predicted essentially correct. Because only two-point correlations are taken into account, it seems incapable to predict collective relaxation or dynamically heterogeneous motion. However, Inhomogeneous mode-coupling theory (IMCT) employs the response of $F(q,t)$ to changes in an external field. This corresponds with using a three-point density correlation function which upon approaching the transition diverges. Such a divergence suggests that the picture of caging in the β-regime, the rattling of particles in local cages, is actually more subtle. In fact, IMCT

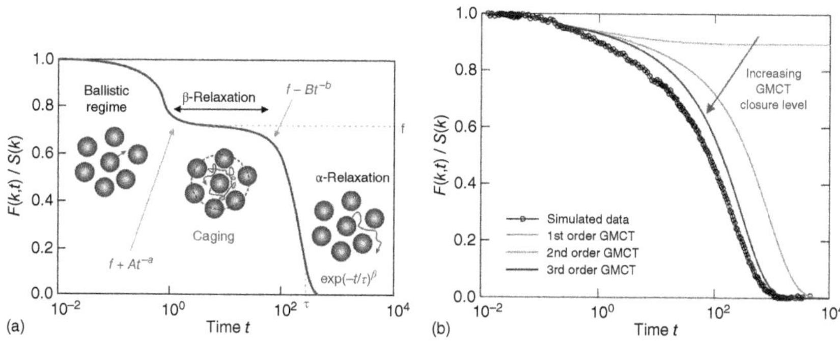

Figure 18.8 (a) Typical MCT prediction for $F(q,t)$ of a supercooled liquid as a function of time, for a wavenumber $q = q_0$ corresponding to the first peak in $S(q)$. Source: Adapted from Janssen [80]. At very short times, particles undergo ballistic motion, while at intermediate times particles become temporarily trapped in cages (β-relaxation) and $F(q,t)$ correspondingly remains approximately constant. Only at sufficiently long times will particles break free and full relaxation takes place (α-relaxation). (b) Microscopic GMCT predictions for $F(q,t)$ [85] compared to numerical data obtained from computer simulations [86] for a system of quasi-hard spheres at packing fraction $\phi = 0.570$ and wavenumber $q\sigma = 7.4$. These results indicate that the GMCT hierarchy apparently converges, and that the theory becomes more quantitatively accurate as the closure level is increased. Source: Adapted from [80].

implies that these cages become more and more collective as the transition is approached, although its predictions are generally not in quantitative agreement with the empirical data.

A route to improve MCT is generalized mode-coupling theory (GMCT) by Szamel [87], which tries to avoid one of the approximations of standard MCT, that is, the factorization of the four-point density correlations appearing in $\mathbb{K}(t)$. A formally exact equation of motion was developed for the four-point correlation functions by applying the Mori–Zwanzig formalism, using as a basis the pair densities $\rho(\boldsymbol{q}_1,t)\rho(\boldsymbol{q}_2,t)$. The resulting equation is governed by another memory function that, to leading order, is controlled by six-point density correlation functions, which in turn are dominated by eight-point correlations, and so on and so forth. Thus, a hierarchy of coupled equations results that can be "closed" at any desired level, akin to the YBG hierarchy for liquids. The approach seems capable of getting systematically improved results (Figure 18.8b).

Apart from GMCT there are several other variations on MCT, such as self-consistent generalized Langevin equation approximation (SCGLE) [88], stochastic beta relaxation theory [89], and extended MCT [90, 91]. Moreover, MCT is part of a wider random first-order transition theory [92], see also Lubchenko [93, 94].

Problem 18.2 ***The Generalized Langevin Equation*** Derive Eq. (18.15).

18.7 Final Remarks

In the classical and technical glass literature, the designations "long" and "short" glasses were invented to indicate that one could work with them in a glass blowing or pressing operation, respectively, relatively long and relatively short before solidification occurs. In other words, they have a long or short working range, related to the behavior of the viscosity as a function of temperature. In 1985, Angell [95] proposed the confusing terms "fragile" and "strong," a labeling that has been taken over by many. Mostly used is the *kinetic fragility index m*, which characterizes the slope of the viscosity η of a material with temperature as it approaches the glass transition temperature T_g from above, and given by

$$m = [\partial \log \eta / \partial (T_g/T)]_{T=T_g}$$

Materials with a higher fragility have a relatively narrow T_g range, while those with low fragility have a relatively broad T_g range, the latter corresponding with a large change in heat capacity at the glass transition. Clearly this relates to the (non)-Arrhenius behavior of glasses and the behavior of fragile glass formers is actively investigated. Analysis of various systems has shown that a steeper interatomic repulsion leads to more fragile liquids, or in other words, soft atoms make strong liquids [96]. It has also become clear that this phenomenon can be linked with the presence of locally heterogeneous dynamics in fragile glass formers, that is, the presence of distinct (if transient) slow and fast regions within the material [97]. It can also be connected to the breakdown of the Stokes–Einstein relation between diffusion and viscosity in fragile liquids [97].

To conclude, it seems clear that gradual freezing in of the structure and dynamics is responsible for the glass transition, contrary to what was largely thought in the past. Many more things could be said, but we just quote several reviews. A broad theoretical overview given by Berthier and Biroli [98]. In a so-called pedagogical review, Lubchenko [93] discussed the physics of the glass transition. Möncke et al. [99] provided a review focusing on the structural relationships between various glass-forming compounds, Zheng et al. [100] reviewed the use of differential calorimetry for glass research, while Mauro et al. [101] discussed various aspects of the fictive temperature concept. Finally, Angell et al. [102] discussed relaxation in glass-forming liquids. In another approach to polymer glassy matter by Ginzburg [103], the material is assumed to be divided into cooperatively rearranging regions where each region can be either in a low-temperature "slow" state or in a high-temperature "fast" state, and which is able to describe the relaxation time together with the *PVT* data. But as Anderson [27] remarked, "Whether it will help make better glass is questionable."

References

1 Zarzycki, 1991.
2 Sadoc, J.-F. and Mosseri, R. (1999). *Geometrical Frustation*. Cambridge: Cambridge University Press.
3 Frost, H.J. (1982). *Acta Metall.* 30: 889.
4 Finney, J.L. (1970). *Proc. R. Soc. A* 319: 479.
5 Bernal, J.D. (1964). *Proc. R. Soc. A* 280: 299.
6 Scholze, H. (1991). *Glas*, 3e. New York: Springer-Verlag (in German).
7 Varsneya and Mauro, 2019.
8 Gracia-Fernández, C.A., Gómez-Barreiro, S., López-Beceiro, J. et al. (2010). *Polym. Test* 29: 1002; Rieger, J. (2001). *Polym. Test.* 20: 199; Achorn, P. and Ferrilo, R.C. (1994). *J. Appl. Polym. Sci.* 54: 2033.
9 Fox, T.G. and Flory, P.J. (1950). *J. Appl. Phys.* 21: 581.
10 Bicerano, J. (2002). *Prediction of Polymer Properties*. New York: Marcel Dekker.

11 Fox, T.G. (1956). *Bull. Am. Phys. Soc.* 1: 123.

12 Fox, T.G. and Loshaek, S. (1955). *J. Polym. Sci.* 15: 371.

13 Hiemenz, P. and Lodge, T. (2007). *Polymer Chemistry*. Boca Raton, FL: CRC Press.

14 Roth, C.B. (ed.) (2021). *Polymer Glasses*. London: CRC Press.

15 Louzguine-Luzgin, D.V. and Inoue, A. (2013). *Bulk Metallic Glasses: Formation, Structure, Properties, and Applications*, Ch. 3 in *Handbook of Magnetic Materials*, Vol. 21 (ed. K.H.J. Buschow), p. 131. Amsterdam: Elsevier.

16 Stachurski, Z.H., Wang, G., and Tan, X. (2021). *An Introduction to Metallic Glasses and Amorphous Metals*. Amsterdam: Elsevier.

17 Chen, M. (2008). *Annu. Rev. Mater. Res.* 38: 445.

18 Böhmer, R.D., Hinze, G., and Rössler, E. (2001). *Prog. Nucl. Magn. Reson. Spectrosc.* 39.

19 Bacon, J.F., Hasapis, A.A., and Whooley, J.W. (1960). *Phys. Chem. Glasses* 1: 90.

20 Kurkjian, C.R. and Douglas, R.W. (1960). *Phys. Chem. Glasses* 1: 19.

21 Mazurin, O.V. (2007). *Glass Phys. Chem* 33: 22.

22 Gutzow and Schmelzer, 2013; (1995). 1e.

23 Simon, F. and Lange, F. (1926). *Z. Phys.* 38: 227.

24 Schultz, A.K. (1954). *Kolloid Z.* 138: 75.

25 Scherer, G.W. (1992). *Relaxation in Glass and Composites*. Malabar, FL: Krieger Publ. Comp.

26 Donth, E.-J. (2001). *The Glass Transition*. Berlin: Springer-Verlag.

27 Anderson, P.W. (1995). *Science* 267: 1615.

28 Dyre, J.C. (2006). *Rev. Mod. Phys.* 78: 953.

29 Kauzmann, W. (1948). *Chem. Rev.* 43: 219.

30 Gutzow, I. and Debrova, A. (1991). *J. Non-Cryst. Solids* 123: 266.

31 Milchev, A. (1983). *Comp. Rend. (Bulgarian Academy of Sciences)* 36: 1415.

32 Wunderlich, B. (1960). *J. Phys. Chem.* 64: 1052.

33 Bartenev, G.M. (1951). *Dokl. Akad. Nauk SSSR* 76: 227; Ritland, H.N. (1954). *J. Am. Chem. Soc.* 37: 370.

34 Schmelzer, J. (2006). *J. Chem. Phys.* 125: 184511.

35 Tropin, T.V., Schmelzer, J.W.P., Gutzow, I. et al. (2012). *J. Chem. Phys.* 136: 124502.

36 Garden, J.-L., Guillou, H., and Richard, J. (2013). *J. Chem. Phys.* 138: 067101; reply Schmelzer, J.W.P. (2013). *J. Chem. Phys.* 138: 067102.

37 Schmelzer, J.W.P. and Tropin, T.V. (2018). *Entropy* 20: 103.

38 Zanotto, E.D. and Mauro, J.C. (2018). *Entropy* 20: 703; reply Schmelzer, J.W.P. and Tropin, T.V. (2018). *Entropy* 20: 704.

39 Leuzzi, L. and Nieuwenhuizen, T.M. (2008). *Thermodynamics of the Glassy State*. New York: Taylor & Francis.

40 Ferry, J.D. (1980). *Viscoelastic Properties of Polymers*, 3e. New York: John Wiley & Sons.

41 van Krevelen, D.W. and te Nijenhuis, K. (2009). *Properties of Polymers*, 4e. Amsterdam: Elsevier.

42 Tobolsky, A.V. and Catsiff, E. (1956). *J. Polym. Sci.* 19: 111.

43 Doolittle, A.K. (1951). *J. Appl. Phys.* 22: 1471; Doolittle, A.K. and Doolittle, D.B. (1957). *J. Appl. Phys.* 28: 901.

44 Cohen, M.H. and Turnbull, D. (1959). *J. Chem. Phys.* 31: 1164; Turnbull, D. and Cohen, M.H. (1961). *J. Chem. Phys.* 34: 120; (1970). *J. Chem. Phys.* 52: 3038.

45 Kumar, P. (2006). *Proc. Natl. Acad. Sci.* 103: 12955.

46 Sengupta, S., Karmakar, S., Dasgupta, C. et al. (2013). *J. Chem. Phys.* 138: 12A548.

47 Wei, S., Evenson, Z., Stolpe, M. et al. (2018). *Sci. Adv.* 4: eaat8632 (1–6).

48 Grest, G.S. and Cohen, M.H. (1981). *Liquids, Glasses, and the Glass Transition: A Free-Volume Approach*, Ch. 6 in *Advances in Chemical Physics*, Vol. 48 (ed. I. Prigogine and S.A. Rice), p. 455. New York: John Wiley & Sons.

49 Cohen, M.H. and Grest, G.S. (1979). *Phys. Rev. B* 20: 1077.

50 Hoffman, J.D., Williams, G., and Passaglia, E. (1966). *J. Polym. Sci. C* 14: 173.

51 Gibbs, J.H. and DiMarzio, E.A. (1958). *J. Chem. Phys.* 28: 373 and 807; Adam, G. and Gibbs, J.H. (1965). *J. Chem. Phys.* 43: 139.

52 Eisenberg, A. and Saito, S. (1966). *J. Chem. Phys.* 45: 1673.

53 Simha, R. and Boyer, R.F. (1962). *J. Chem. Phys.* 37: 1003.

54 Chen, H.S. and Turnbull, D. (1968). *J. Chem. Phys.* 48: 2560.

55 Berry, G.C. and Fox, T.G. (1968). *Adv. Polym. Sci.* 5: 261.

56 Macedo, P.B. and Litovitz, T.A. (1965). *J. Chem. Phys.* 42: 245.

57 Chung, H.S. (1966). *J. Chem. Phys.* 44: 1362.

58 Brummer, S.B. (1965). *J. Chem. Phys.* 42: 4317.

59 Mnyukh, Y. (2014). *Am. J. Condens. Matter Phys.* 4: 1; (2010). *Fundamentals of Solid-State Phase Transitions, Ferromagnetism and Ferroelectricity*. 2e. Farmington, CT: DirectScientific Press; (2020). *Concluding Papers of Yuri Mnyukh on Solid-State Physics*. Farmington, CT: DirectScientific Press.

60 Simon, S.L. and McKenna, G.B. (2009). *J. Non-Cryst. Solids* 355: 672.

61 McKenna, G.B. (2020). *Rubber Chem. Technol.* 93: 79.

62 Gutzow, I. (1962). *Z. Phys. Chem.* 221: 153.

63 Gutzow, I. (1977). *Thermodynamical and Model-Statistical Treatment of the Glassy Solidification*. In *Non-crystalline Solids* (ed. G.H. Frischat), p. 356. Transtech, Aedermansdorf, Switzerland: Proceedings 4th International Conference Clausthal.

64 Milchev, A. and Gutzow, I. (1982). *J. Macromol. Sci. Phys. B* 21: 583.

65 Gutzow, I. (1979). *Wiss. Z. Friedrich-Schiller Univ. Jena, Math.-Naturwiss. Reihe* 28: 243.

66 Milchev, A. and Gutzow, I. (1981). *Phys. Chem. Liq.* 11: 25.

67 Petroff, B., Milchev, A., and Gutzow, I. (1996). *J. Macromol. Sci. B* 35: 763.

68 Gutzow, I. and Schmelzer, J.W.P. (2009). *J. Non-Cryst. Solids* 355: 581.

69 Ritort, F. and Sollich, P. (2003). *Adv. Phys.* 52: 219.

70 Trachenko, K. and Brazhkin, V.V. (2011). *Phys. Rev. B* 83: 014201; Brazhkin, V.V. and Trachenko, K. (2014). *J. Phys. Chem. B* 118: 11417.

71 Trachenko, K. and Brazhkin, V.V. (2009). *J. Phys. Condens. Matter* 21: 425104 (1–9).

72 Frenkel, J. (1946). *Kinetic Theory of Liquids*. Oxford: Oxford University Press (also Dover, 1953).

73 Dexter, A.R. and Matheson, A.J. (1972). *Adv. Mol. Relax. Interact. Processes* 2: 251.

74 Landau, L.D. and Lifshitz, E.M. *Theory of Elasticity*, 3e. Oxford: Butterworth-Heinemann.

75 Trachenko, K. (2007). *Phys. Rev. B* 75: 212201.

76 Brazhkin, V.V. and Trachenko, K. (2014). *J. Phys. Chem. B* 118: 11417.

77 Götze, W. (2009). *Complex Dynamics of Glass-forming Liquids: A Mode-coupling Theory*. Oxford: Oxford University Press.

78 Szamel, G. (2013). *Prog. Theor. Exp. Phys.* 012J01.

79 Reichman, D.R. and Charbonneau, P. (2005). *J. Stat. Mech.* P05013.

80 Janssen, L.M.C. (2018). *Front. Phys.* 6: 97.
81 Hansen, J.-P. and McDonald, I.R. (2006). *Theory of Simple Liquids*, 3e. London: Academic Press; (1976). 1e; (1986). 2e.
82 Zwanzig, R. (1961). *Phys Rev.* 124: 983; Mori, H. (1965). *Prog. Theor. Phys.* 33: 423.
83 Phillies, G.D.J. (2000). *Elementary Lectures in Statistical Mechanics.* New York: Springer.
84 Fuchs, M., Götze, W., Hofacker, I. et al. (1991). *J. Phys. Condens. Matter* 3: 5047; Flenner, E. and Szamel, G. (2005) *Phys. Rev. E* 72: 031508.
85 Janssen, L.M.C. and Reichman, D.R. (2015). *Phys. Rev. Lett.* 115: 205701.
86 Weysser, F., Puertas, A.M., Fuchs, M. et al. (2010). *Phys. Rev. E* 82: 011504.
87 Szamel, G. (2003). *Phys. Rev. Lett.* 90: 228301.
88 Yeomans-Reyna, L., Chávez-Rojo, M.A., Ramírez-González, P.E. et al. (2007). *Phys. Rev.* E76: 041504; Mendoza-Méndez, P., Lázaro-Lázaro, E., Sánchez-Díaz, L.E., et al. (2017). *Phys. Rev. E* 96: 022608.
89 Rizzo, T. and Voigtmann, T. (2015). *Europhys. Lett.* 111: 56008.
90 Das, S.P. and Mazenko, G.F. (2012). *J. Stat. Phys.* 149: 643; *Ibid.* (2013). 152 and 159.
91 Götze, W. and Sjögren, L. (1987). *Z. Phys.* B65: 415; (1992). *Rep. Prog. Phys.* 55: 241.
92 Wolynes, P. (2022). *The random first order transition theory of glasses in real space-time: instantons, strings, flames and flows.* http://dx.doi.org/10.48550/arXiv.2206.00600. This manuscript indicates that it will appear in: *Spin Glass Theory and Far Beyond – Replica Symmetry Breaking after 40 Years* (ed. P. Charbonneau, E. Marinari, M. Mézard, et al.). Singapore: World Scientific, but the chapter by Wolynes in this book appears to be different.
93 Lubchenko, V. (2015). *Adv. Phys.* 64: 283.
94 Lubchenko, V. and Wolynes, P. (2007). *Annu. Rev. Phys. Chem.* 58: 235.
95 Angell, C.A. (1985). *Strong and Fragile Liquids*, Ch. 1 in *Relaxations in Complex Systems* (ed. K.L. Ngai and G.B. Wright), p. 3. Arlington, VA: Office of Naval Research.
96 Krausser, J., Lagogianni, A.E., Samwer, K. et al. (2017). *Phys. Rev. B* 95: 104203.
97 Debenedetti, P.G. and Stillinger, F.H. (2001). *Nature* 210: 259.
98 Berthier, L. and Biroli, G. (2011). *Rev. Mod. Phys.* 83: 587.
99 Möncke, D., Topper, B., and Clare, A.G. (2022). *Rev. Mineral. Geochem.* 87: 1039.
100 Zheng, Q., Zhang, Y., Montazerian, M. et al. (2019). *Chem. Rev.* 119: 7848.
101 Mauro, J.C., Loucks, R.J., and Gupta, P.K. (2009). *J. Am. Ceram. Soc.* 92: 75.
102 Angell, C.A., Ngai, K.L., McKenna, G.B. et al. (2000). *J. Appl. Phys.* 88: 3113.
103 Ginzburg, V.V. (2020). *Soft Matter* 16: 810; (2021). *Macromolecules* 54: 2774; (2021). *Soft Matter* 17: 9094.

Further Reading

Baltă, P. and Baltă, E. (1976). *Introduction to the Vitreous State.* Turnbridge Wells, Kent: Abacus Press. Classical introduction focusing on structure, microphase separation and optical properties.

Gutzow, I.S. and Schmelzer, J.W.P. (2013). *The Vitreous State*, 2e. Berlin: Springer-Verlag (1e, 1995). Dealing mainly with thermodynamics, lattice models and nucleation/crystallization, with a limited introduction to rheology. Unfortunately, in the 2e literature after 1995 is not integrated in the main text (largely unaltered from the 1e) but discussed in an additional chapter.

Vogel, W. (1994). *Glass Chemistry*, 2e. Berlin: Springer-Verlag (1e, 1985). Conform the title, dealing mainly with the chemistry of various inorganic glasses, their microphase separation and crystallization, and optical properties, with a brief introduction to strength.

Varsneya, A.V. and Mauro, J. (2019). *Fundamentals of Inorganic Glasses*, 3e. Amsterdam: Elsevier. A largely descriptive book dealing with many properties, providing a broad overview.

Zarzycki, J. (1991). *Glasses and the Vitreous State*. Cambridge: Cambridge University Press. A broad introduction and overview, covering most topics relevant for glasses, but limited in detail.

19

Irreversibility and the Return to Equilibrium

In this last chapter, we return to some basic aspects that are essential for all types of systems, namely the return to equilibrium and irreversibility. First, we present some general considerations. Thereafter, we briefly review a few approaches to describe the return to equilibrium, which, however, all lead to the same consequences. Finally, we discuss the nature of entropy. For our discussion, many of the concepts introduced in Chapters 2 and 3 are used freely.

19.1 Some Considerations

Without much discussion, in Chapter 7, we loosely used the concept of probability and accepted that equiprobability for equal-energy states via optimizing the (Gibbs) entropy expression $S_G = -k\Sigma_j p_j \ln p_j$ (where p_j is the probability of the system to be in state j) in conjunction with various constraints leads to the various canonical descriptions. We hardly paid any attention as to why the equiprobability procedure is allowed and may be more importantly, after successful application, why it works so well, testifying the results obtained. Furthermore, we note that the basic description of neither classical nor quantum systems, as given in Chapters 2 and 3, explains why a system evolves naturally or irreversibly from one macrostate to another. These two questions touch upon the foundations of statistical thermodynamics.[1]

Let us first recall the concept of probability we use, as briefly described in Appendix E. We use the interpretation, due to Laplace, Jeffreys [5], and Cox [6], which considers probability as a numerical measure of a *reasonable degree of belief*, *degree of plausibility*, or *degree of reasonable expectation*, and which tries to estimate probability distributions consistent with the constraints given.[2] This does not imply that objective probabilities do not exist, but we may be unaware of the proper values and have to make a best guess. Therefore, a probability is taken as the *propensity* of a single event to realize itself in the long run [8]. Probability is thus considered to be a *dispositional property*, that is, a property that may or may not become manifest in a single event. For elaborate considerations, see [6, 9].

To set the scenario further, let us recall a bit of classical particle mechanics (PM) for closed systems, as open systems are trivially irreversible. As briefly discussed in

1 For various views in various depths, see, for example, [1–4].

2 Brief summaries of degree of belief probability theory are given in [7] and chapter 2 of [4].

Phases of Matter and their Transitions: Concepts and Principles for Chemists, Physicists, Engineers, and Materials Scientists, First Edition. Gijsbertus de With.

Section 7.1 – suggested to be read in conjunction with the present section – in PM a system is described by *Hamilton's equations* (Section 2.4). If the system is a molecule, characterized by n generalized coordinates and their associated momenta (degrees of freedom, DoFs), the state of a system is depicted as a point in a $2n$-dimensional Cartesian space – the molecule *phase space* or μ-space – whose axes are labeled by the allowed momenta and coordinates of the DoFs of the system. For a many-molecule assembly, each molecule is thus represented by a point in μ-space.

For a system of interacting molecules, we use the *ensemble* – a large (virtual) collection of identical systems with the same Hamilton function but different initial conditions. By taking a $2nN$-dimensional Cartesian space – the gas *phase space* or Γ-space – where N is the total number of molecules in the system, we can make a representation in which the axes are labeled with the allowed momenta and coordinates of all the molecules. Each system in the ensemble is represented by a point in Γ-space. This point will describe a path – a *trajectory* – as a function of time. From general considerations of first-order differential equations, it follows that through each point of phase space – a *phase point* – can pass one and only one trajectory so that trajectories do not cross. A swirl of points in Γ-space represents the ensemble, and the macroscopic state of a system is described by the average behavior of this swirl. For the density of phase points, we have $\rho(\mathbf{q},\mathbf{p}) \geq 0$ and $\int \rho(\mathbf{q},\mathbf{p})\,\mathrm{d}\mathbf{q}\mathrm{d}\mathbf{p} = 1$, where $\mathbf{q}$ and $\mathbf{p}$ denote the collective of coordinates and momenta, respectively. This density is constant if we follow along with the swirl, a general theorem called *Liouville's theorem* (Section 2.4). A function $F(\mathbf{q},\mathbf{p})$ in phase space representing a certain property is called a *phase function*, the Hamilton function $\mathcal{H}$ providing an important example. The ensemble average of $F(\mathbf{q},\mathbf{p})$ is independent of time if $F(\mathbf{q},\mathbf{p})$ and the boundary conditions are time independent.

Historically, the statistical description started with the Boltzmann approach using μ-space for a single molecule, followed by the Gibbs approach employing Γ-space with the associated ensemble[3] for a set of (interacting) molecules. The literature on which description to prefer is large (and largely confusing), and we cannot deal even with a brief review (see, e.g. [10–12]). Nevertheless, a few remarks seem at place.

The basic idea is that the micro-description, as given by statistical mechanics and based on the dynamical behavior of the system, provides an analogue of the macro-description, as given by phenomenological thermodynamics. The equal a priori probability postulate introduces statistical aspects that, in one way or another, are necessary. As we will see, the Gibbs approach keeps the dynamical behavior unperturbed and introduces the statistics via the initial conditions for the ensemble, while the Boltzmann approach deals with a single system and introduces the statistics via an assumption for the dynamical behavior. As remarked by Grandy [4], it appears that while for most practical (calculation) purposes

3 In classical mechanics, an ensemble represents a (virtual) set of replicas of the system at hand with different initial conditions and probability is interpreted as the fraction of systems in a particular state, that is, the frequency interpretation of probability is used. Providing many replicas of a system generally is prohibitive and Jaynes [9] and Grandy [4] argued that Gibbs used the ensemble terminology as a metaphor for actually employing the degree-of-belief interpretation without having to say so. It will be clear that values of all microscopic variables cannot be determined with only a few macroscopic parameters controlled, giving credit to the degree-of-belief interpretation. Actually, Gibbs himself clearly stated that the ensemble is an artifice to calculate probabilities. In the propensity interpretation of probabilities, an ensemble is not needed as it refers to single events, while quantum mechanics allows the probability that a single system is in a particular state right away.

the Gibbs approach is followed, as far as explanation is concerned, the Boltzmann approach is often called to the rescue. In the course of time, several explanations have been offered for introducing equiprobability. First and foremost, *ergodic theory*, which assumes that time and ensemble averages yield the same result. Another line of attack is *coarse graining*, which denies that an experiment can yield the observer the fine-grained or micro-probabilities but only the coarse-grained probabilities consistent with the scale of the observation. Moreover, there is interventionism and, relatedly, the open system view. *Interventionism* denies the existence of really isolated systems and argues that small influences from outside, considered random, are sufficient to render the system obey the equiprobability assumption. The *open system* view takes explicitly the environment into account and leads, after some approximations, to the master equation $\mathrm{d}p_i/\mathrm{d}t = \Sigma_{i\neq j}\nu_{ij}(p_j-p_i)$ with p_i the probability for state i and $\nu_{ij} = \nu_{ji}$ the jump rate between states i and j. If $p_j > p_i$, $\mathrm{d}p_i/\mathrm{d}t > 0$ and p_i will increase. If $p_j < p_i$, $\mathrm{d}p_i/\mathrm{d}t < 0$ and p_j will increase, which will eventually lead to equiprobability. Here, the external influence is in principle not random, but if two rather different time scales exist for decoherence and thermal equilibration, the influence is essentially random. Finally, we have the *information theory* view, accepting as a basic postulate equiprobability in the form of the principle of maximum entropy (PME). Let us first discuss the Boltzmann and Gibbs approaches[4] and some of their aspects briefly, before turning to discussing equiprobability via the master equation and some further considerations for the information theory approach.

Problem 19.1 ***Liouville Again*** Consider a small volume in phase space, $\mathrm{d}\mathbf{V} = \mathrm{d}\mathbf{q}\mathrm{d}\mathbf{p}$ with surface $\mathrm{d}\mathbf{A}$, and evaluate the number of phase points inside this volume and the number of phase points leaving and entering this volume, that is, calculate $\mathrm{d}R_V/\mathrm{d}t$, with the volume $R_V = \int_V \rho(\mathbf{q},\mathbf{p},t)\,\mathrm{d}\mathbf{V}$ and ρ the probability density of phase points. Using Gauss' theorem (Section A.13) for transforming a surface integral into a volume integral and Hamilton's equations, show that $\mathrm{d}R_V/\mathrm{d}t = 0$. Finally, using that the total number of phase points over the complete phase space Γ is conserved, that is, $\mathrm{d}R_\Gamma/\mathrm{d}t = 0$, and that, since ρ is a probability density, that is, $\rho(\mathbf{q},\mathbf{p},t) \geq 0$, prove *Liouville's theorem* in the form $\mathrm{d}\rho/\mathrm{d}t = 0$.

19.2 The Boltzmann Approach

Soon after Boltzmann [14] in the 1870s elaborated on the necessity of a statistical description for large mechanical systems, strong opposition was given. Boltzmann considered dilute gases, modeled as point-like or hard-sphere molecules, so that intermolecular interactions could be ignored and only binary collisions had to be considered. For these types of configurations, he derived his famous *transport equation*. Although his approach went along a different line, a short derivation is as follows. Consider the density of particles in μ-space, described by $f = f[\mathbf{q}(t),\mathbf{p}(t),t]$, so that the number of particles in volume element $\mathrm{d}\mathbf{q}\mathrm{d}\mathbf{p}$ at time t is $f\mathrm{d}\mathbf{q}\mathrm{d}\mathbf{p}$. At a somewhat later time $t' = t + \mathrm{d}t$, this number will be given by $f' = f[\mathbf{q}(t+\mathrm{d}t), \mathbf{p}(t+\mathrm{d}t), t+\mathrm{d}t]\,\mathrm{d}\mathbf{q}'\mathrm{d}\mathbf{p}'$. The difference $f' - f$ is brought about by changes in

---

4 A classical discussion on Boltzmann versus Gibbs statistical mechanics is the paper by Ehrenfest [13], still worthwhile reading, albeit with a clear bias toward the Boltzmann approach.

f due to interactions (collisions), expressed by a collision term $C[f]$, which is a functional of the function f. Approximating f' by its Taylor expansion, meanwhile realizing that by Liouville's theorem we have $\mathbf{dq}'\mathbf{dp}' = \mathbf{dqdp}$, we obtain

$$\frac{\partial f}{\partial t} + \frac{\partial f}{\partial \mathbf{q}}\frac{\partial \mathbf{q}}{\partial t} + \frac{\partial f}{\partial \mathbf{p}}\frac{\partial \mathbf{p}}{\partial t} = C[f] \tag{19.1}$$

It should be realized that this equation is not exact, as the derivative $\partial f/\partial t$ is approximated by $\Delta f/\Delta t$, where we have $\tau_{\text{col}} \ll \Delta t \ll \tau_{\text{flight}}$ with τ_{col} the duration of a collision and τ_{flight} the time of flight or the time between two collisions. To solve the transport equation, Boltzmann introduced his "*Stosszahlansatz*," essentially saying that (binary) collisions in a gas are completely random, and equivalent to, as we will see, in modern terms, "randomization." More precisely, Boltzmann stated that if $f_2(\boldsymbol{p}_1,\boldsymbol{p}_2,t)$ represents the distribution function of the momenta $\boldsymbol{p}$ of two molecules 1 and 2 in the gas, their momenta behave independently, that is, $f_2(\boldsymbol{p}_1,\boldsymbol{p}_2,t) = f(\boldsymbol{p}_1,t)f(\boldsymbol{p}_2,t)$. He defined his H-function by $H = \int f(\mathbf{p},t) \ln f(\mathbf{p},t)\, \mathbf{dp}$ and showed, using the Stosszahlansatz that H decreases with time, although not necessarily continuously, and hence he associated $-H$ with the thermodynamic entropy S.

The first objection due to Loschmidt, historically known as the "*Umkehreinwand*", or the reversibility objection, is based on the time invariance of classical mechanics. Since quantum mechanics is also time invariant, similar objections can be made there. The second objection due to Zermelo, historically known as the "*Wiederkehreinwand*" or the recurrence objection, is based on a theorem in classical mechanics proven by Poincaré. A similar theorem also exists in quantum mechanics. We discuss each of these objections in a wider context than the diluted gas,[5] while also discussing the abstraction of isolated systems. Finally, we note that it has been shown that $-H$ does not correspond with the thermodynamic entropy in equilibrium and that Boltzmann's H-theorem is violated for real gases [16].

19.2.1 Time Invariance

The laws of classical and quantum mechanics are time invariant, that is, by replacing t by $-t$, the same solution of the evolution equations result. As the negative H-function increases with time, one could object, with Loschmidt [17], that the H-function cannot represent entropy, as with the solutions running backward in time, the negative H-function will also increase, in direct contradiction with entropy, which always increases with forward time (Section 7.1). The actual solution to this objection is that a system is never completely isolated and that minute changes have large consequences. In fact, numerical solutions for a relatively small system using integer arithmetic (so that numerical rounding errors do not occur) [18–20], show (i) that a small disturbance at some time does not change the further behavior significantly and (ii) that time reversal indeed reverses the behavior of the system, including the associated decrease of entropy (Figure 19.1a and b). However, only a minute disturbance of the reversal leads right away to chaotic behavior, essentially blocking the system from returning to its initial configuration and the entropy from decreasing (Figure 19.2a and b). This shows that the development with forward time is highly stable, but the development with reversed time is highly unstable. Example 19.1 shows that completely negligible energetic contributions lead rapidly to complete randomization of

5 For a short, informative narrative, see [15].

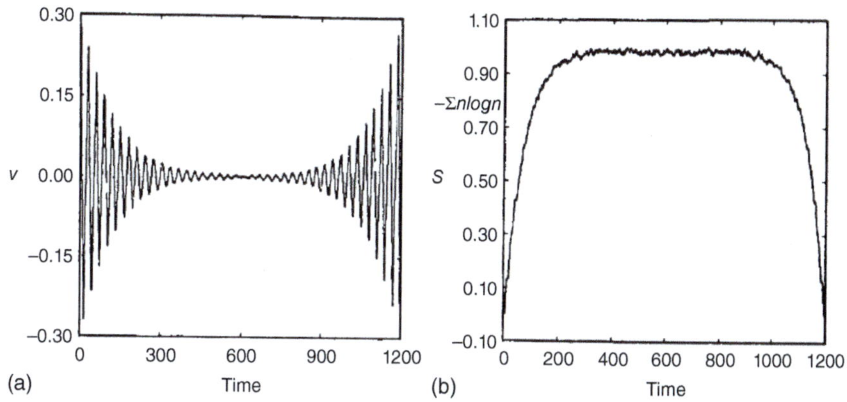

Figure 19.1 Behavior of a system of particles with the positions and momenta under conservation of mass, momentum, and energy, using a cellular automaton with integer calculations, showing (a) the mean velocity v over the box and (b) the entropy S as a function of time t. Initially, v is set to $v \cong 0.3$ and full-time reversal is applied at $t = 600$, from which reversal of both v and S is obtained.

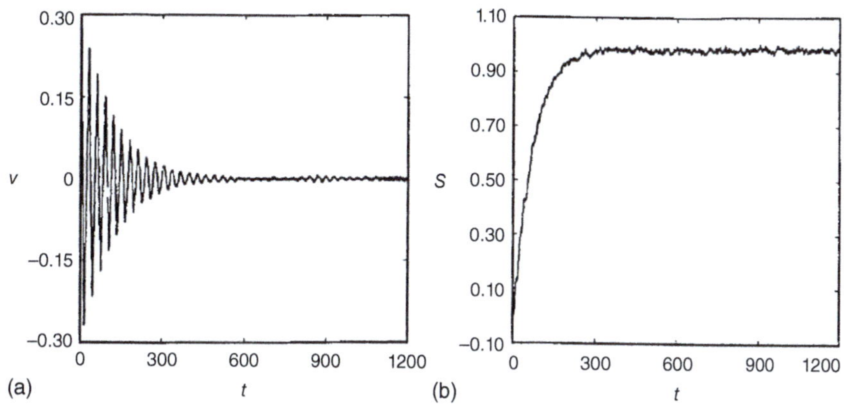

Figure 19.2 The same system as in Figure 19.1, showing (a) v and (b) S as a function of time, with time reversal again applied at $t = 600$ but only for 99.9% of the particles, clearly demonstrating the instability of the reversed trajectory.

the trajectory. In conclusion, the most logical approach is to recognize that a system does interact with its environment and face the consequences in one way or another.

Example 19.1 ***Isolated, or Not?*** Let us estimate the influence of a small perturbation on the trajectory of a particle [21]. Suppose we have two particles with mass m that are initially a distance l apart and are colliding with each other under the influence of a force $\boldsymbol{f}$. Since $\boldsymbol{f}$ varies slightly spatially, it acts in a slightly different way on each particle, and we have $\Delta\boldsymbol{f} = \boldsymbol{f}_1 - \boldsymbol{f}_2$. From Newton's equation for the relative coordinate $\Delta\boldsymbol{x}$, we obtain $m\Delta\ddot{\boldsymbol{x}} = \Delta\boldsymbol{f}$ which leads via $m\Delta\dot{\boldsymbol{x}} \equiv m\Delta\boldsymbol{v} \cong \Delta\boldsymbol{f}t$ to $m\Delta\boldsymbol{x} \cong \frac{1}{2}\Delta\boldsymbol{f}t^2 \cong \frac{1}{2}\Delta\boldsymbol{f}(l/v)^2$. The change in angle after a path of length l is approximately $\Delta\theta \cong |\Delta\boldsymbol{x}|/l \cong \frac{1}{2}|\Delta\boldsymbol{f}|(l/mv^2)$. If the range of the interaction is indicated with r_{int}, the change in deflection angle after one collision $\Delta\theta_1 = (l/r_{\text{int}})\Delta\theta$, while after k collisions it becomes $\Delta\theta_k = (l/r_{\text{int}})^k\Delta\theta$. The perturbed trajectory is no longer correlated to the unperturbed trajectory if $\Delta\theta_k = 2\pi$, that is, if $(l/r_{\text{int}})^k\Delta\theta = 2\pi$. Solving for k we obtain $k = [\ln(2\pi/\Delta\theta)]/[\ln(l/r_{\text{int}})]$. Now, we need to specify a configuration to get a numerical estimate. Suppose we have one mole of He gas molecules, each of mass $m = 6.65 \times 10^{-23}$ g, influenced by the gravitation of an observer with a mass of, say, $M = 80$ kg, and observing the system at a distance of $d = 1$ m. The gravitation potential is given by $W = -GMN_\text{A}m/d$, with the gravitation constant $G = 6.67 \times 10^{-11}$ m^3 kg^{-1} s^{-2} and $N_\text{A} = 6.23 \times 10^{23}$, and results for the numbers given in $W \cong -2 \times 10^{-10}$ J mol^{-1}. This energy is completely negligible as compared to the total kinetic energy at room temperature $E \cong 3RT/2 \cong 3.8$ kJ mol^{-1}. The difference in force $|\Delta \boldsymbol{f}|$ is given by $|\Delta \boldsymbol{f}| = GMm/d^2 - GMm/(d+l)^2 \cong GMml/d^3$, so that the angular change $\Delta\theta$ becomes $\Delta\theta \cong (GMml/d^3)(l/mv^2) = (GM/d^3)(l/v)^2$. This results for the numbers given in $\Delta\theta \cong 4 \times 10^{-28}$. Estimating the distance l as the mean free path, say, $l \cong 140$ nm, and the interaction distance r_{int} as $r_{\text{int}} \cong 0.15$ nm, we obtain $k \cong 10$. Consequently, randomization sets in extremely rapidly.

19.2.2 Recurrence

In 1890, Poincaré proved a theorem [22] that essentially states that, given a few rather general requirements, every closed mechanical system behaves quasi-periodically, that is, after a finite time t_P the system returns to its initial configuration arbitrarily closely. The proof of Poincaré's recurrence theorem is actually not so complex [23] and assumes only that the evolution of a system can be described by a continuous, volume-preserving one-to-one mapping of phase space on itself. We know that in classical mechanics, the evolution of a system is described by a continuous trajectory $\mathbf{q}(t),\mathbf{p}(t)$ in phase space and that, given an initial value $\mathbf{q}(0),\mathbf{p}(0)$, the canonical equations predict $\mathbf{q}(t),\mathbf{p}(t)$ uniquely. These equations also map the evolution of the system one-to-one on phase space, as the map is invertible by replacing t with $-t$, allowed since Hamilton's equations are time invariant. From Liouville's theorem we know further that a volume element $\text{d}\mathbf{V} = \text{d}\mathbf{q}\text{d}\mathbf{p}$ is conserved. Hence Poincaré's theorem applies to classical mechanical systems. Consider now a volume $\text{d}\mathbf{V}$ located at $\mathbf{q},\mathbf{p}$. If, along a trajectory, all volume elements were never intersecting, the total volume of phase space would be infinite. Since we assume that phase space is finite, at some time t_P, $\text{d}\mathbf{q}(t_\text{P})\text{d}\mathbf{p}(t_\text{P})$ must overlap with $\text{d}\mathbf{q}(0)\text{d}\mathbf{p}(0)$. The recurrence time t_P depends on what is considered as a degree of overlap qualifying for representing the initial configuration in phase space again. A similar theorem can be proven in quantum mechanics [24].

The discussion by Boltzmann was couched in terms of his H-function, which he associated with S. Therefore, since entropy is supposed to always increase, Zermelo argued [25] in 1896, that the H-function cannot represent thermodynamic entropy, as a configuration of a system will return arbitrarily closely with time and, therefore, will have the same H-function value. While the Poincaré theorem is undeniably true, it appears that the times involved are excessively long, that is, for reasonably small systems, many times the estimated age t_U of the Universe,[6] as a simple model in Example 19.2 shows. A solution sometimes advocated is that the thermodynamic limit should be taken, thereby avoiding

6 The present best estimate is $t_\text{U} = (13.798 \pm 0.037) \times 10^9$ years or $\ln(t_\text{U}/\text{years}) \cong 23.3$, including the latest estimate of 13.82×10^9 years resulting from observations over 15 months by the European Space Agency's spacecraft Planck obtained in 2013.

essentially the recurrence. However, real systems are finite and hence this argument does not apply (but see Section 19.6). More important, as illustrated before, (minute) external disturbances will influence the future behavior of a system without changing its macroscopic conduct but with significant influence on its microscopic behavior so that recurrence (as derived for undisturbed systems) will not occur. Hence, the conclusion is that for all lengths and time scales relevant to chemistry and physics, the recurrence time is a problem of conceptual nature and not an issue in practice.

Example 19.2 ***Recurrence Indeed, But When?*** As a typical example [26] of a thermodynamic system, we use a set of coupled harmonic oscillators. Since any system of N-coupled harmonic oscillators can be reduced to a system of N-decoupled oscillators, we first focus on a single oscillator. The Hamilton function is given by $\mathscr{H} = ½(p^2/m + aq^2) = ½[(p/m^{1/2})^2 + (a^{1/2}q)^2] \equiv ½(P^2 + \omega^2Q^2) = ½(P + \mathrm{i}\omega Q)(P - \mathrm{i}\omega Q)$ with $P = m^{-1/2}p$, $Q = m^{1/2}q$ and $\omega = (a/m)^{1/2}$. Since $z^* = P - \mathrm{i}\omega Q$ is the complex conjugate of $z = P + \mathrm{i}\omega Q$ and the Hamilton function $\mathscr{H}$ represents the constant energy E, a solution for z is $z = a\exp(\mathrm{i}\omega t)$. This solution represents the uniform motion of a point on a circle of radius a with angular frequency ω in the complex plane with real axis P and complex axis Q (an Argand diagram) as a function of t. The solution z^* represents the same solution but now for $-t$. The probability that z lies within the interval $\Delta\phi$ during one oscillation (rotation) is $\Delta\phi/2\pi$. Suppose now that we have N oscillators for which the ratio of their frequencies is irrational, that is, cannot be represented by a rational number. The probability that these N oscillators are all within their prescribed intervals is $(\Delta\phi/2\pi)^{N-1}$ and hence we need $(2\pi/\Delta\phi)^{N-1}$ oscillations to obtain that configuration again. With an average oscillation period of $1/\omega$, an approximate estimate for the Poincaré time is $t_\mathrm{P} \cong (2\pi/\Delta\phi)^{N-1}/\omega$. Using $N = 10$, $\omega = 10\,\mathrm{s}^{-1}$ and $\Delta\phi = 2\pi/100$, we obtain $t_\mathrm{P} \cong 3.2 \times 10^{12}$ years or $\ln(t_\mathrm{P}/\mathrm{year}) \cong 28.8$. Using $N = 10^{23}$, $\omega = 10^{12}\,\mathrm{s}^{-1}$ and again $\Delta\phi = 2\pi/100$ results in $\ln(t_\mathrm{P}/\mathrm{year}) \cong 4.6 \times 10^{23}$. While for 10 oscillators t_P is already much larger than t_U, for a macroscopic number of oscillators the Poincaré time t_P is excessively long as compared to the age of the Universe t_U.

19.3 The Gibbs Approach

Despite that others, in particular Boltzmann, already used the ensemble approach[3], Gibbs is usually considered as its main advocate. Although he was very careful in his statements about the "analogues" for thermodynamics and did not even mention ergodic theory, one often considers ergodic theory as the basis for the ensemble approach. The idea is to estimate the time average of a property $F[\mathbf{q}(t),\mathbf{p}(t)]$, given by

$$\langle F\rangle_t = \lim_{t\to\infty}\frac{1}{t}\int_0^t F[\mathbf{q}(t'),\mathbf{p}(t')]\,\mathrm{d}t'$$

However, this average cannot be calculated in general since neither the solution $\mathbf{p} = \mathbf{p}(t)$ and $\mathbf{q} = \mathbf{q}(t)$ nor the initial conditions $\mathbf{p} = \mathbf{p}(0)$ and $\mathbf{q} = \mathbf{q}(0)$ are known. To obtain nevertheless estimates of properties of a system, the phase average

$$\langle F\rangle_\Gamma = \int_\Gamma F(\mathbf{q},\mathbf{p})\rho(\mathbf{q},\mathbf{p})\,\mathrm{d}\mathbf{q}\mathrm{d}\mathbf{p}$$

is introduced. The assumption, due to Boltzmann in 1887 for μ-space and known as the *ergodic theorem*, is that over the course of time, independent of the initial microstate, for an isolated system each trajectory will eventually go through every possible microstate for the system given its macroscopic constraints. This implies

$$\langle F \rangle_\Gamma = \langle F \rangle_t \tag{19.2}$$

and essentially says that each trajectory visits each infinitesimal volume element of phase space. This assumption, represented by Eq. (19.2) has been proven false but can be replaced by the *quasi-ergodic theorem*, proven by Birkhoff in 1931 for what he called metrically transitive systems.[7] It states that trajectories approach almost all phase points as closely as desired, given sufficient time to the system. In practice, this means that one accepts the equivalence of the time and phase average, except in exceptional cases, albeit with implicit assuming that a time average over a finite time can be used, so that time-dependent boundary conditions can have an influence.

Nevertheless, despite the wide literature on ergodic theory, many researchers agree that ergodic theory cannot be used as a foundation for statistical thermodynamics. Apart from the (technical) difficulty that ergodicity is extremely difficult to prove, $\langle F \rangle_t$ is a time average, defined as the limit of the behavior over infinite time, ruling out any time dependence. The latter is, evidently, present in real systems, as well as in $\langle F \rangle_\Gamma$ if the ensemble is nonstationary. Further, $\langle F \rangle_t$ also depends on the initial conditions, whereas $\langle F \rangle_\Gamma$ does not. Moreover, nowhere in ergodic theory is reference made to equilibrium configurations [30], which, to say the least, is remarkable. Finally, it can be shown that some systems that are not ergodic still behave according to statistical thermodynamics, while some ergodic systems do not show equiprobability behavior. Examples of non-ergodic systems are due to the so-called KAM theorem[8] which states that, for a system with weak anharmonic coupling, most of the constant energy surface in phase space will consist of confined regions of low-energy modes (invariant tori, in the jargon), even with some perturbation. Increasing the perturbation above a certain limit, the tori will break down and the system will turn to chaotic behavior and, hence, presumably to equiprobability behavior for the total system [24]. An example of an ergodic system not resulting in equiprobability is given in Example 19.3. So, more generally, Thompson [23] states that there is no (mathematical) justification for equiprobability based on ergodicity, even if the ergodic theorems are applicable. Also, after an extensive analysis, Emch and Liu [32] conclude that ergodic theory is an idealization and, in and of itself, is not a justification of statistical mechanics. Similarly, Smale [33] concluded that "the compelling fact is that, with one main exception (a system of elastic spheres) Hamiltonian systems are not ergodic." Hence, ergodic theory has lost its basis to provide a general foundation for statistical thermodynamics.

Still, the Gibbs approach works rather well, witness its success. Of course, in quantum mechanics, probability directly refers to the state of a single system instead of an ensemble.[3] Nevertheless, some further explanation is called for. In the remainder of this chapter, we first briefly outline two possible approaches to rationalizing the master

7 See, for example, [27]. A mechanical system is metrically transitive if the energy surface cannot be divided into two finite regions such that orbits starting from points in one region always remain in that region. This is equivalent to the fact that no constants of the motion are neglected [28]. For a review of classical ergodic theory, see [1, 29].

8 After Kolmogorov (1954), Arnold (1963), and Moser (1962). For a clear narrative, see [31].

equation based on dynamical considerations. These approaches, collectively denoted as the *dynamical approach*, predict eventually for a closed system equiprobability of states and the consequences leading to irreversible behavior. The first variant is the *formal* or *open system approach*,[9] which is in principle straightforward, but loaded with complexity (Section 19.4). The second variant is the *physical approach*, essentially telling the same story but providing some clear insight in the approximations made (Section 19.5). Thereafter, we deal with the *information theory approach*, in which we generously admit that we do not have full knowledge (in whatever sense) of the macrosystem, but that nevertheless we want to make predictions about the future of this macrosystem, given the information available (Section 19.6). Readers who seek only a plausible justification for equiprobability can jump right away to Section 19.6.

Example 19.3 ***Ergodic But Not Equiprobable*** As a simple example of an ergodic system that nevertheless does not show an equiprobable distribution [24], consider a system described by the momentum p $(0<p<1)$ and coordinate q $(0<q<1)$, having periodic boundary conditions and for which the equations of motion read $\mathrm{d}p/\mathrm{d}t = \alpha$ and $\mathrm{d}q/\mathrm{d}t = 1$. The equations of motion are easily integrated to yield $q(t) = q_0 + t$ and $p(t) = p_0 + \alpha t$, so that the trajectory becomes $p = p_0 + \alpha(q-q_0)$. If $\alpha = m/n$ with m and n integers, that is, a rational number, the motion will be periodic, but for irrational numbers the trajectory will not close. Since the phase space is periodic, any integrable function $f(p,q)$ can be written in a Fourier series reading $f(p,q) = \Sigma_{k,l} f_{kl}$ exp[i$(kp + lq)$]. Inserting $p(t)$ and $q(t)$ in the expression for the time average, we obtain $\langle f\rangle_t = f_{00}$, while inserting $f(p,q)$ in the expression for the phase average results in $\langle f\rangle_\Gamma = f_{00}$. As $\langle f\rangle_t = \langle f\rangle_\Gamma$, the system is ergodic. Now consider the distribution $\rho(p_0,q_0,0) = \sin(\pi p_0)\sin(\pi q_0)$, which after time t becomes $\rho(p_0,q_0,t) = \sin[\pi(p_0 - \alpha t)]\sin[\pi(q_0 - t)]$. The shape of this distribution does not change with t but is displaced in phase space and does not represent a uniform distribution.

19.4 The Formal Approach

We now turn to the formal approach to obtain the master equation, which rationalizes equiprobability behavior for an isolated system. As we have seen, the basic trick in thermodynamics to obtain information about an open system is to assume that the system at hand plus the environment (reservoir) is an isolated (overall) system and to proceed from there. For a macroscopic description of quantum systems, conceptually the easiest is to play the same trick. However, complexity sets in rapidly, and, necessarily, the discussion will be condensed.

Suppose we have a system (S) connected to a reservoir (R) and that the state of S + R is described by the density matrix[10] D_{SR}, that is, we describe S + R in a Hilbert space $\mathbf{H} = \mathbf{H}_\mathrm{S}\mathbf{H}_\mathrm{R}$. If we have no further information about the reservoir, we can integrate over the relevant variables and obtain, for the density matrix of the system alone, $\mathrm{D}_\mathrm{S} = \mathrm{tr}_\mathrm{R}\mathrm{D}_{\mathrm{SR}}$, where tr_R

---

9 It will be clear that here "open" refers to being susceptible to outside influences and not to arbitrary changes in energy or number of molecules, as in an "open" thermodynamic system.
10 Recall that a matrix X represents the operator X (Section 3.3.1).

indicates the trace over R. While the dynamic behavior of S + R is completely described by unitary behavior as given by the time-evolution equation (2.156), $D(t) = U_{SR}(t)DU_{SR}{}^{\dagger}(t)$ with $U_{SR} = \exp(-iHt/\hbar)$, the time evolution of the system S behaves, as shown in Justification 19.1, as $D_S(t) = M_{SR}(t)D_S M_{SR}{}^{\dagger}(t) \equiv K(D_S)$. In this so-called *Kraus representation* with K the Kraus operator, $M(t)$ contains all information about the outside world.

Justification 19.1 ***Non-Unitary Development in Time*** To obtain the time evolution, we follow largely the presentation as given by Le Bellac [34] and consider, as a first step, $\mathbf{H} = \mathbf{H}_S\mathbf{H}_R$ using as basis $\{|m_S\mu_R\rangle\} \equiv \{|m_S\rangle|\mu_R\rangle\}$ with for S the set $\{|m_S\rangle\}$ and for R the set $\{|\mu_R\rangle\}$. An arbitrary state $|SR\rangle$ can be written as

$$|\,SR\rangle = \sum\nolimits_{m\mu} c_{m\mu}|m_S\rangle|\mu_R\rangle = \sum\nolimits_m |m_S\rangle\left(\sum\nolimits_\mu c_{m\mu}|\mu_R\rangle\right) \equiv \sum\nolimits_m |m_S\rangle|\widetilde{m}_R\rangle$$

The ket $|\widetilde{m}_R\rangle$ can be seen as containing the information from R relevant to the ket $|m_S\rangle$ of the system S. We chose $|m_S\rangle$ in such a way that it diagonalizes $D_S = \mathrm{tr}_R|SR\rangle\langle SR|$ to $D_S = \sum_m p_m|m_S\rangle\langle m_S|$. But D_S can also be written as

$$D_S = \mathrm{tr}_R D_{SR} = \sum\nolimits_{mn} |n_S\rangle|\widetilde{n}_R\rangle\langle m_S|\langle\widetilde{m}_R| = \sum\nolimits_{mn} \langle\widetilde{m}_R|\widetilde{n}_R\rangle|n_S\rangle\langle m_S|$$

and we obtain $\langle\widetilde{n}_R|\widetilde{m}_R\rangle = p_n\delta_{mn}$. Hence, we normalize the ket $|\widetilde{n}_R\rangle$ by writing $|n_R\rangle = p_n^{-1/2}|\widetilde{n}_R\rangle$ and obtain the *Schmidt decomposition*:

$$|SR\rangle = \sum\nolimits_n p_n^{1/2}|n_S\rangle|n_R\rangle$$

Further, we have $D_R = \mathrm{tr}_S|SR\rangle\langle SR| = \sum_n p_n|n_R\rangle\langle n_R|$, and therefore, D_S and D_R have the same eigenvalues. The number of non-zero eigenvalues (the *Schmidt number*) is 1 for a pure state with $|SR\rangle = |n_S\rangle|n_R\rangle$, while it is N for a random state (entangled state) $|SR\rangle = N^{-1/2}\Sigma_n\exp(i\alpha_n)|n_S\rangle|n_R\rangle$ with $\exp(i\alpha_n)$ a phase factor. It follows that if S and R develop independently, that is, with $U_{SR} = U_S U_R$, that a product state cannot develop in a random state and one needs nonlocal evolution to transform a product state to a random state. Conversely, one needs nonlocal evolution to disentangle a random state in a product state.

In the second step, we describe the time evolution for D_S. We know that the time evolution of D_{SR} is given by $D_{SR}{}' \equiv D_{SR}(t) = U_{SR}D_{SR}(0)U_{SR}{}^{\dagger}$ (Section 3.3.3). Assume that the initial overall state is described by $|m_S0_R\rangle = |m_S\rangle|0_R\rangle$, that is, an arbitrary state for the system $|m_S\rangle$ and a pure state $|0_R\rangle$ for the reservoir. Although this may seem like a rather restricted choice, it appears that it is always possible to choose such a state as a reference, as we will make plausible later. If we take the partial trace over the reservoir, we obtain

$$D_{mn}^{(S)}(t) = \sum\nolimits_\mu U_{m\mu,k0}D_{mn}^{(S)}(0)U_{l0,n\mu} \quad \text{or} \quad D_S{}' \equiv D_S(t) = K(D_S) \equiv \sum\nolimits_\mu M_\mu D_S M_\mu{}^{\dagger}$$

for the time development of the system. The *superoperator* M_μ, that is, an operator operating at an operator, is defined by $M_\mu = \langle\mu_R|U_{SR}|0_R\rangle$. Note that the set $\{M_\mu\}$ is complete, that is, $\sum_\mu M_\mu{}^{\dagger}M_\mu = 1_S$. If we now define $U_{SR}|m_S0_R\rangle = \sum_\mu M_\mu 1_R|m_S\mu_R\rangle$, this operation describes a quantum jump in the time interval $[0,\Delta t]$ from $|m_S0_R\rangle$ at $t = 0$ to the superposition of states $M_\mu|m_S\mu_R\rangle$ at $t = \Delta t$. In this process, the scalar product is preserved so that $\sum_{\mu,\nu}\langle m_S\mu_R|1_R M_\mu{}^{\dagger}M_\nu 1_R|n_S\nu_R\rangle = \langle m_S|n_S\rangle$. Further note that taking the trace over the reservoir yields

$$\mathrm{tr}_R\sum\nolimits_m |m_S0_R\rangle\langle m_S0_R| = \sum\nolimits_\mu\sum\nolimits_m M_\mu|m_S\rangle\langle m_S|M_\mu^{\dagger} = \sum\nolimits_m p_m|m_S\rangle\langle m_S|$$

Time evolution of D_S is thus described by $D_S{}' = K(D_S)$, where $K(\cdot)$ is the *Kraus operator*, which is a linear, Hermitian, and normalized operator ($\text{tr}K(D_S) = 1$). Moreover, $K(D_S)$ is completely positive, that is, $K(D_S)1_C \geq 0$ for any other system C. The Kraus representation is not unique though, since we can also write $K(D_S) = \sum_\nu N_\nu D_S N_\nu{}^\dagger$ with $N_\nu = \sum_\mu U_{\nu\mu} M_\mu$ and U a unitary operator and where the dimension K of M need not to be equal to the dimension L of N. Finally, we note that for any density operator D_S that can be written as $D_S = \sum_m p_m |m\rangle\langle m|$, we can also write $K(D_S) = \sum_\mu M_\mu D_S M_\mu{}^\dagger$ and one can show that any "reasonable" time evolution law for D_S is of this form. So, *for the purpose of describing the time evolution of* D_S, one can always find a (fictitious) environment state with no initial entanglement between the system and the environment. This justifies using an initial overall state $|m_S\rangle|0_R\rangle$.

Next, we consider that, generally, it is impossible to obtain a first-order differential equation in t for D – a *master equation* – describing the evolution as local in time because memory effects will be present and in that case the evolution is described by an integro-differential equation, which is nonlocal in time. Using the memory function $\gamma(t-t')$, we write

$$\text{d}D/\text{d}t = -\int_{-\infty}^{t} \gamma(t-t')D(t')\text{d}t' \cong -D(t)\int_{-\infty}^{t} \gamma(t-t')\text{d}t' = -\Gamma D(t)$$

The second, approximate step can only be made if the characteristic *relaxation time* τ_{rel} is much smaller than the typical *evolution time* τ_{evo} of D. This assumption is known as the *Markov assumption* and essentially states that the time evolution of D depends only on its value at time t and not on its values at earlier times t'. Obviously, this separation of time scales depends on the physical situation at hand, but the approximation appears to be generally applicable. Example 19.4 provides a somewhat detailed calculation for the important example of thermal equilibration.

Assuming that $\tau_{\text{rel}} \ll \tau_{\text{evo}}$ is true, let us introduce a time step Δt for which holds that $\tau_{\text{rel}} \ll \Delta t \ll \tau_{\text{evo}}$. Using the Kraus representation for D, we write

$$\frac{\text{d}D}{\text{d}t} \cong \frac{\Delta D}{\Delta t} = \frac{D(t+\Delta t) - D(t)}{\Delta t} = \frac{K_{t,t+\Delta t}[D(t)] - D(t)}{\Delta t}$$

In order to obtain a master equation, we must (i) from $D_{SR} \cong D_S D_R$ be able to write down the Kraus representation at $t+\Delta t$ and (ii) $K_{t,t+\Delta t}$ must depend only on Δt, but not on t. One can show that both conditions are fulfilled if $|V|\tau_{\text{rel}}/\hbar \ll 1$ with $|V|$, a typical matrix element of the SR interaction. If this is the case, we can write $D(t+\Delta t) = K_{\Delta t}[D(t)] = \sum_\mu M_\mu(\Delta t) D_S(t) M_\mu{}^\dagger(\Delta t)$ and since $D(t+\Delta t) = D(t) + \mathcal{O}(\Delta t)$ is first order in Δt, $K_{\Delta t}[D(t)]$ is first order in Δt. One of the M_μs, which is labeled as M_0, therefore must read $M_0 = 1 + (-\text{i}\hbar^{-1}H - X)\Delta t + \mathcal{O}(\Delta t)^2$, where H and X are Hermitian operators. This term, using the commutator $[A, B] = AB - BA$ and anticommutator $[A, B]_+ = AB + BA$, leads to

$$M_0(\Delta t)D(t)M_0^\dagger(\Delta t) = D(t) - \text{i}\hbar^{-1}\Delta t[H, D] - \Delta t[X, D]_+ + \mathcal{O}(\Delta t)^2 \tag{19.3}$$

The other terms must be of order $(\Delta t)^{1/2}$, that is, $M_\mu(\Delta t) = (\Delta t)^{1/2} L_\mu$, and from the completeness relation $\sum_\mu M_\mu{}^\dagger M_\mu = 1_S$, we obtain

$$1_S = M_0^\dagger M_0 + \sum\nolimits_{\mu>0} M_\mu^\dagger M_\mu = 1_S - 2\Delta t X + \Delta t \sum\nolimits_{\mu>0} L_\mu^\dagger L_\mu \tag{19.4}$$

Hence, $X = ½\sum_{\mu>0} L_\mu{}^\dagger L_\mu$. Combining all the results, we obtain the *Lindblad equation* for D, reading

$$\mathrm{d}D/\mathrm{d}t = -\mathrm{i}\hbar^{-1}[H, D] + ½\sum\nolimits_{\mu>0}\left(\left[L_\mu D, L_\mu^\dagger\right] + \left[L_\mu, DL_\mu^\dagger\right]\right)$$

or, symbolically,

$$\mathrm{d}D/\mathrm{d}t = -\mathrm{i}\hbar^{-1}[H, D] + R(D) \tag{19.5}$$

While the commutator $[H, D]$ leads to oscillatory behavior with constant amplitude, the term $R(D)$ leads to damping.[11] Therefore, the initial oscillations in the elements of D will fade out and, eventually, we obtain $\mathrm{d}D/\mathrm{d}t = 0$, that is, a stationary distribution described by the density matrix D with equal diagonal elements.

Example 19.4 ***Decoherence and Thermal Equilibration*** To illustrate the difference between the decoherence and thermal equilibration time, we consider a simple model [37]. Consider a perfect gas constrained to move in the x-direction only and confined by perfectly reflecting walls at $x = -a$ and $x = a$. At time $t = 0$, all molecules of mass m are located at the center $x = 0$, while having a Maxwell distribution of velocities v. If we introduce the variable $x' \equiv vt$, the relation between x' and the actual position x is given by the many-to-one mapping:

$$\begin{array}{llll} -a < x' < a: & x = x', & a < x' < 3a: & x = 2a - x', \\ 3a < x' < 5a: & x = x' - 4a, & 5a < x' < 7a: & x = 6a - x', \qquad \ldots \end{array}$$

Now we allow for thermal motion of the walls by assuming that each wall has a mass M and a random velocity V. The change in speed δv of the molecules becomes

$$\delta v = \frac{2MV + 2mv}{M + m} \cong 2V$$

since $M \gg m$ and $½M\overline{V^2} \cong ½m\overline{v^2} \cong ½kT$. We denote the speed change at first impact at $x' = a$ by δv_1, at second impact at $x' = 3a$ by δv_2, etc. Then we obtain

$$\begin{aligned} x' &= vt + \delta v_1\left(t - \tfrac{a}{v}\right) + \delta v_2\left(t - \frac{a}{v} - \frac{2a}{v + \delta v_1}\right) + \delta v_3\left(t - \frac{a}{v} - \frac{2a}{v + \delta v_1} - \frac{2a}{v + \delta v_1 + \cdots}\right) + \cdots \\ &\cong vt + \delta v_1\left(t - \tfrac{a}{v}\right) + \delta v_2\left(t - \frac{3a}{v}\right) + \delta v_3\left(t - \frac{5a}{v}\right) + \cdots \end{aligned}$$

where in the last step the denominators $v + \delta v_1 + \delta v_2 + \cdots$ are replaced by v, as all the δvs have zero average value. For the mean value of x', we obtain

$$\overline{(x' - vt)^2} = \overline{(\delta v_1)^2}\left(t - \frac{a}{v}\right)^2 + \overline{(\delta v_2)^2}\left(t - \frac{3a}{v}\right)^2 + \overline{(\delta v_3)^2}\left(t - \frac{5a}{v}\right)^2 + \cdots$$

---

11 Balian [35] discusses the matter in terms of the *Liouville representation*, in which the $n \times n$ Hermitian matrix D is written as a n^2 column matrix **D**. The von Neumann equation then reads $\mathrm{i}\hbar\mathrm{d}\mathbf{D}/\mathrm{d}t = -\mathrm{L}\mathbf{D}$, where L is the *Liouville (matrix) operator*. This change in representation is rather similar to the one often made in solid mechanics, where the stress $\boldsymbol{\sigma}$ and strain $\boldsymbol{\varepsilon}$, second-order tensors represented by a square matrix, become the stress "column" $\boldsymbol{\sigma}$ and the strain "column" $\boldsymbol{\varepsilon}$, respectively. In elasticity, the constitutive relation between $\boldsymbol{\sigma}$ and $\boldsymbol{\varepsilon}$ is given by the fourth-order tensor $\mathbb{E}$, reading $\boldsymbol{\sigma} = \mathbb{E}{:}\boldsymbol{\varepsilon}$. In matrix representation, $\mathbb{E}$ becomes a square matrix $\mathbb{E}$. The constitutive relation can now be written as the matrix equation $\boldsymbol{\sigma} = \mathbb{E}\boldsymbol{\varepsilon}$ (for details, see Chapter 5 or [36]).

The cross-terms vanish on averaging since the speeds at different wall impacts are statistically independent. Using $\overline{\delta v_j^2} = 4\overline{V^2}$ and approximating the series by an integral, we obtain

$$\overline{(x' - vt)^2} = (2vt^3/3a)\overline{V^2}$$

We need to compare this quantity with $(2a)^2$, since, if the uncertainty in x' exceeds $2a$, correlation is effectively lost. By introducing the mean number of impacts on the wall $n = vt/2a$, the condition for attaining equilibrium becomes

$$\overline{(x' - vt)^2}/4a^2 \cong (vt^3/6a^3)\overline{V^2} = 4r^3\overline{V^2}/3v^2 \gg 1 \quad \text{or} \quad n \gg (M/m)^{1/3}$$

The corresponding relaxation time, that is, the *decoherence time*, is given by nT_0, where T_0 is given by $T_0 = 2a/v_0$ with v_0 the average speed.

Another relaxation time describes the relaxation of kinetic energy. If we have $\overline{½MV^2} \cong ½kT_{\text{wall}}$ and $\overline{½mv^2} \cong ½kT_{\text{gas}}$, T_{gas} must approach T_{wall} as the latter is conditioned from outside. The change in kinetic energy of a molecule in a collision with the wall is given by

$$½m(v')^2 - ½mv^2 = [4mM/(M - m)^2][½MV^2 - ½mv^2 - ½(M - m)vV]$$

By averaging over the random velocities V, the last term on the right-hand side vanishes. Since $M >> m$, the final result is

$$\overline{½m(v')^2 - ½mv^2} \cong (4m/M)(½kT_{\text{wall}} - ½mv^2) \quad \text{or} \quad n' \gg M/m$$

where n' denotes the number of impacts on the wall for effective energy transfer between gas and wall, with $n'T_0$ the *thermal equilibration time*. We note that n'/n becomes large for experimental conditions. Suppose $a = 1$ cm, $v \approx 10^5$ cm s^{-1}, so that $T_0 \approx 10^{-5}$ s. Assume that $M/m \approx 10^9$, then we have $nT_0 \approx 10^{-2}$ s, while $n'T_0 \approx 10^4$ s. In conclusion, the decoherence time is much smaller than the thermal equilibration time.

Problem 19.2 ***Independent States*** Show that if the states S and R develop independently, that is, with $U_{\text{SR}} = U_{\text{S}}U_{\text{R}}$, that a product state $|n_{\text{S}}\rangle|n_{\text{R}}\rangle$ remains a product state.

Problem 19.3 ***From Kraus to Lindblad*** Verify Eqs (19.3), (19.4), and (19.5).

19.5 The Physical Approach

In this section, we start afresh and will employ a more physical approach, mixing coarse graining and statistical considerations, largely based on the considerations as given by van Kampen [38] over many years.

Experimentally, a macroscopic system is described by a small number of macro(scopic) variables X, the most important macrovariable probably being the (internal) energy E, but other ones are the pressure P and the number of particles N. Now two uncertainties come into play. First, the experimental uncertainty ΔX as determined by the experimental conditions, and second, the quantum uncertainty δX as given by the Heisenberg relations (Section 3.2). Now, since $\Delta E \gg \delta E$, a macrostate characterized by $E \pm \Delta E$ contains an enormous number of eigenstates of the Hamilton operator H and the associated eigenfunctions span a multidimensional subspace of the total Hilbert space **H** of the system. This implies that an actual measurement does not characterize H, but some coarse-grained version of H, labeled as $\{H\}$. The latter is constructed as follows. If we choose a basis in which H is diagonal, one can arrange all the eigenvalues E_n with increasing value along the diagonal and decompose that sequence in blocks of length ΔE, which we label with ε (Figure 19.1a). For each block, we replace all E_n values with some intermediate value E_ε in such a way that $|E_n - E_\varepsilon| < \Delta E$. The resulting matrix {H} is a representation of the operator $\{H\}$. Experimentally, $\{H\}$ is as good as H, but obviously $\{H\}$ cannot be used in the Schrödinger equation. If one would still do so, the time evolution would contain only rapid fluctuations with periods less than $\hbar/\Delta E$.

Now consider a macrovariable A, represented by the operator A for an observable (quantity) that varies slowly on the time scale $\hbar/\Delta E$, or more precisely, for which the rate of change of the matrix elements $\mathrm{d}|A_{nm}|/\mathrm{d}t = (\mathrm{i}/\hbar)(E_n - E_m)A_{nm} \ll |A_{nm}|/(\hbar/\Delta E)$. Hence, A almost commutes with H. This implies that the matrix elements of A in the energy representation occupy a band along the diagonal, small as compared to ΔE. For the operator A we have also a coarse-grained version $\{A\}$. Consider the representation in which H is diagonal and in which the "blocking" has been done as indicated. There are only very few matrix elements such that E_n and E_m belong to different blocks, and we delete these elements. The remaining matrix is thus decomposed in blocks corresponding to ε. Now apply for each block a unitary transformation for A which diagonalizes the block. In this representation, $\{H\}$ is still diagonal but H is not. Arrange the eigenvalues of A obtained in increasing order and subdivide them in subblocks, labeled α, each having a length ΔA, determined by the experimental uncertainty of A. Again, replace all the eigenvalues for each subblock with an intermediate value A_α. This is the representation of $\{A\}$ since we cannot distinguish experimentally $\{A\}$ from A. This representation is highly degenerate and commutes with $\{H\}$, but not with H. This procedure can be repeated for a second macrovariable B, represented by the operator B, if the commutator $[A,B]$ is small since in that case the matrix of B in the representation constructed before contains again a narrow band of matrix elements along the diagonal. That is, if δA and δB represent the quantum uncertainty for operators A and B, we have $\delta A \delta B \cong |[A,B]|$ and this condition requires that $\delta A \delta B \ll \Delta A \Delta B$.

This process can be repeated for all pertinent macro-operators H, A, B, … that almost commute with each other. The final result is that the Hilbert space is decomposed in subspaces, labeled ε, α, β, …, that are the common eigenspaces of the commuting operators $\{H\}$, $\{A\}$, $\{B\}$, … These subspaces, denoted as phase cells in analogy with classical mechanics and labeled for convenience with the single label J, represent the macrostates of the system. The eigenvalues of $\{H\}$, $\{A\}$, $\{B\}$, … are then E_J, A_J, B_J, … . Since there are many, say g_J, microstates for each macrostate, the macrostates are highly degenerate.

The next step is to choose for each macrostate (phase cell J) an orthonormal set of base vectors $|Jj\rangle$, where $j = 1, 2, \ldots, g_J$. Expanding any wavefunction of the complete system in these base vectors, we have

$$|\Psi\rangle = \sum\nolimits_{Jj} b_{Jj} |Jj\rangle$$

and the probability P_J to be in state J becomes

$$P_J = \sum\nolimits_{j=1}^{j=g_J} |b_{Jj}|^2$$

while the expectation value of $\{A\}$ is

$$\begin{aligned}\langle\Psi|\{A\}|\Psi\rangle &= \sum\nolimits_{Jj}\sum\nolimits_{Kk} b_{Jj}^{*}b_{Kk}\langle Jj|\{A\}|Kk\rangle \\ &= \sum\nolimits_{Jj}\sum\nolimits_{Kk} b_{Jj}^{*}b_{Kk}A_K\langle Jj|Kk\rangle = \sum\nolimits_{Jj} b_{Jj}^{*}b_{Kk}A_K\delta_{JK} \equiv \sum\nolimits_{Jj} P_J A_J\end{aligned}$$

Time development of a system from $t = 0$ to $t = t$ is described by the Schrödinger equation and leads, employing the time-evolution operator U (Section 3.3), to

$$\Psi(t) = \exp(-\mathrm{i}Ht/\hbar)\Psi(0) \equiv U(t)\Psi(0)$$

Using the representation as constructed before, we obtain

$$b_{Jj}(t) = \sum\nolimits_{K,k}\langle Jj|U(t)|Kk\rangle\, b_{Kk}(0)$$

and therefore

$$\begin{aligned}P_J(t) &= \sum\nolimits_j |b_{Jj}(t)|^2 \\ &= \sum\nolimits_j\sum\nolimits_{K,k}\sum\nolimits_{L,l}\langle Jj|U(t)|Kk\rangle\langle Jj|U(t)|Ll\rangle^{*}b_{Kk}(0)b_{Ll}^{*}(0)\end{aligned} \tag{19.6}$$

This expression still contains all the probability amplitudes, and we have to make some assumption if we do not want to (meaning are unable to) solve the full Schrödinger equation. We therefore now assume that there are no unexpected correlations among the terms; this obviously depends on the system and choice of the macrovariables. If this is the case, one might expect that that the sum contains so many terms with varying phases that they cancel by averaging over each phase cell and that only the positive diagonal terms survive, leading to

$$\overline{b_{Kk}b_{Ll}^{*}} = \delta_{KL}\delta_{kl}|b_{Jj}(0)|^2 = \delta_{KL}\delta_{kl}[P_J(0)/g_J] \tag{19.7}$$

This assumption formally corresponds with taking the average for each phase cell over the base vector coefficients (probability amplitudes) b_{Jj} resulting in the absolute square of (probability) P_J. Substitution of Eq. (19.7) in Eq. (19.6) leads to

$$P_J(t) = \sum\nolimits_K T_{JK}(t)P_K(0) \quad \text{with} \quad T_{JK}(t) = g_J^{-1}\sum\nolimits_{j,k}|\langle Jj|U(t)|Kk\rangle|^2 \tag{19.8}$$

The evolution in time is thus described by the *transition probabilities* $T_{JK}(t)$ which involves $U(t)$ and therefore the full solution of the Schrödinger equation. To make further progress, we consider a small time interval Δt for which the $T_{JK}(t)$ can be written as

$$T_{JK}(\Delta t) = (1 - a_J\Delta t)\delta_{JK} + W_{JK}\Delta t + \mathcal{O}(\Delta t^2) \tag{19.9}$$

where W_{JK} are the transition probabilities per unit time and $a_J = \Sigma_K W_{JK}$ is introduced to ensure conservation of probability. Substitution of Eq. (19.9) in Eq. (19.8) results to first order in Δt in

$$\dot{P}_J = \sum\nolimits_{K\neq J}(W_{JK}P_K - W_{KJ}P_J) \tag{19.10}$$

addressed as the (Fermi) *Master Equation* (ME). From Eq. (19.8) we note that

$$T_{KJ} \geq 0 \;\text{ and }\; \sum\nolimits_J T_{JK}(t) = g_K^{-1}\sum\nolimits_k\sum\nolimits_{Jj}|\langle Jj|U(t)|Kk\rangle|^2 = g_K^{-1}\sum\nolimits_k 1 = 1$$

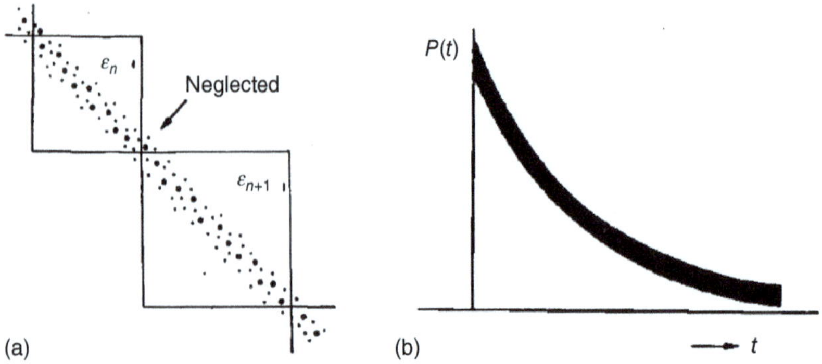

Figure 19.3 (a) Matrix representation of a slowly varying operator in the energy representation. (b) The overall approach to equilibrium. The band indicates that, associated with the overall decrease, significant fluctuations occur, representing the approximate nature of a macroscopic description.

If H is an even function[12] of the momenta p, that is, no magnetic and Coriolis forces are present, we have $\langle n|Jj\rangle = \langle Jj|n\rangle$ and the matrix $\langle Jj|U(t)|Kk\rangle$ will be symmetric. Using $T_{JK} = g_K^{-1}\Sigma_{jk}|\langle Jj|U(t)|Kk\rangle|^2$, we obtain

$$T_{JK}(t)g_K = \sum_J \sum_{Jj} |\langle Jj|U(t)|Kk\rangle|^2 = T_{KJ}(t)g_J \tag{19.11}$$

We conclude from the master equation that the $P_J(t)$ tend to an equilibrium, in which all accessible macrostates (phase cells) are equally occupied (Figure 19.3b).

Sometimes it is beneficial to eliminate the factors g_J and g_K by defining $p_{Jj} \equiv P_J/g_J$ for $j = 1, 2, \ldots, g_J$ and similarly $w_{Jj,Kk} \equiv W_{JK}/g_J$. If we replace the double indices Jj and Kk by the single ones x and y, we can write

$$\dot{p}_x = \sum_y w_{xy}p_y = \sum_{y\neq x}(w_{xy}p_y - w_{yx}p_x)$$

while the symmetry relation Eq. (19.11) becomes $w_{xy} = w_{yx}$. Summing over y while using $\Sigma_J W_{JK} = 0$, we obtain $\Sigma_y w_{xy} = \Sigma_y w_{yx} = 0$, so that Eq. (19.10) becomes

$$\dot{p}_x = \sum_{y\neq x} w_{xy}(p_y - p_x) \tag{19.12}$$

This is actually the form of master equation used at several places. As $w_{xy} \geq 0$, if $p_y > p_x$ its contribution to $\mathrm{d}p_x/\mathrm{d}t$ is larger than zero and p_x will increase with time. If, however, $p_y < p_x$, its contribution is smaller than zero and p_x will decrease with time. Eventually, one reaches the situation where all p_js are equal corresponding to equilibrium (Figure 19.3b). Although in this expression the label x refers to individual states, it is a formal index because these states are coarse grained. In the conventional approach, the coefficients w_{xy} are calculated from individual states using the golden rule (Section 2.4), from which one concludes that $w_{xy} = w_{yx}$. Note that this symmetry does not generally apply because the golden rule assumes that the system is initially in a clearly defined eigenstate, and we noted that this is not the case for a macrostate. Fortunately, the weaker relation $\Sigma_y w_{xy} = \Sigma_y w_{yx} = 0$ suffices [40] to derive Eq. (19.12). Note further that, in order for the master equation to be valid at all times, correlation should be absent, not only at the first time step Δt but after every time step Δt, otherwise all accessible states will be occupied equally in the long run, no

12 The conclusions remain valid when H is an odd function of the various momenta, see [39].

matter how small the transition probability is.[13] This is an unsatisfactory situation [41]. Actually, assuming absence of correlations only at $t = 0$ should suffice and this has been achieved for certain systems [42], though associated with considerable complexity. In the present treatment, at any rate, transitions are limited to the microstates belonging to the same macrostate since $T_{JK} = 0$ if $\{\Delta E\}_J \neq \{\Delta E\}_K$. Of course, this implies that $W_{JK} = 0$ and hence that transitions between states of different energy remain in the long run negligible, consistent with the conservation of energy.

19.6 The Information Theory Approach

The last approach we discuss is actually almost the one practically used in our discussion of statistical thermodynamics in Chapter 7. It is denoted as the information theory approach[14] and based on the PME. We start with a brief summary of the basic procedure followed to obtain the thermodynamic "analogues," and thereafter deal with some other considerations.

19.6.1 A Brief Review

In Chapter 7, the correspondence between statistical and thermodynamic properties was obtained by postulating the expression for the Gibbs or statistical entropy, imposing the relevant constraints, and calculating the optimum using Lagrange multipliers. Let us briefly reiterate this reasoning with two differences. First, the Gibbs entropy in equilibrium was identified right away with the thermodynamic entropy. Here the distinction is important, and we denote the thermodynamic entropy in equilibrium by S_{eq} and the statistical entropy by S_{G}. Second, the average (or expectation) value of any property A was written as $\Sigma_j p_j A_j$. Here, we employ the density operator D, introduced in Chapter 3. The probability p_j of state j is then given by $p_j = D_{jj}$, the diagonal elements of the density matrix D, representing the density operator D, and the value of property A in state j by $A_j = A_{jj}$, the diagonal elements of a matrix $\mathbb{A}$, representing the operator $\mathbb{A}$. Hence, we have $A = \Sigma_j p_j A_j = \langle \mathbb{A} \rangle \equiv \mathrm{tr}\mathrm{D}\mathbb{A}$. This notation has two additional advantages: (i) traces are independent of the QM representation used and (ii) $\mathrm{tr}[\mathrm{d}f(\mathbb{A})/\mathrm{d}t] = \mathrm{tr}[f'(\mathbb{A})\,\mathrm{d}\mathbb{A}/\mathrm{d}t]$, thereby avoiding commutation problems. So, for the Gibbs entropy, we write $S_{\text{G}} = -k\,\mathrm{tr}\mathrm{D}\ln\mathrm{D}$, while normalization is represented by $\mathrm{tr}\mathrm{D} = 1$. For simplicity, we use, apart from normalization, only two constraints. The first deals with the internal energy U, represented by trDH, where H represents the Hamilton operator. The second deals with the number of particles N, represented by trDN, where N represents the number operator. Using the Lagrange multipliers $k\alpha$, $k\beta$ and $k\gamma$, the expression

$$S' = -k\,\mathrm{tr}\,\mathrm{D}\ln\mathrm{D} - k\alpha(\mathrm{tr}\,\mathrm{D} - 1) - k\beta(\mathrm{tr}\,\mathrm{DH} - U) - k\gamma(\mathrm{tr}\,\mathrm{DN} - N)$$

is to be optimized via $\partial S'/\partial\mathrm{D}$, resulting in

$$\mathrm{D}_{\mathrm{R}} = \exp[-(1+\alpha) - \beta\mathrm{H} - \gamma\mathrm{N})] = \Xi^{-1}\exp(-\beta\mathrm{H} - \gamma\,\mathrm{N}) \tag{19.13}$$

13 Essentially, the same problem occurred in the formal approach.
14 *The* advocate of the information theoretical approach to statistical mechanics was E.T. Jaynes (1922–1998). A good summary of that approach has been given by Grandy [4], who has also given an extension to provide a basis for macroscopic (phenomenological) transport equations [43]. We follow largely his arguments.

with $\Xi \equiv \text{tr}(\exp-\beta\, \text{H}-\gamma\, \text{N})$ as $\text{tr}\, \text{D}_\text{R} = 1$ by construction. Here, $\Xi = \exp(1+\alpha)$ is denoted as the grand partition function, essentially acting as the normalization constant for the probability distribution. The Gibbs entropy S_G becomes

$$S_\text{G} = k\beta\langle H\rangle + k\gamma\langle N\rangle + k \ln \Xi$$

which compared with equilibrium thermodynamic entropy

$$S_\text{eq} = T^{-1}(U - \mu N + PV)$$

results in $\beta = 1/kT$, $\gamma = -\mu/kT$, and $\Xi = \exp(PV/kT)$. Here, $\beta = 1/kT$ is identified as essentially the temperature T of the reservoir and the multiplier $\gamma = -\beta\mu$ as essentially the chemical potential μ of the reservoir. For any term $-Aa$ in the entropy expression S_eq, where A is the force corresponding to the displacement a, we can add to S' a constraint $\text{tr}\, \text{DA}-A = 0$ with multiplier δ to yield $\delta = -\beta A$.

It remains to be shown though, that the result is a maximum and we do this by proving that for any pair of density matrices D and D′

$$S_\text{G}(\text{D}) = -k\, \text{tr}\, \text{D} \ln \text{D} \leq S_\text{G}(\text{D}') = -k\, \text{tr}\text{D} \ln \text{D}'$$

To that purpose we consider the concavity of the logarithm yielding the inequality $\ln z \leq z-1$, with equality only holding when $z = 1$. Substituting $z = y/x$, we easily obtain $x \ln y - x \ln x \leq y - x$. If we take D′ for y and D for x and take the trace, the result is $\text{tr}\text{D} \ln \text{D}' - \text{tr}\text{D} \ln \text{D} \leq \text{tr}\text{D}' - \text{tr}\text{D} = 0$ because $\text{tr}D' = \text{tr}\, \text{D} = 1$. Consequently, we have $-\text{tr}\text{D} \ln \text{D} \leq -\text{tr}\text{D} \ln \text{D}'$ or $S_\text{G}(\text{D}) \leq S_\text{G}(\text{D}')$. Now take D_R for D′ and for D any density operator satisfying the constraints $\text{tr}\, \text{DA} = \langle A\rangle = A$. This leads to

$$S_\text{G}(\text{D}) \leq -\text{tr}\text{D} \ln \text{D}_\text{R} = -k \ln \Xi - k\beta\langle H\rangle - k\gamma\langle N\rangle = -\text{tr}\text{D}_\text{R} \ln \text{D}_\text{R} = S_\text{G}(\text{D}_\text{R})$$

and indeed we have, as anticipated, $S_\text{G}(\text{D}_\text{R}) \geq S_\text{G}(\text{D})$.

Having identified the various terms and recognizing that S_G is a Legendre transform of Ξ, we can now consider S_G as a function of $U = \langle H\rangle$ and $N = \langle N\rangle$ from which we obtain $\partial S_\text{G}/\partial U = k\beta = 1/T$ and $\partial S_\text{G}/\partial N = k\gamma = -\mu/T$. Similarly, $\partial\Xi/\partial(k\beta) = U$ and $\partial\Xi/\partial(k\gamma) = -N$. For any extra term $-Aa$ in the entropy expression S_eq, one has $\partial\Xi/\partial(k\delta) = -a$. The properties of entropy are discussed in detail by Wehrl [44].

19.6.2 High and Low Probability Manifolds

At this point, a few further remarks are appropriate. First, re-examining the derivations, as summarized in Section 19.6.1, it will be clear that, following the procedure outlined, not only inferences about statistical mechanics can be made, but also for other problems that can be formulated similarly [45]. Moreover, for statistical mechanics, instead of only constraints on N and U any other constraint can be used for which a term in S_eq can be included. We are only limited by our limited knowledge. For small systems, for example, we could also use, if known, the variance of physical properties as constraint. To denote the constraints collectively, we use $\lambda(\mathbf{r},t)F(\mathbf{r},t)$ with $\lambda(\mathbf{r},t)$ the Lagrange multiplier and $F(\mathbf{r},t)$ the constraint and where the variables indicate that the constraints can be position and time dependent.

Second, although the procedure leads to optimal distributions given the information, the solution is by no means unique. This implies that other distributions than the canonical ones may yield the same average or macroscopic value for a certain property. Under certain conditions one may show though that the micro-canonical distribution is unique. In Chapter 7, we showed that in that case $S = k \ln W$, where W denotes the number of microstates that are thermally accessible for the system compatible with the constraints applied, also called *complexions*. In fact, a rather similar interpretation follows for other distributions [43]. Consider for simplicity the canonical distribution. For this distribution, we derived $U = \mathrm{tr\,DH} = \langle H\rangle = \Sigma_j p_j E_j = -\partial \ln Z(\beta)/\partial\beta$. Formal integration leads to $Z(\beta) = Z(0)\exp(-\beta U)$ which is meaningful only if $Z(0)$ is finite. Now consider the probabilities ordered in such a way that $p_1 \geq p_2 \geq, \ldots,$ and define a number λ such that $0 < \lambda < 1$. Furthermore, add the various p_js until their sum equals or exceeds λ, and define $n(\lambda)$ by

$$\sum\nolimits_{j=1}^{n(\lambda)} p_j = Z(\beta)^{-1} \sum\nolimits_{j=1}^{n(\lambda)} g_j \exp(-\beta E_j) \geq \lambda$$

where g_j denotes the degeneracy of energy level E_j. With this number $n(\lambda)$, there is a corresponding total number of states given by

$$W_\lambda \equiv \sum\nolimits_{j=1}^{n(\lambda)} g_j = Z_\lambda(0) \tag{19.14}$$

That is, W_λ is the finite dimension of a manifold of degenerate states that we label as the *high-probability manifold* (HPM) and its complement is the low-probability manifold (LPM). By varying λ, we obtain a set of nested sequences of HPMs. Since small changes in λ induce large changes in $n(\lambda)$, we use logarithms and write

$$\ln W_\lambda = \ln Z_\lambda(0) = \ln Z_\lambda(\beta) + \beta\langle H\rangle_\lambda$$

where U has been replaced by its expectation value $\langle H\rangle$, allowed since fluctuations are $\mathcal{O}(N^{-1/2})$. Consider now the limit for $N \to \infty$, which we write as

$$N^{-1}\ln W = \lim_{N\to\infty} N^{-1}\ln W_\lambda = \lim_{N\to\infty} N^{-1}[\ln Z_\lambda(\beta) + \beta\langle H\rangle_\lambda] \tag{19.15}$$

In the thermodynamic limit, the third part of Eq. (19.15) appears to be the (finite) entropy per particle [46]. Hence, the limit $N^{-1}\ln W$ exists; it is essentially independent of λ and allows the interpretation of $S = k\ln W$ also for the canonical distribution if we consider W as the number of accessible states in the HPM instead of as all accessible states. Of course, one could object that infinite systems do not exist. Also, at several other places we alluded to the thermodynamic limit. Hence, it may be useful, as is done in Justification 19.2, to assess the effect of surfaces on the density of states. This analysis [24] shows that for macroscopic systems, the boundary terms contribute negligibly and therefore that the thermodynamic limit is a useful abstraction.

Justification 19.2 ***Surface Effects*** To show that the density of states depends only on the volume of the container, consider a particle in a box with mass M [24]. The allowed states for a block with edges a, b, and c are given by

$$k^2 = |\mathbf{k}|^2 = \frac{\pi^2\hbar^2}{2M}\left[\left(\frac{l}{a}\right)^2 + \left(\frac{m}{b}\right)^2 + \left(\frac{n}{c}\right)^2\right]$$

The density of states $g(k)$ is given by $g(k) = \mathrm{d}G(\mathbf{k})/\mathrm{d}k$, where $G(\mathbf{k}) = \sum\nolimits'_{l,m,n} f(l,m,n)$ with $f(l,m,n) = 1$ if $l,m,n = 1,2,3,\ldots$ and $f(l,m,n) = 0$ otherwise. The prime in Σ' indicates that the sum has to be carried out with the restriction $(l/a)^2 + (m/b)^2 + (n/c)^2 \leq 2M(k/\pi\hbar)^2$. The trick is now to write this sum over values of l,m,n including $l,m,n = 0$ and correct for the $l,m,n = 0$ terms. This leads to

$$G(\mathbf{k}) = \frac{1}{8}\left\{\begin{array}{l}\sum'_{l,m,n} f^*(l,m,n) \\ -\left[\sum'_{l,m} f^*(l,m,0) + \sum'_{l,n} f^*(l,0,n) + \sum'_{m,n} f^*(0,m,n)\right] \\ +\left[\sum'_{l} f^*(l,0,0) + \sum'_{m} f^*(0,m,0) + \sum'_{n} f^*(0,0,n)\right] - 1\end{array}\right\}$$

with $f^*(l,m,n) = 1$ for integer values of l,m,n and zero otherwise. The first term counts the number of lattice points within and on an ellipse described by

$$\left(\frac{l}{a}\right)^2 + \left(\frac{m}{b}\right)^2 + \left(\frac{n}{c}\right)^2 = 2M\left(\frac{\mathbf{k}}{\pi\hbar}\right)^2 \quad \text{or} \quad \left(\frac{l}{\alpha}\right)^2 + \left(\frac{m}{\beta}\right)^2 + \left(\frac{n}{\gamma}\right)^2 = 1$$

Similarly, the second set of terms counts the number of lattice points within and on the principal cross-sections of this ellipse and the third set on the principal axes of the ellipse. Defining $V = abc$, $A = 2(ab + ac + bc)$, and $L = 4(a + b + c)$, and approximating the sums with integrals, we obtain

$$\sum'_{l,m,n} f^*(l,m,n) \equiv V = \frac{4}{3}\pi\alpha\beta\gamma = \frac{4}{3}\pi abc(2M)^{3/2}\left(\frac{k}{\pi\hbar}\right)^3$$

$$\sum'_{l,m} f^*(l,m,0) \equiv A_{\alpha\beta} = \pi\alpha\beta = \pi ab(2M)\left(\frac{k}{\pi\hbar}\right)^2$$

$$\sum'_{l} f^*(l,0,0) \equiv L_\alpha = 2\alpha = 2a(2M)^{1/2}\left(\frac{k}{\pi\hbar}\right)$$

with similar expressions for $A_{\alpha\gamma}$, $A_{\beta\gamma}$, L_β, and L_γ. Hence

$$G(\mathbf{k}) = \frac{1}{8}\left\{\begin{array}{l}\frac{4\pi}{3}abc(2M)^{3/2}\left(\frac{k}{\pi\hbar}\right)^3 - \pi(ab + ac + bc)(2M)\left(\frac{k}{\pi\hbar}\right)^2 \\ +2(a + b + c)(2M)^{1/2}\left(\frac{k}{\pi\hbar}\right) - 1\end{array}\right\}$$

$$= \frac{V}{6}\pi(2M)^{3/2}\left(\frac{k}{\pi\hbar}\right)^3 - \frac{A}{16}\pi(2M)\left(\frac{k}{\pi\hbar}\right)^2 + \frac{L}{16}(2M)^{1/2}\left(\frac{k}{\pi\hbar}\right) - 1$$

Consequently,

$$g(\mathbf{k}) = \frac{\mathrm{d}G(\mathbf{k})}{\mathrm{d}k} = V(2M)^{3/2}\frac{k^2}{2\pi^2\hbar^3} - A(2M)\frac{k}{8\pi\hbar^2} + \cdots$$

Therefore, when the dimensions of the box are large, the boundary terms contribute insignificantly.

Irreversibility can now be discussed in terms of the HPM. Suppose we have a system in macroscopic state 1 for which a constraint is released so that we obtain macroscopic state 2. This will increase the size of the HPM enormously. Using $S = k\ln W$, we easily obtain $W_1/W_2 = \exp[-(S_2 - S_1)/k)]$. A simple calculation learns that if $\Delta S = S_2 - S_1 \cong 1 \times 10^{-9}$ cal, $W_1/W_2 \cong \exp(-10^{-12})$ and thus that the number of complexions in the HPM for state 1 is vanishingly small as compared with those for state 2. Thus, to have a significant probability of the reverse being realized, initial complexions of state 2 must lie in the HPM of state 1, but these initial states cannot be realized in many different ways and thus lie in the LPM of state 2. Moreover, this manifold must have about the same dimension as that of state 1, but this dimension is so much smaller than that of state 2 that this renders the return even less probable. In conclusion, for a macroscopic system, returning to the initial state is not impossible but extremely unlikely. Recall further that really small perturbations distort the reverse process to a large extent so that even when a complexion leading to the initial state was "selected", the highly unstable behavior would still lead to the (macroscopic) state 2. For small systems, fluctuations can lead temporarily to a behavior contradicting the second law, but only for a limited time. A similar view has been forwarded by Lebowitz [47] and Balian [35]. The manifold picture assigns essentially the same probability to all states in the HPM. As the HPM is very sharply defined, this leads to highly reliable predictions, while the LPM is responsible for the fluctuations. This *macroscopic homogeneity* leads to experimentally reproducible results and provides isolated systems with a rationale for the "minus-first" law (Section 6.1.1).

To show that this approach leads to an increase in entropy S, consider a closed system in complete thermal equilibrium with density matrix D at time $t = 0$, that is, with no memory of anything done in the past. The microstates of the system are then in its HPM $M(0)$ and we have $S(0) = S(D) = -k\,\mathrm{tr}(D \ln D) = k \ln W(0)$. If the system is now considered at time t, thc system is in its HPM $M(t)$ compatible with the equations of motion, still describing the initial macrostate. Now let us apply an adiabatic change to the system, and thereafter let the system return to its equilibrium macrostate at time t. After completion of the process, the system will have a new HPM $M'(t)$ with dimension $W'(t)$ and entropy $S(t) = k\ln W'(t)$. Because $M'(t)$ contains all the high probability states compatible with the new macrostate, and if the process is to be reproducible, it follows that $M(t)$ must lie in $M'(t)$, but this is only possible if $W'(t) \geq W(t)$. Hence, $S'(t) = k\ln W'(t) \geq k\ln W(0) = S(0)$. For a clear discussion of this so-called "unbelievably short proof of the second law," see [48].

Although these arguments show the increase of S, the objection often made is that, if S is defined by $S_\mathrm{G} = -k\Sigma_j\, p_j \ln p_j$, it cannot increase with time t. Indeed, for proper quantum microstates $|j\rangle$, for which time evolution is described by the unitary transformation $|j(t)\rangle = |j(0)\rangle \exp(-\mathrm{i}E_j t/\hbar)$, one can easily show that the probability p_j remains constant with time and hence that S_G is constant with time (Problem 19.4). Actually, the *microstates used are not proper quantum states* but a kind of average over the accessibility range ΔE, as discussed in some detail in Sections 19.4 and 19.5. This renders their time dependence non-unitary and the p_js will vary with time. In Justification 19.3, we show again in a simplified way, using the (Fermi) master equation, that in that case $\mathrm{d}S/\mathrm{d}t \geq 0$ always. Of course, the master equation has to be rationalized for the argument to be valid, as discussed also in Sections 19.4 and 19.5.

Justification 19.3 *dS ≥ 0 Always* To prove that the coarse-grained statistical entropy increases unless the process is reversible, we first must make the meaning of microstates somewhat more precise. In fact, as discussed (Section 19.4.5), for almost any macroscopic system it is impossible to obtain the exact eigenstates of the system. Moreover, even if we would know them at a certain moment, soon afterwards, because of the interaction between the particles in the system, they would be unknown. For example, for a perfect gas the eigenstates are the described by the momentum eigenfunctions of structureless point particles. During equilibration the particles collide and exchange momentum. The uncertainty in energy E thus becomes $\delta E \geq \hbar/2\delta t$, where δt is mean time between collisions.

As indicated in Section 7.1.1, experimentally we are limited to ΔE with $\Delta E \gg \delta E$. Hence, generally we have groups of approximate eigenstates, often called *coarse-grained states*, associated with a small energy range, the *accessibility range* ΔE, small as compared to E but large as compared to the Heisenberg uncertainty δE and in which the density of states is essentially constant. The probabilities p_i therefore refer to these approximate states and, when ΔE is small enough, can be considered as constant within the accessibility range. It then makes no difference whether we consider transitions from exact state i to exact state j as restricted by δE or transitions from one approximate state (group) i to another state (group) j as restricted by ΔE: in either case, the value of $\mathrm{d}p_i/\mathrm{d}t$ will be the same. Sometimes this result is referred to as the *accessibility* or *equal a priori probability* assumption.

To discuss equilibration, we need the jump rate ν_{ij} from state j to state i. For the rate at which a state i with probability p_i is changing, we have two contributions. First, we have the contribution $\Sigma_j \nu_{ji}\, p_i$ from state i to any state j, and second, the contribution $\Sigma_j \nu_{ij}\, p_j$ from any state j to state i. Quantum mechanics provides us the principle of *jump rate symmetry* (Section 3.4), that $\nu_{ij} = \nu_{ji}$, but it applies to pure quantum states. For coarse-grained states, $\Sigma_j \nu_{ji} = \Sigma_i \nu_{ij}$ applies but this fortunately suffices. Therefore, the total rate is given by the *master equation* $\mathrm{d}p_i/\mathrm{d}t = \Sigma_{i\neq j} \nu_{ij}(p_j - p_i)$.

We now consider the entropy differential $\mathrm{d}S/\mathrm{d}t = -k\ \Sigma_i \ln p_i(\mathrm{d}p_i/\mathrm{d}t)$. For an isolated system, we have for the effect of jumps between two particular states i and j, a contribution to $\mathrm{d}p_i/\mathrm{d}t$ reading $\nu_{ij}(p_j - p_i)$ as well as a contribution $\nu_{ij}(p_i - p_j)$ to $\mathrm{d}p_j/\mathrm{d}t$. The total rate is thus, taking in to account all jumps, $\mathrm{d}S/\mathrm{d}t = k\Sigma_{i,j} \nu_{ij}(p_j - p_i)(\ln p_j - \ln p_i)$. Since the terms in brackets have the same sign, $\mathrm{d}S/\mathrm{d}t$ cannot be negative. Moreover, if $p_i = p_j$, $\mathrm{d}S/\mathrm{d}t = 0$, and so we conclude that $\mathrm{d}S \geq 0$. This result provides also a basis for the equiprobability assumption.

One can now take two different points of view. The first is that, as long as a system can be divided into small subsystems in which local equilibrium prevails, the subsystem properties can be calculated using equilibrium statistical–thermodynamic methods, albeit resulting in time-dependent properties as the system evolves with time. This can be done if the relaxation time of the system, τ_{rel}, is large as compared to the evolution time of the system, τ_{evo}. This is equivalent to the relaxation of the subsystem being fast as compared to the relaxation of the system as a whole (compare Section 6.1.2). Accepting this separation of length and time scales (and with that loss of information, or, equivalently, coarse graining), entropy can be estimated also outside overall equilibrium. For time-dependent properties described by the (Heisenberg) operator F over a domain $R(\mathbf{r},t)$ dependent on time t and position $\mathbf{r}$, the equation of motion becomes $\mathrm{i}\hbar\dot{F} = [\dot{F}, H] + \partial F/\partial t$, where the dot denotes the total time derivative. With the properties varying with time, the constraints become an integral over time with a Lagrange multiplier $\lambda(\mathbf{r},t)$ and a similar maximization procedure as before leads to

$$D = Z^{-1} \exp\left[-\int_R \lambda(\mathbf{r},t) F(\mathbf{r},t)\,\mathrm{d}\mathbf{r}\mathrm{d}t\right] \quad \text{with } Z = \mathrm{tr}\exp\left[-\int_R \lambda(\mathbf{r},t) F(\mathbf{r},t)\,\mathrm{d}\mathbf{r}\mathrm{d}t\right] \tag{19.16}$$

the *partition functional*. The Lagrange multiplier $\lambda(\mathbf{r},t)$ is determined by

$$F(\mathbf{r},t) \equiv \langle F(\mathbf{r},t)\rangle \equiv \mathrm{tr}[\mathrm{D}F(\mathbf{r},t)] = -\delta \ln Z/\delta\lambda(\mathbf{r},t)$$

where $\delta(\cdot)/\delta\lambda(\mathbf{r},t)$ denotes the functional derivative (Section A.8) and $F(\mathbf{r},t)$ denotes the experimental information. Other properties $J(\mathbf{r},t)$ are determined, as before, via $J(\mathbf{r},t) = \mathrm{tr}[\mathcal{D}(t)\mathcal{J}(\mathbf{r})]$. The maximum entropy becomes a functional of $F(\mathbf{r},t)$ obtained from substituting expressions (19.16) in $S_\mathrm{G} = -k\,\Sigma_j p_j \ln p_j$, leading to

$$S_\mathrm{G}[F(\mathbf{r},t)] = k\ \ln Z[\lambda(\mathbf{r},t)] + k\int_R \lambda(\mathbf{r},t)F(\mathbf{r},t)\mathrm{d}\mathbf{r}\mathrm{d}t$$

and an alternative for $\lambda(\mathbf{r},t)$ reading

$$k\lambda(\mathbf{r},t) = \delta S_\mathrm{G}/\delta F(\mathbf{r},t)$$

Grandy developed a theory of time evolution for macroscopic systems based on these expressions leading to a description of transport phenomena [43].

The second point of view does not assume the separability of length and time scale but accepts only that the statistical–thermodynamic expression for the entropy $S_\mathrm{G} = -k\Sigma_j\, p_j \ln p_j$ for the overall equilibrium configuration can be identified with the thermodynamic entropy S_eq. From this point of view, S_eq can be calculated only for the initial and final equilibrium configuration, but not for intermediate times while the process is still ongoing. Obviously, both points of view yield an identical total entropy change for the completed process, given the same information.

Example 19.5 ***The Perfect Gas*** A nice illustration of coarse graining is given by the perfect gas. Suppose one starts with a state with volume V_ini with thermodynamic entropy S_ini, which changes in time, for example, by volume expansion, to V_fin with thermodynamic entropy S_fin. For assessing the time development of the statistical entropy S, one could divide the volume into, say, M subvolumes $V_j = V/M$. Each subvolume should be taken sufficiently large so that it can be considered macroscopic, but also sufficiently small so that we can assume that the energy E_j and density N_j/V_j of each subvolume j can be considered – although time dependent – as constant over each V_j. Let us label the subvolumes in such a way that the first m subvolumes are initially contained within V_ini. For the entropy $S_j(t)$ of each V_j, we have from the Sackur–Tetrode equation, Eq. (7.19),

$$S_j(t) = N_j kT\left(\ln\frac{V_j}{N_j(t)} + \frac{3}{2}\ln\frac{E_j(t)}{N_j(t)}\right) + \mathit{const.}$$

The values of $N_j(t)$ and $E_j(t)$ at each time t obviously depend on how the process is exactly done. Since entropy is additive, we have for the total entropy $S(t) = \Sigma_j S_j(t)$. For the initial and final configuration, we thus have, respectively,

$$S(0) = \sum\nolimits_{j=1}^{j=m} S_j(0) \quad \text{and} \quad S(\infty) = \sum\nolimits_{j=1}^{j=M} S_j(\infty)$$

In both the initial and final configurations, the energy and density are constant so that the equilibrium entropies are $S_\mathrm{ini} = S(0)$ and $S_\mathrm{fin} = S(\infty)$. As long as a proper separation of length and time scale can be made, the entropy, while the process is proceeding, can be considered as given by $S(t)$.

Problem 19.4 ***Time Independence of S_G*** Show that for unitary time evolution of the state $|j\rangle$, the probability p_j is constant, and hence that $S = -k\Sigma_j\, p_j \ln p_j$ is independent of time.

19.7 Closure

From the previous sections, it has hopefully become clear that, whatever approach one prefers, the amount of information (or, more appropriately, lack of information) plays a determining role. The basic concept is that the system at hand plus the reservoir are considered as an isolated overall system. With this in mind, one could say that in the formal, open system approach, the information about the reservoir is integrated out explicitly, while in the physical approach, one uses accessibility and probability assumptions to separate the system from the reservoir. In both cases, one tries to rationalize the equal a priori probability by proving the validity of the master equation. Finally, in the information theory approach, complete knowledge is denied right away, and one accepts the PME directly, which is essentially assuming equiprobability directly. In any case, in all three approaches, information "leaks" from the system to the reservoir and vice versa. Moreover, even if one considers an isolated system, interaction with the outside world is inevitable, and the slightest disturbance leads to randomization, as Example 19.1 shows, while infinitely accurate experimental information is unavailable. In the end, this leads for all three approaches to a system with random probabilities, or a system with no microscopic memory. This "absence of memory" essentially explains the success of statistical thermodynamics.

If, using $S_G = -k\Sigma_j p_j \ln p_j$, each microstate j is considered as a proper quantum state, Problem 19.4 shows that S_G is time independent. This should be no surprise, since in this case the time evolution of each microstate is given by a unitary transformation, so that each probability p_j remains constant. If, however, we consider the microstate j as a state with incomplete or averaged properties over a set of quantum states, that is, coarse-grained states, time evolution becomes non-unitary, and the various p_js change with time to obtain maximum entropy (Justification 19.3).

This all implies that the nature of entropy intrinsically differs from the nature of other extensive properties, say, for example, energy. The latter is not only conserved, also outside equilibrium, but also, given that proper thermodynamic experiments are performed, independent of whether further information is available or not, since its value is completely determined by the work done in an adiabatic process (Section 6.1.1). Entropy, though, depends on what one knows (or assumes) about the system: more information leads to extra terms in the thermodynamic entropy expression

$$T\mathrm{d}S = \mathrm{d}U + P\mathrm{d}V - \sum\nolimits_j A_j \mathrm{d}a_j$$

That is, whether only $-P\mathrm{d}V$ is considered as work, or one also includes other terms $A_j\mathrm{d}a_j$ makes a difference. The entropy S is thus a *contextual property*, that is, a property dependent on the macroscopic, thermodynamic variables chosen to describe the physical system. It is thus essential to distinguish between the *physical system* and its description as a *thermodynamic system*. Given the choice for the thermodynamic variables, S behaves as objectively as any other property. A schematic illustration is given in Figure 19.4.

Let us now briefly address the interpretation of entropy for which there are essentially three interpretative schemes. First, the "disorder" metaphor in which a higher entropy is given to a more disordered state. The problem of this interpretation is that "order" and "disorder" depend on the observer. Hence, two observers can obtain different entropy changes if they have a different perspective of order. Moreover, even if two observers agree on the "disordered" state, there are quite a few examples where the entropy change has

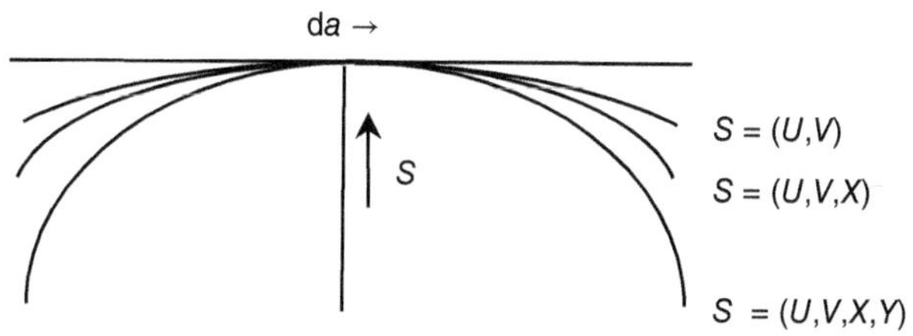

Figure 19.4 Schematic of the behavior of entropy S given a description including only mechanical work $-PdV$ and descriptions containing extra work terms like Xdx or Ydy, showing the dependence of S on the choice of independent variables. Source: Adapted from Woods [49].

opposite sign of what is expected [50]. A clear example of where the "disorder" metaphor leads to nowhere is provided by the free expansion of a (classical) gas. Suppose an isolated volume $2V$ and a partition that keeps the gas in one, say the left, half V. By removing the partition so that the gas is hardly disturbed, it expands to $2V$. The entropy increases but is the disorder for the gas in $2V$ different from that of the gas in V? This metaphor has been labeled, correctly, as a "persistent swindle" [51]. Second, the "missing information" metaphor that relates entropy to information about the system. Given the same information, two observers will obtain the same entropy value. Although well-defined, the picture is rather static and does not address space, time, and energy. The third is the "spreading" metaphor, in which processes are considered in terms of spreading of energy as equally probable over volume and time. Consider again the free expansion mentioned before. If the gas is perfect, the temperature T remains constant. For real gases, T can either decrease or increase, respectively, below or above the inversion temperature T_{inv}. Using the spreading concept, for $T > T_{\text{inv}}$ the intermolecular interactions are mainly repulsive and spatial spreading is enhanced (as compared to the noninteractive configuration) and T increases. This spreading adds to that of the volume increase, so that a larger entropy increase results. For $T < T_{\text{inv}}$, attraction prevails, and spreading is reduced, consistent with a decrease in T. As the temperature change for free expansion is relatively small, all free expansions lead to a net increase in spatial spreading and thus in entropy. The "spreading" metaphor has a dynamic nature (as expressed by the use of a verb instead of a noun) and is a physically motivated, transparent interpretative scheme. It essentially interprets the states and processes quantitatively treated by the information theory (or any other) approach pictorially. Leff [50] has given a highly readable discussion of entropy in these terms for various processes, to which we refer for further information.

In this connection, it may useful to point out that the entropy of an ideal (classical) system, that is, a system for which the molecules experience no interaction, is always larger than (or at best equal to) the entropy for the corresponding nonideal system [52], or $S_{\text{ideal}}(T,V,N) \geq S_{\text{nonideal}}(T,V,N)$ (Problem 19.5). This expression reflects the fact that, for potentials dependent only on position, the degree of spatial and temporal spreading is reduced by the interactions leading to correlations between the positions of the various molecules.

Summarizing in one sentence, entropy is a contextual property that always strives for a maximum, but its precise value depends on the description chosen. The literature on irreversibility is large and we provide here only a few relevant references for the present approach. A lucid discussion has been given by Bricmont [53], while the logic of thermostatistical physics has been reviewed by Emch and Liu [32]. Sklar [11] provided an authoritative, although sometimes not easily readable, philosophical background and a somewhat technical treatment of nonequilibrium and irreversibility is given by Gallavotti [54]. Finally, we refer to a concise review of predictive statistical mechanics and dynamics from the point of view of the PME [55].

Perhaps not surprisingly, we note that not all scientists subscribe to the considerations as given here, as the approaches sketched are based on the conventional *statistical* interpretation of quantum mechanics and other interpretations of quantum mechanics exist. For example, based on a strict belief in *causality*, such as Bohm's interpretation [56–58], or based on the *spontaneous, though infrequent collapse of the wavefunction*, such as the Girardi–Rimini–Weber interpretation [59], for which Albert [60] provides relatively easy reading for its consequences for irreversibility. Zurek [61] has published a highly readable and important paper on decoherence and the transition from quantum to classical mechanics. Noteworthy is also the *cellular automaton* interpretation of quantum mechanics, as advanced by 't Hooft [62], which renders a deterministic interpretation possible meanwhile keeping all well-established results. Other interpretations are the *inherent indeterminism* of the Prigogine school [63, 64], or an interpretation based on *modified basic dynamic laws*, such as those advocated by quantum thermodynamics [65–68]. The latter was originally based on an alternative basis for thermodynamics, as given by Hatsopoulos and Keenan [69] and later considerably extended by Giftopoulos and Beretta [70]. In this formulation, thermodynamics is essentially based on one postulate, the *law of stable equilibrium*, akin to the "minus-first" law (footnote 6, Section 6.1), instead of the usually employed four laws. This brings thermodynamics on a par with classical mechanics and quantum mechanics, both based on one basic postulate, namely, the Hamilton principle (Section 2.2) and the "state vector–operator–property" principle (Section 3.3), respectively. Fortunately, whatever basic considerations one adheres to, their practical implementations all lead to virtually the same results, a comforting thought for any scientist solving practical problems.

Problem 19.5 $S_{ideal}(T, V, N) \geq S_{nonideal}(T, V, N)$ Use the Gibbs–Bogoliubov inequality (Section 7.6.4) to show that $S_{\text{ideal}}(T,V,N) \geq S_{\text{nonideal}}(T,V,N)$. Write the Hamilton function as $\mathscr{H} = \vartheta + \Phi$, where ϑ is the kinetic energy and Φ is the potential energy, and consider the system with $\mathscr{H} = \vartheta$ as the ideal system and that with $\mathscr{H} = \vartheta + \Phi$ as the nonideal system. Use $F = U - TS$ and realize that (classically) the canonical average separates in momentum- and position-dependent terms so that $\langle\vartheta\rangle_{\mathscr{H}} = \langle\vartheta\rangle_{\vartheta}$, where the subscript indicates the energy expression to be used in the average and corresponding partition function.

References

1 Jancel, 1963.

2 Penrose, O. (1979). *Rep. Progr. Phys.* 42: 1937.

3 Tolman, R.C. (1938). *The Principles of Statistical Mechanics*. Oxford: Oxford University Press (also Dover, 1979).

4 Grandy, W.T. (1987). *Foundations of Statistical Mechanics*, Equilibrium Theory, Vol. 1. Dordrecht: D. Reidel Publishing Company.

5 Jeffreys, H. (1961). *Theory of Probability*, 3e. Oxford: Oxford University Press.

6 Cox, R.T. (1961). *Algebra of Probable Inference*. Baltimore: John Hopkins Press.

7 Cox, R.T. (1946). *Am. J. Phys.* 17: 1.

8 Gillies, D. (2000). *Br. J. Phil. Sci.* 51: 807; For a different view by the same author: Gillies, D. (2016). *The Propensity Interpretation*, Ch. 19 in *Oxford Handbook of Probability and Philosophy* (ed. A. Hájek and C. Hitchcock), p. 406. Oxford: Oxford University Press.

9 Jaynes, E.T. (1957). *Phys. Rev.* 106 (620); Brandeis lectures, reproduced as Ch. 4 in Rosenkrantz, R.D. (1989). *E.T. Jaynes: Papers on Probability, Statistics and Statistical Physics*. Dordrecht: Kluwer Academic Publishers.

10 Frigg, R. (2008). *A Field Guide to Recent Work on the Foundations of Statistical Mechanics*, Ch. 3 in *The Ashgate Companion of Contemporary Philosophy of Science* (ed. D. Rickles), p. 99. London: Ashgate.

11 Sklar, 1993.

12 Uffink, J. (2007). *Compendium of the Foundations of Classical Statistical Physics*, Ch. 9 in *Handbook of the Philosophy of Science: Philosophy of Science*, Vol. B (ed. J. Butterfield and J. Earman), p. 923. Amsterdam: North-Holland.

13 Ehrenfest, P. and Ehrenfest, T. (1990). *The Conceptual Foundations of the Statistical Approach in Mechanics*. New York: Dover; (1912). *Enc. Math. Wissenschaften, Teubner, Leipzig* IV(6), Band 2: p. 11.

14 Boltzmann, L. (1868). *Sitz. Ber. Akad. Wiss. Wien.* 58 (517); (1872). *Ibid.* 66: 275.

15 Brown, H.R., Myrvold, W., and Uffink, J. (2009). *Stud. Hist. Phil. Mod. Phys.* 40: 174.

16 Jaynes, E.T. (1965). *Am. J. Phys.* 33: 391; (1971). *Phys. Rev. A* 4: 747.

17 Loschmidt, J. (1876). *Sitz. Ber. Akad. Wiss. Wien.* 73, 128 and 366; (1877). *Ibid.* 75: 287; (1878). *Ibid.* 76: 209.

18 Daniga, B.T., Broadwell, J.E., and Sturtevant, B. (1980). *Study of a Multispeed Cellular Automaton*, Ch. 13 in *Rarefied Gas Dynamics: Theoretical and Computational Techniques*, Vol. 118 (ed. E. Muntz, D. Weaver, and D. Campbell), p. 155. Washington DC: AIAA.

19 Levesque, D. and Verlet, L. (1993). *J. Stat. Phys.* 72: 519.

20 Romero-Rochin, V. and Gonzálex-Tovar, E. (1997). *J. Stat. Phys.* 89: 735.

21 Adapted from Schwabl, F. (2002). *Statistical Mechanics*. Berlin: Springer.

22 Poincaré, J.H. (1890). *Acta Math.* 13: 1.

23 Thompson, C.J. (1972). *Mathematical Statistical Mechanics*. New York: Macmillan Comp.

24 Sachs, I., Sen, S., and Sexton, J.C. (2006). *Elements of Statistical Mechanics*. Cambridge: Cambridge University Press.

25 Zermelo, E. (1896). *Wied. Ann.* 57: 485; *Ibid.* 59: 793.

26 Simplified from Schwabl, F. (2002). *Statistical Mechanics*. Heidelberg: Springer; after his simplification from Hemmer, P.C., Maximon, L.C., and Wergeland, H. (1958). *Phys. Rev.* 111: 689.

27 Khinchin, A.I. (1949). *Mathematical Foundations of Statistical Mechanics*. New York: Dover.

28 van Kampen, N.G. (1971). *Physica* 53: 98.

29 Farquhar, I.E. (1964). *Ergodic Theory in Statistical Mechanics*. London: Interscience Publications.

30 van Lith, J. (2001). *Stud. Hist. Phil. Sci.* 32: 581.

31 Scott Dumas, H. (2014). *The KAM Story - A Friendly Introduction to the Content, History, and Significance of Classical Kolmogorov–Arnold–Moser Theory*. Singapore: World Scientific.

32 Emch, G.G. and Liu, C. (2002). *The Logic of Thermostatistical Physics*. Berlin: Springer.

33 Smale, S. (1980). *Ann. New York Acad. Sci.* 357: 260.

34 Le Bellac, 2006.

35 Balian, R. (1999). *Am. J. Phys.* 67: 1078; (2003). *Séminaire Poincaré* 2: 13; (2005). *Stud. Hist. Phil. Mod. Phys.* 36: 323; Balian, R., Alhassid, Y. and Reinhardt, H. (1986). *Phys. Rep.* 131: 1.

36 de With, G. (2006). *Structure, Deformation, and Integrity of Materials*. Weinheim: Wiley-VCH.

37 Blatt, J.M. (1959). *Progr. Theor. Phys.* 22: 745.

38 van Kampen, N.G. (1988). *Physica A* 153: 97; (1993). *Physica A* 194: 542; A particularly clear exposition, also covering the classical case, is (1956). *Fortschr. Phys.* 4: 405.

39 van Kampen, N.G. (1954). *Physica* 20: 603.

40 Thomsen, J.S. (1953). *Phys. Rev.* 91: 1263; (1955). *Phys. Rev.* 98: 1625.

41 Landsberg, P.T. (1961). *Thermodynamics with Quantum Statistical Illustrations*. London: Interscience Publication.

42 van Hove, L. (1955). *Physica* 21: 517; (1957). *Physica* 23: 441; (1959). *Physica* 25: 268.

43 Grandy, 2008.

44 Wehrl, A. (1978). *Rev. Mod. Phys.* 50: 221.

45 Jaynes, E.T. (2003). *Probability Theory: The Logic of Science*. Cambridge: Cambridge University Press; edited by Bretthorst, G.L.

46 Ruelle, D. (1969). *Statistical Mechanics: Rigorous Results*. New York: Benjamin Inc.; also (1999). London: Imperial College Press.

47 Lebowitz, J.L. (2007). *From Time-symmetric Microscopic Dynamics to Time-asymmetric Macroscopic Behavior: An Overview*. arXiv:0709.0724v1 [cond-mat.stat-mech], 516; (1999). *Physica A* 263: 516.

48 Parker, D. (2011). *Phil. Sci.* 78: 1058.

49 Woods, 1975 and 1986.

50 Leff, H.S. (2007). *Found. Phys.* 37: 1744 and references therein.

51 Bohren, C.F. and Albrecht, B.A. (1998). *Atmospheric Thermodynamics*. Oxford: Oxford Unversity Press and references therein; (2023). 2e. Oxford: Oxford University Press.

52 Baierlein, R. (1968). *Am. J. Phys.* 36: 625; Leff, H.S. (1969). *Am. J. Phys.* 37: 548.

53 Bricmont, J. (1996). *Science of Chaos or Chaos in Science?* Ch. 13 in *The Flight from Science and Reason*, Vol. 775 (ed. P.R. Gross, N. Levitt, and M.W. Lewis), p. 131. New York: The New York Academy of Sciences.

54 Gallavotti, G. (2014). *Nonequilibrium and Irreversibility*. Heidelberg: Springer.

55 Pressé, S., Ghosh, K., Lee, J. et al. (2013). *Rev. Mod. Phys.* 85: 1115.

56 Bohm, D. (1952). *Phys. Rev.* 85: 166.

57 Bohm, D. and Hiley, B.J. (1993). *The Undivided Universe: an Ontological Interpretation of Quantum Theory*. London: Routledge.

58 Dürr, D. and Teufel, S. (2009). *Bohmian Mechanics*. Berlin: Springer.

59 Ghirardi, G.C., Rimini, A., and Weber, T. (1985). *A Model for a Unified Quantum Description of Macroscopic and Microscopic Systems*, Ch. 21 in *Quantum Probability and Applications* (ed. I. Accardi), p. 223. Berlin: Springer.

60 Albert, D.Z. (2000). *Time and Chance*. Cambridge, MA: Harvard University Press.
61 Zurek, W.H. (2002). *Los Alamos Sci.* 27: 86.
62 't Hooft, G. (2016). *The Cellular Automaton Interpretation of Quantum Mechanics*. Heidelberg: Springer.
63 Prigogine, I. and Stengers, I. (1984). *Order Out of Chaos: Man's New Dialogue with Nature*. Toronto: Bantam Books.
64 Prigogine, I. and Stengers, I. (1997). *The End of Certainty*. New York: Free Press.
65 Petruccione, F. and Breuer, H.-P. (2002). *The Theory of Open Quantum Systems*. Oxford: Oxford University Press.
66 Beretta, G.P. (2009). *Nonlinear quantum evolution equations to model irreversible adiabatic relaxation with maximal entropy production and other nonunitary processes*. arXiv:0907.1977v1 [quant-ph].
67 Beretta, G.P. (2009). *What is Quantum Thermodynamics?* and PIAF '09, *New Perspectives on the Quantum State*, http://www.quantumthermodynamics.org/ (accessed 8 February 2016).
68 Gemmer, J., Michel, M., and Mahler, G. (2009). *Quantum Thermodynamics*, 2e. Berlin: Springer.
69 Hatsopoulos, G.N. and Keenan, J.H. (1965). *Principles of General Thermodynamics*. New York: John Wiley & Sons.
70 Gyftopoulos, E.P. and Beretta, G.P. (1991). *Thermodynamics: Foundations and Applications*. New York: Macmillan Publ. Comp.; augmented and corrected, (2005). Dover.

Further Reading

Auletta, G., Fortunato, M., and Parisi, G. (2009). *Quantum Mechanics*. Cambridge: Cambridge University Press. A text discussing basic features and some topics usually not dealt with, such as the measurement problem, quantum information and entanglement issues.

Le Bellac, M. (2006). *Quantum Physics*. Cambridge University Press, Cambridge. A concise book presenting the basics and applications to scattering, atomic physics and open systems.

Jancel, R. (1963). *Foundations of Classical and Quantum Statistical Mechanics*. Pergamon Press. An extensive discussion of the fundamental aspects containing many details.

Grandy, W.T. (2008). *Entropy and the Time Evolution of Macroscopic Systems*. New York: Oxford University Press. A highly readable and recommended booklet providing what the title promises.

van Kampen, N.G. (2007). *Stochastic Processes in Physics and Chemistry*, 3e. Amsterdam: North-Holland. Already a classic treating, amongst others, the master, Langevin and Fokker-Planck equations and applications to chemical reactions, diffusion and unstable systems.

Sklar, L. (1993). *Physics and Chance*. Cambridge: Cambridge University Press. A philosophical treatise on probability with many insights, though not always easy to follow.

Woods, L.C. (1975). *The Thermodynamics of Fluid Systems*. Oxford: Oxford University Press, reprinted and corrected ed. (1986). A gem of a book dealing with classical, internal variable and statistical thermodynamics of (plasma-like) fluids, but, along the lines, elaborating on many issues.

Appendix A

Guide to Mathematics Used

In this appendix, we briefly review some mathematical concepts used as well as some of the operations between them.

A.1 Symbols and Conventions

Often, we will use quantities with subscripts, for example, A_{ij}. With respect to these quantities, a convenient symbol is the *Kronecker delta*, denoted by δ_{ij} and defined by

$$\delta_{ij} = 1 \quad \text{if} \quad i = j \quad \text{and} \quad \delta_{ij} = 0 \quad \text{if } i \neq j \tag{A.1}$$

Note that we can write the identity $\Sigma_i \delta_{ij} a_i = a_j$, and therefore δ_{ij} is also called the *substitution operator*. Applying δ_{ij} to A_{ij} results in $\Sigma_{ij} A_{ij} \delta_{ij} = \Sigma_j A_{jj}$.

We also introduce the *alternator* ε_{ijk} for which it holds that[1]

$$\begin{aligned} &\varepsilon_{ijk} = 1 \text{ for } ijk = 123, 231, 312 \text{ (even permutations)} \\ &\varepsilon_{ijk} = -1 \text{ for } ijk = 132, 213, 321 \text{ (odd permutations) and} \\ &\varepsilon_{ijk} = 0 \text{ otherwise, i.e. if any of the three indices are equal} \end{aligned} \tag{A.2}$$

A single-line expression is given by $\varepsilon_{ijk} = ½(i - j)(j - k)(k - i)$.

A.2 Derivatives, Differentials, and Variations

A single-valued function f, dependent on *variables* x_i and *parameters* p_j, is denoted by $f(x_i;p_j)$ or just $f(x_i)$, omitting the parameters. In practice, reference to the variable is also often omitted by writing f. Often we also write $y = y(x)$. For a single-variable function $y = f(x)$, the *derivative* is another function $f'(x) = \lim_{\Delta x \to 0} [f(x + \Delta x) - f(x)]/\Delta x$. For a function $f(x_i)$, if all variables but one, say x_j, are kept constant during differentiation, the derivative of f with respect to x_j is called the *partial derivative* and is denoted by[2] $\partial f(x_i)/\partial x_j$, also often written as f_j. The function f_j generally is a function of all variables

1 An even (odd) permutation is the result of an even (odd) number of binary interchanges. The character (even or odd) of a permutation is independent of the order of the binary interchanges.

2 Once a choice of independent variables is made, there is in principle no need to indicate which variables are kept constant, and $(\partial f(x_j)/\partial x_1)_{x_j \neq x_1}$ can be written as $\partial f(x_j)/\partial x_k$. Still, in thermodynamics this notation is often used in view of the frequent change of variables.

Phases of Matter and their Transitions: Concepts and Principles for Chemists, Physicists, Engineers, and Materials Scientists, First Edition. Gijsbertus de With.

x_i and, if continuous, may be differentiated again to yield the *second partial derivatives* $f_{ij} = \partial f_i/\partial x_j = \partial^2 f/\partial x_i\partial x_j = \partial^2 f/\partial x_j\partial x_i = \partial f_j/\partial x_i = f_{ji}$.

Example A.1 For $f(x,y) = x^2y^3$, one calculates $\partial f/\partial x = 2xy^3$ and $\partial f/\partial y = 3x^2y^2$. Hence, $\partial^2 f/\partial x^2 = 2y^3$, $\partial^2 f/\partial y^2 = 6x^2y$, and $\partial^2 f/\partial x\partial y = \partial^2 f/\partial y\partial x = 6xy^2$.

For a differentiable function $y = f(x_j)$ with variables x_j, let (x_j,y) and (x_j',y') be two different solutions. We define the *increments* $\Delta x_j \equiv x_j' - x_j$ and $\Delta y \equiv y' - y$ and the *first(-order) differentials* $\mathrm{d}x_j \equiv \Delta x_j$ and $\mathrm{d}y \equiv \Sigma_j(\partial y/\partial x_j)\mathrm{d}x_j$ (see Keisler, 2011). From this definition, we obtain the quotient $\mathrm{d}y/\mathrm{d}x_j = \Sigma_k(\partial y/\partial x_k)(\mathrm{d}x_k/\mathrm{d}x_j)$, which for a single variable reduces to $\mathrm{d}y/\mathrm{d}x = \partial y/\partial x$. Note that the quotient $\mathrm{d}y/\mathrm{d}x_j$ is a function of the x_js and $\mathrm{d}x_j$, and that this quotient equals the derivative $\partial y/\partial x_j$. Keeping $\mathrm{d}x_j$ constant, the second differential is defined as $\mathrm{d}(\mathrm{d}y) = \mathrm{d}^2(y) \equiv [\Sigma_j \mathrm{d}x_j(\partial/\partial x_j)]^2 y$, or, in general, the nth differential as

$$\mathrm{d}^n y \equiv \left[\sum\nolimits_j \mathrm{d}x_j(\partial/\partial x_j)\right]^n y \tag{A.3}$$

Expanding $f(x_j)$ in a Taylor series, we have

$$f(x_i + \mathrm{d}x_i) - f(x_i) = \Delta y = \sum\nolimits_i \frac{\partial f}{\partial x_i}\mathrm{d}x_i + \frac{1}{2!}\sum\nolimits_{i,j}\frac{\partial^2 f}{\partial x_i\partial x_j}\mathrm{d}x_i\mathrm{d}x_j + \cdots \quad \text{or} \tag{A.4}$$

$$f(x_j + \mathrm{d}x_j) = \exp\left[\sum\nolimits_j \mathrm{d}x_j(\partial/\partial x_j)\right] f(x_i) \tag{A.5}$$

Expansion of the exponential yields Eq. (A.4). Using the above we may write

$$\Delta y = \mathrm{d}y + {}^1\!/_2\mathrm{d}^2y + \cdots + (n!)^{-1}\mathrm{d}^n y + \cdots \cong \mathrm{d}y \tag{A.6}$$

if $\Delta r \equiv (\Sigma_j \Delta x_j^{\,2})^{1/2}$ is sufficiently small. Thus, for a single variable $\Delta y \cong \mathrm{d}y = (\partial y/\partial x)\mathrm{d}x$ if $\mathrm{d}x = \Delta x$ is sufficiently small, accounting for the notation $\mathrm{d}y/\mathrm{d}x$ for $f'(x)$.

Example A.2 Consider again the function $f(x,y) = x^2y^3$. A first-order estimate for $f(2.1,2.1)$ is $f(2.1,2.1) = f(2,2) + (\partial f/\partial x)\mathrm{d}x + (\partial f/\partial y)\mathrm{d}y = 32 + 2xy^3\mathrm{d}x + 3x^2y^2\mathrm{d}x = 40.00$, using $f(2,2)$ as the reference value. The actual value is 40.84.

For $\mathrm{d}f$ of a function $f(x_j)$ with independent variables x_j, often called an *exact* (or *total*) *differential*, we thus have $\mathrm{d}f = \Sigma_j f_j \mathrm{d}x_j$ with $f_j \equiv \partial f/\partial x_j$. Since $\mathrm{d}f$ is integrable, $F_b - F_a = \int_a^b \mathrm{d}f$ is path independent or $\oint \mathrm{d}f = 0$. Not all differentials dg are exact differentials. If we have $\mathrm{d}g = \Sigma_j \lambda_j(x_j)\mathrm{d}x_j$, the necessary and sufficient conditions for dg to be an exact differential are that the relations $\partial\lambda_j/\partial x_i = \partial\lambda_i/\partial x_j$ are true for all variables. Else, it is called an *inexact differential*, which we denote by đg. For two variables one can always find an *integrating factor* I (or *integrating divisor* I^{-1}), which transforms an inexact differential đg in an exact differential via $\mathrm{d}g = I$đg. This is not necessarily true for three or more variables. While the label "exact" refers to $\oint \mathrm{d}f = 0$ for any subset of the $\mathrm{d}x_j$s, the label "complete" is sometimes used when all $\mathrm{d}x_j$s are implied.

Example A.3 Consider đ$f(x,y) = \lambda_x\mathrm{d}x + \lambda_y\mathrm{d}y$ with $\lambda_x = y$ and $\lambda_y = -x$. This is an inexact differential because $\partial\lambda_x/\partial y = 1$, while $\partial\lambda_y/\partial x = -1$. Using $I = x^{-2}$, we have $\mathrm{d}f = x^{-2}\lambda_x\mathrm{d}x + x^{-2}\lambda_y\mathrm{d}y$, rendering đ$f$ an exact differential since we have $\partial(\lambda_x/x^2)/\partial y = \partial(y/x^2)/\partial y = x^{-2}$, identical to $\partial(\lambda_y/x^2)/\partial x = \partial(-x^{-1})/\partial x = x^{-2}$.

The differential d$f(x)$ refers to an infinitesimal, actual change in the function $f(x)$ caused by a change dx of the variable x. A *virtual change* δx is any infinitesimal change in x admissible by the constraint(s) applied to $f(x)$. The corresponding infinitesimal change in $f(x)$, given by $\delta f = (\partial f/\partial x)\delta x$, is called the (first) *variation* of f.

A.3 Composite, Implicit, Homogeneous, Complex, and Analytic Functions

If for a function f the independent variables x_i are a function of y_j, it is called a *composite function*. For first-order differentials, the *chain rule* applies so that

$$\mathrm{d}x_i = \sum\nolimits_j \frac{\partial x_i}{\partial y_j}\mathrm{d}y_j \quad \text{and} \quad \mathrm{d}f = \sum\nolimits_{i,j} \frac{\partial f}{\partial x_i}\frac{\partial x_i}{\partial y_j}\mathrm{d}y_j \tag{A.7}$$

If the x_i are not independent, an arbitrary member, say x_1, can be expressed as a function of $x_2, \ldots, x_n$, often resulting in an *implicit function* $f = f(x_i) =$ constant. The differentials can be obtained from $\mathrm{d}f = \Sigma_i(\partial f/\partial x_i)\mathrm{d}x_i = 0$. If x_1 is the dependent variable, putting $\mathrm{d}x_1 = 0$ and division by $\mathrm{d}x_i$ $(i \neq 1)$ yields

$$\left(\frac{\partial f}{\partial x_i}\right)_{x_j,x_1} + \sum\nolimits_j \left(\frac{\partial f}{\partial x_j}\right)_{x_i,x_1}\left(\frac{\partial x_j}{\partial x_i}\right)_{f,x_1} = 0 \tag{A.8}$$

Example A.4 Consider explicitly a function f of three variables x, y, and z, where z is the dependent variable. Taking $x_1 = z$, $x_i = x$ and $x_j = y$, and $x_1 = z$, $x_i = y$ and $x_j = x$, Eq. (A.8) becomes

$$(\partial f/\partial x)_{y,z} + (\partial f/\partial y)_{x,z}(\partial y/\partial x)_{f,z} = 0 \quad \text{and} \quad (\partial f/\partial y)_{x,z} + (\partial f/\partial x)_{y,z}(\partial x/\partial y)_{f,z} = 0$$

Hence, we have the *inverse rule*:

$$(\partial x/\partial y)_{f,z} = 1/(\partial y/\partial x)_{f,z} \tag{A.9}$$

By cyclic permutation of the variables in $(\partial f/\partial x)_{y,z} + (\partial f/\partial y)_{x,z}(\partial y/\partial x)_{f,z} = 0$, we obtain

$$(\partial f/\partial y)_{z,x} + (\partial f/\partial z)_{x,y}(\partial z/\partial y)_{f,x} = 0 \quad \text{and} \quad (\partial f/\partial z)_{x,y} + (\partial f/\partial x)_{y,z}(\partial x/\partial z)_{f,y} = 0$$

resulting, after substitution in each other, in the *minus-one rule*:

$$(\partial x/\partial y)_{f,z}(\partial y/\partial z)_{f,x}(\partial z/\partial x)_{f,y} = -1 \tag{A.10}$$

Now consider x and y to be composite functions of the variables u and v, that is, $x = x(u,v)$ and $y = y(u,v)$, and thus $(\partial f/\partial u)_v = (\partial f/\partial x)_{y,z}(\partial x/\partial u)_v + (\partial f/\partial y)_{z,x}(\partial y/\partial u)_v$. If we have $u = x$, this yields the *substitution rule*:

$$(\partial f/\partial x)_v = (\partial f/\partial x)_{y,z} + (\partial f/\partial y)_{z,x}(\partial y/\partial x)_v \tag{A.11}$$

Moreover, if f is constant, $\mathrm{d}f = 0$, and further taking v as a constant, we have

$$\mathrm{d}f = (\partial f/\partial u)_v = 0 \quad \text{or} \quad (\partial f/\partial x)_{y,z}(\partial x/\partial u)_{f,v} + (\partial f/\partial y)_{z,x}(\partial y/\partial u)_{f,v} = 0$$

Substituting $(\partial f/\partial x)_{y,z} + (\partial f/\partial y)_{x,z}(\partial y/\partial x)_{f,z} = 0$, one obtains the *chain rule*:

$$(\partial y/\partial x)_{f,z} = (\partial y/\partial u)_{f,v}/(\partial x/\partial u)_{f,v} = (\partial y/\partial u)_{f,v}(\partial u/\partial x)_{f,v} \tag{A.12}$$

A function $f(x_i)$ is positively *homogeneous* of degree n if for every value of x_i and for every $\lambda > 0$ we have $f(\lambda x_i) = \lambda^n f(x_i)$. For such a function we have *Euler's theorem*

$$\sum_i x_i(\partial f(x_i)/\partial x_i) = nf(x_i) \tag{A.13}$$

to be proven by differentiation with respect to λ first and taking $\lambda = 1$ afterwards. The chain rule also accounts for why $\mathrm{d}x_j$ is defined as Δx_j in Section A.2.

Example A.5 Consider the function $f(x,y) = x^2 + xy - y^2$. One easily finds $\partial f/\partial x = 2x + y$ and $\partial f/\partial y = x - 2y$. Thus, $x(\partial f/\partial x) + y(\partial f/\partial y) = 2f$ and f is homogeneous of degree 2.

A complex quantity is defined by $z = x + \mathrm{i}y$, where $\mathrm{i}^2 = -1$. The scalars $x = \mathrm{Re}\, z$ and $y = \mathrm{Im}\, z$ are the *real part* and *imaginary part* of z. The (*complex*) *conjugate* z^* is defined by $z^* = x - \mathrm{i}y$. The product $zz^* = z^*z = x^2 + y^2$ is a real quantity and its square root $|z| = (x^2 + y^2)^{1/2}$ is called the *modulus*. The *triangle inequality* states that $|z_1 + z_2| \le |z_1| + |z_2|$. An *Argand diagram* is a graphical representation of z in the *complex plane* (x,y), in which z is a vector connecting the point (0,0) and (x,y). One can also write $z = |z|(\cos\theta + \mathrm{i}\sin\theta)$ with $\tan\theta = y/x$ and where θ, the angle between z and the positive x-axis, is usually called the *phase angle*. From $z_j = |z_j|(\cos\theta_j + \mathrm{i}\sin\theta_j)$, one obtains $z_1z_2 = |z_1||z_2|[\cos(\theta_1 + \theta_2) + \mathrm{i}\sin(\theta_1 + \theta_2)]$. Using *Euler's equation* $\exp(\mathrm{i}\theta) = \cos\theta + \mathrm{i}\sin\theta$, we can alternatively write $z = |z|\exp(\mathrm{i}\theta) = z' + z''$ with $z' = \mathrm{Re}\, z$ and $z'' = \mathrm{Im}\, z$. The operation $\exp(\mathrm{i}\theta)$ represents an anticlockwise rotation in the complex plane.

A real function $f(x)$ is *analytic* at $x = c$ in the open region I $(0 < x < a)$ if $f(x)$ can be written as a Taylor series convergent in some neighborhood of c. A complex function $w(z)$ is analytic if $\partial f(z)/\partial z$ at $z = c$ is independent of the direction of $\mathrm{d}z$ and continuous within the open region I. It satisfies the *Cauchy–Riemann conditions*

$$\partial u/\partial x = \partial v/\partial y \quad \text{and} \quad \partial v/\partial x = -\partial u/\partial y \tag{A.14}$$

where $u(x,y) = \mathrm{Re}\, w(z)$ and $v(x,y) = \mathrm{Im}\, w(z)$. If true, all derivatives of $f(z)$ with respect to z are analytic in I. If u and v have continuous second derivatives, $w(z)$ obeys *Laplace's equation* $\partial^2u/\partial^2x + \partial^2u/\partial^2y = \partial^2v/\partial^2x + \partial^2v/\partial^2y = 0$ and is called *harmonic*.

Many physical properties can be written as a complex quantity. For example, the dielectric permittivity $\varepsilon(t)$ connects the electric field $E(t)$ with the displacement $D(t)$ according to $D = \varepsilon E$. For linear response, $D(t)$ generally can be written as

$$D(t) = \int_{-\infty}^{+\infty} \varepsilon(t - t')E(t')\mathrm{d}t' \quad \text{or as} \quad D(\omega) = \alpha(\omega)E(\omega) \tag{A.15}$$

after taking the Fourier transform (Section A.15) with ω the *angular frequency* and $\alpha(\omega)$ the *complex susceptibility*. However, response can only occur after applying the field, so that we have $D(t) = 0$ for $t < t'$ and $\alpha(\omega) = \int_0^{+\infty} \varepsilon(t)\exp(-\mathrm{i}\omega t)\mathrm{d}t$. Using ω as a complex quantity, it follows from Cauchy's theorem (e.g. Kaplan, 1952) that $\alpha(\xi) = (\mathrm{i}\pi)^{-1}\int_{-\infty}^{+\infty} \alpha(\omega)/(\omega - \xi)\mathrm{d}\omega$. Defining $\mathscr{P} f(x) = \lim_{\delta\to+0}\left[\int_a^{b-\delta} f(x)\mathrm{d}x + \int_{b+\delta}^{a} f(x)\mathrm{d}x\right]$ as the *principal value*, $\alpha'(\omega)$ and $\alpha''(\omega)$ are connected via the *Kramers–Kronig transform*:

$$\alpha'(\omega) - \alpha'(\infty) = \frac{2}{\pi}\mathscr{P}\int_0^{\infty} \frac{x\alpha''(x)}{x^2 - \omega^2}\mathrm{d}x, \qquad \alpha''(\omega) = -\frac{2\omega}{\pi}\mathscr{P}\int_0^{\infty} \frac{\alpha'(x) - \alpha'(\infty)}{x^2 - \omega^2}\mathrm{d}x \tag{A.16}$$

Example A.6 For $w(z) = e^x \cos y + ie^x \sin y = \exp(z)$, $u = e^x \cos y$ and $v = e^x \sin y$. The derivatives are

$$\partial u/\partial x = \partial v/\partial y = e^x \cos y, \quad \partial v/\partial x = -\partial u/\partial y = e^x \sin y, \quad \partial^2 u/\partial x^2 = -\partial^2 v/\partial y^2 = e^x \cos y$$

Hence, the function is analytic and harmonic.

A.4 Extremes and Lagrange Multipliers

For obtaining an extreme of a function $f(x_i)$ of n independent variables x_i, the first variation δf has to vanish. This leads to

$$\delta f = \sum_i (\partial f/\partial x_i) \quad \delta x_i = 0 \tag{A.17}$$

and, since the variables x_i are independent and the variations δx_i are arbitrary, to $\partial f/\partial x_i = 0$ for $i = 1, \ldots, n$. If, however, the extreme of f has to be found when the x_i are dependent and satisfy r constraint functions

$$c_j(x_i) = C_j \quad (j = 1, \ldots, r \text{ and } r < n) \tag{A.18}$$

where the parameters C_j are constants, the variables x_i must also obey

$$\sum_i [\partial c_j(x_i)/\partial x_i] \delta x_i = 0 \quad (j = 1, \ldots, r) \tag{A.19}$$

Of course, the system can be solved in principle by solving Eq. (A.18) for the independent $n - r$ variables x_i as functions of the others, but the procedure is often complex. It can be shown that finding the extreme of f subject to the constraint of Eq. (A.18) is equivalent to finding the extreme of a function g defined by

$$g(x_i, \lambda_j) = f(x_i) - \sum_{j=1}^{r} \lambda_j [c_j(x_i) - C_j] \tag{A.20}$$

where now the original variables x_i and the additional variables λ_j, called the *Lagrange (undetermined) multipliers*, are to be considered independent. Variation of λ_j leads to Eq. (A.18) and variation of x_i to

$$\sum_i \partial g(x_i)/\partial x_i = \sum_i \{[\partial f(x_i)/\partial x_i] - \lambda_j [\partial c_j(x_i)/\partial x_i]\} = 0 \tag{A.21}$$

From Eq. (A.21) the expressions for x_i can be determined. These expressions are still functions of λ_j but they can be eliminated using Eq. (A.18).

Example A.7 For a rectangle with edges by x and y, the circumference $L = 2(x + y)$ and area $A = xy$. To obtain the minimum circumference, the equations to be solved are

$$\partial L/\partial x - \lambda(\partial A/\partial x) = 2 - \lambda y = 0 \text{ and } \partial L/\partial y - \lambda(\partial A/\partial y) = 2 - \lambda x = 0$$

Hence, $y = 2/\lambda$ and $x = 2/\lambda$ and the solution is $x = y$, $\lambda = 2/\sqrt{A}$, and $\min(L) = 4\sqrt{A}$.

A.5 Legendre Transforms

In many problems, we meet the demand to interchange between dependent and independent variables. If $f(x_i)$ denotes a function of n variables x_i, we have

$$df = \sum_j (\partial f/\partial x_j) dx_j \equiv \sum_j X_j dx_j \quad (j = 1, \ldots, n) \tag{A.22}$$

Elimination of x_i from $X_i(x_i)$ and $f(x_i)$ leads to $f = f(X_i)$. However, from $f(X_i)$ it is impossible to uniquely recover $f(x_i)$ by repeating this procedure for $f(X_i)$. Now consider the function $g = f - X_1x_1$. For the differential, we obtain

$$dg = df - d(X_1x_1) = -x_1dX_1 + \sum_j X_jdx_j \quad (j = 2, \ldots, n) \tag{A.23}$$

and we see that the roles of x_1 and X_1 have been interchanged. This transformation can be applied to only one variable, to several variables or to all variables. In the last case we use[3] $g = f - \Sigma_i X_i x_i$ and obtain $dg = -\Sigma_j x_j dX_j$ $(j = 1, \ldots, n)$. This so-called *Legendre transformation*, if applied to the transform, results in the complete original expression.

Example A.8 Consider the function $f(x) = ½x^2$. The dependent variable X is given by $X = \partial f/\partial x = x$, which can be solved to yield $x = X$. Therefore, the function expressed in the variable X reads $f(X) = ½X^2$. For the transform $g(X)$, one thus obtains

$$g(X) = f(X) - Xx = ½X^2 - XX = -½X^2$$

A.6 Coordinate Axes Rotations

Using matrix notation (Section A.12) for a Cartesian coordinate system $\mathbf{e}_j$ and another system $\mathbf{e}_p'$ with the same origin as the initial system, the *direction cosines* C_{pj} are defined as the cosines of the angles between the new axes $\mathbf{e}_p'$ and the old axes $\mathbf{e}_j$ (Figure A.1a). If P denotes a point with coordinates x_j, its coordinates in the second system x_j' are given by the projections of the basis $\mathbf{e}_j$ on the basis $\mathbf{e}_p'$ or

$$x_p' = \Sigma_j C_{pi} x_i \quad \text{or} \quad \mathbf{x}' = \mathbf{Cx} \quad \text{with as inverse} \quad x_j = \Sigma_p C_{pj} x_p' \quad \text{or} \quad \mathbf{x} = \mathbf{C}^{\mathrm{T}}\mathbf{x}' \tag{A.24}$$

Substituting $\mathbf{x}' = \mathbf{Cx}$ in $\mathbf{x} = \mathbf{C}^{\mathrm{T}}\mathbf{x}'$ yields $\mathbf{x} = \mathbf{CC}^{\mathrm{T}}\mathbf{x}$ or $\mathbf{CC}^{\mathrm{T}} = \mathbf{I}$. Thus, the inverse $(C_{kl})^{-1}$ of the matrix C_{kl} is equal to its transpose $(C_{kl})^{\mathrm{T}} = C_{lk}$, which is the definition of an orthogonal matrix. For obvious reasons, the matrix $\mathbf{C}$ is also called a *rotation* matrix.

Example A.9 Consider a rotation of axes in a two-dimensional space of the x_1-axis over an angle θ (Figure A.1a). The original basis is $x_i = \{x_1, x_2\}$ and the rotated one is $x_i' = \{x_1', x_2'\}$. The direction cosines are given by $C_{11} = \cos\theta$, $C_{12} = \cos(\pi/4 - \theta) = \sin\theta$, $C_{21} = \cos(\pi/4 + \theta) = -\sin\theta$, and $C_{22} = \cos\theta$. The matrix $\mathbf{C}$ thus reads

$$C_{ij} = \begin{pmatrix} \cos\theta & \sin\theta \\ -\sin\theta & \cos\theta \end{pmatrix}$$

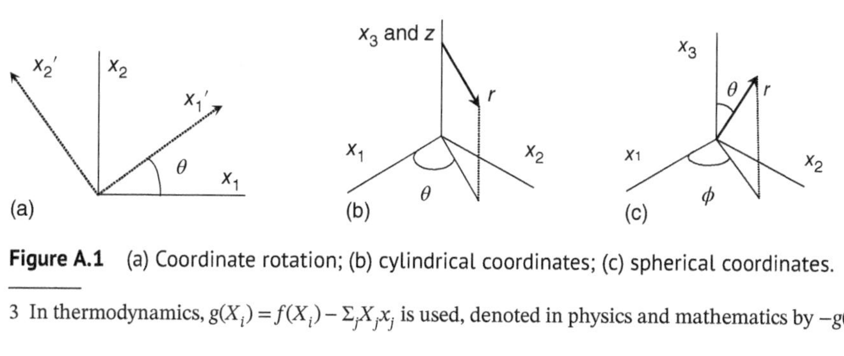

Figure A.1 (a) Coordinate rotation; (b) cylindrical coordinates; (c) spherical coordinates.

3 In thermodynamics, $g(X_i) = f(X_i) - \Sigma_j X_j x_j$ is used, denoted in physics and mathematics by $-g(X_i)$.

For definiteness, consider the point $(x_1, x_2) = (2, 4)$ and a rotation over $\theta = \pi/4$. The new coordinates will be

$$x_1' = C_{11}x_1 + C_{12}x_2 = \tfrac{1}{2}\sqrt{2}\times 2 + \tfrac{1}{2}\sqrt{2}\times 4 \quad \text{and}$$
$$x_2' = C_{21}x_1 + C_{22}x_2 = -\tfrac{1}{2}\sqrt{2}\times 2 + \tfrac{1}{2}\sqrt{2}\times 4$$

yielding $(x_1', x_2') = (3\sqrt{2}, \sqrt{2})$.

So far, we have interpreted the relation $x_p' = \Sigma_j C_{pj}x_j$ as a rotation of "old" coordinate axes x_j over a positive (counterclockwise) angle θ to "new" coordinate axes x_p'. However, we can also suppose that there is only one coordinate system and that the vector x_j is rotated through an angle θ in the positive direction to give a new vector x_p' in the same coordinate system. This is equivalent to a rotation of the coordinate system over an angle θ in the *negative* (clockwise) direction. Since any rotation can be described by the matrix $\mathbf{C}(\theta)$, where θ is the angle of rotation around a proper axis, as illustrated in Example A.9, we obtain in this case $\mathbf{x}' = \mathbf{C}(-\theta)\mathbf{x} = \mathbf{C}^{\mathrm{T}}(\theta)\mathbf{x}$.

A.7 Change of Variables

It is also often required to use different independent variables, in particular in integrals. For definiteness, using direct notation (Section A.12), consider the case of three "old" variables $\boldsymbol{x} = (x,y,z)$ and three "new" variables $\boldsymbol{u} = (u,v,w)$. In this case, we have

$$\boldsymbol{u} = \boldsymbol{u}(\boldsymbol{x}) \quad \text{or} \quad u = u(x,y,z) \quad v = v(x,y,z) \quad \text{and} \quad w = w(x,y,z)$$

where the functions u, v, and w are continuous and have continuous first derivatives in some region R*. The transformations $\boldsymbol{u} = \boldsymbol{u}(\boldsymbol{x})$ are such that a point $\boldsymbol{x}$ corresponding to $\boldsymbol{u}$ in R* lies in a region R and that there is a one-to-one correspondence between the points $\boldsymbol{u}$ and $\boldsymbol{x}$. The Jacobian matrix $J = \partial(\boldsymbol{x})/\partial(\boldsymbol{u}) = \partial(x,y,z)/\partial(u,v,w)$ is defined by

$$J = \frac{\partial(\boldsymbol{x})}{\partial(\boldsymbol{u})} = \frac{\partial(x,y,z)}{\partial(u,v,w)} = \begin{pmatrix} \partial x/\partial u & \partial x/\partial v & \partial x/\partial w \\ \partial y/\partial u & \partial y/\partial v & \partial y/\partial w \\ \partial z/\partial u & \partial z/\partial v & \partial z/\partial w \end{pmatrix} \tag{A.25}$$

The determinant $\det J$ should be either positive or negative throughout the region R*. Consider now the integral

$$I = \int_{\mathrm{R}} F(x,y,z)\,\mathrm{d}x\mathrm{d}y\mathrm{d}z \tag{A.26}$$

over the region R. If the function F is now expressed in u, v, and w instead of x, y, and z, the integral has to be evaluated over the region R* as

$$I = \int_{\mathrm{R}^*} F(u,v,w)\,|\det J|\,\mathrm{d}u\mathrm{d}v\mathrm{d}w \tag{A.27}$$

where $|\det J|$ denotes the absolute value of the determinant of the Jacobian matrix J. The expression is easily generalized to more variables than 3.

Example A.10 ***Cylindrical and Spherical Coordinates*** In many cases the use of cylindrical coordinates is convenient. Here, we consider the Cartesian coordinates x_1, x_2, and x_3 as "new" variables and the cylindrical coordinates r, θ, and z as "old" variables. The relations between the Cartesian coordinates and the cylindrical coordinates (Figure A.1b) and their inverse relations are

$$x_1 = r\cos\theta \qquad x_2 = r\sin\theta \qquad x_3 = z \quad \text{and}$$

$$r = \left(x_1{}^2 + x_2{}^2\right)^{1/2} \qquad \theta = \tan^{-1}(x_2/x_1) \qquad z = x_3$$

The Jacobian determinant is given by $|\det J| = r$. Similarly, for spherical coordinates

$$x_1 = r\cos\phi\sin\theta \qquad x_2 = r\sin\phi\sin\theta \qquad x_3 = r\cos\theta$$

and the corresponding inverse equations

$$r = \left(x_1{}^2 + x_2{}^2 + x_3{}^2\right)^{1/2} \qquad \phi = \tan^{-1}(x_2/x_1) \qquad \theta = \tan^{-1}\left[\left(x_1{}^2 + x_2{}^2\right)^{1/2}/x_3\right]$$

In this case, the Jacobian determinant becomes[4] $|\det J| = r^2\sin\theta$.

A.8 Calculus of Variations

One of the chief applications of the calculus of variations is to find a function for which some given integral has an extreme. We treat the problem essentially as one dimensional, but extension to more than one dimension is straightforward.

Suppose we wish to find a path $x = x(t)$ between two given values $x(t_1)$ and $x(t_2)$ such that the integral $J = \int_{t_1}^{t_2} f(x, \dot{x}, t)\,\mathrm{d}t$ of some function $f(x, \dot{x}, t)$ with $\dot{x} = \mathrm{d}x/\mathrm{d}t$ is an extremum. Such a function of a function is called a *functional*. Let us assume that $x_0(t)$ is the solution we are looking for. Other possible curves close to $x_0(t)$ are written as $x(t,\alpha) = x_0(t) + \alpha\eta(t)$, where $\eta(t)$ is any (reasonable) function that satisfies $\eta(t_1) = \eta(t_2) = 0$. Using this representation, the integral J becomes a function of α,

$$J(\alpha) = \int_{t_1}^{t_2} f[x(t,\alpha), \dot{x}(t,\alpha), t]\,\mathrm{d}t \tag{A.28}$$

and the condition for obtaining the extremum is $(\partial J/\partial\alpha)_{\alpha=0} = 0$. We obtain

$$\frac{\partial J}{\partial\alpha} = \int_{t_1}^{t_2} \left(\frac{\partial f}{\partial x}\frac{\partial x}{\partial\alpha} + \frac{\partial f}{\partial\dot{x}}\frac{\partial\dot{x}}{\partial\alpha}\right)\mathrm{d}t = \int_{t_1}^{t_2}\left(\frac{\partial f}{\partial x}\eta + \frac{\partial f}{\partial\dot{x}}\dot{\eta}\right)\mathrm{d}t \tag{A.29}$$

Through integration by parts, the second term of the integral evaluates to

$$\int_{t_1}^{t_2} \frac{\partial f}{\partial\dot{x}}\dot{\eta}\,\mathrm{d}t = \left.\frac{\partial f}{\partial\dot{x}}\eta\right|_{t_1}^{t_2} - \int_{x_1}^{x_2}\eta\frac{\mathrm{d}}{\mathrm{d}t}\frac{\partial f}{\partial\dot{x}}\,\mathrm{d}t \tag{A.30}$$

At t_1 and t_2, the first or *boundary term* on the rhs vanishes because $\eta(t) = \partial x/\partial\alpha$ vanishes. Hence, we obtain from Eq. (A.29)

$$\delta J = \left[\int_{t_1}^{t_2}\left(\frac{\partial f}{\partial x} - \frac{\mathrm{d}}{\mathrm{d}t}\frac{\partial f}{\partial\dot{x}}\right)\eta\,\mathrm{d}t\right]\delta\alpha = \int_{t_1}^{t_2}\left(\frac{\partial f}{\partial x} - \frac{\mathrm{d}}{\mathrm{d}t}\frac{\partial f}{\partial\dot{x}}\right)\delta x\,\mathrm{d}t = 0 \tag{A.31}$$

4 Unfortunately, in the usual convention for spherical coordinates, the angle ϕ corresponds to the angle θ in cylindrical coordinates.

where the *variations* $\delta J = (\partial J/\partial\alpha)_{\alpha=0}\delta\alpha = (\partial f/\partial\alpha)_{\alpha=0}\delta\alpha$ and $\delta x = (\partial x/\partial\alpha)_{\alpha=0}\delta\alpha$ have been used. Since δx is arbitrary, we obtain the *Euler condition*

$$\frac{\delta J}{\delta x} \equiv \frac{\partial f}{\partial x} - \frac{\mathrm{d}}{\mathrm{d}t}\frac{\partial f}{\partial \dot{x}} = 0 \tag{A.32}$$

with $\delta J/\delta x$ the *functional derivative*. The extremum obtained is not necessarily a minimum but considering the second variation one can establish whether a minimum is obtained. In case the variations at the boundaries do not vanish, that is, the values of η are not prescribed, the boundary term $(\partial f/\partial\dot{x})\eta|_{t_1}^{t_2}$ does not vanish. If we now require $\delta J = 0$, we obtain in addition to Eq. (A.32) also the boundary condition $\partial f/\partial\dot{x} = 0$ at $t = t_1$ and $t = t_2$. Note that if f is independent of $\dot{x}$, $\delta f = \mathrm{d}f$, $\delta x = \mathrm{d}x$, and $\delta f/\delta x = \partial f/\partial x$.

Finally, we note that because

$$\delta \mathrm{d}y = \frac{\partial}{\partial\alpha}\frac{\partial y}{\partial x}\mathrm{d}x\delta\alpha = \frac{\partial}{\partial x}\frac{\partial y}{\partial\alpha}\delta\alpha\mathrm{d}x = \mathrm{d}\delta y \tag{A.33}$$

the operations d and δ *commute*. Similarly, integration and variation commute as

$$\delta\int y\,\mathrm{d}y = \frac{\partial}{\partial\alpha}\int y\,\mathrm{d}y\,\delta\alpha = \int\frac{\partial y}{\partial\alpha}\,\delta\alpha\,\mathrm{d}y = \int\delta y\,\mathrm{d}y \tag{A.34}$$

In case any constraints are present, the Lagrange undetermined multiplier method can be used here as well, again by replacing the function f by the function g, Eq. (A.20).

Example A.11 Let us calculate the shortest distance between points 1 and 2 in a plane. An element of length is $\mathrm{d}s = (\mathrm{d}x^2 + \mathrm{d}y^2)^{1/2}$ and the total length of a curve between the two points is

$$I = \int_1^2 \mathrm{d}s = \int_{x_1}^{x_2} f(\dot{y}, x)\,\mathrm{d}x = \int_{x_1}^{x_2}\sqrt{1+\dot{y}^2}\mathrm{d}x \quad \text{where} \quad \dot{y} = \partial y/\partial x$$

The condition that the curve is the shortest path is

$$\frac{\partial f}{\partial y} - \frac{\mathrm{d}}{\mathrm{d}x}\frac{\partial f}{\partial \dot{y}} = 0$$

Since $\partial f/\partial y = 0$ and $\partial f/\partial\dot{y} = \dot{y}/\sqrt{1+\dot{y}^2}$, we have $\mathrm{d}(\dot{y}/\sqrt{1+\dot{y}^2})/\mathrm{d}x = 0$ or $\dot{y}/\sqrt{1+\dot{y}^2} = c$, where c is a constant. This solution holds if $\dot{y} = a$, where a is given by $a = c/(1+c^2)^{1/2}$. Obviously, this is the equation for a straight line $y = ax + b$, where b is another constant of integration. The constants a and b are determined by the condition that the curve should go through (x_1, y_1) and (x_2, y_2).

A.9 Matrices and Determinants

A *matrix* is an array of numbers (or functions), represented by a roman boldface uppercase symbol, for example, **A**, or by an italic uppercase symbol with indices, for example, A_{ij}. In full we write

$$\mathbf{A} = A_{ij} = \begin{pmatrix} A_{11} & A_{12} & \cdot & A_{1n} \\ A_{21} & & & \cdot \\ \cdot & & & \cdot \\ A_{m1} & \cdot & \cdot & A_{mn} \end{pmatrix} \tag{A.35}$$

The numbers A_{ij} are called the *elements*. The matrix with m rows and n columns is called an $m \times n$ matrix or a matrix of order (m,n). The *transpose* of a matrix, indicated by a superscript T, is formed by interchanging rows and columns. Hence

$$\mathbf{A}^{\mathrm{T}} = A_{ji} \tag{A.36}$$

For square matrices A_{ij}, $m = n$ (order n). A *column matrix* (or *column* for short) is a matrix for which $n = 1$ and is denoted by a lowercase italic symbol with an index, for example, by a_i, or by a lowercase roman bold symbol, for example, $\mathbf{a}$. A *row matrix* (or *row* for short) is a matrix for which $m = 1$ and is the transpose of a column matrix and thus denoted by $(a_i)^{\mathrm{T}}$ or $\mathbf{a}^{\mathrm{T}}$.

Two matrices of the same order are *equal* if all their corresponding elements are equal. The *sum* of two matrices **A** and **B** of the same order is given by the matrix **C** whose corresponding elements are the sums of the elements of **A** and **B** or

$$\mathbf{C} = \mathbf{A} + \mathbf{B} \quad \text{or} \quad C_{ij} = A_{ij} + B_{ij} \tag{A.37}$$

The *product* of two matrices **A** and **B** results in matrix **C** whose elements are given by

$$\mathbf{C} = \mathbf{AB} \quad \text{or} \quad C_{ij} = \sum\nolimits_k A_{ik}B_{kj} \tag{A.38}$$

representing the *row-into-column* rule. Note that, if **A** represents a matrix of order (k,l) and **B** a matrix of order (m,n), the product **AB** is not defined unless $l = m$. For square matrices, we generally have $\mathbf{AB} \neq \mathbf{BA}$ (the matrices do not *commute*), so that the order must be maintained in any multiplication process. The *transpose* of a *product* $(\mathbf{ABC}\ldots)^{\mathrm{T}}$ is given by $(\mathbf{ABC}\ldots)^{\mathrm{T}} = \ldots^{\mathrm{T}}\mathbf{C}^{\mathrm{T}}\mathbf{B}^{\mathrm{T}}\mathbf{A}^{\mathrm{T}}$.

A *real* matrix is a matrix with real elements only while a *complex* matrix is a matrix with complex elements. The *complex conjugate* of a matrix **A** is the matrix $\mathbf{A}^*$ formed by the complex conjugate elements of **A** or

$$\mathbf{A}^* = A_{ij}^* \tag{A.39}$$

If a real (complex), square matrix **A** is equal to its transpose (complex conjugate)

$$\mathbf{A} = \mathbf{A}^{\mathrm{T}}\ (\mathbf{A} = (\mathbf{A}^*)^{\mathrm{T}}) \tag{A.40}$$

then **A** is a *symmetric* (*Hermitian*) matrix. For an *antisymmetric* matrix it holds that

$$\mathbf{A} = -(\mathbf{A}^{\mathrm{T}}) \tag{A.41}$$

and thus **A** has the elements $a_{jj} = 0$ and $a_{ij} = -a_{ji}\ (i \neq j)$. A *diagonal* matrix has only non-zero entries along the diagonal while the *unit* matrix **I** is a diagonal matrix with unit elements, that is, $\mathbf{A} = a_{jj}$ and $\mathbf{I} = \delta_{ij}$. Obviously, $\mathbf{IA} = \mathbf{AI} = \mathbf{A}$, where **A** is any square matrix of the same order as the unit matrix.

The *trace* of a square matrix is defined as the sum of its diagonal elements, that is, $\mathrm{tr}\mathbf{A} = \Sigma_k A_{kk}$. The *determinant* of a square matrix of order n is defined by

$$\det \mathbf{A} = \sum (\pm A_{1i}A_{2j}A_{3k}\ldots A_{np}) \tag{A.42}$$

where the summation is over all permutations of the indices $i, j, k, \ldots, p$. The sign in brackets is positive (negative) when the permutation involves an even (odd) number of permutations from the initial term $A_{11}A_{22}A_{33}\ldots A_{nn}$.

Example A.12 For a matrix **A** of order 3, Eq. (A.42) yields

$$\det \mathbf{A} = A_{11}A_{22}A_{33} + A_{12}A_{23}A_{31} + A_{13}A_{21}A_{32} - A_{12}A_{21}A_{33} - A_{11}A_{23}A_{32} - A_{13}A_{22}A_{31}$$

Alternatively, it can be written as $\det \mathbf{A} = \Sigma_{r,s,t} e_{rst} A_{1r} A_{2s} A_{3t}$.

The determinant of the product **AB** is given by

$$\det \mathbf{AB} = (\det \mathbf{A})(\det \mathbf{B}) \tag{A.43}$$

Further, the determinant of a matrix equals the determinant of its transpose, that is,

$$\det \mathbf{A} = \det \mathbf{A}^{\mathrm{T}} \tag{A.44}$$

The *inverse* of a square matrix **A** is denoted by $\mathbf{A}^{-1}$ and is holds that

$$\mathbf{AA}^{-1} = \mathbf{A}^{-1}\mathbf{A} = \mathbf{I} \tag{A.45}$$

where **I** is the unit matrix of the same order as **A**. Hence, a square matrix *commutes* with its inverse. The inverse only exists if $\det \mathbf{A} \neq 0$. The inverse of the product $(\mathbf{ABC}\ldots)^{-1}$ is given by $(\mathbf{ABC}\ldots)^{-1} = \ldots^{-1}\mathbf{C}^{-1}\mathbf{B}^{-1}\mathbf{A}^{-1}$. The inverse of a transpose is equal to the transpose of the inverse, that is, $(\mathbf{A}^{\mathrm{T}})^{-1} = (\mathbf{A}^{-1})^{\mathrm{T}}$, often written as $\mathbf{A}^{-\mathrm{T}}$.

The *cofactor* α_{ij} of the element A_{ij} is $(-1)^{i+j}$ times the *minor* θ_{ij}. The latter is the determinant of a matrix obtained by removing row i and column j from the original matrix. The inverse of **A** is then found from *Cramers's rule*

$$(\mathbf{A}^{-1})_{ij} = \alpha_{ji} / \det \mathbf{A} \tag{A.46}$$

Note the reversal of the element and cofactor indices.

Example A.13 Consider the matrix $\mathbf{A} = \begin{pmatrix} 1 & 2 \\ 3 & 4 \end{pmatrix}$. The determinant is $\det \mathbf{A} = -2$. The cofactors are

$$\alpha_{11} = (-1)^{1+1}\theta_{11} = (-1)^{1+1}4 = 4 \quad \alpha_{12} = (-1)^{1+2}\theta_{12} = (-1)^{1+2}3 = -3$$

$$\alpha_{22} = (-1)^{2+2}\theta_{22} = (-1)^{2+2}1 = 1 \quad \alpha_{21} = (-1)^{2+1}\theta_{21} = (-1)^{2+1}2 = -2$$

The elements of the inverse $\mathbf{A}^{-1}$ are thus given by

$$[A_{ij}^{-1}] = \begin{pmatrix} \alpha_{11} & \alpha_{21} \\ \alpha_{12} & \alpha_{22} \end{pmatrix} \Big/ \det \mathbf{A} = \begin{pmatrix} 4 & -2 \\ -3 & 1 \end{pmatrix} \Big/ -2 = \begin{pmatrix} -2 & 1 \\ 1.5 & -0.5 \end{pmatrix}$$

For a *diagonal* matrix **A**, the inverse is particularly simple and given by another diagonal matrix with elements $(a_{jj})^{-1}$. For an *orthogonal* (*unitary*) matrix, it holds that

$$\mathbf{A}^{\mathrm{T}} = \mathbf{A}^{-1} \quad \text{or} \quad \mathbf{A}^{\mathrm{T}}\mathbf{A} = \mathbf{I} \quad (\mathbf{A}^{*} = \mathbf{A}^{-1} \quad \text{or} \quad \mathbf{A}^{*}\mathbf{A} = \mathbf{I}) \tag{A.47}$$

implying $\det \mathbf{A} = \pm 1$. With $\det \mathbf{A} = 1$, the matrix **A** is a *proper* orthogonal matrix.

A.10 The Eigenvalue Problem

Consider a square matrix **A** of order n, a column matrix **e** and a scalar λ. It is possible to find column matrices **e** that produce column matrices $\lambda\mathbf{e}$. In formula

$$\mathbf{Ae} = \lambda\mathbf{e} \quad \text{or} \quad (\mathbf{A} - \lambda\,\mathbf{I})\mathbf{e} = 0 \tag{A.48}$$

where **I** is the unit matrix. The scalar λ is called the *eigenvalue*, while the column(s) **e** are called the *eigenvector(s)*. This equation can be solved using det $(\mathbf{A} - \lambda\mathbf{I}) = 0$ or

$$\begin{vmatrix} A_{11}-\lambda & A_{12} & \cdot & A_{1n} \\ A_{21} & A_{22}-\lambda & & \cdot \\ \cdot & & & \cdot \\ A_{n1} & \cdot & \cdot & A_{nn}-\lambda \end{vmatrix} = 0 \tag{A.49}$$

because otherwise the matrix $\mathbf{A} - \lambda\mathbf{I}$ is regular and Eq. (A.48) has only the (trivial) solution $\mathbf{e} = 0$. Expanding the determinant for a matrix (of order 3) yields the *characteristic equation*

$$\lambda^3 - J_{(1)}\lambda^2 - J_{(2)}\lambda - J_{(3)} = 0 \tag{A.50}$$

where the *(principal) invariants* $J_{(1)}$, $J_{(2)}$, and $J_{(3)}$ are given by

$$J_{(1)} = A_{11} + A_{22} + A_{33} = \text{tr}\mathbf{A}$$

$$J_{(2)} = A_{12}{}^2 + A_{23}{}^2 + A_{31}{}^2 - A_{11}A_{22} - A_{22}A_{33} - A_{33}A_{11} = [\text{tr}(\mathbf{A}^2) - (\text{tr}\mathbf{A})^2]/2 \tag{A.51}$$

$$J_{(3)} = \det \mathbf{A}$$

The various Js are called invariants since their numerical value is independent of the coordinate system chosen. The roots of the characteristic equation are the eigenvalues λ_1, λ_2, and λ_3, also independent of the coordinate system. Back substituting λ_1, λ_2, and λ_3 in the eigenvalue equation yields the eigenvectors $\mathbf{e}_1$, $\mathbf{e}_2$, and $\mathbf{e}_3$, respectively. These eigenvectors are indeterminate to the extent of a scalar constant, which can be used to normalize the eigenvectors. Moreover, they are also mutually orthogonal or can be chosen to be so. Hence, we can write

$$\boldsymbol{e}_i \cdot \boldsymbol{e}_j = \mathbf{e}_i{}^{\text{T}}\mathbf{e}_j = \delta_{ij} \tag{A.52}$$

Putting the eigenvectors $\mathbf{e}_i$ as columns of a matrix **E**, the complete set of equations is given by $\mathbf{E}^{\text{T}}\mathbf{A}\mathbf{E} = \boldsymbol{\Lambda}$, where $\boldsymbol{\Lambda}$ is a diagonal matrix with λ_1, λ_2, and λ_3 as elements.

Example A.14 The characteristic equation for a matrix **A** is $(\mathbf{A} - \lambda\mathbf{I})\mathbf{e} = 0$, so that for

$$\mathbf{A} = A_{ij} = \begin{pmatrix} 1 & 0 & 0 \\ 0 & 0 & a \\ 0 & a & 0 \end{pmatrix}, \quad \left|\begin{pmatrix} 1 & 0 & 0 \\ 0 & 0 & a \\ 0 & a & 0 \end{pmatrix} - \lambda\begin{pmatrix} 1 & 0 & 0 \\ 0 & 1 & 0 \\ 0 & 0 & 1 \end{pmatrix}\right| = 0 \quad \text{or} \quad \begin{vmatrix} 1-\lambda & 0 & 0 \\ 0 & -\lambda & a \\ 0 & a & -\lambda \end{vmatrix} = 0$$

Evaluation yields $(1 - \lambda)(\lambda^2 - \text{a}^2) = 0$. The eigenvalues (roots) are $\lambda_1 = 1$, $\lambda_2 = a$ and $\lambda_3 = -a$. Substitution of $\lambda_1 = 1$ in the eigenvalue equation results in

$$\begin{pmatrix} 1-\lambda & 0 & 0 \\ 0 & -\lambda & a \\ 0 & a & -\lambda \end{pmatrix}\begin{pmatrix} e_{11} \\ e_{21} \\ e_{31} \end{pmatrix} = \begin{pmatrix} 0 \\ 0 \\ 0 \end{pmatrix} \quad \text{or} \quad \begin{pmatrix} 0 & 0 & 0 \\ 0 & -1 & a \\ 0 & a & -1 \end{pmatrix}\begin{pmatrix} e_{11} \\ e_{21} \\ e_{31} \end{pmatrix} = \begin{pmatrix} 0 \\ -e_{21} + ae_{31} \\ ae_{21} - e_{31} \end{pmatrix} = \begin{pmatrix} 0 \\ 0 \\ 0 \end{pmatrix}$$

Consequently, we may take $e_{21} = e_{31} = 0$. Since the eigenvectors must be normalized the magnitude $||\mathbf{e}|| = e_{11}{}^2 + e_{21}{}^2 + e_{31}{}^2 = 1$, $e_{11} = 1$. Substitution of $\lambda_2 = a$ results in

$$\begin{pmatrix} 1-a & 0 & 0 \\ 0 & -a & a \\ 0 & a & -a \end{pmatrix}\begin{pmatrix} e_{12} \\ e_{22} \\ e_{32} \end{pmatrix} = \begin{pmatrix} (1-a)e_{12} \\ -ae_{22} + ae_{32} \\ ae_{22} - ae_{32} \end{pmatrix} = \begin{pmatrix} 0 \\ 0 \\ 0 \end{pmatrix}$$

Hence, if $(1-a)\neq 0$, $e_{12}=0$. Further, since $a\neq 0$, $e_{22} = e_{32}$. Normalizing yields $e_{22} = e_{32} = ½\sqrt{2}$ (or $-½\sqrt{2}$). Substitution of $\lambda_3 = -a$ results in

$$\begin{pmatrix} 1+a & 0 & 0 \\ 0 & a & a \\ 0 & a & a \end{pmatrix}\begin{pmatrix} e_{13} \\ e_{23} \\ e_{33} \end{pmatrix} = \begin{pmatrix} (1+a)e_{13} \\ ae_{23}+ae_{33} \\ ae_{23}+ae_{33} \end{pmatrix} = \begin{pmatrix} 0 \\ 0 \\ 0 \end{pmatrix}$$

Hence, if $(1+a)\neq 0$, $e_{13} = 0$. Further, since $a\neq 0$, $e_{23}=-e_{33}$. Normalizing yields $e_{23} = -e_{33} = ½\sqrt{2}$ (or $-½\sqrt{2}$). Collecting results leads to

$$\begin{pmatrix} 1 & 0 & 0 \\ 0 & ½\sqrt{2} & ½\sqrt{2} \\ 0 & ½\sqrt{2} & -½\sqrt{2} \end{pmatrix}\begin{pmatrix} 1 & 0 & 0 \\ 0 & 0 & a \\ 0 & a & 0 \end{pmatrix}\begin{pmatrix} 1 & 0 & 0 \\ 0 & ½\sqrt{2} & ½\sqrt{2} \\ 0 & ½\sqrt{2} & -½\sqrt{2} \end{pmatrix} = \begin{pmatrix} 1 & 0 & 0 \\ 0 & a & 0 \\ 0 & 0 & -a \end{pmatrix}$$

The eigenvalues are also known as *principal values* and the coordinate system in which the matrix is diagonal is known as the *principal axes system*. The principal values are often indicated by a Roman numerical subscript, that is, A_I, A_II, and A_III. In the principal axis system, the invariants become

$$J_{(1)} = \lambda_1 + \lambda_2 + \lambda_3 = A_\text{I} + A_\text{II} + A_\text{III}$$

$$J_{(2)} = \lambda_1\lambda_2 + \lambda_2\lambda_3 + \lambda_3\lambda_1 = A_\text{I}A_\text{II} + A_\text{II}A_\text{III} + A_\text{III}A_\text{I} \tag{A.53}$$

$$J_{(3)} = \lambda_1\lambda_2\lambda_3 = A_\text{I}A_\text{II}A_\text{III}$$

In the above the invariants are given in terms of the principal values λ_1, λ_2, and λ_3, Eq. (A.53), or in terms of combinations of the elements A_{ij} of the matrix $\mathbf{A}$, Eq. (A.51). We can also write them in terms of the *basic invariants* $A_{(1)}$, $A_{(2)}$, and $A_{(3)}$ defined by

$$A_{(1)} = \text{tr}\mathbf{A} = \lambda_1 + \lambda_2 + \lambda_3, \quad A_{(2)} = \text{tr}\mathbf{A}^2 = {\lambda_1}^2 + {\lambda_2}^2 + {\lambda_3}^2 \quad \text{and}$$
$$A_{(3)} = \text{tr}\mathbf{A}^3 = {\lambda_1}^3 + {\lambda_2}^3 + {\lambda_3}^3 \tag{A.54}$$

This results in

$$J_{(1)} = A_{(1)}, \quad J_{(2)} = \left(A_{(2)} - {A_{(1)}}^2\right)/2 \quad \text{and} \quad J_{(3)} = \left(2A_{(3)} - 3A_{(2)}A_{(1)} + {A_{(1)}}^3\right)/6 \tag{A.55}$$

In principal, axes space the matrix $\mathbf{A} = A_{ij}$ and powers of $\mathbf{A}$, in index notation represented by $\Sigma_k A_{ik}A_{kj}$, $\Sigma_{k,l}A_{ik}A_{kl}A_{lj}$, and so on, are represented, respectively, by

$$\mathbf{A} = \begin{pmatrix} A_\text{I} & 0 & 0 \\ 0 & A_\text{II} & 0 \\ 0 & 0 & A_\text{III} \end{pmatrix} \quad \text{and} \quad \mathbf{A}^n = \begin{pmatrix} {A_\text{I}}^n & 0 & 0 \\ 0 & {A_\text{II}}^n & 0 \\ 0 & 0 & {A_\text{III}}^n \end{pmatrix} \tag{A.56}$$

Since the principal values A_I, A_II, and A_III satisfy the characteristic equation, we have

$${A_\text{I}}^3 = J_{(1)}{A_\text{I}}^2 + J_{(2)}A_\text{I} + J_{(3)}$$

Similar equations can be written for A_II and A_III. Collectively, one writes

$$\mathbf{A}^3 = J_{(1)}\mathbf{A}^2 + J_{(2)}\mathbf{A} + J_{(3)}\mathbf{I} \tag{A.57}$$

generally referred to as the *Hamilton–Cayley equation*. Using this equation, it is possible to express any power of $\mathbf{A}$ in terms of $\mathbf{A}^2$, $\mathbf{A}$, and $\mathbf{I}$. Therefore, any power series $\mathbf{B} = a\mathbf{I} + b\mathbf{A} + c\mathbf{A}^2 + d\mathbf{A}^3 + \cdots$ can be reduced to

$$\mathbf{B} = \alpha\mathbf{I} + \beta\mathbf{A} + \gamma\mathbf{A}^2 \tag{A.58}$$

where α, β, and γ are power series in $J_{(1)}$, $J_{(2)}$, and $J_{(3)}$.

Example A.15 Consider the fourth power $\mathbf{A}^4$. This term can be written as

$$\begin{aligned}\mathbf{A}\cdot\mathbf{A}^3 &= \mathbf{A}[J_{(1)}\mathbf{A}^2 + J_{(2)}\mathbf{A} + J_{(3)}\mathbf{I}] = J_{(1)}\mathbf{A}^3 + J_{(2)}\mathbf{A}^2 + J_{(3)}\mathbf{A}\\ &= J_{(1)}[J_{(1)}\mathbf{A}^2 + J_{(2)}\mathbf{A} + J_{(3)}\mathbf{I}] + J_{(2)}\mathbf{A}^2 + J_{(3)}\mathbf{A}\\ &= \left(J_{(1)}{}^2 + J_{(2)}\right)\mathbf{A}^2 + (J_{(1)}J_{(2)} + J_{(3)})\mathbf{A} + J_{(1)}J_{(3)}\mathbf{I}\end{aligned}$$

A.11 Matrix Decompositions

Any matrix $\mathbf{A}$ can be written as the sum of a *symmetric* part $A_{(ij)}$ and an *antisymmetric* part $A_{[ij]}$, defined by

$$A_{(ij)} = (A_{ij} + A_{ji})/2 \quad \text{and} \quad A_{[ij]} = (A_{ij} - A_{ji})/2 \tag{A.59}$$

The additive decomposition is thus given by

$$A_{ij} = A_{(ij)} + A_{[ij]} \quad \text{or} \quad A_{(ij)} = (\mathbf{A} + \mathbf{A}^{\mathrm{T}})/2 \quad \text{and} \quad A_{[ij]} = (\mathbf{A} - \mathbf{A}^{\mathrm{T}})/2 \tag{A.60}$$

It easily follows that for two matrices $\mathbf{A}$ and $\mathbf{B}$, the product $A_{(ij)}B_{[ij]}$ vanishes and that

$$\sum\nolimits_{i,j} A_{ij}B_{ij} = \sum\nolimits_{i,j} (A_{(ij)} + A_{[ij]})(B_{(ij)} + B_{[ij]}) = \sum\nolimits_{i,j} A_{(ij)}B_{(ij)} + \sum\nolimits_{i,j} A_{[ij]}B_{[ij]} \tag{A.61}$$

Example A.16 For matrix $A_{ij} = \begin{pmatrix} 1 & 2 & 3\\ 4 & 5 & 6\\ 7 & 8 & 9\end{pmatrix}$, the symmetric and asymmetric parts are $A_{(ij)} = \begin{pmatrix} 1 & 3 & 5\\ 3 & 5 & 7\\ 5 & 7 & 9\end{pmatrix}$ and $A_{[ij]} = \begin{pmatrix} 0 & -1 & -2\\ 1 & 0 & -1\\ 2 & 1 & 0\end{pmatrix}$. Similarly, for the matrix $B_{ij} = \begin{pmatrix} 1 & 2 & 2\\ 2 & 4 & 4\\ 4 & 8 & 8\end{pmatrix}$, $B_{(ij)} = \begin{pmatrix} 1 & 2 & 3\\ 2 & 4 & 6\\ 3 & 6 & 8\end{pmatrix}$ and $B_{[ij]} = \begin{pmatrix} 0 & 0 & -1\\ 0 & 0 & -2\\ 1 & 2 & 0\end{pmatrix}$. The product $\Sigma_{i,j}A_{ij}B_{ij} = 211$, $\Sigma_{i,j}A_{(ij)}B_{(ij)} = 219$, and $\Sigma_{i,j}A_{[ij]}B_{[ij]} = -8$.

For another way to decompose a symmetric matrix, we use the *trace* of a matrix $\mathbf{A}$ and define the *deviator* $\mathbf{A}'$ by

$$A_{ij}{}' = A_{ij} - (A_{kk}\delta_{ij})/3 = \mathbf{A} - \mathrm{tr}(\mathbf{A})\mathbf{I}/3 \quad \text{where} \quad \mathrm{tr}\,\mathbf{A} = \Sigma_j A_{jj} = \Sigma_{i,j}A_{ij}\delta_{ij} \tag{A.62}$$

The term $\mathrm{tr}(\mathbf{A})\mathbf{I}/3$ is the *isotropic* or *spherical symmetric part* of the matrix $\mathbf{A}$. Solving the deviator expression for A_{ij} yields $A_{ij} = (\Sigma_k A_{kk}\delta_{ij})/3 + A_{ij}{}'$ from which it is clear that every symmetric matrix can be considered as the sum of an isotropic part and a deviator. The trace of the product of two symmetric matrices $\mathbf{A}$ and $\mathbf{B}$, when decomposed into their isotropic and deviatoric parts, can be written as

$$\sum\nolimits_{i,j} A_{ij}B_{ij} = {}^1\!/_3 \sum\nolimits_{k,l} A_{kk}B_{ll} + \sum\nolimits_{i,j} A_{ij}{}'B_{ij}{}' \tag{A.63}$$

The quantity $\text{tr}(\mathbf{A})/3 = \Sigma_k A_{kk}/3$ is the mean value of the diagonal components of the matrix A_{ij}, indicated by A or A_m. The matrix A_{ij} therefore can also be written as $A_{ij} = A_{ij}' + A\delta_{ij}$. This convention is used with the stress matrix σ_{ij} and the strain matrix ε_{ij}.

Example A.17 For a matrix $A_{ij} = \begin{pmatrix} 1 & 2 & 3 \\ 2 & 3 & 4 \\ 3 & 4 & 5 \end{pmatrix}$, $\frac{1}{3}\Sigma_k A_{kk}\delta_{ij} = \begin{pmatrix} 3 & 0 & 0 \\ 0 & 3 & 0 \\ 0 & 0 & 3 \end{pmatrix}$ and $A_{ij}' = \begin{pmatrix} -2 & 1 & 3 \\ 2 & 0 & 4 \\ 3 & 4 & 2 \end{pmatrix}$.

Similarly, for the matrix $B_{ij} = \begin{pmatrix} 1 & 1 & 3 \\ 1 & 3 & 5 \\ 3 & 5 & 5 \end{pmatrix}$, $\frac{1}{3}\Sigma_B A_{kk}\delta_{ij} = \begin{pmatrix} 3 & 0 & 0 \\ 0 & 3 & 0 \\ 0 & 0 & 3 \end{pmatrix}$, while $B_{ij}' = \begin{pmatrix} -2 & 1 & 3 \\ 1 & 0 & 5 \\ 3 & 5 & 2 \end{pmatrix}$. The product $\Sigma_{k,l}(A_{kk}B_{ll})/3 = 27$, while $\Sigma_{i,j}A_{ij}'B_{ij}' = 70$, resulting in $\Sigma_{i,j}A_{ij}B_{ij} = 97$.

Similar to the full symmetric matrix **A**, the invariants and basic invariants of the deviator $\mathbf{A}'$ can be calculated. Note that, since the reduction to the deviator removes only the spherical symmetrical part from **A**, the principal axes of $\mathbf{A}'$ are the same as for **A**, while the principal values are given by $\lambda_\text{I}' = \lambda_\text{I} - \Sigma_j A_{jj}/3$. Obviously, $J_{(1)}' = A_{(1)}' = \text{tr}\,\mathbf{A}' = \Sigma_j A_{jj}' = \Sigma_j(A_{jj} - \Sigma_k A_{kk}\delta_{jj})/3 = 0$. The second invariants are given by

$$A_{(2)}' = \text{tr}\mathbf{A}'^2 = \text{tr}\mathbf{A}^2 - \tfrac{1}{3}\text{tr}^2\mathbf{A} = A_{(2)} - \tfrac{1}{3}A_{(1)}^2 \tag{A.64}$$

$$J_{(2)}' = \tfrac{1}{2}\left(A_{(2)}' - A_{(1)}'^2\right) = \tfrac{1}{2}A_{(2)}' = \tfrac{1}{2}(\text{tr}\mathbf{A}^2 - \tfrac{1}{3}\text{tr}^2\mathbf{A}) \tag{A.65}$$

$$J_{(2)}' = \left[(A_{11} - A_{22})^2 + (A_{22} - A_{33})^2 + (A_{33} - A_{11})^2\right]/6 + A_{12}{}^2 + A_{23}{}^2 + A_{31}{}^2$$
$$= \left[(A_\text{I} - A_\text{II})^2 + (A_\text{II} - A_\text{III})^2 + (A_\text{III} - A_\text{I})^2\right]/6 \tag{A.66}$$

while the third basic invariant reduces to

$$A_{(3)}' = A_{(3)} - A_{(2)}A_{(1)} + 2A_{(1)}{}^3/9 \tag{A.67}$$

Using a multiplicative decomposition, a matrix **A** can be uniquely decomposed into

$$\mathbf{A} = \mathbf{RU} \quad \text{and} \quad \mathbf{A} = \mathbf{VR}$$

where **R** is an orthogonal matrix. Hence, $\mathbf{RR}^\text{T} = \mathbf{R}^\text{T}\mathbf{R} = \mathbf{I}$, and **U** and **V** are symmetric and positive definite matrices, $\mathbf{U} = \mathbf{U}^\text{T}$ and $\mathbf{V} = \mathbf{V}^\text{T}$. These decompositions, conventionally denoted as right and left *polar decompositions*, can only be realized if $\det \mathbf{A} \neq 0$. The actual decompositions are generally quite tedious but fortunately in practice we frequently only make use of the fact that the decomposition always exists.

A.12 Scalars, Vectors, and Tensors

A *scalar* is an entity with a magnitude. It is denoted by an italic, lowercase, or uppercase, Latin or Greek letter, for example, a, A, γ, or Γ.

A *vector* is an entity with a magnitude and direction. It is denoted by a boldface, italic letter, for example, $\boldsymbol{a}$ or $\boldsymbol{A}$. It can be interpreted as an arrow from a point O (origin) to a point P. Let $\boldsymbol{a}$ be this arrow. Its *magnitude* (length), equal to the distance OP, is denoted by $a = |\boldsymbol{a}|$. A unit vector in the same direction as the vector $\boldsymbol{a}$ has a length of 1. Vectors obey the following rules (Figure A.2):

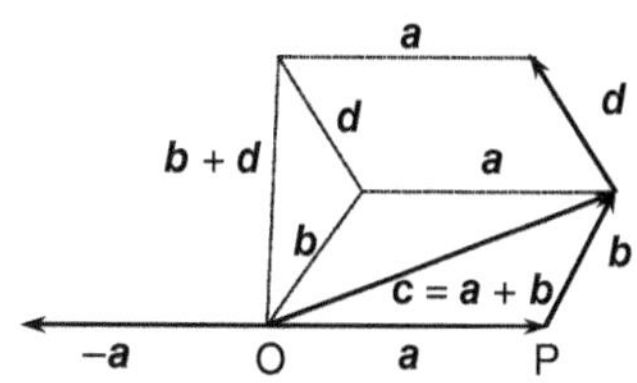

Figure A.2 Scheme of vector properties showing a vector a, its negative $-a$, the sum $a+b$, and the associative rule $(a+b)+d=a+(b+d)$.

- $\mathbf{c}=\boldsymbol{a}+\boldsymbol{b}=\boldsymbol{b}+\boldsymbol{a}$ (commutative rule)
- $\boldsymbol{a}+(\boldsymbol{b}+\boldsymbol{d})=(\boldsymbol{a}+\boldsymbol{b})+\boldsymbol{d}$ (associative rule)
- $(\boldsymbol{a}+\boldsymbol{b})\cdot\boldsymbol{c}=\boldsymbol{a}\cdot\boldsymbol{c}+\boldsymbol{b}\cdot\boldsymbol{c}$ (distributive law)
- $\boldsymbol{a}+(-\boldsymbol{a})=\boldsymbol{0}$ (zero vector definition)
- $\boldsymbol{1}\cdot\boldsymbol{u}=\boldsymbol{u}$ (unit vector definition)
- $\boldsymbol{0}\cdot\boldsymbol{u}=\boldsymbol{0}$
- $\alpha\boldsymbol{a}=\alpha|\boldsymbol{a}|\boldsymbol{e}$, $|\boldsymbol{e}|=1$

Various products can be formed using vectors. The *scalar* (or *dot* or *inner*) *product* of two vectors $\boldsymbol{a}$ and $\boldsymbol{b}$ yields a scalar and is defined as $\boldsymbol{a}\cdot\boldsymbol{b}=\boldsymbol{b}\cdot\boldsymbol{a}=|\boldsymbol{a}|\ |\boldsymbol{b}|\cos(\phi)$, where ϕ is the enclosed angle between $\boldsymbol{a}$ and $\mathbf{b}$. From this definition it follows that $a=|\boldsymbol{a}|=(\boldsymbol{a}\cdot\boldsymbol{a})^{1/2}$. Two vectors $\boldsymbol{a}$ and $\boldsymbol{b}$ are *orthogonal* if $\boldsymbol{a}\cdot\boldsymbol{b}=0$. The scalar product is commutative ($\boldsymbol{a}\cdot\boldsymbol{b}=\boldsymbol{b}\cdot\boldsymbol{a}$) and distributive ($\boldsymbol{a}\cdot(\mathbf{b}+\mathbf{c})=\boldsymbol{a}\cdot\mathbf{b}+\boldsymbol{a}\cdot\mathbf{c}$).

The *vector* (or *cross* or *outer*) *product* of $\boldsymbol{a}$ and $\boldsymbol{b}$ denotes a vector $\mathbf{c}=\boldsymbol{a}\times\mathbf{b}$, defined in three-dimensional (3D) space only. We define a unit vector $\boldsymbol{n}$ perpendicular to the plane spanned by $\boldsymbol{a}$ and $\mathbf{b}$. The sense of $\boldsymbol{n}$ is right-handed: rotate from $\boldsymbol{a}$ to $\mathbf{b}$ over the smallest angle and the direction of $\boldsymbol{n}$ is given by a right-hand screw. It holds that $\boldsymbol{a}\cdot\boldsymbol{n}=\mathbf{b}\cdot\boldsymbol{n}=0$ and $|\boldsymbol{n}|=1$. Explicitly, $\boldsymbol{n}=(\boldsymbol{a}\times\mathbf{b})/|\boldsymbol{a}|\ |\mathbf{b}|$. The vector product $\mathbf{c}$ is equal to $\mathbf{c}=\boldsymbol{a}\times\mathbf{b}=-\mathbf{b}\times\boldsymbol{a}=|\boldsymbol{a}|\ |\mathbf{b}|\sin(\phi)\boldsymbol{n}$. The length $|\mathbf{c}|=|\boldsymbol{a}|\ |\mathbf{b}|\sin(\phi)$ is numerically equal to the area of the parallelogram whose sides are given by $\boldsymbol{a}$ and $\mathbf{b}$. The vector product is anti-commutative ($\boldsymbol{a}\times\mathbf{b}=-\mathbf{b}\times\boldsymbol{a}$) and distributive ($\boldsymbol{a}\times(\mathbf{b}+\mathbf{c})=\boldsymbol{a}\times\mathbf{b}+\boldsymbol{a}\times\mathbf{c}$), but not associative ($\boldsymbol{a}\times(\mathbf{b}\times\mathbf{c})\neq(\boldsymbol{a}\times\mathbf{b})\times\mathbf{c}$).

The *triple product* is a scalar and given by $d=\boldsymbol{a}\cdot\mathbf{b}\times\mathbf{c}=\boldsymbol{a}\times\mathbf{b}\cdot\mathbf{c}$. It yields the volume of the block (or parallelepiped) the edges of which are $\boldsymbol{a}$, $\mathbf{b}$, and $\mathbf{c}$. Three vectors $\boldsymbol{a}$, $\mathbf{b}$, and $\mathbf{c}$ are independent if from $\alpha\boldsymbol{a}+\beta\mathbf{b}+\gamma\mathbf{c}=\boldsymbol{0}$ it follows that $\alpha=\beta=\gamma=0$. This is only the case if $\boldsymbol{a}$, $\mathbf{b}$, and $\mathbf{c}$ are noncoplanar or, equivalently, the product $\boldsymbol{a}\cdot\mathbf{b}\times\mathbf{c}\neq0$.

Finally, we need the *tensor* (or *dyadic*) *product* $\boldsymbol{ab}$. Operating on a vector $\mathbf{c}$, it associates with $\mathbf{c}$ a new vector according to $\boldsymbol{ab}\cdot\mathbf{c}=\boldsymbol{a}(\mathbf{b}\cdot\mathbf{c})=(\mathbf{b}\cdot\mathbf{c})\boldsymbol{a}$. Note that $\boldsymbol{ba}$ operating on $\mathbf{c}$ yields $\boldsymbol{ba}\cdot\mathbf{c}=\mathbf{b}(\boldsymbol{a}\cdot\mathbf{c})=(\boldsymbol{a}\cdot\mathbf{c})\mathbf{b}$.

A *tensor* (of rank 2 or higher), denoted by a special font, that is, $\mathbb{A}$, is a linear mapping that associates a vector $\boldsymbol{a}$ with another vector $\mathbf{b}$ according to $\mathbf{b}=\mathbb{A}\cdot\boldsymbol{a}$. Tensors obey the following rules:

- $\mathbb{A}=\mathbb{A}+\mathbb{B}=\mathbb{B}+\mathbb{A}$ (commutative law)
- $\mathbb{A}+(\mathbb{B}+\mathbb{C})=(\mathbb{A}+\mathbb{B})+\mathbb{C}$ (associative law)
- $(\mathbb{A}+\mathbb{B})\cdot\boldsymbol{u}=\mathbb{A}\cdot\boldsymbol{u}+\mathbb{B}\cdot\boldsymbol{u}$ (distributive law)
- $\mathbb{A}+(-\mathbb{A})=\boldsymbol{O}$ (zero tensor definition)
- $\mathbb{I}\cdot\boldsymbol{u}=\boldsymbol{u}$ ($\mathbb{I}$ unit tensor definition)
- $\mathbb{O}\cdot\boldsymbol{u}=\boldsymbol{0}$ ($\mathbb{O}$ zero tensor, $\boldsymbol{0}$ zero vector)
- $\mathbb{A}\cdot(\alpha\boldsymbol{u})=(\alpha\mathbb{A})\cdot\boldsymbol{u}=\alpha(\mathbb{A}\cdot\boldsymbol{u})$

where α is an arbitrary scalar and $\boldsymbol{u}$ is an arbitrary vector. The simplest example of a tensor is the tensor product of two vectors, for example, if $\mathbf{A} = \mathbf{bc}$, the vector associated with $\boldsymbol{a}$ is given by $\mathbf{A} \cdot \boldsymbol{a} = \mathbf{bc} \cdot \boldsymbol{a} = (\boldsymbol{c} \cdot \boldsymbol{a})\boldsymbol{b}$.

So far, we discussed vectors and tensors using the *direct notation* only, that is using a symbolism, which represents the quantity without referring to a coordinate system and for which a strict ordering is required. It is convenient though to introduce a coordinate system. We use primarily Cartesian coordinates, a rectangular and rectilinear coordinate system with origin O and unit vectors $\boldsymbol{e}_1, \boldsymbol{e}_2$, and $\boldsymbol{e}_3$ along the axes. The set $\boldsymbol{e}_i = \{\boldsymbol{e}_1, \boldsymbol{e}_2, \boldsymbol{e}_3\}$ is called an *orthonormal basis*. It holds that $\boldsymbol{e}_i \cdot \boldsymbol{e}_j = \delta_{ij}$. The vector OP $= \boldsymbol{x}$ is called the position of point P. The real numbers x_1, x_2, and x_3, defined uniquely by the relation $\boldsymbol{x} = x_1\boldsymbol{e}_1 + x_2\boldsymbol{e}_2 + x_3\boldsymbol{e}_3$, are called the (Cartesian) *components* of the vector $\boldsymbol{x}$. It follows that $x_i = \boldsymbol{x} \cdot \boldsymbol{e}_i$ for $i = 1, 2, 3$. Using the components x_i in equations, we use *index notation* in which the scalar product $\boldsymbol{u} \cdot \boldsymbol{v}$ can be written as $\boldsymbol{u} \cdot \boldsymbol{v} = u_1v_1 + u_2v_2 + u_3v_3 = \Sigma_i u_i v_i$. The length of a vector $\boldsymbol{x}$, $|\boldsymbol{x}| = (\boldsymbol{x} \cdot \boldsymbol{x})^{1/2}$, is thus also equal to $(x_1{}^2 + x_2{}^2 + x_3{}^2)^{1/2} = (\Sigma_i x_i x_i)^{1/2}$. Also *matrix notation* is used; in this case, the components x_i are written collectively as a column matrix $\mathbf{x}$ and in which the scalar product $\boldsymbol{u} \cdot \boldsymbol{v}$ is written as $\mathbf{u}^{\mathrm{T}}\mathbf{v}$, while the tensor product $\boldsymbol{ab}$ is given by $\mathbf{ab}^{\mathrm{T}}$.

Using the alternator ε_{ijk} the relations between the unit vectors can be written as $\boldsymbol{e}_i \times \boldsymbol{e}_j = \Sigma_k \varepsilon_{ijk} \boldsymbol{e}_k$. Similarly, the vector product $\mathbf{c} = \boldsymbol{a} \times \mathbf{b}$ can alternatively be written as $\mathbf{c} = \boldsymbol{a} \times \mathbf{b} = \Sigma_{j,k} \varepsilon_{ijk} \boldsymbol{e}_i a_j b_k$. In components, this leads to the following expressions:

$$c_1 = a_2b_3 - a_3b_2 \quad c_2 = a_3b_1 - a_1b_3 \quad c_3 = a_1b_2 - a_2b_1 \tag{A.68}$$

The triple product $\boldsymbol{a} \cdot \boldsymbol{b} \times \mathbf{c}$ in components is given by $\Sigma_{i,j,k} \varepsilon_{ijk} a_i b_j c_k$, while the tensor product $\boldsymbol{ab}$ is represented by elements $a_i b_j$. A useful relation involving three vectors using the tensor product is $\boldsymbol{a} \times \boldsymbol{b} \times \mathbf{c} = (\mathbf{bc} - \mathbf{cb}) \cdot \boldsymbol{a} = (\boldsymbol{c} \cdot \boldsymbol{a})\boldsymbol{b} - (\boldsymbol{b} \cdot \boldsymbol{a})\boldsymbol{c} = [\mathbf{ba} - (\boldsymbol{a} \cdot \boldsymbol{b})\mathbb{I}]\mathbf{c}$.

Recall that the components of a vector $\boldsymbol{a}$ can be transformed to another Cartesian frame by $a_p' = \Sigma_i C_{pi} a_i$ in index notation or $\mathbf{a}' = \mathbf{Ca}$ in matrix notation. Since a tensor $\mathbb{A}$ of rank 2 can be interpreted as the tensor product of two vectors $\boldsymbol{b}$ and $\boldsymbol{c}$, that is,

$$\mathbf{A} = \boldsymbol{bc} \quad \text{or} \quad A_{ij} = b_i c_j \quad \text{or} \quad \mathbb{A} = \mathbf{bc}^{\mathrm{T}}$$

the transformation rule for the components of a tensor $\mathbb{A}$ obviously is[5]

$$A_{pq}{}' = b_p{}' c_q{}' = \Sigma_{i,j} C_{pi} b_i C_{qj} c_j = \Sigma_{i,j} C_{pi} C_{qj} b_i c_j = \Sigma_{i,j} C_{pi} C_{qj} A_{ij} \quad \text{or in matrix notation}$$

$$\mathbb{A}' = \mathbf{b}'\mathbf{c}'^{\mathrm{T}} = (\mathbf{Cb})(\mathbf{c}^{\mathrm{T}}\mathbf{C}^{\mathrm{T}}) = \mathbf{C}\mathbb{A}\mathbf{C}^{\mathrm{T}}$$

If $\mathbb{A}' = \mathbf{A}$ and thus $\mathbb{A}' = \mathbb{A}$, then $\mathbf{A}$ is an *isotropic* (or *spherical*) *tensor*. Further, if the component matrix of a tensor has a property that is not changed by a coordinate axis rotation, that property is shared by $\mathbf{A}'$ and $\mathbf{A}$. Such a property is called an *invariant*. An example is the transpose of a tensor of rank 2: if $\mathbb{A}' = \mathbf{C}\mathbb{A}\mathbf{C}^{\mathrm{T}}$, then $\mathbb{A}'^{\mathrm{T}} = \mathbf{C}\mathbb{A}^{\mathrm{T}}\mathbf{C}^{\mathrm{T}}$. Consequently, we may speak of the transpose $\mathbf{A}^{\mathrm{T}}$ of the tensor $\mathbf{A}$, and we may define the symmetric parts $\mathbf{A}^{(\mathrm{s})}$ and antisymmetric parts $\mathbf{A}^{(\mathrm{a})}$ by

$$\mathbf{A}^{(\mathrm{s})} = (\mathbf{A} + \mathbf{A}^{\mathrm{T}})/2 \quad \text{and} \quad \mathbf{A}^{(\mathrm{a})} = (\mathbf{A} - \mathbf{A}^{\mathrm{T}})/2$$

While originally a distinction in terminology is made for a scalar, a vector, and a tensor, it is clear that they all transform similarly under a coordinate transformation. Therefore,

5 Obviously, if the transformation is interpreted as a rotation of the tensor instead of the frame, we obtain $\mathbb{A}' = \mathbf{C}^{\mathrm{T}}\mathbb{A}\mathbf{C}$. This is the conventional definition of an orthogonal transformation.

a scalar is frequently denoted as a tensor of rank 0 and a vector as a tensor of rank 1. Expressed in components, all tensors obey the same type of transformation rules, that is, $A_{i...j} = \Sigma_{p...q} C_{ip} \ldots C_{jq} A_{p...q}$, where the transformation matrix **C** represents the rotation of the coordinate system. Scalars have no index, a vector has one index, a tensor of rank 2 has two indices, and a tensor of rank 4 has four indices. Their total transformation matrix contains a product of respectively 0, 1, 2, and 4 individual transformation matrices C_{ij}. Obviously, if we define (Cartesian) tensors as quantities obeying the above transformation rules,[6] extension to any order is immediate.

Finally, we have to mention that, like scalars, vectors and tensors can be a function of one or more variables x_k. The appropriate notation is $f(x_k)$, $a_i(x_k)$ and $A_{ij}(x_k)$ or, equivalently, $f(\boldsymbol{x})$, $\boldsymbol{a}(\boldsymbol{x})$, and $\mathbb{A}(\boldsymbol{x})$. If x_k represent the coordinates, $f(\boldsymbol{x})$, $\boldsymbol{a}(\boldsymbol{x})$, and $\mathbb{A}(\boldsymbol{x})$ are referred to as a *scalar*, *vector*, and *tensor field*, respectively.

Example A.18 Consider the vectors $\boldsymbol{a}$, $\boldsymbol{b}$, and $\boldsymbol{c}$, the matrix representations of which are $\mathbf{a}^{\mathrm{T}} = (1,0,0)$, $\mathbf{b}^{\mathrm{T}} = (0,1,1)$, and $\mathbf{c}^{\mathrm{T}} = (1,1,2)$. Then

$$\boldsymbol{a} + \boldsymbol{b} = \mathbf{a} + \mathbf{b} = (1,1,1)^{\mathrm{T}}$$

$$(\boldsymbol{b} \times \boldsymbol{c})^{\mathrm{T}} = (b_2c_3 - b_3c_2,\ b_3c_1 - b_1c_3,\ b_1c_2 - b_2c_1) = \left(1{\cdot}2 - 1{\cdot}1,\ 1{\cdot}1 - 0{\cdot}2,\ 0{\cdot}1 - 1{\cdot}1\right) = \left(1\ 1\ {-1}\right)$$

$$|\boldsymbol{c}| = (\mathbf{c}^{\mathrm{T}}\mathbf{c})^{1/2} = \left[\begin{pmatrix}1 & 1 & 2\end{pmatrix}\begin{pmatrix}1\\1\\2\end{pmatrix}\right]^{1/2} = \sqrt{6} \qquad \boldsymbol{b} \cdot \boldsymbol{c} = \mathbf{b}^{\mathrm{T}}\mathrm{c} = \begin{pmatrix}0 & 1 & 1\end{pmatrix}\begin{pmatrix}1\\1\\2\end{pmatrix} = 3$$

$$\boldsymbol{ab} = \mathbf{ab}^{\mathrm{T}} = \begin{pmatrix}1\\0\\0\end{pmatrix}\begin{pmatrix}0 & 1 & 1\end{pmatrix} = \begin{pmatrix}0 & 1 & 1\\0 & 0 & 0\\0 & 0 & 0\end{pmatrix} \qquad \boldsymbol{a} \cdot \boldsymbol{b} \times \boldsymbol{c} = \begin{pmatrix}1 & 0 & 0\end{pmatrix}\begin{pmatrix}1\\1\\-1\end{pmatrix} = 1$$

$$\boldsymbol{ab} \cdot \boldsymbol{c} = \begin{pmatrix}0 & 1 & 1\\0 & 0 & 0\\0 & 0 & 0\end{pmatrix}\begin{pmatrix}1\\1\\2\end{pmatrix} = \begin{pmatrix}3\\0\\0\end{pmatrix} = 3\boldsymbol{a}$$

A.13 Tensor Analysis

In this section, we consider various differential operators and two important theorems. Consider a_i as a typical representative of tensors of rank 1 and take the partial derivatives

---

6 This transformation rule is only appropriate for proper rotations of the axes. For improper rotations, a rotation followed by a reflection in a plane perpendicular to that axis, there are two possibilities. If the rule applies, the tensor is *polar*; if an additional change of sign occurs, the tensor is *axial*. Hence, generally, $L_{i.j} = (\det \mathbf{C})^p C_{ik} \ldots C_{jl} L_{k...l}$, where $p = 0$ for a polar tensor and $p = 1$ for an axial tensor, or, equivalently, contains an even power and odd power of e_{ijk}, respectively, in case of tensors of second rank. Since det $\mathbf{C} = 1$ for a proper rotation and det $\mathbf{C} = -1$ for an improper rotation, this results in an extra change of sign for an axial tensor under improper rotation. The inner and outer products of two polar or two axial tensors are polar, while the product of a polar and an axial tensor is axial. The permutation tensor e_{ijk} is axial since $e_{123} = 1$. Hence, the vector product of two polar vectors is axial. If one restricts oneself to right-handed systems, the distinction is irrelevant.

$\partial a_i/\partial x_j$. Such a derivative transforms like a tensor of rank 2, is called the *gradient* and denoted by grad a_i in index notation or as $\nabla \boldsymbol{a}$ in direct notation. The gradient can operate on any tensor thereby increasing its rank by 1.

Summing over an index, known as *contraction*, decreases the rank of a tensor by 2. If we apply contraction to a tensor $\mathbb{A} = A_{ij}$ we calculate $\Sigma_i A_{ii}$ and the result is known as the *trace*, written in direct notation as tr $\mathbb{A}$. Contraction of a gradient $\Sigma_i(\partial a_i/\partial x_i)$ yields a scalar, called *divergence*, denoted in direct notation by div $\boldsymbol{a}$ or $\nabla \cdot \boldsymbol{a}$.

The *curl* (or *rot* as abbreviation of rotation) of a vector $\boldsymbol{a}$, in direct notation written as $\nabla \times \boldsymbol{a}$, is a vector with components $\Sigma_{k,j}\varepsilon_{ijk}\partial a_k/\partial x_j$. It is defined for 3D space only. Note that the index j of the derivative coordinate x_j is considered as the second index.

Finally, we have the *Laplace operator*. It can act on a scalar a or on a vector $\boldsymbol{a}$, is denoted by ∇^2 or Δ, and defined by $\nabla^2 a = \Delta a = \Sigma_j \partial a^2/\partial^2 x_j$ or $\nabla^2 \boldsymbol{a} = \Delta \boldsymbol{a} = \Sigma_{i,j}\partial a_j{}^2/\partial^2 x_i$.

Example A.19 If a vector field $\boldsymbol{a}(\boldsymbol{x})$ is represented by $\mathbf{a}^{\mathrm{T}}(\mathbf{x}) = (3x^2 + 2y, x + z^2, x + y^2)$, then

$$\frac{\partial a_i}{\partial x_j} = \nabla \boldsymbol{a} = \begin{pmatrix} \partial/\partial x \\ \partial/\partial y \\ \partial/\partial z \end{pmatrix} \left(3x^2 + 2y,\ x + z^2,\ x + y^2\right) = \begin{pmatrix} 6x & 1 & 1 \\ 2 & 0 & 2y \\ 0 & 2z & 0 \end{pmatrix}$$

$$\nabla \times \boldsymbol{a} = \sum\nolimits_{j,k} \varepsilon_{ijk} \boldsymbol{e}_i \frac{\partial a_k}{\partial x_j} = \begin{pmatrix} 2y - 2z \\ 0 - 1 \\ 1 - 2 \end{pmatrix} = \begin{pmatrix} 2(y - z) \\ -1 \\ -1 \end{pmatrix}, \quad \Delta \boldsymbol{a} = \frac{\partial^2 a_j}{\partial x_i^2} = \begin{pmatrix} 6 \\ 2 \\ 2 \end{pmatrix}$$

and $\nabla \cdot \boldsymbol{a} = \sum_i (\partial a_i/\partial x_i) = 6x + 0 + 0 = 6x$.

For these operations, some general theorems are available. For a *conservative quantity* $\boldsymbol{X} = \nabla x$ with x a scalar, integration yields a path-independent result given by

$$\int_A^B \nabla x \, \mathrm{d}V = x_B - x_A \tag{A.69}$$

Consider now a region of volume V with a piecewise smooth surface S on which a single-valued tensor field A or A_{ij} is defined. The body may be convex or nonconvex. The components of the exterior normal vector $\boldsymbol{n}$ of S are denoted by n_i. The *divergence theorem* or the *theorem of Gauss*, here given without proof, states that

$$\int \sum\nolimits_j \frac{\partial A_{ij}}{\partial x_j} \mathrm{d}V = \int \sum\nolimits_j A_{ij} n_j \, \mathrm{d}S \quad \text{or equivalently} \quad \int \nabla \cdot \mathbb{A} \mathrm{d}V = \int \mathbb{A} \cdot \boldsymbol{n} \, \mathrm{d}S \tag{A.70}$$

The divergence theorem connects a volume integral (integrating over dV) to a surface integral (integrating over $\boldsymbol{n}$dS) and is mainly used in theoretical work. Applying the divergence theorem to a scalar a, a vector $\boldsymbol{a}$ or a tensor $\nabla \times \boldsymbol{a}$, we obtain

$$\int \nabla a \, \mathrm{d}V = \int \boldsymbol{n} a \, \mathrm{d}S, \quad \int \nabla \cdot \boldsymbol{a} \, \mathrm{d}V = \int \boldsymbol{n} \cdot \boldsymbol{a} \, \mathrm{d}S \quad \text{and} \quad \int \nabla \times \boldsymbol{a} \, \mathrm{d}V = \int \boldsymbol{n} \times \boldsymbol{a} \, \mathrm{d}S \tag{A.71}$$

The surface element $\boldsymbol{n}$dS is often written as d$\boldsymbol{S}$. One can also derive *Stokes' theorem*

$$\int \sum_{i,j,k} n_i \varepsilon_{ijk} \frac{\partial a_k}{\partial r_j} \mathrm{d}S = \int \sum_i a_i \mathrm{d}r_i \quad \text{or} \quad \int (\nabla \times \boldsymbol{a}) \cdot \boldsymbol{n} \mathrm{d}S = \int \boldsymbol{a} \cdot \mathrm{d}\boldsymbol{r} \tag{A.72}$$

connecting a surface (integrating over $\boldsymbol{n}\mathrm{d}S$) and line integral (integrating over $\mathrm{d}\boldsymbol{r}$). Some useful relations between vectors $\boldsymbol{a}, \boldsymbol{b}, \ldots$ and scalar ϕ are

$$\begin{aligned}
&\boldsymbol{a}\cdot(\boldsymbol{b}\times\boldsymbol{c}) = \boldsymbol{b}\cdot(\boldsymbol{c}\times\boldsymbol{a}) = \boldsymbol{c}\cdot(\boldsymbol{a}\times\boldsymbol{b}), \qquad &&\boldsymbol{a}\times(\boldsymbol{b}\times\boldsymbol{c}) = (\boldsymbol{a}\cdot\boldsymbol{c})\boldsymbol{b} - (\boldsymbol{a}\cdot\boldsymbol{b})\boldsymbol{c}\\
&(\boldsymbol{a}\times\boldsymbol{b})\cdot(\boldsymbol{c}\times\boldsymbol{d}) = (\boldsymbol{a}\cdot\boldsymbol{c})(\boldsymbol{b}\cdot\boldsymbol{d}) - (\boldsymbol{a}\cdot\boldsymbol{d})(\boldsymbol{b}\cdot\boldsymbol{c}), \qquad &&\nabla\cdot(\phi\boldsymbol{a}) = (\nabla\phi)\cdot\boldsymbol{a} + \phi(\nabla\cdot\boldsymbol{a})\\
&\nabla\times(\phi\boldsymbol{a}) = (\nabla\phi)\times\boldsymbol{a} + \phi(\nabla\times\boldsymbol{a}), \qquad &&\nabla\cdot(\boldsymbol{a}\times\boldsymbol{b}) = (\nabla\times\boldsymbol{a})\cdot\boldsymbol{b} - \boldsymbol{a}\cdot(\nabla\times\boldsymbol{b})\\
&\nabla(\boldsymbol{a}\cdot\boldsymbol{b}) = (\boldsymbol{a}\cdot\nabla)\boldsymbol{b} + (\boldsymbol{b}\cdot\nabla)\boldsymbol{a} + \boldsymbol{a}\times(\nabla\times\boldsymbol{b}) + \boldsymbol{b}\times(\nabla\times\mathbf{a}), \qquad &&\nabla\times(\nabla\phi) = 0\\
&\nabla\times(\boldsymbol{a}\times\boldsymbol{b}) = \boldsymbol{a}(\nabla\cdot\boldsymbol{b}) - (\boldsymbol{a}\cdot\nabla)\boldsymbol{b} + (\boldsymbol{b}\cdot\nabla)\boldsymbol{a} - (\nabla\cdot\boldsymbol{a})\boldsymbol{b}, \qquad &&\nabla\cdot(\nabla\times\boldsymbol{a}) = 0\\
&\nabla\times(\nabla\times\boldsymbol{a}) = \nabla(\nabla\cdot\boldsymbol{a}) - \nabla^2\boldsymbol{a} &&
\end{aligned} \tag{A.73}$$

A.14 Gamma, Dirac, and Heaviside Functions

The *gamma function* is defined by

$$\Gamma(t) = \int_0^\infty x^{t-1}\mathrm{e}^{-x}\mathrm{d}x \tag{A.74}$$

It generally holds that $\Gamma(t+1) = t\Gamma(t)$. For integer n it is connected to the more familiar *factorial function* $n!$ by $\Gamma(n+1) = n! = n(n-1)(n-2)\ldots(2)(1)$. With $x = y^2$ we have

$$\Gamma(t) = 2\int_0^\infty y^{2t-1}\mathrm{e}^{-y^2}\mathrm{d}y$$

which results by setting $t = ½$ in $\Gamma(½) = 2\int_0^\infty \mathrm{e}^{-y^2}\mathrm{d}y = \sqrt{\pi}$. Consequently

$$(-3/2)! = -2\pi 1/2, \quad (-½)! = \pi^{1/2}, \quad (½)! = ½\pi^{1/2} \quad \text{and} \quad (3/2)! = ¾\pi^{1/2}$$

The *Dirac (delta) function* $\delta(x)$ is in one dimension defined by

$$\int_{-a}^{+a}\delta(x)f(x)\,\mathrm{d}x = f(0) \quad \text{or} \quad \int_{-a}^{+a}\delta(x-t)f(x)\,\mathrm{d}x = f(t) \tag{A.75}$$

where $a > 0$ and $a = \infty$ is included and which selects the value of a function f at the value of variable t from an integral expression. Alternatively, $\delta(x)$ is defined by

$$\delta(x-t) = \infty \ \text{ if } x = t \quad \text{and} \quad 0 \quad \text{otherwise and} \quad \int_{-a}^{a}\delta(x-t)\,\mathrm{d}t = 1 \tag{A.76}$$

Indicating differentiation with a prime, some properties of $\delta(x)$ are

$$\delta(-x) = \delta(x), \quad x\delta(x) = 0, \quad \delta(ax) = |a^{-1}|\,\delta(x) \quad \text{and}$$

$$\delta(x^2 - a^2) = \left[\frac{\delta(x-a) + \delta(x+a)}{2\,|a|}\right], \quad \text{or generally} \quad \delta(g(x)) = \sum\nolimits_j \frac{\delta(x-x_j)}{|g'(x_j)|}$$

with x_j the roots of $g(x)$. The derivative $\delta'(x)$ is related to $f'(x)$ according to

$$\begin{aligned}
\int_{-a}^{+a}\delta'(x-t)f(t)\,\mathrm{d}x &= \int_{-a}^{+a}f(t)\,\mathrm{d}\delta(x-t) = -\int_{-a}^{+a}\delta(x-t)\,\mathrm{d}f(t)\\
&= -\int_{-a}^{+a}\delta(x-t)f'(t)\,\mathrm{d}x \equiv -f'(t)
\end{aligned}$$

where the boundary term vanishes due to partial integration. This leads further to

$$\delta'(x) = -\delta'(-x) \quad \text{and} \quad x\delta'(x) = -\delta(x)$$

Related is the *Heaviside (step) function* $h(x)$, defined by

$$h(x-t) = 0 \ \text{if} \ x < t \ \text{ and } \ h(x-t) = 1 \ \text{if} \ x > t \ \text{ or } h(y) = \int_{-\infty}^{x} \delta(y)\,\text{d}y \tag{A.77}$$

Differentiating with respect to x, we get $\delta(x) = h'(x)$. For $a, b > 0$, we have

$$\int_{-a}^{+b} h'(x)f(x)\,\text{d}x = h(x)f(x)|_{-a}^{+b} - \int_{-a}^{+b} h(x)f'(x)\,\text{d}x = f(b) - \int_{0}^{+b} f'(x)\,\text{d}x = f(0)$$

so that the step function can be considered as the integral of the delta function.

A.15 Laplace and Fourier Transforms

The *Laplace transform* transforms a function $f(t)$ into $\hat{f}(s)$ according to

$$L[f(t)] = \hat{f}(s) = \int_{0}^{\infty} f(t)\exp(-st)\,\text{d}t \tag{A.78}$$

where s may be real or complex. The operation is linear, that is,

$$L[c_1 f_1(t) + c_2 f_2(t)] = c_1\hat{f}_1(s) + c_2\hat{f}_2(s)$$

The *convolution theorem* states that the product of two Laplace transforms $L[f(t)]$ and $L[g(t)]$ equals the transform of the convolution of the functions $f(t)$ and $g(t)$

$$L[f(t)]\,L[g(t)] = L\left[\int_{0}^{t} f(t-\lambda)g(\lambda)\,\text{d}\lambda\right] = L\left[\int_{0}^{t} f(\lambda)g(t-\lambda)\,\text{d}\lambda\right] \tag{A.79}$$

Since the Laplace transform has the property

$$L[\text{d}f(t)/\text{d}t] = s\hat{f}(s) - f(0)$$

it can transform differential equations in t to algebraic equations in s. Generalization to higher derivatives is straightforward and reads

$$L[\text{d}^n f(t)/\text{d}t^n] = s^n\hat{f}(s) - s^{n-1}f(0) - \cdots - sf^{n-2}(0) - f^{n-1}(0)$$

Similarly, for integration it is found that

$$L\left[\int_{0}^{t} f(u)\,\text{d}u\right] = s^{-1}\hat{f}(s)$$

Some useful transforms are given in Table A.1.

The *Fourier transform* of a function $f(t)$ and its inverse are defined by

$$F[f(t)] = \tilde{f}(\omega) = N^{(-)}\int_{-\infty}^{+\infty} f(t)\,\exp(-\text{i}\omega t)\,\text{d}t \quad \text{and}$$

$$F^{-1}[\tilde{f}(\omega)] = f(t) = N^{(+)}\int_{-\infty}^{+\infty} \tilde{f}(\omega)\,\exp(\text{i}\omega t)\,\text{d}\omega \tag{A.80}$$

Table A.1 Laplace transform pairs.

$f(t)$	$\hat{f}(s)$	$f(t)$	$\hat{f}(s)$
1	$1/s$	$\exp(-at)\sin(at)$	$a/[(s+a)^2+a^2]$
a	a/s	$\exp(-at)\cos(at)$	$(s+a)/[(s+a)^2+a^2]$
$h(t)$	$1/s$	$t^n\exp(-at)$, $n=0,1,\ldots$	$n!/(s+a)^{n+1}$
$h(t-a)$	$\exp(-as)/s$	$\sin(at)$	$a/(s^2+a^2)$
$\delta(t)$	1	$\cos(at)$	$s/(s^2+a^2)$
$\delta(t-a)$	$\exp(-as)$	$\sinh(at)$	$a/(s^2-a^2)$
t	$1/s^2$	$\cosh(at)$	$s/(s^2-a^2)$
t^n, $n>-1$	$\Gamma(n+1)/s^{n+1}$	$1-\mathrm{erf}[u/2(ct)^{1/2}]$	$\exp(-us^{1/2}/c^{1/2})/s$
$\exp(-at)$	$1/(s+a)$	$\exp(c^2t)[1-\mathrm{erf}(ct)^{1/2}]$	$1/s^{1/2}(s^{1/2}+c)$

The normalization constants $N^{(-)}$ and $N^{(+)}$ can take any value as long as their product is $(2\pi)^{-1}$. If $N^{(-)} = N^{(+)} = (2\pi)^{-1/2}$ is taken the transform is called *symmetric*. Often one takes $N^{(-)} = 1$ and $N^{(+)} = 1/2\pi$. In chemistry and physics, one often encounters $\exp[-\mathrm{i}(\omega t - \boldsymbol{q}\cdot\boldsymbol{r})]$, where the pair (ω,t) represents frequency–time and the pair $(\boldsymbol{q},\boldsymbol{r})$ wavevector–coordinate and in which (ω,t) and $-(\boldsymbol{q},\boldsymbol{r})$ play a similar role. We note that the Fourier transform is a linear operation for which the convolution theorem holds.

For the delta function $\delta(t)$ it holds that

$$\tilde{\delta}(\omega) = \int_{-\infty}^{+\infty} \delta(t)\exp(-\mathrm{i}\omega t)\,\mathrm{d}t = 1 \quad \text{and} \quad \delta(t) = (2\pi)^{-1}\int_{-\infty}^{+\infty} \exp(\mathrm{i}\omega t)\,\mathrm{d}\omega$$

For $a > 0$ and using $\int_{-a}^{+a} \mathrm{e}^{-\mathrm{i}\omega t}\,\mathrm{d}\omega = (\mathrm{i}\omega)^{-1}(\mathrm{e}^{\mathrm{i}\omega a} - \mathrm{e}^{-\mathrm{i}\omega a}) = (2/\omega)\sin(\omega a)$, we can represent $\delta(t)$ also as $\delta(t) = \lim_{a\to\infty}[\sin(at)/\pi t]$. Similarly, we have

$$\tilde{\delta}(\boldsymbol{\omega}) = \int_{-\infty}^{+\infty} \delta(\boldsymbol{t})\exp(-\mathrm{i}\boldsymbol{\omega}\cdot\boldsymbol{t})\,\mathrm{d}\boldsymbol{t} = \boldsymbol{1} \quad \text{and} \quad \delta(\boldsymbol{t}) = (2\pi)^{-3}\int_{-\infty}^{+\infty} \exp(\mathrm{i}\boldsymbol{\omega}\cdot\boldsymbol{t})\,\mathrm{d}\boldsymbol{\omega}$$

for the 3D delta function $\delta(\boldsymbol{t}) = \delta(t_1)\delta(t_2)\delta(t_3)$ for a vector $\boldsymbol{t} = (t_1,t_2,t_3)$.

Defining the *period* $T = 2\pi/\omega_0$, a periodic function $f(t)$ can be approximated by the *Fourier series* $f(t) = \Sigma_n A_n \exp(\mathrm{i}\omega_0 nt)$. For $-\infty \le n \le \infty$ and sufficient "smoothness" of $f(t)$, this expansion is exact, but obviously it is normally applied with n finite. The coefficient A_n is given by $A_n = T^{-1}\int_{-T/2}^{+T/2} f(t)\exp(-\mathrm{i}\omega_0 nt)\,\mathrm{d}t$ and the transform reads

$$\tilde{f}(\omega) = \sum\nolimits_n A_n\delta(\omega - n\omega_0)$$

Finally, we note that by the Gauss theorem applied to a sphere with radius r

$$\int \nabla^2\left(\frac{1}{r}\right)\mathrm{d}V = \int \nabla\left(\frac{1}{r}\right)\cdot\boldsymbol{n}\mathrm{d}S = -\int \frac{\boldsymbol{r}\cdot\boldsymbol{n}}{r^3}\mathrm{d}S = -4\pi \quad \text{or} \quad \nabla^2\left(\frac{1}{r}\right) = -4\pi\delta(\boldsymbol{r})$$

since $\nabla^2(1/r) = 0$ for $r \neq 0$ and $\nabla^2(1/r) = \infty$ for $r = 0$. Therefore, we have

$$\nabla^2 t^{-1} = \nabla^2 F^{-1}[F[t^{-1}]] = \nabla^2(2\pi)^{-3}\int F[t^{-1}]\exp(\mathrm{i}\boldsymbol{\omega}\cdot\boldsymbol{t})\,\mathrm{d}\boldsymbol{\omega}$$

$$= \frac{1}{(2\pi)^3}\int F[t^{-1}](-\omega^2)\exp(\mathrm{i}\boldsymbol{\omega}\cdot\boldsymbol{t})\,\mathrm{d}\boldsymbol{\omega} = \frac{-4\pi}{(2\pi)^3}\int \exp(\mathrm{i}\boldsymbol{\omega}\cdot\boldsymbol{t})\,\mathrm{d}\boldsymbol{\omega} \quad \text{or} \quad F[t^{-1}] = \frac{4\pi}{\omega^2}$$

Applying the inverse transform, we obtain $F^{-1}[F[t^{-1}]] = F^{-1}[4\pi/\omega^2]$ or

$$t^{-1} = (2\pi^2)^{-1} \int_{-\infty}^{+\infty} \omega^{-2} \exp(\mathrm{i}\boldsymbol{\omega} \cdot \boldsymbol{t})\, \mathrm{d}\boldsymbol{\omega} \tag{A.81}$$

A.16 Some Useful Expressions

Quadratic and cubic equations

$ax^2 + bx + c = 0 : x_{1,2} = [-b \pm (b^2 - 4\,ac)^{1/2}]/2a$

$x^3 + bx^2 + cx + d = 0, p = (2b^3 - 9bc + 27d)/54, q = (b^2 - 3c)/9$

$\text{if}\,(p^2 - q^3) > 0 : x_{1,2,3} = -\text{sgn}(p)[(r^2 + q)/r] - b/3, \text{sgn}(p) = p/|p|, r = [(p^2 - q^3)^{1/2} + |p|]^{1/3}$

$\text{if}\,(p^2 - q^3) \le 0 : x_j = -2q^{1/2}\cos(y_j) - b/3, y_1 = \theta/3, y_2 = \theta + 2\pi/3, y_3 = \theta + 4\pi/3, \theta = \text{acos}(p/q^{3/2})$

Integrals

$n = 0 : \quad \int_0^\infty \exp(-ax^2)x^0\,\mathrm{d}x = 2^{-1}(\pi/a)^{1/2}$ $\qquad n = 1 : \quad \int_0^\infty \exp(-ax^2)x^1\mathrm{d}x = (2a)^{-1}$

$n = 2 : \quad \int_0^\infty \exp(-ax^2)x^2\,\mathrm{d}x = (4a)^{-1}(\pi/a)^{1/2}$ $\qquad n = 3 : \quad \int_0^\infty \exp(-ax^2)x^3\,\mathrm{d}x = (2a^2)^{-1}$

$\text{even} : \quad \int_0^\infty \exp(-ax^2)x^{2n}\mathrm{d}x = \dfrac{1\cdot 3\cdot 5\cdot (2n-1)}{2^{n+1}}\dfrac{(\pi/a)^{1/2}}{a^n}$ $\qquad \text{odd} : \quad \int_0^\infty \exp(-ax^2)x^{2n+1}\,\mathrm{d}x = \dfrac{n!}{2a^{n+1}} \cdots$

$\int_0^\infty \exp(-ax)x^n\mathrm{d}x = \Gamma(n+1)/a^{n+1} \quad (n = 0,1,2,\ldots)$

Binomial and multinomial expansions

$(1+x)^{-1} = 1 - x + x^2 - x^3 + x^4 - \cdots$ $\qquad (1+x)^{-2} = 1 - 2x + 3x^2 - 4x^3 + 5x^4 - \cdots$

$(1+x)^{1/2} = 1 + x/2 + x^2/8 + x^3/16 + \cdots$ $\qquad (1+x)^{-1/2} = 1 - x/2 + 3x^2/8 - 5x^3/16 + \cdots$

$(x+y)^n = x^n + nx^{n-1}y + n(n-1)x^{n-2}y^2/2! + n(n-1)(n-2)x^{n-3}y^3/3! + \cdots,$ valid for $y^2 < x^2$

$(x_1 + x_2 + \cdots + x_k)^n = \sum_n (n!/n_1!n_2!\cdots n_k!)x_1^{n_1}x_2^{n_2}\ldots x_k^{n_k}$ with $n = \sum_i n_i$

Trigonometric, exponential, and logarithmic series

$\sin x = x - x^3/3! + x^5/5! - x^7/7! + \cdots$ $\qquad \cos x = 1 - x^2/2! + x^4/4! - x^6/6! + \cdots$

$\tan x = x + x^3/3 + 2x^5/15 + 17x^7/315 + \cdots$ $\qquad \exp x = \lim_{n\to\infty}(1 + x/n)^n = \lim_{m\to 0}(1 + mx)^{1/m}$

$\exp x = 1 + x + x^2/2! + x^3/3! + x^4/4! + \cdots$ $\qquad \ln(1+x) = x - x^2/2 + x^3/3 - x^4/4 + \cdots$

Inversion of series

$y = b_1x + b_2x^2 + b_3x^3 + \cdots \qquad x = c_1y + c_2y^2 + c_3y^3 + \cdots$

$b_1c_1 = 1, b_1{}^3c_2 = -b_2, \quad b_1{}^5c_3 = 2b_2{}^2 - b_1b_3, \quad b_1{}^7c_4 = b_1b_2b_3 - b_1{}^2b_4 - 5b_2{}^3, \cdots$

Euler–McLaurin formula

$$\sum_{i=a}^{i=b} f(i) = \int_a^b f(x)\mathrm{d}x + \frac{1}{2}[f(a) + f(b)] + \sum_{n\geq 1}(-1)^n \frac{B_n}{(2n)!}[f_{2n-1}(a) - f_{2n-1}(b)]$$

Bernoulli numbers: $B_1 = 1/6, B_2 = 1/30, B_3 = 1/42, B_4 = 1/30, \ldots, f_n(a) = \mathrm{d}^n f(x)/\mathrm{d}x^n|_{x=a}$

Further Reading

Adams, R.A. (1995). *Calculus*, 3e. Don Mills, ON: Addison-Wesley.

Jeffreys, H. and Jeffreys, B.S. (1972). *Methods of Mathematical Physics*, 3e. Cambridge: Cambridge University Press.

Kaplan, W. (1952). *Advanced Calculus*. Reading, MA: Addison-Wesley.

Keisler, H.J. (2011). *Elementary Calculus: An Infinitesimal Approach*, 2e. New York: Dover Publ.; Keisler, H.J. (2011). *Foundations of Infinitesimal Calculus*. Boston, MA: Prindle Weber & Schmidt; Both books dealing with so-called nonstandard analysis, more appropriately called infinitesimal analysis, are freely available on the internet.

Kreyszig, E. (1988). *Advanced Engineering Mathematics*, 6e. New York: Wiley.

Margenau, H. and Murphy, G.M. (1956). *The Mathematics of Physics and Chemistry*, 2e. Princeton, NJ: van Nostrand.

Riley, K.F. and Hobson, M.P. (2011). *Foundation Mathematics for the Physical Sciences* and *Essential Mathematical Methods for the Physical Sciences*. Cambridge: Cambridge University Press.

Appendix B

Elements of Special Relativity Theory

In this appendix, we review briefly the basic aspects of the theory of special relativity as relevant for the discussion on the nature of energy. We first discuss Lorentz transformations and subsequently deal with some consequences. We end with a summary of the relativistic Lagrange and Hamilton function for particles.

B.1 Lorentz Transformations

In Chapter 2, we introduced *inertial systems* and the *principle of relativity* (PoR). We accepted *Galilei transformations* to describe transformations from one inertial frame to another, in which the spatial coordinates were taken as relative but time as absolute. The usual description of interaction in terms of potential energy (PE) thus supposes instantaneous propagation of interaction, as PE does not involve time but only changes in the configuration of the particles that constitute the system.

Actually, such instantaneous interaction does not exist and any change in configuration at location $\boldsymbol{x} = 0$ at time $t = 0$ needs some time t before it affects other particles at location $\boldsymbol{x} = \boldsymbol{x}$, that is, the interaction propagates with a maximum velocity $\boldsymbol{v} = \boldsymbol{x}/t$, or actually, a maximum speed $v = |\boldsymbol{v}|$, as required by homogeneity of space and time. Accepting the PoR, this implies that v is an invariant (a universal constant in all inertial frames). It experimentally appears to be the speed of light in vacuum c [1], the constancy of which is verified by experiments (e.g. Tolman, 1934; Møller, 1952). This excludes the possibility of a unique reference frame (but the debate continues, see, e.g. [2]). While in classical mechanics using Galilei transformations coordinates are taken as relative but time as absolute, in (special) relativistic mechanics both the coordinates and time are taken as relative, implying that a definite time interval can only be assigned to events described in the same inertial frame.

It is convenient to think therefore in $(3+1)$-dimensional Euclidean space or *Minkowski space* using the *Galilei coordinates* $(\boldsymbol{x},\tau) = (\boldsymbol{x},ct) = (x,y,z,ct)$. An *event* is described by the position $\boldsymbol{x}$ at time t, where it occurred and is represented by a point, called a *world point*, in this $(3+1)$ space. The history of a particle is described by a line, called a *world line*, in $(3+1)$ space. A freely moving particle moving in a straight line is described by a straight world line (Figure B.1b).

Consider now an inertial frame I in which two events occur: one at $\boldsymbol{x} = 0$ at time $t = 0$ and another at $\boldsymbol{x} = \mathrm{d}\boldsymbol{x}$ at $t = \mathrm{d}t$. The distance $\mathrm{d}\boldsymbol{x}$ is also given by $c\mathrm{d}t$ and we have $\mathrm{d}\boldsymbol{x} = c\mathrm{d}t$

Phases of Matter and their Transitions: Concepts and Principles for Chemists, Physicists, Engineers, and Materials Scientists, First Edition. Gijsbertus de With.

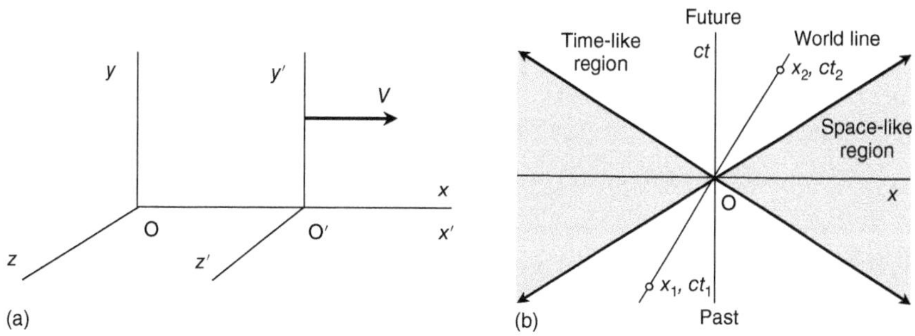

Figure B.1 (a) The relationship between inertial frames I and I′ having a relative velocity V with respect to I. (b) The light cone showing the world line and time-like and space-like intervals.

or $\mathrm{d}s^2 \equiv c^2\mathrm{d}t^2 - \mathrm{d}\boldsymbol{x}^2 = 0$. For another frame I′, moving with relative velocity V with respect to I in the positive x-direction, we similarly have $\mathrm{d}s'^2 \equiv c^2\mathrm{d}t'^2 - \mathrm{d}\boldsymbol{x}'^2 = 0$ (Figure B.1a). For the same events, we thus have that if $\mathrm{d}s^2 = 0$, also $\mathrm{d}s'^2 = 0$. To obtain a complete relation, we apply a linear homogeneous transformation to transform $\mathrm{d}s^2$ in $\mathrm{d}s'^2$. That means $\mathrm{d}s^2 = \alpha\mathrm{d}s'^2$, but also $\mathrm{d}s'^2 = \alpha\mathrm{d}s^2$, and thus $\mathrm{d}s^2 = \alpha^2\mathrm{d}s^2$ or $\alpha^2 = 1$. Using that $\mathrm{d}s' \to \mathrm{d}s$ if $(\boldsymbol{x},t) \to (\boldsymbol{x}',t')$, we see that $\alpha = 1$, that is, the *interval* $\mathrm{d}s = \mathrm{d}s'$ is an invariant. Thus, both world points and world lines are *invariants*.

To obtain a general transformation, we use the Galilei coordinates $(\boldsymbol{x},\tau)$ and write for motion along the x-axis ($y = y'$, $z = z'$)

$$x' = \alpha x + \beta\tau \quad \text{and} \quad t' = \gamma x + \delta\tau \tag{B.1}$$

Taking x as fixed, we have $\mathrm{d}x' = \beta\mathrm{d}\tau$ and $\mathrm{d}\tau' = \delta\mathrm{d}\tau$. The velocity $V = \mathrm{d}x'/\mathrm{d}\tau'$ becomes

$$V = \beta/\delta \quad \text{or} \quad \beta = \delta V \tag{B.2}$$

If we take $x' = \tau'$, we also have $x = \tau$, and thus

$$x' = \alpha x + \beta\tau = \alpha x + \beta x = \tau' = \gamma x + \delta\tau = \gamma x + \delta x \quad \text{or} \quad \alpha + \beta = \gamma + \delta \tag{B.3}$$

If we take $x' = -\tau'$, we also have $x = -\tau$, and thus

$$x' = \alpha x + \beta\tau = -\alpha\tau + \beta\tau = -\tau' = -\gamma x - \delta\tau = +\gamma\tau - \delta\tau \quad \text{or} \quad \alpha - \beta = \delta - \gamma \tag{B.4}$$

Combining Eqs. (B.3) and (B.4), we easily find $\alpha = \delta$ and $\beta = \gamma$. Therefore,

$$x' = \alpha x + \beta\tau = \delta(x + V\tau) \quad \text{and} \quad \tau' = \gamma x + \delta\tau = \delta(Vx + \tau) \tag{B.5}$$

as also could have been obtained by noticing that we have 4 equations and 3 constraints.

Inverting leads to

$$x = (x' - V\tau')/[\delta(1 - V^2)] \quad \text{and} \quad \tau = (t' - Vx')/[\delta(1 - V^2)] \quad \text{and thus to} \tag{B.6}$$

$$\delta = [\delta(1 - V^2)]^{-1} \quad \text{or} \quad \delta = (1 - V^2)^{-1/2} \tag{B.7}$$

Consequently, we obtain

$$x' = \delta(x + V\tau) \qquad t' = \delta(Vx + \tau) \tag{B.8}$$

$$x = \delta(x' - V\tau') \qquad \tau = \delta(-Vx' + \tau') \tag{B.9}$$

as could have been expected on basis of the PoR. The equations could also have been solved for γ and, in fact, γ is the conventionally used symbol instead of δ, and is the symbol that we use from now on. Returning to physical coordinates ($\boldsymbol{x}$,ct), we must further replace τ by ct and V by V/c and obtain

$$x' = \gamma(x + Vt) \qquad t' = \gamma[(V/c^2)x + t] \tag{B.10}$$

$$x = \gamma(x' - Vt') \qquad t = \gamma[-(V/c^2)x' + t') \quad \text{with} \quad \gamma = (1 - V^2/c^2)^{-1/2} \tag{B.11}$$

Equations (B.10) and (B.11) are denoted as the *Lorentz transformations*. A similar result is obtained for general motion (Møller, 1952). We note that for an infinite speed c, the Galilei transformations are obtained. One can show that, apart from a trivial solution, the Galilei and Lorentz transformations are the only two possible transformations satisfying the PoR [3, 4].

Now consider again $\mathrm{d}s^2 = c^2\mathrm{d}t^2 - \mathrm{d}\boldsymbol{x}^2$. If two events occur at the same points for some position in I′, we have $\mathrm{d}\boldsymbol{x}'^2 = 0$, $\mathrm{d}s^2 = c^2\mathrm{d}t^2 - \mathrm{d}\boldsymbol{x}^2 = c^2\mathrm{d}t'^2 > 0$, and such an interval is called a *time-like interval*. If two events occur at the same time for some moment in I′, we have $c^2\mathrm{d}t'^2 = 0$, $\mathrm{d}s^2 = c^2\mathrm{d}t^2 - \mathrm{d}\boldsymbol{x}^2 = -\mathrm{d}\boldsymbol{x}'^2 < 0$, and the interval is called a *space-like interval*. In the case $\mathrm{d}s^2 = 0$, we have a *singular interval* (Figure B.1b).

B.2 Velocities, Contraction, Dilatation, and Proper Quantities

One of the starting points was that the relative velocity between two systems never can be larger than c. To verify that, let us use $\mathrm{d}x' = \gamma(\mathrm{d}x + V\mathrm{d}t)$ and $\mathrm{d}t' = \gamma(V\mathrm{d}x/c^2 + \mathrm{d}t)$ to obtain in I for the velocity $u' = \mathrm{d}x'/\mathrm{d}t' = (\mathrm{d}x + V\mathrm{d}t)/(V\mathrm{d}x/c^2 + \mathrm{d}t) = (u + V)/(1 + uV/c^2)$ and $u' = \mathrm{d}x'/\mathrm{d}t'$ for the velocity in I′. Similarly, we obtain $v' = \mathrm{d}y'/\mathrm{d}t' = v/(1 + uV/c^2)$ and $w' = \mathrm{d}z'/\mathrm{d}t' = w/(1 + uV/c^2)$. Contrary to the classical results, the components v' and w' do change as time intervals change. These expressions show that u' can never exceed c and reduce at low velocity to the expressions $u' = u + V$, $v' = v$ and $w' = w$.

The length of a rod at rest in frame I and lying parallel to the x-axis is given by $\Delta x = x_2 - x_1$, but is measured in frame I′ moving with relative velocity V with respect to I as $\Delta x' = x_2' - x_1'$. Using Eq. (B.10a), we have $x_1' = \gamma(x_1 + Vt)$ and $x_2' = \gamma(x_2 + Vt)$. Eliminating t using Eq. (B.11b), we obtain $\Delta x' = \Delta x/\gamma = (1 - V^2/c^2)^{1/2}\Delta x$. Note that using Eq. (B.11a) would result in x'-values at different times. The length of the rod Δx as measured in the moving frame I′ is thus shorter than Δx in the stationary frame I and the moving observer will find the rod shorter. This shortening is known as the *Lorentz contraction*. The length of the rod measured in its own stationary frame is called its *proper length*.

Similarly, but now considering time, we have a clock at rest in frame I and therefore moving with velocity V in frame I′. The time interval between two events in I is given by $\Delta t = t_2 - t_1$. The corresponding time interval in I′ can be obtained from Eq. (B.11b) as $t_1' = \gamma(t_1 - Vx/c^2)$ and $t_2' = \gamma(t_2 - Vx/c^2)$, so that for a moving observer, $\Delta t' = \gamma\Delta t = (1 - V^2/c^2)^{-1/2}\Delta t$. Note that using Eq. (B.10b) would result in t'-values measured at different positions. The time interval Δt as measured in the moving frame I′ is thus shorter than Δt in the stationary frame I and the moving observer will find the stationary clock running slow. This phenomenon is called *time dilatation*. The time t as measured with a clock in its own stationary frame is called the *proper time*, for which we thus have $c\,dt' = c\gamma\,dt = ds$, corresponding to a straight world line.

B.3 Relativistic Lagrange and Hamilton Functions

In Chapter 2, we showed that mechanics can be derived from Hamilton's principle, stating that the time integral over the Lagrange function $\mathscr{L}$ with (generalized) coordinates $\boldsymbol{q}$, $\dot{\boldsymbol{q}}$, and t, that is, the *action*

$$S \equiv \int_{t_1}^{t_2} \mathscr{L}\,dt = \int_{t_1}^{t_2} \mathscr{L}(\boldsymbol{q}, \dot{\boldsymbol{q}}, t)\,dt \tag{B.12}$$

should be minimal. For a freely moving particle having mass m and moving with velocity $\boldsymbol{v}$, we obtained $\mathscr{L} = \vartheta = ½mv^2$ with ϑ the kinetic energy. The PoR states that the expression for S should be covariant, but, as time has become relative, we should consider the situation anew. The action S can be considered as the integral along a world line but should be covariant and thus a scalar, because Lorentz transformations do not preserve (the components of) a vector, that is, neither their length, nor their direction. This scalar integrand must also be a differential. The only scalar that satisfies these conditions is the interval ds. We thus have

$$S = k\int_a^b ds \tag{B.13}$$

where a and b denote two world points and the constant $k < 0$, as S should be minimal, while the integral in Eq. (B.13) is maximal for straight world line using a stationary clock. From $ds = (c^2dt^2 - d\boldsymbol{x}^2)^{1/2}$, we have $ds = c(1 - v^2/c^2)^{1/2}dt = c\gamma^{-1}dt$, resulting in

$$k\,ds = kc(1 - v^2/c^2)^{1/2}\,dt \cong kc - ½kc(v/c)^2 + \ldots \tag{B.14}$$

If we take $k = -mc$, we obtain

$$\mathscr{L} = -mc(ds/dt) = -mc^2(1 - v^2/c^2)^{1/2} \cong -mc^2 + ½mv^2 + \ldots \tag{B.15}$$

consistent with the low-speed result $\mathscr{L} = \vartheta = ½mv^2$, as mc^2 is a constant.

In Chapter 3, we also saw that a transformation to the Hamilton function $\mathscr{H}$ was beneficial. The momentum $\boldsymbol{p}$ was defined as $\boldsymbol{p} = \partial\mathscr{L}/\partial\boldsymbol{v}$ and results here in the expression $\boldsymbol{p} = (1 - v^2/c^2)^{-1/2}m\boldsymbol{v} = \gamma m\boldsymbol{v}$. This yields for the energy E

$$E = \boldsymbol{p}\cdot\boldsymbol{v} - \mathscr{L} = (1 - v^2/c^2)^{-1/2}mc^2 = \gamma mc^2 \quad \text{or} \tag{B.16}$$

$$E^2 = m^2c^4/(1 - v^2/c^2) = m^2c^4 + [m^2c^2v^2/(1 - v^2/c^2)] = m^2c^4 + c^2p^2 \tag{B.17}$$

Equation (B.17) is often called the *energy–momentum relation*. The Hamilton function, being the energy expression in terms of momenta and coordinates, thus becomes (taking the positive root to agree with the limiting nonrelativistic limit case)

$$\mathscr{H} = c(m^2c^2 + p^2)^{1/2} \cong mc^2 + p^2/2m \tag{B.18}$$

The *rest energy* mc^2 is obtained from $E = \gamma mc^2$ for $\gamma = 1$, while $\vartheta = p^2/2m$ is the *kinetic energy*. From $E = \gamma mc^2$, we have for $m \neq 0$, if $v \to c$, $p \to \infty$, and $E \to \infty$. This implies that for particles with mass m, the speed of light c cannot be reached as an infinite amount of energy is required. For a particle with $m = 0$, for example, a photon, we obtain $\boldsymbol{p} = \gamma m\boldsymbol{v}$, or $p = E/c$.

Finally, consider the potential energy Φ and its contribution to the action given by $-\int \Phi\,\mathrm{d}t$. If we consider that $\Phi = \Phi(\boldsymbol{x},\tau)$, we can obtain a covariant expression by using the proper time, resulting in $-c^{-1}\int \Phi\,\mathrm{d}s$. The total action including ϑ thus becomes $-\int (mc+\Phi/c)\,\mathrm{d}s = -c\int (m+\Phi/c^2)\,\mathrm{d}s$. We note that the mass of the particle increases by an amount Φ/c^2 and there is *no conservation of mass*. This consideration applies to all potential energy terms as well as heat, but as c^2 is huge, in all low-energy physics conservation of mass still applies approximately. Some examples of the effects involved are given in Chapter 1.

References

1 Lee, A.R. and Kalotas, K.M. (1975). *Am. J. Phys.* 43: 434.
2 Múnera, H.A. (1998). *Apeiron* 5: 37; Cahill, R.T. (2005). *Progr. Phys.* 1: 25 and (2006). *Progr. Phys.* 2: 73; Seaver, J.R. (2016). *Progr. Phys.* 4: 332 and references therein.
3 von Ignatowsky, W. (1911). *Arch. Math. Phys.* 17: 1.
4 Pelissetto, A. and Testa, M. (2015). *Am. J. Phys.* 83: 338.

Further Reading

Landau, L.D. and Lifshitz, E.M. (1996). *The Classical Theory of Fields*, 4e with corrections, Oxford: Butterworth-Heinemann.
Lanczos, C. (1970). *The Variational Principles of Mechanics*, 4e. Toronto: University of Toronto Press (also Dover, 1986).
Møller, C. (1952). *The Theory of Relativity*. Oxford: Oxford University Press (also Clarendon Press, 1972); For a gentle introduction to special relativity, see W.G.V. Rosser (1991). *Introductory Special Relativity*, Taylor and Francis, London, or the first 80 pages of R.J.A. Lambourne (2010). *Relativity, Gravitation and Cosmology*, Cambridge University Press, Cambridge.
Taylor, E.F. and Wheeler, J.A. (1992). *Spacetime Physics: Introduction to Special Relativity*, 2e. W. H. Freeman & Company.
Tolman, R.C. (1934). *Relativity, Thermodynamics and Cosmology*. Oxford: Oxford University Press.

Appendix C

The Lattice Gas Model

In this appendix, we briefly review the lattice gas model [1], which is widely used in chemical physics and known under different names in various fields, for example, as the regular solution model for liquid solutions, as the order–disorder model for alloys [2], or as the Ising model [2] for magnetic problems [3]. This broad range of applications explains its importance. For pure liquids, the model was introduced by Frenkel [4] in 1932, and in 1936 independently by Eyring and Cernuschi [5]. It is one of the simplest models, if not the simplest model, showing a phase transition, but we note that, although the model looks deceptively simple, its solution is complex.

C.1 The Lattice Gas Model

In the *lattice gas model*, the volume available to material is divided into *cells* of molecular size. Usually, for simplicity, the cells are arranged in a regular lattice with coordination number z. They are occupied by one molecule at most, representing repulsion, and only nearest-neighbor cell attractive interactions are considered.

The lattice gas model can be solved exactly for one-dimensional (1D), and in this case, the model does not show a phase transition. It can also be solved exactly for two-dimensional (2D) without an applied field, as was achieved for the first time by Onsager [6]; in this case, the model does show a phase transition. A complete exact solution (i.e. with applied field) is neither known for 2D nor for three-dimensional (3D), so that one has to approximate. We will not deal with the exact 2D solution [2] but discuss here the *zeroth* or *mean field approximation*[1] in terms of a model for mixtures, and thereafter the *first* or *quasi-chemical approximation*. We follow largely the treatment as given by Guggenheim [7] and reinforced by Tompa [8]. Benigni [9] provided a compact overview.

C.2 The Zeroth or Mean-Field Approximation

For a lattice model of a mixture, we distribute molecules of type 1 and type 2 over the lattice (without any further holes). In Chapter 9, we showed that if we have N_1 molecules of type 1

1 Also denoted as the Bragg–Williams approximation.

Phases of Matter and their Transitions: Concepts and Principles for Chemists, Physicists, Engineers, and Materials Scientists, First Edition. Gijsbertus de With.

and N_2 molecules of type 2, so that the total number becomes $N = N_1 + N_2$, we obtain the general relations

$$N_{11} = ½zN_1 - ½N_{12} \quad \text{and} \quad N_{22} = ½zN_2 - ½N_{12} \tag{C.1}$$

while the energy is given by

$$E = \varepsilon_{11}N_{11} + \varepsilon_{22}N_{22} + \varepsilon_{12}N_{12} = (\varepsilon_{12} - ½\varepsilon_{11} - ½\varepsilon_{22})N_{12} + ½z\varepsilon_{11}N_1 + ½z\varepsilon_{22}N_2 \tag{C.2}$$

Here N_{11} (N_{22}, N_{12}) represent the number of type 1–1 (2–2, 1–2) pairs, ε_{11} ($\varepsilon_{22}, \varepsilon_{12}$) represent the energy of 1–1 (2–2, 1–2) pairs, and the quantity $w \equiv \varepsilon_{12} - ½\varepsilon_{11} - ½\varepsilon_{22}$ is denoted as *interchange energy*. We use $zX \equiv N_{12}$, so that

$$E = zwX + ½z\varepsilon_{11}N_1 + ½z\varepsilon_{22}N_2 \tag{C.3}$$

The canonical partition function is easily constructed, and we obtain, using $\beta = 1/kT$,

$$Q(T,N,V) = f_1^{N_1}f_2^{N_2}\sum\nolimits_{\{X\}} \exp[-\beta(zwX + ½z\varepsilon_{11}N_1 + ½z\varepsilon_{22}N_2)] \tag{C.4}$$

Recognizing that the term $B \equiv ½(z\varepsilon_{11}N_1 + z\varepsilon_{22}N_2)$ is constant, Q reduces to

$$Q(T,N,V) = f_1^{N_1}f_2^{N_2}\exp(-\beta B)\sum\nolimits_{\{X\}} \exp(-\beta zwX) \tag{C.5}$$

where the sum $\Sigma_{\{X\}}$ is over all configurations having zX type 1–2 pairs and f_j is the internal partition function of molecule j. Equivalently, we write

$$Q(T,N,V) = f_1^{N_1}f_2^{N_2}\exp(-\beta B)\sum\nolimits_X g(N_1,N_2,X)\exp(-\beta zwX) \tag{C.6}$$

where the sum Σ_X is now over the number of configurations having zX type 1–2 pairs with the degeneracy factor $g(N_1,N_2,X)$ for each configuration labeled with (N_1,N_2,X).

In the zeroth approximation, we consider that molecules are distributed randomly over the lattice despite their interaction with others, which we restrict to their z nearest neighbors. For the number of type 1–2 pairs zX, we get

$$zX = N_1 \cdot z(N_2/N) = N_1 \cdot z[(N - N_1)/N] = zN_1 - zN_1^2/N = zNx_1x_2 \tag{C.7}$$

The next step is to calculate $g(N_1,N_2,X)$. Considering a random distribution of N_1 molecules of type 1 and N_2 of type 2 over N cells, we have

$$g(N_1,N_2,X) = N!/(N_1!N_2!) \tag{C.8}$$

Therefore, the partition function becomes

$$Q(T,N,V) = f_1^{N_1}f_2^{N_2}\exp(-\beta B)[N!/(N_1!N_2!)]\exp(-\beta Nzwx_1x_2) \tag{C.9}$$

so that for the Helmholtz energy the result is

$$F = -kT\ln Q = -kT\ln\left(f_1^{N_1}f_2^{N_2}\right) + B - kT\ln(N!/N_1!N_2!) + Nzwx_1x_2 \tag{C.10}$$

$$= -kT\ln\left(f_1^{N_1}f_2^{N_2}\right) + B + NkT(x_1\ln x_1 + x_2\ln x_2) + Nzwx_1x_2 \tag{C.11}$$

Consequently, the mixing Helmholtz energy reads, assuming that f_j is not changing,

$$\Delta_{\text{mix}}F = Nzwx_1x_2 + NkT(x_1\ln x_1 + x_2\ln x_2) \tag{C.12}$$

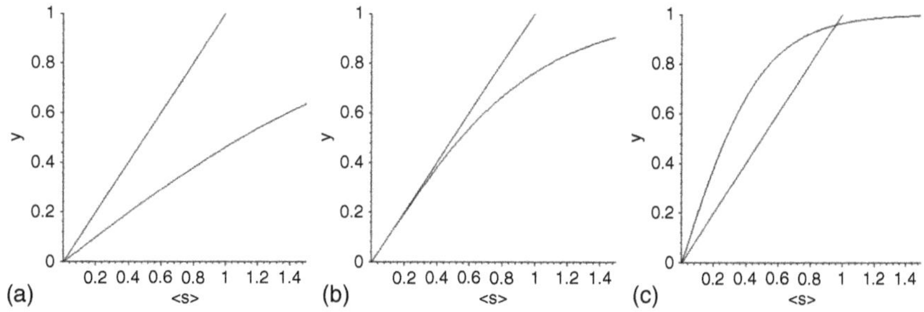

Figure C.1 The curves $y = s$ and $y = \tanh(as)$ for (a) $a = 0.5$, (b) $a = 1.0$, and (c) $a = 2.0$, showing for the zeroth-order approximation, respectively, the homogeneous state ($T > T_{cri}$), the critical state ($T = T_{cri}$), and the phase-separated state ($T < T_{cri}$). All curves are plotted for $s \geq 0$ only.

As a check, we calculate the internal energy of mixing $\Delta_{mix}U$ energy from $\Delta_{mix}F$ via $U = \partial(F/T)/\partial(1/T)$, which leads, as expected, to

$$\Delta_{mix}U = Nzwx_1x_2 \tag{C.13}$$

The coexistence line between a phase-separated system and a homogeneous system can be calculated from $\partial\Delta_{mix}G/\partial x_2 = \partial\Delta_{mix}F/\partial x_2 = 0$. This leads to

$$\ln(x_2/x_1) = \beta\, zw(2x_2 - 1) \tag{C.14}$$

Using the transformation $s = 2x_2 - 1$, the final expression is

$$\beta zw = s^{-1}\ln[(1+s)/(1-s)] = 2s^{-1}\tanh^{-1}s \quad \text{or} \quad s = \tanh(½\beta zws) \tag{C.15}$$

The *critical temperature* T_{cri} is the temperature where the molecules start to order, and we determine T_{cri} by considering that $\tanh(½\beta zws)$ is a continuously increasing function of s passing through $s = 0$. Hence, there is always the solution $s = 0$ but for $d[\tanh(½\beta zws)]/ds|_{s=0} > 1$, two other solutions occur (Figure C.1). Because this derivative equals $½\beta zw$, we find $T_{cri} = zw/2k$. The composition as a function of T can be found by solving Eq. (C.15), either numerically or graphically (Figure C.1a).

The associated heat capacity C_V can be found from $C_V = \partial U/\partial T$, with $U = -Q_N{}^{-1}\partial Q_N/\partial\beta$ resulting in (after some calculation and plotted in Figure C.2b)

$$C_V = \frac{2Nks^2(T_{cri}/T)^2}{\cosh^2(sT_{cri}/T) - (T_{cri}/T)} \tag{C.16}$$

Note that C_V shows a finite jump at $T = T_{cri}$ with a maximum of $3Nk$.

C.3 The First or Quasi-Chemical Approximation

Improvement over the zeroth approximation is possible by randomly distributing pairs of molecules over the lattice instead of single molecules. This solution is known as the *first* or *quasi-chemical approximation*.[2]

2 Also denoted as the Bethe–Peierls approximation. For further details, see Refs. [7] or [10].

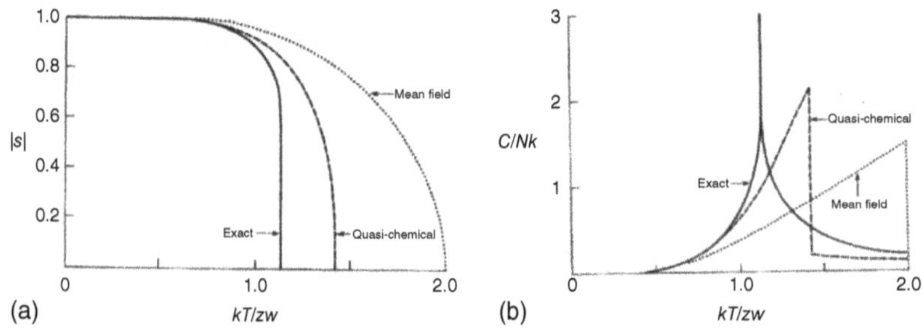

Figure C.2 The composition $|s| = |2x_2 - 1|$ where phase separation occurs (a) and the associated heat capacity C/Nk (b) as a function of kT/zw for the 2D square lattice, showing the zeroth order (mean field) approximation, the first order (quasi-chemical) approximation, and the exact solution.

Let us consider an independent distribution of pairs 1–1, 1–2, 2–1, and 2–2 over the lattice. Any such distribution is characterized by the numbers N_1, N_2, and $N_{12} = zX$, and this leads in the usual way to the number of possible configurations:

$$h(N_1, N_2, X) \equiv \frac{(\tfrac{1}{2}zN)!}{[\tfrac{1}{2}z(N_1 - X)]!(\tfrac{1}{2}zX)!(\tfrac{1}{2}zX)![\tfrac{1}{2}z(N_2 - X)]!} \tag{C.17}$$

where we have used a separate entry for the 1–2 and 2–1 pairs in order to avoid the use of a symmetry number. This result cannot be correct, as the sum over all configurations zX should yield $N!/N_1!N_2!$. However, we can remedy this defect – at least approximately – by introducing a normalization factor $c(N_1,N_2)$, and write

$$g(N_1, N_2, X) = c(N_1, N_2)h(N_1, N_2, X) \qquad \text{so that} \tag{C.18}$$

$$G(N_1, N_2) \equiv \sum\nolimits_X g(N_1, N_2, X) = c(N_1, N_2)\sum\nolimits_X h(N_1, N_2, X) \tag{C.19}$$

We evaluate the sum G by taking its largest term (Justification 7.4) and this term, labeled X^*, is obtained by setting $\partial G/\partial X = 0$. The result is

$$X^{*2} = (N_1 - X^*)(N_2 - X^*) \quad \text{or} \quad X^* = N_1N_2/(N_1 + N_2) \tag{C.20}$$

This implies that the sum G for X^*, labeled $G^*(N_1,N_2)$, becomes

$$G^*(N_1, N_2) = \frac{c(N_1, N_2)(\tfrac{1}{2}zN)!}{[\tfrac{1}{2}z(N_1 - X^*)]!(\tfrac{1}{2}zX^*)!(\tfrac{1}{2}zX^*)![\tfrac{1}{2}z(N_2 - X^*)]!} \tag{C.21}$$

If we sum $g(N_1,N_2,X)$ over all values of X, we must regain the total number of ways of placing N_1 molecules of type 1 and N_2 of type 2 over $N = N_1 + N_2$ sites. In other words, we also have $G = N!/N_1!N_2!$, and we obtain after quite some algebra

$$c(N_1, N_2) = G(N_1, N_2)/\sum\nolimits_X h(N_1, N_2, X) = N!/N_1!N_2!h(N_1, N_2, X^*) \qquad \text{or} \tag{C.22}$$

$$c(N_1, N_2) = \left(\frac{N!}{N_1!N_2!}\right)^{1-z} \quad \text{and} \quad g(N_1, N_2, X) = \left(\frac{N!}{N_1!N_2!}\right)^{1-z} h(N_1, N_2, X) \tag{C.23}$$

The configurational partition function Q is still given by Eq. (C.6). The maximum term trick is now played again and hence we seek the maximum term of Q, labeled $\overline{X}$, by setting

$\partial Q/\partial X = 0$, or equivalently, by setting $\partial \ln Q/\partial X = 0$. After a straightforward calculation, we obtain

$$\overline{X}^2 = (N_1 - \overline{X})(N_2 - \overline{X})\exp(-2\beta w) \tag{C.24}$$

For an explicit solution, we define $\eta^2 \equiv \exp(2\beta w)$ and the parameters α and δ by

$$\overline{X} = \frac{N_1 N_2}{N_1 + N_2}\frac{2}{\alpha + 1} \quad \text{and} \quad \delta = ½(\alpha - 1) \tag{C.25}$$

so that we can transform Eq. (C.24) to $\alpha^2 - (1 - 4x_1x_2) = [\alpha^2 - (1 - 2x_2)^2] = 4\eta^2 x_1 x_2$, which has the solution $\alpha = [1 + 4x_1x_2(\eta^2-1)]^{1/2}$. The energy of mixing becomes $\Delta_{\text{mix}}U = 2x_1x_2zwN/(\alpha + 1)$. For $\alpha = 1$ (or $\delta = 0$), we regain the zeroth approximation.

The configurational Helmholtz energy F is given by $F = -kT\ln Q$ and becomes

$$\begin{aligned} F &= -kT \ln g(N_1, N_2, \overline{X}) + (½z\varepsilon_{11}N_1 + ½z\varepsilon_{22}N_2 + zw\overline{X}) \\ &= -kT \ln\left[\frac{N!h(N_1, N_2, \overline{X})}{N_1!N_2!h(N_1, N_2, X^*)}\right] + (½z\varepsilon_{11}N_1 + ½z\varepsilon_{22}N_2 + zw\overline{X}) \end{aligned} \tag{C.26}$$

The evaluation of F is complex, and it is convenient to calculate first the chemical potential $\mu_j = \partial F/\partial N_j$. If this calculation is handled without thinking, it is a straightforward but tedious task. However, $\overline{X}$ is obtained from $\partial F/\partial \overline{X} = 0$, implying that all terms resulting from $\partial F/\partial \overline{X}$ taken together cancel. Moreover, X^* is obtained from $\partial \ln g/\partial X = 0$, implying that all terms from $\partial \ln g/\partial X^*$ taken together also cancel. This renders the differentiation of F a relatively simple task, leading to

$$\beta\mu_j = \beta\partial F/\partial N_j = \ln\left(\frac{N_j}{N_1 + N_2}\right) + ½z \ln\left(\frac{N_j - \overline{X}}{N_j - X^*}\right) + ½z\varepsilon_{jj} \tag{C.27}$$

To obtain $\mu_j - \mu_j{}^\circ$, we realize that $\mu_j{}^\circ$ is obtained from Eq. (C.26), again via $\mu_j = \partial F/\partial N_j$, but now with $N_2 = 0$, leading to $\mu_j{}^\circ = ½z\varepsilon_{jj}$. Writing $F = N_1\mu_1 + N_2\mu_2 = N(x_1\mu_1 + x_2\mu_2)$, we obtain for the molar Helmholtz energy of mixing

$$\frac{\Delta_{\text{mix}}F_{\text{m}}}{RT} = x_1 \ln x_1 + x_2 \ln x_2 + ½z\left[x_1 \ln\left(\frac{N_1 - \overline{X}}{N_1 - X^*}\right) + x_2 \ln\left(\frac{N_2 - \overline{X}}{N_2 - X^*}\right)\right] \tag{C.28}$$

Using the absolute activity $\lambda_j = \exp(\beta\mu_j)$ and, as before, assuming that the activity can be replaced by the partial pressure, the activity coefficient γ_j becomes

$$\lambda_j/(\lambda_j{}^\circ x_j) = P_j/(P_j{}^\circ x_j) = \gamma_j = [(N_j - \overline{X})/(N_j - X^*)]^{½z} \tag{C.29}$$

In all these expressions X^* and $\overline{X}$ are given by Eqs. (C.20) and (C.25), respectively.

The above expression for $\Delta_{\text{mix}}F_{\text{m}}$ can be made more explicit by writing the various contributions in terms of α, $x_1 = N_1/(N_1 + N_2)$ and $x_2 = N_2/(N_1 + N_2)$. We find

$$\overline{X}/N = 2x_1x_2/(\alpha + 1) \qquad X^*/N = x_1x_2 \tag{C.30}$$

$$(N_1 - \overline{X})/N = x_1(\alpha + 1 - 2x_2)/(\alpha + 1) \qquad (N_1 - X^*)/N = x_1^2 \tag{C.31}$$

$$(N_2 - \overline{X})/N = x_2(\alpha - 1 + 2x_2)/(\alpha + 1) \qquad (N_2 - X^*)/N = x_2^2 \tag{C.32}$$

so that $\Delta_{mix}F_m$ in terms of α and δ becomes

$$\frac{\Delta_{mix}F_m}{RT} = x_1 \ln x_1 + x_2 \ln x_2 + \tfrac{1}{2}z\left[x_1 \ln\left(\frac{\alpha+1-2x_2}{x_1(\alpha+1)}\right) + x_2 \ln\left(\frac{\alpha-1+2x_2}{x_2(\alpha+1)}\right)\right] \quad \text{and}$$

$$\frac{\Delta_{mix}F_m}{RT} = x_1 \ln x_1 + x_2 \ln x_2 + \tfrac{1}{2}z\left[x_1 \ln\left(\frac{1-x_2-\delta}{(1-x_2)(1+\delta)}\right) + x_2 \ln\left(\frac{x_2+\delta}{x_2(1+\delta)}\right)\right] \quad (C.33)$$

A similar substitution can be made for $\lambda_j = (\lambda_j^\circ/P_j^\circ)P_j$ and P_j, leading to

$$\frac{P_1}{P_1^\circ x_1} = \gamma_1 = \left[\frac{\alpha+1-2x_2}{x_1(\alpha+1)}\right]^{1/2z} \quad \text{and} \quad \frac{P_2}{P_2^\circ x_2} = \gamma_2 = \left[\frac{\alpha-1+2x_2}{x_2(\alpha+1)}\right]^{1/2z} \quad (C.34)$$

The critical point, due to symmetry, still occurs at $x_1 = x_2 = ½$. To obtain the expression for T_{cri}, we first derive the expression for the coexistence line, determined by $P_1/P_1^\circ = P_2/P_2^\circ$. Upon substitution of the expression for P_j/P_j°, we obtain

$$(\alpha-1+2x_2)/(\alpha+1-2x_2) = (x_2/x_1)^{(z-2)/z} \equiv r^{(z-2)/z} \equiv \gamma \quad (C.35)$$

where the *molecular ratio* $r \equiv x_2/x_1$ and the abbreviation γ are introduced. Solving for α yields $\alpha/(1-2x_2) = (1+\gamma)/(1-\gamma)$, and thus leads to

$$\alpha-1+2x_2 = 2\gamma(1-2x_2)/(1-\gamma) \quad \text{and} \quad \alpha+1-2x_2 = 2(1-2x_2)/(1-\gamma) \quad (C.36)$$

Multiplying these two expressions leads to

$$\alpha^2-(1-2x_2)^2 = 4\gamma(1-2x_2)^2/(1-\gamma)^2 \quad (C.37)$$

Comparing with the transform of Eq. (C.24), $\alpha^2-(1-2x_2)^2 = 4\eta^2 x_1x_2$, we obtain

$$\eta = (\gamma/x_1x_2)^{1/2}(1-2x_2)/(1-\gamma) \quad \text{or} \quad \eta = e^{\beta w} = (1-r)/(r^{1/z}-r^{(z-1)/z}) \quad (C.38)$$

Note that if $r = r_1$ is a solution of Eq. (C.38), the other one is $r_2 = 1/r_1$.

The critical temperature T_{cri} is now obtained by putting $r = 1+\xi$ in Eq. (C.38), expanding in powers of ξ and taking the limit $\xi \to 0$. We thus obtain

$$\eta_{cri} = \exp(\beta_{cri}w) = z/(z-2) \quad \text{or} \quad zw/kT_{cri} = z\ln[z/(z-2)] \quad (C.39)$$

For the SC lattice ($z = 6$) we have $zw/kT_{cri} = 2.433$, while for the BCC lattice ($z = 8$) and FCC lattice ($z = 12$) we obtain $zw/kT_{cri} = 2.301$ and $zw/kT_{cri} = 2.188$, respectively. Note that if we take $z = \infty$, we regain the zeroth approximation $zw/kT_{cri} = 2$. Provided that $\eta^2 < 1$, we may expand the logarithm in Eq. (C.33) in terms of δ, then δ in terms of η^2-1, and thereafter η^2 in terms of $2\beta w$, leading to $F^E/N = x_1x_2zw[1 - ½x_1x_2 \cdot 2\beta w + \mathcal{O}(2\beta w)^2]$. The maximum effect occurs for $x_1 = x_2 = 0.5$ near T_{cri}. For the FCC lattice with $\beta_{cri}w = 0.365$, this leads to a not quite negligible correction of ≅5%.

Calculating, as before, the internal energy via $U = \partial(F/T)/\partial(1/T)$ is considerably more complex as for the zeroth approximation and leads to

$$\Delta_{mix}U = 2Nzwx_1x_2/(\alpha+1) \quad (C.40)$$

So far, w was assumed to be temperature independent, but taking w as a temperature-dependent parameter presents no specific problems, and by using the definition $u \equiv w - T(\partial w/\partial T)$, one can show that in this case one obtains $\Delta_{mix}U = 2Nzux_1x_2/(\alpha+1)$.

As stated in Section C.1, only a (partial) exact solution exists for 2D lattices. For a square lattice, this solution reads $zw/kT_{cri} = 3.523$, to be compared with $zw/kT_{cri} = 2.773$ for the first and $zw/kT_{\text{cri}} = 2.00$ for the zeroth approximation. The results from the exact solution for s and C_V are shown, as are the results for the zeroth and first approximations, in Figure C.2b. From these plots it becomes clear that considerable improvement is obtained by using the first approximation, but also that the exact solution is still not approached. Finally, the higher the number of dimensions, the better the mean-field solution (Section 15.4), which implies that the situation for 3D is somewhat better than that for 2D.

C.4 Athermal Entropy for Chain-Like Molecules

So far, we have discussed the lattice gas model for components of equal size, so that each molecule occupies only one lattice site. For chain-like molecules, like polymers, the difference in size with solvent molecules is large, and to consider that we use a reference volume V_0 representing the molar volume of a lattice site and a molecular reference volume $v_0 = V_0/N_A$. Assuming no volume change, for a volume V_j of component j the volume of a solution is $V = V_1 + V_2$, and the number of sites is $N = V/v_0$. Of these sites a volume fraction $\phi_j = V_j/V$ is occupied by component j. Now, consider a molecule which, in terms of an *equivalent chain*, consists of r segments. If the molecular volume v_j of component j is given by $v_j = V_j/N_A$, the *degree of polymerization* (DoP) is defined by $r_j = v_j/v_0 = V_j/V_0 = M_j/\rho_j V_0$, where M_j is the molar mass, and ρ_j and V_j, respectively, are the density and volume in the amorphous state at the solution temperature. Because $v_j = r_j v_0$, the number of molecules N_j of component j is $N_j = V_j/v_j = N\phi_j/r_j$. Hence, on the lattice, r_1N_1 sites are occupied by solvent molecules and r_2N_2 sites by polymer molecules, and the total number of lattice sites is $N = r_1N_1 + r_2N_2$. When a molecule consisting of connected segments is positioned on a lattice, each segment is connected either *internally* (to a segment of the same molecule) or *externally* (to a segment of another molecule). Each molecule j thus occupies r_j sites, each site having z neighbors. The number of pairs of neighboring sites of which one site is occupied by molecule j, and the other is not, is indicated by zq_j. The parameter q_j thus represents the molecular area, similarly as r_j represents the molecular volume. For *open-chain molecules*, that is, molecules containing no rings and no nonbonded internal contacts, r_j and q_j are related via $zq_j = r_j(z-2)+2$. A solvent molecule is denoted as a monomer, and a polymer molecule containing r segments as an r-mer. We limit the discussion to solutions, indicating the number of monomers by N_1 and the number of r-mers by N_r, so that $r_1 = q_1 = 1$ and relabel r_2 to r and q_2 to q. Moreover, we use *volume fractions* $\phi_1 = 1-\phi = N_1/(N_1 + rN_r)$ and $\phi_2 = \phi = rN_r/(N_1 + rN_r)$, and *surface fractions* $\theta_1 = 1-\theta = N_1/(N_1 + qN_r)$ and $\theta_2 = \theta = qN_r/(N_1 + qN_r)$.

Assuming that all configurations have the same energy and thus the arrangement of molecules is random, we define α as the ratio of the probability that a group of sites is wholly occupied by an r-mer to the probability that the group is occupied only with monomers. The quantity α is key as it relates on the one hand to thermodynamic expressions, such as ΔF, and on the other hand to the composition, characterized by either N_r/N_1 or ϕ. Using grand partition functions, we obtain $\alpha = \lambda_r f_r/(\lambda_1 f_1)^r$ with the activity $\lambda = \exp(\beta\mu)$. At constant T and P, we have for a lattice without expansion $\text{d}G = \text{d}F = \mu_1\text{d}N_1 + \mu_r\text{d}N_r$ or, equivalently, $\beta\text{d}F = \ln\lambda_1\text{d}N_1 + \ln\lambda_r\text{d}N_r$. Changing variables from (N_1,N_r) to (N_s,ϕ), where $N_s = N_1 + rN_r$ denotes the total number of sites, we obtain $N_1 = (1-\phi)N_s$ and $N_r = (\phi/r)N_s$. Hence, $\beta\text{d}F = [(1-\phi)\ln\lambda_1 + \phi r^{-1}\ln\lambda_r]\text{d}N_s + N_s(-\ln\lambda_1 + r^{-1}\ln\lambda_r)\text{d}\phi$, reducing to, as $\text{d}N_s = 0$,

$\beta \mathrm{d}F = (N_s/r)\ln[\lambda_r/(\lambda_1)^r]\mathrm{d}\phi$. Inserting α results in $\beta \mathrm{d}F = r^{-1}\ln[\alpha(f_1)^r/f_r]\mathrm{d}\phi$. The Helmholtz energy of mixing for N_1 monomers and N_r r-mers, $\Delta F = F(N_s,\phi) - (1-\phi)F(N_s,0) - \phi F(N_s,1)$, becomes

$$\beta \Delta_{\text{mix}} F = N_s \left[r^{-1} \int_0^\phi \ln \alpha \, \mathrm{d}\phi - (\phi/r) \int_0^1 \ln \alpha \, \mathrm{d}\phi \right] \tag{C.41}$$

The total amount of molecules being $(1-\phi+\phi/r)N_s$, we obtain for the molar Helmholtz energy of mixing, recalling that we consider athermal mixtures ($\Delta_{\text{mix}}U = 0$),

$$\Delta_{\text{mix}} F_{\text{m}}/RT = -\Delta_{\text{mix}} S_{\text{m}}/R = [r-(r-1)\phi]^{-1} \left(\int_0^\phi \ln \alpha \, \mathrm{d}\phi - \phi \int_0^1 \ln \alpha \, \mathrm{d}\phi \right) \tag{C.42}$$

Using the degeneracy factor $g(N_1,N_r)$ for putting N_1 monomers and N_r r-mers on the lattice, $-\beta F = \ln g(N_1,N_r) + N_1 \ln q_1 + N_r \ln q_r$, integrating $\beta \mathrm{d}F$ and comparing, we obtain

$$\ln g(N_1,N_r) - \ln g(N_s,0) = \ln g(N_1,N_r) = -N_s r^{-1} \int_0^\phi \ln \alpha \, \mathrm{d}\phi \tag{C.43}$$

where the first step can be made since for $N_r = 0$, the lattice is fully covered with monomers and therefore $g(N_s,0) = 1$.

We now turn to an approximate calculation for $g(N_1,N_r)$ for open-chain molecules. The number of ways in which a polymer can be placed after one of its segments has been placed is denoted by ρ. For a monomer ($r = 1$), $\rho = 1$, for a dimer ($r = 2$), $\rho = z$, and for a flexible trimer ($r = 3$), $\rho = z(z-1)$. For more than three segments, the maximum value for flexible molecules is $\rho \cong z(z-1)^{r-2}$. This expression is somewhat in error as it includes also configurations for which two segments occupy the same site; however, eventually, knowledge of ρ is not required. The probability of a site being occupied by a monomer is $f(\text{M}) = 1-\phi$ and for an r-mer is $f(\text{R}) = \phi$. If a site is occupied by a monomer, the probabilities that a neighboring site is occupied by another monomer or an r-mer are as zN_1 to zqN_r, or $f(\text{MM}) = (1-\phi)(1-\theta)$ and $f(\text{MR}) = (1-\phi)\theta$. Similarly, $f(\text{RM}) = \theta(1-\theta)$, $f(\text{R}_1\text{R}_2) = \theta^2$ if the two sites are occupied by different r-mers and $f(\text{R}_1\text{R}_1) = (r-q)\theta/r$ if they are occupied by the same r-mer. Now consider a set of r sites congruent with the r-mer. The probability that these r sites are occupied by monomers is $f_1 = (1-\phi)(1-\theta)^{r-1}$. The probability of one of these sites to be occupied by any segment of the r-mer is ϕ, so that the probability of that site to be occupied by a particular segment is ϕ/r. Hence, the probability that the set of r sites is occupied by a single r-mer is $f_r = (\sigma/\rho)(\phi/r)$, with ρ as given before and σ the symmetry number of the set. Therefore, we obtain

$$\alpha = \frac{f_r}{f_1} = \frac{(\sigma/\rho)(\phi/r)}{(1-\phi)\theta^{r-1}} = \frac{\sigma}{r\rho} \frac{\phi}{1-\phi} \left[\frac{1-(r-q)\phi/r}{1-\phi} \right]^{r-1} \tag{C.44}$$

Inserting α in $\Delta_{\text{mix}} S_{\text{m}}/R$ and changing back to (N_1,N_r) yields, after quite some algebra,

$$-(N_1+N_r)\Delta_{\text{mix}} S_{\text{m}} = \ln \frac{(N_1)!(rN_r)!}{(N_1+rN_r)!} + \frac{r-1}{r-q} \ln \frac{(N_1+rN_r)!(qN_r)!}{(N_1+qN_r)!(rN_r)!} \tag{C.45}$$

Differentiation with respect to N_1 and N_r, using $\mathrm{d}N! = \ln N \, \mathrm{d}N$, leads to

$$\frac{\lambda_1}{\lambda_1{}^\circ} = \frac{N_1}{N_1+rN_r} \left(\frac{N_1+rN_r}{N_1+qN_r} \right)^{\frac{(r-1)}{(r-q)}} \quad \text{and} \quad \frac{\lambda_r}{\lambda_r{}^\circ} = \frac{N_r}{N_1+rN_r} \left(\frac{N_1/r+N_r}{N_1/q+N_r} \right)^{q\frac{(r-1)}{(r-q)}} \tag{C.46}$$

Evaluating Eq. (C.46), meanwhile reverting to fractions and substituting the expression $(r-1)/(r-q) = z/2$, leads, again after some algebra, to the expression first derived by Guggenheim:

$$\frac{\Delta_{\text{mix}}S_{\text{GS}}}{R} = -N_1 \ln(1-\phi) - N_r \ln\phi - \tfrac{1}{2}z\left(N_1 \ln\frac{1-\theta}{1-\phi} + qN_r \ln\frac{\theta}{\phi}\right) \tag{C.47}$$

A simplification, known as the *Flory–Huggins entropy*, is obtained if $q \sim r$, leading to

$$\Delta_{\text{mix}}S_{\text{FH}}/R = (1-x)\ln(1-\phi) + x\ln\phi \tag{C.48}$$

Finally, by substituting Eq. (C.45) in Eq. (C.44), one obtains after some calculation

$$g(N_1, N_r) = \left(\frac{\rho}{\sigma}\right)^{N_r} \frac{(N_1 + rN_r)!}{(N_1)!(N_r)!} \left[\frac{(N_1 + qN_r)!}{(N_1 + rN_r)!}\right]^{(r-1)/(r-q)} \tag{C.49}$$

Staverman [11] indicated that f_1, written as $f_1 = (1-\phi)^r[(1-\theta)/(1-\phi)]^{r-1}$, can be interpreted as follows. The probability that one site is occupied by a monomer is $1-\phi$ and the probability that r sites are occupied by r monomers is $(1-\phi)^r$. The factor $[(1-\theta)/(1-\phi)]^{-1}$ corrects for the $r-1$ links in open-chain molecules. It is thus reasonable to write for molecules with l links, whether internally bonded or nonbonded contacts, $f_1 = (1-\phi)^r[(1-\theta)/(1-\phi)]^l$ and thus formally Eqs. (C.46) and (C.47) remain valid if the r-mer contains rings or nonbonded contacts if one considers r and q as independent parameters. Hence, Eq. (C.47) is known as the *Guggenheim–Staverman entropy*.

References

1. Yang, C.N. and Lee, T.D. (1952). *Phys. Rev.* 87: 404 and 410.
2. See e.g., Huang, K. *Statistical Mechanics*, 2e. New York: Wiley.
3. Brush, S.G. (1967). *Rev. Mod. Phys.* 39: 883.
4. Frenkel, J. (1946). *Kinetic Theory of Liquids*. Oxford: Oxford University Press (also Dover, 1953).
5. Eyring, H. (1936). *J. Chem. Phys.* 4: 283; Cernuschi, F. and Eyring, H. (1939). *J. Chem. Phys.* 7: 547.
6. Onsager, L. (1944). *Phys. Rev.* 65: 117.
7. Guggenheim, E.A. (1944). *Proc. Roy. Soc. A* 183: 203; (1952). *Mixtures*. Oxford: Clarendon Press.
8. Tompa, H. (1956). *Polymer Solutions*. London: Butterworth.
9. Benigni, P. (2020). *J. Non-cryst. Sol.* 534: 119942.
10. Fowler, R.H. and Guggenheim, E.A. (1939). *Statistical Thermodynamics*. London: Cambridge University Press.
11. Staverman, A.J. (1950). *Rec. Trav. Chim. Pays-Bas* 69: 163.

Further Reading

Teraoka, I. (2002). *Polymer Solutions*. New York: John Wiley & Sons.

Kamide, K. and Dobashi, T. (2000). *Physical Chemistry of Polymer Solutions*. Amsterdam: Elsevier.

Appendix D

Elements of Electrostatics

In this appendix, we briefly review some aspects of electrostatics that are relevant to a number of discussions in this book. Free use has been made of the literature indicated in the section "Further Reading."

D.1 Coulomb, Gauss, Poisson, and Laplace

In vacuum, a charge q_1 at a distance $r = |\boldsymbol{r}|$ of another charge q_2 experiences a force $\boldsymbol{f}$ given by *Coulomb's law*, using $\mathscr{C} \equiv (4\pi\varepsilon_0)^{-1}$ with ε_0 the permittivity of vacuum,

$$\boldsymbol{f} = \mathscr{C}\, q_1 q_2 \boldsymbol{r}/r^3 \tag{D.1}$$

The *field* $\boldsymbol{E}$ is the force a unit test charge experiences from a charge q at position $\boldsymbol{r}$ so that we have

$$\boldsymbol{E} = \mathscr{C}\, q\boldsymbol{r}/r^3 \tag{D.2}$$

The energy increment $\mathrm{d}w_{12}(r)$ of a charge q_1 in the field of a charge q_2 is given by $\mathrm{d}w_{12}(r) = -q_1 \boldsymbol{E}_2 \cdot \mathrm{d}\boldsymbol{r}$ and accordingly the (potential or) *Coulomb energy* reads

$$w_{12}(r) = -q_1 \int_\infty^r \boldsymbol{E}_2 \cdot \mathrm{d}\boldsymbol{r} = -\mathscr{C}\, q_1 q_2 \int_\infty^r r^{-3}\boldsymbol{r} \cdot \mathrm{d}\boldsymbol{r} = \mathscr{C}\, q_1 q_2 r^{-1} \equiv q_1 \phi_2 \tag{D.3}$$

where the *potential* $\phi = \mathscr{C} q/r$ – the potential energy per unit charge as defined by $\boldsymbol{E} \equiv -\nabla\phi$ – is introduced. As conventional, the force is thus $\boldsymbol{f} = -\nabla w$.

A pair of opposite and equal charges, q_1 and $q_2 = -q_1$, separated by a (small) vector $\boldsymbol{d}$ pointing from q_2 to q_1, is called a *dipole*, and characterized by the dipole moment $\boldsymbol{\mu} = q_1\boldsymbol{d}$. The potential of such a dipole fixed in space at large distance r from the center of the dipole is (Figure D.1)

$$\phi = \phi_1 + \phi_2 = \mathscr{C} q_1/|\boldsymbol{r} - ½\,\boldsymbol{d}| + \mathscr{C} q_2/|\boldsymbol{r} + ½\,\boldsymbol{d}| \cong \mathscr{C}\,\boldsymbol{\mu} \cdot \boldsymbol{r}/r^3 \tag{D.4}$$

The field $\boldsymbol{E}$ is then given by

$$\boldsymbol{E} = -\nabla\phi = -\mathscr{C}\,\boldsymbol{\mu} \cdot \nabla\frac{\boldsymbol{r}}{r^3} = -\frac{\mathscr{C}\,\boldsymbol{\mu}}{r^3}\nabla\boldsymbol{r} - \boldsymbol{\mu} \cdot \boldsymbol{r}\nabla\frac{1}{r^3} = \frac{\mathscr{C}}{r^5}\,[3(\boldsymbol{\mu} \cdot \boldsymbol{r})\boldsymbol{r} - (\boldsymbol{r} \cdot \boldsymbol{r})\boldsymbol{\mu}] \tag{D.5}$$

The Coulomb energy of a dipole in a field with potential ϕ can be calculated to be $w = \boldsymbol{\mu} \cdot \nabla\phi = -\boldsymbol{\mu} \cdot \boldsymbol{E}$.

Phases of Matter and their Transitions: Concepts and Principles for Chemists, Physicists, Engineers, and Materials Scientists, First Edition. Gijsbertus de With.

Figure D.1 Schematic for the potential of a dipole $\mu = qd$ at distance r.

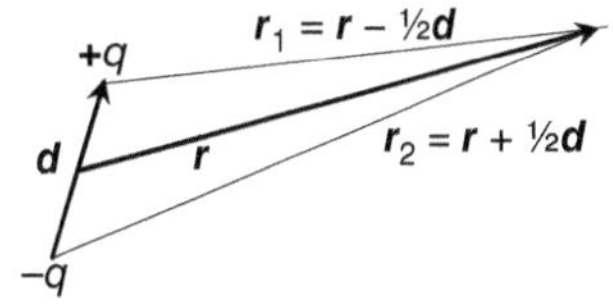

Consider now an arbitrary closed surface S with surface elements $\mathrm{d}\boldsymbol{S} = \boldsymbol{n}\mathrm{d}S$, located at position $\boldsymbol{r}$ from the origin, where $\boldsymbol{n}$ is the positive outward normal of the surface element and $\mathrm{d}S$ its area. Because the scalar product $\boldsymbol{r}\cdot\mathrm{d}\boldsymbol{S}/r$ is the projection of $\mathrm{d}S$ on a sphere of radius $\boldsymbol{r}$, the product $\boldsymbol{r}\cdot\mathrm{d}\boldsymbol{S}/r^3$ is the projection on a unit sphere centered at the origin. Hence, if S encloses the origin, we have

$$\int \boldsymbol{E}\cdot\mathrm{d}\boldsymbol{S} = -\int \nabla\phi\cdot\mathrm{d}\boldsymbol{S} = -\int \mathscr{C}\,q\nabla\frac{1}{r}\cdot\mathrm{d}\boldsymbol{S} = \mathscr{C}\,q\int\frac{1}{r^3}\boldsymbol{r}\cdot\mathrm{d}\boldsymbol{S}$$

$$= \mathscr{C}\,q\int\frac{1}{r^3}\boldsymbol{r}\cdot\boldsymbol{n}\mathrm{d}S = \mathscr{C}\,q\int_0^1\frac{r}{r^3}\,4\pi r^2\mathrm{d}r = 4\pi\mathscr{C}\,q = \frac{q}{\varepsilon_0} \tag{D.6}$$

If S does not enclose the origin, $\boldsymbol{r}$ cuts S at two points, and the contributions from these two intersections cancel and the integral is zero. It follows that the integral $\int\boldsymbol{E}\cdot\mathrm{d}\mathbf{S}$ is q/ε_0 if S does contain q and zero otherwise. For a general charge distribution ρ within the closed surface S having a volume V, we obtain *Gauss' law*:

$$\int \boldsymbol{E}\cdot\mathrm{d}\boldsymbol{S} = \varepsilon_0^{-1}\int\rho\,\mathrm{d}V \tag{D.7}$$

In order to go from this global description to a local one, we use the Gauss divergence theorem and find *Poisson's equation*:

$$\int \boldsymbol{E}\cdot\mathrm{d}\boldsymbol{S} = \int\nabla\cdot\boldsymbol{E}\mathrm{d}V = \varepsilon_0^{-1}\int\rho\,\mathrm{d}V \quad \text{or} \quad \nabla\cdot\boldsymbol{E} = \varepsilon_0^{-1}\rho \tag{D.8}$$

Using $\boldsymbol{E} = -\nabla\phi$ we obtain

$$\nabla^2\phi = -\varepsilon_0^{-1}\rho \tag{D.9}$$

reducing to *Laplace's equation* $\boldsymbol{\nabla}^2\phi = 0$ if $\rho = 0$, which is the usual case for a dielectric material. For a distribution of volume charge $\boldsymbol{\rho}$ and surface charge σ, the potential, using the dummy variable ξ, becomes

$$\phi(\boldsymbol{r}) = \mathscr{C}\int\frac{\rho(\xi)}{|\,\boldsymbol{r}-\xi\,|}\,\mathrm{d}V + \mathscr{C}\int\frac{\sigma(\xi)}{|\,\boldsymbol{r}-\xi\,|}\,\mathrm{d}S \tag{D.10}$$

Molecules are polarizable and therefore a field induces in a molecule a dipole moment $\boldsymbol{\mu}_{\text{ind}} = \alpha\boldsymbol{E}$, where the proportionality factor α is a material-dependent parameter called the *polarizability*. For macroscopic matter this occurs also, and we have the *polarization* $\boldsymbol{P} = \varepsilon_0\chi\boldsymbol{E}$ with the *susceptibility* χ. Since $\boldsymbol{P}$ represents a volumetric distribution of induced dipoles, we have for the associated potential

$$\phi(\boldsymbol{r}) = -\int\frac{\mathbf{P}(\xi)\cdot\boldsymbol{x}}{x^3}\,\mathrm{d}V = \int\mathbf{P}(\xi)\cdot\nabla x^{-1}\,\mathrm{d}V = \int\nabla\cdot[\mathbf{P}(\xi)r^{-1}]\,\mathrm{d}V - \int r^{-1}\nabla\cdot\boldsymbol{P}\,\mathrm{d}V$$

$$= \int x^{-1}\mathbf{P}(\xi)\cdot\mathrm{d}\boldsymbol{S} - \int x^{-1}\nabla\cdot\mathbf{P}(\xi)\,\mathrm{d}V \tag{D.11}$$

where $\boldsymbol{x} = \boldsymbol{r} - \xi$ and $x = |\boldsymbol{x}|$ is used. The second line states that the potential is analogous to one caused by a surface charge density $\boldsymbol{n}\cdot\boldsymbol{P}$ and a volume charge density $-\nabla\cdot\boldsymbol{P}$. Using

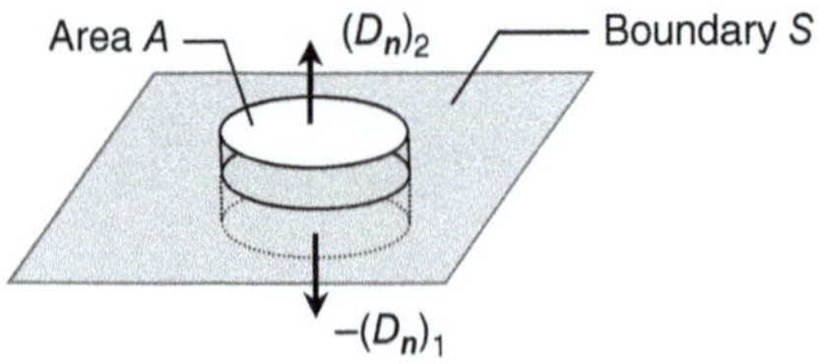

Figure D.2 The pillbox construction to derive the continuity of $\boldsymbol{D}$ at the boundary S.

the same reasoning as before, and going again from a global to a local description via the Gauss divergence theorem using the volume charge density $\rho - \nabla \cdot \boldsymbol{P}$, we have within the dielectric, since the surface charges do not enter Poisson's equation,

$$\nabla \cdot \boldsymbol{E} = \varepsilon_0^{-1}(\rho - \nabla \cdot \boldsymbol{P}) \quad \text{or} \quad \nabla \cdot (\varepsilon_0 \boldsymbol{E} + \boldsymbol{P}) = \rho \quad \text{or} \quad \nabla \cdot \boldsymbol{D} = \rho \tag{D.12}$$

with the *dielectric displacement* $\boldsymbol{D} \equiv \varepsilon_0 \boldsymbol{E} + \boldsymbol{P} = (1 + \chi)\varepsilon_0 \boldsymbol{E} \equiv \varepsilon_r \varepsilon_0 \boldsymbol{E} \equiv \varepsilon \boldsymbol{E}$, thus introducing the *relative permittivity* (or *dielectric constant*) $\varepsilon_r = 1 + \chi$, and *permittivity* $\varepsilon \equiv \varepsilon_r \varepsilon_0$. Outside the dielectric, $\boldsymbol{P} = 0$ and $\nabla \cdot \boldsymbol{D} = \rho$ reduces to $\varepsilon_0 \nabla \cdot \boldsymbol{E} = \rho$.

So, the dielectric displacement satisfies the condition $\nabla \cdot \boldsymbol{D} = \rho$, and $\varepsilon_0 \nabla \cdot \varepsilon_r \nabla \phi = -\rho$ is the governing equation for the resulting potential ϕ. From this equation, we can derive the requirement for the continuity of $\boldsymbol{D}$ across the boundary between media by writing the integral equivalent of $\nabla \cdot \boldsymbol{D} = \rho$ to obtain

$$\int \nabla \cdot \boldsymbol{D} \, \mathrm{d}V = \int \boldsymbol{D} \cdot \mathrm{d}\boldsymbol{S} = \int \rho \, \mathrm{d}V \tag{D.13}$$

Applying this relation to an infinitesimally small pillbox-shaped volume with a bottom and top surface with area A parallel to the boundary on each side of the boundary (Figure D.2), we have

$$A[(D_n)_2 - (D_n)_1] = \int \rho \, \mathrm{d}V \tag{D.14}$$

where D_n is the component of $\boldsymbol{D}$ normal to the boundary (with the direction from medium 1 to 2 as positive). If the charge density at the boundary remains finite, the integral tends to zero when volume of the pillbox V goes to zero. Hence,

$$(D_n)_2 - (D_n)_1 = 0 \tag{D.15}$$

If a surface charge is present, we obtain

$$(D_n)_2 - (D_n)_1 = \sigma \tag{D.16}$$

In conclusion, the normal component D_n of $\boldsymbol{D}$ across a boundary between two dielectrics is continuous in the absence of surface charge, but if a surface charge σ is present, this component jumps by σ.

D.2 A Dielectric Sphere in a Dielectric Matrix

To illustrate the use of the above theory, we now calculate the potential field and some associated relations outside (domain 1) and inside (domain 2) a sphere with radius a and relative permittivity ε_2 embedded in a matrix with relative permittivity ε_1 to which, by a set of external charges, a field $\boldsymbol{E}$ along the z-axis is applied that would be uniform if $\varepsilon_1 = \varepsilon_2$.

Generally, it is most convenient to use a coordinate system associated with the boundary conditions and for spherical entities these are the spherical coordinates r, θ, and ϕ defined in terms of the Cartesian coordinates x, y, and z by

$$x = r\sin\theta\cos\phi, \qquad y = r\sin\theta\sin\phi, \qquad z = r\cos\theta \tag{D.17}$$

In these coordinates, the Laplace operator, operating on a scalar, reads

$$\nabla^2(\cdot) = \frac{1}{r^2}\frac{\partial}{\partial r}\left(r^2\frac{\partial(\cdot)}{\partial r}\right) + \frac{1}{r\sin\theta}\frac{\partial}{\partial\theta}\left(\sin\theta\frac{\partial(\cdot)}{\partial\theta}\right) + \frac{1}{r^2\sin^2\theta}\frac{\partial^2(\cdot)}{\partial\phi^2} \tag{D.18}$$

Since we interested in an axially symmetric system, we have $\phi = \phi(r,\theta)$ and use a trial solution in the form $\phi(r,\theta) = R(r)\Theta(\theta)$ from which we obtain

$$\frac{1}{R}\frac{\partial}{\partial r}\left(r^2\frac{\partial R}{\partial r}\right) + \frac{1}{\Theta\sin\theta}\frac{\partial}{\partial\theta}\left(\sin\theta\frac{\partial\Theta}{\partial\theta}\right) = 0 \tag{D.19}$$

As the two parts involve different spatial variables, this expression can be only satisfied if each part equals a constant, say λ, and therefore

$$\frac{\partial}{\partial r}\left(r^2\frac{\partial R}{\partial r}\right) - \lambda R = 0 \quad \text{and} \quad \frac{1}{\sin\theta}\frac{\partial}{\partial\theta}\left(\sin\theta\frac{\partial\Theta}{\partial\theta}\right) + \lambda\Theta = 0 \tag{D.20}$$

By substituting $x = \cos\theta$ and $\Theta(\theta) = y(x)$, the equation for Θ transforms to *Legendre's equation*, of which the generic solutions are $y(x) = A_\nu P_\nu(x) + B_\nu Q_\nu(x)$, where P_ν and Q_ν are the Legendre functions of the first and second kind, respectively. The label ν is determined by $\lambda = \nu(\nu+1)$, and A_ν and B_ν are arbitrary constants. The full generic solution must be used when the z-axis ($x = \cos\theta = \pm 1$) is excluded from the region where the potential has to be calculated. In our case, however, the axis has to be included. It appears that the $P_\nu(x)$ for fractional ν and the $Q_\nu(x)$ for all values of ν have a singularity at $x = \pm 1$, that is, at $\theta = 0$ and $\theta = \pi$. The consequence is that all B_ν coefficients must be zero because all $Q_\nu(x)$ functions lead to singularities. Moreover, ν can only take integer values n, so $\lambda = n(n+1)$ and $\Theta(\theta) = \Sigma_n A_n P_n(\cos\theta)$. The first few Legendre polynomials $P_n(\cos\theta)$ are

$$P_0 = 1, \qquad P_1 = \cos\theta, \qquad P_2 = (3\cos^2\theta - 1)/2, \qquad P_3 = (5\cos^3\theta - 3\cos\theta)/2 \tag{D.21}$$

With $\lambda = n(n+1)$, the solution for the equation for R becomes

$$R = A_n r^n + B_n/r^{n+1} \tag{D.22}$$

Hence, the generic solution for the Laplace operator for axial symmetry, recalling that outside the sphere is domain (i) and inside the sphere is domain (2), becomes

$$\phi_1 = \sum_{n=0}^{n=\infty}\left(A_n r^n + \frac{B_n}{r^{n+1}}\right)P_n(\cos\theta) \quad \text{and} \quad \phi_2 = \sum_{n=0}^{n=\infty}\left(C_n r^n + \frac{D_n}{r^{n+1}}\right)P_n(\cos\theta) \tag{D.23}$$

To find the specific solution, we must add the boundary conditions reading:

I: $(\phi_1)_{r\to\infty} = -Er\cos\theta$ because the field gradient far removed from the origin should approach the applied field $\boldsymbol{E}$.
II: $(\phi_1)_{r=a} = (\phi_2)_{r=a}$ as ϕ should be continuous across the boundary.
III: $\varepsilon_1(\partial\phi_1/\partial r)_{r=a} = \varepsilon_2(\partial\phi_2/\partial r)_{r=a}$ because at the boundary the normal dielectric displacement component D_n should be continuous.
IV: ϕ_2 is finite at $r = 0$ since the potential may not have a singularity at $r = 0$.

Satisfying boundary condition IV can only be achieved if $D_n = 0$ and condition I requires $A_1 = -E = -|\boldsymbol{E}|$, while all other $A_n = 0$. Conditions II and III say that, except for $n = 1$,

$$B_n/a^{n+1} = C_n a^n \quad \text{and} \quad -\varepsilon_1(n+1)B_n/a^{n+2} = \varepsilon_2 n C_n a^{n-1} \tag{D.24}$$

From these equations, it follows that $B_n = C_n = 0$ for all $n \neq 1$. For $n = 1$ it is found that

$$B_1/a^2 - Ea = C_1 a \quad \text{and} \quad \varepsilon_1(2B_1/a^3 + E) = -\varepsilon_2 C_1 \tag{D.25}$$

Therefore,

$$B_1 = \frac{\varepsilon_2 - \varepsilon_1}{\varepsilon_2 + 2\varepsilon_1} a^3 E \quad \text{and} \quad C_1 = -\frac{3\varepsilon_1}{\varepsilon_2 + 2\varepsilon_1} E \tag{D.26}$$

and ϕ_1 and ϕ_2 become

$$\phi_1 = \left(\frac{\varepsilon_2 - \varepsilon_1}{\varepsilon_2 + 2\varepsilon_1}\frac{a^3}{r^3} - 1\right) Ez \quad \text{and} \quad \phi_2 = -\frac{3\varepsilon_1}{\varepsilon_2 + 2\varepsilon_1} Ez \tag{D.27}$$

Because the potential due to the external charges is $\phi = -Ez$, the contributions $\phi_1' = \phi_1 - \phi$ and $\phi_2' = \phi_2 - \phi$ due to the polarization become

$$\phi_1' = \frac{\varepsilon_2 - \varepsilon_1}{\varepsilon_2 + 2\varepsilon_1}\frac{a^3}{r^3} Ez \quad \text{and} \quad \phi_2' = \frac{\varepsilon_2 - \varepsilon_1}{\varepsilon_2 + 2\varepsilon_1} Ez \tag{D.28}$$

The expression for ϕ_1' is identical to that of an equivalent ideal dipole $\boldsymbol{\mu}_{\text{equ}}$ at the center of the sphere, surrounded by vacuum with the dipole vector directed along the external field $\boldsymbol{E}$ (the z-axis), and is given by

$$\boldsymbol{\mu}_{\text{equ}} = \frac{\varepsilon_2 - \varepsilon_1}{\varepsilon_2 + 2\varepsilon_1}\frac{a^3}{\mathscr{C}} \boldsymbol{E} \tag{D.29}$$

The field $\boldsymbol{E}_2'$ inside the sphere due to the polarization and the total field $\boldsymbol{E}_2$ inside the sphere are described by, respectively,

$$\boldsymbol{E}_2' = -\frac{\varepsilon_2 - \varepsilon_1}{\varepsilon_2 + 2\varepsilon_1} \boldsymbol{E} \quad \text{and} \quad \boldsymbol{E}_2 = \frac{3\varepsilon_1}{\varepsilon_2 + 2\varepsilon_1} \boldsymbol{E} \tag{D.30}$$

In case the sphere is a cavity so that $\varepsilon_1 = \varepsilon$ and $\varepsilon_2 = 1$, the total field $\boldsymbol{E}_2$ becomes the *cavity field* $\boldsymbol{E}_{\text{cav}} = 3\varepsilon\boldsymbol{E}/(2\varepsilon + 1)$.

D.3 A Dipole in a Spherical Cavity

If in a spherical cavity in an infinite matrix a dipole $\boldsymbol{\mu}$ is placed at the center, this dipole will polarize the surrounding matter and the resulting polarization will give rise to a field at the dipole, called the reaction field $\boldsymbol{R}$. The reaction field will be proportional to $\boldsymbol{\mu}$, having the same direction as $\boldsymbol{\mu}$ so that $\boldsymbol{R} = f\boldsymbol{\mu}$. We choose the direction of the z-axis along the dipole vector so that we can employ again the axisymmetric solutions (Eq. (D.23)) to the Laplace equation. In this case, the boundary conditions are, using $\varepsilon_1 = \varepsilon$ and $\varepsilon_2 = 1$:

$$\text{I: } (\phi_1)_{r\to\infty} = 0, \quad \text{II: } (\phi_1)_{r=a} = (\phi_2)_{r=a}, \quad \text{III: } \varepsilon(\partial\phi_1/\partial r)_{r=a} = (\partial\phi_2/\partial r)_{r=a} \tag{D.31}$$

The D-terms in ϕ_2 are due to the charges inside the cavity. Because the potential of an ideal dipole along the z-axis is $\phi = \mathscr{C}\mu r^{-2}\cos\theta$, the contribution $D_n = 0$ for all n except $n = 1$, which reads $D_1 = \mathscr{C}\mu$. For ϕ_2, we thus have

$$\phi_2 = \sum\nolimits_{n=0}^{n=\infty} C_n r^n P_n(\cos\theta) + \mathscr{C}\mu r^{-2}\cos\theta \tag{D.32}$$

From condition I, we conclude that $A_n = 0$ for all n, so that ϕ_1 becomes

$$\phi_1 = \sum\nolimits_{n=0}^{n=\infty} \frac{B_n}{r^{n+1}} P_n(\cos\theta) \tag{D.33}$$

From conditions II and III, we obtain for all $n \neq 1$,

$$B_n/a^{n+1} = C_n a^n \quad \text{and} \quad -\varepsilon(n+1)B_n/a^{n+2} = nC_n a^{n-1} \tag{D.34}$$

which can be satisfied only if $B_n = C_n = 0$ for $n \neq 1$, while for $n = 1$ the result is

$$B_1/a^2 = C_1 a + \mathscr{C}\mu/a^2 \quad \text{and} \quad -2\varepsilon B_1/a^3 = C_1 - 2\mathscr{C}\mu/a^3 \tag{D.35}$$

Solving for B_1 and C_1 leads to

$$B_1 = \frac{3}{2\varepsilon+1}\mathscr{C}\mu \quad \text{and} \quad C_1 = -\frac{2(\varepsilon-1)}{2\varepsilon+1}\frac{\mathscr{C}\mu}{a^3} \tag{D.36}$$

Substitution in Eqs. (D.32) and (D.33) results in

$$\phi_1 = \frac{3}{2\varepsilon+1}\frac{\mathscr{C}\mu}{r^2}\cos\theta \quad \text{and} \quad \phi_2 = \frac{\mathscr{C}\mu}{r^2}\cos\theta - \frac{2(\varepsilon-1)}{2\varepsilon+1}\frac{\mathscr{C}\mu}{a^3}r\cos\theta \tag{D.37}$$

Because the potential due to the dipole is given by $\phi = \mathscr{C}\mu r^{-2}\cos\theta$, we see from the first part of Eq. (D.37) that formally the potential in the dielectric matrix can be described as the field of an effective (or virtual) dipole at the center of the cavity given by

$$\mu_{\text{eff}} = \frac{3\varepsilon}{2\varepsilon+1}\mu \tag{D.38}$$

From the second part of Eq. (D.37), we see that the field in the cavity is the field of the dipole in vacuum plus the uniform reaction field $\boldsymbol{R}$, given by

$$\boldsymbol{R} = f\boldsymbol{\mu} \quad \text{with} \quad f \equiv \frac{2(\varepsilon-1)}{2\varepsilon+1}\frac{\mathscr{C}}{a^3} \tag{D.39}$$

Using again the potential due to the dipole given by $\phi = \mathscr{C}\mu r^{-2}\cos\theta$, the contributions $\phi_1{}' = \phi_1 - \phi$ and $\phi_2{}' = \phi_2 - \phi$ due to the surface charges become

$$\phi_1{}' = -\frac{2(\varepsilon-1)}{2\varepsilon+1}\frac{\mathscr{C}\mu}{r^2}\cos\theta \quad \text{and} \quad \phi_2{}' = -\frac{2(\varepsilon-1)}{2\varepsilon+1}\frac{\mathscr{C}\mu}{a^3}r\cos\theta \tag{D.40}$$

These potentials can be described by a dipole $\boldsymbol{m}$ given by

$$\boldsymbol{m} = -\frac{2(\varepsilon-1)}{2\varepsilon+1}\boldsymbol{\mu} \tag{D.41}$$

The results of Eqs. (D.30) and (D.39) are used in Chapter 8.

Further Reading

Böttcher, C.J.F. (1973). *Theory of Electric Polarization*, 2e, Vol. 1. Amsterdam: Elsevier.

Jackson, J.D. (1999). *Classical Electrodynamics*, 3e. New York, NY: John Wiley & Sons.

Panofsky, W.K.H. and Phillips, M. (1955). *Classical Electricity and Magnetism*. Reading, MA: Addison-Wesley Publ. Comp.

Robinson, F.N.H. (1973). *Electromagnetism*, Oxford Physics series. Oxford: Clarendon Press.

Stratton, J.A. (1941). *Electromagnetic Theory*. McGraw-Hill Book Comp.

Appendix E

Elements of Probability and Statistics

In this appendix, some basic aspects of probability and statistics are briefly introduced, mainly for setting the terminology. More details can be found in the textbooks quoted in the section "Further Reading."

E.1 Probability

For probability, an elusive concept quantifying our degree of certainty, the syntax is generally accepted, but its semantics is not (see, e.g. [1]). That is, one accepts the rules, but does not accept the interpretation, and, consequently, various interpretations exist. Here, first, we enumerate the rules and thereafter indicate the interpretation.

An experiment is considered to yield reproducible and enumerable outcomes, which are conveniently described by set theory. A *set* A is a collection of objects, called *elements*. If ω is (is not) an element of A, we write $\omega \in A$ ($\omega \notin A$). A *subset* B is another set whose elements are also elements of A and one writes $B \subset A$ (or $A \supset B$). An *empty* set $\emptyset$ is a set containing no elements. The total collection of sets considered is called the *sample space* S. Thus, for any A, $\emptyset \subset A \subset A \subset S$. Elements are *transitive*, that is, if $C \subset B$ and $B \subset A$, then $C \subset A$, and are equal if and only if $A \subset B$ and $B \subset A$. The *sum* or *union* $A + B$ (or $A \vee B$ or $A \cup B$) of two events A and B is the event that occurs when A or B or both occur. The *product* or *intersection* AB (or $A \wedge B$ or $A \cap B$) is the event that occurs when both A and B occur. These operations are *commutative*, that is, $A + B = B + A$ and $AB = BA$, *associative*, that is, $A + (B + C) = (A + B) + C$ and $A(BC) = (AB)C$, and *distributive*, that is, $A(B + C) = (AB) + (AC)$. The elements A and B are *mutually exclusive* if the occurrence of one excludes the occurrence of the other, that is, if $AB = \emptyset$. A *partition* P of a set S is a collection of mutually exclusive subsets A_j whose union equals S, that is, $A_1 + A_2 + \ldots A_n \equiv \bigcup_{i=1}^{i=n} A_i = S$. The *complement* $\overline{A}$ of a set A with respect to space S is the set of all elements in S but not in A, that is, $A + \overline{A} = S$ and $A\overline{A} = \emptyset$. Finally, we mention *De Morgan's law* reading $\overline{A + B} = \overline{A}\,\overline{B}$ and $\overline{AB} = \overline{A} + \overline{B}$. An *event* $\mathcal{A}$ (no order in time implied) is a subset of S, which obeys the rules of a (sigma) *algebra*, that is, has properties (i) $S \in \mathcal{A}$, (ii) if $A \in \mathcal{A}$, then $\overline{A} \in \mathcal{A}$, and (iii) if $A_i \in \mathcal{A}$, then $\bigcup_{i=1}^{i=\infty} A_i \in \mathcal{A}$. For countable sets, the *event space* contains all subsets of the sample space S, and if S is finite with n elements, it contains in total 2^n subsets. For a noncountable infinite set, the above rules guarantee that we can assign probabilities to all events. An *elementary event* is an event consisting of a single element ω, while a *sure event* is the event that occurs in every trial, that is, for S or $\mathcal{S}$.

Phases of Matter and their Transitions: Concepts and Principles for Chemists, Physicists, Engineers, and Materials Scientists, First Edition. Gijsbertus de With.

The *probability* $p(\mathcal{A})$ of event $\mathcal{A}$ is interpreted as a number assigned to the event, as based on three axioms:

(1) The probability $p(\mathcal{A})$ is a positive number, that is, $p(\mathcal{A}) \geq 0$.
(2) The probability of a sure event $p(\mathcal{S})$ equals 1, that is, $p(\mathcal{S}) = 1$.
(3) The probabilities of n mutually exclusive events, where n can be ∞, are additive, that is, $p\left(\bigcup_{j=1}^{j=n} \mathcal{A}_j\right) = \sum_{j=1}^{j=n} p(\mathcal{A}_j)$.

In this *axiomatic definition* of probability, due to Kolmogorov [2], the numerical values for the probabilities can be based on empirical information, for example, using repeated experiments, or, if unavailable or impossible to obtain, on the *principle of insufficient reason* (Bernoulli), also called the *principle of indifference* (Bayes, Laplace), essentially stating that in the absence of information, we must assign equal probabilities $1/n$ to an exhaustive set of n mutually exclusive events. As shown in Chapters 7 and 19, this leads to the *principle of maximum entropy* (PME). To explicitly indicate which information is used, for the probability p of event $\mathcal{A}$ given the information $\mathcal{H}$, the notation $p(\mathcal{A}|\mathcal{H})$ is often used. In this axiomatic definition, one considers probability as a numerical measure of a *reasonable*[1] *degree of belief* [3], *degree of plausibility* [4], or *degree of reasonable expectation*, and tries to estimate probability distributions consistent with the constraints given. This system can be rationalized, primarily using the intuitively reasonable arguments $p(\overline{A}) = f[p(A)]$ and $p(AB|C) = F[p(B|C), p(A|BC)]$, where f and F denote functions, and consistency arguments based on Boolean algebra, as advanced by Cox [5] and Jaynes [4] (see also [3]). We essentially use this interpretation of probability throughout.[2]

Using $A + \overline{A} = S$, we have

$$p(\mathcal{A}|\mathcal{H}) + p(\overline{\mathcal{A}}|\mathcal{H}) = 1 \tag{E.1}$$

Furthermore, these definitions and axioms lead to the product and sum rules (see, e.g. [4]), reading, respectively,

$$p(\mathcal{AB}|\mathcal{H}) = p(\mathcal{A}|\mathcal{BH})\,p(\mathcal{B}|\mathcal{H}) = p(\mathcal{B}|\mathcal{AH})\,p(\mathcal{A}|\mathcal{H}) \quad \text{and} \tag{E.2}$$

$$p(\mathcal{A} + \mathcal{B}|\mathcal{H}) = p(\mathcal{A}|\mathcal{H}) + p(\mathcal{B}|\mathcal{H}) - p(\mathcal{AB}|\mathcal{H}) \tag{E.3}$$

For a set $\{\mathcal{A}\}$ of n mutually exclusive and exhaustive events (given the information $\mathcal{H}$), that is, a *partition*, we will have $\Sigma_j p(\mathcal{A}_j|\mathcal{H}) = 1$. In case no further information is available, the

1 Reasonable implies here both "rational" as well as "fair."

2 Other definitions of probability exist. The *relative frequency* definition, associated with Venn, von Mises, and Feller [6], considers probability as the *frequency* of an event, that is, the limiting fraction with which an event occurs in a series of $m \to \infty$ repetitions of the experiment. This approach assumes an underlying probability distribution, prescribed by outside means, and tries to determine this distribution. Obviously, the problem is that an infinite repetition, and in many cases even a large number of repetitions, is impossible, so that the fraction cannot be, even approximately, equated to a limit. The frequency interpretation also uses a priori information, while the degree-of-belief interpretation cannot avoid frequency arguments (chapter 2 in Ref. [7]) and Kendall [8] has attempted to reconcile both approaches. The *classical definition*, associated with Bernoulli, Bayes, and Laplace, equates the ratio of favorable outcomes to the total number of outcomes, provided that all outcomes are equally likely. In this case, the problem is that, apart from that "likely" actually means "probable," and thus a recursive definition is used. This definition can be applied only to equally probable outcomes. Moreover, for continuum distributions, this definition leads to ambiguities. For a more detailed but still brief introduction, see chapters 1 and 2 of Ref. [9]. Finally, a similar but distinctive treatment of probability based on degree of belief is given by de Finetti [10].

principle of indifference states that in that case $p(\mathcal{A}_j|\mathcal{H}) = 1/n$. The product rule also easily leads to *Bayes' theorem*, reading

$$p(\mathcal{A}|\mathcal{B}\,\mathcal{H}) = p(\mathcal{A}|\mathcal{H})\,p(\mathcal{B}|\mathcal{A}\,\mathcal{H})/p(\mathcal{B}|\mathcal{H}) \tag{E.4}$$

The term $p(\mathcal{A}|\mathcal{H})$ is often denoted as the *prior* or *conditional probability* (that event $\mathcal{A}$ occurs). When new (prior) information with probability $p(\mathcal{B}|\mathcal{H})$ becomes available, the rule leads to the updated or *posterior probability* $p(\mathcal{A}|\mathcal{B}\,\mathcal{H})$ for event $\mathcal{A}$. The probability $p(\mathcal{B}|\mathcal{A}\mathcal{H})$ is denoted as the *likelihood*.

E.2 Single Variable

A *random variable* (rv) $X(x)$ is a function with a definitive value and associated *cumulative distribution function* (cdf) $F(x)$ for every element (or *sample point*) x of the sample space S (i.e. for each individual observation). The cdf, defined by $F(x_0) = p(x|x \le x_0)$, is continuous except perhaps at a denumerable number of points. If F is continuous everywhere, F is monotonically increasing and $\mathrm{d}F = (\mathrm{d}F/\mathrm{d}x)\mathrm{d}x \equiv f(x)\mathrm{d}x$ with $f(x)$ the *probability density function* (pdf). Hence, with $F(-\infty) = 0$ and $F(+\infty) = 1$,

$$F(x) = \int_{-\infty}^{x} \mathrm{d}F(x') = \int_{-\infty}^{x} f(x')\,\mathrm{d}x' \quad \text{and} \quad F(\infty) = \int_{-\infty}^{+\infty} f(x)\,\mathrm{d}x = 1 \tag{E.5}$$

The *expectation value* $\varepsilon\{Y\}$ for any rv Y, possibly a function of x, is defined by

$$\varepsilon\{Y\} \equiv \int_{-\infty}^{+\infty} Y(x)f(x)\,\mathrm{d}x \tag{E.6}$$

If $Y = x$, the expectation value is usually denoted as the *mean* μ. It is a measure for the center of the distribution. More generally, the pdf may be characterized by its expansion in a set of *moments* m_n or *central moments* μ_n of the order n given by

$$m_n = \varepsilon\{x^n\} = \int_{-\infty}^{+\infty} x^n f(x)\,\mathrm{d}x \quad \text{and} \quad \mu_n = \varepsilon\{(x-\mu)^n\} = \int_{-\infty}^{+\infty} (x-\mu)^n f(x)\,\mathrm{d}x \tag{E.7}$$

The first moment m_1 is equal to the *mean* μ, while the first central moment μ_1 is zero. The second central moment $\mu_2 = V(x) = \varepsilon\{x^2\} - [\varepsilon\{x\}]^2 \equiv \sigma^2$ is called the *variance* of the distribution. Its square root, the *standard deviation* σ, is a measure for the width of the distribution. Using $m_n(a) \equiv \varepsilon\{(x-a)^n\}$, the relation $m_n(a) = [m(b)+c]^n$ with $c = b - a$, exists, in which after expanding the rhs binomially, $m(b)^n$ is to be identified as $m_n(b)$. In particular, if either $(a = \mu, b = 0)$ or $(a = 0, b = \mu)$, we have

$$\mu_2 = m_2 - \mu^2, \quad \mu_3 = m_3 - 3\mu m_2 + 2\mu^3, \quad \mu_4 = m_4 - 4\mu m_3 + 6\mu^2 m_2 - 3\mu^4, \ldots \tag{E.8}$$

$$m_2 = \mu_2 + \mu^2, \quad m_3 = \mu_3 + 3\mu\mu_2 + \mu^3, \quad m_4 = \mu_4 + 4\mu\mu_3 + 6\mu^2\mu_2 + \mu^4, \ldots \tag{E.9}$$

For any symmetrical pdf, $\mu_n = 0$ if n is odd. The *skewness* $\mu_{\text{ske}} = \mu_3{}^2/\mu_2{}^3$ measures the asymmetry of a pdf and the *kurtosis* $\mu_{\text{kur}} = \mu_4/\mu_2{}^2$ its degree of peakedness.

For many distributions, a *characteristic function* (cf) $C_x(k)$ may be defined by the Fourier transform (Section A.15):

$$C_x(k) = \varepsilon\{\exp(-ikx)\} = \int_{-\infty}^{+\infty} \exp(-ikx)f(x)\,\mathrm{d}x = \sum_{n=0}^{n=\infty} \frac{(-ik)^n}{n!}\varepsilon\{x^n\} \tag{E.10}$$

Random variables with the same cf have the same pdf. The cf is a *moment-generating function* from which the moment m_n can be calculated using

$$m_n = \varepsilon\{x^n\} = \mathrm{i}^n[\mathrm{d}^n C_x(k)/\mathrm{d}k^n]_{k=0} \tag{E.11}$$

that is, the nth moment is the nth derivative of $C_x(k)$, evaluated at $k = 0$. Taking the inverse Fourier transform of Eq. (E.10), the result is

$$f(x) = \frac{1}{2\pi}\int_{-\infty}^{+\infty} C_x(k)\mathrm{e}^{\mathrm{i}kx}\,\mathrm{d}k = \frac{1}{2\pi}\int_{-\infty}^{+\infty}\sum_{n=0}^{n=\infty}\frac{(-\mathrm{i}k)^n}{n!}\varepsilon\{x^n\}\mathrm{e}^{\mathrm{i}kx}\,\mathrm{d}k \tag{E.12}$$

In Eqs. (E.10) and (E.12), the last step can only be made if all moments of $f(x)$ exist.

Other measures for the central tendency of the rv x are:

- The *median* x_{50} defined by the value of x which divides the area beneath the $f(x)$ curve into half. The median is the so-called 50 *percentile*. For the percentile, other percentages z may be used leading to $F(x_z) + F(x_{100-z}) = 1$. Hence, for x_{50}

$$\int_{-\infty}^{x_{50}} f(x)\,\mathrm{d}x = \int_{x_{50}}^{+\infty} f(x)\,\mathrm{d}x \quad \text{or} \quad F(x_{50}) = 1/2 \tag{E.13}$$

- The *mode* x_{mod}, defined as the most frequently occurring value of x, given by

$$[\mathrm{d}f(x)/\mathrm{d}x]_{x=x_{\mathrm{mod}}} = 0 \tag{E.14}$$

For distributions skewed to the right (tail to the right) we have $\mu - x_{\mathrm{mod}} > 0$, while for distributions skewed to the left, $\mu - x_{\mathrm{mod}} < 0$. In both cases, the value of x_{50} is between those of μ and x_{mod}. For unimodal curves of moderate asymmetry, the empirical relationship $\mu - x_{\mathrm{mod}} = 3(\mu - x_{50})$ exists [11].

E.3 Multiple Variables

For multivariable problems with $\mathbf{x} = (x_1, x_2, \ldots, x_n)$, we have the pdf $f_n(\mathbf{x})\mathrm{d}\mathbf{x} = f_n(\mathbf{x})\mathrm{d}x_1\,\mathrm{d}x_2\ldots\mathrm{d}x_n$ as the probability of finding $\mathbf{x}$ between $\mathbf{x}$ and $\mathbf{x}+\mathrm{d}\mathbf{x}$. The mth moment of any function $G(\mathbf{x})$ of $\mathbf{x}$ obviously is

$$\varepsilon\{G^m(\mathbf{x})\} = \int_{-\infty}^{+\infty} G^m(\mathbf{x})f_n(\mathbf{x})\,\mathrm{d}\mathbf{x} \tag{E.15}$$

Since the function $G(\mathbf{x})$ itself is a rv, it has a pdf dependent on a rv which we denote by g, so that we have $f_G(g)$. From the multivariable generalizations of Eqs. (E.10) and (E.12), we have

$$f_G(g) = \frac{1}{2\pi}\int_{-\infty}^{+\infty}\varepsilon\{\mathrm{e}^{-\mathrm{i}kg}\}\mathrm{e}^{\mathrm{i}kg}\,\mathrm{d}k = \frac{1}{2\pi}\int_{-\infty}^{+\infty}\sum_{n=0}^{n=\infty}\frac{(-\mathrm{i}k)^n}{n!}\varepsilon\{G^n(\mathbf{x})\}\,\mathrm{e}^{\mathrm{i}kg}\,\mathrm{d}k \tag{E.16}$$

Inserting Eq. (E.15) and summing, we obtain

$$f_G(g) = (2\pi)^{-1}\int_{-\infty}^{+\infty} f_n(\mathbf{x})\int_{-\infty}^{+\infty}\exp[\mathrm{i}k(g - G(\mathbf{x})]\,\mathrm{d}k\,\mathrm{d}\mathbf{x} \tag{E.17}$$

which, using $\delta(y) = (2\pi)^{-1}\int_{-\infty}^{+\infty}\exp(\mathrm{i}ky)\,\mathrm{d}k$, results in

$$f_G(g) = \int_{-\infty}^{+\infty} f(\mathbf{x})\delta[g - G(\mathbf{x})]\,\mathrm{d}\mathbf{x} = \varepsilon\{\delta[g - G(\mathbf{x})]\} \tag{E.18}$$

Moreover, we have now correlation coefficients K_{ij} between x_i and x_j, defined by

$$K_{ij} = \varepsilon\{(x_i - \varepsilon\{x_i\})(x_j - \varepsilon\{x_j\})\} \tag{E.19}$$

If the pdf is given by $f(\mathbf{x}) = f_i(x_i) f_j(x_j)$ with $j \neq i$, so that $f_j(x_j)$ does not depend on x_i, $K_{ij} = 0$ for $j \neq i$, and the variable x_i is uncorrelated with the remaining variables. In case $f(\mathbf{x}) = f_1(x_i)$ $f_2(x_j) \ldots f_n(x_n)$, the variables $\mathbf{x}$ are completely uncorrelated.

If the pdf $f_n(\mathbf{x}) = f(x_1, x_2, \ldots, x_n)$ is integrated over a subset of $\mathbf{x}$, say x_n, or, more generally, over $x_{m+1}, \ldots, x_n$ with $m < n$, we have the *marginal distributions*, given by

$$f_{n-1}(x_1, \ldots, x_{n-1}) = \int_{-\infty}^{+\infty} f_n(x)\,\mathrm{d}x_n \quad \text{and} \quad f_m(x_1, \ldots, x_m) = \int_{-\infty}^{+\infty} f_n(x)\,\mathrm{d}x_{m+1} \ldots \mathrm{d}x_m \tag{E.20}$$

respectively. These marginal distributions are used for the probability of event a, given that event b is certain. For that we encountered the product rule, in the present notation, $p(ab) \equiv p(a|b)p_B(b) = p(b|a)p_A(a)$, leading to *Bayes' theorem*

$$p(a|b) = p(ab)/p_B(b) = p(b|a)p_A(a)/p_B(b) \tag{E.21}$$

If we identify probabilities with statistical frequencies, the pdf $f_n(x_1, \ldots, x_n)$ also determines the frequencies of $x_1, \ldots, x_m$ for given, fixed values of $x_{m+1}, \ldots, x_n$. Using

$$f_{n-m}(x_{m+1}, \ldots, x_n) = \int_{-\infty}^{+\infty} f_n(x)\,\mathrm{d}x_1 \ldots .\mathrm{d}x_m$$

we obtain for the *conditional probability*

$$p_{m|n-m}(x_1, \ldots, x_m | x_{m+1}, \ldots, x_n) = p_n(x_1, \ldots, \mathrm{d}x_n)/p_{n-m}(x_{m+1}, \ldots, \mathrm{d}x_n) \tag{E.22}$$

describing the probability of the random variables $x_1, \ldots, x_m$ given those of $x_{m+1}, \ldots, x_n$.

E.4 The Normal Distribution and the Central-Limit Theorem

Many distribution functions are known. We only deal here with the *normal* (or *Gauss*) pdf, in view of its particular importance. Let us denote the mean and standard deviation of this distribution by μ and σ, respectively. Introducing a standardized variable z by $z = (x - \mu)/\sigma$, the normal pdf $g(x)$ and cdf $G(x)$ are given by

$$g(z) = (2\pi)^{-1/2} \exp(-z^2/2) \quad \text{and} \quad G(z) = (2\pi)^{-1/2} \int_{-\infty}^{z} \exp(-t^2/2)\,\mathrm{d}t \tag{E.23}$$

The cf for the standardized and original normal distribution are, respectively,

$$C_z(t) = \exp(-½\sigma^2 t^2) \quad \text{and} \quad C_x(t) = \exp(-\mathrm{i}\mu\, t - ½\sigma^2 t^2) \tag{E.24}$$

with t a dummy variable. All odd moments are equal to zero and the even central moments are given by $\mu_{2n} = (2n)!\sigma^{2n}/2^n n!$. So, for a normal pdf, $\mu_{\text{kur}} = 3$. Generally but not always, when for a certain pdf $\mu_{\text{kur}} > 3$ ($\mu_{\text{kur}} < 3$), this pdf is more (less) peaked than a normal pdf.

Finally, we briefly indicate the central-limit theorem. Suppose, we have $\mathbf{x} = (x_1, \ldots, x_n)$ with the various x_j independent rvs (measurements) having the same pdf. A *statistic* is a function of these rvs only, not containing unknown parameters. To calculate the pdf for the statistic $Y = \Sigma_j x_j$, we first introduce the random function

$$Z = \sum_{j=1}^{j=n}(x_j - \mu_x)/n^{1/2} = (Y - n\mu_x)/n^{1/2} \tag{E.25}$$

where $\mu_x \equiv \varepsilon\{\Sigma_j x_j/n\}$. From Eq. (E.18), the pdf $f_Z(z)$ for Z is

$$\begin{aligned} f_Z(z) &= \int_{-\infty}^{+\infty} f(\mathbf{x})\delta[z - Z(\mathbf{x})]\,\mathrm{d}\mathbf{x} = \int_{-\infty}^{+\infty} f(\mathbf{x})\delta\left(z - n^{-1/2}\sum_i x_i + n^{1/2}\mu_x\right)\mathrm{d}\mathbf{x} \\ &= (2\pi)^{-1}\int_{-\infty}^{+\infty}\exp(\mathrm{i}kz)\int_{-\infty}^{+\infty} f(\mathbf{x})\exp\left[-\mathrm{i}k\left(n^{-1/2}\sum_i x_i\right) + \mathrm{i}kn^{1/2}\mu_x\right]\mathrm{d}\mathbf{x}\,\mathrm{d}k \\ &= (2\pi)^{-1}\int_{-\infty}^{+\infty}\exp[\mathrm{i}kz + \mathrm{i}kn^{1/2}\mu_x][C_x(n^{-1/2}k)]^n\mathrm{d}k \end{aligned} \tag{E.26}$$

To make further progress, we expand $\ln[C_x(q)] = \ln[\Sigma_{n=0}(n!)^{-1}(-\mathrm{i}q)^n\varepsilon\{x^n\}]$, yielding

$$\ln[C_x(q)] = -\mathrm{i}qc_1 - ½q^2c_2 + \ldots \quad \text{or} \quad C_x(q) = \exp\sum_{n=1}^{n=\infty}(n!)^{-1}(-\mathrm{i}q)^n c_n \tag{E.27}$$

The coefficients c_n are called *cumulants* and a general expression of cumulants c_n in terms of moments μ_n can be derived [12]. Generally, if we have the expression $\Sigma_{n=0}(n!)^{-1}\mu_n\alpha^n = \exp[\Sigma_{k=1}(k!)^{-1}c_k\alpha^k]$, expand the exponentials and collect terms, we find

$$\mu_1 = c_1,\quad \mu_2 = c_2 + c_1^{\,2},\quad \mu_3 = c_3 + 3c_2c_1 + c_1^{\,3},\ldots. \tag{E.28}$$

Similarly, if we write $\Sigma_{k=1}(k!)^{-1}c_k\alpha^k = \ln[\Sigma_{n=0}(n!)^{-1}\mu_n\alpha^n]$, expand the logarithm and collect terms, we obtain

$$c_1 = \mu_1,\quad c_2 = \mu_2 - \mu_1^{\,2} = \sigma^2,\quad c_3 = \mu_3 - 3\mu_2\mu_1 + 2\mu_1^{\,3},\ldots. \tag{E.29}$$

A cumulant is zero if one of its variables is independent of the others [13]. Similarly as the moments μ_n can be calculated from C_x, the cumulants c_n can be calculated from $\ln(C_x)$. Substituting Eq. (E.27b) in Eq. (E.26) results in

$$f_Z(z) = (2\pi)^{-1}\int_{-\infty}^{+\infty}\mathrm{e}^{\mathrm{i}kz - ½k^2\sigma_x^2 + \ldots + \mathrm{i}k^3n^{-1/2} + \ldots}\,\mathrm{d}k = \left(2\pi\sigma_x^2\right)^{-1/2}\mathrm{e}^{-z^2/2\sigma_x^2} \tag{E.30}$$

where the last step can be made since for large n, terms in $n^{-1/2}$ and the like are negligible, and in $\exp(-½k^2\sigma_x^{\,2})$ we recognize the cf for the normal distribution. Using $f_Y(y)\mathrm{d}y = f_Z(z)\mathrm{d}z$, we finally obtain for the pdf of Y with variable y

$$f_Y(y) = \left(2\pi n\sigma_x^2\right)^{-1/2}\exp\left[-(y - n\mu_x)^2/2n\sigma_x^2\right] \tag{E.31}$$

This expression represents the *central-limit theorem*, which states that, if we have several independent stochastic variables distributed according to some (not necessarily normal) pdf with finite mean and variance, the pdf of the sum of the variables is approximately normal for sufficiently large number of variables. We have for the mean $\mu_Y = n\mu_x$, for the standard deviation $\sigma_Y = n^{1/2}\sigma_x$, and for the relative deviation $\sigma_Y/\mu_Y = n^{1/2}\sigma_x/n\mu_x = \sigma_x/n^{1/2}\mu_x$. The theorem remains true if the variables x_i have different distributions, in which case μ_x and σ_x are replaced by their individual means and standard deviations. Unbiased estimates for the μ and σ^2 using n data x_j are given by the average

$m = \Sigma_j x_j/n$ and sample variance $s^2 = \Sigma_j(x_j - m)^2/(n-1)$. The corresponding variances for m and s^2 are $V(m) = s^2/n$ and $V(s^2) = 2s^4/(n-1)$ [11]. Because for any statistic x with mean θ, $V[f(x)] = (\mathrm{d}f/\mathrm{d}x)^2_{x=\theta} V(x)$, an approximate estimate for the variance of s is $V(s) = [\mathrm{d}(s^2)^{1/2}/\mathrm{d}s^2]^2_{s=\sigma} V(s^2) = [2(n-1)]^{-1} s^2$.

References

1 Emch, G.G. and Liu, C. (2002). *The Logic of Thermostatistical Physics*. Berlin: Springer.
2 Kolmogorov, A. (1933). *Grundbegriffe der Wahrscheinlichkeitsrechnung*. Berlin: Springer; (1956). *Foundations of the Theory of Probability*. New York: Chelsea.
3 van Horn, K.S. (2003). *Int. J. Approx. Reason.* 34: 3; (2004). *Ibid.* 35: 107; Shafer, G. (2004). *Ibid.* 35: 97.
4 Jaynes, E.T. (2003). *Probability Theory: the Logic of Science* (ed. G.L. Bretthorst). Cambridge: Cambridge University Press.
5 Cox, R.T. (1946). *Am. J. Phys.* 17: 1; Cox, R.T. (1961). *Algebra of Probable Inference*. Baltimore: John Hopkins Press.
6 von Mises, R. (ed.) (1957). *Probability, Statistics, and Truth*, 2e, revised by Geiringer, H., Allen and Unwin Ltd, London (also Dover, 1981); Feller, W. *An Introduction to Probability Theory and its Applications*, 3e, Vol. 1 (1968) and 2e, Vol. 2 (1971). New York: John Wiley & Sons.
7 Grandy, W.T. (1987). *Foundations of Statistical Mechanics, Equilibrium Theory*, Vol. 1. Berlin: Springer.
8 Kendall, M.G. (1949). *Biom.* 36: 101.
9 Mood, Graybill and Boes, 1974; Papoulis, (1991).
10 de Finetti, B. (1972). *Probability, Induction and Statistics*. New York: John Wiley & Sons.
11 Kendall and Stuart, 1963, Vol. 1.
12 Zwanzig, R.W. (1954). *J. Chem. Phys.* 22: 1420.
13 Kubo, R. (1962). *J. Phys. Soc. Jpn.* 17: 1100.

Further Reading

Kendall, M.G. and Stuart, A. *The Theory of Advanced Statistics*, Vol. 1 (1963) *Distribution Theory*, Vol. 2 (1961) *Inference and Relationship*, Vol. 3 (1966) *Planning and Analysis, and Time-Series*. London: Charles Griffin.

Mood, A.M., Graybill, F.A., and Boes, D.C. (1974). *Introduction to the Theory of Statistics*, 3e. Tokyo: McGraw-Hill Book Comp.

Papoulis, A. (1991). *Probability, Random Variables, and Stochastic Processes*, 3e. New York: McGraw-Hill.

Appendix F

Selected Data

Table F.1 Lennard-Jones parameters ε and σ for some molecules from simulations.

Compound	ε/k (K)	σ (10^{-12} m)	Compound	ε/k (K)	σ (10^{-12} m)
Ar	111.84	362.3	CH_4	140.42	401.5
Kr	154.87	389.5	CD_4	138.16	402.3
Xe	213.96	426.0	C_2H_6	216.12	478.2
O_2	113.27	365.4	nC_3H_8	255.18	547.1
N_2	91.85	391.9	nC_4H_{10}	287.20	608.1
F_2	104.29	357.1	nC_5H_{12}	309.75	670.9
Cl_2	296.27	448.5	nC_6H_{14}	327.47	731.9
CF_4	155.31	491.1	nC_7H_{16}	340.97	790.2
CCl_4	378.86	624.1	cC_3H_6	277.46	509.2
SiH_4	193.65	453.9	cC_5H_{10}	346.11	610.0
SiF_4	140.14	669.2	C_6H_6	377.46	617.4
$SiCl_4$	338.01	668.2	C_2H_2	209.11	463.5
$TiCl_4$	417.49	672.5	C_2H_4	200.78	458.9
SF_6	206.85	578.3	CO_2	201.71	444.4
UF_6	323.61	637.9	CS_2	388.81	498.3

Adapted from Cuadros et al. [1], obtained by fitting molecular dynamics (MD) simulation results of the pressure at low density and the Soave equation of state.

Phases of Matter and their Transitions: Concepts and Principles for Chemists, Physicists, Engineers, and Materials Scientists, First Edition. Gijsbertus de With.

Table F.2 Lennard-Jones parameters ε and σ for some molecules from experimental data.

Compound	ε/k (K)	σ (10^{-12} m)	Compound	ε/k (K)	σ (10^{-12} m)
Ne	35.60	274.9	CH_4	148.2 (137*)	381.7 (382.2*)
Ar	119.8	340.5	C_2H_6	230*	441.8*
Kr	171	360	nC_3H_8	254*	506.1*
Xe	221	410	nC_4H_{10}	410*	499.7*
O_2	117.5	358	nC_5H_{12}	345*	576.9*
N_2	95.05	369.8	nC_6H_{14}	413*	590.9*
F_2	112*	365.3*	CH_3OH	507*	358.5*
Cl_2	357*	411.5*	C_2H_5OH	391*	445.5*
Br_2	520*	426.8*	C_6H_6	440*	527*
I_2	550*	498.2*	C_2H_2	185*	422.1*
CH_3Cl	855	337.5	C_2H_4	205*	423.2*
CH_2Cl_2	406	475.9	CO_2	189	448.6
$CHCl_3$	327	543.0	CS_2	488*	443.8*
CCl_4	327*	588.1*	SiF_4	200.9	551
CF_4	152.5	470	SF_6	200.9	551

Adapted from Hirschfelder et al. [2], obtained from the viscosity (*) or from the second virial coefficient.

Table F.3 Thermodynamic vapor pressure constants a', b', and c' for some compounds.

Molecule	a'	b'	c'	L_0 (cal mol^{-1})
Ar	14.059	2.814	498	2278
CS_2	15.9206	2.90	1844	8444
H_2O	20.9586 (17.443)	4.0843 (3.868)	2825.3 (2795)	12 926 (13 611)
H_2S	11.778	1.510	1145	5248
NH_3	17.718	3.406	1612.5	7377
CH_4	14.9640	3.283	598.4	2739
CH_3Cl	22.0114	5.133	1701	7786
CH_2Cl_2	20.2795	4.2857	2098.1	9601
$CHCl_3$	12.2999	3.9158	2179.1	9974
CCl_4	24.3085	5.669	2452.6	11 220
C_2H_6	16.1506	3.332	1055	4831
nC_3H_8	15.772	3.003	1325	6066
nC_4H_{10}	19.877	4.376	1729	7912
nC_5H_{12}	22.752	5.253	2099	9611
nC_6H_{14}	25.281	6.013	2465	11 284
nC_8H_{18}	30.040	7.396	3162	14 473

(Continued)

Table F.3 (Continued)

Molecule	a'	b'	c'	L_0 (cal mol^{-1})
CH_3OH	22.43	4.634	2661	12 180
nC_4H_9OH	40.21	10.35	4100	18 750
$nC_5H_{11}OH$	46.49	12.42	4580	20 940
$nC_6H_{13}OH$	51.00	13.80	5068	23 170
$nC_7H_{15}OH$	56.20	15.41	5580	25 510
$nC_8H_{17}OH$	65.21	18.40	6190	28 300
C_6H_6	20.818 (26.075)	4.7793(6.203)	2388 (2610)	10 927 (11 940)
$(CH_3)_3N$	30.094	7.978	2142	10 030
C_6H_5COOH	30.172	6.720	4714	21 567
$(CH_3)_3CO$	19.557	3.966	2203	10 078

$\log P_{sat}$ (mm Hg) $= a' - b'\log T - c'/T$, where T (K) and $L_0 = 2.303Rc'$ with R the gas constant.
Source: Adapted from Moelwyn-Hughes [3].

Table F.4 Trouton constant C for some molecules.

Molecule	C	Molecule	C	Molecule	C
Ar	9.0	CH_4	8.8	H_2S	10.6
Kr	9.1	C_2H_6	9.6	SO_2	11.4
Xe	9.2	n-C_4H_{10}	9.9	CH_3Cl	10.4
Rn	9.4	i-C_4H_{10}	9.8	$CHCl_3$	10.6
N_2	8.7	n-C_7H_{16}	10.3	$(CH_3)_2O$	10.4
O_2	9.1	c-C_6H_{12}	10.2	NH_3	11.7
Cl_2	10.3	C_6H_6	10.5	H_2O	13.1
CO	8.9	HCl	10.3	CH_3OH	12.6
N_2O	10.8	HBr	10.3	C_2H_5OH	13.2

Source: Rowlinson and Swinton [4]/with permission of Elsevier.

Table F.5 Dipole moment μ and polarization volume $\alpha' = \alpha/4\pi\varepsilon_0$ for some compounds.

Compound	μ (D)	α' (10^{-30} m^3)	Compound	μ (D)	α' (10^{-30} m^3)
He	0	0.201	$(CH_3)_2CO$	2.84	6.33
Ar	0	1.63	CH_3OCH_3	1.29	5.16
C_6H_6	0	10.3	CH_3OH	1.71	3.23
$C_6H_5CH_3$	0.36	12.3	C_2H_5OH	1.69	5.11
C_6H_5Cl	1.70	12.3	HF	1.91	0.80
1,2-$C_6H_4(CH_3)_2$	0.62	—	HCl	1.03	2.63

(Continued)

Table F.5 (Continued)

Compound	μ (D)	α' (10^{-30} m^3)	Compound	μ (D)	α' (10^{-30} m^3)
CH_4	0	2.60	HBr	0.79	3.61
CH_3Cl	1.86	4.56	HI	0.38	5.45
CH_2Cl_2	1.58	6.48	H_2	0	0.82
$CHCl_3$	1.02	8.23	N_2	0	1.76
CCl_4	0	10.5	O_2	0	1.60
H_2O	1.84	1.49	Cl_2	0	4.61
NH_3	1.47	2.26	CO	0.10	1.95
$N(CH_3)_3$	0.60	8.29	CO_2	0	2.65

Dipole moment data are mainly from Landolt et al. [5]/Springer Nature.

Table F.6 Density ρ, compressibility κ, and expansivity α for some liquids at 20 °C.

Liquid	ρ (g cm^{-3})	κ_S (10^{11} Pa^{-1})	α_P (10^3 K^{-1})	Liquid	ρ (g cm^{-3})	κ_S (10^{11} Pa^{-1})	α_P (10^3 K^{-1})
n-C_6H_{14}	0.654	130	1.38	H_2O	0.998	45.6	0.207
n-C_7H_{16}	0.684	108	1.24	CH_3OH	0.792	100	1.49
n-C_8H_{18}	0.703	99.3	1.11	C_2H_5OH	0.798	91.1	1.40
CCl_4	1.595	71.4	1.16	C_3H_7OH	0.804	83.2	1.22 (0 °C)
CH_3Cl	1.489	66.5	1.32	C_4H_9OH	0.810	76.9	1.12 (0 °C)
C_6H_6	0.878	64.9	1.14	$CH_3CO_2C_2H_5$	0.900	79.3	1.32
$C_6H_5CH_3$	0.866	65.5	1.08	$CONH_2$	1.139	36.5	0.779
$C_6H_5C_2H_5$	0.868	64.4	1.01	$(CH_3)_2CO$	0.792	89	1.48
$C_6H_5NH_2$	1.022	35.7	0.843	$(C_2H_5)_2O$	0.714	138	1.66
Br_2	3.10	64	1.10	CS_2	1.26	92.7	1.19

Data for ρ and κ_S are mainly from Bergmann [6]. Data for κ_S can be converted from κ_T using Eq. (6.23). Data for α_P are mainly from Kayode Coker [7].

Table F.7 Kihara parameters for some molecules with nondimensional core size $a^* \equiv 2a/(\sigma - 2a)$.

Compound	a^* (–)	σ (10^{-12} m)	ε/k (K)	Compound	a^* (–)	σ (10^{-12} m)	ε/k (K)
Ar	0.121	331.7	146.5	C_2H_6	0.359	350.4	496.7
Kr	0.144	353.3	213.7	nC_3H_8	0.470	461.1	501.9
Xe	0.173	388.0	298.2	nC_4H_{10}	0.661	471.7	701.2
O_2	0.308	310.9	194.3	nC_5H_{12}	0.818	502.9	837.8
N_2	0.250	352.6	139.2	$C(CH_3)_4$	0.551	576.2	557.8
CH_4	0.283	356.5	227.1	CO_2	0.615	376.0	424.2
C_6H_6	0.750	533.5	832.0	CF_4	0.500	431.9	289.7

Source: Adapted from Prausnitz et al. [8].

Table F.8 Critical data, acentric factor ω, refractive index n_D, and normal boiling point T_n for some organic compounds.

Name	Molar mass	ω (–)	T_{cri} (K)	P_{cri} (bar)	V_{cri} ($cm^3\ mol^{-1}$)	n_D (–)	T_n (K)
Acetone	58.080	0.307	508.2	47.01	209	1.3590	329.4
MEK[a)]	72.11	0.329	535.6	41.5	267	1.3788	267
Dimethylether	46.07	0.192	400.0	53.7	178	—	249
Diethylether	74.123	0.281	466.7	36.40	280	1.353	307.6
Tetrafluoride	88.00	0.191	227.6	37.4	140	1.0005	145
Tetrachloride	153.822	0.193	556.4	45.60	276	1.4601	349.8
Chloroform	119.377	0.222	536.4	54.72	239	1.4459	334.3
Water	18.015	0.345	647.1	220.55	55.9	1.3330	373.2
Methanol	32.042	0.564	512.6	80.97	118	1.33	337.9
Ethanol	46.069	0.645	513.9	61.48	167	1.36	351.4
1-Propanol	60.096	0.622	536.8	51.75	219	1.387	370.4
i-Propanol	60.096	—	508.3	47.6	220	1.3776	356
1-Butanol	74.123	0.594	563.1	44.23	275	1.399	390.8
Benzene	78.114	0.210	562.2	48.98	259	—	353.2
Toluene	92.141	0.262	591.8	41.06	316	1.399	383.8
Ethylbenzene	106.167	0.303	617.2	36.06	374	—	409.4
Aniline	93.13	0.384	699.	53.1	274	—	457
Pyridine	79.1	0.24	620.0	56.3	254	1.5093	254
o-Xylene	106.167	0.310	630.3	37.34	369	1.505 45	417.6
m-Xylene	106.167	0.326	617.1	35.36	376	1.497 22	412.3
p-Xylene	106.167	0.322	616.2	35.11	379	1.495 82	411.5
Phenol	94.113	0.444	694.3	61.30	229	—	455.0
Methane	16.04	0.011	190.6	45.19	98.7	—	111.6
Ethane	30.07	0.100	305.4	48.83	148	—	184.6
Propane	44.1	0.153	369.85	42.48	200.0	—	231.1
Cyclopropane	42.08	0.264	397.8	54.9	210	—	240
Butane	58.123	0.200	425.1	37.96	255	—	272.7
Cyclobutane	56.107	0.209	459.9	49.8	210	—	286
Pentane	72.150	0.252	469.7	33.70	313	1.358	309.2
Cyclopentane	70.134	0.196	511.8	45.02	258	—	322.4
Neopentane	72.15	0.197	433.8	31.96	311.1	—	282.6
Hexane	86.177	0.301	507.6	30.25	371	1.375	341.9
Cyclohexane	84.161	0.210	553.6	40.73	308	1.4262	353.9
Heptane	100.204	0.350	540.2	27.40	428	1.387	371.6
Octane	114.231	0.400	568.7	24.90	486	1.398	398.8
Nonane	128.258	0.444	594.6	22.90	544	1.405	424.0

(Continued)

Table F.8 (Continued)

Name	Molar mass	ω (–)	T_{cri} (K)	P_{cri} (bar)	V_{cri} (cm^3 mol^{-1})	n_D (–)	T_n (K)
Acetaldehyde	44.05	0.303	461	56	154	—	293
Acetonitrile	41.05	0.309	508.1	47.0	209	1.344	354
Chlorobenzene	112.56	0.249	632.4	45.2	308	—	404
Trimethylamine	59.11	0.195	433.2	40.7	254	—	270

a) Methyl-ethyl ketone.
Source: Adapted from Smith et al. [9] and Pitzer [10].

Table F.9 Physical data for some liquids at 20 °C (25 °C).

Liquid	ε	μ (D)	R_M (cm^3)	γ (mN m^{-2})	ρ (g cm^{-3})	t_n (°C)
Acetic acid	6.19	1.74	13	26.9	1.0497	117.9
Acetic anhydride	21	2.8	22.4	31.9	1.075	140
Acetone	20.7	2.88	16.2	22.9	0.792	56.3
Acetonitrile	3.44	3.44	0	28.45	0.7768	81.6
Aniline	6.8	1.55	—	42.8	1.022	184.4
Benzene	2.28	0	26.2	28.2	0.878	184.4
Benzyl alcohol	13.1	1.71	32.5	39	1.042	205.2
Bromobenzene	5.4	1.7	34	35.7	1.495	156
Bromoform	4.3	1.3	0	31.7	2.8776	149.6
Butanol	17.5	1.66	0	24.3	1.0269	117.7
i-Butanol	17.7	1.64	22.4	22.5	0.798	107.7
t-Butanol	10.9	1.66	22.2	20	0.7812	82.4
Carbon disulfide	2.6	0	—	31.5	1.2566	46.26
Carbon tetrachloride	2.23	0	26.7	26.2	1.5842	76.8
Chlorobenzene	5.62	1.69	31.1	32.7	1.11	131
Chloroform	4.81	1.04	21.5	26.5	1.4799	61.2
Cyclohexane	2.02	0	27.8	24.4	0.7739	80.7
Dichloromethane	8.93	1.60	16.3	27.3	1.3168	39.8
Diethyl ether	4.34	1.15	22.5	16.5	0.7078	34.6
N,*N*-Dimethylacetamide	37.8	3.81	24.4	33.3	0.937	165
N,*N*-Dimethylformamide	36.7	3.86	20	35.2	0.9445	153
Dimethyl sulfoxide	49	3.96	20.1	42.8	1.0958	189
1,4-Dioxane	2.21	0	21.7	32.9	1.0269	101.3
Ethanol	24.3	1.69	14.9	21.8	0.798	78.3
Ethyl acetate	6.02	1.78	22.1	23.2	0.8945	77.1

(Continued)

Table F.9 (Continued)

Liquid	ε	μ (D)	R_M (cm^3)	γ ($mN\,m^{-2}$)	ρ ($g\,cm^{-3}$)	t_n (°C)
Ethylene glycol	37.7	2.28	14.5	48.1	1.1097	1.042
Formamide	110	3.7	10.7	58.5	1.1292	210.5
Formic acid	58	1.41	8.6	37.1	1.2131	100.7
Glycerol	42.5	2.56	20.5	62.5	1.2582	290
n-Heptane	1.92	0	34.6	19.8	0.684	98.5
n-Hexane	1.89	0.08	29.9	17.9	0.654	68.7
Methanol	32.6	1.7	8.2	22.4	0.792	64.7
Methyl acetate	6.7	1.72	17.5	24.1	0.932	56.9
n-Methylpyrrolidone	33	4.1	—	—	0.703	202
Nitrobenzene	35	4.22	32.9	42.8	1.1987	210.8
Nitromethane	38.6	3.46	12.5	50.7	1.1312	101.2
n-Pentane	1.84	0	25.3	15.5	0.626	36.2
Propanol	20.1	1.68	17.5	23.4	0.804	97.5
i-Propanol	18.3	1.66	17.6	20.8	0.781	82.3
Pyridine	12.3	2.19	24.1	36.3	0.982	115.3
Tetrahydrofuran	7.32	1.63	19.9	26.9	0.8844	66
Toluene	2.38	0.36	31.1	27.9	0.866	110.6
Water	80.2	1.84	3.7	72.8	0.998	100

Source: Adapted from Connors [11]. t_n represents the normal boiling point.

Table F.10 van der Waals constants *a* and *b* for some compounds.

Compound	*a* ($atm\,l^2\,mol^{-2}$)	*b* ($10^{-2}\,l\,mol^{-1}$)	Compound	*a* ($atm\,l^2\,mol^{-2}$)	*b* ($10^{-2}\,l\,mol^{-1}$)
He	0.03457	2.370	C_6H_{14}	24.71	17.35
Ne	0.2135	1.709	H_2O	5.536	3.049
Ar	1.363	3.219	H_2S	4.490	4.287
Kr	2.349	3.978	CH_3OH	9.649	6.702
Xe	4.250	5.105	C_2H_5OH	12.18	8.407
C_6H_6	18.24	11.54	CH_3Cl	7.570	6.483
$C_6H_5CH_3$	24.38	14.63	CCl_4	19.75	12.81
C_6H_5F	20.19	12.86	$SiCl_4$	4.377	5.786
C_6H_5Cl	25.77	14.53	$(CH_3)_2C{=}O$	14.09	9.94
C_6H_5Br	28.94	15.39	CO	1.505	3.985
C_6H_5I	33.52	16.56	CO_2	3.640	4.267

(Continued)

Table F.10 (Continued)

Compound	a (atm l^2 mol^{-2})	b (10^{-2} l mol^{-1})	Compound	a (atm l^2 mol^{-2})	b (10^{-2} l mol^{-1})
CH_4	2.283	4.278	H_2	0.2476	2.661
C_2H_6	5.562	6.380	O_2	1.378	3.183
C_3H_8	8.779	8.445	N_2	1.408	3.913
C_4H_{10}	14.66	12.26	Cl_2	6.579	5.622
C_5H_{12}	19.26	14.6	NH_3	4.225	3.707

Source: Adapted from Weast [12].

Table F.11 Critical data, acentric factor ω, and normal boiling point T_n for some inorganic compounds.

Name	Molar mass	ω (–)	T_{cri} (K)	P_{cri} (bar)	V_{cri} (cm^3 mol^{-1})	Z_{cri} (–)	T_n (K)
Ne	20.180	−0.016	44.44	27.605	41.70	0.312	27.1
Ar	39.948	−0.002	150.9	48.98	74.57	0.291	87.3
Kr	83.800	−0.002	209.4	55.00	91.20	0.288	119.7
Xe	131.29	0.002	289.7	58.40	118.0	0.286	165.0
N_2	28.014	0.037	126.2	33.98	91.10	0.289	77.4
O_2	31.999	0.022	154.6	50.43	73.4	0.288	90.2
F_2	37.997	0.051	144.3	52.2	66.2	0.288	85.0
Cl_2	70.905	0.069	417.2	77.00	124	0.275	239.1
Br_2	159.81	0.119	584.1	103.0	135	0.269	331.9
I_2	253.81	0.229	819	117.0	155	0.266	457.6
CO	28.01	0.045	132.9	34.94	93.10	0.292	81.7
CO_2	40.01	0.225	304.1	73.74	94.07	0.247	—
CS_2	76.14	0.115	552	79.0	170	0.293	319
HF	20.01	0.329	461.0	65.0	69.0	0.117	292.7
HCl	36.46	0.133	324.7	83.1	81.0	0.249	188.2
HBr	80.91	0.069	363.2	85.1	—	—	206.4
HI	127.91	0.038	423.9	90.00	132.7	0.303	237.6
NH_3	17.031	0.255	405.4	113.53	72.47	0.255	239.8
H_2S	34.08	0.090	373.4	89.63	98.00	0.283	212.8

Source: Adapted from Poling et al. [13].

Table F.12 Antoine constants *A*, *B*, and *C* for the vapor pressure P_{sat} of some compounds.

Name	Formula	A	B	C	Range (°C)	$\Delta_{vap}H_n$ (kJ mol^{-1})	t_n (°C)
Acetone	C_3H_6O	14.3145	2756.22	228.060	−27–77	29.10	56.2
MEK	C_4H_8O	14.1343	2838.4	218.690	−8–103	31.30	79.6
Ether	$C_4H_{10}O$	14.0735	2511.29	231.200	−43–55	26.52	34.4
Tetra	CCl_4	14.0572	2914.23	232.148	−14–101	29.82	76.6
Chloroform	$CHCl_3$	13.7324	2548.74	218.552	−23–84	29.94	61.1
Dichloromethane	CH_2Cl_2	13.9891	2463.912	223.240	−38–60	28.06	39.7
Water	H_2O	16.3872	3885.70	230.170	0–200	40.66	100.0
Methanol	CH_4O	16.5785	3638.27	239.500	−11–83	35.21	64.7
Ethanol	C_2H_6O	16.8958	3795.17	230.918	3–96	38.56	78.2
1-Propanol	C_3H_8O	16.1154	3483.67	205.807	20–116	41.44	97.2
2-Propanol	C_3H_8O	16.6796	3640.20	219.610	8–100	39.85	82.2
1-Butanol	$C_4H_{10}O$	15.3144	3212.43	182.739	37–138	43.29	117.6
2-Butanol	$C_4H_{10}O$	15.1989	3026.03	186.500	25–120	40.75	99.5
Iso-butanol	$C_4H_{10}O$	14.6047	2740.95	166.670	30–128	41.82	107.8
Tert-butanol	$C_4H_{10}O$	14.8445	2658.29	177.650	10–101	39.07	82.3
Benzene	C_6H_6	13.7819	2726.81	217.572	6–104	30.73	80.0
Toluene	C_7H_8	13.9320	3056.96	217.625	13–136	33.18	110.6
Ethylbenzene	C_8H_{10}	13.9726	3259.93	212.300	33–163	35.57	136.2
Phenol	C_6H_6O	14.4387	3507.80	175.400	20–208	46.18	181.8
o-Xylene	C_8H_{10}	14.0415	3358.79	212.041	40–172	36.24	144.4
m-Xylene	C_8H_{10}	14.1387	3381.81	216.120	35–166	35.66	139.1
p-Xylene	C_8H_{10}	14.0579	3331.45	214.627	35–166	35.67	138.3
Phenol	C_6H_6O	14.4387	3507.80	175.400	80–208	46.18	181.8
Butane	C_4H_{10}	13.6608	2154.70	238.789	−73–19	22.44	−0.5
Iso-butane	C_4H_{10}	13.8254	2181.79	248.870	−83–7	21.30	−11.9
Pentane	C_5H_{12}	13.7667	2451.88	232.014	−45–58	25.79	36
Cyclopentane	C_5H_{10}	13.9727	2653.90	234.510	−35–71	27.30	49.2
Hexane	C_6H_{14}	13.8193	2696.04	224.317	−19–92	28.85	68.7
Cyclohexane	C_6H_{12}	13.6568	2723.44	220.618	9–105	29.97	80.7
Heptane	C_7H_{16}	13.8622	2910.26	216.432	4–123	31.77	98.4
Octane	C_8H_{18}	13.9346	3123.13	209.635	26–152	43.41	125.6
Iso-octane	C_8H_{18}	13.6703	2896.31	220.767	2–125	30.79	99.2
Nonane	C_9H_{20}	13.9854	3311.19	202.694	46–178	36.91	150.8
Decane	$C_{10}H_{22}$	13.9748	3442.76	193.858	65–203	38.75	174.1
Acetonitrile	C_2H_3H	14.8950	3413.10	250.253	−27–81	30.19	81.6
Acetic acid	$C_2H_4O_2$	15.0717	3580.80	224.650	24–142	23.70	117.9

$\ln P_{sat}$ (kPa) $= A - B/(t\,(°C) + C)$, where t is the temperature in degree Celsius, $\Delta_{vap}H_n$ is the enthalpy of evaporation, and t_n is the normal boiling temperature. Source: Adapted from Smith et al. [9].
MEK = methyl ethyl ketone.

Table F.13 Structural contributions for the parachor $\mathscr{P}$.

Carbon–hydrogen		Special groups			
C	9.0	–COO– (esters)	63.8	N (if not noted above)	17.5
H	15.5	–COOH (acids)	73.8	S	49.1
CH_3	55.5	–OH	29.8	P	40.5
CH_2 in $-CH_2-$		$-NH_2$	42.5	F	26.1
$n < 12$	40.0	–O–	20.0	Cl	55.2
$n > 12$	40.3	$-NO_2$	74	I	90.3
Alkyl groups		$-NO_3$ (nitrate)	93	Ethylenic bonds =C<	19.1
1-Methylethyl	133.3	$-CO(NH_2)$	91.7	*Terminal*	
1-Methylpropyl	171.9	*R–[–CO–]–R′ for CO in ketones*		2,3 position	17.7
1-Methylbutyl	211.7	$R + R' = 2$	51.3	Triple bond	40.6
2-Methylpropyl	173.3	$R + R' = 3$	49.0	Ring closure	
1-Ethylpropyl	209.5	$R + R' = 4$	47.5	Three membered	12
1,1-Dimethylethyl	170.4	$R + R' = 5$	46.3	Four membered	6.0
1,1-Dimethylpropyl	207.5	$R + R' = 6$	45.3	Five membered	3.0
1,2-Dimethylpropyl	207.9	$R + R' = 7$	44.1	Six membered	0.8
1,1,2-Trimethylpropyl	243.5	–CHO	66		
C_6H_5	189.6	O (if not noted above)	20		

Adapted from Sugden [14] and Quayle [15], to be used with γ (dyn cm^{-1} = 10^{-3} N m^{-2}) and M/ρ (cm^3 mol^{-1}). For polymers, M is the molecular mass of the repeating unit.

Table F.14 Ionic radii according to Pauling r_P and Shannon–Prewitt r_{SP} for IV and VI coordination.

Ion	r_P (pm)	$^{IV}r_{SP}$ (pm)	$^{VI}r_{SP}$ (pm)	Ion	r_P (pm)	$^{IV}r_{SP}$ (pm)	$^{VI}r_{SP}$ (pm)
Li^+	60	73	90 (88)	B^{3+}	20	25 (26)	41
Na^+	95	113	116	Al^{3+}	50	53	67.5 (67.0)
K^+	133	151	152	Sc^{3+}	81	—	88.5 (87.0)
Rb^+	148	—	166 (163)	Ga^{3+}	62	61	76
Cs^+	169	—	181 (184)	Y^{3+}	93	—	104 (103.2)
Be^{2+}	31	41	59 (45)	In^{3+}	81	76	94 (3)
Mg^{2+}	65	71 (63)	86.0	Fe^{3+}		63 HS	69 LS
Ca^{2+}	99	—	114			—	78.5 HS
Zn^{2+}	74	74	88.0 (88.5)	F^-	136	117	119
Sr^{2+}	113	—	132 (130)	Cl^-	181	—	167

(*Continued*)

Table F.14 (Continued)

Ion	r_P (pm)	$^{IV}r_{SP}$ (pm)	$^{VI}r_{SP}$ (pm)	Ion	r_P (pm)	$^{IV}r_{SP}$ (pm)	$^{VI}r_{SP}$ (pm)
Ba^{2+}	135	—	149 (150)	Br^-	195	—	182
Fe^{2+}		77 HS	75 LS	I^-	216	—	206
		—	92 HS (91)	O^{2-}	140	124	126
Mn^{2+}		80 HS	81 LS	S^{2-}	184	—	170
			97 HS (96)	Se^{2-}	198	—	184
Ni^{2+}		69	84 (83)	Te^{2-}	221	—	207

Data are from Pauling [16], Shannon and Prewitt [17], and Shannon [18]. For VI coordination, values not in parenthesis indicate that the same values were reported in 1969 and 1976; data in parenthesis refer to the 1969 paper. LS = low spin, HS = high spin.

Table F.15 Pauling EN, IP, EA[a)] and Sanderson data[b)] for some elements.

Element	X_P	IP (eV)	EA (eV)	X_S	$\Delta X_{S,1}$	r_{cov} (Å)	E_{ato} (kcal mol^{-1})	E''' (kcal mol^{-1})
H	2.10	13.598	0.754	2.592	2.528	0.320	52.1	104.2
Li	0.98	5.392	0.618	0.670	1.285	1.336	38.6	24.6
Be	1.57	9.323	−0.2[c)]	1.810	2.112	0.887	77.5	67.6
B	2.04	8.298	0.280	2.275	2.368	0.822	133.8	76.7
C	2.55	11.260	1.262	2.746	2.602	0.772	171.3	85.4[d)]
N	3.04	14.534	−0.07[c)]	3.194	2.806	0.734	113	94.9
O	3.44	13.618	1.461	3.654	3.001	0.702	59.6	104.0
F	3.98	17.423	3.401	4.000	3.140	0.681	18.9	113.1
Na	0.93	5.139	0.548	0.560	1.175	1.539	25.8	16.4
Mg	1.31	7.646	−0.2[c)]	1.318	1.802	1.373	35.2[e)]	43.3
Al	1.61	5.986	0.433	1.714	2.055	1.258	78.8	48.2
Si	1.90	8.152	1.390	2.138	2.296	1.169	107.6	54.1
P	2.19	10.487	0.747	2.515	2.490	1.107	75.6	60.0
S	2.58	10.360	2.077	2.957	2.700	1.049	66.2	65.9
Cl	3.16	12.968	3.613	3.475	2.927	0.994	29	71.8
K	0.82	4.341	0.501	0.445	1.047	1.962	21.4	13.1
Ca	1.00	6.113	0.025	0.946	1.527	1.740	42.6	30.8
Cu	1.90	7.726	1.236	2.033	2.239	1.331	80.7	31.3
Zn	1.65	9.394	−0.49[c)]	2.223	2.341	1.292	31.2	35.8
Ga	1.81	5.999	0.301	2.419	2.442	1.256	65.4	40.3
Ge	2.01	7.899	1.232	2.618	2.540	1.223	89.5	44.8
As	2.01	9.789	0.804	2.816	2.635	1.194	72.3	49.3

(Continued)

Table F.15 (Continued)

Element	X_P	IP (eV)	EA (eV)	X_S	$\Delta X_{S,1}$	r_{cov} (Å)	E_{ato} (kcal mol^{-1})	E''' (kcal mol^{-1})
Se	2.55	9.752	2.021	3.014	2.726	1.167	54.3	53.8
Br	2.96	11.814	3.364	3.219	2.817	1.142	26.7	58.3
Rb	0.82	4.177	0.486	0.312	0.877	2.160	19.6	12.4
Sr	0.95	5.695	0.052	0.721	1.333	1.910	39.1	24.6
Ag	1.93	7.576	1.304	1.826	2.122	1.533	68.1	27.7
Cd	1.69	8.994	−0.33[c)]	1.978	2.208	1.493	26.7	30.4
In	1.78	5.786	0.404	2.138	2.296	1.455	58	33.1
Sn	1.96	7.344	1.112	2.298	2.380	1.420	72	35.8
Sb	2.05	8.608	1.047	2.458	2.461	1.389	63.2	38.5
Te	2.10	9.010	1.971	2.618	2.540	1.360	47	41.2
I	2.66	10.451	3.059	2.778	2.617	1.333	25.5	43.9
Cs	0.79	3.984	0.472	0.220	0.736	2.350	18.7	10.8
Ba	0.89	5.212	0.145	0.651	1.267	1.980	42.5	22.2
Hg	1.9	10.438	−0.63[c)]	2.195	2.326	1.500	14.7	8.6
Tl	1.8	6.108	0.320	2.246	2.353	1.490	43.6	16.6
Pb	1.8	7.417	0.357	2.291	2.376	1.480	46.7	24.2
Bi	1.9	7.286	0.942	2.342	2.403	1.470	50.1	32.2

a) Rumble [19], https://hbcp.chemnetbase.com/faces/contents/ContentsSearch.xhtml.
b) Sanderson [20].
c) Sanderson [21].
d) For *n*-alkanes 82.8.
e) Experimental value, empirical 50.3 preferred.

Table F.16 Unweakened (E'''), half weakened (E''), and fully weakened bond energies (E') according to Sanderson.[a)]

Element	E''' (kcal mol^{-1})	E'' (kcal mol^{-1})	E' (kcal mol^{-1})	Element	E''' (kcal mol^{-1})	E'' (kcal mol^{-1})	E' (kcal mol^{-1})
N	94.9	66.9	38.8	As	49.3	45.1	40.9
O	104.0	68.8	33.6	Se	53.8	46.0	38.2
F	113.1	76.8	40.5	Br	58.3	52.2	46.1
P	60.0	56.9	53.7	Sb	38.5	35.8	33.0
S	65.9	60.4	54.9	Te	41.2	39.2	37.2
Cl	71.8	64.9	58.0	I	43.9	40.0	36.1

In the polar covalence scheme, molecules containing N, O, F, P, S, and Cl atoms need special attention. For this, Sanderson introduced unweakened, half weakened, and fully weakened bond energies on the basis of his empirical considerations. For details, we refer to the references quoted.

a) Sanderson [22]. More extensive data are given in Sanderson [21]. Note that the data given in *Simple Inorganic Substances* differ from those given in *Polar Covalence*.

Table F.17 *B*-values for some elements according to Sanderson.[a)]

Element	B_{gas} (pm e^{-1})	B_{solid} (pm e^{-1})	Element	B_{gas} (pm e^{-1})	B_{solid} (pm e^{-1})	Element	B_{gas} (pm e^{-1})	B_{solid} (pm e^{-1})
H	97.2	—	Be	61.2	—	F	52.0	94.1
Li	115.8	72.0	Mg	62.7	—	Cl	68.8	115.7
Na	92.5	74.4	Ca	63.0	—	Br	66.8	116.7
K	106.0	91.5	Sr	59.0	44.1	I	67.0	128.0
Rb	112.7	100.5	Ba	48.1	35.9			
Cs	117.9	94.1						

a) Sanderson [22]. More extensive data are given in Sanderson [23]. Note that the data given in *Polar Covalence* differ from those given in *Chemical Bonds and Bond Energies*.

Table F.18 Data for the orbital electronegativity model as given by Hinze.[a)]

Element	IP (eV)	EA (eV)	b^0 (V)	b^1 (V e^{-1})	c (V e^{-1})	X_M (eV)	X_P
H	13.599	0.755	20.021	0.000	6.422	7.18	2.17
Li (s)	5.392	0.620	7.778	0.000	2.386	3.01	0.91
Be (di)	8.552	1.018	12.319	6.307	3.767	4.79	1.45
B (tr)	11.520	1.130	16.310	8.823	5.060	6.19	1.88
C (te)	14.423	1.719	20.775	10.741	6.352	8.07	2.45
C (tr)	15.447	2.315	22.013	10.677	6.566	8.88	2.69
C (di)	17.277	3.677	24.077	10.596	6.800	10.48	3.17
C (π)	11.0.36	0.400	16.354	11.026	5.318	5.72	1.73
N (15%s)	16.974	2.390	24.266	12.900	7.292	9.68	2.93
O (10%s)	20.166	3.644	28.427	14.792	8.261	11.91	3.61
F (5%s)	22.853	4.469	32.045	0.000	9.192	13.66	4.14
Na (s)	5.144	0.560	7.436	0.000	2.292	2.85	0.86
Mg (di)	7.029	0.929	10.079	5.070	3.050	3.98	1.21
Al (tr)	8.860	1.806	12.387	6.118	3.527	5.33	1.62
Si (te)	11.384	2.580	15.786	7.225	4.402	6.98	2.12
P (15%s)	13.431	2.785	18.754	8.816	5.323	8.11	2.46
S (10%s)	14.049	3.377	19.385	10.524	5.336	8.71	2.64
Cl (5%s)	15.921	4.183	21.790	0.000	5.869	10.05	3.05
K (s)	4.431	0.501	6.261	0.000	1.920	2.42	0.73
Ca (di)	5.747	0.997	8.122	4.108	2.375	3.37	1.02
Cu (s)	7.726	1.226	10.976	0.000	3.250	4.48	1.36
Zn (di)	8.254	1.407	11.678	5.781	3.424	4.83	1.46
Ga (tr)	9.762	1.926	13.680	6.474	3.918	5.84	1.77

(Continued)

Table F.18 (Continued)

Element	IP (eV)	EA (eV)	b^0 (V)	b^1 (V e^{-1})	c (V e^{-1})	X_M (eV)	X_P
Ge (te)	11.544	2.564	16.034	7.397	4.490	7.05	2.14
As (15%s)	12.350	2.534	17.258	9.572	4.908	7.44	2.25
Se (10%s)	12.831	3.401	17.546	9.833	4.715	8.12	2.46
Br (5%s)	14.653	4.057	19.951	0.000	5.298	9.36	2.83
Rb (s)	4.176	0.484	6.022	0.000	1.846	2.33	0.71
Sr (di)	5.369	0.979	7.564	3.837	2.195	3.17	0.96
Ag (s)	7.576	1.303	10.713	0.000	3.137	4.44	1.35
Cd (di)	7.790	1.091	11.140	5.645	3.350	4.44	1.35
In (tr)	9.073	1.707	12.756	6.017	3.683	5.39	1.63
Sn (te)	10.346	3.680	13.679	5.986	3.333	7.01	2.12
Sb (15%s)	11.983	2.217	16.866	7.628	4.883	7.10	2.15
Te (10%s)	11.483	3.607	15.421	8.712	3.938	7.54	2.29
I (5%s)	13.197	3.741	17.925	0.000	4.728	8.47	2.57

a) Hinze [24]. Various valence state values for carbon are given: s = monovalent, di = divalent, tr = trivalent, te = tetravalent. For other elements, data corresponding to the most probable valence state are given. More extensive data are given by Bergmann and Hinze [25]. Note that IP, EA and X_P values deviate slightly from those in Table F.15.

Table F.19 Covalent radii r_{cov} and bond energy D_{AA} according to the orbital electronegativity model as given by Hinze.[a)]

Element	r_{cov} (Å)	Element	r_{cov} (Å)	Element	D_{AA} (kcal mol^{-1})
H (s)	0.343	Li (p)	1.502	H (s)	107.5
B^- (te)	0.880	B (tr)	0.831	Li (s)	33.0
B (di)	0.831	B (di π)	0.619	C (te)	85.6
C (te)	0.763	C (tr)	0.747	C (tr)	(103.0)
C (di)	0.710	C (tr π)	0.664	C (di)	139.2
C (di π)	0.617	C (di ππ)	0.603	C (tr π)	174.9
N (te)	0.695	N (tr)	0.686	C (di ππ)	235.8
N^+ (te)	0.767	N (tr π)	0.601	N (te)	78.2
N (di ππ)	0.562	O (te)	0.686	O (p)	46.9
O (tr π)	0.552	F (p)	0.569	F (p)	49.7
F (te)	0.726	Na (p)	1.793	Na (s)	16.2
Si (te)	1.106	Si (p p)	1.002	Si (te)	75.9
P (p)	1.097	P (te)	1.018	P (p)	57.3
P (pe)	1.006	P^+ (te)	1.022	S (p)	63.2

(Continued)

Table F.19 (Continued)

Element	r_{cov} (Å)	Element	r_{cov} (Å)	Element	D_{AA} (kcal mol^{-1})
P (te π)	0.905	S (p)	1.023	Cl (p)	62.2
S (te)	1.016	S (oh)	1.029	K (s)	10.6
S (p π)	0.944	S (te π)	0.883	Ge (te)	62.4
Cl (p)	0.988	Cl (te)	0.974	Br (p)	44.6
K (p)	2.198	Ge (te)	1.192	Rb (s)	7.1
Ge (p π)	1.082	As (p)	1.220	Sn (te)	45.7
Se (p)	1.157	Se (p π)	1.074	Sb (te)	23.2
Br (p)	1.135	Br (te)	1.084	I (p)	37.1
Rb (p)	2.374	Sn (te)	1.379		
Sn (p π)	1.278	Sb (p)	1.394		
Te (p π)	1.275	I (p)	1.338		
I (te)	1.284				
c_0 (Å e^{-1})	0.075			c_3 (kcal mol^{-1} eV^{-1})	18.36
c_1 (V Å e^{-2})	−0.844			c_4 (kcal Å mol^{-1} e^{-2})	−751.98
c_2 (V Å^2 e^{-3})	14.558				

a) Hinze [24]. The various valence state values are given: s = monovalent, di = divalent, tr = trivalent, te = tetravalent. More extensive data are given by Bergmann and Hinze [25].

References

1 Cuadros, F., Cachadina, I., and Ahamuda, W. (1996). *Mol. Eng.* 6: 319.

2 Hirschfelder, J.O., Curtiss, C.F., and Bird, R.B. (1954). *Molecular Theory of Gases and Liquids*. New York: John Wiley & Sons.

3 Moelwyn-Hughes, E.A. (1961). *Physical Chemistry*, 2e. Oxford: Pergamon Press.

4 Rowlinson, J.S. and Swinton, F.L. (1982). *Liquids and Liquid Mixtures*, 3e. London: Butterworth.

5 Landolt, H.H. and Börnstein, R. (1951). *Zahlenwerte und Functionen, Atom- und Molekularphysik*, 6e, Vol. 1, Part 3. Berlin: Springer.

6 Bergmann, L. (1954). *Der Ultraschall*, 6e. Stuttgart: Hirzel Verlag.

7 Kayode Coker, A. (2007). *Ludwig's Applied Process Design for Chemical and Petrochemical Plants*, 4e, Vol. 1. Amsterdam: Elsevier.

8 Prausnitz, J.M., Lichtenthaler, R.N., and de Azevedo, E.G. (1986). *Molecular Thermodynamics of Fluid-phase Equilibria*, 2e. Englewood Cliffs, NJ: Prentice-Hall.

9 Smith, J.M., Van Ness, H.C., and Abbott, M.M. (2005). *Introduction to Chemical Engineering Thermodynamics*, 7e. New York: McGraw-Hill Book Comp.

10 Pitzer, K.S. (1995). *Thermodynamics*, 3e. New York: McGraw-Hill Book Comp.

11 Connors, K.A. (1990). *Chemical Kinetics*. Weinheim, Germany: VCH.

12 Weast, R.C. (ed.) (1972). *Handbook of Chemistry and Physics*, 53e. Cleveland: Chemical Rubber Company.

13 Poling, B.E., Prausnitz, J.M., and O'Connell, J.P. (2001). *The Properties of Gases and Liquids*, 5e. New York: McGraw-Hill Book Comp.

14 Sugden, S. (1924). *J. Chem. Soc.* 125: 32 and 1177.

15 Quayle, O.R. (1953). *Chem. Rev.* 53: 439.

16 Pauling, L. (1960). *The Nature of the Chemical Bond*, 3e. Ithaca: Cornell University Press.

17 Shannon, R.D. and Prewitt, C.T. (1969). *Acta Cryst. B* 25: 925.

18 Shannon, R.D. (1976). *Acta Cryst. A* 32: 751.

19 Rumble, J.R. (ed.) (2022). *CRC Handbook of Chemistry and Physics*, Online, 103e.

20 Sanderson, R.T. (1983). *J. Am. Chem. Soc.* 105: 2259.

21 Sanderson, R.T. (1989). *Simple Inorganic Substances*. Malabar, FL: R.E. Krieger Publishing Company.

22 Sanderson, R.T. (1983). *Polar Covalence*. NY: Academic Press.

23 Sanderson, R.T. (1976). *Chemical Bonds and Bond Energies*. New York: Academic Press.

24 Hinze, J. (1999). *The Concept of Electronegativity of Atoms in Molecules*, Ch. 7 in *Pauling's Legacy: Modern Modelling of the Chemical Bond* (ed. Z.B. Maksić and W.J. Orville-Thomas), p. 189. Amsterdam: Elsevier.

25 Bergmann, D. and Hinze, J. (1987). *Electronegativity and Charge Distribution*, Ch. 7 in No. 66 of *Structure and Bonding* (ed. K.D. Sen and C.D. Jørgensen), p. 145. Berlin: Springer.

Appendix G

Answers to Selected Problems

2.1 Consider first the *linear momentum* $\boldsymbol{p}$ for a closed system with constant mass m and velocity $\boldsymbol{v} = \dot{\boldsymbol{r}}$, defined by $\boldsymbol{p} = \Sigma_j \boldsymbol{p}_j = \Sigma_j m_j \boldsymbol{v}_j$. The rate of change $\mathrm{d}\boldsymbol{p}/\mathrm{d}t = \dot{\boldsymbol{p}}$ is, with acceleration $\boldsymbol{a} = \dot{\boldsymbol{v}}$, given by $\dot{\boldsymbol{p}} = m\dot{\boldsymbol{v}} \equiv m\boldsymbol{a} = \boldsymbol{f}$. For $\boldsymbol{f} = 0$, $\mathrm{d}\boldsymbol{p}/\mathrm{d}t = 0$ or $\boldsymbol{p}$ = constant, expressing the *conservation of linear momentum.*

Consider next the *angular momentum* $\boldsymbol{l}$ with respect to the origin for such a system with particles at position $\boldsymbol{r}_i$, defined by $\boldsymbol{l} = \Sigma_j \boldsymbol{l}_j = \Sigma_j \boldsymbol{r}_j \times \boldsymbol{p}_j = \Sigma_j \boldsymbol{r}_j \times [m_j \mathrm{d}\boldsymbol{r}_j/\mathrm{d}t]$. Introducing the *moment of force* or *torque*, defined as $\boldsymbol{q} = \Sigma_j \boldsymbol{q}_j = \Sigma_j \boldsymbol{r}_j \times \boldsymbol{f}_j$, the rate of change $\mathrm{d}\boldsymbol{l}/\mathrm{d}t$ is given by $\dot{\boldsymbol{l}} = \dot{\boldsymbol{r}} \times m\dot{\boldsymbol{r}} + \boldsymbol{r} \times m\ddot{\boldsymbol{r}} = 0 + (\boldsymbol{r} \times \boldsymbol{f}) = \boldsymbol{q}$. Hence, if $\boldsymbol{q} = 0$, $\mathrm{d}\boldsymbol{l}/\mathrm{d}t = 0$. Likewise, for linear motion $\boldsymbol{f} = 0$ leads to $\boldsymbol{p}$ = constant, and for angular motion $\boldsymbol{q} = 0$ leads to $\boldsymbol{l}$ = constant, expressing *conservation of angular momentum.*

Consider now the motion of such a system under the influence of a conservative force $\boldsymbol{f}$. A force $\boldsymbol{f}$ is conservative if the work done by that force, when moving a particle from one point to another, is independent of the path taken. Equivalently, the force $\boldsymbol{f}$ is conservative if it can be derived from a function and is called the *potential energy*, $\Phi(\boldsymbol{r})$ by $\boldsymbol{f} = -\nabla\Phi(\boldsymbol{r})$. We evaluate the integral I of $\boldsymbol{f} \cdot \mathrm{d}\boldsymbol{r} = \boldsymbol{f} \cdot \boldsymbol{v}\,\mathrm{d}t$. Here, $\boldsymbol{f} \cdot \boldsymbol{v}$ is the *power*, that is, work done by the force per unit time. On the one hand, we have

$$I = \int_{t'}^{t''} \boldsymbol{f} \cdot \mathrm{d}\boldsymbol{r} = \int_{t'}^{t''} (m\ddot{\boldsymbol{r}} \cdot \dot{\boldsymbol{r}})\mathrm{d}t = {}^1\!/\!_2 m(\dot{\boldsymbol{r}} \cdot \dot{\boldsymbol{r}})|_{t'}^{t''} = \vartheta(t') - \vartheta(t')$$

where the *kinetic energy* $\vartheta = {}^1\!/\!_2 m\boldsymbol{v}^2$ is introduced. We also have

$$I = \int_{t'}^{t''} \boldsymbol{f} \cdot \mathrm{d}\boldsymbol{r} = -\int_{t'}^{t''} \nabla\Phi \cdot \mathrm{d}\boldsymbol{r} = -\Phi(t'') + \Phi(t')$$

For a conservative force, the (total mechanical) *energy* $E = \vartheta(\dot{\boldsymbol{r}}, \boldsymbol{r}) + \Phi(\boldsymbol{r})$ is thus constant during the motion of the system, denoted as *conservation of energy.*

3.5 From $\langle\Phi|\Phi\rangle = 1$, we obtain $\partial\langle\Phi|\Phi\rangle/\partial\alpha = 0$. Thus, from $\partial\langle\Phi|H|\Phi\rangle/\partial\alpha = \langle\partial\Phi/\partial\alpha|H|\Phi\rangle + \langle\Phi|\partial H/\partial\alpha|\Phi\rangle + \langle\Phi|H|\partial\Phi/\partial\alpha\rangle = \langle\Phi|\partial H/\partial\alpha|\Phi\rangle + \partial\langle\Phi|\Phi\rangle/\partial\alpha = \langle\Phi|\partial H/\partial\alpha|\Phi\rangle$.

Phases of Matter and their Transitions: Concepts and Principles for Chemists, Physicists, Engineers, and Materials Scientists, First Edition. Gijsbertus de With.

4.4 Assuming that the plane of the electron orbit with radius a_0 displaces by l from the proton, the induced dipole moment is $\mu_{\text{ind}} = \alpha E = el$, where α is the polarizability, E is the electric field, and e is the unit charge. Denoting the angle between proton, electron, and center of the orbit by θ, the required internal force $F_{\text{int}} = e^2\sin\theta/4\pi\varepsilon_0 {a_0}^2 \cong e^2 l/4\pi\varepsilon_0 {a_0}^3 = e\mu_{\text{ind}}/4\pi\varepsilon_0 {a_0}^3$. Equating F_{int} to the external force $F_{\text{ext}} = eE$, one has $\mu_{\text{ind}} = 4\pi\varepsilon_0 {a_0}^3 E$ or $\alpha = 4\pi\varepsilon_0 {a_0}^3$.

4.9 Taking $\alpha_1/\alpha_2 = 2$, we obtain the following results rendering the Berthelot approximation doubtful:

σ_{11}/σ_{22}	$\varepsilon_{11}/\varepsilon_{22}$	Berthelot	Eq. (4.17)	σ_{11}/σ_{22}	$\varepsilon_{11}/\varepsilon_{22}$	Berthelot	Eq. (4.17)
1	1	1.00	0.80	2	1	1.0	0.33
1	2	1.41	1.33	2	2	1,41	0.34

4.10 $U_{\text{coh}} = 14.3\,\text{kJ mol}^{-1}$ (and $H_{\text{coh}} = U_{\text{coh}} + RT = 16.8\,\text{kJ mol}^{-1}$) and $U_{\text{coh}} = 28.9\,\text{J mol}^{-1}$ (and $H_{\text{coh}} = U_{\text{coh}} + RT = 31.4\,\text{kJ mol}^{-1}$).

6.1 Limiting ourselves to thermal interaction only, in general, a system receives heat from a number of reservoirs. Using T for the subsystem of interest,

$$dS = T^{-1}\Sigma_j\delta Q_j = \Sigma_j {T_j}^{-1}\delta Q_j + \Sigma_j(T^{-1} - {T_j}^{-1})\delta Q_j$$

For a two-part system, using T_{res} for the reservoir, we have

$$dS = {T_{\text{res}}}^{-1}\delta Q + \left(T^{-1} - {T_{\text{res}}}^{-1}\right)\delta Q$$

For $T_{\text{res}} > T$, $\delta Q > 0$ and for $T_{\text{res}} < T$, $\delta Q < 0$. Hence, $(T^{-1} - {T_{\text{res}}}^{-1})\delta Q \geq 0$ always. Therefore, we can write

$$dS = {T_{\text{res}}}^{-1}\delta Q + \left(T^{-1} - {T_{\text{res}}}^{-1}\right)\delta Q \equiv dS^{(r)} + dS^{(i)} \quad \text{with} \quad dS^{(i)} \geq 0$$

6.2 From $dU = TdS - PdV$, we have $(\partial U/\partial V)T = T(\partial S/\partial T)_T - P = T(\partial P/\partial T)_V - P$. From $P = nRT/V$, we have $(\partial P/\partial T)_V = nR/V = P/T$, resulting in $(\partial U/\partial V)T = 0$.

6.3 Using the substitution rule for partial differential quotients, we have

$$(\partial S/\partial T)_P = (\partial S/dT)_V + (\partial S/dV)_T(\partial V/dT)_P$$

Using $(\partial S/dV)_T = (\partial P/dT)_V = \alpha/\kappa_T$ and multiplying by T, we arrive at

$$T(\partial S/\partial T)_P = T(\partial S/dT)_V + \alpha^2 VT/\kappa_T \quad \text{or} \quad C_P - C_V = \alpha^2 VT/\kappa_T$$

Similarly,

$$(\partial V/\partial P)_T = (\partial V/\partial P)_S + (\partial V/\partial S)_P(\partial S/\partial P)_T \text{ or, using Maxwell relations,}$$

$$V^{-1}(\partial V/\partial P)_T = V^{-1}(\partial V/\partial P)_S + V^{-1}(\partial T/\partial P)_S(\partial V/\partial T)_P \quad \text{or} \quad \kappa_T = \kappa_S + (\partial T/\partial P)_S\alpha$$

Using the minus-one rule, we have

$$(\partial T/\partial P)_S = -(\partial S/\partial P)_T(\partial T/\partial S)_P = (\partial V/\partial T)_P T/C_P = V\alpha T/C_P \text{ and thus}$$

$$\kappa_T - \kappa_S = \alpha^2 VT/C_P$$

Solving leads to $C_P/C_V = \kappa_T/\kappa_S$. Alternatively, $\kappa_T/\kappa_S = (\partial V/\partial P)_T/(\partial V/\partial P)_S$. Using the minus-one rule, we obtain

$$(\partial V/\partial P)_T = -(\partial V/\partial T)_P/(\partial P/\partial T)_V \quad \text{and} \quad (\partial V/\partial P)_S = -(\partial S/\partial P)_V/(\partial S/\partial V)_P$$

Substitution leads to

$$\kappa_T/\kappa_S = (\partial V/\partial T)_P(\partial S/\partial V)_P/(\partial P/\partial T)_V(\partial S/\partial P)_V = (\partial S/\partial T)_P/(\partial S/\partial T)_V = C_P/C_V$$

6.4 From $\mathrm{d}U = T\mathrm{d}S - P\mathrm{d}V$, we have $(\partial U/\partial V)_T = T(\partial S/\partial V)_T - P$. Inserting the Maxwell relation $(\partial P/\partial T)_V = (\partial S/\partial V)_T$ leads to $(\partial U/\partial V)_T = T(\partial P/\partial T)_V - P$. From $P = ⅓u$ with $u = U/V$, we have $(\partial P/\partial T)_V = ⅓(\partial u/\partial T)_V$, and thus $(\partial U/\partial V)_T = T \cdot ⅓(\partial u/\partial T)_V - ⅓u$. Hence,

$$T(\partial u/\partial T)_V = u + 3(\partial U/\partial V)_T = u + 3u = 4u \quad \text{or} \quad u = aT^4 \quad \text{and} \quad P = \tfrac{1}{3}aT^4$$

Now $S(V,T) = \int_0^V (\partial S/\partial V)_T \mathrm{d}V = \int_0^V (\partial P/\partial T)_V \mathrm{d}V = 4aT^3V/3$ or

$$T = (3S/4aV)^{1/3} \quad \text{and} \quad P = \tfrac{1}{3}a(3S/4aV)^{4/3}$$

With $u = aT^4$ and $T = (3S/4aV)^{1/3}$, we find $U/V = a(3S/4aV)^{4/3}$ or

$$S = 4a^{1/4}\, U^{3/4}V^{1/3}/3$$

6.5 For the entropy we write $S = N_1s_1 + N_2s_2$ with $s = s(u,v)$, while the constraints are $C_N = N_1 + N_2 - N = 0$, $C_U = N_1u_1 + N_2u_2 - U = 0$, and $C_V = N_1v_1 + N_2v_2 - V = 0$. Using the Lagrange technique, we thus have to solve $\delta(L) = 0$, where $L = S + \alpha C_N + \beta C_U + \gamma C_V$, which leads, using $\delta(\cdot) = \Sigma_j(\partial\cdot/\partial x_j)\mathrm{d}x_j$, to

$$\begin{aligned}\delta(L) = {} & (s_1 + \alpha + \beta u_1 + \gamma v_1)\mathrm{d}N_1 + (s_2 + \alpha + \beta u_1 + \gamma v_1)\mathrm{d}N_2 \\ & + [N_1(\partial s_1/\partial u_1) + \beta N_1]\mathrm{d}u_1 + [N_2(\partial s_2/\partial u_2) + \beta N_1]\mathrm{d}u_2 \\ & + [N_1(\partial s_1/\partial v_1) + \gamma N_1)]\mathrm{d}v_1 + [N_2(\partial s_2/\partial v_2) + \gamma N_1)]\mathrm{d}v_2 = 0\end{aligned}$$

Using $\partial s/\partial u = 1/T$ and $\partial s/\partial v = P/T$, the final result becomes

$$\begin{aligned}\delta(L) = {} & (s_1 + \alpha + \beta u_1 + \gamma v_1)\mathrm{d}N_1 + (s_2 + \alpha + \beta u_1 + \gamma v_1)\mathrm{d}N_2 \\ & + [(N_1/T_1) + \beta N_1)]\mathrm{d}u_1 + [(N_2/T_2) + \beta N_1)]\mathrm{d}u_2 \\ & + [(N_1P_1/T_1) + \gamma N_1)]\mathrm{d}v_1 + [(N_2P_2/T_2) + \gamma N_1)]\mathrm{d}v_2 = 0,\end{aligned}$$

where $\mathrm{d}N_1$, $\mathrm{d}N_2$, $\mathrm{d}u_1$, $\mathrm{d}u_2$, $\mathrm{d}v_1$, and $\mathrm{d}v_2$ can now be considered as independent. Hence, first, we have $\beta = -1/T_1 = -1/T_2$, or $T_1 = T_2 = T$. Second, we have $\gamma = -P_1/T = -P_2/T$, or $P_1 = P_2 = P$. Finally, we have $\alpha = -s_1 - (-u_1/T) - (-Pv_1/T) = -s_2 - (-u_2/T) - (-Pv_2/T)$ and, because $\mu = u - Ts + Pv$, $\mu_1 = \mu_2$.

7.2 $Z = 1 + \exp(-\varepsilon/kT)$, $p_1 = 0.73$, $p_2 = 0.27$, $U = \varepsilon/[\exp(\varepsilon/kT) + 1]$,

$$C_V = [\exp(\varepsilon/kT) + 1]^{-2}(\varepsilon^2/kT^2)\exp(\varepsilon/kT)$$

7.9 Given $\phi = \phi(\mathbf{r},\boldsymbol{\omega})$ and $\overline{\phi^n} = \int \phi^n \mathrm{d}\boldsymbol{\omega}$, we have

$$\langle\phi\rangle = \int \phi \exp(-\beta\phi)\mathrm{d}\mathbf{r}\mathrm{d}\boldsymbol{\omega} \Big/ \int \exp(-\beta\phi)\mathrm{d}\mathbf{r}\mathrm{d}\boldsymbol{\omega} \quad \text{with} \quad Q = \int \exp(-\beta\phi)\mathrm{d}\mathbf{r}\mathrm{d}\boldsymbol{\omega}$$

$$= \int \phi(1-\beta\phi+½\ldots)\mathrm{d}\mathbf{r}\mathrm{d}\boldsymbol{\omega} \Big/ \int (1-\beta\phi+½\ldots)\mathrm{d}\mathbf{r}\mathrm{d}\boldsymbol{\omega}$$

$$\cong \left[\int \overline{\phi}\mathrm{d}\mathbf{r} - \beta\int \overline{\phi^2}\mathrm{d}\mathbf{r} + ½\ldots\right]\left[\int \overline{1}\mathrm{d}\mathbf{r} + \beta\int \overline{\phi}\mathrm{d}\mathbf{r} - \ldots\right] \cong -\beta\int \overline{\phi^2}\mathrm{d}\mathbf{r}$$

where the last step is made by $(1-x)^{-1} = 1+x-\ldots$ and noting that $\overline{\phi} = \int \phi \mathrm{d}\boldsymbol{\omega} = 0$ as this is an odd function.

8.3 Expanding the vdW equation $(P+an^2/V^2)(V-nb) = 0$ in a virial form, we obtain

$$P = (nRT/V)[1+(bn/V)+(bn/V)^2+\ldots] - a(n/V)^2 \quad \text{or}$$

$$P \cong (nRT/V)[1+B] \quad \text{with} \quad B = (nRT/V)(b-an/RTV)$$

From $B = 0$ we thus obtain $b-an/RTV$ or $T_\mathrm{B} = a/Rb$, from $\partial BT^{-1}/\partial T = 0$ we have $T^{-2}(-b+2a/RT) = 0$, and discarding $T_\mathrm{JT} = 0$, the result is $T_\mathrm{JT} = 2a/Rb = 2T_\mathrm{B}$.

9.10 Using $\mathrm{g}(\omega) = 3\omega^2/\omega_\mathrm{D}{}^3$, the nth moment of $\mathrm{g}(\omega)$ reads $\mu_n = \int_0^{\omega_\mathrm{D}} \omega^n \mathrm{g}(\omega)\mathrm{d}\omega = 3\omega_\mathrm{D}{}^n/(n+3)$, reducing for $n = 0$ to $\mu_0 = \int_0^{\omega_\mathrm{D}} \mathrm{g}(\omega)\ln\omega\,\mathrm{d}\omega$. Defining $\omega(n) = \mu_n{}^{1/n}$, we have $\omega(n) = [3/(n+3)]^{1/n}\omega_\mathrm{D}$. Now, two options arise to calculate $\omega(0)$. The first is to use $\mathrm{e} = \lim\limits_{N\to\infty}(1+N^{-1})^N = \lim\limits_{n\to 0}(1+n)^{1/n}$. Hence, $\lim\limits_{n\to 0}[3/(n+3)]^{1/n} = \mathrm{e}^{-1/3}$ and thus $\omega(0) = \varpi = \mathrm{e}^{-1/3}\omega_\mathrm{D}$. The second is to use the standard integral $\int \omega^n \ln\omega\,\mathrm{d}\omega = \omega^{n+1}\ln\omega/(n+1)+C$, which leads for $n = 3$ to $\ln\varpi = \ln\omega_\mathrm{D} - 1/3$.

9.11 Using the minus-one rule for $V(P,T) = 0$, we have $(\partial V/\partial P)_T(\partial P/\partial T)_V(\partial T/\partial V)_P = -1$ or $\mathrm{d}T = (\partial T/\partial V)_P\mathrm{d}V + (\partial T/\partial P)_V\mathrm{d}P$. Substitution of $\mathrm{d}T$ in $\mathrm{d}\ln\omega = (\partial\ln\omega/\partial T)_V\mathrm{d}T + (\partial\ln\omega/\partial V)_T\mathrm{d}V$ leads, after some rearrangement, to $(\partial\ln\omega/\partial T)_V = \alpha[(\partial\ln\omega/\partial V)_T - (\partial\ln\omega/\partial V)_P] = \alpha(\gamma_T - \gamma_P)$.

9.12 For $Z = Z_\mathrm{vib}Z_\mathrm{dif}$, we have

$$S = -(\partial F/\partial T)_V = k[\partial\ln Z_\mathrm{vib}/\partial T + \ln Z_\mathrm{vib} + \partial\ln Z_\mathrm{dif}/\partial T + \ln Z_\mathrm{dif}]$$

Hence, in view of $U_\mathrm{dif} = kT^2\partial\ln Z_\mathrm{dif}/\partial T \ll U_\mathrm{vib}$,

$$S \cong k[\partial\ln Z_\mathrm{vib}/\partial T + \ln Z_\mathrm{vib} + \ln Z_\mathrm{dif}]$$

For S, $\ln Z_\mathrm{dif}$ is not negligible and for a quantitative assessment we need a model for diffusion. For C_V we have

$$C_V = T(\partial S/\partial T)_V = T\left\{\partial\left[T(\partial\ln Z_\mathrm{vib}/\partial T)_V + \partial(\ln Z_\mathrm{vib} + \ln Z_\mathrm{dif})/\partial T\right)_V\right\}$$
$$\cong T\left\{\partial\left[T(\partial\ln Z_\mathrm{vib}/\partial T)_V + \partial(\ln Z_\mathrm{vib})/\partial T\right)_V\right\}$$

11.1 Initially, $S = 72.8-(28.9+35.0) = 8.9\,\mathrm{mJ\,m^{-2}}$, so that benzene spreads on water. After some time, the liquids become mutually saturated, $S = 62.2-(28.8+35.0) = -1.6\,\mathrm{mJ\,m^{-2}}$, and the benzene retracts in lens-shaped droplets.

11.6 If we take $zw = a + bT + cT\ln T$ to be the Helmholtz energy f, we have

$$s = -\partial f/\partial T = -(b + c + c\ln T) \quad \text{and} \quad u = f + Ts = a - cT$$

From $u = u_0 + \int_0^T C_V\,dT$ and $s = s_0 + \int_0^T (C_V/T)dT$ we find

$$a = u_0, -(b + c) = s_0 \text{ and } -c = C_V \text{ if we take } C_V = const.$$

Hence, using $zw = a + bT$ means taking $C_V = 0$.

12.4 From $F - G = -PV$, we have $f - g = -P$, where $g = G/V$. Next, we rewrite $\mathrm{d}F = -S\mathrm{d}T - P\mathrm{d}V + \mu\mathrm{d}n$ to $\mathrm{d}F = -S\mathrm{d}T + (f - g)\mathrm{d}V + \mu\mathrm{d}n$, and further to $V^{-1}\mathrm{d}F - (F/V^2)\mathrm{d}V = -(S/V)\mathrm{d}T - (G/V^2)\mathrm{d}V + (\mu/V)\mathrm{d}n$ or $\mathrm{d}(F/V) = -s\mathrm{d}T + \mu n\,\mathrm{d}(1/V) + (\mu/V)\mathrm{d}n = -s\mathrm{d}T + \mu\mathrm{d}\rho$, where in the last step $G = \mu n$ and $\rho = n/V$ are used.

13.1 (a) $L = 31.93\ \mathrm{kJ\ mol^{-1}}$; (b) $L = 30.87\ \mathrm{kJ\ mol^{-1}}$. In spite of the relatively low temperature (80 °C) and pressure (1 atm), the difference $\Delta = (31.93 - 30.87)/31.93 = 3.6\%$, of which only about 0.3% is due to setting $\Delta V = V_V$.

15.1 If $f(x) \sim x^a \leq g(x) \sim x^b$, we also have $\ln f(x) \leq \ln g(x)$. For $x < 1$, we have $\ln x < 0$ and $\ln f(x)/\ln x \geq \ln g(x)/\ln x$. Hence, for $x \to 0$ it follows from $a = \lim \mathrm{d}\ln f(x)/\mathrm{d}\ln x$ and a similar expression for b that $a \geq b$.

18.1 $\rho_{\mathrm{rel}} = 0.984 \cdot \rho_{\mathrm{glass}}/\rho_{\mathrm{FCC}} = 0.984 \cdot 0.637/0.741 = 0.85$. Figure 18.2 provides $0.9 < V^* < 1.1$, hence $0.77 < \rho_{\mathrm{rel}} < 0.95$.

19.1 Consider a small volume V in phase space located at $\mathbf{q}$,$\mathbf{p}$. The probability for the phase point to be in this arbitrary volume is $R_V = \int_V \rho(\mathbf{q}, \mathbf{p}, t)\,\mathrm{d}\mathbf{V}$, where $\mathrm{d}\mathbf{V} = \mathrm{d}\mathbf{q}\mathrm{d}\mathbf{p}$ with surface $\mathrm{d}\mathbf{A}$. For $\mathrm{d}R_V$ we obtain

$$\frac{\mathrm{d}R_V}{\mathrm{d}t} = \int \frac{\partial \rho}{\partial t}\,\mathrm{d}\mathbf{V} + \int \rho\dot{\mathbf{q}}\mathrm{d}\mathbf{A} + \int \rho\dot{\mathbf{p}}\,\mathrm{d}\mathbf{A} = \int \frac{\partial \rho}{\partial t}\,\mathrm{d}\mathbf{V} + \int \frac{\partial \rho\dot{\mathbf{q}}}{\partial \mathbf{q}}\,\mathrm{d}\mathbf{V} + \int \frac{\partial \rho\dot{\mathbf{p}}}{\partial \mathbf{p}}\,\mathrm{d}\mathbf{s}$$

$$= \int \left(\frac{\partial \rho}{\partial t} + \rho\frac{\partial \dot{\mathbf{q}}}{\partial \mathbf{q}} + \dot{\mathbf{q}}^{\mathrm{T}}\frac{\partial \rho}{\partial \mathbf{q}} + \rho\frac{\partial \dot{\mathbf{p}}}{\partial \mathbf{p}} + \dot{\mathbf{p}}^{\mathrm{T}}\frac{\partial \rho}{\partial \mathbf{p}}\right)\mathrm{d}\mathbf{V} = \int \dot{\rho}\,\mathrm{d}\mathbf{V} \quad \text{with}$$

$$\dot{\rho} = \frac{\mathrm{d}\rho}{\mathrm{d}t} = \frac{\partial \rho}{\partial t} + \dot{\mathbf{q}}^{\mathrm{T}}\frac{\partial \rho}{\partial \mathbf{q}} + \dot{\mathbf{p}}^{\mathrm{T}}\frac{\partial \rho}{\partial \mathbf{p}} \quad \text{and} \quad \frac{\partial \dot{\mathbf{q}}}{\partial \mathbf{q}} + \frac{\partial \dot{\mathbf{p}}}{\partial \mathbf{p}} = 0$$

In the first line, Gauss' theorem for transforming a surface integral in a volume integral has been used. In the last line, the first relation provides the total derivative of ρ and the second relation follows from Hamilton's equations. Since the number of phase points over the complete phase space Γ is conserved, that is, $\mathrm{d}R_\Gamma/\mathrm{d}t = 0$, and since ρ being a probability density, $\rho(\mathbf{q},\mathbf{p},t) \geq 0$, it follows that $\mathrm{d}\rho/\mathrm{d}t = 0$. Hence, we have *Liouville's theorem*, which states that for a closed, conserved system the probability density of phase points is constant.

19.5 Using a subscript to indicate the Hamilton function used, we obtain from $S_{\mathcal{H}} = -\partial F_{\mathcal{H}}/\partial T$ by inserting $F_{\mathcal{H}} = -kT\ln Z_{\mathcal{H}}$ and $F_{\mathcal{H}} = U_{\mathcal{H}} - TS_{\mathcal{H}} = \langle \mathcal{H} \rangle_{\mathcal{H}} - TS_{\mathcal{H}}$ that

$$S_{\mathcal{H}} = k\ln Z_{\mathcal{H}} + \langle \mathcal{H} \rangle_{\mathcal{H}}/T = k\ln Z_{\mathcal{H}} + \langle \vartheta \rangle_{\mathcal{H}}/T + \langle \Phi \rangle_{\mathcal{H}}/T$$

For the ideal gas with $\mathscr{H} = \mathscr{H}_0 = \vartheta$, we have thus $S_\vartheta = k\ln Z_\vartheta + \langle\vartheta\rangle_\vartheta/T$. Using the Gibbs–Bogoliubov inequality $F_0 + \langle\mathscr{H}_1\rangle_1 \leq F$, we obtain

$$F_\vartheta + \langle\Phi\rangle_\mathscr{H} \leq F_\mathscr{H} \quad \text{or} \quad U_\vartheta - TS_\vartheta + \langle\Phi\rangle_\mathscr{H} \leq U_\mathscr{H} - TS_\mathscr{H}$$

Employing $U_\vartheta = \langle\vartheta\rangle_\vartheta = \langle\vartheta\rangle_\mathscr{H}$ and $U_\mathscr{H} = \langle\vartheta\rangle_\mathscr{H} + \langle\Phi\rangle_\mathscr{H}$, the result is $-S_\vartheta \leq -S_\mathscr{H}$ or $S_\vartheta \geq S_\mathscr{H}$.

Index

Phases of Matter and their Transitions: Concepts and Principles for Chemists, Physicists, Engineers, and Materials Scientists, First Edition. Gijsbertus de With.

i

j

k

q

r

t